Field Trip Guidebook on Chinese Sedimentary Geology

activities (Wang et al. 2021). The decrease in Lake Daihai's inflow runoff has led to the shrinkage of the lake area. In recent years, the water level of Lake Daihai has dropped from 1,220.08 m in 2006 to 1,214.54 m in 2018, with an average annual decline of about 0.43 m/year (Fig. 3) (Wang et al. 2021).

2.3 Source Rocks

The source rock to the north of the lake basin is dominated by ~1.93–1.89 Ga garnet-rich Liangcheng charnockites (magmatic orthopyroxene granitic rocks) and subordinate Liangcheng S type granite, which constitute the Manhan Mountain (Fig. 4; Peng et al. 2012). The charnockites developed during the peak formation of the gabbronorites, whereas the S-type granites were generated during late-stage with lower temperature melting (Peng et al. 2012). Charnockites have a mineral assemblage of plagioclase, perthite, K-feldspar, garnet, orthopyroxene (hypersthene) and quartz, with or without clinopyroxene and biotite (Le Maitre 2002). S-type granite are mainly composed of plagioclase, perthite, K-feldspar, quartz, and garnet with or without cordierite and biotite.

To the south of the lake, the source rock contains predominantly Fengzhen khondalite (~2.0–1.96 Ga), which refers to a quartz–feldspar–sillimanite paragneiss (Peng et al. 2012). The Fengzhen mobile belt sits along a paleo collisional belt (Zhao et al. 2005).

2.4 Rivers

More than ten rivers feed the lake basin. The longitudinal rivers are Muhua River and Gongba River with relatively longer-distance (20–30 km) sediment transport. The transverse rivers that are sourced from the north Manhan Mountain are characterized by short distance transport (<10 km), and include Yuanzigou River, Bantanzi River, Jiucaigou River, Jinggouzi River and Sandao River. The rivers that are originated from the southern Matou Mountain comprise Buliang River and Tiancheng River. About 262 mm precipitation occurs from June to August (Tan et al. 2021; Tan and Plink-Björklund 2021). This results in seasonal or even ephemeral rivers surrounding Daihai Lake.

Gongbahe River and Wuhaohe River were perennial rivers 10 years ago (Yu et al. 2013). However, due to anthropogenic effect, the rivers became ephemeral in recent years (Fig. 5).

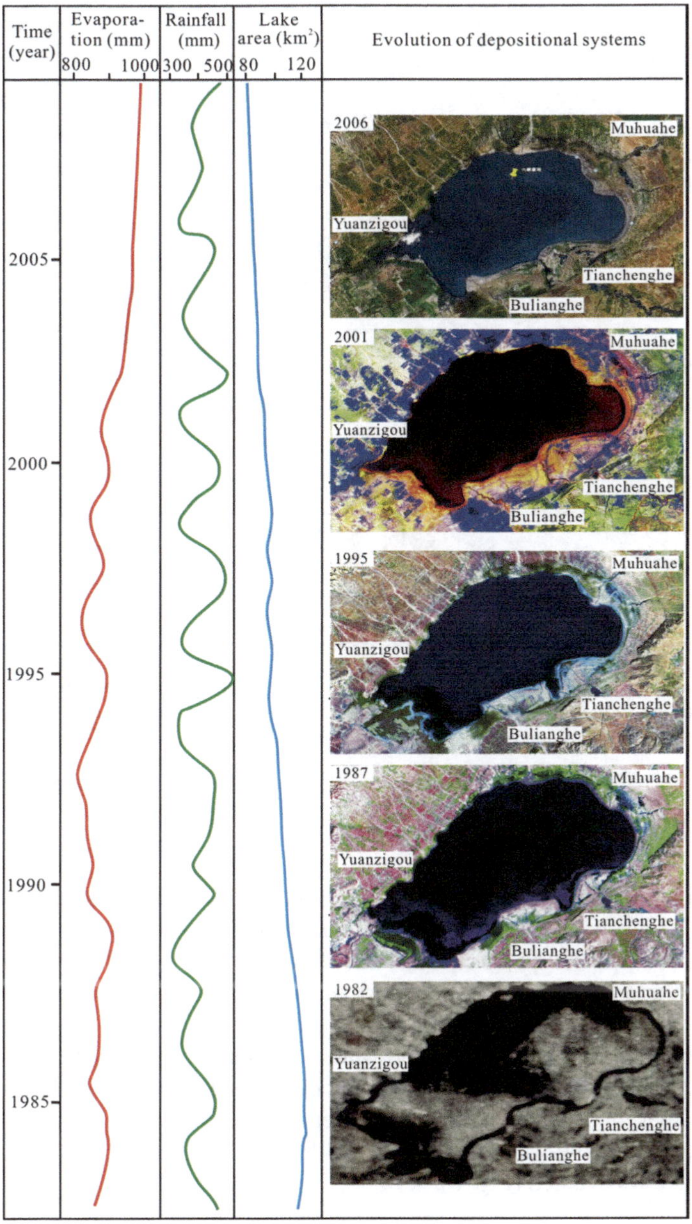

Fig. 3 Change of climate (evaporation and rainfall) and lake area of the Daihai Lake in the past 30 years. The lake shrinked significantly as a result of increased evaporation and anthropogenic activities

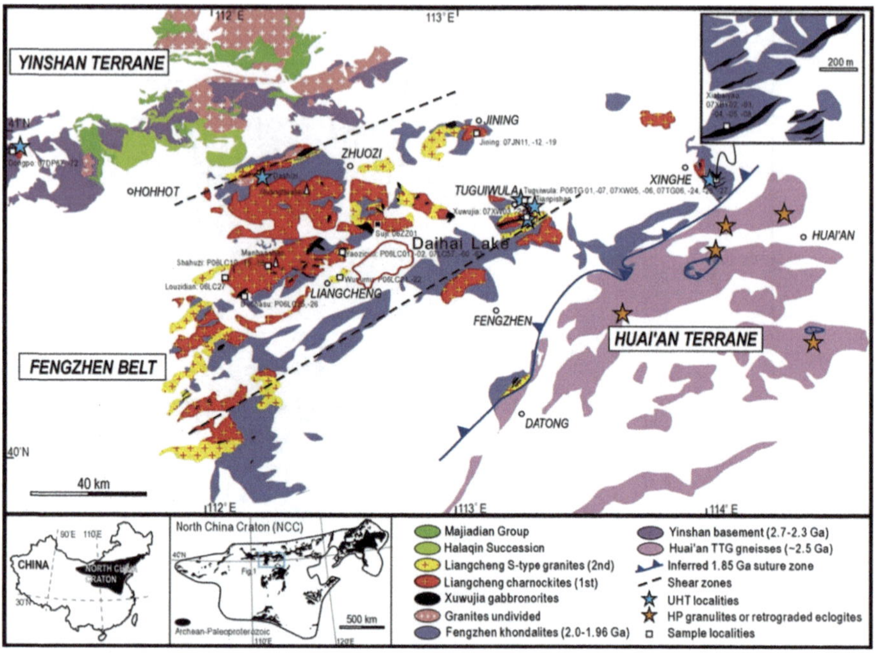

Fig. 4 Source rocks of the Daihai Lake (Peng et al. 2012). The Manhan mountain (northern provenance) lies north of the Lake, which consists of charnockites and S type granite. The southern source rocks from the Matou mountain is composed of Fengzhen khondalites

3 Excursion Schedules

The field investigation was mainly carried out around Daihai Lake in Liangcheng County, Wulanchabu City, Inner Mongolia Autonomous Region. The excursion includes 17 stops that is scheduled for 3 days. The stops on the first day consists of the northern provenance (Manhan Mountain), alluvial fan and its associated fan delta system on the north part of Daihai Lake were observed (4.1.1–4.2.3). On the second day, the source rocks of the Matou Mountain, Buliang River, Tiancheng River and their corresponding braided river delta systems in the gentle slope area on the south margin of Daihai Lake were observed (4.3.1–4.4.3). On the third day, the Muhua River shallow-water delta system (4.5.1–4.5.4) along the longitudinal axis of Daihai Lake was visited (Fig. 6).

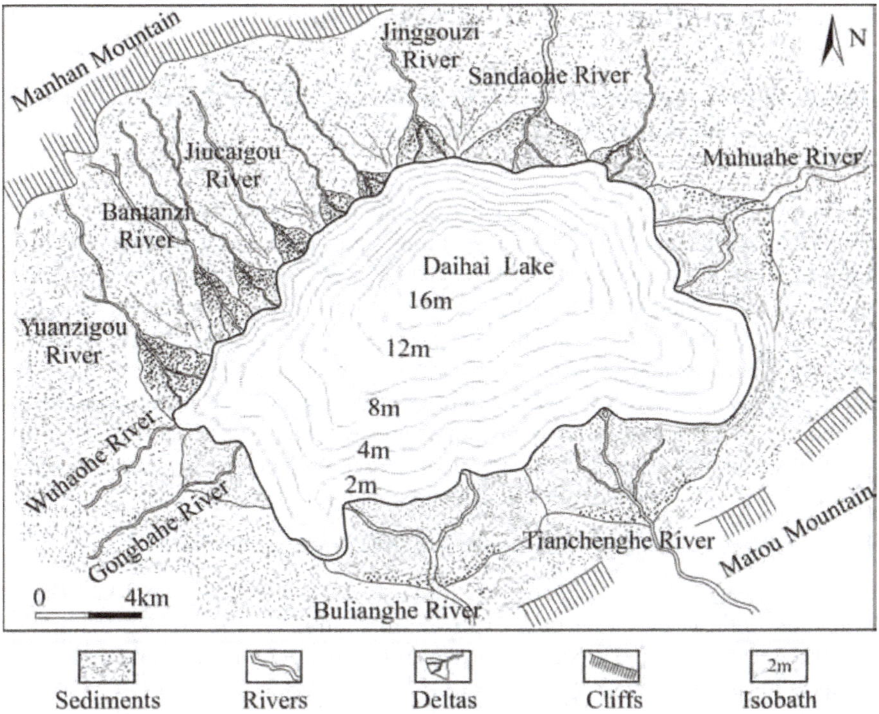

Fig. 5 The map that shows the locations of rivers and deltas surrounding the Daihai Lake. The deltas on the northern margin of the Lake is small, whereas the deltas on the southern margin are much larger

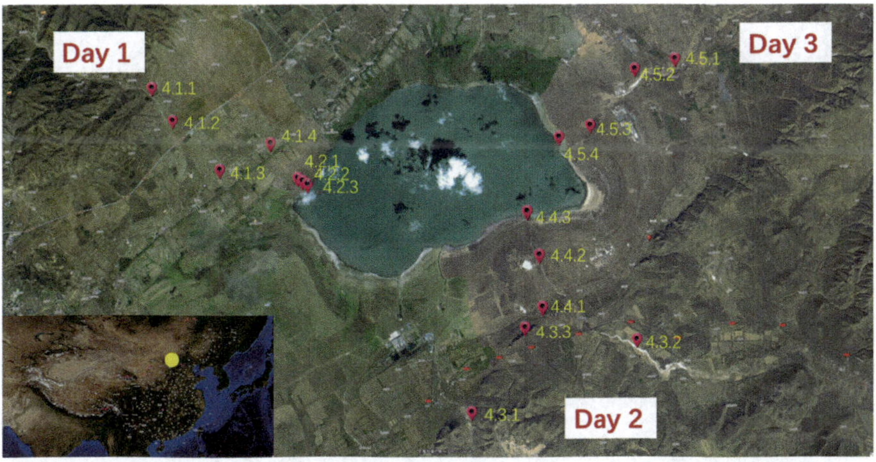

Fig. 6 The google map that shows stops of the field trip. The stops lies along the Yuanzigou River (4.1.1–4.1.3), the Bantanzi River to its delta (4.1.4–4.2.3), the Muhua River to its delta (4.5.1–4.5.4) and Bulianghe and Tianchenghe River to delta (4.3.1–4.4.3)

4 Geological Itinerary and Excursion Stops

4.1 Provenance (Manhan Mountain) and Alluvial Fans on the Northern Steep-Slopes

This branch of the field trip focuses on the sediment sources derived from the northern mountains, and the primary sediment accumulations as alluvial fans in the proximal realm of steep slopes, northern Daihai Lake (Fig. 7). The surficial deposits in the catchment and stream bed, and older deposits exposed along cutbanks are the dominant target of observation. From the perspective of sedimentary evolution, we start from the piedmont and walk along the arroyo of ephemeral streams to the distal part of alluvial fans. This excursion will present a complete view about the sediment production, transportation, and deposition from catchment to alluvial fans. There are 4 stops that are set to show the features of source rocks and catchment, proximal alluvial fan, middle to distal alluvial fan and ephemeral streams.

Stop 1: Source Rocks and Catchments on the Northern Slopes

The first stop (E 112° 34′ 2″, N 40° 37′ 53″) is located at the foothill of the Manhan Mountain, with an elevation of 1490 m. At this site, we observe the features of bedrock, and interpret the processes involved in sediment production and their initial transport in catchment.

The bedrock (i.e., granite gneiss) is well exposed in the source area. The exhumation of bedrocks is resulted from physical weathering with dense jointed fractures. But less chemical weathering due to limited precipitation and sparse vegetation cover

Fig. 7 Google map that shows stops on the northern margin of the Daihai Lake

(Fig. 8). This weathering pattern in the source area partially determine that how the sediment produced from the source and transported to the sink at mountain front.

This stop represents a confluence point of tributaries. Standing at this spot and looking at the mountains, we get a view of the steep drainage basin mantled with coarse and angular rockfall talus and debris-flow colluvium (Fig. 8), which indicate the primary processes in the catchment.

Stop 2: Proximal Alluvial Fan

The second stop (E 112° 34′ 29″, N 40° 37′ 21″, Elevation: 1380 m) is located at about 1.15 km southeast of the first stop. The slope gradient of the area is ~0.1°. The subject of this stop is to observe and describe the sedimentary characteristics of the proximal alluvial fan based on exposed sections.

Coarse particles, up to cobbles and even boulders, were accumulated in the proximal alluvial fan. The coarse gravels are commonly matrix- or clast-supported without organization (Fig. 9). These coarse-grained facies were wedge-shaped and flat based, which were abruptly overlain by coarse gravels (Fig. 10). The second facies are diffusely stratified gravels which were integrated with more sharp bases (Fig. 11). These two dominated facies indicate sediments that are accumulated by debris flows and river floods in the proximal alluvial fan (Blair and McPherson 1994; Blair 1999; Went 2005).

Stop 3: Middle-to-Distal Alluvial Fan

The third stop (E 112° 34′ 59.33579″, N 40° 37′ 5.90969″, Elevation: 1338 m) is at the transition from middle to distal part of the alluvial fan, which located at about 1 km southeast of the second stop. This stop is set for understanding the sedimentary processes of the middle to distal alluvial fan.

The dominant facies in the middle to distal alluvial fan are centimeter-thick, planar-stratified couplets alternating with gravel and gravelly sand or sandy clay. Most gravel particles show the form of a–b planes that are parallel with the bedding planes, although some grains show imbricated structures (Fig. 12). Scattered coarse cobbles and boulders are present (Fig. 12a). The overlying fine-grained layers consists

Fig. 8 Field photographs showing the bedrocks and landscape of the catchment. **a** The bowl-like hollow and steep hillslopes are mantled by unconsolidated rockfall talus and fine-grained colluvium. The whole catchments are planted by sparse mountainous meadows, grasses, and herbs. **b** The bedrock (granite gneiss) with dense jointed fractures. **c** The rock avalanches containing brecciated sediments along loess-draped hillslopes

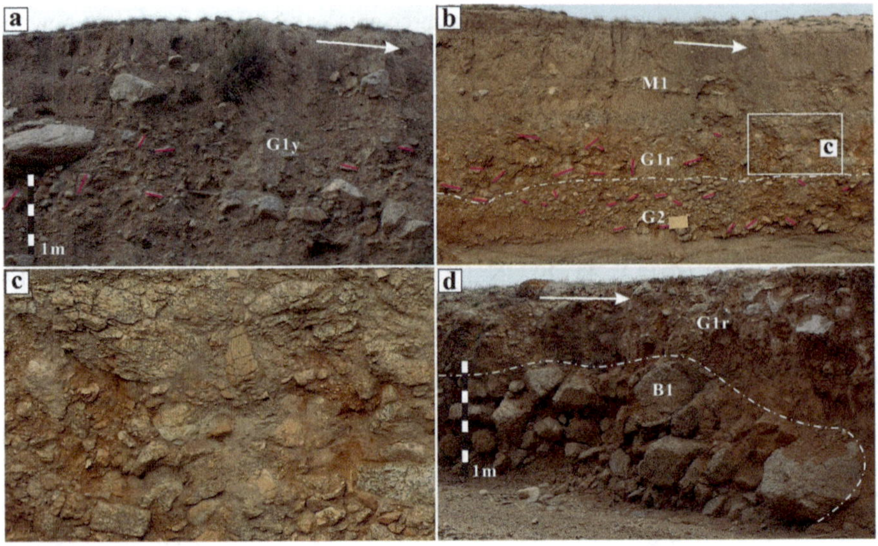

Fig. 9 Representative photographs and interpretations of debris-flow deposits in the proximal alluvial fan. **a** Light-yellowish debris-flow wedges of disorganized, sharp-based angular to angular, unsorted, clay and sandy clay matrix-supported bouldery, very coarse pebble-cobble-sized clasts overlain by loess (L1) with dark varnish surface in the proximal fan. Some elongate boulders are aligned parallel to the fan slope. **b** Brown-reddish tabular beds of unstratified, angular to subangular, poorly sorted, matrix to clast-supported coarse pebble to fine cobble gravel (G2) and overlying brown-reddish clay matrix-supported bouldery, pebble-cobble gravel (G1r) in the middle fan. **c** Close-up view of facies G1r of brown-reddish, matrix-supported bouldery, very coarse pebble-cobble breccias. **d** Angular, matrix-supported boulder gravel (B1) with inverse grading is overlain by sharp angular, clay matrix-supported bouldery, pebble-cobble gravel (G1r)

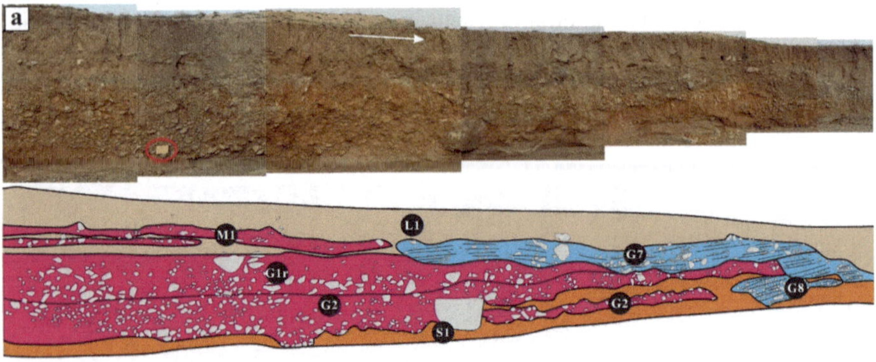

Fig. 10 Representative photographs and interpretations of proximal alluvial fan deposits. Brown-reddish debris-flow lobe deposits (G2, G1r) are cut by light-yellowish fluvial deposits (G7). The debris-flow deposits consist of matrix to clast-supported fine bouldery cobble (G2) and overlying clay matrix-supported bouldery, pebble-cobble (facies G1r). Tabular beds with erosional base may be abruptly halted by coarse boulders and then pinch out. The notebook for scale is ca. 30 cm long and 20 cm wide

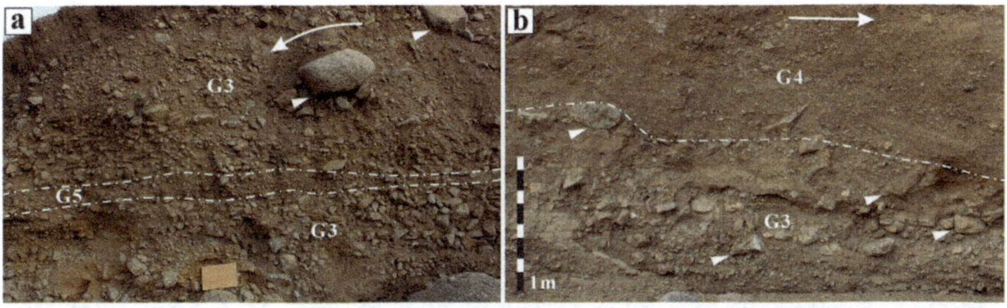

Fig. 11 Representative photographs and interpretations of channel deposits in the proximal alluvial fan. **a** Diffusely-stratified, disorganized, subangular to subrounded, poorly to moderately sorted, clast-supported cobbly coarse pebbles to fine pebbly cobbles (G3) separated by a thin interlayer of parallel-stratified sandy pebbles (G5). Some scattered coarse cobbles and boulders are locally distributed (arrow). The clipboard for scale is ca. 30 cm long and 20 cm wide. **b** The lower incised-channel deposits are overlain by sheetflood deposits. Outsized boulders (arrow) scatter within facies G3

of poorly sorted pebbly sand or sandy clay. This rhythmically stacked couplets form depositional units that are 2.5–4.5 m thick and laterally continuous over tens of to hundreds of meters long (Fig. 12b, c). This facies in the distal fan were deposited by sheetfloods as a result of torrential precipitation (Blair 1999; Went, 2005; Muravchik et al., 2014; Suresh et al., 2007).

Stop 4: Ephemeral Stream

The fourth stop (E 112° 35′ 22″, N 40° 36′ 33.7″, Elevation: 1297 m) is at the middle reach of Bantanzi ephemeral stream. This stop aims to provide the sedimentary and geomorphic features of the ephemeral stream, which is primary conduit for sediment transport from source to lake.

The fluvial deposits are dominated by amalgamated lenses of stratified gravel and sand, and planar parallel-stratified gravel (Fig. 13), which is the product of vertical accretion of longitudinal bars within partially confined channel.

At this site, we investigate the variation of the present channel dimension. The channel dimension exhibits an overall increasing longitudinal trend in the upland and a general decreasing trend beyond this site, but locally, the lowland channel morphology varies quasi-periodically (Fig. 14). Variations in the upland channel morphology are directly influenced by discharge/drainage area, composition of substrate, sediment flux, and slope gradient (Tan et al. 2021). In the lowland, variation in channel morphology is an integrated result of the local flow regimes, vegetations, and anthropogenic activities (Tan et al. 2021).

Fig. 12 Representative photographs and interpretations of middle to distal alluvial fan deposits. **a** Massive gravel lenses (G6) of disorganized, very poorly sorted, open-framework to sandy and clay matrix, clast-supported sandy cobble to pebble gravel interbedded with planar couplets of cobbly, pebble gravel and gravelly sand (G4) overlain by lenticular beds of subangular to subrounded, poorly to moderately sorted, clast-supported cobbly coarse pebbles (G3). Scattered coarse cobbles and boulders (arrow) are found locally. **b** Wedge-shaped bed of light-yellowish clay matrix-supported boulders (B1) and overlying planar couplets of pebble gravel and pebbly sand (G4). The underlying channel deposits are composed of facies G3 of angular to subrounded, poorly to moderately sorted, clast-supported cobbly coarse pebbles. **c** The couplets of alternating fine cobbly, coarse pebble and sandy clay (G4) are overlain by lenticular beds of subangular to subrounded, moderately sorted, clast-supported cobbles (G3). **d** Close-up view of sheetflood deposits. The a–b planes of elongated, angular boulders are parallel to the bedding planes

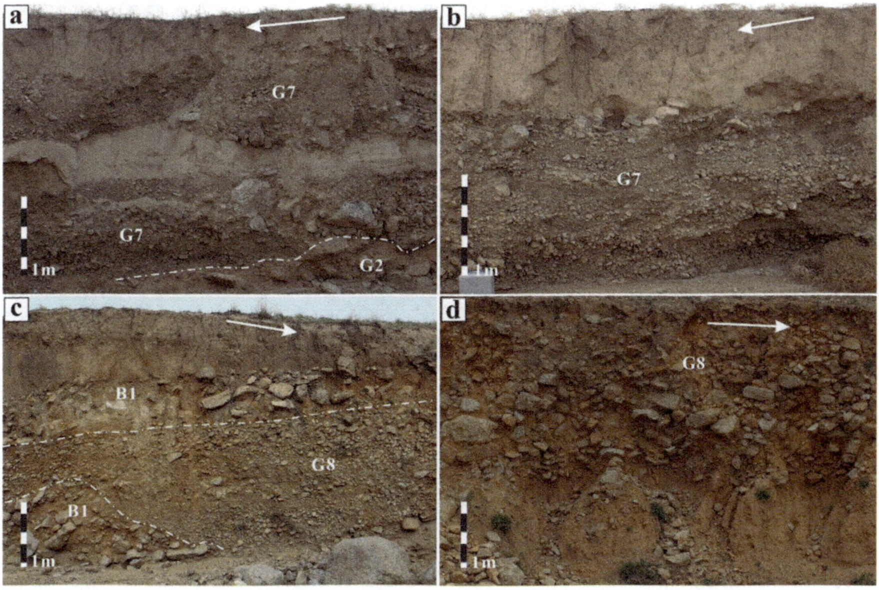

Fig. 13 Representative photographs and interpretations of Bantanzi river deposits. **a** Matrix to clast-supported coarse pebble to fine cobble gravel (G2) overlain by amalgamated lenses of stratified, moderately to well-sorted, matrix-poor to no matrix, clast-supported cobble gravel and gravelly sand (G7). Above G7, the finer grained layers likely indicate inter-flood deposition. **b** Exposure of superimposed lenses of stratified gravel and gravelly sand (G7). The basal part of stratified, subangular to subrounded, moderately to well-sorted, matrix-poor to no matrix, clast-supported coarse pebble to cobble gravel is sharply overlain by massive to faintly stratified, poorly sorted, gravelly sand. **c** Planar parallel-stratified clast-supported coarse pebbles and pebbly sand (G8) have an erosional contact with underlying matrix-supported boulder gravel (B1). **d** Thick beds of parallel-stratified, clast-supported cobbles that have a matrix of light-yellowish mud (G8)

4.2 Bantanzi Fan Delta on Steep Slope

This field route aims to explore a representative coarse-grained delta on the steep slope of the Daihai Lake—the Bantanzi delta, which is fed by a ephemeral stream. The Bantanzi deltaic deposits are continuously exposed along the riverbanks that are incised by newly formed channels, which is favorable to observe the sedimentary characteristics of deltaic deposits. We will start from the apex of this delta and walk through the arroyo of Bantanzi stream down to the present shore of Daihai Lake (Fig. 15). Longitudinal variation of deltaic deposits from the delta plain to pro-delta are observed and interpreted. This excursion is about 1.6 km in distance, and there are 3 stops where characteristic of delta-plain, delta-front, and pro-delta deposits are highlighted.

Fig. 14 Images and interpretations of the ephemeral channel morphology (i.e., Bantanzi River). **a**, **b** Ground views of the upstream reach showing gravelly stream bed and rills developing in the active channel (**b**). **c**–**e** Aerial views showing wide and shallow channels alternating with narrow and deep channels in the moniliform reaches. **f** Ground view of the downstream reach showing alternation of gravelly and sandy stream bed. The number with unit of km in the bottom right of each photo refers to the distance from the source to the spot where the photo was taken. The white arrows indicate flow direction. Encircled person in A is ~1.75 m tall and encircled red shovel in F is ~0.6 m long for scale

Stop 1: Fan Delta Plain Deposits

The first stop (E 112° 36′ 59″, N 40° 35′ 2″) is located near the apex of the Bantanzi delta, which is about 1.5 km away from the shore of Daihai Lake, at an elevation of 1230 m. This stop is to observe typical delta-plain deposits in terms of architecture and facies.

Facies in the Bantanzi delta plain are dominated by clast-supported, normally graded gravels (Fig. 17e) and low-amplitude, convex-up laminated pebbly sands

Fig. 15 Field route and stops for the Bantanzi delta on steep slope

(Figs. 16b and 17g), commonly interbedded with low-angle cross-stratified gravelly sands (Fig. 17c) and gradational planar laminated pebbles (Fig. 17a). Crudely trough cross-stratified pebbles (Fig. 17e) and sigmoidal cross-stratified gravelly sands (Fig. 17b) occur occasionally. Gravelly or sandy backsets occurring as upstream-dipping cross-sets at the downstream margin of scours (Fig. 16b) are diagnostic features of Froude supercritical flow in the delta plain (Cartigny et al. 2014; Slootman and Cartigny 2020; Tan and Plink-Björklund 2021).

Architectures of the Bantanzi delta plain are characterized by irregular low-angle downstream accretion sets that transition downstream into upstream-dipping strata (backsets) across broad scours (Fig. 16). Regular downstream accretion sets also occur in places (Fig. 17i). They are commonly 0.6–1.0 m thick and can be traced laterally for 15–20 m. The set boundaries are undulatory and sharp, and generally overlied with pebbles to cobbles, and even boulder lenses and layers (Fig. 17h).

Stop 2: Fan Delta Front Deposits

The second stop (E 112° 37′ 17″, N 40° 34′ 41″) is located at a paleo-shoreline of the Daihai Lake, which is about 800 m away from the present shoreline of Daihai Lake, with an elevation of 1222 m. Striking delta foresets can be observed in this stop. Along the clinoforms basinward, the variation of grain size, structures, geometries can be identified.

The delta front deposits are characterized by gentle (0.3°) to steep (4°) clinoforms that are ca. 1–2 m high (Figs. 18, 19, and 20). Sedimentary facies in gentle clinoforms are dominated by sands with symmetrical and asymmetrical cm-scale sedimentary structures that are superficially ripple-like, but lack internal cross laminae foresets, and rather are structureless or contain low-angle cross laminae that dip down- or upstream, commonly with mud interlayers. In addition to the cm-scale structures, gentle clinoforms contain wedge-shaped up to 2 m long and 0.4 m thick sands with

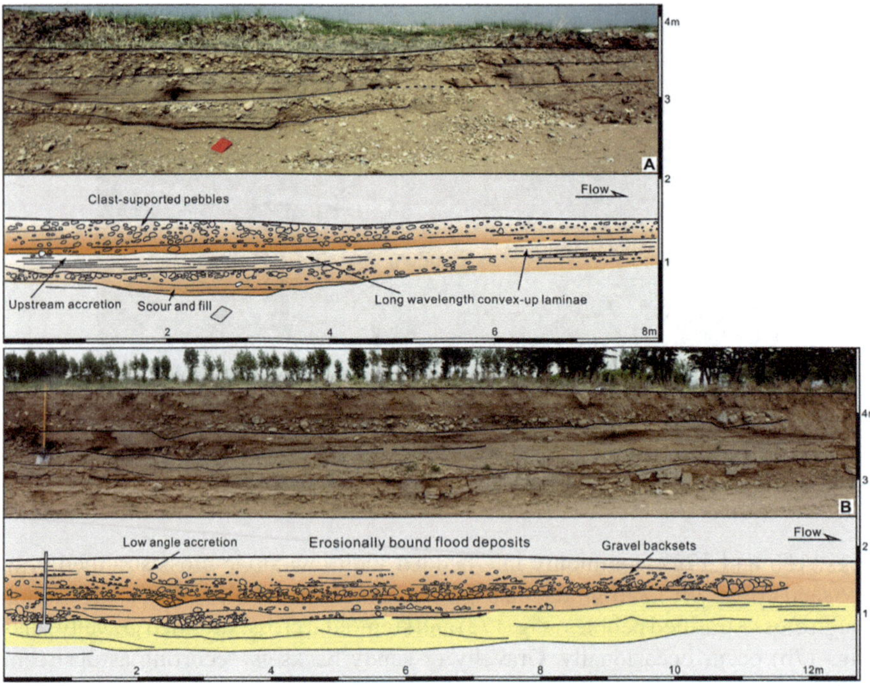

Fig. 16 Photographs and line drawings illustrating architectures of fluvial deposits in the delta plain. **a** Coarse-grained deposits of clast-supported pebbles and low-amplitude, convex-up pebbly sands. **b** Fining-upward successions of conglomerates and sands with basal cobbles, gravel backsets and low angle accretion

internal upstream-dipping strata up to 1.2 m long and 5–20 cm thick, or internally structureless strata.

The steep clinoforms consist of upward coarsening heterolithic deposits composed of dark-gray muds and very fine-grained sands up to 1.6 m thick. The dominant facies consist of fine-grained sand and mud deposits that have the appearance of flaser, lenticular or wavy bedding (Fig. 20).

Stop 3: Pro-delta and Lake Deposits

Moving forward to the present lake shore, the riverbank profile becomes progressively less exposed. The third stop (E 112° 37′ 31″, N 40° 34′ 28″) is only about 200 m away from the present lake shore, with an elevation of ~1217.7 m which is close to the present lake level elevation of 1217 m. Although the exposed profiles are less than 1 m high, the fine-grained pro-delta and lacustrine deposits are well shown.

The pro-deltaic deposits occur lakeward or below the delta front deposits. Although it is hard to obtain a complete picture of the pro-deltaic architectures, individual flat beds interbedded as tabular form could be revealed from the limited exposed profile (Fig. 21). The main facies are fine-grained sand and argillaceous

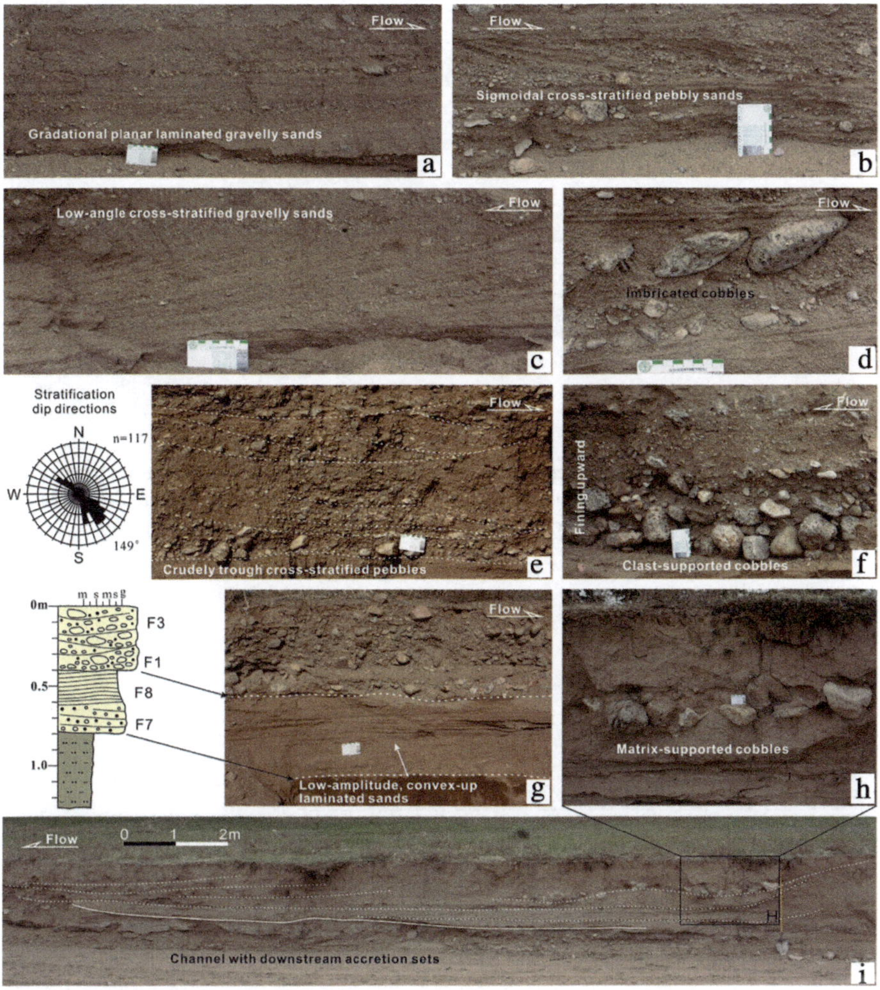

Fig. 17 Representative measured section and illustrations of fluvial deposits in the Bantanzi delta plain. The scale card is 8 cm long and the shovel is 1.4 m long. **a** Gradational planar laminated gravelly sands; **b** sigmoidal cross-stratified gravelly sands; **c** low-angle cross-stratified gravelly sands; **d** imbricated cobbles; **e** crudely trough cross-stratified pebbles; **f** finning-upward motifs of channel deposits; **g** multiple erosion surfaces (dashed lines); **h** fining-upward bed with a sharp base; **i** wedge-shaped geometry and downstream accretions sets of the channel deposits

deposits, showing flaser-like, lenticular or wavy bedding (Figs. 20i and 21b–e). Inside, the lenticular, flaser and wavy bedding are composed of convex and concave structures. The interior is inconsistent with the shape of the structure (Fig. 21b–e) or symmetrical (Fig. 21c). Complete or fragmented helicoidal fossil shells were found in sediments (Fig. 21i).

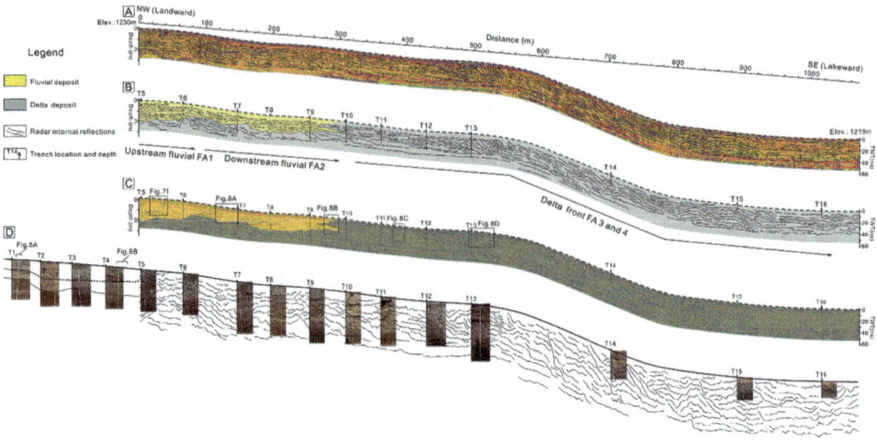

Fig. 18 Original and interpreted ground penetrating radar profile. **a** original ground penetrating radar profile; **b** interpreted profile based on (**a**); **c** the original profile that highlight fluvial deposit; **d** the images of trenches were shown. The trenches lies on the profile without separation

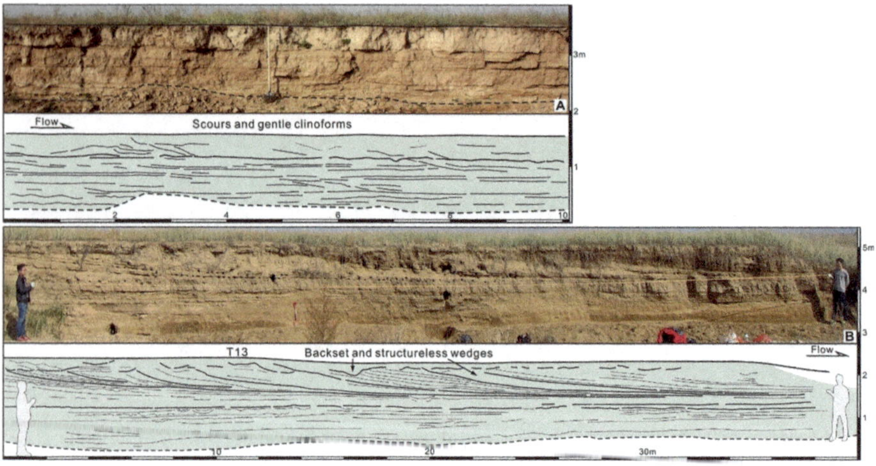

Fig. 19 Photographs and line drawings showing architecture of delta front deposits. In clinoforms, channel-shaped sediments were absent. **a** Low angle (ca 0.3°) delta front clinoforms; **b** high angle (ca 4°) delta front clinoforms

4.3 *Provenance (Matou Mountain) on the Southern Gentle Slope*

The south side of Daihai Basin is Matou Mountain, which is relatively flat, with the highest elevation of 1600 m. The bedrock is mainly Miocene-Pliocene basalt, that intercalated a thin layer of loess and lignite, and a small amount of Archaean granitic

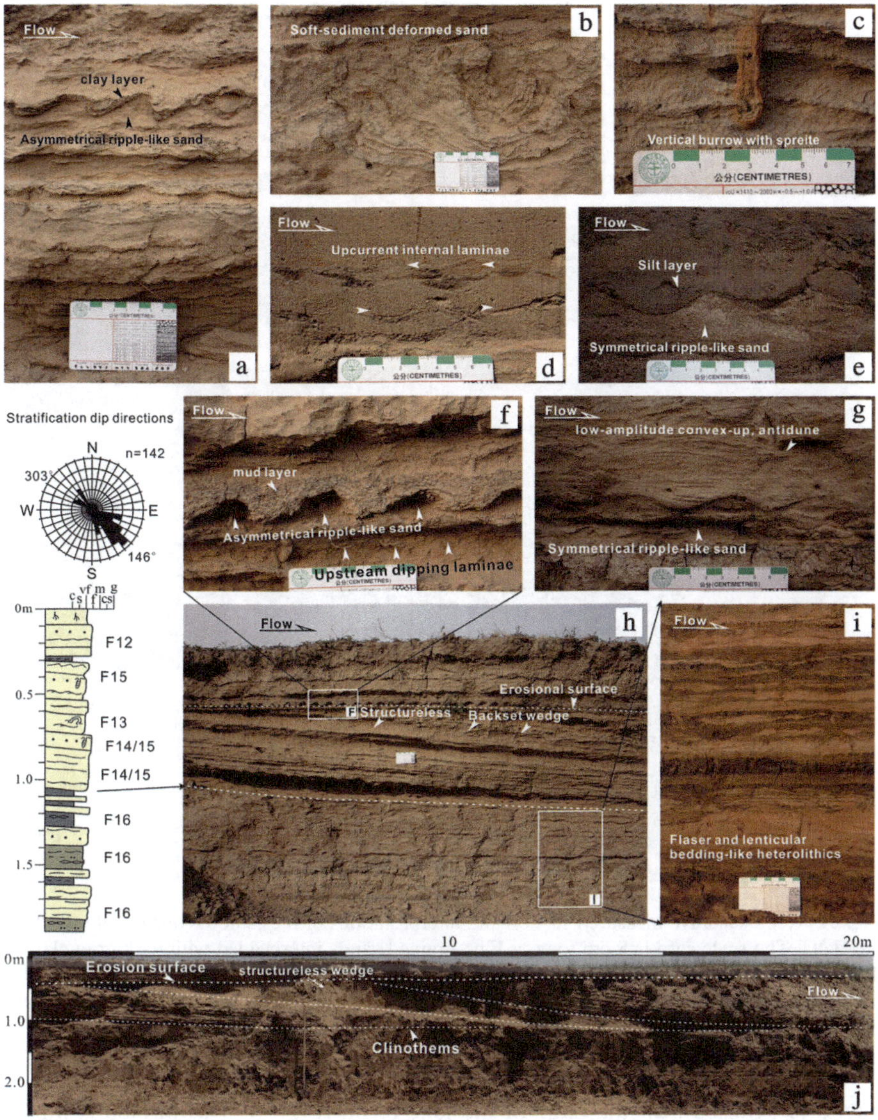

Fig. 20 Representative measured section and interpretations of facies in the Bantanzi delta front deposits. The scale card is 8 cm long. Asymmetrical and symmetrical cm-scale structures (**a**, **d**–**g**) predominate in the facies association. Clay (**a**, **f**) and silt (**e**) layer on the cm-scale sandy structures. Flaser and lenticular bedding-like heterolithic deposits (**i**) are ubiquitous in the delta front. High-angle concave clinoforms with backsets and structureless wedges are diagnostic features of parts of the delta front (**h**, **j**). Soft sediment deformation (**b**) occurs in places. Vertical burrows (**c**) with orange staining commonly occurred

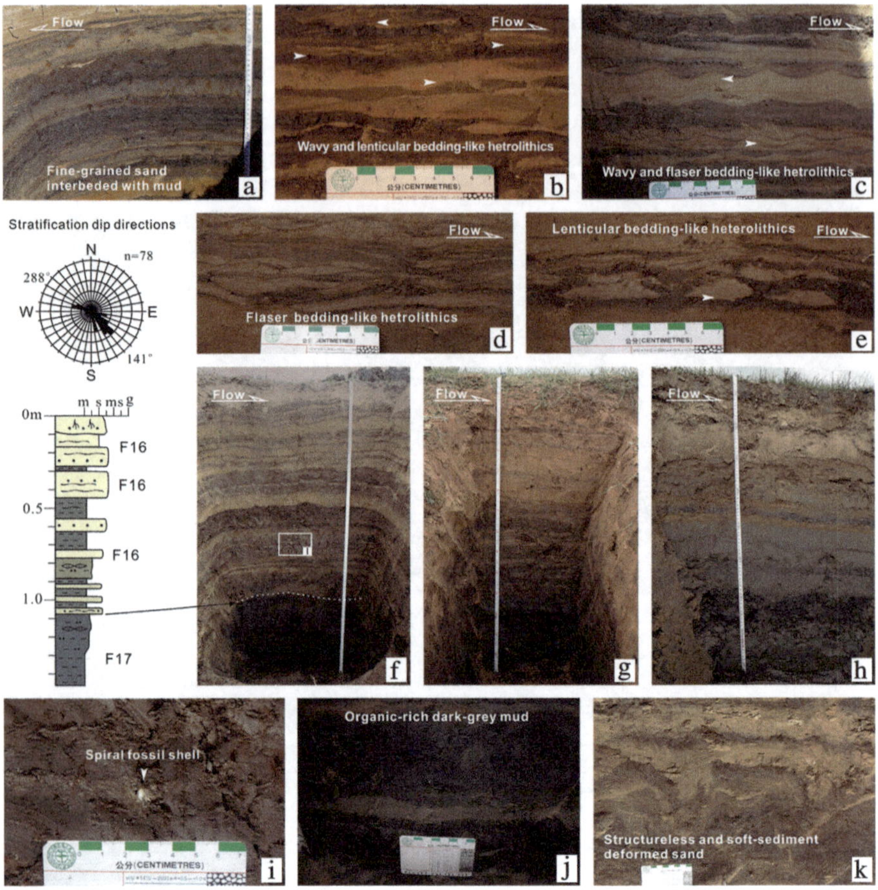

Fig. 21 Representative measured section and interpretations of facies in the distal delta front and pro-delta. The scale card is 8 cm long. **a** Fine-grained sands interbedded with dark gray clay are dominant. Flaser, lenticular and wavy bedding-like heterolithic deposits are also found (**b**–**e**). The succession typically coarsens upward from organic-rich dark gray (**j**) mud to fine-grained sands (**f**–**h**, **j**). Spiral shell fossil (**i**) and soft-sediment deformation (**k**) occur in places

gneiss is exposed near the edge of the lake basin (Fig. 2). The surface is mostly covered by yellow map, and a small amounts of vegetation is developed.

Stop 1: Source Rocks and Catchments on the Southern Slopes

The first terrace of the Tiancheng River Delta on the southern Daihai is well exposed. Uncompacted sediments overlay the metamorphic basement, reflecting changes in early water levels and evolution of the sediment.

Fig. 22 Bulianghe and Tianchenghe Rivers on southern margin of the Daihai Lake

Stop 2: Buliang and Tiancheng Rivers

Buliang River Delta and Tiancheng River Delta are located on the south side of Daihai (Fig. 22). These deltas have a finer grain size than the northern slope deltas due to longer sediment transport distances (15–30 km). The average slope of the southern sedimentary slope is 2°–3°. The rivers feeding the delta are also short-lived streams with a wide grain size distribution; these rivers dry up during the dry season.

Stop 3: Fault Scarps and Transformation Zones on the Southern Slopes

The south coast of Daihai Lake is characterized by gentle slope topography, but there are still many faults exposed on the surface (Fig. 23). These faults have great differences in plan view, and some of the faults are exposed on earth surface. The lithology is mainly Archean granitic gneiss, and the dip angle of the fault is greater than 60°. Several tectonic regulation zones are formed by different faults, which control the flow direction of the Buliang River and the Tiancheng River on the southern slope of the Daihai Lake basin.

4.4 Tianchenghe Braided-River Delta

Tianchenghe delta is located in the southern slope of Daihai Lake. The grain size of these deltas is coarser than Muhuahe delta due to the shorter distance of sediment supply of 15–30 km. The average slope gradient of south depositional slope is 2°–3°. The feeding river of the delta is an ephemeral stream with a wide range of grain size distribution, and they just dried up in arid seasons.

Fig. 23 Fault scarp on the southern slope of the Daihai Lake

Stop 1: Delta Deposits at the Early Periods

In this study, eight deltaic beds were subdivided (Fig. 24). During the early stage of the Tiancheng River delta (probably early Holocene), the metamorphic basement was eroded by the River; thus, mafic gravels were generated (bed ①). The sediments are medium rounded in shape and the grain size ranges from 0.5 to 5 cm. Coarse-grained sand layer and thin gravel layer are interbedded; in addition, there are wave ripples in the coarse-grained sand layer, indicating a delta in a shallow lake environment. Above bed 1 (beds ②–⑤), four layers of upward coarsening fine-grained sand layers are present, of which ripples are the main structures. The thickness of the layers gradually decreases, reflecting the gradual rise of the lake level, and retrogradation of the delta. After the period of flooding, the lake area began to shrink and the delta entered the lake again. At that time, the climate was in a short-term cold period. A freeze slip structure can be observed in beds ⑥. ⑦ A large number of holess were formed between the layers; these caves are vertical, indicating that the water depth was very shallow at that time. The climate here is warm and humid, so there was abundant shells and trace fossils during the deposition of the strata. The bed 8 is composed of fine-grained sand; wavy ripples rarely developed. In short, the section shows a complete depositional cycle from retrogradation to progradation (Fig. 24).

Stop 2: Delta Plain Deposits of the Tianchenghe Delta

The delta plain develops braided distributary channels, containing gravel, gravelly sand and coarse sand. The pebbles in the channel are moderately rounded, with a diameter between 0.05 and 0.1 m. Massive bedding, planar bedding and cross bedding are the main sedimentary structures in the sediments (Fig. 25).

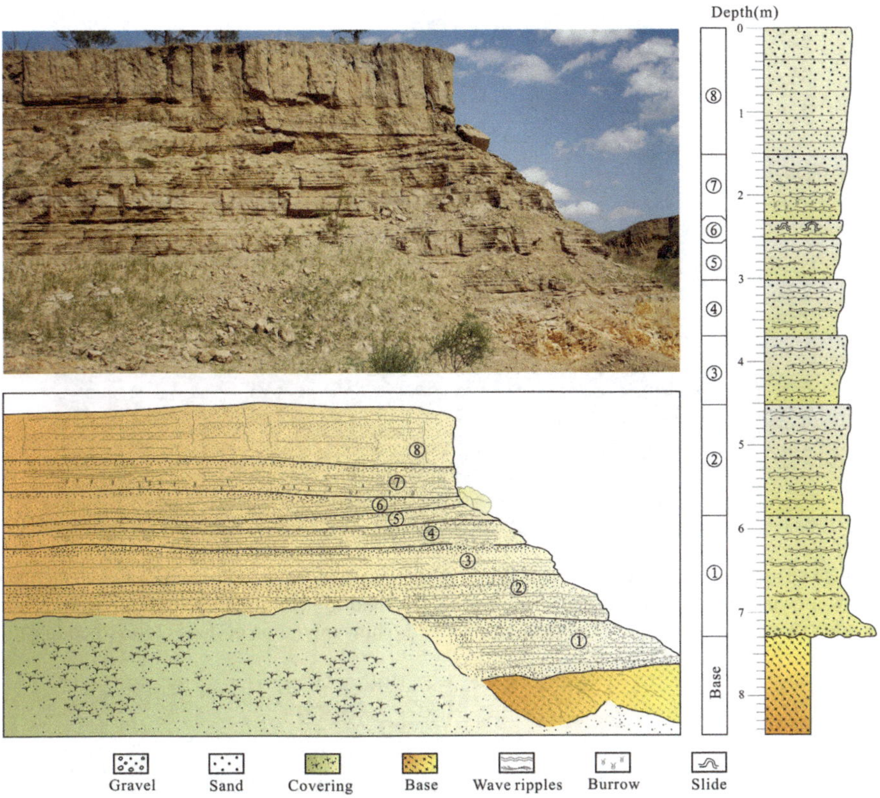

Fig. 24 Basement rocks and the early stage of the Tiancheng delta on southern margin of the Daihai Lake

Stop 3: Delta Front Deposits of the Tianchenghe Delta

The delta front developed silty and muddy deposits. The deposits are characterized by corrugated beds. The Tiancheng River Delta has a large-scale channel with moderate degree of sinuosity. The sedimentary sequence in the channel fines upwards, with erosional surfaces and coarse-grained lags. Delta front deposits are characterized by mud interbeds and rippled fine sand layers. The dark gray mud layer at the bottom of the sequence indicates a coarsening upward sequence from the delta front. Sand bodies in the delta front are widely distributed, showing good connectivity (Fig. 26).

Stop 4: Pro-delta Deposits of the Tianchenghe Delta

As the lake level drops year by year, the prodelta deposits are exposed. The predominant lithofacies includes hydroplastically deformed mud beds. Individual beds have thickness ranging from decimeters to meters and commonly show coarsing-up trend. The element is typical wave reworked littoral sediments of delta front and the sediments features indicate river mouth processes and waves (Fig. 27).

Fig. 25 Trenches of the river bank and dunes in the river bed (of the Tiancheng River delta plain) show a coarsening upward sequence from pro-delta mud to heterolithic delta front sediments to sandier delta plain sediments. The modern river beds show migration of dunes (cross-bedding)

Fig. 26 Delta front deposits showing coarsening-upward feature of the Tianchenghe Delta. The delta front deposits are often heterolithic

Fig. 27 Muddy and silty sediments with mud diapirs of the prodelta in the Tianchenghe Delta

4.5 Muhuahe Meandering-Delta on Eastern Axial Slope

Muhua River Delta is located on the east side of Daihai Lake, close to Maihutu Town. Due to the long-distance (more than 50 km) transport from the source area, the delta sediments are fine and rounded grains. The Muhua River developed on a gentle slope with moderate degree of sinuosity. Fine to coarse-grained sands with trough cross-bedding, planar cross-bedding and hummocky cross-stratification bedding are common in delta plain channels. Rippled silt or mud layers are developed at the delta front (Fig. 28).

Stop 1: Meandering River and Delta Progradations at the Early Periods

The upper reaches of Muhua River delta are supplied by meandering river, and the width of the river channel can reach 1 km. The river channel migrates in the upper reaches. It is controlled by the lake level fall and has incised dramatically. The river channel cuts the early delta lobes, exposing the foreset of the early Muhua River delta. The length of Muhua River is more than 100 km, and shows characteristics of meandering river in the lower reaches, forming sedimentary architectures such as point bar and oxbow lake. On the eastern side of rift axis, Muhuahe River, the biggest one in this area with a large drainage system, is an ephemeral river in the strict sense, which generated a cuspate-shape delta, implying that the delta is influenced by wave processes significantly.

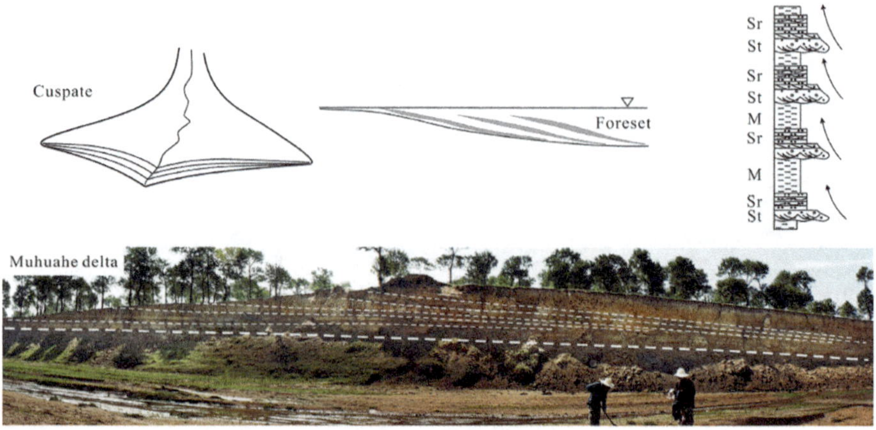

Fig. 28 The planform, foresets and sedimentary sequences of the Muhua Delta. The delta is cuspate in shape, implying a wave-influenced delta

Stop 2: Delta Plain Deposits of the Muhuahe Delta

Sedimentary successions of the Muhuahe delta were coarsening upward with some isolated fining-upward channel beds. Erosional surfaces and lag sediments were common in channel sedimentation in the delta plain (Fig. 29).

Fig. 29 Point bar deposits of a distributary channel in the Muhuahe delta plain

Fig. 30 Sedimentary log (number in centimeter) and trench image of the Muhuahe delta front deposit

Stop 3: Delta Front Deposits of the Muhuahe Delta

The wave ripples in the delta front suggest that coastal waves reworked the sediments and formed distal bar along the shoreline with high connectivity. However, from 1988 to 1993, the evaporation of Daihai Lake was low while discharge of the Muhuahe River was high. The delta kept prograding into the Daihai Lake, forming low angle, imbricated foresets of the delta front (Fig. 30).

Stop 4: Pro-delta and Wave-Dominated Beach Deposits of the Muhuahe Delta

The prodelta was dominated by muddy and silty sediments. The Muhua River delta is affected by the action of the river more than the action of the wave, and it is in the shape of a bird's foot. Subsequently, due to the high evaporation rate and strong wave action, the Muhua River delta showed a cuspate shape. Therefore, when the evaporation decreases, the delta front is dominated by distributary mouth bars due to strong fluvial action. In contrast, as evaporation increases, the delta is dominated by waves.

5 Discussion and Conclusion

The Daihai Lake is a rifted lake basin that is strongly affected by East Asian Summer Monsoon. The formation of the Daihai Lake and the evolution of the sedimentary system are driven by a variety of complex geological factors. This paper aims to demonstrate the influence of tectonics, climate change and hydrology on delta development. Tectonic movement controls the structure pattern of Daihai Lake and determines the formation and size of the fluvial and deltaic system. In addition, due to changes in climate and hydrology, the sedimentary processes and resultant products changes dramatically.

The alluvial fans on the northern steep slope are mainly constructed by debris-flow and river floods, which are dominated in the last interglacial and glacial periods, respectively. The last interglacial debris-flow-dominated unit is characterized by coarse-grained (bouldery coarse pebbles to cobbles or boulders), thick (0.5–1 m, except for facies pebble mud), elongated (from fan apex to distal fan) sedimentary units. The last glacial flood-dominated unit with well-developed supercritical-flow sedimentary structures is relatively fine-grained, sheet to tubular sedimentary bodies. The evolution of these alluvial fans is controlled by climate change rather than tectonic activities due to relatively stable tectonics since the late Pleistocene and synchrony of the alluvial units and eolian deposits that have definite interglacial-glacial climate indication. During the interglacial, stronger extreme rainstorms and more precipitation may trigger widespread colluvial slope failure to generate debris flows due to a combination of the warmer climate and northwestern advancement of the East Asian Summer Monsoon rainfall belt. On the contrary, the sediment-charged flash floods converged along the feeding channels and then transformed into supercritical flows across the unconfined active lobe surfaces due to the retreating East Asian Summer Monsoon towards the southeast during the last glacial period. In addition, catchment size and geographic settings should be taken into consideration in the understanding of the sensitivity of alluvial fans to orbital-scale climate change. To better discuss the effect of orbital-scale climate variation on the alluvial fan evolution, the chronology of the alluvail fan deposits should be established in the future via correlation with the loess-paleosol sequences supplemented by AMS 14C dating and optical luminescence dating.

The Bantanzi Delta (ephemeral stream) in northern Daihai Lake documents two distinct types of clinoforms, which are dominated by bedding structures formed by supercritical flow. One of these clinoform geometries is consistent with deltaic processes controlled by hydraulic jumps at the river mouth, and the other with hydraulic jumps at the base of delta foresets. Comparison to flume experiments and quantitative assessment show that ephemeral rivers, like the Bantanzi stream are likely to produce supercritical flow also on their delta fronts, and link the supercritical flow occurrences to the highly seasonal and intense monsoon precipitation in this semi-arid region. This implies that delta-front deposits composed of supercritical flow structures may indicate precipitation variability and occur in specific hydroclimate conditions, similar to rivers. Such hydroclimates with highly seasonal, intense and intermittent precipitation promote marked seasonal and inter-annual hydrological changes leading to highly variable discharge regimes and exceptional flood events. Intense floods and the steep topography are likely to deliver high sediment concentrations to river mouth and ensure formation and preservation of supercritical flow bedforms. Bantanzi Delta is an unique modern analog, which provides insights into studies on sedimentary processes of other deltas fed by ephemeral streams.

The deltas on the southern gentle slope fed by Buliang and Tiancheng Rivers are much larger than the northern deltas as a result of tectonic control. The terraces developed on the southern slope may have been formed during the rapid drop of lake level. However, the specific timing of lake level drop is still unknown, and dating work is needed. On the basis of dating, the effects of climate and lake level change on the delta evolution can be explored more fully. In addition, the well exposed fluvial deposits profiles along the upper reach of the Buliang River can be finely dissected in order to establish a fluvial-deltaic linkage in the future.

The Muhua delta is fed by a river with sediment transport of a long distance. The sediments are finer than deltas on the northern slope (e.g., Bantanzi delta). The Muhua Delta is cuspate in shape, which is indicative of strong wave influences on delta shape. The clinorms that we investigated are apparently older than the present Muhua delta. The clinoforms should be accumulated when lake level was much higher than the present. Future works should investigate depositional age of the clinoforms and evolution of the Muhua delta.

Acknowledgements We greatly appreciate the field work done by Dr. Mingxuan Gao. We thank Prof. Xiumian Hu and Dr. Hanting Zhong for their help.

References

Blair TC (1999) Sedimentology of the debris-flow-dominated Warm Spring Canyon alluvial fan, Death Valley, California. Sedimentology 46:941–965

Blair TC, McPherson JG (1994) Alluvial fans and their natural distinction from rivers based on morphology, hydraulic processes, sedimentary processes, and facies assemblages. J Sediment Res 64:450–489

Cartigny MJB, Ventra D, Postma G, van Den Berg JH (2014) Morphodynamics and sedimentary structures of bedforms under supercritical-flow conditions: new insights from flume experiments. Sedimentology 61:712–748

Chen X, Dong S, Shi W, Zuza AV, Li Z, Chen P et al (2021) Magnetostratigraphic ages of the Cenozoic Weihe and Shanxi Grabens in North China and their tectonic implications. Tectonophysics 813:228914

Le Maitre RW (ed) (2002) Igneous rocks: a classification and glossary of terms. Cambridge Univ. Press, Cambridge, UK, p 235

Muravchik M, Bilmes A, D'Elia L, Franzese JR (2014) Alluvial fan deposition along a rift depocentre border from the Neuquén Basin, Argentina. Sed Geol 301:70–89

Peng P, Guo JH, Windley BF, Liu F, Chu Z, Zhai MG (2012) Petrogenesis of Late Paleoproterozoic Liangcheng charnockites and S-type granites in the central-northern margin of the North China Craton: implications for ridge subduction. Precambr Res 222:107–123

Slootman A, Cartigny M (2020) Cyclic steps: review and aggradation-based classification. Earth Sci Rev 201:102949

Suresh N, Bagati TN, Kumar R, Thakur VC (2007) Evolution of quaternary alluvial fans and terraces in the intramontane Pinjaur Dun, Sub-Himalaya, NW India: interaction between tectonics and climate change. Sedimentology 54:809–833

Tan C, Plink-Björklund P (2021) Morphodynamics of supercritical flow in a linked river and delta system, Daihai Lake, Northern China. Sedimentology 68:1606–1639

Tan C, Feng S, Zhao X, Shan X, Feng S (2021) Longitudinal variations in channel morphology of an ephemeral stream from upland to lowland, Daihai Lake basin, North China. Geomorphology 372:107450

Wang Q, Chen X, Peng W, Liu X, Dong F, Huang A, Wang W (2021) Changes in runoff volumes of inland terminal lake: a case study of Lake Daihai. Earth Space Sci 8(11):e2021EA001954

Went DJ (2005) Pre-vegetation alluvial fan facies and processes: an example from the Cambro-Ordovician Rozel Conglomerate Formation, Jersey, Channel Islands. Sedimentology 52:693–713

Yu X, Li S, Tan C, Xie J, Chen B, Yang F (2013) The response of deltaic systems to climatic and hydrological changes in Daihai Lake rift basin, Inner Mongolia, northern China. J Palaeogeogr 2(1):41–55

Zhao G-C, Sun M, Wilde SA, Li S-Z (2005) Late Archean to Paleoproterozoic evolution of the North China Craton: key issues revisited. Precamb Res 136:177–202

Core Investigation of the International Continental Scientific Drilling Program from the Cretaceous Songliao Basin (SK-1/SK-2/SK-3), NE China

Yuan Gao, Youfeng Gao, and Daniel E. Ibarra

1 Introduction

The exploration of the universe, of the earth's interior and of the oceans are three great endeavors of humanity challenging nature in order to improve living conditions for humans. Scientific drilling is a significant exploration tool in current earth science research. We can directly observe the lithosphere which we know little about and obtain the rock record from its deeper parts through scientific drilling programs. Drilling provides important information about natural resources, disaster prediction and a variety of environmental problems currently facing modern human society. The Deep Sea Drilling Project (DSDP) in 1968, followed by the Ocean Drilling Program (ODP) and the Integrated Ocean Drilling Program (IODP), led early to a major scientific breakthrough and geosciences revolution which has changed our views of the Earth (Wang 2007). Since the International Continental Scientific Drilling Program (ICDP) was jointly initiated in 1996 by Germany, the United States and China, many continental scientific drilling projects have been completed. As a result, discoveries have been made in five key research themes: the physical and chemical processes responsible for earthquakes and volcanic eruptions; the manner in which the earth's climate has changed in the recent and deep past and the reasons for such changes; the effects of major impacts on climate and mass extinctions; how

Y. Gao (✉)
School of Earth Sciences and Resources, China University of Geosciences, Beijing 100083, China
e-mail: huxm@nju.edu.cn

Y. Gao
Key Lab for the Evolution of Past Life and Environment in Northeast Asia, Ministry of Education, Jilin University, Changchun 130026, China

D. E. Ibarra
Department of Earth, Environmental and Planetary Science, Institute at Brown for Environment and Society, Brown University, Providence, RI 02912, USA

X. Hu (ed.), *Field Trip Guidebook on Chinese Sedimentary Geology*,
https://doi.org/10.1007/978-981-99-6936-4_2

sedimentary basins and hydrocarbon resources originate and evolve; how ore deposits are formed in diverse geologic settings (Harms et al. 2007).

In the past hundred years, the global climate has experienced a great change represented by significant warming. Human civilization is faced with an increasing urgency to understand global climate change and its potential effects on ecosystems, resources and habitability. This research has so far been focused on the modern and Quaternary climate record. The way to understanding the earth's climate system demands a fully integrated exploration of the earth and its geological history, especially of the greenhouse climate (Wang et al. 2017; Tierney et al. 2020). Cretaceous is the most recent long period of greenhouse climate. The study of the Cretaceous greenhouse climate is of great significance for understanding the past and predicting future climate change (Hay 2011).

Geological records indicate that the sea surface temperatures in the equatorial area during the hottest period of the Cretaceous can be as high as 36–40 °C, which is approximately 8–10 °C higher than the present day. The temperature gradient from the equator to the poles was only 24 °C in Cretaceous, which is approximately 25 °C lower than today (O'Brien et al. 2017). Cretaceous atmospheric carbon dioxide concentration varies from 800 to 2400 ppmv, which is 2–6 times that of present day (Wang et al. 2014). There were no ice sheets at the North and South poles in the Cretaceous except for a few brief periods, and the sea level was approximately 50 m higher than that of present day (Miller et al. 2005). Several major geological-environmental-biological events have been occurred in the Cretaceous, such as the Ocean Anoxic Events (OAEs) and the mass extinction at the Cretaceous-Paleogene boundary (K-Pg boundary). The OAE1a event at approximately 120 Ma and the OAE2 event at approximately 94 Ma are two global ocean anoxic events. Driven by large-scale volcanism, a series of processes of warming, hypoxia, mass extinction, and land weathering have occurred on the earth's surface, resulting in a large amount of organic matter buried on the seabed and creating the most important source rocks in modern petroleum industry (Jenkyns 2010). The mass extinction at the K-Pg boundary at approximately 66 Ma, with the extinction of non-avian dinosaurs as a typical representative, is one of the five mass extinctions in geological history. The relationships between mass extinction and meteorite impact, volcanic eruption and climate change are still controversial (Hull et al. 2020).

Marine sedimentary records reveal the close relationships between Cretaceous climate changes and high atmospheric carbon dioxide concentration, tectonic-volcanic activity, and major mass extinction-recovery events (O'Brien et al. 2017; Hull et al. 2020). However, terrestrial climate in the Cretaceous is barely known because of the lack of continuous, high-resolution terrestrial paleoclimate records. The Songliao Basin in China was located on the eastern margin of Eurasia continent, which was the largest landmass in the world in the Cretaceous. A nearly complete Cretaceous continental sedimentary record was formed in the Songliao Basin (Wang et al. 2013a, b). This basin contains abundant oil and gas resources. After 60 years of development, Daqing Oilfield has produced 2.4 billion tons crude oil and 138.6 billion cubic meters of natural gas, which makes outstanding contributions to China's industry (Wang et al. 2021). The implementation of continental scientific drilling in the Songliao Basin has turned the possibility of obtaining a continuous Cretaceous continental sedimentary record into reality.

The International Continental Scientific Drilling Project of Songliao Basin is the world's first continental scientific drilling to drill through the Cretaceous continental strata. This project is also the only scientific drilling project aiming at the entire terrestrial Cretaceous under the framework of the International Continental Scientific Drilling Program (ICDP) (Wang et al. 2013a, b; Gao et al. 2019). The Continental Scientific Drilling Project of the Cretaceous Songliao Basin has two major objectives: one is to correlate the oceanic and continental records and to determine the principal drivers of climate change in order to assist in future climate change predictions and its influence on the earth's environment; the other is to further test the theory of terrestrial genesis of hydrocarbons, and to explore the formation of intra-continental sedimentary basins, as many are major hydrocarbon provinces. In addition, this project provides the scientific basis for exploration of the Daqing Oilfield.

The Continental Scientific Drilling Project of Songliao Basin includes three phases, "SK-1", "SK-2" and "SK-3", and achieves "three wells with four drill-holes and eight kilometers of continuous coring" (Wang et al. 2016a, b, c, d; Gao et al. 2019). Scientific research on the core provides evidences for the reconstruction of Cretaceous terrestrial paleo-climate and sedimentary environment, the terrestrial sedimentary response to major geological events, the formation of large-scale terrestrial source rocks, the evolution of terrestrial biota, and the rapid climate change under greenhouse climate. Such research also provides new scientific support for the construction of the century-old Daqing Oilfield.

The International Sedimentological Conference in 2022 will be held in Beijing. Investigation on the core of the Continental Scientific Drilling of Songliao Basin (SK-1, SK-2, SK-3) is an important part of the intra-conference trips. This guide book systematically introduces the geological background of the Songliao Basin, the general information of the Continental Scientific Drilling of Songliao Basin, and characteristics and scientific research of the cores. Trip visitors are suggested to use this guide book when investigating the core, so that they will systematically understand the achievements of the Continental Scientific Drilling of Songliao Basin.

2 Geological Background of the Songliao Basin

The Songliao basin is 750 km long, 330–370 km wide and has a total area of 26 $\times$ 10^4 km^2. The basin is divided into six structural units: northern plunge, central downwarp, northeastern uplift, southeastern uplift, southwestern uplift, and western slope (Fig. 1). The main oil and gas-producing province is the central downwarp, which includes the Daqing anticline, Qijia-Gulong, Sanzhao and Changling sags and Chaoyanggou terrace.

The basin contains Jurassic, Cretaceous, Palaeogene and Neogene clastic deposits that are about 10 km thick in the basin center. Sediment fill thins toward the basin

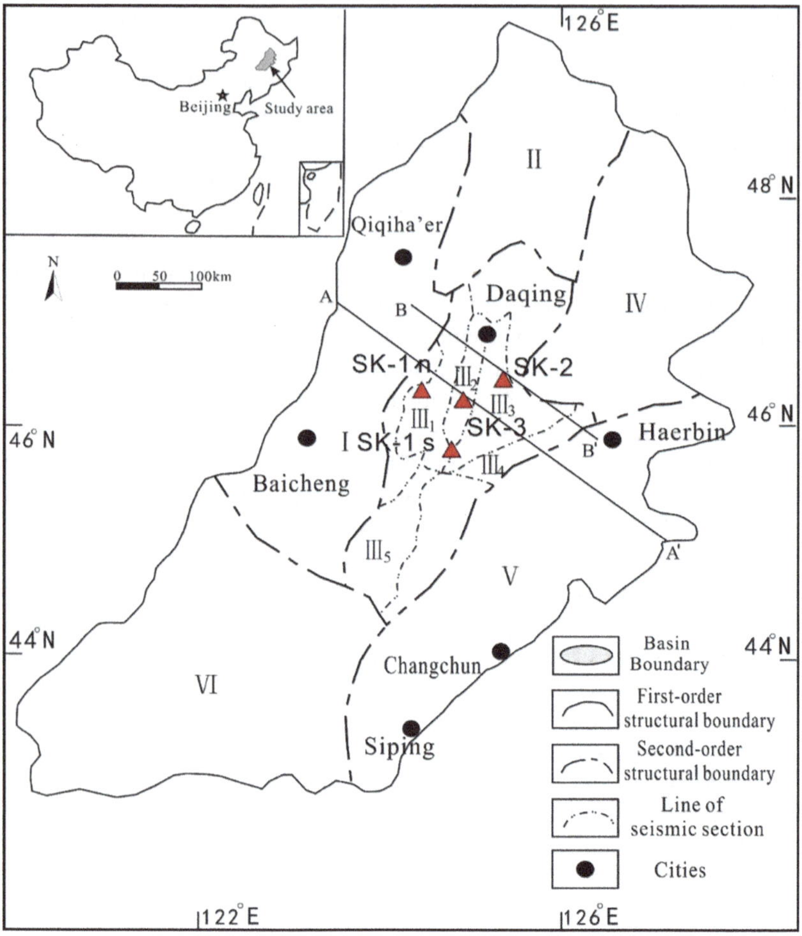

Fig. 1 Division of tectonic units in Songliao Basin and sites of scientific drilling. Location map of the Songliao basin showing major structural divisions referred to in the text. Lines of cross section and seismic sections show locations of Fig. 2 (A-A′). I, Western slope; II, Northern plunge; III, Central downwarp; III1, Qijia-Gulong sag; III2, Daqing anticline; III3, Sanzhao sag; III4, Chaoyanggou terrace; III5, Changling sag; IV, Northeastern uplift; V, Southeastern uplift; VI, Southwestern uplift zone. Red triangles represent sites of Continental Scientific Drilling of Songliao Basin. Modified from Wang et al. (2016a, b, c, d)

margins resulting in a cross-sectional steer's-head geometry (Fig. 2). These sediments are underlain by Palaeozoic metamorphic, plutonic and volcanic rocks. Upper Jurassic to Lower Cretaceous pre-rift and syn-rift deposits (pre-rift and syn-rift tectonostratigraphic units) occur only within isolated fault blocks with a maximum thickness of 7000 m in the Shiwu and Dehui sags. Lower to Upper Cretaceous post-rift deposits (post-rift tectonostratigraphic unit), with a thickness of 3000–4000 m (maximum 6000 m), unconformably overlie syn-rift strata and extend beyond the

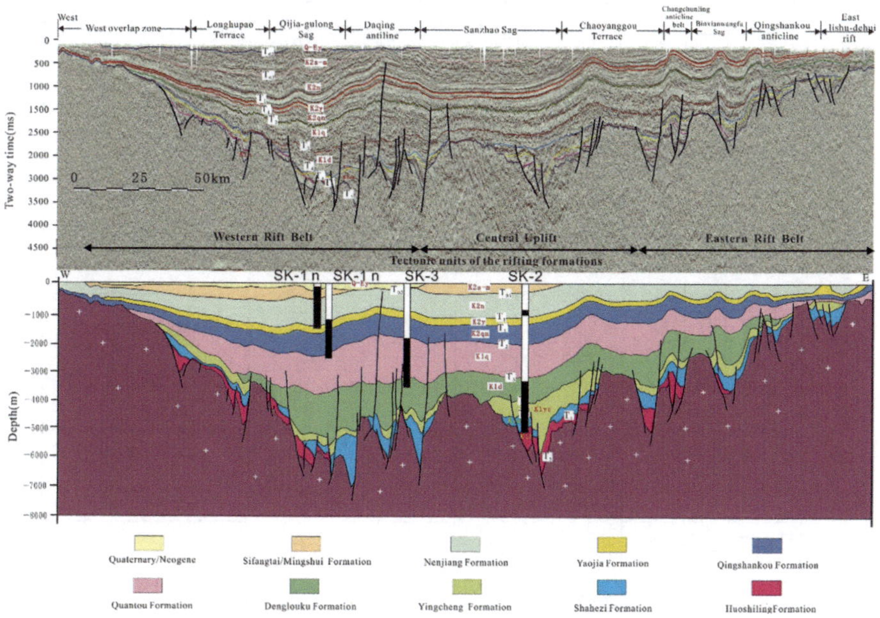

Fig. 2 Regional seismic section and structural cross-section across the central part of the Sangliao Basin. Lithologic information derived from geophysical logs and cores are tied to seismic sections. Location of the section is shown in Fig. 1 (A-A′). Black segments in drilling holes indicate cored intervals. Modified from Wang et al. (2016a, b, c, d)

fault blocks to cover the whole basin. Palaeogene and Neogene deposits (structural inversion tectonostratigraphic unit) with a thickness of 0–510 m occur only locally in the western part of the basin (Feng et al. 2010; Fig. 3).

2.1 Tectonic Evolution of the Songliao Basin

The tectonic evolution of the Songliao Basin can be separated into four stages: pre-rift doming, syn-rift extensional faulting, downwarping (driven by thermal subsidence) and structural inversion.

The pre-rift doming stage: During the middle and late Jurassic, upper mantle perturbation caused crustal extension and thinning leading to the development of a series of fault-controlled depressions in the basin (Fig. 3). This was accompanied by magmatism along the faults. Magmatic activity was stronger in the western part of the Songlia Basin, while in the eastern part subsidence along crustal-scale faults was associated with the volcanic-sedimentary deposition with little magmatic activity (Wang et al. 2006a, b).

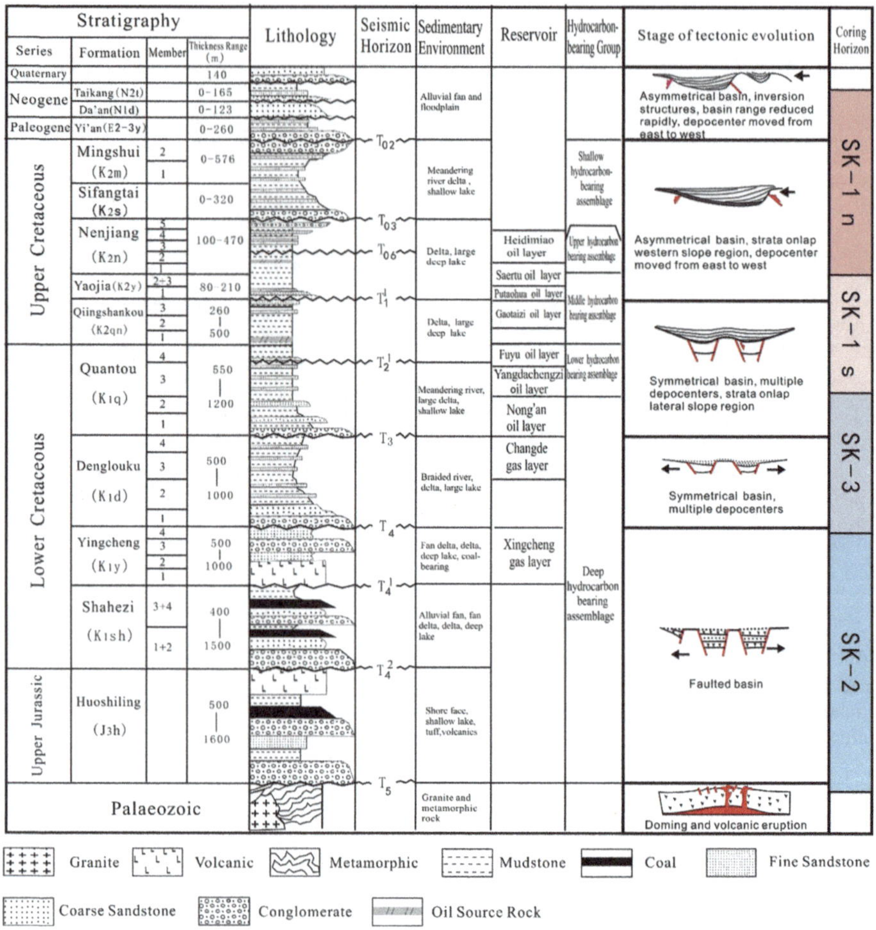

Fig. 3 Generalized stratigraphic column of the Songliao Basin (modified from Wang et al. 2016a, b, c, d)

Syn-rift extensional faulting stage: During the Early Cretaceous mantle doming caused further crustal extension. The Sunwu-Shuangliao fault became active, with uplift of the central horst. Sediments deposited during this time were mainly coarse-grained flysch (Feng et al. 2010). The final shape of the Songliao Basin was formed at this time. During deposition of the Shahezi Formation, the Songliao Basin was undergoing extension accompanied by a widespread volcanism. A series of NNE or NE-trending faults were formed. Deposition of the Yingcheng Formation was influenced by westward compression from the Pacific Plate subduction to the east, which terminated the rifting processes. As a result, previously formed depressions began to contract and the rate of tectonic subsidence rate decreased. Areas surrounding the Songliao Basin began to rise.

Downwarping stage: This stage was driven by thermal subsidence. During the Early Cretaceous, lithosphere underlying the Songliao Basin gradually cooled and contracted resulting in large scale subsidence. The rate of subsidence accelerated to form a central deep depression in the area that had been the locus of maximum doming of the mantle. Two large transgressions resulted in deposition of the Qingshankou and Nenjiang formations when up to 3000 m of sediments were deposited (Wang et al. 2013a). The sediments comprised fluvial, lacustrine, and deltaic lithofacies that contain hydrocarbon. Because of lateral tectonic variability during the early stages of the Songliao Basin development two deposition centers developed-one in the eastern and other one in the central part of the basin. However, during the middle and late stages of basin development, the eastern depocenter disappeared, which caused faulted depressions to develop in eastern part, sags in the middle part, and a long-lived slope zone to develop in the west of the basin (Feng et al. 2010).

Structural inversion stage: During the Late-Cretaceous, the crust of Songliao Basin as a whole rose and the lake basin contracted under the influence of the Pacific plate and Eurasian plate extrusion (Feng et al. 2010). Previously deposited sediments were folded. The eastern part of the basin underwent continuous uplift, causing the westward migration of the depocenter. The rate of subsidence was slow. The structural movement in the eastern and central parts of the basin affected deeper strata, while shallow deformation structures were formed in the west. During the Palaeogene and Neogene, the deposition of alluvial fans and small-scale lakebeds imply the cessation of major tectonic activity (Feng et al. 2010).

2.2 *Tectono-stratigraphic Framework of the Songliao Basin*

2.2.1 Pre-rift and Syn-rift Tectonostratigraphic Units

The pre-rift and syn-rift tectonostratigraphic units comprise the Upper Jurassic Huoshiling (J_3h), and the Early Cretaceous Shahezi (K_1sh) and Yingcheng (K_1y) formations (Fig. 3; Feng et al. 2010; Wang et al. 2013a). The Upper Jurassic Huoshiling Formation, with a thickness of 500–1600 m, was deposited during the doming and initial rifting stage of the basin.

The Lower Cretaceous Shahezi Formation was deposited in the middle of extensional faulting stage. With the accelerated of tectonic subsidence, a large lake was formed in the central part of the rift belt. The formation is usually 400–1500 m thick, and is thickest in the Xujiaweizi and Changling fault blocks, where it is 2500 m thick in the centre part. Strata consist of grey to black lacustrine and floodplain mudstone and siltstone interbedded with grey sandstone and conglomerate. A thin layer of felsic tuff and tuff breccia occurs at the base of the formation, and five coal seams in the middle. Black lacustrine mudstone hundreds of meters thick, with an average total organic carbon (TOC) of 2.1%, forms the second and third members of the Shahezi Formation. These mudstone units are highly mature because of deep burial, and are the most important hydrocarbon source rocks in the Songliao Basin.

The Lower Cretaceous Yingcheng Formation was deposited in down-faulted depressions. The formation is typically 500–1000 m thick, but in the Xujiaweizi Fault block, it has a maximum thickness of 2900 m. The formation comprises mainly felsic volcanics, which are the main hydrocarbon reservoirs, interbedded with clastic sediments and several discontinuous coal beds. During the early sedimentary period of Yingcheng Formation, with the strong volcanic eruption, widely distributed volcanic eruptive rocks were formed, and the sedimentary center tended to migrate northward. The middle and upper Yingcheng Formation is generally a set of sandy mudstone conglomerate with lake as the main body.

2.2.2 Stratigraphic Sequences of the Early Thermal Subsidence Stage

The tectonostratigraphic units deposited during the post-rift thermal subsidence period and the structural inversion period comprise the Denglouku (K_1d), Quantou (K_1q), Qingshankou (K_2qn), Yaojia (K_2y), Nenjiang (K_2n), Sifangtai (K_2s) and Mingshui (K_2m) formations (Fig. 3; Feng et al. 2010; Wang et al. 2013a). This unit is bounded at the base by an unconformity, equivalent to seismic horizon T_4, and at the top by another unconformity, equivalent to seismic horizon T_{02}. The upper unconformity is between the Upper Cretaceous Mingshui (K_2m) and Palaeogene Yi'an (E*y*) formations. Early post-rift strata are divided into three sequences (II_1, II_2 and II_3) separated by unconformities. The sediments comprise alluvial fan, fluvial, floodplain, shallow lacustrine and deltaic facies.

Sequence II_1, which corresponds to the Denglouku Formation (K_1d), was deposited during the transition from rift-induced fault-block basins to cooling-induced regional subsidence. During this period the basin had multiple depositional and subsidence centres. The thickness of sequence II_1 is usually 500–1000 m, with a maximum of 2000 m. Strata comprise interbedded grey to white structureless sandstone, dark sandy mudstone, various coloured sandstone and mudstone, conglomerate and a thick argillite.

Sequence II_2, which corresponds to the first, second and third members of the Quantou Formation, is characterized by red-brown, purple and purple-brown mudstone, coarse greyish-white sandstone and conglomerate of fluvial and floodplain origins deposited in arid or semi-arid conditions. The thickness of sequence II_2 is usually 550–1200 m, with a maximum of 1650 m. Strata of Sequence II_2 progressively overlap the basin margins. N-S trending fluvial channel belt sandstones are common in the first and second members. The third member contains fewer thin fluvial channel sandstones within fine floodplain deposits. Fluvial systems change from braided to meandering channels through the Quantou Formation. Shallow lacustrine and delta deposits occur in the central area of the third member (Fig. 4).

Sequence II_3 consists of the fourth member of the Quantou Formation (K_1q^4) and the Qingshankou Formation (K_2qn), and is 260–500 m thick with a maximum thickness of 635 m. The Quantou Formation is characterized by greyish-green or black mudstone, siltstone and sandstone of fluvial and lacustrine origin (Fig. 4). The Qingshankou Formation consists of grey, dark-grey and black mudstone interbedded

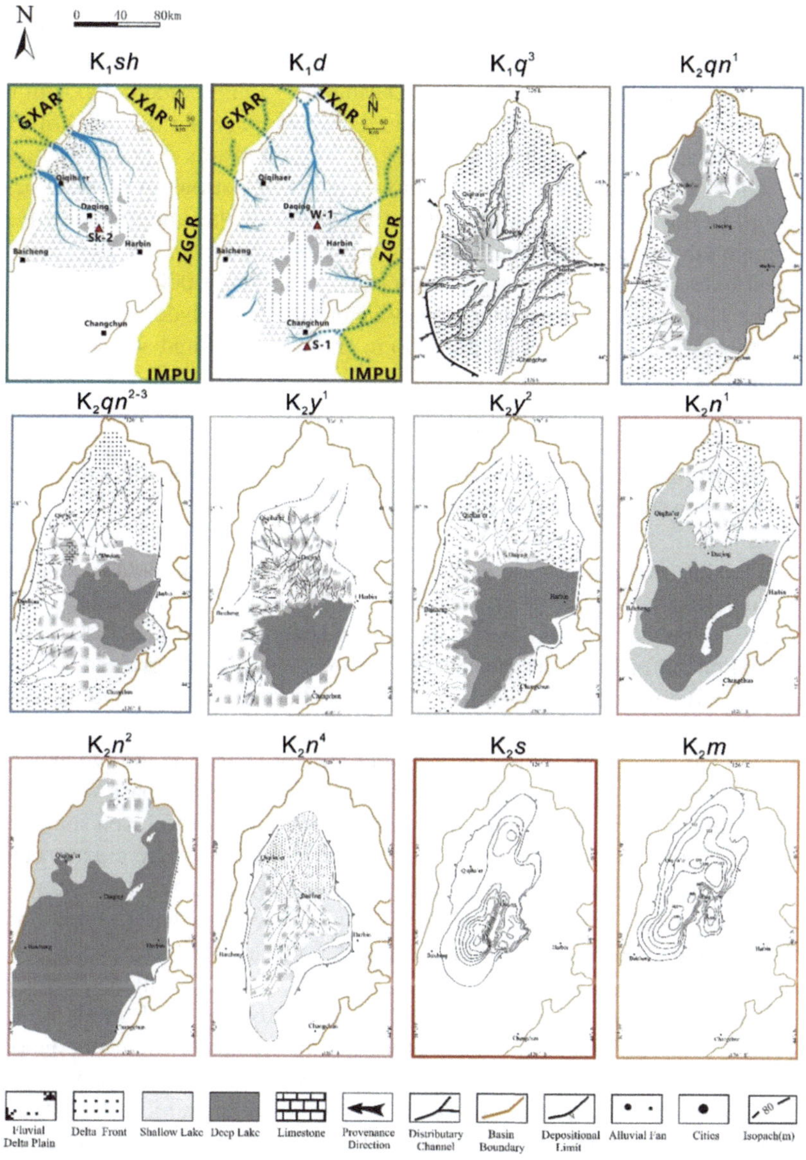

Fig. 4 Sedimentary facies of stratigraphy sequences during the evolution of the Songliao Basin (modified from Feng et al. 2010)

with oil shale and grey sandstone and siltstone. Deltaic and shallow lacustrine facies dominate both formations. Strata near the base of the sequence contain transitional fluvial to lacustrine facies environments. During early Qingshankou deposition, the lake reached its maximum depth and extent, covering an area of 87,000 km^2. Deep lacustrine black mudstone with a thickness of 60–100 m was deposited across the entire central downwarp, forming the most important petroleum source rocks in the Songliao Basin. The second and third members of the Qingshankou Formation contain large deltaic sediment systems that prograded from the north and west. The areal extent of lacustrine strata diminishes to about 41,000 km^2 in the upper part of the Qingshankou Formation. Deep-water lacustrine shale in the second and third members has a thickness of 200–300 m and was deposited across most of the downwarp.

2.2.3 Stratigraphic Sequences from the Late Thermal Subsidence Stage to the Structural Inversion Stage

Strata of the late thermal subsidence stage are divided into three sequences separated by unconformities, sequences II_4, II_5 and II_6. The fundamental attribute of these sequences is that they contain facies that fine and deepen upward and are capped by a subaerial unconformity. Late post-rift strata are generally fine grained.

Sequence II_4, which corresponds to the Yaojia Formation (K_2y) and the first and second members of the Nenjiang Formation (K_2n^{1+2}), comprises red, grey, greyish-green and black mudstone, siltstone and sandstone of lacustrine, fluvial and deltaic origins. Fluvial sandstone and red floodplain mudstone at the base of the Yaojia Formation onlap the unconformity surface on top of the grey to black shales of the underlying Qingshankou Formation. Mudstone rip-up clasts, mud cracks, calcrete, rhizo concretions and bioturbation on the unconformity surface in the central part of the basin indicate a period of subaerial exposure and weathering. The remainder of the 80–210 m thick Yaojia Formation contains a fining-upward succession of fluvial to deltaic to lacustrine facies (Fig. 4). The 100–470 m thick Nenjiang Formation is dominated by deep-water lacustrine grey to black mudstone, marl, shelly limestone and oil shale interbedded with grey siltstone and fine sandstone. During deposition of the first member of the Nenjiang Formation (K_2n^1), the lake expanded rapidly and reached its maximum extent of 20×10^4 km^2, covering almost the entire basin (Fig. 4). Lake expansion was accompanied by a significant increase in subsidence rate. This is the second phase of large-scale lake expansion, following that of the first member of the Qingshankou Formation. In the first member of the Nenjiang Formation, deep-water lacustrine facies were deposited in the basin centre, turbidite deposits occur in the northern part of the basin and shallow lake-margin facies in the south. In the second member, deep-water lacustrine, organic-rich mudstones with a thickness of 200 m covered nearly the entire basin and form a regional barrier to fluid flow. Shallow lacustrine deposits occur only in the north, west and southeast parts of the basin.

Sequence II_5, which corresponds to the third, fourth and fifth members of the Nenjiang Formation, also comprises grey, grey-green and black mudstone, siltstone and sandstone of lacustrine and deltaic origins. Both the lake and the area of sediment accumulation contracted after deposition of the third member, from 110,000 to 10,000 km^2 and from 151,000 to 40,000 km^2 respectively.

Sequence II_6 consists of the Sifangtai (K_2s) and Mingshui (K_2m) formations. Differential subsidence across the basin during Sifangtai deposition created thickness variations from 0 to 320 m (Fig. 4). Sifangtai Formation consists of brick-red pebbly sandstone and shale interbedded with brown, grey and grey–green sandstone and muddy siltstone, purple-red mudstone. These facies represent meandering river, delta and shallow lacustrine environments. Sediment supply was from the north and northwest. The thickness of Mingshui Formation varies from 0 to 576 m (Fig. 4). The formation is composed of grey–green, grey, black and brown–red shale and grey–green sandstone. Sediment was supplied from the south, north and east, and this created NNE-aligned facies tracts of braided river, delta, and shallow lacustrine environments from east to west.

2.3 *Previous Paleoclimatic Research in the Songliao Basin*

Previous reconstruction of Cretaceous paleoclimate in Songliao Basin mainly relies on spore/ pollen and plant fossils, oxygen isotope data, paleoecology, and climatically sensitive deposits.

Researchers have long employed percent ratios of pollen and spore taxa, to estimate temperature, humidity, as well as ecological environment (White et al. 1997; Barron and Bukry 2006). In this way, Gao et al. (1999) reconstructed the climate history of the Songliao Basin on the basis of more than 20,000 samples from more than 500 cores. The vegetation landscape of the basin in Cretaceous time was mainly a conifer forest and steppe (Fig. 5). The Cretaceous atmospheric temperature changed relatively frequently in the Songliao Basin, but was mainly humid to semi-humid and subtropical. During the Cretaceous Period, four cooling events and three warming events occurred in the Songliao Basin (Fig. 5).

Chamberlain et al. (2013) provide the first reported oxygen isotopic results from the SK-1 core of the Songliao Basin (Fig. 5). The oxygen isotopic data are from ostracods collected that cover the interval from the Qingshankou Formation through to the Mingshui Formation. These data record distinct oxygen isotope shifts. There is a negative oxygen isotope shift in the Qingshankou Formation, with $\delta^{18}O$ values changing from ~$-10‰$ to as low as $-18‰$ (Fig. 5). Following this negative shift, oxygen isotopes generally tend to increase from the Qingshankou through the Nenjiang formations. In the Sifangtai Formation, $\delta^{18}O$ values decrease and for the most part are more positive thereafter. In general, the trends in the oxygen temperature records from the Far East sites are similar to the oxygen data analyzed in the Songliao Basin (Zakharov et al. 2011a, b). In the Far East marine record, Cretaceous climate changed relatively frequently, and mainly in temperate.

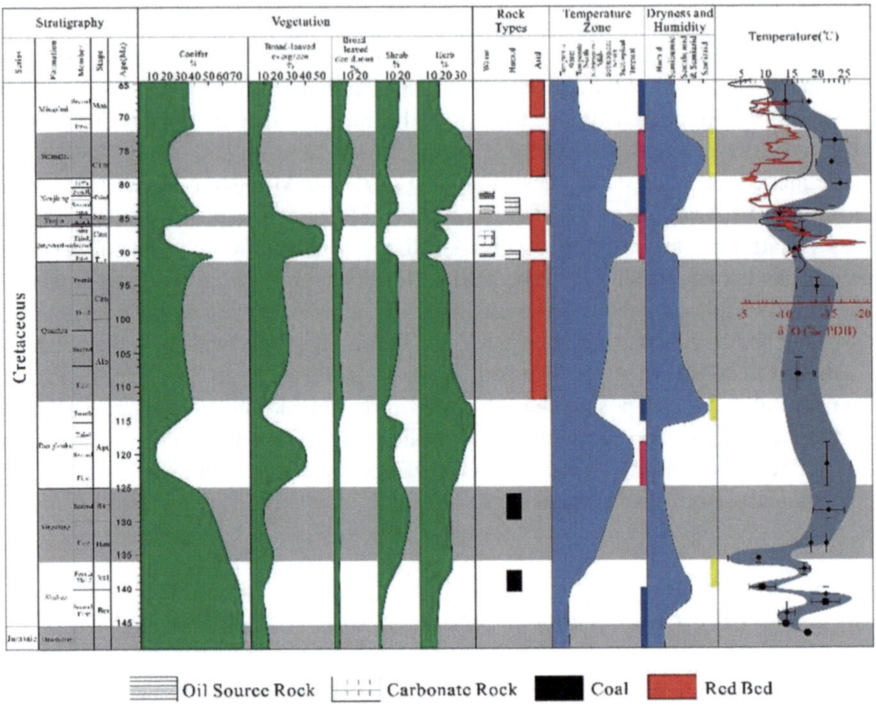

Fig. 5 Cretaceous paleoclimate evolution in the Songliao Basin. Spore/pollen relative abundances, paleotemperature zones and paleohumidity are derived from Gao et al. (1999). Climate-sensitive rock types are from Wang et al. (2013a). Temperatures (black curve, black dots, diamonds) are from Zakharov et al. (1999, 2009, 2011a, b). Oxygen isotope data (red curve) are from Chamberlain et al. (2013). Classification of temperature zones is based on the species of spore/pollen spectra that define tropical, tropical–subtropical, subtropical, tropical-temperate, and temperate conditions. Dryness and humidity are based on species of parent plants of spore/pollen fossils subdivided into xerophyte, mesophyte, hygrophyte, helophyte, and hydrophyte types, which correspond to arid, semi-arid, semi-humid, and semi-arid, semi-humid, humid. The red bar = warming event; blue bar = cooling event; yellow bar = semiarid event

Plant fossils and spore/pollen are abundant and coal is widely distributed (Fig. 5) in Lower Cretaceous of Songliao Basin, indicating a relatively well-developed plant ecosystem (Gao et al. 1999). Floras reflect a warm and humid temperate climate, as does lacustrine Jehol biota (Huang et al. 1999) the middle-late Cretaceous of the Songliao Basin. At that time the diversity of genera and species reached a maximum, reflecting well-developed plant and aquatic ecosystems. Many species of foraminifera, chlorophyta and fish first appeared during this period. During late Cretaceous, the fossils are mainly chlorophyta and ostracoda (Ye et al. 2002). Compared with before, the fossil diversity of the Mingshui biota decreased markedly in numbers of classes and species, which implies an associated shrinkage of the basin and a suppression of the entire ecosystem.

The distribution of coal indicates a range of humid climates under different temperature conditions (Meyerhoff and Teichert 1971; Boucot et al. 2009). Coal beds occur widely in the Lower Cretaceous Shahezi and Yingcheng formations and indicate Early Cretaceous humid climates in the Songliao Basin (Fig. 5). Coal beds are not present in the Denglouku to Mingshui formations, indicating that the paleoenvironment of Songliao Basin changed to a dryer climate with deposition of fluvial to lacustrine facies sediments. Terrestrial red beds are considered to be indicative of a climate that is warm and dry or seasonal with respect to rainfall (Parrish 1998; Du et al. 2011). In the Songliao Basin four large-scale red bed depositional events are evident in the Quantou Formation, second to third members of the Qingshankou and Yaojia Formations, the Sifangtai Formation, and second member of the Mingshui Formation.

In summary, based upon previous paleoclimatic studies, the Huoshiling Formation represents a humid temperate environment. Shahezi Formation coniferous forest decreased, shrubs and herbaceous plants increased, coal seam appeared in large quantities, indicating humid climate conditions. The Yingcheng Formation indicates a semi-humid to humid subtropical environment, in which the temperature decreased and conditions became more humid. Coal beds occur in this formation and there was a decrease in conifers and an increase in shrubs and herbs. The climate of the Denglouku Formation changed rapidly and sharply. There was a rapid decrease in conifers and a sharp increase in broad-leaved evergreens and herbs. Above the Denglouku Formation is the Quantou Formation, that indicates relatively stable temperature and humidity conditions, and a semi-humid subtropical environment. The presence of red beds and absence of coal beds suggests a more arid environment than the lowest stratigraphical unit and a transition from fluvial to lacustrine facies. The Qingshankou Formation was a humid subtropical-tropical environment; there was a sharp decrease in conifers and a sharp increase in broad-leaved evergreens and herbs. The occurrence of red beds suggests an abrupt changes in the climate. The climate during deposition of the Yaojia Formation was cooler and more arid than the Qingshankou Formation. The Nenjiang Formation was characterized by a warm and cool climate. The Sifangtai Formation represents a semi-humid subtropical environment, with the occurrence of red beds. There was a decrease in conifers and an increase in broad-leaved deciduous trees and herbs. The Mingshui Formation represents a semihumid temperate environment.

2.4 Development History of the Daqing Oilfield

The large-scale petroleum geological survey in Songliao Basin began in 1956. On September 26, 1959, the first high production well (called Songji-3 gusher) was drilled successfully, which started the history of Daqing Oilfield. The Daqing Oilfield Company has produced a total of 23.99×10^8 t crude oil and 1386.16×10^8 m^3 natural gas in the last 60 years. The annual crude oil output of Daqing Oilfield Company reached more than 5000×10^4 t from 1960 to 1975. It continued to produce high and stable production for 27 years from 1976 to 2002. Subsequently, Daqing Oilfield Company adjusted its annual crude oil production strategy to 4000×10^4 t, achieving continuous stable production for 12 years (2003–2014). The annual crude oil output of Daqing Oilfield Company is more than 3000×10^4 t by 2020, creating a miracle in the history of similar oilfield development in the world. This company has made outstanding contributions to safeguarding national energy security (Cui et al. 2018; Wang 2019; Sun and Wang 2019; Wang et al. 2021). The Daqing Oilfield Company's exploration and development process is generally divided into five major stages based on the principle of "exploration to achieve block discovery and development to continuously increase production" (Wang et al. 2021).

Discovery of the Daqing Oilfield (1959–1964). Petroleum exploration in the Songliao Basin is guided by the continental oil generation theory to explore the oil and gas in the basin. With the discovery of Well Songji-3, the light spot seismic technology has been used to quickly discover the large-scale structural oil reservoirs in Lamadian, Saertu, and Xingshugang of the Daqing Placanticline.

Structural oil reservoir exploration and development (1965–1985). On the basis of continental oil generation theory, guided by source control theory and anticline reservoir theory, simulated earthquakes and reservoir reconstruction technologies have been developed based on spot seismic technology to explore the Sanzhao depression and the Qijia-Gulong depression on the east and west sides of the Daqing Placanticline. A number of oil fields have been discovered.

Structural-lithologic reservoir exploration and development (1986–2000). With the continuous discovery of structural oil reservoirs and the gradual complexity of exploration targets, a large-scale continental depression lake basin petroleum exploration theory and a thin interbedded low-permeability lithological reservoir exploration supporting technology have been created based on the previous exploration results. Three-dimensional seismic exploration technology was created based on two-dimensional seismic technology. This technology explored the lithologic oil reservoir dominated by Putaohua and Fuyu-Yangdachengzi (Fu-Yang oil layer for short, the main stratum for exploration at present) oil layers in the northern Songliao Basin. The breakthroughs were made in the low part of the Sanzhao and Qijia-Gulong depression.

Lithological reservoir exploration and development (2001–2010). As the exploration targets become more and more complex and covert, the geological understanding has been continuously deepened. The understanding of synclinal reservoir formation has gradually formed, which has enriched the theory of petroleum exploration in continental depression lake basins. Based on theoretical and technological innovation, the Putaohua oil layer in the northern Songliao Basin has achieved a full oil-bearing scene and the Fu-Yang oil layer has achieved oil-bearing contiguous areas in the main part of the Sanzhao depression.

Exploration and development for various types of oil and gas (2011–present). Facing the deteriorating quality of new reserves, low recovery rate, low single-well production, and short stable production period in old areas, the petroleum exploration direction has changed from conventional petroleum to unconventional conventional petroleum in the Daqing Oilfield Company. It has officially entered the stages of detailed conventional petroleum exploration and the exploration stage of tight oil and gas, shale oil and new areas.

In the course of more than 60 years of development, Daqing Oilfield has perfected the continental oil generation theory and source control theory. It also developed exploration theories for continental depression lake basins, complex fault depressions, volcanic rocks, and tight oil and gas, which formed a complete suite of technology for exploration and development. With the continuous expansion and extension of the field of petroleum exploration and development, combined with a series of development problems faced by Daqing Placanticline Oilfield during the ultra-high water cut period and the production of difficult-to-recover oil and gas reserves in the Placanticline oilfields, the Daqing Oilfield has made it clear that it should speed up shale oil and carbonate exploration and development supporting technical research and greatly improve the recovery rate, and effectively use supporting technical research for difficult-to-recover reserves, etc. Based on exploration and development technology, which aims to achieve the effective replacement of reserve resources and the goal of increasing oil and gas production.

3 Progress of Continental Scientific Drilling Project of Cretaceous Songliao Basin

The International Continental Scientific Drilling Project of Songliao Basin is the world's first continental scientific drilling to drill through the Cretaceous continental strata. This project is also the only scientific drilling project aiming at the entire terrestrial Cretaceous under the framework of the International Continental Scientific Drilling Program (ICDP) (Wang et al. 2013a, b; Gao et al. 2019). The implementation of the scientific drilling project provides key information to understand terrestrial climate change in the Cretaceous greenhouse period, and to predict future climate warming. This project also provides scientific support for exploring mechanism of massive organic carbon burial in terrestrial basins, and for stable petroleum production in the Daqing Oilfield.

3.1 Technical Progress of Continental Scientific Drilling of Cretaceous Songliao Basin

The Continental Scientific Drilling Project of Cretaceous Songliao Basin consists of three stages: SK-1, SK-2, and SK-3. A total of 8186.85 m cores were recovered, and the cumulative coring recovery ratio reached 97.03%."Three wells, four holes, eight kilometers of continuous coring" have been realized (Figs. 6 and 7, Table 1).

SK-1 drilling was started in Daqing City, Heilongjiang Province in 2006 and was completed in 2007. According to stratigraphy and engineering requirements, and

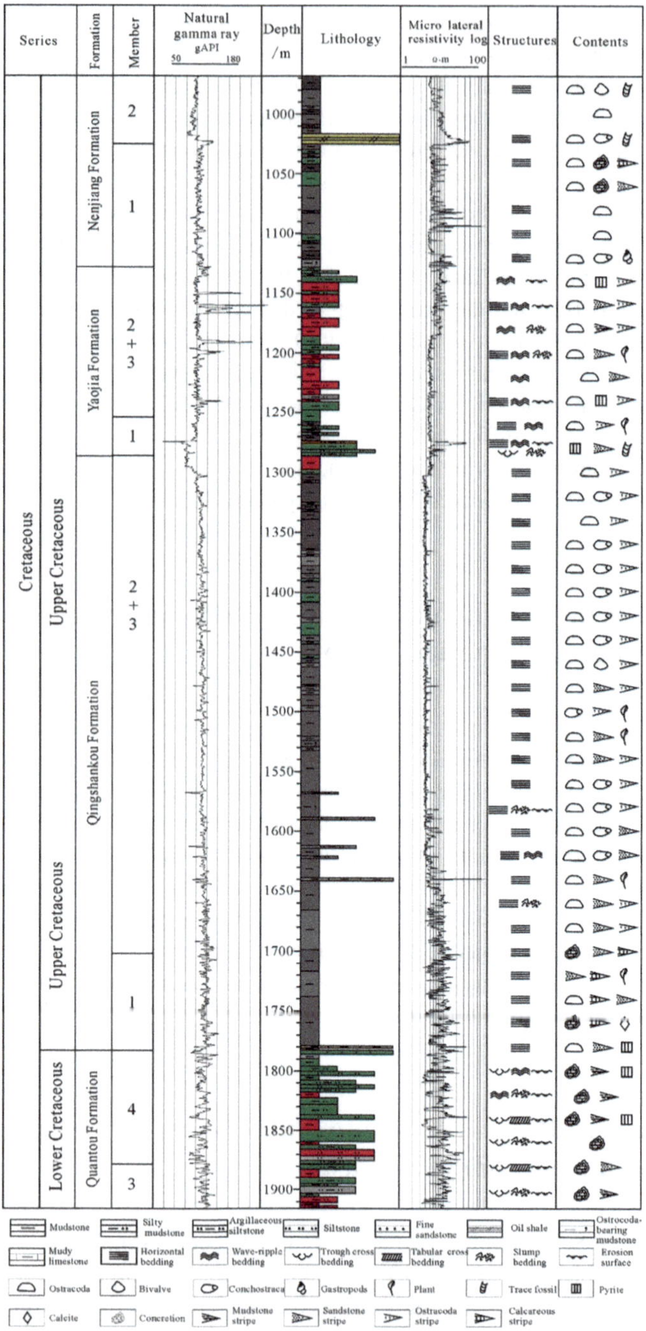

Fig. 6 Stratigraphic column of SK-1s core (modified from Wang et al. 2016a, b, c, d)

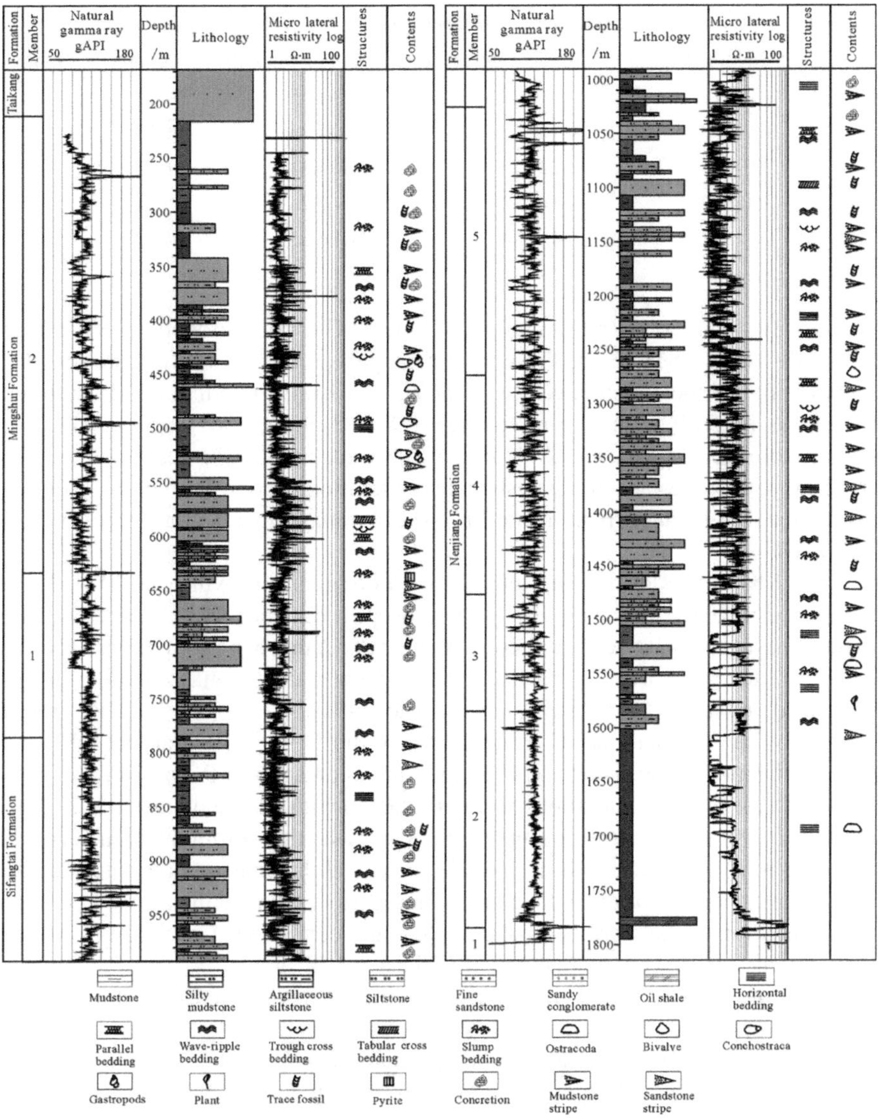

Fig. 7 Stratigraphic column of SK-1n core (modified from Wang et al. 2016a, b, c, d)

based on extensive experience in scientific drilling initiatives, a complete design concept was applied to SK-1 and this included the four systems/techniques of "economical and flexible one well-two holes design", "advancement by combined drilling technique", "ultra-long high-recovery integrated coring technique and management system," and "core preservation in crystal resin plastic seals-organic glass brackets", organically integrated into the drilling plan of Cretaceous terrestrial strata. SK-1

Table 1 Drilling and coring statistics of the Contiental Scientific Drilling of Songliao Basin

	Drilling depth (m)	Coring footage (m)	Core length (m)	Core recovery (%)
SK-1n	1811.18	1630.14	1541.55	94.55
SK-1s	1915.00	946.83	944.23	99.73
SK-2	7018.00	4279.73	4134.81	96.61
SK-3	3600.00	1580.34	1566.26	99.11
Total	–	8437.04	8186.85	97.03

contains two boreholes (north and south holes). The drilling depth of SK-1 north hole (SK-1n) is 1811.18 m, the coring footage is 1630.41 m, the core length is 1541.65 m, and the core recovery is 94.56%. The drilling depth of SK-1 south hole (SK-1s) is 1915.00 m, the coring footage is 946.83 m, the core length is 944.23 m, and the core recovery is 99.73%. A total of 2486-m continuous cores were recovered from the Upper Cretaceous Qingshankou Formation, Yaojia Formation, Nenjiang Formation, Sifangtai Formation and Mingshui Formation with a total recovery ratio of 96.46%.

The "SK-2" was drilled in Anda City, Heilongjiang Province in 2014. It was completed in 2018. This well was 7018 m deep. It obtained 4134.81 m from basement of basin to the Lower Cretaceous Shahezi and Yingcheng formations. The core recovery rate was 96.41%. The SK-2 is the deepest continental scientific drilling organized and implemented by Asian countries. It is also the deepest drilling implemented by the International Continental Scientific Drilling Program (ICDP) since its establishment 22 years ago. During the implementation of the project, the world record for the longest continuous coring (1650 m) with a large diameter of 311 mm and the longest single-barrel coring (30, 41, 33 m) with three different diameters (311, 216, 152 mm) was created. It has greatly improved the construction efficiency and core recovery rate, shortened the construction period, lowered the cost, and ensured the acquisition of the world's most complete and continuous Cretaceous terrestrial stratigraphic record and the realization of the scientific goals.

The "SK-3" was drilled in Nong'an County, Changchun City, Jilin Province in September 2020. It was completed in February 2021 with a depth of 3600 m. The Denglouku Formation and Quantou Formation in the mid-Cretaceous period of 1566.26 m was obtained for coring. The recovery rate was 99.1%. The SK-3 adopts the conventional three-open well structure, and rationally optimizes the designs of the technical casing depth and expansion hole, shortens the soak period of well wall, avoids complicated well conditions, and ensures the smooth progress of coring construction. The coring tools were optimized, which strictly implementing anti-blocking and anti-jamming technical measures. Combined with analysis of core crushing conditions and the reasonable selection of coring parameters, there is no plugging, which affects the single core drilling footage. In the light of loose and broken strata, the plugging force of the drilling fluid and adjusting the rheological parameters of the drilling fluid were strengthened, which to make sure that the rock strength is improved, the scouring effect is reduced, and the smooth coring construction is ensured.

3.2 *Core Lithology of Continental Scientific Drilling of Cretaceous Songliao Basin*

3.2.1 Core Lithology of SK-1

In the SK-1 boreholes, continuous coring recovered strata from the top of Quantou Formation (Late Cretaceous) to the base of Taikang Formation (Paleogene). Among them, the coring horizon of borehole SK-1n is from the bottom of the second member of Nenjiang Formation to the bottom of Taikang Formation, and borehole SK-1s is from the top of the third member of Nenjiang Formation to the bottom of the second member of Nenjiang Formation. Lithologic characteristics of each formation are summarized below and the formations are described in stratigraphically descending order (Fig. 8).

Taikang Formation: Only the basal part of the formation was cored. It is represented by light grey sandy conglomerate and pebbly sandstone.

Mingshui Formation: Strata are dominated by siltstone and mudstone with subordinate argillaceous siltstone and silty mudstone, a few thin-layers of sandy conglomerate, fine-grained and medium-grained sandstone. Color of the sandstone and some of the mudstones is mainly grey. Mudstones are grey, grayish-black, black, mauve, greenish-gray and or grayish-green. Sedimentary structures in fine and medium-grained sandstones are parallel bedding, wave-ripple lamination, cross bedding, minor small-scale trough bedding, and graded bedding. Slump features are common. Mudstones are massively bedded. In grey, grayish-black and black mudstone the bedding is horizontal and ripple laminations occur. Sandy conglomerates have massive structure and deformation structures are present. Sandstones enclose abundant charcoal and plant debris. Sparse clam and spiro fossils have been found. This formation is interpreted to have been deposited in a shallow lake and near-shore lake environment.

Sifangtai Formation: Middle to lower part of the formation is mainly siltstone. The upper part is dominated by mudstone with subordinate argillaceous siltstone and silty mudstone. Color of the sandstone and conglomerate is gray. Mudstone colors are predominantly mauve and greenish-grey. In siltstone and fine-grained sandstone, sedimentary structures are parallel bedding, wave ripple bedding and wave cross bedding. Slump features are common. Mudstone are massively bedded, with occasional massive laminae, horizontal bedding and ripple lamination. Sandstones enclose abundant charcoal and plant debris. Sparse clam and spiro fossils have been found. This formation is interpreted to have been deposited in a shallow lake and near-shorelake environment.

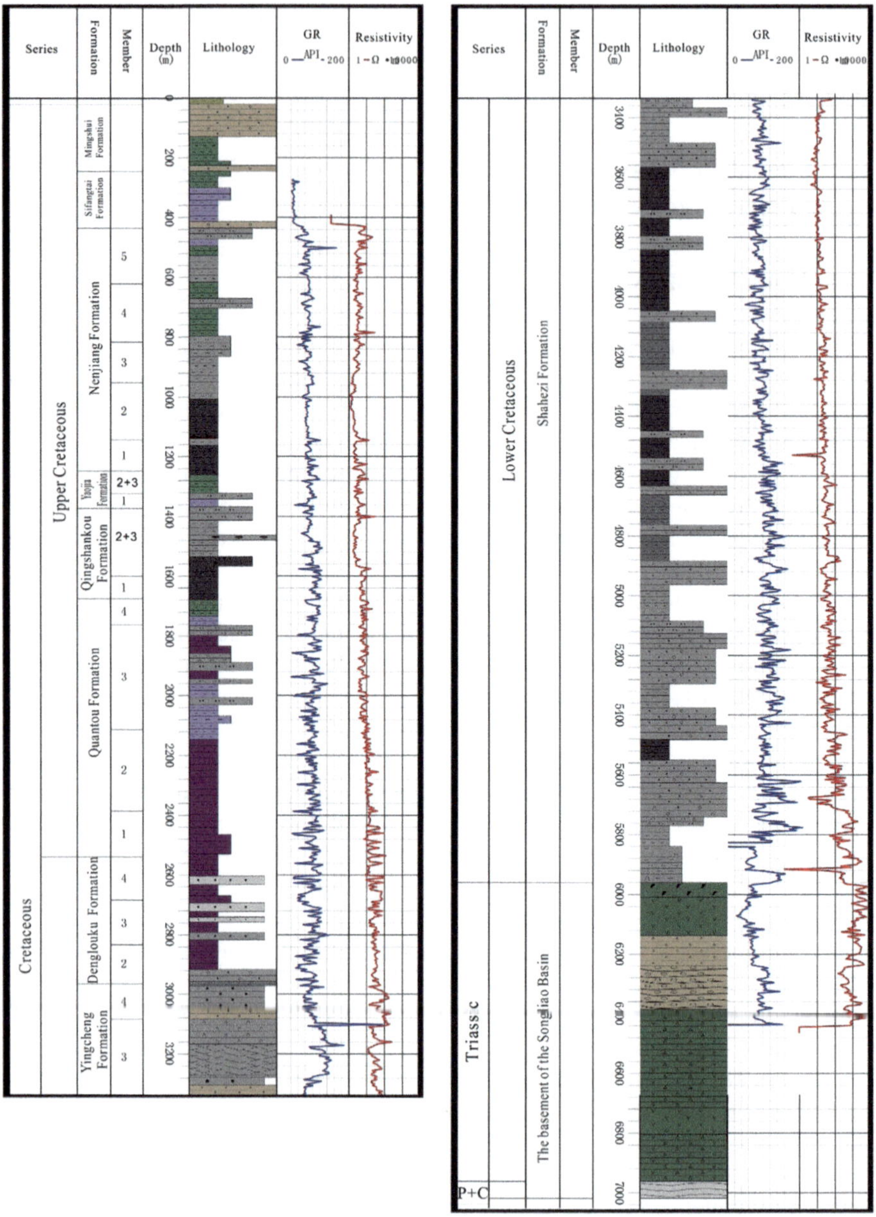

Fig. 8 Stratigraphic column of SK-2 core

The fifth member of Nenjiang Formation: Mauve mudstone-mottled mauve mudstone and grayish-green mudstone dominate, with subordinate grey siltstone, fine-grained sandstone, argillaceous siltstone, some grey mudstone and grayish-green mudstone. Ripple lamination and some charcoal debris occur in the sandstone. Mudstone is massively bedded. At sandstone mudstone contacts deformation features occasionally occur. The sediments of this formation were deposited in a shallow lake environment.

The fourth member of Nenjiang Formation: Unit 4 is mainly comprised of grey-colored fine-grained sandstone, siltstone, argillaceous siltstone and silty mudstone. Grey mudstone and greenish-grey mudstone are subordinate lihtologies. In sandstone, foreset cross-bedding, wavy bedding are present, with some trough bedding in siltstone and fine-grained sandstone. Local slumps and erosion surfaces occur. Sandstones enclose sparse charcoal debris. Sediments show several cycles with reversed grain-size gradation. The sedimentary environment was deltaic, specifically delta front.

The third member of Nenjiang Formation: In the middle to lower part of Unit 3, black and grey mudstone dominate, while in the upper part grey fine-grained sandstone, siltstone, argillaceous siltstone and silty mudstone are dominant. In the middle to lower part of the formation sedimentary structures present in sandstone are trough bedding, wave bedding, climbing ripples and deformation bedding. In the upper part, ripple lamination and wavy bedding occur in the sandstone. Dark mudstone is horizontally bedded. The deposition environment changed from semi-deep lake in the lower part to shallow lake in the upper part of Unit 3.

The second member of Nenjiang Formation: This is mainly black, horizontally bedded grayish-black mudstone interbeded with sandstone. At the base of Unit 2 is inferior oil shale, which is laminated and has an oily smell. In black mudstone, ostracoda, conchonstracan and other fossil fragments are present. In sandstone, charcoal debris is occasionally seen. Unit 2 is interpreted as being deposited in a semi-deep to deep lake environment.

The first member of Nenjiang Formation: The dominant lithology is dark-grey, medium-dark grey, olive-grey, dark greenish-grey mudstone, interbeded with thin-layered oil shale and ostracod-bearing mudstone. The depositional environment was deep lake.

The second and third member of Yaojia Formation: These units are dominated by greenish-grey, dark greenish-grey, grayish-brown mudstones with interbeded thin layers of dark greenish-grey, greenish-grey argillaceous siltstone and silty mudstone. The depositional environment was semi-deep lake to deep lake.

The first member of Yaojia Formation: The upper part of Unit 1 is comprised of greenish- grey and dark greenish-grey mudstone. Mudstone in the middle to lower part is mainly grayish-brown interbedded with dark greenish-grey argillaceous siltstone, and grayish-brown oil-bearing sandstone. Deposition occurred in a shallow lake with delta-front subfacies. The contact with underlying strata is disconformable.

The second and third member of Qingshankou Formation: The dominant lithology of these units is a dark grey, medium-dark grey, greenish-grey and darkgreenish-grey mudstone, interbedded with thin-layers of olive-grey oil shale and ostracod-bearing mudstone. Deposition occurred in a deep lake environment.

The first member of Qingshankou Formation: This unit is mainly comprised of dark grey, olive-grey and olive-black mudstone. In the lower part are thin layers of dark grey, medium dark-grey silty mudstone and argillaceous siltstone. The sediments were deposited in a deep lake environment. A disconformity at the base of Unit 1 separates the Qingshankou and Quantou formations.

The fourth member of Quantou Formation: This unit is mainly comprised by greenish-grey, dark greenish-grey, brownish-grey and grayish-brown mudstone interbeded with argillaceous siltstone, and some oil-bearing brownish-grey siltstone and fine-grained sandstone. Sediments were deposited on a flood plain with meandering river subfacies.

The third member of Quantou Formation (not fully penetrated): This unit is dominated by greenish-grey, dark greenish-grey, brownish-grey, grayish-brown mudstone and silty mudstone, with thin layers of greenish-grey fine-grained sandstone. The depositional environment was a meandering river flood plain.

Figures 6 and 7 are Stratigraphic columns of SK-1n and SK-1s boreholes.

3.2.2 Core Lithology of SK-2

In the SK-2 borehole, continuous coring recovered strata from the basement of the Songliao Basin (early Mesozoic to Paleozoic) to the base of the Denglouku Formation (Early Cretaceous). The first and second member of Nenjiang Formation were also cored in SK-2. Lithologic characteristics of each formation are summarized below and the formations are described in stratigraphically descending order.

The first and second member of Nenjiang Formation: The main lithology of these units is gray black mudstone, dark gray mudstone, gray mudstone, olive gray oil shale, light olive gray dolomite and gray white volcanic ash.

The base of Denglouku Formation: The main lithology of this unit is brownish gray mudstone, red gray mudstone, silty mudstone interbedded with light gray argillaceous siltstone, green gray argillaceous siltstone, green gray siltstone and green gray sandstone.

Yingcheng Formation: In this unit, the upper part is conglomerate, sandstone interbedded with siltstone. The middle part is rhyolite with rhyolitic breccia lava and a small amount of andesitic breccia lava. The lower part is conglomerate mixed with glutenite and sandstone.

Shahezi Formation: In this unit, the upper part is mainly silty mudstone and mudstone, interbedded with fine sandstone and medium sandstone, showing a coarsening upward trend. The middle and lower parts are dominated by dark gray mudstone and silty mudstone, interbedded with sandstone, conglomerate and coal.

Huoshiling Formation: In this unit, the upper part is mainly composed of sedimentary pyroclastic rocks and pyroclastic sedimentary rocks, the middle part is dominated by intermediate-basic volcanic rocks, and the lower part is composed of a set of conglomerate.

The basement of the Songliao Basin (early Mesozoic to Paleozoic): In this unit, the upper part is characterized by shallow metamorphic glutenite, the middle and lower part is composed of volcanic breccia with lava, and the base is composed of dynamic metamorphic rocks.

3.2.3 Core Lithology of SK-3

In the SK-3 borehole, continuous coring recovered strata from the Denglouku Formation to the third member of Quantou Formation. Lithologic characteristics of each formation are summarized below and the formations are described in stratigraphically descending order.

The third member of Quantou Formation: In this unit, the lithology is mainly siltstone, argillaceous siltstone, silty mudstone and mudstone. Colors of mudstone include purple red, purple gray, gray purple; siltstone is mainly gray in color; argillaceous siltstone is mainly in gray; silty mudstone is mainly in gray purple, purple gray (Fig. 9).

The second member of Quantou Formation: In this unit, the lithology is mainly siltstone, fine sandstone, argillaceous siltstone, sandy mudstone and mudstone. Mudstone is mainly purple gray and gray purple in color, siltstone in gray, argillaceous siltstone in gray, and silty mudstone in gray purple and dark purple.

The first member of Quantou Formation: In this unit, the lithology is mainly siltstone, fine sandstone, argillaceous siltstone, sandy mudstone and mudstone. Mudstone is mainly purplish red in color, siltstone and argillaceous siltstone in gray, and silty mudstone mainly in gray purple.

Denglouku Formation: In this unit, the lithology is mainly glutenite, fine sandstone, pebbly fine sandstone, silty mudstone and mudstone. Mudstone is mainly purplish red in color, glutenite in gray and varying colors, fine sandstone and pebbly fine sandstone in gray, and silty mudstone in gray and grayish purple.

3.3 Chronological Framework of SK Cores

The study of geochronology is the basis for solving various geological problems. Multiple chronostratigraphic studies including biostratigraphy, magnetostratigraphy, radioisotope geochronology, and cyclostratigraphy were carried out on the continental scientific drilling cores of the Songliao Basin, and a high-precision chronostratigraphic framework was established. The comprehensive chronological framework of scientific drilling in the Songliao Basin shows that age span of the core is ~242–65 Ma, whose age of the continuous core in Cretaceous is ~118–65 Ma, and the chronological resolution based on cyclostratigraphy has reached the level of 100,000 years. According to the age constraints of paleontology, magnetostratigraphy and cyclic stratigraphy, the K-Pg boundary is located in the member 2 of the Mingshui Formation.

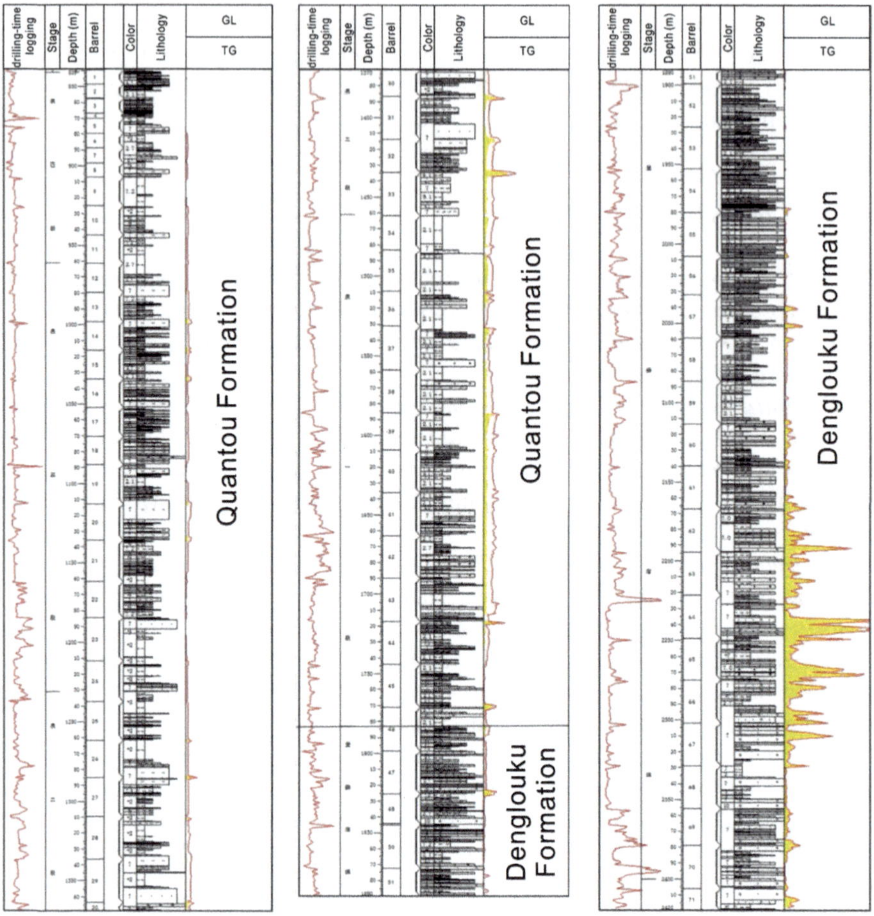

Fig. 9 Stratigraphic column of SK-3 core

3.3.1 Biostratigraphy

The research of scientific drilling biostratigraphy in Songliao Basin is mainly concentrated in SK-1, and the research involves ostracods, sporopollen, chara, foraminifera, calcareous nannofossil and other biological categories. The fossils of SK-2 is poorly preserved due to the large burial depth. The biostratigraphic research of SK-3 is underway. This section summarizes the biostratigraphic research of SK-1.

Ostracods: The core of the SK-1s was processed and analyzed according to the standard of 1 m spacing and 100 g samples. Abundant and well-preserved ostracod fossils were obtained. More than 87 species of fossils were identified in 25 genera, and whose distribution in stratums are: The member 3 of Quantou Formation to the member 2 of Mingshui Formation (Wan et al. 2013). According to the combination of fossils and vertical changing characteristics, it is divided into 21 fossil assemblages, and the age is from Late Cretaceous to Early Paleocene. The evolutionary law of ostracods reflects that with the changes in the salinity, turbulence, and water depth of the lake, the degree of differentiation, individual size, and number of preserved fossils of ostracods have also changed. The overall conclusion is that during the development period of the basin, the lake was wide and the lives were prosperous. Especially during the sedimentary period of the member 1 of the Qingshankou Formation and the members 1 and 2 of the Nenjiang Formation, three largest lake transgression events occurred in the history of basin evolution, which triggered the appearance of anoxic environment in the whole basin, and led to the renovation and prosperity of subsequent lacustrine bios.

Palynology: The palynological fossil assemblages are important indicators of biostratigraphic division. The palynological samples were collected at intervals of 15–25 m between mudstone and siltstone, and about 5–10 m between mudstone and shale. A total of 154 samples have been analyzed. Powder analysis (Li et al. 2011). Synthesizing morphology, research level and other factors, through the production of some angiosperm pollen, gymnosperm pollen, and fern palynological, the SK-1 reveals the formation of 8 palynological zones. After comparing the palynological zones of SK-1 with the international terrestrial Cretaceous palynological strata, the following conclusions can be drawn: (1) The age of the upper members 3 and 4 of the Quantou Formation is from the Late Turonian to the Coniacian, Qingshankou Formation and Yaojia Formation are the main component that should have been formed during the Coniacian to Santonian; (2) A suspecting end-Cretaceous extinction event was discovered. A large number of Cretaceous palynological groups disappeared in this layer and replaced by very monotonous Cenozoic palynological groups, which is located in the upper part of the member 2 of Mingshui Formation.

Charophyte: The core in the SK-1n was sampled at an average of 1 m intervals and more than 1100 samples were prepared for the study of Charophytes. Li et al. (2019) redefined four charophyte biozones and one superzone in the SMF. *Atopochara trivolvis ulanensis* Zone (964.57–813.57 m) in the Sifangtai Formation is defined between the FAD (first appearance datum) of *Atopochara trivolvis ulanensis* and the FAD of *Microchara gobica*, with totally six species. *Microchara cristata* Superzone includes three biozones: the *Microchara gobica*, *Microchara prolixa* and *Peckichara sinuolata* zones. *Microchara gobica* Zone is located in the upper part of Sifangtai formation and lower part of Mingshui Formation (813.57–521.71 m), which is defined between the FAD of *Microchara gobica* and the FAD of *Microchara prolixa*, with totally seven species. *Microchara prolixa* Zone (521.71–329.06 m) in the middle part of Mingshui Formation is defined between the FAD of *Microchara prolixa* and the FAD of *Peckichara sinuolata*, with totally eleven species. The base of *Peckichara sinuolata* Zone in the upper part of Mingshui Formation is defined by the FAD of *Peckichara sinuolata*, with totally ten species, the top of this zone is undefined. The first appearance of *Peckichara sinuolata* at 329.06 m in chron C29 has been proposed as the basal marker of the Paleocene (Li et al. 2019).

Foraminifera: A total of 9 species in 8 genera of foraminiferal fossils have been found in the members 1 and 2 of the Nenjiang Formation in the SK-1s. The habits of foraminifera are mainly benthic and planktonic, whose age is from the Late Cretaceous Coniacian period to Santonian period, and indicate that seawater invaded the Songliao Basin in the early depositional stage of the Nenjiang Formation (Xi et al. 2011).

3.3.2 Radioisotope Geochronology

Yin et al. (2019) used (LA-ICP-MS) U–Pb geochronology to carry the U–Pb isotope dating on the samples of andesite at 6031.9 m, sandstone at 6286.2 m and andesite at 6373.9 m from SK-2. The sample at 6031.9 m shows that the formation age of this layer is 242.4 ± 2.1 Ma; the age-weighted average of zircon samples at 6373.9 m is 242.6 ± 1.5 Ma; the peaks of several ages of sandstone at 6386.2 m show that its age is between 240–2546 Ma. The youngest peak age is 242.2 Ma. Therefore, it is inferred that the volcanic-sedimentary strata in the basement of the Songliao Basin were formed in the Triassic.

A layer of sedimentary tuff with a thickness of nearly 1m was found at the bottom of the Shahezi Formation at 5943.19 m in SK-2, and a layer of rhyolite crystal tuff with a thickness of about 6 mm was found at 5958.62 m. Liu et al. (2021) used LA-ICP-MS U–Pb dating of the zircon samples in it, which shows that the time when the Shahezi Formation began to deposit is 118.2 ± 1.5 Ma. A 5–7 mm thick tuff layer was found at 3961 m in SK-2. Yu et al. (2020) conducted SIMS U–Pb dating analysis on zircon samples in it, and found that the eruption time of tuff was 113.9 ± 0.9/1.5 Ma, the sedimentary age of the upper part of the Shahezi Formation is about 114 Ma.

He et al. (2012) collected 4 bentonite layers from the 1780, 1705, 1673, and 1019 m volcanic ash interlayers of the SK-1s for SIMS U–Pb zircon analysis, and obtained the weighted average age of the sample at 1780 m is 90.2 ± 0.6 Ma, the weighted average age of the sample at 1673 m is 90.4 ± 0.6 Ma, and the weighted average age of the sample at 1019 m is 83.7 ± 0.5 Ma. Wang et al. (2016d) conducted ID-TIMS high-precision U–Pb. Wang et al. (2016d) used the zircon grains that had been analyzed by He et al. (2012) through SIMS U–Pb dating to analyze the samples of 1780, 1705, 1673, and 1019 m through ID-TIMS for high-precision U–Pb isotope dating and gained more accurate results. The weighted average age of the sample at 1780 m is 91.886 ± 0.11 Ma, the weighted average age of the sample at 1705 m is 90.974 ± 0.12 Ma, and the weighted average of the sample at 1673 m. The age is 90.536 ± 0.12 Ma, and the weighted average age of the sample at 1019 m is 83.269 ± 0.11 Ma.

3.3.3 Magnetostratigraphy

Through the high-resolution magnetic stratigraphic study of scientific drilling in the Songliao Basin, a magnetic polarity sequence can be established, and compare it with the geomagnetic polarity chronology to establish a core chronology framework and achieve an accurate comparison of the continental Cretaceous strata in the Songliao Basin with the global marine strata. At present, the research results of magnetostratigraphy are concentrated in the SK-1. The north broehole of SK-1 recorded 11 magnetic polarity zones and the south borehole recorded 8 polarity zones. According to the characteristics of the Cretaceous paleomagnetic field and the anchor point of the zircon U–Pb chronology of SK-1, the north borehole recorded the deposition between the C29r negative polarity and the C34n positive polarity (65.6–83 Ma), the south borehole recorded the deposition between the negative polarity of C33r and the positive polarity of C34n (79.1–92 Ma), and the K/Pg boundary is located in the upper part of the member 2 of the Mingshui Formation (Deng et al. 2013).

3.3.4 Cyclostratigraphy

Cyclostratigraphy is an important component of stratigraphy, aiming to decipher the rhythmic features, which controlled by the orbital forcing, recorded in the sedimentary sequence. The SK-1 is the first scientific drilling core recovered the complete terrestrial Cretaceous strata. It is of great scientific significance to carry out high-precision astronomical chronology research and establish the Cretaceous astronomical time scale in the Songliao basin.

Wu et al. (2013) conducted a detailed cyclostratigraphic study of natural gamma-ray (GR) log, thorium (Th) log, and magnetic susceptibility (MS) data from the SK-1s core. According to the robust spectral analysis, all Milankovitch cycles have been recognized. An astronomical time scale (ATS) was established by tuning interpreted 405 kyr cycles to a 405 kyr orbital eccentricity target curve, and anchored to four SIMS U–Pb zircon radioisotope ages. The ATS provides a continuous 'absolute time framework' from 83.40 to 92.08 Ma (Early Turonian–Early Campanian). Afterwards, Wang et al. (2016d) provided new high-precise ID-TIMS U–Pb ages for the tuff layers, and recalibrated the ATS for the SK-1s core, which yield the SK-1s age framework range from 83.90 to 92.58 Ma, and simultaneously estimating the age of the termination of the Cretaceous Normal Superchron is expected to be 83.07 $\pm$ 0.15 Ma.

The SK-1n borehole recovered a continuous, 1541.66 m Late Santonian–Early Danian terrestrial succession in Songliao Basin (SB), northeastern China. Wu et al. (2014) used Th logging data as a paleoenvironmental and paleoclimatic proxy to conduct a detailed cyclostratigraphic study on the SK-1n core. Based on the spectral analysis, all Milankovitch cycles have been recognized. The ATS of the SK-1n, which is tuned to the astronomical solution La2010d, provides a high-resolution time framework from Late Santonian to Early Danian ($\sim$84–65 Ma) and estimates the positions of the Cretaceous-Paleogene (K–Pg) boundaries at depths of 318 m with an age of 66 Ma.

3.3.5 Comprehensive Chronological Framework of Continental Scientific Drilling of Cretaceous Songliao Basin

In summary, the establishment of a comprehensive chronological framework for scientific drilling in the Songliao Basin is based on multiple chronostratigraphic studies (Fig. 10). The age of the basement and Huoshiling Formation is Triassic or even older. The bottom of the Huoshiling Formation reaches 242 Ma. The age of the Shahezi Formation is ~118–112 Ma in Early Cretaceous, and the age of Yingcheng Formation is ~112–102 Ma of Early Cretaceous. The age of the Denglouku Formation and Quantou Formation is estimated to be ~102–92 Ma. The age of Qingshankou Formation is ~92–86 Ma. The age of Yaojia Formation is 86–84 Ma. The members 1 and 2 of Nenjiang Formation are 84 to ~82 Ma, the members 3–5 of Nenjiang Formation are ~82–79 Ma. The age of Sifangtai Formation is 79–72 Ma, and the age of Mingshui Formation is 72 to ~65 Ma.

3.4 Research Progress of Continental Scientific Drilling of Cretaceous Songliao Basin

Through scientific research of more than ten years, significant progress has been made in terrestrial climate and environment in Cretaceous, tectonic evolutions and petroleum exploitation. Detailed information on these scientific progresses is omitted here but will be introduced in the following chapters.

4 Basement of the Songliao Basin from SK-2 Core

The basement of the Songliao Basin represents strata unconformably underlying the Mesozoic-Cenozoic strata, which includes Carboniferous to Permian meta-sedimentary strata, Paleozoic to Mesozoic plutonic rocks, and pre-Cambrian crystalline rocks (Wang et al. 1995, 2009a, b; Zhang et al. 2008). In the Songliao Basin, the existence of lower Mesozoic strata remains a debatable issue. Previous studies indicated the absence of Triassic to Lower and Middle Jurassic strata in northeastern China because of uplift and erosion events associated with the return of geo-synclinal folds and orogenic movement during the Late Permian–Early Jurassic. To date, geochronological studies of intrusive and metamorphic rocks in the basement of the Songliao Basin have also confirmed Carboniferous, Permian, and Late Jurassic ages for the basement formations in general (Gao et al. 2009). Based on radiometric dating on core samples of SK-2 scientific drilling, Triassic volcanic-sedimentary strata in the basement of the Songliao Basin have been discovered (Yin et al. 2019).

Triassic volcanic-sedimentary strata revealed by the SK-2 east borehole consist of andesitic volcanic breccias at the bottom; andesites, sandstones, and conglomerates

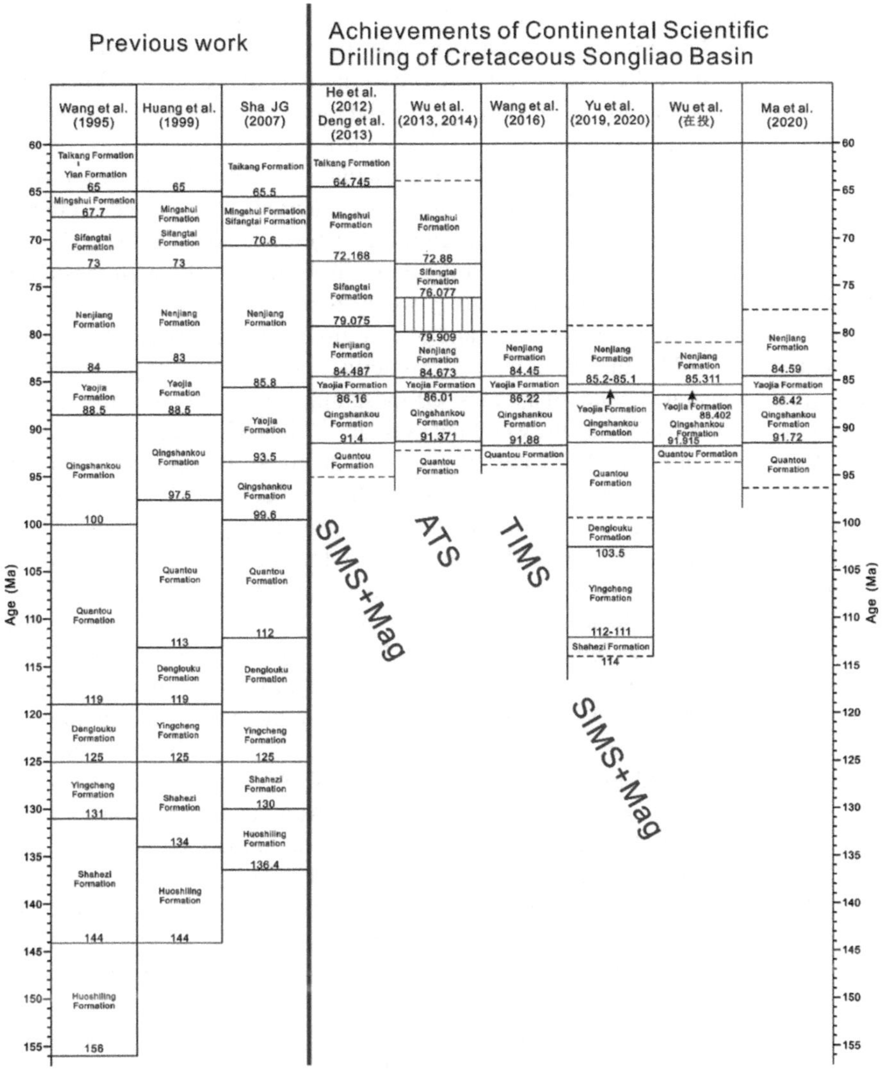

Fig. 10 Evolution of chronological framework of SK cores

in the middle; and andesites at the top. The total thickness of these strata is over 500 m. The main mineral components of the andesite at 6031.9 m are magnetite (5% of the total minerals), amphiboles (10% of the total minerals) and plagioclases (85% of the total minerals). The main mineral components of the andesite at 6373.9 m are magnetite (10% of the total minerals), amphiboles (altered) (15% of the total minerals), and plagioclases (75% of the total minerals). Porphyritic and hyalopilitic textures are developed in the andesites at 6031.9 and 6373.9 m. The main components of the sandstone at 6286.2 m are feldspars (10% of the total minerals), lithic fragments

(15% of the total minerals), and quartz (85% of the total minerals). The structure is medium-grained with moderate sorting and poor roundness (Yin et al. 2019).

The zircons from the sandstone at 6286.2 m are euhedral crystals with various diameters from 60 to 150 μm and a range of axial ratio from 1:1 to 2:1. The zircons from the andesites at 6031.9 and 6373.9 m are euhedral crystals or subhedral crystals with assorted diameters from 45 to 100 μm and a range of axial ratios from 1:1 to 3:1. Zircons from the two andesite samples show clear internal structure and oscillating growth rings under cathode luminescence (Fig. 11d). Based on the relatively high Th/U ratios of the zircon samples and the above information, these zircons are of magmatic origin. For U–Pb isotope dating, twenty zircon crystals were selected from the andesite at the depth of 6031.9 m, and twenty zircon crystals were selected from the andesite at 6373.9 m. In the 6031.9 m interval, nine zircon samples from this andesite demonstrated concordant ages. Of these nine zircon samples, the isotopic ages of seven zircons showed the weighted average $^{206}Pb/^{238}U$ age of 242.4 ± 2.1 Ma (MSWD = 0.06, n = 7). The $^{207}Pb/^{206}Pb$ ages of the other two zircons revealed the concordant ages of 1917 ± 36 Ma and 2463 ± 33 Ma, respectively (Fig. 11d). Among the reliable isotopic age data above, the dominant isotopic age (n = 7) of 242.4 ± 2.1 Ma is inferred to represent the formation age of this interval. The remaining two isotopic age data characterize the ages of zircons captured by the andesite. In the 6373.9 m interval, eighteen zircon samples from the andesite displayed concordant ages, which showed a weighted average of $^{206}Pb/^{238}U$ age 242.6 ± 1.5 Ma (MSWD = 1.02, n = 18) (Fig. 11d). This date is inferred to represent the formation age of this interval. The isotopic analysis results demonstrate that these two andesite intervals formed during the Triassic. One hundred and ten zircon crystals were randomly selected from the sandstone of the 6286.2 m interval. All the zircon samples from this sandstone show concordant ages, with an age distribution between 240 and 2546 Ma (young zircons (<10 Ga) show $^{206}Pb/^{238}U$ ages and old zircons (>10 Ga) show $^{207}Pb/^{206}Pb$ ages in Fig. 11d) with several age peaks. The youngest peak age is 242.2 Ma, which indicates that the Triassic source rocks were the youngest sediment supplier during the development of the sandstone (Fig. 11d). Because the sandstone is sandwiched between the two sets of andesite, we infer that the chronologic age of this interval is likely Triassic.

Studies from SK-2 core demonstrate that in the Songliao Basin, there are not only Carboniferous and Permian strata, but also a Triassic volcanic-sedimentary succession in the basement of the basin. The SK-2 drilling core reveals that this volcanic-sedimentary sequence has great thickness. These Triassic volcanic-sedimentary strata provide new clues for the study of the origin and development of the Songliao Basin. As both volcanic and sedimentary rocks can be oil and gas reservoirs, this discovery also provides a new target for oil and gas exploration deep in the Songliao Basin.

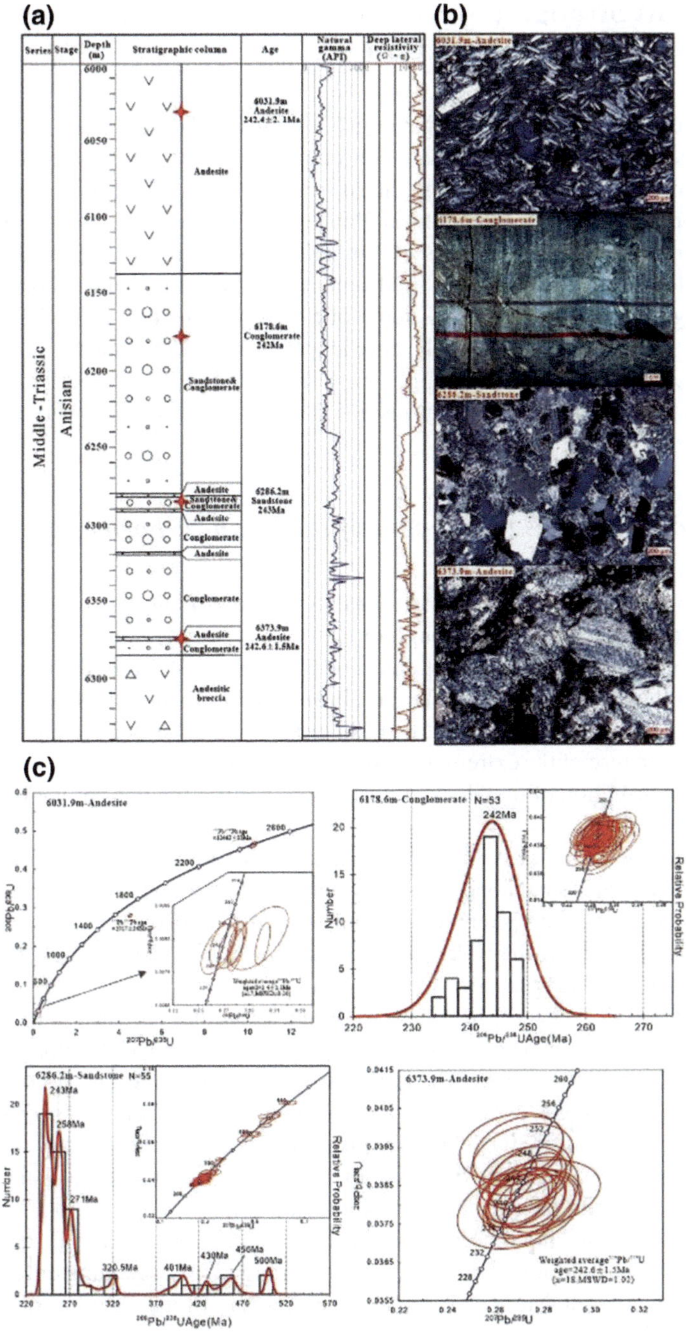

Fig. 11 Triassic volcanic-sedimentary strata and age data of the Songliao Basin revealed by the SK-2 borehole. **b** Triassic volcanic-sedimentary strata revealed by the SK-2 east borehole. **c** Photomicrographs (cross-polarized light) of zircon U–Pb dating samples. **d** Zircon U–Pb dating concordia diagrams and probability density diagrams with histograms. Modified from Yin et al. (2019)

5 Syn-rift Stratigraphy (Shahezi Formation-Yingcheng Formation) from SK-2 Core

The Shahezi Formation and Yingcheng Formation were deposited in the syn-rift extensional faulting stage. The Shahezi Formation is a thick continental clastic sedimentary sequence deposited during the rift period of Songliao Basin. It is underlain by conglomerate and volcanic rocks of Huoshiling Formation and overlying by sedimentary rocks and volcanic rocks of Yingcheng Formation (Wang et al. 2016a, b, c, d). Drilling of SK-2 has recovered continuous core of Shahezi Formation and Yingcheng Formation. The coring depth of Shahezi Formation is 3335.99–5695 m, the coring footage is 2358.41 m, the core length is 2296.17 m, and the core recovery is 97.36%. The coring depth of Yingcheng Formation is 2965–3335.99 m, the coring footage is 370.99 m, the core length is 365.32 m, and the core recovery is 98.47%.

5.1 Sedimentary Environment of the Shahezi Formation

SK-2 is located in the northern part of the Xujiaweizi fault depression zone in the central depression area of the Songliao Basin. It is mainly controlled by the Xuxi fault among the three NW-trending faults (Chen et al. 2014). The Shahezi Formation belongs to the filling deposit of the fault basin. The top burial depth of the Shahezi Formation has a large range (between 1000–6000 m, and the average top burial depth is 3900 m) because of the extremely developed structural fault depressions. The layer thickness distribution of the Shahezi Formation in different areas of the basin is also different.

The stratum granularity of the Shahezi Formation generally becomes finer upwards. The coarse sandstone and variegated conglomerate are at the bottom, which transforms upwards into interbedded sandstone and mudstone with unequal thickness, black mudstone with black coal line, thin gray fine sandstone, dark gray powder sandstone and argillaceous siltstone. The upper strata are in unconformable contact with the underlying strata (Liu 2020a, b). Previous studies revealed that the Shahezi Formation in Xujiaweizi Sag has gone through the entire rift development period during the deposition process. The upper part of the strata presents an overall trend of reverse cycles. However, the lower part presents an overall trend of positive cycles, during which several small cycles have developed. For the sedimentary facies of the Shahezi Formation, previous conclusions mainly include fan delta facies, braided river delta facies and lacustrine facies (Cai et al. 2015).

An (2018) divided the lithofacies of the upper part of the Shahezi Formation. The proportions of each lithological thickness are arranged in descending order. They are mudstone, coarse sandstone, silty mudstone, fine sandstone, argillaceous siltstone, medium sandstone, fine conglomerate, medium conglomerate, coarse conglomerate and coal seams (Figs. 12 and 13). The section of 4510.01–4359.96 m belongs to shore-shallow lacustrine deposits, with thick mudstone and silty mudstone. The

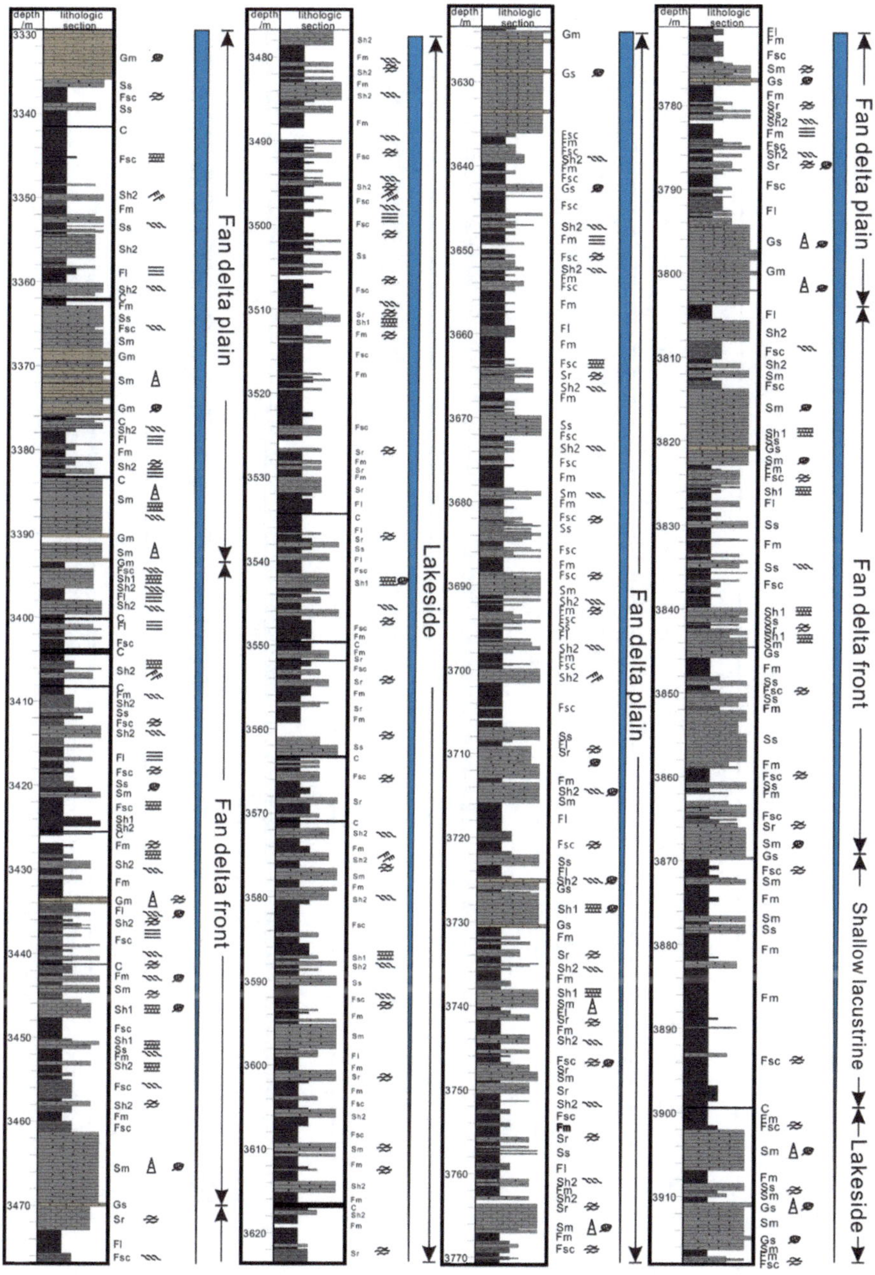

Fig. 12 Stratigraphic column of Shahezi Formation in SK-2 (3915–3330 m)

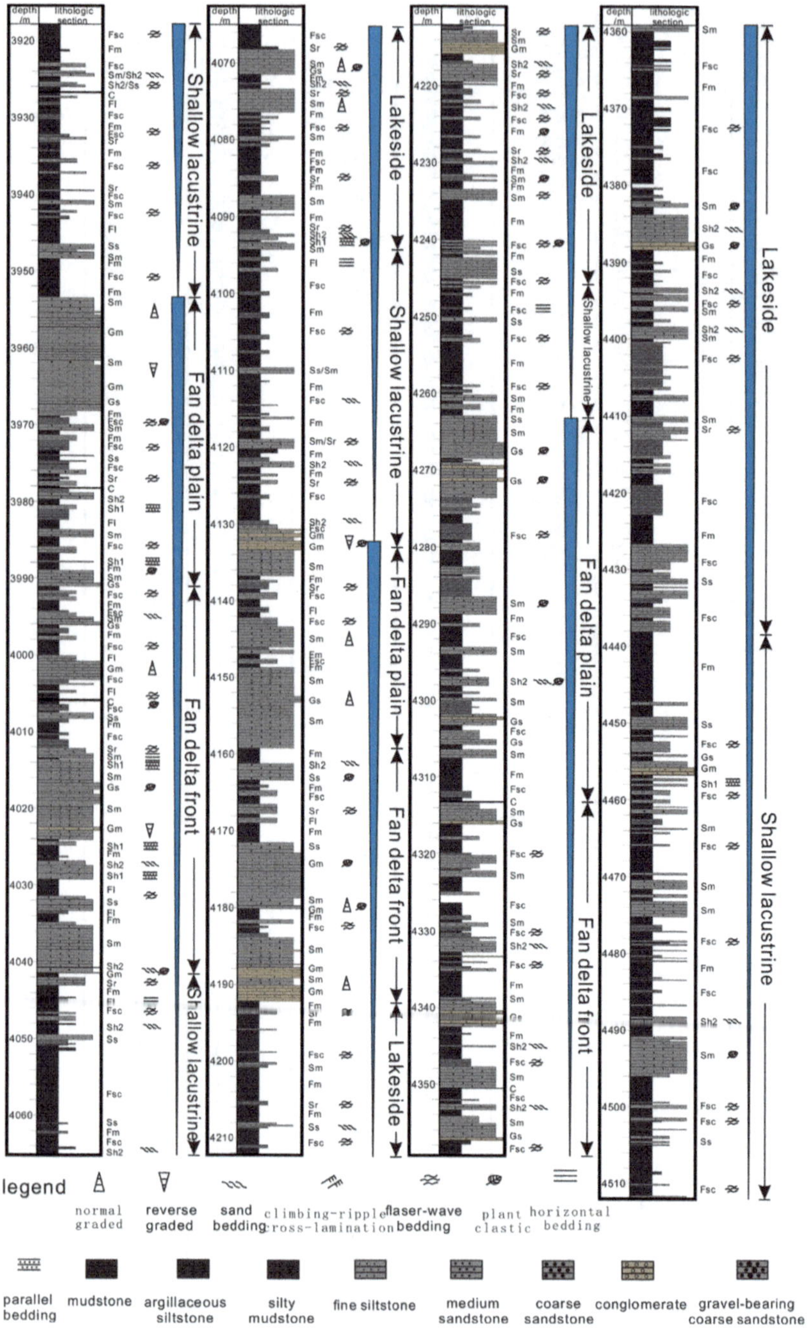

Fig. 13 Stratigraphic column of Shahezi Formation in SK-2 (4510–3915 m)

average thickness of the sandstone layer increases significantly with the increase in depth. The conglomerate occurs locally as a lens, which is a relatively deeper extension from underwater distributary channel to the lake water body (containing plant debris). The section of 4359.96–4262.98 m belongs to the fan delta facies deposits, with the development of conglomerate caused by grain flow and the interlayered mudstone and sandstone, which belong to underwater distributary channel and fine-grained sediments between channels. The section of 4313.29–4262.98 m is a subfacies of delta plain, with relatively thick conglomerate and gravel-bearing coarse sandstone, and fine-grained mudstone constitutes a braided distributary channel on the delta plain. Furthermore, the microfacies of the floodplain, in which the coal seams still developed, may be swamps or abandoned river microfacies. The section of 4262.98–4192.23 m belongs to shore-shallow lacustrine facies. The mudstone in the low-energy fluid environment accounts for a large proportion. However, the sandstone accounts for the proportion is relatively small. Some conglomerates can also be seen, which are sublacustrine fan deposits. The section of 4192.23–4130.52 m belongs to fan delta facies deposits, with finer grain size from bottom to top. It is an underwater distributary channel and finer sediments (contain plant debris) between channels. The thick layers of sandstone and conglomerate develop upward with relatively small proportion of fine-grained sediments. The braided distributary channels are relatively developed. The wavy bedding and positive-grain bedding can be seen in the braided distributary channels. The section of 4130.52–4041.36 m belongs to shore-shallow lacustrine deposits, with thick mudstone intercalated layered sandstone. The section of 4041.36–3953.44 m belongs to fan delta facies deposition. In the lower shore-shallow lacustrine facies deposition of this section, the grain size of the sediments is obviously thicker. The proportion of sandstone and conglomerate is obviously larger, which showing the development of thin coal seams. The section of 3953.44–3869.84 m belongs to shore-shallow lacustrine facies deposits, with thick mudstone intercalated with thin fine-grained sandstones, well-developed coal seams, visible wavy bedding and sand-line bedding, which belonging to lake marsh deposits. The 3869.84–3623.95 m section is fan delta facies deposits, with a relatively low proportion of fine-grained sediments, mainly developed with multiple thick layers of glutenite. The section of 3623.95–3469.74 m belongs to lakeside deposits, mainly fine-grained sediments. The sandstone with coarser grains accounts for a relatively small proportion, and no conglomerate facies is seen in this section. The wave-vein bedding, climbing sand bedding, and parallel bedding can be seen in this section. The section of 3469.74–3330.90 m belongs to the fan delta facies deposits. The grain size has a relatively obvious upward trend of coarsening as a whole, with positive-grained sequence bedding, climbing sand bedding, wavy-vein bedding, and parallel bedding. Horizontal bedding is rare.

The lower part of the Shahezi Formation is dominated by dark colors. The bottom part of the Shahezi Formation is dominated by coarse sandstone and variegated conglomerate. This formation transformed upward into interbedded sandstone and mudstone with unequal thickness, black mudstone interposed with black coal lines, interbedded gray with thin fine sandstone layers, dark gray siltstone, argillaceous siltstone (Zhao 2020). The lithology includes coarse conglomerate, medium conglomerate, fine conglomerate, coarse sandstone, medium sandstone, fine sandstone, silty mudstone, argillaceous siltstone, black mudstone and thin coal seams. Among them, conglomerate accounted for approximately 5%, coarse sandstone accounted for approximately 13%, medium sandstone accounted for 6%, fine sandstone accounted for 7%, and silty mudstone, argillaceous siltstone and mudstone accounted for approximately 64%. The conglomerate in the lower part of the Shahezi Formation can be divided into two major categories. One is the massive conglomerate dominated by mud and gravel. The other is the massive conglomerate dominated by quartz conglomerate with bad sorting. The roundness is sub-angular to sub-circular, without obvious directional arrangement. The maximum gravel diameter can reach 200 * 100 mm. The sedimentary rocks in the lower part of the Shahezi Formation drilled by SK-2 mainly developed wave-like bedding, deformed bedding, climbing wave-mark bedding, parallel bedding, and well-developed erosion surfaces. Thin coal seams and biological disturbance structures are found in many places. Among them, it can be found that the liquefied sandstone layer structure may be caused by paleo-seismic activity. It is shown that the sedimentary facies are mainly fan deltas facies, shore-shallow lake facies, and deep lake facies based on the tectonic evolution stage of the Shahezi Formation and the characteristics of lithological combination (Fig. 14).

5.2 *Chronology and Provenance of the Shahezi Formation*

Some results have been obtained by biostratigraphy, chronostratigraphy and magnetic stratigraphy to constrain the age of the Shahezi Formation. Gao et al. (1999) showed that fossils of *Klukisporites, Cicatricosisporites* and *Aequitriradites* from the Shahezi Formation belong to the Early Cretaceous. An (2018) obtained an LA-ICP-MS U–Pb age of 111.13 ± 0.84 Ma as the maximum depositional age of sandstone at the top of the Shahezi Formation. Afterwards, Yu et al. (2020) obtained a SIMS U–Pb age of 113.9 ± 0.9 Ma from the 3961 m tuff layer in the upper part of the Shahezi Formation in SK-2. Liu et al. (2021) obtained detrital zircon ages from two tuff layers at the bottom of Shahezi Formation at 5943.19 m and 5,958.62 m, which indicates maximum depositional ages 117.9 ± 1.6 Ma and 118.2 ± 1.5 Ma respectively (Fig. 15).

An (2018) analyzed the upper source of the Shahezi Formation through detrital zircon U–Pb dating method. The 105–140 Ma detrital zircon ages in the upper part of the Shahezi Formation, obtained from samples in SK-2 core, can be matched with the central Daxing'an Range granites. The 155–200 Ma ages can be matched with the central Daxing'an Range Middle Jurassic granites. Ages of 220–260 Ma and a few around 300 Ma correspond to the Daxing'an Range central granites. It is inferred that the central part of the Daxing'an Range provided the source for the upper part of the Shahezi Formation clastic sandstone (Fig. 16).

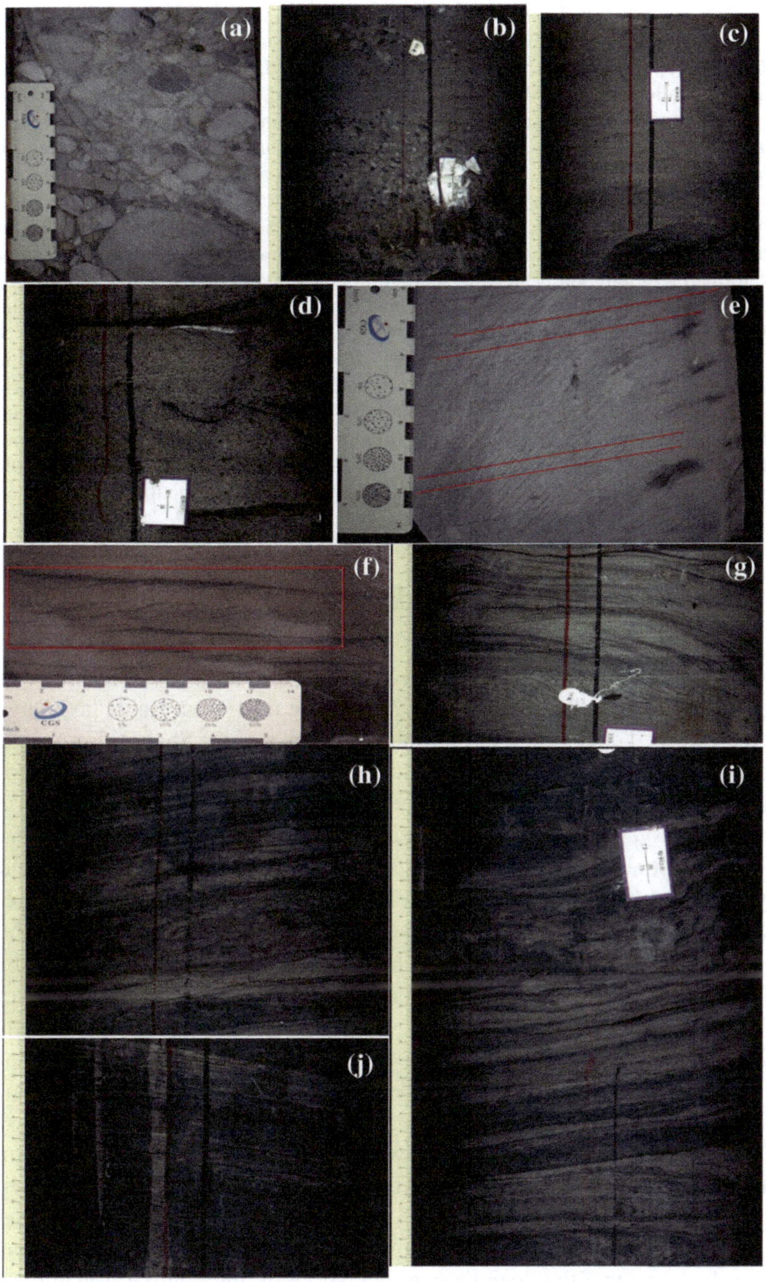
(a)
(b)
(c)
(d)
(e)
(f)
(g)
(h)
(i)
(j)

◀**Fig. 14** Core photos of the Shahezi Formation. **a** Clastic grains supports conglomerate, not obviously directional, poorly sorted, rounded sub-angular rounded, generally formed in braided divergent channels or alluvial fans. **b** Miscellaneous base-supported, thinly bedded conglomerate, poorly sorted, rounded to sub-angular-sub-rounded, occurring in thin layers, sandwiched between sandstone layers, as an extension of the submerged diversion channe. **c** Very fine-coarse sandstone with gravel. **d** Fine-coarse sandstone, poorly sorted, fine-coarse sandstone, clastic flow deposits, visible with plant debris (black). **e** Fine-coarse sandstone, parallel laminated, formed in relatively hydrodynamic environments, such as riverbeds or lakeside environments, with high flow regimes characterized by faceted bottom-flow deposits. **f** Very fine-medium sandstone, sand grain laminations, formed by migration of flowing sand grain under weak traction flow conditions. **g** Very fine-medium sandstone with creeping sand laminae, where the ripple textures appear superimposed on each other as the water ripples migrate forward and grow upward at the same time. **h** Very fine-coarse sandstone, vein-like-wavy laminations, dark muddy sediments are distributed in vein-like bodies among gray sandy sediments, formed under alternating strong-weak hydrodynamics, lenticular sandy bodies or muddy bands are visible. **i** Siltstone, mudstone, undulating laminations, sandy in mudstone, formed under alternating strong–weak hydrodynamic action, quasi-syngenetic deformation is visible. **j** Siltstone, mudstone, grainy laminations, small corrugated laminations, dark mudstone, formed under weak hydrodynamic conditions, developed in floodplains, abandoned rivers and lakes with relatively deep water bodies

5.3 *Oil and Gas Resources of the Shahezi Formation*

SK-2 encountered dark mudstone in the Shahezi Formation and Huoshiling Formation with a cumulative thickness of 1359.37 m. The maximum single layer thickness reached 23.21 m. The fine-grained sedimentary rocks of the Shahezi Formation with the most content mainly include silty mudstone, argillaceous siltstone and mudstone, most of which are darker in color. This is also the reason why the Shahezi formation is a potential oil-generating layer in the deep part of the Songliao Basin. The samples tested by Hou et al. (2018) showed that the organic carbon value of dark mudstone mostly exceeded 1.0% and the highest temperature (T_{max}) of rock pyrolysis reached 505 °C. Based on above, it is inferred that the strata rich in organic matter in the deep depression zone of the fault depression has a large cumulative thickness and a high abundance of organic matter. It is in a high-mature to over-mature thermal evolution stage.

SK-2 reveals that the deep mud shale strata have good gas-bearing properties. These strata have good mechanisms of deep heat conduction and shallow heat accumulation structure. The geothermal resources are well-occurring. Hou et al. (2018) reveals that 118 gas logging anomalies were encountered in the Shahezi Formation and Huoshiling Formation (3350 m). The cumulative thickness is 268 m with the average gas logging anomaly ratio is 8.33. The percentage of methane accounts for 96.8% of total hydrocarbon hydrocarbons. The lithology is mainly mudstone and glutenite. It shows that the deep source rock strata in the Songliao Basin have self-generation and self-accumulation conditions. The results of multiple temperature loggings of SK-2 showed that this basin has a large geothermal gradient, with a terrestrial heat flow value of 84 mW/m^2. The geothermal abnormality is obvious. At

Fig. 15 Chronology of the Shahezi Formation (Liu et al. 2021)

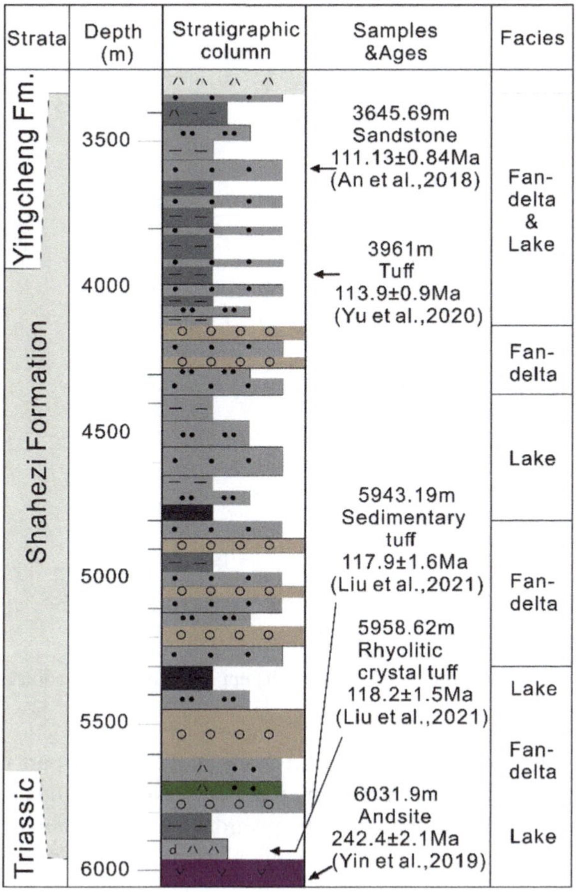

the depth of 4400 m, the temperature exceeds 150 °C, which reaches the standard of dry hot rock.

5.4 *Lithology and Chronology of the Yingcheng Formation*

Three suites of sedimentary rocks and three suites of volcanic rocks were developed in the Yingcheng Formation from bottom to top. The boundary between the third and fourth sections of the Yingcheng Formation is 3102.37 m, which is marked by the transition from the third large suite of volcanic rocks to the fourth large suite of conglomerates. Gao et al. (2017) analyzed the sedimentary environment and climate of the Yingcheng Formation based on the fine description of the core of the Yingcheng Formation and the core color, lithology, and lithofacies: The section I of sedimentary rock is 3278.96–3335.99 m long. The medium conglomerates with variegated

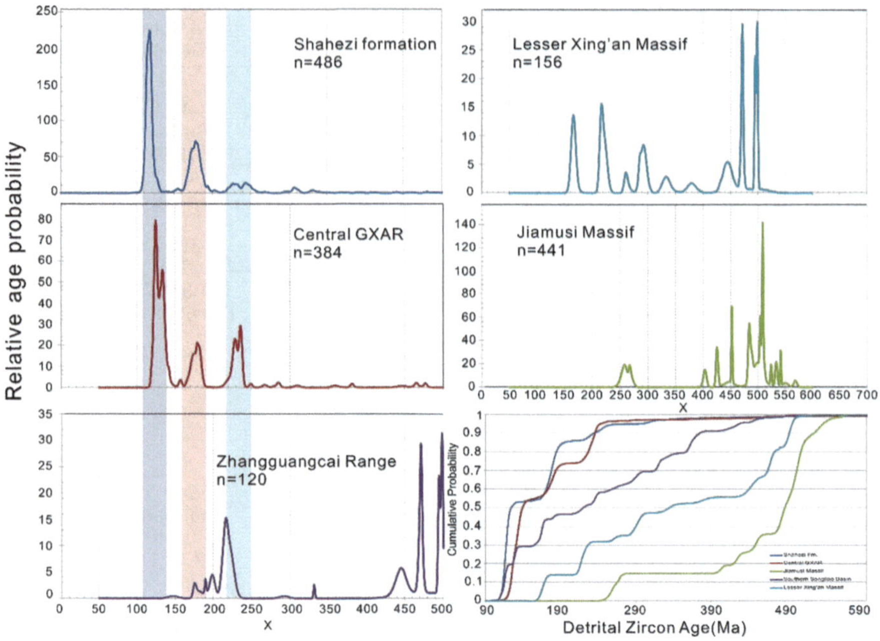

Fig. 16 Detrital zircon ages of the upper Shahezi Formation (An 2018)

colors are developed at the bottom. The lower part mainly develops light-medium gray coarse sandstone intercalated with silty mudstone, calcareous bands and coal lines. Wormholes develop upwards in the Yingcheng Formation. The color gradually becomes darker. The lithological grain size becomes finer until it becomes dark gray mudstone, indicating the water gradually deepening. The climate mainly transitioned from a relatively dry climate to a humid climate. The section I of volcanic rock is 3275.87–3278.96 m long. The gray andesitic agglomerated lava was developed in section I. The overlying and underlying lithology of section I is a transition type of volcanic-sedimentary rock, containing certain volcanic clastics, and breccias and aggregates. The section II of sedimentary rock is 3268.11–3275.87 m. Light-medium gray tuffaceous sandstone, variegated medium conglomerate, and black-gray tuffaceous siltstone with breccia-bearing aggregates were developed from bottom to top. The color gradually deepens and the particle size gradually decreases. The initial tuffaceous sandstone was developed instead of tuff, which indicates that the initial transport is intense. Then, black-gray breccia-bearing agglomerated tuffaceous siltstone is formed because of the rise of the water surface. The section II of the volcanic rock is 3102.37–3268.11 m, all of which are acidic eruptive rocks. These rocks are mainly eruptive rhyolite tuff lava, rhyolite breccia agglomerate and eruptive rhyolite. The core color is mainly light gray and light green gray. The section III of sedimentary rock is 2972.33–3102.37 m. This section mainly developed light gray medium-coarse sandstone and variegated medium conglomerate. The overall hydrodynamic force is relatively intense. The water surface is shallow. The channel of the fan delta plain and a few inter-fan swamps are generally developed. The climate began to change into arid climate. Volcanic Rock section III is a suite of blue-gray rhyolite tuff, in which the tuff (3.12 m) can be well consolidated and diagenetic. The underlying lithology of the volcanic rock in this period is the dark gray silty mudstone, which represents the inter-fan swamp depositional environment. This indicates that the background of deposition environment is stable. After the deposition of Volcanic Rock section III, the hydrodynamic force gradually strengthened, the water surface deepened. The climate then transitioned to a humid and rainy climate (Figs. 17 and 18).

Large-scale volcanic rocks in the Yingcheng Formation provide abundant materials for isotope dating research. Data of the Yingcheng Formation before 2008 indicate the ages range from 110 to135 Ma. However, the ages of the Yingcheng Formation reported after 2011 were later than those of previous studies, yielding ages of 101–112 Ma. Liu (2020a, b) considered that the ages of 110–112 Ma belonged to the stratigraphic age of the Yingcheng Formation. Wang et al. (unpublished data) also obtained the age of 112 Ma by ID-TIMS high-precision dating of the volcanic rocks in the Yingcheng Formation of the SK-2. The isotopic ages of Yingcheng Formation are estimated approximately as 105–112 Ma based on the age of 105 Ma at the bottom of the Denglouku Formation overlying Yingcheng Formation (Wang et al. 2016d; Liu 2020a, b) and the age of 112 Ma at the top of the lower Fushahezi Formation (An 2018).

5.5 Volcanism and Paleoclimate, Paleoenvironment of the Yingcheng Formation

The Aptian-Albian volcanism is widely distributed in the Songliao Basin and Northeast Asia. It is represented by the volcanic rocks of the Yingcheng Formation (covering area of approximately 2.3×10^6 km^2), which is similar to the flooded basalts (2.0×10^6 km^2) of the Ontong Java Plateau (OJP) in the Western Pacific. The average thickness of the volcanic rock calculated by drilling and outcrops is 0.35–1.2 km. The total volume of eruption is approximately 1.8×10^6 km^3. Its volume of eruption should be larger if the underlying magma system is considered. The violent and explosive volcanic eruption lasted from 121 to 109 Ma. It is estimated that at least a total of approximately 1.89×10^{14} t of volcanic volatiles entered the atmosphere during this period (Wang et al. 2016a, b, c, d).

Volcanism in the Songliao Basin and the Northeast Asia produced a large amount of volcanic ash and outgassed products that were discharged into the upper atmosphere in the early Aptian period (119.9–120.2 Ma). This is consistent with the short-term global cooling that occurred during the OAE1a. Therefore, the abrupt decrease in temperature may have been caused by the increase in sunlight reflection caused by the increase of volcanic ash and aerosol in the atmosphere. The greenhouse gases produced by volcanic degassing in Songliao and Northeast Asia from

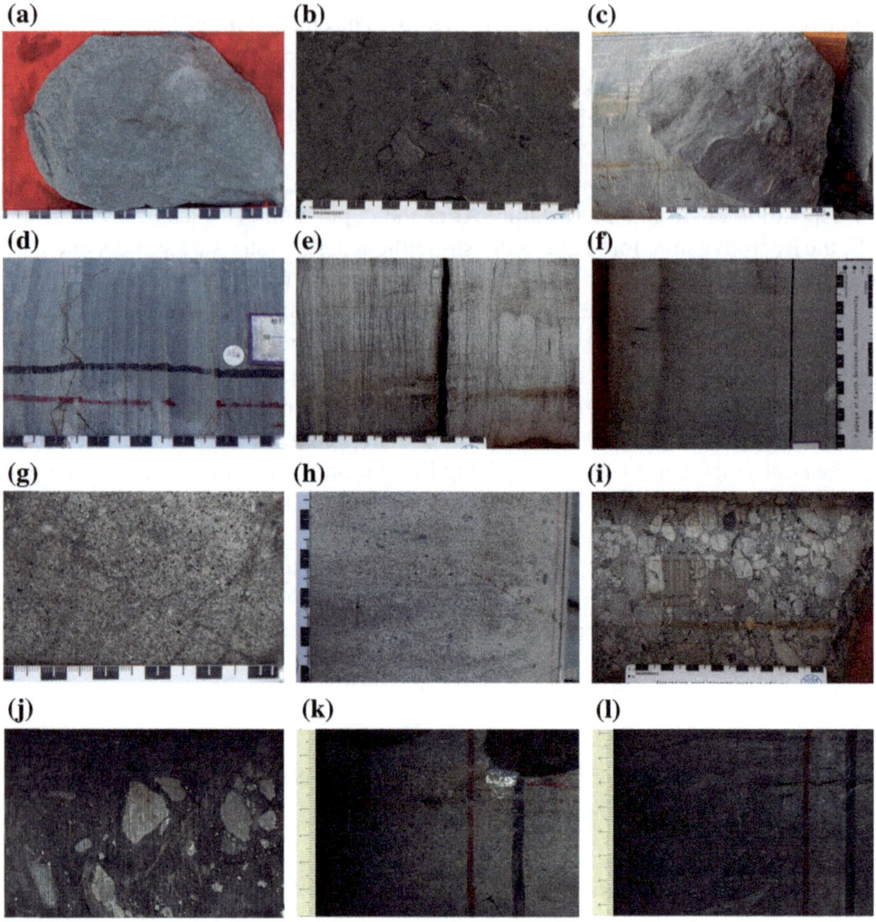

Fig. 17 Core photos of the Yingcheng Formation. **a** Mudstone. **b** Siltstone. **c** Siltstone. **d** Fine sandstone. **e** Medium sandstone. **f** Coarse sandstone. **g** Conglomeratic coarse sandstone. **h** Fine conglomeratic coarse sandstone. **i** Complex conglomerate. **j** Conglomeratic tuffaceous siltstone. **k**, **l** Tuffaceous sandstone. Modified from Gao et al. (2017)

the Aptian period to the early Albian period (121–109 Ma) may have a certain impact on the subsequent long-term climate warming and the increase of atmospheric CO_2 (Wang et al. 2016a, b, c, d).

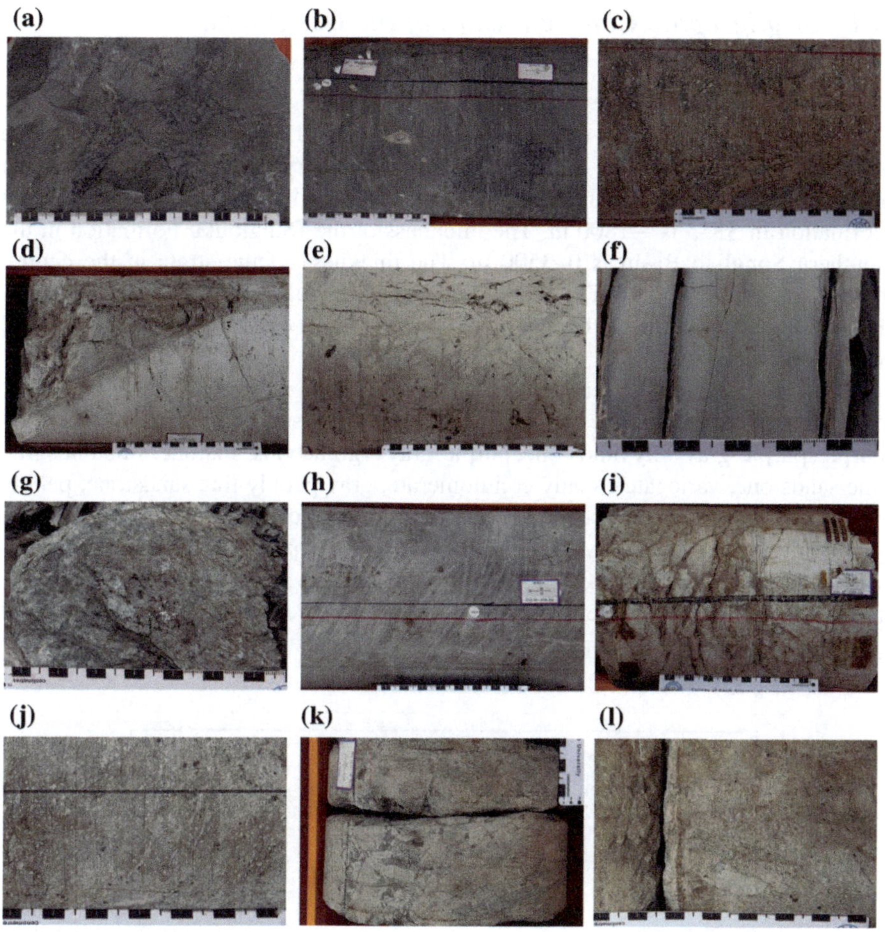

Fig. 18 Core photos of the Yingcheng Formation. **a**, **b** Andesitic agglomerate lava. **c** Rhyolitic brecciated tuff lava. **d** Rhyolitic brecciated agglomerate lava. **e**, **h** Rhyolite. **f**, **g** Rhyolitic tuffs. **i** Rhyolitic brecciated agglomerate lava. **j** Rhyolitic felsic tuff lava. **k** Rhyolitic breccia **l** Cryptoblastic breccia. Modified from Gao et al. (2017)

6 Stratigraphy of the Early Thermal Subsidence Stage (Denglouku Formation-Quantou Formation) from SK-3 Core

Stratigraphy in the early thermal subsidence stage includes Denglouku and Quantou formations. SK-3 has a drilling depth of 3600 m, coring footage of 1580.34 m, total core length of 1566.26 m, and a recovery ratio of 99.1%. The Quantou Formation and the Denglouku Formation were continuously cored for 1580.34 m with a core diameter of 100 mm.

6.1 Stratigraphic Characteristics of the Denglouku Formation

The maximum burial depth of the top boundary of the Denglouku Formation in the southern Songliao Basin is −4065 m. The top boundary of Dehui fault depression in the southeast uplift is −1200 to −2600 m. The top boundary of the Denglouku Formation in SK-3 is −1800 m. The thickness of the Denglouku Formation in the southern Songliao Basin is 0–3500 m. The thickness of the strata in the central depression is the largest. The thickness of the southeast uplift is relatively thin. However, the formation of the Denglouku Formation is continuously deposited and well preserved (Fig. 19).

The coring section of Denglouku Formation in SK-3 is 1783.03–2383.48 m. The core length is 600.45 m. The upper part of the lithology is purple-red mudstone, gray-purple, purple-gray silty mudstone, purple-gray argillaceous siltstone, siltstone, gray fine sandstone, variegated sandy conglomerate, gray pebbly fine sandstone, pebbly coarse sandstone, and interbedded coarse sandstone. The middle part is composed of gray-white, variegated sandy conglomerate, purple-red, purple-grey mudstone, gray-violet, gray to gray-white silty mudstone, gray pebbly fine sandstone, gray-white medium sandstone, and interbedded fine sandstone. The lower part is composed of gray sandy conglomerate, gray fine sandstone, silty mudstone, purple and interbedded

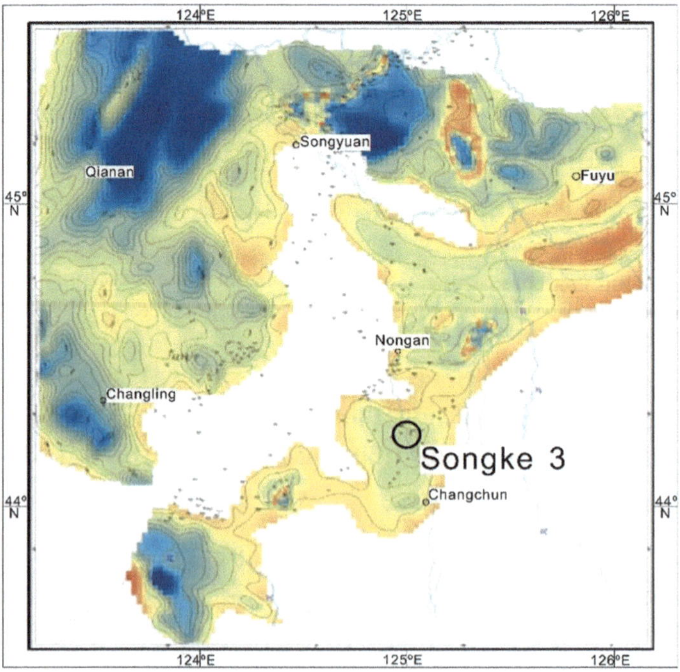

Fig. 19 Thickness of the Denglouku Formation in southern Songliao Basin

gray mudstone, with gray sandy conglomerate at the bottom. Occasionally, a thin layer of conglomerate and argillaceous conglomerate was developed, which is in unconformity contact with the underlying strata.

6.2 Sedimentary Environment of the Denglouku Formation

The Denglouku Formation was mainly deposited in a fan delta sedimentary system. The sedimentary facies include alluvial fan facies, fan delta facies, braided river facies, braided river delta and lake facies. The rock composition of alluvial fan facies is mostly variegated conglomerate and sandy conglomerate with brown mudstone. The gravel sorting and rounding are extremely poor. The branch channels of the fan delta plain have relatively coarse lithology and low component maturity. They are mainly gravel-bearing coarse sandstone, coarse sandstone and gray fine sandstone. Occasionally, some gray muddy were deposited on the top, which are poor in sorting and rounding. The sedimentary grain size of fan delta front is relatively fine and the color is mainly dark gray in the reduction environment. The sand is relatively pure, and the sorting is good. The logging lithology shows that the fossils, carbon chips and other materials are rich. The bottom of braided river facies is usually glutenite with coarser lithology. It transitions to fine-grained sandstone at the top. The floodplain is the product of flooding period. The lithology is mainly red-brown mudstone in oxidized environment, accompanied by deposition of calcareous, silty mudstone. Braided river deltas are developed in the second part of the Denglouku Formation, which are finer in lithology and grain size than fan deltas. Lacustrine deposits are mostly black mudstone, which is darker in color but purer in lithology (Figs. 20 and 21).

6.3 Stratigraphic Characteristics of the Quantou Formation

The maximum burial depth at the top of the Quantou Formation in the southern Songliao Basin is −2430 m. The burial depth of central depression belt is the deepest. The burial depth of southeast uplift area is shallower, with a burial depth of −300 to −1000 m. The top surface of the Quantou Formation is buried at a depth of −860 m from the SK-3 in the Dehui fault depression. The thickness of the Quantou Formation in the Dehui Fault Depression in the southern Songliao Basin is 100–1800 m. The Quantou Formation in the SK-3 is deposited continuously and the stratum is intact, with a thickness of approximately 940 m (Fig. 22).

In the SK-3, the fourth member of the Quantou Formation has a depth of 800.00–926.05 m with a core length of 126.05 m. The upper part of the lithology is composed of brown–red mudstone, gray silty mudstone, argillaceous siltstone,

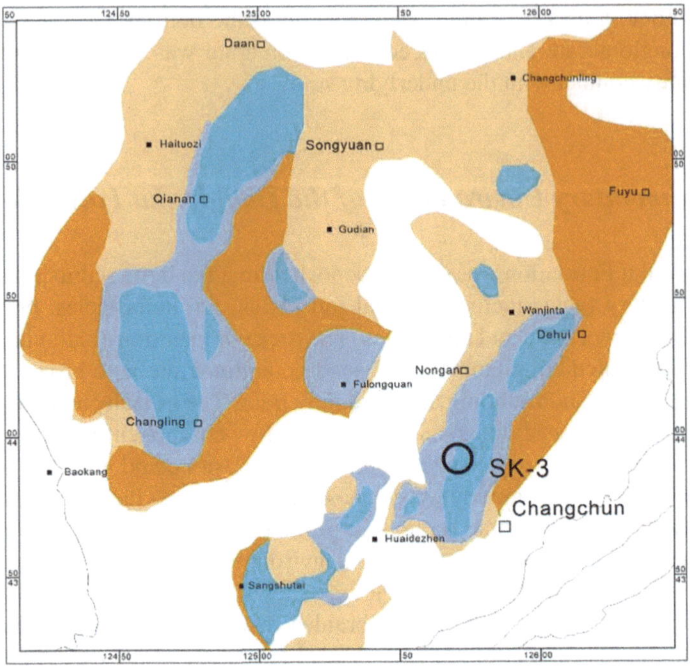

Fig. 20 Sedimentary facies of the Denglouku Formation

and siltstone. The middle part is purple-red and dark purple mudstone, gray—gray-purple—green-gray—dark-gray silty mudstone, gray—green-gray argillaceous siltstone, interbedded siltstone, occasionally gravelly coarse sandstone. The lower part is mainly gray-purple, purple-gray mudstone, green-gray mud qualitative siltstone and gray siltstone, with dark purple mudstone bands, gray-purple mudstone with silty mudstone bands. Occasionally, a layer of gray fine sandstone is in integrated contact with the underlying strata.

The third member of Quantou Formation has a depth of 926.05–1191.56 m with a core length of 265.51 m. The upper part of the lithology is interbedded with dark purple, purple-gray mudstone and gray siltstone. The middle part is composed of gray siltstone, purple-gray silty mudstone, and dark gray mudstone. The lower part is the interbedded layer of gray argillaceous siltstone, dark purple mudstone, and gray-purple silty mudstone. The bottom is gray fine sandstone, which is in integrated contact with the underlying strata.

The second member of the Quantou Formation has a depth of 1191.56–1512.94 m with a core length of 321.38 m. The upper part of the lithology is dark purple mudstone, gray siltstone, and fine sandstone interbedded. The middle part is gray siltstone, gray fine sandstone, dark purple mudstone, and gray purple interbedded argillaceous sandstone. The lower part is gray siltstone, gray fine sandstone, dark purple mudstone, gray-purple argillaceous siltstone, purple-red mudstone interbeds.

Fig. 21 Core photos of the Denglouku Formation. 1-Scouring surface, dark green-grey muddy siltstone at the bottom, gray coarse-sand microconglomerate at the top, 2-bioturbated texture, 3-scouring surface, green-grey siltstone at the bottom, gray gravel-bearing gritstone at the top, 4-trough cross bedding, green-grey coarse-sand microconglomerate; 5-concretion; 6-soft sedimentary structure; 7-root trace, brown-gray muddy siltstone; 8-scouring surface, gray-white polymictic fine conglomerate; 9-paleosols, visible redox spots, brown-gray silty mudstone

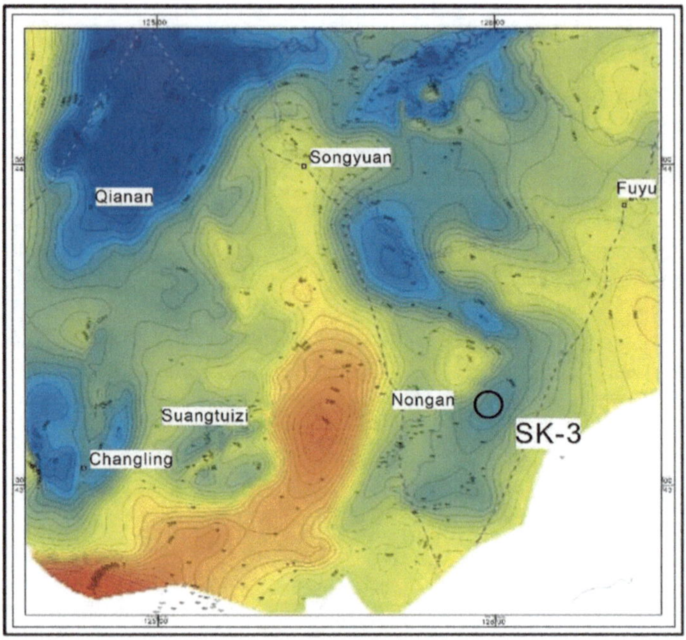

Fig. 22 Thickness of the Quantou Formation in southern Songliao Basin

The purple-red mudstone is at the bottom, which is in integrated contact with the underlying strata.

The first member of the Quantou Formation has a depth of 1512.94–1783.03 m with a core length of 270.09 m. The upper part of the lithology is interbedded with purple-red mudstone, gray siltstone, and gray medium sandstone. The middle part is composed of purple-red mudstone, gray siltstone, gray medium sandstone, and gray-purple silty mudstone, interbedded sandy mudstone and gray argillaceous siltstone. The lower part is purple-red mudstone, purple-gray mudstone, gravel-bearing medium sandstone. The bottom is purple-red mudstone. Occasionally, thin layers of glutenite and boulderstone are integrated with the underlying strata.

6.4 Sedimentary Environment of the Quantou Formation

The fluvial facies, delta facies, and lacustrine facies were mainly developed in the Quantou Formation in the southern Songliao Basin (Fig. 23). The channel and flood subfacies of the meandering river facies are developed. The channel subfacies are dominated by channel retention sediments, which are light gray coarse sandy fine conglomerate, gray white gravel coarse sandstone, light brown red/light brown medium sandstone, brown–red fine sandstone. The scouring-filling structure

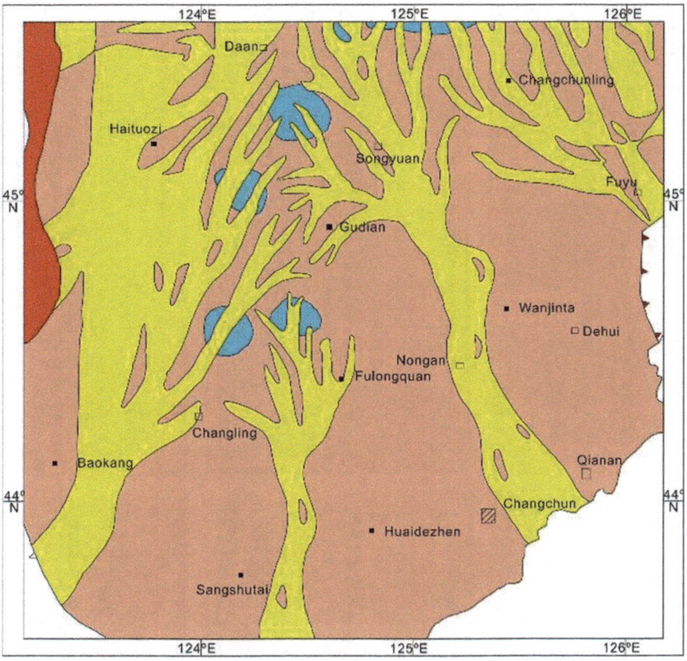

Fig. 23 Sedimentary facies of the Quantou Formation in southern Songliao Basin

was generally at the bottom. The visible mud and gravel and some of the mud and gravel are arranged in a directional arrangement (Fig. 24), trough-like cross bedding (Fig. 24). The structures of parallel bedding, wedge-like cross bedding, etc. is common. The flood subfacies include three sedimentary microfacies: floodplain, flood lake, and flood lake and rupture fan. The floodplain is composed of brown–red mudstone topped with light brown–red argillaceous siltstone. Calcareous nodules are developed in the mudstone. Wormholes can be found in the argillaceous siltstone. The delta facies only develops the delta front subfacies, in which the underwater distributary channel deposits are represented by light brown–red/light brown-gray medium sandstone, light brown–red fine sandstone. The mud is very little and there are scour surface, mud gravel, parallel bedding and scouring-filling structures such as wavy cross bedding a developed. The underwater bays are dominated by brown–red mudstone with a small amount of thin sandstone layers or lenses. The bedding structure is not developed. The wavy cross bedding and deformation are only seen in sandstone interbeds. Bedding, visible calcareous nodules and wormholes are developed. The underwater distributary sand dams are light brown–red or light brown-grey fine sandstone, light gray medium sandstone or fine sandstone, climbing bedding, wave marks formed by waves, parallel bedding. Wavy cross bedding and deformed bedding are developed. Wormholes and biological disturbances are common. The lacustrine facies mainly develops lakeside subfacies, which includes four sedimentary microfacies of muddy lake beach, sandy lake beach, gravel lake beach and

①
②
③
④
⑤
⑥
⑦
⑧
⑨

◂**Fig. 24** Core photos of Quantou Formation. 1-Light gray medium grain debris arkose, tempestite deposit sequence, convolution structure; 2-Gy fine sandstone, wave ripple cross bedding at the bottom, parallel bedding in the middle, wave ripple cross bedding at the top; 3-light gray fine sandstone, strident layering; 4-light gray fine sandstone, wave liquefaction deformation bedding; 5-light gray medium grain debris arkose, parallel bedding at the bottom, wave ripple cross bedding at the top, local deformation bedding; 6-light gray fine sandstone, wave ripple cross bedding; 7-concretion, 8-parallel bedding; 9-deformation bedding

adjacent shore sand bar. The muddy lake beach deposits are dominated by brown–red mudstone with a small amount of light brown-gray mudstone intercalated with thin layers of fine sandstone. In the interlayer, the exposed structures such as wavy cross bedding, wormholes, and calcareous nodules can be seen.

6.5 Petroleum Resources in the Denglouku and Quantou Formations

The source rocks of the Denglouku Formation in the Songliao Basin were formed in the shallow lake environment. The mudstone is gray-dark gray in color, with low organic carbon content. The parent material type is Type III, which belongs to poor quality source rocks. Although they are mature-highly mature, the potential for generation and expulsion of hydrocarbons is limited. They are mainly generation and exhaust and are almost impossible to expel liquid hydrocarbons. The reservoir lithology of the Denglouku Formation is dominated by glutenite, argillaceous siltstone, gravel-bearing argillaceous siltstone, and gravel-bearing fine sandstone, occasionally siltstone. The argillaceous siltstone and gravel-bearing fine sandstone are argillaceous cements and are relatively loose. The glutenite is argillaceous cementation and denser, which can be used as a good reservoir. The second and third members of Quantou Formation develop thick mudstone, which can be used as a good regional caprock.

7 Stratigraphy of the Late Thermal Subsidence Stage (Qingshankou-Yaojia-Nenjiang Formations) from SK-1 Cores

The Qingshankou, Yaojia and Nenjiang formations are mainly deposited in the late thermal subsidence stage of the Songliao basin. These three sets of strata are deposited in river-lacustrine facies environment, in which Qingshankou formation in semi-deep lake facies, Yaojia formation in shallow lake facies to fluvial facies, and Nenjiang Formation in deep lake to semi-deep lake facies. The three stratigraphic units were completely recovered from SK-1 cores. The coring depth of Qingshankou Formation is 1285.91–1782.93 m in SK-1s core, with a core recovery ratio nearly 100%. The coring section of Yaojia formation is from 1128.17 to 1285.91 m in SK-1s core, and the coring length is 157.74 m. The Nenjiang Formation is recovered in both SK-1 cores, of which the SK-1s obtains the strata of first and second member of the Nenjiang Formation, and the coring section is from 1128.17 to 963 m. The SK-1n starts coring from the bottom of the second member of Nenjiang Formation, and the coring section is from 1582.93 to 1783.16 m. The north and south holes of Nenjiang Formation were connected through the oil shale at the bottom of the second member. The North core of SK-1 coring the third to fifth members of Nenjiang Formation, the coring depth is 1811.18 m, the footage is 1630.41 m, the core length is 1541.66 m, and the recovery rate is 94.56%.

7.1 Lithology and Sedimentary Environment of the Qingshankou Formation

Fine grained sedimentary rocks dominate the Qingshankou Formation. Nomenclature scheme of fine-grained sedimentary rocks proposed by Lazar et al. (2015) is applied for detailed description of the Qingshankou Formation in the SK-1s core. Nine rock types, including dark gray laminated mudstone, dark gray massive mudstone, brownish gray massive mudstone, medium gray silty mudstone, light gray argillaceous siltstone, olivine gray dolostone, olivine gray marl, ostracoda limestone and volcanic ash, were identified. Among them, laminated dark grey silty mudstone, massive dark grey mudstone and olivine grey dolostone are the main lithology of the First Member of Qingshankou Formation, which occur in large thickness and high frequency. The second and third member of Qingshankou Formation are mainly composed of dark gray laminated mudstone, dark gray mudstone containing ostracoda, medium gray silty mudstone and dolostone. Other rock types occur as thin layers or intercalations, some of which are very thin and occur infrequently (Figs. 25 and 26).

Sedimentary environment of the Qingshankou Formation is dominated by deep lake to shallow lake. Sedimentary facies in the Qingshankou Formation can be divided into 3 sedimentary subfacies and 9 sedimentary microfacies. Among them, the First Member of Qingshankou Formation was deep lake subfacies which has a large lake basin with deep lake water. There were 5 sedimentary microfacies including deep lake mudstone, deep lake turbidity, dolostone, marl and volcanic ash deposit. During the second and third member of Qingshankou Formation, the lake basin area shrank and the lake water became shallower, and the whole lake was semi-deep lake subfacies, and the shallow lake subfacies appeared at the top of stratum. There were 7 sedimentary microfacies including semi-deep lake mudstone, semi-deep lake turbidity, dolomite, marl, shallow lake mudstone, shallow lake turbidity and volcanic ash deposition.

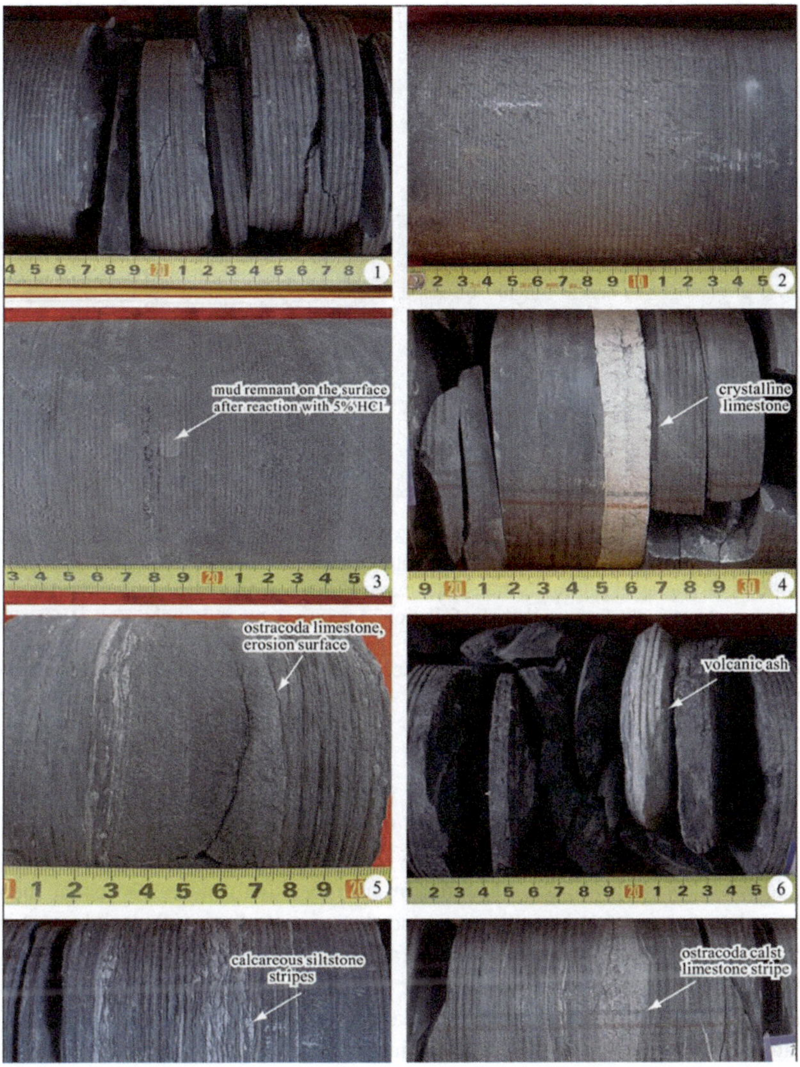

Fig. 25 Core photos of Qingshankou Formation Unit 1. 1-Dark gray mudstone, horizontal bedding; 2-light olive gray dolostone, massive structure; 3-light olive gray marl; 4-recrystallized limestone; 5-dark grey ostracoda limestone, ostracoda relatively complete, see scouring surface at the bottom; 6-volcanic ash; 7-bands of light gray calcareous siltstone in dark gray mudstone; 8-bands of light grey ostracoda clastic rock in dark grey mudstone (Gao et al. 2009)

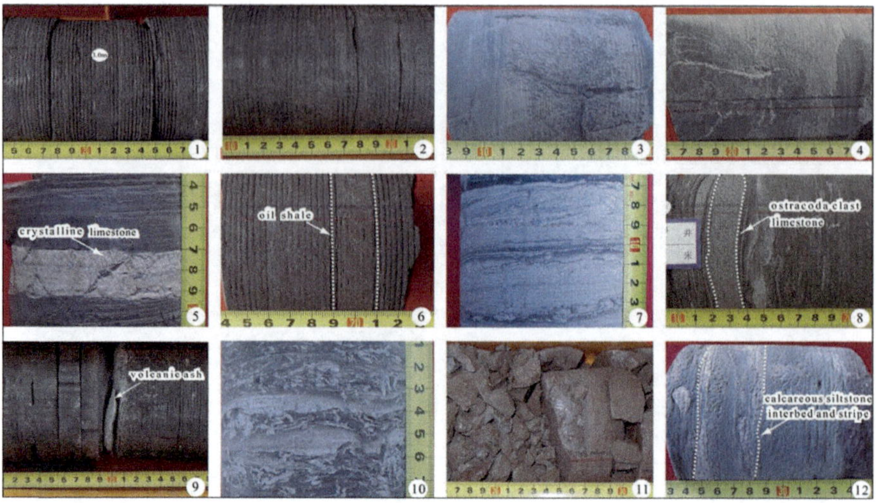

Fig. 26 Core photos of Qingshankou Fm Unit 2 + 3. 1-Dark gray mudstone, horizontal bedding; 2-dark gray ostracod bearing mudstone, horizontal bedding; 3-olive gray dolomite, massive structure; 4-light olive gray argillaceous limestone, massive structure; 5-white recrystallized limestone, massive structure; 6-olive gray oil shale, foliation structure; 7-medium light gray calcareous siltstone, horizontal ripple bedding, locally see convolute bedding; 8-light gray ostracod clastic rock; 9-volcanic ash; 10-light gray calcareous siltstone; 11-grayish brown mudstone, massive structure; 12-light gray calcareous siltstone thin layer and strip in green gray mudstone (Wang et al. 2009a)

(1) Deep lake subfacies

Deep lake subfacies occurred in the whole first member of Qingshankou Formation. Dark gray massive mudstone, dark gray laminated mudstone without calcareous silty lens and with continuous lamination and less disturbance represent the deep lake mudstone microfacies as background deposits, and other deep-lake turbidites, dolostone, marl and volcanic ash microfacies appear as event deposits. Deep-lake turbidites and dolostone deposits occur frequently, indicating that the lithology is olivine gray nodular dolomite and dark gray laminated mudstone with many calcareous silty lens and bands, discontinuous laminae and more disturbance. Many calcite veins and pyrite can be observed, which may indicate a relatively anoxic environment.

(2) Semi-deep lake subfacies

The semi-deep lake subfacies occurred in most of the second and third member of Qingshankou Formation except the top. Similar to the first member of Qingshankou Formation, the semi-deep lake mudstone deposits represented the dark gray laminated mudstone with continuous and less disturbed lithology without calcareous silty lens as the background deposits, and other semi-deep lake turbidity, dolostone, marl deposits as the event deposits. Turbidity deposits occur frequently, and the typical lithology is dark gray laminated mudstone with many calcareous silty lens and discontinuous bands and laminations with more disturbance. Pyrite and calcite veins rarely appear in there, while biodisturbance, sandy bands, phyllolimia and ostracoda fossils occur frequently, which may indicate a relatively less anoxic environment, and the water flow is more turbulent than the first member of Qingshankou Formation.

(3) Shallow lake subfacies

The shallow lake subfacies occurred at the top of the second and third member of Qingshankou Formation, including the shallow lake mudstone and shallow lake turbidity deposits. Among them, shallow lake mudstone deposits as the background deposits, and the typical lithology is brown-gray massive mudstone. Turbidite deposits occur as event deposits, and the typical lithology is light gray calcareous siltstone interbeddings and bands.

In general, according to the core characteristics of SK-1, the Songliao Basin experienced a process of gradually shrinking lake basin and shallower water body during the deposition of Qingshankou Formation. The subfacies transition from deep lake in the first member of Qingshankou Formation to semi-deep lake in most of the second and third member of Qingshankou Formation, and then to shallow lake at the top of the third member of Qingshankou Formation. The sedimentary environment of the basin is transitional from lake to delta.

7.2 Chronology of the Qingshankou Formation

The Qingshankou Formation contains abundant fossils, including ostracods, conchostraca, dinoflagellate and sporopollen. The bottom of Qingshankou Formation was characterized by the initial emergence and large-scale prosperity of the *Triangu-licypris torsuosus,* and followed by the ostracod fossils, such as *Cypridea dekhoinensis, C. adumbrate, C. vicina, C. panda, Limnocypridea copiosa, L. bucerusa, L. inflata.* Conchostraca fossils mainly occur in the middle and upper part of Qingshankou Formation, mainly consists of *Nemestheria lineata, N. qingshankouensis, N. robusta, N. jiutaiensis, N. furcate, Jilinestheria nonganensis, J. qianguoqiensis, J. insculpta, Orthestheria bayanensis, Cratostracus merus* (Cui et al. 1988). Dinoflagellates are characterized by the emergence and relative prosperity of *Kiokansium declinatum, K. regulatum, Dinogymniopsis spinulosa, Sentusidinium bifidum* (Gao et al. 1992). Furthermore, at the bottom of Qingshankou Formation, some angiosperm pollen with important age significance were also found, such as *Cranwellia, Lythraites, Complexiopollis* etc. *Borealis, buttinia, areicipites, sabalpollenites* etc. appear in the middle of Qingshankou Formation. These paleontological and plant pollen records indicate that the Qingshankou Formation may have been deposited in the Turonian period of the late Cretaceous.

Three thin volcanic tuff layers were found in the lacustrine dark mudstone of Qingshankou Formation in SK-1s core, which are located at 1780.1 and 1705.8 m of the first member of Qingshankou Formation and 1673.1 m of the second member of Qingshankou Formation, respectively. The U–Pb SIMS dating results of these three layers of volcanic ash are 91.4 ± 0.5 Ma, 90.1 ± 0.6 Ma and 90.4 ± 0.4 Ma, indicating that the bottom of Qingshankou Formation is equal to Turonian stage of Late Cretaceous. However, the ages of 90.1 ± 0.6 Ma and 90.4 ± 0.4 Ma are contradictory and do not accord with the law of superposition. Wang et al. (2016d) re-dated the three layers of volcanic ash of Qingshankou Formation in SK-1s, and obtained high-precision U–Pb ages which in line with the stratigraphic law. The ages of three layers are 91.886 ± 0.11 Ma, 90.974 ± 0.12 Ma and 90.536 ± 0.12 Ma respectively. The new dating results show that the depositional ages of the first member and the bottom of the second and third members of Qingshankou Formation are in the period of Turonian, which extends the span of Qingshankou Formation to the extent of 0.5 Ma, and recalibrates the U–Pb chronological framework of SK-1s. Deng et al. (2013) analyzed the magnetostratigraphy of SK-1. Based on the zircon ages and polarity conversion as anchor points, the Qingshankou Formation is located in the Cretaceous Normal Superchron. Therefore, the magnetic stratigraphy cannot accurately constrain the sedimentary ages of the internal stratigraphic units of the Qingshankou Formation.

Wu et al. (2013) conducted a detailed cyclostratigraphic study of natural gamma-ray (GR) log, thorium (Th) log, and magnetic susceptibility (MS) data from the SK-1s core. According to the robust spectral analysis, all Milankovitch cycles have been recognized. An astronomical time scale (ATS) was established by tuning interpreted 405 kyr cycles to a 405 kyr orbital eccentricity target curve, and anchored to four SIMS U–Pb zircon radioisotope ages. The ATS provides a continuous 'absolute time framework' for the SK-1. Among them, about 13.5 long eccentricity cycles and about 54 short eccentricity cycles are identified in the Qingshankou Formation. Thus, the sedimentary duration of the Qingshankou Formation is about 5.4 Ma. Afterwards, Wang et al. (2016d) adopt the new high-precise ID-TIMS U–Pb ages as the anchored points, and recalibrated the ATS for the SK-1s core, which yields the sedimentation duration of Qingshankou Formation ranges from 86.55 to 91.906 Ma (Fig. 27).

7.3 *Cretaceous OAE3 Recorded in the Qingshankou Formation*

Ocean Anoxic Events (OAEs) are global Anoxic Events that occur in the Ocean. The bottom water of the ocean was anoxic at the time of these events. Therefore, organic-rich black shale appears in all marine basins, indicating a high productivity and high rate of carbon burial in the marine environment. During the Cretaceous period, there were three OAEs called OAE1 (Aptian-Albian), OAE2 (Cenomanian–Turonian) and OAE3 (Coniacian-Santonian), and OAE1 could be divided into OAE1a, OAE1b, OAE1c and OAE1d. Previous studies suggested that OAE1a marked the beginning of the mid-Cretaceous extreme greenhouse climate, OAE2 was the peak period of the greenhouse climate, and OAE3 occurred with the greenhouse climate weakening.

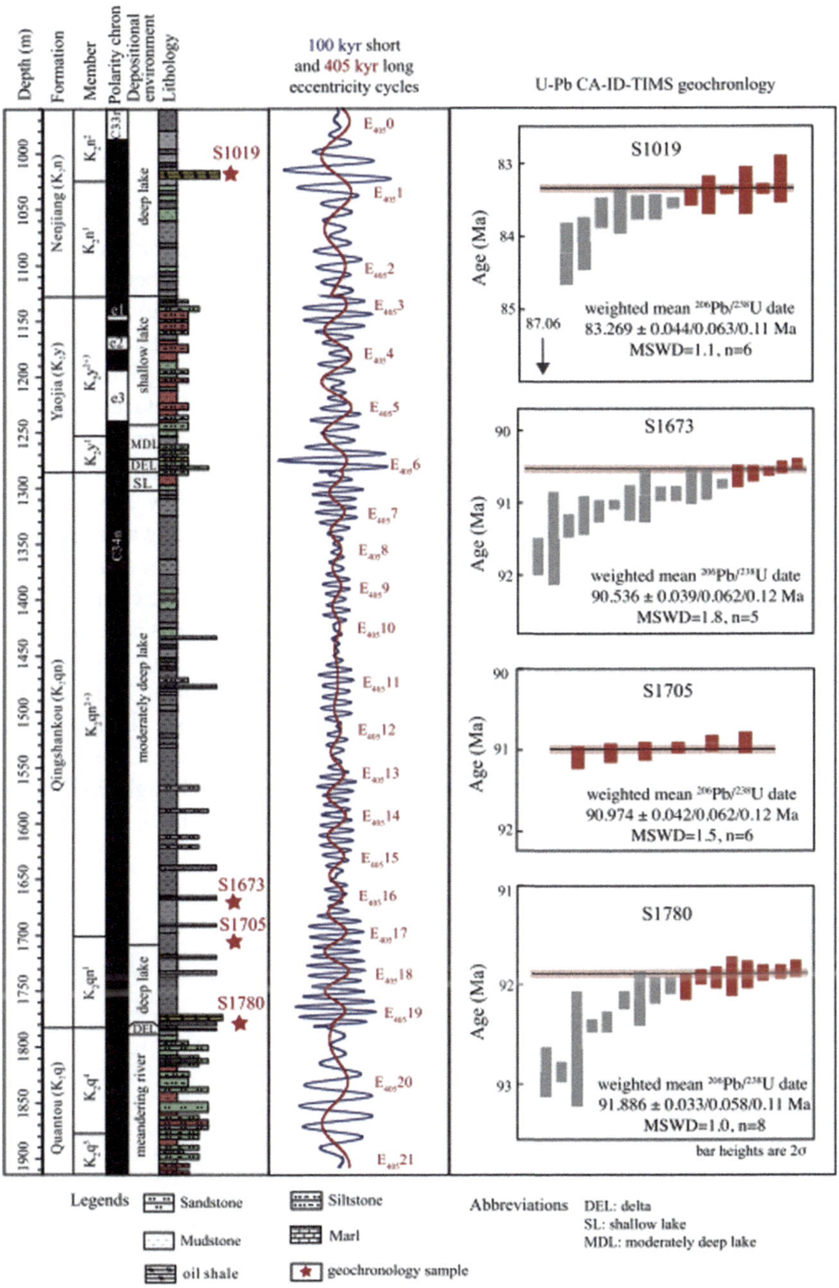

Fig. 27 Chronology of the Qingshankou Formation and adjacent stratigraphic layers (Wang et al. 2016d)

OAE3 is also an important transition in the evolution of the long-term Cretaceous climate system, representing a transition from Cretaceous greenhouse climate to Paleocene cooling (Friedrich et al. 2012). The large scale organic carbon burial caused by the OAEs has a profound impact on the atmosphere–ocean system, which mainly leads to the decrease of CO_2 and the increase of O_2 in the atmosphere–ocean system.

The sedimentation duration of Qingshankou Formation ranges from 86.55 to 91.906 Ma, which is equivalent to the Turonian to the Coniacian, during which OAE3 occurred. The response of Songliao Basin to OAE3 can be better understood by studying the lake sedimentary records of Qingshankou Formation. Jones et al. (2018) have studied bulk organic carbon isotopes ($\delta^{13}C_{org}$), elemental concentrations (XRF), and initial osmium ratios ($^{187}Os/^{188}Os$, Os_i) from the Qingshankou Formation (Fig. 28). Through geochemical analyses, a regional Late Cretaceous paleoclimate records and lacustrine carbon burial dynamics of the Qingshankou Formation have been reconstructed. Correlation of Turonian Coniacian $\delta^{13}C_{org}$ records from the Songliao Basin to the WIB confirms the presence of OAE3 in a low TOC interval of the second and third member of Qingshankou Formation, providing a unique record of the event in a lake system. The chemostratigraphic results from the Songliao Basin demonstrate that OAE2 and OAE3 did not trigger elevated organic carbon burial in an expansive East Asian lake. Furthermore, the significant decreases in marine and lacustrine ^{13}C can be attribute to a drawdown of pCO_2 and cooling through OAE3 and decreased soil productivity in the drainage of Songliao Basin. This finding is consistent with enhanced burial of organic carbon on a global scale and is analogous to interpretations for other prominent Cretaceous OAEs. However, Os_i stratigraphy records no evidence for significant changes in global volcanism though OAE3, which suggests an event trigger unique from OAE2 (i.e., LIP volcanism).

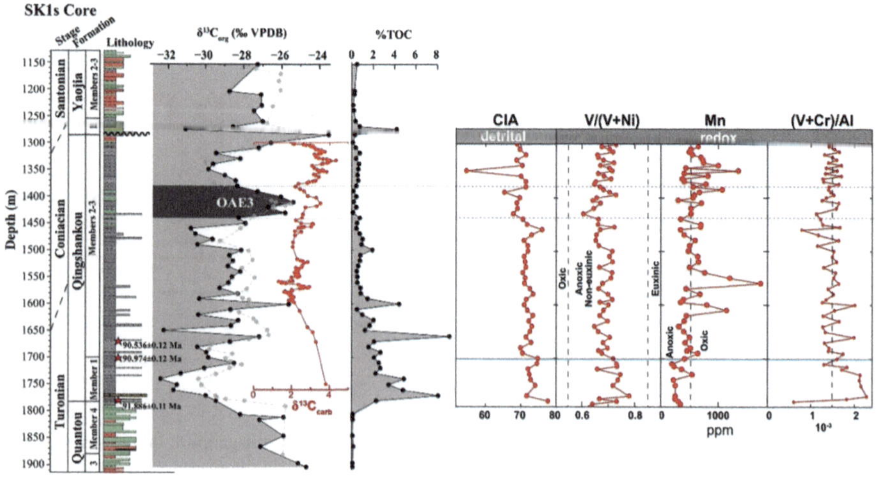

Fig. 28 Organic and elemental geochemistry of the Qingshankou Formation (Jones et al. 2018)

7.4 Sedimentary Cycle in the Qingshankou Formation

The Milankovitch theory provides a remarkable tool to investigate climatic response to latitudinally varying top-of-atmosphere insolation changes at ten thousand to million-year time scales (Zachos et al. 2001; Lourens et al. 2005; Palike et al. 2006). Previous cyclostratigraphic studies have revealed the Qingshankou Formation recorded Milankovitch cycles, which shows the sensitive response of large lake sedimentary system to astronomical force during the Cretaceous period. In combination with other absolute dating methods, Wu et al. (2013) used astronomical tuning to obtain a high-resolution astronomical time scale, and limited the depositional age of Qingshankou Formation range from 91.9 to 86.5 Ma. Additionally, three layers of volcanic ash were found in first and second members of Qingshankou Formation, which effectively refined the sedimentation rate of this interval to about 8–9 cm/ka, which is consistent well with the average sedimentation rate obtained by cyclostratigraphic analysis based on GR logging data (Wu et al. 2013). Therefore, under the constraints of high-precision chronology, we can carry out orbital-scale climatic and environmental research.

Huang et al. (2021) collected organic geochemical samples with equal spacing (1 m) from the Qingshankou Formation, and obtained high-resolution total organic carbon content (TOC), carbon nitrogen ratio (C/N) and total organic carbon isotopes ($\delta^{13}C_{org}$) dataset. Spectral analysis of the untuned TOC data from the Qingshankou Formation reveals significant cyclicity at wavelengths of 34, 13.8, 9, 4.5, and 3.4 m. Combined with the sedimentation rate from absolute ages constraint and estimated by ASM method (8.3cm/ka), these cycles roughly represent ~408 ka, ~166 ka, ~108 ka, ~54 ka, and ~40.8 ka, respectively. Among them, 408 ka may represent the long eccentricity cycle, 108 ka may represent the short eccentricity cycle, 54 and 40.8 ka may represent the obliquity cycle. In addition, the TOC data also recorded a significant ~170 ka orbital cycle. Astronomical calculation indicates that the obliquity modulation cycles not only have ~1.2 Ma cycle, but also have a ~170 ka short modulation cycle (Hinnov 2013). Recently, it was prominently detected in the sedimentary and geochemical proxies of the Middle Eocene and Late Cretaceous as a result of AM of the ~40 ka obliquity cycle. This obliquity AM cycle has been used as a new geochronometer to construct the geological time scale for closing the "Middle Eocene time scale gap" (Boulila et al. 2018) and to refine the duration of Cretaceous Oceanic Anoxic Event 2 (OAE2) (Charbonnier et al. 2018). This ~170 ka cycle in the Qingshankou Formation is the first time found in terrestrial strata, which is of great significance to study the universality of this obliquity modulation cycle in geological records. Therefore, Huang et al. (2021) summarized the records of various high-resolution paleoclimate proxies from the middle and high latitudes in the past 200 million years, and found that there is also a ~170 ka "unconventional" obliquity amplitude modulation cycle (Fig. 29).

It is unexpected that these theoretically low-AM cycles which have little effect on the variation of the Earth's insolation, but produced significant and identifiable signals in the geological records. These orbital quasi-periodic cycles are considered to

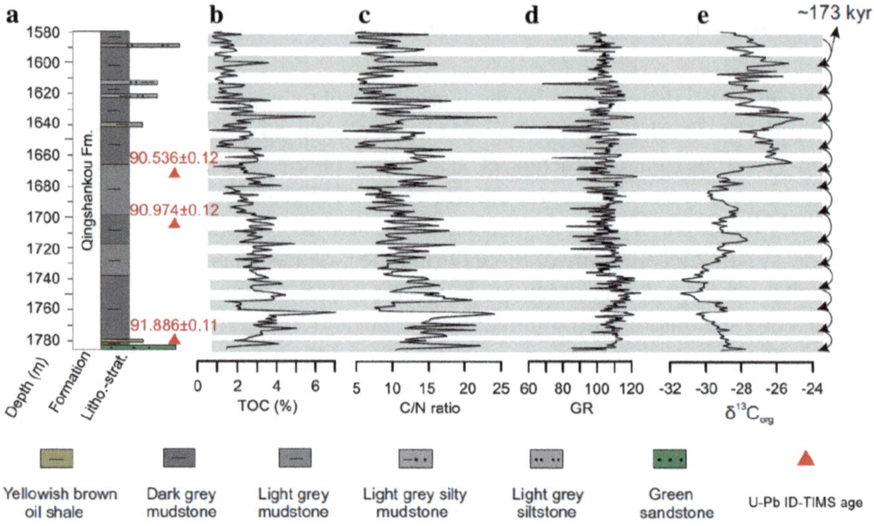

Fig. 29 Geochemical records and ~173 ka cycle recorded in the Qingshankou Formation from Songliao Basin

have been amplified through nonlinear response feedbacks between external (orbital) forcing and the internal climate and sedimentary systems, reflecting the nonlinear characteristics of the Earth's climate system. Therefore, the main reason why they can be recorded in the stratum may lie in the complex orbit signal frequency reduction mode, signal transmission process, Earth climate system feedback mechanism and so on (Laurin et al. 2015; Huang et al. 2021).

According to the spectral results of the three TOC time series, the ~173 ka signal is not only expressed in the obliquity envelope series but also clearly detected in the original TOC datasets. This feature indicates that the obliquity AM cycle is not the only contributor for the ~173 ka cycle; there is a specific process to enhance the ~173 ka cycle itself. Here, we provide a possible mechanism related to threshold responses to the obliquity forcing. The ~173 ka AM cycle is enhanced in a scenario of obliquity forcing across a critical geochemical threshold within sedimentary basins (Fig. 30).

The equilibrium of the lake biogeochemistry is a considerable "tipping point" capable of clipping the obliquity signal. When the system falls below the threshold, the net OC burial flux is decreased in lacustrine settings owing to oxygenated bottom waters and reduced terrestrial input, resulting in OC-poor sediments. Conversely, as obliquity exceeds a critical threshold, the net carbon burial flux is increased through increased carbon export to the bottom and/or enhanced preservation through anoxic bottom conditions (Fig. 31). Therefore, the lake biogeochemical processes can cause a threshold for burial or oxidation of OC, and thus, this threshold response could transfer the basic high-frequency ~40-ka obliquity cycle to the longer low frequency AM periods.

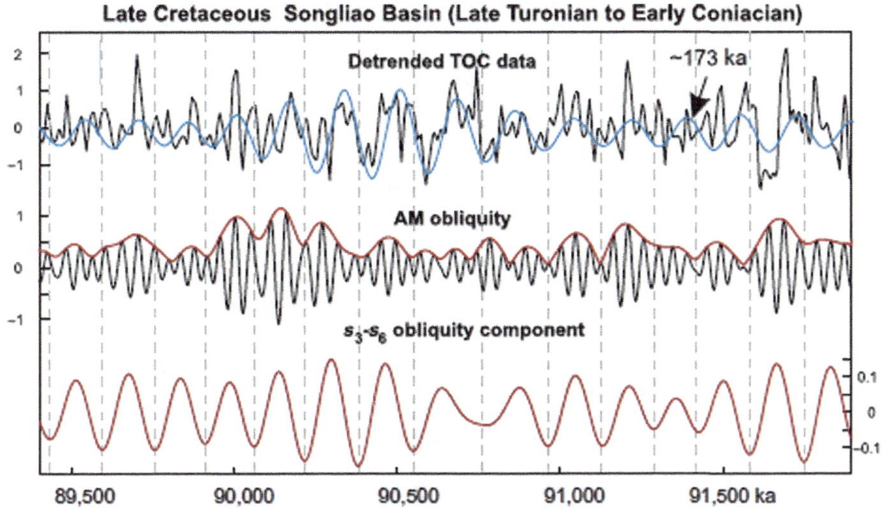

Fig. 30 ~173 ka cycle recorded in TOC data from the Qingshankou Formation

7.5 Unconventional Oil and Gas Resources in Qingshankou Formation

The Songliao Basin is one of the most promising continental shale oil resource basins in China, with rich oil and gas resources and the maximum depth of more than 10,000 m. The Qingshankou Formation, as the main source rock of Gulong Sag, is mainly composed of fine-grained sedimentary rocks deposited in deep lacustrine to semi-deep lacustrine environment, and the sedimentary thickness is 260–500 m. The distribution area is 8×10^4–9×10^4 km, and the geological resources are 20–25 billion tons.

TOC of shale in Qingshankou Formation is 0.13–6.12%, with an average of 2.32%. TOC of shale in first member of Qingshankou Formation is 1.01–6.12%, with an average of 3.44%. TOC of shale in second and third members of Qingshankou Formation is 0.13–3.60%, with an average of 1.14%. TOC in first member of Qingshankou Formation is much higher than that in second and third members of Qingshankou Formation, that is, organic matter abundance in first member of Qingshankou Formation is higher (Fig. 32).

The hydrocarbon potential (S1 + S2) of shale in Qingshankou Formation is 0.23–31.55mg HC/g, with an average of 11.64 mg HC/g. The hydrocarbon potential (S1 + S2) of shale in first member of Qingshankou Formation (an average of 17.51 mg HC/g) is much larger than that of shale in second and third members of Qingshankou Formation (an average of 5.47 mg HC/g). The hydrocarbon potential of Qingshankou Formation shale increases with the increase of depth (Fig. 31). The variation regularity of the hydrocarbon potential and TOC is similar with depth, which is coupled. The hydrocarbon (S1) (an average of 1.95 mg HC/g) and the pyrolyzed hydrocarbon

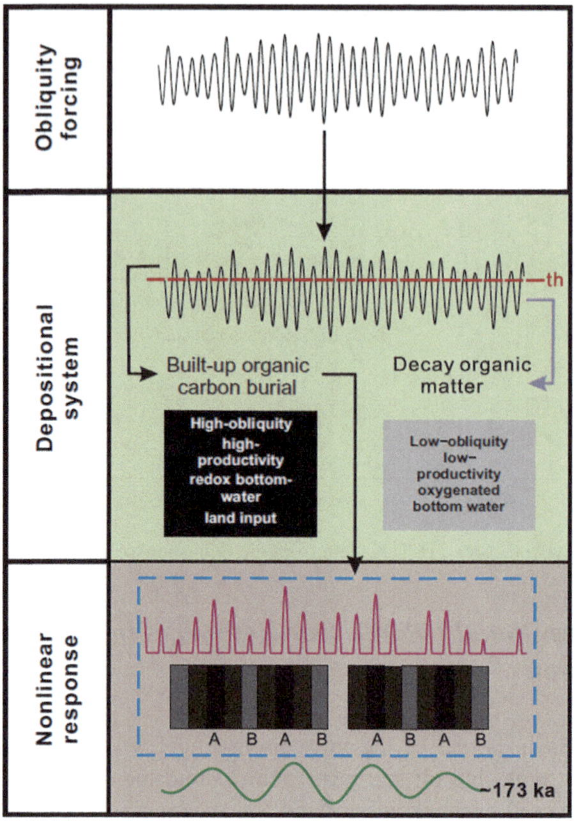

Fig. 31 A simplified possible sedimentary process of nonlinear climate feedback linked to the ~173 ka amplitude of obliquity

(an average of 15.56 mg HC/g) of shale in first member of Qingshankou Formation is significantly higher than The hydrocarbon (S1) (an average of 0.44 mg HC/g) and the pyrolyzed hydrocarbon (an average of 5.03 mg HC/g) of shale in second and third members of Qingshankou Formation.

Different organic matter types can be distinguished by hydrogen index (HI). The HI of I kerogen (Oil-generating type, usually lacustrine facies) is greater than 600 mg HC/g TOC. The HI of II kerogen (Oil-generating type, usually Marine) is 300–600 mg HC/g TOC, The HI of II/III kerogen (Raw oil and gas type) is 200–300 mg HC/g TOC, The HI of III kerogen (gas-generating type) is 50–200 mg HC/g TOC, The HI of IV kerogen (Dry gas type) is less than 50 mg HC/g TOC. The hydrogen index (HI) of Qingshankou Formation varies widely, ranging from 127 to 656 mg HC/g TOC, with an average of 411 mg HC/g TOC. The organic matter type of Qingshankou Formation is II, belonging to oil-generating kerogen.

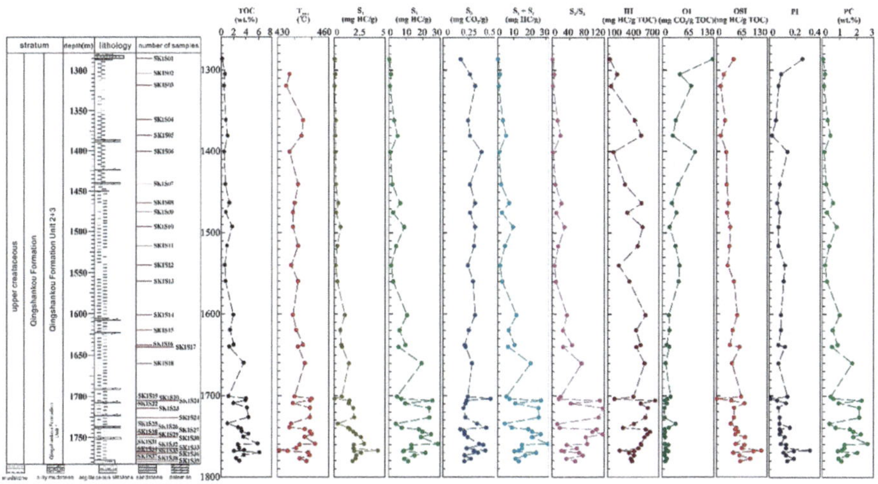

Fig. 32 Shale porosity of the Qingshankou Formation

The oxygen index (OI) of Qingshankou formation is 5–127 mgCO_2/g TOC, with an average of 23 mg CO_2/g TOC. The oxygen index (OI) of first member of Qingshankou Formation (an average of 10 mg CO_2/g TOC) was lower than that of shale in second and third members of Qingshankou Formation (an average of 37 mg CO_2/g TOC). On the whole, the oxygen index (OI) of Qingshankou Formation decrease sharply with the increase of depth.

The oil saturation index (OSI) of Qingshankou Formation varies widely, ranging from 2 to 116 mg HC/g TOC, with an average of 47 mg HC/g TOC. The oil saturation index of first member of Qingshankou Formation (an average of 59 mg HC/g TOC) is higher than that of second and third members of Qingshankou Formation (an average of 35 mg HC/g TOC).

The maximum pyrolysis temperature (Tmax) can be used to determine the maturity of organic matter. The Tmax between 430 and 455 °C belongs to the oil generation stage. The Tmax of Qingshankou Formation is 431–452 °C, with an average of 443 °C. The whole is in the stage of low maturity to maturity, the maturity of first member of Qingshankou Formation is slightly higher than that of second and third members of Qingshankou Formation. Tmax generally increase with depth.

The Productivity Index of Qingshankou Formation ranged from 0.01 to 0.32, with a wide range. The Productivity Index of first member of Qingshankou Formation (an average of 0.12) is slightly higher than that of second and third members of Qingshankou Formation (an average of 0.09). It shows that the hydrocarbon conversion rate of first member of Qingshankou Formation is higher, and the Productivity Index is positively correlated with the oil saturation index.

7.6 *Yaojia Formation*

The Yaojia Formation recovered by CCSD SK-1s borehole is 157.67 m long with the core recovery rate of 99.96%. The span of the Formation corresponds to a range from the Santonian to the early Campanian. The sequences and processes of lithology, lithofacies and cyclic stratigraphy were revealed by a detailed core description. Eleven rock types and three kinds of sedimentary subfacies including shallow lake, deep lake and delta front were recognized from the drilling core. There are 11 sedimentary microfacieses including dolostone, argillaceous limestone, shallow lake turbidite, deep lake turbidite, estuary dam, distal bar, sheet sandstone, subaqueous distributary bay, slump deposits, shallow lake mudstone and deep lake mudstone. The Yaojia Formation represents 151 m scale cycles (sixth order cycle), 45 fifth order cycles, 9 fourth order cycles and 3 third order cycles. Meticulous depiction (centimeter level) of the whole formation provides an approach to study its sedimentary process with high precision. The red beds with thickness of 43% in the Yaojia Formation, being one of fewer continental red beds depicted meticulously in the world, will become a standard potential formulation section for the studies of Cretaceous global oxygen enriched events (Fig. 33).

7.7 *Chronology of the First and Second Members of Nenjiang Formation*

Studies on the age of the Nenjiang Formation can dated back to the 1970s and 1980s. Zhang and Zong (1984) have reported ostracod fossils from the Nenjiang Formation, and meanwhile listed many fossils including sporopollen, conchostraca, bivalve, gastropod, fish, reptiles and plants. Chen (2012) have reported seven assemblage zones of *Euestherites* fauna which widely developed in Songhua Lake Basin. Among them, the Nenjiang Formation have developed four assembleage zones which consist of *Dictyestheria ovata, Halysestheria yui, Euestherites bifurcates, Calestherites sertus* and *Mesolimnadiopsis anguangensis.* These fossil assembleages have refined the sedimentation duration of the Nenjiang Formation ranges from Turonian to Santonian stage. Wan et al. (2013) have conducted a detail study on the Nenjiang Formation of the SK-1s well, finding the *Borealipollis* and *Aquilapollenites* sporopollen assemblage zones, and *Pediastrum-Botryococcus, Dinogy mniopsisminor Balmula, Dinogymniopsis-Chlamydophrella-Vesperopsis bifurcata* and *Pedias trum* Phytoplankton communities, *Cypridea anonyma-Candona fabiforma, Cypridea gracila-Cypridea gunsulinensis, Mongolocypris magna-Mon golocypris heilun tsziane nsis, Cypridea liaukhenensis-Cyprideabella, Periaca nthella portentosa-Limnocypridea subscalariformis, Limnocypridea dilinensis-Limnocypridea nova* and *Strumosia inandita-Cypridea spangvosa* Ostracod assemblages. Besides, the foraminiferal fossil including *Archaeoglobigerina blowi, Archaeoglobigerina cretacea, Hedbergella*

Fig. 33 Core photos of the Yaojia Formation (Cheng et al. 2009)

flandrini, *Karrorulina hokkaidoana*, *Gavlinella* sp. and *Anomalinoides* sp. have been found in the first member of the Nenjiang Formation. These fossils have constrained the sedimentation duration of Nenjiang Formation to Early Santonian to Campanian stage of Late Cretaceous. Afterwards, Xi et al. (2011) have detected the foraminiferal fossil by studying the SK-1s borehole, these fossil assemblages including the benthic foraminifera (*Gavlinella* sp., *Anomalinoides* sp., *Haplophragmoides* sp., *Pullenia* sp., *Karrorulina hokkaidoana, Clavulinoides* sp.) and planktonic foraminifera (*Archaeoglobigerina blowi, Archaeoglobigerina cretacea, Hedbergella flandrini*), which indicated that the sedimentation duration of Nenjiang Formation was equal to Late Turonian to Late Santonian stage.

Previous studies have provided two U–Pb ages for the Nenjiang Formation in SK-1s core by using different dating methods (He et al. 2012; Wang et al. 2016d). He et al. (2012) used the SIME U–Pb method to measure the age sample of Nenjiang Formation (S1019 m) and obtained the age was 83.7 ± 0.8 Ma. Afterwards, Wang et al. (2016d) adopted the CA-ID-TIMS U–Pb method to re-measure the same age sample of the Nenjiang Formation, which yielding the U–Pb age equal to 83.269 ± 0.044 Ma. Deng et al. (2013) and Wu et al. (2013) used the magnetic stratigraphy and cyclostratigraphy to calculate the boundary age of the Nenjiang and Yaojia Formations. However, there are no direct isotopic ages, and there is a lack of chronological constraints on important events in the lower part of the Nenjiang Formation and the chronological data of international stratigraphic correlation. Yu et al. (2019) has taken the SK-2 core as a reference, combining with Yaojia, Houjinggou, and Yuewangcheng outcrops, and further established a typical profiles for Nenjiang Formation. Yu et al. (2019) provided the boundary age of Nenjiang and Yaojia Formations (85.1–85.5 Ma), which shows that the boundary between Nenjiang Formation and Yaojia Formation is located in Santonian stage.

As discussed above, the boundary age between Yaojia and Nenjiang Formation was estimated 85.1–85.5 Ma, which is older than previous study (Deng et al. 2013; Wu et al. 2013). This new age also confirm that the sedimentation rate of Nenjiang Formation is much slower than previous study. The Nenjiang Formation was deposited in a deep lake environment, and the lithological associations were dominated by the dark-mudstone, all these information have indicated the Nenjiang Formation has a relative low sedimentation rates. Therefore, the isotopic age at the boundary between Nenjiang Formation and Yaojia Formation is 85.1–85.5 Ma, the boundary age between the first member of Nenjiang Formation and second member of Nenjiang Formation is 83.1–83.5 Ma, and the core length of Nenjiang Formation is about 101 m which calculated that the sedimentation rate from the first member of Nenjiang Formation to second member of Nenjiang Formation is 42.1–63.1 m/Ma.

7.8 Sedimentary Environment of the First and Second Members of Nenjiang Formation

The Nenjiang Formation is mainly composed of deep-water lacustrine gray to black mudstone, marl, limestone, oil shale and lenticular dolomite with intercalation of gray sandstone and fine sandstone. According to the lithologic associations and paleontological fossil assemblages, the Nenjiang Formation can be divided into five members, of which the first and second members of the Nenjiang Formation are the main source rocks, and the third and fourth members of the Nenjiang Formation are the important Heidimiao oil layer, which is also the shallowest oil layer in the Songliao Basin. High quality source rocks were developed in the first and second members of Nenjiang Formation in the Songliao Basin. Massive studies have been carried out on the controlling factors and sedimentary environment of continental source rocks. Previous studies indicated that the first and second members of Nenjiang Formation was a warm and humid climate environment.

Wang et al. (2010) and Song et al. (2013) conducted a comprehensive geochemical analysis of SK-1 in Songliao Basin to reveal the controlling factors of ancient lake water environment and TOC content change in the first and second members of Nenjiang Formation. According to the change patterns of TOC data, the first and second members of Nenjiang Formation are divided into four parts for discussion. It is considered that in the high TOC intervals, the lake water body is characterized by partial reduction and hypoxia, and there will be water stratification (salinity stratification). During the low TOC intervals, the bottom water environment is characterized by partial oxidation and low salinity, and there is no water stratification.

Biomarkers also further indicate the geochemical characteristics of water during deposition of the Nenjiang Formation. Among them, Pr/Ph is widely used to reflect the redox characteristics of water bodies. It is found that Pr/Ph at the bottom of the first member of Nenjiang Formation indicates a partial oxidation water characteristic, and this period also corresponds to a lower TOC value. However, the first member of Nenjiang Formation has a lower Pr/Ph value, reflecting a reduction environment. The second member of Nenjiang Formation shows different water characteristics and organic matter enrichment mechanism from the first member of Nenjiang Formation. Specifically, at the position of high TOC value at the bottom of the second member of Nenjiang Formation, Pr/Ph value indicates the partial oxidation water environment.

Previous studies have pointed out that the Nenjiang Formation have developed two sets of oil shale layers, located in the middle of K_2n^1 and bottom of K_2n^2. At present, many geological evidences show that the two sets of oil shale were formed in different sedimentary environments. According to the hydrogen index (HI), kerogen, carbon isotope value of n-alkanes and other indicators, the organic matter in K_2n^1 mainly comes from the photosynthesis of algae in the lake. According to the evidences from paleontology and biomarkers, the Songliao Basin has experienced intermittent transgression in the K_2n^1, which promoted the salinity stratification of lake water and the hypoxia of bottom water, resulting in the enrichment and preservation of organic matter. The organic matter of K_2n^2 mainly comes from the lake algae and terrestrial higher plants input. The K_2n^2 is a partial oxidation water environment, so the formation mechanism of oil shale at the bottom of K_2n^2 is different from that of K_2n^1. A recent work on hydrogen isotopes of n-alkanes found an obvious negative isotope excursion in the upper part of K_2n^1. It is considered that this excursion event represents an abnormal hydrological process and may indicate the change of hydrological conditions around the Songliao Basin. It is considered that the new hydrological processes provide a large amount of nutrients, promotes the productivity of the lake itself, and leads to the formation of oil shale at the bottom of K_2n^2 (Fig. 34).

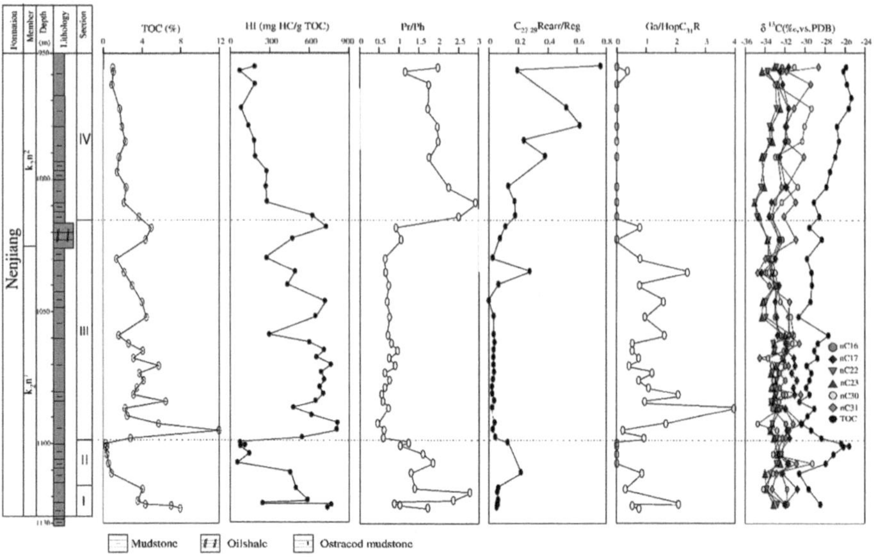

Fig. 34 Organic geochemistry of the first and second members of Nenjiang Formation (Song et al. 2013)

Additionally, Niu et al. (2021) discussed the formation mechanism and spatial distribution of oil shale in the Nenjiang Formation in combination with different drilling and profile records in Songliao Basin. According to the TOC variation characteristics of different wells, stratigraphic correlation was carried out and three potential oil shale layers were identified (a 5–10 m thick oil shale layer was newly identified at the bottom of K_2n^1), but the newly identified oil shale layer was only recorded in SK-1 and LY-1 wells, which may indicate the spatial difference of oil shale distribution in this layer. Meanwhile, combined with different paleolake environmental indicators (redox, salinity and stratification), the sedimentary environment and oil shale enrichment mechanism of the first and second members of Nenjiang Formation are discussed. It is considered that the organic matter enrichment of the first member of Nenjiang Formation is mainly dominated by the "preservation" model related to transgression, and the formation mechanism of oil shale in the second member of Nenjiang Formation is mainly controlled by the change of hydrological conditions in Songliao Basin. The "productivity" model is the main factor for the enrichment of organic matter in K_2n^2 (Fig. 35).

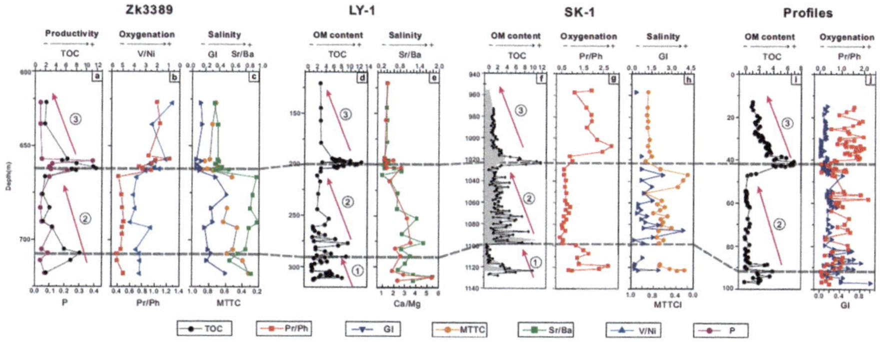

Fig. 35 Basin scale correlation of Nenjiang Formation Unit 1–2 (Niu et al. 2021)

7.9 Fine-Grained Sedimentary Rocks and Varves in the First and Second Members of Nenjiang Formation

The laminations of the first member of the Nenjiang Formation are well developed. The laminations are mainly couplets composed of silt- and clay-rich sublamina (marked in the Fig. 36), which are structures of annual laminations/varves. The lamina couplets have a good continuity in the cores, which has a diameter of 21.6 cm (Zou et al. 2016). According to the classification criteria of fine-grained sedimentary rocks by Lazar et al. (2015), we classified the mudstones of the first member of the Nengjiang Formation into coarse mudstone, medium mudstone and fine mudstone according to grain size. For example, in the fine mudstone, the color is darker due to the high organic content, while in the medium mudstone, the color is lighter, and laminations are extremely developed in the fine mudstone (Fig. 36).

Based on sedimentological studies, we found the first member of the Nengjiang Formation was influenced by muddy hyperpycnal flows (Tian et al. 2020), which is relatively dense land-derived gravity flow enters into a marine or lacustrine water reservoir, thus carrying land-derived debris (Mulder et al. 2003; Zavala et al. 2011; Zavala and Pan 2018). Some of the laminations in the first member of the Nengjiang Formation tend to be varves influenced by hyperpycnal flows (Fig. 37). The strong rhythm of the laminated mudstone (Fig. 37) suggests that this laminated mudstone in the first member of the Nengjiang Formation is likely to be climate-driven, in other words, climate could drive river flux in the Late Cretaceous Northeast China and thus control the deposition of varves in the Songliao paleo-lake (Fig. 37). This analog to some modern hyperpycnal rivers, for example, the Yellow River is influenced by modern monsoon, leading to obvious seasonal characteristics, and the hyperpycnal flows at Xiaolangdi Reservoir will be released during flood season (Yang et al. 2019).

Climate simulations by previous work, for instance, Chen et al. (2009, 2013) found the Late Cretaceous climate in East Asia was influenced by the monsoon, indicates the presence of seasonal precipitation at that time, and this provides a good basis for the interpretation of seasonal characteristics of our annual laminations. As shown in

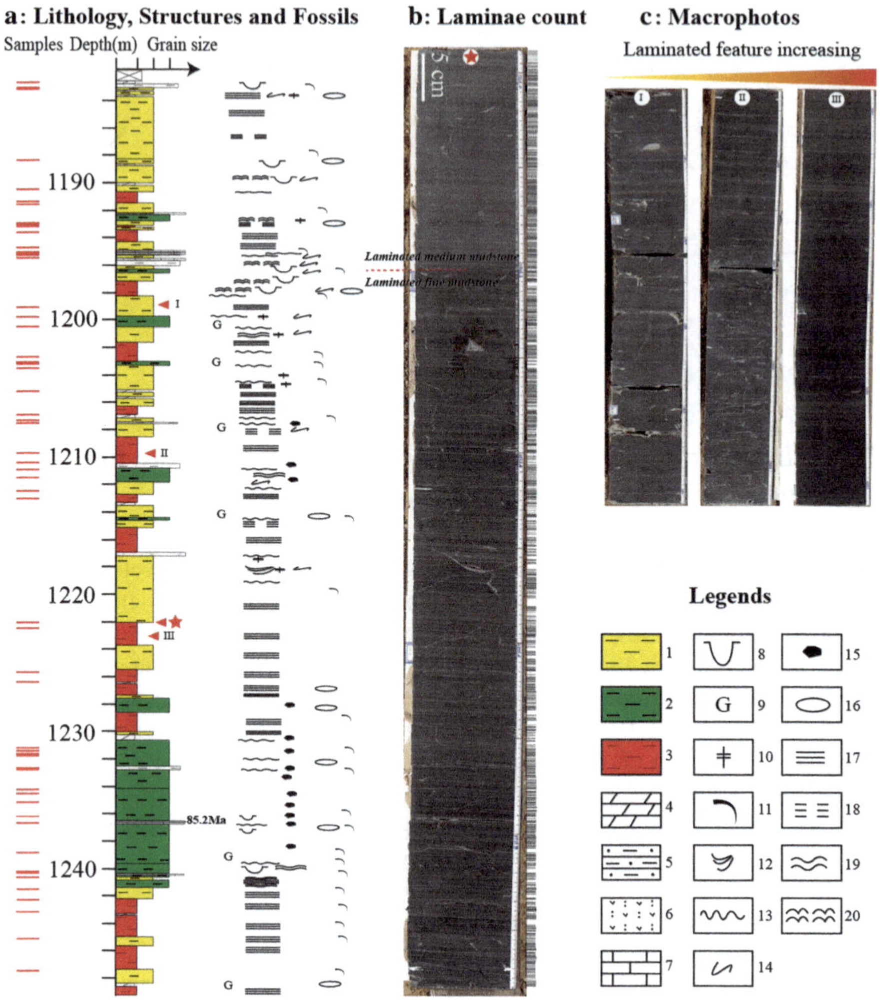

Fig. 36 Sedimentary log showing samples, lithology, structures, fossils, and selected core photos of the first member of the Nenjiang Formation of SK-2 from Songliao Basin. 1. Medium mudstone; 2. coarse mudstone; 3. fine mudstone; 4. dolostone; 5. muddy sandstone; 6. tuff; 7. limestone; 8. burrow; 9. graded bed; 10. crack; 11. bioclast; 12. curved lamination; 13. scour surface; 14. deformed structure; 15. mudstone intraclast; 16. dolomitic lens; 17. continuous planar parallel lamination; 18. di continuous planar parallel lamination; 19. continuous wavy parallel lamination; 20. dis continuous wavy parallel lamination

Fig. 36 the silt-rich sublamina corresponding to a relatively wet season, during which rivers carry a large amount of sediments into the Songliao paleo-lake, while the clay-rich sublamina corresponding to a relatively dry season, and suspended sedimentation dominants. Therefore, the thickness of this clay-rich sublamina is thinner than the silt-rich sublamina (Fig. 37), and the thickness variation of the varve couplets reflects

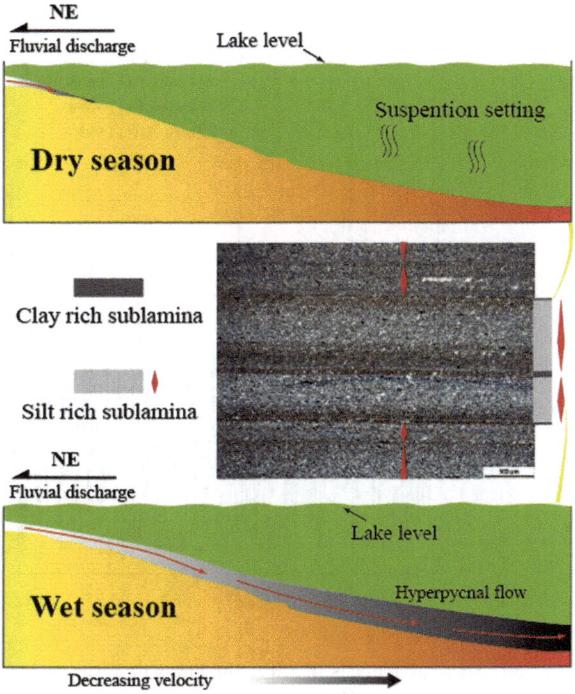

Fig. 37 Influences on the laminated sediments between dry and wet seasons

the rainfall variation during the Late Cretaceous period in the Songliao Basin. The vares are tend to be thicker in years of abundant rainfall and thinner in the opposite.

As described above, it is likely that the lamina collected from the first member of the Nenjiang Formation is annual lamination, and it is necessary to do some further research. In this paper, the varve couplet thicknesses of the samples in Fig. 36b (depth range: 1221.977–1222.961 m) were counted, and a total of 1929 couplets, and the average thickness was 510 μm. The thickest varve couplet is 8598 μm, while the thinnest is 87 μm, and this means large interannual variation of precipitation might exist in the Late Cretaceous. Spectral results (Fig. 38) show significant 13.1-year and 2–4-year cycles (with >99% confidence), which belong to the ENSO cycle. Therefore, the varves of the Nenjiang Formation recorded the Late Cretaceous ENSO signal, which provide information for further study on the ENSO characteristics (e.g., ENSO's frequency or amplitude).

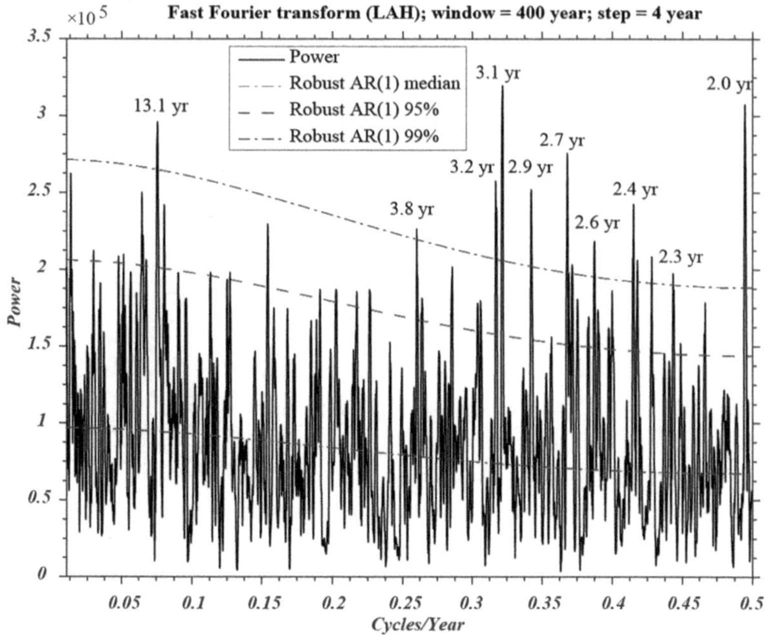

Fig. 38 Spectral analysis of varve thickness time-series from the First Member of the Nenjiang Formation of Songliao Basin

7.10 Marine Incursion (Transgression) and Organic Matter Enrichment in the Nenjiang Formation Unit 1–2

In the 1970s, when studying the sedimentary evolution of Late Cretaceous in Songliao Basin, Daqing Oilfield put forward the view that "transgression" occurred in Qingshankou Formation and Nenjiang Formation. Since then, the debate on "transgression" in Songliao basin has not stopped in China's petroleum geological communities. Previous studies have found dinoflagellate sterane and dinoflagellate fossils from the Nenjiang Formation, which are considered to represent the marine fossil record. Therefore, it is considered that marine transgression occurred during the Nenjiang Formation (Gao et al. 1992). Because of dinoflagellasterane was widely existing in Songliao Basin, Hou et al. (1999) Proposed that there were two obvious transgressive events in Songliao Basin during the Middle Cretaceous, which occurred in the first member of Qingshankou Formation and the first and second members of Nenjiang Formation. In addition, based on the study of biomarkers, Feng et al. (2009) found that the lake water body during the Qingshankou Formation and Nenjiang Formation were characteristed with high salinity and stratification, which was also considered to be the result of seawater transgression. However, there are many studies against the seawater transgression. They believe that paleontological and biomarker evidences are not enough to explain transgression events. Similar biological fossils and geochemical characteristics can also be found in terrestrial brackish water environment (Fig. 39).

With the completion of SK-1 drilling project, complete cores of Qingshankou Formation and Nenjiang Formation were obtained, which provided continuous and high-resolution materials for the study of Songliao transgression events. Xi et al. (2011) found foraminiferal fossils from the Qingshankou Formation and Nenjiang Formation in SK-1, which are considered to be reliable indicators of marine environment. In addition, Cao et al. (2016) also found calcareous nannofossils in the layer corresponding to foraminifera fossils, which are considered to be derived from the marine environment, and considered the calcareous ultramicro plankton in the Northwest Pacific was brought into the Songliao Basin when the sea water incursion.

In addition, the results of biomarkers in SK-1 show that the lower part of the second member and the lower part of the first member of the Nenjiang Formation

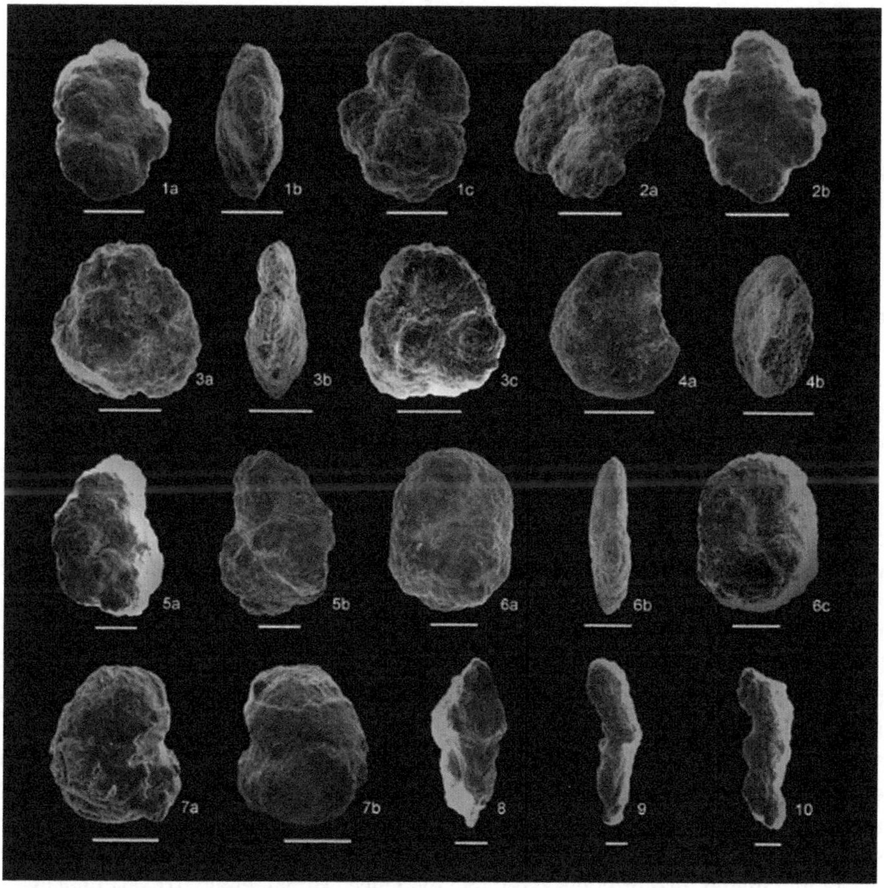

Fig. 39 Foraminifera fossils in Nenjiang Formation of SK-1 cores (Xi et al. 2011)

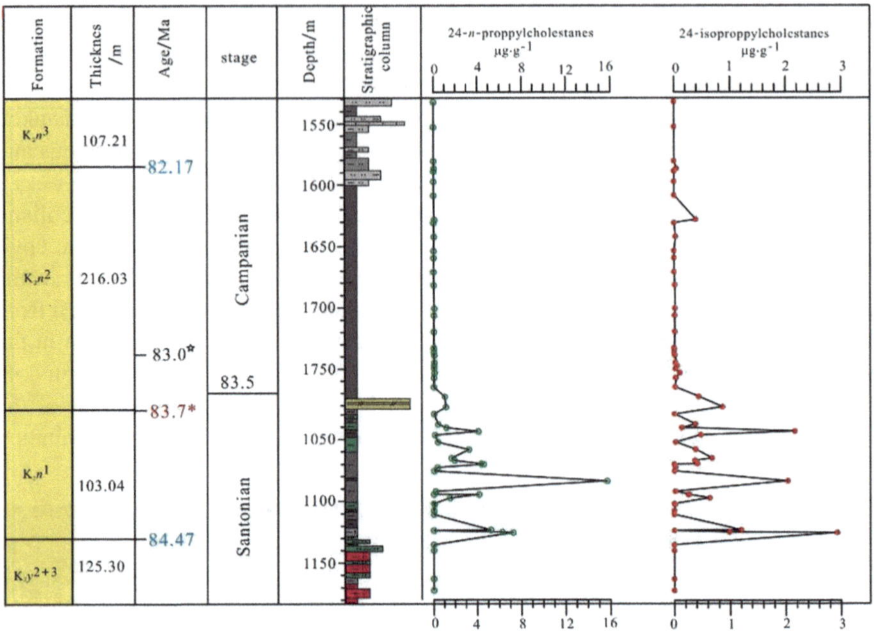

Fig. 40 Marine biomarkers from Nenjiang Formation of SK-1 cores indicating marine incursion events (Hu et al. 2015)

show high and serrated C30 sterane content, suggesting that there are large-scale and multi-stage transgression events in these two periods, which is similar to the previous research results (Hou et al. 2004). According to the results of SK-1, transgression events occurred in the first and second members of Nenjiang Formation and the first and second members of Qingshankou Formation, especially in the lower part of the first and second members of Nenjiang Formation (Fig. 40).

In conclusion, the evidence of marine incursion (transgression) in the Songliao Basin mainly comes from the Cretaceous mudstone shale development sections, such as the mudstone of the first and second members of Qingshankou Formation and the mudstone of the first and second members of Nenjiang Formation. At present, some scholars believe that the formation of Cretaceous organic rich shale in Songliao basin is related to the salinity stratification of lake water and hypoxia at the bottom caused by periodic seawater intrusion (Wang et al. 2013a, b; Hu et al. 2015). This view holds that the periodic intrusion of seawater into the lake basin is the fundamental reason for the enrichment of organic matter. Seawater entering the lake basin causes water stratification and hypoxia at the bottom of the lake, so as to promote the preservation of organic matter. Meanwhile, the salinization of the lake basin caused by seawater intrusion will also improve the primary productivity of the lake and increase the supply of organic matter. The sulfur isotope of transgression affecting the enrichment of organic matter also provides a reliable evidence (Huang et al. 2013) (Fig. 41).

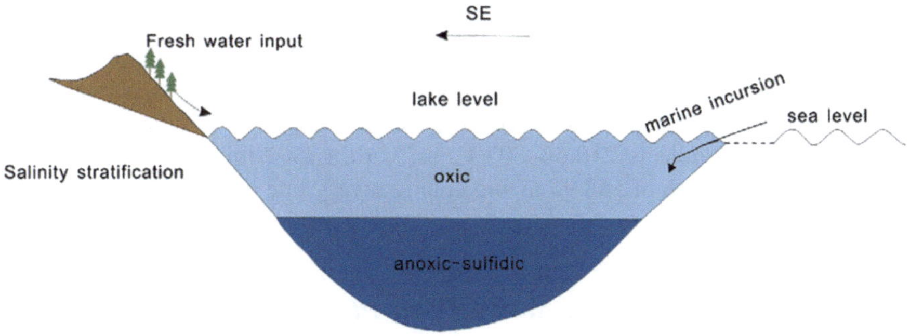

Fig. 41 Model of lake water stratification caused by marine incursion in Late Cretaceous Songliao Basin (Wang et al. 2013a, b)

7.11 Nenjiang Formation Unit 3–5

SK-1n core has collected 532.07 m of members 3–5 of the Upper Nenjiang Formation which correspond to the middle Campanian. The borehole gets a highly continuous core section with recovery rate of 94.8%. Nine litho-types are recognized in the section. Sedimentary facies are predominantly lacustrine, delta and fluvial, which can be subdivided into fifteen microfacies including channel-lag deposit, point bar, natural levee, crevasse splay, crevasse channel, flood plain, flood lake, distributary mouth bar, subaqueous distributary channels, subaqueous natural levee, subaqueous inter-distributary bay, distal bar, mudstone of still water, adjacent shore sand bar and lake turbidite. The whole sequence is composed of 417 m scale cycles (sixth-order cycle), 94 fifth-order cycles, 26 fourth-order cycles and 3 third-order cycles. Centimeter-scale depiction of the core-section revealed some previously unknown horizons such as special kind of mudstone, dolomite and volcanic ash of the formation. The red beds in the fifth member of Nenjiang Formation, being an important case study of mid-Cretaceous continental red beds found in the northeast Asia, will become a standard of potential formulation section for the studies of late Cretaceous global oxygen-enriched events. The detailed description of the third and fourth members of Nenjiang Formation is significant for their further exploration in the Songliao Basin.

8 Stratigraphy of the Structural Inversion Stage (Sifangtai and Mingshui Formations) from the SK-1 Cores

The structural inversion strata are mainly comprised of the Sifangtai and Mingshui Formations. The Sifangtai and Mingshui Formations (SMF) are a set of lacustrine, deltaic and fluvial sedimentary sequences, underlain by sedimentary rocks with high organic matter content of the Nengjiang Formation and overlain by sedimentary rocks of the Paleogene Taikang Formation (Wang et al. 2016a). The SMF were continuously cored in the SK-1n borehole. The cored interval of the Sifangtai Formation in the SK-1n borehole is 807.12–1021.60 m with a total thickness of 214.48 m, accumulative core length of 209.02 m and core recovery rate of 97.5%. The Mingshui Formation in the SK-1n borehole is 210.66–807.12 m, with a total thickness of 596.46 m, cumulative core length of 558.94 m and core recovery rate of 93.71%.

8.1 Lithology and Sedimentary Environment in the Sifangtai and Mingshui Formations

The assignation of lithofacies codes and their environmental interpretations follow the methods of DeCelles et al. (2007) and Miall (2010). Overall, the SMF of SK-1n core is a predominantly fine grain-sized sedimentary sequence (Fig. 42), in which fine-grained lithofacies comprise more than 80% of the total thickness, sandstone lithofacies comprise ~15%, and conglomerate lithofacies comprise less than 5%. The sedimentary environments in the SMF are reconstructed based on an integrated analysis of lithofacies associations, contacts and erosional surfaces between beds, typical sedimentary structures, and fossil occurrences. The SMF contains three major facies associations, deposited in fluvial, deltaic, and lacustrine environments. The fluvial facies are characterized by fining-upward trends, with erosional surfaces at their bases, conglomerates and cross-bedded sandstones in the lower portions, and fine-grained sediments commonly with pedogenic structures in the upper portions. The lacustrine facies are lithologically characterized by greenish to grayish siltstones and mudstones with horizontal laminations and the frequent occurrence of ostracod/conchostraca fossils. The deltaic facies represent transitional environments between fluvial and lacustrine deposits and are characterized by both coarsening upward and fining upward trends as well as the presence of soft-sediment deformation and burrowing. Accordingly, the SMF is subdivided into five discrete fining-upward depositional units, each of which is marked by relatively coarse-grained fluvial facies at the base and transitions upward to generally finer-grained deltaic and lacustrine facies. The lower boundary of each depositional units is defined at an unconformity (Gao et al. 2021).

Paleosols are well-developed from the SK-1n core. Based on the characteristics of plant root traces, calcareous nodules, redoximorphic mottling and slickensides, the paleosols can be classified as Aridisol, Inceptisol, Vertisol and Alfisol. Aridisols are usually developed under arid to semi-arid climates, in which rainfall is insufficient to leach soluble salts, and calcareous (Bk) horizons are commonly formed (Retallack 2001; Soil Survey Staff 2014). Inceptisols are characterized by weakly developed paleosol horizons and pedogenic features, and the presence of relict primary sedimentary structures (Retallack 2001; Soil Survey Staff 2014). Vertisols are clayey soils with deep, wide cracks and abundant slickensides, formed through repeated strong shrinking and swelling of smectitic clays during wet and dry seasons of the year

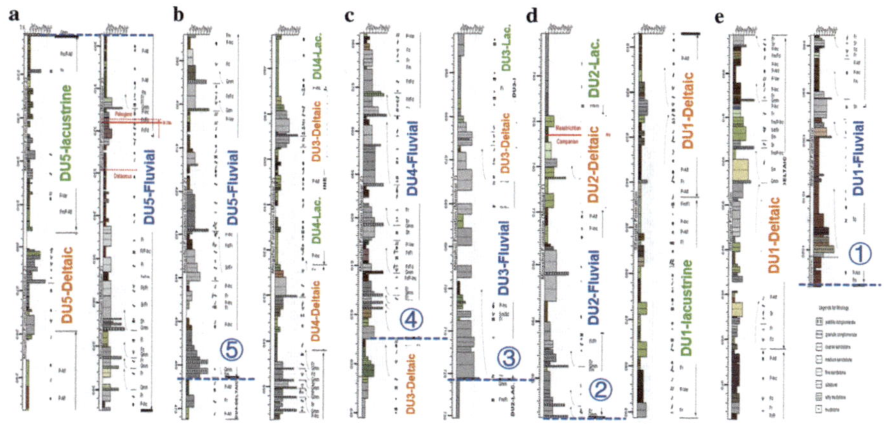

Fig. 42 Logs of lithofacies and paleosol types in the SMF of SK-1n core

(Retallack 2001; Soil Survey Staff 2014). Alfisols are formed in wetter climates and develop redoximorphic mottling, with a few calcareous nodules at depths of more than 1 m (Retallack 2001; Soil Survey Staff 2014). The development of paleosol type in the SMF is fluctuating, such as the significant development of Alfisols in the depth intervals of ~850–800 m, ~550–450 m and ~350–300 m, and the occurrence of Vertisols and Aridisols in the depth intervals of ~1000–850 m and ~700–550 m (Fig. 42).

8.2 *The Chronostratigraphic Framework in the Sifangtai and Mingshui Formations*

The SMF possesses a uniquely chronostratigraphic framework, which provides excellent material for the study of terrestrial climate and environment of the Cretaceous to Paleogene (Fig. 43; Feng et al. 2010; Wang et al. 2013a, b). Based on magnetostratigraphy, biostratigraphy, and cyclostratigraphy, the age of the SMF is constrained as late Campanian to early Danian (~76.1 to 65.1 Ma).

The study of ostracod biostratigraphy has divided the SMF into two ostracod assemblages: the *Talicypridea amoena—Metacypris kaitunensis—Ziziphocypris simakovi* assemblage (970–335.09 m) and the Ilyocypris assemblage (335.09–317 m), with the Ilyocypris assemblage crossing the K-Pg boundary (Fig. 43; Qu et al. 2014). Five palynological biozones were established that constrain the SMF mostly from late Campanian to early Danian, with only a small number of spores and pollen were discovered within the depth interval of 360.6–233.9 m that probably reflect extinction of some key elements of the Cretaceous vegetation in East Asia (Fig. 43; Li et al. 2011). The study of charophyte redefined four charophyte biozones and

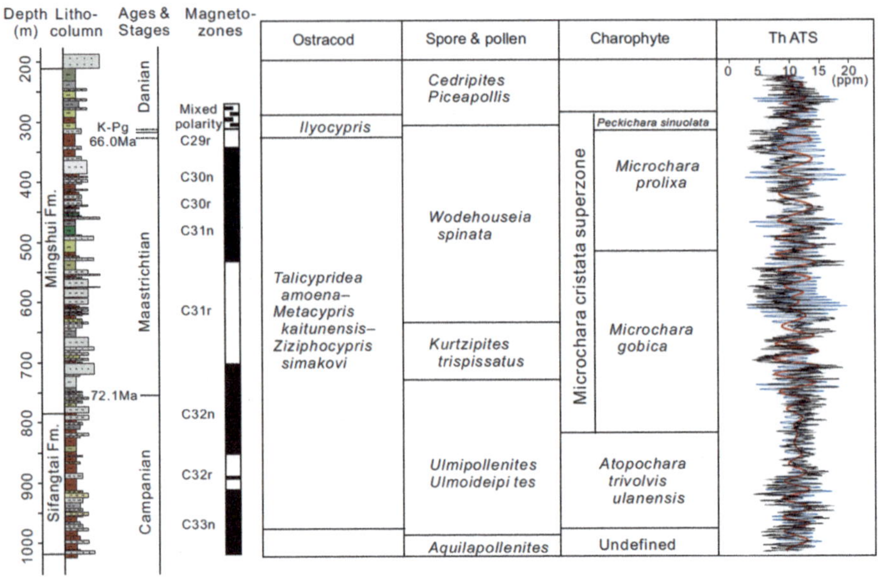

Fig. 43 Integrated chronostratigraphic framework of the SMF in the SK-1n core. Lithostratigraphic column after Wang et al. (2015a, b); magnetostratigraphic zones after Deng et al. (2013); ostracod zones after Qu et al. (2014); spore and pollen zones after Li et al. (2011); charophyte zones after Li et al. (2019); astronomical timescale (ATS) based on thorium (Th) logging data (black curve) with 405 kyr (red curve) and 100 kyr (blue curve) filter outputs (Wu et al. 2014). Thin dotted lines indicate possible range of K-Pg interval inferred by biostratigraphy

one superzone in the SMF, with the Peckichara sinuolata belt constraining the K-Pg boundary between 329.06–328.77 m (Fig. 43; Li et al. 2019; Wan et al. 2013).

Paleomagnetic studies subdivided the SMF into nine magnetozones (Deng et al. 2013). Based on the U–Pb dating results (from 91.4 ± 0.9 Ma to 83.7 ± 0.8 Ma) of four Bentonite beds in the underlying Nengjiang and Qingshankou Formations that correspond to C34n, the polar zones of the Sifangtai-Mingshui Formation can be assigned to C33n (910.2–1020.4 m), C32r (852.6–910.2 m), C32n (700.9–852.6 m), C31r (530.78–700.88 m), C31n-C30n (342.1–530.78 m), C29r (317.0–342.1 m), and one chron with mixed polarity (175–317.03 m) (175–317.03 m). C30r and C32n.1r are missing in the SMF, possibly due to their short time durations (Fig. 43; Deng et al. 2013).

An astronomical time scale (ATS) was established by tuning filtered 405 kyr eccentricity cycles of thorium logging data to a target curve of the astronomical solution La2010d (Fig. 43; Wu et al. 2014). The ATS is anchored by the boundary between paleomagnetic chrons C29r and C30n at 317.0 m, with an age of 66.3 Ma, and was tested by comparison to a series of age anchors at boundaries of geomagnetic chrons (Gradstein et al. 2012). According to the ATS, the hiatus within the unconformity at the bottom of the Sifangtai Formation is resolved to be ~3.8-million-year-long, and the Campanian/Maastrichtian boundary and the K-Pg boundary are placed at 752.8 m and 318 m, respectively.

8.3 *Climatic and Environmental Changes of the Songliao Basin in the Late Cretaceous*

The benthic foraminifera oxygen isotope records ($\delta^{18}O_{benthic}$) show that the global climate was in a cooling process in the late Campanian and Maastrichtian, but with multiple short-term climate fluctuations on a million-year scale (Barrera and Savin 1999; Friedrich et al. 2012; Linnert et al. 2014; Hull et al. 2020). In the late Campanian (~76–74 Ma), pedogenic carbonate $\delta^{13}C$ ($\delta^{13}C_{pc}$), $\delta^{18}O$ ($\delta^{18}O_{pc}$) and soil water $\delta^{18}O$ ($\delta^{18}O_w$) records in the Songliao Basin indicate warmer and drier climate, probably caused by the presence of coastal mountains along the eastern margin of the East Asia continent prevented Pacific moisture from reaching into the Songliao Basin. During ~74–72 Ma, a broad negative $\delta^{18}O_{benthic}$ excursion indicates marine warming. In the Songliao Basin, a shift from fluvial to lacustrine sedimentary facies in the SMF occurs in concurrence with changes in paleosol types from presence of Aridisols and Vertisols to dominance of mottled-Alfisols at ~74 Ma, which is presumed to be caused by a combination of coastal mountain range weathering and erosion, altitude loss, and global warming (Fig. 44).

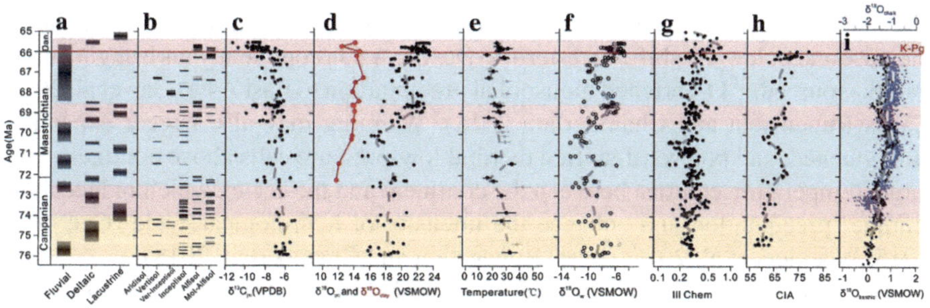

Fig. 44 Correlation of paleoclimatic records between the Songliao Basin and the marine realm in the latest Cretaceous. Records in the SMF of the SK-1n core are presented as **a** sedimentary facies, **b** paleosol types, **c** $\delta^{13}C_{pc}$ values, **d** $\delta^{18}O_{pc}$ (black dots) and $\delta^{18}O_{clay}$ (red dots) values, **e** Δ_{47}-temperatures derived from pedogenic carbonates in fine (black dots) and coarser (gray dots) paleosol matrix, **f** calculated $\delta^{18}O_{water}$ values based on pedogenic carbonates, **g** illite chemistry index, and **h** CIA. The pink, blue, and orange bands indicate warm-wet, cool-dry, and warm-dry climates, respectively. Marine records (**i**) are $\delta^{18}O_{benthic}$ values from the tropical Pacific (black dots, ODP 1210B, Jung et al. 2013) and $\delta^{18}O_{chalk}$ values from the mid-latitude Chalk Sea (blue dots, Stevns-1, Thibault et al. 2016). Kernel smoothed timeseries (gray and yellow dashed lines) were calculated using a fixed bandwidth (250 or 50 kyrs) using an Epanechnikov kernel centered on each datapoint (e.g., Gao et al. 2015). Non-mudstone paleosol carbonate clumped isotope data are shown in gray and are not included in the kernel smoothed timeseries (see discussion in Gao et al. (2021))

A significant global cooling event occurred in the early Maastrichtian (EME event, early Maastrichtian Event, ~70 Ma), which is characterized by a positive $\delta^{18}O_{benthic}$ excursion of ~1‰. During ~71.5–69.5 Ma, a striking $\delta^{18}O_{pc}$ negative excursion of ~3‰ in the SMF with contemporaneous occurrences of fluvial sediments and Vertisols, increasing $\delta^{13}C_{pc}$ values (higher than ~7‰), and decreasing $\delta^{18}O_w$ (by ~3‰) and CIA (by ~8) values are considered as the terrestrial climate response within the Songliao Basin to this global cooling (Fig. 44; Gao et al. 2015). This cooling event in the Songliao Basin are likely caused by equator-ward movement of westerlies over the mid-latitude Asian continent, whereas strengthening of westerlies at the expense of monsoons caused decreases in precipitation, soil respiration and weathering intensity in the Songliao Basin (Gao et al. 2015; Caves Rugenstein and Chamberlain 2018) (Fig. 44).

A warming event occurred in the mid-Maastrichtian (MME Event, Mid-Maastrichtian Event, ~69.5–68.5 Ma). In the Songliao Basin, a warmer and wetter climate at ~69.5–68.5 Ma is recorded by presence of lacustrine and deltaic facies, development of Alfisols and especially mottled-Alfisols, decreasing$\delta^{13}C_{pc}$ values (as low as ~10‰) indicative of increased soil respiration, and positive excursions by ~4‰ in both $\delta^{18}O_{pc}$ and δ^{18}Owater values in support of more Pacific-sourced precipitation (Fig. 44). Increases illite chemistry index (as high as ~0.5) CIA (by more than 5) values during the MME indicate intensified terrestrial chemical weathering in response to a warmer and more humid climate (Fig. 44g, h). The cause of MME warming could have been an intense phase of Ninety East Ridge volcanism, which is a basaltic volcanic chain that formed by the northward migration of the Indian plate over the Kerguelen hotspot (Keller et al. 2016; Mateo et al. 2017). Greenhouse warming triggered by volcanism may have induced land temperature increase and poleward-shifted westerlies over the Asian continent, thus bringing more Pacific-sourced, ^{18}O-enriched monsoonal precipitation to East Asia (Gao et al. 2015; Caves Rugenstein and Chamberlain 2018). Warming may also trigger expanded, strengthened, and poleward-shifted thermal low-pressure cells above continents and larger temperature contrast between the continent and the ocean, which induce more summer precipitation and enhance the intensity of hydrological cycle (Gao et al. 2015; Woelders et al. 2017; Caves Rugenstein and Chamberlain 2018).

The temperature returned to cold after the MME event, and this cooling trend has persisted for around two million years (Petersen et al. 2016). Occurrence of fluvial facies, development of more Inceptisols, Vertisols and even Aridisols, as well as observations of more negative $\delta^{18}O_{pc}$ and $\delta^{18}O_w$ values (decreases by ~2‰ in both), more positive $\delta^{13}C_{pc}$ values (higher than ~7‰), combined with lower values of illite chemistry index (~0.3–0.4) and CIA (~65–70) in the Songliao Basin at ~68.0–66.5 Ma all indicate a cooler and drier climate with decreased intensity of chemical weathering (Fig. 44). Enhanced silicate weathering during the MME greenhouse condition, which is also suggested by weathering indices in the SMF, and waning volcanism may have contributed to lowering atmospheric pCO_2 levels and global cooling (Dessert et al. 2001; Mateo et al. 2017).

8.4 Climatic and Environmental Changes of the Songliao Basin Across the K-Pg Boundary

The K-Pg boundary biological mass extinction involved the extinction of multiple phyla of organisms (e.g. non-avian dinosaurs), which was one of five catastrophic mass extinction events in Earth history. The marine records indicate significant fluctuations of global climate across the K-Pg boundary, manifested by multiple cooling and warming events that affected biological extinction and recovery. However, the changes in terrestrial climatic environments across the K-Pg boundary and their linkages with major geological-biological events are still poorly understood. Climatic evolution in the Songliao Basin across this critical time interval has been heavily investigated through sedimentological and geochemical methods and correlated to marine counterparts (Gao et al. 2015, 2021; Zhang et al. 2018).

A warming event occurred during the Late Maastrichtian (LME Event, ~66.4–66.1 Ma), and the release of large amounts of CO_2 from the Deccan basalt may have led to a warming climate (Hull et al. 2020). In the Songliao Basin, a Δ_{47}-temperature rise of ~6 °C during the LME provides further evidence for terrestrial warming (Zhang et al. 2018; Fig. 44f). Moreover, the occurrence of mottled-Alfisols and the negative $\delta^{13}C_{pc}$ excursion of ~2‰ at the LME interval in the SMF suggest more humid climate contemporaneous with warming (Fig. 45c, d). Similar to the MME event, global greenhouse warming in the late Maastrichtian caused an increase in land temperatures, poleward shifted westerlies over mid-latitude East Asia, and a regional precipitation increase in the Songliao Basin (Gao et al. 2015; Caves Rugenstein and Chamberlain 2018).

Following the LME warming, a transient cooling at ~100 kyr timescale occurred immediately preceding the K-Pg boundary, featuring a temperature decrease to pre-LME conditions in the marine realm and an up to ~8 °C decrease in the terrestrial realm (Li and Keller 1998; Wilf et al. 2003; Tobin et al. 2014). Weathering of Deccan basalts has been proposed as a possible reason for this cooling event, with disputable evidence from osmium isotopes in marine sediments (Schoene et al. 2019; Sprain et al. 2019), but the records in the SMF strongly support this hypothesis. Elevated values of CIA (increase by ~10, Fig. 45i) and illite chemistry index (increase by ~0.2, Fig. 45j) occur at ~66.2–66.1 Ma, right in the trend of decreasing temperatures after the peak warmth of LME as indicated by marine and terrestrial isotopes (Fig. 45f, j). At ~66.1–66.0 Ma values of CIA and illite chemistry index drop down to pre-LME levels, interpreted as decreased weathering due to a cooler and drier climate, which provides additional support for preservation of the transient cooling before K-Pg in the Songliao Basin.

After the K-Pg boundary event, the Earth continued to warm during the first 500,000 years of the Paleogene (Hull et al. 2020), consistent with the warming, wetting, and weathering enhancement trends in the Songliao Basin, which may be related to the continued CO_2 emissions from the Deccan basalt.

How climate change affects biological extinction and recovery across the K-Pg boundary remains controversial and is at the forefront and hotspots of international

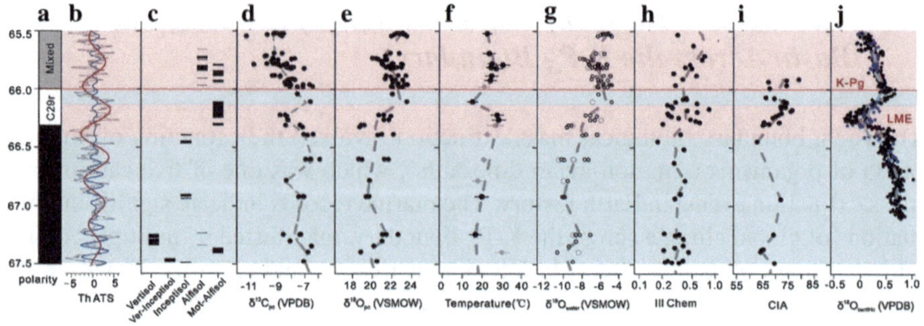

Fig. 45 Correlation of paleoclimatic records of the Songliao Basin (**a**–**i**) and the marine realm (**j**) across the K-Pg boundary. Records in the SMF of the SK-1n core are presented as **a** magnetostratigraphic chrons, **b** Th logging data (gray curve) and astronomical time scale at 400 ka (blue curve) and 100 ka (red curve), **c** paleosol types, **d** $\delta^{13}C_{pc}$ values, **e** $\delta^{18}O_{pc}$ values, **f** Δ_{47}-temperature from pedogenic carbonates in mudstone (black dots) and coarser (gray dots) paleosol matrix, **g** calculated $\delta^{18}O_{water}$ values based on carbonates in mudstone (black dots) and coarser (gray dots) paleosol matrix, **h** illite chemistry index, and **j** CIA values. Marine records (**j**) are $\delta^{18}O_{benthic}$ values tropical Pacific (black dots) and South Atlantic (gray dots) with data from Barnet et al. (2018). Kernel smoothed timeseries (gray and dark blue lines) were calculated using a fixed bandwidth (250 or 50 kyrs) using an Epanechnikov kernel centered on each datapoint (e.g., Gao et al. 2015a). Non-mudstone or silty mudstone paleosol carbonate clumped isotope data are shown in gray (**g**) and are not included in the kernel smoothed timeseries

research. The fossil record of marine foraminifera suggests that MME events promote increased biodiversity, while LME events lead to ecological stress and foraminiferal miniaturization that directly affect extinctions (Keller et al. 2016). However, the study of terrestrial biological evolution is still deficient.

Palaeoecological studies based on SK-1n and their linkages with the East Asia continent suggest that EME cooling events lead to a decrease in microbial biodiversity in East Asia (e.g. Charophytes), while MME and LME warming events correspond to an increase in biodiversity (Li et al. 2021). The results of a comparative study of terrestrial fossils and climate change across the K-Pg boundary provide preliminary evidence that terrestrial climate has significant influence on biological evolution by similar and/or different ways to the marine environment, but further research is needed.

9 Logistics

The whole trip will be in the Physical Geological Data Center, China Geological Survey, about 45 km away from the Conference center. The address is No. 245 Yanling Road, Yanjiao Economic and Technological Development Zone, Sanhe City, Hebei Province. Duration of the trip is one day. Trip will be indoor with a temperature of ~20–30 °C. Short walk is needed. Shuttle bus will be used for transportation and lunch will be provided in the Physical Geological Data Center. In case of emergency, contact info of trip leaders are: +8,613,466,721,378 (Yuan Gao); + 8,613,504,707,690 (Youfeng Gao).

Acknowledgements Trip leaders greatly appreciate the Physical Geological Data Center, China Geological Survey for help on logistics. We thank Prof. Xiumian Hu in Nanjing University, Prof. Pujun Wang in Jilin University, and Prof. Xiaoqiao Wan in China University of Geosciences Beijing for their help on this guide.

References

An D (2018) Early Cretaceous sedimentary records and provenance analysis of the upper member of Shahezi formation in Songliao Basin. China University of Geoscience (Beijing) (in Chinese)

Barnet JSK, Littler K, Kroon D et al (2018) A new high-resolution chronology for the late Maastrichtian warming event: establishing robust temporal links with the onset of Deccan volcanism. Geology 46:147–150

Barrera E, Savin SM (1999) Evolution of late Campanian-Maastrichtian marine climates and oceans. Special Paper of the Geological Society of America, pp 245–282

Barron JA, Bukry D (2006) Development of the California current during the past 12,000 yr based on diatoms and silicoflagellates. Palaeogeogr Palaeoclimatol Palaeoecol 248(3):313–338

Boucot AJ, Chen X, Scotese CR et al (2009) Reconstruction of global paleoclimate in Phanerozoic. Science Press, Beijing (in Chinese)

Boulila S, Vahlenkamp M, De Vleeschouwer D et al (2018) Towards a robust and consistent middle Eocene astronomical timescale. Earth Planet Sci Lett 486:94–107

Cai L, Lu S, Li A et al (2015) Sequence stratigraphy characteristics and its control action on sedimentary and course rock in Shahezi Formation of Zhaozhou area in Songliao Basin. J Jilin Univ: Earth Sci Ed 45(3):724–735

Cao W, Xi D, Huang Q (2016) Seawater incursion event in Songliao Basin: new evidence from calcareous nannofossils of SK-1. Geol Bull China 35(6):866–871

Caves Rugenstein JK, Chamberlain CP (2018) The evolution of hydroclimate in Asia over the Cenozoic: a stable-isotope perspective. Earth Sci Rev 185:1129–1156

Chamberlain CP, Wan XQ, Graham SA et al (2013) Stable isotopic evidence for climate and basin evolution of the Late Cretaceous Songliao basin, China. Palaeogeogr Palaeoclimatol Palaeoecol 385:106–124

Charbonnier G, Boulila S, Spangenberg E et al (2018) Obliquity pacing of the hydrological cycle during the Oceanic Anoxic Event 2. Earth Planet Sci Lett 499:266–277

Chen P (2012) Cretaceous conchostracan biostratigraphy of China. J Stratigr 36(2):300–313 (in Chinese)

Chen J, Zhao P, Wang C, Huang Y (2009) Modeling the East Asian climate during the late Cretaceous (80 Ma). Earth Sci Front 16(6):226–239

Chen J, Zhao P, Wang CS et al (2013) Modeling East Asian climate and impacts of atmospheric CO_2 concentration during the Late Cretaceous (66 Ma). Palaeogeogr Palaeoclimatol Palaeoecol 385:190–201

Chen J, Zhao Z, Jiang L et al (2014) The establishment of sequence stratigraphic framework and its models of the Shahezi Formation in Xujiaweizi Fault Depression, Songliao Basin. Sci Technol Eng 14(10):183–190. (in Chinese)

Cheng R, Wang G, Wang P et al (2009) Description of Cretaceous sedimentary sequence of the Yaojia Formation recovered by CCSD-SK-I s borehole in Songliao Basin: lithostratigraphy, sedimentary facies and cyclic stratigraphy. Earth Sci Front 16(02):272–287 (in Chinese)

Cui T (1988) Discovery of Eosestheria in Songliao Basin and its significance. P.G.O.D.D 7(1) (in Chinese)

Cui B, Meng Q, Bai X et al (2018) Petroleum exploration progress and suggestions for north Songliao Basin. Pet Geol Oilfield Dev Daqing 37(3):1–9 (in Chinese)

DeCelles PG, Kapp P, Ding L et al (2007) Late Cretaceous to middle Tertiary basin evolution in the central Tibetan Plateau: changing environments in response to tectonic partitioning, aridification, and regional elevation gain. Geol Soc Am Bull 119:654–680

Deng C, He H, Pan Y et al (2013) Chronology of the terrestrial Upper Cretaceous in the Songliao Basin, Northeast Asia. Palaeogeogr Palaeoclimatol Palaeoecol 385:44–54

Dessert C, Dupre B, François LM et al (2001) Erosion of Deccan Traps determined by river geochemistry: impact on the global climate and the 87Sr/86Sr ratio of seawater. Earth Planet Sci Lett 188:459–474

Du X, Xie X, Lu Y et al (2011) Distribution of continental red paleosols and their forming mechanisms in the Late Cretaceous Yaojia Formation of the Songliao Basin, NE China. Cretac Res 32:244–257

Feng Z, Fang W, Wang X et al (2009) Microfossils and molecular records in oil shales of the Songliao Basin and implications for paleo-depositional environment. Sci China Ser D-Earth Sci 52(10):1559–1571 (in Chinese)

Feng Z, Jia C, Xie X et al (2010) Tectonostratigraphic units and stratigraphic sequences of the nonmarine Songliao basin, northeast China. Basin Res 22:79–95

Friedrich O, Norris RD, Erbacher J (2012) Evolution of middle to late Cretaceous oceans—a 55 m.y. record of Earth's temperature and carbon cycle. Geology 40:107–110

Gao F, Song Y, Liang Z et al. Development characteristics of organic pore in the continental shale and its genetic mechanism: a case study of Shahezi Formation shale in the Changling fault depression of Songliao Basin. Acta Petrolei Sinica 40(9) (in Chinese)

Gao R, He C, Qiao X et al. A new genus and species of Cretaceous dinoflagel lates from two transgressive beds in Songliao Basin, NE China. Acta Palaeontologica Sinica 31:17–29 (in Chinese)

Gao R, He C, Qiao X et al (1992) Cretaceous non-marine dinoflagellates, green algae and acritarchs in Songliao Basin. Nanjing University Press, Nanjing (in Chinese)

Gao R, Zhao C et al (1999) Cretaceous oil strata palynology from Songliao basin. Geological Publishing House, Beijing (in Chinese)

Gao Y, Wang P, Cheng R et al (2009) Description of Cretaceous sedimentary sequence of the first member of the Qingshankou Formation recovered by CCSD-SK-Is borehole in Songliao Basin: lithostratigraphy sedimentary facies and cyclic stratigraphy. Earth Sci Front 16(02):314–323 (in Chinese)

Gao Y, Ibarra DE, Wang C et al (2015) Mid-latitude terrestrial climate of East Asia linked to global climate in the late Cretaceous. Geology 43:287–290

Gao X, Gao Y, Qu X et al (2017) Volcanic-sedimentary succession description of the Lower Cretaceous Yingcheng Formation based on the ICDP scientific drilling borehole in the Songliao Basin(SK-2). Earth Sci Front 24(1):265–275 (in Chinese)

Gao Y, Wang C, Wang P et al (2019) Progress on continental scientific drilling project of Cretaceous Songliao Basin (SK-1 and SK-2). Sci Bull 64(2):73–75

Gao Y, Ibarra DE, Rugenstein JKC et al (2021) Terrestrial climate in mid-latitude East Asia from the latest Cretaceous to the earliest Paleogene: a multiproxy record from the Songliao Basin in northeastern China. Earth Sci Rev 216:103572

Gradstein FM, Ogg JG, Schmitz M et al (2012) The geologic time scale 2012 2-volume set. Elsevier, Amsterdam, 1172 pp

Harms U, Koeberl C, Zoback MD (eds) (2007) Continental scientific drilling: a decade of progress, and challenges for the future, vol 252. Springer Science & Business Media

Hay WW (2011) Can humans force a return to a 'Cretaceous' climate? Sed Geol 235:5–26

He H, Deng C, Wang P et al (2012) Toward age determination of the termination of the Cretaceous Normal Superchron. Geochem Geochem Geophys Geosyst 13:Q02002

Hinnov LA (2013) Cyclostratigraphy and its revolutionizing applications in the earth and planetary sciences. Geol Soc Am Bull 125:1703–1734
Hou D, Huang Q, Huang F et al (1999) The characteristics of molecular geochemistry of marine transgression strata in Songliao Basin. Acta Petrolei Sinica 20(2):30–34 (in Chinese)
Hou G, Feng D, Wang W et al (2004) Reverse structures and their impacts on hydrocarbon accumulation in Songliao basin. Oil Gas Geol 25(1):49–53 (in Chinese)
Hou H, Wang C, Zhang J et al (2018) Deep continental scientific drilling engineering in Songliao Basin: progress in earth science research. Geol China 45(4):641–657 (in Chinese)
Hu J, Peng P, Liu M et al (2015) Seawater incursion events in a Cretaceous Paleo-lake revealed by specific marine biological markers. Sci Rep 5:9508
Huang Q, Zheng Y, Yang M et al (1999) On Cretaceous paleoclimate in the Songliao Basin. Acta Micropalaeontologica Sinica 16(1):95–103 (in Chinese)
Huang Y, Yang G, Gu J et al (2013) Marine incursion events in the Late Cretaceous Songliao Basin: constraints from sulfur geochemistry records. Palaeogeogr Palaeoclimtol Palaeoecol 385:152–161
Huang H, Gao Y, Ma C et al (2021) Organic carbon burial is paced by a ~173-ka obliquity cycle in the middle to high latitudes. Sci Adv 7(eabf9489):2375–2548
Hull PM, Bornemann A, Penman DE et al (2020) On impact and volcanism across the Cretaceous-Paleogene boundary. Science (New York, N.Y.) 367(6475):266–272
Jenkyns HC (2010) Geochemistry of oceanic anoxic events. Geochem Geophys Geosyst 11(3):Q03004
Jones MM, Ibarra DE, Gao Y et al (2018) Evaluating Late Cretaceous OAEs and the influence of marine incursions on organic carbon burial in an expansive East Asian paleo-lake. Earth Planet Sci Lett 484:41–52
Jung C, Voigt S, Friedrich O et al (2013) Campanian-Maastrichtian Ocean circulation in the tropical Pacific. Paleoceanography 28:562–573
Keller G, Punekar J, Mateo P (2016) Upheavals during the late Maastrichtian: volcanism, climate and faunal events preceding the end-Cretaceous mass extinction. Palaeogeogr Palaeoclimatol Palaeoecol 441:137–151
Laurin J, Meyers SR, Uličný D et al (2015) Axial obliquity control on the greenhouse carbon budget through middle- to high-latitude reservoirs. Paleoceanography 30:133–149
Lazar OR, Bohacs KM, Macquaker JHS et al (2015) Capturing key attributes of fine-grained sedimentary rocks in outcrops, cores, and thin sections: nomenclature and description guidelines. J Sediment Res 85(3):230–246
Li L, Keller G (1998) Abrupt deep-sea warming at the end of the Cretaceous. Geology 26:995–998
Li J, Batten DJ, Zhang Y (2011) Palynological record from a composite core through late Cretaceous–early Paleocene deposits in the Songliao Basin, Northeast China and its biostratigraphic implications. Cretac Res 32:1–12
Li S, Wang Q, Zhang H et al (2019) Charophytes from the Cretaceous–Paleocene boundary in the Songliao Basin (north-eastern China): a Chinese biozonation and its calibration to the geomagnetic polarity time scale. Paleontology 5(1):47–81
Li S, Sanjuan J, Wang Q et al (2021) Response of the lacustrine flora in East Asia to global climate changes across the K/Pg boundary. Global Planet Change 197:103400
Linnert C, Robinson SA, Lees JA et al (2014) Evidence for global cooling in the late Cretaceous. Nat Commun 5:1–7
Liu J (2020a) Establishment of chemical stratigraphic framework and climate cycle analysis on the upper section of Shahezi Formation of Songke 2 well in Songliao Basin, China. China University of Geoscience (Beijing) (in Chinese)
Liu S (2020b) Demarcation of the boundary between Yingcheng formation and Denglouku formation of lower Cretaceous in the north of Songliao Basin and paleoclimate change—based on continuous coring data of SK2. Jilin University (in Chinese)

Liu H, Wang P, Gao Y et al (2021) New data from ICDP borehole SK2 and its constraint on the beginning of the lower Cretaceous Shahezi Formation in the Songliao basin, NE China. Sci Bull 66:411–413
Lourens LJ, Appy S, Dick K et al (2005) Astronomical pacing of late Palaeocene to early Eocene global warming events. Nature 435:1083–1087
Mateo P, Keller G, Punekar J et al (2017) Early to late Maastrichtian environmental changes in the Indian Ocean compared with Tethys and South Atlantic. Palaeogeogr Palaeoclimatol Palaeoecol 478:121–138
Meyerhoff AA, Teichert C (1971) Continental drift, III: late Paleozoic glacial centers, and Devonian-Eocene coal distribution. J Geol 79(3):285–321
Miall A (2010) Alluvial deposits. In: James NP, Dalrymple RW (eds) Facies models 4. Geol Assoc, Canada, pp 105–138
Miller KG, Kominz MA, Browning JV et al (2005) The Phanerozoic record of global sea-level change. Science 310:1293–1298
Mulder T, Syvitski JPM, Migeon S et al (2003) Marine hyperpycnal flows: initiation, behavior and related deposits. A review. Mar Pet Geol 20(6–8):861–882
Niu L, Gao Y, Huang H et al (2021) Controlling factors for organic carbon burial in the Late Cretaceous Nenjiang Formation of the Songliao Basin, NE China. Energies 14:4783
O'Brien CL, Robinson SA, Pancost RD et al (2017) Cretaceous sea-surface temperature evolution: constraints from TEX86 and planktonic foraminiferal oxygen isotopes. Earth Sci Rev 172:224–247
Palike H, Norris RD, Herrle JO et al (2006) The heartbeat of the Oligocene climate system. Science 314:1894–1898
Parrish JT (1998) Interpreting pre-Quaternary climate from the rock record. Columbia University Press, New York
Petersen SV, Dutton A, Lohmann KC (2016) End-Cretaceous extinction in Antarctica linked to both Deccan volcanism and meteorite impact via climate change. Nat Commun 7:1–9
Qu H, Xi D, Li S et al (2014) Late Cretaceous–early Paleocene ostracod biostratigraphy of Scientific Drilling Sk1(N) in the Songliao Basin, Northeast China. J Paleontol 88:786–798
Retallack GJ (2001) Soils of the past: an introduction to paleopedology. Blackwell Science Ltd., Oxford
Schoene B, Eddy MP, Samperton KM et al (2019) U-Pb constraints on pulsed eruption of the Deccan Traps across the end-Cretaceous mass extinction. Science 363:862–866
Soil Survey Staff (2014) Keys to soil taxonomy. Natural Resources Conservation Service, Washington, 600 pp
Song Z, Qin Y, George SC et al (2013) A biomarker study of depositional paleoenvironments and source inputs for the massive formation of Upper Cretaceous lacustrine source rocks in the Songliao Basin, China. Palaeogeogr Palaeoclimtol Palaeoecol 385:137–151
Sprain CJ, Renne PR, Vanderkluysen L et al (2019) The eruptive tempo of Deccan volcanism in relation to the Cretaceous-Paleogene boundary. Science 363:866–870
Sun L, Wang G (2019) Revitalization and development strategy of Daqing Oilfield. Pet Geol Oilfield Dev Daqing 38(5):1–7 (in Chinese)
Thibault N, Harlou R, Schovsbo NH et al (2016) Late Cretaceous (late Campanian–Maastrichtian) sea-surface temperature record of the Boreal Chalk Sea. Clim Past Discuss 12:429–438
Tian X, Gao Y, Li Z et al (2020) Fine-grained gravity flow deposits and their depositional processes: a case study from the Cretaceous Nenjiang Formation, Songliao Basin, NE China. Geol J 56(3):1496–1509
Tierney JE, Poulsen CJ, Montanez I et al (2020) Past climates inform our future. Science (New York, N.Y.) 370(6517)
Tobin TS, Wilson GP, Eiler JM et al (2014) Environmental change across a terrestrial Cretaceous-Paleogene boundary section in eastern Montana, USA, constrained by carbonate clumped isotope paleothermometry. Geology 42:351–354

Wan X, Zhao J, Scott RW et al (2013) Late Cretaceous stratigraphy, Songliao Basin, NE China: SK1 cores. Palaeogeogr Palaeoclimatol Palaeoecol 385(1):31–43
Wang P (2007) View of the Earth from the seabed—third platform for the earth system. China J Nat 29(3):125–131 (in Chinese)
Wang C (2016) Late Cretaceous climate changes recorded in Eastern Asian lacustrine deposits and North American Epieric sea strata. Sci Technol Innov Herald 126(126):275–299
Wang Y (2019) Analysis and countermeasure on the exploration situation of Daqing Oilfield. Pet Geol Oilfield Dev Daqing 38(5):23–33 (in Chinese)
Wang X, Liu L, Liu Z et al (1994) Basement structure and sedimentary/tectonic evolution of Meso-Cenozoic basins in Manzhouli-Suifenhe Geoscience Section, China. (in Chinese)
Wang P, Du X, Wang J et al (1995) The chronostratigraphy and stratigraphic classification of the cretaceous of the songliao basin. Acta Geol Sin 69(4):372–381 (in Chinese)
Wang X, Cai J, Bao Y (2006a) Catalysis of clay mineral to organic matter in hydrocarbon genesis. Mar Orig Pet Geol 11(3):27–38 (in Chinese)
Wang P, Chen F, Chen S et al (2006b) Geochemical and Nd-Sr-Pb isotopic composition of Mesozoic volcanic rocks in the Songliao basin, NE China. Geochem J 40:1–11
Wang P, Gao Y, Cheng R et al (2009a) Description of Cretaceous sedimentary sequence of the second and third member of the Qingshankou Formation recovered by CCSD-SK-I s borehole in Songliao Basin: lithostratigraphy, sedimentary facies and cyclic stratigraphy. Earth Sci Front 16(02):288–313 (in Chinese)
Wang P, Gao Y, Ren Y et al (2009b) $^{40}Ar/^{39}Ar$ age and geochemical features of mugearite from the Qingshankou Formation: significances for basin formation, hydrocarbon generation and petroleum accumulation of the Songliao Basin in Cretaceous. Acta Petrologica Sinica 25(05):1178–1190 (in Chinese)
Wang P, Chen S, Li W et al (2010) Chronology, petrology and geochemistry of the Cretaceous crypto-explosive breccia-bearing volcanic rocks: implications for volcanic reservoir and tectonics of the Songliao Basin, NE China. Acta Petrologica Sinica 26(1): 33–46 (in Chinese)
Wang C, Feng Z, Zhang L et al (2013a) Cretaceous paleogeography and paleoclimate and the setting of SK1 borehole sites in Songliao Basin, northeast China. Palaeogeogr Palaeoclimatol Palaeoecol 385:17–30
Wang C, Scott RW, Wan X et al (2013b) Late Cretaceous climate changes recorded in eastern Asian lacustrine deposits and North American Epeiric sea strata. Earth Sci Rev 126:275–299
Wang Y, Huang C, Sun B et al (2014) Paleo-CO_2 variation trends and the Cretaceous greenhouse climate: Earth-Sci Rev 129:136–147
Wang P, Zhao R, Meng Q et al (2015a) The Cretaceous Songliao Basin: dynamic background from volcanic rift to interior sag basin. Earth Sci Front 03:99–117 (in Chinese)
Wang G, Cheng R, Wang P et al (2015b) High resolution continuous sedimentary records of Upper Cretaceous obtained from the continental drilling (SK-1) borehole in Songliao Basin: Sifangtai and Mingshui formations. Geosci Front 6:895–912
Wang C, Feng Z, Wang P (2016a) Initial report of continental scientific drilling project of the Cretaceous Songliao Basin (SK-1). Science Press, Beijing (in Chinese)
Wang C, Feng Z, Chen X (2016b) Paleoclimate implications for future climate change. Earth Sci Front 24(01):1–17 (in Chinese)
Wang P, Mattern F, Didenko NA et al (2016c) Tectonics and cycle system of the Cretaceous Songliao Basin: an inverted active continental margin basin. Earth-Sci Rev 159:82–102
Wang T, Ramezani J, Wang C et al (2016d) High-precision U-Pb geochronologic constraints on the Late Cretaceous terrestrial cyclostratigraphy and geomagnetic polarity from the Songliao Basin, Northeast China. Earth Planet Sci Lett 446:37–44
Wang P, Liu H, Ren Y et al (2017) How to choose a right drilling site for the ICDP Cretacedus continental scientific drilling in the Songliao Basin (SK2), Northeast China. Earth Sci Front 24(1):216–228 (in Chinese)
Wang G, Wang F, Zhao B et al (2021) Exploration and development situation and development strategy of Daqing Oilfield Company. China Pet Explor 26(1):19 (in Chinese)

White JM, Ager TA, Adam DP et al (1997) An 18 million year record of vegetation and climate change in northwestern Canada and Alaska: tectonic and global climatic correlates. Palaeogeogr Palaeoclimatol Palaeoecol 130(1):293–306

Wilf P, Johnson KR, Huber BT (2003) Correlated terrestrial and marine evidence for global climate changes before mass extinction at the Cretaceous-Paleogene boundary. Proc Natl Acad Sci USA 100:599–604

Woelders L, Vellekoop J, Kroon D et al (2017) Latest Cretaceous climatic and environmental change in the South Atlantic region. Paleoceanography 32:466–483

Wu H, Zhang S, Jiang G et al (2013) Astrochronology of the Early Turonian-Early Campanian terrestrial succession in the Songliao Basin, northeastern China and its implication for long-period behavior of the Solar System. Palaeogeogr Palaeoclimatol Palaeoecol 385:55–70

Wu H, Zhang S, Hinnov LA et al (2014) Cyclostratigraphy and orbital tuning of the terrestrial upper Santonian–lower Danian in Songliao Basin, northeastern China. Earth Planet Sci Lett 407:82–95

Xi D, Wan X, Feng Z et al (2011) Discovery of Late Cretaceous foraminifera in the Songliao Basin: evidence from SK-1 and implications for identifying seawater incursions. Chin Sci Bull 56:253–256

Yang R, Jin Z, van Loon AJ et al (2019) Climatic and tectonic controls of lacustrine hyperpycnite origination in the Late Triassic Ordos Basin, central China: Implications for unconventional petroleum development: reply. AAPG Bull 103(2):511–514

Ye D, Huang Q, Zhang Y et al (2002) Cretaceous ostracoda biostratigraphy in Songliao Basin. Petroleum Industry Press, Beijing (in Chinese)

Yin Y, Gao Y, Wang P et al (2019) Discovery of Triassic volcanic-sedimentary strata in the basement of Songliao Basin. Sci Bull 64:644

Yu Z, He H, Deng C et al (2019) New geochronological constraints for the Upper Cretaceous Nenjiang Formation in the Songliao Basin, NE China. Cretac Res 102:160–169

Yu Z, He H, Deng C et al (2020) New SIMS U-Pb geochronology for the Shahezi Formation from CCSD-SK-IIe borehole in the Songliao Basin, NE China. Sci Bull 65(13)

Zachos JC, Shackleton NJ, Revenaugh JS et al (2001) Climate response to orbital forcing across the Oligocene-Miocene Boundary. Science 292:274–278

Zakharov YD, Boriskina NG, Ignatyev AV et al (1999) Palaeotemperature curve for the Late Cretaceous of the northwestern circum-Pacific. Cretaceous Res 20(6):685–697

Zakharov YD, Sha J, Popov AM et al (2009) Permian to earliest Cretaceous climatic oscillations in the eastern Asian continental margin (Sikhote-Alin area), as indicated by fossils and isotope data. GFF 131(1–2):25–47

Zakharov YD, Shigeta Y, Popov AM et al (2011a) Campanian climatic change: isotopic evidence from far East, North America, North Atlantic and Western Europe. Acta Geologica Sinica-Engl Ed 81(6):1049–1069

Zakharov YD, Shigeta Y, Popov AM et al (2011b) Cretaceous climatic oscillations in the Bering area (Alaska and Koryak Upland): isotopic and palaeontological evidence. Sed Geol 235(1–2):122–131

Zavala C, Pan S (2018) Hyperpycnal flows and hyperpycnites: origin and distinctive characteristics. Lithol Reserv 30(1):1–27

Zavala C, Arcuri M, Di Meglio M et al (2011) A genetic facies tract for the analysis of sustained hyperpycnal flow deposits. AAPG Stud Geol 61:31–51

Zhang X, Zhou J, Chi X et al (2008) Late Paleozoic Tectonic Sedimentation and Petroleum Resources in Northeastern China. J Jilin Univ (Earth Science Edition) 38(5):719–725 (in Chinese)

Zhang X, Zong J (1984) On the age of Nenjiang Formation in Songliao Basin based on ostracoda. Oil Gas Geol 5(1):73–76 (in Chinese)

Zhang L, Wang C, Wignall PB et al (2018) Deccan volcanism caused coupled pCO_2 and terrestrial temperature rises, and pre-impact extinctions in northern China. Geology 46:271–274

Zhao X (2009) The impact of clay minerals on oil-gas reservoir. Xinjiang Pet Geol 2009(04):139–142 (in Chinese)
Zhao Y (2020) A dissertation submitted to China University of Geosciences for mast degree. China University of Geoscience (Beijing) (in Chinese)
Zou C, Xiao L, Niu Y et al (2016) General design of geophysical logging of the CCSD-SK-2 East Borehole in the Songliao Basin of Northeast China (in Chinese with English abstract). Earth Sci Front 279–287

Cambrian Tempestites in the Xiaweidian Area, Western Beijing City

Bo Liu, Kaibo Shi, Hangyu Liu, Chun Wu, and Qing He

1 Introduction

This field trip is an invitation to observe the Cambrian outcrops at the Xiaweidian Section located in the Miaofengshan Town, western Beijing. The outcrop completely records the sedimentary process of the North China Craton (NCC), one of the three cratons in China, in the sedimentary environment of the epeiric sea at equatorial paleolatitude in the Cambrian. The Cambrian lithologic association occurred in the section includes siliciclastic rocks deposited in the supratidal zone, mixed siliciclastic-carbonate rocks deposited in the intertidal zone, and carbonates deposited in the subtidal zone. The most remarkable phenomenon in this section is the abundant storm-originated wormkalks, also named flat-pebble conglomerates, occurred frequently in Cambrian sequences. Besides, large-scale oolitic limestones, diverse microbial mounds, marl-limestone high-frequency cycles are well-developed. This section very well outcrops representative Cambrian sedimentary systems which could help us to understand the Cambrian sedimentary evolution of the NCC.

The field trip is scheduled to last about six hours, and 13 stops are set, which are only tens of meters away from each other. The stops mainly focus on (1) The sedimentary textures and origins of different types of wormkalk tempestites, including in-situ tempestites and allochthonous tempestites, and (2) Characteristics of tempestites between supratidal to subtidal environments and their associated facies sequence with oolitic limestones and microbial mounds.

In addition to observation of various tempestites, the sedimentary characteristics of oolitic limestones with different textures (from fine ooid, coarse ooid to giant ooid), the internal structure, growth and extinction mechanisms of different forms of microbial mounds, and the rhythmic characteristics of marl-limestone high-frequency cycles will be observed on the stops. This field trip is close to the downtown of Beijing,

B. Liu (✉) · K. Shi · H. Liu · C. Wu · Q. He
School of Earth and Space Sciences, Peking University, Beijing, China
e-mail: bobliu@pku.edu.cn

X. Hu (ed.), *Field Trip Guidebook on Chinese Sedimentary Geology*,
https://doi.org/10.1007/978-981-99-6936-4_3

with convenient transportation and a beautiful environment. The planned field route is along an abandoned railway, nestling under a mountain and riverside with high security and short walking distance. It is believed that scholars and colleagues who are interested in the Cambrian tempestites and other above-mentioned sedimentary phenomena will make a worthwhile trip.

2 Geological Background

The Xiaweidian section is located in Miaofengshan Town, Mentougou District, Beijing. It is about 33 km away from the downtown of Beijing. Tectonically, it belongs to the northeast of NCC, one of the three major cratons in China (Fig. 1a, b), where the Qingbaikou, Cambrian, Ordovician, Carboniferous, Permian, Triassic, Jurassic strata are exposed (Fig. 1c). The Cambrian on Xiaweidian section are typical Cambrian sedimentary strata of the NCC with complete outcrops and mainly carbonate rocks.

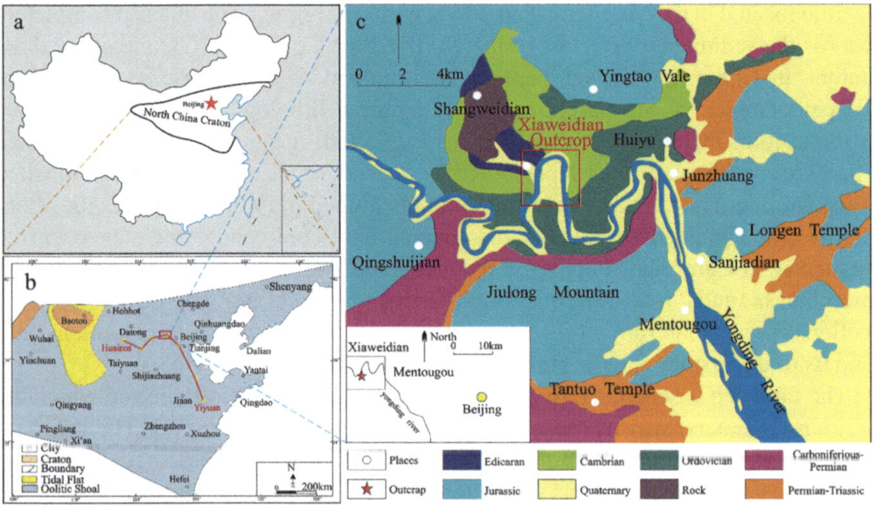

Fig. 1 The position of the Xiaweidian section and the regional strata outcrops (modified from Liu 2020). **a** The location of NCC; **b** The NCC Paleogeography of the Zhangxia period; **c** The regional geologic map and location of the Xiaweidian section

3 Tectono-Sedimentary Evolution of the Early Paleozoic Epeiric Sea in North China

3.1 Tectonic Setting and Evolution

The NCC is one of the oldest blocks in China, including most of the North China, the southern Northeast China, the Inner Mongolia, the Bohai Bay, and the northern Yellow Sea. It is about 1500 km long from east to west and 1000 km wide from north to south, with a total area of about 140×10^4 km^2 (Fig. 1a). The NCC is bounded by the Central Asian orogenic belt in the north, the Qinling-Dabie orogenic belt in the south, the Qilian Orogenic belt in the southwest, the Tanlu fault zone in the east and the Yangtze Plate in the south (Fig. 2). It consists of an Archaean to the Paleoproterozoic metamorphic crystalline basement and overlying lightly metamorphic or metamorphosed sedimentary covers and igneous rocks of Mesoproterozoic to Mesozoic.

During the Cambrian-Ordovician period, the world's major plates including the Gondwana supercontinent, Laurentian, Baltic and Siberian plates convergent together in the southern hemisphere as the Panthalassa. Whereas the NCC and some other small blocks are located between the equator and 15 °N latitude (Fig. 3). The Caledonian Movement resulted in the NCC lack of the Late Ordovician, Silurian, Devonian, and early to middle Carboniferous. With the influences of the Indonesian Movement and the Yanshan Movement, the NCC experienced intensive destruction and transformation in Mesozoic. During the Cenozoic, the Himalayan Movement creates its today's geologic and geographic features.

3.2 The Sedimentary Infilling

The Cambrian NCC developed as an epeiric sea on the tectonic setting formed at the end of the Proterozoic, in which carbonate deposits extent thousands of kilometers. Due to the combined influence of differential subsidence and sea-level fluctuation, the environment of the NCC Cambrian epeiric sea changed rapidly (Yu 2015; Wang et al. 2018). The frequent transgression and regression deposited the alternation of supratidal, intertidal and subtidal sedimentary rocks, in which the clasts of trilobites, gastropods, brachiopods, cephalopods and echinoderms, etc. could be found (Chen et al. 2012).

In the Early Cambrian, with the relative sea level rose, the large area of the NCC deposited terrigenous fine clastic rocks. In the Middle Cambrian, seawater covered almost all of the NCC, the deposition was dominated by carbonate rocks of intertidal and subtidal facies. The largest transgression occurred in the Cambrian Zhangxia Period with the characterized deposition of thick-bedded oolites almost covering the entire craton (Fig. 1b) (Su et al. 2016). In the Late Cambrian, the further rising and fluctuation of relative sea-level caused different sedimentary assemblages to alternate

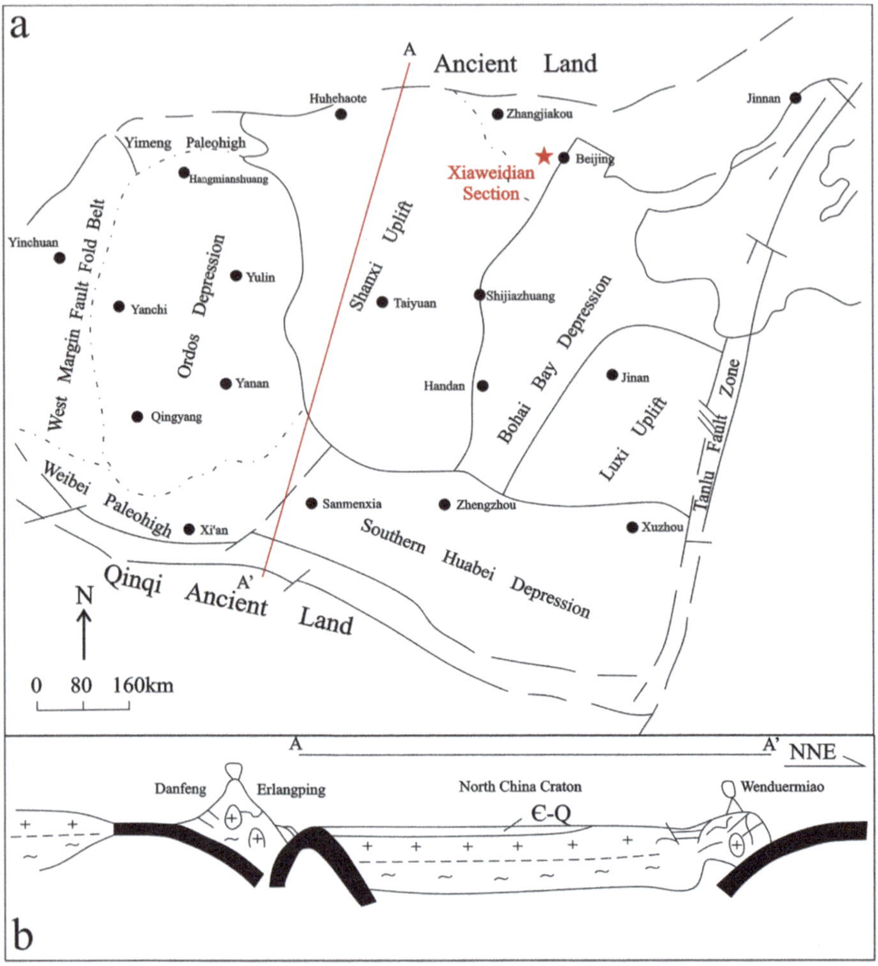

Fig. 2 Proterozoic tectonic division of the NCC (after Wang 2002)

frequently (Sang et al. 2002). During the Middle and Late Cambrian, the NCC was located in the tropical climate belt near the equator where the epeiric sea developed and frequent storms reworked the shallow-water carbonate sediments and deposited many flat-pebble conglomerates in the strata (Wang et al. 2018).

The Xiaweidian section represents the typical early Paleozoic deposition in the northern NCC, where the Cambrian strata were traditionally divided into eight formations, from bottom to top, are Changping Formation, Mantou Formation, Maozhuang Formation, Xuzhuang Formation, Zhangxia Formation, Gushan Formation, Changshan Formation and Fengshan Formation, which are equivalent to the Second Series to Furong Series of the World Stratigraphic Chronology (Fig. 4). The Terreneuvian series of Cambrian is absent in this area due to the Jixian Movement uplifting. The

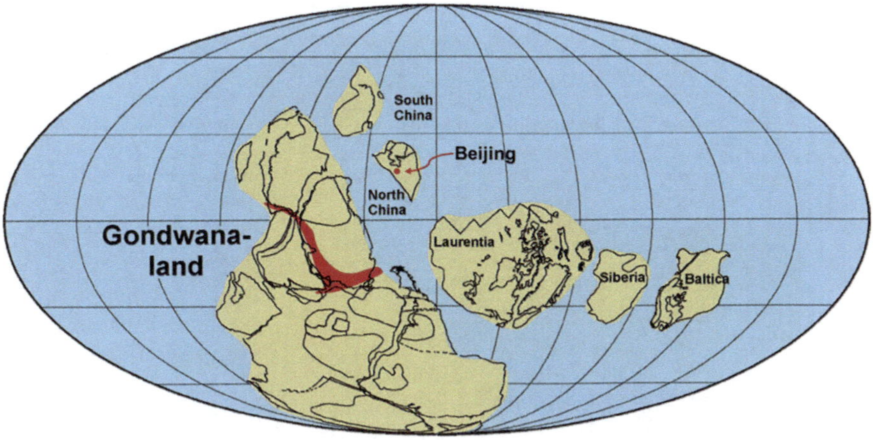

Fig. 3 Global Plate reconstruction in Early Cambrian (530 Ma), the location of Beijing in the NCC is indicated by a red arrow (after Li et al. 2008)

strata of each formation are stable and comparable in most areas of North China. The section characteristics of each formation are as follows:

(1) The Changping Formation

Changping Formation is the earliest deposit of Cambrian in northern North China, and only developed transgressive system tract and high system tract. There is an unconformity between the Changping Formation and the underlying Proterozoic Jingeryu Formation, on which the yellow–brown paleo weathering crust developed. The formation is dominated by micrite-silty dolostone with a thickness of 75.22 m. Some breccia derived from the Jingeryu dolostone can be observed at the bottom of the Changping Formation.

(2) The Mantou Formation

The Mantou Formation is mainly composed of mudstone and dolomitic shale with a thickness of 61.36 m. This formation is characterized by gypsum dissolved breccia and a reddish appearance, such as the yellowish to brownish-yellow of the lower member and the purplish-red of the upper. It reflects that the sedimentary environment is an intertidal zone under oxidation conditions.

(3) The Maozhuang Formation

The thickness of the Maozhuang Formation is 228.22 m, in which the purple-red massive mudstone is mainly developed in the lower and the greyish-green siltstone is gradually increased and thickened upward. The upper part sometimes has interbeds of light gray lenticular or thin-bedded bioclastic limestone. From bottom to top, the overall sedimentary environment became shallower and the general environment was transformed from a supratidal zone to an intertidal zone.

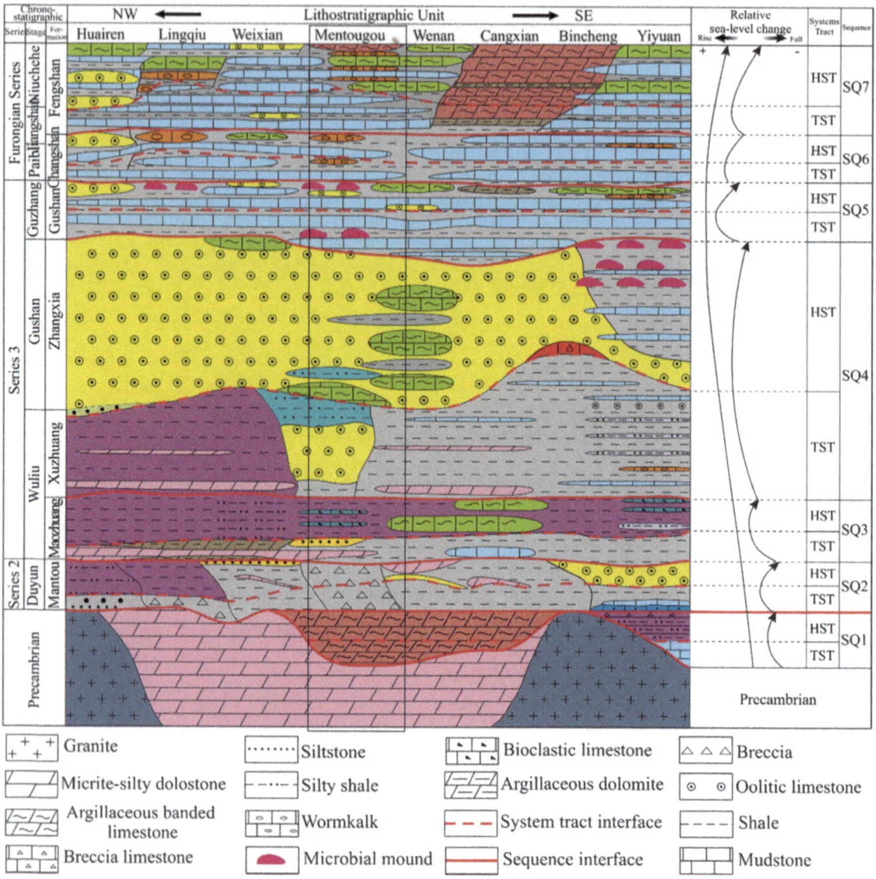

Fig. 4 Strata occurrences and lithological association of Cambrian in the middle-eastern NCC (locations as in Fig. 2a), Xiaweidian Section is shown in the black box (after Xiao 2020)

(4) The Xuzhuang Formation

The thickness of the Xuzhuang Formation is 74.99 m. The bottom is greyish-green mudstone with fine sandstone. It is deposited in the subtidal to the intertidal sedimentary environment. Dark gray massive fine oolites are developed upward. The characteristics of wavy bedding and parallel bedding could be observed.

(5) The Zhangxia Formation

The thickness of the Zhangxia Formation is 148.39 m. Its lower part is characterized by greyish-green mudstone, marl and calcareous siltstone. Thick massive oolitic limestone is widely developed in the middle and upper Zhangxia Formation. Compared with Xuzhuang Formation, the thickness of oolitic limestone in the Zhangxia Formation is larger and the oolitic grain size is larger. It's a deposit of high-energy area. The formation is characterized by many tempestites and at least two microbial mounds.

(6) The Gushan Formation

The thickness of the Gushan Formation is 51.46 m. Oolitic limestone and mudstone develop at the bottom. The hydrodynamic conditions of the forming environment are weak. There are several sets of argillaceous banded limestone and wormkalks developed upward. There's still oolitic limestone at the top. It reflects the sedimentary environment of the subtidal zone below the fair-weather wave base and above the storm wave base.

(7) The Changshan Formation

The thickness of the Changshan Formation is 17.7 m. The sedimentary environment is the subtidal zone below the fair-weather wave base and above the storm wave base. Many cycles are composed of glaucous argillaceous banded limestone and wormkalk. It reflects frequent storm activity during the period.

(8) The Fengshan Formation

The thickness of the Fengshan Formation is 96.35 m. At the bottom is the greyish-green mudstone in the deepwater reduction environment with several sets of wormkalk. The middle and upper is thin-bedded mudstone. Bioclastic limestone is developed at the top.

3.3 Cambrian Tempestite in Epeiric Sea

In Owen's classical epeiric sea model, there existed three energy zones: the low energy zone X, the high energy zone Y and the low energy zone Z. Laporte (1969) divided the tidal epeiric sea into four zones: deep subtidal zone X, shallow subtidal zone Y, intertidal and supratidal zone Z. The epeiric sea of the NCC during the Cambrian was controlled by tides. Its carbonate sedimentary stage can be divided into three sedimentary facies zones: supratidal zone, intertidal zone and subtidal zone (Fig. 5). The supratidal zone is the nearshore facies of the epeiric sea, where evaporation is the most intense. Dolomitic mudstone is developed and dolomitization is generally developed, reflecting the arid climate. The intertidal zone is the seaward sedimentary part of the supratidal zone. It is a region with strong hydrodynamic force, where bioclastic limestone, bioclastic mudstone, stromatolites and biological disturbances structures are deposited. The subtidal zone is an underwater open platform. The upper is a high energy zone (Zhao 1986) with middle-thick sparry oolitic limestone and sparry bioclastic limestone (Sang et al. 2002). The middle-lower is mudstone and argillaceous banded limestone.

Storm deposition refers to sediment assemblages formed by sediment deposition of suspended materials in peak and decay periods by agitation and transport of seafloor sediments by storm surge and backflow (Kumar et al. 1978; Allen 1982; Wang 2011). Tempestite refers to the sedimentary rock formed by the rapid accumulation of sediments formed by storms (Kelling et al. 1975; Aigner et al. 1982; Liu et al. 1986; Brenchley et al. 2010).

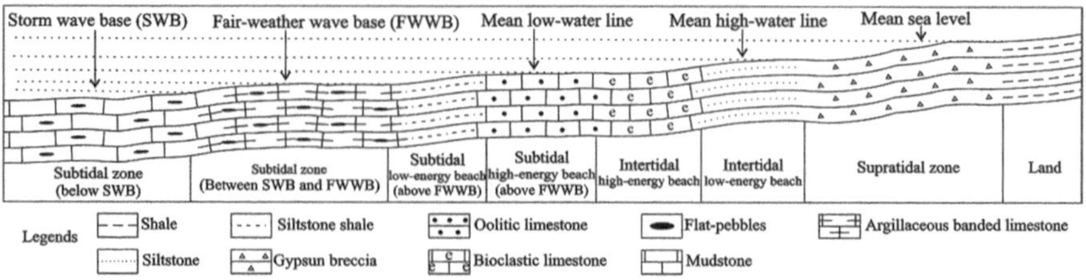

Fig. 5 The epeiric sea sedimentary model in Cambrian of the NCC

In the 1890s, Gilbert (1899) first discovered storm products. He reported that dome-shaped three-dimensional mound-like stratification and trough cross-stratification were the products of storm waves in shallow sea zones. Scholars around the world began studying tempestites in the 1950s. They have found records of different forms of storm deposits in different parts of the globe (Mckee 1959; Ball 1971; Howard 1971, 1972; Ager 1973). Since the 1980s, Chinese scholars have also begun to pay extensive attention to tempestites in many places in China. There are many reports in western Beijing, eastern Hebei and Xuzhou of Jiangsu province (Meng et al. 1986; Jin et al. 1997) and northwest Hunan, northeast Guizhou, eastern Hubei, southwest Jiangxi, etc. (Kuang et al. 2008; Zhou et al. 2006; Yang et al. 2007; Wang et al. 1995; Jin et al. 1995) about the Cambrian tempestite in North and South China. In these studies, some scholars have studied the conditions for the occurrence of storms in the Cambrian period. During the Cambrian, North China was located in the storm zone at the lower latitude of the equator (5° ~ 20°). The climate was tropical. It is located in the favorable latitude area, which was prone to storm action (Dott 1974; Ma 2011).

A complete storm process can be divided into pre-storm, storm intensification, storm peak, storm decay and post-storm (Allen 1982), or simplified into three stages: storm peak, storm decay and storm abatement (Kreisa and Bambach 1982). Controlled by the energy and time of storm periods, the hydrodynamic force directly affects the accumulation characteristics of sediments. From bottom to top, there are bottom erosion surface, shell concentration, parallel bedding, hummocky cross-bedding and mudstone. The shells show the characteristics of graded bedding (Aigner 1982; Liu et al. 1987; Wu et al. 1987; Yao et al. 2013; Wang et al. 1998; Friedman 1994). And the deposition of the shell is formed by the instantaneous action of storms (Dattilo 2008). In different sections, different storm energy will show different feature combinations in different facies (Song 2016).

When there is a certain correlation between the sedimentary textures, it can be combined into a set of tempestite assemblages with different lateral division schemes. In the background of a tidal flat in the North China epeiric sea, tempestite can be divided into backflow tempestite, agitation tempestite, proximal tempestite and rudaceous shale (Liu et al. 1987). These tempestites are typical wormkalks with different structures and morphological characteristics.

4 Field Description of Xiaweidian Section in Western Beijing

The outcrops in western Beijing belong to typical epicontinental deposits in northern North China Craton during the Early Paleozoic, which recorded a large number of storms sediments. Based on the study of this section, a series of results have been obtained on the sedimentary model of the epeiric sea, sea-level change and sediment response, origin of oolitic limestones, tempestite and origin of microbial carbonates (Meng et al. 1986; Mei et al. 2011; Xing et al. 2015; Jin et al. 2016; Ma et al. 2017; Xiao et al. 2017). Tempestite is the main observation object in this field study, and 13 stops are arranged as follows.

Stop 1: Thick-Bed Fine-Grained Oolitic Limestone in the Upper Xuzhuang Formation

Location: west side of the mountain; Coordinates: N 39°59.8117'; E 116° 01.3141'.

The Xuzhuang Formation records a slow rise in relative sea level. Before the deposition of the Xuzhuang Formation, the Cambrian mainly deposited terrigenous clastic rocks in a tidal flat environment. Several sets of oolitic limestone began to develop in Xuzhuang Formation. From this stop, we can see the upward shallower meter-scale cycles in the upper Xuzhuang Formation. The lower of the cycle is thin-bedded micrite intraclastic limestone interbedded with thin-bedded banded siltstone, the thickness of which is generally less than 1 m. Gray thick-bedded oolitic limestone is developed in the upper of the cycle (Fig. 6). Oolitic limestone shows obvious biological disturbance structure and tabular cross-bedding, and the bioturbation part is darker than the undisturbed part. The development of oolitic limestone in the Xuzhuang Formation reflects high energy oolitic shoals in the intertidal zone of the epeiric sea (Feng et al. 1990; Mei et al. 2011).

The oolitic limestone in Xuzhuang Formation is mainly sparry with fine grain size, which generally less than 1 mm. The ooids content are more than 50%. Ooids types include radioactive ooids, single-crystal ooids, micrite ooids, compound ooids, etc. and some oolitic grains have the cores of bioclasts (Fig. 7). In addition to the traditional "snowball" model, the diagenetic mechanism of the oolitic limestone in the Xuzhuang Formation is explained based on the presence of special fabrics such as dark pellets, organic residue, calcite and dolomite condensed grains, pyrite in the oolitic limestone, and bioclastic attachment. Xing and Feng (2015) believed that the formation of oolitic limestone in the Xuzhuang Formation may be related to microbial action.

Stop 2: L-M (Limestone-Marl) Meter-Scale Cycles in the Lower Zhangxia Formation

Location: west side of the mountain; Coordinates: N 39°59.7632'; E 116°01.2998'.

This stop is in the lower Zhangxia Formation with multiple L-M (limestone-marl) type meter-scale cycles developed (Fig. 8). The lower unit of the cycle is argillaceous siltstone with marl nodules, silty shale or argillaceous siltstone with marl bands. The

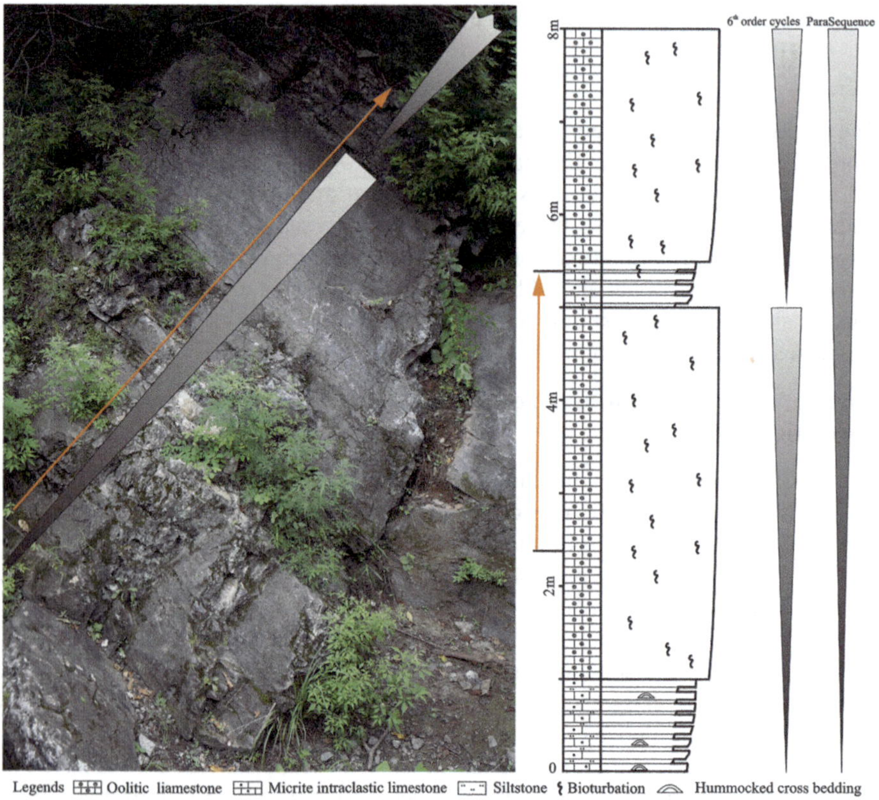

Fig. 6 Thick fine-grained oolitic limestone in the Upper Xuzhuang Formation

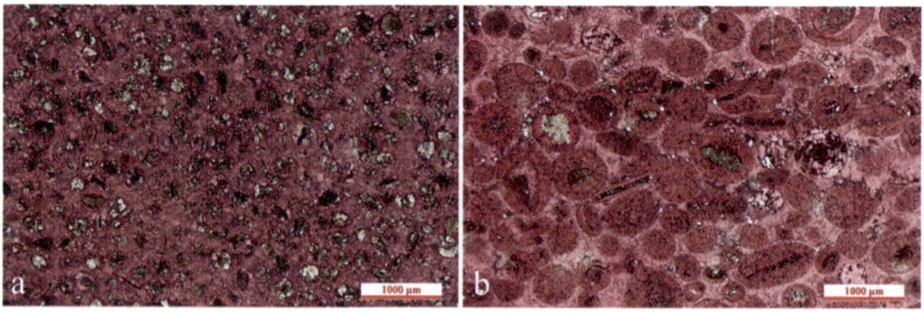

Fig. 7 Microscopic characteristics of the fine-grained oolitic limestone in the Xuzhuang Formation. **a** Dolomite composes of fine-grained ooids with diameters less than 100 μm; **b** Fine-grained ooids with diameters less than 500 μm on average

thickness of this unit is more than 2 m. The marl nodules or bands are eroded to form the concave due to weathering. The upper unit of the cycle is bioclastic limestone with flat-pebbles, which are dozens of centimeters thick. It contains flat-pebbles ranging from 10 to 25%. Bioclasts include trilobites, ostracods and other fragments, as well as a small number of ooids.

Wormkalk appeared many times in this stop but the proportion of flat-pebbles was not high. It did not belong to the typical wormkalk (Fig. 9). The composition of poor sorting flat-pebbles is mudstone. The diameter of pebbles is 2 mm to 1 cm. The shape of pebbles is mainly flat, but also subequiaxed. The lack of mudstone in the surrounding rocks and the good roundness of flat-pebbles indicate that the tempestite is deposited after allochthonous transport. The occurrence of storm-induced flat-pebbles and exposed markers in the bioclastic limestone indicates that the bioclastic limestone was deposited in the subtidal zone above the storm wave base but below the fair-weather wave base.

The overall sedimentary environment of the lower Zhangxia Formation belongs to the subtidal zone of the epeiric sea. In addition to large sections of grayish-green low-energy deposition as indicators, small microbial mounds developed in argillaceous siltstones are also typical indicators (Fig. 8). Microbial mounds are located in banded argillaceous siltstone and marl. They are several centimeters high and more than 10 cm wide. Some microbial mounds develop continuously to form a wave-like appearance.

Stop 3: Silty Shale-Marl-Oolitic Limestone Cycle Sequence in the Lower Zhangxia Formation

Location: west side of the mountain; Coordinates: N 39°59.7216'; E 116°01.2957'.

The stop is an outcrop of the middle and lower Zhangxia Formation. It consists of multiple sets of L-M meter-scale cycles (Fig. 10). However, the lithologic composition is different from the L-M meter-scale cycle of the lower Zhangxia Formation (Fig. 8). The lower units of the cycle are argillaceous siltstone with marl, siltstone with marl and gradually transitioning to mudstone, reflecting the decreasing trend of terrigenous material input. The upper units of the cycle are oolitic limestones. At the top of one of the lower cycles, giant ooids with very coarse grain sizes are developed.

The stop reveals the lower part of the highstand systems tract above the condensation of SQ4 in the third-order cycle. The 3rd relative sea-level gradually lowers. The development of sparry oolitic limestone in the upper part of the cycle indicates that the sedimentary environment was shallow water in the subtidal zone or high-energy oolitic beach in the intertidal zone. It indicates that the maximum high-frequency relative sea level has fallen to the intertidal zone. In general, the environment represented by this outcrop is a subtidal low energy environment and an intertidal high energy environment occasionally occurs. The upward shallower sea level also responds to the longitudinal variation of cycle thickness. The thickness of the cycle gradually decreases, which is caused by the gradual decrease of the relative sea-level leading to the gradual decrease of the sedimentary space.

In this investigation, the giant ooids only developed at this stop (Fig. 11). The diameter of the giant ooids is nearly 4 cm with Oolites abundance of about 50%. The

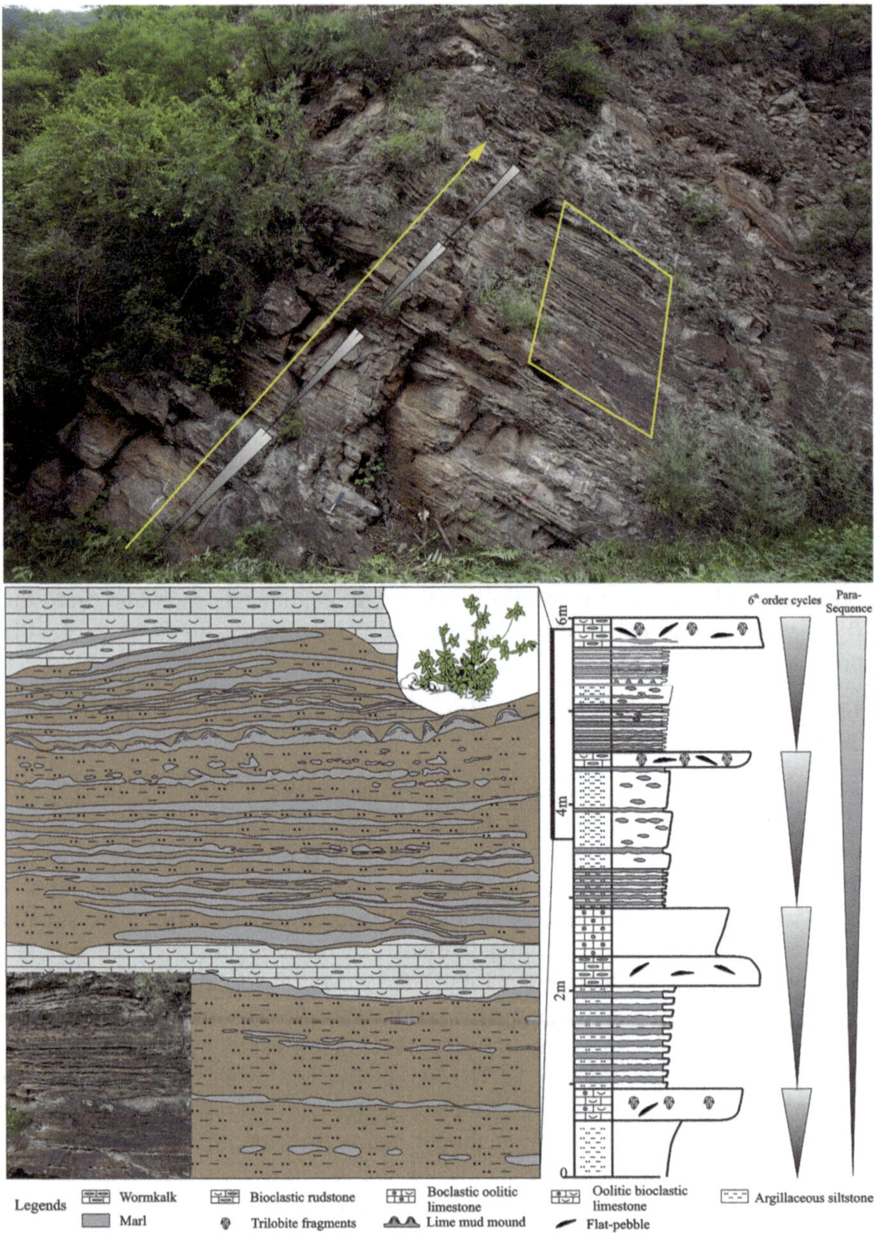

Fig. 8 Sedimentary characteristics of the marl-limestone meter-scale cycles in the Lower Zhangxia Formation

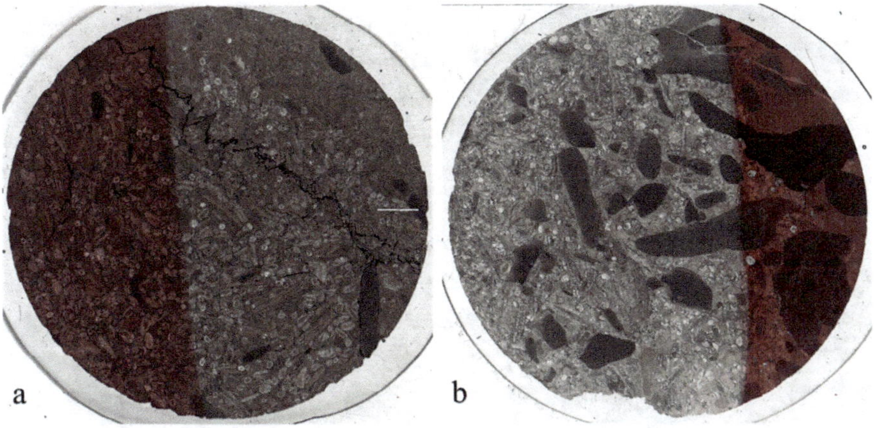

Fig. 9 Microscopic characteristics of the limestone units of L-M meter-scale cycles in the Lower Zhangxia Formation (diameter is 2.5 cm). **a** bioclastic packstone contains bit flat-pebbles; **b** bioclastic packstone contains flat-pebbles with the percentage of 35%

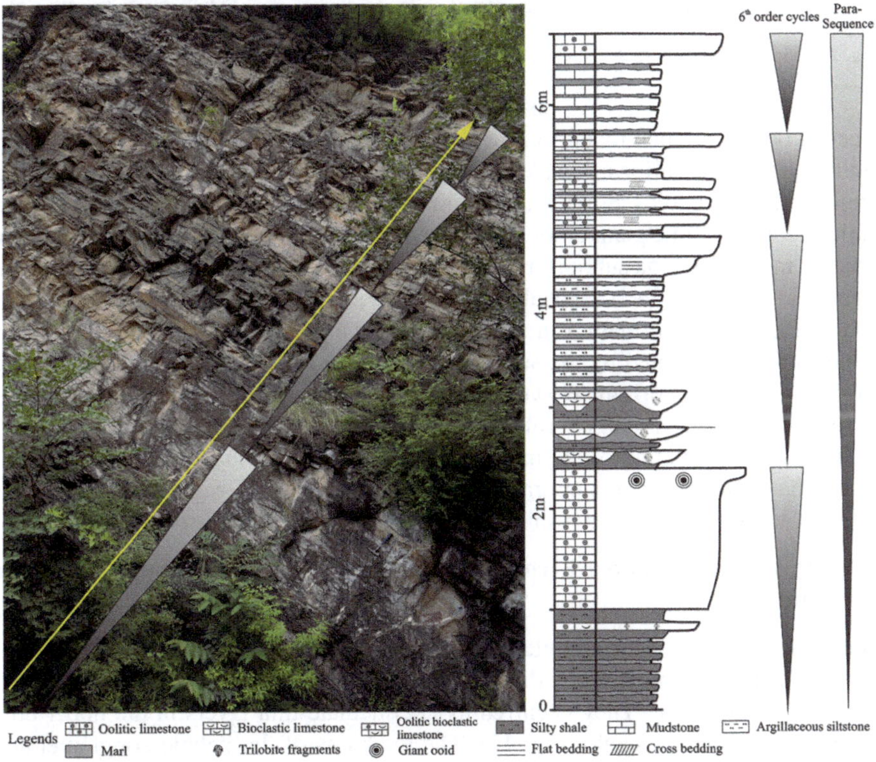

Fig. 10 The silty shale-marl-oolitic limestone cycles in the Lower Zhangxia Formation

Fig. 11 Macroscopic characteristics of the giant oolitic limestone in the Lower Zhangxia Formation

types of oolites are mainly concentrically layered oolites and single-crystal oolites. Oolitic grains are uniform in size of 2-5 mm with good roundness. Regarding the origin of the giant ooids, Sumner and Grotzinger (1993) inferred the inorganic origin of the Neoproterozoic giant ooids. However, the giant ooids discovered in the Lower Triassic in South China support more microbial mechanisms in the formation of the giant ooids (Mei 2012a, b; Li et al. 2013). Dai et al. (2014) found two kinds of giant ooids in Zhangxia Formation in North China. The giant ooids without Cirvanella protonema were formed by the superposition of calcified microbial mineralization based on inorganic origin. However, the giant ooids with Cirvanella protonema were formed by mudstone spherules with microbiological calcification around them, without the participation of the inorganic genetic process. Duan et al. (2015) interpreted the giant ooids contemporaneous with the Zhangxia Formation in the Shizhu area of Chongqing as the microbial origin.

Stop 4: Wormkalk and Microbial Mounds in the Middle-Upper Zhangxia Formation

Location: west side of the mountain; Coordinates: N 39°59.7216'; E 116°01.2957'.

At this stop, the most typical flat-pebble conglomerates of the Cambrian of the NCC began to appear. They occurred as intermediate-thin layers in the upper unit of the high-frequency cycle (Fig. 12). The thickness of flat-pebble conglomerates is about 20 cm (Fig. 13). The contact with the lower dark gray oolitic limestone is the wavy erosion surface. The upper lithology is glaucous silty shale with gray mudstone bands. Light gray flat-pebbles are intersected in long bands, accounting

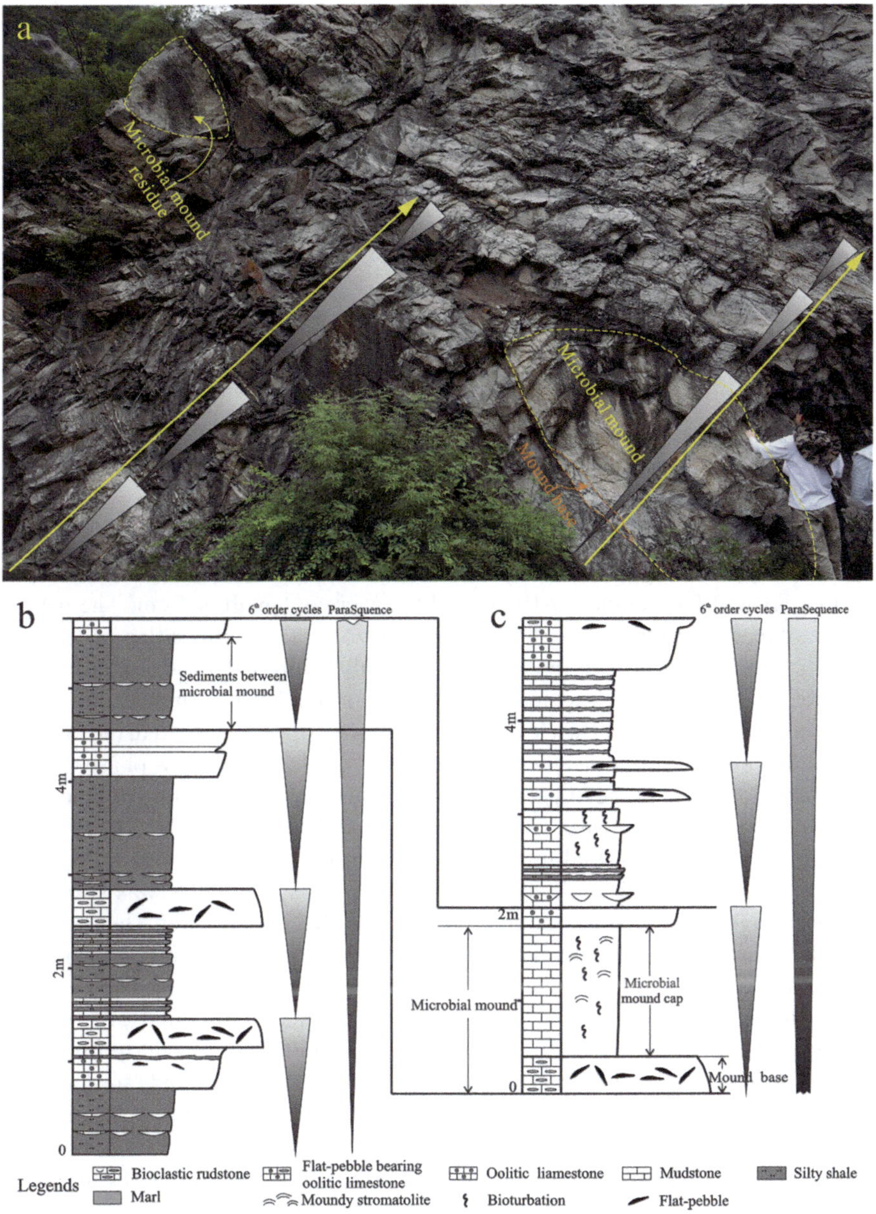

Fig. 12 Argillaceous siltstone or mudstone-oolitic limestone-wormkalk cycles in the Middle Zhangxia Formation

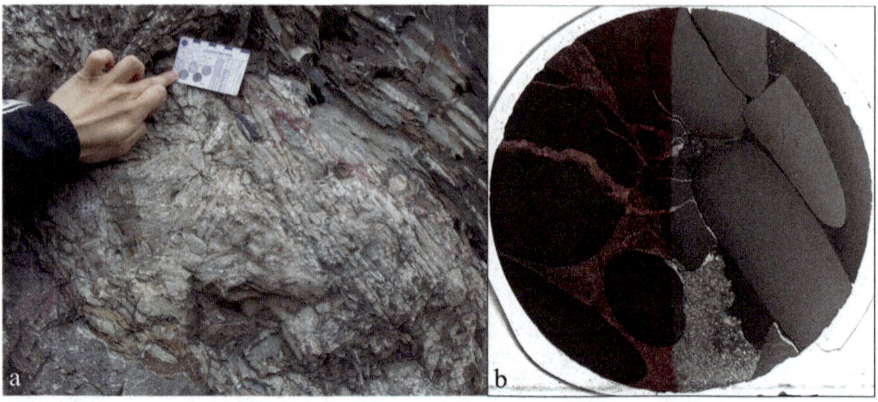

Fig. 13 Macroscopic and microscopic characteristics of wormkalk in the Middle Zhangxia Formation. **a** The macroscopic characteristics of wormkalk. **b** The wormkalk is flat-pebble-supported. The component of flat-pebbles is mudstone

for 70% content. They are scattered in a dark gray matrix. The maximum length of the long axis of flat-pebbles is 10 cm and the average length is 4 cm. The pebbles are distributed along the purplish-red oxidizing zone. The composition of the well-sorted and rounded pebbles is mudstone. It is mainly composed of calcite and contains cluttered floating trilobite fragments. The set of wormkalk was formed in the subtidal zone with a relatively rising sea level. The oolitic limestone deposited in the turbulent environment near the fair-weather wave base was eroded by the new phase of storm action and deposited in the oxygen-rich environment between the fair-weather wave base and the storm wave base with the backflow of the storm.

Multiple sets of microbial mounds were developed in the Zhangxia Formation on the Xiaweidian section. A complete set of microbial mounds was exposed at this stop (Fig. 14). Microbial mounds are 1.8 m high and 3 m wide. The weathered surface is gray-brown. Microbial mounds are composed of mound base of wormkalk and mound cover. The grain size of flat-pebbles is 3–10 cm. They have weathered to brown. From the perspective of the structure of microbial mounds, the mounds grow upwards from the core of the mound bases and gradually grow into ellipses. The laminar structure of the microbial mound is obvious. Microbial mounds are composed of microbial bonded limestone. Microscopically, there are bioclasts of microbiological origin. The mound-building microbes are mainly cyanobacteria, gorvania, etc., accompanied by trilobites, ostracods and other bioclasts. Bioclasts are mainly concentrated near the mound base, forming mound-like geological bodies through calcification and bonding of lime mud during growth.

The underlying strata of the mound are sparry oolitic limestones. The thickness is about 40 cm, indicating a high energy sedimentary environment. There are grayish-green silty shales between the mounds, indicating that the mounds grew in the subtidal zone of the epicontinental sea. The overlying stratum of the mounds is sparry oolitic

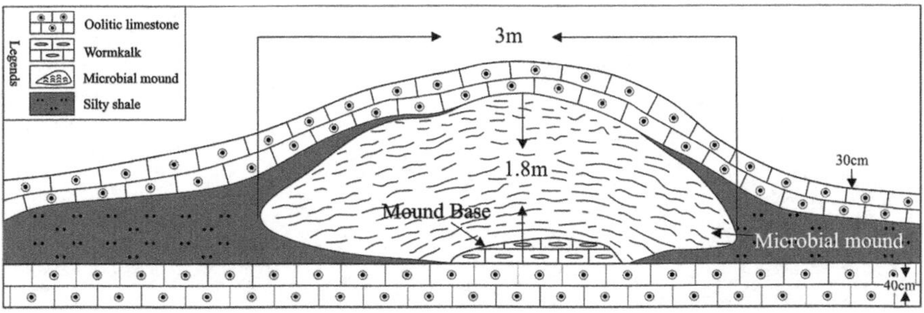

Fig. 14 The microbial mound architecture in the Zhangxia Formation (microbial mounds are shown in Fig. 12a)

limestone with a thickness of 30 cm, which represents that the water becomes shallower to the intertidal zone due to the lowering of relative sea level, and then the intertidal oolitic beach develops again. Microbial mounds and the relationship between their upper and lower strata represent the process of relative sea-level rise and drop.

The lower unit of the high-frequency cycle below the mounds is argillaceous siltstone with lenticular and thin-bedded mudstone. The overall color is grayish-green. The lower unit of the high-frequency cycle above the mounds turns into gray mudstone. The above changes of lithologic assemblages indicate that the development period of mounds is the turning point of the significant changes in the sedimentary environment.

Stop 5: Stromatolitic Mounds in the Lower Gushan Formation

Location: west side of the mountain; Coordinates: N 39°59.6396'; E 116°01.3082'.

This stop shows us the lower Gushan Formation, in which meter-scale high-frequency cycles developed (Fig. 15). In the lower part of the cycle, gray thin-bedded micrite intraclastic limestone with irregular veined siltstone can be observed. The thickness of a single layer of thin-bedded micrite intraclastic limestone is about 2–5 cm and the thickness of siltstone vein is even thinner with a thickness of 0.5–2 cm. Due to its iron content (Fig. 16), the siltstone veins are earthy yellow and show strong weathering resistance. It protrudes outwardly on outcrop compared to the thin-bedded micrite intraclastic limestone. Wormkalk about 30 cm thick can be seen in the upper unit of the cycle, which is composed of mudstone. The matrix between flat-pebbles is composed of intraclasts and bioclasts. The longitudinal distribution of flat-pebbles has the characteristic of graded bedding. The grain size of flat-pebbles below the siltstone vein is 5–10 cm and about 1–2 cm above, which represents two successive stages of storm action. The intensity of the first stage is large while the intensity of the second stage is small. The siltstone was deposited in the period of storm silence between the two stages.

In the other cycle, there is an erosion surface in the lower part of wormkalk, which is lenticular and only a few centimeters thick. There are long axial massive gravels on

Fig. 15 Mudstone with interlayered banded siltstone—wormkalk cycles in the Lower Gushan Formation

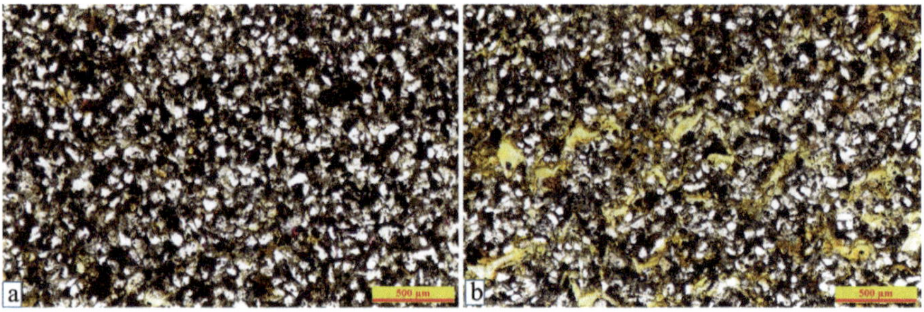

Fig. 16 Microscope characteristics of siltstone in the Lower Gushan Formation. **a** The siltstone contains ferruginous material which shows a dark color. **b** Some clay minerals shown yellow color are unevenly distributed in the siltstone

it. The composition is mudstone. The gravels are somewhat rounded, similar to flat-pebbles magnified several times like soles. The matrix between the gravels is earthy yellow iron-bearing siltstone (Fig. 15). It is speculated that the pebbles are caused by the slump. There is a certain transport distance during the slumping process, which makes the pebbles have the characteristics of roundness.

The lithologic assemblage of the lower Gushan Formation is mainly composed of dark micrite intraclastic limestone and siltstone which represent low energy. Wormkalk in the upper part of the cycle indicates storm events in the subtidal environment. The collapse of gravels indicating a subtidal environment is also developed. It lacks coarse deposits indicating high-energy facies and exposure surfaces indicating shallow-water deposits. It belongs to the SQ5 transgressive system tract and the relative sea level rises gradually.

In the transgressive system of SQ5 in the lower Gushan Formation, another set of microbial mounds was developed over the cycle of micrite intraclastic limestone-wormkalk (Fig. 17). The mound is 2 m high and 3 m wide with a grayish-white weathering surface and a mound base. The base is stable normal topography and composed of storm-origin wormkalk. It is mainly flat-pebbles with some ooids.

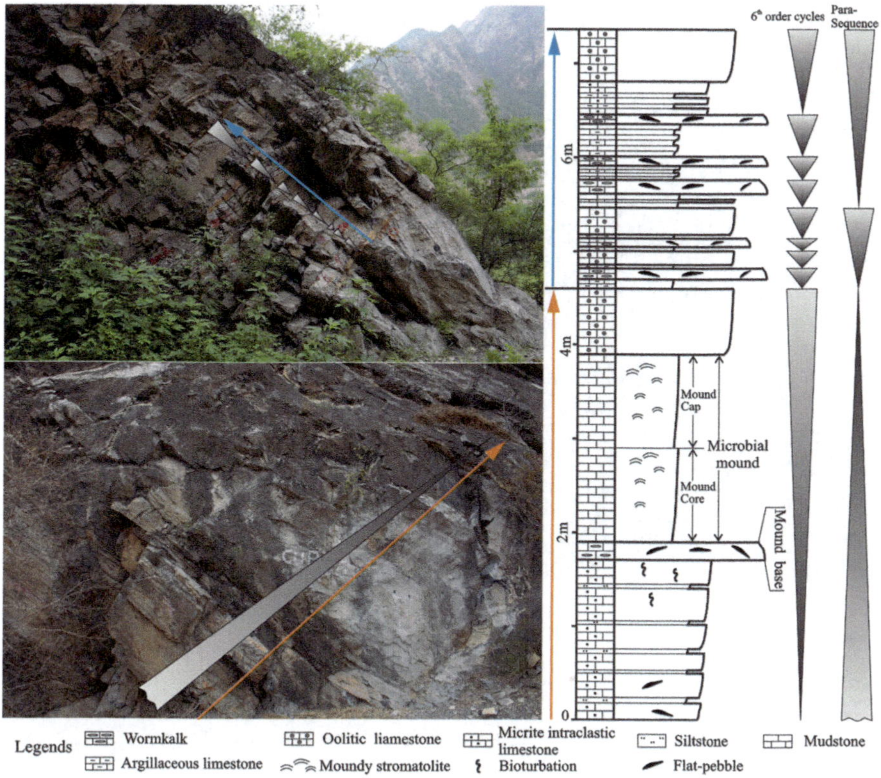

Fig. 17 Microbial mounds and sets of micrite intraclastic limestone intercalated with siltstone-wormkalk cycles in the Lower Gushan Formation

The flat-pebbles have good roundness, which reflects that the base was formed in a high-energy environment. The mound is mainly composed of columnar stromatolites accompanied by bioclasts, ooids, flat-pebbles, clumps and spherules, which belong to the stromatolite mound. The detritus and clumps contained in the mounds were brought by high-energy water during the growth process. The organisms involved in the construction are mainly cyanobacteria, epidophytes and a small number of algae such as gewanalgae. Trilobites, crinoids, ostracodes and lamellibranchs can also be seen.

According to the section observation of the microbial mounds, the mounds experienced six growth stages during the deposition process, namely the six layers evident in Fig. 18. The mounds were all in a high-energy environment during the six growth stages and the energy of the second and fourth layers was relatively higher. Due to the strong water dynamic brings rich nutrients, the biological community thrives. Columnar stromatolites have thick laminae. Individual stromatolites also have large widths. There are flat-pebbles captured between stromatolites. The flat-pebbles captured in the second layer are larger. They are more than 2 cm and the

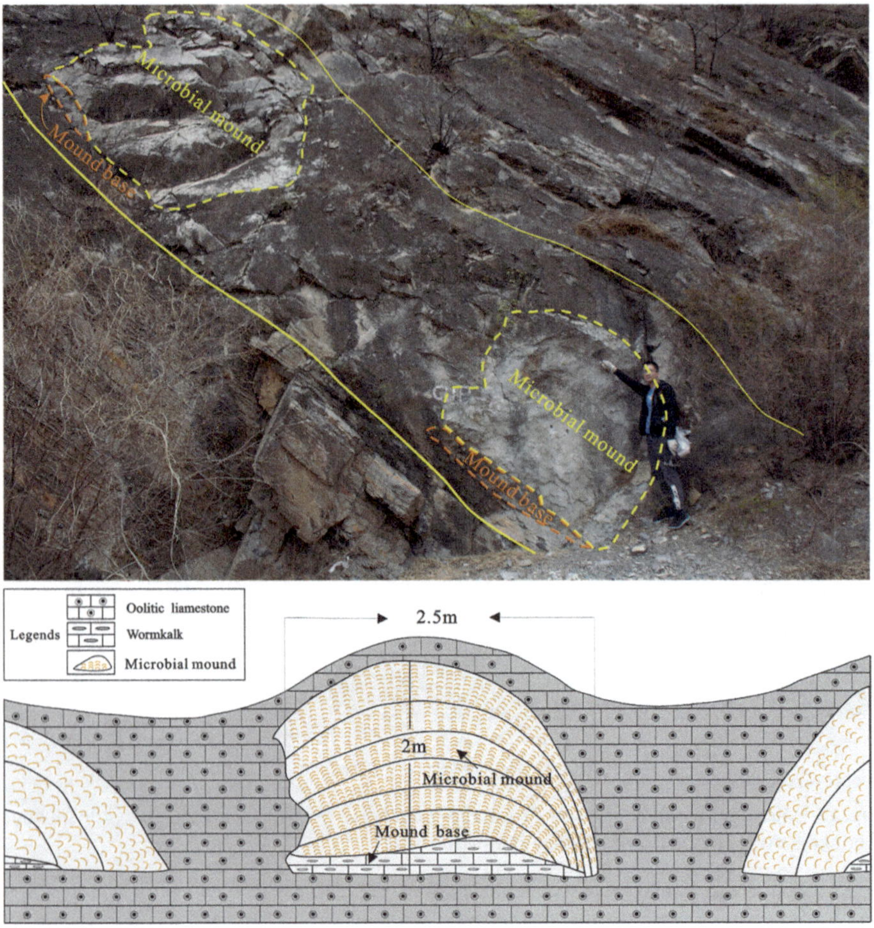

Fig. 18 Architecture of the microbial mound in the Lower Gushan Formation

largest flat-pebbles can reach 10 cm. The flat-pebbles captured by the fourth stromatolites are relatively small, generally less than 2 cm. The lamina of other growth stages of stromatolites is fine. They are columnar or undulating, in which the bottom of the first layer and the third layer of the columnar stromatolites are more obvious. From the height, internal structure and sedimentary structure of the mounds, stromatolite mounds are more resistant to waves than the clotted bioherms at the bottom.

The rocks near the top and bottom of the microbial mounds are oolitic limestones. The oolitic limestones of the same deposition contact the underlying rocks of the mounds. The oolitic limestones were uniform in size and layered with sparry calcite cementation. Due to recrystallization, the internal structure of some oolitic limestones was not obvious, indicating that the seawater energy was high during the formation of the second mound. One side of the mound was damaged (Fig. 18), which may be caused by strong hydrodynamic conditions such as waves or tides during the growth of the mound.

Oolitic limestones formed in the upper part of the subtidal zone and the intertidal zone under the condition of relatively weak deepwater dynamics began to appear near this set of mounds. Oolitic limestone occurs in the upper part of the high-frequency upward shallowing cycle, while the lower part of the cycle mainly consists of micrite intraclastic limestone, marl and siltstone, which represent the low energy subtidal zone.

Stop 6: Tempestite Types in Gushan, Changshan and Fengshan Formations

Location: west side of the mountain; Coordinates: N 39°59.5843'; E 116°01.3391'.

The stop indicates the top of the Gushan Formation, Changshan Formation and Fengshan Formation. Lithology association is relatively simple, consisting of dozens of upward shallowing high-frequency cycles. The lower part of the cycle is interbedded by banded silty shale with argillaceous siltstone or thin-bedded mudstone with calcareous siltstone.

The tempestite can be divided into in-situ tempestite and allochthonous tempestite according to its sedimentary process. On this outcrop, in-situ tempestite is lenticularly distributed in the brown argillaceous siltstone (Fig. 19). The length and width of the lens are about 30 and 8 cm. The light gray flat-pebbles are floating in the brown-yellow matrix with different sizes, most of which are long bands and some are irregular ellipses. The flat-pebbles account for 70% of the textural composition of the host rock. The longest axis length of the largest flat-pebbles is 4 cm and the average length is 1 cm. The flat-pebbles are distributed along with the bedding and locally overlapped by bending deformation. They are well-sorted and moderately rounded mudstone with 30% of micrite. It is mainly calcite, containing trilobite fragments.

Allochthonous tempestite is located at the bottom of the Fengshan Formation (Fig. 20), which is layered with a thickness of about 35 cm. The upper and lower lithology is light gray thin-bedded mudstone mixed with brown banded silty shale. The lower interbedded layer is about 20 cm thick. The bottom is an allochthonous tempestite with the same lithology. Light gray flat-pebbles floated in the yellow matrix with different sizes. Most of them were long bands and some were irregular, accounting for 40%. The longest axis length of the largest flat-pebble was 10 cm and the average length of flat-pebbles is 4 cm. The long axis of them distributed in the middle of the layer was larger than that on both sides. They are scattered along with the bedding with well sorted and rounded. The composition is mudstone. Matrix accounted for 60%, mainly microcrystalline dolomite.

Stop 7: Thick Bioturbated Oolitic Limestone in the Upper Xuzhuang Formation

Location: east side of the mountain; Coordinates: N 39°59.5843'; E 116°01.3391'.

The distance between the east and west sides of the mountain is about 800 m, providing a comparative observation of the spatial changes of the relevant sedimentary systems. This stop exposes oolitic limestone with bioturbation characteristics in the middle and upper part of the Xuzhuang Formation (Fig. 21a, b). The bioturbated

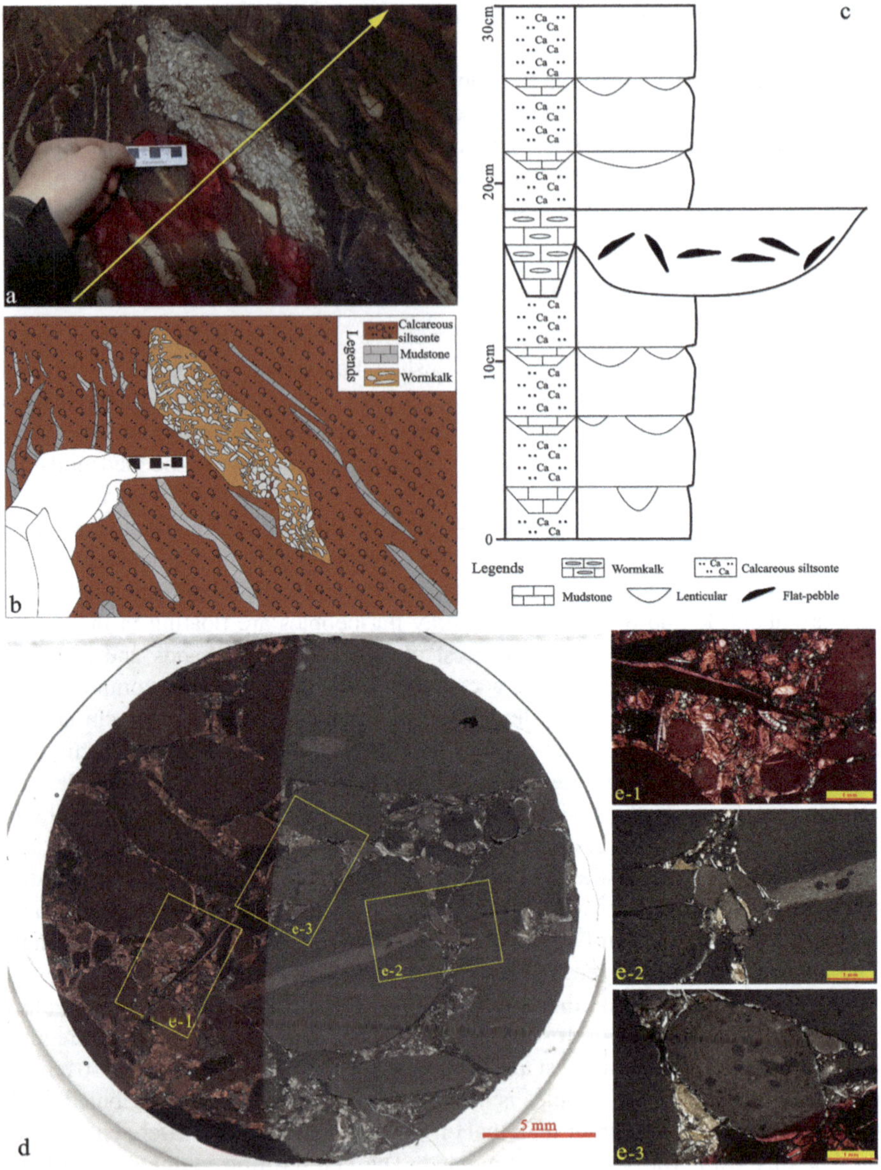

Fig. 19 Macroscope and microscope characteristics of in-situ wormkalk tempestite in the Changshan Formation

portion appears as dark patches with the same oolitic texture as the host, but the boundary is blurry (Fig. 21c, d).

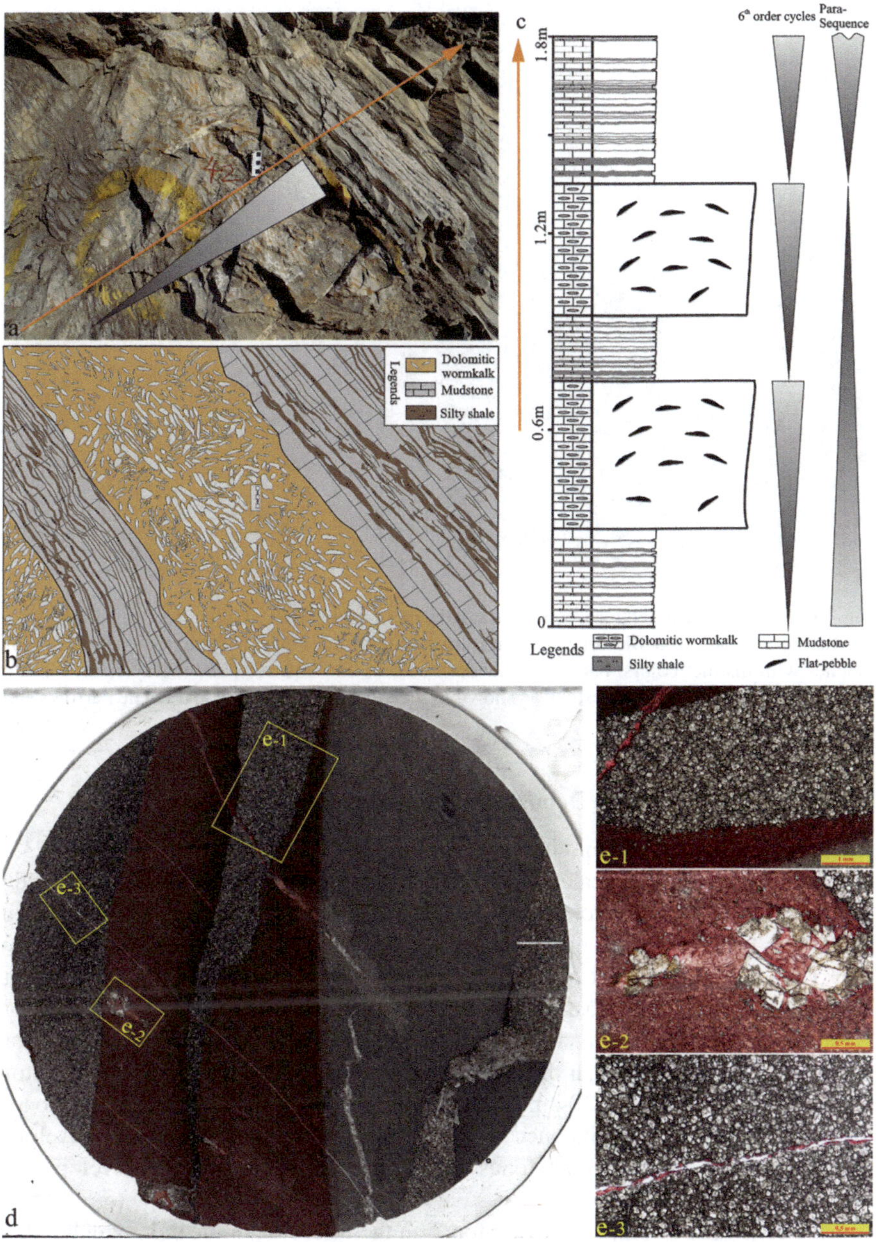

Fig. 20 Macroscope and microscope characteristics of allochthonous wormkalk tempestite in the Changshan Formation

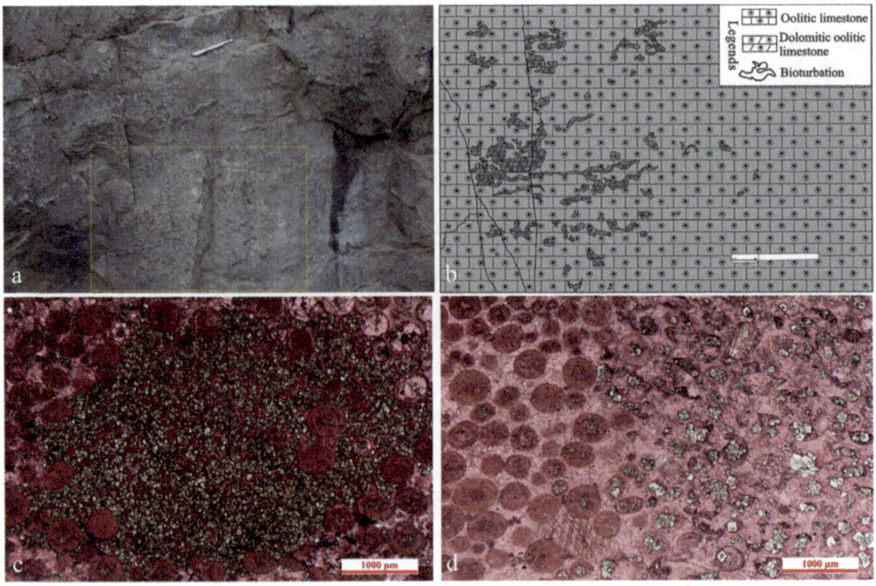

Fig. 21 Macroscope and microscope characteristics of bioturbated oolitic limestone in the Upper Xuzhuang Formation. **a** Bioturbated oolitic limestone in the Upper Xuzhuang Formation shows patches with distinct colors, burrows show dark color while host rock show gray color. **b** Sketch map of bioturbated oolitic limestone as shown in the yellow box in (**a**). **c** and **d** The burrows show selective dolomitization

Stop 8: Mixed Tidal Flat Deposits in the Lower Zhangxia Formation

Location: east side of the mountain; Coordinates: N 39°59.5843'; E 116°01.8933'.

The L-M type meter-scale cycle of the lower Zhangxia Formation is exposed at this stop. The lower portion of the cycle is thin-bedded calcareous siltstone with banded or lenticular marl, and the upper part is oolitic limestone. Figure 22 shows 1.5 cycles. The lower part of the first cycle is gray yellow thin-bedded calcareous siltstone with gray-white banded or bedded marl, ripple occurred in the calcareous siltstone. The difference in weathering resistances between the marl and siltstone makes it clear to distinguish them. The upper part of the cycle is oolitic limestone with a thickness of about 20 cm. In the lower part of the second cycle, marl bands or lenticular calcareous siltstones with a thickness of about 50 cm were developed and thin-bedded calcareous siltstones interbedded with marl began to develop above them.

This stop can be compared with Stop 2 on the west side of the mountain but there are fewer storm-originated wormkalks here. It indicates that though the development of tempestite in the epeiric sea is very frequent, its lateral continuity is poor.

Stop 9: Microbial Mounds in the Lower Zhangxia Formation

Location: east side of the mountain; Coordinates: N 39°59.8753'; E 116°01.8970'.

Fig. 22 Thin argillaceous siltstone and marl interbed-oolitic limestone-siltstone cycle in the Lower Zhangxia Formation

The biological mounds in this stop and those in Stop4 are located in the middle and lower part of the Zhangxia Formation. They are comparable horizontally. Outcrops show a layer of ellipse mounds growing adjacent to each other. The base of mounds is not observed clearly. The exposed part is about 2.3 m high and 3.2 m wide (Fig. 23).

Stop 10: Microbial Mound and Ooids-Bearing Tempestite in the Lower Gushan Formation

Location: east side of the mountain; Coordinates: N 39°59.8753'; E 116°01.8970'.

This stop indicates that the microbial mounds of the lower Gushan Formation are similar to that of Stop5. However, their morphology and structural characteristics are different. The mounds here are 70 cm high and 1.1 m wide with a base of wormkalk and a cover of microbial bonded mudstone (Fig. 24).

The deposit between mounds is argillaceous siltstone, indicating a subtidal zone environment. The overlying beds are mudstone with bands of silty mudstone, indicating that the growth stagnation of microbial mounds is caused by the change of sedimentary environment.

Above the overlying mudstone with bands of argillaceous siltstone, the upper beds became grayish-green argillaceous siltstone with a thickness of about 30 cm. The argillaceous siltstone or mudstone developed interbed of 8–10 cm thick oolitic packstone with small pebbles. The pebbles are mudstone with 1 to 2 cm long axis

Fig. 23 Microbial mounds and overlying silty shale in the Lower Zhangxia Formation

Fig. 24 Characteristics of microbial mounds in the subtidal zone of the Lower Gushan Formation

Fig. 25 Thin-bedded packstone deposited in subtidal zone of the Lower Gushan Formation. **a** The interbed of packstone with small pebbles. **b** The pebbles are matrix-supported

being parallel to the bed plane. They are distributed in the matrix, which is mainly filled with ooids and mud (Fig. 25).

Stop 11: Cone Lime Mud Mound in the Lower Fengshan Formation

Location: east side of the mountain; Coordinates: N 39°59.7574'; E 116°01.9265'.

In the lower Fengshan Formation, there are large sets of thick mudstone (Fig. 26) with thin-bedded and banded dolomitic limestones (Fig. 27). In the mudstone, there are obvious lime mud mounds. Several cone lime mud mounds are developed laterally. They are small spires or cones about 30 to 40 cm high and a maximum width of 30–50 cm at the base. There is no obvious sedimentary structure inside them, and the filling lithology between the mounds is the same as the surrounding rocks above and below, which is mudstone with dolomitic limestones interbeds.

It is generally believed that the shape of the lime mud mound is related to hydrodynamic conditions. The nearly horizontal gently sloping lime mud mounds are usually formed in the supratidal zone with weak hydrodynamics. The gently sloping lime mud mounds are mainly distributed in the intertidal zone with weak hydrodynamics.

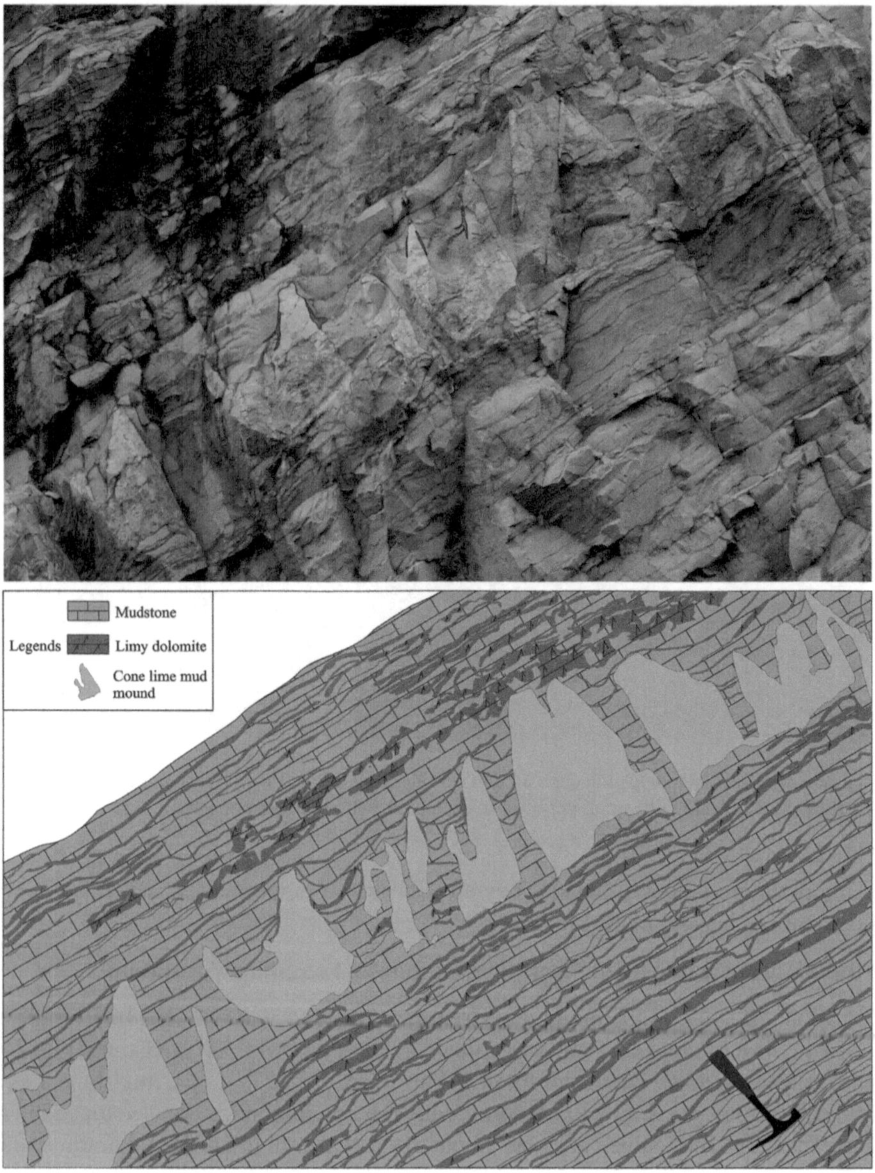

Fig. 26 Macroscopic characteristics of cone lime mud mounds in the Lower Fengshan Formation

Fig. 27 Microscope characteristics of intercalated limy dolostone in thick mudstone (The macroscope characteristics are shown in Fig. 26). **a** Thinly interbedded mudstone and limy dolostone. **b** There is a clear boundary between mudstone and limy dolostone. **c** Fractures occur in the limy dolostone. **d** The dolomite in the limy dolostone is less than 0.03 μm in crystalline size and dirty appearance

Cones and hemispheres lime mud mounds form in intertidal zones where the hydrodynamics are strong (Cao and Yuan 2006, 2009; Bosak et al. 2013). However, the actual situation is much more complicated. Small spires and large cone lime mud mounds developed in Antarctica's perennial ice-covered Lake Untersee, are recognized as a result of biological behavior for the morphology of such lime mud mounds (Andersen et al. 2011). The sedimentary environment of the lime mud mounds can be inferred from the contemporaneous deposition of the cone mounds. The surrounding rocks between the mounds and above and below the mounds are all mudstone, indicating the subtidal sedimentary environment under the input background of intermittent terrigenous argillaceous silts.

Stop 12: Microbial Mounds and Wormkalk Tempestite Between Mounds in the Upper Fengshan Formation

Location: east side of the mountain; Coordinates: N 39°59.7212'; E 116°01.9214'.

In the thinly interbedded mudstone and argillaceous siltstone of the upper Fengshan Formation, a set of gray and dark purple interbedded strata developed (Fig. 28) with a thickness of about 2 m. The dark purple columns are narrow with a width less than 50 cm. They are trapped in the wider gray and white rectangular microbial mounds. The gray rectangular microbial mounds show the growth of multiple columnar stromatolites. The lower part of the mound contains more flat-pebbles

Fig. 28 Microbial mound and wormkalk filling in the Upper Fengshan Formation

with red oxidation circles, which constitute the base for the growth of the microbial mounds. The purple-red filling between the rectangular mounds is wormkalk with poorly sorting and orientation pebbles in micritic limestone.

Beneath the microbial mounds are thin-interbedded mudstone and argillaceous siltstone, indicating the sedimentary environment of the subtidal zone. The mound

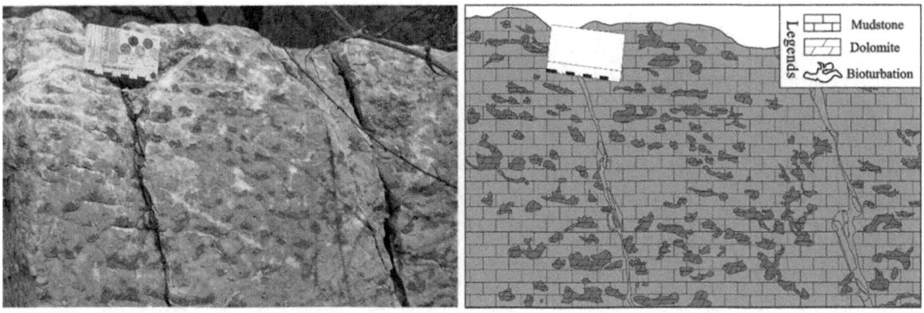

Fig. 29 Bioturbation-associated dolomitization of micrite limestone in the Ordovician Yeli Formation

is composed of multiple high-frequency cycles. The lower part of the single cycle is argillaceous siltstone with marl bands or lenses and the upper part is wormkalk. There is no high-energy oolitic grainstone deposition and lack of exposure markers. Stromatolite microbial mounds grow in subtidal environments. Based on storm-caused wormkalk, multiple mounds gradually grow upwards. The lack of terrigenous input and very slow chemical deposition rates resulted in unusually slow deposition rates between mounds. When the mounds grow to 1.5–2 m high, the storm develops and carries flat-pebbles and shallow water carbonate materials to fill between mounds. The flat tops of microbial mounds with tempestite accumulation and the oxidizing coloration of storm-originated wormkalk suggest that sea level limits the upward growth of microbial mounds and the possibility of exposure or top erosion of the mounds. Above it is the next transgressive sedimentary cycle.

Stop 13: Bioturbated Micrite Limestone in the Ordovician Yeli Formation

Location: east side of the mountain; Coordinates: N 39°59.6069'; E 116°01.9110'.

The stop exposes the strata of the Ordovician Yeli Formation which is continuously deposited with the Cambrian. Bioturbated limestone is developed at the stop. The bioturbated limestone traces are *Thalassinoides* (Marine trace) (Fig. 29). The burrows formed by bioturbation are dark patches with a long axis of 1–2 cm. The dark patches are subhedral fine-mesocrystalline dolomites with a light matrix of micrite limestone (Fig. 30). The special burrow microenvironment caused by bioturbation may be the cause of dolomitization.

5 Description and Interpretation of Tempestite

The NCC is an old tectonic stable block on which widely developed epeiric sea carbonate during the Cambrian and Ordovician. In the Middle and Late periods of Cambrian, the NCC underwent periodically affection of strong storms and deposited

Fig. 30 Microscopic characteristics of bioturbation-associated dolomitization in micrite limestone of Ordovician Yeli Formation. **a** The burrow filling was dolomitized in the host mudstone. **b** and **c** The burrow shows coarse crystalline dolomites

many sets of wormkalks in the Zhangxia-Gushan Formations. The Xiaweidian section provides us perfect outcrops for observation of the sedimentary process.

Based on detailed field observation and intensive studies, 59 sets of tempestites were identified from the outcrop of Cambrian strata in the Xiaweidian section (Fig. 31). Vertically, the storm-originated tempestites (wormkalks) are well developed in the bottom and middle of Zhangxia Formation, the bottom and top of Gushan Formation, the bottom of Changshan Formation and the bottom of Fengshan Formation. They are mainly in-situ tempestites and proximal tempestites. Distal tempestites are very few and are mostly distributed at the bottom of the Zhangxia Formation. Laterally, the tempestites are continuous beds or discontinuous lenses. Accompanied the tempestites are deform bedding, wave bedding and other sedimentary structural characteristics.

The thicknesses of tempestites are mostly less than 20 cm with an average of 18 cm. Thick tempestites are only found in Zhangxia and Fengshan Formations (Fig. 32). The Zhangxia Formation has the largest number of tempestites and has a corresponding proportion in each thickness. All the tempestites in Gushan Formation are thin layers. Changshan Formation is mainly composed of thin to medium beds with rarely thick beds.

The Zhangxia Formation is an oolitic shoal and tidal flat sedimentary system characterized by oolitic limestone formed during the most extensive transgressive period of the NCC. In the middle and lower part of the Zhangxia Formation, the distal tempestites developed while the proximal tempestites are observed in the middle and upper part. The distal tempestites are suggested formed by the storm flow transportation the shallow sediments from tidal flat to the lower subtidal zone and deposited

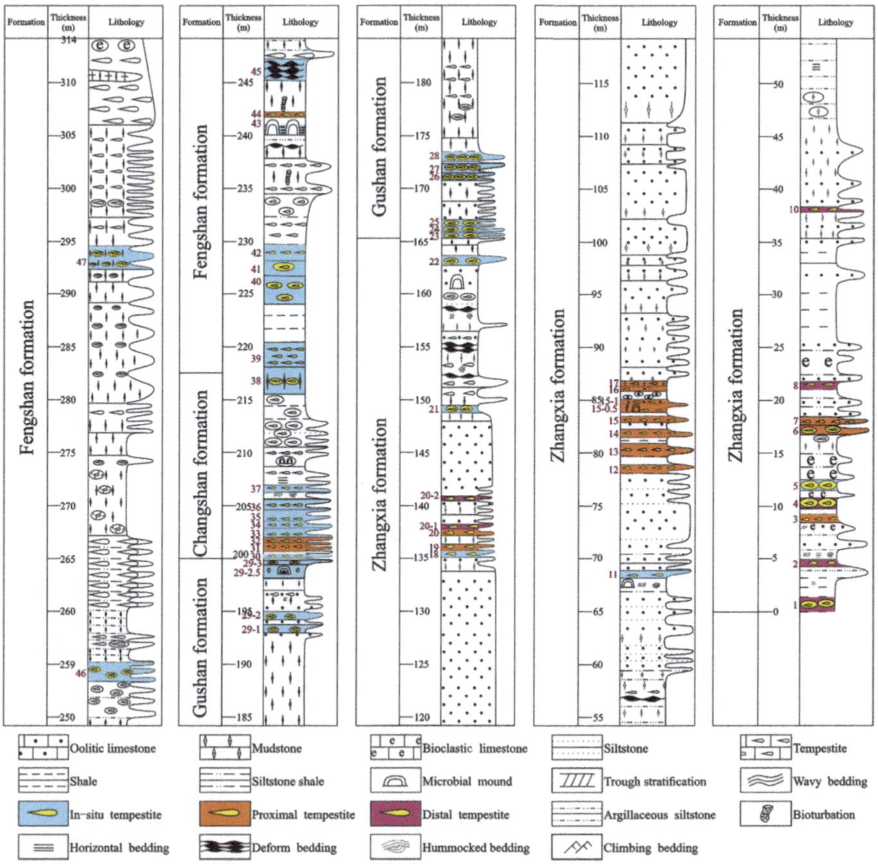

Fig. 31 Histogram of wormkalk tempestites in the Zhangxia to Fengshan formations

below the storm wave base, which experienced the longest transportation distance, formed the smallest grain size pebbles and the best degree of sorting and rounding. The proximal tempestite is deposited between the storm wave base and the fair-weather wave base. Its flat-pebble size and the degree of sorting and rounding were between the in-situ tempestites and the distal tempestites. The transition from the distal type to the proximal type of tempestites in the Zhangxia Formation reflects the transition of the sedimentary environment from relatively deep water to shallow water.

There is gray thin-bedded mudstone with oolitic limestone in the bottom of the Gushan Formation, and dark gray thin-bedded argillaceous mudstone with glaucous silty veins in the middle and upper part of the Gushan Formation. The bottom oolitic limestone interlayers reflect the subtidal high-energy environment. In this condition, the storm current is deposited in situ above the fair-weather wave base to form the in-situ tempestites. The frequent occurrence of glaucous silty veins in the middle

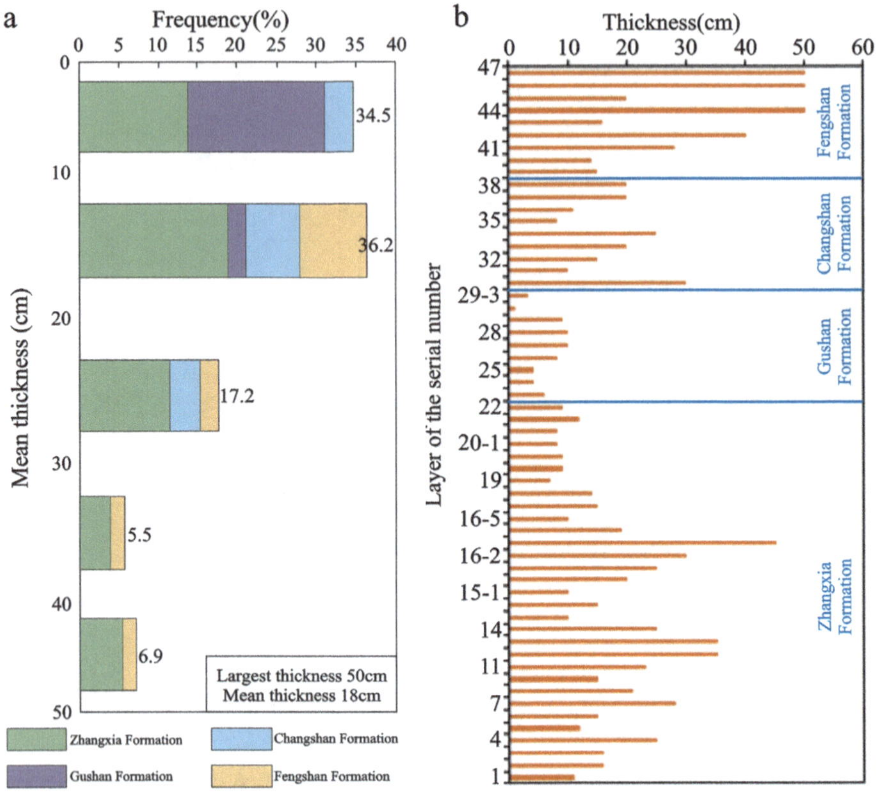

Fig. 32 Statistical chart of wormkalk tempestite thickness from the Cambrian outcrop of the Xiaweidian section. **a** Frequency distribution histogram of wormkalk tempestites thickness; **b** Variation of wormkalk tempestites thickness from the Zhangxia to Fengshan formations

and upper part reflects the increase of water depth and multiple layers of distal tempestites were deposited near the storm wave base. The in-situ tempestite has the highest energy and the strongest erosion to the basement. Most of them are distributed lenticular in the subtidal environment above the fair-weather wave base. Due to high oxygen content and suitable temperature, the Gushan Formation developed many mounds bonded by microorganisms, on which another storm period might stop its growth or even destroy it by deposition of the tempestite with the appearance of a purplish-red oxidation circle under the condition of oxygen enrichment.

There are interbedded gray thin-bedded mudstone and oolitic limestone in the bottom of Changshan Formation and glaucous silty shale and gray thin-bedded mudstone in the middle and upper part of Changshan Formation. The extensive development of tempestites indicates that during the Changshan period the NCC experienced extensive transgression and intensive storm activities. The occurrence of oolitic limestone reflects the environment of shoal and the proximal tempestites indicate their backflow and short-distance transport. The widespread occurrence of glaucous silty shale reveals the higher relative sea level and deeper environment.

From bottom to top of Fengshan Formation, the glaucous silty shale becomes thinner and the gray thin-bedded mudstone becomes thicker. The tempestite and bioclastic limestone with the purplish-red oxidizing appearance developed at the top, reflecting the lowering of relative sea level. The storm flow might transport the sediments over a long distance to form the distal tempestite, which was distributed in silty shale and mudstone in a deeper environment. The occurrence of bioclastic limestone at the top indicates the development of the intertidal zone caused by rapid shallowing.

Generally speaking, extensive transgression occurred in the northern part of the NCC during the Cambrian, which resulted in the formation of multiple carbonate sedimentary cycles in the epeiric sea. During the early Cambrian, the deposition of tidal flat shale and gypsum breccia transformed into subtidal banded shale, argillaceous siltstone and mudstone. In the middle Cambrian of maximum transgression, huge thick oolitic limestone was formed, and finally, the tidal flat mudstone and silty shale were formed in the late period. In the middle and upper part of the carbonate rocks in Cambrian, many different kinds of wormkalks could be observed. The tempestite wormkalks deposition was the result of the combination of the sedimentary background of the carbonate epeiric sea and the tropical climate belt of hurricanes near the equator.

6 Field Logistics

6.1 Travel Arrangements

(1) Date

Beijing time on August 24, 2022.

(2) Scheduling

All participants gather at the gate of China National Convention Center at 8:30 Beijing time to start. It is expected to end the field trip and return to China National Convention Center at 17:00.

(3) Transportation

A bus will be arranged to take all participants to the section and back. The one-way driving time is about 1 h (Fig. 33).

Fig. 33 Route map to field section

6.2 Lunch

A picnic lunch will be provided by trip organizers, such as pizza, hamburgers. If anyone to be diet or avoiding particular food, please inform the trip organizers in advance.

6.3 Climate

In August, the average temperature in Beijing is ~20 °C ~30 °C. The weather is generally sunny, warm and dry. Please pay close attention to the weather forecast of the Mentougou district, Beijing on August 24 in advance to prepare appropriate excursion equipment.

6.4 Safety Instruction

1. It is recommended to wear long-sleeve shirts and long trousers. Sun hat is needed.
2. Field routes are mainly on an abandoned train rail. Hiking boots is needed.

3. Participants should inform the leader of any physical or mental condition which may affect performance in the field (e.g. asthma, diabetes, epilepsy, vertigo, heart condition, back problem, ear disorder, lung disease, allergies, etc.).
4. To prevent property and personal safety accidents caused by all kinds of emergencies, please read carefully and strictly follow the following precautions for emergencies:

    (1) Earthquake

    ① Choose open ground and stay far away the cliff.

    (2) Rainstorm and flood

    ① Just stay on the base of the railway to avoid standing on the steel track.

    (3) Taking the Bus

    ① Do not chat with the driver or interfere with the driver's operation, do not open the door at will, do not throw items outside

5. Emergency treatment of field trips and accidents

① contusion: cold compress treatment as soon as possible.
② sprain: light symptom treated by anti-inflammatory pain cream; serious cases should be sent to the nearest hospital for treatment.
③ trauma: should be sent to the nearest hospital for treatment. If there is respiratory failure, artificial respiration should be done immediately, until the situation is stable.
④ Dysentery (diarrhea): light symptom should end the trip and return to hotel. And serious cases should be sent to the nearest hospital for treatment.
⑤ Fracture: use both hands to stabilize and support the injury, limit the activity of the fracture, and place soft pads, with bandages, splints, or substitutes to properly fix the injured limb.
⑦ Food poisoning: to induce vomiting and sending to the nearest hospital.
⑧ Harmful insects: Most insect venoms are acidic and can be neutralized with an alkaline liquid. Use dilute alkali water, soapy water to clean or soak the wound.

6. Emergency call

Mobile phone coverage is good in the field area. In case of emergency, please dial the following numbers:

The emergency telephone number for an ambulance is (008,610) + 120.
The emergency telephone number for police is (008,610) + 110.
The emergency telephone number for the fire alarm is (008,610) + 119.
The emergency telephone number for traffic police is (008,610) + 122.

Acknowledgements The preparation of this guidebook is directed by the China Sedimentology and Sedimentary Geology Professional Committee. The gathering of digital data was completed with the help of the team of Dr. Zheng Yanwei of the China University of Geosciences (Beijing). The graduate students of the Carbonate workshop at Peking University participated in the fieldwork. The laboratory work was carried out in the Carbonate Laboratory, PKU. Thanks to the foundation from the National Natural Science Foundation of China (No. U19B6003) and the RIPED cooperative project (No. RIPED-2020-JS-50751).

References

Ager DV (1974) Storm deposits in the Jurassic of the Moroccan High Atlas. Palaeogeogr Palaeoclimotol Palaeoecol 15(2):83–93

Aigner T (1982) Calcareous tempestites: storm-dominated stratification in upper Muschelkalk limestones (Middle Trias, SW-Germany). Springer, Berlin Heidelberg, Berlin

Allen JRL (1982) Sedimentary structures: their character and physical basis. Dev Sedimentol 30:137–177

Andersen DT, Sumner DY, Hawes I et al (2011) Discovery of large conical stromatolites in Lake Untersee, Antarctica. Geobiology 9(3):280–293

Ball SM (1971) The Westphalia Limestone of the northern Midcontinent: a possible ancient storm deposit. J Sediment Petrol 41(1):217–232

Bosak T, Knoll AH, Petroff AP (2013) The meaning of stromatolites, vol 41. Annual Review of Earth and Planetary Sciences, pp 21–44

Brenchley PJ, Pickerill RK, Stromberg SG (2010) The role of wave reworking on the architecture of storm sandstone facies, Bell Island Group (Lower Ordovician), eastern Newfoundland. Sedimentology 40(3):359–382

Cao R, Yuan X (2006) Stromatolites. University of Science and Technology Of China Press, Hefei, pp 1–383

Cao R, Yuan X (2009) Advances of stromatolite study in China. Acta Palaeontol Sinica 48(3):314–321

Chen J, Lee JH, Woo J (2014) Formative mechanisms, depositional processes, and geological implications of Furongian (late Cambrian) reefs in the North China platform. Palaeogeogr Palaeoclimatol Palaeoecol 414:246–259

Dattilo BF, Brett CE, Tsujita CJ et al (2008) Sediment supply versus storm winnowing in the development of muddy and shelly interbeds from the Upper Ordovician of the Cincinnati region, USA. Can J Earth Sci 45(2):243–265

Dai MY, Qi YA, Chen Y et al (2014) Giant ooids and their genetic analysis from the Zhangxia formation of Cambrian Series 3 in Mianchi area, western Henan province. J Palaeogeog 16(5):726–734

Duan X, Shi Z, Jin X et al (2015) Microbial cause for giant ooids: evidence from the lower Cambrian in Shizhu area, Chongqing. J Palaeog 17(2):241–248

Dott RH Jr (1974) Cambrian tropical storm waves in Wisconsin. Geology 2(5):243–246

Friedman GM (1994) Upper Cambrian-lower Ordovician (Sauk) platform carbonates of the northern Appalachian (Gondwana) passive margin. Carbon Evapor 9(2):143–150

Feng Z, Wang Y, Zhang J (1993) Lithofacies paleogeography of the early Paleozoic of north China platform. Petroleum Industry Press, Beijing, pp 28–48

Gilbert GK (1899) Ripple-marks and cross-bedding. Bull Geol Soc Am 10(1):135–140

Howard JD (1972) Nearshore sedimentary processes as geologic studies. In: Swift DJP, Duane DB, Pilkey OH (eds) Shelf sediment transport: process and pattern. Pa, Stroudsburg, pp 645–648

Jin K (1997) Tempestites in the Cambrian Strata in Dabeiwang, Xuzhou, Jiangsu. Sedim Facies Palaeog 17(1):34–38

Jin Z, Jiang P, Feng Z (1995) Cambrian storm deposits on carbonate slope at east Guizhou and west Hunan provinces. J Univ Petrol China 19(3):1–6

Kelling G, Mullin PR (1975) Graded limestones and limestone-quartzite couplets: possible storm-deposits from the Moroccan Carboniferous. Sed Geol 13(3):161–190

Kreisa RD (1981) Storm-generated sedimentary structures in subtidal marine facies with examples from the Middle and upper Ordovician of southwestern Virginia. J Sediment Res 51(3):823–848

Kumar N, Sanders JE (1978) Storm deposits. Springer, Berlin Heidelberg, Berlin, pp 1139–1142

Kuang W, Yang S, Liu X (2008) Significance of tempestite from the Cambrian Qingxudong formation in Yutang area of Northwestern Hunan province. J Jilin Univ: Earth Sci Ed 38(2):225–232

Laporte LF (1967) Recognition of transgressive carbonate sequence within Epeiric sea: helderberg group (Lower Devonian) of New York State: Abstract. AAPG Bull 51(3):473–473

Li F, Yan J, Algeo T et al (2013) Paleoceanographic conditions following the end-Permian mass extinction recorded by giant ooids (Moyang, South China). Global Planet Change 105:102–120

Li ZX, Bogdanova SV, Collins AS et al (2008) Assembly, configuration, and break-up history of Rodinia: a synthesis. Precambr Res 160(1):179–210

Li Z, Yu J, Guo J (2002) Study on coal accumulation under marine transgression events in the epicontinental basin of North China. Coal Geol Explorat 30(5):1–5

Liu B, Zhang J, Xu X (1986) On the calcareous tempestites in the lower permian of Silong, Xingwen, Sichuan. Acta Geological Sinica 01:57–124

Liu H, Jia Y, Long Y et al. (1987) The features of the barrier island systems of the Epeiric sea and their event deposits of coal-bearing formations in carboniferous of north China. Acta Sedimentologica Sinica 03:78–85+193

Ma R, Zhang L, Du Y et al (2011) Sedimentary characteristics and its geological implications of Cambrian tempestite in Northern Henan province. Geol Sci Technol Inform 30(04):15–20

Ma Y, Mei M, Zhou R et al (2017) Forming patterns for the oolitic bank within the sequence-stratigraphic framework: an example from the Cambrian series 3 at the Xiaweidian section in the Western Suburb of Beijing. Acta Petrologica Sinica 33(4):1021–1036

Mckee ED (1959) Storm sediments on a pacific atoll. J Sediment Res 29(3):354–364

Mei M, Guo R, Hu Y (2011) Sedimentary fabrics for the stromatolitic bioherm of the Cambrian Gushan Formation at the Xiaweidian section in the western suburb of Beijing. Acta Petrologica Sinica 27(8):2473–2486

Mei M (2012a) Organomineralization derived from the biomineralization: an important theme within the framework of geobiology. Geol Rev 58(5):937–951

Mei M (2012b) Brief introduction on new advances on the origin of Ooids. Acta Sedimentol Sin 30(1):20–32

Meng X, Qiao X, Ge M (1986) Study on ancient shallow sea carbonate storm deposits (Tempestite) in North China and Dingjiatan model of facies sequences. Acta Sedimentologica Sinica 02:1–18+130–133+142

Mou C, Zhou K, Chen X (2016) Lithofacies and paleogeography of China: Ediacaran-Silurian. Geological Publishing House, Beijing, pp 1–154

Sang S, Lei X, Zhang H (2002) Depositional model of lime sand barriers of Cambrian in Xuzhou Area. J China Univ Min Technol 06:21–26

Song J, Liu S, Zhao Y et al (2016) Characteristics and sedimentary geological significances of lower-middle Cambrian tempestites in central Sichuan Basin. Acta Petrolei Sinica 37(01):30–42

Su DC, Loon AV, Sun AP (2016) How quiet was the epeiric sea when the middle Cambrian Zhangxia formation was deposited in SW Beijing, China? Mar Pet Geol 72:209–217

Sumner DY, Grotzinger JP (1993) Numerical modeling of ooid size and the problem of Neoproterozoic giant ooids. J Sediment Res 63(5):974–982

Wang B, Wang H (1998) The middle and late carboniferous depositional characteristics of Shanxi Epeiric Sea. J Taiyuan Univ Technol 05:37–41+45

Wang J (2002) The geochemical and potential of hydrocarbon-producing of likely source rocks in the north China. Master's dissertation, Lanzhou: Lanzhou Institute of Geology, Chinese Academy of Science, pp 1–65
Wang L, Wu H, Zhang R et al (2018) The types, characteristics and depositional models of carbonate platform: implications for Cambrian sedimentary patterns of Epeiric-drowned carbonate platform in north China. Geol Rev 64(01):62–76
Wang W (2011) Storm deposits of shallow lake in lower submember of member 2 of Shahejie formation of Wennan-Liuzhuang area, Dongpu Sag. J Earth Sci Environ 33(4):384–389
Wang X, Zhou G (1995) The discovery of storm rock in the upper Cambrian Shuishi formation and its significance in Southwest Jiangxi. Jiangxi Geol 9(1):21–27
Wu X, Hu B, Wang G (1987). Storm deposits in upper Carboniferous shallow marine carbonates of Jiaozuo, Henan. Acta Sedimentol Sinica 04:1–13+164
Xiao F (2020) Sequence stratigraphy and lithofacies paleogeography of the Cambrian in the Middle-eastern North China Platform. Doctoral dissertation, China University of Geosciences (Beijing)
Xing Y, Feng L (2015) A study on ooids in limestones of the Cambrian Xuzhuang Formation at Xiaweidian outcrop in Western Hill of Beijing. J Palaeogeog 17(04):517–528
Yang B, Yang K, Xia W (2007) Discovery of middle-upper Cambrian Tempestites and its significance in Huangshi, eastern Hubei province. Geol Sci Technol Inform 26(3):33–36
Yao X, Zhou Y, Li S (2013) Sedimentary environments and geochemical characteristics of early permian carbonate platform in the western Henan province. Geoscience 27(06):1340–1347
Yu K, Chang T, Qiu L et al (2015) Research development of flat-pebble conglomerate characteristics and their origin in early Paleozoic north china platform. Acta Sedimentol Sin 33(06):1111–1125
Zhao Z (1988) A sedimentary model of the tidal flat of epicontinetnal sea. Acta Sedimentol Sinica 02:68–76+135
Zhou Q, Yan J, Zhang M (2006) Calcareous Tempestite from the Cambrian Qingxudong formation in northeastern Guizhou province and their geological implications. Geol Sci Technol Inf 25(2):25–28

Lower Cretaceous Lacustrine Sedimentary Sequences of a Small Intracontinental Rift Basin in Luanping, China

Ping Luo, Chuanmin Zhou, Xingyang Zhang, Zhijie Zhang, Xuanjie Zhang, and Xiufen Zhai

1 Objectives and Strategy

The Luanping Basin is a typical continental rift basin that developed during the Middle Jurassic to Early Cretaceous in the Yanshan fold-thrust belt in Northeast China. The Yanshan fold-thrust belt, which is characterized by terrestrial sedimentation, magmatism and deformation—including multiple phases of folding and contractional, extensional, and strike-slip faulting throughout China during the Jurassic–Cretaceous Cretaceous—was mentioned in the Chinese literature as a result of deformation episodes named the Yanshan (or Yenshan) movement, Yanshanian movement, or Yanshanian deformation (Cope, 2003; Davis et al. 1998, 2001; Dou et al. 2020; Li et al. 2003, 2007). The Luanping Basin is a small half-graben and elongated in a NE–SW direction, approximately 40 km long and 20 km wide. Controlled by multiepisodic Yanshanian movements, the basin experienced extensional and transtensional stresses leading to a complex rifting process. Three rifting episodes were divided: the first episode occurred in the early Middle Jurassic, the second episode occurred in the middle to late Middle Jurassic, and the third episode took place during the late Jurassic to early Cretaceous. During these rifting episodes, thick nonmarine terrestrial volcanic and clastic synrift successions (~5000 m) accumulated. The synrift strata are composed of five lithologic stratigraphic units: the Jiulongshan, Tiaojishan, Houcheng, Zhanjiakou and Xiguayuan Formations.

Outcrops of the Lower Cretaceous Xiguayuan Formation found throughout Luanping County provide well-exposed stratigraphic transects that record the final rifting

P. Luo (✉) · C. Zhou · X. Zhang · Z. Zhang · X. Zhai
Research Institute of Petroleum Exploration & Development, PetroChina, Beijing, China
e-mail: 19537974@qq.com

X. Zhang
China Aero Geophysical Survey and Remote Sensing Center for Land and Resource, Beijing, China

X. Hu (ed.), *Field Trip Guidebook on Chinese Sedimentary Geology*,
https://doi.org/10.1007/978-981-99-6936-4_4

episode of the Luanping basin. The Xiguayuan Formation is approximately 2000 m thick and records a typical overall fining-upward to coarsening-upward lacustrine synrift succession. Sedimentary facies are defined, and facies associations are divided based on detailed descriptions of well-exposed outcrops. Four facies associations have been identified from alluvial to lake deposits: volcanogenic sedimentary facies association, alluvial facies association, littoral to sublittoral facies association, and profundal facies association. The volcanogenic sedimentary facies association comprises two main lithofacies—extrusive volcanic rocks (mainly andesite) and pyroclastic deposits. The alluvial facies association is composed mainly of massive conglomerates with variable mudstones, siltstones and sandstones. The littoral to sublittoral facies association is primarily characterized by fan delta sandy conglomerates and sandstones separated by lacustrine mudstone and sandstones. The profundal facies association consists dominantly of laminated shales interbedded with turbidite conglomerates.

The aim of this field trip is to further develop our understanding of the facies, facies association, stratigraphic architecture and evolution of nonmarine half-graben-type rift basins. This guide contains 4 road/section logs (Fig. 1) and a stop description for a field trip in Luanping Basin/County. Section 1 (Xingzhouhe Section) involves a walk along the Xingzhouhe River to examine sedimentary successions from alluvial fans and fan deltas to turbidites intercalated with profundal lake sediments. Section 2 (Caoying Section) is along a road cut near the basin margin where both alluvial fan packages and pyroclastic deposits can be observed. Section 3 (Sangyuan Section) is characterized by alternating shales/mudstones and conglomerates that dominate in fan delta front deposits and sediment gravity flow deposits. Section 4 (Wangjiagou Section) is a cliff section where profundal sedimentary units intercalated with mass-transported deposits can be inspected.

2 Introduction

2.1 Traffic and Geographic Locations

Luanping County, which belongs to Chengde city, is located in northern Hebei Province, where four provinces and cities, including Beijing, Tianjin, Liaoning and Inner Mongolia, meet. Being in this "Golden Triangle" position, Luanping is called the North Gate for the capital city. Luanping County is approximately 165 km to the northwest of Beijing and 65 km to the east of Chengde city, which is a famous place of interest (Fig. 1). Luanping County is situated in the middle section of the Yanshan Mountains. The topography is high in the middle and low-lying in the east and west. The Luan River and Chao River are the main rivers that flow through this area. The total area is 3212 km^2, with a total population of 333,000. Among them, 196,000 are minorities. Luanping County is also an ethnic community dominated by Manchurians. There are more than 30 kinds of ore deposits in this area, including gold,

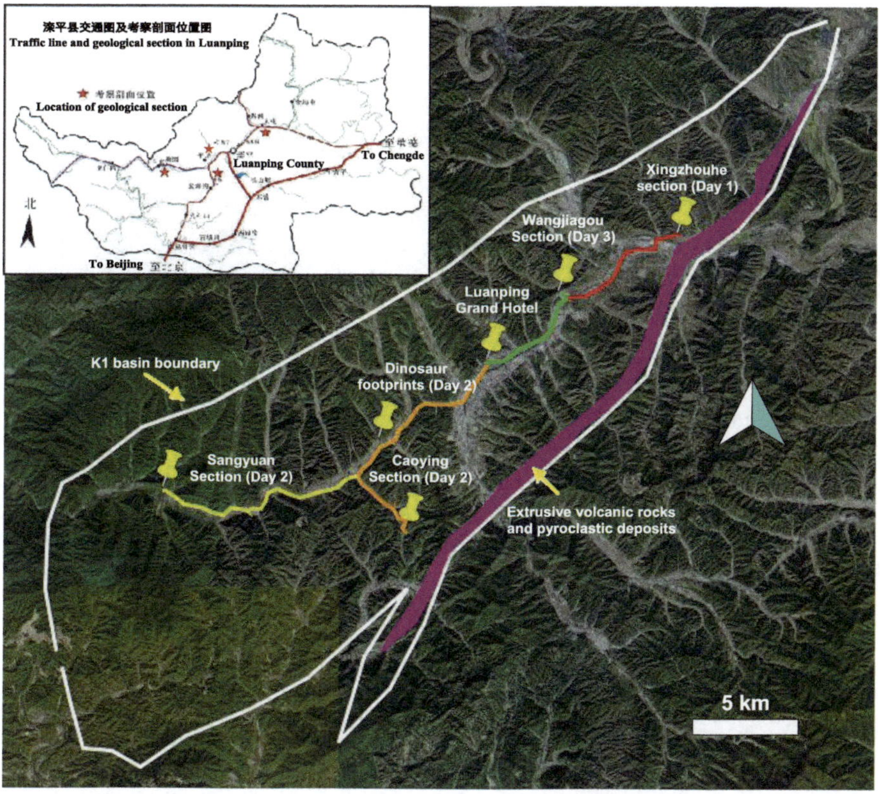

Fig. 1 Geological sections/stops and traffic lines in the Luanping Basin

iron, silver, vermiculite, K feldspar, granite, mineral water, and others. Sightseeing resources in this area are unique, including Jinshanling Great Wall, Baicaowa Forest Park, Zhuanshan Lake, and Bixia Geopark distributed along the Beijing-Chengde Highway. Traveling to Luanping is quite easy and convenient with the construction of the Beijing-Chengde tourist highway, Beijing-Shenyang 101 national road, and Beijing-Tongliao Railway (Fig. 1).

2.2 General Geology and Outcrop Locations

The Luanping rifted basin (Fig. 2) belongs to the Bohai Bay J_3-K_1 basin groups, which underwent less rifting during the Late Jurassic and intense rifting during the Early Cretaceous.

The basement of the Luanping Basin is composed of Neoarchean metamorphic rocks and Paleoproterozoic intrusive rocks. The Middle Jurassic to Early Cretaceous was the main basin-forming time when thick pyroclastic rocks and terrigenous clastic

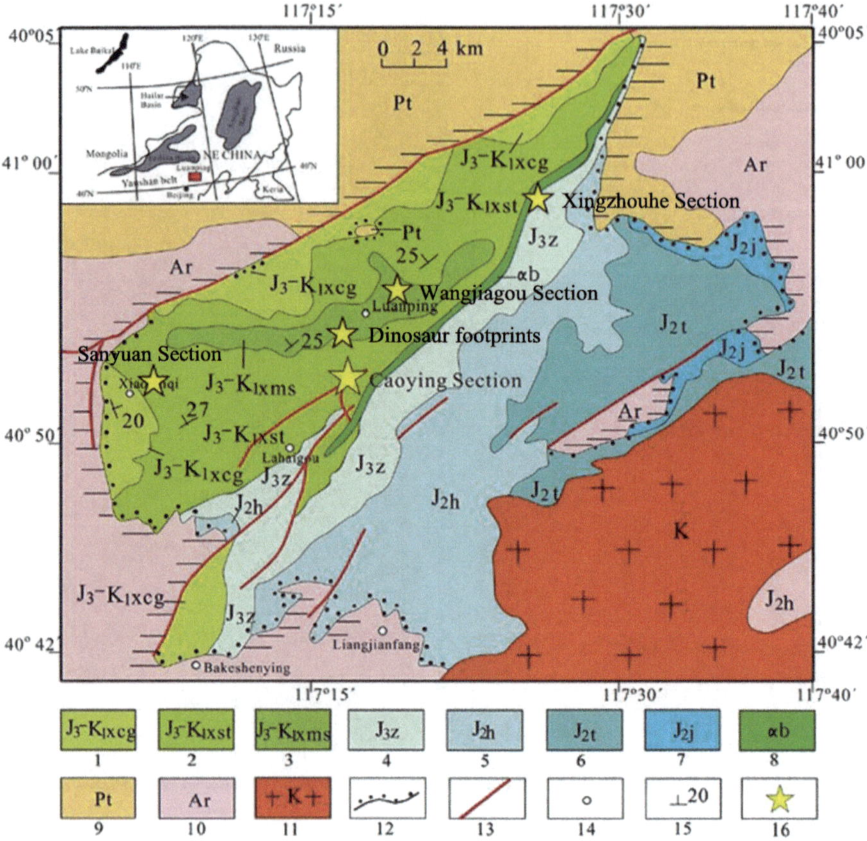

Fig. 2 Sketch geological map of the Luanping Basin (Dou et al. 2020) with the locations of outcrops marked as Sangyuan, Caoying, Xingzhouhe and Wangjiagou (situated in Luanping County). (1) Conglomerate deposits of the Upper Jurassic Xiguayuan Formation; (2) siltstone deposits of the Upper Jurassic Xiguayuan Formation; (3) mudstone deposits of the Upper Jurassic Xiguayuan Formation; (4) Zhangjiakou Formation; (5) Houcheng Formation; (6) Tiaojishan Formation; (7) Jiulongshan Formation; (8) Andesite; (9) Proterozoic intrusive rocks; (10) Archean metamorphic rocks; (11) Cretaceous intrusive rocks; (12) unconformity boundary; (13) normal faults; (14) geographic names; (15) dip and strike; and (16) location of the studied outcrops

rocks of the alluvial system mainly filled into the basin (Fig. 3). The Xiguayuan Formation developed during the third rifting episode of the tectonic evolution of the basin. The fault-bounded lake basin in the Early Cretaceous experienced three phases (i.e., the prerifting stage, the rifting stage and the contracting stage), which correspond to different nonmarine sedimentary packages (Fig. 3). The alluvial fan, fan delta, and lacustrine depositional systems in this basin are observed during this field excursion (Fig. 4).

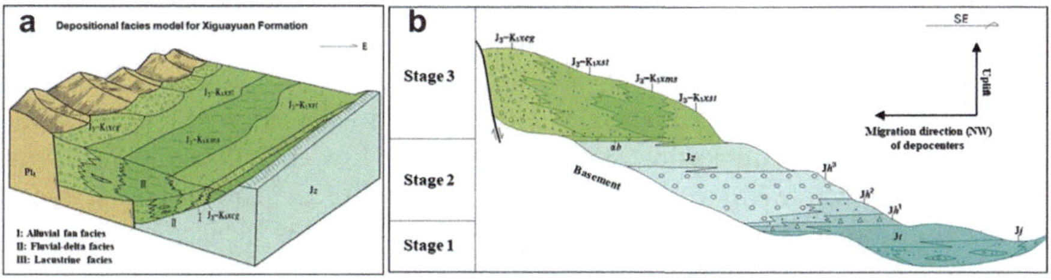

Fig. 3 Depositional facies model (**a**) and stratigraphic framework (**b**) of the Luanping Basin

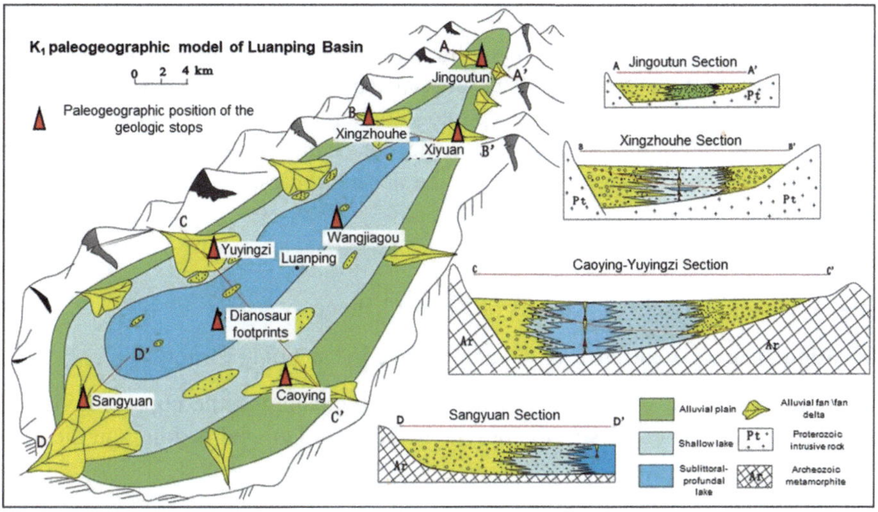

Fig. 4 Conceptual paleogeographic model of the Xiguayuan Formation in the Luanping Basin

3 Field Itinerary and Stops

Four sections included in this three-day field trip are the Xingzhouhe, Caoying, Sangyuan and Wangjiagou sections (Figs. 1 and 4) to investigate the sedimentary characteristics of alluvial fan, fan delta and lacustrine depositional systems, including sub-lacustrine gravity flow deposits. The first day will be devoted to the Xingzhouhe section, where a typical sedimentary succession of alluvial fans, fan deltas and turbidites intercalated with profundal lake sediments is well exposed. On Day 2, we will visit the Caoying Section and Sangyuan Section. Both the Caoying Section and Sangyuan Section are characterized by fan delta to profundal lake sediments, with the latter abundant with thick debris flow conglomerates of sub-lacustrine channel-lobe systems sandwiched between deep-lake shales. On the last day, we will examine mass transported deposits (MTD, including slide and slump blocks) sandwiched in profundal lake shales in the Wangjiagou Section, where hydroclastic deposit-like breccias can be observed.

3.1 Xingzhouhe Section and Stops

The Xingzhouhe Section is approximately 10 km to the northeast of Luanping County. The outcrops dip northwest and belong to the Lower Cretaceous Xiguayuan Formation. The whole strata corresponding to a second-order sequence cycle consist of three third-order sequences (Fig. 5). The field excursion is focused on two third-order sequences in the middle and lower parts of the Xingzhouhe Section. The outcrops are along the bank of the Xingzhouhe River from east to west. Five stops have been included in this guide to reflect our current state of understanding of the geologic evolution of this region (Fig. 5). In addition, a number of optional stops can be observed in the field.

This field trip starts at the Xiyuan alluvial fan deposits, which developed in the initial rifting stage of the basin. The alluvial fan deposits show a complete retrograding sequence that is upward composed of upper, middle and lower alluvial fans. Typical depositional microfacies, such as feeder channels, debris flows and sheet-flow conglomerate/sandstone facies, can be observed closely in the eastern part of this section.

Stop Descriptions

Stop 1: Upper alluvial fan deposits in Xiyuan Village.

Location: The west entrance of Xiyuan Village in Luanping County.

Content: Lithology, sedimentary structure, main channel in the upper alluvial fan deposits, the characteristics of the pluvial bedding and graded bedding on the flank

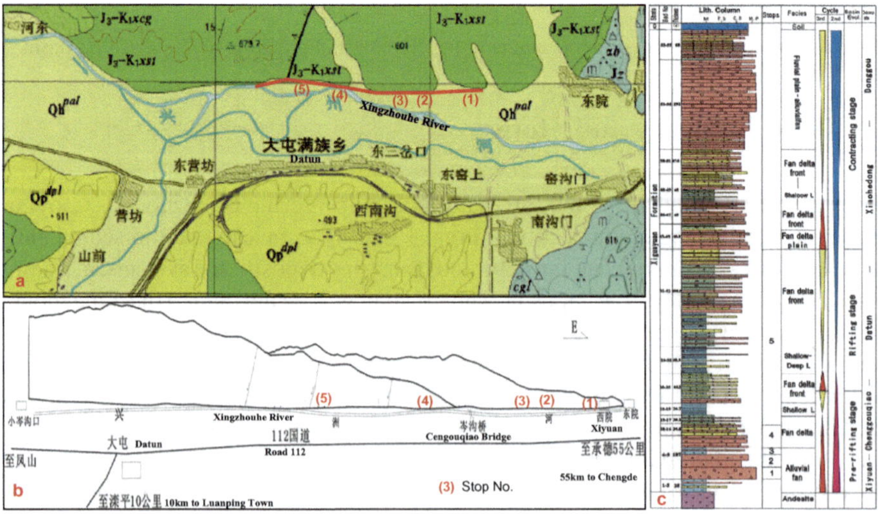

Fig. 5 Road log (**a**, **b**), lithology column and sequence stratigraphy framework (**b**) of the Xingzhouhe Section

of the main channel, the sedimentary characteristics of the braided channels within the middle fan deposits.

Stop 2: Middle alluvial fan deposits in Xiyuan Village.

Location: The west entrance of Xiyuan Village in Luanping County.

Content: Reverse-graded bedding, erosional surfaces, rounded gravels and coarse-grained sandstone in braided channel deposits.

Stop 3: Lower alluvial fan deposits in Xiyuan Village.

Location: The west entrance of Xiyuan Village in Luanping County.

Content: Red massive mudstones with conglomerate lenses deposited by debris flow.

Stop 4: Retrograding fan delta plain succession beside Cengouqiao Bridge.

Location: Cengouqiao Bridge in Datun Village of Luanping County.

Content: Amalgamation of debris flow deposits and sheet flow conglomerates of the braided channel in the fan delta plain (middle fan) and lateral facies change from fan-delta plain to delta front.

Stop 5: Fan delta front deposits and the flooding surfaces to the west of the Cengouqiao Bridge.

Location: West of Cengouqiao in Datun Village of Luanping County.

Content: Retrogradation and progradation of the fan delta front, light-colored conglomerates, the first and the maximum flooding surfaces of the third-order sequence.

Stop Figures

Stop 1: Upper alluvial fan deposits in Xiyuan Village

See Figs. 6, 7 and 8.

Stop 2: Middle alluvial fan deposits in Xiyuan Village

See Figs. 9 and 10.

Stop 3: Lower alluvial fan deposits in Xiyuan Village

See Fig. 11.

Stop 4: Retrograding fan delta plain succession to the north of the Cengouqiao Bridge

See Figs. 12, 13, 14 and 15.

Stop 5: Fan delta front deposits and the first flooding surface to the west of the Cengouqiao Bridge

See Figs. 16, 17, 18, 19 and 20.

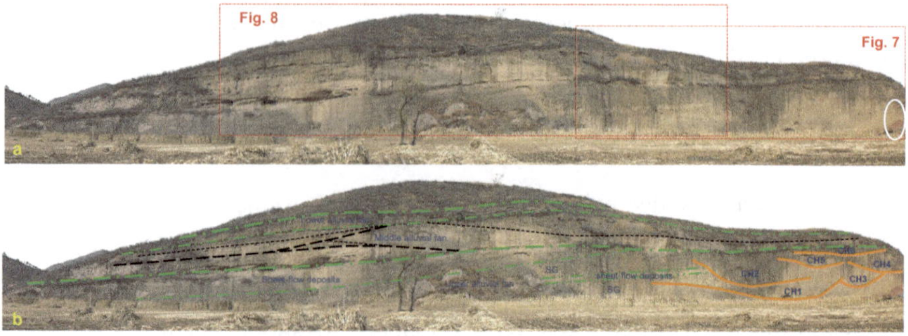

Fig. 6 Panoramic view (**a**) and interpretation (**b**) of the Xiyuan Section. The annotated lower photograph shows conglomerate beds consisting of sheet-flow elements, sediment gravity flow elements (SG), and channel elements (CH). The building circled height is approximately 3.5 m

Fig. 7 Feeder channels developed in the upper alluvial fan succession in Xiyuan Village. Note, amalgamated channel element (CH) passes laterally into sheet flow and sediment gravity flow element (SG)

Fig. 8 Lateral stacked channel bodies within the middle alluvial fan succession. Please note the laterally restricted thick channel element (CH), laterally extensive thin sheet flow and sediment gravity flow element (SG)

Fig. 9 Close view of middle alluvial fan deposits, which consist dominantly of structureless, poorly sorted and matrix-supported sandy conglomerates with large volcaniclastic boulders floating in the matrix. Scale bar, 1 m

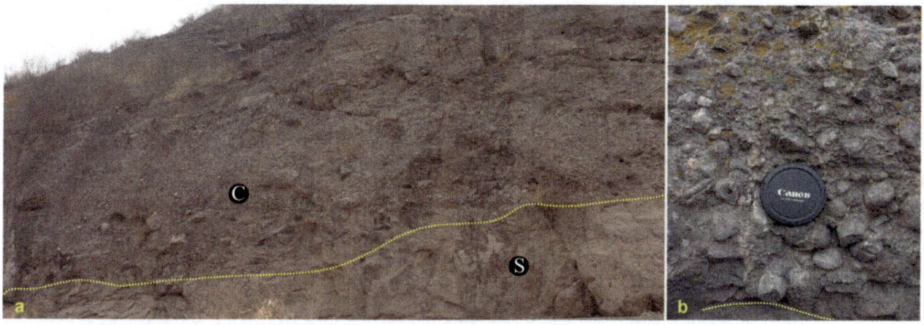

Fig. 10 Details of channel deposits in the middle alluvial fan (as in Fig. 9). **a** Coarse-grained granular sandstone (S) overlayed by sandy conglomerate (C). Note the slightly normally graded conglomerate with erosion scour (dashed line) at the base (**b**). Scale bar, 1 m. Lens cap diameter is 6 cm

3.2 Caoying Section and Stops

The Caoying-Yuyingzi Section is composed of four outcrops: Caoying, Pianling, Yuyingzi and Yinwogou (Fig. 21). It shows a series of lacustrine-basin facies successions of the Lower Cretaceous Xiguayuan Formation, including retrograding fan delta, shallow lake to semi-deep lake, fan delta front, and prograding fan delta. A symmetrical cycle consisting of one second-order sequence cycle and three third-order sequences was recognized in this section (Fig. 21). The field excursion is mainly focused on the Caoying section, which corresponds to the lower two third-order sequences of the Caoying-Yuyingzi section (Fig. 21).

The Caoying Section is composed of an overall thinning-upward and fining-upward lithogenetic unit that includes two parts: a lower retrograding fan delta succession deposited at a gentle slope in the initial rifting stage and an upper profundal lake succession formed during the peak-rifting stage.

The Caoying section is approximately 20 km to the southwest of Luanping County, along County Road No. 354 from Pingfang to Bakeshiying, 1 km long between

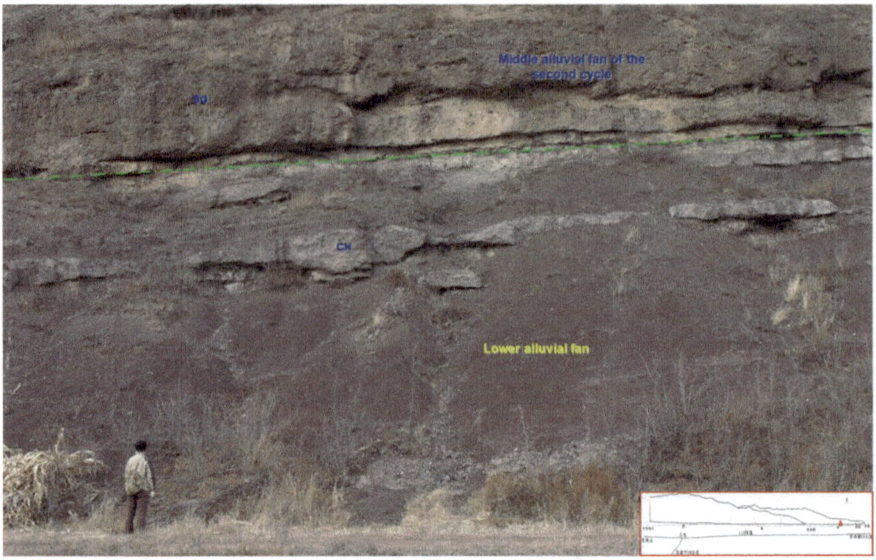

Fig. 11 Lower alluvial fan deposits composed of overbank fines element and lenticular channel element (CH). The upper part of the view is middle alluvial fan conglomerate dominated by sediment gravity flow element (SG). Person for scale

Fig. 12 Sedimentary sequence of fan delta plain to fan delta front around the Cengouqiao Bridge. Note that the lower part of the view is middle fan/fan delta plain conglomerate mainly composed of sediment gravity flow channel elements (CH). The upper part of the view is fan delta front deposits, dominated by subaqueous sediment gravity flow channel elements (SCH). The lithological column on the left shows upward retrogradation of the fan delta. Person circled for scale

Fig. 13 Partly matrix-supported debris flow conglomerates in fan-delta plain deposits of the middle alluvial fan. **a** Amalgamated gravity flow channel-lobe complexes. **b** Detail of amalgamated channels. Scale bar, 1 m. **c** Detail of partly matrix supported conglomerates. Hammer is 25 cm long

Fig. 14 Gigapan photograph of the retrograding fan delta developed around Cengouqiao. Note the deepening-upward trend in lake evolution from the middle fan/fan delta plain and the fan delta front to the pro-fan delta/shallow lake

Caoying Village and Yaozigouliang. Nine stops have been included in this guide to reflect our current state of understanding of the geologic evolution of this region (Fig. 21). Both terrigenous clastic sedimentary packages and pyroclastic packages will be observed in the field (Fig. 21).

Stop Descriptions

Stop 6: Stratal packages of fan delta to delta plain deposited in the initial rifting stage.

Fig. 15 Detail of the retrograding fan delta succession (as in Fig. 9), showing that sandy conglomerate beds gradually transition upward into grayish shallow lake shales and mudstones. Note normal faults (red solid lines) near the middle and right end of the view

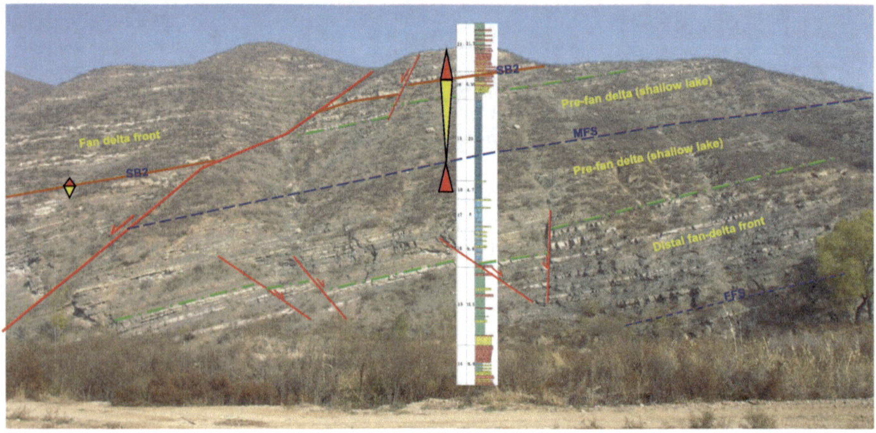

Fig. 16 Gigapan photograph shows fan delta front debris flow deposits, pro-fan delta deposits, sequence boundary (SB), the first flooding surface (FFS) and maximum flooding surfaces (MFS) in dark gray–greenish mudstones. The lithology column shows an overall fining-upward to coarsening-upward rift cycle. Note that normal faults (red solid lines) are present near the lower and upper parts of the section

Location: Lingdangmugou Valley in Luanping County.

Content: Depositional cycles of fan delta plain and unconformity surface.

Stop 7: Stratal packages of fan delta and delta front.

Location: Lingdangmugou Valley in Luanping County.

Content: Fan delta front packages, parasequence set, mouth bar, and the MFS.

Stop 8: Turbidite sequences within the prodelta succession.

Location: Lingdangmugou Valley in Luanping County.

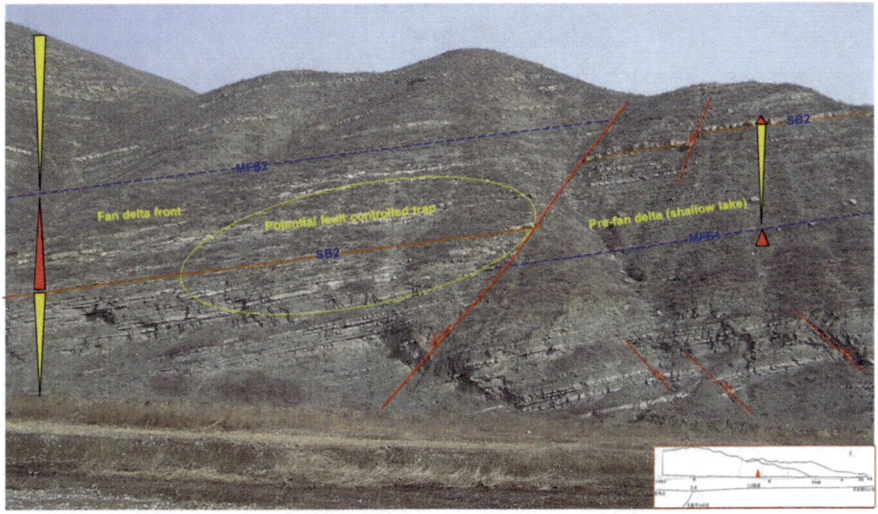

Fig. 17 A notated fan delta front succession showing third-order sequences, maximum flooding surfaces (MFS), sequence boundaries (SB) and normal faults (red solid line). Note that fault-controlled traps (circled area near middle of view) may be present with sandstone beds juxtaposed across the fault against impervious greenish shales

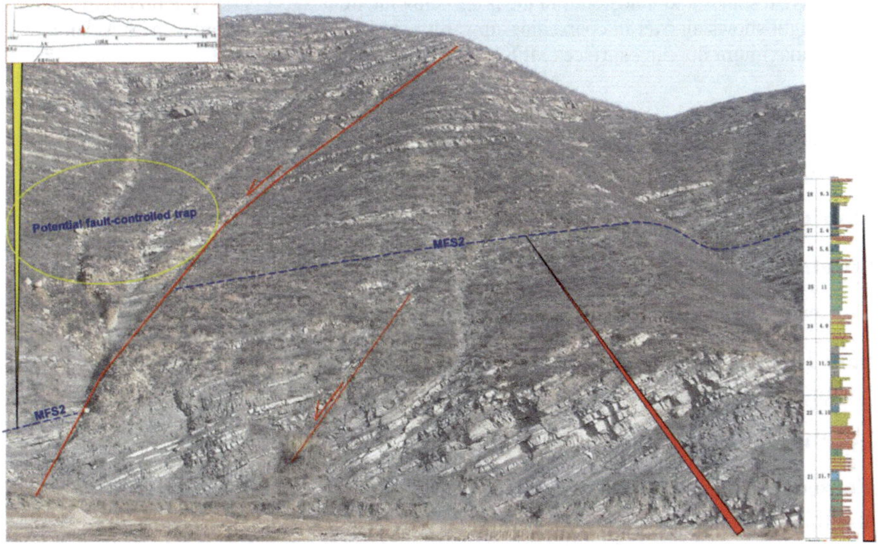

Fig. 18 Detail of potential fault-controlled trap beyond the normal fault (as in Fig. 8). The lithological column on the right shows an overall fining-upward and deepening-upward transgressive sequence of the lower part section below the second maximum flooding surface (MFS2)

Fig. 19 Distal part of the Xingzhouhe section showing sediment gravity flow sandbodies alternating with grayish shales and mudstones in the prograding fan delta succession. The lithological column on the right shows an overall coarsening-upward trend in lake evolution after the formation of the second maximum flooding surface (MFS2)

Content: Turbidite packages, slide/slump blocks, and soft-sediment deformation structures.

Stop 9: Third-order sequence boundary formed in the late initial rifting stage.

Location: Lingdangmugou Valley in Luanping County.

Content: 3rd SB and associated sedimentary packages and slump deposits.

Stop 10: Sediment gravity flow lenticular bodies and thin-bedded marlstones within profundal succession.

Location: Lingdangmugou Valley in Luanping County.

Content: Thin-bedded mudstone and marlstone, sedimentary gravity flow deposits, and the MFS.

Stop 11: Thin layered hyperpycnites in profundal lake deposits.

Location: Lingdangmugou Valley in Luanping County.

Content: Lacustrine rhythmic packages, layered and laminated shales, hyperpycnite sequences, and the MFS.

Fig. 20 Gigapan photographs of the distal Xingzhouhe Section moving from east (**a**) to the west (**b**). Note the channel and sheet sandy conglomerate elements (CH, SH) of the sub-lacustrine fan system sandwiched in profundal lake shales and mudstones, also a reverse fault (red solid line) with minor displacement through sub-lacustrine fan lobe deposits

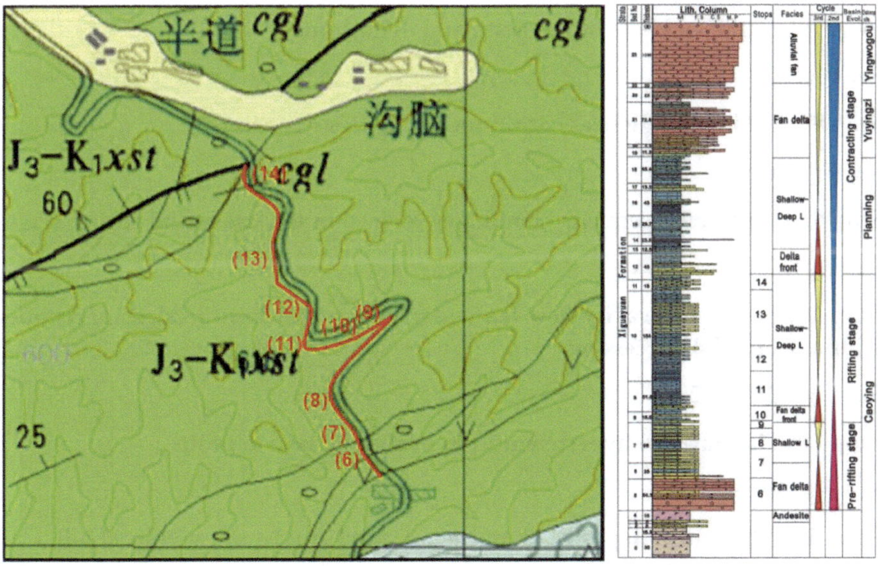

Fig. 21 Road log, lithology column and sequence structures of the Caoying Section

Stop 12: Rhythmic deposits of marlstone and mudstone within shallow to profundal lake succession.

Location: Lingdangmugou Valley in Luanping County.

Content: Rhythmic beds of marlstone, mudstone and siltstone.

Stop 13: Medium- to thin-bedded siltstone packages in shallow lake succession.

Location: Lingdangmugou Valley in Luanping County.

Content: Thin-bedded siltstone packages, climbing ripples, and laminations, including lenticular, flaser and wavy laminations.

Stop 14: Syn-depositional fault and associated soft-sediment deformation structure within shallow lake succession.

Location: Lingdangmugou Valley in Luanping County.

Content: Climbing ripples, trace fossils, syn-depositional faults and associated soft-sediment deformation structures.

Stop Figures

Stop 6: Stratal packages of fan delta to delta plain deposited in the initial rifting stage

See Figs. 22, 23 and 24.

Stop 7: Stratal packages of fan delta and delta front

See Fig. 25.

Stop 8: Turbidite sequences within the prodelta succession

See Fig. 26.

Stop 9: Third-order sequence boundary formed in the late initial rifting stage

See Figs. 27.

Stop 10: Sediment gravity flow lenticular bodies and thin-bedded marlstones within profundal succession

See Figs. 28 and 29.

Stop 11: Thin layered hyperpycnites in profundal lake deposits

See Fig. 30.

Stop 12: Rhythmic deposits of marlstone and mudstone within shallow to profundal lake succession

See Fig. 31 and 32.

Fig. 22 Gigapan photograph of the retrograding fan delta deposits developed on a gentle slope (**a**). Note that the sequence boundary (SB1) outlines the base of the fan delta succession. Lateral stacked channel element (CH) in the middle fan/fan delta plain succession (**b**). The lithology column on the right shows an overall fining-upward and deepening-upward transgressive sequences. Stratigraphic interval (**b**) corresponds roughly to the lower part (red rectangle) of the lithology column on the right

Fig. 23 Detail of the sequence boundary (SB1) that developed on extrusive volcanic rocks/ pyroclastic deposits (as in Fig. 22), outlining the base of the Lower Cretaceous strata. Width of view ~4 m

Fig. 24 Detail of amalgamated channel deposits (as in Fig. 22). Note the common occurrence of solid bitumen (red arrows) along the boundaries between conglomerates and siltstones. Hammer 25 cm

Fig. 25 **a** Retrograding fan delta succession developed on a gentle slope with the first maximum flooding surface (FFS1) outlining the boundary between proximal fan delta front and distal fan delta front deposits. **b** Parasequence set produced by rapid progradation of fan deltas, with local mouth bars. Stratigraphic interval (**a**) corresponds roughly to the middle part (red rectangle) of the lithology column on the right. Person for scale

Fig. 26 Gigapan photograph of prodelta succession with the maximum flooding surface (MFS1) marked (**a**). **b**, **c** Slide/slump blocks and associated soft-sediment deformation structures in deep-water sediments. Stratigraphic interval (**a**) corresponds roughly to the upper part (red rectangle) of the lithology column is on the right. Scale bar, 1 m

Fig. 27 Gigapan photograph of the profundal lake deposits. Note the sequence boundary (SB2) presented at the base of the rift climax stage succession. Scale bar, 1 m

Stop 13: Medium- to thin-bedded siltstone packages in shallow lake succession

See Fig. 33.

Stop 14: Syn-depositional fault and associated soft-sediment deformation structure within shallow lake succession

See Fig. 34.

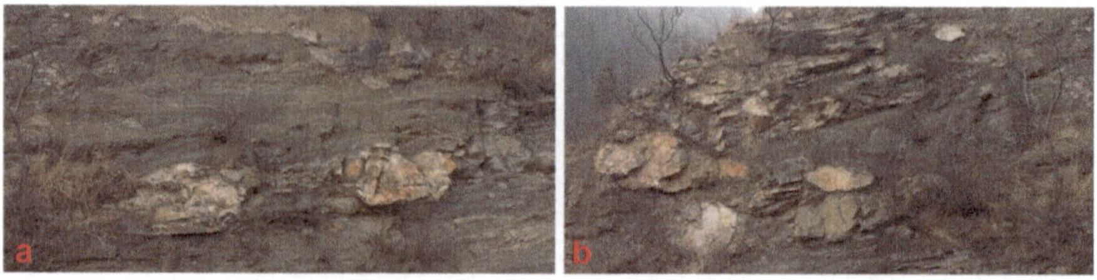

Fig. 28 Detail of lens-like slump blocks of deep-water channels produced by sediment gravity flows (as in Fig. 27)

Fig. 29 Examples of lenticular channel bodies in deep water sediment produced by sediment gravity flow. Hammer 25 cm; coin diameter is 2.5 cm

Fig. 30 Profundal lake succession showing alternation of siltstone, mudstone and marlstone intervals. The width of view is approximately 8 m

3.3 Sangyuan Section and Stops

The Sangyuan Section is 22 km to the west of Luanping County, on the east entrance of Sangyuan village (Fig. 35). The section is mainly composed of an overall fining-upward and deepening-upward transgressive sequence that grades upward into a coarsening-upward retrogressive sequence. This fining-upward to coarsening-upward rift cycle corresponds to the late rifting phase (equivalent to the contracting stage in Fig. 5) of the third rifting episode during the Middle Jurassic to Early

Fig. 31 Thin-layered hyperpycnites/turbidites in the profundal lake succession (**a**) deposited by long-lived river-flood-generated hyperpycnal flows. Note that the maximum flooding surface (MFS) developed during the rift climax stage (**b**). Person for scale

Fig. 32 Marlstone and shale deposited in sublittoral (**a**) to profundal (**b**) lakes during the rift climax stage. Note that marlstone beds are characterized by thin bedding and are laterally extensive across the outcrop. Person for scale

Fig. 33 Laminated, thin- to medium-bedded siltstone within the deltaic-coastal succession (**a**). **b** Small-scale (ripple) cross-laminated siltstone with lenticular and flaser laminations (FL) in parts. Coin diameter is 2.5 cm. **c** The siltstone bed shows climbing-ripple cross-laminae (CR) overlying planar parallel laminae (PL). Lens cap diameter is 6 cm

Cretaceous. The Sangyuan Section showing an overall fining-upward and deepening-upward transgressive sequence consists of two parts: a lower aggradational sequence set of fan delta deposits on a steep slope and an upper profundal lake succession characterized by sub-lacustrine channel-lobe packages (Fig. 36). The lower part of the section locally shows progradational fan delta sequences, which may be related to lateral deltaic lobe shifting.

The trip is from Luozigoumen to Sangyuan Village, covering a distance of 1.1 km, with five stops included in this guide to reflect our current state of understanding of the geologic evolution of this region (Fig. 35). The sandstones and conglomerates on these stops are rather thick, and the lateral trends and geometry of channel bodies in the delta front are obvious.

Stop Descriptions

Stop 15: Depositional cycle of the fan delta front influenced by sediment influx from the lower alluvial fan to the littoral lake.

Location: The east entrance of Sangyuan Village in Luanping County.

Content: Characteristics of channel deposits and the first flooding surface.

Fig. 34 **a** Laminated, thin- to medium-bedded siltstone succession with syn-depositional faults and associated soft-sediment deformation structure (upper left). Note that the upper bed is chaotically deformed, but the underlying beds show no internal deformation, with thin- to medium-bedded sheet-like sand bodies (SH) extending across the outcrop. **b** Intense deformation of strata showing slump folds and deformed subaqueous channel bodies (SCH). **c** Detail of laminated, thin- to medium-bedded siltstone. **d** Small-scale (ripple) cross-laminated siltstone with climbing ripples in parts. Scale bar, 1 m

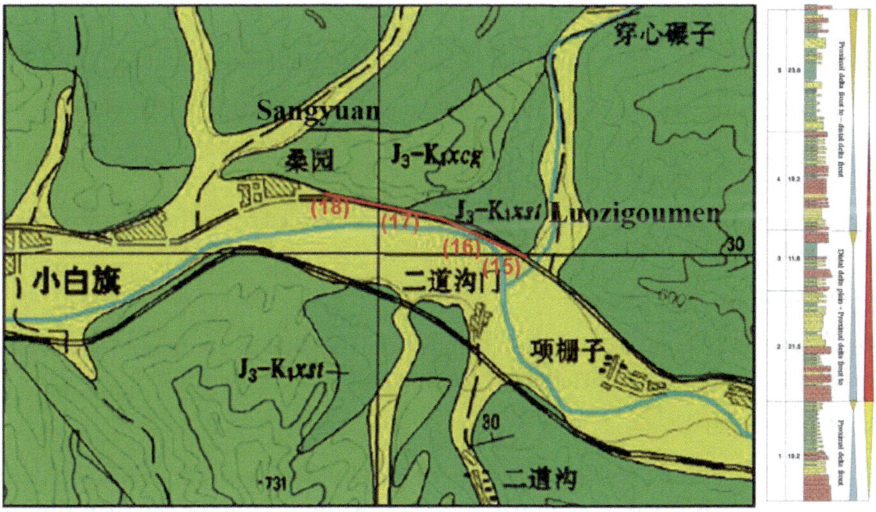

Fig. 35 Road log, lithologic column and sequence structure of the Sangyuan Section

Fig. 36 Panoramic view of the Sangyuan Section, annotated with stops, the first flooding surface (FFS) and maximum flooding surface (MFS)

Stop 16: Depositional cycle of the fan delta front deposited by sediment influx from the lower alluvial fan to the sublittoral-profundal lake.

Location: The east entrance of Sangyuan Village in Luanping County.

Content: Shales below and above the maximum flooding surface, principal features of subaqueous distributary channel sandstone bodies.

Stop 17: Depositional cycle and sediment gravity flow deposits within the fan delta front sequence formed by sediment influx from the middle alluvial fan to the profundal lake.

Location: The east entrance of Sangyuan Village in Luanping County.

Content: Characteristics of deep-water channel-lobe complexes, feeder channels of sub-lacustrine lobes, and dark-gray shale within sublittoral to profundal lake facies.

Stop 18: Depositional cycle and sediment gravity flow deposits of fan delta front formed by the influx of upper alluvial fan to profundal lake.

Location: The east entrance of Sangyuan Village in Luanping County.

Content: Characteristics of debris flow deposits and channel-lobe complexes.

Stop Figures

Stop 15: Depositional cycle of fan delta front influenced by sediment influx from lower alluvial fan to littoral lake

See Figs. 37 and 38.

Stop 16: Depositional cycle of fan delta front deposited by sediment influx from lower alluvial fan to sublittoral-profundal lake

See Figs. 39 and 40.

Stop 17: Depositional cycle and sediment gravity flow deposits within fan delta front sequence formed by sediment influx from middle alluvial fan to profundal lake

See Fig. 41.

Fig. 37 Retrogradation of the fan delta recorded in the Sangyuan Section with the first flooding surface (FFS3) marked. Note the alternation of deltaic deposits (e.g., channel sandstone) and coastal deposits (e.g., sheet-like sandstone) within the delta front succession. Car for scale

Fig. 38 Detail of distributary channels present within the delta front succession with the first flooding surface (FFS3) marked (as in Fig. 37). Note the lenticular channel bodies in (**a**) and (**b**), (**c**) channel, channel margin and interdistributary deposits within delta front sequences; channel axis deposit is characterized by massive, sandy conglomerate. Person for scale

Stop 18: Depositional cycle and sediment gravity flow deposits of fan delta front formed by the influx of upper alluvial fan to profundal lake

See Fig. 42.

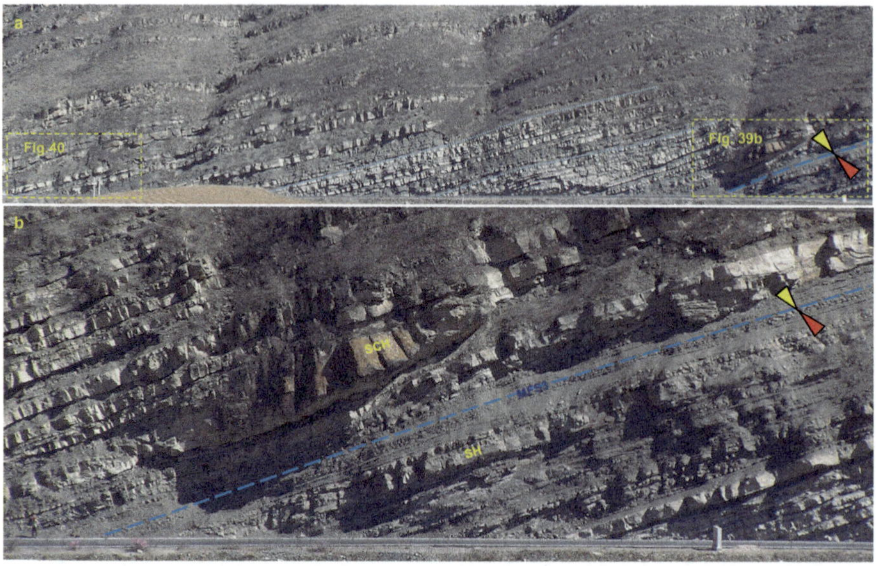

Fig. 39 Panoramic view of the distal fan-delta front sequence with the maximum flooding surface (MFS3) marked (**a**). **b** Detail of fan-delta front deposits below and above the maximum flooding surface (MFS3). Note the lenticular sand body (center of view) formed by a subaqueous channel cutting into a thin-bedded turbidite succession

Fig. 40 Detail of subaqueous channel-lobe complexes sandwiched in distal fan-delta front succession (as in Fig. 39). Note thick bedded sandstone turbidites (T, center of view) with several thin sandstone beds grading laterally into a single thick, amalgamated bed (A). Person for scale

Fig. 41 Deep-water channel-lobe complexes within the distal fan-delta front shale succession. Note feeder channels (CH) of sub-lacustrine lobes (LO) filled with massive sandy conglomerates. Person for scale

Fig. 42 Panoramic view of the sub-lacustrine channel-lobe complexes within profundal lake succession. Note sub-lacustrine lobes are characterized by convex-upward shapes and are mainly composed of thick to very thick sandy conglomerate beds that commonly connect to incised feeder channels and grade laterally into flat lying, unchannelized profundal deposits. Circled part is a car for scale

Fig. 43 Dinosaur footprints (arrows) preserved on the surface of muddy siltstone in a sub-littoral lake sequence (**b**, **c**, **d**). Note that the tracks lining the bed surface indicate a relatively shallow water depth. Persons for scales

3.4 Dinosaur Footprints in Pingfang Village

On the way back to the Luanping Grand Hotel from the Sangyuan Section, we can also find dinosaur footprints in Pingfang, Luanping, near the Caoying section (Fig. 43).

3.5 Wangjiagou Section and Stops

The Wangjiagou section, where mass transport deposits (MTD or slide and slump) crop out perfectly, is located 3500 m to the east of the Luanping Grand Hotel (Fig. 44). The Wangjiagou section is mainly composed of profundal facies associations, which consist dominantly of laminated shales interbedded with turbidite conglomerates.

The MTD bearing section is exposed approximately 2000 m in width and 8–12 m in thickness. Hydroclastic deposit-like breccias can be observed at the west end of the Wangjiagou Section. Three stops have been included in this guide to reflect our current state of understanding of the geologic evolution of this region (Fig. 44).

Stop Descriptions

Stop 19: Intensely deformed beds within the profundal lake shale/mudstone succession.

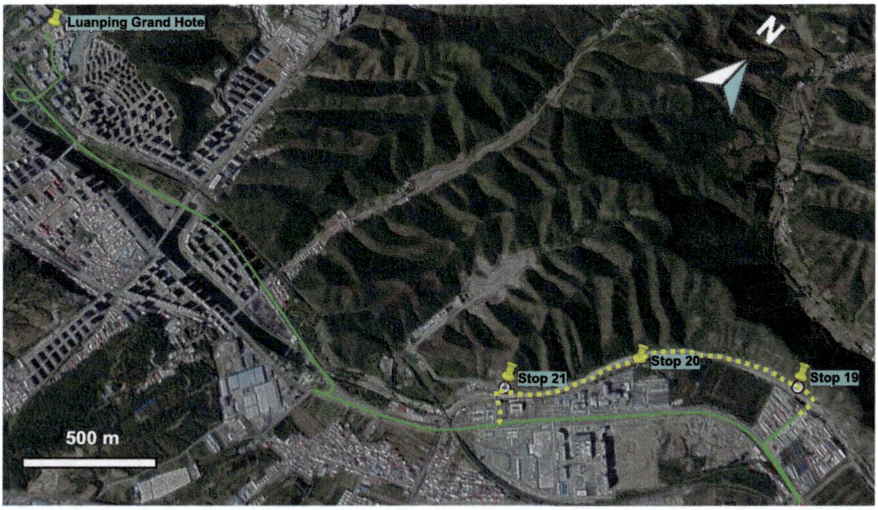

Fig. 44 Road log of the Wangjiagou Section, showing the locations of stops and the Luanping Grand Hotel

Location: North bank of the Mangniuhe River.

Content: Characteristics of mass transport deposits, slides, slumps and deformed beds.

Stop 20: Debris flow deposits of slump block.

Location: North bank of the Mangniuhe River, ~500 m to the west of Stop 19.

Content: Principal features of the deformed bed and debrites.

Stop 21: Pyroclastic/Hydroclastic deposits within profundal lake strata.

Location: On a hill slope approximately 50 m north of the Government Affairs Services Center of Luanping County.

Content: Angular boulder of andesite/basalt, massive conglomerates and breccias of hydroclastic deposits.

Stop Figures

Stop 19: Intensely deformed beds within profundal lake shales/mudstones succession

See Figs. 45 and 46.

Stop 20: Debris flow deposits of slump block

See Figs. 47, 48 and 49.

Fig. 45 The slump block sandwiched between profundal lake deposits above and under dark gray and black shale and siltstones. Its thickness is approximately 8 to 12 m, and it extends approximately 2 km in the east–west direction. The slump is characterized by a chaotic mix lithology, composed of partly folded beds of gravelly sandstone and sandy conglomerate and partly debris breccia with matrix-supported texture. The lithic composition is mostly crystalline, such as granite, diorite and/or gneiss. Person for scale

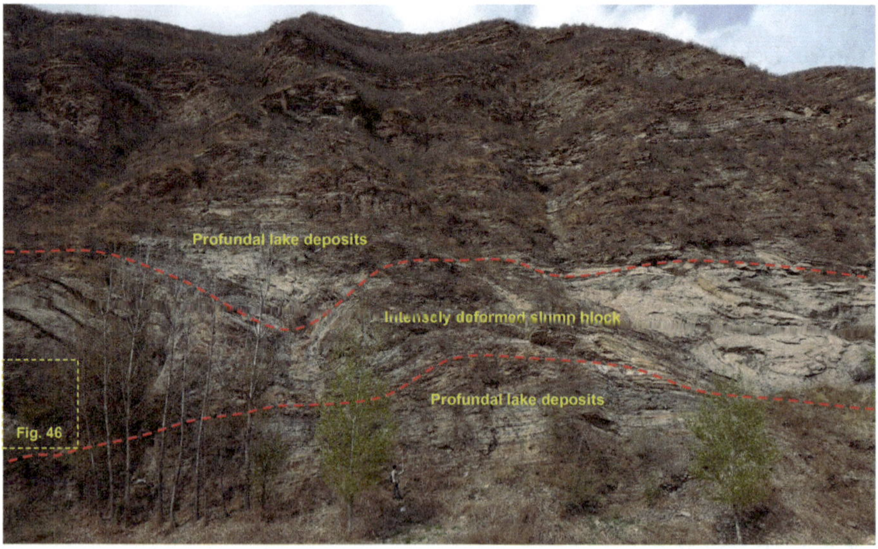

Fig. 46 Close view of intensely deformed beds. It shows overturned folds but still exhibits a sequence of shale, siltstone, sandstone and conglomerate (coarsening-upward). This implies that the sequence was possibly deposited in a delta front environment. Scale bar, 1 m

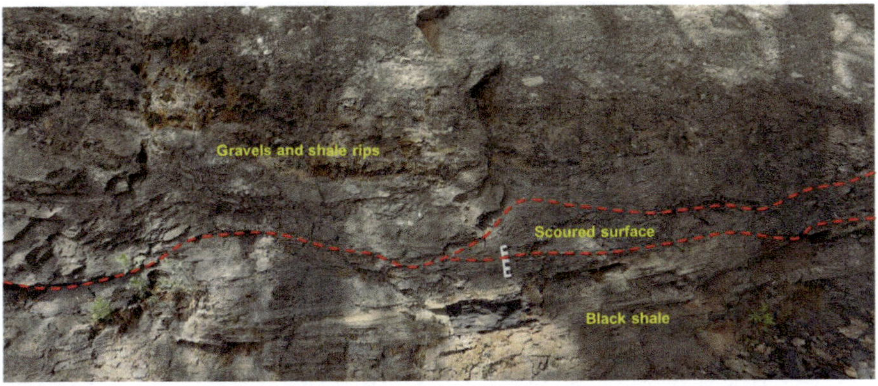

Fig. 47 Thick bed slump block with broad erosional scours at the base. Note that the upper bed is chaotically deformed, but the underlying beds show slight internal deformation

Fig. 48 Deformed slump block and associated debris flow conglomerates sandwiched between profundal lake shales/mudstones. The lower left slump is pinched out and composed of several small slumps. Sharp margins of debrite indicate cutting out of underlying thin-bedded turbidites and laminated shales by slides

Stop 21: Pyroclastic/hydroclastic deposits in profundal lake strata

See Figs. 50, 51 and 52.

Summary of the field trip

The itinerary concerns a Cretaceous sedimentary succession of a small intracontinental rift basin in Luanping County, Hebei Province. The lacustrine successions show a complete rift cycle (equivalent to the third rifting episode during the Middle Jurassic to Early Cretaceous) that includes (1) alluvial fan succession deposited in the initial rifting stage, (2) retrogradational delta succession deposited in the waxing rifting stage, (3) lacustrine sediment gravity flow deposits accumulated in the climax rifting stage, and (4) progradational delta succession, which was not included in this guide and was deposited in the waning rifting stage. Such a complete stratigraphic

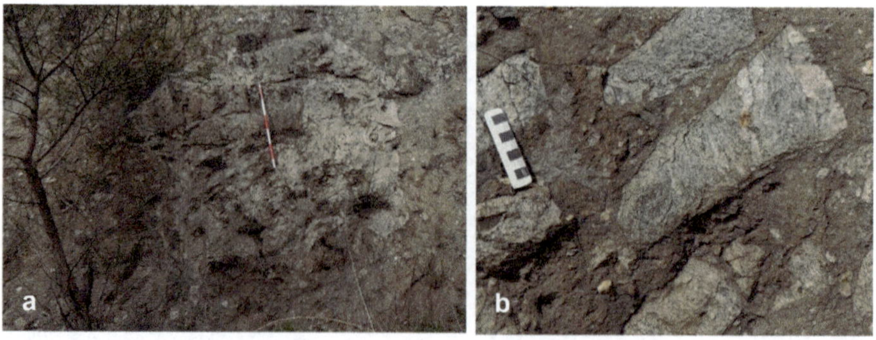

Fig. 49 Photo **a** shows a giant boulder with a diameter greater than 2 m, and photo **b** shows a matrix-supported breccia with poor sorting. The gravels are gneiss and granite or plagiogranite, implying that the sediment source of the Proterozoic comes from fault blocks located at the northwestward margin of the basin. Scale bar, 1 m

Fig. 50 An annotated drone photograph illustrating a debrite-hydroclastic complex sandwiched in profundal lake deposits. Note the color change between MTD and the debrite-hydroclastic complex

record of a small nonmarine rift basin can be used as an analog for larger petroliferous rift basins to predict sedimentary facies in underground.

4 Proposed Investigation Schedule

The 21st International Sedimentological Congress will be held in Beijing, China, from August 22 to August 26, 2022. The conference includes oral presentations, poster displays, field investigation and short-term courses. The "Lower Cretaceous

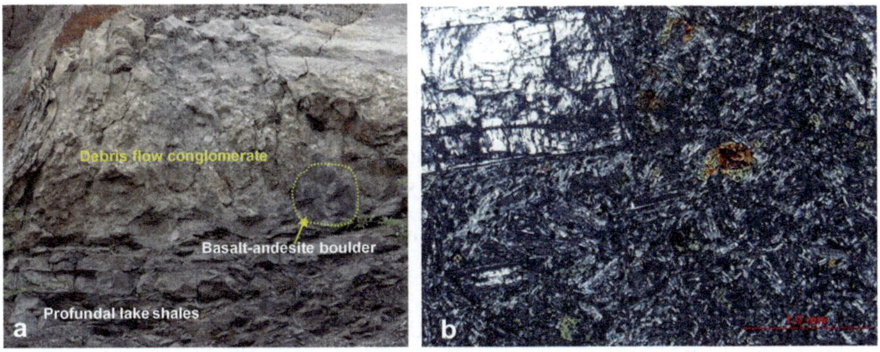

Fig. 51 Debris flow conglomerate sandwiched in profundal lake shale. **a** Normal graded debris flow conglomerate with broad erosional scour at the base; ~ 2 m of view. **b** Thin section photomicrograph of the basalt boulder (**a**). Note a feldspar phenocryst in a finely crystalline groundmass

Fig. 52 Outcrop photos of hydroclastic breccia-like deposits. **a** Massive, matrix-supported breccia with angular tuffaceous pebbles. Lens cap diameter is 6 cm. **b** Massive, clast-supported breccia with very angular tuffaceous pebbles. The very angular pebbles may be explosion breccia or pyroclastic breccia deposited during hydroclastic eruptions. Hammer is 25 cm long

Lacustrine Sedimentary Sequences of a Small Intracontinental Rift Basin in Luanping, Hebei" investigation route, as one of the preconference investigation routes, is scheduled to begin on the morning of August 19, 2022, Beijing time. Please set your clock to Beijing time in advance (UTC/GMT + 08:00).

4.1 Dates and Locations

Investigation dates: August 19–21, 2022.

Departure place: From conference venue in Beijing to Luanping County, Hebei Province.

Ending place: Conference venue in Beijing.

4.2 Travel Arrangements

The participants are requested to arrive at the gathering place near the conference venue in Beijing before 8:30 AM on August 19, 2022, take the TOYOTA COASTER minibus to Luanping County, and check in to the Luanping Grand Hotel at approximately 12:00 AM. The following schedule is:

(1) August 19: (Afternoon) Take bus to Xingzhouhe Section, and investigate sedimentary successions from alluvial fan, fan delta to turbidites intercalated with profundal lake sediments.
(2) August 20: (Morning) Field investigation in Caoying Section of fan delta to lacustrine sedimentary successions; (Afternoon) Field examination in Sangyuan Section of fan delta to gravity-flow deposits sedimentary successions.
(3) August 21: (Morning) Visit Wangjiagou Section of profundal sedimentary successions, where mass-transported deposits (i.e., slide, slump and debris flow deposits) were widely exposed in profundal lake shales; (Afternoon) gather at the parking lot in front of Luanping Grand Hotel on 13:30 PM, then go back to Beijing by bus.

Table 1 shows the specific schedule of "Lower Cretaceous Lacustrine Sedimentary Sequences of a Small Intracontinental Rift Basin in Luanping, Hebei" investigation route.

Table 1 Schedule of the investigation route in the Luanping Basin

Date	Time	Trip	Transportation mode	Contents	Note
August 19 (day 1)	08:30–11:30	Beijing-Luanping Grand Hotel	Bus	Journey	Lunch: Luanping Grand Hotel
	13:30–17:30	Hotel-Xingzhouhe Section	Bus	Field investigation	Accommodation: stay at the Luanping Grand Hotel
August 20 (day 2)	08:30–11:30	Hotel-Caoying Section	Bus	Field investigation	Lunch: Luanping Grand hotel
	13:30–17:30	Hotel-Sangyuan Section	Bus	Field investigation	Accommodation: stay at the Luanping Grand Hotel
August 21 (day 3)	08:30–11:30	Hotel-Wangjiagou Section	Bus	Field investigation	Lunch: Luanping Grand hotel
	13:30–17:30	Hotel-Beijing	Bus	Return journey	Accommodation: stay at a hotel near conference venue

5 Logistics

The following arrangements for this field trip are provided for your reference.

5.1 Accommodation and Meals

Participants in this investigation will be arranged with accommodation from the night of August 19 to the night of August 20 (staying at the Luanping Grand Hotel, Luanping County, Hebei Province). Staying a night in a Beijing hotel near a conference venue on August 18 is recommended. All the reserved hotels are safe, clean and comfortable. You will have breakfasts and dinners at the hotels where you stay; lunches may be available at restaurants along the way or packed prior and brought with you. Necessary meals/drinks will be arranged during the field investigation.

5.2 Transport

The investigation stops are all located within 20 km of your hotels, and the TOYOTA COASTER minibus will be used for transportation.

5.3 Weather

Luanping County of Hebei Province is located in northern China, which belongs to the warm temperate zone and has a half moist continental monsoon climate. During the investigation period, mid-August, the climate was usually mild and pleasant, with an average temperature of ~25 °C during the day and ~15 °C at night.

Moreover, the participants are recommended to prepare caps for shade, sunscreen, antiskid outdoor shoes, and a thin to prevent any possible sun exposure and temperature drops at night.

5.4 Safety

(1) Be careful to avoid vehicles and bad road conditions, especially when walking along road cuts.
(2) Be careful during investigation at high and steep locations to avoid any possible injury by falling rocks. Observe the signals of geological disasters, such as abnormal noise, loosening and collapse. Wear the helmet at or near the crags.

(3) The investigation arrangement will be delayed or cancelled in the case of rainstorms, gales, thunderstorms, hail and other adverse weather conditions or the possibility of natural disasters.
(4) Follow the directions of the organizer during the investigation. DO NOT leave the team and act alone.
(5) Inform the organizer before the trip if you have any health problems (e.g., hypertension, heart disease, asthma, diabetes, etc.). Common medicines will be available at the vehicles.
(6) When necessary, call 110 for police support, 120 for emergency medical assistance, or 112 where there is no mobile phone signal for field emergency help.

Acknowledgements This work is funded by the China National Science and Technology Major Project (2017ZX05001) and the PetroChina Science and Technology Major Project (2021DJ0401). Most of the materials for this guide are from the geology field works of RIPED and the accumulation of the years' graduate practice in this teaching base of field geology. Particular due to the reviewer, Prof. Xiumian Hu, for his helpful comments. Special thanks are also due to Prof. Rukai Zhu and Prof. Lianhua Hou for reading and commenting on the proof stage.

References

Cope T (2003) Sedimentary evolution of the Yanshan fold-thrust belt, Northeast China. Stanford University, pp 93–163

Davis GA, Wang C, Zheng Y, Zhang J, Zhang C, Gehrels GE (1998) The enigmatic Yinshan fold-and-thrust belt of northern China: new views on its intraplate contractional styles. Geology 26(1): 43–46

Davis GA, Zheng Y, Wang C, Darby BJ, Zhang C, Gehrels G (2001) Mesozoic tectonic evolution of the Yanshan fold and thrust belt, with emphasis on Heibei and Liaoning provinces, northern China. Geol Soc Am Memoir 194:171–197

Dou L, Hou J, Song S, Zhang L, Liu Y, Sun S, Li Y, Wang X, Ren X, Tang Y, Tian H, Yang Y (2020) Sedimentary characteristics of hyperpycnites in a shallow lacustrine environment: a case study from the Lower Cretaceous Xiguayuan Formation, Luanping Basin, Northeast China. Geol J 55(5):3344–3360

Li W, Li X (2003) Petrogenesis and its implication for the middle-late Jurassic intermediate-acidic volcanic rocks in the middle section of Yanshan orogeny. Acta Petrologica Sinica 20(3):501–510

Li W, Zhao Y, Li X, Lu F, Liang X, Tu X (2007) Genesis of the middle-later Jurassic volcanic rocks of Yiaojishan (Lanqi) period and its geodynamics implication, Yanshan Orogen, east China. Acta Petrologica Sinica 23(3):557–564

Songliao Basin to Changbaishan Mountain Volcanic and Sedimentary Successions and Oil Shale Sedimentary Sequence of the Mesozoic and Cenozoic

Youfeng Gao, Pujun Wang, Frank Mattern, Jian Yi, Jianguang Zhang, Xuejiao Qu, and Weidong Xu

Itinerary and Schedule

See Fig. 1 and Table 1.

1 Oil Shale Sedimentary Sequence of Nenjiang Formation in the Songliao Basin

1.1 Geological Setting

The Songliao Basin (SB), situated on the Mongol-North China Plate, contain China's largest oil field, the Daqing Field, which is one of the very few supergiant oil fields in the world that produces lacustrine oil (e.g., Yang 1985; Li et al. 1995). During the Late Jurassic and Early Cretaceous, hundreds of rift basins developed in Northeast Asia (Ren et al. 2002), covering an area of more than 4 million km^2 from Lake Baikal to the Sikhote-Alin Region in E-W direction and from the Mongol-Okhotsk Fold Belt to the Xilamulun River-Yanji Suture Zone in N-S direction, as shown in Fig. 2. The

Y. Gao (✉) · P. Wang · J. Yi · W. Xu
College of Earth Sciences, Jilin University, Changchun 130061, China
e-mail: huxm@nju.edu.cn

F. Mattern
Department of Earth Sciences, Sultan Qaboos University, Muscat, Oman

J. Zhang
Technische Universität Darmstadt, Schnittspahnstraße 9, 64287 Darmstadt, Germany

X. Qu
Chongqing University of Science and Technology, Chongqing 401331, China

X. Hu (ed.), *Field Trip Guidebook on Chinese Sedimentary Geology*,
https://doi.org/10.1007/978-981-99-6936-4_5

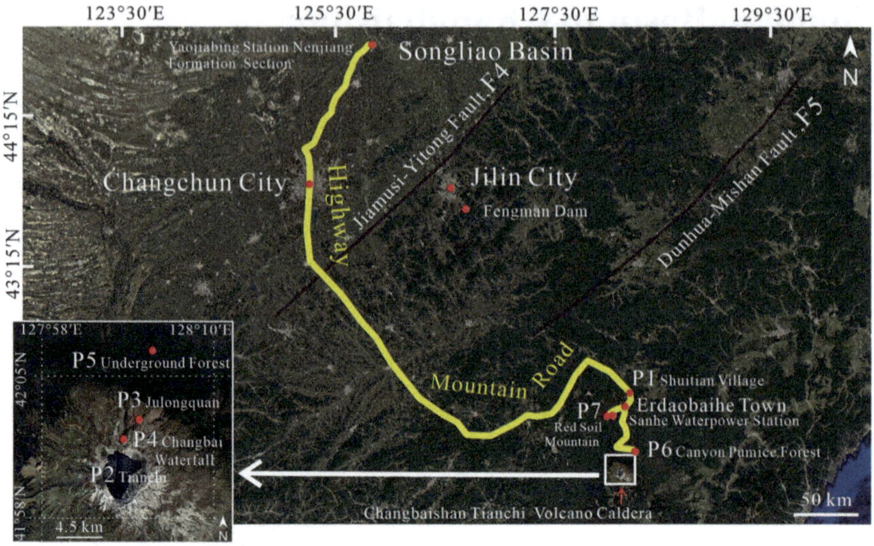

Fig. 1 Satellite images showing route and sites of the field excursion from the Songliao Basin to the Changbaishan Tianchi Volcano (Google Earth 2021)

Table 1 Itinerary schedule with topics of the Field Trip

Day	Time	Topics	Destination
1	08:00–17:00		Changchun
2	08:00–10:30		Yaojia Station
	10:30–14:30	The Nenjiang Formation section	
	14:30–17:00		Changchun
3	8:00–12:30		Shuitian Village
	12:30–15:00	Holocene Lahar deposits	
	15:00–15:30		Erdaobaihe Town
4	8:00–8:40		The North Scenic Spot of the Changbaishan Mountain
	8:40–15:40	The modern volcano and sedimentary	
	15:40–16:00		The Canyon Pumice Forest
	16:00–17:30	Pyroclastic flow deposits	
	17:30–18:30		Town of Erdaobaihe
5	8:00–8:40		Red Soil Mountain
	8:40–11:40	Pyroclastic flow deposits	
	11:40–16:00		Changchun

SB is a unique case among them as it evolved into the largest rift basin and because it is the only one with a sizable Upper Cretaceous basin fill. These Upper Cretaceous deposits may reach a thickness of up to 6000 m and represent the most important oil and gas generating sequence in China (Feng et al. 2010).

1.2 Introduction to the Section of the Upper Cretaceous Nenjiang Formation at the Yaojia Military Station

The section of the Upper Cretaceous Nenjiang Formation at the Yaojia Military Station is located at the south bank of Songhua River Bridge of Yaojia Military Station, Caiyuanzi Town, Dehui City, Jilin Province, where the strata of the first member of Nenjiang Formation are exposed (Fig. 3). The section is in good condition with a length of 320 m. The bedding attitude is 120°/4° and the total thickness measures 42 m. The rocks mainly consist of dark gray, grayish-green shale, grayish-white silty sandstone, gray ostracod limestone, and dark gray oil shale, representing interbedded shallow lacustrine facies, moderately deep, and deep lacustrine facies. This section can be correlated with the first member of the Nenjiang Formation revealed by the south borehole of SK1 in the Songliao Basin in terms of lithology, sedimentary facies, and cyclicity. The south borehole of SK1 is an ICDP scientific exploration borehole with full coring from the top of the third member of the Quantou Formation to the bottom of the second member of the Nenjiang Formation, located in the south of Qijiagulong Sag, in the central depression area of the Songliao Basin. The well has completely revealed the formation of the first member of Nenjiang Formation with a total coring length of 103 m and a coring recovery of 100% (Gao et al. 2008). Previous studies on the chronostratigraphy, magnetic stratigraphy, and biostratigraphy of the first member of the Nenjiang Formation of the south borehole of SK1 determined that the age of the first member of the Nenjiang Formation is 83.7–84.5 Ma (Late Santonian) (Wan et al. 2013), while that of the south borehole of SK1, in the Nenjiang Formation section of at the Yaojia Military Station is 84.0–84.5 Ma.

Liu and Wang (1997) proved that the isotopic composition of the sedimentary basin in the Nenjiang Formation during formation of dolomite concretions was derived from seawater through the identification of the origin of carbon and oxygen isotope in carbonate rocks and their initial $^{87}Sr/^{86}Sr$ ratio. Several dolomite layers were also found in the cores of members 1 and 2 of the Nenjiang Formation taken from the south borehole of SK1. Wang et al. (2008) suggested that formation of the dolomite concretions was facilitated (1) by marl brought in by suspension and (2) by Late Cretaceous regional metasomatism of Mg^{2+} which also caused ostracod extermination during the Late Cretaceous transgression in the Songliao Basin. Xi et al. (2011) found a large number of foraminifers in the first member of the Nenjiang Formation of well SK1, which accurately indicated the existence of lake-sea connection events during deposition of this formation.

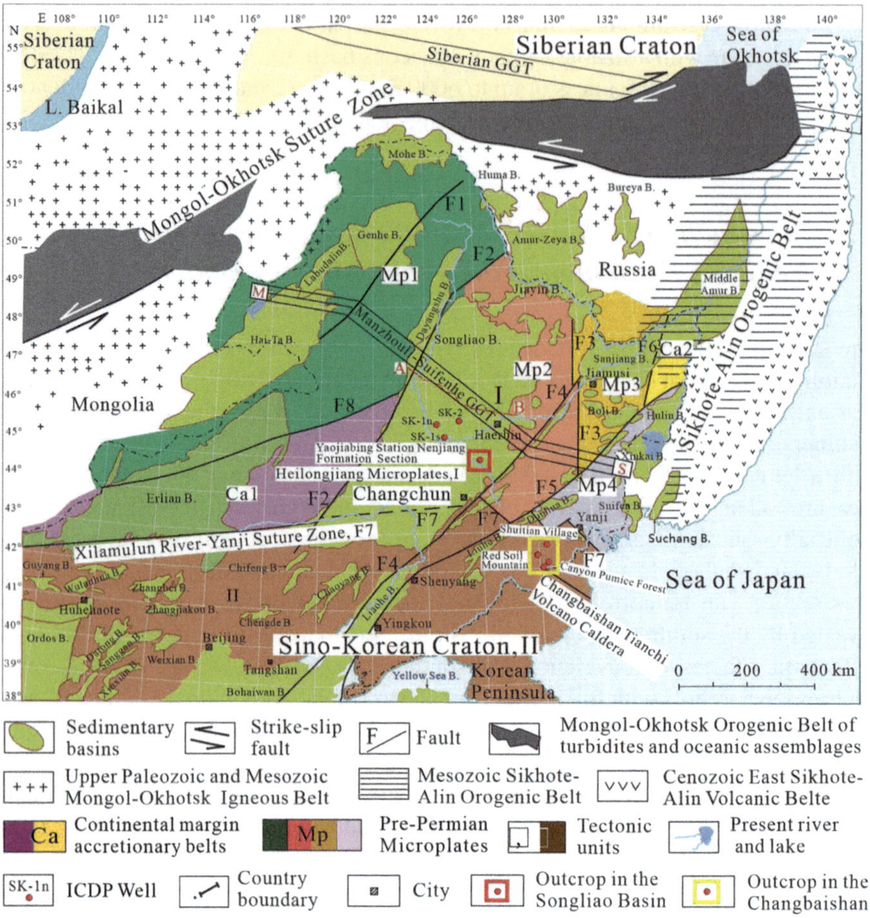

Fig. 2 Tectonic collage from the Songliao Basin to the Changbaishan Volcanoes and their surroundings (adapted from Wang et al. (2016), Zhang et al. (1998), Karsakov and Zhao (2001), Li et al. (2012), Xu et al. (2014), Khanchuk et al. (2015); Wang et al. (2015)). *Note* The faults are the Tayuan-Xijiatu Suture Zone (F1); Nenjiang-Balihan Fault (western boundary fault of the SB) (F2); Mudanjiang Suture Zone (F3); Jiamusi-Yitong Fault (eastern boundary fault of the SB) (F4); Dunhua-Mishan Fault (F5); Lower Heilongjiang or Dahezhen Fault (F6); Xilamulun River-Changchun-Yanji Suture Zone (F7); and Heihe-Hegenshan Suture Zone (F8). The tectonic units are the Mongol-Great Xing'an Microplate (Mp1); Songliao-Zhangguangcailing Microplate (Mp2); Jiamusi Massif of Archean age (Mp3); Khanka Massif of Early Palaeozoic (Mp4); Wenduermiao-Hegenshan Continental Margin Accretion Belt of Paleozoic age (Ca1); and Nadanhada Continental Margin Accretion Wedge of Jurassic age (Ca2). The assemblage of Mp2, Mp3, Mp4, and Ca1 are collectively termed "Heilongjiang Microplates" (I). Block-II indicates the Sino-Korean Craton. The so-called "Mongol-Jiamusi Block" (MJB) was constructed by the assembly of Block-I and Mp1 before the Devonian (Wang et al. 2009a, b). The Mongol-North China Plate was constructed when the Sino-Korean Craton and the MJB were welded together along the Xilamulun River-Changchun-Yanji Suture Zone at the end of the Late Permian (Wang and Fan 1997). By then, the pre-Triassic basement of the SB had been built up. GGT indicates Global Geosciences Transect

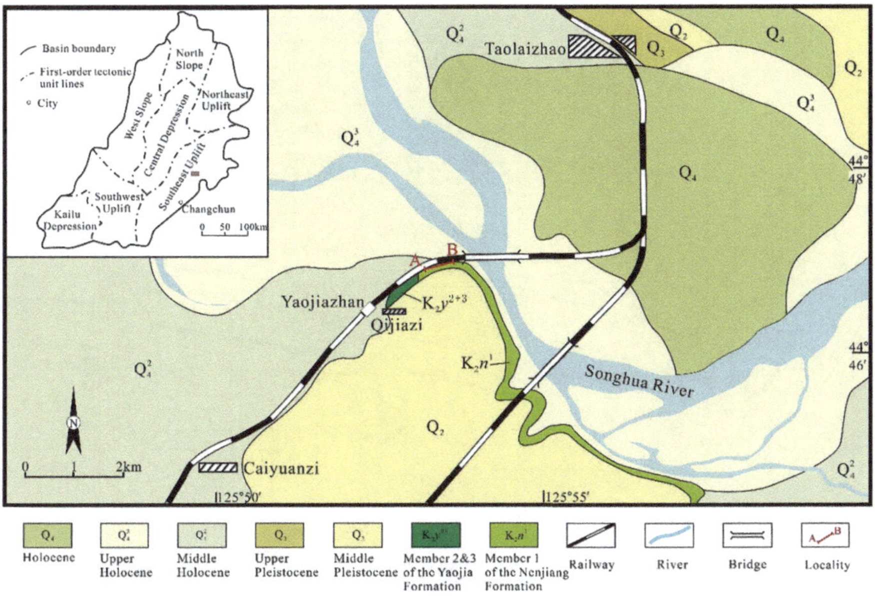

Fig. 3 Geological map of Yaojiazhan area showing outcrop locations

As the dip angle of the strata measures only 4°, they are oriented nearly horizontally, and the section is incompletely exposed at different locations. Four observation points are designated for their presentation. See Fig. 3 for the specific observation positions.

1.3 The Oil Shale Sedimentary Sequence Outcrops Description of the Nenjiang Formation

Stop 1 Overview site at the boundary between Yaojia Formation and Nenjiang Formation (44° 47′ 11.50″ N, 125° 53′ 39.10″ E)

Stop 1 is located at the western end of the section, showing the top of the Yaojia Formation and the overlying basal strata of the first member of the Nenjiang Formation and (Figs. 4, 5 and 6). The boundary of the two formations is located at the interface of olive-gray, brownish-gray mudstone, and greenish-gray mudstone. The delineation of the stratigraphy is based on the research results of Li et al. (2009) and Xi et al. (2009) on this section and the correlation with the lithologic stratigraphy of the boundary between Nenjiang Formation and Yaojia Formation in well SK 1 (Gao et al. 2008). The section is divided into 11 layers (layers 1–11 in Figs. 4 and 5).

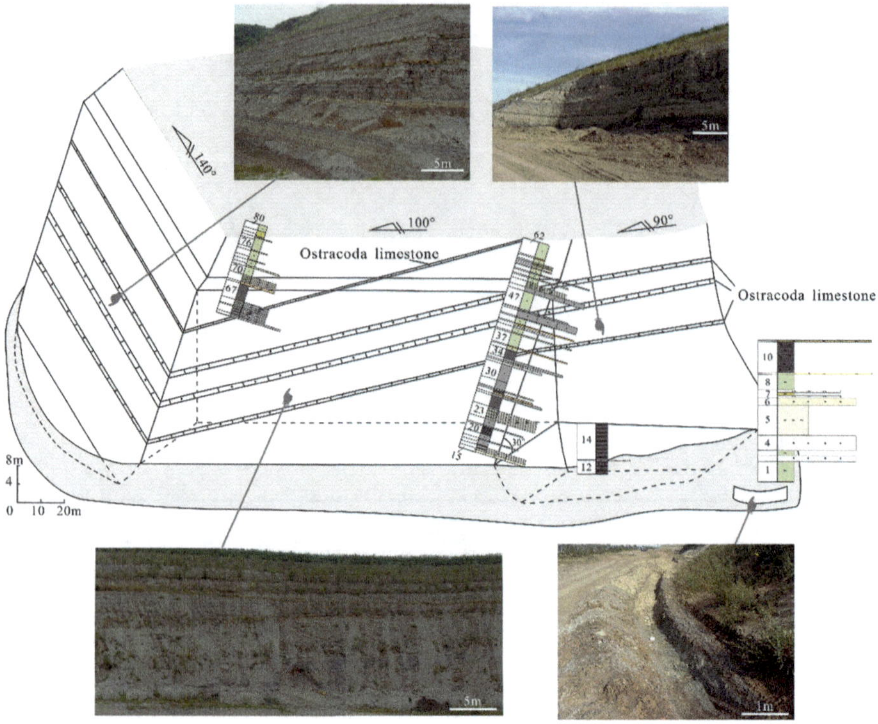

Fig. 4 Sketch of measuring the position of A–B section

Stop 2 Oil shale observation site

The rocks of Stop 2 are located in an artificial pit in the eastern part of the trench. It comprises dark gray oil shale (Fig. 7) from the lower part of the first member of the Nenjiang Formation alternating with marl and turbidite siltstone, which can be divided into three layers (layers 12–14 in Figs. 4 and 5). The oil shale displays a fissility, severe surface weathering, weak oil taste, and is regarded as an inferior oil shale. Marl usually occurs in the form of a thin conformable intercalation. Most of the marl is yellowish gray and with massive structures, showing residual argillaceous material after acid bubbling. Turbidite deposits are embedded in dark oil shale as thin layers of gray siltstone (Fig. 9-6).

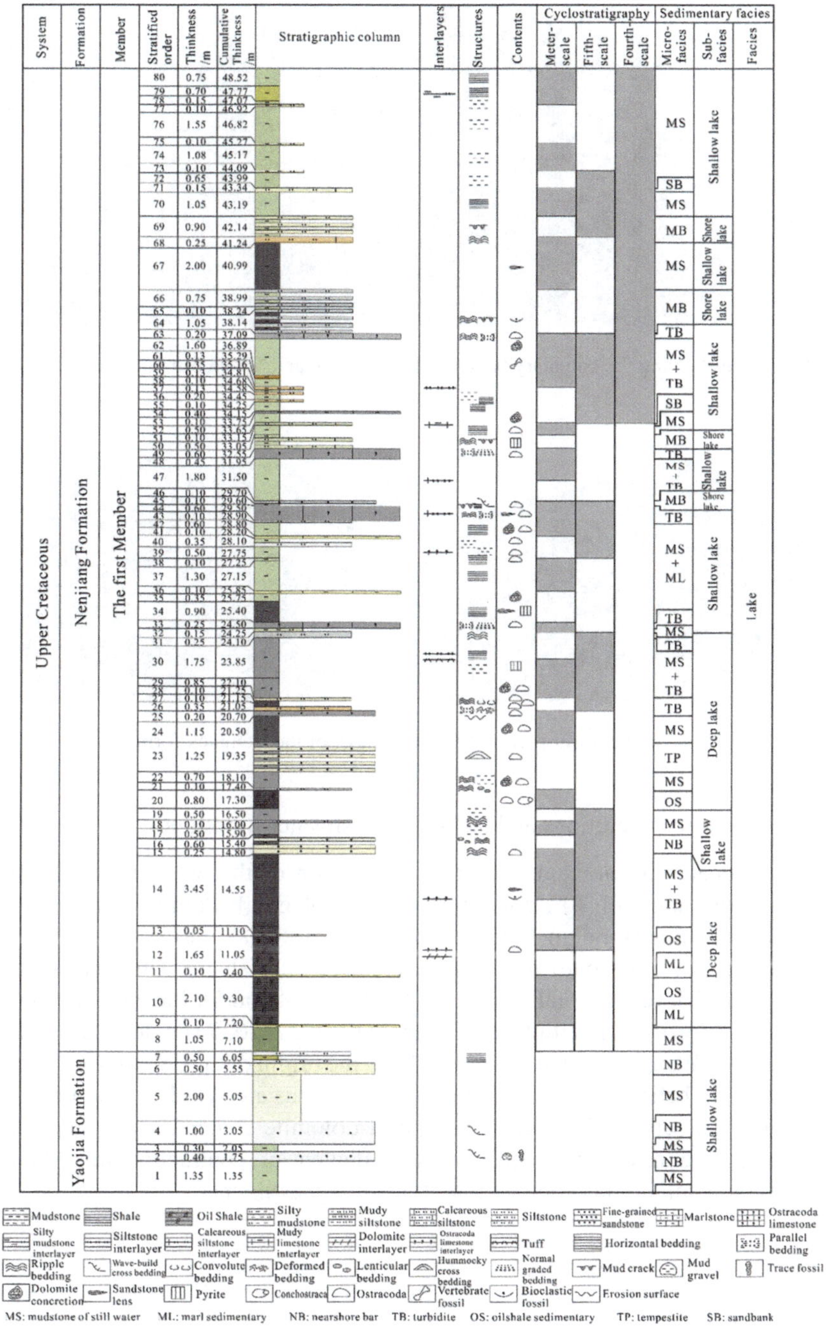

Fig. 5 Comprehensive stratigraphic column of the Yaojia and Nenjiang formations

Fig. 6 Trench showing the boundary between the Nenjiang and Yaojia formations

Fig. 7 The oil shale in the lower part of the first member of the Nenjiang formation

Stop 3 Dolomite concretions and lacustrine freshwater limestone

Stop 3 is located in the central part of the section of the west end of the slope, with an updip angle of 30°, consisting of a set of interbedding of greenish-gray, dark gray, gray mudstone (including dolomite concretions), grey ostracod limestone (including oolitic limestone), and light gray, gray, yellowish gray siltstone, which are divided into 47 layers in total (layers 15–61 in Figs. 4 and 5). Most of the dolomite concretions (Fig. 9-3) are elliptic, and a single nodule is 20–50 cm long and 5–15 cm thick. Liu and Wang (1997) discussed in detail the origin of the dolomite concretions in this section, believing that the dolomite concretions are syn-depositional or early diagenetic concretions. The elliptic shape of dolomite concretions is the result of complex interations of syngenetic crystalline diagenesis, contraction, microlamellar fractures, and continuous compaction of overlying sediments.

Lacustrine freshwater limestone (including biological limestone, bioclast limestone oolitic limestone, and oncolitic limestones etc.) were recognized in the Qingshankou Formation and Nenjiang Formation in the Songliao Basin.

In the northwest of the Songliao Basin, dozens of square kilometers reef facies carbonate rocks developed in the Qingshankou Formation, includes ostracoda limestone, oolitic limestone (Fig. 8-1), marl and oncolitic limestones (Liu et al. 2008; Wang et al. 2009a, b). Drilling data shows that the oolites are dominated by superficial ooids and true ooids with 1–4 cortices. The cores of the oolites are micrite-filled, smooth and thin ostracod valves, intraclasts and various types of detritus (Wang et al. 1983; Wang et al. 2009a, b). Herringbone cross-bedding and trough cross-bedding are developed in the oolitic limestone (Zhuo et al. 2007). Moreover, there are three kinds of oncolites in the Qingshankou Formation, annular oncolite, stalagmite oncolite and ginger-shaped oncolite. Most oncolites are composed of a core and multi ring lamina, but minor amount doesn't have distinct core or monolayer lamina (Fig. 8-2–4) (Liu et al. 2008). Wang et al. (2009a, b) suggested that the formation of the limestones in the Qingshankou Formation was related to the volcanic activites caused by lithosphere tension-shear fracture.

On the other hand, the oolites also discovered in the Nenjiang Formation in the southeast of the Songliao basin, especially in the Yaojia Military Station section (Cao 2016). There are four layers of ostracoda limestone (Figs. 8-5–8 and 9-4), with ostracods as rock formers (>90% of ostracod). Oolites with ostracods as the core are found in some layers, with individual sizes ranging from 1 to 3 mm. The rocks show wavy and parallel bedding. The 33rd layer displays obvious graded bedding (Fig. 5). And further more, the high-resolution paleoclimatic and palaeoenvironmental change and mechanisms will be discussed by Dr. Xuejiao Qu based on the lacustrine ooids in the Nenjiang Formation (Late Santonian), which supported by the National Natural Science Foundation of China (Grant Numbers. 42102135).

Stop 4 Top of the section

Stop 4 is located at the eastern end of the section. The section displays 19 layers (layers 62–80 in Figs. 4 and 5) starting from the top mudstone as seen at Stop 3, mainly consisting of greenish-gray mudstone interbedded with light yellowish gray silty mudstone and siltstone. The section description is in detail. Generally, the minimum thickness of the identified lithology measures 5 cm. A thickness of <5 cm is considered as interlayer. The minimum layer thickness of special lithology is 2 cm, and less than 2 cm is described as interlayer.

Fig. 8 Photographs showing features of the lacustrine freshwater limestone in the Qingshankou Formation and Nenjiang Formation. *Note* 1. Oolitic limestone with ostracoda in the Qingshankou Formation, Well Jin 6, 1787.0 m (from Wang et al. 2009a, b); 2. Oncolitic limestones in the Qingshankou Formation, Well La8-PS1413, 1239.2 m (from Liu et al. 2008); 3. The laminations of oncolitic limestones in the Qingshankou Formation, Well La8-PS1413, 1239.4 m (from Liu et al. 2008); 5, 6, 7, Oolitic limestone in the Nenjiang Formation, the 33rd layer; 8, Oolitic limestone in the Nenjiang Formation, the 63rd layer

Fig. 9 Photos showing features of sedimentary structure and special layers in the Nenjiang section. *Note* 1. Ripple marks, the 64th layer; 2. Mud crack impression, the 46th layer; 3. Dolomite concretions, the 52nd layer; 4. Graded bedding, the 33rd layer; 5. Erosion surface in the bottom of sandstone, the 25th layer; 6. Parallel bedding and deformation bedding, the 25th layer; 7. Hummocky cross-bedding, the 23rd layer; 8. Hummocky cross-bedding, the 23rd layer; 9. Pyrite layer with limonitization, the 49th layer

2 Field Excursion to the Quaternary Volcanogenic Successions at the Changbaishan Tianchi Volcano

2.1 Introduction of the Changbaishan Tianchi Volcano

The Changbaishan Tianchi Volcano is located in the eastern part of Jilin Province (Figs. 1 and 2). Its geographic coordinates are 41° 20′–42° 40′ N/127° 00′–129° 00′ E. Hundreds of volcanic cones are situated in the boundary region between China and North Korea (Liu et al. 1998). This is one of the largest active volcano groups in the world. They include three main sub-volcanoes (known as the sub-eruptive centers) which are the Tianchi Volcano (2755 m a.s.1.), the Wangtian'e Volcano (2438 m a.s.1.), and the South Paotaishan Volcano (2434 m a.s.1.). The three eruptive centers have similar magmatic characteristics despite their different ages. They formed from the Early Miocene to the present day, producing basaltic lava flow, volcanic cones and pyroclastic deposits of trachyte and comendite[1] composition.

The name of the volcanic field surrounding Tianchi is 长白山 in Chinese characters, which means "perpetually white mountain". These characters are Romanized as Baegdu-san or Paektu-san in Korea, Hakuto-san in Japan, and Changbaishan in China. The field has >100 Cenozoic volcanic centers (Wei et al. 2013). Tianchi ("Heavenly Lake") is the youngest and highest cone in this field (Jin et al. 1994). Baitoushan (or Baitoufeng) was a former name for the highest point of Tianchi at 2750 m, now called Jiangjunfeng (Wei et al. 2013).

The eruptive activities of the Changbaishan volcano can be divided into three main stages: (1) central vent and fissure eruptions of basaltic magmas started approximately in Pliocene and culminated in Early Pleistocene (ca. 5–1 Ma), forming a shield-like lava plateau; (2) multi-stage eruptions of voluminous silicic (and minor intermediate) magmas constructed cones of the polygenetic volcanoes between Late Pliocene and Pleistocene (3.14–0.01 Ma); and (3) explosive silicic eruptions during Holocene dominated the Tianchi Volcano and led to the formation of its summit caldera (Fig. 10). Small-scale eruptions of basaltic magmas from monogenetic scoria cones (and minor fissures) were coeval with the Tianchi cone-construction stage (ca. 1–0.01 Ma) (Zhang et al. 2018).

The latest and largest eruption of the Tianchi Volcano is named the "Millennium Eruption" (ME). The volume of the volcanic ejecta from this eruption measures approximately 40–98 km^3 (Yang et al. 2021). Chen et al. (2021) presented radiocarbon chronology and grain-specific glass chemistry data, including for the first time, single grain trace element data, for a composite proximal sequence and a distal tephra from Lake Kushu, northern Japan (ca. 1100 km away from Changbaishan). They demonstrated the interaction of two compositionally distinct magma batches (with end-member compositions) during the caldera forming eruption. Zhao et al. (2020) examined the proximal to distal variations in the pyroclastic density current (PDC) facies during the ME by analyzing the topography, stratigraphy, granulometry, and petrography. They found the ME PDCs form a pyroclastic apron that is thin at both ends and thick in the middle. Yin et al. (2012) used wiggle-match dating from carbonized logs collected from fallout deposits on the southern and northern flanks of the Tianchi crater to yield a high-precision wiggle-match date of 921–941 cal CE (95.4%) for the Millennium eruption. Xu et al. (2013) used high-precision Accelerator Mass Spectrometry (AMS) obtained the wiggle-match (WM) of sequences of ^{14}C ages for a 264-year-old tree trunk with bark buried in pyroclastic flow deposits yielding a well-constrained ME calendar age of 946 $\pm$ 3 CE. Cheong et al. (2019) presented U-Th disequilibrium ages and Hf isotopic compositions of zircons from trachydacitic pumice ejected during the ME. Taking together with the previous U-Th isotope data, their laser ablation-multicollector ICP-MS analyses of the zircons

[1] Originally described as a variety of porphyritic leucocratic alkali rhyolite containing phenocrysts of quartz, alkali feldspar, aegirine, arfvedsonite or riebeckite and minor biotite. It is now defined and distinguished from pantellerite as a variety of peralkaline rhyolite of TAS field R in which Al,0, > 1.33 $\times$ total iron as FeO + 4.4 (Le Maitre et al. 2002).

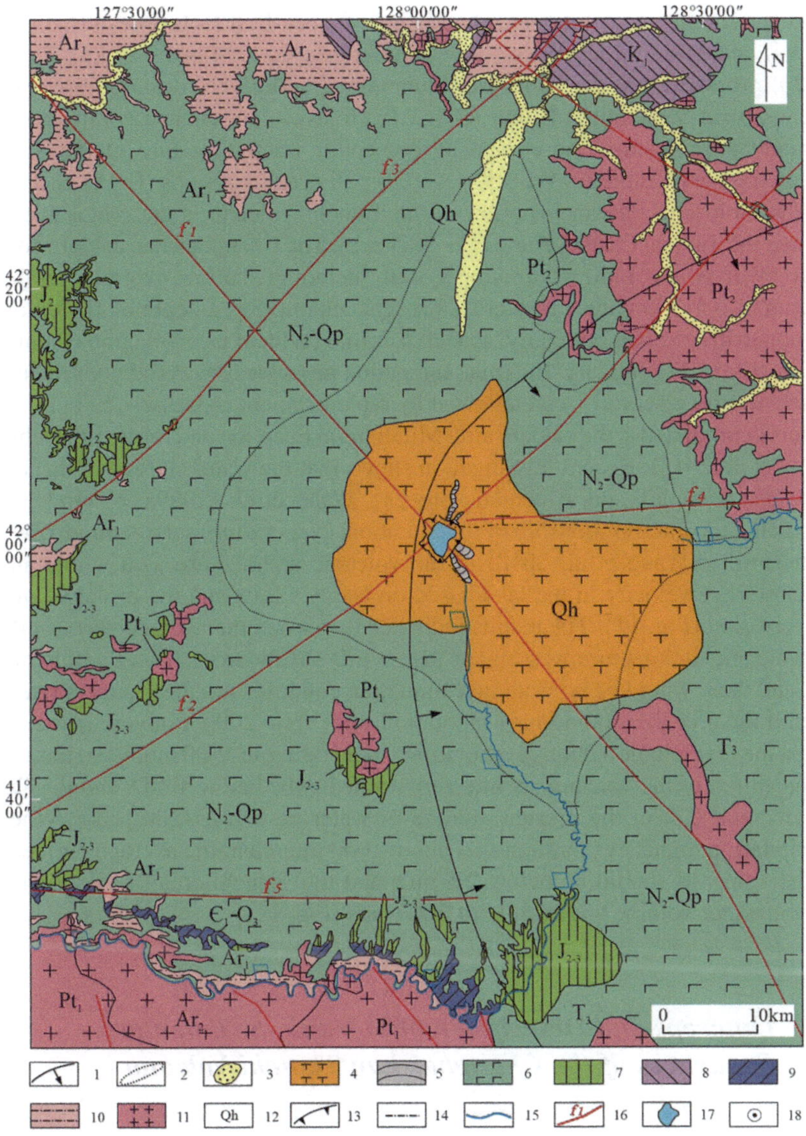

Fig. 10 Geological map of the Changbaishan Tianchi Volcano and adjacent region (Compilation based on the Geological survey report of Manjiang and Changbaishan area (1:200,000) (1963); Geological survey report of Fusong area (1:200,000) (1971); Geological survey report of Baitoushan area (1:200,000) (1976); the International Geological Map of Asia (1:5,000,000) (2013); Jin et al. (1994); Liu et al. (1997) and Wei et al. (2013)). *Note* 1. Fallout deposit distribution region and transport direction; 2. Pyroclastic flow deposit distribution circle; 3. Lahar deposits; 4. Holocene trachyte; 5. Alkali rhyolite coupled with comendite blocks; 6. Neogene and Quaternary basalt; 7. Jurassic andesite; 8. Cretaceous shale and marlstone; 9. Cambrian and Ordovician shale, limestone, and marlstone; 10. Proterozoic metamorphic rocks; 11. Proterozoic and Archean granite; 12. Quaternary; 13. Caldera; 14. Reference line of country boundary; 15. River; 16. Fault; 17. Tianchi, Heavenly Lake; 18. Village and town

yielded multiple age components consisting of an earliest Holocene cluster (11.3 ± 1.3 ka) and Middle-Late Pleistocene populations (ca. 230 and 110 ka). Pan et al. (2020) proposed that the 946 CE Millennium Eruption, be formally re-named the "Generalized Millennium Eruption" (GME) to avoid entrenched confusion and to formally acknowledge the composite comendite-trachyte nature of the eruption. Following the ME, there were still several small-scale eruptions (Liu et al. 1999).

The Holocene Plinian Eruption of the Tianchi Volcano is the major event that formed the Tianchi Crater (Wang et al. 2020a, b). The Changbaishan caldera-forming stage involved a complex plumbing system including shallow magma chambers at a depth of 0.5–3 km and a mid-crustal magma chamber at a depth of 10–15 km and the upper-lower crust boundary, as well as some residual basic magma chambers at the crust-mantle boundary. Magma segregation processes of crystal poor melts from crystal mushy reservoirs were allowed by replenishment processes (Yi et al. 2021). Water accumulated at the caldera after Millennium Eruption and formed Tianchi Lake (Yang et al. 2018). Recent investigation shows that the water storage of the Tianchi Lake is a complex and systematic process. Wang et al. (2020a, b) modeled and calculated the interrelationship of precipitation, evaporation, surface runoff, average permeability (Wang et al. 2018) of the volcanic architecture as a whole and the water storage velocity in the lake based on Darcy's Law and the principle of mass conservation (Fig. 11). The results indicated that when the average permeability of the volcanic architecture was higher than 6 mD (Millidarcy). Water would all leak out, and there would be no water accumulation conditions in the Tianchi crater lake. Even if the precipitation minus evaporation went up to 2000 mm/year in the region. When the precipitation minus evaporation would be about 1500 mm/year, the average permeability of the volcanic architecture needed to be below 4 mD, and the Tianchi crater could collect the present volume of water in at least 200 years. When the precipitation minus evaporation was about 1000 mm/year, the average permeability of the volcanos should be below 2.5 mD, and the Tianchi crater could collect the present water volume within at least 200 years (Fig. 12).

2.2 *Outcrops and Description of Volcanic-Sedimentary Sequences of the Changbaishan Tianchi Volcano*

Stop 1 The Shuitian Village Lahar Sequence (42° 30′ 50.74″ N, 128° 09′ 40.93″ E; P1 in Fig. 1)

Lahars generated from the Tianchi crater lake in the Changbaishan area include coarse-grained lithic types and fine-grained ones, characterized by pumice. They were transported northwest along the Erdaobaihe and Songhua rivers from the nearest Sanhe Waterpower Station being 40 km away from the crater, passed by Fengman Dam (360 km) and Jilin City (380 km) to Xiaobaiqitun 450 km away from the crater lake (Fig. 1). All the way along, the lahar flows left behind widespread lahar deposits

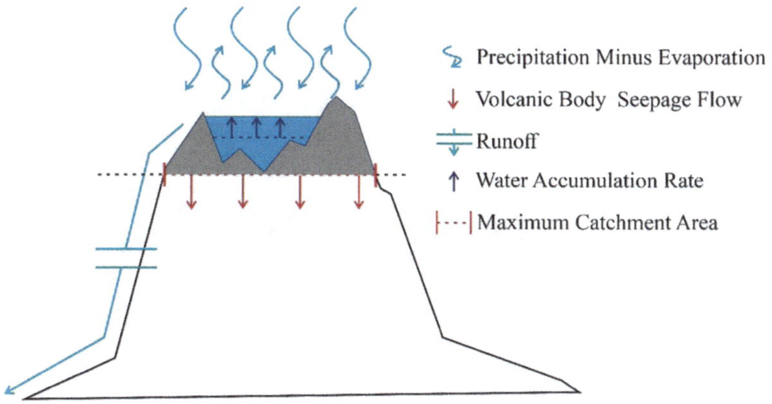

Fig. 11 Schematic diagram of Tianchi Lake water supply-savings-leakage model based on the conservation of mass

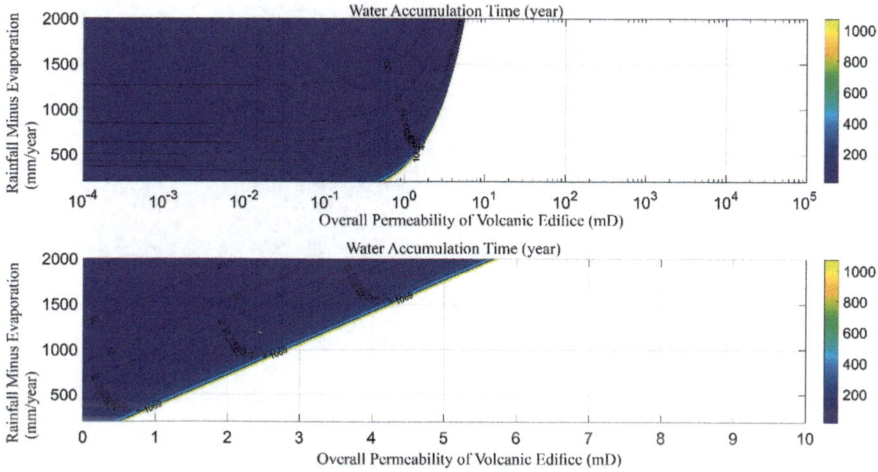

Fig. 12 Time required for the Tianchi Lake to fill up with water

in the region. The Shuitian Village outcrop of the Lahar Sequence in a distance of 57.73 km NW of the Tianchi crater, with a cumulative thickness of 12 m. It is a relatively complete geological record of the volcanic lahar deposition and of the fluvial background deposition of the Changbaishan Sequence since the ME (Figs. 13 and 15).

There are two explanations concerning the formation of these volcanic lahar deposits. Firstly, they formed by a sudden water discharge of lake water caused by dam-break during the ME, which occurred about 946 ± 3 CE at the Changbaishan Tianchi crater lake, and pyroclastic materials were transported and deposited by the same off flood event (Liu et al. 2000). The second is that the lithic and pumice lahar

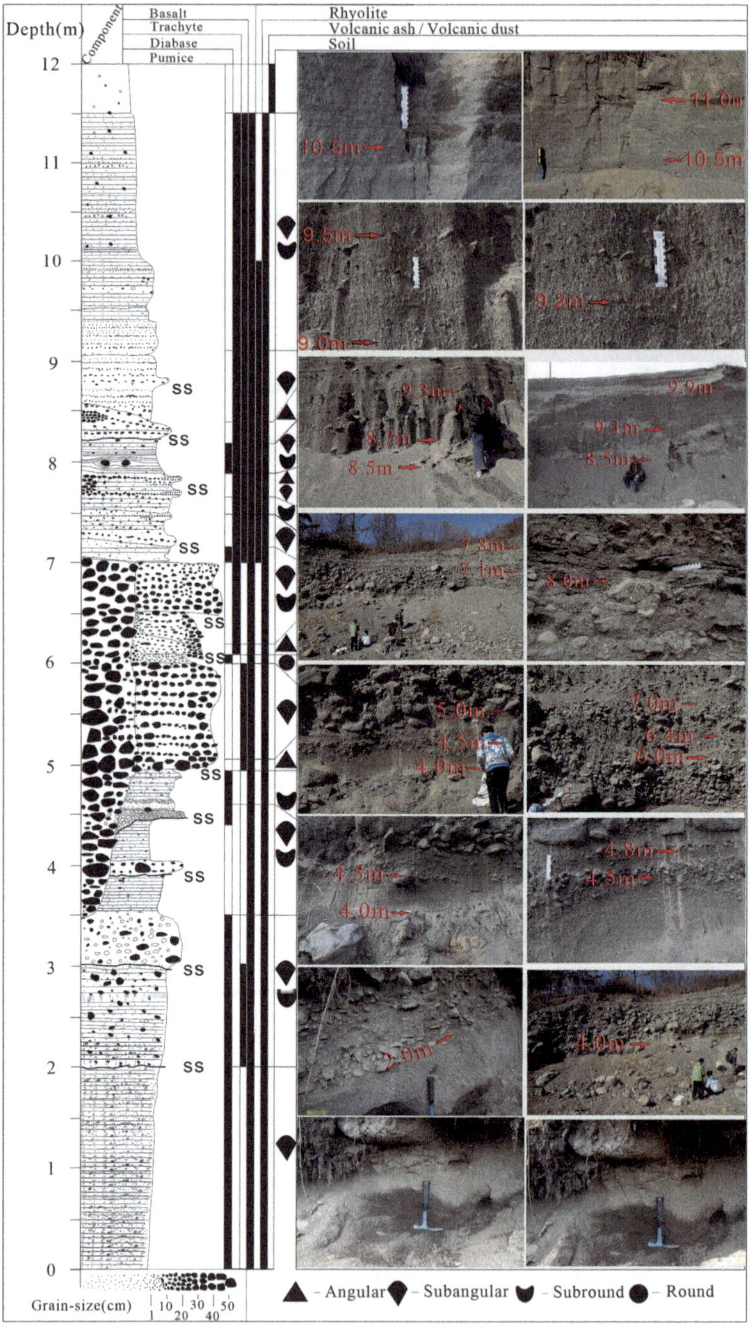

Fig. 13 The field photographs on the right depict what is recorded in the drawing on the left. *Note* SS, scoured surface

deposits were separately formed by two episodes of transportation and deposition following the ME (Yi et al. 2019). The problem with the two models is that the water was not considered, i.e. whether or not there was any water in the crater lake at the respective time and how the water could have accumulated. Moreover, the latter scheme potentially avoids, to some extent, key aspects of the sediment source, hydrodynamic conditions, and the mechanisms controlling the transition that led to frequent interbedding of the lithic and pumice lahars. Wang et al. (2020a, b) challenged the above explanations and researched in detail the well-known examples of the so-called "classic lahar successions" exposed in Shuitian Village. They found that there is a steady-state of the inversely back-stripping relationship between the source denudation area and the sedimentary accumulation process.

That is to say, changes in sequence about the components of both coarse debris in diameter 32–500 mm and fine debris of 0.0625–16 mm are as follows. The fragments in the top part of the source area are mainly pumices. Going down stratigraphically a significant increase in trachyte and basalt fragments occurs. Dike diabase and basement rhyolite are situated in the deeper horizon in the source area. The fragments in the lower part of the section are mainly pumices. Going up stratigraphically a significant increase in trachyte and basalt fragments occurs. And the detrital components in the upper part are dike diabase and basement rhyolite. In addition, sedimentary structures and lithofacies associations can be seen in the pumiceous lahar, which is relatively fine-grained, well-sorted, well stratified, and horizontally bedded, indicating fluvial and terminal fan facies having formed in a relatively stable environment of downslope flow runoff. The coarse-grained lithic lahars in the middle part are poorly stratified and are diluvial event deposits that were most likely caused by sudden flooding. Frequently occurring scours indicate that water currents may have been rapid and periodically increasing. From base to top, the scale of the scour surfaces changes from small to large and then back to small, indicating paleo-current fluctuations of the water flow regime (Figs. 14 and 15).

By comparing the sedimentary sequence of the lahar deposits with the calculated results of water storage of the Tianchi crater lake, Wang et al. (2020a, b) arrived at the following suggestions: (1) The lahar deposits are a post-ME interbedded sequence representing fluvial background sediments and diluvial event deposits. The fluvial deposits are downflow runoff or river deposits and the event deposits represent deposits related sudden release of the Tianchi crater-lake water. (2) The background sediments of the lower part may correspond to the continuous water accumulation process in the Tianchi Lake, following the ME, which may last over 200 years. The upper background sedimentation may correspond to the stable stage of the river system and landform which were relatively similar to the present situation. Taking into account the growing age of ***Abies*** *holophylla* Maxim. (1749–1768 CE) recorded by tree rings on the North and the West slopes of the volcano, Wang et al. (2020a, b) inferred that the stable state of the environment in the Tianchi region may be no later than the 1700s after the ME because coniferous plants are fairly sensitive to the stability of the water system and geomorphology.

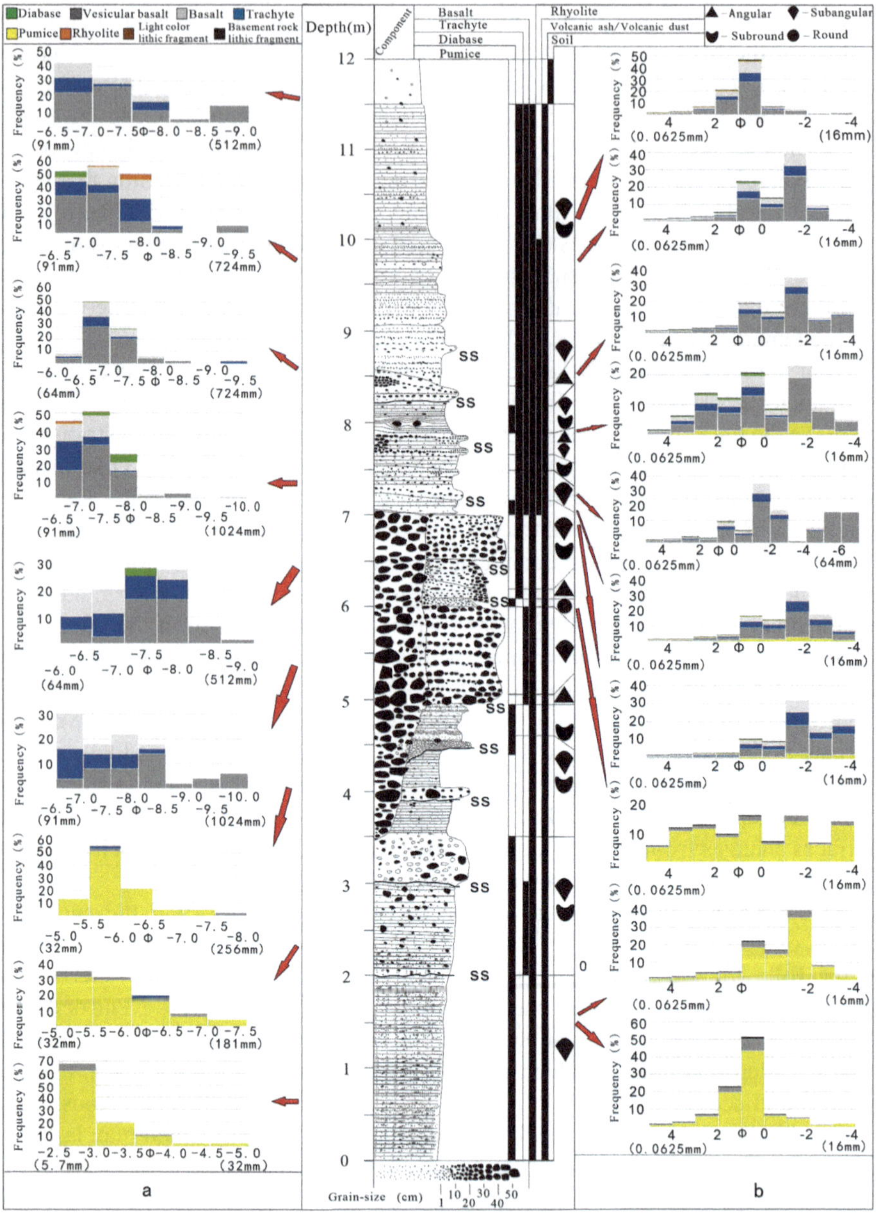

Fig. 14 Stratigraphic sequence and vertical changes of the lahar deposits in composition, particle size, and roundness of the fragments. **a** Coarse components, **b** Fine components. For legend see Fig. 13

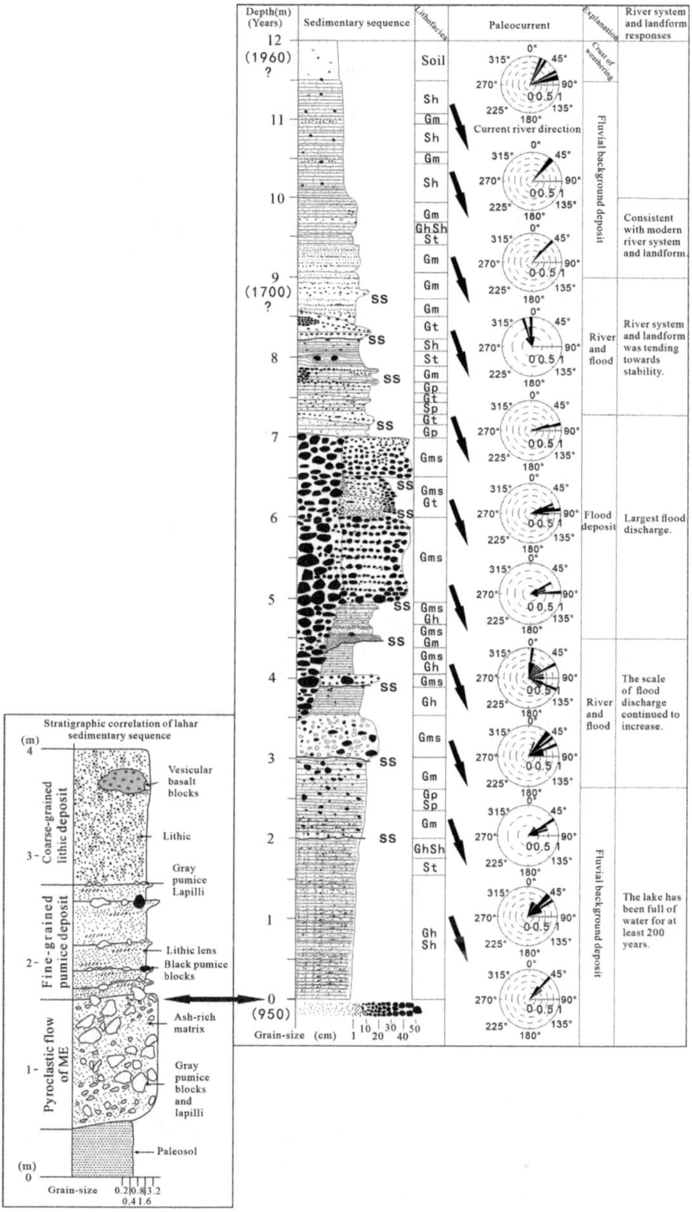
Depth(m) (Years)
Sedimentary sequence
Lithofacies
Paleocurrent
Explanation
River system and landform responses
12
(1960)
?
11
10
9
(1700)
?
8
7
6
5
4
3
2
1
0
(950)
Grain-size (cm)
1 10 20 30 40 50
Soil
Sh
Gm
Sh
Gm
Sh
Gm
GhSh
St
Gm
Gm
Gm
Gt
Sh
St
Gm
Gp
Gt
Sp
Gt
Gp
Gms
Gms
Gt
Gms
Gms
Gh
Gms
Gm
Gms
Gh
Gms
Gh
Gms
Gm
Gp
Sp
Gm
GhSh
St
Gh
Sh
SS
Current river direction
Crust of weathering
Fluvial background deposit
River and flood
Flood deposit
River and flood
Fluvial background deposit
Consistent with modern river system and landform.
River system and landform was tending towards stability.
Largest flood discharge.
The scale of flood discharge continued to increase.
The lake has been full of water for at least 200 years.
Stratigraphic correlation of lahar sedimentary sequence
(m)
Coarse-grained lithic deposit
Fine-grained pumice deposit
Pyroclastic flow of ME
Vesicular basalt blocks
Lithic
Gray pumice Lapilli
Lithic lens
Black pumice blocks
Ash-rich matrix
Gray pumice blocks and lapilli
Paleosol
Grain-size
0.2 0.4 0.8 1.6 3.2

◂**Fig. 15** Stratigraphic column of lahar deposits, lithofacies, paleocurrents, water and landform response (the lower-left figure is the stratigraphic correlation of lahar sequences overlying the ME pyroclastic flow deposits. Cited from Yi et al. 2019, their Fig. 3b). *Note* Gms, gravel with massive matrix (sand and mud) supported gravel, debris flow deposit. Gm, massive or crudely bedded gravel, longitudinal bars, lag deposits, sieve deposits. Gt, trough cross-bedded, clast-supported gravel, minor channel fills. Gp, planar cross-bedded gravel and/or matrix-supported gravel, linguid bars, or deltaic growth. Gh, horizontally stratified fine gravel or gravel-bearing coarse sand, lahar-runout facies. Sh, horizontally stratified sand, upper flow regime. St, trough cross-stratified sand, lower flow regime. Sp, planar cross-stratified sand, transverse bars, lower flow regime. Sedimentary interpretation, fluid properties, and geomorphological modification are based on Einsele (2000, p. 31), Sigurdsson et al. (2000, p. 601–611) and Scott (1988). Paleocurrent directions were determined by cross-stratification and imbrication of gravel

Stop 2 The Changbaishan Tianchi Volcano (42° 00′ 22.41″ N, 128° 03′ 20.92″ E; P2 in Fig. 1)

Stop 2 is located in the Southeast of Jilin Province of northeast China, the Changbaishan Tianchi, the lake is the border lake between China and North Korea. The lake is the source of the three rivers in northeast China and North Korea. They include the Songhua River, the Tumen River, and the Yalu River. These are main river systems nourishing a population of several tens of millions. The lake is called "Tianchi" (meaning "Heavenly Lake") because of its occurrence at a high elevation of 2189.7 m above sea level (Liu et al. 2000). Yang et al. (2018) used a multibeam sounding system to explore the basic parameters and lakebed topographic features in the lake of Tianchi. The results showed that the crater lake has a maximum depth of 373.2 m, a water boundary circumference of 13.44 km, a water area of 9.4 km^2, and total water volume of 19.88 × 10^8 m^3 (Fig. 16)

.

Stop 3 Julongquan—Tufa Deposition (42° 02′ 42.69″ N, 128° 03′ 29.64″ E; P3 in Fig. 1)

Julongquan (聚龙泉) is a group of hot springs. Ju (聚) means "Get Together". Long (龙) means "Dragon". Quan (泉) means "Spring". It is 750 m north of the Changbaishan Waterfall (Fig. 17). Quaternary trachyte, trachy-breecia and basalt are distributed in and around hot springs. There are three groups of hot springs in the upper reaches of Erdaobaihe River. The area of springs is 2817.5 m^2 on the east bank of the river. Two small groups springs with an area of nearly 500 m^2 are on the west bank of the river. The number of springs in Julongquan is 103, and the total flow is 6465.48 t/d. The highest temperature of hot springs is 82 °C (Jin et al. 2003). The intense hydrothermal activity reflects the continuous heating and baking of the deep high-temperature magma chamber and the magma degassing (Guo et al. 2010).

Stop 4 Changbaishan Waterfall—Pyroclastic Alluvial Deposits and Large Talus Fans (42° 02′ 20.85″ N, 128° 03′ 23.30″ E; P4 in Fig. 1)

The Changbaishan Waterfall is located on the north side of the Tianchi Lake, at the end of Chengcha River (Fig. 18). The water of the Chengcha River drops straight

Fig. 16 Photographs of the Tianchi Lake/Volcano (Field photographs were taken by Pujun Wang in September 2019. Field participants included Pujun Wang, John Charles Eichelberger, Alan Patrick Bischoff, Hanfei Wang, Jiannan Guo and Weidong Xu)

Fig. 17 Photographs of the Julongquan (Field photographs were taken by Pujun Wang in September 2019. Field participants included Pujun Wang, John Charles Eichelberger, Alan Patrick Bischoff, Hanfei Wang, Jiannan Guo and Weidong Xu)

Fig. 18 Photographs of the Changbaishan Waterfall (Field photographs were taken by Pujun Wang in September 2019. Field participants included Pujun Wang, John Charles Eichelberger, Alan Patrick Bischoff, Hanfei Wang, Jiannan Guo and Weidong Xu)

down from the cliff at 1250 m and forms a huge waterfall of 68 m height. The rock wall on both sides of the waterfall is influenced by weathering and gravity to cause rock collapse and debris flows (Wang et al. 1989).

Stop 5 Underground Forest (42° 05′ 58.99″ N, 128° 04′ 51.79″ E; P5 in Fig. 1)

The Underground Forest is the name of a valley. During the ME, pyroclastic flows of high speed and temperature were generated from eruption column collapse. With great power, the pyroclastic flows cut and widened a gully into a broad valley. Upon cooling, vertical joints formed by contraction. Then the joints evolved into many open cracks, along which rocks were broken down under retrogressive erosion from downstream to upstream. Finally, a 3 km-long, 100 m-deep valley with a width of 300 m width formed. Trees started to grow forming a spectacular "Underground Forest" (Fig. 19). Taking into account the growing age of ***Abies*** *holophylla* Maxim. (1749–1768 CE) recorded by the tree rings on the North and the West slopes of the volcano (Fig. 20), Wang et al. (2020a, b) inferred that the stable state of the environment in the Tianchi region may date no later than the 1700s after the ME

Fig. 19 Photographs of the Underground Forest (Field photographs were taken by Pujun Wang in September 2019. Field participants included Pujun Wang, John Charles Eichelberger, Alan Patrick Bischoff, Hanfei Wang, Jiannan Guo and Weidong Xu)

because coniferous plants are fairly sensitive to the stability of the water system and geomorphology.

Stop 6 Canyon Pumice Forest (42° 08′ 2.76″ N, 128° 11′ 59.14″ E; P6 in Fig. 1)

The Canyon Pumice Forest, located about 18.5 km northeast of Tianchi Lake, is a natural landscape formed by the accumulation of pyroclastic flows after the ME (Sun et al. 2017). It appears as the Canyon-filled pyroclastic flow facies (Yi et al. 2020). Cooling/contraction joints are well developed, and water erosion along them formed a spectacular Pumice Forest (Shi et al. 2005). The columnar joints have a rhythm of alternating soft and hard layers. The top of the columnar joints is developed by a layer of earth yellow debris mixed with some dark and light-colored debris, and the

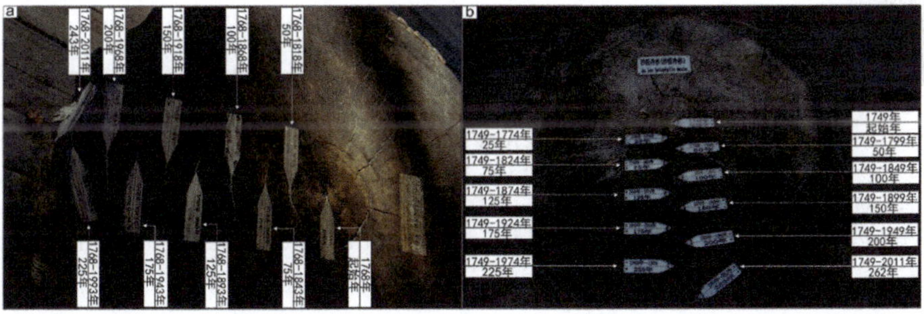

Fig. 20 Photographs of ancient tree rings from trees of the Changbaishan Tianchi Volcano (Field photographs were taken by Pujun Wang in September 2019. Field participants included Pujun Wang, John Charles Eichelberger, Alan Patrick Bischoff, Jiannan Guo, Hanfei Wang and Weidong Xu). *Note* **a** ***Abies*** *holophylla* Maxim. (living from 1749 to 2011 CE) from the Underground Forest scenic spot on the north slope of the Tianchi Volcano, 128° 4′ 51.79″ E/42° 5′ 58.99″ N and 10.49 km north to the Tianchi Lake. **b** ***Abies*** *holophylla* Maxim. (living from 1768–2011 CE) from the Jinjiang Grand Canyon scenic spot on the west slope of the Tianchi Volcano, 127° 55′ 31.45″ E/41° 54′ 33.33″ N and 15.44 km southwest to the lake

Fig. 21 Pyroclastic flows in the Canyon Pumice Forest (Field photographs were taken by Pujun Wang in October 2019. Field participants included Pujun Wang, Haichao Chen, Zhoulong Yang and Weidong Xu)

lower part is composed of a 16-m-thick sedimentary layer dominated by dark debris (Fig. 21).

Stop 7 Red Soil Mountain—Scoria and Pyroclastic-Flow Deposits (42° 21′ 29.70″ N, 127° 56′ 20.17″ E; P7 in Fig. 1)

The Red Soil Mountain cinder cone is located on the Changbaishan volcanic shield, 38 km northwest of the Changbaishan Tianchi crater (Fig. 22). The color variation between red and black is due to different oxidation. It is a cinder cone which formed in the Tianchi cone-construction stage between 1 and 0 Ma (Fig. 23). The Red Soil Mountain formed by the accumulation of pyroclasts from bottom to top by the eruption of a typical Strombolian volcano with different energies (Guo et al. 2020).

Fig. 22 Photographs of the Red Soil Mountain (Field photographs were taken by Pujun Wang in October 2019. Field participants included Pujun Wang, Haichao Chen, Zhoulong Yang and Weidong Xu)

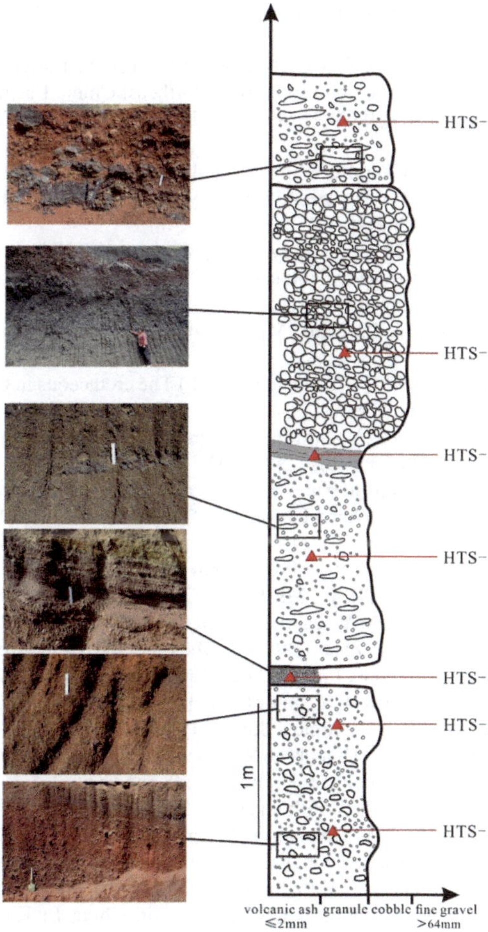

Fig. 23 The accumulation of pyroclasts in the Red Soil Mountain (based on Guo et al. 2020)

References

Cao HS (2016) Biochemical cycles of C-N-S and paleoenvironmenttal evolution in the Nenjiang Formation Songliao Basin. Dissertation, Jilin University (in Chinese with English abstract)

Chen ZQ, Chen Z (2021) Identifying references to volcanic eruptions in Chinese historical records. Geological Society, London, Special Publications 510:271–289

Cheng RH, Wang, GD, Wang PJ (2008) Sedimentary cycles of Quantou-Nenjiang, Cretaceous and Milankovitch cycles from the south hole of CCSD-SK in Songliao Basin. Acta Geol Sin 82(1):55–64 (in Chinese with English abstract)

Chen XY, Blockley SPE, Tarasov PE et al (2016) Clarifying the distal to proximal tephrochronology of the Millennium (B-Tm) eruption, Changbaishan Volcano, northeast China. Quat Geochronol 33:61–75

Cheong ACS, Jeong YJ, Jo HJ et al (2019) Recurrent quaternary magma generation at Baekdusan (Changbaishan) volcano: New zircon U-Th ages and Hf isotopic constraints from the Millennium Eruption. Gondwana Res 68:13–21

Einsele G (2000) Sedimentary basins: evolution, facies, and sediment budget, 2nd edn. Springer, Berlin

Feng ZQ, Jia CZ, Xie XN et al (2010) Tectonostratigraphic units and stratigraphic sequences of the nonmarine Songliao Basin. Northeast China. Basin Res 22(1):79–95

Fischer AG (1964) The lofter cyclothems of the alpine Triassic. Kansas Geol Surv Bull 169(1):107–149

Gao YF, Wang PJ, Wang CS et al (2008) Well site selecting, core profile characteristics and distribution of the special lithology in CCSD-SKII. Acta Geol Sin 82(5):669–675 (in Chinese with English abstract)

Guo JN (2020) A study on the structure of mafic magma system in the lower crust of the late Pleistocene to Holocene in Changbaishan volcano. Dissertation, Jilin University (in Chinese with English abstract)

Guo ZF, Li XF, Zhang MF (2010) Volcanic activities and deep carbon cycle. Quat Sci 30(03):497–505

Hao YC, Su DY, Yu JX et al (1986) The cretaceous in China. Geological Publishing House, Beijing (in Chinese)

Harms JC (1975) Stratification and sequence in prograding shoreline deposits. Soc Econ Paleontol Min Short Course 2:81–102

Heilongjiang Bureau of Geology and Mineral Resources (1997) Lithostratigraphic of Heilongjiang Province. China University of Geosciences Press, Wuhan (in Chinese)

Jin BL, Zhang XY (1994) Researching volcanic geology in Mount Changbai. Northeast Korea Ethnic Education Press, Yanji (in Chinese)

Jin DC, Cui TR, Jin ML (2003) Observation and research on hot and warm springs in ChangbaishanTianchi Volcano. Seismol Geol 25(A1):121–127 (in Chinese with English abstract)

Karsakov LP, Zhao CJ (2001) Tectonic map of the Central Asia-Pacific Belts junction area, scale 1:1,500,000. In: Yu A (eds) Kosygin Institute of Tectonics and Geophysics, Far Eastern Branch of Russian Academy of Sciences and Shengyang Institute of Geology and Mineral Resources of China Geological Survey. Printed at 65 Kim Yu Chen St., Khabarovsk, 680000

Khanchuk AI, Didenko AN, Popeko LI et al (2015) Structure and evolution of the mongol-okhotsk orogenic belt. In: Kröner A (ed) The Central Asia orogenic belt-geology, evolution, tectonics, and models. Borntraeger Sci, Stuttgart, pp 211–234

Le Maitre RW, Streckeisen A, Zanettin B et al (2002) Igneous rocks: a classification and glossary of terms, 2nd edn. Cambridge University Press, New York

Li DS, Jiang RQ, Katz BJ (1995) Petroleum Generation in the Non-Marine Qingshankou Formation (Lower Cretaceous), Songliao Basin, China. In: Katz BJ (ed) Petroleum Source Rocks. Springer, Berlin Heidelberg, pp 131–148

Li G, Chen PJ, Wan XQ et al (2004) Stratotype of the basal boundary of the Nenjiang Stage, Cretaceous. J Stratigr 28(4):297–299, 335 (in Chinese with English abstract)

Li G, Wan XQ, David JB et al (2009) Spinicaudatans from the upper Cretaceous Nenjiang Formation of the Songliao Basin, Northeast China: taxonomy and biostratigraphy. Cretac Res 30:687–698
Li SZ, Zhao GC, Dai LM et al (2012) Cenozoic faulting of the Bohai Bay Basin and its bearing on the destruction of the eastern North China Craton. J Asian Earth Sci 47:80–93
Liu BJ, Zeng YF (1985) Foundation and working methods of lithofacies paleogeography. Geological Publishing House, Beijing (in Chinese)
Liu JQ (1999) Volcanos in China. Science Press, Beijing (in Chinese)
Liu RX, Wei HQ, Chou SH (1997) Study on the age of the last major eruption of Tianchi Volcano in Changbai Mountain and its significance. Sci China 27(5):437–441 (in Chinese)
Liu RX, Wei HQ, Li JT (1998) Modern eruption of Changbaishan-Tianchi volcano. Science Press, Beijing (in Chinese)
Liu WZ, Wang PJ (1997) Genesis and engironmental significance of the dolomite concretions from the Nenjiang Formation in the Songliao Basin, Northeastern China. Lithofacies Palaeogeogr 17(1):22–26 (in Chinese with English abstract)
Liu WZ, Wang PJ, Gao YF et al (2008) Discovery of oncolites in the Qingshankou Formation of Cretaceous, Songliao Basin and its environmental significance. Acta Geol Sin 82(5):594–600 (in Chinese with English abstract)
Liu X, Sui WG, Wang XK (2000) Lahar deposits of 1000 a B.P. eruption at Changbaishan volcano and their hazards. J Changchun Univ 30(1):14–17 (in Chinese with English abstract)
Pan B, De Silva SL, Xu JD et al (2020) Late Pleistocene to present day eruptive history of the Changbaishan-Tianchi Volcano, China/DPRK: new field, geochronological and chemical constraints. J Volcanol Geoth Res 399:106–870
Park SC, Kim TU, Cho IW et al (2016) Scoriae magma evolution at Paekdu volcano, Democratic People' s Republic of Korea. Acta Petrol Sinica 32(10):3214–3224 (in Chinese with English abstract)
Ren JS, Niu BG, Wang J (2013) International geological map of Asia. Geological Publishing House, Beijing (in Chinese)
Ren JY, Tamaki K, Li ST et al (2002) Late Mesozoic and Cenozoic rifting and its dynamic setting in Eastern China and adjacent areas. Tectonophysics 344(3):175–205
Scott KM (1988) Origins, behavior, and sedimentology of lahars and lahar-runout flows in the Toutle-Cowlitz River system: Lahars and lahar-runout flows in the Toutle-Cowlitz River system, Mount St. Helens, Washington. United States Government Printing Office, Washington
Shi LB, Chen XD, Yang QF et al (2005) Petrochemistry of pumices of various colors produced by the eruption of Changbaishan Tianchi volcano at 1000 years ago. Seismol Geol 27(1):73–82
Sigurdsson H, Houghton BF, McNutt SR et al (2000) Encyclopedia of volcanoes. Academic Press, San Diego
Specialized comprehensive Prospecting Brigade directly under Geological Bureau of Jilin Province (1963) Geological survey report of Manjiang and Changbai area (1:200000). Jilin Provincial Bureau of Geology and Mineral Resources, Changchun (in Chinese)
Specialized comprehensive Prospecting Brigade directly under Geological Bureau of Jilin Province (1971) Geological survey report of Fusong area (1:200000). Jilin Provincial Bureau of Geology and Mineral Resources, Changchun (in Chinese)
Specialized comprehensive Prospecting Brigade directly under Geological Bureau of Jilin Province (1976) Geological survey report of Baitoushan area (1:200000). Jilin Provincial Bureau of Geology and Mineral Resources, Changchun (in Chinese)
Sun CQ, Liu JQ, You HT et al (2017) Tephrostratigraphy of Changbaishan volcano, northeast China, since the mid-Holocene. Quatern Sci Rev 177:104–119
Tan XC, Wang HS (1929) Geology along the valley of Nengjiang River and Heilongjiang River. Bull Geol Surv China 13:32–41 (in Chinese)
Wang JP (1989) Changbai mountain chronicles. Jilin Literature and History Press, Jilin (in Chinese)
Wan XQ, Zhao J, Scott R et al (2013) Late Cretaceous stratigraphy, Songliao Basin, NE China: SK1 cores. Palaeogeogr Palaeoclimatol Palaeoecol 385:31–43

Wang CW, Sun YW, Li N et al (2009a) Tectonic implications of Late Paleozoic stratigraphic distribution in Northeast China and adjacent region. Sci China Earth Sci 52(5):619–626

Wang PJ, Gao YF, Ren YG et al (2009b) $^{40}Ar/^{39}Ar$ age and geochemical features of mugearite from the Qingshankou Formation: significances for basin formation, hydrocarbon generation and petroleum accumulation of the Songliao Basin in Cretaceous. Acta Petrol Sinica 25(5):1178–1190 (in Chinese with English abstract)

Wang GD, Cheng RH, Wang PJ et al (2008) The forming mechanism of dolostone of Nenjiang Formation in Songliao Basin-example from CCSD-SKII. Acta Geol Sin 82(1):48–54 (in Chinese with English abstract)

Wang HJ, Zhou SX, Teng YH et al (1983) A study of freshwater carbonates in the Cretacous Qingshankou and Nenjiang Formation in western Songliao basin (I). Pet Explor Dev 5:1–6 (in Chinese with English abstract)

Wang N, Li S, Hu R et al (2020a) Geomorphology of the underwater caldera of the Changbaishan Tianchi volcano using 3D virtual visualization. Geol J 55:5186–5196

Wang PJ, Xu WD, Chen HC et al (2020b) Characteristics and source-sink response of the Lahar deposits in Changbaishan-Tianchi volcano and adjacent area following the Millennium Eruption. Acta Petrol Sinica 36(9):2893–2911 (in Chinese with English abstract)

Wang PJ, He KL, Yi J et al (2018) Volcanostratigraphy, volcanic architecture and reservoir of intermediate-mafic volcanic rocks: a comparison between buried Cretaceous volcanoes in the Songliao Basin and the modern volcanoes in Changbai Mountain. Geophys Prospect Petrol 57(5):775–787 (in Chinese with English abstract)

Wang PJ, Liu WZ, Shan XL et al (2001) Depositional events: introduction example application. Jilin Science & Technology Press, Changchun, pp 83–91 (in Chinese with English abstract)

Wang PJ, Mattern F, Didenko NA et al (2016) Tectonics and cycle system of the Cretaceous Songliao Basin: an inverted active continental margin basin. Earth Sci Rev 159:82–102

Wang PJ, Zhao RL, Meng QA et al (2015) The Cretaceous Songliao Basin: dynamic background from volcanic rift to interior sag basin. Earth Sci Front 22(3):99–117 (in Chinese with English abstract)

Wang YJ, Fan ZY (1997) Discovery of Permian radiolarians in ophiolite belt on northern side of Xar Moron River, Inner-Mongolia and its geological significance. Acta Palaeontol Sin 36(1):58–69 (in Chinese with English abstract)

Wei HQ, Liu GM, Gill J (2013) Review of eruptive activity at Tianchi volcano, Changbaishan, northeast China: implications for possible future eruptions. Bull Volcanol 75(4):706

Xi DP, Li G, Wan XQ, Jing X et al (2009) Yaojia-Lower Nenjiang Formation and lake evolution of Sourtheastern Songliao Basin, NE China. Acta Palaeontol Sinica 48(3):259–271 (in Chinese with English abstract)

Xi DP, Wan XQ, Feng ZQ et al (2011) Discovery of Late Cretaceous foraminifera in the Songliao Basin: evidence from SK-1 and implications for identifying seawater incursions. Chin Sci Bull 56(3):253–256

Xu B, Zhao P, Bao QZ et al (2014) Preliminary study on the pre-Mesozoic tectonic unit division of the Xing-Meng Orogenic Belt (XMOB). Acta Petrol. Sin 30(7):1841–1857 (in Chinese with English abstract)

Xu JD, Pan B, Liu TZ et al (2013) Climatic impact of the Millennium eruption of Changbaishan volcano in China: new insights from high-precision radiocarbon wiggle-match dating. Geophys Res Lett 40(1):54–59

Yang QF, Yuan XJ, Wu CZ et al (2018) The multibeam sounding exploration of the Tianchi caldera lakebed topography at the China-North Korea border. Acta Petrol Sinica 34(1):185–193 (in Chinese with English abstract)

Yang QY, Jenkins SF, Lerner GA et al (2021) The Millennium Eruption of Changbaishan Tianchi Volcano is VEI 6, not 7. Bull Volcanol 83(11):1–10

Yang WL (1985) Daqing oil field, People's Republic of China: a giant field with oil of nonmarine origin. AAPG Bull 69(7):101–1111

Yi J, Wang PJ, Shan XL et al (2020) Facies variations in the pyroclastic density currents(PDCs)produced by the Millennium Eruption of the Changbaishan Tianchi volcano, NE China. Acta Petrol Sinica 36(11):3346–3362

Yi J, Wang PJ, Shan XL et al (2021) Modeling the multi-level plumbing system of the Changbaishan caldera from geochemical, mineralogical, Sr-Nd isotopic and integrated geophysical data. Geosci Front 12(5):23–42

Yi J, Wang PJ, Shan XL et al (2019) Lahar deposits generated after the Millennium eruption of the Changbaishan Tianchi volcano in the Erdaobaihe River system, China. J Volcanol Geoth Res 380:1–18

Yin JH, Jull AJT, Burr GS et al (2012) A wiggle-match age for the Millennium eruption of Tianchi volcano at Changbaishan, Northeastern China. Quat Sci Rev 47(4):150–159

Zhang ML, Guo ZF, Liu JQ et al (2018) The intraplate Changbaishan volcanic field (China/North Korea): a review on eruptive history, magma genesis, geodynamic significance, recent dynamics and potential hazards. Earth Sci Rev 187:19–52

Zhang SQ, Ren YG (2003) The study of base level changes of the Songliao Basin in Mesozoic. Journal of Chang'an University (earth Science Edition) 25(2):1–5 (in Chinese with English abstract)

Zhang YX, Sun YS, Zhang XZ et al (1998) Geotraverse from Manzhouli to Suifenhe, China (Map & Guidebook). Geological Publishing House, Beijing (in Chinese)

Zhao B, Xu JD, Yu HM et al (2020) Pyroclastic density current facies in the Millennium Eruption of Tianchi Volcano, Northeast China: insights from topography, stratigraphy, granulometry, and petrography. Front Earth Sci 8:323

Zheng RC, Peng J, Wu CR (2001) Grade division of Base-level cycles of terrigenous basin and its implications. Acta Sedimentol Sin 19(2):250–255 (in Chinese with English abstract)

Zhuo HC, Lin CM, Li YL et al (2007) Characteristics of sedimentary facies and sequence boundary in Upper Cretacous Qinghsankou-Yaojia Formation of northern Songliao basin. Acta Sedimentol Sin 25(1):29–38 (in Chinese with English abstract)

Field Logistics

Safety in the field is closely related to awareness of potential difficulties, fitness and use of appropriate equipment. Safety is a personal responsibility and all participants should be aware of the following issues. The participants will be carried with buses. All participants require comfortable walking boots. Trainers or running shoes are unsuitable footwear in the field. A waterproof coat/jacket is essential. In summer weather conditions are good. Anyway a waterproof coat/jacket is recommended as well as a bottle of water and a backpack. Sun protection can be useful as well a hat or a headscarves. Participants should inform the excursion leaders (in confidence) of any physical or mental condition, which may affect performance in the field (e.g. asthma, diabetes, epilepsy, vertigo, heart condition, back problem, ear disorder, lung disease, allergies etc.), and of problems related to particular diets or food. Each vehicle will carry one basic first aid kit.

Mobile/cellular phone coverage is good but could be limited in same place.

The emergency telephone number for police is **110**.

The emergency telephone number for ambulance is **120**.

Fieldtrip Leaders

Prof. Dr. Pujun Wang

Geologist on Oil & Gas Sedimentary Basins
Research Fellow of the AvH Germany
ICDP Geologist in-Chief, Songliao Basin of NE China
School of Earth Sciences, Jilin University, China
E-mail: wangpj@jlu.edu.cn
Website: http://www.petrorevol.com

Prof. Dr. Frank Mattern

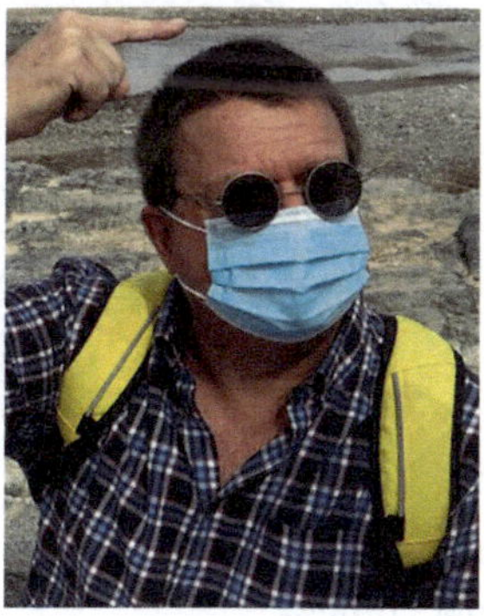

Diplom, FU Berlin; Ph.D., FU Berlin; Habilitation, FU Berlin
Sedimentologist, Basin Analyst, Tectonicist
Department of Earth Sciences, Sultan Qaboos University, Oman
E-mail: frank@squ.edu.om;fmattern@zedat.fu-berlin.de
Website: https://www.squ.edu.om/science/Departments/Earth-Sciences/Staff

Dr. Youfeng Gao

Geologist on Oil & Gas
Sedimentary Basins
School of Earth Sciences, Jilin University, China
E-mail: gaoyoufeng@jlu.edu.cn
Website: http://www.petrorevol.com

Dr. Jianguang Zhang

Palynologist on the Mesozoic
Paleoenvironment and Paleoclimate
Technische Universität Darmstadt, Germany
E-mail: zhangjianguang108@126.com
Website: http://www.sporopollen.com

Dr. Jian Yi

Volcanologist
School of Earth Sciences, Jilin University, China
E-mail: yijian@jlu.edu.cn
Website: http://www.petrorevol.com

Dr. Xuejiao Qu

Geologist on Oil & Gas
Sedimentary Basins
Chongqing University of Science and Technology, China
E-mail: quxuejiao2008@aliyun.com

Coal Accumulation Patterns and Paleoclimates in a Carboniferous–Permian Cratonic Basin, North China

Longyi Shao, Christopher Fielding, Guangyuan Mu, Jing Lu, Peixin Zhang, Ya-nan Li, Ye Wang, Yanwen Shao, and He Wen

The North China Plate was at a paleolatitude of ~10–20° N, near the paleo-equator during the Carboniferous and Permian. During this period, land plants flourished in the low latitudes of the northern hemisphere, resulting in large-scale coal accumulation in the North China Plate, while extensive glaciation occurred in the supercontinent Gondwana of the southern hemisphere. In the middle and late Permian, the global icehouse climate system gradually changed to a greenhouse mode, coal accumulation on the North China Plate gradually weakened and continental red-bed deposition began to dominate paleocontinental landscapes. Therefore, the North China Plate was characterized by a coal-bearing series in the lower part and continental red-bed strata in the upper part of the Carboniferous and Permian succession. The coal-bearing series is composed of offshore carbonate rocks, and terrigenous clastic facies of shallow marine, shoreline and coastal plain origin. The continental red-bed strata mainly consist of terrigenous clastic rocks formed in alluvial plain environments. The Carboniferous and Permian strata contain rich coal resources and paleoenvironmental information.

The term "cyclothem" was first used to describe the repetitive strata of Pennsylvanian age in the Eastern Interior Basin (or Illinois Basin) of the USA, and this terminology was then applied to similar cyclical strata across the paleo-equatorial regions of the world. The cyclic records in these strata were found to correspond to orbital frequencies in the Milankovitch band. Identification of orbital frequencies is of great significance in constructing a high-resolution chronostratigraphic framework and in understanding controls on paleoclimate evolution. Cycolothemic coal-bearing successions have also been recognised in the North China region. However, due to the

L. Shao (✉) · G. Mu · J. Lu · P. Zhang · Y.-n. Li · Y. Wang · Y. Shao · H. Wen
China University of Mining and Technology, Beijing, China
e-mail: ShaoL@cumtb.edu.cn

C. Fielding
University of Connecticut, Storrs, CT, USA

X. Hu (ed.), *Field Trip Guidebook on Chinese Sedimentary Geology*,
https://doi.org/10.1007/978-981-99-6936-4_6

diversity of sedimentary facies in the Late Carboniferous to early Permian successions of the North China Plate, the cyclic strata and their recorded Milankovitch orbital signals need to be further studied. Furthermore, relationships among stratal cyclicity and deep time paleoclimate changes also need to be explored.

Environments of coal formation in the North China region during the Carboniferous and Permian have been widely studied by many institutions and scholars in China (e.g., Shang 1997; Cheng 1992). Chinese geologists have put forward numerous theories on coal-accumulating models, such as (1) the tidal flat coal-accumulation model, (2) the delta plain coal-accumulation model, (3) the barrier-lagoon coal-accumulation model, and (4) the carbonate platform coal accumulation model. The coal-bearing succession in North China was formed in an epicontinental sea during the Carboniferous and Permian. The shallow water delta depositional system and coal accumulation model is explored in this excursion, based on field outcrop sections of Carboniferous and Permian strata in the study area.

The End-Permian Mass Extinction represents the largest mass extinction event in Earth history. Studies of the End-Permian Mass Extinction with well-preserved strata and precise age control have to date mainly focused on marine successions, especially the Global Boundary Stratotype Sections and Points (GSSPs) in southern China. As the probable source of contaminants flushed into the marine environment, time-equivalent records from continental settings are becoming more intensively studied, including those in North China, in order to establish the relationship between the continental stratal record and the marine mass extinction at the end of the Permian. As part of a comprehensive evaluation of Permian to Triassic continental sedimentary records from the North China region, changes in fluvial facies recording changes informative river types on ancient alluvial plains are investigated, and their relationships to paleoclimate change are addressed on this excursion.

The field excursion route takes in the Carboniferous and Permian strata of Shuiyuguan section in Jiaocheng City and Palougou section in Baode City, Shanxi Province as the main investigation sites, and also includes the Pingshuo open-pit coal mine and the Yungang Grottoes in Datong City as additional points of interest. The purpose of this excursion is to study scientifically significant geological phenomena preserved in the late Paleozoic record of North China, including (1) Depositional environments and coal accumulation model of the Carboniferous–Permian in the North China Plate; (2) Deep-time paleoclimate and coal accumulation patterns; (3) Geological events and sedimentary characteristics near the Permo-Triassic boundary; (4) Fluvial styles of Carboniferous and Permian coastal and alluvial plain deposits in North China; and (5) Milankovitch Cycle signals recorded in the Carboniferous–Permian coal-bearing strata of the North China Plate.

1 Introduction

1.1 Geological Setting

1.1.1 Regional Tectonics

The North China Plate is a paleo-plate with a history of 3.8 billion years, and also the main tectonic unit in China (Fig. 1; Zhai 2019). It is bounded by Yinshan structural belt to the north, the Qinling and Dabie Mountains to the south, the Helan Mountain and Liupan Mountain to the west and Liaodong Peninsula, Shandong Peninsula and the wide areas of the Huanghuai plain to the east (Fig. 2; Shen et al. 2019b).

There are two ancient continental blocks in the North China Plate, namely, the Ordos continental block and the Jilu continental block (Fig. 2). In the Early Proterozoic, the two continental blocks and the parts between and around them were consolidated into a unified North China Plate through the Luliang movement (Meng 2004). The North China Plate began to enter the plate tectonic development stage in the Mesoproterozoic and Neoproterozoic. The positional changes of the North China Plate gradually stabilized, and the plate deformation stage gradually came to an end

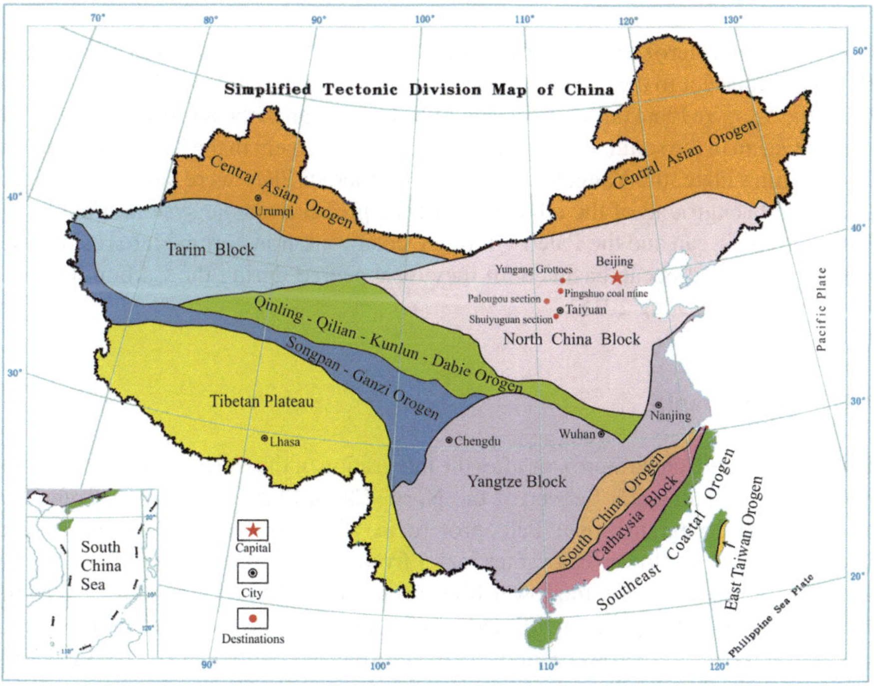

Fig. 1 Simplified tectonic division map of China (modified after Geng et al. 2011). The locations of field trip routes were shown in the North China Plate (North China Block)

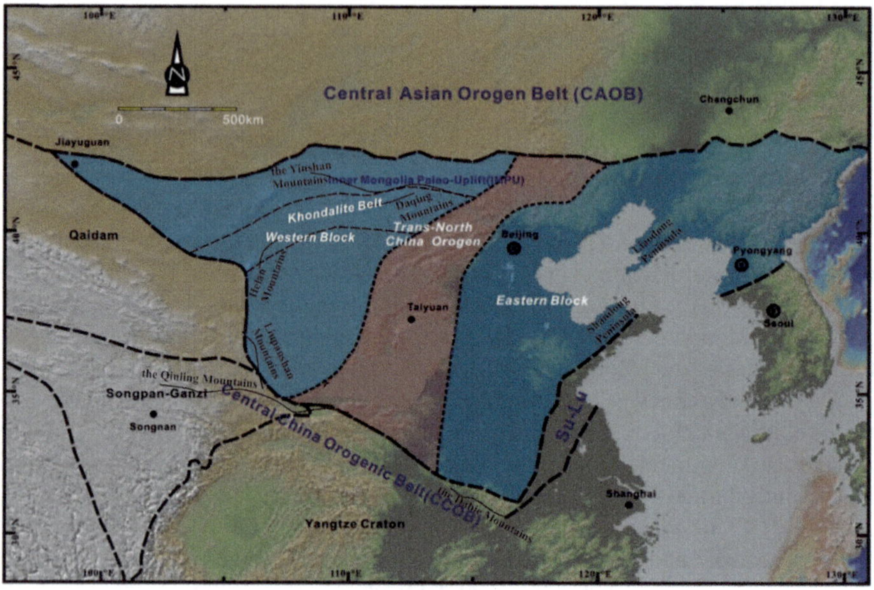

Fig. 2 Tectonic subdivision of the North China Craton (modified after Zhu et al. 2014)

from the Neoproterozoic to the Early Paleozoic. In the Paleozoic, the North China Plate was adjacent to the Paleo-Asian Ocean and the Siberian block in the north, and faced the Yangtze Plate through the Paleo-Qinling Ocean of the northern branch of the Tethys Ocean in the south. During the Hercynian movement in the Late Paleozoic, the North China Plate, the Yangtze Plate and the Siberian Plate were slowly combined, under the background of the adjustment of the global dynamic system. Finally, the Paleo-Asian Ocean and the Paleo-Qinling Ocean were transformed into residual sea basins or residual ocean basins until they disappeared during the Carboniferous or Permian period (Dong 2018).

1.1.2 Stratigraphy

In the middle and late Proterozoic after the integration of the North China Plate (1850–850 Ma), the sedimentary cover of the North China Plate was formed, composed of the Changchengian (upper Paleoproterozic), Jixianian (lower Mesoproterozic) and Qingbaikouan (lower Neoproteroozic) from bottom to top. There is an unconformable contact between the lower Mesoproterozoic (Jixianian) and the overlying Neo-Proterozoic (Qingbaikouan), and an unconformable contact between Neoproterozoic and overlying Lower Paleozoic (Cambrian and Ordovician) offshore marine carbonates. In the middle and late Ordovician, and Early Carboniferous, the North China Plate was continuously uplifted due to the extrusion of oceanic crust on the north and south sides of the plate, and was subjected to weathering and denudation for a long time, forming a regional weathering crust and becoming the basement to the coal-bearing succession from the late Carboniferous to late Permian (Li 2021).

An epicontinental sea was formed over the Ordovician paleo-weathering crust following a widespread transgression in the Late Carboniferous and early Permian. During this time, the entire tectonic plate was located in a low latitude region north of the equator (Fig. 3), with humid climate and prolific vegetation facilitating extensive and thick coal accumulation in coastal and alluvial plain settings. The Late Carboniferous to early Permian is the most important coal accumulation period in the North China Plate. In the middle-late Permian, the sedimentary environment changed to entirely continental, forming river and lake deposits with abundant paleosol nodules, and coal accumulation gradually ended with regional uplift of the plate and transformation of the climate to dry and hot state. Late Permian coal accumulation was restricted in the southern zone of the North China Plate. The Carboniferous and Permian strata in the North China Plate preserve both a record of diverse depositional environments ranging from epicontinental shallow marine to continental settings, and a record of deep time paleoclimate change with a variety of associated geological events. The succession is an important reference point for the study of the Carboniferous and Permian systems in China and elsewhere.

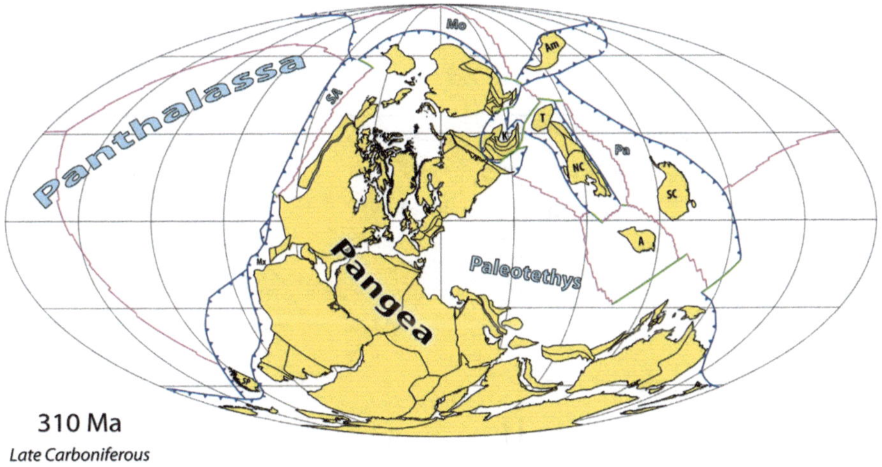

Fig. 3 310 Ma (late Carboniferous) paleogeographic reconstruction showing simplified plate boundaries and labels of some major features (Domeier and Torsvik 2014). A, Annamia; Am, Amuria; K, Kazakhstania; Mx, Mixteca-Oaxacan; NC, North China; SC, South China; SP, South Patagonia; T, Tarim; oceanic domains: Mo, Mongol-Okhotsk Ocean; SA, Slide Mountain-Angayucham Ocean; Pa, Paleoasian Ocean

1.2 Regional Stratigraphy

The Carboniferous–Permian strata of the North China Plate were divided into three stratigraphic sub-regions, namely the northern, central and southern zones, seen in Fig. 4 (Shang 1997). Geographically the northern zone mainly includes Jilin, Liaoning, Hebei provinces, and part of the Inner Mongolia Autonomous Region. The central zone mainly includes part of Inner Mongolia Autonomous Region, Gansu, Ningxia, Shaanxi, Shanxi, Beijing, Hebei and Shandong. The southern zone mainly includes Henan, Jiangsu and Anhui. Although stratigraphic nomenclature for the different regions differs greatly, and the stratigraphic nomenclature for individual regions has also evolved over time, the currently used stratigraphic correlation has become unified (Fig. 5). The Carboniferous–Permian succession of the North China Plate can be correlated to the Pennsylvanian to Lopingian in the International Chronostratigraphic Chart (Cohen et al. 2013).

The principal formations of Carboniferous–Permian succession, including Benxi, Taiyuan, Shanxi, Xiashihezi, Shangshihezi, and Sunjiagou Formations, are described below, in ascending stratigraphic order. The mark layers of the succession are shown in Fig. 6.

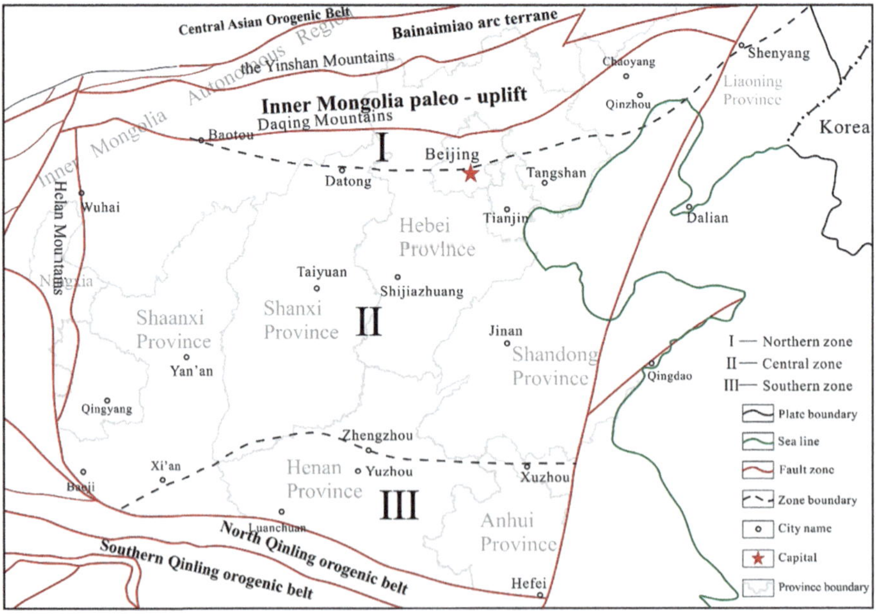

Fig. 4 Stratigraphic zonation of Carboniferous–Permian in North China Plate (modified after Shang 1997). I—northern zone; II—central zone; III—southern zone

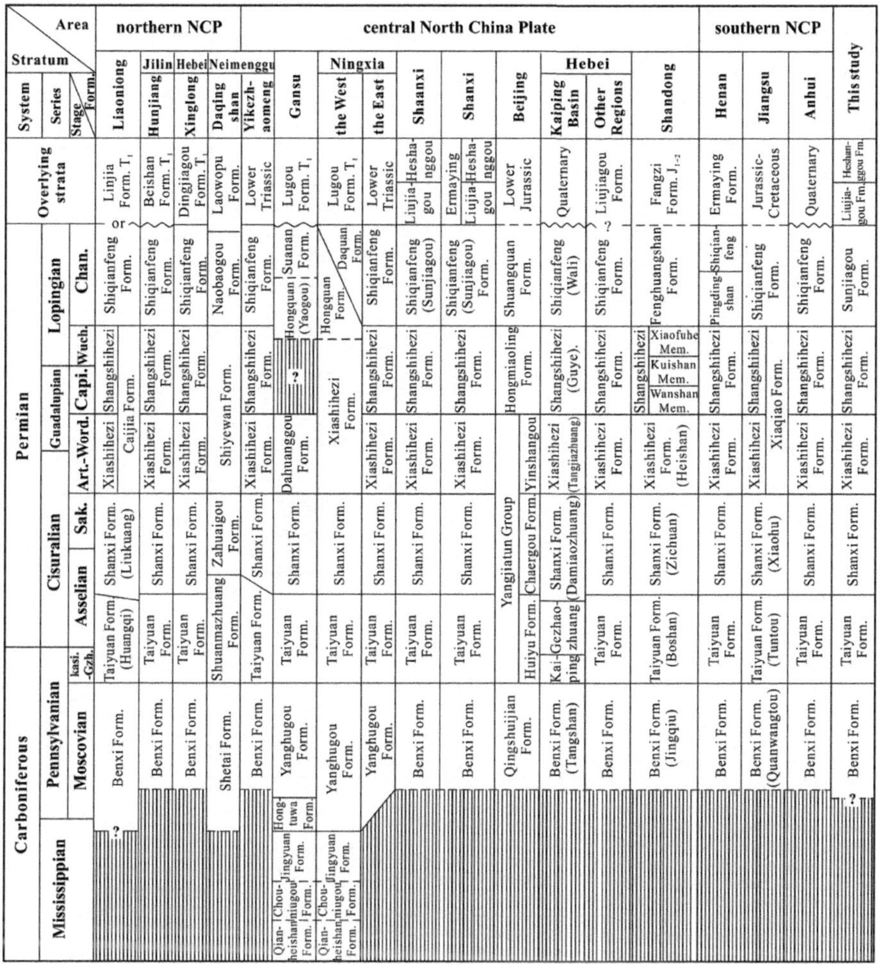

Fig. 5 Stratigraphic division and correlation of Carboniferous–Permian in North China Plate (revised after Li 2012; Cohen et al. 2013)

(1) Benxi Formation

The Benxi Formation corresponds to the Pennsylvanian Subseries, unconformably overlies Ordovician limestone at its base, and is conformably overlain by the Taiyuan Formation. The formation is developed in almost the entire area of the North China Plate, and its thickness varies greatly. Only discontinuous coal traces or thin seams are developed in this formation.

In the northeastern region of the North China Plate, a typical section is located in the Niumaoling area of Benxi, Liaoning Province, and is characterized by rhythmic alternation of (1) purple mudstone (Ferralic horizon) and G-layer bauxite developed at the base and (2) mudstone, limestone, sandstone, thin coal seams or carbonaceous

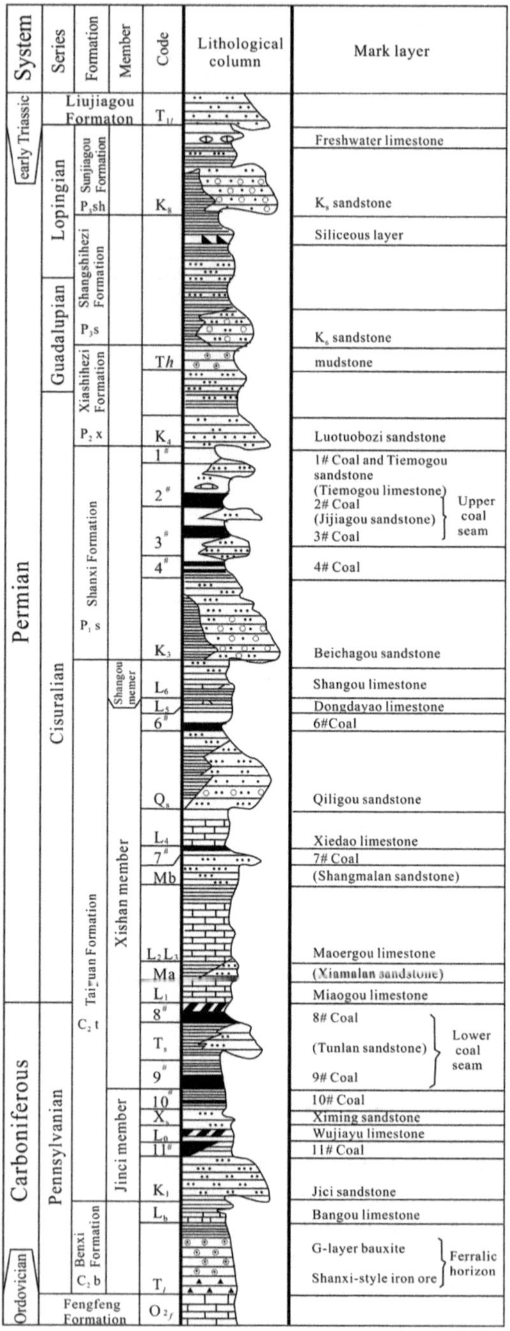

Fig. 6 Schematic diagram showing Late Paleozoic stratigraphic succession with typical marker layers in Shanxi Province (modified after Cheng 1992; Dong 2018)

mudstone at the top. Two fossil horizons preserving *Fusulina-Fusulinella* assemblage and *Eostaffella-subsolana* assemblage have been recognised in this area, and marine fossils such as crinoids, brachiopods, bivalves, and gastropods can be seen, while plant fossils are rare. In places such as Gansu and Ningxia provinces in western North China Plate, this unit is named the Hongtuwa Formation or the Yanghugou Formation. The Yanghugou Formation in Ningxia contains marine invertebrate fossils such as brachiopods, bivalves, gastropods and also some plant fossils. In the central zone of the North China Plate, the Benxi Formation is characterised by a set of sedimentary sequences, with Shanxi-style iron ores, claystone and G-layer bauxite in the lower part, and sandstone, mudstone, tuff and coal in the upper part (Cheng 1992). There are *Fusulina-Fusulinella* fossil assemblage in this area, as well as marine invertebrate fossils such as gastropods, bivalves, brachiopods and bryozoans, with abundant plant fossils.

(2) Taiyuan Formation

The Taiyuan Formation is one of the major coal-bearing units of North China, corresponding to the lower Pennsylvanian Subseries to the Upper Cisuralian Series. The base is a conformable contact with the Benxi Formation. In the south and parts of the northwest, the strata of the Taiyuan Formation directly overlie the Ordovician, Cambrian or Precambrian. The top of the Taiyuan Formation is a conformable contact with the overlying Shanxi Formation. The total stratal thickness in the east is greater than that in the west.

The Taiyuan Formation in the northern zone of North China Plate consists of sandstones, mudstones, coals and limestones, while in the central zone the deposits are mainly mudstones and sandstones, interbedded with thin layers of limestones and coal beds. The standard section of the central zone of the North China Plate, located in Xishan, Taiyuan, is composed of sandstones, mudstones, limestones and coal, with the Bangou Limestone taken as the basal boundary and the Dongdayao Limestone as the top boundary. From the north to the south of the area, the thickness of the carbonate rocks gradually increases, while the thickness of the coal seams decreases, accompanied by the transformation of sedimentary environment from terrestrial, through coastal, to shallow marine. The formation contains fossils of fusulinids, gastropods, echinoderms, brachiopods and abundant trace fossils of *Zoophycos*, *Planolites*, *Nereites*, *Teichichnus* and *Rhizocorallium* in the western part of Henan Province (Guo et al. 1991) and the southeastern part of Shanxi Province (Cheng 1992).

(3) Shanxi Formation

The Shanxi Formation is one of the major coal-bearing units in North China, corresponding to the Central Cisuralian Series. This formation has conformable contacts with underlying Taiyuan Formation and overlying Xiashihezi Formation. Its thickness varies from greatest in the east and north to least in the west and south, within the North China Plate.

In the northern zone of the North China Plate, sedimentary sequences consisting of sandstones, mudstones and coals, representing fluvial and deltaic environments, are mainly developed. In the central zone of the region, paleogeographical zonation is evident, with a transition from a fluvial clastic-dominated facies association to a deltaic coal-bearing clastic-dominated facies association, interbedded with several thin marine carbonate intervals. The southern zone of the plate is characterised by carbonate rocks, mudstones and coal seams at the base, and mudstones and coal seams in the upper part. The thicknesses of coal seams decreases from north to south. The Yuzhou area in Henan Province is rich in fossil plants such as *Gigantopterids* and *Lobatannularia* (Shang 1997; Lu et al. 2020).

(4) Xiashihezi Formation

The Xiashihezi (Lower Shizhezi) Formation is the most economically important coal-bearing unit in the southern zone of the North China Plate, corresponding to the Upper Cisuralian Series to the Lower Guadalupian Series. This formation has conformable contacts with underlying Shanxi Formation and overlying Shangshihezi Formation. It varies in thickness from greatest in the north to least in the south.

The lithology of the Xiashihezi Formation is highly variable across the area, and its type section is located in the Shihezigou of Dongshan area (eastern mountains) of Taiyuan city (Kong et al. 1996). The sedimentary sequence in the northern zone of the North China Plate is characterised by coarse clastic deposits such as conglomerate, coarse-grained and medium-grained sandstone, while in the central zone it is composed of interbedded sand and mudstone, and in the southern zone is dominated by mudstone interbedded with thin sandstones. The rock colour also varies, with yellow-green and purple predominating in the north-central region and grey in the south. A layer of bauxite is commonly developed at the top of the formation, which is named differently in different areas, such as A-layer bauxite in Jilin, Liaoning and the southern foot of Yanshan Mountains in Hebei, Taohua mudstone in Shanxi and the eastern foot of Taihang Mountains in Hebei, B-layer bauxite in Shandong province, Dazi mudstone in Henan, and purple bauxitic mudstone in Huainan and Huabei. The thickness of coal seams gradually increases from north to south. Only a few, thin coal beds are developed in the middle and northern zones of the plate, and the thickness of coal seams is the largest in Lianghuai and southeast Henan province. The brachiopod *Lingula* and plant fossils of the Middle Cathaysian Flora were found in this unit.

(5) Shangshihezi Formation

The Shangshihezi (Upper Shizhezi) Formation corresponds to the Upper Guadalupian to the Lower Lopingian (Wang et al. 2014). This formation is in conformable contact with the underlying Xiashihezi Formation and with the overlying Sunjiagou Formation. The formation thickness is generally greater in the east and less in the west. This formation doesn't contain coal seams in central and northern zones of North China Plate, although it is one of the main coal-bearing strata in the southern zone, mainly in Henan and Anhui provinces (Shang 1997; Shao et al. 2014 and 2018).

The formation has a pronounced paleogeographical zonation across the region, with a type section located in the Shihezigou in Dongshan, Taiyuan City. In the northern and central zones of the North China Plate, the formation is characterized by a suite of purplish-red, mottled or yellow-green, coal-free nonmarine clastic rocks, interspersed with several layers of siliceous mudstones. The southern zone is characterized by a set of coal-bearing clastic facies of broadly deltaic origin. Moving south, the color of the rocks changes from purple to yellow-green, while the grain-size gradually becomes finer. The thin coal seams were mainly restricted in the southern zone of the North China Plate, with the thickness gradually thickening from north to south. The brachiopod *Lingula* and the plant fossils of the late Cathaysian flora were found in some areas (Shang 1997).

(6) Sunjiagou Formation

The Sunjiagou Formation corresponds to the Upper Lopingian Series. This formation is in conformable contact with the underlying Shangshihezi Formation and with the overlying Triassic Liujiagou Formation. The stratigraphic thickness is highly variable and can be greater than 1000 m in Huainan and Huabei.

This formation is characterized by a series of purplish red, sandy matrix conglomerates, and mudstones of fluvial origin, intercalated with lacustrine, argillaceous limestone and thin layers of gypsum. Southward, the color of this rocks changes from purple to color-mottled. There is no coal in the Sunjiagou Formation throughout the entire North China Plate. The plant assemblage *Ullmannia bronnii–Yuania magnifolia* is found in Sunjiagou Formation, and ostracod fossils are occasionally found in the argillaceous limestones of this formation in the southern North China region.

1.3 Tectonics and Magmatism

The North China Plate gradually entered the development stage of plate tectonics in the Meso-Neoproterozoic. During the Neoproterozoic to early Paleozoic, the stage of North China Plate changed from plate deformation to fixed plate position. At the end of that period, the North China Plate faced the Paleo-Asian Ocean in the north, the Siberian block in the east, the Paleo-Qinling Ocean (the northern branch of the Tethys Ocean) in the south, and then faced the Yangtze block across the ocean in the south (Fig. 1). During the late Paleozoic, the global plate tectonics was dominated by Hercynian movement (Fig. 7), and the North China block began to merge with the Yangtze block in the South and the Siberian block in the north; the Paleo-Asian Ocean and the Paleo-Qinling Ocean gradually evolved into residual ocean basins and eventually disappeared in the Carboniferous and Permian, respectively (Chen et al. 2000).

The growth of Late Paleozoic sedimentary basins in North China was controlled by the construction of orogenic belts on its northern and southern sides. After the Caledonian movement during early Paleozoic in the northern margin of the North China Plate, an island arc chain-accretionary fold zone was formed in the northern margin. The paleo-Asian Ocean was subducted and compressed the North China

Stratigraphy			Mark	Tectonic evolution stage
Erathem	System	Series		
Cenozoic	Quatrnary		Q	
	Neogene		N	Himalayan tectonic evolution stage
	Paleogene		E	
Mesozoic	Cretaceous	Upper	K_2	Yanshan tectonic evolution stage
		Lower	K_1	
	Jurassic	Upper	J_3	
		Middle	J_2	
		Lower	J_1	
	Triassic	Upper	T_3	Indosinian tectonic evolution stage
		Middle	T_2	
		Lower	T_1	
Palezoic	Permian	Upper	P_3	Hercynian tectonic evolution stage
		Middle	P_2	
		Lower	P_1	
	Carbonife-rous	Upper	C_2	
		Lower	C_1	
	Silurian and Devonian		S-D	
	Ordovician	Upper	O_3	Caledonian tectonic evolution stage
		Middle	O_2	
		Lower	O_1	
	Cambrian	Upper	$Є_3$	
		Middle	$Є_2$	
		Lower	$Є_1$	
Neoproterozoic			Pt_3	Jinning tectonic evolution stage
Paleo- to Meso-proterozoic			Pt_{1-2}	

Fig. 7 Tectonic evolution of North China Plate (modified after Liang 2007)

Plate, collided during the late Devonian to the Early Carboniferous, and forming the Hercynian accretionary fold belt (Shao 1991; Tang 1992). The northern zone of the North China Plate is in a compressional depression tectonic state, with the nature of a foreland basin (Meng and Ge 2001). In the late Paleozoic, the southern side of the north China plate was affected by a collision orogeny process. In the early Permian, the southern margin of the North China Plate had changed from a tectonic highland to an epicontinental sea sedimentary basin, and the North Qinling belt had also begun to rise and gradually transformed into a provenance area to supply clastic materials to the southern zone of the North China Plate. In the middle and Late Permian, due to the collision, compression and uplift on both sides of the plate, the sea gradually and finally completely withdrew from the North China Plate, leading to a full transformation into a continental sedimentary environment. The transformation of the tectonic system from the Late Carboniferous to the Early Permian caused a "tilting" tectonic movement in the North China Plate, and the tectonic paleogeographic pattern changed from "southern uplift and northward tilt" to "northern uplift and southward tilt" (Shang 1997; Chen et al. 2000).

Magmatism occurred in North China from Archean and Proterozoic, through the Paleozoic, Mesozoic and Cenozoic. The North China Plate experienced strong lithospheric thinning, magmatism and mineralization during the Mesozoic and Cenozoic (Mao et al. 2003; Zhao et al. 2004; Zhai 2019). In recent years, late Paleozoic to early Mesozoic intrusions have been found in North China (Luo et al. 2001; Ma et al. 2004; Wang et al. 2007). During the late Paleozoic to early Mesozoic, there were at least four distinct magmatic episodes: Devonian, late Early Carboniferous, middle Permian, and late Permian to Triassic. The Devonian magmatism was probably related to the Bainaimiao arc belt and extension after arc continent collision in North China Craton, and its time range was about 400–360 Ma. The lithology is mainly alkaline syenite and monzonite, followed by basic to ultrabasic rocks, monzodiorite, alkaline granite and rhyolite, and these rocks are revealed by limited outcrops. The late Early Carboniferous to middle Permian magmatism was related to the subduction of the paleo Asian Ocean to the North China Plate, the time range of which was about 330–265 Ma. The lithologies are mainly diorite, quartz diorite, granodiorite and granite, with lesser gabbro and tonalite. The Late Permian to Triassic magmatism has a time limit from 250 to 200 Ma. The lithologies are mainly potassium feldspar granite, monzogranite and complexes of alkaline rocks, followed by basic—ultrabasic rocks and a small amount of intermediate-felsic volcanic rocks (Zhang et al. 2010).

1.4 Lithofacies and Paleogeography

1.4.1 Facies and Facies Associations

Based on the borehole cores and outcrops data, 14 facies were recognized in the Carboniferous–Permian strata, with lithologies including sandstone, mudstone, coal, and limestone, and these were arranged into 5 associations, namely (A) alluvial plain, (B) delta plain, (C) delta front and pro-delta, (D) barrier coast, and (E) offshore marine. Detailed descriptions and interpretations of these facies are summarized in Table 1.

Facies association A includes alluvial plain channel (Facies 1), floodplain (Facies 2) and incised valley fill (Facies 3) deposits. Facies 1 is characterized by fining-upward succession of conglomerate, coarse- to medium-grained sandstone with trough cross-bedding interbedded with thin partings of mudstone with ripple cross-lamination. Facies 2 is composed of thick-bedded mudstone with purple-red mottling interbedded with thin sandstone partings. Facies 3 is characterized by erosionally-based thick multi-storey sandstone with trough cross-bedding interbedded with siltstone.

Facies association B includes distributary channel (Facies 4), levee (Facies 5), coastal floodbasin (Facies 6) and coastal mire (Facies 7) deposits. Facies 4 is characterized by fine-grained, and grayish sandstone bodies with coaly traces and mudstone partings with ripple cross-lamination. Facies 5 is characterized by regularly interbedded sandstone and laminated mudstone. Facies 6 is characterized by thick-bedded mudstone deposits with rhythmic bedding, wavy bedding and flaser bedding or horizontal lamination and containing plant leaf fossils, the inarticulate brachiopod *Lingula* and sponge spicules. Facies 7 is dominated by coal and carbonaceous mudstone.

Facies association C includes delta front (Facies 8) and pro-delta (Facies 9) deposits. Facies association C comprises a coarsening-upward succession of pinstripe laminated mudstone (Facies 9) with rhythmic bedding, fine-grained sandstone with wavy and flaser bedding, and fine- to medium-grained sandstone with coaly traces (Facies 8).

Facies association D includes tidal sand bar (Facies 10), tidal channel (Facies 11), and lagoon (Facies 12) deposits. Facies 10 is characterized by Coarsening-upward sandstone, gravelly coarse sandstone and conglomerate. Facies 11 is characterized by erosionally-based sandstone with bidirectional/bimodal cross-bedding. Facies 12 is dominated by thick dark grey mudstone with horizontal lamination, plant root and siderite concretion.

Facies association E includes transgressive, offshore black marine mudstone (Facies 13) and clear-water, offshore marine carbonates (Facies 14). Facies 13 is composed of a succession of dark grey mudstone containing bivalve, brachiopod, and gastropod shells. Facies 14 comprises bioclastic limestone with abundant fossils including corals, foraminifers, crinoids, brachiopods, and bryozoans, and trace fossils identified in western Henan Province including *Zoophycos*, *Nereites*, *Ophiomorpha*, *Rhizocorallium*, and *Teichichnus* (Hu et al. 2015).

Table 1 Summary of the description and interpretation of the facies associations for the upper Carboniferous–Permian strata in the North China Plate

Facies association/ facies	Interpretation	Lithology	Sedimentary structure/fossils
A	Alluvial plain		
1	Alluvial plain channel	5–17 m thick fine-upward succession of erosionally-based conglomerate, coarse- to medium-grained sandstone interbedded with thin partings of mudstone	Trough cross-bedding and ripple cross-lamination
2	Floodplain	4–40 m thick succession of mudstone with purple-red mottling interbedded with thin sandstone partings	Ripple cross-lamination, horizontal lamination, siderite concretion, pedogenic nodular carbonate, and nodular gypsum
3	Incised valley fill	4–10 m erosionally-based thick multi-storey sandstone interbedded with siltstone	Trough cross-bedding, load casts, low angle planar lamination, horizontal lamination, clast imbrication, plant debris, coaly traces, trace fossils (*Diplocraterion*, *Thalassinoides*)
B	Delta plain		
4	Distributary channel	3–10 m thick fining-upward succession of single-story, erosionally-based medium- to fine-grained sandstone and mudstone	Trough cross-bedding, ripple cross-lamination, coaly traces
5	Levee	1–10 m thick successions of regularly interbedded sandstone and laminated mudstone	Ripple cross-lamination, horizontal lamination, muscovite fragments, plant debris

(continued)

Table 1 (continued)

Facies association/ facies	Interpretation	Lithology	Sedimentary structure/fossils
6	Coastal floodbasin	Thick-bedded mudstone interbedded with thin-bedded fine-grained sandstone	Rhythmic bedding, wavy and flaser bedding, ripple cross-lamination, horizontal lamination, coaly traces, and siderite concretion, plant leaf *Lepidodendron*, *Pecopteris*, *Calamites*, *Taeniopteris*, *Cordaites* fragments, *Lingula*, and sponge spicules (Ma 2013)
7	Coastal mire	0.2–19.9 m thick coal and carbonaceous mudstone	Pyrite staining
C	Delta front and pro-delta		
8	Delta front	Coarsening-upward succession composed of mudstone, fine- to medium-grained sandstone deposits	Flaser and wavy bedding, rhythmic bedding, convolute bedding, trough cross-bedding, ripple cross-lamination, and coaly traces
9	Pro-delta	Pinstripe laminated mudstone	Siderite concretion, plant debris, and pyrite staining
D	Barrier coast		
10	Tidal sand bar	Coarsening-upward sandstone, gravelly coarse sandstone and conglomerate	Tabular cross-bedding, trough cross-bedding, bidirectional/bimodal cross-bedding
11	Tidal channel	Erosionally-based sandstone	Tabular cross-bedding and bidirectional/bimodal cross-bedding
12	Lagoon	Dark grey mudstone	Horizontal lamination, plant root and siderite concretion
E	Offshore marine		
13	Offshore black marine mudstone	Dark grey mudstone	Bivalve, brachiopod, and gastropod shell
14	Clear-water offshore marine carbonates	0.6–9 m thick bioclastic limestone	Corals, foraminifers, crinoids, brachiopods, and bryozoans, trace fossils: *Zoophycos*, *Nereites*, *Ophiomorpha*, *Rhizocorallium*, and *Teichichnus* identified in western Henan province (Hu et al. 2015)

1.4.2 Sequence Stratigraphic Framework

According to the identification of the sequence stratigraphic boundary of Carboniferous–Permian in North China, combined with the previous research of sequence stratigraphy, the Carboniferous and Permian succession as a whole that which corresponds to a first order sequence. During the Late Carboniferous, the entire basin was mainly epicontinental sea and barrier coastal system, and in the early Permian, it was dominated by river, delta, and lake sedimentary system. Over time, an overall regressive, progradational trend is preserved. According to the regional unconformity surface and regional tectonic transformation surface, the whole succession is subdivided into two second-order sequences: The earlier second-order sequence extends from the Late Carboniferous to the early Early Permian (corresponding to the top of the Taiyuan Formation). The base of this sequence was paleo-weathering surface from which the epicontinental sea sediments started to develop and then turned into a river, delta, and lake sediments. This second-order sequence includes 2 third-order sequences, SQ1 and SQ2 (Fig. 8). The later second-order sequence extended from the early Permian (base of the Shanxi Formation) to the late Permian (top of the Sunjiagou Formation). This second-order sequence can be further subdivided into 5 third-order sequences, including SQ3, SQ3, SQ5, SQ6, and SQ7 in ascending order (Fig. 8).

The first second-order sequence contains two third-order sequences, labelled SQ1 and SQ2 in Fig. 8. The base of SQ1 is marked by the angular unconformity between the underlying Middle Ordovician (Majiagou or Fengfeng Formation) and the overlying Upper Carboniferous (Benxi Formation), and the top boundary of SQ1 is marked by the paleosol horizon beneath a regionally correlatable coal seam known as the No.9 coal seam in the Taiyuan Xishan coalfield, the No.1 coal seam in the Henan area, and the No.17 coal seam in the Shandong area. The age range of SQ1 is from Moscovian to Gzhelian (Late Carboniferous). SQ2 extends from the top boundary of SQ1, representing a change to a transgressive trend (Shang 1997), to the base Beichagou sandstone of the Shanxi Formation, and the age range is from Asselian to the late Artinskian (early Permian).

The second second-order sequence contains 5 third-order sequences labelled SQ3 to SQ7 (Fig. 8). SQ3 is equivalent to the regionally widespread Shanxi Formation. The basal boundary of SQ3 is marked by the erosional bases of the fluvial-channel or distributary-channel sandstones ("Beichagou Sandstones"), and the top boundary is represented by the erosional bases of regionally-correlatable sandstone formations ("Luotuobozi Sandstones") of the Xiashihezi Formation (Fig. 8). The age roughly ranged from the late Artinskian to the late Kungurian. SQ4 is equivalent to the Xiashihezi Formation and spans from the base of the Luotuobozi Sandstones to the top of the Taohua Mudstone (a regionally correlatable reddish mottled mudstone). The age range of SQ4 is from late Kungurian to early Wordian. SQ5 is equivalent to the lower part of the Shangshihezi Formation, spanning from the base of this formation to the base of thick-bedded sandstones in the lower middle part of the formation (e.g. the K7 Sandstones in the Taiyuan Xishan coalfield), and the age range is equivalent to the late Wordian and Capitanian of the middle Permian. SQ6

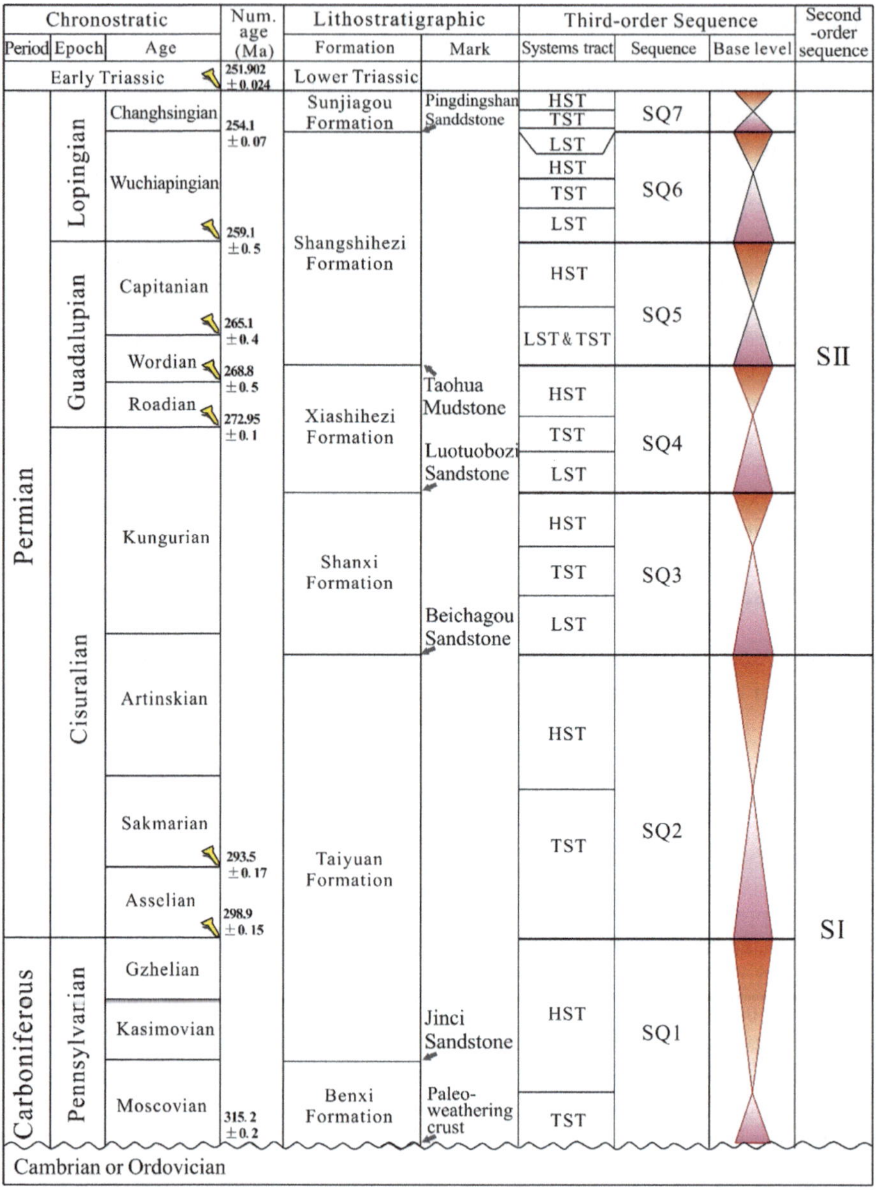

Fig. 8 Sequence stratigraphic framework for Carboniferous–Permian coal-bearing strata in the North China coalfield. The upward-facing triangle represents the base-level rise within the third-order sequence, and the downward-facing triangle represents the base-level fall within the third-order sequence (revised after Cohen et al. 2013; The Stratigraphic Chart of China 2014; Shao et al. 2020)

is equivalent to the middle-upper part of the Shangshihezi Formation and is defined at its base by a thick-bedded sandstone and its top by the base of the Sunjiagou Formation, and the age range is within the Wuchiapingian stage. SQ7 is equivalent to the or Sunjiagou Formation, and its top corresponds to the P–Tr boundary. The age range of SQ7 is equivalent to the Changhsingian stage.

1.4.3 Facies Composition and Paleogeography of Sequences

(1) Sequence SQ1

The bottom surface of SQ1 corresponds to the weathering surface at the top of Ordovician and the basal contact of the Benxi Formation. The top surface of SQ1 corresponds to the No.9 coal seam in Taiyuan Xishan, the base of No.1 coal seam in southern North China, and the bottom of No.17 coal seam in the Shandong region. The time range of SQ1 corresponds to Moscovian to Gzhelian. The total stratal thickness of SQ1 is generally no more than 100 m, and the maximum thickness is more than 130 m only near the Fuzhou Bay section on Liaodong Peninsula. Overall, it is thicker in north central region and thinner toward to north and south edges (Fig. 9). The main sedimentary centers are distributed in Fuzhou Bay in East Liaoning, Northwest Ordos Basin, Beijing, northern Lishi in Shanxi, eastern Henan, and western Anhui. Sandstone is not well developed in SQ1. There are two areas where thick sandstones are recorded: (1) generally 37–70 m in the northwest of the Ordos Basin, almost all of which is sandstone with subordinate conglomerate, and (2) 10–33 m in West Beijing. The total thickness of sandstone in other areas is mostly <20 m. However, the thickness of mudstone is <125 m, and the average mudstone thickness over the whole region is 25 m. The thickness of mudstone in the Pingdingshan, Luoyang, and Sanmenxia areas of the western Henan and the Weibei coalfields in the south of the Ordos Basin is significantly lower than the average level. The high limestone content is mainly concentrated in the western and eastern parts of the North China Plate, which could up to 60%. The limestone content in the central part of North China is generally less than 20%, while the limestone content in the northern margin of the basin is less than 5% (Fig. 10). In this period, the coal accumulation was poor, and five coal accumulation centers were developed: (1) Tongchuan-Hancheng, (2) Jungar Banner-Taiyuan, (3) Zhoukou-Huaibei, (4) South Yulin, and (5) Etuoke Banner (Fig. 11).

During SQ1 period of the Late Carboniferous, the northern zone of the North China Basin was dominated by a fluvially dominated delta complex, which developed delta plain facies and delta front facies and transitioned distally into a shallow marine environment to the south. Several tidal sand bars developed between the tidal flat and offshore carbonate shelf, which played a certain barrier role (Fig. 12). The alluvial fan sedimentary system was developed in the Chengde area in the north. Shizuishan and Yinchuan in the West were dominated by offshore carbonate shelf sedimentary system. Due to the influence of the central uplift of the Ordos Basin, the carbonate offshore marine in the south of the Uxin Banner (County) and the offshore carbonate

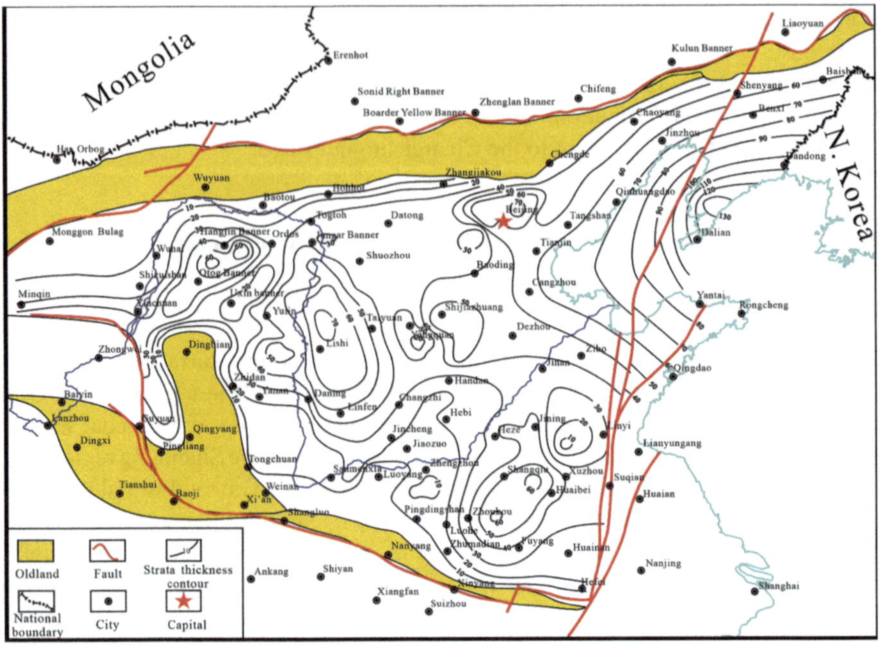

Fig. 9 Strata thickness contour map of SQ1 in North China Plate

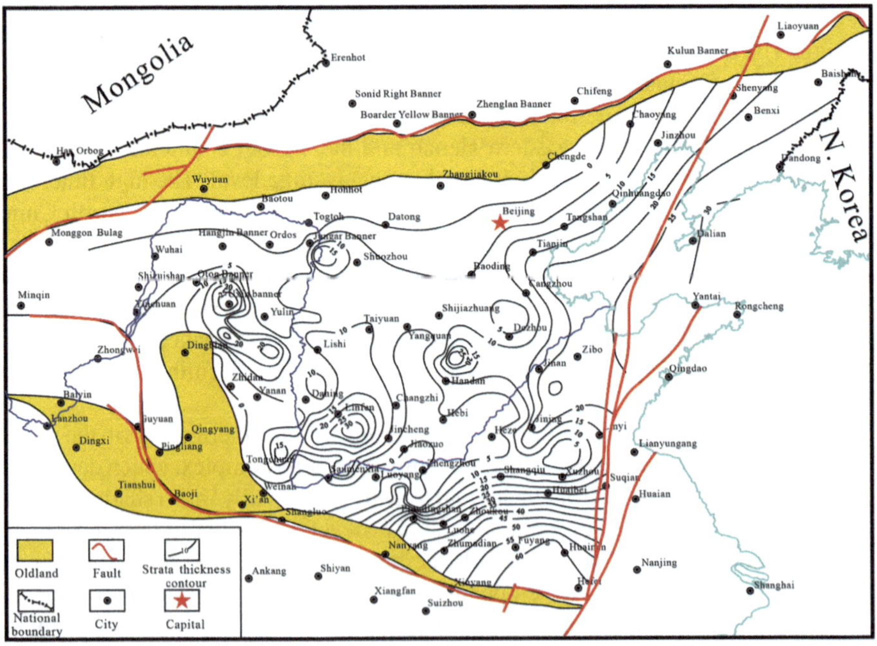

Fig. 10 Limestone percentage content contour map of SQ1 in North China Plate

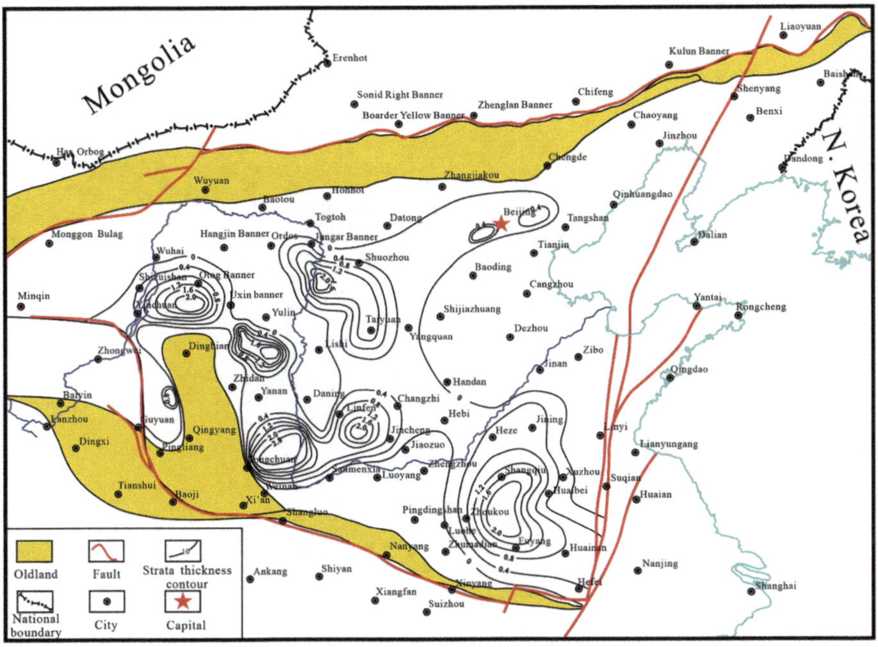

Fig. 11 Coal thickness contour map of SQ1 in North China Plate

shelf in the west edge were separated by a tidal flat zone. The Liaodong Peninsula in northeastern North China was dominated by a offshore carbonate shelf sedimentary system, which graded into the tidal flat and fluvially dominated delta complex to the north. In the southern part of North China, Eastern Henan and southern Anhui were also dominated by a offshore carbonate shelf sedimentary system and transitioned to the lagoon-tidal flat sedimentary system to the northwest.

(2) Sequence SQ2

SQ2 incorporates the strata between the top surface of SQ1 and the basal boundary of the Beichagou sandstone (Shanxi Formation), and its time range is Asselian to the late Artinskian. The total stratal thickness of SQ2 is thicker than SQ1, ranging from 10 to 200 m, with an average value of about 80 m. The thickness near Tianjin is more than 200 m. The total stratal thickness in the east is greater than that in the west. The sequence thickness is greatest in Beijing, Tianjin, Tangshan, Cangzhou, Shijiazhuang, Linyi, Jining and Heze, with an average value of 129 m. Due to the continuous strengthening of transgression in the Western Ordos Basin, the blocking effect of the central uplift in the central region on seawater was gradually weakened, leading to uniformity in facies on both sides of the central ridge. Similarly, the sequence thickness on both sides also tends to be consistent, generally 20–100 m, with an average of 50 m, which is thicker than that in the middle of Ordos Basin. In addition, the stratal thickness in Shanxi and Western Hebei, in the middle of the

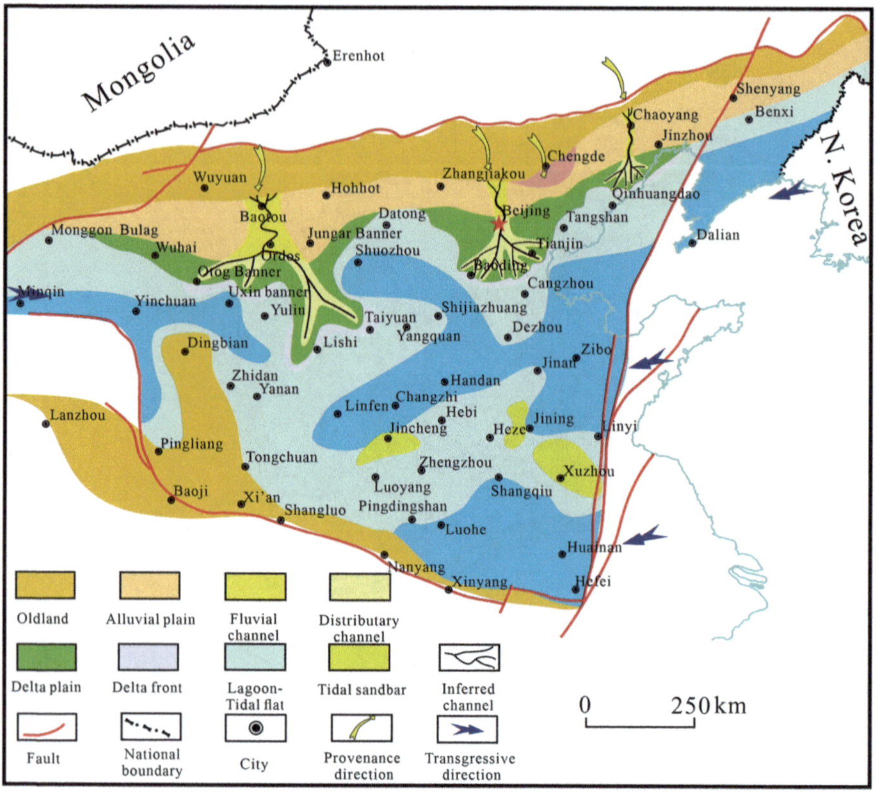

Fig. 12 Lithofacies paleogeography of SQ1 in North China Plate

basin, is also relatively uniform, within the range 25–115 m, with an average of 79 m. It can be seen that there are two depocenters in North China Basin during formation of SQ2, one near the belt of Beijing-Tianjin-Tangshan, Cangzhou and Shijiazhuang, while the other is the belt of southwest Shandong to North Anhui (Fig. 13). In the SQ2, sandstone is a subordinate constituent. There are two principal high-value areas: Hangjin Banner and Jungar Banner in the north of Ordos Basin with 30–40 m, Chengde and West Beijing with 20–95 m. The sandstone thickness in other areas is relatively small. Comparatively, the thickness of mudstone is relatively high, generally <130 m. The depocenters during formation of SQ2 included Dezhou (the total thickness of mudstone 130 m) and southwest Shandong (the average thickness 101 m). The limestone content shows a decreasing trend from south to north. The limestone content in the northern part of the North China is generally less than 5% (Fig. 14), indicating that the transgression occurred in the southeast, while the sediments in the north are dominated by coarse-grained terrigenous debris, which is the main provenance area. In the SQ2 period, the coal accumulation in North China Plate was strong, and thicker coal seams was formed in the whole area. The total thickness of coal seams in the whole area is 0–38m, with the average thickness

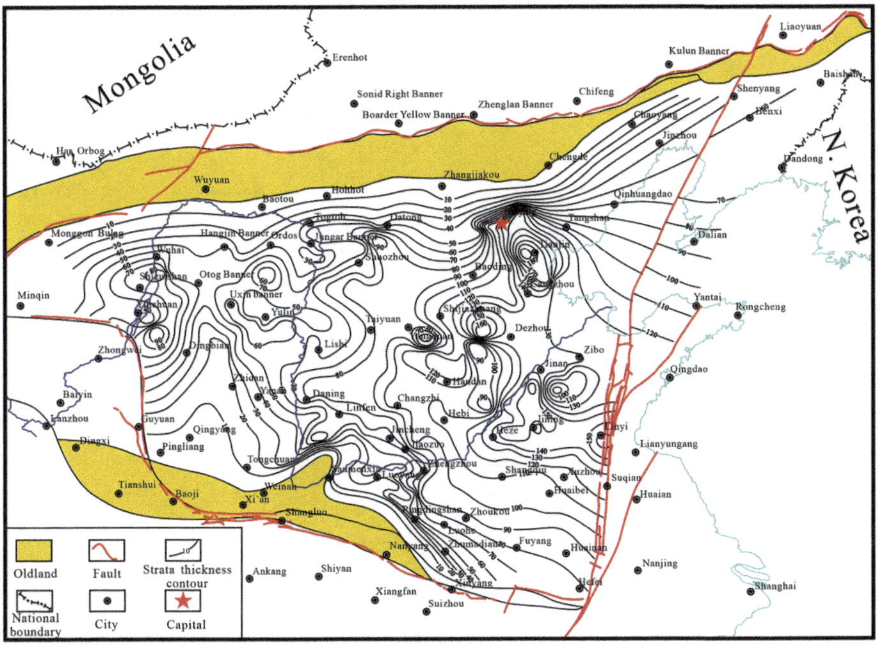

Fig. 13 Strata thickness contour map of SQ2 in North China Plate

of 7.6m. The average total thickness of coal seams in the north is 10m, which is obviously higher than that of the southern area. The thickest coal seam is located at Jungar Banner, followed by Datong, Pingshuo, Beijing, and Tianjin-Cangzhou area (Fig. 15).

During the early Permian formation of SQ2, the northern zone of the North China Basin was dominated by a fluvially-dominated delta complex, transiting to the lagoon-tidal flat sedimentary system southward, and several tidal sand bars were developed. Offshore carbonate shelf sedimentary systems were widely developed in the central and southern regions. Shizuishan, Yinchuan, and other places in the West study area were still dominated by a offshore carbonate shelf system. The Liaodong Peninsula in northeastern North China was dominated by a offshore carbonate shelf, which transited to lagoon-tidal flat and delta environments to the north. At this time, the Zhongtiao massif in the Qinling Mountains had begun to provide a modest amount of sediment. Fluvially dominated deltas were developed in Pingliang, Tongchuan and Sanmenxia at the south margin of the Ordos Basin (Fig. 16).

(3) Sequence SQ3

The basal surface of SQ3 is the base of the Beichagou sandstone, and the top surface is the base of the Luotuobozi sandstone. The sequence records from the late Artinskian to the late Kungurian stage. The total stratal thickness of SQ3 is greater than SQ2, generally 25–220 m, and the average thickness is about 82 m. The contour

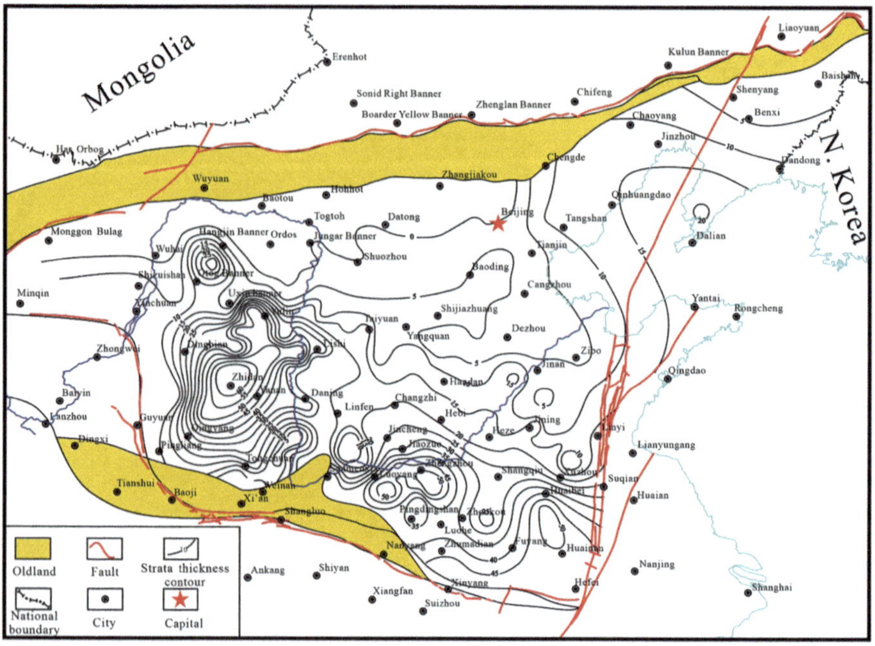

Fig. 14 Limestone percentage content contour map of SQ2 in North China Plate

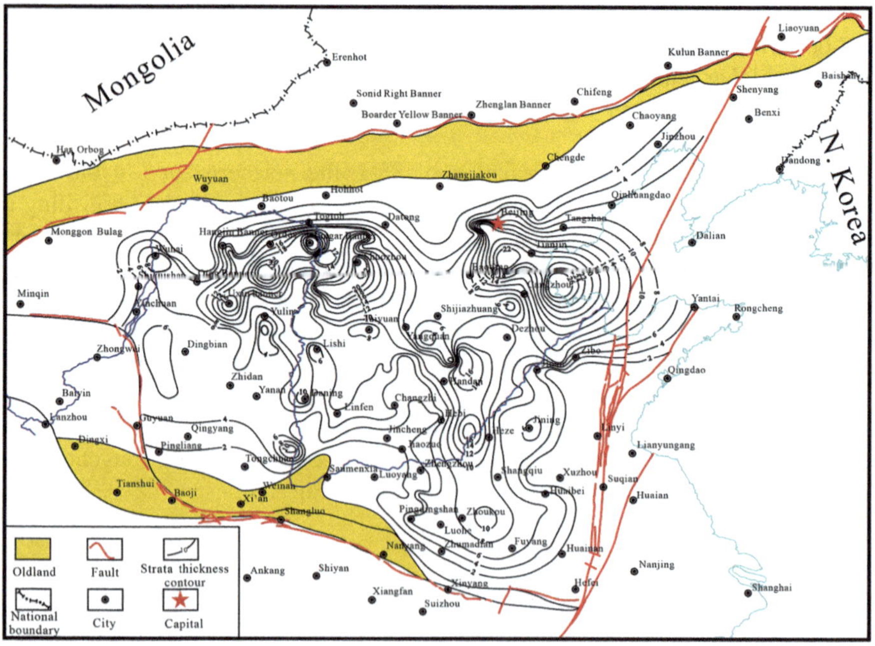

Fig. 15 Coal thickness contour map of SQ2 in North China Plate

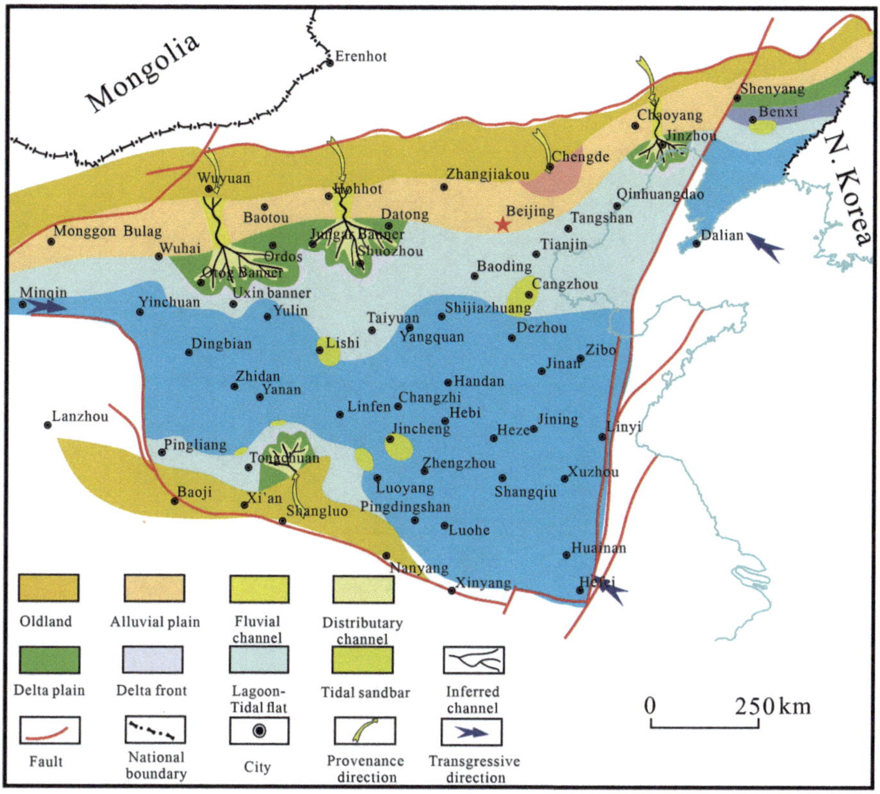

Fig. 16 Lithofacies paleogeography of SQ2 in North China Plate

map of stratal thickness reflects three depocenters: the first is the line from Baodi in Tianjin, Cangzhou and Bohai Bay to Hebi-Anyang-Puyang, with a total stratal thickness of 110–160 m. The second is a belt joining Tai'an-Jining-Linyi-Xuzhou, where the total thickness is 110–150 m. The third is the belt connecting Yulin-Dingbian-Zhidan-Yan'an-Daning-Liulin in the Ordos Basin, with a total thickness of 100–150 m (Fig. 17). The overall trend is that the succession is thicker in the east and west, whereas the central zone is characterized by a relatively thin succession in most areas of Shanxi (except in the Hedong coalfield). Sandstone is well developed, with a thickness range of 10–217 m. Except for western Henan, and the zone connecting Daning-Yan'an-Zhidan-Dingbian-Jixian-Qingyang. The thickness of mudstone is also relatively high, ranging from 5 to 160 m, with an average thickness of 43 m. The coal accumulation was still higher in the SQ3 period. Except for the northern and southern margins, coal seams were developed in the whole area, but the thickness of coal seams were lower than that in the SQ2 period. The total thickness of coal seams in the whole area is 0–31m, with the average thickness of 4.7m (Fig. 18). The coal thickness become thicker from south to north.

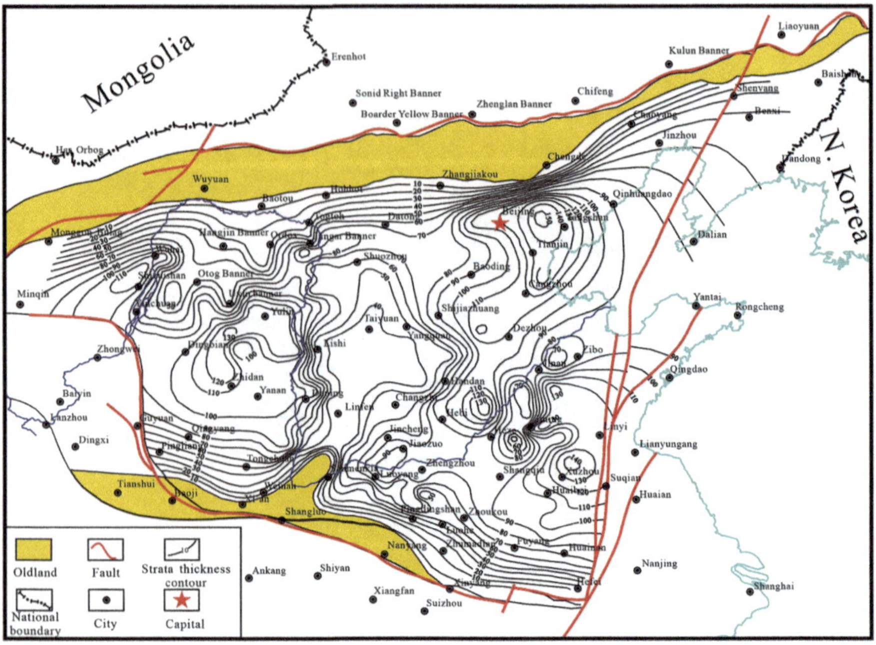

Fig. 17 Strata thickness contour map of SQ3 in North China Plate

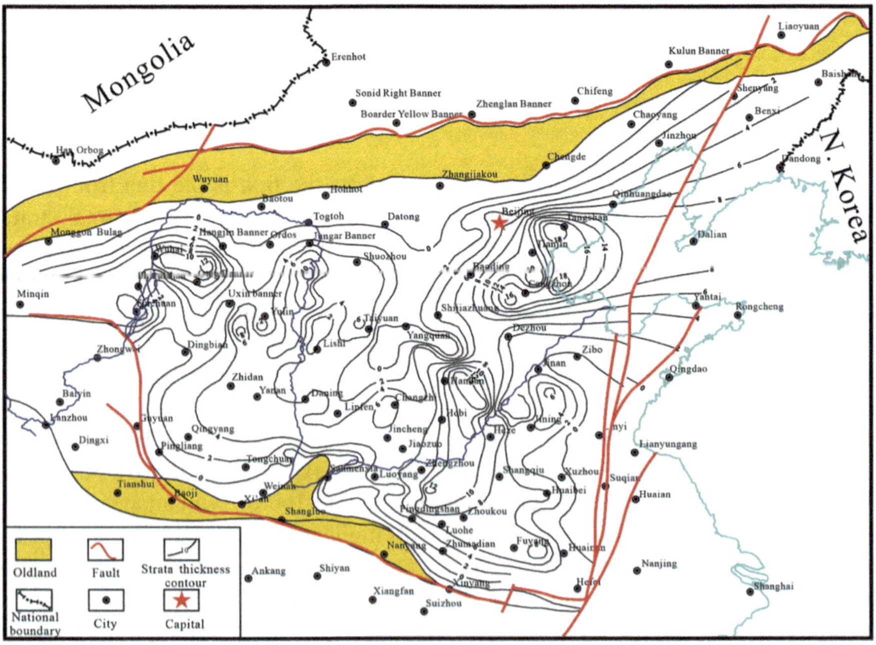

Fig. 18 Coal thickness contour map of SQ3 in North China plate

As SQ3 was forming in the Early Permian, the North China basin was mostly occupied by a fluvially-dominated delta complex. There was a large lake in the Zhidan-Yan'an-Daning-Jixian area of the Ordos Basin, and the southeast Zibo-Jinan-Hebi-Jincheng areas transited southward into the lagoon-tidal flat sedimentary system. The Zhongtiao upland in the southern Qinling Mountains continued to be a source of sediment. Weibei coalfield and Sanmenxia in Eastern Henan continued to be occupied by a fluvial-dominated delta system (Fig. 19).

(4) Sequence SQ4

The base of SQ4 is placed at the base of the Luotuobozi sandstone, and its top boundary is the top surface of the Taohua mudstone, and is equivalent to the entire Xiashihezi Formation, which age range is from late Kungurian to early Wordian. The total thickness of SQ4 is generally 11–370 m, with an average thickness of about 118 m. The contour map of stratal thickness shows three depocenters: the first is between Tangshan and Baodi in Tianjin, with a thickness of 180–230 m. The second is a narrow and elongate zone, located in the Dezhou-Handan-Hebi-Jiaozuo-Zhoukou region,

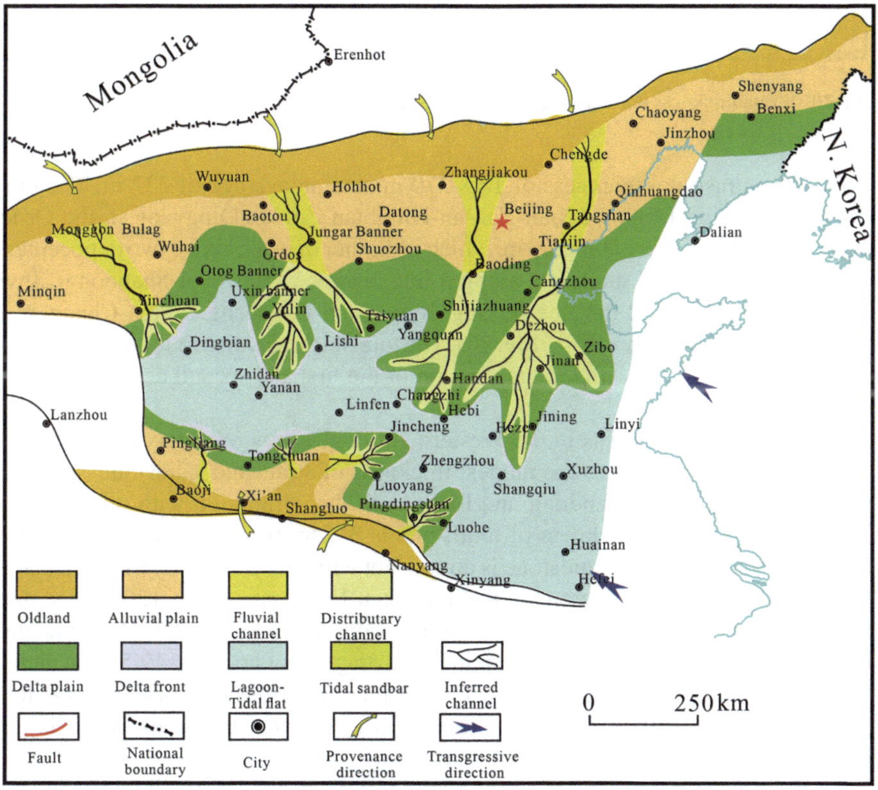

Fig. 19 Lithofacies paleogeography of SQ3 in North China Plate

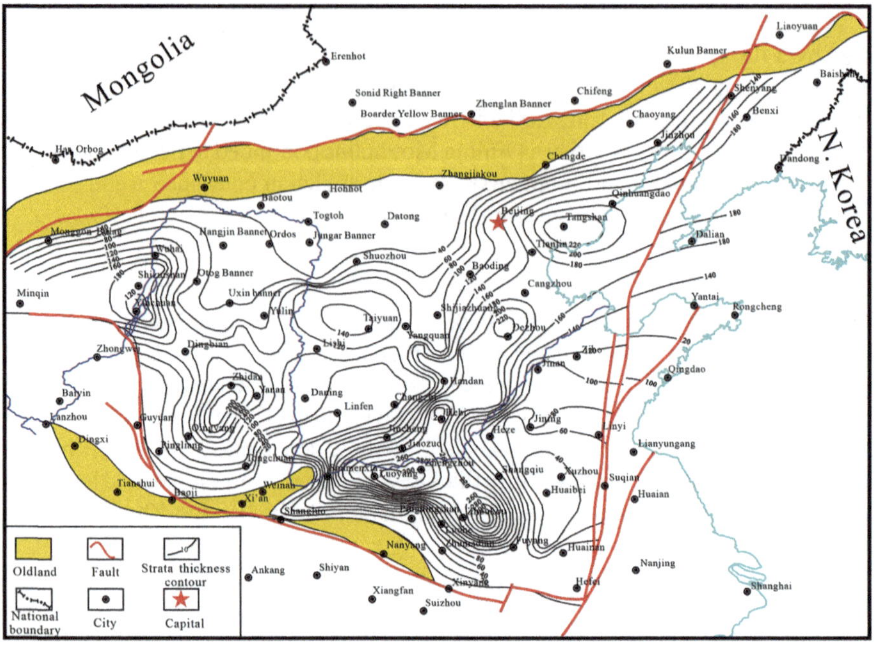

Fig. 20 Strata thickness contour map of SQ4 in North China plate

with a range instratal thickness of 180–370 m, and an average of 237 m. The third is the belt linking Daning-Jixian-Yan'an-Dingbian-Zhidan-Qingyang in the Ordos Basin, with a thickness of 90–200 m, which is thinner than the above two depocenters. In addition, the stratal thickness along a line between southwest Shandong-Linyi-Xuzhou-Lianghuai and the line linking Tongchuan-Hancheng-Linfen-Changzhi is relatively thin, generally <80 m (Fig. 20). The total thickness distribution of sandstone is basically coincident with the distribution of the first and second depocenters, and is 70–170 m.

The sandstone thickness in the Western Ordos Basin is relatively uniform, with an average of 37 m. Due to the modest thickness of strata and sandstone in eastern Shandong, Southwest Shandong and Lianghuai, the correlation with other regions is not clear. The total thickness of mudstone is 3–246 m, with an average value of 73 m. The depocenter for mudstone is coincident with that defined by total thickness. The total thickness of mudstone in the Tangshan-Baodi area is 80–120 m, and dark argillaceous limestone can be seen in some wells; Dezhou-Handan-Hebi-Jiaozuo-Zhoukou area preserves 62–246 m, with an average thickness of 149.5 m, which is the thickest mudstone accumulation in the study area; The thickness of mudstone in Daning-Jixian-Yan'an-Dingbian-Zhidan-Qingyang-area of Ordos Basin is 38–207 m, with an average thickness of 84 m.

During the Middle Permian accumulation of SQ4, the North China Basin was dominated by fluvial sedimentary systems, with well-developed channels, alluvial

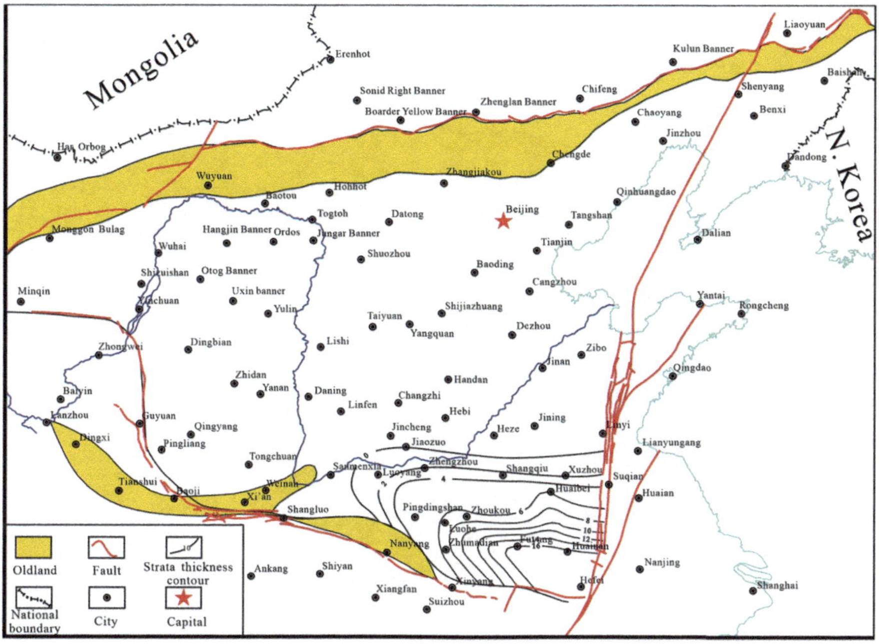

Fig. 21 Coal thickness contour map of SQ4 in North China Plate

plains and floodbasins, secondly by a lacustrine sedimentary system and associated deltas. The Dingbian-Yan'an-Zhidan-Daning-Jixian area in the middle of the Ordos Basin, southern Yulin and Tianjin-Cangzhou area were dominated by the lacustrine basin, and the lakes gradually changed around into lacustrine delta plain and front facies, with little peat accumulation in general (Fig. 21). A fluvially-dominated delta system was developed in the south of the Jining-Hebi-Jiaozuo-Zhengzhou-Luohe-Zhumadian area in the south of the North China Basin, forming coal seams with industrial value (Fig. 22).

(5) Sequence SQ5

SQ5 is equivalent to the lower part of the Shangshihezi Formation, spanning from the bottom of that formation to the base of thick-bedded sandstones in the lower middle part of the formation (e.g. the K7 Sandstones in the Taiyuan Xishan coalfield), and the age range is equivalent to the late Wordian and Capitanian stage of the late middle Permian. The total thickness of the SQ5 strata varies considerably from north to south from 46 to 450 m, with an average of 126 m. The principal depocenter was located in the Xuzhou-Lianghuai area in the southeast. In the north, the strata are generally <100 m thick, but are thicker in the southeast, ranging from 140 to 450 m. The average thickness in Huainan and Huaibei areas was 302 m, and the thickness in the western Ordos Basin is relatively uniform, generally 62–207 m, with an average of 105 m (Fig. 23). The distribution of sandstone across the whole area is relatively uniform,

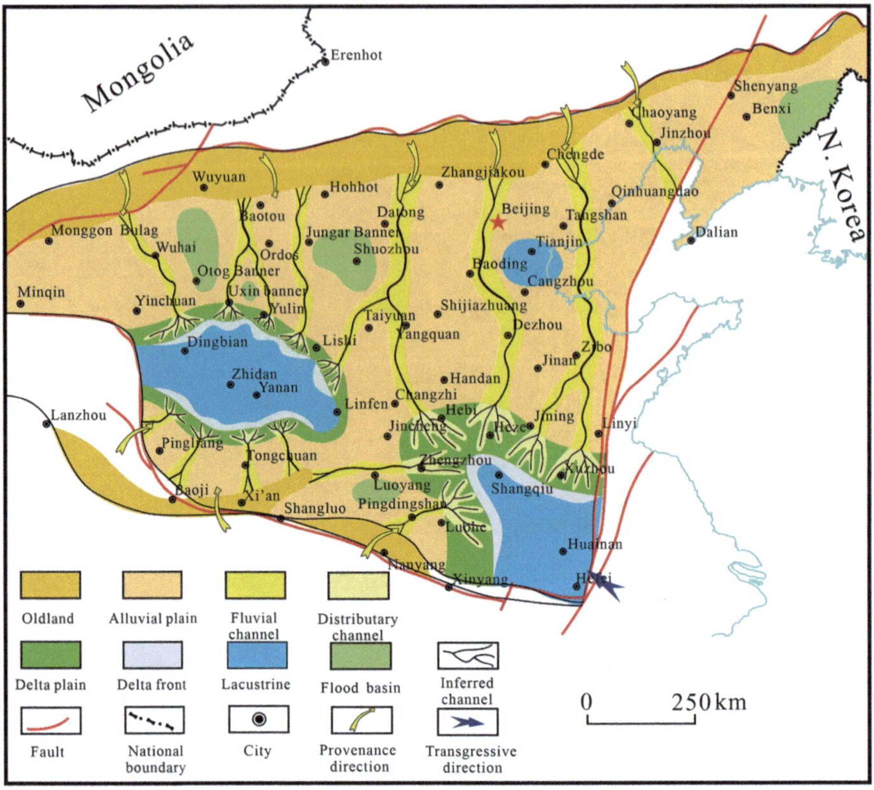

Fig. 22 Lithofacies paleogeography of SQ4 in North China Plate

with modest changes in thickness. The thickest sandstone accumulation includes the Palougou section near Baode, Shanxi Province, with a thickness of 152 m, and in the Lianghuai-Xuzhou area, where a borehole preserves 151 m of SQ5. The ShaBaGou section south of Handan, Hebei is 122 m thick. The total thickness of sandstone reaches a maximum of 100 m near Daning and Jixian. It is considerably thinner in other areas, with an average value of about 50 m. The total thickness of mudstone is 3–246 m, with an average value of 73 m. The mudstone depocenter was located in Xuzhou and Lianghuai areas, generally preserving thicknesses of 110–354 m, with an average value of 212 m. In addition, the total thickness of mudstone in the Shabagou section in the south of Handan, and the Daning-Jixian area was also relatively thick, which is consistent with the depocenter defined by total thickness of strata and sandstone. The thickness of mudstone in other areas is generally 20–80 m.

In the middle and late Permian interval when SQ5 was formed, the North China Basin was dominated by an alluvial plain, with widespread development of fluvial channel, and floodbasin environments, and secondly by a lacustrine system with associated deltas. Lacustrine systems were developed in the south of Dingbian-Yan'an-Zhidan-Yulin in the central Ordos Basin and the southern Liaodong Peninsula, and

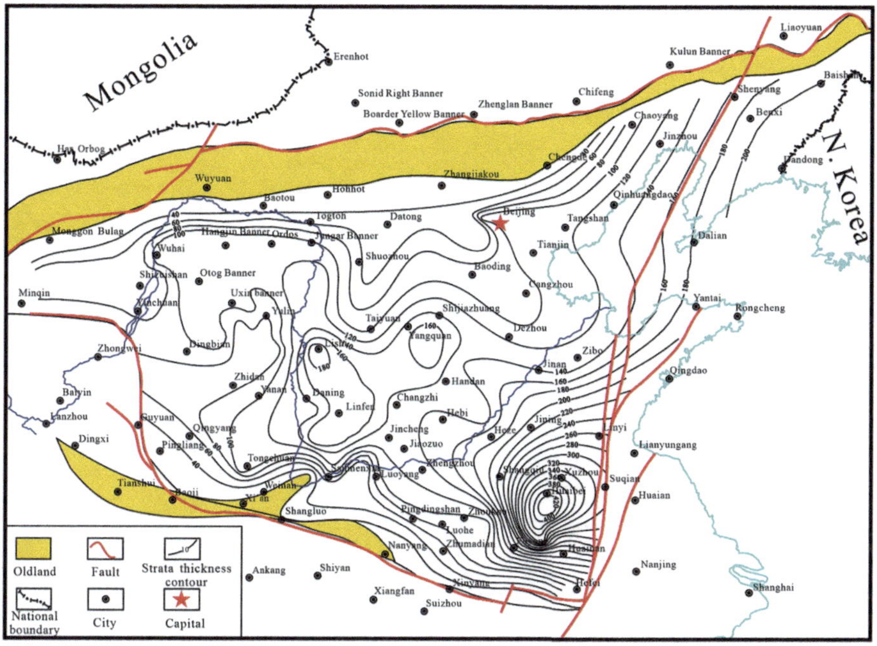

Fig. 23 Strata thickness contour map of SQ5 in North China Plate

the surrounding lakes gradually transformed over time into floodbasin facies and channel facies. A lacustrine system was also developed in the east of Dongying-Binzhou-Jinan-Jining-Linyi area in eastern North China, accompanied by lacustrine deltas, but generally did not for and preserve coal. A delta system is preserved south of East Henan-Xuzhou-the Lianghuai in the southeast of North China, where delta plain and delta front environments were developed, and peat accumulation formed coal seams with industrial value (Fig. 24). Most other areas in North China were dominated by the alluvial plainsystem, with channel, floodplain, and floodbasin facies developed, and the Hangjinqi-Ordos area preserves widespread floodbasin facies (Fig. 25).

(6) Sequence SQ6

SQ6 is equivalent to the middle-upper part of the Shangshihezi Formation and is bounded at its base by a thick-bedded sandstone and its top by the base of the Sunjiagou Formation. Its age range is within the Wuchiapingian stage. The total thickness of strata ranges from 47 to 417 m, with an average of 159 m. The overall trend is of thin accumulations in the north and thick in the south, with two depocenters. The primary depocenter was located east of the Zibo-Jining-Shangqiu-Huainan belt, with a stratigraphic thickness of 300–417 m and an average thickness of 367 m. The second, subsidiary depocenter is located in the Daning-Jixian-Yuncheng-Jincheng-Changzhi Handan-Taiyuan-Lishi area, with thickness ranging from 148 to 337 m, and

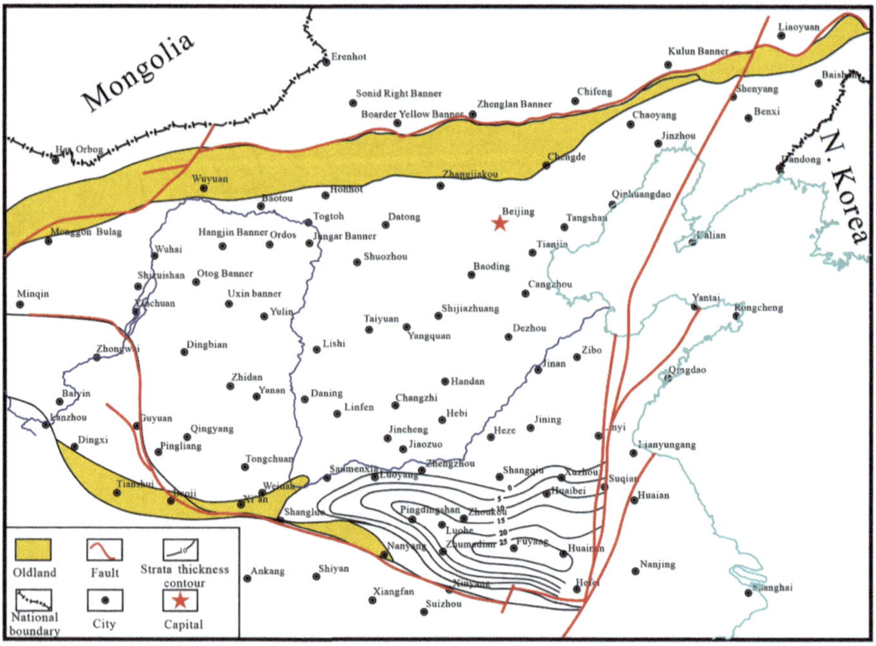

Fig. 24 Coal thickness contour map of SQ5 in North China Plate

an average of 264 m (Fig. 26). The distribution of sandstone is relatively uniform across the entire region, generally 5–195 m, with an average thickness of 50 m. There are thick sandstone deposits only in Daning-Jixian and Handan. The thickness in Daning is 195 m and the latter is 115 m. Mudstone was relatively thick across the entire region, with an average of 109 m, showing three depocenters. The first was located in the Mayicun section near Benxi, Liaodong Peninsula, which is the thickest accumulation in the whole area, mainly composed of mudstone with minor fine-grained sandstone. The second depocenter is located in the Linyi-Jining-Xuzhou-Lianghuai area in the southeast, with an average thickness of 304 m. The third depocenter is located in the Lishi-Liulin area, with an average thickness of 150 m.

During the late Permian accumulation of SQ6, the North China Basin was dominated by fluvial deposits, with fluvial channels, alluvial plain, and floodbasins, and secondly by lacustrine environments. Large-scale lacustrine sedimentation occurred in the south of Benxi in the Liaodong Peninsula in the northeast of North China, in the east and south of Cangzhou-Dezhou-Heze-Luohe-Fuyang, and the west of Otog Banner-Shizuishan-Dingbian-Zhidan-Yan'an-Lishi belt. Small lacustrine basins were developed in the Jincheng-Jiaozuo-Luoyang area, and floodbasin facies and channel facies were gradually developed around the lake. Other areas in North China were alluvial plains, with well-developed fluvial channels and floodbasins. Floodbasin facies of different scales were developed in the Ordos-Jungar Banner-Shuozhou-Baode area in the north of North China and some areas in the west of

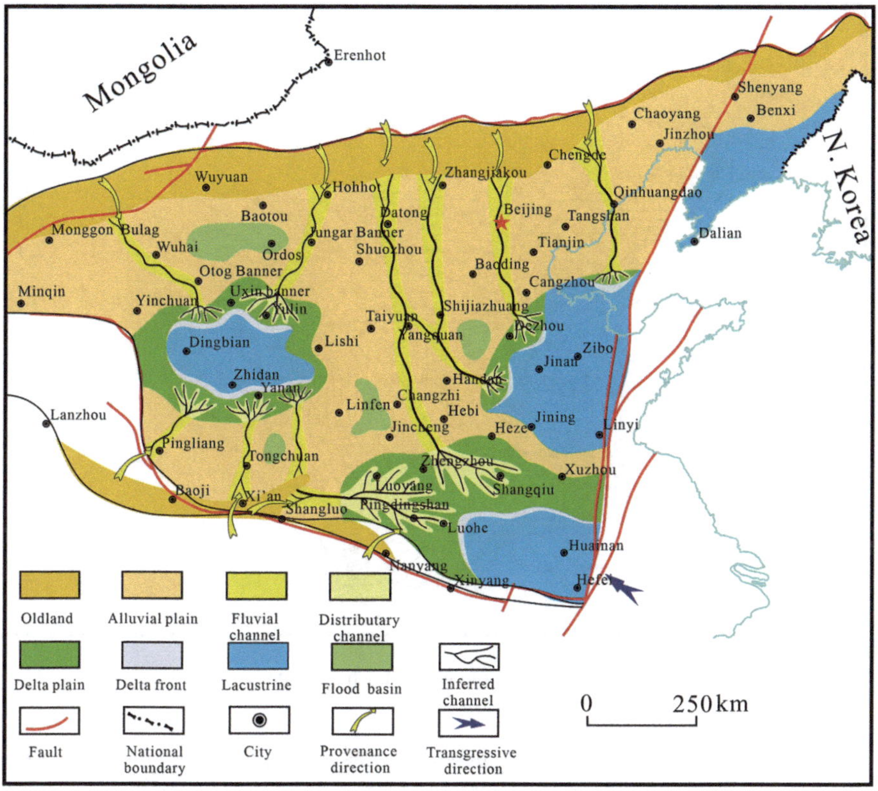

Fig. 25 Lithofacies paleogeography of SQ5 in North China Plate

Tongchuan. The provenance of the sediment forming SQ6 sedimentary was the Yinshan basement in the north margin and Zhongtiao basement in the South Qinling Mountains, and the lake transgression direction was from southeast to northwest (Fig. 27).

(7) Sequence SQ7

SQ7 is equivalent to the Sunjiagou Formation, and its top corresponds to the Permian–Triassic boundary. The age range of SQ7 is equivalent to the Changhsingian stage. The strata are composed of terrestrial redbed deposits. The climate was warm and humid with a well-drained continental environment, and the total thickness of the strata ranges from 85 to 367 m, with an average of 226 m. The stratal thickness in the Ordos Basin is somewhat uniform, ranging from 137 to 346 m, with an average thickness of 249 m. The strata in the Luoyang-Zhengzhou area in the west of Henan Province, south of North China were relatively thick, and in the Yaoling section reached 367 m, constituting a depocenter (Fig. 28). Sandstone in SQ7 is unevenly distributed from east to west across the region. In the west, the sandstone is generally thicker, especially in the Ordos Basin, ranging from 51 to 247 m, with an average

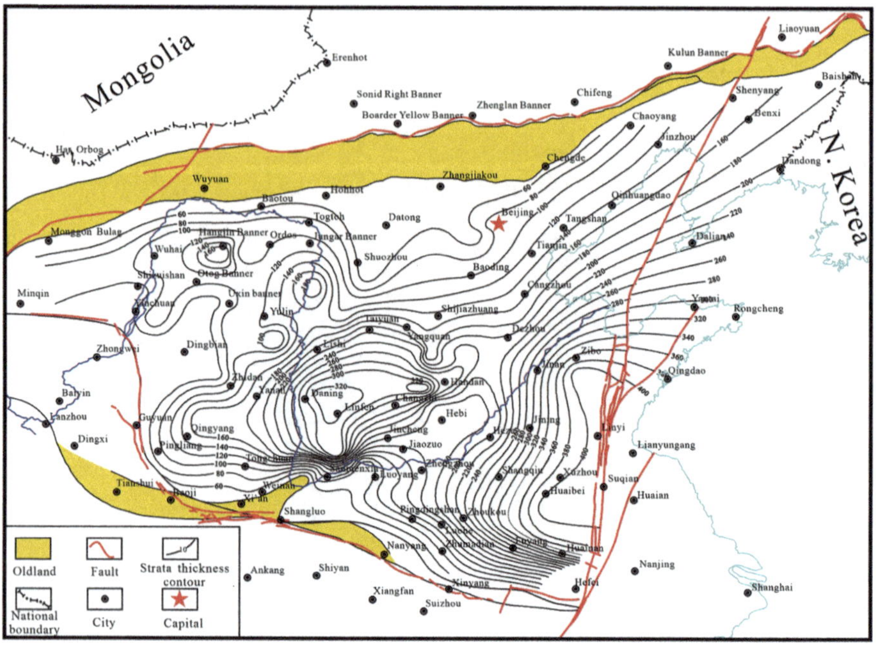

Fig. 26 Strata thickness contour map of SQ6 in North China Plate

thickness of 110 m. In the east, the range is 10–175 m, with an average value of 66 m, which is significantly thinner than in the west. Mudstone is relatively thick across the entire region, with an average of 127 m, and the depocenter is located in the Ordos Basin and Luoyang in western Henan. The thickness of mudstone in the Ordos Basin is 30–230 m, with an average thickness of 138 m, and the thickness in the Yaoling section in Luoyang, western Henan is 216 m. Further, mudstone thickness is lower in other areas than in the above two locations.

In the Late Permian during accumulation of SQ7, the North China Basin was dominated by fluvial and lacustrine sedimentary environments, and fluvial channels, alluvial plain, and floodbasins were extensively developed. Large-scale lacustrine basins were formed south of Benxi-Dandong in the northeast belt of North China, east of the Zibo-Dezhou-Handan-Changzhi-Jincheng-Jiaozuo-Shangqiu-Fuyang-Hefei belt in the southeast of North China, and in the Dingbian-Zhidan-Yulin-Yan'an area in the center of the Ordos Basin. The lacustrine sedimentary environment gradually passed into floodbasin and fluvial settings and in other areas, fluvial channel, floodplain, and floodbasin facies were developed. Floodbasin facies of different scales were developed in the south of Hangjin Banner-Ordos-Otok Banner area and east of Datong in North China. The provenance of the SQ7 sediments is principally Yinshan massif in the north, Zhongtiao massif in southern Qinling, and Funiu massif, and the Pingliang-Qingyang area in the southwest. The lake transgression direction was from southeast to northwest. During this period, the North China Plate was completely transformed

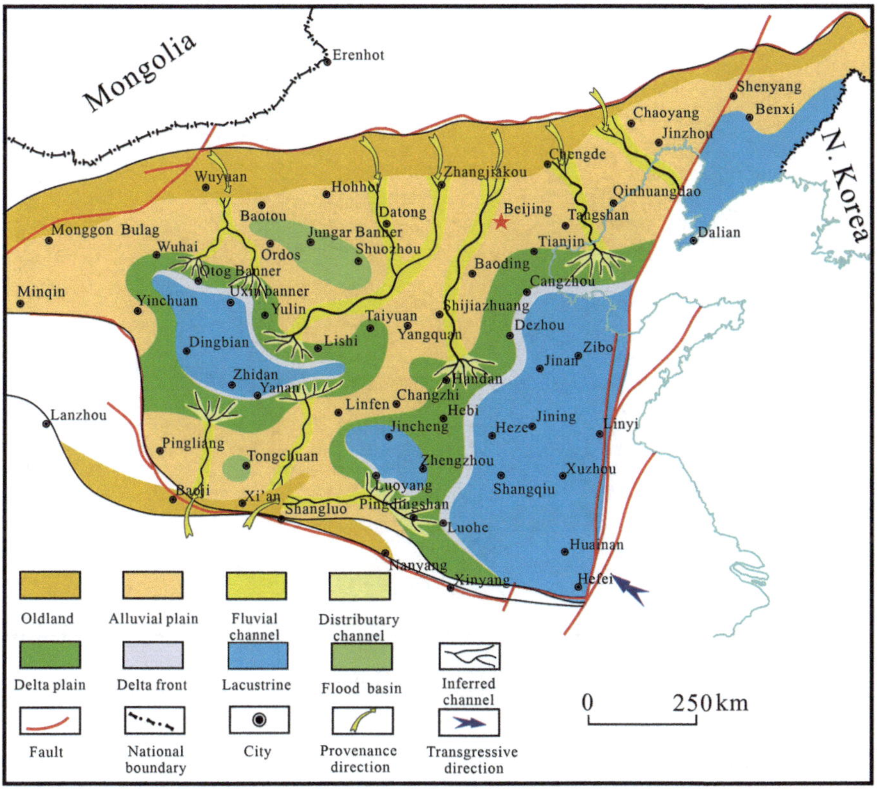

Fig. 27 Lithofacies paleogeography of SQ6 in North China Plate

into a continental setting, with a warm and humid climate. However, compared with the previous SQ1–SQ6 period, landscapes were well-drained, precluding the accumulation of peat that could be buried to form coal. The sediments are mainly redbeds, and lack coal (Fig. 29).

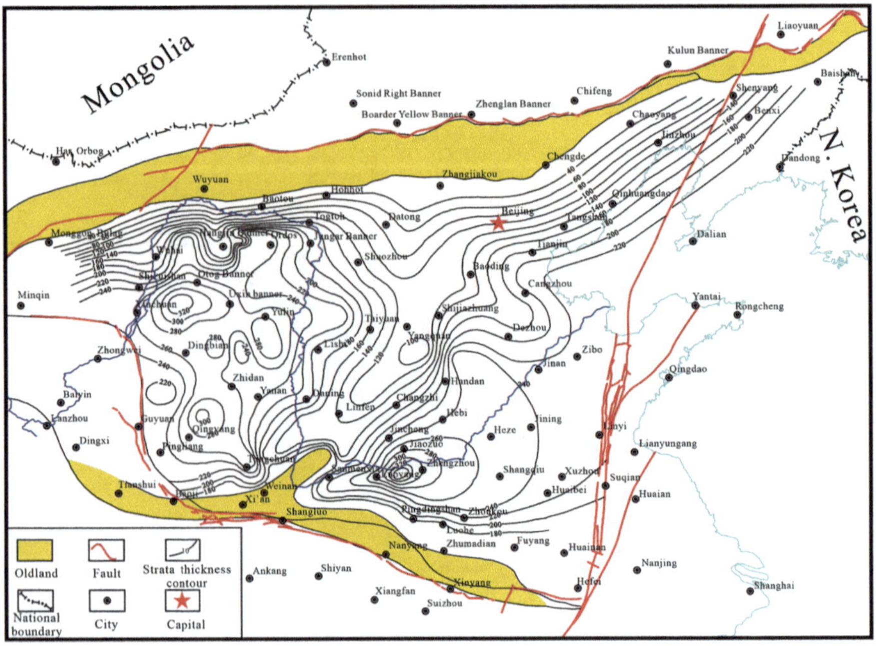

Fig. 28 Strata thickness contour map of SQ7 in North China plate

2 Scientific Significance

2.1 *Paleoclimate Changes During the Icehouse to Greenhouse Transition*

2.1.1 Low-Latitude Sedimentary Environment Changes Linked High-Latitude Glacial Cycles During Late Carboniferous

The Late Paleozoic glacial age (LPIA) of the Carbonifero-Permian (c. 360–260 Ma) is the only recorded transition from the glacial to greenhouse period on Earth (Montañez et al. 2007). This paper studies the relationship between the Pennsylvanian sediment characteristics and glacial-interglacial climatic cycles in Palougou section, Ordos Basin. The combination of three tuff zircon radiometric ages (315.4–299.3 Ma, Fig. 30) and biostratigraphy has established a high-resolution stratigraphic framework for the target strata, which defines the Benxi Formation as the Late Bashkirian–Middle-Late Moscovian stage of the Carboniferous. Taiyuan Formation is the Middle and late Moscovian–Permian Sakmarian stage of Carboniferous.

Based on the analysis of the sedimentary environment, the sea level evolution of the North China Plate during the Late Carboniferous has been recovered. The relatively high sea level of the Late Bashkirian–Early Moscovian interglacial, the

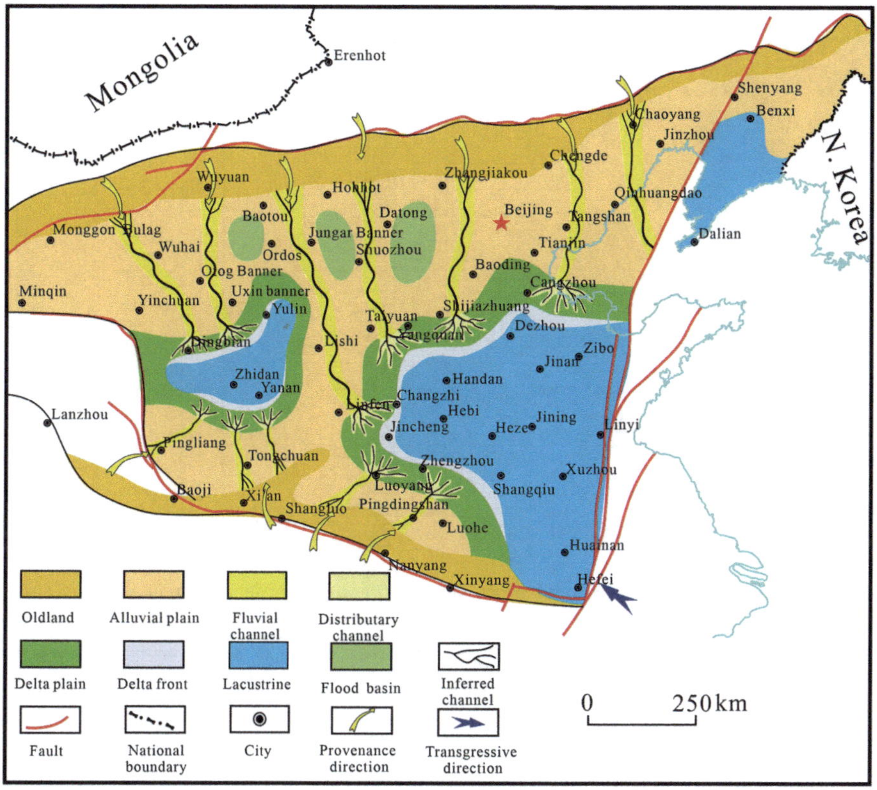

Fig. 29 Lithofacies paleogeography of SQ7 in North China plate

large sea level decline of the Middle Moscovian interglacial, the sea level rise of the Moscovian-Gzhelian interglacial, and until the emergence of Maoergou limestone reached a fixed point. The extensive development of the Asselian glaciation led to a large drop in sea level, which provided a good record for the deposition of the high-latitude Gondwana glacier in the low-latitude region. In addition, we find that the interglacial sedimentary features are extensive development of ferric nodular mudstone and limestone, and coal accumulation is widespread in the glacial period (Fig. 31).

2.1.2 Driving Mechanism of Volcanism on Glacial Cycles

The late Paleozoic ice age (LPIA) during the Carboniferous and Permian (ca. 360–260 Ma) was a vegetated Earth's only recorded icehouse-to-greenhouse transition (Gastaldo et al. 1996; Montañez et al. 2007) and provides a deep-time perspective for climate-glaciation-environment co-evolution and future climate change. During the

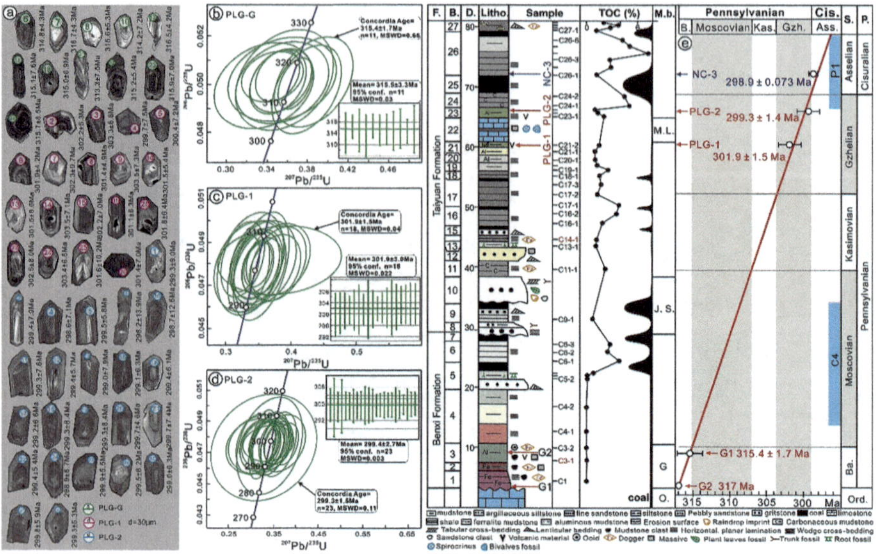

Fig. 30 Zircon $^{206}Pb/^{238}U$ age distribution diagrams (unpublished data by Lu et al.)

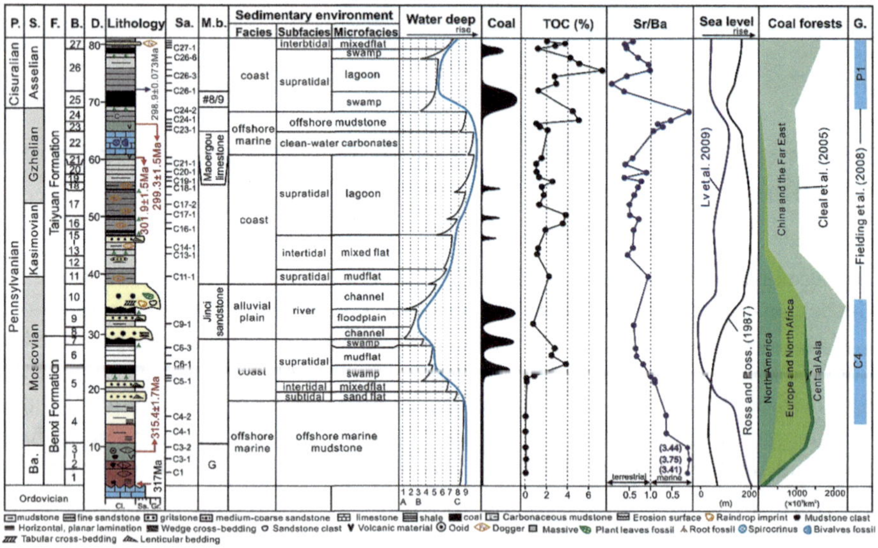

Fig. 31 Pennsylvanian-early Cisuralian paleoclimate changes indicated by coal seams and bauxitic mudstone from the Palougou section in the North China Plate. Abbreviations: P.—Period; S.—Stage; F.—Formation; B.—Bed; Litho.—Lithofacies; M.b.—Marker bed; S./L.—Sample/ Limestone. Glacial intervals from Gastaldo et al. (1996), Fielding et al. (2008) and Lu et al. (2021) (unpublished data by Lu et al.)

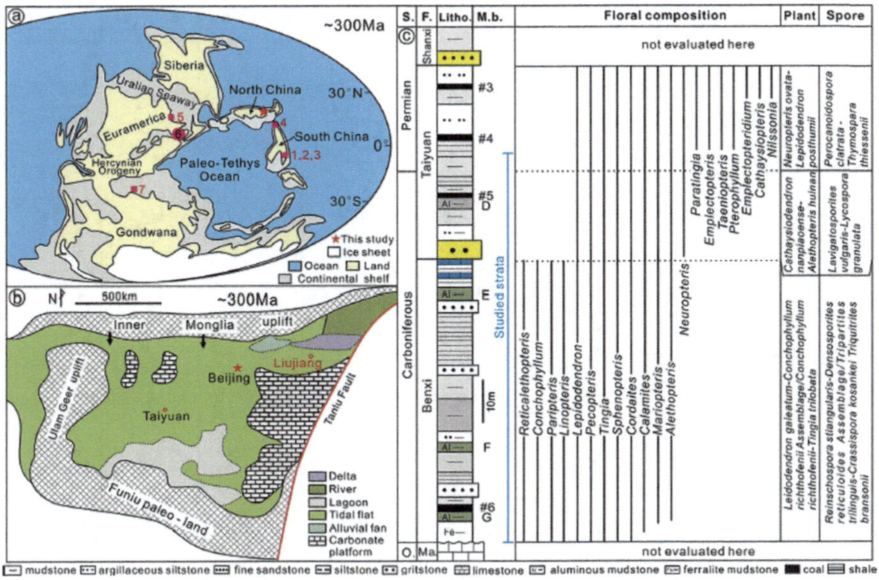

Fig. 32 Location and geological context for the research area (Lu et al. 2021)

LPIA, changes in glacial-interglacial cycles have been closely linked to atmospheric pCO_2 (Lu et al. 2021).

We investigate the relationship between volcanism and glacial to nonglacial intervals during the Pennsylvanian to earliest Cisuralian in the Liujiang Coalfield of the North China Plate (NCP) (Fig. 32). We use integrated data from Zircon U–Pb dating, Hg/TOC ratios and $\delta^{13}C_{org}$ to recover temporal trends in volcanic intensity and relate these to fluctuations in the global carbon cycle and atmospheric pCO_2. The main results are summarised in Fig. 33.

Four new radiometric ages from tuffaceous claystones provide a precise chronostratigraphic framework and constrain the succession to ~322–301 Ma. We interpret the weighted mean ages from the younger peaks as the depositional ages of the tuffaceous claystone horizons for the following reasons. Firstly, the Th/U ratios of the zircon crystals vary from 0.41–2.70 (average = 0.96) and they have euhedral crystal morphologies with clear oscillatory zoning that are indicative of volcanism-sourced zircons. Secondly, during the Pennsylvanian frequent volcanic activity in Inner Mongolia uplift (IMU) source area provides the possibility for the frequent inputs of tuffaceous and pyroclastic materials into the depositional basin, and the ages of the younger peaks are consistent with a Pennsylvanian age (Fig. 34f). Finally, the distribution of the four younger peak weighted mean ages is consistent with the Pennsylvanian ages determined from biostratigraphy and regional lithostratigraphy, unlike the older peaks that provide ages that are consistently older than the Pennsylvanian.

Results from Hg/TOC ratios record four peaks of volcanic intensity from the Bashkirian to the Carboniferous–Permian (C–P) transition period, and two low

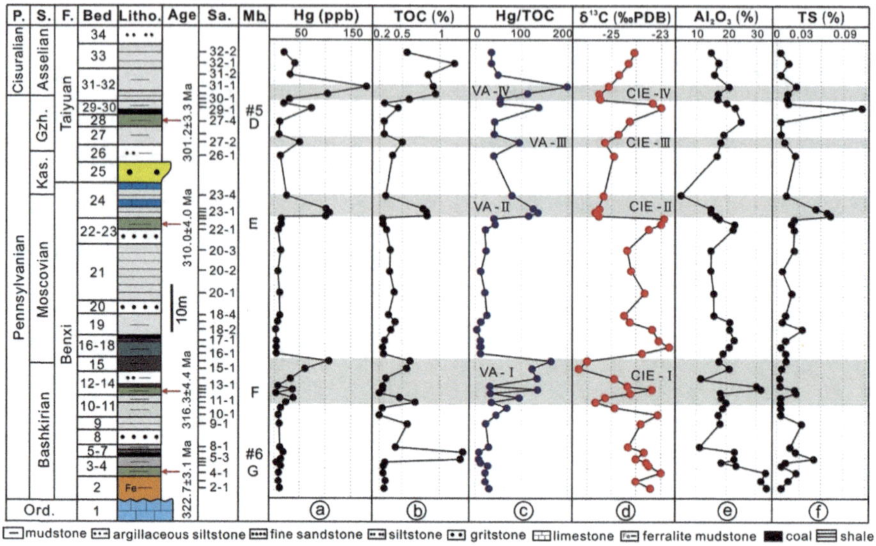

Fig. 33 Hg concentration, TOC values, Hg/TOC (ppb/%) ratios, $\delta^{13}C_{org}$ values, Al_2O_3 concentrations and total sulfur (TS) concentrations from the Shimenzhai section in the North China Platform. Abbreviations: P.—Period; S.—Stage; F.—Formation; Ord.—Ordovician; Litho.—Lithofacies; Sa.—Sample; M.B.—Marker bed

ebbs during the early-middle Bashkirian and the early-middle Moscovian. Peaks in volcanic intensity coincide with the four negative excursions of $\delta^{13}C_{org}$ and correlate with volcanism in the North China Plate and NW Europe, inferring volcanic drivers for the carbon isotope excursions and perturbations in the global carbon cycle. Periods of higher atmospheric pCO_2 correlate to interglacial intervals, the two lower ebbs in pCO_2 correlate with two positive isotope plateaus, while periods of lower atmospheric pCO_2 and cooling correlate with glacial intervals C3 and C4 recognized from other sections globally. Results suggest volcanism to be a significant driver for late Pennsylvanian to earliest Cisuralian shifts in global climates from glacial to nonglacial through greenhouse gas emissions and changes to the global carbon cycle. Our results represent an important contribution to the development of an accurate chronostratigraphy for the continental facies of the NCP during the Pennsylvanian and earliest Permian, enabling correlation of volcanic and climate events including glacial and nonglacial intervals (Fig. 35).

2.1.3 Coal Accumulation Patterns in the Epicontinental Basin Affected by Glacioeustatic Sea Level Cycles

To study the stratal stacking patterns and controls on coal accumulation in a lowland alluvial plain, coastal plain, and shallow-marine setting, the depositional facies, sequence stratigraphy, and paleogeography of the Carboniferous–Permian strata in

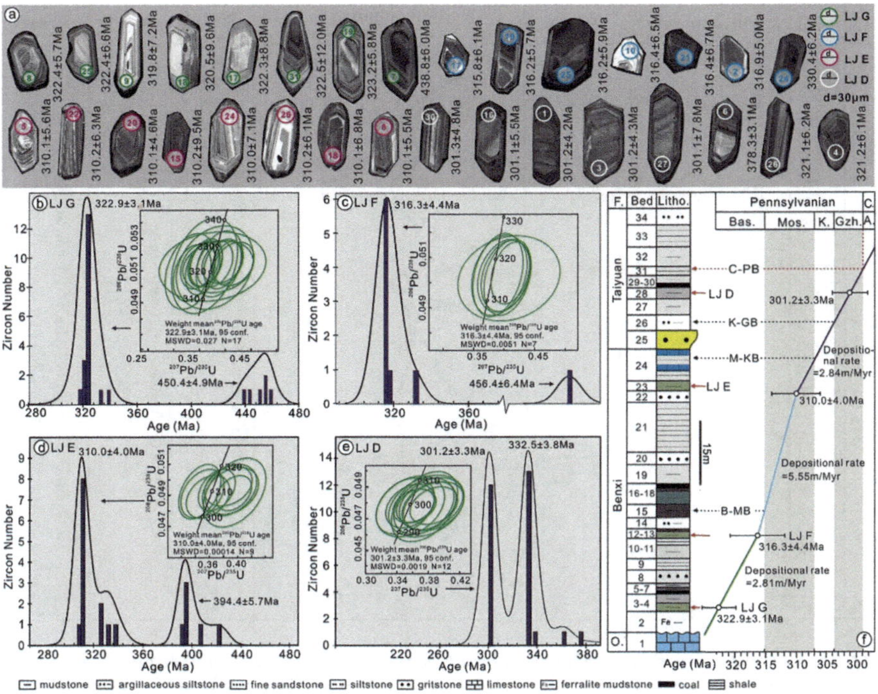

Fig. 34 Zircon $^{206}Pb/^{238}U$ age distribution diagrams. **a** Example zircon crystals from each sample. **b** Sample LJ #G. **c** Sample LJ #F. **d** Sample LJ #E, **e** Sample LJ #D. **f** Age model through the Shimenzhai profile succession. Abbreviations: F.—Formation, O.—Ordovician, Litho.—Lithofacies, P.—Permian, Bas.—Bashkirian, Mos.—Moscovian, K.—Kasimovian, Gzh.—Gzhelian, A.—Asselian. *Note* Position of C–P boundary is from the age dating developed in this paper (Lu et al. 2021)

the Anyang-Hebi (Anhe) coalfield, Henan Province, in the southern zone of NCP (Fig. 36), were analyzed based on borehole cores.

Ten facies were identified in the Carboniferous–Permian strata. These facies were grouped into five facies associations, which represent alluvial plain, delta plain, delta front and pro-delta, barrier coast, and offshore marine depositional environments (Table 1). Based on the vertical variation of facies, ten sequences (S1–S10) were recognized, and their internal Lowstand Systems Tract (LST), Transgressive Systems Tract (TST), and Highstand Systems Tract (HST) were determined. From S1 to S10, the Anhe coalfield experienced an evolution of depositional environment from offshore marine, pro-delta and delta front through delta plain to alluvial plain (Fig. 37).

(1) Interbedded siliciclastic and carbonate depositional succession and coal accumulation model for S2

Cyclothems of coal, siliciclastic and carbonate strata typically represent 10^4–10^5 years, and these successions are widely attributed to glacioeustasy (Wanless and

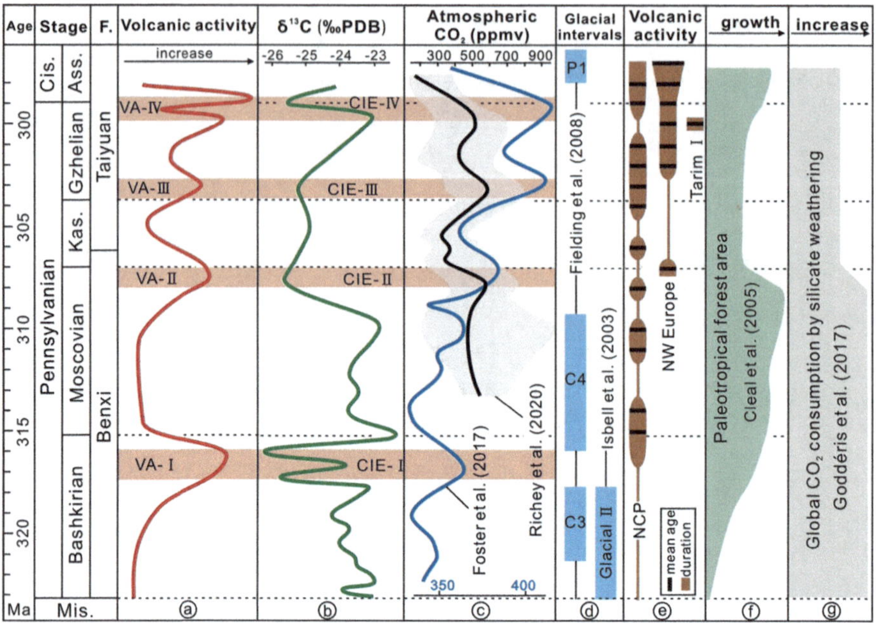

Fig. 35 Correlation of Pennsylvanian-earliest Cisuralian global events. **a** Volcanic activity from this study. **b** $\delta^{13}C$ variation from this study. **c** Atmospheric CO_2 from Foster et al. (2017) and Richey et al. (2020). **d** Glacial intervals from Fielding et al. (2008) and Isbell et al. (2003). **e** Timing of volcanic activity. Data from the northern NCP from Zhang et al. (2009 and the references therein), Europe (Germany) from Torsvik et al. (2008) and Tarim I from Xu et al. (2014). **f** Paleotropical forest extent from Cleal and Thomas (2005). **g** Global CO_2 consumption by silicate weathering from Goddéris et al. (2017). Abbreviations: Cis.—Cisuralian, Mis.—Mississippian, Ass.—Asselian, Kas.—Kasimovian (diagram from Lu et al. 2021)

Shepard 1936; Crowell 1978; Heckel 1986; Veevers and Powell 1987). These cyclothems are grouped into successions of 10^6 years duration, which are generally thought to be influenced by sea-level change, tectonic evolution, and local variations in climate and sedimentation (Calder and Gibling 1994). S2, constrained from 301.13 to 299.32 Ma and comparable to a cyclothem, is characterized by a series of parasequences influenced by repetitive relative sea-level changes, most likely driven by local sediment supply variations (see Fig. 38). Each parasequence is bounded top and base by flooding surfaces and is composed of regressive and transgressive deposits (Ainsworth et al. 2020; Colombera and Mountney 2020). In a parasequence, the offshore carbonate deposits are preserved at the bottom (Fig. 38a), then the offshore carbonate deposits are overlain by pro-delta and delta front deposits, or coastal plain deposits at the top (Fig. 38b–d).

The coal distribution in S2 indicates that the coal mainly developed during the TST in the western Anhe coalfield (Fig. 38). The presence of parasequences in S2 indicates a more distal location and more profound transgressions, which may be

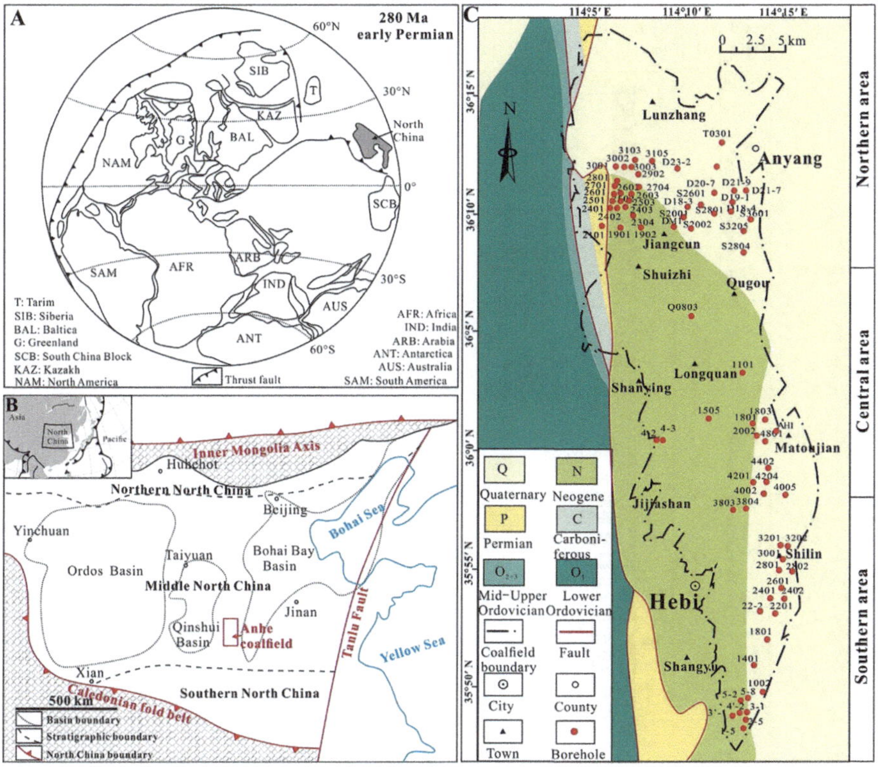

Fig. 36 **a** Global plate location map during the early Permian (modified from Torsvik and Cocks 2004). **b** Tectonic context of the Anhe coalfield (modified from Shang 1997). **c** Regional geological map and borehole distribution in the Anhe coalfield (modified from Li et al. 2020a, b)

related to the cyclothems of the Taiyuan Formation record glacial-interglacial cycles that are part of glaciation P1 in eastern Australia, identified by Fielding et al. (2008).

(2) Siliciclastic depositional succession and coal accumulation model for S4

An SB was developed at the base of S4 (Fig. 39). During the LST, distributary channel fine-grained sandstone deposits were preserved in the western Anhe coalfield, and no deposits were preserved in the eastern Anhe coalfield (Fig. 39a). During the TST, the coastal mire deposits firstly occupied the eastern Anhe coalfield, and later were preserved in the western Anhe coalfield. The coal seams display a thinning trend from the western to the eastern part of the Anhe coalfield (Fig. 39b, c). During the HST, the Anhe coalfield was widely occupied by coastal floodbasin mudstone deposits (Fig. 39d). Therefore, thick coals in S4 were mainly developed during the late TST in the western Anhe coalfield.

The coal accumulation model of S4 in the Anhe coalfield is similar to the paralic coal formation model within a depositional sequence proposed by Bohacs and Suter (1997). However, the paralic coals in Bohacs and Suter (1997) are interpreted to form

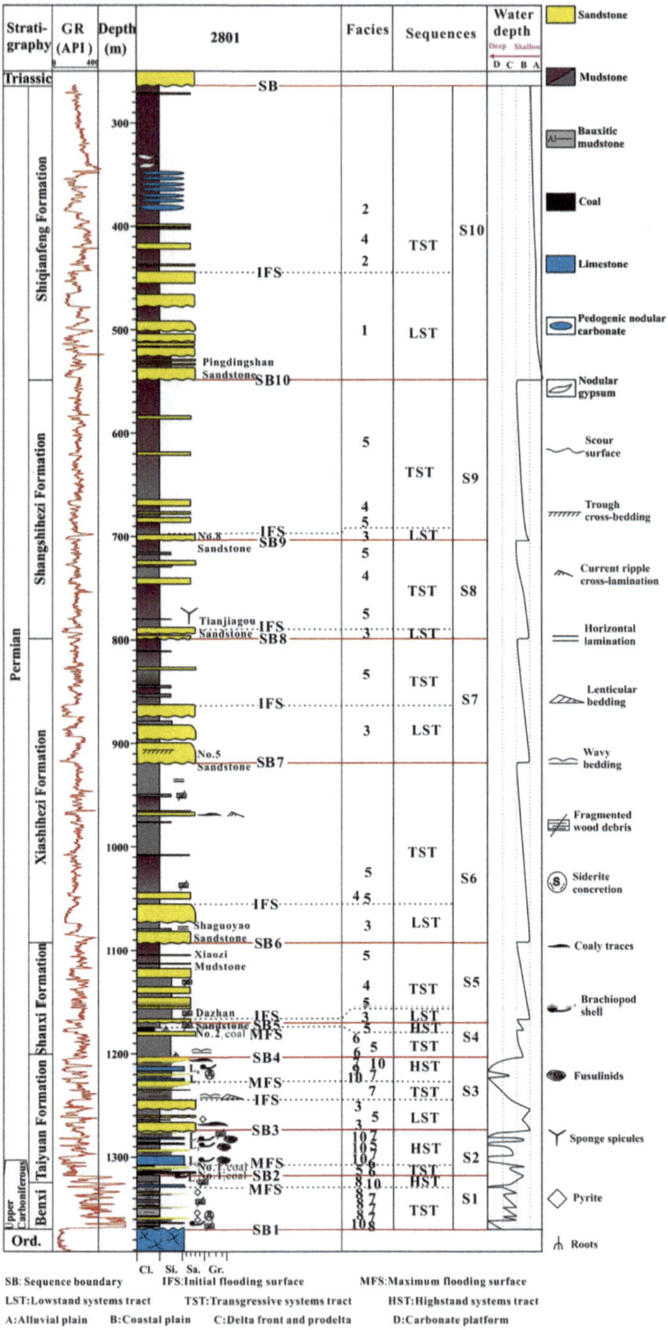

Fig. 37 A columnar section showing vertical facies variation and sequence stratigraphy of the upper Carboniferous–Permian in the Anhe coalfield (borehole 2801). The water depth is based on the facies associations A, B, C, and D recognized in the well cores. GR = Gamma ray; Ord = Ordovician (Li et al. 2021)

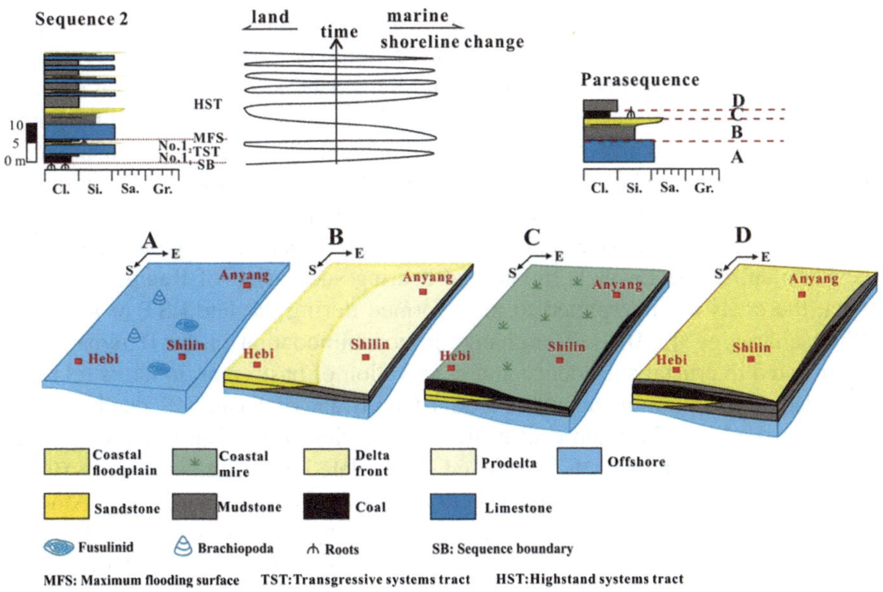

Fig. 38 Interbedded siliciclastic and carbonate depositional succession and coal accumulation model for S2 (Li et al. 2021)

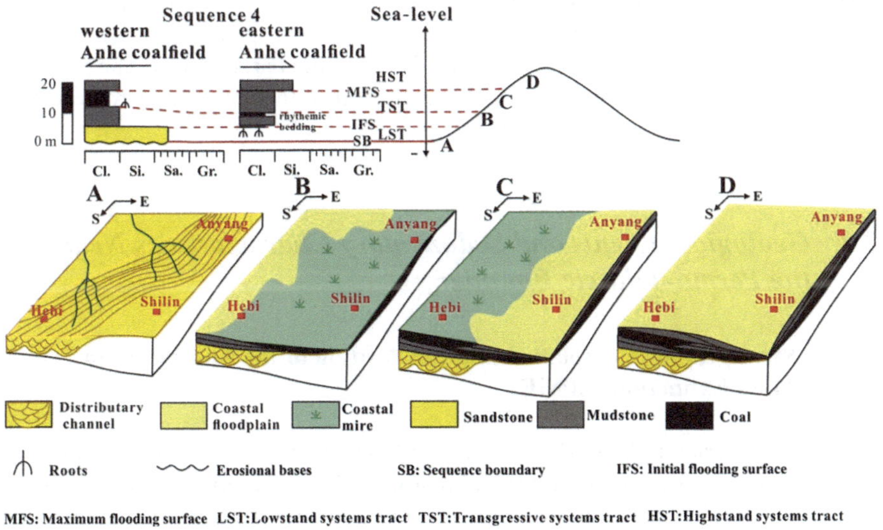

Fig. 39 Siliciclastic depositional succession and coal accumulation model for S4 (Li et al. 2021)

during the late LST in seaward areas and early HST in landward areas. The coals in the TST are believed to be restricted and scattered because of increased overbank flow and clastic sedimentation (Bohacs and Suter 1997). Whether the paralic coals in the seaward area are assigned to the late LST or early TST and whether the paralic coals in the landward area are assigned to the late TST or early HST seem largely semantic arguments. The debate may arise from different recognition of stratigraphic evidence for marine flooding in these paralic successions.

Thick, laterally extensive coals were widely preserved across Gondwana during the Permian. In the lower Permian coal-bearing succession of Raniganj Basin in India, the coals are interpreted to have formed during the late LST of sequences (Bhattacharya et al. 2018). The change in accommodation in the Raniganj Basin is attributed to crustal subsidence and reactivation of basin-margin faults related to post-glacial isostatic adjustment in Permian Gondwana rift basins (Bhattacharya et al. 2018). In the lower Permian coal-bearing succession of the Karoo Basin in South Africa, formation of the coals is believed to be related to post-glacial events (Cairncross 2001). In the lower Permian succession of the Paraná Basin, South America, glacial diamictites were replaced by thick coals during a post-glacial stage (Limarino et al. 2014). The ages of Sequence 4 in the Anhe coalfield are constrained from 295.65 to 293 Ma. Similarly, the significant sea-level fall recorded at the base of S4 (late Asselian) in the Anhe coalfield may be related to Gondwanan ice dynamics during the P1 glacial interval dating from 299 to 291 Ma identified by Fielding et al. (2008) in eastern Australia. The beginning of early Permian glaciations was newly dated using U–Pb zircon chemical abrasion thermal ionizing mass spectrometry (CA-TIMS) analysis to >298 Ma by Griffs et al. (2018, 2019) in the Paraná basin. The S4 shows a weaker glacioeustatic control compared to S2, indicating that S4 may represent the early post-glacial stage following P1.

2.2 *Geological Events and Sedimentary Characteristics Near the Permo-Triassic Boundary*

2.2.1 Stratigraphic Correlation and the Position of the End-Permian Mass Extinction (EPME)

The Permian–Triassic Mass Extinction (PTME) is the largest mass extinction event in Earth history with as many as 90% of marine and 70% of terrestrial species wiped out (Erwin 1994; Stanley 2016). To reveal the terrestrial records and the potential drivers of the End-Permian Mass Extinction (EPME), we investigate continental $\delta^{13}C_{org}$, nickel concentrations, wildfire, and climate change proxies from the late Permian Changhsingian stage to Early Triassic Induan stage in the ZK21-1 borehole core of the Yuzhou coalfield, Henan Province, southern NCP.

Previous studies of continental weathering in the Yima and Shichuanhe sections in the North China Plate (NCP) during the Permo-Triassic transition have shown that

CIA values tends to increase first and then decrease, with the maximum CIA values occurring at the top of the Sunjiagou Formation and approximately correspond to the End-Permian Plant Extinction (EPPE) (Cao et al. 2019). This provides a timeline for the position of the EPPE in the NCP. We follow this conclusion, using the peak in CIA as a marker for the EPPE and place the EPPE at the horizon coincident with C isotope excursion (CIE)-II in the Yuzhou coalfield (Fig. 40).

Furthermore, chemostratigraphy can provide evidence for the correlation between the EPPE in NCP and marine settings. Investigations on $\delta^{13}C$ distribution patterns from stratigraphically well-constrained Lopingian to early Triassic profiles have been undertaken in marine (Meishan, Niushan) (Shen et al. 2011, 2013; Liao et al. 2016, 2020) and terrestrial (Dalongkou, Lubei, Guanbachong, Chahe, Longmendong, Bunnerong^{-1}) strata (e.g., Zhang et al. 2016; Fielding et al. 2019; Shen et al. 2019b). These reveal $\delta^{13}C$ is relatively stable during the early Changhsingian, followed by a gradual and slow decrease during the late Changhsingian prior to a globally significant excursion with an average negative offset of 3–5‰ shortly before the P–T

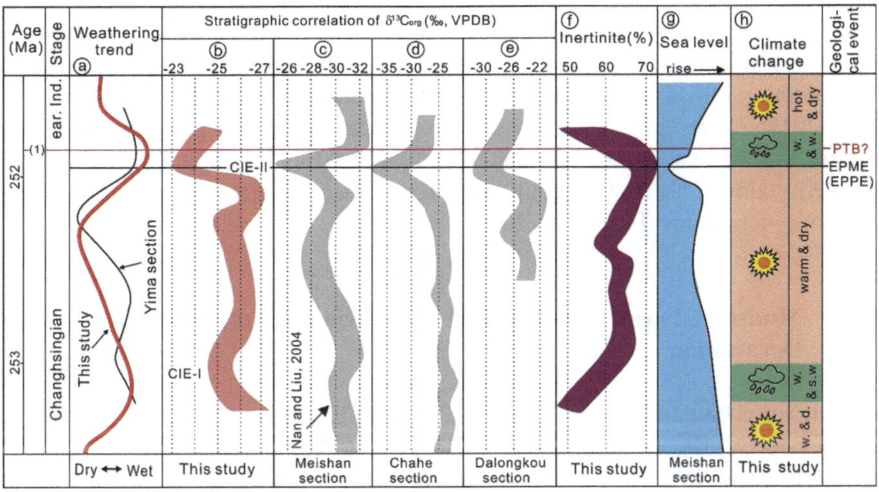

Fig. 40 Comparison among weathering trend, carbon isotope records, inertinite content, sea-level, and the paleoenvironment events from Changhsingian to early Induan. Age stratigraphic framework from Shen et al. (2019b); (1) Represents P–T boundary age: 251.902 ± 0.024 Ma; *a* red curve represent a weathering trend by MIA and CIA in the study area, and the black curve represent a weathering trend by CIA in Yima section (southern NCP) near the study area (Cao et al. 2019); *b* vertical change trend of $\delta^{13}C_{org}$ in the study area showing two negative isotope excursions near the middle and end of the Changhsingian; *c–e* $\delta^{13}C_{org}$ from the marine (Meishan) and terrestrial (Chahe and Dalongkou) sections by Nan and Li (2004), Zhang et al. (2016), and Shen et al. (2019a); *f* inertinite content from the present study; *g* sea-level curve (relative to current sea level) of the Meishan section from (Cao et al. 2009); *h* paleoclimate change inferred from clay mineral component, CIA and MIA. Abbreviations: ear. Ind. = Early Induan; w. & d. = Warm and dry; w. & s. w. = Warm and slight wet; w. & w. = Warm and wet; CIE-I to CIE-II = From the first organic CIE to the second organic CIE; PTB = Permian–Triassic boundary; EPME = End Permian mass extinction; EPPE = End Permian plant extinction (Lu et al. 2020)

boundary (Shen et al. 2019b; Fig. 40). The end-Changhsingian negative excursion of $\delta^{13}C$ represents a major reorganization of the global carbon cycle associated with the EPME interval and is a global phenomenon (Korte and Kozur 2010; Shen et al. 2011, 2013, 2019b; Frank et al. 2021).

In the study area, the $\delta^{13}C_{org}$ trend is very similar to that in the Meishan Changhsingian stratotype section at Changxing, South China (Nan and Li 2004). CIE-I occurs in mudstones on top of the Pingdingshan Sandstone, but it may be a regional negative excursion, since while it is present in the Meishan section in South China, it is absent in many other sections globally (Yin et al. 2007). In our study, no changes in plant species composition occur at this level (Lu et al. 2020). CIE-II occurs near the top of the Sunjiagou Formation, and it is coincident with a significant floral extinction event just below the P–T boundary that occurs across the NCP. Moreover, peaks in nickel concentration and Ni/Al ratio during the P–T transition period also are within the extinction interval of the EPME (Rampino et al. 2017; Fielding et al. 2019). As a result, we interpret CIE-II as correlating with the end-Changhsingian negative excursion associated with the EPME in Meishan Changhsingian stratotype section. This represents the first time the EMPE has been demonstrated in the NCP based on combined evidence from negative carbon isotope excursion, concurrent weathering trends, Ni/Al ratio and biotic extinctions. Our study supports the hypothesis that the extinction is synchronous in both terrestrial and marine successions (Shen et al. 2011, 2013; Zhang et al. 2016), but some recent researches show that the extinction of terrestrial life was earlier than that of marine life (Fielding et al. 2019; Chu et al. 2020; Gastaldo et al. 2020). This might suggest that the terrestrial extinction event was not synchronous (Feng et al. 2020).

2.2.2 Mudstone Facies, Wildfire, and End-Permian Terrestrial Devastation

Fusinite, or charcoal, is largely fire-derived and provides evidence for wildfires in the rock record (Scott 2000; Glasspool and Scott 2010). In the Yuzhou coalfield, inertinite (fusinite) is the most abundant kerogen maceral group ($\overline{x} = 63.1\%$). The high fusinite content and its vertical variation pattern indicate that the paleo-fires prevailed in the southern NCP during the middle Changhsingian-early Induan and reach their peak near the Permo–Triassic boundary. This increasing frequency of continental paleo-fires appears to be a global phenomenon, with similar records recorded in other parts of NCP as well as in South China, Australia, and Canada (Wang and Chen 2001; Grasby et al. 2011; Shen et al. 2011; Yan et al. 2019; Chu et al. 2020; Vajda et al. 2020).

Factors affecting wildfire include availability of combustible fuel, atmospheric oxygen concentration to enable burning, a suitable climate lacking high moisture, and an ignition mechanism (Scott 2000). The Zechstein-type flora present across the NCP during the Changhsingian was adapted for dry climates and would have been an appropriate source of fuel. Atmospheric oxygen concentration at the end-Permian has been estimated as 21–27%, far in excess of the minimum oxygen requirement of 15% for plant combustion (Glasspool and Scott 2010). Dry and hot climates favor the prevalence of wildfires, and water-limited conditions persisted during the Changhsingian in the study area before the EPME. However, prior to this, the climate tended to be relatively humid thereby reducing the likelihood of wildfire. Therefore, the prevalence of wildfire in the late Changhsingian may have been controlled by ignition factors. Under natural conditions, ignition is caused by lightning, volcanic eruption and less probably meteor impact (Glasspool et al. 2015). Of these, there was no volcanic activity in proximity to the Yuzhou coalfield in Henan Province (southern NCP) suitable to ignite wildfires, nor is there any evidence for meteor impact as an ignition mechanism at this time. Lightning would have been the main ignition source of wildfire in the run-up to the EPME, the occurrence of which was related to the climate and atmospheric pCO_2 (Glasspool and Scott 2010; Glasspool et al. 2015).

In the NCP, wildfire may have been the direct cause for both the destruction of terrestrial ecosystems and the rapid decline of plant biodiversity at the EPME, also playing an important role in the extinction of marine organisms (Shen et al. 2011; Zhang et al. 2016). Damage to the land surface vegetation system by frequent wildfire during the EPME interval would have led to increased soil erosion as well as exposing bedrock and increasing continental weathering leading to siltation (Shen et al. 2011).

In the study area, many greyish-green and purplish red mudstone facies formed in lake settings at the time of the CIE-II and persisted into the early Triassic in the ZK21-1 borehole core in the Yuzhou coalfield (southern NCP). This is a common phenomenon in the NCP and similar mudstone was also observed in the uppermost Sunjiagou Formation in a borehole core profile from the Liujiang area in Hebei province (central NCP), and the Shuiyuguan section in Shaanxi Province (central NCP) (Fig. 41). These mudstones may indicate the increased soil erosion after the collapse of terrestrial vegetation systems. The increased rates of soil erosion did not have a significant effect on chemical weathering, because it promoted erosion and transportation of the surface soil (Cao et al. 2019). However, the decline in CIA values following the EPPE may reflect loss of weathered soils through physical erosion (Cao et al. 2019; Frank et al. 2021).

As a result of wildfire, large amounts of organic matter (including charcoal and un-charred matter) and nutrients (including phosphorus and potassium) produced by plant combustion and weathering of parent rock would later enter the oceans through surface runoff (Algeo et al. 2013; Glasspool et al. 2015). These inert organic particles would float in ocean for some time, increasing oceanic turbidity through siltation, and affecting the penetration of light and the photosynthesis of marine organisms (Glasspool et al. 2015).

Large nutrient inputs may be one of the main reasons for prospering cyanobacteria and algae in oceanic surface waters (Meyer et al. 2011; Shen et al. 2015). In marine settings, bacterial (including cyanobacterial) communities thrived in the aftermath of the EPME (Grice et al. 2005; Kershaw et al. 2012). Similarly, marine algae experienced a near-global abundance "spike" in strata immediately succeeding the EPME (Thomas et al. 2004; Schneebeli-Hermann et al. 2012; Spina et al. 2015; van Soelen and Kürschner 2018). The initial post-EPME increase of algae in nearshore environments has been interpreted as an in situ bloom linked to decreased salinity due

Fig. 41 Abundant mudstone clasts in uppermost Sunjiagou Formation in the NCP. **a** Abundant greyish-green and purplish-red mudstone clasts in lake mudstone associated with CIE-II and lasting into the early Triassic in core ZK21-1, Henan province (southern NCP). *Note* The colors on the log mimic actual colors of rocks, and the numbers (e.g., *(1)*–*(3)*) represent the vertical order of lithologies. Abbreviations: M = Mudstone; S = Sandstone; **b** Mudstone clasts observed in the uppermost Sunjiagou Formation in the borehole core profile in Liujiang area, Hebei province (middle NCP). **c** Photograph of the Shuiyuguan section, Shaanxi province (middle NCP) showing the boundary of the Sunjiagou and Liujiagou Formations with highlighted box enlarged in **d. d** Enlargement from **c** showing details of mudstone clasts in the uppermost Sunjiagou Formation (diagram from Lu et al. 2020)

to enhanced input of freshwater (van Soelen et al. 2018) and/or soil nutrient (Kaiho et al. 2016). Eutrophication of seawater during the Permo-Triassic transition was considered as a localized phenomenon (Algeo et al. 2013) while Sun et al. (2019) considered the transformation of nitrogen to ammonium the main reason for the cyanobacterial proliferation at this time. However, a recent study (Mays et al. 2021) shows that the primary constituents of the continental algal assemblages are similar to the most common algae in several shallow marine records during the PTME intervals, suggesting that algae from freshwater blooms were transported to the marine realm en masse.

In this context, oxygen circulation between seawater and atmosphere would have been inhibited by floating inert organic particles, cyanobacteria and algae in surface waters, increasing the consumption of dissolved oxygen by the decomposition of dead cyanobacteria and algae remains (Algeo et al. 2013; Glasspool et al. 2015; Sun et al. 2019). Meanwhile, these cyanobacteria and algae blooms also produced secondary metabolites toxic to marine/lake animals (Paerl and Otten 2013). This would have further contributed to oceanic anoxia and the extinction of aerobic marine organisms (Lu et al. 2020) and impeding the recovery of Early Triassic freshwater ecosystems (Mays et al. 2021).

2.3 *Fluvial Style Changes During the Icehouse to Greenhouse Transition*

The coefficient of interannual peak discharge variance (COV; standard deviation of the peak flood discharge over the mean peak flood discharge) has been proposed to measure the inter-annual variability in the water discharge through a fluvial system (Fielding et al. 2018). These authors found a strong correlation between COV and the sedimentological characteristics of preserved alluvium in modern rivers, and proposed that this may form the basis for an alternative classification of alluvial deposits. Modern rivers can be divided into two end-members (low and high COV), with intermediate COV characterizing the patterns between two-end members (Fig. 42). The rivers with low COV tend to be reworked by multiple events of comparable magnitude and are dominated by cross-bedding at various scales with ripple cross-lamination. The rivers with high COV tend to preserve event deposits and are dominated by flat and low-angle bedding, mud drapes and in situ vegetation. The rivers with intermediate COV preserved a variety of styles of cross bedding, together with a record of critical and supercritical flow in the form of flat and low-angle bedding (Fielding et al. 2018). The alluvial features mentioned above form a more reliable basis for alluvial classification than attempting to identify planform from preserved geological characteristics.

In the Palougou and Shuiyuguan sections, the lower Permian to early upper Permian (Taiyuan Formation to Shangshihezi Formation) successions are characterized by sand and gravel-sized coastal fluvial and alluvial channel bodies with

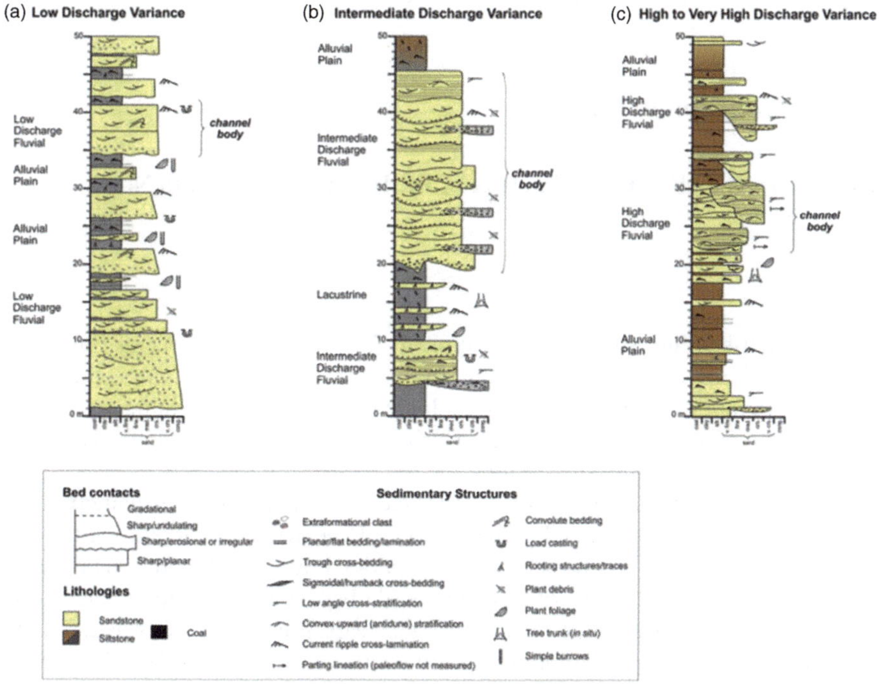

Fig. 42 Suite of three schematic graphic logs summarizing the sedimentological characteristics that define **a** low, **b** intermediate, and **c** high and very high peak discharge variance alluvial channel bodies. Diagram is based on intervals within the thick, stratigraphically complete Pennsylvanian succession of the Maritimes Basin Complex of Atlantic Canada (Fielding et al. 2018)

abundant cross-bedding and transported wood debris. The fluvial styles in the successions can be characterized as low discharge variance deposits. The low COV fluvial deposits are overlain by coal-bearing coastal floodbasin deposits and fine-grained floodplain deposits.

The latest Permian (Sunjiagou Formation) to Lower Triassic (Liujiagou Formation) successions are characterized by sand-rich fluvial channel bodies with some cross-bedding, but abundant flat and low-angle bedding (Fig. 43). The fluvial styles in the successions are characteristic of systems of moderate discharge variance. Critical and supercritical flow structures within fluvial channel bodies gradually increase upwards in the succession. A concomitant increase in the abundance of pedogenically modified floodplain facies is also observed. Coal deposits are absent from these successions. The double mud drapes and reverse paleocurrent orientations at several stratigraphic intervals indicate tidal modulation of fluvial outflow, and thus a location within the coastal, backwater zone. Coastal indicators die out upwards showing that the depositional environment became more inland fluvial over time (Fig. 43).

During the icehouse to greenhouse period, the transition from coastal fluvial and fluvial channel deposits with low discharge variance in the lower to initial upper Permian to fluvial channel deposits with moderate discharge variance in the late upper Permian to Lower Triassic is observed in North China.

2.4 *Milankovitch Cycles*

2.4.1 Milankovitch Cycle Analysis of Early Permian Strata in North China

Introduction

Paralic coal-bearing succession is typical for the intercalation of sandstone, mudstone, limestone, and coal, tending to display an obvious lithological cyclicity Cyclostratigraphy is a useful tool for geological age scaling, providing an excellent opportunity for improving resolution of Carboniferous–Permian geological time with these paralic coal-bearing successions.

Cyclostratigraphy mainly involves three earth orbital parameters, namely eccentricity (E, e), obliquity (o) and precession (p). Eccentricity is the ellipticity of the earth's elliptical orbit around the sun, with periods of 400,000 and 100,000 years. Obliquity is measured as the intersection angle between the ecliptic plane and the equatorial plane, with a period of about 40,000 years. Precession refers to the oscillation (wobble) of the earth's rotation axis around the vertical axis of the ecliptic plane, with a period of about 20,000 years. These three types of earth's orbital parameters of periodic change will cause the earth to experience cyclical changes in insolation leading to consequent cycles in biological productivity, glacier development, sea level change, and finally recorded in changes in sedimentary systems. Cycles forced by these orbital parameters are termed Milankovitch cycles. Once Milankovitch

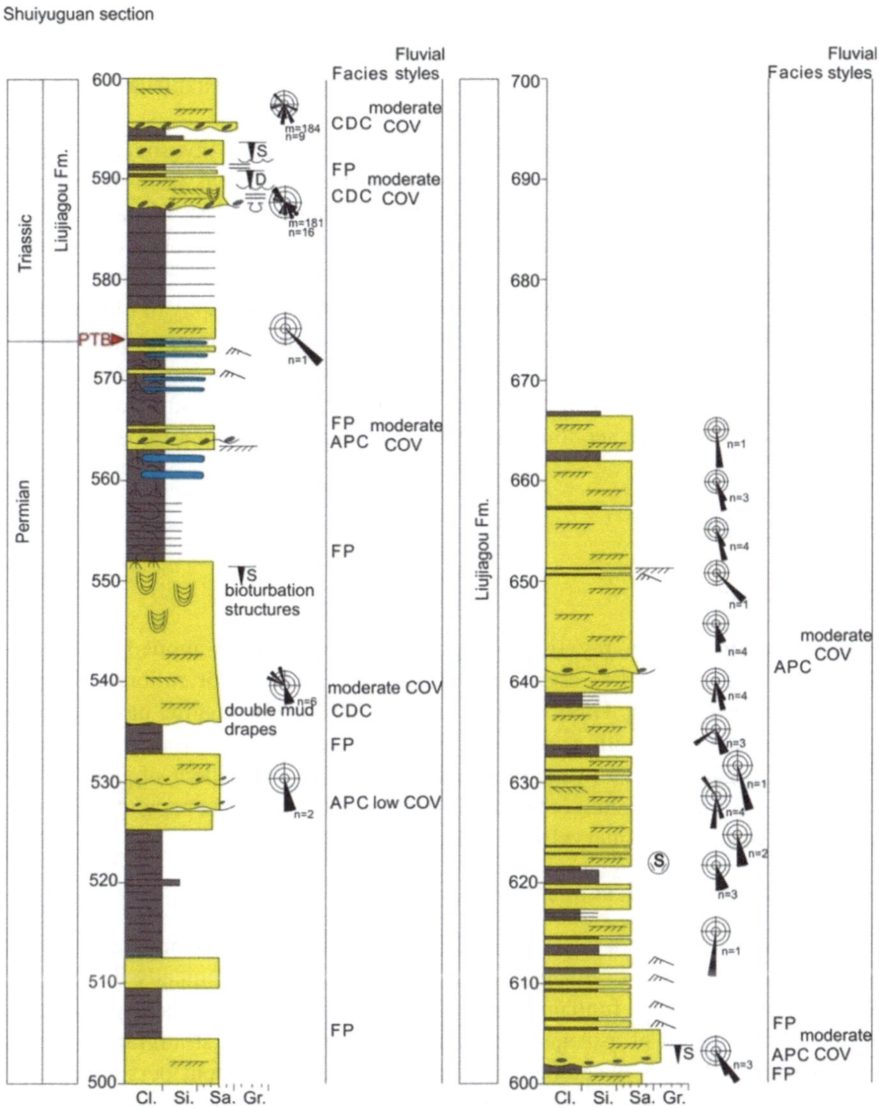

Fig. 43 Graphic log of the Sunjiagou Formation to Liujiagou Formation at Shuiyuguan section illustrating the progressive change in Facies and fluvial styles. APC—Alluvial plain channel; FP—Floodplain; CDC—Coastal distributary channel; low COV—The river with low coefficient of interannual peak discharge variance; moderate COV—The river with moderate coefficient of interannual peak discharge variance

cyclicity has been identified from strata, it is possible to establish a "floating" age scale with resolution of 0.4–0.02 Myr (Hinnov 2004). With radioisotope ages or established magnetic zone boundary ages as "anchors", the established age scale then has absolute age significance.

Because there are no data suitable for cyclostratigraphic investigation on the Palougou and Shuiyuguan sections, this study has used sections from the Shenmu region as the research object. Its geographical location is close to the Palougou section, the stratigraphic classification and tectonic characteristics are similar, and it has continuous Gamma Ray logging data from boreholes. By analogy to the study of Shenmu strata and logging, we can understand the cyclostratigraphy of the Palougou section and Shuiyuguan section. In future studies, magnetic susceptibility (MS) of the section can be measured to provide data support for cyclostratigraphic study. The Shanxi Formation in the Shenmu region which is adjacent at its east to the Baode area where the Palougou section is located is the target formation of this study.

Dataset and Methods

(1) Data selection

High-resolution gamma ray (GR) logging has been widely applied to analyze Milankovitch cycles and to estimate sedimentation rates and/or calibrate the ages of geological events (e.g. Hinnov 2013; Li et al. 2018a). GR logging has the advantages of easy data acquisition, good continuity, and sensitivity to reflect the content of mud and organic matter in sediments, thus reflecting the changes of paleoclimate and paleoenvironment (Chen et al. 2004; Wu et al. 2011). In this paper, GR logging with 0.125 m sampling interval were selected for analysis.

(2) Correlation coefficient method

The correlation coefficient method (COCO) is an astronomical test based on multiple statistical algorithms to evaluate whether the strata are astronomically driven. This function addresses two fundamental issues in cyclostratigraphy and paleoclimatology: identification of astronomical forcing in sequences of stratigraphic cycles, and accurate evaluation of sedimentation rates. The technique considers these issues part of an inverse problem and estimates the product-moment correlation coefficient between the power spectra of astronomical solutions and paleoclimate proxy series across a range of test sedimentation rates. The number of contributing astronomical parameters in the estimate is also considered. The estimation procedure tests the hypothesis that astronomical forcing had a significant impact on proxy records. The null hypothesis of no astronomical forcing is evaluated using a Monte Carlo simulation approach (Li et al. 2018b). This paper uses COCO method to perform astronomical test on *Acycle* software platform.

(3) Time-series analysis

Time-series analysis has been widely applied in many fields. In the field of geoscience, the application of this method has rapidly promoted the development of cyclostratigraphy (Wu et al. 2011). The identification of potential astronomical signals in stratigraphic data series using *Acycle* involves selection and sampling of paleoclimate proxy, data pre-processing, spectral analysis, filtering, tuning, and establishment of age scale (Li et al. 2019).

① Data pre-processing

Data pre-processing can eliminate the influence of non-orbit factors on the analysis results, including sort, interpolation, unique, detrending, pre-whitening, etc. (Fang 2015). The detrending operation is very important for time-series analysis. Due to the influence of tectonic activities of sedimentary basins and other factors, paleoclimate proxy often retain some long-period trends (Xue et al. 2020). Removing these long-period trends can avoid the influence of low frequency components in spectral analysis, thus improving the accuracy of spectral analysis results. In this study, the "locally weighted regression (LOWESS)" method provided by *Acycle* software was used to remove the long-period trend in GR logging data.

② Power spectral analysis

Power spectral analysis can identify the period information of earth orbit parameters preserved in paleoclimate proxy. In this study, the multi-taper method (MTM) estimators provided by *Acycle* software is used to carry out spectral analysis. This method is a low variance and high resolution spectral analysis method, which can provide the best trade-off between spectral resolution and variance (Jiang et al. 2001). Meanwhile, the "Robust AR" model was used to estimate the confidence level of spectral analysis results (Mann and Lees 1996).

③ Evolutionary power spectral analysis

The spectral power can vary with depth or time, especially for paleoclimatic and cylostratigraphic series, being a natural property of most geoscientific data and leading to potential distortion of paleoclimate signals in depth domain (Kodama and Hinnov 2014; Li et al. 2018b; Weedon 2003). When the length of the time series is longer than the cyclicity of interest, then it is possible to generate power spectra from many segments of the time series, given that each segment is long enough for detection of all cycle frequencies of interest. This is variously known as evolutionary (evolutive) spectral analysis, and sliding-window (running-window) spectral analysis (Kodama and Hinnov 2014; Li et al. 2018c; Meyers and Hinnov 2010; Meyers et al. 2011; Weedon 2003).

④ Filtering

Filtering is the operation of extracting periodic signals of target earth orbit parameters from paleoclimate proxy, including high-pass filtering, low-pass filtering and band-pass filtering, among which band-pass filtering is widely used (Fang 2015). In this paper, Gaussian band-pass filtering provided by *Acycle* software was used to extract 405 kyr eccentricity cycle signals from GR logging data (Kodama and Hinnov 2014).

⑤ Age model and tuning

Tuning is the operation of comparing cycle records (filter curves) in sediment or paleoclimate proxy to theoretical target curves of eccentricity, obliquity and precession (Huang 2014), aiming to establish high-resolution age scales on this basis. However, due to the lack of a reliable astronomical model in Paleozoic, there is no suitable target curve for filtering curve comparison, and the filtering curve can only be transformed from the depth domain to the time domain relatively simply (Fang 2015). In this paper, the "Build Age Model" function of *Acycle* software was used to assign a value of 405 kyr to each long eccentricity filtering cycle, and the data in the depth domain was converted to the data in the time domain. Then, use the "Age Scale" function to generate a time scale.

Calculation of Optimal Sedimentation Rate

The COCO method was used to test the hypothesis, and the sedimentation rate of the Shanxi Formation in the S-8 well was estimated statistically. Due to the lack of theoretical earth orbit parameters in the Paleozoic, the Berger94 astronomical solution (Berger and Loutre 1994) is used as the reference value. The early Permian astronomical solutions are 405 kyr (E), 95–123 kyr (e), 34.3–42.9 (o) and 17.4–20.7 (p).

According to previous biostratigraphy or high-precision zircon U–Pb geochronological studies, the sedimentary time of the Shanxi Formation varies widely, ranging from 2.5 to 5 Ma (Wang 2010; Liu et al. 2015; Shen et al. 2019b; Wu 2020). The depositional thickness of the Shanxi Formation in the S-8 well is 124 m. Therefore, a reasonable average deposition rate of Shanxi Formation is 2.5–4.9 cm/kyr.

Firstly, the range of sedimentation rate was limited to 0–20 cm/kyr with a step of 0.2 cm/kyr. The optimal sedimentation rate (the sedimentation rate corresponding to the larger value of ρ and H_0) in the reasonable range were 1.4, 2.2 and 3 cm/kyr, respectively. In order to obtain the optimal deposition rate with higher accuracy, the range of sedimentation rate was again limited to 1–4 cm/kyr with a step of 0.01 cm/kyr. The optimal sedimentation rate was obtained as 2.17 and 2.94 cm/kyr. When the sedimentation rate is 2.17 cm/kyr, the number of astronomical terms is small, so 2.94 cm/kyr is selected as the optimal sedimentation rate of Shanxi Formation in the S-8 well (Fig. 44).

Identification of Astronomical Periods

The MTM estimators of the GR logging stratigraphic series in well S-8 shows significant peaks above 99% confidence interval at 11.9, 3.6, 3.4, 2.8, 1.12, 1.10 and 1.09 m

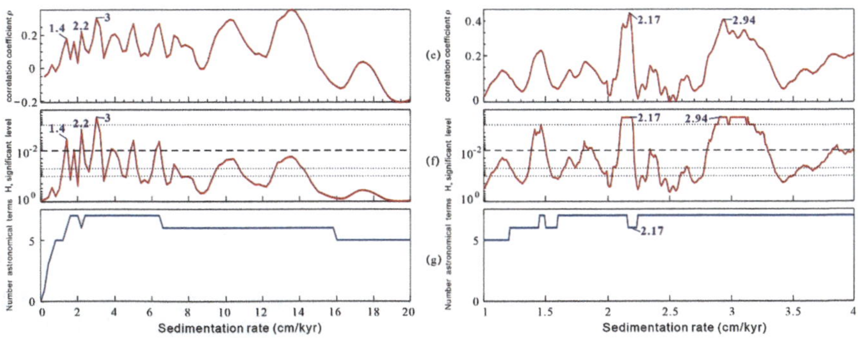

Fig. 44 COCO analysis of S-8 well. **a–c** Initial test of sedimentation rates range from 0 to 20 cm/kyr with a step of 0.2 cm/kyr; **e–f** Refined test of sedimentation rates range from 1 to 4 cm/kyr with a step of 0.01 cm/kyr; Monte Carlo simulation of 2000 iterations

and above 90% confidence interval at 0.94 m (Fig. 45). Previous work suggests that the optimal sedimentation rate is 2.94 cm/kyr for the Shanxi Formation. Therefore, the above stratigraphic cycles probably represent 405 kyr (long orbital eccentricity), 96.2–123.2 kyr (short orbital eccentricity), 31.8–38.1 kyr (obliquity). The precession period could not be identified in the S-8 well, either because the resolution of GR logging data sampling was insufficient, or because short-period information was easily lost due to various geological effects. Except for the absence of the precession period, other earth orbit parameters are highly consistent with the theoretical parameters given by Berger94 astronomical solution (Xue et al. 2020; Yang et al. 2021), and according to the above analysis, the Shanxi Formation in the Shenmu region retains a clear Milankovitch cycle signal.

Evolutionary MTM estimators (Fig. 45) shows the distribution of depositional cycle signals in the formation. Long orbital eccentricity signal distribution is continuous and stable, short orbital eccentricity signal is discontinuous, the obliquity signal is the weakest, and a precession signal cannot be recognized, consistent with MTM estimators.

Build Age Scale

The establishment of the "floating" age scale is achieved through 405 kyr astronomical correction in *Acycle* software. According to the previous spectral analysis, band-pass filtering was applied to Shanxi Formation in the S-8 well, and the stratigraphic cycles representing 405 kyr were filtered out, resulting in 10 long orbital eccentricity cycles (Fig. 46a). The duration of 405 kyr was assigned to each long orbital eccentricity cycle, thus converting GR logging from the depth domain to the time domain (Fig. 46).

According to the "floating" age scale, the time taken to accumulate the Shanxi Formation was 4.14 Ma (Fig. 46d). In the S-8 well, the thickness of the Shanxi

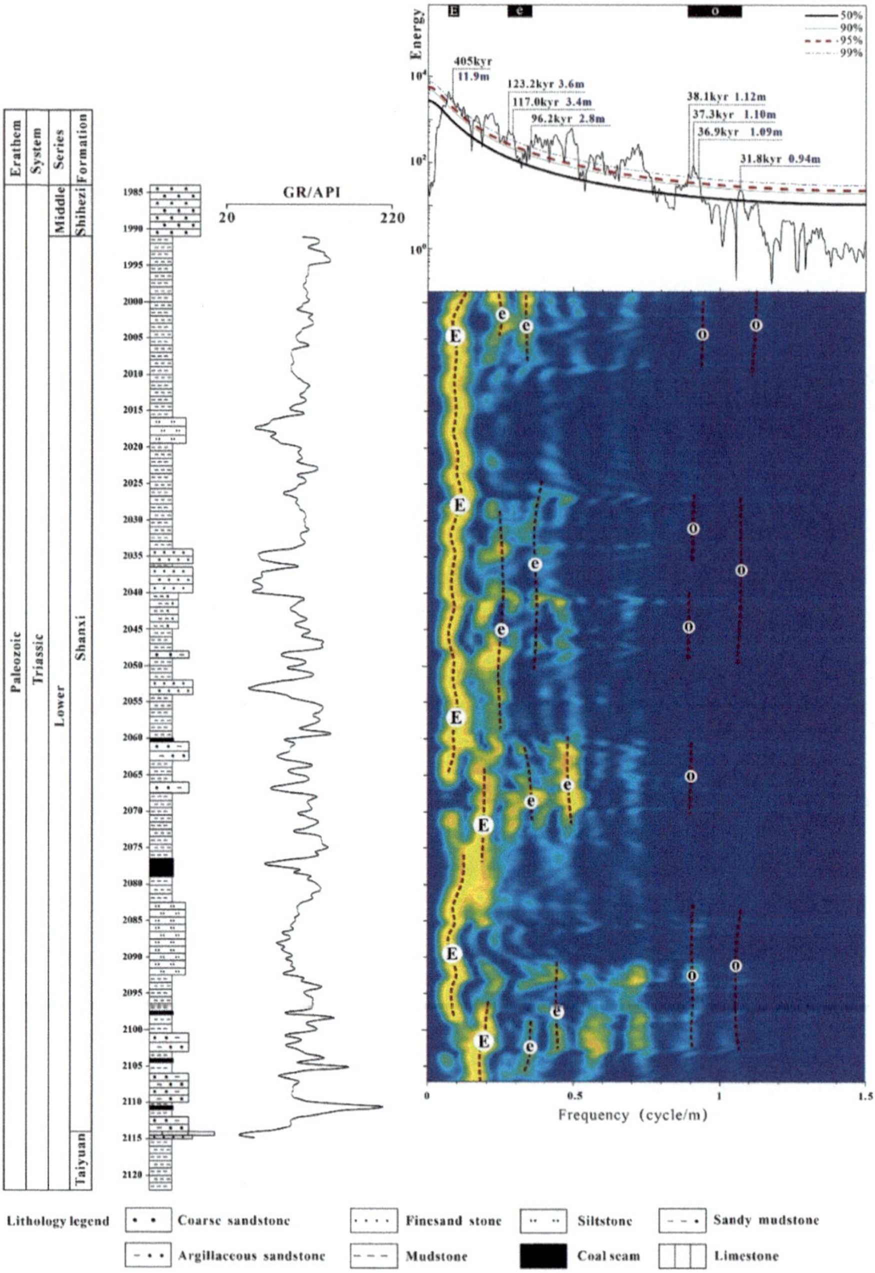

Fig. 45 MTM power spectrum of the S-8 well GR data. The black, green, red, and blue lines (from bottom to top) are 50, 90, 95, and 99% confidence level, respectively

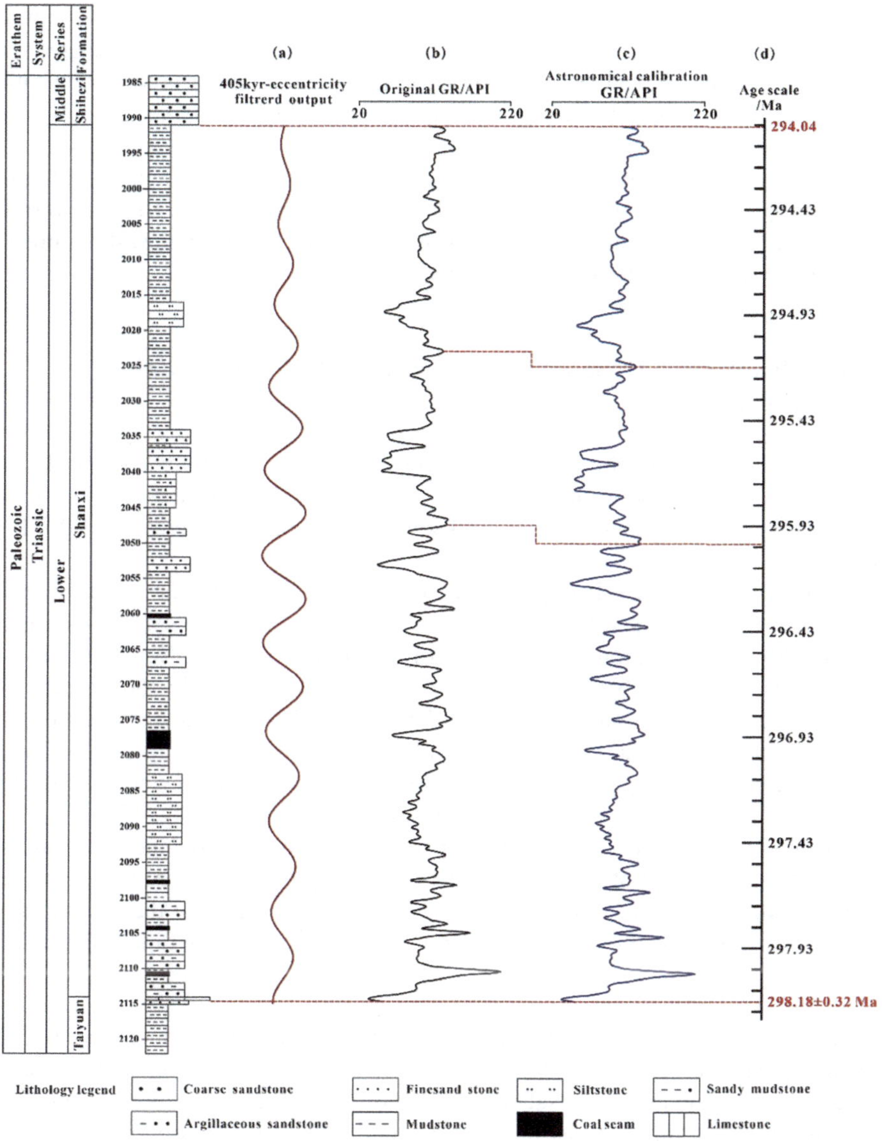

Fig. 46 "Floating" age scale of Shanxi Formation in S-8 well. **a** 405 kyr-eccentricity with the Gauss bandpass-filtered outputs (passband: 0.083787 ± 0.007787 cycles/m); **b** Original GR log, ordinate represents depth, corresponding to lithology column; **c** Astronomical calibration GR log, the ordinate represents time, corresponding to the age scale (**d**)

Formation is 124 m, so the average sedimentation rate was 2.995 cm/kyr, which is highly consistent with the optimal sedimentation rate of 2.94 cm/kyr calculated by COCO. In addition, the sedimentation duration age of the formation is consistent with the high-precision zircon U–Pb geochronological studies (Wu 2020), confirming the reliability of the age scale.

The zircon U–Pb dates of Wu (2020) has provided the most reliable and high-precision age constrains the Shanxi Formation at 298.18 ± 0.32 Ma, which confirms the "anchor" of the "floating" astronomical age scale and establishes the astronomical age scale of the Shanxi Formation (Fig. 46d).

2.4.2 Milankovitch Cycle Signals Recorded in Thick Coal Seams of Paralic Facies

Thick coal seams in paralic settings in North China have recently been considered as successions of stacked mires developed by several base-level oscillations and separated by multiple hiatal surfaces (Glikson and Fielding 1991; Jerrett et al. 2011a, b).

To study the possible role of base-level oscillations in the formation of thick coal seams in coastal plain (paralic) environments, the No.21 thick coal seam of the early Permian Shanxi Formation in the Anhe (Anyang-Hebi) coalfield, Henan Province, in southern zone of the North China Plate, was studied. Regionally, the No.21 coal in the Anhe coalfield is subcorrelated to the No.4 coal in the Palougou section or the No.4 coals in the Shuiyuguan section of Shanxi Province (Shang 1997). Three borehole cores spanning a distance of 29 km in the Anhe coalfield were investigate using sedimentology, sequence stratigraphy, and coal petrology (Fig. 47).

The No.21 coal seam was developed as part of a TST and displays maximum thickness in the central area of the Anhe coalfield, thinning both landward and seaward. The splitting of the No.21 coal seam by siliciclastic sediments in the central area can be mostly attributed to locally accelerated subsidence, and the rapid subsidence in the lower delta plain led to the development of thick coal seams there (Fielding 1984, 1987).

Six types of key stratigraphic surfaces are recognized in the thick coal seam, including terrestrialization surface, paludification surface, give-up transgressive surface, accommodation reversal surface, exposure surface, and flooding surface. The No.21 coal seam consists of at least three drying-up and wetting-up cycles defined by these surfaces, with each cycle spanning from 20 to 30 ka (Fig. 48). These cycles arrived at a period close to that of climatic precession for Milankovitch cycles during the early Permian, and therefore the precession cycles are recorded in paralic thick coal seams formed in cratonic settings. The oscillating high and low summer isolation may have caused the alternating wet and dry climate periods that were then recorded as the wet-dry cycles in the coal seams (Li et al. 2020a, b).

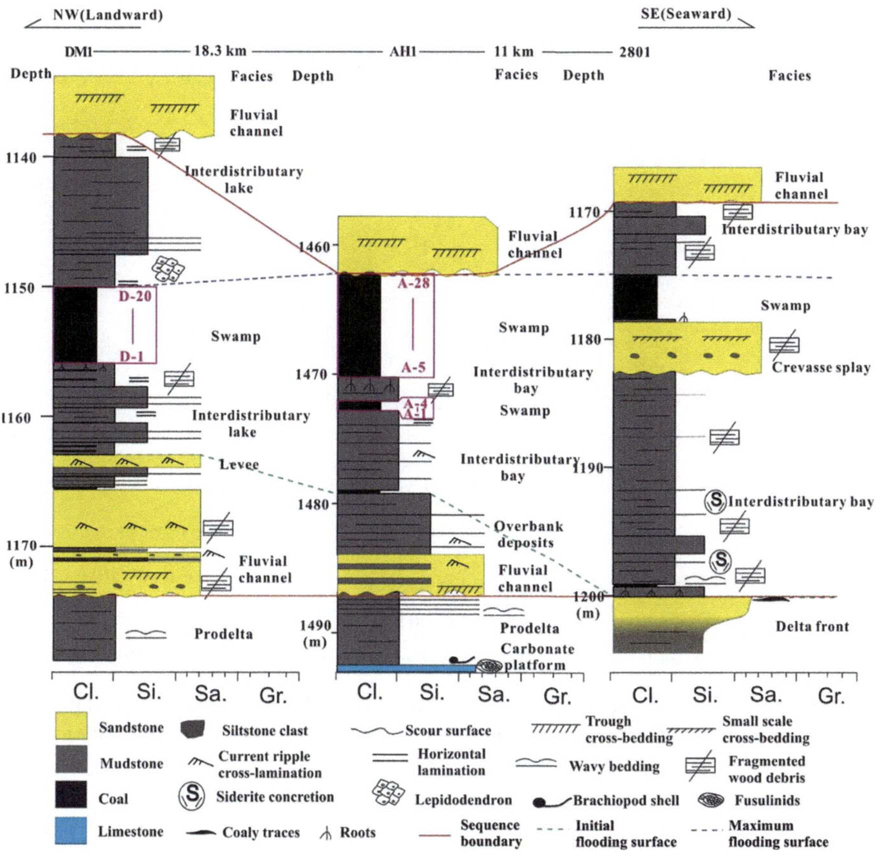

Fig. 47 Schematic cross section showing vertical and lateral variation of the No.21 coal member in the Anhe coalfield (Li et al. 2020a, b). The pink box represents the coal samples in boreholes DM1 and AH1

3 Field Trip Routes

Taiyuan is located in the center of Shanxi Province, 400 km southwest of Beijing, at the north end of Taiyuan Basin, and the intersection of North–South Tongpu and Shijiazhuang Taiyuan railway lines (Fig. 49). Jiaocheng County is under the jurisdiction of Lvliang City, Shanxi Province. It is located at the eastern foot of Lvliang Mountain, in the middle of Shanxi Province, 51 km away from Taiyuan. Baode county is subordinate to Xinzhou, Shanxi Province, 180 km northwest of Taiyuan. Pingshuo mining area is located in Shuozhou, Shanxi Province, about 100 km away from Datong.

Shuiyuguan section is located in the middle of Lvliang Mountain and is southwest of Taiyuan Xishan coalfield, at 111° 52′–112° 00′ E and 37° 37′–37° 45′ N. The

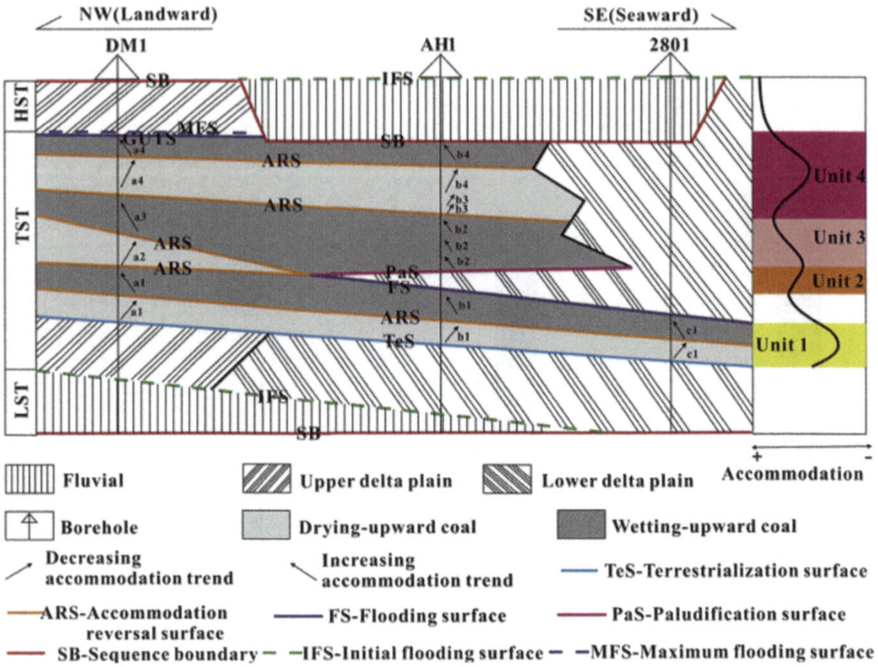

Fig. 48 A sequence stratigraphic model showing the spatial and temporal correlation of the No.21 coal member with interpreted sequence stratigraphic surfaces. The interfingering black line indicates a lateral facies transition (Li et al. 2020a, b)

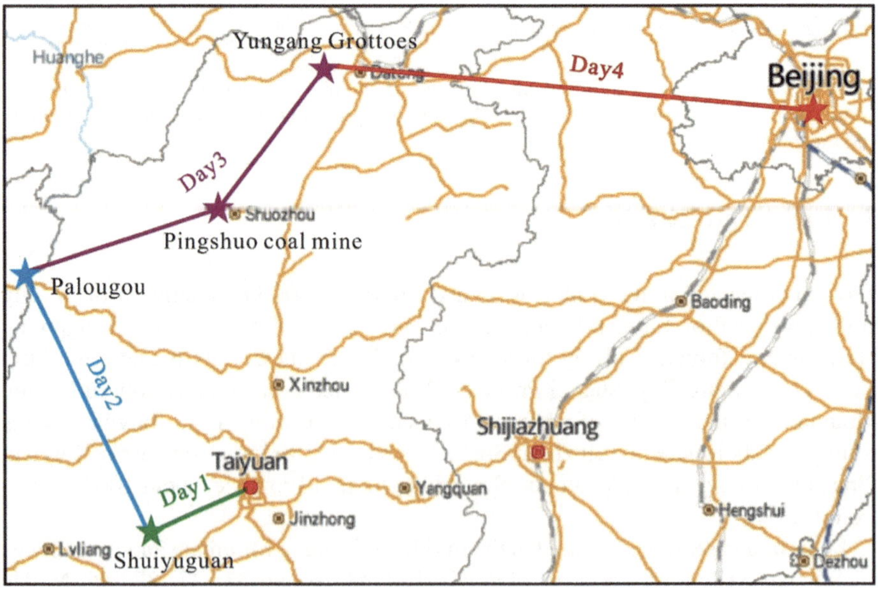

Fig. 49 Traffic location map showing the field trip routes

Fig. 50 Location map of Paluogou and Shuiyuguan sections in Shanxi Province

straight-line distance from Taiyuan is 56.5 km, and provincial highway S219 runs through the area, with convenient transportation (Fig. 50).

The Palougou section is located in Palougou village, Nanhegou Township, south of Baode County, Shanxi Province, 46 km away from the county seat. The geographical coordinates are 111° 8′ 41.55″ E and 38° 45′ 34.78″ N. A road in the village leads to the coal mine in the village (Fig. 50).

Pingshuo mining area is located in Shuozhou, Shanxi Province. It belongs to the northern area of the Ningwu coalfield, about 100 km from Datong, with an area of 380 km^2, and geological reserves of 12.75 billion tonnes. The large-scale open-pit coal mine is divided into three coal fields, namely Antaibao open-pit coal mine, Anjialing open-pit coal mine and East open-pit coal mine (Figs. 51 and 52).

The Yungang Grottoes are located at the south foot of Wuzhou (Zhou) mountain and the north bank of Wuzhouchuan, about 16 km west of Datong, Shanxi Province, China. It is an undulating, low mountain and hilly area. Its geographical location is 113° 20′ E and 40° 04′ N, and its topographic elevation is 1133–1178 m. The grottoes were excavated along the mountainside, on a grand scale, extending for about 1 km from east to west. The grottoes are divided into three areas according to the natural mountain from east to west. There are 45 main caves and 209 affiliated caves, with a carving area of more than 18,000 m^2 (Fig. 53). The highest statue is 17 m and the smallest is 2 cm. There are about 1100 Buddhist niches and more than 51,000 statues. Yungang Grottoes has a history of 1500 years. It is the first Buddhist art treasure house carved by a nation in one dynasty after Buddhist art spread to China. It is a historical monument of the integration of Chinese and Western cultures in the fifth century. In March 1961, it was announced by the State Council as the first batch

Fig. 51 Location map of Pingshuo open-pit coal mine and Yungang Grottoes in Shanxi Province

of national key cultural relic protected units; In December 2001, it was approved by UNESCO to be included in the list of "world cultural heritage"; In May 2007, it became one of the first 5A scenic spots in China.

4 Itinerary

4.1 Section Investigation Routes

The Shanxi field section investigation route includes the Shuiyuguan section near Jiaocheng and the Palougou section near Baode. In the Carboniferous–Permian strata of these two sections, 14 sedimentary facies are identified, which are channel, distributary channel, incised valley fill, floodplain, levee, swamp, floodbasin, delta front, pro-delta, tidal channel, tidal sand bar, lagoon, offshore shale and offshore carbonate facies. The sedimentary characteristics of these sedimentary facies are shown in Table 2.

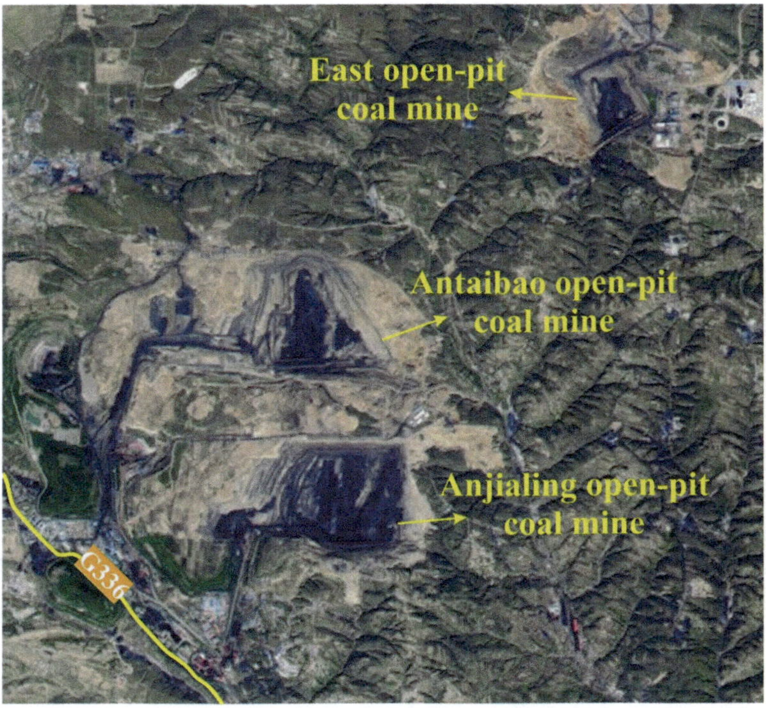

Fig. 52 An aerial view showing the three coal fields of Pingshuo open-pit coal mine, Shanxi Province

4.1.1 Day One: Shuiyuguan Section in Jiaocheng County

The Shuiyuguan section is located in Shuiyuguan Town, Jiaocheng County, Luliang City, Shanxi Province (Fig. 55). The strata exposed in this section range from the Carboniferous Benxi Formation to the Triassic Heshanggou Formation from bottom to top. The main investigation objectives are the sedimentary characteristics and important geological phenomena of Carboniferous–Permian strata. The investigated strata include Benxi, Taiyuan, Shanxi, Xiashihezi, Shangshihezi, Sunjiagou and Liujiagou formations (Figs. 56 and 57). See Table 3 for detailed stratum description.

Stop 1.1: Unconformable contact between Ordovician limestone and overlying Carboniferous ferro-aluminous mudstone of the Benxi Formation

GPS co-ordinates: 37° 40′ 55.52″ N, 111° 56′ 46.40″ E

Objectives: Ordovician Fengfeng Formation limestone and overlying Carboniferous Benxi Formation ferro-aluminous mudstone

Description and interpretation: The gray limestone of the Fengfeng Formation is in unconformable contact with the reddish brown ferrous mudstone of Benxi Formation, and the gray-yellow aluminous mudstone overlying the ferrous mudstone is rich in plant root fossils and *Planolites* trace fossils. The limestone of Fengfeng Formation

Fig. 53 Cave 20, Yungang Grottoes, Datong City, Shanxi Province

is interpreted as shallow sea carbonate deposition, and the ferro-aluminous mudstone of Benxi Formation is interpreted as offshore marine mudstone deposition (Figs. 58, 59, and 60).

Stop 1.2: Carboniferous coastal distributary channel Jinci Sandstone

GP co-ordinates: 37° 40′ 53.07″ N, 111° 56′ 46.93″ E

Objectives: Sedimentary characteristics of distributary channel Jinci Sandstone at the base of the Carboniferous Taiyuan Formation

Description and interpretation: Thick Jinci Sandstone is developed at the base of the Taiyuan Formation, and the lower member is grayish brown medium-coarse sandstone with tabular cross-bedding; The middle member is gray white medium-coarse grained quartz sandstone with trough cross-bedding and fossil plant axes; The upper member is grayish brown gravelly coarse sandstone and conglomerate, with tabular cross-bedding. Figure 61 shows the fining-upward sedimentary sequence of the middle and upper member of Jinci Sandstone, which is interpreted as distributary channel deposition.

Stop 1.3: Carboniferous tidal channel Ximing Sandstone

GPS co-ordinates: 37° 40′ 50.18″ N, 111° 56′ 48.76″ E

Objectives: Sedimentary characteristics of tidal channel Ximing Sandstone of the Carboniferous Taiyuan Formation

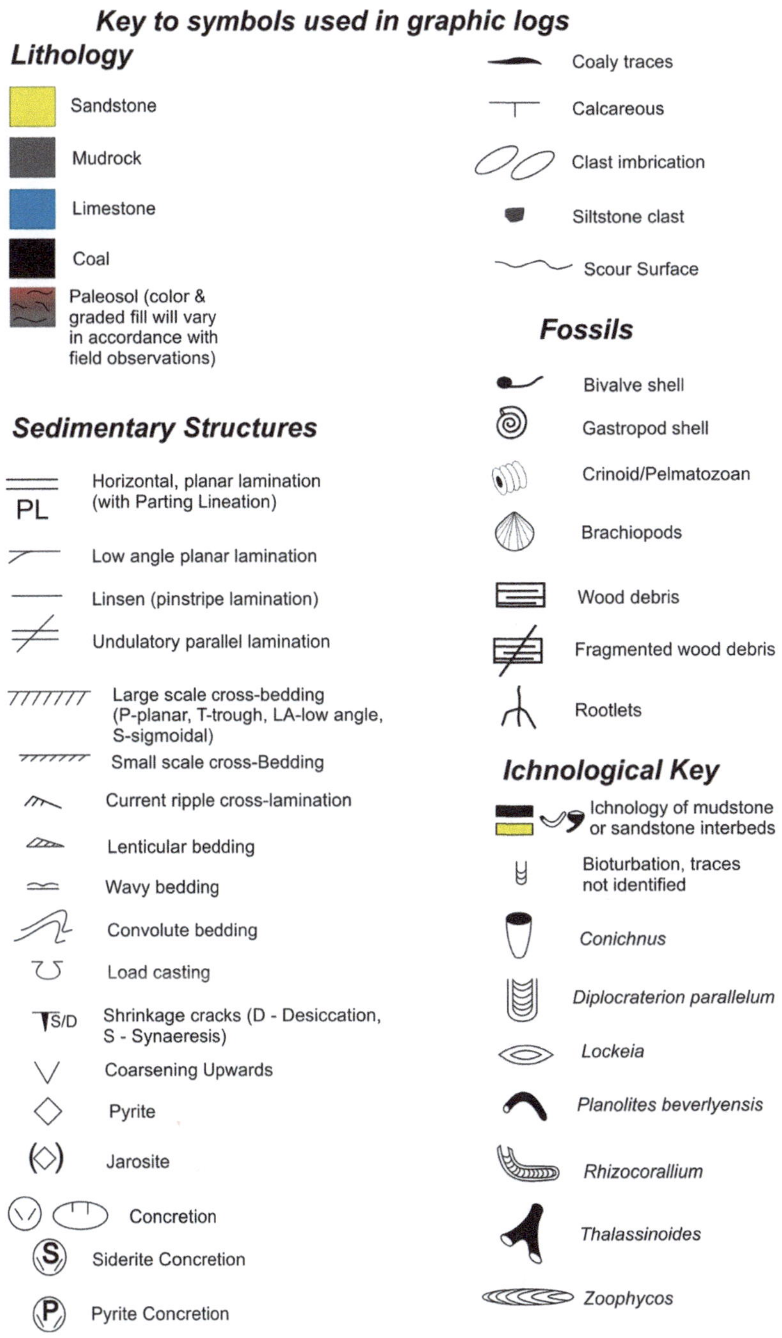

Fig. 54 Key to symbols used in graphic logs of Figs. 56 and 57

Table 2 Summary of the characteristics and interpretation of the facies within the Permian and Lower Triassic successions in the Palougou and Shuiyuguan sections

Facies	Interpretation	Lithology	Sedimentary structures/fossils	Geometry
1	Alluvial plain channel (APC)	Erosionally-based sandstone	Cross-bedding, clast imbrication, ripple cross-lamination	Extensive sheets
2	Coastal distributary channel (CDC)	Erosionally-based sandstone	Cross-bedding, bidirectional/bimodal cross-bedding, clast imbrication, double mud drapes, convolute bedding, bioturbation, undulatory parallel lamination, ripple cross-lamination, horizontal lamination, load casts, plant debris, tree trunks and coaly traces	Extensive sheets
3	Incised valley fill (IVF)	Erosionally-based thick multi-storey sandstone bodies interbedded with siltstone	Trough cross-bedding, bidirectional/bimodal cross-bedding, double mud layers, load casts, low angle planar lamination, horizontal lamination, clast imbrication, plant debris, Jarosite stain, coaly traces, trace fossils (*Diplocraterion*, *Thalassinoides*)	Sheets
4	Floodplain (FP)	Siltstone, and mudstone with color mottling and pedogenic nodular carbonate interbedded with thin sandstone partings	Horizontal lamination and pedogenic modification (coloration, mottling, plant root penetration)	Sheets
5	Levee	Regularly interbedded siltstone and sandstone	Ripple cross-lamination and horizontal lamination	Sheets and lenses
6	Coastal mire (CM)	Coal and carbonaceous mudstone	Clastic partings, pyrite stain	Sheets, <4m

(continued)

Table 2 (continued)

Facies	Interpretation	Lithology	Sedimentary structures/fossils	Geometry
7	Coastal floodbasin (CF)	Stratified and root-penetrated siltstone and mudstone, and thin sandstone	Pinstripe lamination, ripple cross-lamination, plant debris, bivalve shells, jarosite stain and siderite concretions	Sheets and lenses
8	Delta front (DF)	Coarsening-upward siltstone and sandstone	Lenticular bedding, wavy bedding, pinstripe lamination, horizontal lamination, low angle planar lamination, current ripple cross-lamination, large scale cross-bedding, plant debris, pyrite and siderite concretions	Sheets and lobate aprons
9	Pro-delta (PD)	Siltstone interbedded with thin sandstone parting	Pinstripe lamination, plant debris and siderite concretion	Sheets
10	Transgressive, offshore black marine shale (OBMS)	Dark grey shale	Bivalve shells, brachiopods, gastropods	Sheets
11	Clear-water, offshore marine carbonates (OMC)	Bioclastic limestone	Invertebrate fossils (brachiopods, bivalves, gastropods, crinoids), trace fossils (*Conichnus*, *Diplocraterion*, *Lockeia*, *Planolites*, *Rhizocorallium*, *Thalassinoides*, *Zoophycos*)	Sheets and lenses
12	Tidal sand bar (TSB)	Coarsening-upward sandstone, gravelly coarse sandstone and conglomerate	Tabular cross-bedding, trough cross-bedding, bidirectional/bimodal cross-bedding	Sheets

(continued)

Table 2 (continued)

Facies	Interpretation	Lithology	Sedimentary structures/fossils	Geometry
13	Tidal channel (TC)	Erosionally-based sandstone	Tabular cross-bedding and bidirectional/bimodal cross-bedding	Extensive sheets
14	Lagoon	Dark grey mudstone	Horizontal lamination, plant root and siderite concretion	Sheets

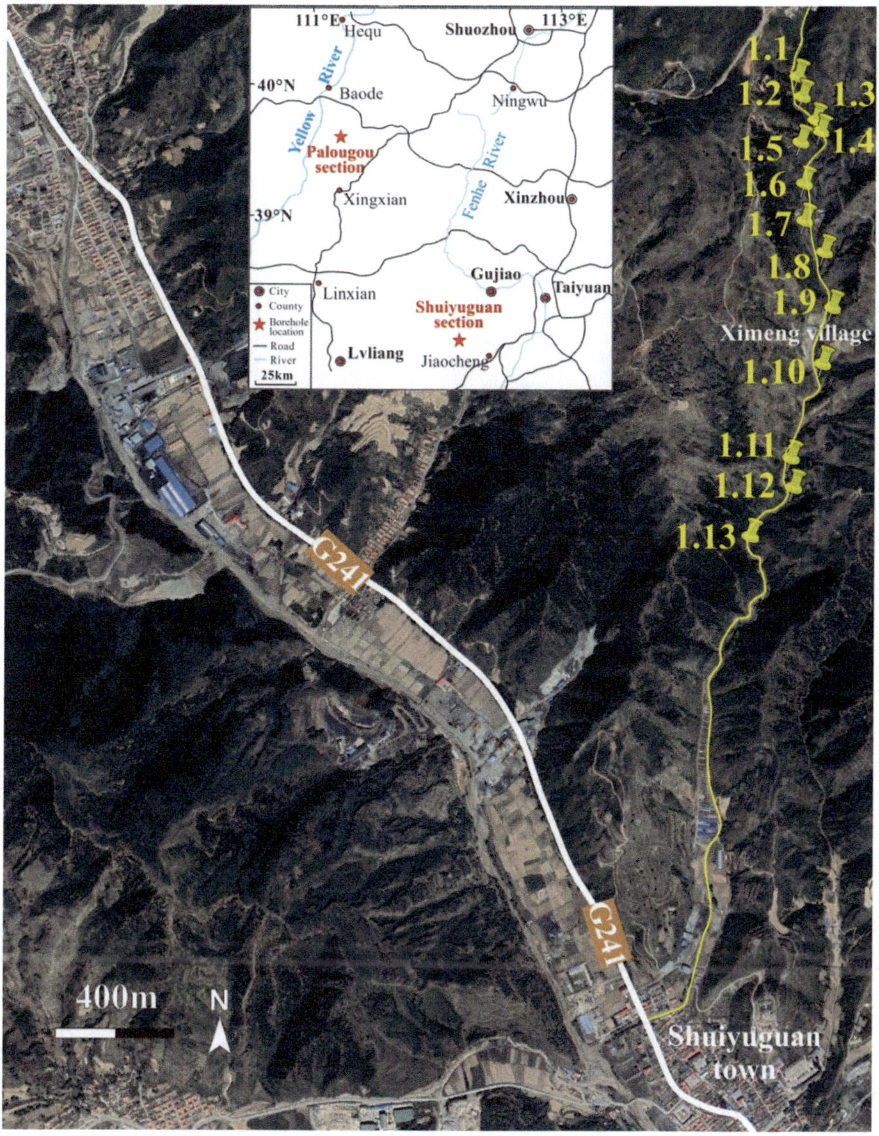

Fig. 55 Location map of observation points in Shuiyuguan section, Shanxi Province

Description and interpretation: The Ximing Sandstone in the lower part of the Taiyuan Formation is gray to white medium to coarse-grained sandstone. The basal surface is erosional, and trough cross-bedding (including bidirectional–herringbone–cross-bedding) are developed in the sandstone. The sedimentary sequence is interpreted as recording tidal channel deposition (Figs. 62 and 63).

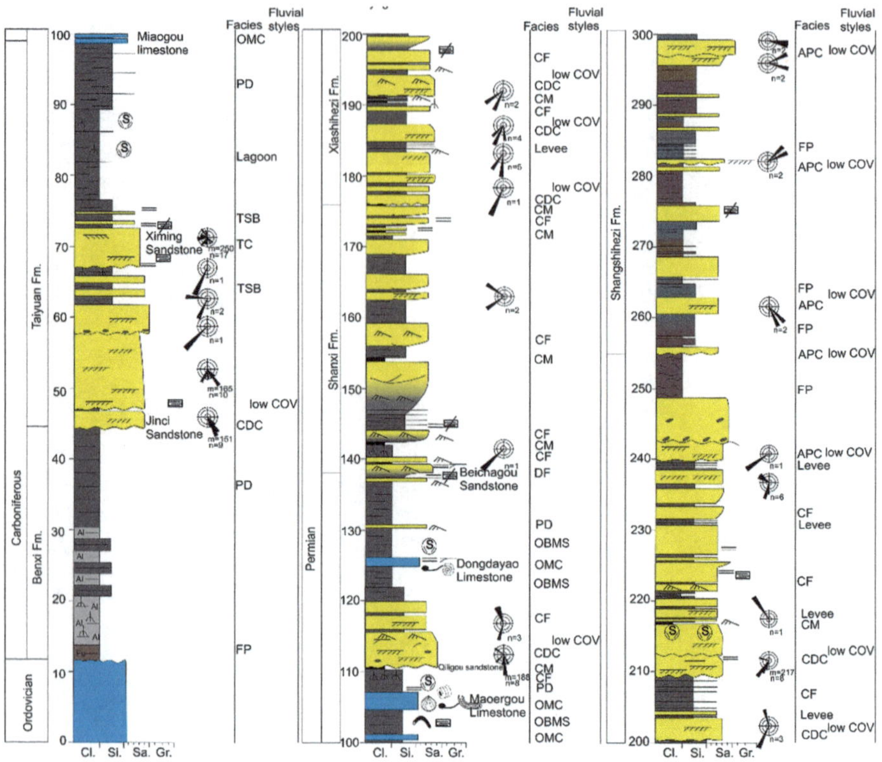

Fig. 56 Graphic log of the Benxi Formation to Shangshihezi Formation in the Shuiyuguan section illustrating the progressive change in Facies and fluvial styles. Low COV-low coefficient of discharge variance; The abbreviations of Facies are listed in the Table 2. See Fig. 54 for symbol key

Stop 1.4: Sedimentary characteristics of dark gray lagoon mudstone of the Taiyuan Formation in the Late Carboniferous

GPS co-ordinates: 37° 40′ 48.79″ N, 111° 56′ 49.50″ E

Objectives: Dark gray mudstones of the Late Carboniferous Taiyuan Formation

Description and interpretation: Thick dark gray mudstone with a thin layer of siltstone is developed in the upper part of the Upper Carboniferous section, with a thickness of 12.9 m. The mudstone contains siderite concretions, plant root fossils and horizontal bedding in the lower part, but is thinly laminated in the upper part. The sedimentary sequence is interpreted as lagoon deposits (Fig. 64).

Stop 1.5: Sedimentary characteristics of epicontinental sea deposits in the Early Permian

GPS co-ordinates: 37° 40′ 47.51″ N, 111° 56′ 46.54″ E

Objectives: Sedimentary characteristics of shallow sea carbonate limestone and underlying offshore shelf argillaceous rock of Taiyuan Formation in the Early Permian

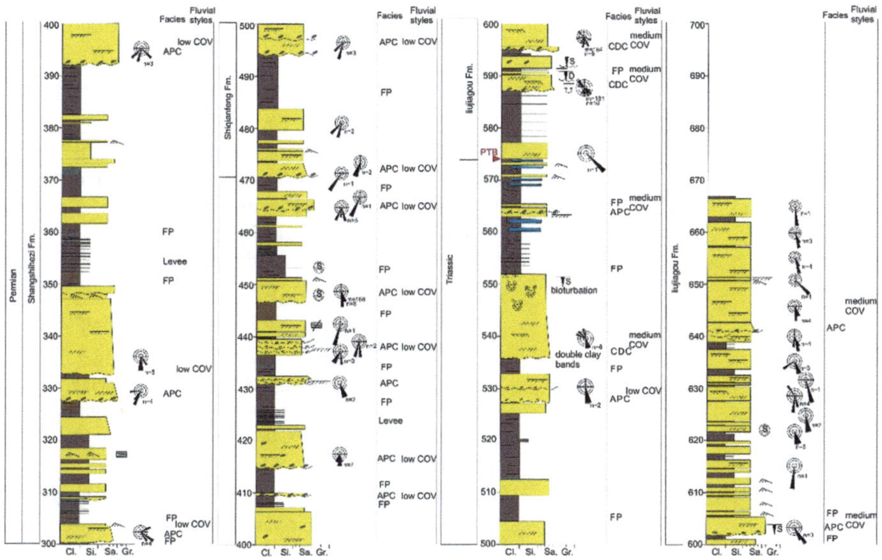

Fig. 57 Graphic log of the Shangshihezi Formation to Liujiagou Formation in the Shuiyuguan section illustrating the progressive change in Facies and fluvial styles. Low COV—low coefficient of discharge variance; Medium COV—medium coefficient of discharge variance; The abbreviations of Facies are listed in the Table 2. See Fig. 54 for symbol key

Description and interpretation: Early Permian Gray to white Miaogou Limestone, dark gray mudrock and gray to white Maoergou Limestone are exposed from base to top (Fig. 65). Horizontal bedding is developed in argillaceous rocks, and limestone contains a large number of brachiopod, bivalve, and crinoid fossils (Fig. 66), and *Rhizocorallium* trace fossils. The limestone in the sedimentary sequence is interpreted as shallow open marine carbonate deposits, and the argillaceous rock layer is interpreted as offshore muddy shelf deposit. The sequence of limestone and siltstone interbedded with thin fine sandstones represents vertical alternating cyclothem of marine and nonmarine facies.

Stop 1.6: Delta deposits at the boundary between the Permian Taiyuan and Shanxi Formations

GPS co-ordinates: 37° 40′ 42.55″ N, 111° 56′ 46.90″ E

Objectives: Sedimentary characteristics of delta deposits at the boundary between the Permian Taiyuan Formation and Shanxi Formation

Description and interpretation: The base of the Beichagou Sandstone is the basal boundary of the Shanxi Formation. The Beichagou Sandstone is gray to white, medium to fine-grained sandstone, containing detrital plant remains, and preserves horizontal bedding, low angle planar lamination and current ripple cross-lamination. The Beichagou Sandstone is overlain by a thin layer of carbonaceous mudstone and coal, and then transits upward to grayish brown fine-grained sandstone and dark

Table 3 Stratigraphic characteristics of Benxi Formation to Liujiagou Formation in the Shuiyuguan section, Shanxi Province

Stratigraphy	Layer no.	Thickness (m)	Description	Interpretation
Liujiagou Fm.	361–374	27.81	Erosional-based grey and greyish white fine- to medium-grained sandstone, intercalated with multi-layer purplish red siltstones, developed with cross-beddings and ripple cross-beddings	Fluvial channel
	360	1.60	Interbedded purplish red siltstones and mudstones	Floodplain
	347–359	15.15	Grey fine-grained sandstones, intercalated with multi-layer purplish red siltstones, developed with cross-beddings	Fluvial channel
	327–346	15.36	Interbedded grey fine-grained sandstones, purplish red siltstones and mudstones, developed with ripple cross-beddings	Floodplain
	326	3.40	Erosional-based greyish purple fine-grained conglomerate, developed with cross-beddings	Fluvial channel
	325	0.70	Purplish red mudstones	Floodplain
	323–324	6.10	Erosional-based grey fine- to medium-grained sandstone, developed with cross-beddings and channel lag deposits	Distributary channel
	317–322	4.73	Interbedded purplish red fine-grained sandstones, siltstones and mudstones, developed with horizontal beddings	Floodplain
	315–316	2.30	Erosional-based purplish red fine- to coarse-grained sandstone, developed with cross-beddings and parallel beddings	Distributary channel
	314	10.00	Purplish red mudstones, intercalated with multi-layer thin fine-grained sandstones	Flood basin
	313	3.08	Purplish grey fine-grained sandstones, developed with cross-beddings	Distributary channel

(continued)

Table 3 (continued)

Stratigraphy	Layer no.	Thickness (m)	Description	Interpretation
Sunjiagou Fm.	304–312	8.34	Purplish red mudstones, intercalated with multi-layer thin grey fine-grained sandstones, developed with ripple cross-beddings and plant root fossils	Floodplain
	303	1.60	Greyish white fine-grained sandstones, developed with gravel layer and cross-beddings	Fluvial channel
	301–302	11.30	Interbedded purplish red mudstones and siltstones, developed with plant root fossils	Floodplain
Sunjiagou Fm.	298–300	16.50	Erosional-based purplish grey fine- to medium-grained sandstone, developed with bioturbation and double mud drapes	Distributary channel
	297	3.20	Purplish red mudstones	Floodplain
	294–296	7.50	Erosional-based purplish grey fine- to medium-grained sandstone, developed with gravel layer and cross-beddings	Fluvial channel
	291–293	12.90	Purplish red and greyish green mudstones, intercalated with thin siltstone	Floodplain
	290	3.00	Greyish white medium-grained sandstone	Distributary channel
	289	5.10	Purplish red mudstones	Floodplain
	287–288	10.00	Purplish red fine-grained sandstones, developed with cross-beddings and channel lag deposits	Fluvial channel
	286	10.80	Purplish red mudstones	Floodplain
	283–285	6.60	Purplish red fine-grained sandstones, developed with cross-beddings	Distributary channel
	280–282	3.00	Purplish red mudstones, intercalated with thin fine-grained sandstone	Floodplain
	274–279	5.25	Erosional-based purplish red fine- to medium-grained sandstone, developed with cross-beddings and channel lag deposits	Fluvial channel

(continued)

Table 3 (continued)

Stratigraphy	Layer no.	Thickness (m)	Description	Interpretation
Shangshihezi Fm.	273	3.00	Purplish red mudstones	Floodplain
	269–272	4.65	Erosional-based greyish green medium- to coarse-grained sandstone, developed with cross-beddings	Fluvial channel
	263–268	11.65	Purplish red mudstones and siltstones, intercalated with greyish white thin fine-grained sandstones, developed with siderite concretions	Floodplain
	261–262	4.80	Erosional-based greyish green fine- to medium-grained sandstone, developed with cross-beddings and siderite concretions	Fluvial channel
	260	3.60	Brown yellow mudstones	Floodplain
	255–259	6.36	Erosional-based greyish white fine- to medium-grained sandstone, developed with cross-beddings and fossil plant debris	Fluvial channel
	251–254	4.10	Purplish red mudstones	Floodplain
	249–250	0.80	Erosional-based greyish white medium- to coarse-grained sandstone, developed with cross-beddings	Fluvial channel
	248	5.00	Purplish red mudstones	Floodplain
Shangshihezi Fm.	244–247	3.98	Interbedded yellowish brown siltstones and dark grey mudstones	Levee
	240–243	8.10	Erosional-based greyish white medium- to coarse-grained sandstone, developed with cross-beddings and channel lag deposits	Fluvial channel
	237–239	1.90	Interbedded yellowish brown siltstones and purplish red mudstones	Floodplain
	235–236	15.30	Erosional-based greyish white medium- to coarse-grained sandstone, developed with cross-beddings and channel lag deposits	Fluvial channel

(continued)

Table 3 (continued)

Stratigraphy	Layer no.	Thickness (m)	Description	Interpretation
	223–234	25.28	Purplish red mudstones, intercalated with grey thin fine-grained sandstones and siltstones	Floodplain
	217–222	8.10	Purplish red mudstones, intercalated with yellowish brown fine-grained sandstones	
	211–216	8.77	Interbedded yellowish brown siltstones and purplish red mudstones	Levee
	203–210	21.60	Erosional-based greyish white fine- to coarse-grained sandstone, developed with cross-beddings and channel lag deposits	Fluvial channel
	200–202	3.13	Purplish red and greyish green mudstones	Floodplain
	199	3.28	Greyish white fine- to medium-grained sandstone	Fluvial channel
	191–198	10.50	Interbedded purplish red siltstones and brown yellow fine-grained sandstones	Floodplain
	184–190	6.05	Interbedded brown yellow siltstones and purplish red/greyish green mudstones	Levee
	182–183	3.47	Erosional-based grey fine- to medium-grained sandstone, developed with cross-beddings	Fluvial channel
	181	1.20	Greyish green siltstones	Floodplain
	178–180	3.57	Erosional-based purplish red fine- to coarse-grained sandstone, developed with cross-beddings	Fluvial channel
	166–177	12.42	Greyish green siltstones, intercalated with fine-grained sandstones and mudstones	Floodplain
	161–165	1.65	Erosional-based greyish green fine- to medium-grained sandstone, developed with cross-beddings	Fluvial channel
	159–160	4.65	Purplish red mudstones	Floodplain
	157–158	2.25	Purplish red/greyish green fine-grained sandstones	Fluvial channel
Shangshihezi Fm.	149–156	5.22	Brown red siltstones and purplish red/greyish green mudstones	Floodplain
	148	2.80	Brown yellow fine-grained sandstones	Fluvial channel

(continued)

Table 3 (continued)

Stratigraphy	Layer no.	Thickness (m)	Description	Interpretation
	145–147	3.01	Greyish green siltstones and purplish red mudstones	Floodplain
	144	2.10	Purplish red fine-grained sandstones with cross-beddings	Fluvial channel
	139–143	3.41	Purplish red siltstone intercalated with thin mudstones	Floodplain
	138	0.72	Erosional-based grey fine-grained sandstone	Fluvial channel
Xiashihezi Fm.	135–137	6.30	Purplish red/greyish green mudstones	Floodplain
	133–134	8.90	Erosional-based greyish green medium- to coarse-grained sandstone, developed with cross-beddings and channel lag deposits	Fluvial channel
	132	1.50	Interbedded greyish green siltstones and yellowish brown fine-grained sandstones	Levee
	126–131	7.43	Interbedded greyish green siltstones and fine- to medium-grained sandstones	Flood basin
	125	0.40	Interbedded yellowish green siltstones and fine-grained sandstones	Levee
	117–124	8.80	Interbedded greyish brown siltstones and greyish white fine-grained sandstones	Flood basin
	108–116	4.48	Interbedded yellowish green siltstones and greyish brown fine-grained sandstones	Levee
	106–107	6.90	Erosional-based greyish white fine- to medium-grained sandstone, developed with cross-beddings and siderite concretions	Distributary channel
	105	5.00	Interbedded dark grey siltstones and mudstones	Flood basin
	103–104	1.30	Interbedded grey siltstones and greyish brown fine-grained sandstones	Levee
	102	2.90	Erosional-based greyish white medium-grained sandstone with cross-beddings	Distributary channel
	93–101	6.75	Interbedded grey siltstones and greyish yellow fine-grained sandstones	Flood basin

(continued)

Table 3 (continued)

Stratigraphy	Layer no.	Thickness (m)	Description	Interpretation
	91–92	2.83	Erosional-based greyish white fine- to medium-grained sandstone, developed with cross-beddings	Distributary channel
	86–90	1.47	Interbedded dark grey siltstones, mudstones and coals	Swamp
	83–85	3.35	Greyish green mudstones intercalated with fine-grained sandstone	Flood basin
	81–82	2.70	Greyish brown fine- to medium-grained sandstone with cross-beddings	Distributary channel
	80	0.90	Interbedded grey mudstones and medium-grained sandstones	Levee
	71–79	7.55	Erosional-based greyish white fine- to medium-grained sandstone, developed with cross-beddings	Distributary channel
Shanxi Fm.	64–70	3.31	Erosional-based greyish white fine- to medium-grained sandstone, developed with cross-beddings	
	60–63	1.85	Coals and dark grey mudstones	Swamp
	53–59	17.20	Dark grey siltstones intercalated with greyish brown fine-grained sandstones	Flood basin
	52	0.80	Coals	Swamp
	48–51	11.75	Dark grey siltstones intercalated with greyish brown fine-grained sandstones	Flood basin
	46–47	2.02	Coals and dark grey mudstones with fossil plant roots	Swamp
	41–45	3.20	Greyish brown fine- to medium-grained sandstone (Beichagou sandstone), intercalated with dark grey mudstones and siltstones, developed with cross-beddings	Delta front

(continued)

Table 3 (continued)

Stratigraphy	Layer no.	Thickness (m)	Description	Interpretation
Taiyuan Fm.	39–40	6.40	Dark grey mudstones	Pro-delta
	35–38	10.70	Dark grey mudstones and grey Dongdayao limestones with brachiopod fossils	Offshore carbonate shelf and muddy shelf
	31–34	4.16	Interbedded dark grey siltstones and greyish white fine-grained sandstones	Flood basin
	29–30	4.50	Qiligou sandstone, erosional-based greyish white medium- to coarse-grained sandstone, developed with cross-beddings	Distributary channel
	28	1.10	Carbonaceous mudstones with fossil plant roots and siderite concretions	Swamp
	27	2.25	Dark grey siltstones	Pro-delta
	26	2.48	Maoergou limestone, developed with brachiopods, bivalves, crinoids fossils and *Rhizocorallium* trace fossils	Offshore carbonate shelf
	25	3.42	Dark grey siltstones, developed with tree trunk fossils and *Planolites* trace fossils	Offshore muddy shelf
	24	0.70	Miaogou limestone	Offshore carbonate shelf
	22–23	9.60	Dark grey mudstones and siltstones	Pro-delta
	21	12.90	Dark grey mudstones with siderite concretions	
Taiyuan Fm.	17–20	4.27	Interbedded dark grey siltstones and brown fine grained sandstones	Tidal sand bar
	15–16	5.14	Ximing sandstone, erosional-based greyish white medium- to coarse-grained sandstone, developed with parallel beddings, bidirectional beddings and tree trunk fossils	Tidal channel
	12–14	5.40	Interbedded dark grey siltstones and fine- to coarse-grained sandstones	Tidal sand bar
	6–11	16.75	Erosional-based brown medium- to coarse-grained sandstone, developed with cross-beddings and tree trunk fossils	Distributary channel

(continued)

Table 3 (continued)

Stratigraphy	Layer no.	Thickness (m)	Description	Interpretation
Benxi Fm.	5	16.20	Grey mudstones	Pro-delta
	4	10.00	Interbedded grey siltstones and greyish white aluminous mudstone	
	3	7.10	Greyish white aluminous mudstones with plant root fossils	
	2	2.20	Reddish brown ferrous mudstones	
Ordovician	1	>11.80	Greyish white limestone	Offshore carbonate platform

Fig. 58 Unconformable contact between Ordovician limestone and overlying Carboniferous ferro-aluminous mudstone, Shuiyuguan section, Shanxi Province (Stop 1.1, layer 1–3)

gray mudstone. The coal seam is developed in the distributary bay sedimentary environment and belongs to the coal accumulation model of siliciclastic depositional succession. The mudstone contains plant root fossils and horizontal bedding. The Beichagou Sandstone is interpreted as mouth bar deposits, and the overlying thick mudstone with thin layer of fine sandstone and coal is interpreted as interdistributary bay deposits (Fig. 67).

Fig. 59 Ferrous mudstone of Benxi Formation, Shuiyuguan section, Shanxi Province (Stop 1.1, layer 2)

Fig. 60 Aluminous mudstone of the Benxi Formation, Shuiyuguan section, Shanxi Province (Stop 1.1, layer 3)

Fig. 61 Distributary channel sandstone (Jinci Sandstone) at the base of the Taiyuan Formation, Shuiyuguan section, Shanxi Province (Stop 1.2, layer 8–11)

Fig. 62 Trough cross-bedding developed in tidal channel Ximing Sandstone in the lower part of the Taiyuan Formation, Shuiyuguan section, Shanxi Province (Stop 3, layer 15)

Fig. 63 Bidirectional (herringbone) cross-bedding developed in tidal channel Ximing Sandstone in the lower part of the Taiyuan Formation, Shuiyuguan section, Shanxi Province (Stop 1.3, layer 15)

Fig. 64 Abundant plant root fossils are developed in the lower part of the Late Carboniferous dark gray mudstone, Shuiyuguan section, Shanxi Province (Stop 1.4, layer 21)

Fig. 65 Maoergou Limestone and underlying argillaceous rocks of Taiyuan Formation of the lower Permian, Shuiyuguan section, Shanxi Province (Stop 1.5, layer 25–26)

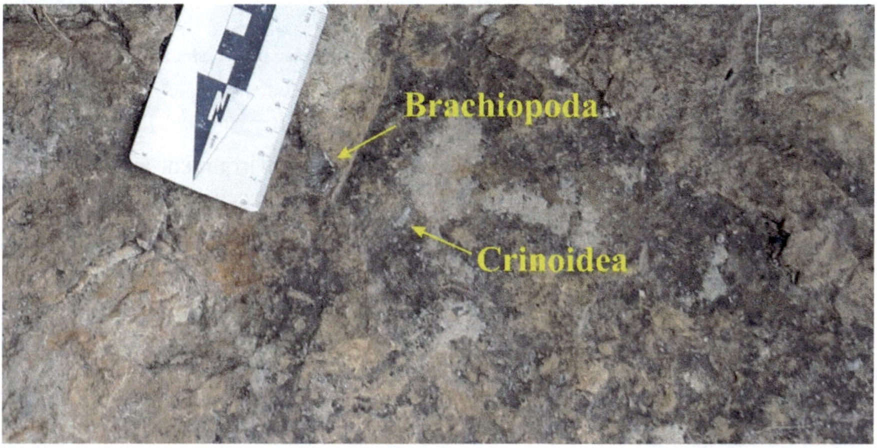

Fig. 66 Brachiopod and crinoid fossils in the Maoergou Limestone, Shuiyuguan section, Shanxi Province (Stop 1.5, layer 26)

Stop 1.7: Sedimentary sequence of delta plain deposits at the boundary between the Permian Shanxi and Xiashihezi Formations

GPS co-ordinates: 37° 40′ 38.10″ N, 111° 56′ 46.79″ E

Objectives: Sedimentary characteristics of delta plain deposits at the boundary between the Permian Shanxi Formation and Xiashihezi Formation

Description and interpretation: The base of the Xiashihezi Formation is a sedimentary sequence of interbedded grayish brown, fine to medium-grained sandstone and grayish black mudstone, with grain-size coarsening upward overall, and the mudstone is intercalated with thin coal layers. An erosional surface is developed at the base of the sandstone, and small-scale cross bedding is developed in the sandstone. The sedimentary sequence is interpreted as delta plain deposits, the sandstone as distributary channel deposits and mudstone as interdistributary bay deposits. The

Fig. 67 Mouth bar deposition of Beichagou Sandstone at the base of the Shanxi Formation (Stop 1.6, layer 43–50)

coal seam is developed in the floodbasin sedimentary environment and belongs to the coal accumulation model of siliciclastic depositional succession (Fig. 68).

Stop 1.8: Delta front and Pro-delta deposits of shallow water delta system in the Permian Shanxi Formation

GPS co-ordinates: 37° 40′ 34.33″ N, 111° 56′ 49.52″ E

Objectives: Sedimentary characteristics of shallow water delta mouth bar and pro-delta facies in the upper part of the Shanxi Formation

Fig. 68 Sedimentary sequence of delta plain deposits at the boundary between the Shanxi Formation and Xiashihezi Formation, Shuiyuguan section, Shanxi Province (Stop 1.7, layer 66–75)

Fig. 69 Mouth bar and pro-delta deposits in the upper part of the Shanxi Formation, Shuiyuguan section, Shanxi Province (Stop 1.8, layer 132–133)

Description and interpretation: A thick interval of grayish brown, fine-grained sandstone is developed in the upper part of the Shanxi Formation, and a thin layer of grayish black mudstone is preserved under it. The sandstone is lenticular and contains small-scale cross-bedding. The mudstone is intercalated with thin siltstone seams and contains plant root fossils. Lenticular sand bodies are interpreted as delta front mouth bar deposits, and thin mudstone and siltstones are interpreted as pro-delta deposits (Fig. 69).

Stop 1.9: Sedimentary characteristics at the boundary between the Permian Xiashihezi and Shangshihezi Formations

GPS co-ordinates: 37° 40′ 27.73″ N, 111° 56′ 50.47″ E

Objectives: Sedimentary characteristics at the boundary between the Xiashihezi Formation and the Shangshihezi Formation

Description and interpretation: Thick, purple mottled mudstone is developed at the top of the Xiashihezi Formation, containing plant root fossils and horizontal bedding. The mudstone of the Xiashihezi Formation is conformably overlain by thick fine-grained sandstone of the Shangshihezi Formation, and trough cross-bedding is developed in the sandstone. The mudstones are interpreted as floodplain deposits and sandstones as crevasse splay deposits (Fig. 70).

Stop 1.10: Seasonally exposed purplish red and grayish green floodplain deposits, upper Permian Shangshihezi Formation

GPS co-ordinates: 37° 40′ 21.20″ N, 111° 56′ 48.80″ E

Fig. 70 Crevasse splay and floodplain deposits at the boundary between the Xiashihezi Formation and Shangshihezi Formation, Shuiyuguan section, Shanxi Province (Stop 1.9, layer 136–142)

Objectives: Sedimentary sequence of seasonally exposed floodplain deposits of the upper Permian Shangshihezi Formation

Description and interpretation: The base of the exposure is gray green medium-grained sandstone with an erosional base, and trough cross-bedding is developed. The upper part comprises rhythmically interbedded purplish red and grayish green siltstone and fine-grained sandstone, with horizontal bedding in siltstone, and current ripple cross-lamination in fine-grained sandstone. The alternation of red and green in sediment color indicates that the deposition was affected by seasonal exposure, so the sedimentary sequence is interpreted as seasonal exposure floodplain deposits. The distinctive rhythmic character is reminiscent of some levee deposits (Fig. 71).

Stop 1.11: Channel deposits showing characteristics of with low coefficient of discharge variation, and abandoned channel deposits in the transition period from icehouse to greenhouse

GPS co-ordinates: 37° 40′ 10.36″ N, 111° 56′ 43.95″ E

Objectives: Sedimentary characteristics of channel facies with low coefficient of discharge variation properties and abandoned channel facies in the upper Permian Shangshihezi Formation

Description and interpretation: The lower part of the exposure is thick gray medium-grained sandstone, with tabular cross-bedding and trough cross-bedding developed, with a basal scour surface and gravel lag. The overlying grayish black mudstone is interbedded with yellow siltstone. The sandstone is interpreted as

Fig. 71 Sedimentary sequence of seasonally exposed floodplain deposits in the Shangshihezi Formation, Shuiyuguan section, Shanxi Province (Stop 1.10, layer 165–168)

channel deposits, and the interbedded mudstone and siltstone is interpreted as abandoned channel deposits (Fig. 72).

Stop 1.12: Sedimentary sequence at the boundary between the upper Permian Shangshihezi and Sunjiagou Formations

GPS co-ordinates: 37° 40′ 7.01″ N, 111° 56′ 44.22″ E

Objectives: Channel and floodplain deposits at the boundary between the late Permian Shangshihezi Formation and Sunjiagou Formation

Description and interpretation: In the upper Permian Shangshihezi Formation, purple red thick mudstone and grayish green gravelly coarse-grained sandstone are interbedded. The planar cross-beddings and trough cross-beddings are developed in grayish green sandstone, and a scour surface is developed at the base. Thick mudstone is interpreted as floodplain deposits and sandstone is interpreted as channel deposits. The gray-green sandstone at the base of the Sunjiagou Formation is conformable with the purplish red mudstone at the top of Shangshihezi Formation (Fig. 73).

Stop 1.13: Characteristics of sedimentary facies at the Permian to Triassic boundary

GPS co-ordinates: 37° 40′ 1.50″ N, 111° 56′ 38.06″ E

Objectives: Sedimentary characteristics near the boundary between the Permian and Triassic successions

Description and interpretation: The Sunjiagou Formation and Liujiagou Formation are in conformable contact, and no obvious sedimentary discontinuities are observed. The color at the top of the Sunjiagou Formation is mottled, and siltstone and

Fig. 72 Deposits of channels with low coefficient of discharge variation characteristics and abandoned channel facies in the Shangshihezi Formation, Shuiyuguan section, Shanxi Province (Stop 1.11, layer 241–247)

claystone with diagenetic carbonate concretions are intercalated with thin layers of sandstone, which are collectively interpreted as floodplain deposits. The sandstone at the base of the overlying Liujiagou Formation contains low angle cross-stratification and horizontal bedding, and is interpreted as channel deposits (Figs. 74 and 75).

4.1.2 Day Two: Palougou Section in Baode County

The Palougou section is located in Palougou village, Baode County, Xinzhou City, Shanxi Province. The strata exposed in this section are mainly the Upper Carboniferous Benxi Formation to the upper Permian Shangshihezi Formation from base to top. The main investigation objectives are sedimentary characteristics and important geological phenomena of Carboniferous–Permian strata. The investigated strata include the Benxi Formation, Taiyuan Formation, Shanxi Formation, Xiashihezi Formation and Shangshihezi Formation from base to top (Figs. 76 and 77). See Table 4 for detailed stratum description.

Fig. 73 Channel and floodplain deposits at the boundary between the Shangshihezi Formation and Sunjiagou Formation, Shuiyuguan section, Shanxi Province (Stop 1.12, layer 273–281)

Fig. 74 The boundary between the Permian and Triassic succession in the Shuiyuguan section, Shanxi Province (Stop 1.13, layer 306–314)

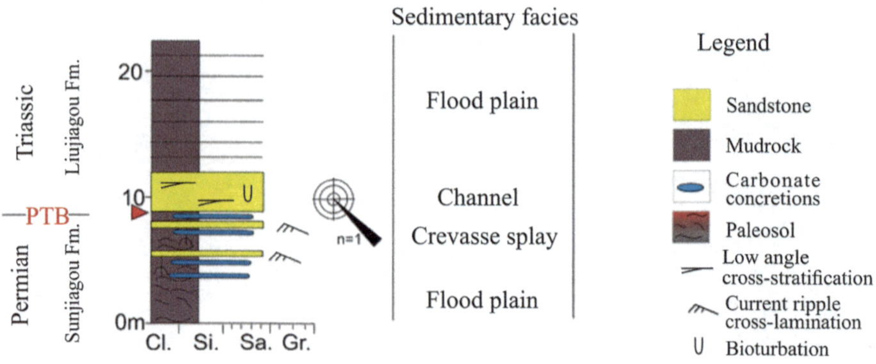

Fig. 75 Graphic log of the boundary between the Permian and Triassic succession in the Shuiyuguan section, Shanxi Province (Stop 1.13, layer 306–314)

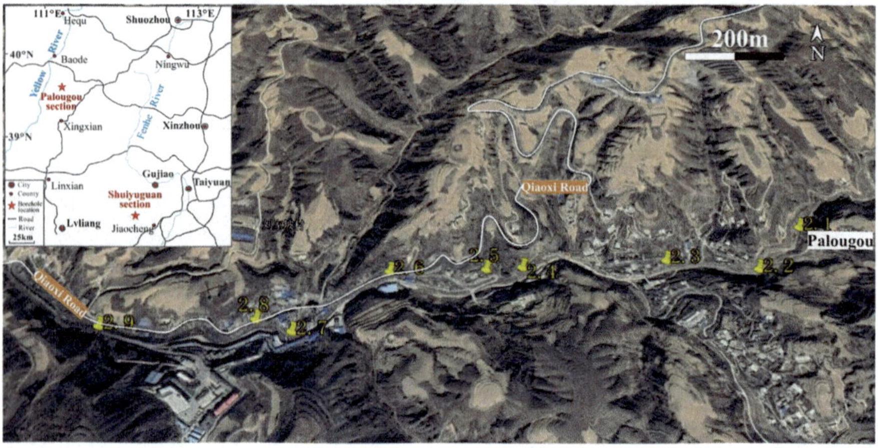

Fig. 76 Location map of observation points in the Palougou section, Shanxi Province

Stop 2.1: Sedimentary characteristics at the boundary between the Majiagou and Benxi Formations

GPS co-ordinates: 38° 45′ 32″ N, 111° 8′ 14″ E

Objectives: Contact relationship between limestone of the Middle Ordovician Majiagou Formation and iron/aluminum-rich mudrocks of the Benxi Formation

Description and interpretation: The limestone of the Majiagou Formation at the base of the measured section is flesh red, filled with calcite veins, containing siderite nodules, and surface fractures are developed. Above this is red ferrous mudstone of the Carboniferous Benxi Formation with horizontal bedding and red oolite. Discontinuous aluminous argillite is developed on the ferrous mudstone. The limestone is interpreted as a shallow sea carbonate deposit, and the ferro-aluminous mudstone is interpreted as lagoon argillaceous deposits (Fig. 78).

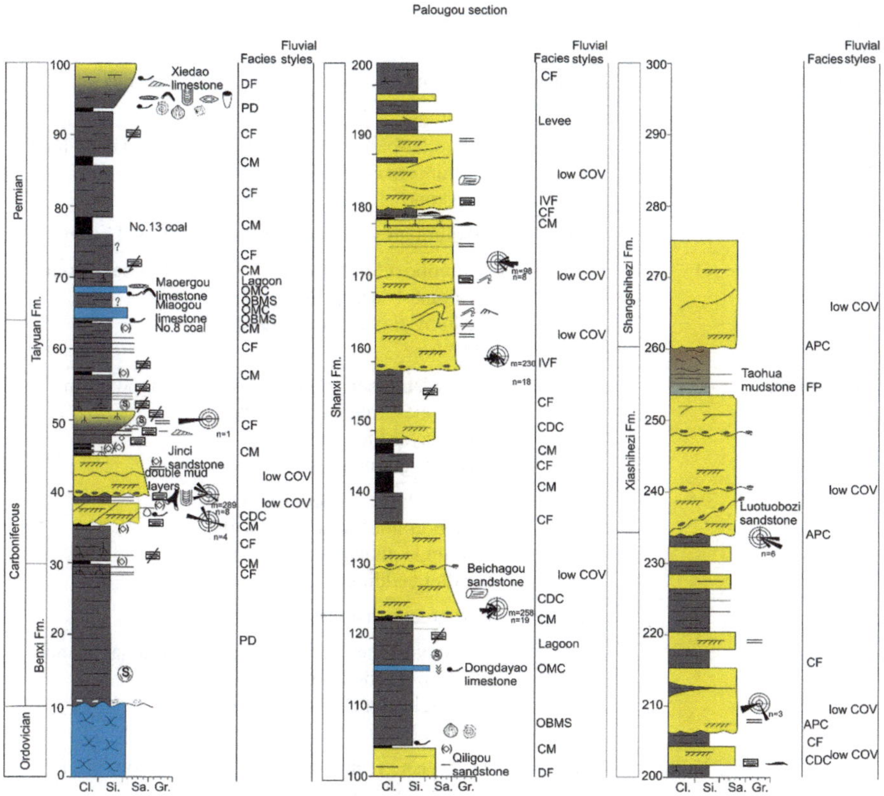

Fig. 77 Graphic log of the Benxi Formation to Shangshihezi Formation at Palougou section illustrating the progressive change in facies and fluvial styles. See Table 2 for the abbreviation of sedimentary facies. The legend is the same as Fig. 55

Stop 2.2: Distributary channel Jinci Sandstone of the Carboniferous

GPS co-ordinates: 38° 45′ 29″ N, 111° 8′ 1″ E

Objectives: Superimposition of Jinci Sandstone sand body at the base of the Upper Carboniferous Taiyuan Formation

Description and interpretation: The base is medium to coarse-grained sandstone, laterally variable and containing gravel lags. A scour surface is developed at the bottom of sandstone, trough cross-bedding is preserved, and the sandstone is interbedded with gray to black claystone, gray siltstone and thin coal layers, and overlain by thick carbonaceous mudstone. The grayish black mudstone contains plant clastic fossils and siderite nodules, and horizontal bedding is developed. Sandstone is interpreted as distributary channel sediments, gray black mudstone and gray siltstone as interdistributary bay sediments, and coal seam and carbonaceous mudstone as swamp sediments (Figs. 79 and 80).

Table 4 Stratigraphic characteristics of Benxi Formation to Shangshihezi Formation in the Palougou section, Shanxi Province

Stratigraphy	Layer no.	Thickness (m)	Description	Interpretation
Shangshihezi Fm.	43	15.10	Erosional-based greyish white medium- to coarse-grained sandstone	Fluvial channel
Xiashihezi Fm.	42	7.06	Medium grey Taohua mudstone	Flood basin
	41	19.65	Luotuobozi Sandstone, erosional-based yellowish green medium- to very coarse-grained sandstone	Fluvial channel
Shanxi Fm.	40	13.57	Yellowish green siltstones and mudstones, intercalated with multi-layer fine- to medium-grained sandstone	Flood basin
	39	2.50	Yellowish green medium- to coarse-grained sandstone with low angle planar beddings and parallel beddings	
	38	3.30	Red/green siltstones	
	37	8.00	Erosional-based greyish green medium-grained sandstone with low angle planar beddings and parallel beddings	Fluvial channel
	33–36	2.29	Yellowish green siltstones intercalated with greyish white fine-grained sandstone	Flood basin
	32	2.60	Greyish white medium- to coarse-grained sandstone with coaly plant axes and coaly traces	Distributary channel
	31	5.20	Grey siltstones with plant root fossils	Flood basin
	30	5.59	Interbedded greyish brown siltstones and sandstones	Levee
	24–29	9.06	Erosional-based greyish white coarse-grained sandstones intercalated with greyish brown siltstone, developed with low angle planar bedding and large scale fossil plant axes	Incised valley fill
	23	1.30	Dark grey siltstones intercalated with a thin coal seam, coaly traces are developed	Flood basin

(continued)

Table 4 (continued)

Stratigraphy	Layer no.	Thickness (m)	Description	Interpretation
	22	21.00	Erosional-based greyish white coarse-grained sandstone with low angle planar beddings, parallel beddings and convolute beddings	Incised valley fill
	21	6.20	Grey mudstones with plant debris fossils	Flood basin
	20	3.80	Greyish white fine-grained sandstones with low angle planar beddings	Distributary channel
	19	2.50	Dark grey siltstones intercalated with a thin coal seam	Flood basin
	18	2.00	Coal seam	Swamp
	17	5.50	Dark grey mudstones	Flood basin
	16	12.80	Beichagou Sandstone, erosionally-based greyish yellow medium- to coarse-grained sandstone with low angle planar beddings and large scale fossil plant axes	Distributary channel
Taiyuan Fm.	15	6.80	Dark grey siltstone, with thin coal seam at the top, developed with siderite concretions and plant debris fossils	Lagoon
	14	0.80	Greyish yellow Dongdayao Limestone, with cone-in-cone structure and brachiopod fossils	Offshore carbonate shelf
Taiyuan Fm.	13	9.00	Thick dark grey siltstone, intercalated with thin coal seams, rich in brachiopod fossils	Lagoon and offshore muddy shelf
	12	10.40	Qiligou Sandstone, greyish white, medium-grained, intercalated with a thin limestone	Delta front
	11	20.00	Dark grey siltstone intercalated with multiple coal seams, and plant debris fossils are developed	Flood basin or swamp
	10	0.80	Maoergou Limestone, greyish yellow, with brachiopod fossils and trace fossils	Offshore carbonate shelf
	9	2.00	Grey siltstone	Lagoon and Offshore muddy shelf

(continued)

Table 4 (continued)

Stratigraphy	Layer no.	Thickness (m)	Description	Interpretation
	8	1.50	Miaogou Limestone, greyish yellow, with brachiopod fossils	Offshore carbonate shelf
	7	13.00	Siltstones intercalated with thin layer of fine-grained sandstone, containing plant debris and root fossils	Lagoon and swamp
	6	2.80	Grey siltstones, with yellowish green fine-grained sandstones on the top, with plant debris and root fossils	Interdistributary bay
	5	1.30	Coals and dark grey carbonaceous mudstones are interbedded, and plant debris and root fossils are developed	Swamp
	4	9.85	Jinci Sandstone, erosional-based greyish yellow coarse-grained sandstones interbedded with thin siltstones, and cross-beddings are developed	Distributary channel
Benxi Fm.	3	7.70	Yellowish green siltstone and mudstones, intercalated with thin coal seams, with plant debris and root fossils	Pro-delta
	2	18.00	Reddish brown ferrous mudstones and greyish yellow aluminous mudstones, interbedded with siltstone, with siderite concretions	Pro-delta
Ordovician	1	>10.00	Light grey limestone	Offshore carbonate platform

Stop 2.3: Sedimentary characteristics of early Permian epicontinental sea deposits

GPS co-ordinates: 38° 45′ 33″ N, 111° 7′ 59″ E

Objectives: Offshore carbonate shelf deposits of the Miaogou limestone and Maoergou limestone

Description and interpretation: The Miaogou Limestone and Maoergou Limestone are mainly developed in the middle and upper part of the Taiyuan Formation. They are mainly dark gray argillaceous limestone and limestone, with *Zoophycos* and *Planolites* ichnofossils. The Miaogou Limestone was formed in a relatively deep, open marine environment, deeper than that of other limestones in the Taiyuan Formation. The formation environment of the Miaogou and Maoergou limestone in general

Fig. 78 Superimposition relationships between limestone of the Majiagou Formation, ferro-aluminous mudstone of the Benxi Formation and Jinci Sandstone of the Taiyuan Formation in Palougou section, Shanxi Province (Stop 2.1, layer 1–4)

Fig. 79 Coal-bearing sedimentary sequence of the deltaic Jinci Sandstone, Taiyuan Formation, Palougou section, Shanxi Province (Stop 2.2, layer 4)

represents the maximum transgression period of the Late Paleozoic in North China (Figs. 81, 82, and 83)

Fig. 80 Plant root and stem fossils in mudstone of the Taiyuan Formation, Palougou section, Shanxi Province (The length of the bottom edge of the card is about 8 cm for scale)

Fig. 81 The offshore carbonate shelf deposits of the Miaogou limestone, Taiyuan Formation, Palougou section, Shanxi Province (Stop 2.3, layer 8)

Fig. 82 *Zoophycos* trace fossils developed in the Maoergou Limestone, Palougou section, Shanxi Province (Stop 2.3, layer 8)

Fig. 83 The offshore carbonate shelf deposits of the Maoergou limestone, Taiyuan Formation, Palougou section, Shanxi Province (Stop 2.3, layer 9–11)

Stop 2.4: Offshore marine deposits at the top of the Taiyuan Formation in Permian

GPS co-ordinates: 38° 45′ 28″ N, 111° 7′ 46″ E.

Objectives: Cone-in-cone structure is developed in offshore carbonate shelf deposits at the top of the Taiyuan Formation

Description and interpretation: The base of the interval is thickly bedded, gray black carbonaceous mudstone, rich in siderite nodules, *brachiopods* and *bivalve* fossils. This facies is interpreted as offshore muddy shelf deposits. Above the black mudstone is the Dongdayao Limestone formed at the end of a transgression, and

Fig. 84 Sedimentary characteristics at the boundary between the Shanxi and Taiyuan Formations, Palougou section, Shanxi Province (Stop 2.4, layer 13–16)

defines the top of the Taiyuan Formation. It is mainly argillaceous banded limestone and bioclastic limestone, in which cone-in-cone structure is developed. The limestone is interpreted as offshore carbonate shelf deposits; The upper part is gray black Tumen Shale, containing siderite nodules, and is interpreted as lagoon deposits (Figs. 84, 85, 86, and 87). The topmost of this interval is distributary channel Beichagou Sandstone.

Stop 2.5: Distributary channel deposits of the Shanxi Formation with low discharge variance characteristics

GPS co-ordinates: 38° 45′ 21″ N, 111° 7′ 41″ E

Objectives: Sedimentary characteristics of distributary channel with low discharge variance characteristics

Photographs: Figs. 81, 82, and 83

Description and interpretation: The base of the Beichagou Sandstone is the lower boundary of the Shanxi Formation. The lithology is mainly gray to white thickly bedded, medium to coarse-grained quartz-lithic sandstone and quartz sandstone. The sandstone body is multistorey and distributed in a lenticular shape. Planar cross-bedding, trough cross-bedding and pseudostratification are developed, beds and nodules of ironstone are developed, and fossils are preserved in the nodules. The sandstone is interpreted as distributary channel deposits formed under a low coefficient of discharge variance. See the following figures for specific analysis—Figs. 88 and 89.

Fig. 85 Carbonaceous mudstone with siderite nodules in the Taiyuan Formation (Tumen Shale), Palougou section, Shanxi Province (Stop 2.4, layer 15)

Fig. 86 Cone-in-cone structure in limestone in the Taiyuan Formation (Tumen Shale), Palougou section, Shanxi Province (Stop 2.4, layer 14)

Fig. 87 *Dictyoclostus* fossil mold in offshore shelf mudstone deposits of the Taiyuan Formation, Palougou section, Shanxi Province (Stop 2.4, layer 13)

Stop 2.6: Thick, multistorey distributary channel sandstone superposition of the Permian Shanxi Formation

GPS co-ordinates: 38° 45′ 21″ N, 111° 7′ 25″ E

Objectives: Thick layered distributary channel sand body superposition

Description and interpretation: The main lithology is grayish yellow, thickly bedded, fine to medium-grained arkose sandstone, containing basement gravel and mud clasts locally. The convolute bedding, soft-sedimentary deformation structure and pseudostratification are observed in this succession. The sand body shows a multistorey stacking pattern, with a total thickness of 20–30 m, and the thickness of a single layer storey is 5–6 m. Horizontal bedding and trough cross bedding are extensively developed in the sandstones. The multi-layered superimposed thick sandstone is interpreted as a distributary channel deposit (Fig. 90).

Stop 2.7: Distributary channel deposits and swamp deposits of the Permian Shanxi Formation

GPS co-ordinates: 38° 45′ 21″ N, 111° 7′ 22″ E

Objectives: Thick sandstone distributary channel deposits and carbonaceous mudstone and thin coal seam swamp deposits of the Permian Shanxi Formation

Description and interpretation: The main lithology is interbedded thick, gray to white, fine to medium-grained sandstone, carbonaceous mudstone and thin coal beds. The thickness of sandstone in the lower part is about 3.6 m, and trough cross-bedding and water escape structures are developed. The middle part is composed of thin coal seams and carbonaceous mudstone. The 10 cm thick coal seam is developed in the

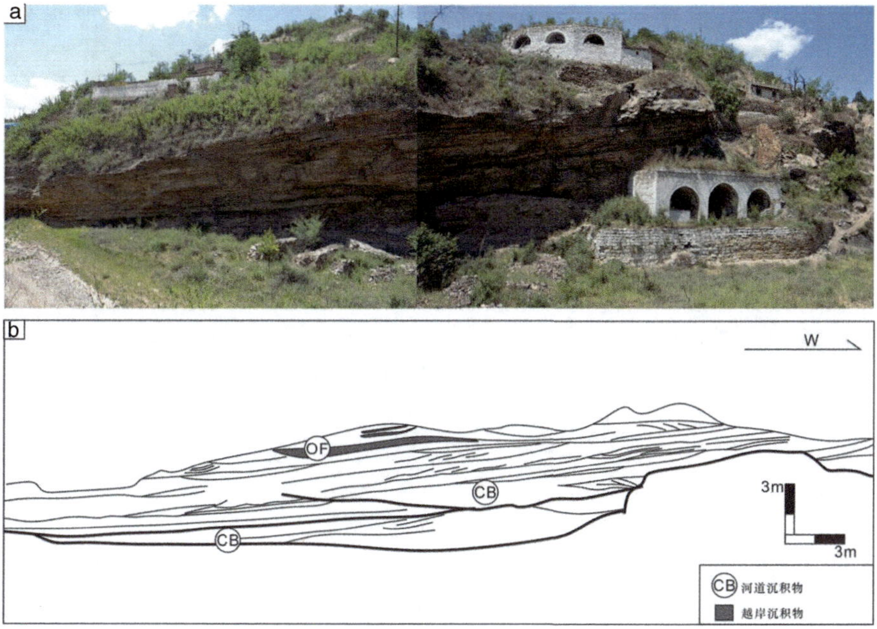

Fig. 88 Illustrations of channel bodies from the rock record representative of low discharge variance. **a** Photomosaic and **b** line drawing with interpretation of a large cliff face exposing part of the Beichagou Sandstone, Shanxi Formation, Palougou section. The cliff is composed of a series of fluvial channel bodies (CB) and overbank fines (OF). The fluvial channel bodies are dominated by cross-bedded sandstone (Stop 2.5, layer 16)

swamp sedimentary environment and belongs to the coal accumulation model of siliciclastic depositional succession. It is in direct contact with the underlying sandstone. The thickness of carbonaceous mudstone is about 1.2 m, and plant clastic fossils are developed. The carbonaceous mudstone is overlain by thick, medium-grained sandstone, about 8.4 m thick, with a scour surface at its base. The thick sandstone is interpreted as a distributary channel deposit, and the carbonaceous mudstone and thin coal seams are interpreted as swamp deposits (Fig. 91).

Stop 2.8: Coastal plain succession, Permian Shanxi Formation

GPS co-ordinates: 38° 45′ 18″ N, 111° 7′ 14″ E

Objectives: Sedimentary characteristics of a coastal plain succession at the top of the Permian Shanxi Formation.

Description and interpretation: The lower part is a thick layer of grayish brown mudstone with a thin layer of fine sandstone, showing horizontal bedding. The upper part is thick, medium to coarse-grained sandstone with small-scale cross-bedding. The lower mudstone is interpreted as interdistributary bay deposits, the thin fine sandstone interlayer is interpreted as a crevasse splay deposit, and the upper thick sandstone is interpreted as distributary channel deposits (Fig. 92).

Fig. 89 Illustrations of channel bodies from the rock record representative of low discharge variance systems. **a** Channel bodies dominated by trough cross-bedded sandstone, the cliff is 2 m high, Shanxi Formation, Palougou section. **b** Channel bodies with significant internal erosional surfaces, dominated by basal conglomerate and cross-bedded sandstone, the hammer is 25 cm high, Shanxi Formation, Palougou section. **c** Channel bodies dominated by large-scale cross-bedded sandstone, the hammer is 25 cm high, Shanxi Formation, Palougou section. **d** Channel bodies dominated by trough cross-bedded sandstone are overlain by levee deposits composed of siltstone interbedded with sandstone, the truck is 2 m high, Shanxi Formation, Paolougou section. **e** Incised valley fill with trough cross-bedded sandstone, the cliff is 10 m high, Shanxi Formation, Palougou section. **f** Several distributary fluvial channel bodies each composed of cross-bedded sandstone interbedded with floodbasin siltstone and claystone, the cliff is 12 m high, Shanxi Formation, Palougou section

Stop 2.9: Sedimentary characteristics at the boundary between the Permian Shanxi, Xiashihezi, and Shangshihezi Formations

GPS: 38° 45′ 20″ N, 111° 7′ 11″ E

Fig. 90 Thick layered distributary channel sandstone of Shanxi Formation, Palougou section, Shanxi Province and its sketch (Stop 2.6, layer 22)

Objectives: The contact relationship between mudstone of the Shanxi Formation and Luotuobozi Sandstone of the basal Xiashihezi Formation; Characteristics of Taohua Mudstone

Description and interpretation: Luotuobozi sandstone is earthy yellow-colored at the surface (grayish green on fresh surfaces), thickly bedded medium-grained sandstone with discordant major bedding surfaces. The base of the Luotuobozi sandstone is defined as the contact between the Shanxi and Xiashihezi Formations. Above the contact are the Luotuobozi Sandstone and Taohua Mudstone, in ascending order. The base of the Luotuobozi Sandstone is erosional, and both planar cross-bedding and trough cross-bedding are developed within, with calcareous cementation. The sandstone is interpreted to record river channel deposition. The sandstone is overlain

Fig. 91 Distributary channel and swamp deposits of Shanxi Formation, Palougou section, Shanxi Province (Stop 2.7, layer 23–24)

Fig. 92 Coastal plain sedimentary sequence of Shanxi Formation, Palougou section, Shanxi Province (Stop 2.8, layer 32–37)

Fig. 93 The boundary between the Shanxi and Xiashihezi Formations, Palougou section, Shanxi Province (Stop 2.9, layer 41)

by the purplish red Taohua Mudstone, containing plant fossils, which is interpreted as floodplain sediments (Figs. 93, 94 and 95).

4.2 Other Investigation Routes

4.2.1 Day Three: Pingshuo Open-Pit Coal Mine

Location

Pingshuo coal mine is owned by the China National Coal Group Corporation and is the core production enterprise of the group. Pingshuo mining area is located in Shuozhou City, Shanxi Province (Fig. 96). Its geographical coordinates are 112.15°–112.35° E and 39.22°–39.42° N, covering an area of 380 km^2 and the geological reserves are 127.5 billion tons. The actual output of Pingshuo coal mine averages close to 20 million tons. The national large-scale open-pit coal mine is divided into three separate fields, namely Antaibao open-pit coal mine, Anjialing open-pit coal mine and East open-pit coal mine (Figs. 52 and 97).

Fig. 94 The boundary between the Shangshihezi Formation and Xiashihezi Formation, Palougou section, Shanxi Province (Stop 2.9, layer 41–43)

Tectonic Setting

Pingshuo depression is the northernmost depression in the Ningwu coalfield, located between Hongtao mountain and Heituoshan mountain. It is bounded by the Danshuigou fault in the south and adjacent to the Shuozhou fault in the Datong rift basin. The overall structural morphology is controlled by NNW trending folds and NE trending faults. The fold structure mainly developed as the Maguanhe syncline, Luzigou anticline, Baijiaxinyao syncline and Erpu anticline. Fault structures mainly include Anjialing reverse fault.

Lithology and Coal Accumulation Characteristics of Coal-Bearing Strata

The coal-bearing strata in Pingshuo mining area include the Upper Carboniferous Benxi Formation, and the lower Permian Taiyuan Formation and Shanxi Formation. The base of the Benxi Formation is generally composed of purplish red ferruginous deposits and aluminium deposits, and it unconformably overlies the Shuoxian Group of Early Ordovician age. The upper Benxi Formation is composed of mudstone, marl, limestone and thin coal seams. The entire Benxi Formation is about 40 m thick. Two divisions are recognized in the overlying Taiyuan Formation. The lower section is mainly composed of mudrocks with thick coal seams. The upper section is mainly composed of sandstones and preserves no important minable coal seam.

Fig. 95 The large scale plant root fossil developed in Taohua Mudstone, Xiashihezi Formation, Palougou section, Shanxi Province (Stop 2.9, layer 42). The width of the book is 18 cm

The total thickness of the Taiyuan Formation is 78 m on average. The lower part of the Shanxi Formation is the principal coal-bearing member and is mainly composed of sandstones with three claystone interlayers and 1–3 thin coal beds (Fig. 98). The coal seam numbers are No.1 to No.12 from top to base. The total aggregate thickness of mineable seam and locally minable seam is 34 m, and the total thickness of main mineable seams 4, 9 and 11 is 26.14 m on average. The main mineable coal seams of Antaibao Coal Mine are No.4, No.9 and No.11 coal seams with continuous distribution and small thickness change (Figs. 99 and 100).

(1) Benxi Formation

The lowermost Benxi Formation is mainly composed of ferruginous sediments and aluminium-rich mudstone, and its thickness is 3–7 m. The K1 layer is relatively pure limestone with a uniform distribution. The thickness of K1 limestone in Pingshuo mining area is 3.5 m. Four layers of sandstone or pebbly sandstone are developed in the mining area, with great variability in thickness. Coal seams are not developed.

(2) Taiyuan Formation

The sandstone is well-represented in this formation. The thickest sandstone is in Jingping and Pingfancheng mines, with a thickness of more than 34 m, and the

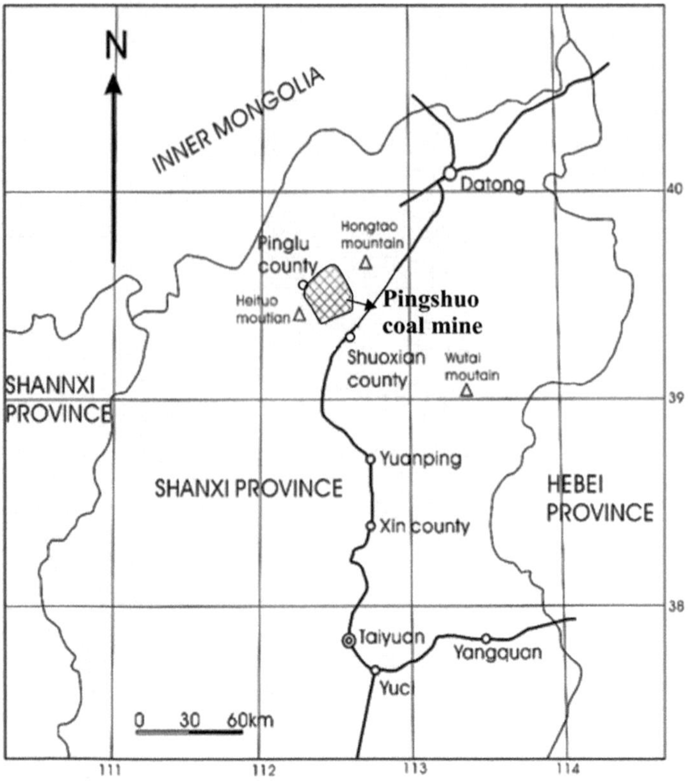

Fig. 96 The location map of Pingshuo Open-pit Coal Mine, Shanxi Province (Geological exploration branch of Academy of Coal Sciences 1978)

thinnest is in Taixi mine, with a thickness of 16.14 m. Limestone is mainly developed in the roof of No.11 coal seam, thinning to the northwest. The thickest coal seams are in Anjialing, Taocun, Luzigou, Antaibao and Maguanhe areas, with a general thickness of more than 22 m. The coal seams in Jingping and Pingfancheng mines are thinner, with a thickness of less than 17 m. The mudstone in Taixi, Taocun and Yangjian is thicker, with a thickness of more than 26 m, and the local thickness is 36.05 m. The mudstone with the minimum thickness appears in Jingping, about 14.5 m.

(3) Shanxi Formation

Sandstone is best developed in Jingping and Pingfancheng mining areas, with a thickness of about 44 m, thinning to the southeast, and the thickness in Yangjian area is 29.2 m. The thickness of mudstone and siltstone in Luzigou and Antaibao mines is greatest, at 41 m. The thickness in Yangjian, Jingping and Pingfancheng mines is less than 27 m. Coal is widely developed in this group, especially in the Hedong, Luzigou, Antaibao and Shiyu mines.

Fig. 97 A photograph showing Pingshuo Open-pit Coal Mine, Shuozhou City, Shanxi Province

Paleogeography of Coal-Bearing Strata

Shanxi Formation and Taiyuan Formation are the main coal-bearing strata in Pingshuo mining area.

(1) Taiyuan Formation

Three transgressions occurred in the study area during the Late Carboniferous. The first one occurred after the deposition of No.11 coal seam, and argillaceous limestone containing marine fossils was deposited. The second time occurred after No.9-1 coal seam, and calcareous mudstone containing phytoclasts and wavy bedding was deposited. The sedimentary environment of the mining area was mainly coastal plain. The last transgression occurred following formation of No.7 coal seam. During the sedimentation of lower Taiyuan Formation, No.9-1 and No.11 coal seams were formed. Because of the long-term balance between siliciclastic input and sedimentation, coal swamps were widely developed in the area at this time (Fig. 101).

(2) Shanxi Formation

The deposition of the Shanxi Formation made the sea retreat in this area and a broad coastal plain environment developed. The main sedimentary environments were lakes, rivers, floodplains and swamps. In the early deposition stage of the Shanxi Formation, the swamp environment was well developed, and the No.4-1 and No.4-2 coal seams were accumulated over a wide area. The Datong Ningwu area was a wide alluvial plain with anastomosing channels. The swamp environment behind the floodplain was an ideal place for coal accumulation. At the end of the upper member

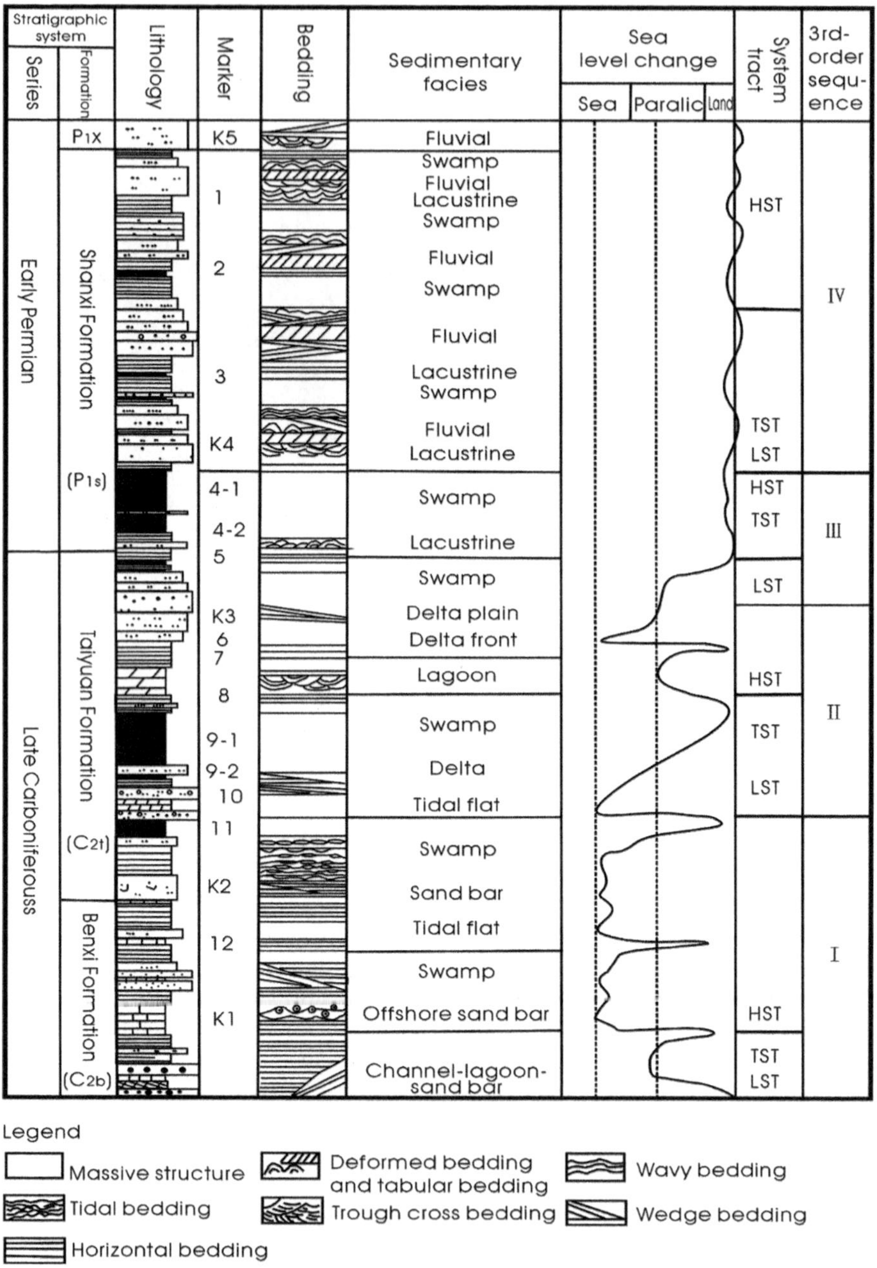

Fig. 98 Graphic log and sequence division of coal-bearing strata in Pingshuo Coal Mine (Geological exploration branch of Academy of Coal Sciences 1978)

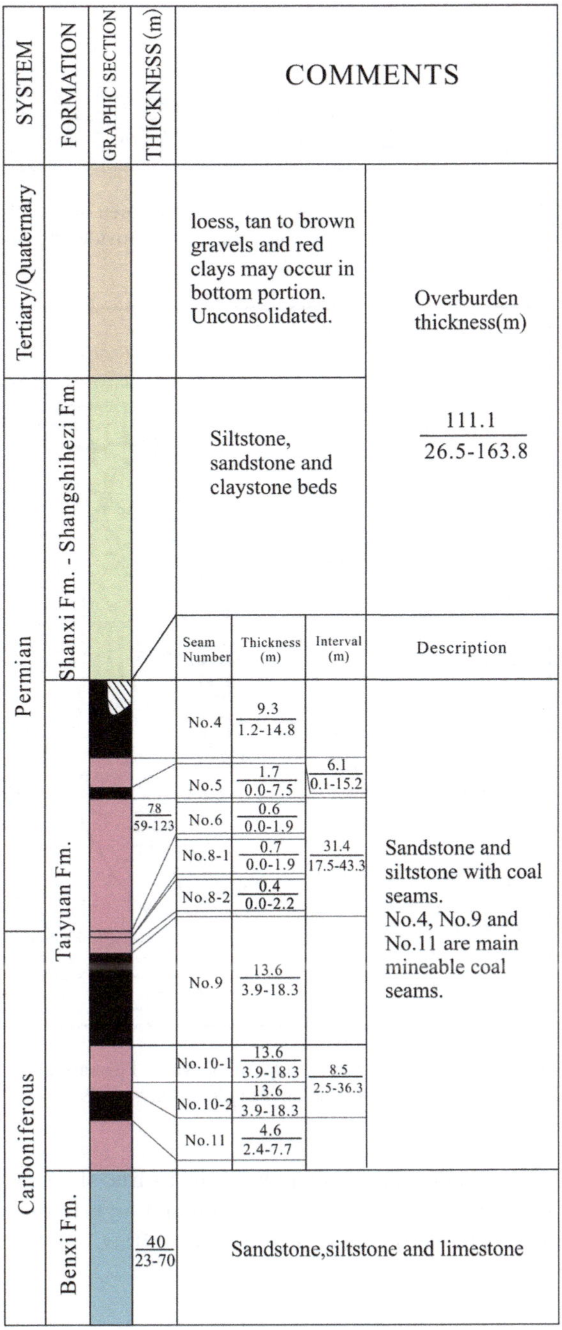

Fig. 99 Graphic log of main mining seams in Antaibao mining area (provided by Antaibao Coal Mine)

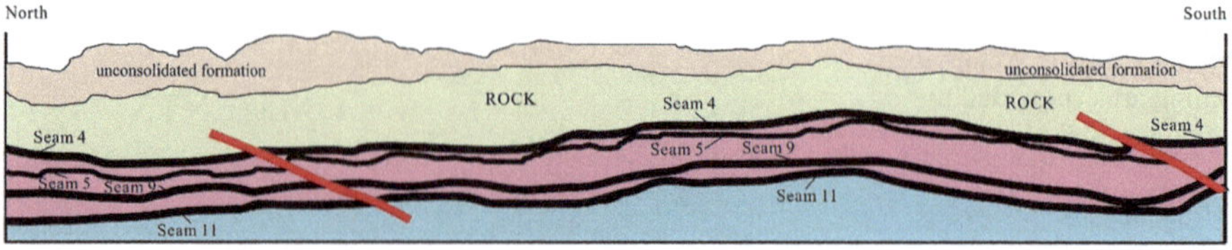

Fig. 100 The typical section of No.2 mining area of Antaibao Coal Mine shows the distribution characteristics of main mineable coal seams (provided by Antaibao Coal Mine)

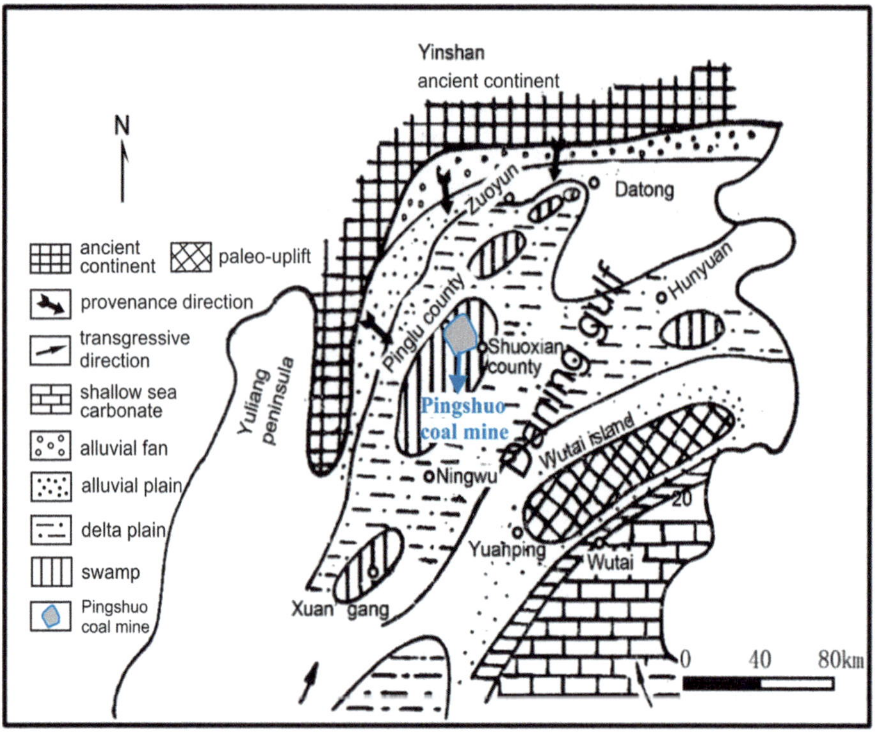

Fig. 101 Paleogeographic map of Taiyuan Formation in Pingshuo mining area (Geological Exploration Branch of Academy of Coal Sciences 1978)

of Shanxi Formation, there was a lacustrine-fluvial-swamp facies sedimentary cycle. The enhancement of water energy and high-frequency facies alteration limited the development of coal at this time (Fig. 102).

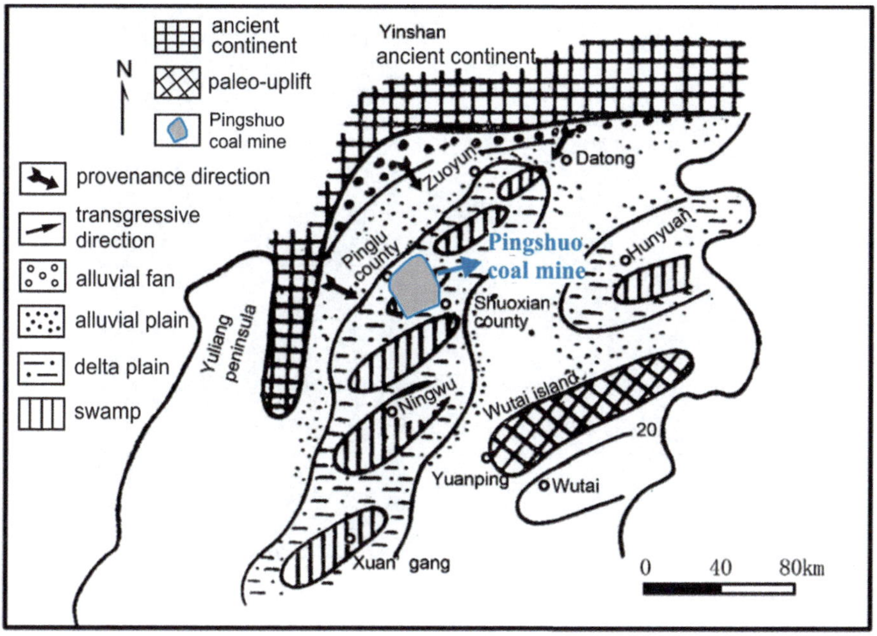

Fig. 102 Paleogeographic map of Shanxi Formation in Pingshuo mining area (Geological Exploration Branch of Academy of Coal Sciences 1978)

4.2.2 Day Four: Datong Yungang Grottoes

The Yungang Grottoes are one of the three major grotto sites in China. The site is located on the southern part of Wuzhou Mountain on the northern bank of Ten-miles River, about sixteen kilometers west of Datong City, Shanxi Province. The grottos extend about one kilometer from east to west, and were all carved out of the mountain. The 254 extant grottoes consist of 45 major grottoes and more than 51,000 sculptures (Fig. 103). Representing the highest level of world carved art in the fifth century, the Yungang Grottos is considered to be exemplary Chinese Buddhist art work at its finest. The site was listed on the first batch of Chinese key cultural relic preservation units in March, 1961 by the State Council. In December, 2001, Yungang Grottos was listed as a UNESCO World Heritage Site and became one of the nation's earliest 5A tourist areas in May, 2007.

As a large-scale cultural relic of grotto temple during the period when the capital Pingcheng (now Datong, Shanxi) was established in the Northern Wei Dynasty, construction took nearly 70 years from the beginning by Emperor Wen Cheng of the Northern Wei Dynasty (AD 452) to the end of the Zhengguang period of the Northern Wei Dynasty (AD 520). The grottoes are rich in art and exquisitely carved. They were carved by the royal family of the Northern Wei Dynasty who ruled northern China at that time with the concentration of national skills, human and material resources. They are a Buddhist temple created by generations and groups of skilled craftsmen.

Fig. 103 The first cave (Shigu cave) of the Yungang Grottoes, Datong City, Shanxi Province

They are the first large Grottoes east of Xinjiang. They are different from other early Grottoes with magnificent and typical Royal statues. The Buddhist culture and art displayed involves many aspects such as history, architecture, music and so on. It is not only the soul of Oriental stone carving art, but also a model of the integration of Chinese and Western cultures. It also represents the highest achievement of Buddhist art in the fifth–sixth century. It occupies an important position in the art history of China and even the world. It is known as one of the three major Grottoes in China together with Dunhuang Mogao Grottoes and Luoyang Longmen Grottoes. It is also known as one of the three major stone carving art treasures in the world together with the Indian Ajanta grottoes and the Afghanistan Bamiyan grottoes.

The strata exposed in Yungang Grottoes belong to the Middle Jurassic Yungang Formation, which were formed by sandy braided fluvial systems. The Yungang Formation is divided into three sections from bottom to top, including lower sand-conglomerate member, middle pebbly sandstone (grottoes) member and upper mudstone member (Fig. 104). The plant root fossils are developed in the mudstone (Fig. 105).

Yungang Grottoes are located on the steep sandstone wall on the northern bank of Shilihe river. South of the grottoes are the first-class terrace, floodplain and river bed of Shilihe river. There are large gullies (Xigou) in the west separated from Wuguantun coal mine. The top and north of the grottoes belong to low mountains and hills with little topographic fluctuation. Elevation is about 1170 m in the South and 1180 m in the north. There are two relatively well-developed gullies in the distribution area of the grottoes. One is located between Grottoes 4 and 5. The gully is about 300 m long and the maximum incised depth of the gully is 30 m; The other is located

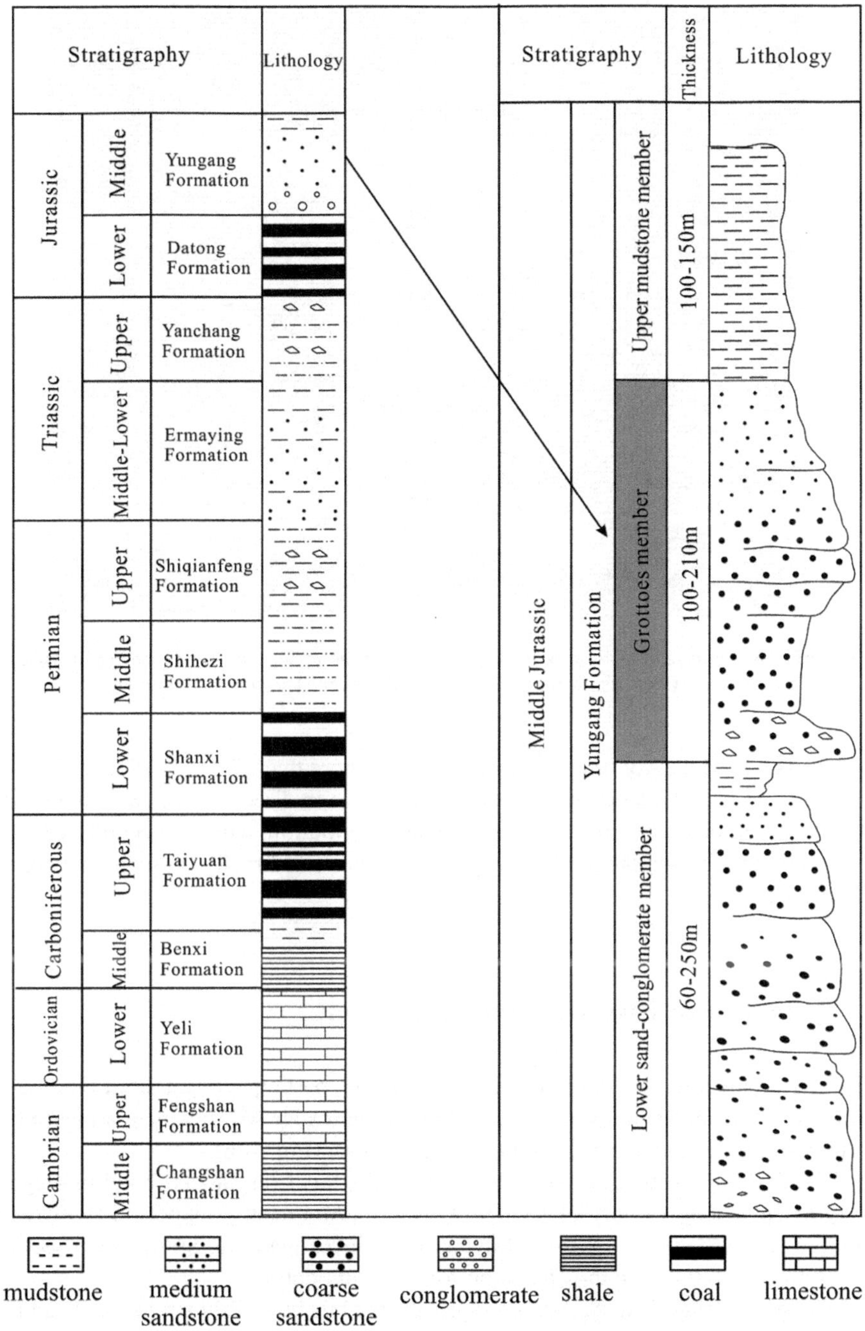

Fig. 104 Stratigraphic histogram of grottoes member in Datong basin (revised after Li 1992)

Fig. 105 The large scale plant root fossil developed in the upper mudstone member of Yungang Formation, Yungang Grottoes, Datong City, Shanxi Province. The pen is 15 cm long

near cave 10. The gully runs north–south and is more than 100 m long. The gravel at the bottom of the Quaternary System in the area is in direct contact with the coarse sandstone of the underlying Jurassic Yungang Formation, which has good hydraulic connection with each other. Shallow phreatic water can directly supplement bedrock fissure water.

5 Schedule of Investigation Suggestions

Before the meeting (Wednesday, August 17): The participants gather at the Shanxi hotel in Taiyuan, Shanxi Province, and the route leader will organize a meeting to briefly introduce and discuss the field investigation (Fig. 106).

The first day (Thursday, August 18): Take the bus from Shanxi hotel in Taiyuan City to Jiaocheng County of Lvliang City at 8:00 a.m. It is expected to arrive at Shuiyuguan village in 2 h from Taiyuan City, investigate the Shuiyuguan section, and return to Taiyuan in the afternoon for accommodation;

The second day (Friday, August 19): Take the bus from Taiyuan to Baode county at 8:00 a.m. It is expected to arrive at the section in 6 h, observe the section of Palougou in Baode County on the eastern edge of Ordos Basin, and stay at hotel in Fugu County in the evening;

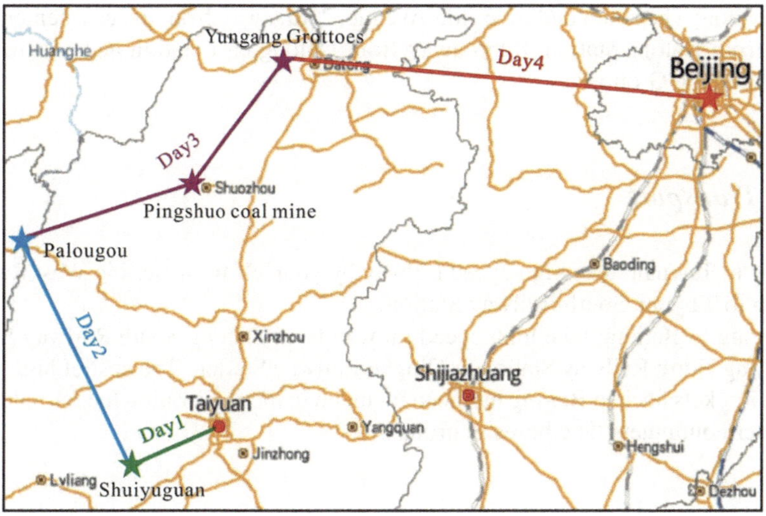

Fig. 106 Investigation route and schedule

The third day (Saturday, August 20): Take the bus from Fugu County to Shuozhou City at 8:00 a.m. It is expected to arrive in 3 h, visit Pingshuo coal mine, take a bus to Datong City in the afternoon, and stay at the hotel of Datong in the evening;

The fourth day (Sunday, August 21): Take the bus from hotel to Yungang Grottoes at 8 a.m., visit the grottoes and observe Jurassic fluvial sandstone. Return to Datong in the afternoon, take the high-speed railway to Beijing, and take the bus to the meeting place after arriving in Beijing.

6 Logistics

6.1 Accommodation

Taiyuan City

Room type: Twin room/single room

Location: The intersection of Yingze street and Xinjian South Road, 5 km away from the city center, 13 km away from Taiyuan South Station, and 2 h' drive from the Shuiyuguan section.

Fugu County

Room type: Twin room/single room

Location: No.1, Hebin Road, Fugu County, located in the city center, 1.5 h' drive from the Palougou section, with convenient transportation and complete facilities.

Datong City

Room type: Twin room/single room

Location: The south end of Weidu Avenue, 2 km away from the city center, 6 km away from Datong station, 8 km away from Datong east station and 16 km away from Yungang Grottoes.

6.2 Transport

Beijing to Taiyuan: take high-speed railway by yourself from Beijing West Railway Station to Taiyuan South Railway Station.

Datong to Beijing: take high-speed railway from Datong South Railway Station to Beijing North Railway Station or Qinghe Railway Station. The cost of high-speed railway tickets back to Beijing has been included in the registration fee for field trips.

Other commuters take buses in groups.

6.3 Clothes

Shanxi Province is located in the middle latitude of the inland, belonging to the temperate continental monsoon climate. The sections for this expedition are located in mountainous areas and require outdoor sports shoes or hiking boots, long sleeves and trousers, sun caps and sunscreen, as well as backpacks to carry field work supplies, food and drinking water.

6.4 Meals

Meals will be arranged uniformly during the excursion, and meals will be served in the hotel in the morning and evening. When the field site is remote, we will purchase portable food and drinking water as lunch for participants.

6.5 Safety

(1) Road safety: there is a direct provincial road from the lodging place to the section, and there are expressways between each investigation site, which is in good condition;
(2) Dehydration, heat stroke and sunburn: the highest temperature in summer can reach 36 °C, so it is necessary to replenish water in time and do a good job in sunscreen protection;

(3) Road conditions of geological sections: the survey sections are all in mountainous areas with steep and narrow terrain, so it is necessary to wear comfortable anti-skid shoes and pay attention not to climb high;
(4) Unexpected situation: If there is an unexpected situation, you can go to the local hospital for emergency treatment. Emergency number: 120.

Acknowledgements The study has been supported by the National Natural Science Foundation of China (41572090). We thank Liu Chao and Sun Beilei of Taiyuan University of Technology, Wu Peng and Fang Chao of China United Coalbed Methane, and all those who helped with preparation of these field trip routes.

References

Ainsworth RB, McArthur JB, Lang SC, Vonk AJ (2020) Parameterizing parasequences: importance of shelf gradient, shoreline trajectory, sediment supply, and autoretreat. AAPG Bull 104(1):53–82

Algeo TJ, Henderson CM, Tong J, Feng Q, Yin H, Tyson RV (2013) Plankton and productivity during the Permian-Triassic boundary crisis: an analysis of organic carbon fluxes. Glob Planet Change 105:52–67

Berger A, Loutre MF (1994) Astronomical forcing through geological time. Blackwell Scientific Publications, Oxford, pp 15–24

Bhattacharya B, Bhattacharjee J, Bandyopadhyay S, Banerjee S, Adhikari K (2018) Early Permian transgressive-regressive cycles: sequence stratigraphic reappraisal of the coal-bearing Barakar Formation, Raniganj Basin, India. J Earth Syst Sci 127, 29 pp.

Bohacs KM, Suter JR (1997) Sequence stratigraphic distribution of coaly rocks: fundamental controls and paralic examples. AAPG Bull 81(10):1612–1639

Calder JH, Gibling MR (1994) The Euramerican Coal Province: controls on late Paleozoic peat accumulation. Palaeogeogr Palaeoclimatol Palaeoecol 106(1–4), 1–21

Cairncross B (2001) An overview of the Permian (Karoo) coal deposits of southern Africa. J Afr Earth Sci 33:529–562

Cao C, Love GD, Hays LE, Wang W, Shen S, Summons RE (2009) Biogeochemical evidence for euxinic oceans and ecological disturbance presaging the end-Permian mass extinction event. Earth Planet Sci Lett 281:188–201

Cao Y, Song H, Algeo TJ, Chu D, Du Y, Tian L, Wang Y, Tong J (2019) Intensified chemical weathering during the Permian-Triassic transition recorded in terrestrial and marine successions. Palaeogeogr Palaeoclimatol Palaeoecol 519:166–177

Chen ZH, Cha M, Jin Q (2004) Application of natural gamma ray logging and natural gamma spectrometry logging to recovering paleoenvironment of sedimentary basin. Chin J Geophys 47(06):1145–1150 (in Chinese with English abstract)

Chen SY, Xu YF, Xu HJ (2000) Late Paleozoic sequence stratigraphy and coal accumulation in North China. China University of Petroleum Press, Dongying, 162 pp. (in Chinese)

Cheng BZ (1992) Late Paleozoic sedimentary environments and coal accumulation in Shanxi. Shanxi Science and Technology Press, Taiyuan (in Chinese with English abstract)

Chu D, Grasby SE, Song HJ, Dal Corso J, Wang Y, Mather TA, Wu Y, Song H, Shu W, Tong J, Wignall PB (2020) Ecological disturbance in tropical peatlands prior to marine Permian-Triassic mass extinction. Geology 48:288–292

Crowell JC (1978) Gondwanan glaciation, cyclothems, continental positioning, and climate change. Am J Sci 278(10):1345–1372

Cleal CJ, Thomas BA (2005) Palaeozoic tropical rainforests and their effect on global climates: is the past the key to the present? Geobiology 3:13–31

Cohen KM, Finney SC, Gibbard PL, Fan J-X (2013) The ICS international chronostratigraphic chart. Episodes 36:199–204

Colombera L, Mountney NP (2020) On the geological significance of clastic parasequences. Earth Sci Rev 201:103062

Domeier M, Torsvik TH (2014) Plate tectonics in the late Paleozoic. Geosci Front 5(3):303–350

Dong DX (2018) Sequence-palaeogeography and coal-seams change law of Benxi-Shanxi formation in North China platform. PhD thesis, China University of Mining & Technology (Beijing), 150 pp. (in Chinese with English abstrct)

Erwin DH (1994) The Permo-Triassic extinction. Nature 376:231–236

Fang Q (2015) Milankovitch cycles from south China and the environmental responds at the end of the late paleozoic ice age. PhD thesis, China University of Geosciences (Beijing), 141 pp. (in Chinese with English abstrct)

Feng Z, Wei HB, Guo Y, He XY, Sui Q, Zhou Y, Liu HY, Gou XD, Lv Y (2020) From rainforest to herbland: new insights into land plant responses to the end-Permian mass extinction. Earth Sci Rev 204:103153

Fielding CR (1984) A coal depositional model for the Durham coal measures of NE England. J Geol Soc Lond 141(5):919–931

Fielding CR (1987) Coal depositional models for deltaic and alluvial plain sequences. Geology 15(7):661–664

Fielding CR, Frank TD, Birgenheier LP, Rygel MC, Jones AT, Roberts J (2008) Stratigraphic imprint of the late Paleozoic ice age in eastern Australia: a record of alternating glacial and nonglacial climate regime. J Geol Soc Lond 165:129–140

Fielding CR, Alexander J, Allen JP (2018) The role of discharge variability in the formation and preservation of alluvial sediment bodies. Sed Geol 365:1–20

Fielding CR, Frank TD, McLoughlin S, Vajda V, Mays C, Tevyaw AP, Winguth A, Winguth C, Nicoll RS, Bocking M, Crowley JL (2019) Age and pattern of the southern high-latitude continental end-Permian extinction constrained by multiproxy analysis. Nat Commun 10:385

Foster GL, Royer DL, Lunt DJ (2017) Future climate forcing potentially without precedent in the last 420 million years. Nat Commun 8:14845

Frank TD, Fielding CR, Winguth AME, Savatic K, Tevyaw A, Winguth C, McLoughlin S, Vajda V, Mays C, Nicoll R, Bocking M, Crowley JL (2021) Pace, magnitude, and nature of terrestrial climate change through the end-Permian extinction in southeastern Gondwana. Geology 49:1089–1095

Gastaldo RA, Dimichele WA, Pfefferkorn HW (1996) Out of the icehouse into the greenhouse: a late Paleozoic analogue for modern global vegetational change. GSA Today 10:1–7

Gastaldo RA, Kamo SL, Neveling J, Geissman JW, Looy CV, Martini AM (2020) The base of the Lystrosaurus Assemblage Zone, Karoo Basin, predates the end-Permian marine extinction. Nat Commun 11:1–8

Geng SF, You GQ, Wang ZY (2011) Simplified tectonic division map of China. China Geol Surv

Geological Exploration Branch of Academy of Coal Sciences, Ministry of Coal (1978) Sedimentary environment of coal bearing strata in Pingshuo mining area. Shaanxi People's Education Press, China

Glasspool IJ, Scott AC (2010) Phanerozoic concentrations of atmospheric oxygen reconstructed from sedimentary charcoal. Nat Geosci 3:627–630

Glasspool IJ, Scott AC, Waltham D, Pronina N, Shao L (2015) The impact of fire on the late Paleozoic earth system. Front Plant Sci 6:1–13

Glikson M, Fielding CR (1991) The Late Triassic Callide coal measures, Queensland, Australia: coal petrology and depositional environment. Int J Coal Geol 17(3–4):313–332

Goddéris et al., 2017Goddéris Y, Donnadieu Y, Carretier S, Aretz M, Dera G, Macouin M (2017) Onset and ending of the late Palaeozoic ice age triggered by tectonically paced rock weathering. Nat Geosci 10(5):382–38

Grasby SE, Sanei H, Beauchamp B (2011) Catastrophic dispersion of coal fly ash into oceans during the latest Permian extinction. Nat Geosci 4:104–107
Griffs NP, Mundil R, Montañez IP, Isbell J, Fedorchuk N, Vesely F, Iannuzzi R, Yin QZ (2018) A new stratigraphic framework built on U-Pb single-zircon TIMS ages with implications for the timing of the penultimate icehouse (Paran´ a Basin, Brazil). Geol Soc Am Bull 130:848–858
Griffs NP, Montañez IP, Mundil R, Richey J, Isbell J, Fedorchuk N, Linol B, Iannuzzi R, Vesely F, Mottin T, da Rosa E, Keller B, Yin QZ (2019) Coupled stratigraphic and U-Pb zircon age constraints on the late Paleozoic icehouse-togreenhouse turnover in south-central Gondwana. Geol 47:1146–1150
Grice K, Cao C, Love GD, Böttcher ME, Twitchett RJ, Grosjean E, Summons RE, Turgeon SC, Dunning W, Jin Y (2005) Photic Zone Euxinia during the Permian-Triassic Superanoxic Event. Sci 307:706–709
Guo XN, Tang ZL, Li WC (1991) The Late Palaeozoic coal-accumulating laws in Henan Province. China University of Geosciences Press, Wuhan, 314 pp. (in Chinese)
Heckel PH (1986) Sea-level curve for Pennsylvanian eustatic marine transgressive-regressive depositional cycles along midcontinent outcrop belt, North America. Geology 14(4):330–334
Hinnov LA (2004) Earth's orbital parameters and cyclestratigraphy. Cambridge University Press, Cambridge, pp 55–62
Hinnov LA (2013) Cyclostratigraphy and its revolutionizing applications in the earth and planetary sciences. Geol Soc America Bulletin 125:1703–1734
Hu B, Song F, Chen SM, Song HB, Jin Y (2015) Sedimentary environment and Lithofacies paleogeography of coal-bearing upper Paleozoic in Henan Province. China University of Mining and Technology Press, Xuzhou, 244 pp. (in Chinese with English abstract)
Huang CJ (2014) The current status of cyclostratigraphy and astrochronology in the mesozoic. Earth Sci Front 21(02):48–66 (in Chinese with English abstract)
Isbell JL, Miller MF, Wolfe KL, Lenaker PA (2003) Timing of late Paleozoic glaciation in Gondwana: was glaciation responsible for the development of Northern Hemisphere cyclothems? Extreme depositional environments: mega end members in geological time, Special Paper 370, pp 5–24
Jerrett RM, Davies RC, Hodgson DM, Flint SS, Chiverrell RC (2011a) The significance of hiatal surfaces in coal seams. J Geol Soc Lond 168(3):629–632
Jerrett RM, Flint SS, Davies RC, Hodgson DM (2011b) Sequence stratigraphic interpretation of a Pennsylvanian (Upper Carboniferous) coal from the central Appalachian Basin, USA. Sedimentology 58(5):1180–1207
Jiang ZH, Tu QP, Shi N (2001) Multi-taper method of spectral analysis and applications in global warming study. Acta Meteor Sin 59(04):480–490 (in Chinese with English Abstract)
Kaiho K, Saito R, Ito K, Miyaji T, Biswas R, Tian L, Sano H, Shi Z, Takahashi S, Tong J, Liang L, Oba M, Nara FW, Tsuchiya N, Chen ZQ (2016) Effects of soil erosion and anoxic-euxinic ocean in the Permian-Triassic marine crisis. Heliyon 2:e00137
Kershaw S, Crasquin S, Li Y, Collin PY, Forel MB, Mu X, Baud A, Wang Y, Xie S, Maurer F, Guo L (2012) Microbialites and global environmental change across the Permian-Triassic boundary: a synthesis. Geobiol 10:25–47
Kodama KP, Hinnov LA (2014) Rock magnetic cyclostratigraphy. Rock Magnetic Cyclostratigraphy, Oxford, p 176
Kong XZ, Xu HL, Li RL et al (1996) Late Paleozoic coal bearing strata and paleobiota in Shanxi. Shanxi Science and Technology Press, Taiyuan, 340 pp. (in Chinese)
Korte C, Kozur HW (2010) Carbon-isotope stratigraphy across the Permian-Triassic boundary: a review. J Asian Earth Sci 39:215–235
Li YN (2021) Icehouse to greenhouse paleoenvironmental records of Carboniferous-Permian strata in North China. China University of Mining & Technology, Beijing (in Chinese with English abstract)
Li MS, Hinnov LA, Huang CJ, Ogg JG (2018a) Sedimentary noise and sea levels linked to land–ocean water exchange and obliquity forcing. Nat Commun 9:1004

Li MS, Kump LR, Hinnov LA, Mann ME (2018b) Tracking variable sedimentation rates and astronomical forcing in phanerozoic paleoclimate proxy series with evolutionary correlation coefficients and hypothesis testing. Earth Planet Sci Lett 501:165–179

Li MS, Huang CJ, Hinnov LA, Chen WZ, Ogg JG, Tian W (2018c) Astrochronology of the Anisian stage (Middle Triassic) at the Guandao reference section, South China. Earth Planet Sci Lett 482:591–606

Li MS, Hinnov L, Kump L (2019) Acycle: time-series analysis software for paleoclimate research and education. Comput Geosci 127:12–22

Li MP (2012) Sequence stratigraphy and lithofacies and paleogeography of the Carboniferous-Permian in North China Platform. Master's thesis, China University of Mining & Technology (Beijing), 102 pp. (in Chinese with English abstract)

Li MP, Shao LY, Li ZX, Dong DX, Li JQ (2020a) Lithofacies palaeogeography of lower coal group accumulation period of Carboniferous Permian in North China. J China Coal Soc 45(07):2399–2410 (in Chinese with English abstract)

Li YN, Shao LY, Fielding CR, Wang DW, Mu GY, Luo HH (2020b) Sequence stratigraphic analysis of thick coal seams in paralic environments—a case study from the Early Permian Shanxi Formation in the Anhe coalfield, Henan Province, North China. Int J Coal Geol 222:103451

Li YN, Shao LY, Fielding CR, Wang DW, Mu GY (2021) Sequence stratigraphy, paleogeography, and coal accumulation in a lowland alluvial plain, coastal plain, and shallow-marine setting: upper Carboniferous-Permian of the Anyang-Hebi coalfield, Henan Province, North China. Palaeogeogr Palaeoclimatol Palaeoecol 567:110287

Li ST (1992) Analysis of sequence stratigraphy and sedimentary system in northeast Ordos Basin. Geological Publishing House, Beijing, 194 pp. (in Chinese)

Liang JW (2007) Study on Jurassic sedimentary system and sequence stratigraphy in Ordos Basin. Xi'an University, Xi'an, 175 pp. (in Chinese)

Liao Z, Hu W, Cao J, Wang X, Yao S, Wan Y (2016) Permian-Triassic boundary (PTB) in the lower Yangtze Region, southeastern China: a new discovery of deep-water archive based on organic carbon isotopic and U-Pb geochronological studies. Palaeogeogr Palaeoclimatol Palaeoecol 451:124–139

Liao Z, Hu W, Cao J, Wang X, Fu X (2020) Oceanic anoxia through the late Permian Changhsingian stage in the Lower Yangtze region, South China: evidence from sulfur isotopes and trace elements. Chem Geol 532:119371

Limarino CO, C´esari SN, Spalletti LA, Taboada AC, Isbell JL, Geuna S, Gulbranson EL (2014) A paleoclimatic review of southern South America during the late Paleozoic: a record from icehouse to extreme greenhouse conditions. Gondwana Res 25:1396–1421

Liu F, Zhu HC, Ouyang S (2015) Late pennsylvanian to wuchiapingian palynostratigraphy of the baode section in the ordos basin, north china. J Asian Earth Sci 111:528–552

Lu J, Zhang P, Yang M, Shao L, Hilton J (2020) Continental records of organic carbon isotopic composition ($\delta^{13}C_{org}$), weathering, paleoclimate and wildfire linked to the end-Permian mass extinction. Chem Geol 558:119764

Lu J, Wang Y, Yang M, Shao L, Hilton J (2021) Records of volcanism and organic carbon isotopic composition ($\delta^{13}C_{org}$) linked to changes in atmospheric pCO2 and climate during the Pennsylvanian icehouse interval. Chem Geol 570:120168

Luo ZK, Miao LC, Guan K, Qiu YS, Qiu YM, McNaughton NJ, Groves DI (2001) SHRIMP chronological study of Shuiquangou intrusive body in Zhangjiakou Area, Hebei Province and its geochemical significance. Geochimica 30(002):116–122. (in Chinese with English abstrct)

Ma DD (2013) Analysis on stratigraphic and tectonic characteristics in Dangzhonggang exploration district of Anhe coalfeld. Zhongzhou Coal 209(5):26–28 (in Chinese with English abstract)

Ma F, Mu ZG, Liu YL (2004) Geochronology and geologic significance of the orbicular Dioritic rocks in Luanping, Hebei Province. Geol Rev 2004(04):360–364 (in Chinese with English abstract)

Mann ME, Lees JM (1996) Robust estimation of background noise and signal detection in climatic time series. Clim Change 33(3):409–445

Mays C, McLoughlin S, Frank TD, Fielding CR, Slater SM, Vajda V (2021) Lethal microbial blooms delayed freshwater ecosystem recovery following the end–Permian extinction. Nat Commun 12:5511

Mao JW, Zhang ZH, Yu JJ, Wang YT, Niu BG (2003) Geodynamic background of Mesozoic large-scale mineralization in North China and its adjacent areas: enlightenment from precise dating of metal deposits. Science in China (series d) 04:289–299 (in Chinese with English abstract)

Meng XH, Ge X (2001) Discovery and evidence of the foreland basin in the North China platform in the Permian period. Geol Sci Technol Inf 20(1):8–14 (in Chinese with English abstract)

Meng XH (2004) Sino-Korean plate sequence, event, evolution. Science Press, Beijing, 542 pp. (in Chinese)

Meyer KM, Yu M, Jost AB, Kelley BM, Payne JL (2011) $\delta^{13}C$ evidence that high primary productivity delayed recovery from end-Permian mass extinction. Earth Planet Sci Lett 302:378–384

Montañez IP, Tabor NJ, Niemeier D, DiMichele WA, Frank TD, Fielding CR, Isbell JL, Birgenheier LP, Rygel MC (2007) CO2-forced climate and vegetation instability during Late Paleozoic deglaciation. Science 315:87–91

Nan J, Li Y (2004) Organic and inorganic carbon-isotope shift and paleoenvironment at the P-T boundary section in Meishan. Geochimica 33:9–19

Paerl HW, Otten TG (2013) Harmful cyanobacterial blooms: causes, consequences, and controls. Microb Ecol 65:995–1010

Rampino MR, Rodriguez S, Baransky E, Cai Y (2017) Global nickel anomaly links Siberian Traps eruptions and the latest Permian mass extinction. Sci Rep 7:12416

Richey JD, Montañez IP, Goddéris Y, Looy CV, Griffis NP, DiMichele WA (2020) Influence of temporally varying weatherability on CO_2-climate coupling and ecosystem change in the late Paleozoic. Clim past 16:1759–1775

Schneebeli-Hermann E, Kürschner WM, Hochuli PA, Bucher H, Ware D, Goudemand N, Roohi G (2012) Palynofacies analysis of the Permian–Triassic transition in the Amb section (Salt Range, Pakistan): implications for the anoxia on the South Tethyan Margin. J Asian Earth Sci 60:225–234

Scott AC (2000) The pre-quaternary history of fire. Palaeogeogr Palaeoclimatol Palaeoecol 164:281–329

Shang GX (1997) Late Paleozoic coal geology of North China platform. Shanxi Science and Technology Press, Taiyuan (in Chinese with English abstract)

Shao JA (1991) Crustal evolution in the middle of the northern margin of Sino-Korean plate. Peking University Press, Beijing (in Chinese)

Shao LY, Dong DX, Li MP, Wang HS, Wang DD, Lu J, Zheng MQ, Cheng AG (2014) Sequence-paleogeography and coal accumulation of the Carboniferous-Permian in the North China Basin. J China Coal Soc 39(8):1725–1734 (in Chinese with English abstract)

Shao LY, Wang XT, Wang DD, Li MP, Wang S, Li YJ, Shao K, Zhang C, Gao GX, Dong DX, Cheng AG, Lu J, Ji CW, Gao D (2020) Sequence stratigraphy, paleogeography, and coal accumulation regularity of major coal-accumulating periods in China. Int J Coal Sci Technol 7:240–262

Shao LY, Zheng MQ, Hou HH, Dong DX, Wang HS (2018) Characteristics sequence-palaeogeography and coal accumulation of Permo-Carboniferous coal measures in Shanxi Province. Coal Sci Technol 46(02):1–8+34. (in Chinese with English abstract)

Shen SZ, Crowley JL, Wang Y, Bowring SA, Erwin DH, Sadler PM, Cao CQ, Rothman DH, Henderson CM, Ramezani J, Zhang H, Shen Y, Wang XD, Wang W, Mu L, Li WZ, Tang YG, Liu XL, Liu LJ, Zeng Y, Jiang YF, Jin YG (2011) Calibrating the end-Permian mass extinction. Science 334:1367–1372

Shen S, Cao C, Zhang H, Bowring SA, Henderson CM, Payne JL, Davydov VI, Chen B, Yuan D, Zhang Y, Wang W, Zheng Q (2013) High-resolution $\delta^{13}C$ carb chemostratigraphy from latest Guadalupian through earliest Triassic in South China and Iran. Earth Planet Sci Lett 375:156–165

Shen J, Schoepfer SD, Feng Q, Zhou L, Yu J, Song H, Wei H, Algeo TJ (2015) Marine productivity changes during the end-Permian crisis and Early Triassic recovery. Earth Sci Rev 149:136–162

Shen J, Yu J, Chen J, Algeo TJ, Xu G, Feng Q, Shi X, Planavsky NJ, Shu W, Xie S (2019a) Mercury evidence of intense volcanic effects on land during the Permian-Triassic transition. Geology 47:1117–1121

Shen SZ, Zhang H, Zhang YC, Yuan DX, Chen B, He WH, Mu L, Lin W, Wang WQ, Chen J, Wu Q, Cao CQ, Wang Y, Wang XD (2019b) Permian integrative stratigraphy and timescale of China. Sci China-Earth Sci 62(1):154–188

Spina A, Cirilli S, Utting J, Jansonius J (2015) Palynology of the Permian and Triassic of the Tesero and Bulla sections (Western Dolomites, Italy) and consideration about the enigmatic species Reduviasporonites chalastus. Rev Palaeobot Palynol 218:3–14

Stanley SM (2016) Estimates of the magnitudes of major marine mass extinctions in earth history. Proc Natl Acad Sci (u.s.a.) 113:E6325–E6334

Sun YD, Zulla MJ, Joachimski MM, Bond DPG, Wignall PB, Zhang ZT, Zhang MH (2019) Ammonium ocean following the end-Permian mass extinction. Earth Planet Sci Lett 518:211–222

Tang KD (1992) Tectonic evolution and metallogenic regularity of north fold belt in Sino-Korean plate. Peking University Press, Beijing (in Chinese)

The stratigraphic chart of China (Illustration 1) (2014) Acta Geoscientica Sinica 35(03)

Thomas BM, Willink RJ, Grice K, Twitchett RJ, Purcell RR, Archbold NW, George AD, Tye S, Alexander R, Foster CB, Barber CJ (2004) Unique marine Permian-Triassic boundary section from Western Australia. Aust J Earth Sci 51:423–430

Torsvik TH, Cocks LRM (2004) Earth geography from 400 to 250 Ma: a palaeomagnetic, faunal and facies review. J Geol Soc Lond 161(4):555–572

Torsvik TH, Smethurst MA, Burke K, Steinberger B (2008) Long term stability in deep mantle structure: evidence from the ~300 Ma Skagerrak-Centered Large Igneous Province (the SCLIP). Earth Planet Sci Lett 267:444–452

Vajda V, McLoughlin S, Mays C, Frank TD, Fielding CR, Tevyaw A, Lehsten V, Bocking M, Nicoll RS (2020) End-Permian (252 Mya) deforestation, wildfires and flooding—an ancient biotic crisis with lessons for the present. Earth Planet Sci Lett 529:115875. https://doi.org/10.1016/j.epsl.2019.115875

van Soelen EE, Kürschner WM (2018) Late Permian to Early Triassic changes in acritarch assemblages and morphology in the Boreal Arctic: new data from the Finnmark Platform. Palaeogeogr Palaeoclimatol Palaeoecol 505:120–127

van Soelen EE, Twitchett RJ, Kürschner WM (2018) Salinity changes and anoxia resulting from enhanced run off during the late Permian global warming and mass extinction event. Clim Past 14:441–453

Veevers JJ, Powell CMcA (1987) Late Paleozoic glacial episodes in Gondwanaland reflected in transgressive-regressive depositional sequences in Euramerica. Geol Soc Am Bull 98(4):475–487

Wang ZQ, Chen AS (2001) Traces of arborescent lycopsids and dieback of the forest vegetation in relation to the terminal Permian mass extinction in North China. Rev Palaeobot Palynol 117:217–243

Wang HC, Zhao FQ, Li HM, Sun LX, Miao LC, Ji SP (2007) Zircon SHRIMP U-Pb age of the dioritic rocks from northern Hebei: the geological records of late Paleozoic magmatic arc. Acta Petrologica Sinica 03:597–604 (in Chinese with English abstract)

Wang J (2010) Late paleozoic macrofloral assemblages from weibei coalfield, with reference to vegetational change through the late paleozoic ice-age in the north china block. Int J Coal Geol 83(2–3):292

Wanless HR, Shepard FP (1936) Sea level and climatic changes related to late Paleozoic cycles. Geol Soc Am Bull 47(8):1177–1206

Weedon GP (2003) Time series analysis and cyclostratigraphy: examining stratigraphic records of environmental cycles. Cambridge University Press, p. 259

Wu HC, Zhang SH, Feng QL, Fang NQ, Yang TS, Li HY (2011) Theoretical basis research advancement and prospects of cyclostratigraphy. Earth Sci J China Univ Geosci 36(03):409–428 (in Chinese with English abstract)

Wu Q (2020) High-precision zircon u-pb geochronological studies of the permian ash beds from china and north america. PhD thesis, University of Science and Technology of China

Xu YG, Wei X, Luo ZY, Liu HQ, Cao J (2014) The early Permian Tarim large igneous province: main characteristics and a plume incubation model. Lithos 204:20–35

Xue NH, Wang W, Guan CG, Fang YN, Zhao XD, Zhang HC (2020) Cyclostratigraphy of the lower-middle Ediacaran Doushantuo formation in Yichang, South China. J Stratigr 44(03):250–259 (in Chinese with English abstract)

Yan Z, Shao L, Glasspool IJ, Wang J, Wang X, Wang H (2019) Frequent and intense fires in the final coals of the Paleozoic indicate elevated atmospheric oxygen levels at the onset of the End-Permian mass extinction event. Int J Coal Geol 207:75–83

Yang Y, Fu WZ, Yu JF, Ning ZG, Cui JJ, Li Q, Zhou XZ, Chen JP (2021) Cyclostratigraphical analysis of continental red beds below k/pg boundary in the jiaolai basin. Acta Sedimentol Sin 39(04):942–952 (in Chinese with English abstract)

Yin H, Feng Q, Lai X, Baud A, Tong J (2007) The protracted Permo-Triassic crisis and multi-episode extinction around the Permian-Triassic boundary. Glob Planet Change 55:1–20

Zhai MG (2019) Tectonic evolution of the North China craton. J Geomech 25(05):722–745 (in Chinese with English abstract)

Zhang SH, Zhao Y, Kroener A, Liu XM, Xie LW, Chen FK (2009) Early Permian plutons from the northern North China block: constraints on continental arc evolution and convergent margin magmatism related to the Central Asian Orogenic Belt. Int J Earth Sci 98(6):1441–1467

Zhang SH, Zhao Y, Liu JM, Hu JM, Song B, Liu J, Wu H (2010) Geochronology, geochemistry and tectonic setting of the Late Paleozoic-Early Mesozoic magmatism in the northern margin of the North China block: a preliminary review. Acta Petrologica Et Mineralogica 29(06):824–842 (in Chinese with English abstract)

Zhang H, Cao CQ, Liu XL, Mu L, Zheng QF, Liu F, Xiang L, Liu LJ, Shen SZ (2016) The terrestrial end-Permian mass extinction in South China. Palaeogeogr Palaeoclimatol Palaeoecol 448:108–124

Zhao Y, Zhang SH, Xu G, Hu JM (2004) The Jurassic major tectonic events of the Yanshanian intraplate deformation belt. Geol Bull China 2004(Z2):854–863

Zhu XQ, Zhu WB, Ge RF, Wang X (2014) Late Paleozoic provenance shift in the south-central North China Craton: implications for tectonic evolution and crustal growth. Gondwana Res 25(1):383–400

Cretaceous Fluvio-Lacustrine Stratigraphy, Sedimentology and Paleontology in the Laiyang, Zhucheng and Lingshan Island Areas of Eastern Shandong Province, North China

Han Zuozhen, Zhou Yaoqi, and Li Shoujun

Organizing Committee

Han Zuozhen; Li Shoujun; Yang Renchao; Han Chao; Zhou Changfu; Wang Zhaopeng; Chen Lei; Cheng Yanjun; Du Qingxiang; Fan Aiping; Zhao Yanyang; Pang Yumao

Shandong University of Science and Technology, Qingdao 266590, China

Zhou Yaoqi; Lin Chengyan; Wu Zhiping; Qiu Longwei; Zhou Tengfei

China University of Petroleum (East China), Qingdao 266590, China

Sun Yejun; Cui Weisheng; Chen Shuqing

Laiyang Cretaceous National Geological Park, Yantai 265216, China

Liang Zhaohui; Wang Zhengdong

Shandong Zhucheng National Dinosaur Geopark, Weifang 262299, China

H. Zuozhen (✉) · L. Shoujun
Shandong University of Science and Technology, Qingdao, China
e-mail: hanzz@163.com

Z. Yaoqi
China University of Petroleum (East China), Dongying, China

X. Hu (ed.), *Field Trip Guidebook on Chinese Sedimentary Geology*,
https://doi.org/10.1007/978-981-99-6936-4_7

1 Introduction

The Jiaolai Basin is a typical rift basin located at eastern Shandong Province in Eastern China on the edge of the East Asian Continent (Fig. 1). It is formed in the Cretaceous and bordered by large-scale normal faults. To the west, the Yishu fault, which is the northeastern branch of the Tanlu fault, separates it from Luxi uplift. To the south and southeast, the Wulian-Yantai fault divided it from the Sulu orogen. To the east, it is connected with the Sulu ultra-high-pressure metamorphic belt.

The metamorphic basement for Jiaolai basin was comprised of Archaean Jiaodong Group, Lower Proterozoic Jingshan Group, Fenzishan Group and Upper Proterozoic Penglai Group, while the overlying strata on this basement consist of Cretaceous terrestrial clastic sediments and volcanics comprised of Laiyang Group (K1l), Qingshan Group (K1q), Wangshi Group (K2w) (Fig. 2). Eocene and Oligocene systems are distributed in a limited way, being exposed only in Pingdu area.

The Lower Cretaceous Laiyang Group consists primarily of fluvial and lacustrine facies siltstone and sandstone interbedded with shale and volcaniclastic rocks, yielding abundant fossils including gastropods, ostracods, clam shrimps, insects, fish and plants (Zhang et al. 2010; Wang et al. 2016). The Lower Cretaceous Qingshan Group is composed of volcanic and volcaniclastic complexes with minor terrestrial successions, yielding bivalves, gastropods, ostracods, clam shrimps, turtles, lizards, psittacosaurids, pterosaurs, plant spores and pollen (Wang et al. 2016).

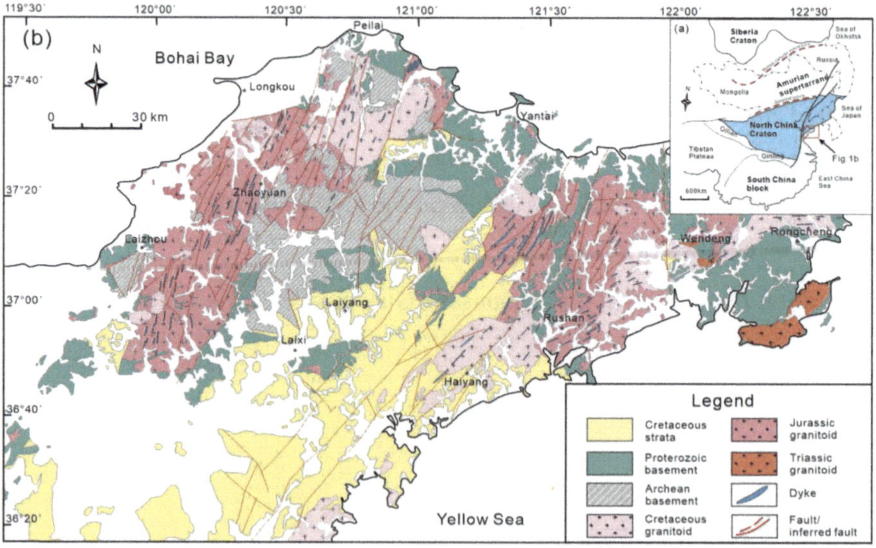

Fig. 1 **a** Tectonic sketch map displaying the position of the NCC in East Asia (modified from Meng et al. 2019). **b** Simplified geological map of the Jiaodong Peninsula including the Neoarchean-Proterozoic ancient crystallized basement, Mesozoic magmatism, and sedimentary covers

Fig. 2 Simplified cretaceous stratigraphic column in the Jiaolai Basin

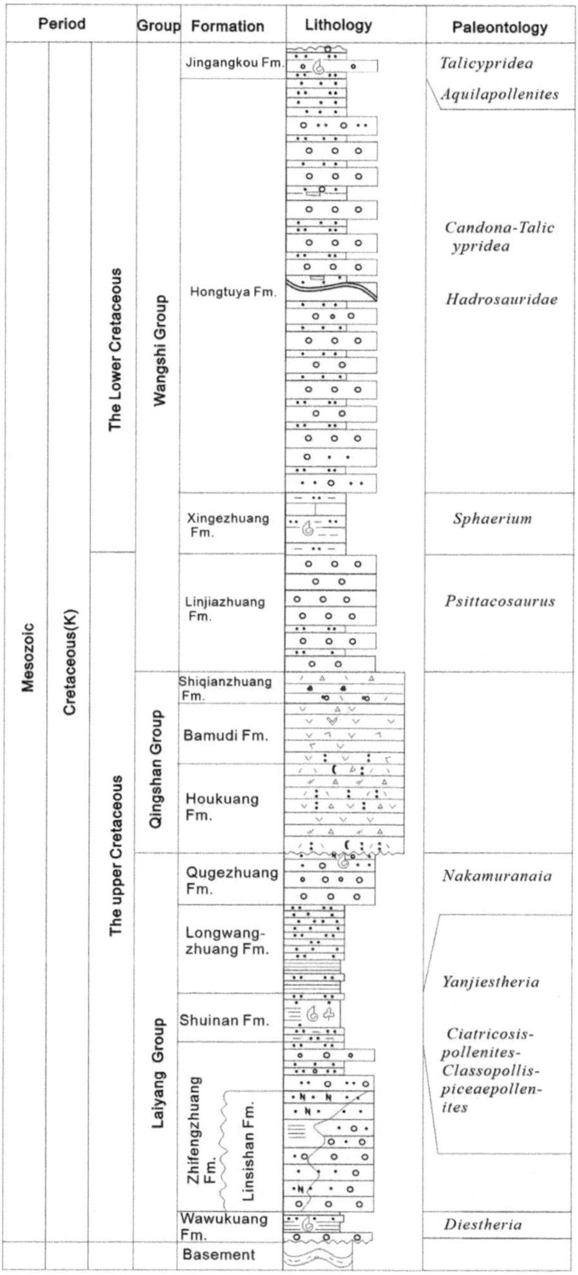

2 Stratigraphy of Jiaolai Basin

2.1 Laiyang Group

The Lower Cretaceous Laiyang Group in the Laiyang Sag developed continuously with the Wawukuang (K_1Lw), Zhifengzhuang (K_1Lz), and Shuinan Formations (K_1Ls), and the Longwangzhuang (K_1Ll) and Qugezhuang Formations (K_1Lq), from the lower to upper positions.

2.1.1 Wawukuang Formation (K_1Lw)

Wawukuang Formation unconformably covers the basement, including Mesozoic metamorphic and intrusive rocks and Precambrian lithosphere units. This formation consists of several rhythmic formations from conglomerate to mudstone. The grain size of the conglomerate in the bottom rhythmic layer is obviously larger than those of the conglomerates in the upper rhythmic layers. Yellow is the dominant color of these rocks, except that the conglomerate at the bottom is yellowish green. The presence of abundant mud cracks in the field and gypsum in mudstone units suggest that a strong evaporation under dry climatic conditions occurred at that time. The Wawukuang Formation is characterized by a set of thin conglomerates at the base of strata, fine-grained arkosic arenites in the middle, and coarse-grained arkosic arenites in the top representing a depositional cycle with fan-delta facies deposits to restricted lacustrine facies deposits. Samples from different levels of the strata show variable features in feldspathic grains and lithic fragments.

2.1.2 Zhifengzhuang Formation (K_1Lz)

This formation is comprised of conglomerate, pebbly sandstone and sandstone with interbeds of siltstone and mudstone. Its bottom is dominated by purple to reddish conglomerate and sandstone, and the lithology then fines upwards and changes to grey-to-greenish sandstone and mudstone. The rocks particle size of Zhifengzhuang Formation moves from coarse to fine, with low compositional maturity and structural maturity. The main sedimentary structure is block bedding, followed by wedge-shaped staggered bedding, grain sequence bedding and rhythmic bedding. The main sedimentary facies include alluvial fan, fan delta and shallow lake.

2.1.3 Shuinan Formation (K1Ls)

This stratigraphic unit is a set of fine-grained sediments composed of argillaceous siltstone, sandstone and shale interbedded with some marlstones. Abundant sedimentary structures, such as horizontal bedding, symmetrical wavy ripples and soft-sediment deformation structures, characterize the basal Shuinan Formation, which also yields abundant organic-rich black shale. It is interpreted as a lake facies deposit. There are distal sub-lacustrine turbidites primarily in the Shuinan Member and located mostly in the eastern Laiyang Subsag. There are also fine-grained fan deposits interbedded within thick, dark-gray, deep-lacustrine mudstone intervals.

2.1.4 Longwangzhuang Formation (K_1Ll)

This formation is dominated by conglomerate and sandstone mixes with thin-layered muddy argillaceous siltstone and shale. Sedimentary structures of horizontal bedding, crossbedding, mud cracks and ripple marks are pronounced in the field.

2.1.5 Qugezhuang Formation (K_1Lq)

This formation is composed of conglomerate and sandstone with a mixture of the sedimentary rhythms formed by medium-to-fine-grained sandstone, muddy siltstone and mudstone, which unconformably underlies the pyroclasticrocks of the Qingshan Group. The clastic rocks of the Qugezhuang Formation are mostly lithic sandstone or lithic complex sandstone. The mineral composition is dominated by quartz and rock cuttings, with a particle size ranging from 0.03 to 0.6 mm, and the degree of sorting and roundness is average. It is mostly supported by particles, and the main type of cementation is pore cementation. It reflects that this set of rocks should be formed in the piedmont filling or intermountain basin near the denudation zone of the strong tectonic deformation zone. It is a set of rapid accumulation products with strong physical denudation and close transportation distance.

2.2 *Qingshan Group*

The Qingshan Group in the Jiaolai Basin is characterized by the development of volcanic rocks. According to the volcanic activity and rock types, it is divided into four cycles from bottom to top, corresponding to the Houfen Formation, the Bamudi Formation, and the Shiqianzhuang group and Fanggezhuang group.

Within the basin, the Houfen, Shiqianzhuang and Fanggezhuang formations are mainly distributed in the northern boundary of the Wulian-Rongcheng fault on the southern margin of the Zhucheng Sag. There is also a small amount of Houkuang and Shiqianzhuang formations at the margins of other secondary structural units. The

Bamudi Formation is widely distributed in the basin, and it is exposed on a large scale in the northern Laiyang Sag, the Muping-Jimo fault zone, and the southern margin of the Zhucheng Sag. The stratum of the Houkuang Formation is exposed in the east of Baichihe Town. The entire rock layer is a grayish-purple set of volcanic clastic rocks. The Bamudi Formation in the basin is relatively developed, and its lower part is a cycle of medium-thick gray-white, purple-red coarse sandstone and medium-thick purple-red shale. Parallel bedding is occasionally seen in the sandstone. Sedimentary rocks with similar characteristics can be seen in the Bamudi Formation in the north of Wandi Town. The bottom is gray-black intermediate-basic volcanic rock; the middle is purple-red mudstone wrapped gray-black volcanic bomb; the rock formation has irregular quartz veins, and the upper part is gray-black intermediate-basic volcanic rock with irregular intrusive veins.

There is an outcrop of the Shiqianzhuang Formation in the southwest of Wadian Town, Zhucheng Depression. The rock formations are thick blue-gray volcanic clastic rocks interbedded with middle-thick brick red and blue-gray siltstones, with thick layers of conglomerate on the interior. The Fanggezhuang Formation exposed in the southeast of Zhigou Town is a set of neutral volcanic rocks with a grayish-yellow tone.

2.3 Wangshi Group

The Wangshi Group is composed of a set of brick-red lacustrine facies clastic deposits, with volcanic interbedded layers developed in some areas. Its main lithology is clastic rocks such as conglomerate, sandstone, siltstone and mudstone. It is exposed in the south-central Laiyang Sag, the middle part of the Muping-Jimo fault zone, the middle eastern part of the Gaomi Sag, and the mid-north part of the Zhucheng Sag. From bottom to top, the Wang's Group in the Jiaolai Basin are the Linjiazhuang Formation, the Xingezhuang Formation, the Hongtuya Formation, the Jingangkou Formation and the Jiaozhou Formation. The top strata of the Wangs Group divided into different regions in the basin have certain differences. The top strata of the Wang Group in the Laiyang Sag is called the Jingangkou Formation, and the top strata of the Wang Group in the Gaomi Sag is called the Jiaoxian Formation.

2.3.1 Linjiazhuang Formation

The Linjiazhuang Formation is the bottom strata of the Wangs Group. It is not only found in the northeast of Laiyang City, but also in the north of Taocun Town, the northwest of Facheng Town, the south of Zhigou Town and the north of Junan County. The Linjiazhuang Formation is composed of a set of purple-red, light-green conglomerate and gravel-bearing coarse sandstone and mudstone. The Linjiazhuang Formation in the north of Taocun Town is mainly composed of purple-red conglomerate, gravel-bearing coarse sandstone with coarse sandstone interbedded. The gravels in the conglomerate layer are sorted and rounded. The gravel composition is volcanic rock and is oriented in several layers. The Linjiazhuang Formation in the northeastern part of Laiyang City is mainly composed of purple-red and light green conglomerate and coarse sandstone interbedded layers. In the northwest of Facheng Town, it is composed of interbedded purple-red conglomerate and gravelly coarse sandstone, and transitioned upward into interbedded purple-red siltstone, mudstone and mid-sandstone. In the south of Zhigou Town, it is composed of purple-red massive gravel-bearing coarse sandstone, fine conglomerate and coarse sandstone interbedded, transitioning upward to purple-red mudstone with volcanic breccias in the mudstone layer. In the north of Junan County, it is a thick layer of purple-red medium sandstone and fine gravel-bearing coarse sandstone interbedded.

2.3.2 Xingezhuang Formation

The Xingezhuang Formation is distributed in the north-central part of Laiyang Sag, around Facheng Town, Yuanshang Town-Lingshan Town, West Jiayue Town and Zhigou–Zhangying Town. The formation is composed of brick red-gray-green siltstone, mudstone, gray-white-purple red coarse sandstone, and gravel-bearing coarse sandstone. The coarse sandstone is often in scouring contact with the bottom silt mud, and the sandstone rock mass is produced in the form of a lens.

2.3.3 Hongtuya Formation

The Hongtuya Formation is widely distributed in the basin. It is developed in the central-southern Laiyang Sag, the central-northern Zhucheng Sag, and the southeast of the Gaomi Sag. The Hongtuya Formation in the Laiyang Depression is mainly developed in the southern part of Laiyang City. The strata are mainly composed of purple-red and brick-red pebbly coarse sandstone, basic coarse sandstone and mudstone. The pebbly coarse sandstone often contains gray-white fine conglomerate bands. The gravel in the conglomerate is dominated by volcanic rocks. The coarse sandstone and the underlying mudstone are in scouring contact, and the gravel-bearing coarse sandstone gradually transitions to coarse sandstone, while the sandstone layer contains boulders. The Hongtuya Formation in the southern part of the basin has similar characteristics with the Hongtuya Formation in the Laiyang Depression, but purple-red sandstones are found in some points with blue-grey volcanic rocks.

2.3.4 Jingangkou Formation

The stratum distribution of the Jingangkou Formation is relatively limited, and the development of the Jingangkou Formation is only found in the area of Jingangkou Village in the southern margin of the Laiyang Depression. It is composed of gray-white coarse sand, gravel-bearing coarse sandstone and purple-red mudstone interbedded, in which the sandstone is in a lens shape and in contact with the underlying mudstone.

3 Field Trip Introduction

Four stops are provided to investigate the Cretaceous strata in Jiaolai basin, Shandong province, include Lingshan Island, Laiyang Cretaceous National Geopark, Shandong Zhucheng National Dinosaur Geopark, Jimo Mashan Geopark (Fig. 3).

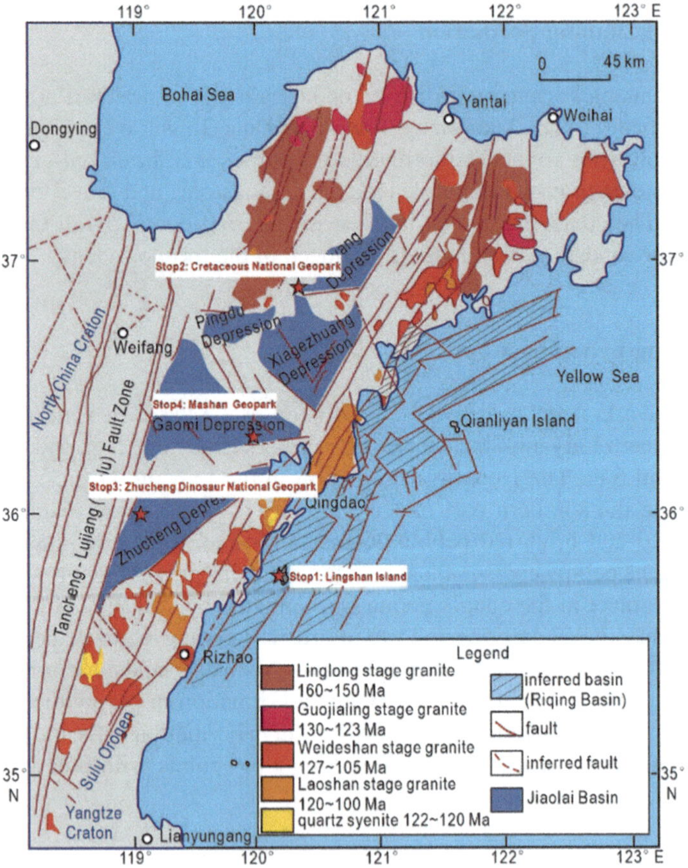

Fig. 3 Location map of four Cretaceous strata stops (Modified from Yang and van Loon 2016)

3.1 Stop 1: Lingshan Island

Lingshan Island is located in the western Yellow Sea, 37 km south of Qingdao City, Shandong Province, East China (Fig. 3). The surface of the island is only 7.66 km^2, and the highest altitude is 513.6 m. There are two main rock units on the island (Fig. 4). The Old Formation is composed of ~90 m gray-yellow thin sandstone interbedded with dark gray to black shale (the lowest part of the Lingshandao Formation is not exposed because it is below sea level). This unit is part of the Early Cretaceous Lingshandao Formation (Zhang et al. 2013), and SBGMR (1991) named it "Fajiaying Formation". The unit is covered by a 15-m-thick rhyolitic lava unconformity, overlaid with more than 150 m of pyroclastic and spewing rock. These together constitute the Bamudi Formation of the Qingshan Group. This formation also belongs to the Early Cretaceous. There are several dolomite dike intrusions on the sedimentary rocks throughout the profile. The stratigraphic age is definitely Early Cretaceous, as weighted average 206Pb/238U measurements of zircons from upper rhyolite samples using LA-ICP-MS gave an age estimate of 119.2 ± 2.2 Ma (Middle Aptian) (Zhou et al. 2015), while the maximum sedimentary age of the Lingshandao Formation is about 121 Ma (Lower Aptian) (Wang et al. 2014). The marine sediments of the Lingshandao Formation correspond in age to the uppermost part of the Early Formation Laiyang Group in the Jiaolai Basin (Wang et al. 2014). Previously considered to be the terrestrial lacustrine deposits of the Laiyang Group in the Jiaolai Basin (SBGMR 1991), it is now considered a separate lithostratigraphic unit that does not belong to the Laiyang Group; this decision was based on knowledge of marine plant fossils and turbidites (Lu et al. 2011, 2012). Due to the existence of the Sulu orogenic belt (Fig. 3) between Lingshan Island and the Jiaolai Basin depression, the Late Mesozoic magmatic rocks intruded and the sedimentary backgrounds are different, the Lingshan Island Formation must be separately divided into a stratigraphic entity.

Lingshan Island was formed in the Early Cretaceous, initially as part of a sedimentary basin and later as part of a volcanic arc. Uplifted by the tectonic tilting related to later compression, the formation as a whole dips toward ENE. Despite the simplicity of the overall construction, the construction background has been debated for many years. This is because Lingshan Island is located in the Sulu orogenic belt (extending from Jiangsu to Shandong), an important structural suture (Zhu et al. 2009), which is the eastern belt connecting the Yangtze Craton and the North China Craton (Lu et al. 2011).

Detrital zircon U–Pb ages of the Lingshandao Formation and field data indicate that the sediments of the Yangtze Craton and the North China Craton are composed of a mixture of materials from the basement of the North China and Yangtze Blocks (Wang et al. 2014).

On the other hand, Lingshan Island is located in the large NE-SW trending rift basin developed in the Late Mesozoic off the coast of Shandong (Fig. 3). This setting was obtained by Zhou et al. (2015) after investigating the tectonic and magmatic activities in this vast area and after basin development and reconstruction. The lithosphere in eastern China experienced significant thinning during the Late Mesozoic

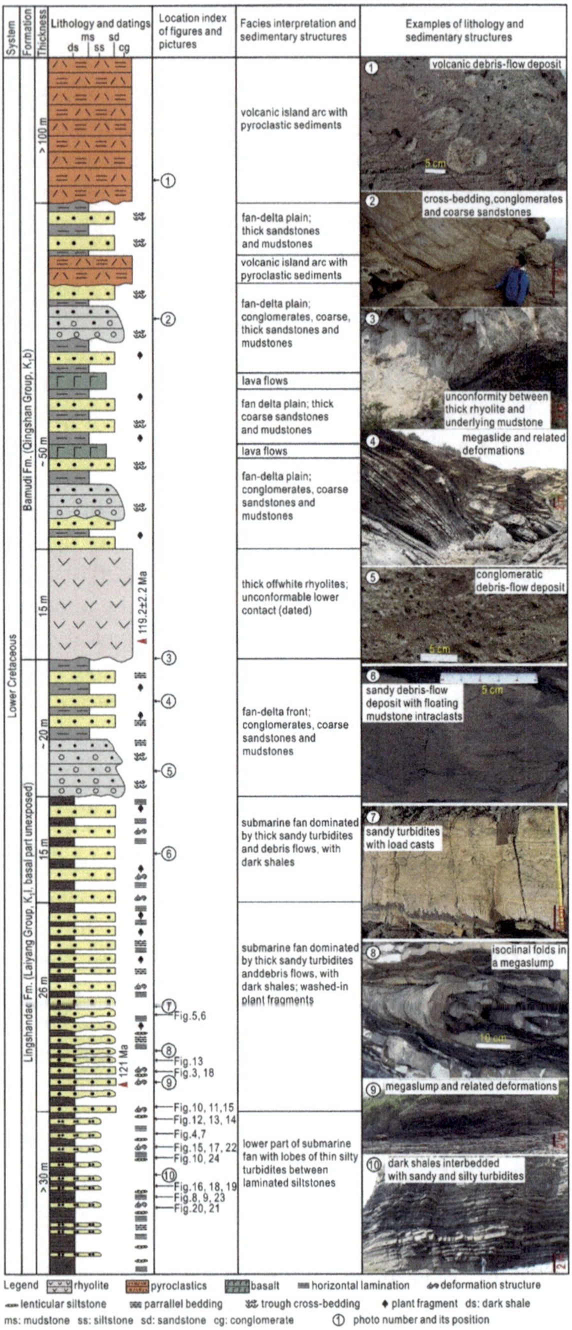

Fig. 4 Lithostratigraphy of the Lingshandao Fm. and the overlying Bamudi Fm., with some field examples

(Zhou et al. 2003a, b; Zhai et al. 2004; Wu et al. 2005). In the extensional environment of eastern China during the late Yanshanian (125–106 Ma) tectonic period, intrusions and frequent volcanic eruptions were particularly active (Zhou et al. 2015).

3.1.1 Stratigraphy

The Lower Cretaceous Laiyang Group

The lower part of Lingshan Island is a set of turbidites (the part below sea level is possible much thicker), with a huge layer of gray-white rhyolite (Fig. 4). The surface is built of a set of terrestrial molasse topped by massive volcanic rock. The upper body of the exposed formation on Lingshan Island may also be mixed with it, and the clastic rock requires further investigation.

It is mainly exposed in the western and southern part of Lingshan Island with a small amount in the back stone attractions at the northern end of the island. It is easy to observe at the southern and western ends of the island due to stratigraphic tendencies and topography. The lower part of this formation is submerged in seawater, the exposed surface thickness over 100 m, manifested in gray (weathered surface is soil yellow) thin sandstone (occasionally huge thick sandstone with granular sequence), siltstone and black mudstone constitute good rhythm layer. The sandstone layer is mostly composed of massive silty and fine sand, while the mud unit is grey-black mudstone, which partially contains black and small plant fragments. Sandstone thin sections show that the debris particles are feldspar, rock fragment and quartz, and the filling material is mainly calcite and mud (Zhang et al. 2013).

The Lower Cretaceous Qingshan Group

According to field observation, the bottom and top of the rock layer have obvious magma flow lines, and the overall appearance is pale gray-white cryptocrystalline, indicating that it is volcanic lava overlying the sandstones with angular unconformity (Fig. 4). This thick layer of rhyolite emerges in the south, southwest and southeast of Lingshan Island, with the best near the southernmost Laohuzui/Tiger mouth, the highest location with an overall thickness of about 15–20 m, thick south to east and thin north to west. The thickness extending eastward to the south of the central reef is slightly thinner; westward to the Ferry Yard near the south of the pier to more than 2 m. After a LA-ICP-MS zircon U–Pb dating of a rhyolite sample from the southern of Lingshan Island, data were obtained as 125 $\pm$ 1.4 Ma.

The upper part of the Lingshan Island is a set of molasse architecture. The lower part of this section is clastic rock about 40 m thick, and the upper part is volcanic rock, pyroclastic rock, and occasionally clastic rock. Volcanic rocks above the molasse cannot be observed due to the steep terrain, but the collapsed rocks are visible.

The uppermost of Lingshan Island is composed of volcanic lava, mainly various volcanic lava with huge thickness, complex lithology and variable occurrence. Among these, lava flows of crater phases were found on several hills.

3.1.2 Sedimentology Environment

The Lingshan Island sediments have long been regarded as the continental sedimentary environment at the edge of the Jiaolai Basin (SBGMR 1991). However, the discovery of marine fossil Dasycladales and distal turbidites (Lu et al. 2011) clearly indicate that the depositional environment must have been marine. The continuous shallowing of the Early Cretaceous basin resulted in the facies transition of the present-day Lingshan Island from an abyssal basin with slumps and turbidites to a shallow basin with turbidite channels and turbidites. From there to the volcanic basin, the volcanic overflow flows into the fan delta, to the volcanic deposits and overflow (Zhou et al. 2015).

The lower part of the Lingshandao Formation is composed of transported sediments and turbidites (Lu et al. 2011; Shao et al. 2014). The upper part of the Lingshandao Formation is now interpreted as a shallow-water continental delta front developed in an intracontinental setting, without any possible association with plate margins or collisions (Shao et al. 2014; Zhou et al. 2015). This interpretation is based on evidence of short-distance sliding, well-developed slump deposits, imbricated structures supported by sand grains, well-developed troughs, abundant carbonized plant debris, and xylose vitreous bands (Zhong 2012).

Although there is still controversy over the depositional environment of the Lingshandao Formation, namely continental (SBGMR 1991; Zhong 2012; Shao et al. 2014) and marine debates (Lu et al 2011, 2012), for slumps and turbidites Consensus exists for the source rocks: these are deltaic deposits, probably located in the southwest of Lingshan Island (Lu et al. 2012; Zhong 2012).

The Lingshandao Formation turbidites are only exposed on Lingshan Island, and there is no similar deposition in nearby stratigraphic units (Zhang et al. 2013). It is therefore difficult to determine the extent of the sedimentary facies, as this can only be investigated within a 1.6 km lateral exposure on the islet.

Evidence to distinguish turbidites from the Lingshandao Fm include bottom structures and sedimentary sequences. Bottom structures, such as load structures, groove casts (Fig. 5a) and flute casts (Fig. 5b) are well developed at the contact of sandstone and mudstone. The Lingshandao Formation turbidites generally have massive bedding, normal gradation (Fig. 6), climbing lines (cross-stratification) and horizontal stratification. There is also a complete Bouma sequence (Ta – Te) (Fig. 7). The sandy turbidites are poorly sorted, the quartz grains are horn-like to sub-horn-like, and the matrix is more (Fig. 6). Intramudstone clastics (Fig. 8) are common in some thick sandstones, most of which are deposited from sandy mudflows or eroded from dike systems by turbidity currents. Detrital sandstones are usually overlaid by normally graded siltstones and argillaceous siltstones, implying that debris flows and turbidity flows occur simultaneously and may accompany each other.

Fig. 5 Lower surfaces of fine-grained sandstones with load casts and sole marks (**a**), such as groove casts and flute casts, and numerous well developed flute casts (**b**)

From bottom to top, the thick turbidites of the Lingshandao Fm commonly consist of fine sandstone, siltstone, argillaceous siltstone, silty mudstone and shales. The turbidites that are a few centimeters thick are mainly composed of siltstone, argillaceous siltstone, silty mudstone and shale.

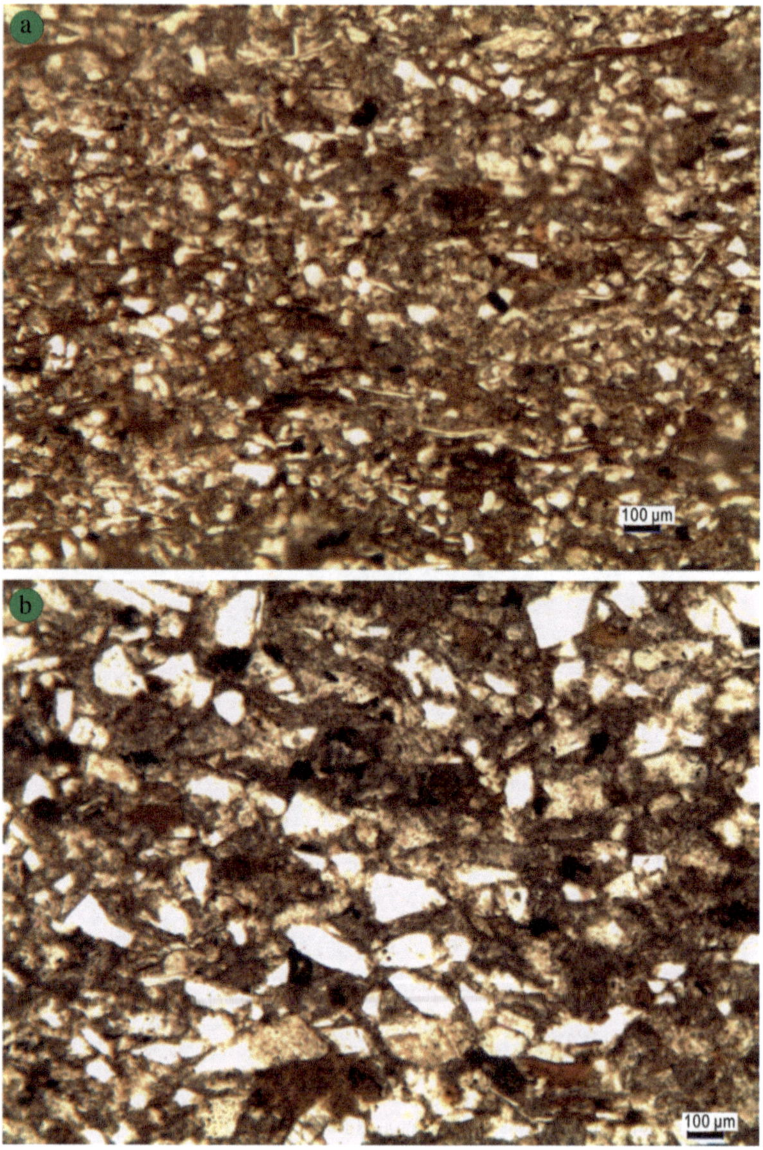

Fig. 6 Thin sections of a turbidite sample (5 cm thick) from the lower part of the Lingshandao Fm. The lower part **b** consists of fine sandstone; the quartz grains are angular and poorly sorted. The upper part **b** consists of argillaceous siltstone and contains considerably more matrix than the lower part

Fig. 7 Two well developed Bouma sequences in a sandstone succession of 51 cm thick. With load casts and flame structures at the base, both turbidites show, from bottom to top, a massive part, normal grading, parallel lamination, cross-lamination, and horizontal lamination

3.1.3 Some Important Phenomena of Sedimentology

Several well-exposed gravity flow deposits in the Late Cretaceous Lingshandao Formation provide a rare opportunity to study deep-water processes. The main aims of this presentation are, consequently, to establish the spatial and temporal relationships between the various gravity flows and to establish the co-existence of these deep-water deposits and their sequence stratigraphic patterns and sedimentary environments.

Field observations, large-scale measurements (1:20), sketches of field profiles and thin-section analysis were performed for the study of the gravity flow deposits in the Late Cretaceous Lingshandao Fm, western Yellow Sea.

Dark shales (DS), hyperpycinites (HP), turbidites (TB), debrites (DB), slumps (SP), etc. build 5 types of lithofacies and 5 middle-term base-level cycles (MSC) that were identified in a 16.2 m high field section on Lingshan Island. Horizontal lamination is well developed in dark shales and argillaceous siltstones (Fig. 9a and b). Load casts are common at the bottom of sandstones and siltstones (Fig. 9c). Various soft-sediment deformation structures are well developed in sandstones and siltstones (Fig. 9). Normal and inverse grading is common in both sandstones and siltstones. The lithology of the Late Cretaceous Lingshandao Fm. consists of dark shales, grey

Fig. 8 Floating mudstone intraclasts in a graded sandstone of 35 cm thick (diameter of coin 2 cm). This sandstone is interpreted as a sandy debris-flow deposit. It overlies greyish mudstone and is overlain by numerous thin-bedded laminated turbidites

siltstones and light grey fine sandstones. Sandstone textures are characterized by poor sorting, a high matrix content (~20%), and angular to sub-angular particles (Fig. 9d).

The identification of the short-term base-level cycles was based on the alternations of dark shales, siltstones and sandstones. The middle-term base-level cycles could subsequently be constructed on the basis of asymmetric short-term cycles and upward changes in thickness. Each of the middle-term cycles starts with a fairly short rising base-level, followed by a longer lowering base-level (Fig. 9). Couplets of beds with upward coarsening and upward fining are separated by micro-scale erosion surfaces; these couplets are interpreted as hyperpycnites originated from estuarine floods (Fig. 9a and b). Combined occurrences of sole marks, graded bedding in siltstones/sandstones and alternations with dark shales are interpreted as deep-sea turbidites (Fig. 9e). Sharp basal and top boundaries of sandstones/siltstones with floating mudstone intraclasts most commonly typify deep-sea debris flow deposits (Fig. 9f and g). Large-scale convolutions and other sedimentary deformations, combined with mixing of various lithologic components, are interpreted as resulting from slumping over depositional slopes (Fig. 9h, i and j), mostly on delta fronts.

The lithofacies change from deep-sea dark shales at the base to hyperpycnites, turbidites, debrites and slumps at the top in each middle-term base level cycle. The turbidites, debrites and slumps tend to become ever thicker from MSC1 to MSC5.

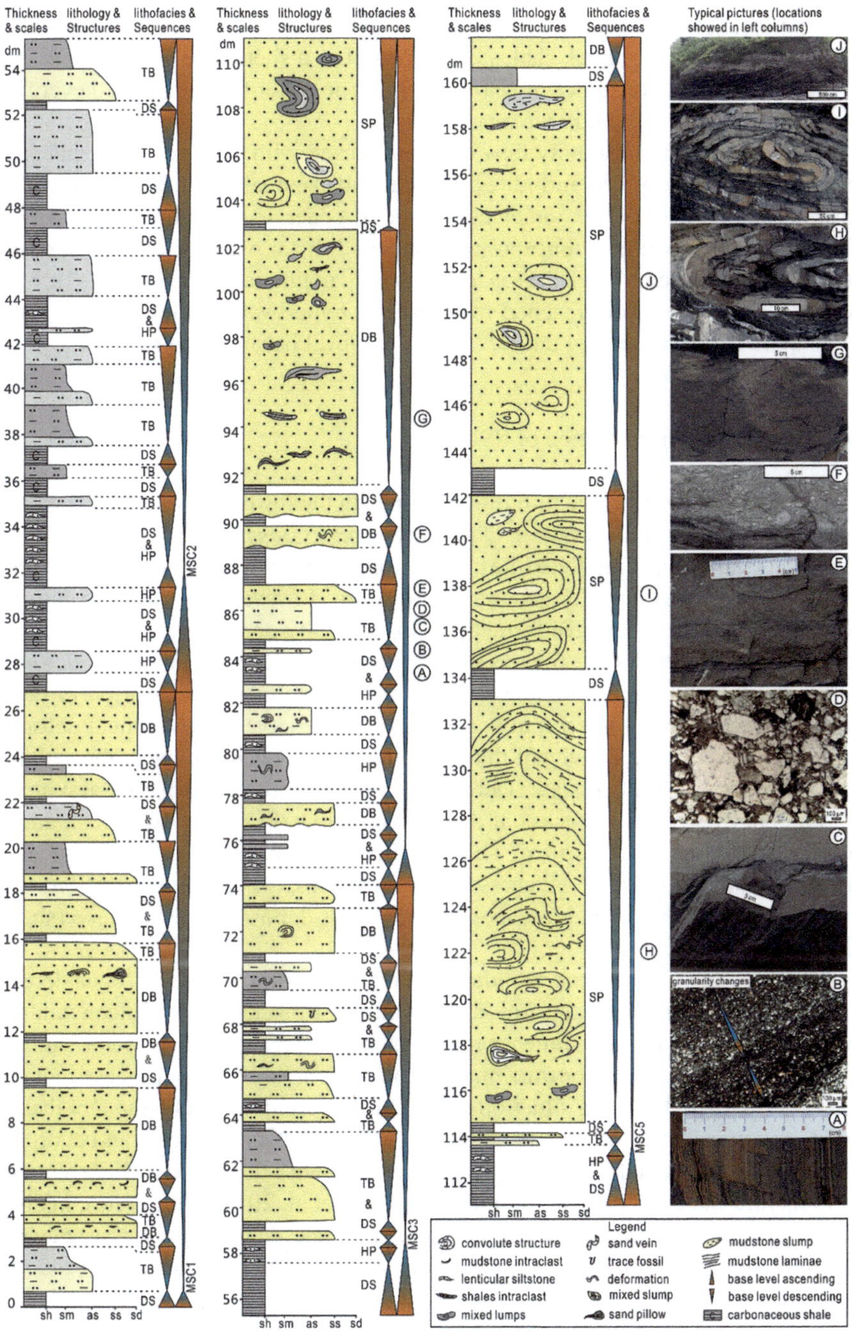

Fig. 9 Lithological log and base-level cycles of the Lingshandao Fm, Ferry Yard profile. DS: deep sea; HP: hyperpycnites; TB: turbidites; DB: debrites; SP: slumps; sh: shales; sm: silty mudstones; as: argillaceous siltstones; ss: siltstones; sd: sandstones

This indicates that the delta fronts prograded more extensively in successive middle-term base-level cycles. Dark shales developed after a short time when the sea level rose due to episodic tectonics or sea-level rise. The water then shallowed ever more with ongoing sedimention in the basin.

The development of gravity flow deposits started with deep-sea dark shales and hyperpycnites, followed by turbidites, debrites and eventually slumps in each middle-term base-level cycle. This indicates that flood-generated hyperpycnal flows with high suspended-sediment concentrations reached distal basin floors and deposited fine-grained sediments when the sea level rose and maintained a high stand. Turbidity currents and debris flows reached to where the hyperpycnal flows were deposited when delta fronts prograded farther and the sea level became ever shallower.

Description and Interpretation of SSDS

The SSDS were found in the turbidite sequence in the lower part of the Lingshandao Formation. The geological setting of the early Cretaceous island arc formation in the Lingshan Island area is related to the internal forces (structures) of the island arc; these structures were obviously formed after deposition. However, there are also deformation structures that must have formed synsedimentally or while the sediments were still in a non-lithified state (penecontemporaneous deformation). These deformed structures, which can in principle develop during continuous deposition or during periods of sedimentary interruption, are often referred to as soft sediment deformed structures (SSDS) if it is evident that the deformation occurred when the sediment was still in a non-lithified state. Deformation in the loose state can be assumed, if not proven, by whatever process, if the deposit below the deformed deposit and the deposit above it is not deformed. This situation may be caused by three processes: (1) rework of new sediment layers (such as slumping); (2) unstable conditions (such as load shedding, pore water escape, etc.); (3) post-deposition disturbance (such as earthquake shock).

In this context, neither consolidation nor lithification should be considered to represent a well-defined state (cf. Davies 2005; Einsele 2013). Unconsolidated, initially water-saturated sediments that become stronger and less water-saturated consolidated sediments over time may initiate a later lithification process that typically It is also gradual. The situation can be further complicated by the fact that in a range of layers of varying composition and grain size, some layers can be strongly consolidated and even lithified, while others (even those deposited earlier) remain in an immature State of consolidation (cf. Van Loon et al. 2013). Thus, one can distinguish between brittle deformation structures and ductile deformation structures as final members, with the middle group filling the complete spectrum of soft sediment deformation structures (SSDS).

The lower part of the Lingshandao Formation is solid sediments. Represented by brittle, ductile and intermediate deformation forms, as well as deformation induced by liquefaction and/or fluidization, depending on precise sedimentary, diagenetic and tectonic conditions. Under the conditions of sedimentary slopes accumulated in the lower part of the Lingshandao Formation, block waste, tectonic activity, high deposition rate and underwater environment, various types of SSDS were formed. Most are well known and frequently described and analyzed.

We therefore focus here on some rare SSDS, some very well developed types, and some structures that provide additional information on the early Cretaceous tectonic setting of the region.

Faults

The study area has been affected by tectonic activity leading to fractures in already lithified rocks (see Sect. 1), which may appear insignificant since fractures are generally considered to be manifestations of brittle deformation. However, fracturing in water-saturated alithic sediments is also evident. It is well known that faulting is not restricted to petrified or strongly consolidated sediments, but can even occur in unconsolidated sediments saturated with water (see Van Loon and Wiggers 1975, 1976). In this case, faults occur if instability occurs suddenly, such as overloading (e.g. storm surges, sedimentary overburden) or shocks (e.g. earthquakes, volcanic eruptions). This penecontemporaneous fault does exist in the lower Lingshandao Formation.

- **Description of penecontemporaneous faults**

Some relatively small faults may have developed when the sediments studied were not yet lithified. This can be inferred from the fact that these faults affect only one or a limited number of layers and not the layers below or above, although the characteristics of the sediments involved suggest continuous deposition. Almost all of these penecontemporaneous faults formed normal faults (Fig. 10a), indicating extension, although there were some rare examples (reverse faults), which seemed to indicate compression (upthrust faults) (Fig. 10b). A process that may lead to simultaneous forward and reverse faults is proposed.

Synsedimentary and penecontemporaneous faults are also frequently present in this sequence of landslides. The internal stress field in and around the slumped rock mass is often quite chaotic, which is reflected in the complex system of small-scale faults in the slumped rock mass. Interestingly, in this case, there are massive landslides, composed of turbidites that together can measure several meters thick. However, at the head of such a landslide, a compressive environment tends to dominate, while further back, a tensional environment may dominate. This provides evidence that not all features indicative of compression or tension are due to endogenous conformation.

Finally, a special type of fault can be found at several places in the turbidite profile, similar in some respects to faults in the "fault-graded layers" described by

Fig. 10 Examples of small penecontemporaneous faults in the Lingshandao Formation. **a** The faults shown here are, just like the great majority of fields in the turbidite succession, normal faults. Note the stepwise displacement (graded faults). See also Fig. 11a and b. **b** One of the few synsedimentary upthrusts in the Lingshandao Formation

Seilacher (1969). These faults are treated separately due to their special properties and implications for their genetic interpretation.

- **Interpretation of the penecontemporaneous faults**

Almost only the presence of normal faults indicates extension rather than compression. Typically, states of tension leading to fractures in unlithified sediments may only be localized due to immediately adjacent movements such as slumps; however, states of tension appear more likely to be caused by broader conditions, suggesting regional tectonics, may be a putative rift valley in the region. This is consistent with the geological background.

It cannot be completely ruled out that some of the uplift is not the result of a local stress field, but may be due to conditions occurring over a wide area. This implies a compressed state. Considering that most of the syngenetic soft-sedimentary faults are normal faults, this may indicate that the rifting is not completely regular expansion, and some irregular rift block movements may also occur, resulting in local compression deformation.

For example, if this extensional tectonics is accompanied by a slight rotation of some blocks, one can imagine that local compression phases could occasionally lead to features such as upthrusts. This may explain the ongoing debate in the region about compressional or tensional tectonic regimes.

Faults Resulting in Domino Boudins

Some (times) vertical small-scale faults in the research area affect specific turbidity rocks at a relatively long horizontal distance (number of rice). These turbid rocks are usually composed of a relative sand layer with a thickness of only a few centimeters.

- Description of the domino boudins

The turbidity rock affected by the faulting action described here is decomposed by a small decisive layer at a considerable a few centimeters away, so a rectangular block of several centimeters wide to one point is formed on the horizontal section (Fig. 11a) Essence Due to the form of broken, these faults have gone through slight rotation (Fig. 11a and b). This characteristic is called a Domino card. According to what we can trace in the literature, we have described the border of Domino, which completely comes from GosCombe et al. 2004; Dabrowski and Grasemann 2014) caused by the Zhang sex structure system. (Alsop and Marco 2011) are mainly calcium collapsed blocks, but there is no silicon fragmented sediment that has not yet been rocky during Boudinal.

Figure 11c shows more Mimino borders in the research area. In some places, that is, a certain layer of local disturbances, the tension characteristics of the strong field have become more obvious, because the "block" is indeed separated from each other (Fig. 11d).

- **Interpretation of the boudinage structures**

In the literature, the description of the boundary structure almost exactly comes from the hard rock unit that is exposed to the sexual structure (lloyd et al. 1982; He et al. 2015; Peters and Others 2015). The mud in loose deposits is rarely described, and there are usually no specific analysis and genetic interpretation (Burg and Harris 1982; OGIWARA and ITO 2011). The debris formed in the still soft sediment is considered to be caused by the landslide caused by gravity and the tension generated by it; the boundary generated by the sagging limbs of the stretch (Altermann 1986). In Israel's evening update of the Liman Lisan group, some similar structures are considered to be a fracture, rather than the real Domino bone border (soft depression deformation structure) (Alsop and Marco 2011).

GosCombe et al. (2004) and Dabrowski and Grasemann (2014) believed that the formation of the Domino bone boundary in the rocky rocky rock was made due to the Zhang sex structure mechanism. However, experiments have proven that the binding

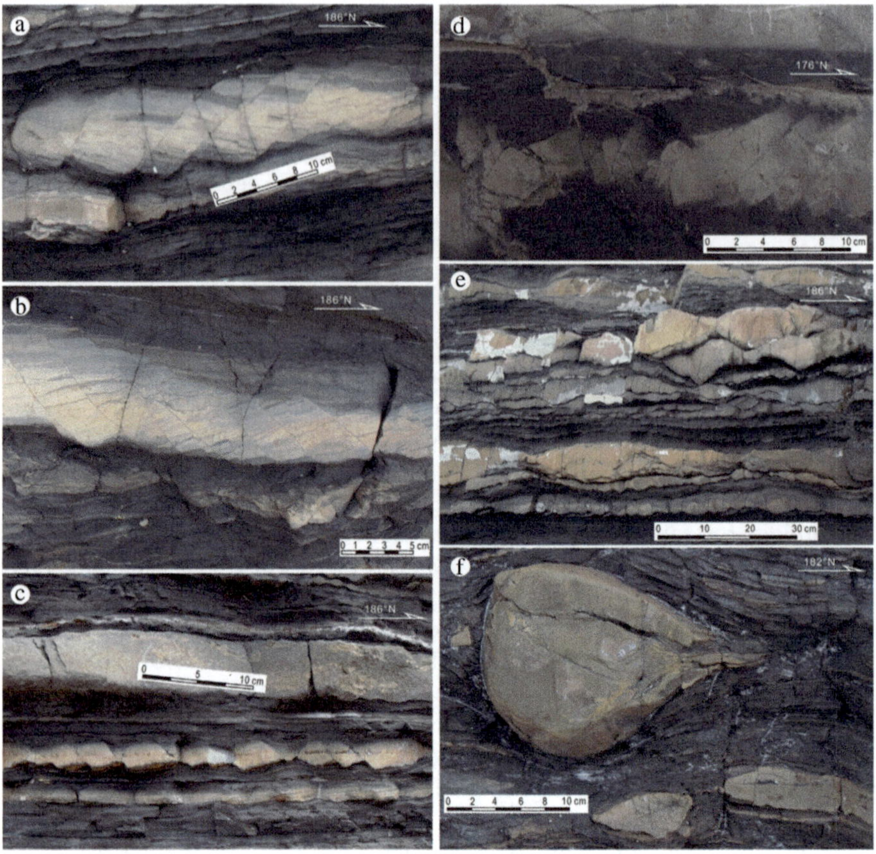

Fig. 11 Boudins and associated SSDS indicating extension. **a** and **b** Domino boudins. More or less parallel penecontemporaneous normal faults broke up the respective layers due to tension. **c** Domino boudinage. The sandstones reacted in a brittle way during tension, whereas the shales were still ductile. Note the rotation that the individual domino 'blocks' show as an unavoidable consequence of the stretching. **d** Pulling apart of a sandstone that had already underwent domino boudinage. This indicates continued, or more likely, repeated tensional conditions. **e** Pinch-and-swell structures in turbidites and dark, possibly autochthonous, shales. **f** Isolated, fairly rounded sandstone masses resulting from boudinage while the sediments were not yet lithified, but when the sand was already much more consolidated than the clay

part can also be formed in unsteady stubborn (Zulauf et al. 2011; Marques et al. 2012). All research shows that nervousness is the reason for flesh; there is no reason to assume that the structural conditions of the areas studied here are different.

Considering that the border does not appear in all rocks that seem to be appropriate, but the layer with boundary also shows other types of soft depression deformation. For example, a half-ground rigid structure must be inferred that the layer affected by tension has been solidly condensed. It is manifested as crispy. This means that the influential layer has been compacted under the influence of the upper cover, or the early ripple junction has been precipitated from the porosity water.

One of the main questions in this case is why tension is not a continuous feature. This can be explained by the geological background of the research area. The area is considered a crack or arc basin. The earthquake must occur frequently, and it is well known that earthquake activities can cause damage (such as Dugué 1995; Fossen et al. 2013). Therefore, the break on the study of the study is likely to be caused by a short stretch stage. During this period, the area experienced stretching.

Common Boudinage Structures

Not all Boudinals form a Domino bone brand Boudins: there are more common types: squeezing structure, and more extreme forms: Boudins; Altermann (1986) describing similar SSDS combinations.

- **Description of the pinch-and-swell structures and boudins**

The turbidtest rock section with a Dimino bone pattern also contains a layer with SSDS, called the bloating structure (Fig. 11e); these structures show the remnants of the affected layer with thicker neck connection. The same deformation process must also be considered an extreme form of squeezing-expansion structure, that is, more or less isolated sandstones surrounded by shale, thereby forming a characteristic boundary (Fig. 11f). The size of the sandstone boundary is generally 1–2 mm.

The border sometimes shows vague layers, which is not always continuous, but it is usually curved and roughly parallel to the outer border of the boundary, but occasionally the layer combination may be twisted. Generally, the original sedimentation structure in the wrapping shale cannot be recognized because they may be covered by the shear effect during the deformation process. In isolation, it is also embedded in more tough shale.

- **Interpretation of the pinch-and-swell structures and boudins**

The expansion structure should be considered as the result of extreme continuous neck deflation (Schmalholz et al. 2008), which can be formed during the parallel stretch of the rigid viscosity layer of the weaker sticky layer (Schmalholz and Maeder 2012), and a long time ago (Ramberg 1955) was discovered as the result of tension, which has also been proven by experiments (such as Mandal et al. 1992). Related tension must also affect the lower and higher rock formations in the strata, but the nature of the adhesive structure between each layer (the length of the neck or neck, the length of the neck, the size of the bonding, the "block" of the Domino or the non-broken layer or the non-broken layer Wait) different (as mentioned earlier, may be due to the difference in ability and/or grain size), so it is unlikely that there are multiple layers of bonding at the same time.

Theoretically, the difference between the types of various layers may be due to the differences in each layer of rock, but the replacement is composed of turbid rocks (and collapsed rocks) with a small amount of rock. Some people may say that the differences in the division of boundaries in the serial level may be the difference in compact state, or even a slight rocky role. In this case, there should be a constant layer with boundary, and the type of boundary has regular changes; however, this gradual change does not exist. This shows that when the tension is large enough, it is enough to affect the single layer, rather than too tight or even petrochemical (from silicon dioxide; rare square decomposite glue knots obviously have no effect), although it cannot be ruled out, the tension becomes strong enough In the follow-up stage, the layer of bonding has gone through further bonding.

Load Casts and Related Structures

At the contact between the light (clay) layer and the relative (sand) layer, deformation may occur due to the opposite density gradient (CF. Anketell et al. 1970). This situation often leads to load casting (Fig. 12). This type of tough deformation may be the most common in all SSDS and has a wealth of description and genetic explanation. Macar (1948) is one of the first research methods using sedimentary methods. Since then, about tens of thousands of studies have been published, describing and analyzing the causes of load castings (such as Moretti and Others 1999; Mokhtari Fard and Van Loon 2004; Moretti and Sabato 2007), pseudo-nodules, etc. Related structures (such as Van Loon and Bronzikowski 1987; Owen et al. 2011), ball pillow structure (such as Allen 1986; Wang et al. 2011; Van loon and Pisarska-Jamrozy 2014) and flame structure (such as Mills 1983; Cojan and Thing 1992; He et al. 2015), also from Zhenjiyan (such as Alfaro et al. 1999; Plaziat and Ahmamou 1998). Here, it seems that it is enough for them to describe them.

The delay sediments under the development load casting and the side pressure are under pressure (Fig. 12a), so it has also become deformed. Therefore, they are partially bent and partially compressed along the external shape of the load casting to form a flame structure. During a similar process, more corrosive (sandy) substances push away more tough (clay) sediments, causing complex deformation of the fold structure (Fig. 12b and c). When the sandy load casting is finally depressed in the mud sediment below, they eventually transform into a fake tumor, because the clay material from the bottom layer is pushed too far, and the load casting is completely embedded.

Sand ball-pillow-shaped structure also exists (Fig. 12d). Their size ranges from 1 to 13 cm. The sediment is bent or folded in certain ball pillow structures, and even mixed with each other.

Water-Escape and Associated Structures

When the sediment is liquefied or more specifically, many SSDs are formed when they are darling. This process can occur in several cases. In order to prove the existence of the Lingshan Island group. It is also applicable to liquefaction and duty. We described

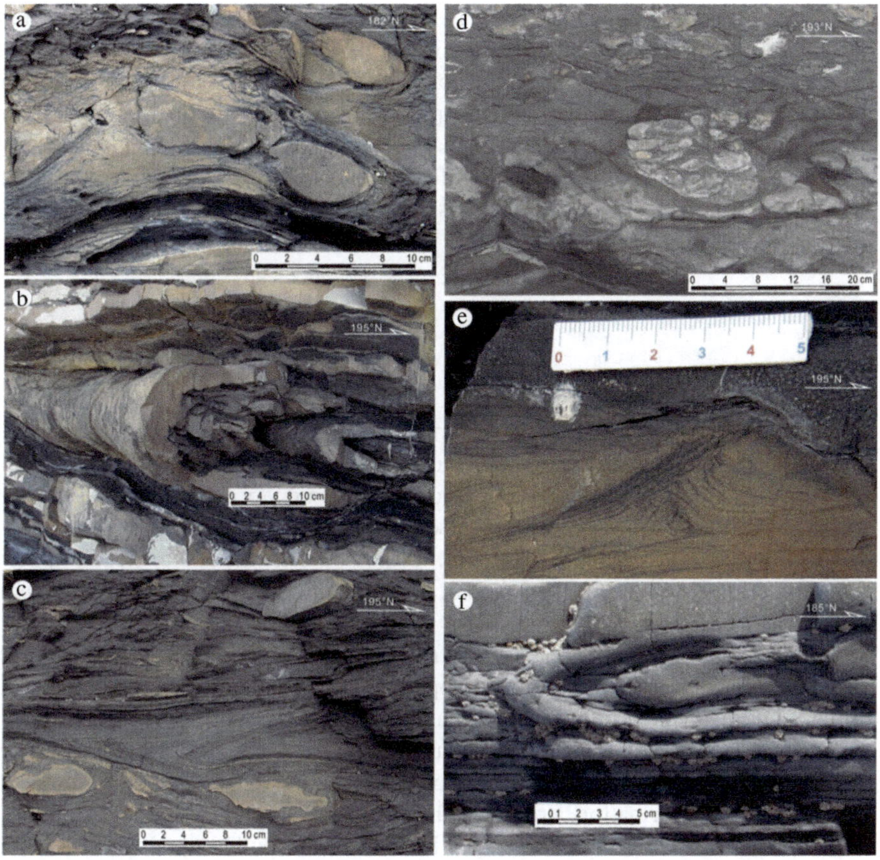

Fig. 12 Soft-sediment deformation structures. **a** Squeezing where a sandy layer overlying a clayey unit underwent irregular loading. The parent sand layer itself disappeared and eventually pseudo nodules were formed with irregular squeezed remnants of clay layers in between. **b** Slump fold with space-problem-induced deformations in which ductile sediments were squeezed between more resistant sandstones. **c** Ductile folded layer squeezed between more resistant material. **d** Ball-and-pillow structures. **e** Escape structure of a water/sediment mixture (see the preserved lamination in the lower part of the structure) with a characteristic flame-like shape. **f** Isoclinally folded layer. This is ascribed to dragging over a water-saturated sediment

and briefly analyzed the most common SSDS types. The origin of the demand of the darrorization, that is, the water escape structure.

Many water avoidance structures are relatively small; the biggest example is 20 cm high. It can be seen from many other descriptions that their shapes are different, that is, from vein (Etienne et al. 2012) or funnel-shaped (Chen et al. 2009) to "classic" 1994) (Fig. 12e). Similar structures were found in the arc environment similar to Lingshan Island group. Deposition (for example, Di Celma and Cantalassa 2007).

The origin of the water avoidance structure has been clarified a long time ago (Lowe 1975; Allen 1982). When the pore water pressure of the sediment exceeds the threshold of the sedimentary coverage layer weight, they are generated. Because the slow sedimentation provides the possibility of slowly accumulating the accumulated upward sediment, the possibility of "burst" is not suddenly "explosive", the fluid renewal structure usually reflects rapid sedimentation or sudden sedimentation events, such as turbidity deposits. This happened in the Lingshan Island Group.

The water saturation characteristics of sediments allow the development of water reassers and related structures, and it is also shown that these layers must be attributed to dragging.

Slumps

Landscapes appeared in some places (Fig. 13a). Their thickness ranges from one point to a few meters. They are the most common sets of turbidity, which have obviously become unstable after sedimentation, and begin to decline, forming a quality body composed of a pair of multi-layers, which shows complex deformation (Fig. 13b) locally. Several slips may be superimposed (Fig. 13c). If it is multi-layer sliding, it may form more or less oblique folds (Fig. 13d). Especially the sliding head can show complex wrinkle mode (Fig. 13e), and at the same time, there are many small faults due to space problems.

Only when the sediment is easily deformed, a large-scale slump will occur (PIPER 2005; Weiss et al. 2016). Therefore, the appearance of the sliding of the murmur group of the thick meter means that when a large number of previously deposited turbid rocks are not petrified, this turbid rock sequence may accumulate. This clearly proves that at least most of the above deformation structures were formed before rocky.

Small-scale deformation structures usually exist around large slippery slopes (Fig. 13d and f). The tough sand layer is distorted or pulled open to form a thin graded fault. Some harder sand layers are broken by small fractures (see arrows in Fig. 13e). Therefore, there must be different stress fields in the inside and outside of large landslides.

People often describe the origin of economic recession (such as Clennell 1991; Festa et al. 2014; Ogata et al. 2014), we can refer to the nearest literature and textbooks. The relatively mild impact may cause the landslide and sliding and sliding of saturated water sinks (Van Loon et al. 2012), even if the slope is very flat; study the slope is <1° (García-Tortosa et al. 2011), even <0.25, (Field et al. 1982). Very low-angle exercise has also been proven by experiments (Owen 1996).

Large-scale landslides and landslides must occur on the sedimentary slope. In this case, it may be the far-end part of the front edge of the Delta (CF. Elliott 2005; Shanmugam 2009). When the slope of the slope is not enough to move them further down, they stop, which usually means that the bottom is actually horizontal. Considering that slipping always exists below the head of large-quality transportation materials, thereby reducing Shear stress and friction.

Fig. 13 Slumps and slump structures. **a** Complex deformation in a slump, with large-scale folds and numerous small-scale normal faults, upthrusts and folds. **b** Multilayer large-scale slump with isolated, broken and/or folded pieces of layers in front of it, indicating a chaotic stress field surrounding the slump. **c** Large-scale deformations in stacked, thin-bedded, slumped siltstones and shales showing interconnected soft-sediment folds. **d** Large slump mass with a relatively simply folded head. **e** Slump head with complex folding. **f** Different deformations inside and outside a large-scale slump, indicating a chaotic stress field

3.1.4 Division of SSDS

As a special type of sediment structure, soft-sediment deformation structure has attracted wide attention and has conducted in-depth research on its morphological characteristics, origin mechanism, and classification scheme. Earthquakes are a key factor in the deformation of soft sediment, while storm, tsunami, collapse, volcanic activity, freezing or ice melting, rapid sediment accumulation, extraterrestrial object impact cannot be ignored. Since different events can induce the same deformations, the classification scheme according to induced events is more applicable to the same event. In deformations with "multicause and same phenomenon", it is difficult to distinguish true causes, thus, the classification based on the trigger mechanism often has limitations.

SSDS is a series of structures formed by the transformation of the plastic sediment after the sedimentation of the rocky front of the rock. In different sizes, different shapes, different periods, and different causes, solid deposits are often developed. There is no doubt that earthquakes are one of the most common reasons for the deformation of soft sediments, but it is not limited to this. SSD is not a diagnosis of any specific trigger mechanism. According to the classification scheme of induced events, it is more suitable for a single event. It is difficult to distinguish the real reasons in the deformed "due to the same phenomenon". Therefore, the classification of origin is often limited. In fact, the driving force is the most direct factor that causes the deformation of soft sediments, and directly controls the specific form of the deformation structure.

In order to clarify its specific types and development, we advocate SSD classification according to different drivers. According to the literature and exposure surveys, Lingshan Island has developed a variety of gravity flows such as multi-scale, multi-form, multi-layered, multi-stage, multi-phase, and multiple types of hemp rocks in Lingshan Island. Depending on the driving force and form, SSDS can be divided into three types: SSDS driven by slope gravity; due to reverse density gradients, SSD driven by gravity.

In addition, considering the development of the internal SSDS development of the Baiji Tunxia Belt under the Lingshan Island, the SSDS is further divided into three types: compressed, tensile, and shear type. The type is the SSDS on the slope gravity. Under Lingshan Island, Baiji Tong has divided three categories of SSDS (Figs. 14, 15, 16, 17 and 18) with more than ten kinds of slope gravity drives: SSDS driven by slope gravity, including compression types (the same deposition reverse layer and irregular compression deformation structure, etc.), Tension type (with the sedimentary boundary, step jump layer, etc.) and shear type (shear folds, shear cracks, soft dual structures, etc.); due to the reverse density gradient, SSDS driven by gravity, including load structure,, including load structure, including load structure,, including load structures, including load structures, including load structures, including load structures, including load structures, including load structures, including load structures, including load structures, including load structures, including load structures. Flame structure and ball pillow structure, etc.; SSDS driven by liquefied diaper, water

Fig. 14 Soft-sediment deformation structures of compression types driven by gravity acting on slopes

seepage structure, curl layer, sandstone dam and other pore fluids. From the perspective of the type, development and distribution of SSDS, the early earthquake activities of Lingshan Island are frequent. The driving force of earthquake activities directly acts on soft sediments, and its type, strength, role and duration affects the type, shape and scale of soft sediments. Through the power classification of the soft sedimentary deformation structure under the Lingshan Island, it can distinguish the soft sink and deformation structure in the sliding layer and the soft sinking deformation structure in the normal sedimentation layer, as well as The soft sink and deformation structure

Fig. 15 Soft-sediment deformation structures of tension types driven by gravity acting on slopes

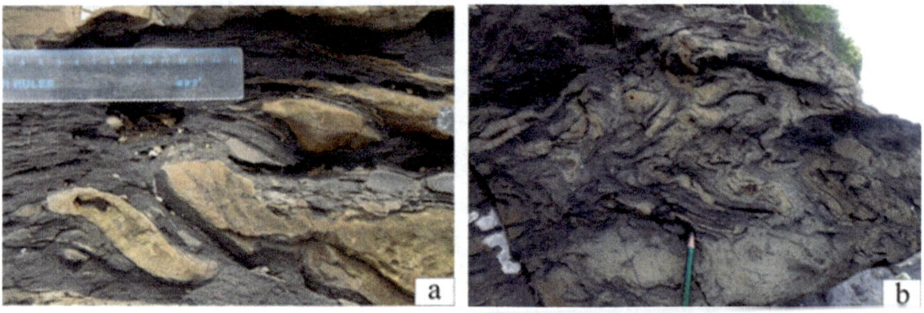

Fig. 16 Soft-sediment deformation structures of shear types driven by gravity acting on slopes

during the earthquake. This classification can clarify the dynamic mechanism of soft sand deformation. Therefore, the division of SSD based on power is scientific, applicable, and feasible. Discussing its classification is conducive to its field identification and cause analysis, and provides theoretical basis for geological research.

3.1.5 Gravity-Flow Deposition

Four rock phases were divided into the Lingshan Island group. Because this article only focuses on the ancient salinity of one of the rock phase (Rock phase 4), the 1–3

Fig. 17 Soft-sediment deformation structures of driven by gravity due to a reverse density gradient

Fig. 18 Soft-sediment deformation structures driven by shear due to pore fluid

phase will only be discussed briefly. For more detailed descriptions, see Lu et al. (2011), Zhong (2012), Shao et al. (2014), Yang and Van Loon (2016) and Yang and Others (2016).

Phase 1 consists of deformation sandstone, powder sandstone and shale. Transformers are concentrated in the layer that is inserted from the undesirable layer. The thickness of the deformation layer ranges from several centimeters to 6 m (Fig. 19). The deformed layer is explained as a slippery, which is consistent with the sedimentation of the slope or near the slope (no matter how small; because it is located in the earthquake activity area, it must be noted that on the slope of <0.25°, the earthquake will cause the saturated water sink accumulation Slope destroy; Field et al. 1982).

Phase 2 consists of thick sandstone and powder sandstone, which contains block or spherical chips with a diameter of more than 0.3 m (Fig. 19c). The sandstone is manifested as the tip of the bottom, and the downward deposition of the rock phase 1. Sandstone and powder sandstone are interpreted as debris caused by slope destruction.

Phase 3 consists of a grading layer with <1 mm to more than 50 cm thick. In some strata at a vertical distance of only 1 cm (Fig. 20a), it shows the rapid and normal classification of rough sand to powder mudstone. The water avoidance structure,

Fig. 19 Debrites (DB) and slumps (SL) in the lower part of the Lingshandao Formation. **a** Overview of an outcrop showing superimposed slumps. The positions of **b–c** are indicated. **b** Detail of a large strongly deformed slump built mainly of turbidites. **c** Large sandstone ball/lumps floating in argillaceous sandstone

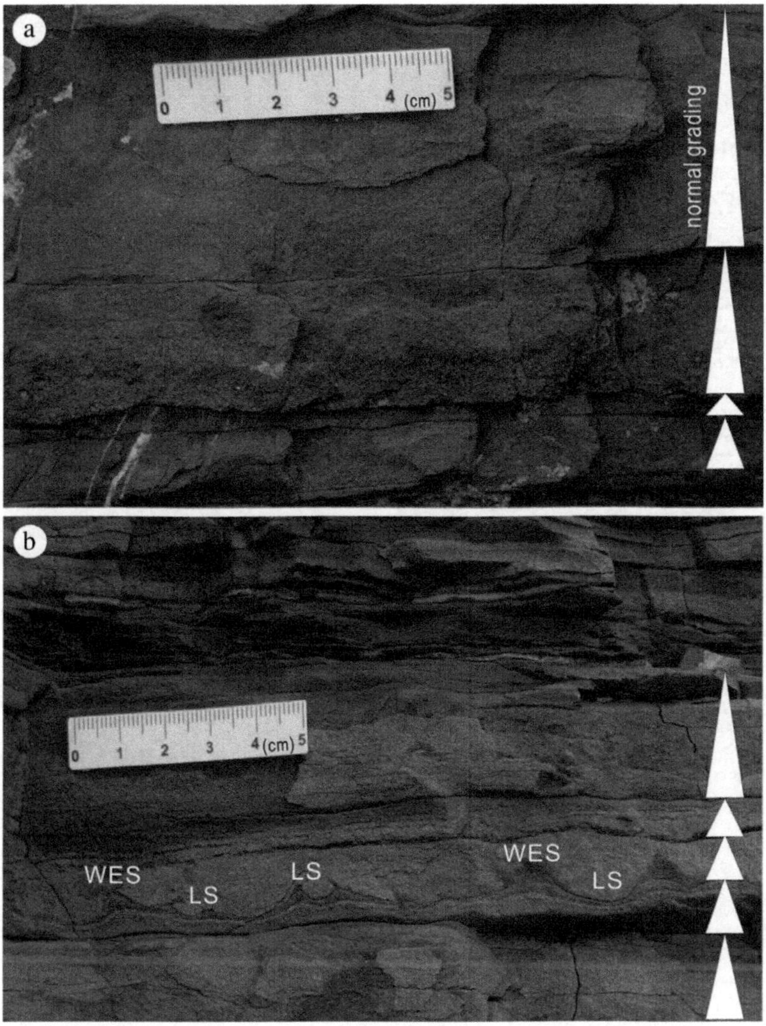

Fig. 20 Turbidite characteristics. **a** Normally graded thin sandstone turbidites. **b** Normal grading in a thin turbidite, showing flames and water-escape structures (WES) and load structures (LS)

flame structure, and load structure of some layers develop well (Fig. 20b). The sedimentation of the phase 3 is explained as a turbidity in the pot.

Phase 4 consists of a sandstone layer that is opposite to the positive sequence, which is mixed with black shale, mud powder sandstone and powder mudstone. These dark sediments show the horizontal layer (Fig. 21a), most of which are less than one millimeter thick. Powder mudstone layers are usually connected, reverse and normal. The continuous layers are separated by sharp straight lines or wave

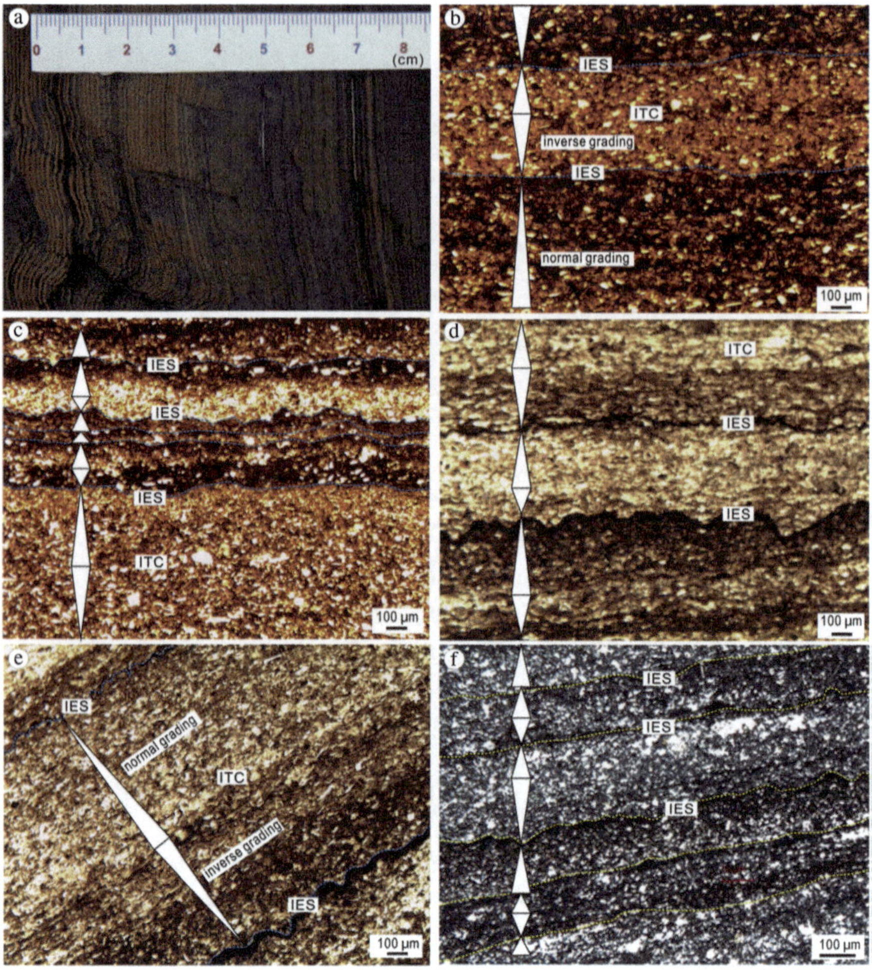

Fig. 21 Very thin hyperpycnites in the Lingshandao Formation. **a** Hyperpycnites approx. 1mm thick. **b–f**: Microscopic details of the thin hyperpycnites, showing inter-layer erosion surface (IES) and intra-layer contacts (ITC). Note the inverse to normal grading, which is considered typical of deposition from hyperpycnal flows

erosion, while internal contact in the layer is transition, which is usually represented by the thickest particle layer (Fig. 21b–f).

The above features show that the phase 4 mine is mainly high manganese ore. High-inclined streams can form huge scale. The dense freshwater flow is tilted downward in a turbid stream (Fig. 21), mixing with the surrounding water. These high manganese ore rocks are different from the turbid rocks that also exist in the Lingshan Island group. They show a lower anti-level accumulation, which originated from the increase in traffic (waxy), covering a normal level accumulated during the weakened traffic period. Plants, fish, shells and rock fragments may be banded into the basin, and the existence of plant debris in deep water sediment has been proposed as recognition of high manganese ore The main standards. The thickness, color, grain size and composition of the particles in the continuous stream reflect the fluctuation fluid dynamic conditions, the fluid dynamics of the particles provided in the suspension, the volume and deposition rate of the high-quality flowing feathers (Yang et al. 2016).

The rock description and rock phase analysis cannot be used as a strong evidence of determining the sedimentation environment, because these four kinds of rock phase may also be deposited in the lake environment (Yang et al. 2014). The combination of chemical evidence may better avoid vulnerabilities directly on-site and rock-phase observation (Fig. 22).

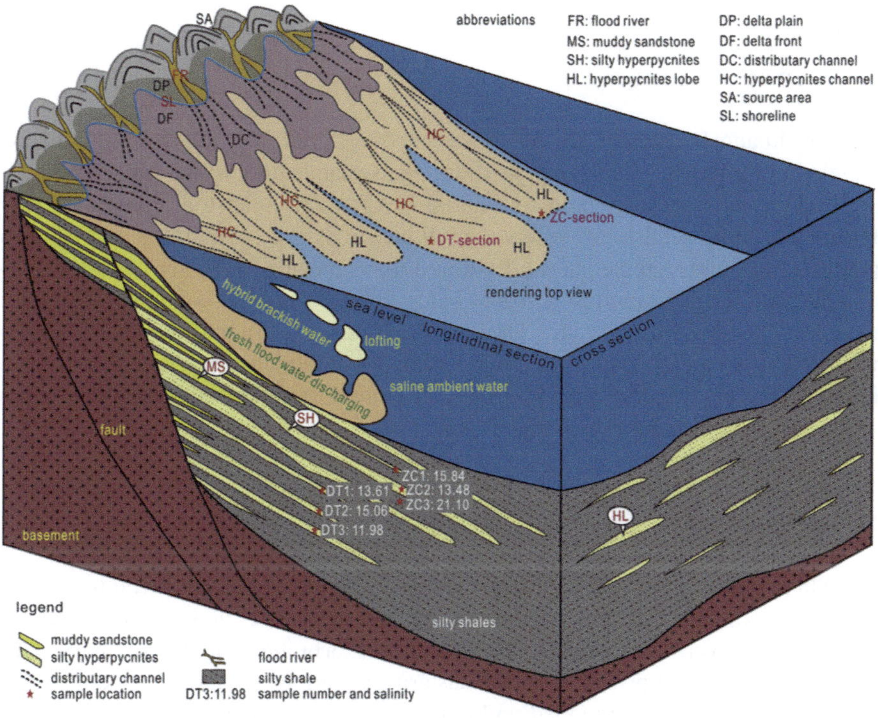

Fig. 22 Depositional model for the Lingshandao Formation, with a key role for hyperpycnites. The locations of all samples (all from hyperpycnites) are indicated

3.2 Stop 2: Laiyang Cretaceous National Geological Park

3.2.1 Sedimentology Context

Wawukuang Formation

The Wawukuang Formation is characterized by a set of thin conglomerates at the base of strata, fine-grained arkosic arenites in the middle, and coarse-grained arkosic arenites in the top representing a depositional cycle with fan-delta facies deposits to restricted lacustrine facies deposits (Fig. 23a, c, e). Samples from distinct levels of the strata show variable features in feldspathic grains and lithic fragments (Fig. 23b, d, f).

Zhifengzhuang Formation

Zhifengzhuang Formation are composed of purple red and yellow conglomerate, conglomerate-bearing sandstone, medium-coarse sandstone, with a small amount of argillaceous siltstone, silty mudstone and shale (Fig. 24). Sandstone types are mainly feldspathic lithic sandstone, followed by lithic sandstone, a small amount of feldspathic lithic quartz sandstone, feldspathic lithic sandstone. The rocks particle size of Zhifengzhuang Formation is from coarse to fine, with low compositional maturity and structural maturity are. The main sedimentary structure is block bedding, followed by wedge-shaped staggered bedding, grain sequence bedding and rhythmic bedding. The main sedimentary facies include alluvial fan, fan delta and shallow lake (Xie et al. 2012; Zhang et al. 2019).

Shuinan Formation

The Shuinan Formation (K_1Ls) is mainly composed of yellowish-green to grayish-green fine-grained to silty sandstone, shale, and marlstone, and is interpreted as a lake facies deposit. There are distal sub-lacustrine turbidites primarily in the Shuinan Member and located primarily in the eastern Laiyang Subsag. There are also fine-grained fan deposits interbedded within thick, dark-gray, deep-lacustrine mudstone intervals (Fig. 25).

- **Soft-sediment deformation structures in Shuinan Formation**

Load structures: Load structures are recognized at the interface between sandstone on top of muddy siltstone or mudstone, with contrasting density (Fig. 26).

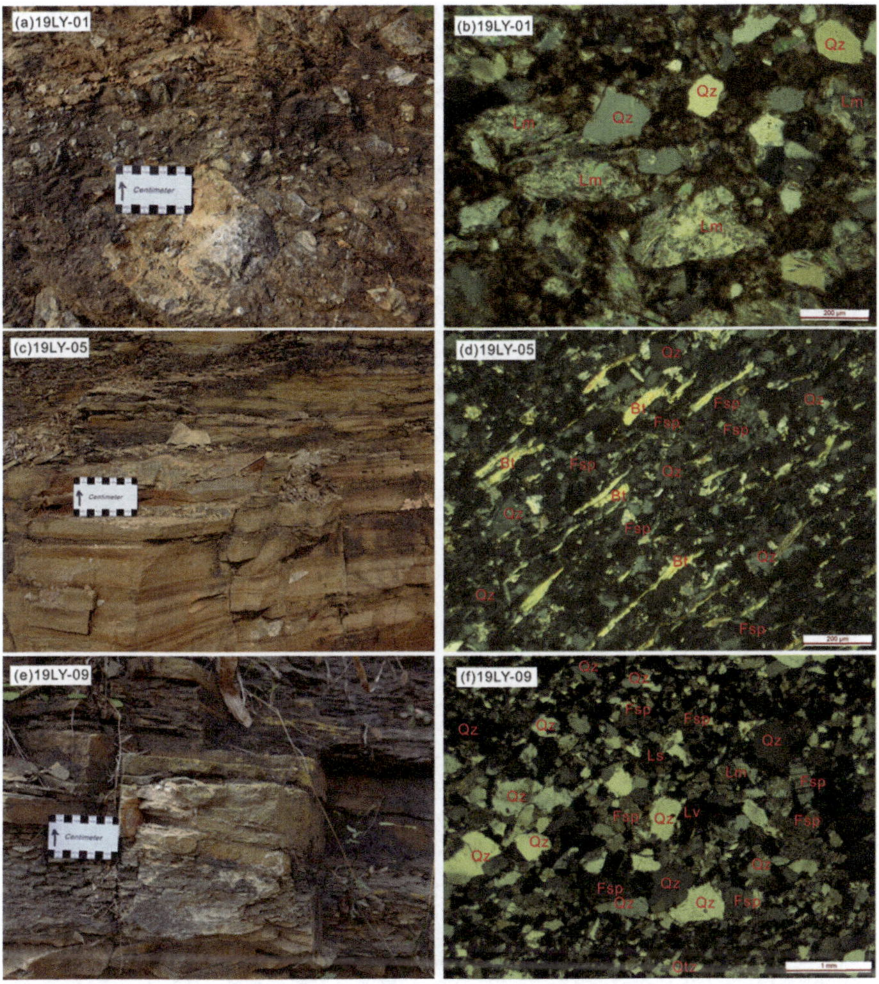

Fig. 23 Field photographs and microphotographs of three representative sandstone samples from the Wawukuang Formation. Abbreviations: Qz = quartz; Fsp = feldspar; Bt = biotite; Lv = volcanic lithic fragments; Lm = metamorphic lithic fragments; Ls = sedimentary lithic fragments

Soft-sediment deformation structure (SSDS) refers to the deformation induced by the sudden loss or decrease of shear strength, which can be induced by processes of liquefaction, fluidization, and thixotropy in water-saturated, unconsolidated sediments in various depositional environments.

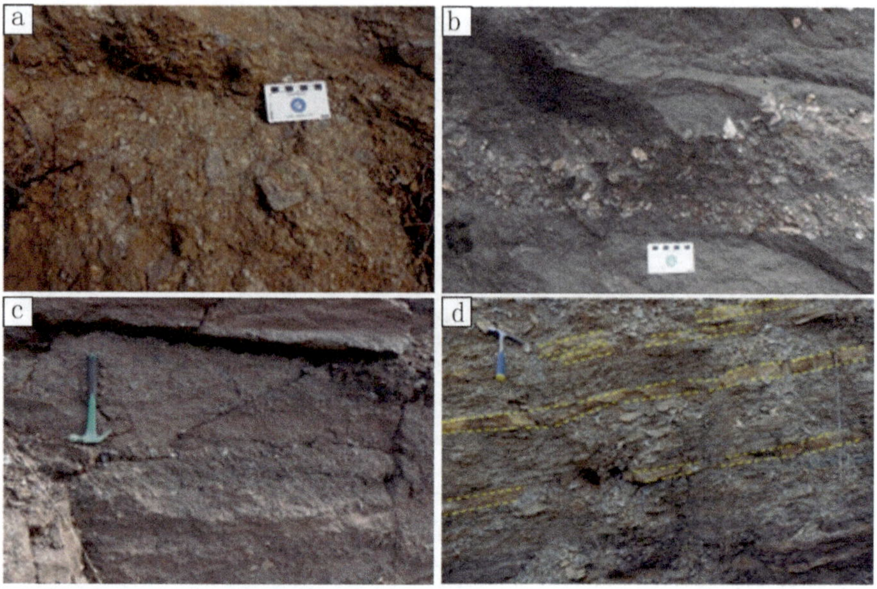

Fig. 24 Typical rock beds and sedimentary structures of Zhifengzhuang Formation

Fig. 25 Typical rock beds of Shuinan Formation

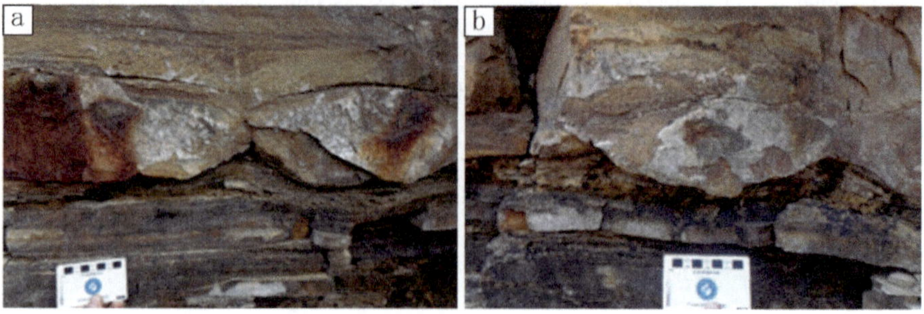

Fig. 26 Load structures of Shuinan Formation

Wangshi Group Hongtuya Formation

The Wangshi group is composed of a set of brick-red lacustrine facies clastic deposits, with volcanic interbedded layers developed in some areas. Its main lithology is clastic rocks such as conglomerate, sandstone, siltstone and mudstone (He et al. 2015; Wang et al. 2016). The Wangs Group stratum is mainly exposed in the south-central Laiyang Sag, the middle part of the Muping-Jimo fault zone, the middle eastern part of the Gaomi Sag, and the middle-north part of the Zhucheng Sag. From bottom to top, the Wang's Group in the Jiaolai Basin are the Linjiazhuang Formation, the Xingezhuang Formation, the Hongtuya Formation, the Jingangkou Formation and the Jiaozhou Formation (Han et al. 2020; Tan et al. 2020). The top strata of the Wangs Group divided into different regions in the basin have certain differences. The top strata of the Wang Group in the Laiyang Sag is called the Jingangkou Formation, and the top strata of the Wang Group in the Gaomi Sag is called the Jiaoxian Formation.

The Hongtuya Formation exposed in the Jiangjunding section of Laiyang develops a set of red clastic rocks, which are integrated and covered on the Xingezhuang Formation, and the degree of rock cementation is poor (Fig. 27a). According to the lithology, the Hongtuya Formation can be divided into three parts: the lower part is composed of purple-red fine conglomerate and gray-green conglomerate (An et al. 2016; Zhang et al. 2020). Above the red silty mudstone, the gravel is composed of quartz, feldspar and volcanic debris; the middle part mainly develops fine conglomerate, sandstone and siltstone; the grain size of the conglomerate is smaller than that of the lower part, and the grain sequence is developed locally; the upper part is comprised of thick layers of mudstone and conglomerate interbedded with mudstone. The section of Hongtuya Formation in Jiangjunding is divided into 12 layers.

- **lithology characteristics**

Conglomerate: most are complex conglomerates, which can be seen from fine to boulder structures. The distribution of gravels is disorderly, and the content is maintained at about 70–80%. Some long axes are distributed in parallel with the plane at 180°, and the arrangement is disorderly (Fig. 27a, b).

The gravel composition is dominated by sandstone and volcanic debris, most of which contain granite and acid lava, as well as some metamorphic rocks with complex composition. The gravel is generally rounded, and the sorting is also poor, mostly in the shape of sub-round-sub-angular, indicating that its deposition is close to the provenance area and the deposition process is short. The lower limit of the diameter of conglomerate can be 2 mm, the medium is 50–100 mm, and the upper limit can reach 150 mm. Under normal circumstances, the diameter of fine gravel is about 2–8 mm.

Siltstone: The main composition of the rock is compound siltstone. In addition to quartz and feldspar, the clastics also contain other rock debris. Mostly block structure. The argillaceous cement and silt dust constitute the main components of the mineral composition. The main part of the particles contain feldspar and quartz, with a small amount of mica, apatite and tourmaline. The particle size is small, between 0.01 and

Fig. 27 **a** Panoramic view of Hongtuya Formation at Jiangjunding section. **b** Purple-red conglomerate. **c** Dinosaur skeleton fossil

0.1 mm, the roundness is average, the sorting is good, while the degree of cementation is poor. It is a porous-basal cement. The cement is mud cement and part calcium, quality calcite, etc. (Fig. 27c).

Miscellaneous sandstone: The mixed sandstone is more common, and it has a rhythmic interbedded relationship with the mudstone. Plagioclase, quartz and some rock fragments constitute the clastic part. The composition of the rock fragments is more complex, including not only acidic volcanic rock fragments and quartzite, but also granite and argillaceous rocks. Some cuttings and a small amount of feldspar particles are replaced by carbonate minerals such as calcite. The particles are mostly sub-angular-sub-round, and their sortability and roundness are poor. Among them, it is rich in clay matrix, which can reach about 20%. The cement in the mixed sandstone is much less than that in the sandstone. This is because the impermeability of the matrix is not conducive to the passage of the solution, and the resulting precipitation is correspondingly less. Cementation type multi-heterobase support-substrate cementing structure (Fig. 28a, b).

Detrital feldspar sandstone: The structure is mostly sandy and unequally grained. The clastic composition is feldspar and quartz, each accounting for about 40%, followed by intermediate-acid lava and other debris. It also contains components such as mica, greystone and opaque minerals. The shape of the debris is angular to sub-circular, followed by an angular shape. The lower limit of the particle diameter range is 0.05 mm, and the upper limit is 1.5 mm. Few can reach 2–3 mm. It is composed of clay minerals and calcareous glue, and the composition forms a cement.

Fig. 28 **a** Conglomerate greywacke. **b** Greywacke

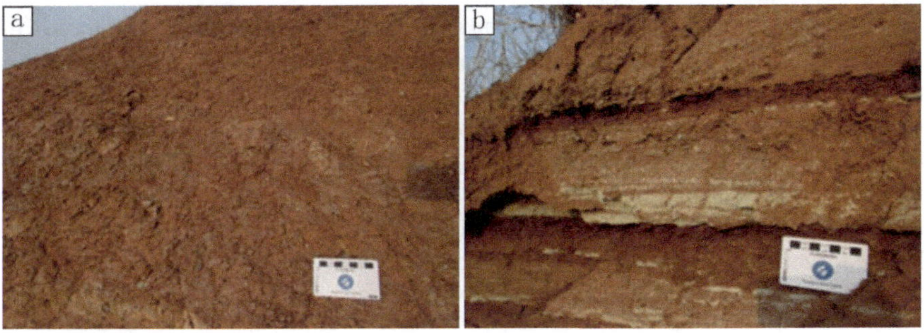

Fig. 29 **a** Brick red mudstone. **b** Interbedded mudstone and sandstone

Mudstone: Silt-like structure, mud-like structure sand and mud-like structure, block structure, and bedding is not obvious. The main components are weakly consolidated clay. The mudstones of the Hongtuya Formation include silt mudstone, silty mudstone, and some calcareous mudstone (Fig. 29a, b). Among them, clay minerals such as hydrous mica and montmorillonite. It is characterized by a low degree of consolidation and a soft texture.

- **Sedimentary environment of Hongtuya Formation**

The Hongtuya Formation in the Jiangjunding section of the Laiyang Sag is divided into lower, middle and upper sections according to the characteristics of the sedimentary environment, and the lithological characteristics, sedimentary facies and typical probability accumulation curves of each section are sorted into a comprehensive histogram.

In the lower part of Hongtuya Formation, there are interbeds of purple-red fine conglomerate and gray-green conglomerate, partially interbedded with brick-red siltstone, and integrated and developed on the purple-red silty mudstone of the Xingezhuang Formation with a thin layer of gray-white limestone. The probability accumulation curve belongs to the wide and gentle upward arch, reflecting that under the high-energy condition of the water body, the sand, gravel and mud in the block are too late to be sorted, so the particles rely on the support of the miscellaneous base and are transported in a whole block shape. The detrital material is similar to the gravity flow deposition of debris flow and detrital flow formed by simultaneous deposition. It reflects the deposition of clastic channels in the fan delta plain.

The middle section develops fine conglomerate, sandstone and siltstone. The grain size of the conglomerate is smaller than that of the lower part, and the granular sequence bedding is developed locally. The probability accumulation curve type is a two-stage type. The transportation method is skip transportation, indicating the turbidity deposition supported by turbulence and transported in the high-speed turbulent flow state, specifically the high-density turbidity current type. It reflects the deposition of the braided river channel of the nearshore underwater fan or the water distributary channel of the fan delta plain.

Thick mudstone and interbedded conglomerate and mudstone are developed in the upper part. The probability accumulation curve is a suspended one-stage type, and the overall transportation method is suspended transportation. Compared with the two-stage type, the particle size is finer, indicating that the hydrodynamic conditions change, and the reduction of fluid energy reduces its load capacity for particles, indicating high-density turbidity current deposition. Reflects the sedimentation of the water distributary channel in the fan delta plain or the branch channel of the braided river in the front edge of the fan delta.

3.2.2 Provenance Analysis

- **Wawukuang Formation**

Detrital zircon ages of the Wawukuang Formation encompass a wide range (Fig. 30) and are characterized by distinctive age modes which can be used in combination with associated Hf isotopes to plausibly infer the detrital zircon sources.

The results of three detrital zircon U–Pb dating and Hf isotope samples as well as sediment compositions of fourteen sandstones constrain the maximum depositional age of the Laiyang Group and construct the provenance evolution trend of the Wawukuang Formation in the Laiyang Sag. A population of the youngest detrital zircons from the basal sandstone of strata gives a maximum depositional age estimation of the Laiyang Group to be 125 $\pm$ 1 Ma close to the true depositional age (Huo et al. 2015; Ren et al. 2021) (Fig. 31). Integrated detrital zircon U–Pb dates and Hf isotopes, as well as sediment compositions of sandstones, demonstrate that the dominant source region of the Wawukuang Formation transited from the Sulu Orogen to the Jiaobei Terrane. The provenance evolution of the Wawukuang Formation over time constrains two significant tectonic events. The first event refers to the exhumation of the Sulu Orogen to an extent to the surface before sedimentation of the Laiyang Group. The second event refers to a rapid uplift of the Jiaobei Terrane after ca. 125 Ma. The later event is interpreted to be driven by the removal of the

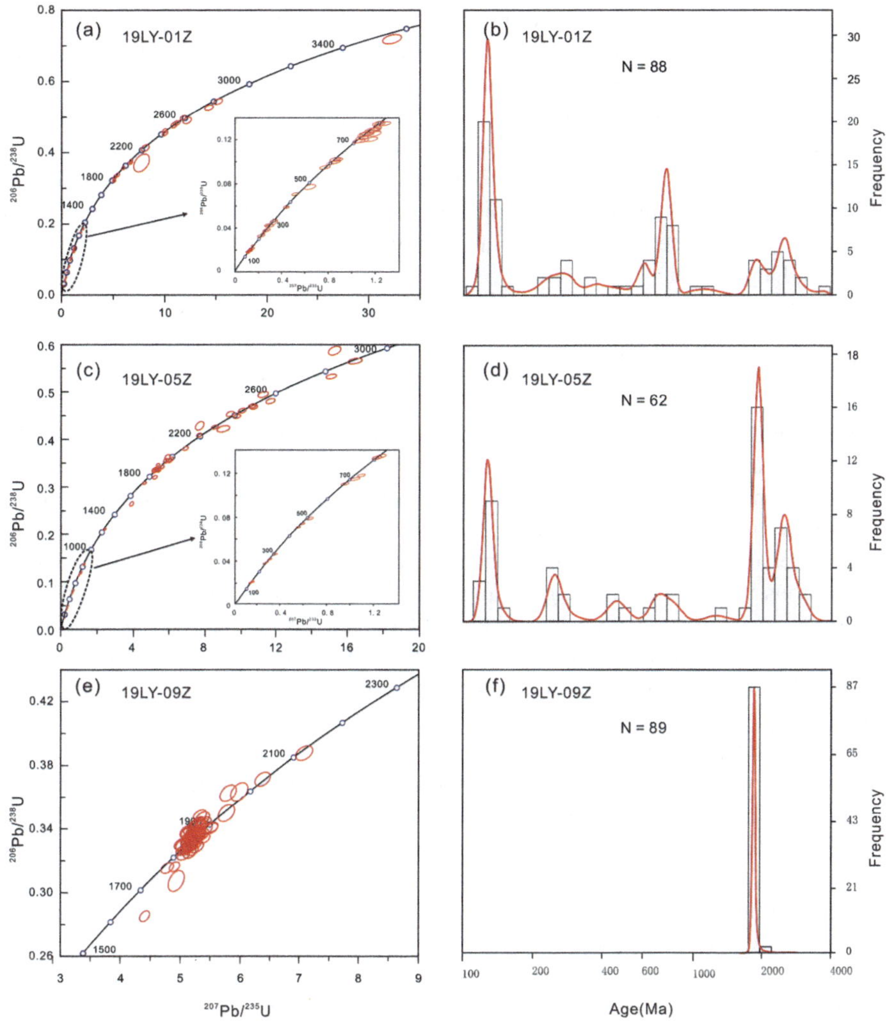

Fig. 30 Zircon U–Pb concordia diagrams for three sandstone samples (19LY-01Z, 19LY-05Z, and 19LY-09Z) from the Wawukuang Formation (**a**, **c**, **e**); Kernel density estimation plots of the sandstones (**b**, **d**, **f**). *Note* Age axis is logarithmic

lithosphere due to the rollback of the Paleo-Pacific slab in the Early Cretaceous (Ren et al. 2021).

- **Hongtuya Formation**

According to the above-mentioned observation of gravels in the study area under the microscope, the gravels of the Hongtuya Formation are mainly intermediate-basic volcanic rocks, pyroclastic rocks, acidic volcanic rocks, and small amounts of sedimentary rocks and metamorphic rocks. The granite of the body is far related and closely related to the volcanic strata of the Qingshan Group: the Qingshan Group is mainly a set of volcanic rock layers with volcanic clastic rock layers in between (Zhao et al. 2018). The Houkui Formation is a set of acidic volcanic clastic rocks at the bottom of the Qingshan Group. The main lithology is rhyolite vitreous tuff, breccia tuff, and dacian volcanic breccia; the Bamudi Formation is the first eruption product of the strongest volcanic activity in the Qingshan Period. The lithology is dominated by andesite, andesitic volcanic breccia, andesitic agglomerate breccia. Therefore, the source of the Hongtuya Formation may be the Qingshan Group in the Laiyang Depression.

3.2.3 Paleontology of Cretaceous Deposits of Laiyang

Starting in the early 1920s, paleontologists began to discover a wide range of Cretaceous terrestrial fossils in the Laiyang, Qingshan, and Wangshi Groups of the Jiaolai Basin which resides in the eastern part of the Shandong Peninsula of northern China. As it is a major fossil site, Laiyang area is remarkable with the Jehol Biota from lower Cretaceous Laiyang and Qingshan Groups, and the dinosaur fauna of the Upper Cretaceous Wangshi Group (e.g., Wang et al. 2010; Zhang et al. 2017; Fig. 32).

Shandong Laiyang Cretaceous National Geopark (SLCNGP) was established in 2015. Its main park is located on the outcrop of the Upper Cretaceous Wangshi Group. Most fossils are exhibited in the museum of SLCNGP, which is open to the public (Fig. 33). The dinosaur fossils are situated in the museum where they are exhibited under two fossil protection zones (Fig. 34).

The Early Cretaceous Jehol Biota is widely distributed throughout central and eastern Asia. Its typical area in northeastern China is well known for its excellently preserved fossils, including feathered dinosaurs, basal birds, primitive mammals, and early angiosperms (Zhou et al. 2003a, b). Based on the hypothesis of the Jehol radiation and migration, the fossils from the Laiyang and Qingshan Groups of Laiyang represent a second or final phase of the Jehol Biota development (Chen and Jin 1999; Zhang et al 2010). Abundant fossils have been recovered from the Laiyang Group, including plants, insects, conchostracans, various other invertebrates, and dinosaur tracks (Chow 1923; Grabau 1923; Hong and Wang 1990; Zhang 1992; Young 1960; Li and Zhang 2000, 2001; Chen et al. 2019). The Qingshan Group consists of volcanic rocks interbedded with sedimentary rocks, and contains the ceratopsian dinosaur *Psittacosaurus* (Young 1958; Zhao 1962), pterosaurs (Young 1958; Zhou 2010), lizards (Gao and Cheng 1999) and the turtle *Peishanemys* (Chow 1954).

The dinosaur fauna of the Upper Cretaceous Wangshi Group is dominated by the hadrosaurids *Tsintaosaurus* and *Tanius*, and dinosaur-egg assemblages (e.g., elongatoolithids, ovaloolithids, spheroolithids and doctyoolithids) (e.g., Wang et al. 2010; Zhang et al. 2017). Other vertebrate fossils include pachycephalosaur, ankylosaurs, stegosaurs, theropods, and sauropods, and lizards (e.g., Wiman 1929; Chow 1951; Young 1958; Buffetaut 1995; Poropat and Kear 2013). Dinosaur eggs (Fig. 35) from the Wangshi Group show that dinosaur species represented by elongatoolithids were present from the middle to later Late Cretaceous, whereas those represented by the

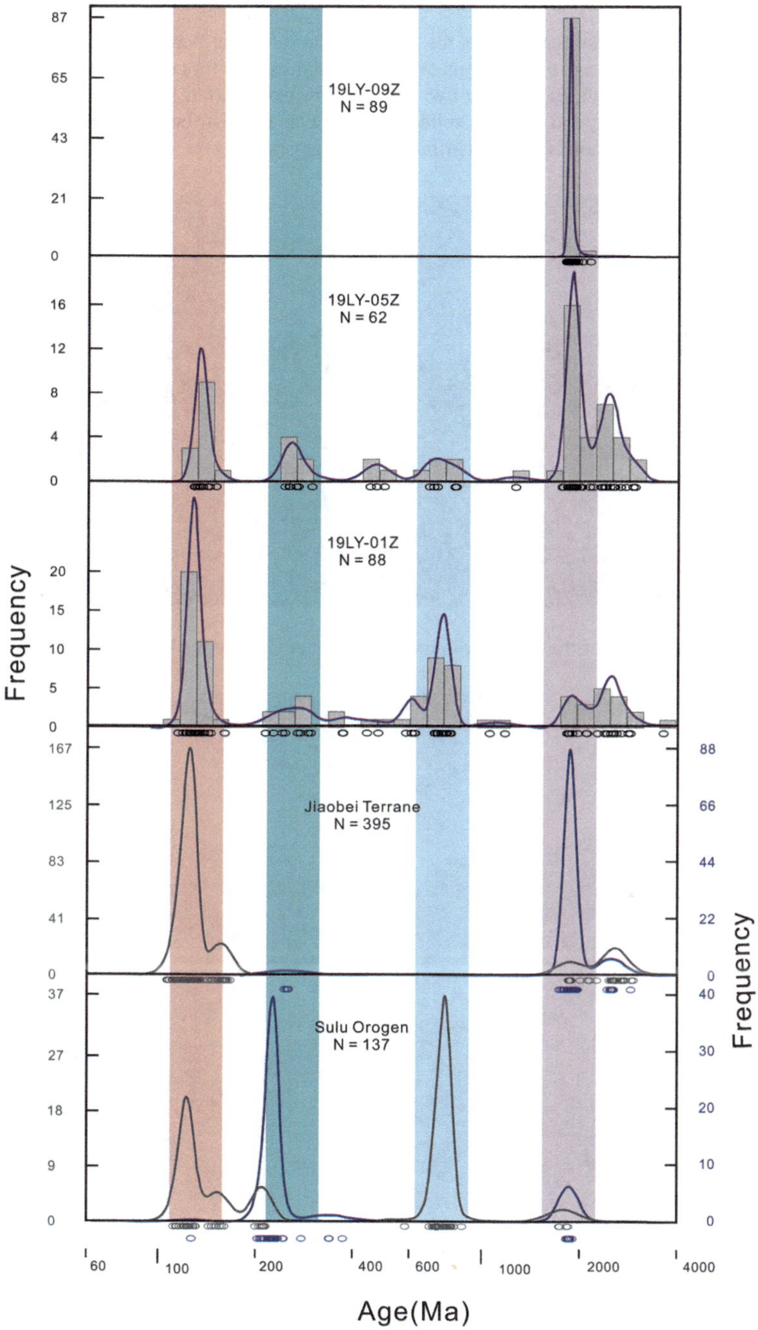
19LY-09Z
N = 89
19LY-05Z
N = 62
19LY-01Z
N = 88
Jiaobei Terrane
N = 395
Sulu Orogen
N = 137
Frequency
Age(Ma)
60
100
200
400
600
1000
2000
4000

◂**Fig. 31** Kernel density estimation plots for three sandstones of the Wawukuang Formation. Also shown in the diagram are age distributions of two potential sources for comparison. *Note* X-axis in this diagram is logarithmic. Each of the two panels below exists two different Y-axis scale shown in way of assorted colors, and the blue solid line represents the distribution of metamorphic age; the gray solid line corresponds to the distribution of magmatic ages

Fig. 32 Reconstruction of the dinosaur fauna of the Upper Cretaceous Wangshi Group

Fig. 33 Dinosaur Museum of Shandong Laiyang Cretaceous National Geopark (SLCNGP)

Fig. 34 One fossil protection of Shandong Laiyang Cretaceous National Geopark (SLCNGP)

Fig. 35 Dinosaur eggs of Shandong Laiyang Cretaceous National Geopark (SLCNGP)

dictyoolithids and spheroolithids became extinct by the middle Late Cretaceous. Estimated eggshell conductance of water vapor in spheroolithids and dictyoolithids exceeded that of elongatoolithids and ovaloolithids by factors of over 4–115 (respectively) indicating that eggs of the first two oofamilies required higher humidity during incubation.

3.3 Stop 3: Shandong Zhucheng National Dinosaur Geopark

3.3.1 Introduction

The "Shandong Zhucheng National Dinosaur Geopark" is located on Longdu Street in Zhucheng, Shandong Province, 10 km from downtown Zhucheng, and is a one-hour drive from Weifang, Qingdao, or Rizhao. The Geopark consists of two scenic areas, Dinosaur Valley and Zangjiazhuang, and has a total area of 9.45 km^2.

There are many dinosaur-bearing beds in Shandong Zhucheng National Dinosaur Geopark, which looks like a natural field museum in which people can learn about dinosaurs and the geosciences. The Geopark is a key that can unlock the study of the history of the Earth along with a window into the world of the dinosaurs. The Geopark not only provides direct and reliable evidence of the origin and evolution of the dinosaurs, but also presents reconstructions of the local paleoenvironment, paleogeography, and paleoclimate and is dedicated to the Earth's protection.

3.3.2 Stratum

According to stratigraphic subdivision, the Geopark is located within the east Shandong partition of the Shanxi–Shandong–Hebei–Henan Region of north China. The Haogezhuang fault zone separates the Lai Yang and Jiaonan–Donghai subregions. The Paleoproterozoic Jingshan and Fenzishan Groups mostly outcrop in Taolin, Taiyuan, Shimen and southwest Haogezhuang, south of the fault zone. The Cretaceous Laiyang, Qingshan and Wangs Groups, consisting of volcanic and clastic rocks, are widespread in a Mesozoic depressed basin that lies north of the Haogezhuang fault zone. Late Neogene sediments are known to occur on the alluvial plain of the Weihe and Quhe Rivers, as well as in mountain basins.

The Geopark is located in the Jiaolai Basin where the Cretaceous strata are well exposed. In particular, red beds of the Wangs Group are extremely widespread. Paleontological and stratigraphic work has made it possible to find abundant vertebrate fossils, including the bones of hadrosaurs and other dinosaurs (Xu et al. 2010; He et al. 2015; Yu et al. 2018), and to establish a standard section for Upper Cretaceous strata in China. Most of the dinosaur fossils preserved in the Geopark were found in Dinosaur Valley and Zangjiazhuang. Dinosaur Valley is actually a gully which is well-known for its beautiful landscape and geological relics. The Valley is situated to the east of the Juan River, a branch of the Weihe River. The dinosaur footprints found in Huanglonggou, Huang Hua, Zhucheng extend over an area of up to 5000 m^2. More than 11,000 footprints are present, making this perhaps the world's largest assemblage of dinosaur footprints exposed at a single site.

3.3.3 Some Important Fossils

Zhucheng dinosaur fossils are characterized by large size, great abundance, high density, and good preservation with prominent distinction, rarity and representativeness. The Zhucheng dinosaur fauna is a significant source of information for research on the distribution, evolution, migration, and extinction of Late Cretaceous dinosaurs, as well for paleoclimatic, paleogeographic, and paleoenvironmental studies.

① **Dinosaur Ravine fossil gallery**

The world's largest dinosaur fossil coenosis—Dinosaur Ravine fossil gallery is about 500 m long and 30 m deep, exposing nearly 10 thousand of pieces of fossils. It is believed by UNESCO to be the world's rare geological wonder (Fig. 36).

Fig. 36 Dinosaur Ravine fossil gallery

② Huanglonggou dinosaur footprint coenosis

The world's largest dinosaur footprint coenosis—Huanglonggou dinosaur footprint coenosis has more than 11,000 dinosaur footprints exposed within 5000 m^2, demonstrating multi-layer distribution (Fig. 37).

③ *Ornithischians*

Ornithischians found in ZhuchengMajor ornithischian groups include the Ornithopoda, Stegosauria, Ceratopsia, Ankylosauria, and Pachycephalosauria (e.g., Xu et al. 2010; He et al. 2015; Yu et al. 2018). All, other than the Stegosauria, are known from Zhucheng.

④ *Shantungosaurus giganteus*

In May 1966, the famous paleontologist Hu Chengzhi was assigned by the Ministry of Geology to lead a dinosaur excavation in Dinosaur Valley. By 1968, some thirty tons of hadrosaur fossils had been collected. From these fossils a dinosaur skeleton

Fig. 37 Footprints of dinosaurs

measuring 15 m long and 8 m high was reconstructed and mounted, representing a species that was named *Shantungosaurus giganteus* by Mr. Hu. At the time, it was the world's largest known hadrosaur and largest known ornithischian.

⑤ ***Zhuchengosaurus maximus***

In May 1989 a team led by Zhao Xijin, a researcher from the CAS Institute of Vertebrate Paleontology and Paleoanthropology, carried out excavation work in Dinosaur Valley. In May 1992, the material they collected was used to reconstruct and mount another hadrosaur skeleton, measuring 9.1 m in height and 16.6 m in length. This mount represented a new species that Zhao Xijin named *Zhuchengosaurus maximus*, and which was recognized by "Guinness World Records".

⑥ ***Huaxiaosaurus aigahtens zhao, gen, et sp, nov.* (Zhao et al.** 2011*)*

Huaxiaosaurus aigahtens Zhao gen. et sp. nov. was found at Zangjiazhuang, Zhucheng, Shandong, and a mount representing this species was reconstructed under the direction of Zhao Xijin of the CAS Institute of Vertebrate Paleontology and Paleoanthropology. The Huaxiaosaurus mounted skeleton is 18.7 m long and 11.3 m high, making this species the world's largest known hadrosaur (Fig. 38).

⑦ ***Zhuchengceratops inexpectus* (Xu et al.** 2010)

Zhuchengceratops inexpectus, from Zhucheng's Dinosaur Valley, is an advanced leptoceratopsid species because some of its characteristics are similar to those of

Fig. 38 Huaxiaosaurus aigahtens Zhao, gen, et sp, nov

Ceratopsidae. Xu Xing from the Institute of Vertebrate Paleontology and Paleoanthropology has noted that the discovery of *Zhuchengceratops inexpectus* was an important addition to the known diversity of leptoceratopsids, suggesting that Leptoceratopsidae exhibited an entirely different evolution process from Ceratopsidae in the Late Cretaceous.

⑧ ***Sinoceratops zhuchengensis* (Xu et al.** 2010)

Sinoceratops zhuchengensis was discovered in Zangjiazhuang, Zhucheng, Shandong, and is the first ceratopsid species known from outside North America. Representatives of other major dinosaur groups known from the Upper Cretaceous of North America had previously been found in Asia before, so the discovery of *Sinoceratops zhuchengensis* filled an important gap in dinosaur paleobiogeography. Among ceratopsids, *Sinoceratops zhuchengensis* is classified as a primitive centrosaurine, but resembles primitive chasmosaurines in various characteristics including large body size. This species blurs the distinction between the two major types of ceratopsid. *Sinoceratops zhuchengensis* also provides vital information regarding the origin of ceratopsids, supporting the hypothesis that large ceratopsians first appeared in Asia.

⑨ ***Zhuchengtyrannus magnus***

Zhuchengtyrannus magnus was discovered in Zangjiazhuang, Zhucheng, Shandong, and was a top predator in a trophic pyramid within the Late Cretaceous terrestrial food web. *Zhuchengtyrannus magnus* is one of the largest carnivorous dinosaurs known from Asia, showing affinities with Tarbosaurus from Mongolia and Tyrannosaurus rex from North America. Zhuchengtyrannus is an important addition to the known diversity of Asian tyrannosaurids (Fig. 39).

⑩ **Coelurosaur fossil**

The fossil thigh bone of a small coelurosaur was found in the Upper Cretaceous Wangshi Group. It was suggested by an American dinosaur researcher, Mark Loewen, that this fossil is likely the thigh bone of an advanced maniraptoran theropod within the Coelurosauria. This specimen may contribute new information about the origin of birds (Fig. 40).

3.4 Stop 4: Jimo Mashan Geopark

Mashan nature reserve is located on the southeast edge of the Jiaolai basin. Its basement comprises Archaean Jiaodong Group, Proterozoic Jingshan group, Fenzishan group, and Penglai group. The caprock is lower Cretaceous Laiyang group, lower Cretaceous Qingshan group, and Upper Cretaceous Wangshi group from bottom to top, and the Paleogene is the Wutu or Huangxian group. The stratum in reserve is single, the exposed stratum is the Cretaceous Laiyang group, and the volcanic rock mass in the upper part is classified as the Qingshan group.

Fig. 39 Sinoceratops zhuchengensis & Zhuchengtyrannus magnus

Fig. 40 Dinosaur skeleton fossil

Fig. 41 The boundary between Laiyang Group and Qingshan Group

3.4.1 The Boundary Between Laiyang Group and Qingshan Group

At Jimo Mashan, the Qugezhuang (QGZ) Formation on the top of the Laiyang Group and the overlying Qingshan Group of Bamudi (BMD) Formation are exposed. The lithology of the QGZ Formation is dominated by yellow-green and gray-green sandstones, which are in unconformable contact with the columnar joint andesite of the upper Bamudi Formation (Fig. 41).

3.4.2 Laiyang Group Qugezhuang Group

The clastic rocks of the Qugezhuang Formation are mostly lithic sandstone or lithic complex sandstone. The mineral composition is dominated by quartz and rock cuttings, with a particle size ranging from 0.03 to 0.6 mm, with an average degree of sorting and roundness. It is mainly supported by particles, and the main type of cementation is pore cementation. It reflects that this set of rocks should be formed in the piedmont filling or intermountain basin near the denudation zone of the strong tectonic deformation zone. It is a set of rapid accumulation products with strong physical denudation and close transportation distance.

Features of Sedimentary Structure

The Qugezhuang Formation is well exposed in Jimo Mashan, which is divided into 11 layers. The sedimentary structure includes horizontal bedding, parallel bedding, cross bedding, wave mark, drainage structure and imbricate structure. Horizontal bedding occurs in fine-grained silt and fine sandstone, and the laminae are arranged horizontally, reflecting weak hydrodynamic conditions. Parallel bedding mainly appears in the gray-green coarse sandstone of the second layer. The laminae are parallel to each other and almost horizontal, and the associated peeling lineage structure can be seen when peeling off along the bedding surface. Cross bedding is more developed in this set of sandstones, and a large trough-like cross bedding is developed in the 9th layer. Huge, silicified wood can be found in the sandstone (Fig. 42). The upper and lower interfaces of the cross-layer unit are concave downwards. The crests and troughs of the corrugated structure are smooth and asymmetrical and are formed by directional flow of water. The drainage structure is formed by loose sediment that accumulates quickly under compaction. Due to the outflow of pore water, the supporting relationship of the original particles was destroyed, and the particles were rearranged. Some sandstones have invaded the upper sandstone layer. The imbricate structure is the phenomenon that flat gravels are arranged obliquely in the same direction along the direction of the water flow under the action of the water flow. The inclination direction of the flat surface is opposite to the direction of the water flow, so it can be used to judge the flow direction of the ancient water. This also serves as a distinguishing mark of fluvial deposits. Mud cracks are caused by unconsolidated sediments exposed to the water surface, water evaporation, dry cracks, and volume shrinkage, which are typical signs of exposure (Fig. 43).

Sedimentary Facies Analysis

The color of this set of terrigenous clastic rocks is yellow and purple-red warm tones, which can determine the iron oxidation state as trivalent iron, reflecting that the sedimentary environment is an arid-semi-arid terrestrial oxidation or a strong oxidation environment. From the perspective of mineral composition, it is a near-source rapid accumulation with a shorter transport distance and a faster deposition rate. Combined with the analysis of its sedimentary occurrence, this set of sandstone bodies is longitudinally represented as lenticular sand bodies distributed in belts, dominated by large-scale cross bedding, and no rhythmic bedding is seen, and the top and bottom of the sandstone body and the overlying boundary is obvious, the lithology is dominated by sand and gravel, the content of shale is less, and the upper part of a single cycle has a binary structure with the upper part being coarse and the upper part being fine. In the wild, oriented gravel and mud cracks can be seen (Fig. 43), which may be indicators of river facies. In addition, an incomplete semi-rhythmic structure that becomes thinner further upward can be seen. The coarser sandstone at the bottom gradually becomes finer in grain size upward, then the grain

Fig. 42 Silicified wood buried in sandstone of Qugezhuang Formation of Laiyang Group

Fig. 43 A typical mud crack

size of the sediment becomes coarser again, and the bottom scour phenomenon can be seen at the bottom of the single rhythm.

In summary, it can be judged that this set of sandstones of the Qugezhuang Formation has well-developed riverbank deposits. The sedimentary environment should be a piedmont alluvial basin, a set of braided river sediments with strong hydrodynamic conditions and high oxygen content in the water body.

3.4.3 Volcanic Rocks of the Qingshan Group

The stratum of the Qingshan Group is mainly a set of volcanic rock series with volcanic clastic rock layers in between (Fig. 44). The Qingshan Group stratum is mainly exposed in the west and north of Laiyang Sag, the middle part of Muping-Jimo fault zone and the middle part of Zhucheng Sag inside the basin. The Qingshan Group in the Jiaolai Basin is characterized by the development of volcanic rocks. According to the volcanic activity and rock properties, the Qingshan Group in the basin can be divided into four cycles from bottom to top, corresponding to the Houkuang Formation, the Bamudi Formation, and the Shijiazhuang Formation. Qianzhuang group and Fanggezhuang group. Within the basin, the Houkuang, Shiqianzhuang and Fanggezhuang formations are mainly distributed on the northern boundary of the Wulian-Rongcheng fault on the southern margin of the Zhucheng Sag. There are also a small number of Houkuang and Shiqianzhuang formations at the margins of other secondary structural units. The Bamudi Formation is widely distributed inside the basin and is exposed on a large scale in the northern Laiyang Sag, the Muping-Jimo fault zone, and the southern margin of the Zhucheng Sag.

The Bamudi group in Mashan area is well exposed. The Bamudi Formation is the eruption product of the first cycle with the strongest volcanic activity in the Qingshan Period. It is widely distributed and has a large stratum thickness. The lithology is dominated by andesite, andesitic basalt, andesitic volcanic breccia, andesitic agglomerated breccia, with purple and gray tuffaceous sandstone and conglomerate interbedded. The Bamudi Formation is in integrated contact with the underlying Houkuang Formation, and it is in angular unconformity contact with the overlying Wang Clan Linjiazhuang Formation.

- **Lithology characteristics of volcanic rocks**

The results under the microscope show that this is a set of neutral hornblende andesite, with higher quartz content than other neutral extruded rocks, yellow-green, gray-green in color, and chlorite development. It has a porphyritic structure, most of the phenocrysts are hornblende, which appear brown under the microscope, with a poor degree of self-shape. The matrix is composed of plagioclase crystallites and is arranged in a semi-directional arrangement. Stomatal structures and rhyolite structures are developed, and the pores are often filled with materials such as calcite and chlorite. So far, there is no definite conclusion about the location of the crater in this area. The appearance of rhyolite structure indicates that the location of the crater should be farther away. Columnar joints are developed. The joint columns

Fig. 44 **a**, **b** Stratigraphy of the Bamudi Formation of Baqingshan Group in Jimo Mashan; **c** rock xenoliths; **d** stomatal structure; **e** rhyolite structure; **f** amphibole phenocryst

are pentagonal or hexagonal, with a diameter of about 1 m. The joint surfaces are parallel to each other. The hot lava gushes from the ground and cools rapidly, and the particles shrink quickly and are arranged in parallel.

4 Proposed Investigation Schedule

The 21st International Sedimentological Congress will be held in Beijing, China, from August 22 to August 26, 2022. The conference includes oral presentations, poster displays, field investigation and short-term courses. The "Cretaceous fluvio-lacustrine stratigraphy, sedimentology and paleontology in Laiyang, Zhucheng and Lingshan island, east of Shandong province" investigation route, as one of the post-conference investigation routes, is scheduled to begin in the morning of August 27, 2022, Beijing time. Please set your clock to Beijing time in advance (UTC/GMT+08:00).

4.1 Dates and Location

Investigation dates: August 27–31, 2022.
Departure place: Beijing City.
Ending place: Qingdao City.

4.2 Travel Arrangements

The participants are requested to take flight or train from Beijing to Qingdao in August 27, 2022. The following schedule is Table 1.

5 Logistics

The following arrangements for this field trip are provided for your reference.

5.1 Accommodation and Meals

Participants in this investigation will be provided with accommodations from August 27 to August 30 in Qingdao City. All reserved hotels are safe, clean and comfortable. Breakfasts and dinners will be provided at the hotels, while lunches will be at restaurants along the way to the investigation site or by bento. Necessary meals/drinks will be provided in the field during the investigation.

Table 1 Schedule of investigation route of the Cretaceous successions in Eastern Shandong Province, North China

Date	Time	Trip	Transportation mode	Contents	Note
August 27 (DAY1)	09:00–17:00	Beijing-Qingdao	Plane/train	Inbound journey	Accommodation: Qingdao
August 28 (DAY2)	07:30–09:30	Qingdao-Laiyang	Bus	Journey	
	09:30–16:00	Laiyang Cretaceous National Geopark		Field investigation	Lunch: Laiyang
	16:00–18:00	Laiyang-Qingdao	Bus	Journey	Accommodation: Qingdao
August 29 (DAY3)	07:30–09:00	Qingdao -Lingshan Island	Bus & Ferry	Journey	
	09:00–16:00	Lingshan Island		Field investigation	Lunch: Lingshan Island
	16:00–18:00	Lingshan Island -Qingdao	Bus & Ferry	Journey	Accommodation: Qingdao
August 30 (DAY4)	07:30–09:00	Qingdao -Zhucheng	Bus	Journey	
	09:00–12:00	Zhucheng National Dinosaur Geopark		Field investigation	Lunch: Zhucheng
	12:00–14:00	Zhucheng-Jimo	Bus	Journey	
	14:00–16:00	Jimo Mashan Geopark		Field investigation	
	16:00–18:00	Lingshan Island -Qingdao	Bus	Journey	Accommodation: Qingdao
August 31 (DAY5)	09:00–17:00	Qingdao -	Plane/train	Return journey	

5.2 *Transport*

The investigation stops are all located within 150 km of the hotels, and the TOYOTA COASTER minibus will be used for the transportation.

5.3 *Weather*

Qingdao is located in the south of Shandong Peninsula, southeast of the Yellow Sea, a warm temperate sub-humid monsoon climate. In summer, it is humid, hot and rainy, and the average temperature is 25 °C. Please prepare appropriate tools in the event of rainfall.

Moreover, to guard against sun and rain exposure, their is recommended that participants come prepared with hats/visors, sunscreen, antiskid, rainproof outdoor shoes, as well as a thin coat.

5.4 *Safety*

(1) Be careful to avoid vehicles, bad road conditions, rivers, and falls.
(2) Be extra cautious at steep locations to avoid falling rocks. Observe the signals of potential geological disasters, such as abnormal noises, loosening or collapse of rocks. Wear a helmet at or near the crags.
(3) The investigation arrangement will be delayed or cancelled in the event of a rainstorm, gale, thunderstorm, hail or other adverse weather conditions, or the possibility of a natural disaster.
(4) Follow the organizer during the investigation and do not stray from the group on your own.
(5) Inform the organizer before the trip if you have any health considerations (e.g. hypertension, heart disease, asthma, diabetes, etc.). Common medicines will be available.
(6) When necessary, call 110 for police support, 120 for emergency medical assistance, or 112 where there is no mobile phone signal for field emergency help.

Acknowledgements We appreciate Professor Hu Xiumian of Nanjing University for his assistance in scheduling the investigation and, Professor Lin Changsong of China University of Geosciences (Beijing) for guidance and suggestions on the arrangement of the investigation stops. The preparation of this guidebook has referenced the research results of many scholars, all of whom will be credited with their work. Thanks to all the researchers.

References

Alfaro P, Estévez A, Moretti M, Soria J-M (1999) Structures sédimentaires dedéformation interprétées comme séismites dans le Quaternaire du Bassin du Bas Segura (Cordillère Bétique Orientale). CR l'Acad Sci Ser IIA Earth Planet Sci 328:17–22

Allen JRL (1986) Earthquake magnitude-frequency, epicentral distance, and soft sediment deformation in sedimentary basins. Sed Geol 46:67–75

Allen JRL (1982) Sedimentary structures, their character and physical basis. Developments in sedimentology. Elsevier, Amsterdam, p 2

Alsop GI, Marco S (2011) Soft-sediment deformation within seismogenic slumps of the Dead Sea Basin. J Struct Geol 33:433–457

Altermann W (1986) The Upper Paleozoic pebbly mudstone facies of peninsular Thailand and western Malaysia—continental margin deposits of Paleoeurasia. Geol Rundsch 75:371–381
Ambraseys N (1988) Engineering Seismology. Earthq Eng Struct Dyn 17:1–105
Amorosi A (1995) Glaucony and sequence stratigraphy: a conceptual framework of distribution in siliclastic sequences. J Sediment Res B65:419–425
Amorosi A (1997) Detecting compositional, spatial, and temporal attributies of glaucony: a tool for provenance research. Sed Geol 109:135–153
An W, Kuang HW, Liu YQ (2016) Detrital zircon dating and tracing the provenance of dinosaur bone beds from the Late Cretaceous Wangshi Group in Zhucheng, Shandong, East China. J Palaeogeogr 5–1:72–99
Anketell J, Cegla J, Dzulynski S (1970) On the deformational structures in systems with reversed density gradients. Ann Soc Géol Pologne 40:3–29
Atkinson G (1984) Simple computation of liquefaction probability for seismic hazard applications. Earthq Spectra 1:107–123
Audemard FA, De Santis F (1991) Survey of liquefaction structures induces by moderate recent earthquakes. Bull Int Assoc Eng Geol 44:5–16
Awramik SM, Sprinkle J (1999) Proterozoic stromatolites: the first marine evolutionary biota. Hist Biol 13:241–253
Buffetaut E (1995) An ankylosaurid dinosaur from the Upper Cretaceous of Shandong (China). Geol Mag 132(6):683–692
Burg JP, Harris LB (1982) Tension fractures and boudinage oblique to the maximum extension direction: an analogy with lüders' bands. Tectonophysics 83:347–363
Chen P, Jin F (eds) (1999) The Jehol Biota. Palaeoworld 11:1–342 (in Chinese with English summary)
Chen J, Van Loon AJ, Han Z, Chough SK (2009) Funnel-shaped, breccia-filled clastic dykes in the Late Cambrian Chaomidian Formation (Shandong Province, China). Sed Geol 221:1–6
Chen T, Liu S, Lei X, Chen L (2019) A new cockroach (Instecta: Blattaria: Blattulidae) from the Lower Cretaceous Laiyang Formation of China. Cretac Res 101:17–22
Chow TH (1923) A preliminary note on some younger Mesozoic plants from Shantong. Bull Geol Soc China 5(2):81–141
Chow MC (1951) Notes on the Late Cretaceous dinosaurian remains and the fossil eggs from Laiyang, Shantung. Bull Geol Soc China 31:89–96
Chow MC (1954) Cretaceous turtles from Laiyang, Shantung. Acta Palaeontologica Sinica 2(4):395–408
Clennell B (1991) The origin and tectonic significance of mélanges in Eastern Sabah, Malaysia. J SE Asian Earth Sci 6:407–429
Cojan I, Thiry M (1992) Seismically induced deformation structures in Oligocene shallow-marine and aeolian coastal sands (Paris Basin). Tectonophysics 206:79–89
Dabrowski M, Grasemann B (2014) Domino boudinage under layer-parallel simple shear. J Struct Geol 68A:58–65
Davies RJ (2005) Differential compaction and subsidence in sedimentary basins due to silica diagenesis: a case study. Geol Soc Am Bull 117:1146–1155
Di Celma C, Cantalamessa G (2007) Sedimentology and high-frequency sequence stratigraphy of a forearc extensional basin: the Miocene Caleta Herradura Formation, Mejillones Peninsula, northern Chile. Sed Geol 198:29–52
Dong ZM (1978) A new genus of Pachycephalosauria from Laiyang, Shantung. Vertebrata PalAsiatica 16(4):225–228
Dugué O (1995) Séismites dans le Jurassique supérieur du Bassin anglo-parisien (Normandie, Oxfordien supérieur, Calcaire gréseux de Hennequeville). Sed Geol 99:73–93
Einsele G (2013) Sedimentary basins—evolution, facies and sediment budget, 2nd edn. Springer, New York, p 792
Elliott T (2005) Sedimentary environments—deltas encyclopedia of geology. Elsevier, Amsterdam, pp 528–539

Etienne S, Mulder T, Bez M, Desaubliaux G, Kwasniewski A, Parize O, Dujoncquoy E, Salles T (2012) Multiple scale characterization of sand-rich distal lobe deposit variability: examples from the Annot Sandstones Formation, Eocene-Oligocene, SE France. Sed Geol 273(274):1–18
Faure M, Lin W, Monié P, Le Breton N, Poussineau S, Panis D, Deloule E (2003) Exhumation tectonics of the ultrahigh-pressure metamorphic rocks in the Qinling orogen in east China: new petrological-structural-radiometric insights from the Shandong Peninsula. Tectonics 22:1018
Festa A, Dilek Y, Gawlick H-J, Missoni S (2014) Mass-transport deposits, olistostromes and soft-sediment deformation in modern and ancient continental margins, and associated natural hazards. Mar Geol 356:1–4
Field ME, Gardner V, Jennings AE, Edwards BD (1982) Earthquake-induced sediment failures on a 0.25 slope, Klamath river delta. Calif Geol 10:542–546
Fossen H, Teyssier C, Whitney DL (2013) Transtensional folding. J Struct Geol 56:89–102
Galli P (2000) New empirical relationships between magnitude and distance for liquefaction. Tectonophysics 324:169–187
Gao K, Cheng Z (1999) A new lizard from the lower Cretaceous of Shandong, China. J Vertebr Paleontol 19:456–465
García-Tortosa FJ, Alfaro P, Gilbert L, Scott G (2011) Seismically induced slump on an extremely gentle slope (<1) of the Pleistocene Tecopa paleolake (California). Geology 39:1055–1058
Ghosh SK, Pandey AK, Pandey P, Ray Y, Sinha S (2012) Soft-sediment deformation structures from the Paleoproterozoic Damtha Group of Garhwal Lesser Himalaya, India. Sed Geol 261(262):76–89
Goscombe BD, Passchier CW, Hand M (2004) Boudinage classification: end member boudin types and modified boudin structures. J Struct Geol 26:739–763
Grabau AW (1923) Cretaceous fossils from Shantung. Bull Geol Surv China 5(2):143–182
Guiraud M, Plaziat J-C (1993) Seismites in the fluviatile Bima sandstones: identification of paleoseisms and discussion of their magnitudes in a Cretaceous synsedimentary strike-slip basin (Upper Benue, Nigeria). Tectonophysics 225:493–522
Han F, Sun JP, Qin HF (2020) Magnetostratigraphy of the Upper Cretaceous and Lower Paleocene terrestrial sequence, Jiaolai Basin, eastern China. Palaeogeogr Palaeoclim-atol Palaeoecol 538
He B, Qiao X, Zhang Y, Tian H, Cai Z, Chen S, Zhang Y (2015) Soft-sediment deformation structures in the Cretaceous Zhucheng Depression, Shandong Province, East China; their character, deformation timing and tectonic implications. J Asian Earth Sci 110:101–122
Hong YC, Wang WL (1990) Insects fossil from the Laiyang Basin, Shandong Province. In: The Regional Geological Surveying Team, Shandong Bureau of Geology and Mineral Resources (eds) The stratigraphy and palaeontology of Laiyang Basin, Shandong Province. Geological Publishing House, Beijing, pp 44–189
Huo TF, Yang DB, Xu WL et al (2015) U-Pb ages and Hf isotope compositions of detrital zircons from the sand stone in the Early Cretaceous Wawukuang Formation in the Jiaolai Basin, Shandong Province and its Tectonic Implications. Geotectonicaet Metallogenia 39:355–368
Kuenen PH (1958) Experiments in geology. Trans Geol Soc Glasgow 23:1–26
Li R, Lockley MG (2011) Masaki Matsukawa et al. An unusual theropod track assemblage from the Cretaceous of the Zhucheng area, Shandong Province, China. Cretaceous Res
Li RH, Zhang GW (2000) New dinosaur ichotaxon from the early Cretaceous Laiyang Group in the Laiyang Basin, Shandong Province. Geol Rev 46(6):605–610
Li RH, Zhang GW (2001) A preliminary study of nonmarine trace fossils from the Laiyang Group (Early Cretaceous), Laiyang Basin, eastern China. Acta Palaeontologica Sinica 40(2):252–261
Ling WL, Duan RC, Xie XJ (2009) Contrasting geochemistry of the Cretaceous volcanic suites in Shandong province and its implications for the Mesozoic lower crust delamination in the eastern North China craton. Lithos 113(3–4):640–658
Liu C, Wang J, Zhao H, Zhang D, Deng Y, Zhao X (2015) The classification of sedimentary basins and discussion on relevant issues. Earth Sci Front 22:1–26 (in Chinese)
Lloyd GE, Ferguson CC, Reading K (1982) A stress-transfer model for the development of extension fracture boudinage. J Struct Geol 4:355–372

Lowe DR (1975) Water escape structures in coarse-grained sediments. Sedimentology 22:157–204
Lu H, Wang J, Zhang H (2011) Discovery of the Late Mesozoic slump beds in Lingshan Island, Shandong, and a pilot research on the regional tectonics. Acta Geol Sin 85:938–946 (in Chinese)
Lu H, Zhang H, Wang J, Dong X, Zhang X (2012) A giant olistolith discovered in the Late Mesozoic turbidites in Lingshan Island, Jiaonan, Shandong. Geol Rev 58:80–81 (in Chinese)
Ma Q, Zhou YQ, Mu HY et al (2020) Geochemistry of the Laiyang Group from outcrops and Lingke-1 core on Lingshan Island, Shandong Province, Eastern China: implications for provenance, tectonic setting and palaeo-environment. Geol Mag 159–1:37–54
Macar P (1948) Les pseudo-nodules du Famennien et leur origine. Ann Soc Géol Belgique 72:47–74
Mandal N, Khan D, Deb SK (1992) An experimental approach to wide-necked pinch-and-swell structures. J Struct Geol 14:395–403
Marques FO, Fonseca PD, Lechmann S, Burg J-P, Marques AS, Andrade AJM, Alves C (2012) Boudinage in nature and experiment. Tectonophysics 526(529):88–96
Meng QR, Wu GL, Fan LG, Wei HH (2019) Tectonic evolution of early Mesozoic sedimentary basins in the North China Block. Earth-Science Reviews 190:416–438. https://doi.org/10.1016/j.earscirev.2018.12.003
McCabe AM, Carter RWG, Haynes JR (1994) A shallow marine emergent sequence from the northwestern sector of the last British ice sheet, Portballintrae, Northern Ireland. Mar Geol 117:19–34
Mills PC (1983) Genesis and diagnostic value of soft-sediment deformation structures—a review. Sed Geol 35:83–104
Mokhtari Fard A, Van Loon AJ (2004) Deformation of an early Preboreal deposit at Nykvarn (SE Sweden) as a result of the bulldozing effect of a grounding iceberg. Sed Geol 165:355–369
Montenat C, Barrier P, Ott d'Estevou P, Hibsch C (2007) Seismites: an attempt at critical analysis and classification. Sed Geol 196:5–30
Moretti M, Sabato L (2007) Recognition of trigger mechanisms for soft-sediment deformation in the Pleistocene lacustrine deposits of the Sant'Arcangelo Basin (Southern Italy): seismic shock vs. overloading. Sed Geol 196:31–45
Moretti M, Alfaro P, Caselles O, Canas JA (1999) Modelling seismites with a digital shaking table. Tectonophysics 304:369–383
Ogata K, Pogačnik Ž, Pini GA, Tunis G, Festa A, Camerlenghi A., Rebesco M (2014) The carbonate mass transport deposits of the Paleogene Friuli Basin (Italy/Slovenia): internal anatomy and inferred genetic processes. Mar Geol 356:88–110
Ogiwara H, Ito M (2011) Origin and internal organization of widespread composite soft-sediment deformation units in a deep-water forearc basin: the lower Pleistocene Kazusa Group on the Boso Peninsula, Japan. Sed Geol 237:209–221
Owen G (1996) Experimental soft-sediment deformation: structures formed by the liquefaction of unconsolidated sands and some ancient examples. Sedimentology 43:279–293
Owen G, Moretti M, Alfaro P (2011) Recognising triggers for soft-sediment deformation: current understanding and future directions. Sed Geol 235:133–140
Peters M, Veveakis M, Poulet T, Karrech A, Herwegh M, Regenauer-Lieb K (2015) Boudinage as a material instability of elasto-visco-plastic rocks. J Struct Geol 78:86–102
Piper DJW (2005) Sedimentary processes—deep water processes and deposits. Encyclopedia of geology. Elsevier, Amsterdam, pp 641–649
Plaziat J-C, Ahmamou M (1998) Les différents mécanismes à l'origine de ladiversité des séismites, leur identification dans le Pliocène du Saïss de Fès et deMeknès (Maroc) et leur signification tectonique. Geodin Acta 11:183–203
Poropat SF, Kear BP (2013) Reassessment of coelurosaurian (Dinosauria, Theropoda) remains from the Upper Cretaceous Wangshi Group of Shandong Province, China. Cretac Res 45:103–113
Ramberg H (1955) Natural and experimental boudinage and pinch and swell structures. J Geol 63:512–526

Ren X, Han ZZ, Dong YP et al (2021) Early Cretaceous uplift of the Jiaobei Terrane: evidence from the detrital zircon and sediment compositions of sandstones in the Jiaodong Peninsula, China. Geol J 1–22
Rodríguez-Pascua MA, Calvo JP, De Vicente G, Gómez-Gras D (2000) Soft-sediment deformation structures interpreted as seismites in lacustrine sediments of the Prebetic Zone, SE Spain, and their potential use as indicators of earthquake magnitudes during the Late Miocene. Sed Geol 135:117–135
SBGMR (Shandong Bureau of Geology and Mineral Resources) (1991) Shandong regional geological annals. Geological Publishing House, Beijing, 639 pp (in Chinese)
Schmalholz SM, Maeder X (2012) Pinch-and-swell structures and shear zones in viscoplastic layers. J Struct Geol 37:75–88
Schmalholz SM, Schmid DW, Fletcher RC (2008) Evolution of pinch-and-swell structures in a power-law layer. J Struct Geol 30:649–663
Seilacher A (1969) Fault-graded beds interpreted as seismites. Sedimentology 13:155–159
Shanmugam G (2009) Slides, slumps, debris flows, and turbidity currents. Encyclopedia of ocean sciences, 2nd edn. Elsevier, Amsterdam, pp 447–467
Shao Z, Zhong J, Li Y, Ni L, Liu S, Fan L, Chen B (2014) The sedimentary characteristics and environmental analysis of Late Mesozoic gravity flows in Lingshan Island. Geol Rev 60:555–566 (in Chinese with English abstract)
Tan J, Zhang L, Wang C et al (2020) Late Cretaceous provenance change in the Jiaolai Basin, East China: implications for paleogeographic evolution of East Asia. J Asian Earth Sci 194:104188
Van Loon AJ, Brodzikowski K (1987) Problems and progress in the research on soft-sediment deformations. Sed Geol 50:167–193
Van Loon AJ, Pisarska-Jamroży M (2014) Sedimentological evidence of Pleistocene earthquakes in NW Poland induced by glacio-isostatic rebound. Sed Geol 300:1–10
Van Loon AJ, Wiggers AJ (1975) Holocene lagoonal silts (formerly called "sloef") from the Zuiderzee. Sed Geol 13:47–55
Van Loon AJ, Wiggers AJ (1976) Metasedimentary "graben" and associated structures in the lagoonal Almere Member (Groningen Formation, The Netherlands). Sed Geol 16:237–254
Van Loon AJ, Han Z, Han Y (2012) Slide origin of breccia lenses in the Cambrian of the North China Platform: new insight into mass transport in an Epeiric sea. Geologos 18:223–235
Van Loon AJ, Han Z, Han Y (2013) Origin of vertically oriented clasts in brecciated shallow-marine limestones of the Chaomidian Formation (Furongian, Shandong Province, China). Sedimentology 60:1059–1070
Wang P, Zhang B, Qiu W, Wang J (2011) Soft-sediment deformation structures from the Diexi paleo-dammed lakes in the upper reaches of the Minjiang River, east Tibet. J Asian Earth Sci 40:865–872
Wang J, Chang S, Lu H, Zhang H (2014) Detrital zircon U-Pb age constraints on Cretaceous sedimentary rocks of Lingshan Island and implications for tectonic evolution of Eastern Shandong, North China. J Asian Earth Sci 96:27–45
Wang J, Chang SC, Lu HB et al (2016) Detrital zircon provenance of the Wangshi and Laiyang groups of the Jiaolai basin: evidence for early Cretaceous uplift of the Sulu orogen, Eastern China. Int Geol Rev 58–6:719–736
Wang XL, Wang Q, Wang JH, Zhang JL, Cheng X, Jiang SX, Pan R (2010) An overview on the Cretaceous dinosaurs and their eggs from Laiyang, Shandong Province, China. In: Dong W (ed) Proceedings of the twelfth annual meeting of the Chinese society of vertebrate paleontology. China Ocean Press, Beijing, pp 293–316
Weiss BJ, Hübscher C, Lüdmann T, Serra N (2016) Submarine sedimentation processes in the southeastern Terceira Rift/São Miguel region (Azores). Mar Geol 374:42–58
Wiman C (1929) Die Kreide-Dinosaurier aus Shantung. Paleontologica Sinica, Ser C 6(1):1–67
Wu F, Lin J, Wilde S, Zhang X, Yang J (2005) Nature and significance of the early Cretaceous giant igneous event in eastern China. Earth Planet Sci Lett 233:103–119

Xie S, Wu Y, Zhang Z et al (2012) U-Pb ages and trace elements of detrital zircons from early Cretaceous sedimentary rocks in the Jiaolai Basin, north margin of the Sulu UHP terrane: Provenances and tectonic implications. Lithos 154:346–360
Xu X, Wang K, Zhao X et al (2010) A new Leptoceratopsid (Ornithischia: Ceratopsia) from the Upper Cretaceous of Shandong, China and Its Implications for Neoceratopsian evolution. PLoS One 5(11):e13835
Yang R, van Loon AJ (2016) Early Cretaceous slumps and turbidites with peculiar soft-sediment deformation structures on Lingshan Island (Qingdao, China) indicating a tensional tectonic regime. J Asian Earth Sci 129:206–219
Yang R, Fan A, van Loon AJ, Han Z, Wang X (2014) Depositional and diagenetic controls on sandstone reservoirs with low porosity and low permeability in the eastern Sulige Gas field, China. Acta Geologica Sinica (English Ed.) 88:1513–1534
Yang R, Jin Z, van Loon AJ, Han Z, Fan A (2017) Climatic and tectonic controls of lacustrine hyperpycnite origination in the Late Triassic Ordos Basin, central China: implications for unconventional petroleum development. AAPG Bull 101:97–117
Yang R, Fan A, van Loon AJ, Zavala C (2018) The influence of hyperpycnal flows on the salinity of deep-marine environments, and implications for the interpretation of marine facies. Mar Pet Geol 98:1–11
Yang R, Fan A, Han Z, van Loon AJ (2016) An upward shallowing succession of gravity flow deposits in the early Cretaceous Lingshandao Formation, western Yellow Sea. Acta Geologica Sinica 90:1553–1554
Young CC (1958) The dinosaurian remains of Laiyang, Shantung. Palaeontologica Sinica, New Ser C 16:1–138
Young CC (1960) Fossil footprints in China. Vertebrata PalAsiatica 4(2):53–66
Yu Y, Wang K, Chen S et al (2018) A new caenagnathid dinosaur from the Upper Cretaceous Wangshi Group of Shandong, China, with comments on size variation among oviraptorosaurs. Sci Rep 8(1)
Zhai M, Zhu R, Liu J, Meng Q, Hou Q, Hu S, Liu W, Li Z, Zhang H, Zhang H (2004) Time range of Mesozoic tectonic regime inversion in eastern North China block. Sci Chin Ser D: Earth Sci 47:151–159
Zhang JF (1992) Late Mesozoic insects and ecological characteristics in Laiyang, Shandong. Chin Sci Bull 37(5):431–434
Zhang H, Wang B, Fang Y (2010) Evolution of insect diversity in the Jehol Biota. Sci China Earth Sci 53:1908–1917
Zhang H, Lu H, Li J, Wang J, Chang S, Dong X, Zhang X, Huang Z, Shu Y, Ren X (2013) The Lingshandao Formation: a new lithostratigraphic unit of the early Cretaceous in Qingdao, Shandong, China. J Stratigr 37:216–222 (in Chinese with English abstract)
Zhang JL, Wang Q, Jiang SX, Cheng X, Li N, Qiu R, Zhang XJ, Wang XL (2017) Review of historical and current research on the Late Cretaceous dinosaurs and dinosaur eggs from Laiyang, Shandong. Vertebrata PalAsiatica 55(2):187–200
Zhang B, Liu S, Lin C et al (2019) Source-to-sink system reconstruction in the northern Jiaolai Basin, eastern China, by multiproxy provenance methods and implications for exhumation of the Sulu Orogen. Tectonophysics 754:18–32
Zhang B, Liu S, Lin C et al (2020) Provenance of the Late Cretaceous sediments in Jiaolai Basin, Eastern China, and its tectonic implications. Int Geol Rev 63–8:973–991
Zhao XJ (1962) A new genus of Psittacosauridae from Laiyang, Shantung. Vertebrata PalAsiatica 6(4):349–364
Zhao R, Wang Q, Deng J, et al (2018) Late Mesozoic magmatism and sedimentation in the Jiaodong Peninsula: new constraints on lithospheric thinning of the North China Craton. Lithos 322
Zhao X, Wang K, Li D (2011) "Huaxiaosaurus aigahtens." Geol Bullet China 30(11):1671–1688
Zhong J (2012) The Mesozoic sedimentary rocks on Lingshan Island: are they deep-water far-sourced turbidites or intra-continental deltaic sediments?—a discussion with Prof. Lu Hongbo. Geol Rev 58:1180–1182 (in Chinese)

Zhou CF (2010) A possible azhdarchid pterosaur from the Lower Cretaceous Qingshan Group of Laiyang, Shandong, China. J Vertebr Paleontol 30(6):1743–1746
Zhou X, Yang J, Zhang L (2003a) Metallogenesis of super-large gold deposits in Jiaodong region and deep processes of subcontinental lithosphere beneath North China Craton in Mesozoic. Sci Chin Ser D: Earth Sci 46:14–25
Zhou ZH, Barrett PM, Helton J (2003b) An exceptionally preserved Lower Cretaceous ecosystem. Nature 421:807–814
Zhou Y, Zhang Z, Liang W, Li S, Yue H (2015) Late Mesozoic tectono-magmatic activities and prototype basin restoration in Eastern Shandong Province, China. Earth Sci Front 22:137–156 (in Chinese with English abstract)
Zhou JB, Han W, Song MC et al (2019) Zircon U-Pb ages of the Cetaceous sedimentary rocks in the Laiyang Basin, eastern China and their tectonic implications. J Asian Earth Sci 194:103956
Zhu G, Liu G, Niu M, Xie C, Wang Y, Xiang B (2009) Syn-collisional transform faulting of the Tan-Lu fault zone, East China. Int J Earth Sci 98:135–155
Zulauf J, Zulauf G, Hammer J, Zanella F (2011) Tablet boudinage of an anhydrite layer in rock-salt matrix: results from thermomechanical experiments. J Struct Geol 33:1801–1815

Discover the Glacial World on the Eve of the Cambrian Life Explosion: The Ediacaran-Early Cambrian Glaciation in the West of Henan Province, North China

Hongwei Kuang, Xiaoshuai Chen, Yongqing Liu, Yuchong Wang, Nan Peng, Daniel Paul Le Heron, Thomas Matthew Vandyk, and Kening Qi

1 Schedule and Logistics

1.1 Schedule

The field trip in western Henan Province, North China lasts for 5 days. Please see Table 1 and Fig. 1 for the details of the arrangement, location of sections, transportation, and accommodation.

H. Kuang (✉) · X. Chen · Y. Liu · Y. Wang · N. Peng · K. Qi
Institute of Geology, Chinese Academy of Geological Sciences, Beijing, China
e-mail: kuanghw@126.com

Y. Wang
College of Geosciences and Surveying Engineering, China University of Mining and Technology Beijing, Beijing, China

D. P. Le Heron
Department of Geology, University of Vienna, Vienna, Austria

T. M. Vandyk
Department of Geography, Royal Holloway University of London, Egham, UK

K. Qi
School of Resources and Environmental Engineering, Hefei University of Technology, Hefei, China

X. Hu (ed.), *Field Trip Guidebook on Chinese Sedimentary Geology*,
https://doi.org/10.1007/978-981-99-6936-4_8

1.2 Logistics

The investigation area is located in the western mountainous area of Henan Province. It is midsummer in late August, with an average temperature of about 21–33 °C. During this period, it is mostly sunny with thunderstorms. We recommend wearing thinner clothes and long shirts, and to come equipped with sunscreen and rainwear. The routes are generally along simple roads or flat sand roads, but it often requires short-distance walking. Meanwhile, to prevent mosquito bites and spiny plants, it is best to wear scratch-resistant long-sleeved tops and trousers, and hiking shoes or outdoor shoes are recommended.

2 Glacial Epochs in Earth History and Snowball Earth Event

2.1 Glacial Epochs in Earth History

The Earth has experienced at least six major ice ages since its birth (Eyles 2008, Fig. 2). Glaciations are not only closely related to continental breakup (Eyles 2008; Young 2019) but also related to biological evolution. Unique erosion, transport and deposition records preserved in different glaciers periods are important evidence for reconstructing palaeogeography, palaeoclimate and palaeoenvironment of pre-glacial period. The continental glaciers on the modern Earth have retreated to the polar regions, global warming and the rapid rise of ice-melting sea level induced by the intensification of greenhouse climate bring a great threat to the survival of human beings, which is a serious issue that all mankind must face together. This also makes us deeply realize the importance of studying ice ages, the formation and evolution, kinematic law, sedimentary characteristics of glaciers, as well as the correlation with climate and environment, which are exactly the key content of glacial sedimentology, palaeogeography and other disciplines.

The earliest ice age in the geological record occurred 2.9–2.8 Ga ago is named Pongola glacial period, which can be only found in South Africa (Young 1998). Viewpoints on the existence of Pongola Glacier are different due to the lack of geological records. The existing evidence suggests that the glaciation may be transient and local, and it is difficult to define an ice age strictly.

The second glaciation occurred in the Paleoproterozoic (2.4–2.1 Ga), which is called the Huronian glacial period. It lasted for more than 300 Ma years and was the most severe and longest glaciation in geological history. It was named after Lake Huron in North America. Previous studies (Young 2019) pointed that the Huronian glaciation represented the first Snowball Earth event. Glaciers extended from the Poles to the equator and the whole Earth was covered by thick ice. The thick ice in Huronian glaciation reflects all the sunlight into space; this turned the Earth into a snowball. After nearly 300 Ma years of cold climate, the heat inside the Earth

Table 1 Schedule of the field trip (August 27th–31st, 2022)

Schedule		Route	Vehicle	Outcrops	Investigation contents	Time	AC
27th	a.m.	BWRS to LLRS	HSR	–	–	8:00–12:00	Luoyang
	p.m.	Longmen-Shangxuma	Bus	Shangxuma	Glacial pavement and Luoquan Fm.-Xinji Fm.	14:30–17:30	
28th	Whole day	Luoyang-Yiyang-Ruzhou	Bus	Zhoucun	Mesoproterozoic basement-Luoquan Fm.-Xinji Fm.	08:30–17:00	Ruzhou
29th	Whole day	Ruzhou-Lushan-Ruzhou	Bus	Shimengou	Glacial pavement and Luoquan Fm.-Xinji Fm.	08:30–17:00	
30th	a.m.	Ruzhou-Luoquan	Bus	Luoquancun	Luoquan Fm.	08:30–12:00	
	p.m.	Luoquan-Yangpo-Ruzhou	Bus	Yangpo	Mesoproterozoic basement-Luoquan Fm.-Xinji Fm.	13:00–17:00	
31st	a.m.	Ruzhou-Liuzhanggou-Ruzhou	Bus	Liuzhanggou	Luoquan Fm.-Xinji Fm., sand wedges	08:30–11:30	
	p.m.	Ruzhou-Return trip	At one's sole discretion End of the trip, head for the station			13:00–17:00	

Note BWRS—Beijing West Railway Station, LLRS—Luoyang Longmen Railway Station; HSR—high-speed railway; AC—accommodation city

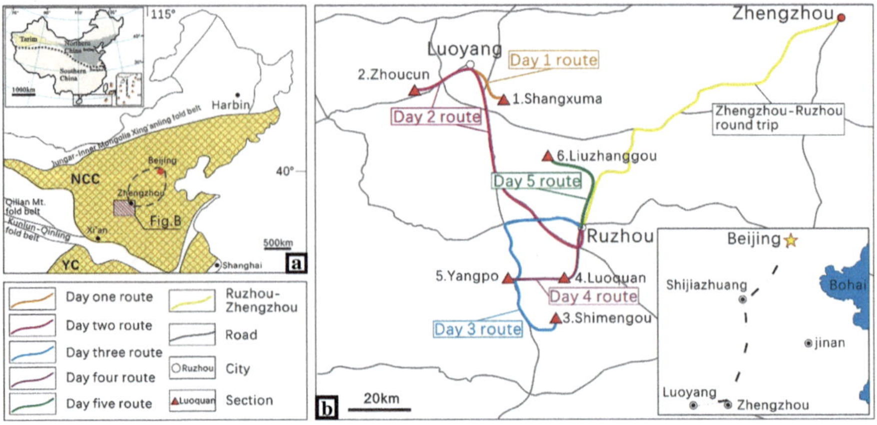

Fig. 1 General routes for the field trip of the Ediacaran Luoquan Formation in Western Henan Province, North China. Abbreviations: NCC—North China Craton; YC—Yangtze Craton

was finally unable to be released. Massive volcanic eruptions released hundreds and millions of tons of carbon dioxide into the atmosphere. Carbon dioxide is a typical greenhouse gas, covered the Earth with a thick "quilt", making the surface heats up slowly and eventually released from the ice cover.

The third ice age occurred during the Cryogenian or called Sturtian-Varangian (720–635 Ma), encompassed the Sturtian and Marinoan glaciations in Australia, and extended over more than 100 million years. The concept of "Snowball Earth" (Hoffman et al. 1998) was derived from this glaciation. Kirschvink (1992) proposed the hypothesis of "Snowball Earth", however, it's much shorter than the first glaciation in the Palaeoproterozoic. This hypothesis has been a hot topic in Earth science for more than 20 years. According to this hypothesis, the breakup of the Rodinia supercontinent located near the equator and low latitude rapidly increased the area of marginal sea, and a large number of basalts formed by large-scale volcanic eruptions have undergone strong chemical weathering, which intensified the consumption of CO_2, resulting in a sharp decline of carbon dioxide content in atmospheric and the emergence of the ice house effect. With the continuous increase of the ice cover on the earth, the positive feedback of ice-albedo made the ice sheets at the poles expand and gradually advance to lower latitudes, eventually leading to global ice cover and the formation of "snowball".

Under the background of transgression after Marinoan glaciation, the global climate turned warm rapidly. Under the warm and clean shallow water environment, the Marinoan glacial deposits were covered by the Ediacaran cap dolostone of varied thickness, which was reported almost on the margin of every main continent in the world (North America: James et al. 2001; Corsetti and Grotzinger 2005; Creveling et al. 2016; South America: Font et al. 2006; India: Etienne et al. 2011; Oman: Allen et al. 2004; Australia: Kennedy 1996; Rose and Maloof 2010; Namibia: Hoffman et al. 1998; Hoffman et al. 2007; Hoffman and Macdonald 2010; Central Africa:

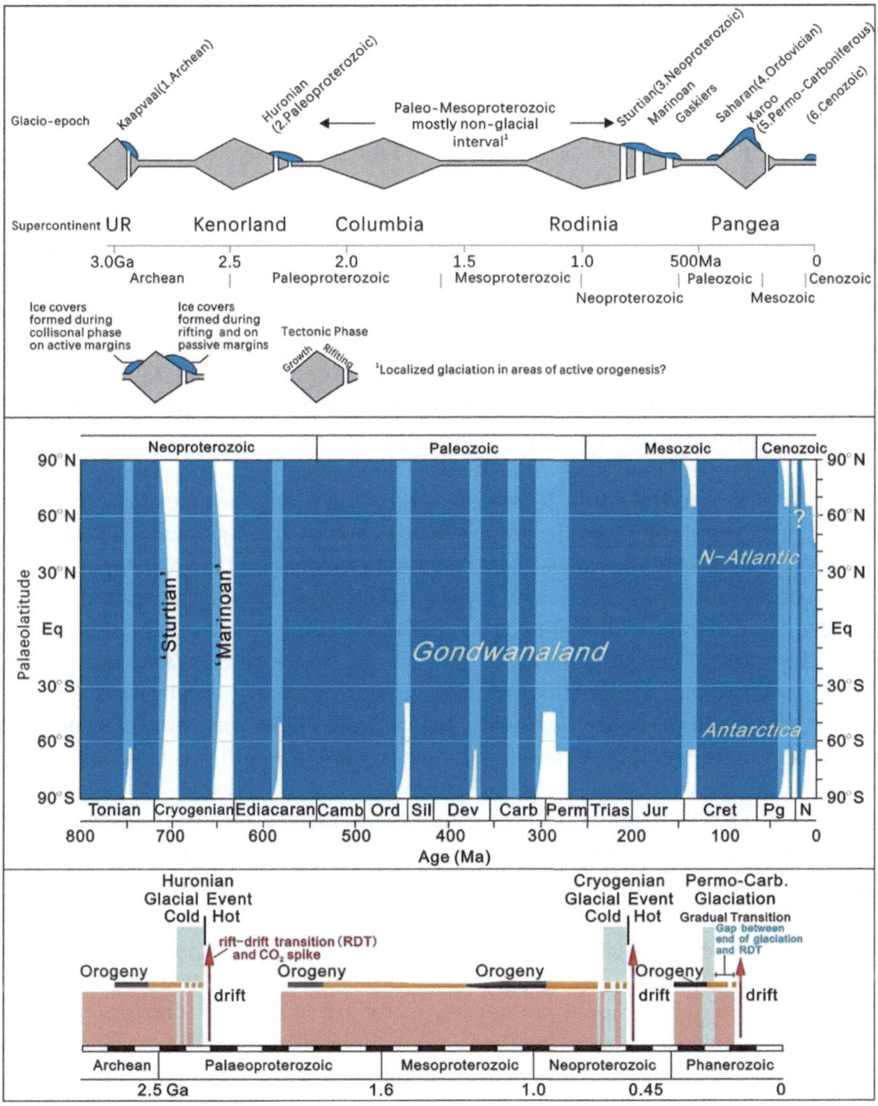

Fig. 2 Glaciers in the geological history (Modified after Eyles 2008; Hoffman 2009; Young 2019)

Mickala et al. 2014; Tarim: Xiao et al. 2004; South China: Jiang et al. 2006; Kuang et al. 2019, 2022).

However, the "Snowball Earth" hypothesis has been challenged by increasing sedimentological evidence, such as the development of wavy bedding, hummocky cross-bedding, subglacial deformation structures, dolomite lenses, red beds, and dropstone-bearing intervals in glacial sedimentary succession (Allen and Etienne

2008; Le Heron et al. 2011, 2013; Bai et al. 2020). In addition to the aspects of sedimentary facies and sedimentary structure mentioned above, the evolution of glacial sedimentary succession, such as the deposition at the far and near ends of glaciers, can also represent that during the Snowball Earth period, the glacier has a dynamic cyclic evolution process at the periglacial and the grounding area (Lang et al. 2018; Hu et al. 2020a; Chen et al. 2021).

After the Cryogenian glaciations, life flourished and developed further in the Ediacaran (Young 2017). During "Snowball Earth", cyanobacteria in the ocean proliferated and a large amount of oxygen was released from the ocean, changing the structure of the atmosphere. Large quantities of carbon dioxide were consumed and oxygen concentration increased continuously, followed by the Shuram event (the largest global negative carbon isotope drift) (Fike et al. 2006; Le Guerroué 2010; Grotzinger et al. 2011), and the Ediacaran glaciation, which is typically represented by the Luoquan glaciation we will investigate.

The fourth glaciation known as the Andean-Saharan glacial period (Eyles and Young 1994), occurred 450–420 Ma ago (end of the Early Paleozoic). With a duration of just 30 million years, it's called the most short-lived glacier. The glaciation occurred in the Late Ordovician (mainly the Ashgillian Stage) in West Africa and the Tamadjert Formation in the Sahara Desert, Morocco (Tindouf Basin), and central-western Saudi Arabia, all belonged to the high-latitude polar region at that time. From the late Ordovician to the early Silurian, the center of the glacier moved from North Africa to southwestern South America. This ice age is short in duration, and the origin of it is even stranger. The black hole from 6,000 light-years away collided with a neutron star, producing a burst of gamma rays, and the earth was unfortunately hit by one. Gamma rays destroyed a third of the ozone layer and shattered atmospheric molecules. These broken atmospheric molecules formed nitrogen dioxide (brown gas) and block the atmosphere, inducing rapid cooling and a massive ice age, marked by a mass extinction which is the second to the extinction at end of the Permian.

The fifth glaciation is known as the Karoo Glaciation (Haldorsen et al. 2001), occurred between 360 and 260 million years ago at the end of the oxygen-rich Carboniferous period to the Permian (Late Paleozoic) and lasted 100 Ma years. Climate was mild and humid in the early Carboniferous, huge forests can be seen everywhere. However, plants absorbed carbon dioxide and released oxygen, causing lower carbon dioxide level and weaker greenhouse affection in the late Carboniferous. As oxygen exploded, temperatures began to fall, and the fifth glaciation began.

The sixth glaciation is the Quaternary glaciation (Hewitt 2000), which is closely related to the evolution of human beings. The Quaternary glaciation occurred two million to three million years ago and ended about 10,000 years ago, lasting only a few million years. Abundant Quaternary glaciation remnants are preserved in the world due to the close association to the time human beings live in now. We are now in a warm period between the two small ice ages of the Quaternary glaciation, also known as an interglacial period, and may soon enter the next ice age.

2.2 *Ediacaran Glaciation*

2.2.1 Spatial Distribution

The Ediacaran-Cambrian transition period (635–540 Ma) after the Snowball Earth is recognized as a global greenhouse climate period. However, Ediacaran glacial deposits have been reported from more than a dozen different paleocontinents around the world (Hoffman and Li 2009; Li et al. 2013), e.g., Northwest Australia (Corkeron and George 2001), North China, Tarim, South Africa (Germs and Gaucher 2012), Congo (Hofmann et al. 2015), West Africa (Vernhet et al. 2012), Central Asia (Chumakov 2004), Arabian-Nubian (Vickers-Rich and Fedonkin 2007; Miler et al. 2008), Europe (Bingen et al. 2005), Baltica (Chumakov 2011), Greenland (Stouge et al. 2011), North America (Pu et al. 2016), and South America (de Alvarenga et al. 2007; Figs. 3 and 4).

Ediacaran glacial deposits are extensive and crop out over 5000 km in Asia, including a batch of diamictite localities in Southern North China (south Anhui; west Henan; Shaanxi; Ningxia; Qilian Mountain of central Gansu, Luoquan Glaciation, North China), Fengtai/Luoquan /Shangzhangwan/Zhengmuguan formations; Northwest China (Quanjishan, Qaidam, central Qinghai; Tarim, NE Xijiang, NW China), Luoquan Glaciation, Hongtiegou/Hankalchough formations (Mu 1981; Guan et al. 1986; Wu and Guan 1988a, b; Xiao et al. 2004; Zhan et al. 2007; Xu et al. 2009; Shen et al. 2010; Zhu and Wang 2011; Dong et al. 2017; Yang et al. 2017, 2019; Le Heron et al. 2018, 2019; Li et al. 2020; Wang et al. 2022; Liu et al. 2023) and those in Central Asia and West Asia (as mentioned above) as well (Fig. 4).

2.2.2 Sediments and the Coeval Products

Globally, the Ediacaran glacial sediments either directly overlie the Precambrian basement on the so-called Great Unconformity (Keller et al. 2019; Li et al. 2020), or rest on underlying marine deposits formed in the Ediacaran greenhouse climate. The glacial sediments themselves are mostly several meters to tens of meters, e.g., in West Africa, Australia, North China, and Iran, or even more than hundred or tens of hundred meters thick, e.g., in Tarim, Newfoundland of Canada, Brazil, Kazakhstan and Kyrgyzstan, and mainly consist of stratified and massive diamictite, conglomerates, laminated sandstone along with dropstone-bearing siltstones which are originated from subglacial and periglacial fluvial, lacustrine environments, and ice-rafting in open water.

The vertical evolution of facies indicates a deglaciation sequence typically comprising (1) subglacial erosional surface (e.g., a striated pavement); (2) diamictites or finer-grained sediments with evidence for ice-rafted dropstones. These glaciogenic deposits may correspond to the marginal ice-sheet sediments deposited during the ice-sheet retreat; (3) stratified diamictites are proglacial outwash and marine deposits (Chen et al. 2023). Striated pebbles may be found. Additionally, facies commons to

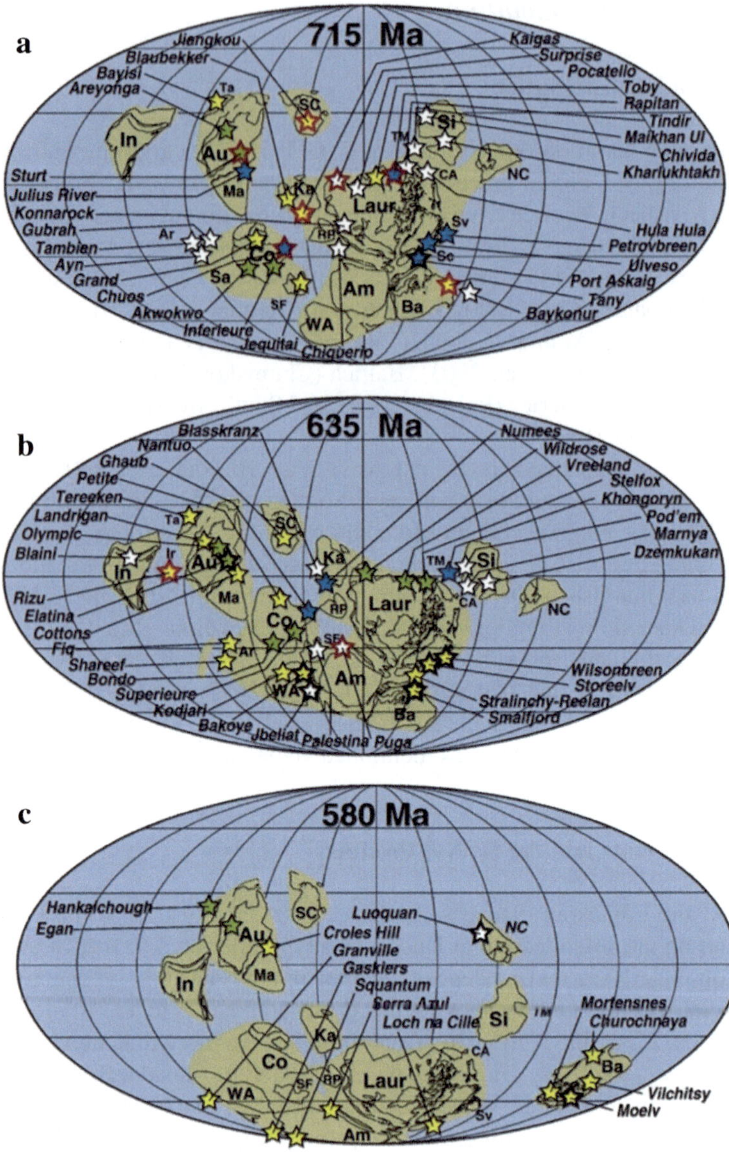

Fig. 3 Global glaciations from Cryogenian (**a**, **b**) to Ediacaran-Cambrian transition period (**c**) (Hoffman and Li 2009)

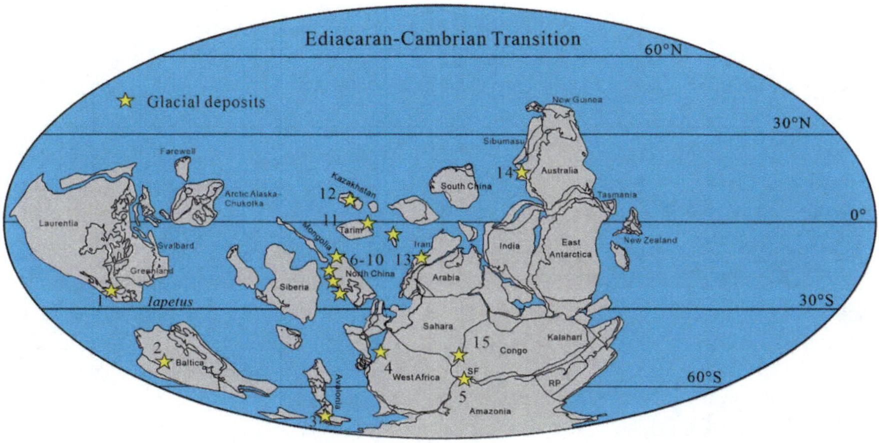

Fig. 4 Global distributions of the major glacial diamictites in the Ediacaran (Map is modified from Torsvik and Cocks 2013; Location of glacial deposits is modified from Hoffman and Li 2009; Liu et al. 2023)

many Neoproterozoic glacial succession is the iron formation ("BIF") corresponding to ferruginous tillites or the underlying glacial sequence in NW China (Fig. 5). "BIF" were dated as 593–532, 520–516 Ma, and more akin to Superior-type, which constitutes an additional line of evidence for the widespread return of BIFs in the Ediacaran (Xu et al. 2015; Hu et al. 2017, 2020b).

2.2.3 Geochronology

Critical tests of the idea of global glaciations require robust, very precise, and accurate time constraints on the deposition of glacial deposits in different palaeogeographical and environmental settings on different palaeocontinents. Precise U-Pb zircon dating of interbedded multiple volcanic ash layers of the Gaskiers Formation at the type locality in Newfoundland, Canada, has shown it (the Gaskiers glaciation) to be 580 Ma and to have lasted less than 1.0 Ma (Pu et al. 2016). However, age constraints for assumed Gaskiers-equivalent strata elsewhere are poorly defined, and several Ediacaran glaciations postdating the Gaskiers Glaciations have been identified worldwide (Hebert et al. 2010).

For example, a glacial succession ~565 Ma old is extensive on the margin in the Ediacaran of peri-Gondwanan West Africa, and the Fauquier glaciation is ~571 Ma in Virginia (USA). Detrital zircon age for sandstones below and above the Ediacaran Kahar diamictite intervals in Iran indicates the youngest glaciation of ~550–560 Ma. The probable density distribution of U-Pb ages through time for the Ediacaran (Youbi et al. 2020) shows that a glaciation of ~580 Ma accounts for more than 49.5% of the entire Ediacaran glacial record.

Fig. 5 Ediacaran glacial deposits in the northern part of China and Newfoundland, Canada. 1. Hongtiegou Fm., Qaidam, Qinghai; 2. Zhengmuguan Fm., Helanshan, Ningxia; 3. Luquan Fm., Zhoucun, west Henan; 3–4. Gaskiers Fm., Newfoundland, Canada; 5. Shangzhangwan Fm., Luonan, south Shaanxi; 6. Luquan Fm., Yangpo, west Henan; 7. Fengtai Fm., Huainan, south Anhui; 8. Madian Fm., Huoqiu, south Anhui. Gravels in diamictite are mainly dolomite (1, 2, 5, 6, 7, 8), and multiple compositions (3, 4). Purple-red iron ribbons and dropstone are similar to the Cryogenian BIF (5), and deep water glacial or rafting deposits (4, 8). Len cover of camera is 6 cm in diameter and pencil is 20 cm long for scale

2.2.4 Post-glacial Cap Carbonate

Thin cap dolostones blanket glacial sediments and show distinct sedimentological, biological, and textural features, are increasingly used for global correlation and understanding the evolution of the atmosphere and biosphere during the Precambrian (Kaufman and Knoll 1995; Kirschvink et al. 2000). Although the Cryogenian cap dolostones of the Sturtian and Marinoan glaciations are recognized worldwide, the counterparts related to the Ediacaran glaciation are less studied, assumed to be less extensive, and typically do not have an associated cap carbonate. In fact, cap dolostones with idiosyncratic sedimentary structures, negative carbon isotopic signature, and varied thickness occurred worldwide in the aftermath of the Ediacaran glaciation as well, such as northern Virginia, America, Avalon Peninsula, Newfoundland, Kazakhstan and Kyrgyzstan, and North China.

2.2.5 Subglacial Erosional Landscapes

Globally, well-preserved glacial pavements with multi-scale subglacial erosional landforms occur on the surface of the Great Unconformity, e.g., in Australia (Corkeron and George 2001), Africa (Germs and Gaucher 2012; Vernhet et al. 2012; Letsch et al. 2018), Iran (Najmeh Etemad-Saeed et al. 2016) and North China (Mu 1981; Guan et al. 1986; Wu and Guan 1988a, b; Le Heron et al. 2018, 2019; Chen et al. 2020; Zhang et al. 2021). This may be the unique feature of the Ediacaran glaciation, and significantly different from the physical nature of the Cryogenian glaciation, and seems most similar to the Gondwana continental glaciation in the Paleozoic. An outstanding record of glacial erosion, and extensively glacial pavements with varied subglacial erosional landforms, are well-preserved in North China (Fig. 6).

2.2.6 Tectonic Background

Supercontinent breakup has been considered to be a causal mechanism for the Neoproterozoic glacial events, and each breakup of the supercontinent would have a significant glacial event associated with it. The Ediacaran glaciations were concomitant with the final breakup of the Rodinia supercontinent. The ~615–583 Ma mafic-ultramafic intrusions, mafic dykes, continental flood basalts with different sizes in the underlying or interbedded Ediacaran diamictites are well developed in Australia, West, and Central Asia (Xu et al. 2015). The central Iapetus magmatic province event, in Laurentia, Baltica, and NW Africa, consisting of multiple pulses of mafic-ultramafic LIP ~620–520 Ma, were all interpreted to be correlated with the final fragmentation of Rodinia. In addition, palaeogeographic reconstruction from the Neoproterozoic to the Early Cambrian shows that those continental blocks preserved the Ediacaran glacial diamictites, e.g., North China, the microblocks of Central Asia, Siberia, Baltic and Laurencia, etc., were originally part of the Rodinia supercontinent.

Fig. 6 Composite aerial image and Digital Elevation Model from the Shimengou outcrop, West Henan Province (Courtesy of Prof. D.P. Le Heron)

2.2.7 Glaciations and Early Metazoan Evolution

Glaciations have a major impact on the evolution of life and affect the entire planet Earth (Hoffman 2009), and the Ediacaran glaciation opens a wider window into the relationship between glaciation and early metazoan evolution (Fig. 7). *Tianzhushania*-dominated assemblage and Doushantuo embryos of the Pre-Ediacaran biotas are succeeded by the assemblages of large complex Ediacaran fossils, which evolved only two million years after the Ediacaran ice sheets melted. In Newfoundland, the cosmopolitan taxa *Charnia* is preserved immediately below the 579 Ma ash layers (Hoffman and Li 2009) and followed by the fossiliferous 565 Ma ash layers (Benus 1988). The Ediacaran glacial diamictites in Quanjishan, Qinghai, NW China, were immediately covered by the fossiliferous siltstone containing *Charnia* and *Shannxilithes* as well (Shen et al. 2007; Pang et al. 2021). It appears that Earth's earliest complex metazoans evolved amid strong climatic oscillations of the Ediacaran glaciation and intermittent greenhouse conditions.

3 Geological Background

3.1 Brief Introduction to the Study Area

West Henan Province is located on the southern margin of the North China Craton. Rocks of Archean, Palaeoproterozoic, and Meso-Neoproterozoic age are widely developed. Among them, metamorphic rock series such as Archean Dengfeng Group, Taihua Group, and Lower Proterozoic Songshan Group comprises the basement of the North China Craton (Hu et al. 2012a). The Mesoproterozoic is composed of volcanic-sedimentary rocks of Xiong'er Group, carbonate rocks, and clastic rocks of Ruyang Group and Luoyu Group; The Neoproterozoic Luanchuan Group comprises of quartz sandstone, siltstone, dolostone, and volcanic rock series, which are widely distributed in the Xiaoqinling area at Shaanxi and the junction of Henan Provinces from west to east, such as Luonan, Lingbao, and Lushi. To the east, passing through Xiaoshan, Xiong'er mountain, and Songshan, to Lushan, Yexian, Wuyang, Queshan, and Wulou in the south of Henan Province (Guan et al. 1988) (Fig. 8a, b).

The three-armed rift system represented by the volcanic rock series such as Xiong'er Group and Xiyanghe Group developed in the southern margin of North China Craton, and above Mesoproterozoic and Neoproterozoic sediments are developed. According to the lithologic association and sedimentary characteristics, the Meso-Neoproterozoic strata in this area are divided into Xiong'er stratigraphic district, Mianchi-Queshan district, and Songji district. The stratigraphic system and correlation relationship of each district are established too (Guan et al. 1988; Zhao et al. 2015; Fig. 8c).

I—Xiong'er stratigraphic district: Guandaokou Group and Luanchuan Group are mainly distributed in Lushi, Luoning, Luanchuan, Fangcheng, and other places in

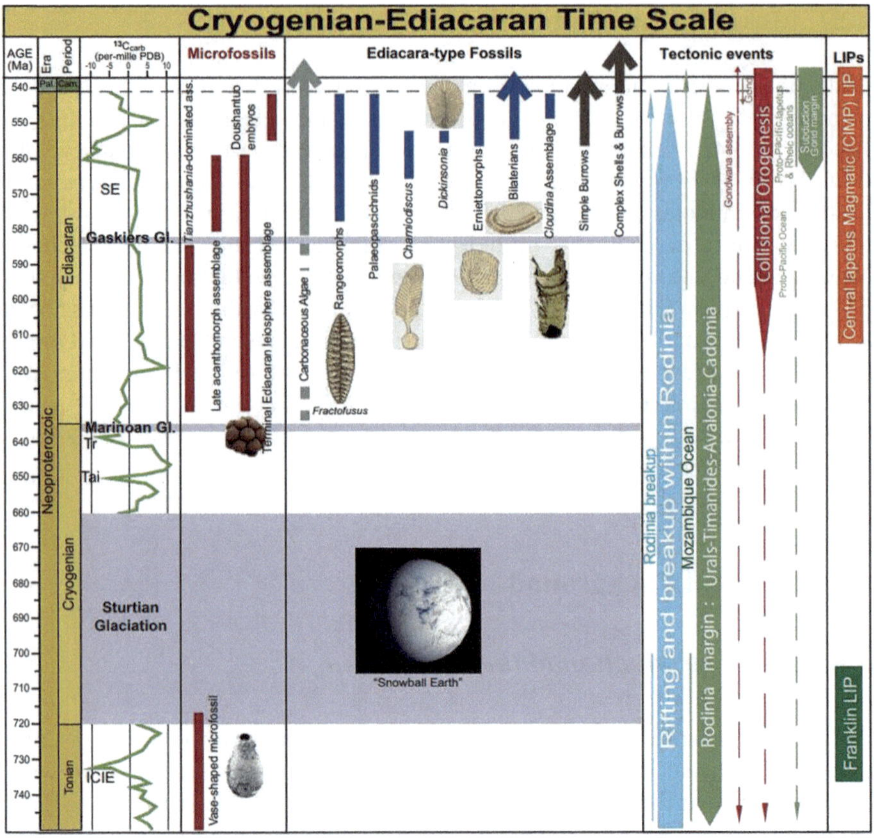

Fig. 7 Selected major trends in Cryogenian and Ediacaran geologic history (Ogg et al. 2016). The carbon-isotope curve is a smoothed version modified by Ogg et al. (2016) from the synthesis for the late Proterozoic by Cohen and Macdonald (2015) calibrated by them to the Cryogenian time scale of Rooney et al. (2015). Ranges and images of organic-walled microfossils, Ediacaran metazoans, and bioturbation styles are from Narbonne et al. (2012). Timing of events associated with continental collision and convergent and divergent plate-margin activities related to assembly and dispersal of Rodinia and subsequent assembly of Gondwana is from Cawood et al. (2016). The timing of large igneous provinces (LIPs) is from Ernst (2014), Ernst and Youbi (2017), and Macdonald and Wordsworth (2017). Additional geochemical trends, biostratigraphic ranges, regional stages, and details on calibrations were compiled in Narbonne et al. (2012). Abbreviations: SE, Tr, Tai and ICIE-Shuram, Trezona, Taishir, and Islay carbon-isotope excursions, respectively; Gond-Gondwana; Pal-Paleozoic; Cam-Cambrian; Gl.-glaciation; CIMP-Central Iapetus magmatic province; PDB-PeeDee belemnite. Figure modified from Youbi et al. (2020)

western Henan Province. Guandaokou Group is mainly composed of quartz sandstone and dolostone, which is a set of terrigenous clastic rock-carbonate sedimentary succession. From bottom to top, Guandaokou Group is divided into Gaoshanhe Formation, Longjiayuan Formation, Xunjiansi Formation, Duguan Formation, and

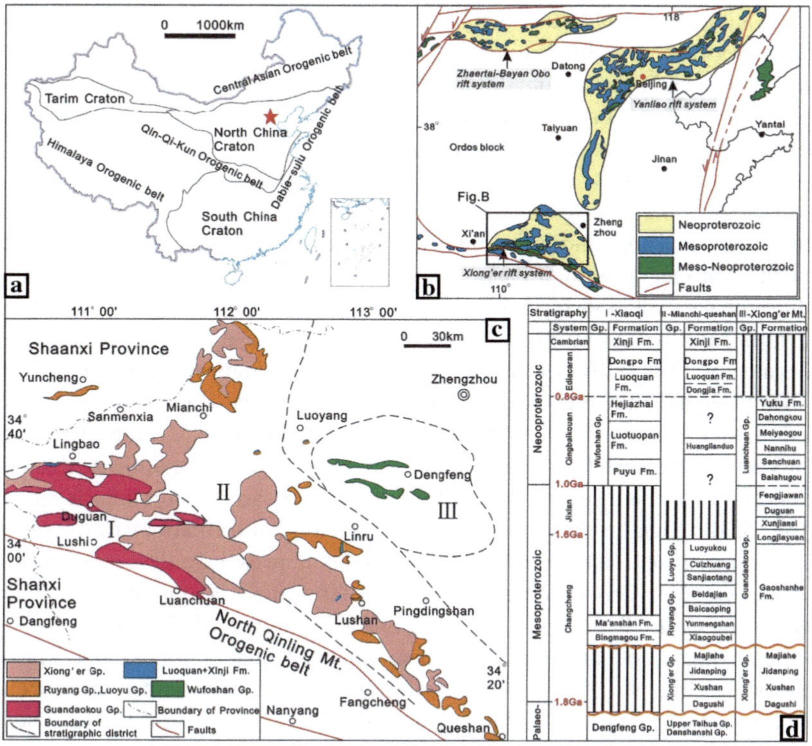

Fig. 8 **a** Schematic tectonic map of China with the major Precambrian blocks (modified from Zhao et al. 2001), **b** The location of the study area within the North China Craton (NCC) and North Qinling Terrane (NQT) (modified from Hu et al. 2016; Zuo et al. 2019), **c**, **d** Division of stratigraphic districts of Meso-Neoproterozoic strata in southern North China Craton and corresponding stratigraphic chart (After Zhao et al. 2015; Wang et al. 2015); I—Xiong'er stratigraphic district; II—Mianchi-Queshan stratigraphic district; III—Songji stratigraphic district

Fengjiawan Formation, which are covered by Luanchuan Group or Luoquan Formation, Dongpo Formation and Xinji Formation, etc.

II—Mianchi-Queshan stratigraphic district: the Mesoproterozoic and Neoproterozoic are distributed in Mianchi, Ruyang, Lushan, Fangcheng, Biyang, Queshan in western Henan Province, and Yangcheng, Yuanqu in Southern Shanxi Province (Fig. 8). Ruyang Group overlies on Xiong'er Group unconformably, according to lithologic characteristics, sedimentary cycle, isotopic age data, and paleontology (including stromatolites), Guan et al. (1988) divided Ruyang Group into Xiaogoubei, Yunmengshan, Baicaoping, and Beidajian formations in ascending order. Ruyang Group is mainly composed of terrigenous clastic rocks in the lower part and only a small amount of carbonate rocks in the top. Luoyu Group is mainly distributed in the area of Chaoyangguan-Jiunvdong-Shixiakou-Luanshipan in Xiatang. Generally, it extends in a NEE arc and parallel unconformity contact with the underlying Ruyang Group and the overlying Ediacaran strata. From bottom to top, it can be divided into

Cuizhuang, Sanjiaotang, and Luoyukou formations, and there is a conformity contact relationship between each Formation. Luoyu Group is mainly composed of shale, quartz sandstone, and stromatolite dolomite. In addition, as a set of clastic rock and carbonate rock combinations, the position of Huanglianduo Formation and Dongjia Formation in the stratigraphic chart is not clear.

III—Songji stratigraphic district: the Mesoproterozoic and Neoproterozoic are mainly distributed in the Songshan-Jishan area (short for the Songji area) in Yanshi, Dengfeng, Yichuan, Linru and Yuxian, Henan Province. It is distributed on the landwards of the continental margin basin in the south of the North China Craton and has a phase change relationship with the continental margin basin sedimentation-Mesoproterozoic Ruyang Group. Based on the comprehensive study of the previous work, Guan et al. (1988) divided the Wufoshan Group in Songshan area into Bingmagou, Ma'anshan, Xiamen (Menwai), Puyu, Luotuopan, and Hejiazhai formations from bottom to top, above which is the Hongling Formation of late Neoproterozoic. In recent years, there have been many studies on Bingmagou Formation in western Henan, and most of them regard the clastic rocks below the volcanic interlayers at the bottom of Bingmagou, Xiaogoubei, and Yunmengshan formations in the south of Ruyang as synchrono-heteropic facies, and regard them as the oldest set of clastic rocks above Xiong'er Group and under the overlying Meso-Neoproterozoic cover deposits (Zheng et al. 2016, 2017; Yue and Liu 2018; Meng et al. 2018) (Fig. 8c).

The Ediacaran Luoquan Formation rests disconformable above the Meso-Neoproterozoic (Fig. 9). The residual stratigraphic succession under the unconformity changes significantly. Within a range of more than 100 km, the thickness and integrity of the underlying Meso-Neoproterozoic show a pattern of gradually increasing from north to south (Fig. 9). In general, in Shangxuma-Liuzhanggou-Gandanchang in the north of the investigation area, the degree of glacier erosion on the underlying basement rocks is the strongest, so the residual thickness of basement rocks under the Luoquan Formation is also the smallest; Accordingly, Jiunvdong, Lushan in the south of the investigation area, the thickness of the basement rocks under the Luoquan Formation is the largest, the residual strata are relatively complete, and the glacier erosion is the weakest, Luoquan Formation is in parallel unconformity (or conformity) contact with the overlying Cambrian Xinji Formation.

In the northernmost part of the investigation area, Shangxuma, Luoyang, Liuzhanggou, and Gandanchang sections, under the Luoquan Formation, only 170 m thick Ma'anshan Formation/Yunmengshan Formation-Bingmagou Formation of the lowermost part of the Mesoproterozoic remains, and the top surface of the quartz sandstone of Ma'anshan Formation is the glacial pavement; However, there are exceptions. In Zhoucun, Yiyang, west of Shangxuma, there is no glacial pavement preserved while Luoquan Formation may overlie the shale of Cuizhuang Formation of Luoyu Group or the dolomite of Luoyukou Formation of Luoyu Group within a very small range (<1 km), and the thickness of the basement rocks is much greater than the remaining 170 m thick Ma'anshan Formation and Bingmagou Formation in Shangxuma. In contrast, in the middle part of the investigation area, for example in Yangpo, it is the dolomite of Luoyukou Formation of Luoyu Group and the rest of Mesoproterozoic basement below Luoquan Formation. In the adjacent Luoquan

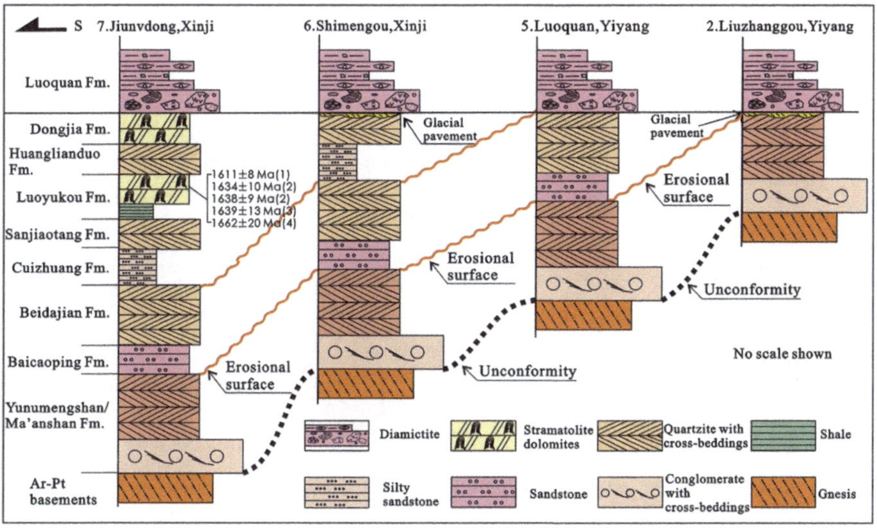

Fig. 9 The stratigraphic framework of Mesoproterozoic residual basement rocks under Luoquan Formation in Western Henan Province. References for the age data: (1) Su et al. (2012), Zircon U-Pb LA-ICP-MS; (2) Li et al. (2017), Zircon U-Pb LA-ICP-MS; (3) Peng et al. (2018), Zircon U-Pb SHRIMP; (4) Wang et al. (2015), Zircon U-Pb LA-ICP-MS

village, it is the quartz sandstone of the Beidajian Formation of the top Ruyang group, which is more complete than that in the north of the investigation area. What's more, in the Jiunvdong section of Lushan in the southernmost part of the investigation area, the strata under the Luoquan Formation are the most completely preserved, and the total thickness is more than 1000 m (Chen 1990). So far, we tend to consider that the unconformity between Luoquan Formation and underlying Meso-Neoproterozoic is caused by Ediacaran glacial erosion, which belongs to macro glacial erosional landform. At present, this great unconformity and different scale glacial pavements on it can be observed (e.g., Li et al. 2020; Chen et al. 2020).

3.2 Luoquan Glaciation: History of Research and State of the Art

The Chinese geologist Prof. Yuelun Wang discovered glacial erratic boulders in South Shaanxi and West Henan Provinces and mentioned that there could be a Precambrian glacier preserved in the northern slope of Qinling Mountain. In the same year of 1958, Yang Zhijian clarified the glacial origin of these deposits, and were named as "Linru tillite bed". In 1959, the phosphate team led by the very famous academician Lianjun Ye from the Chinese Academy of Sciences also demonstrated the development of tillite in Western Henan Province. Later, Chang'an Liu and Weixing Lin confirmed

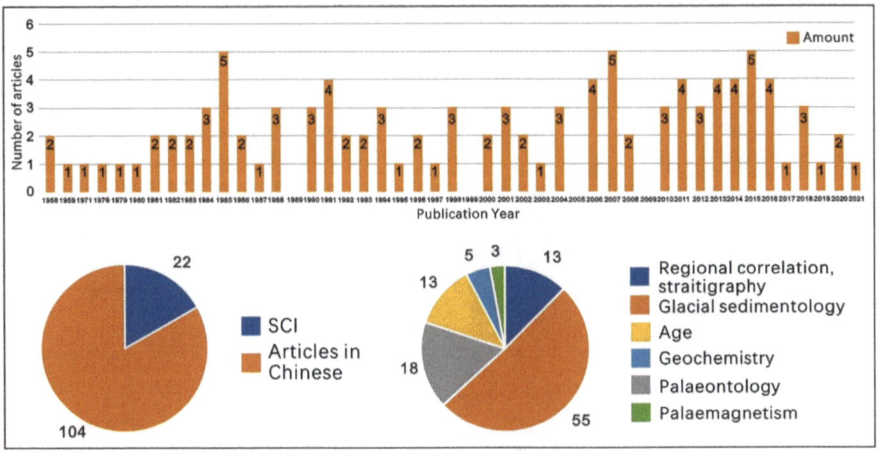

Fig. 10 Research history and related achievements of the Luoquan Formation (up to 2021)

the existence of the tillite bed, which they named the Luoquan Formation (Ye 1984). Since then, it has attracted broad attention at home and abroad (Guan et al. 1988), and has carried out in-depth research in glacial sedimentology, subglacial erosional structure, and glacial landform, detrital zircon chronology, and stratigraphic paleontology (Fig. 10).

From the 1970s to the 1980s, a large number of geologists studied the properties and characteristics of the diamictite of the Luoquan Formation. Generally, there are two view points. One thought the succession records non-glacial gravity flow deposition (Hong 1984) or subaqueous alluvial fan sedimentation (Li et al. 1990). Alternatively, some have emphasized turbidite deposition (Wang 1996); However, most geologists prefer a glacial origin. For example, gelogists Hongzhen Wang, Xiufu Qiao, Ziqiang Wang, Ruiji Cao, and Wenjie Zhao compared the Luoquan Formation with Nantuo Formation and regarded it as the deposits of a mountain glacier (Ye 1984); Wu and Guan (1988a, b) further put forward the concept of glacigene gravity flow, summarized five glacial sedimentary markers and the grain size law of the tillite of Luoquan Formation (Wu and Yang 1991), meanwhile, according to the soft sediments deformation structure in the diamictite, they further considered these deposits belong to periglacial sedimentary environment, and two types of diamictite are divided, in-situ massive diamictite and stratified diamictite of glacial fluvial origin (Yang and Wu 1994). Ge and Cui (1993) proposed that some cross-bedding structures support the viewpoint of glacial outwash deposits; Wang and Wang (2015) also identified and divided the types of ice raft and glacial lacustrine sedimentary facies through the characteristics of sediment fabric. It is worth mentioning that Mu (1983), Guan et al. (1986), Wu and Guan (1988a, b), Yang and Wu (1992) have all identified glacial pavements on the surface of the underlying bedrock in western Henan. A systematic study of the subglacial erosional forms further supports the glacial origin of the Luoquan Formation. In particular, based on the sedimentary characteristics of the Luoquan Formation in Yiyang, Luoyang, Henan Province and the development and distribution of the underlying basement strata, Wang and Yang (1996) believed that the diamictite of the Luoquan Formation was the product of continental glaciers, and further studied the erosion conditions of the basement by glaciers, and speculated the ancient landforms and glacier movement during the glaciation, laying a good foundation for subsequent research. Pioneers such as Baode Guan and Ruitang Wu took the lead in introducing China's Ediacaran glaciers to the world and made outstanding contributions to Henan's basic geological research.

In the first 10 years of the twenty-first century, there were only sporadic studies on Luoquan glacial deposits (Zhang et al. 2001, 2008). Only in the past 10 years has it set off a wave of research on Luoquan glaciers. Miao and Zhu (2014), Bai et al. (2018), Wang et al. (2021), and Pang et al. (2021) have made breakthroughs in paleontological research, helping to determine the age of the Luoquan Formation from the perspective of biostratigraphy; Hu et al. (2015) and Li et al. (2020) used detrital zircon data to propose that the Luoquan Formation should belong to the early or late Neoproterozoic; More progress is reflected in the in-depth study of glacial sedimentology and paleogeography (Zhou et al. 2010; Zheng and Liu 2012; Cai et al. 2013, 2014; Wang 2015; Wang and Wang 2015; Le Heron et al. 2018, 2019; Zheng et al. 2019; Zhang and Zheng 2019; Zhang 2019; Zhang et al. 2019; Chen et al. 2020; Yue et al. 2020; Zhang et al. 2021; Chen et al. 2023; Liu et al. 2023). Le Heron et al. (2018) further systematically studied the sedimentary succession, sedimentary facies, and environment of the Luoquan Formation in western Henan and eastern Shaanxi. At the same time, they also conducted a small-scale study of the local glacial pavements and subglacial erosional forms, re-studied and summarized the continental ice sheet, sedimentary facies, and environmental evolution model of the Luoquan Formation in the study area. Based on previous studies, Chen et al. (2020) also reinforced the interpretation of the Luoquan Formation as the deposits of a typical continental ice sheet and developed from in-situ massive, periglacial stratified diamictite and ice-distal glacial lacustrine, shallow seas, and ice rafts. In this model, the advance was from highlands in the north to a marine environment in the south. Recently, Zhang et al. (2021) filled a gap in the study of the glacial sedimentology of the Luoquan Formation in the Zhoucun section of Yiyang in the northwest of the study area and discussed the provenance and climate background of the Luoquan Formation based on geochemical and detrital zircon data. Liu et al. (2023) systematically reviewed the spatial distribution, stratigraphy and sedimentary environment and sedimentary facies of the Ediacaran diamcitites (later than 580 Ma) in northern China, and provided geological evidences for the global paleogeographic reconstruction and tectonic setting restore of the Ediacaran period. Based on the sedimentary characteristics and succession of eight sections of the Luoquan and Dongpo formations in western Henan (six of them are mentioned in this guidebook), Chen et al. (2023) confirmed the Luoquan glaciation was a large ice sheet that extended from land to the ocean. In addition, there are a small number of geochemical results (Zhang et al. 2008; Shen et al. 2010; Yang et al. 2013, 2019), which provided good materials and research results for this investigation.

3.3 Basic Characteristics of the Luoquan Formation and Associated Strata

3.3.1 Stratigraphy and Paleontology and Geochronology

Stratigraphy

Since the positions of the Huanglianduo Formation and Dongjia Formation in the stratigraphic chart have not yet been determined, the Ediacaran in Western Henan Province in this guide only includes the Luoquan Formation and the Dongpo Formation. As mentioned above, Luoquan Formation is in parallel or angular unconformity contact with several different strata of the underlying Mesoproterozoic basement, and above is in conformable contact with the sandstone or siltstone of the Dongpo Formation, or is directly covered by the Xinji Formation of Cambrian third stage.

The thickness of the strata of the Luoquan Formation varies greatly. Affected by the differential erosion of glaciers, later structural transformation, and preservation conditions, the thickness of the strata exposed varies from tens of meters to hundreds of meters. The sedimentary filling succession has obvious two-fold characteristics: the lower part is dominated by massive and weakly stratified diamictite, and its interior can be subdivided into multiple secondary cycles according to the rhythmic changes of gravel diameter and gravel content, interbedded with pebble-bearing sandstone and sandy conglomerate lenses; The upper part is dominated by dropstone-bearing siltstone, silty mudstone, or shale (Fig. 11).

The distribution range of the Dongpo Formation is smaller than that of the Luoquan Formation and can only be seen in the Lushan-Queshan area in the southern part of the investigation area. It was named in the section of Luoquan Village, Mangchuan Town, Ruzhou City, which refers to a set of siltstone and shale assemblages deposited above the Luoquan Formation without glacial character, and have the cement of glauconite, the bottom is interbedded lenticular calcareous quartz sandstone, pebbly shale, and glutenite, to show that it is different from the glacial deposits of the Luoquan Formation. The Dongpo Formation is famous for the large number of late Ediacaran standardized fossil-*Shannxilites* Sp. The Cambrian Xinji Formation above the Luoquan Formation or the Dongpo Formation has phosphorous sandstones in the lower part that can be compared with the world (Runnegar 1985; Brasier 1990; Li et al. 2016). The Xinji Formation develops horizontal cylindrical burrows (Miao and Zhu 2014), fossils developed at the bottom of the Xinji Formation, such as spicules, bivalves, trilobites, and other metazoans (He and Pei 1985; Pan et al. 2018), and a large number of hummocky cross-bedding and parallel bedding are developed in the middle part, while the upper part is argillaceous dolostone and marl, containing rock salt pseudocrystals and gypsum breccias. What is particularly rare is that in recent years, we have discovered large-scale and spectacular sand wedges at the interface between the Luoquan Formation and the Xinji Formation. It is likely to be a product of the melting process of post-glacial permafrost, adding new materials to decipher the evolution and properties of the Luoquan glacier (Wang et al. 2022).

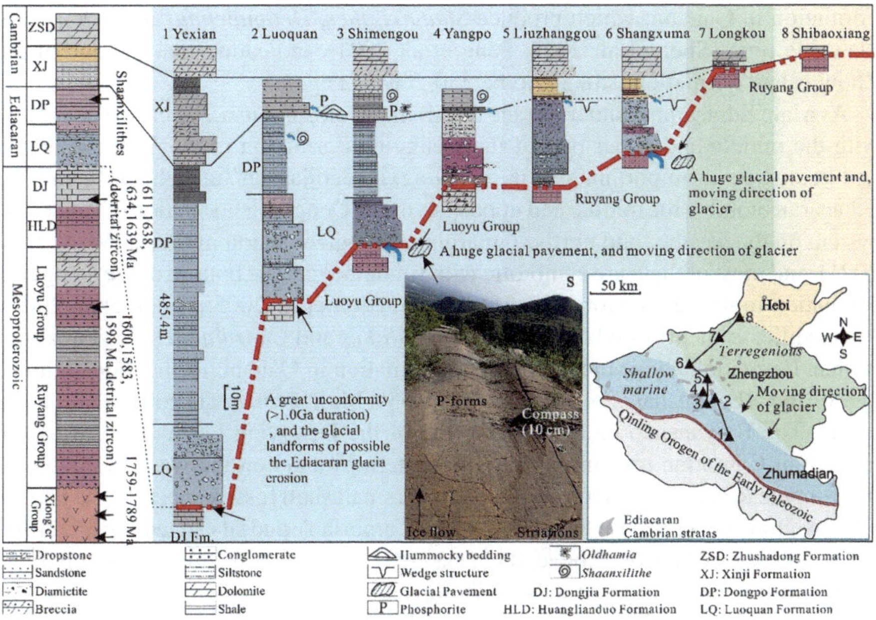

Fig. 11 Glacial successions and subglacial pavements with the varied subglacial erosional landforms, west Henan, North China

Paleontology

Although fossils in the Ediacaran-Cambrian transition diamictite from East Asia to Central Asia are generally rare, acritarchs are common. For example, the *Micrhystridium*-like spinose acritarchs preserved in the Dongpo Formation in West Henan can be classified into the acritarchs AHC assemblage (*Asteridium-Heliosphaeridium-Comasphaeridium*) of lower Meishucunian Stage originated the southern margin of the Yangtze Craton (Yin and Gao 1995; Yin and Guan 1999; Yao et al. 2005; Ahn and Zhu 2017), previous studies pointed out that the latter is also widely seen in the Lower Cambrian strata in Avalonia, South China, Spain, Tarim, India and South Australia (Högström et al. 2013; Ahn and Zhu 2017). At the same time, a series of trace fossils are developed in the coastal or lacustrine sediments above the Luoquan Formation (and equivalents) in WesternL487 Henan Province, such as vertical burrow *Skolithos* and vertical U-shaped burrow *Arenicolites*, oblique burrow system *Treptichnus*, horizontal burrow *Palaeophycus* and *planolites* indicate that the Dongpo Formation is not higher than the first trace fossil biozone of the Early Cambrian and is a Cambrian deposit later than the biozone of *Treptichnus* (Phycodes) pedum.

In contrast, *Shannxilithes* are more common in this layer, such as western Henan, Ningqiang in Shaanxi, Helan Mountain in Ningxia, Qaidam, etc., especially the shallow sea sediments of the Zhoujieshan Formation overlying the Hongtiegou

Formation in Qaidam, which produce *Shannxilithes*, *Helanoichnus* and *Charnia* at the same time (Shen et al. 2007; Pang et al. 2021), indicating that the age of the Zhoujieshan Formation is likely to be 550–539 Ma ago.

As a late Ediacaran standard (trace fossil) molecule, *Shannxilites* can be compared with the middle and upper part of the Dengying Formation in the Yangtze Craton and belong to an important specie of Xilingxia/Gaojiashan/Jiangchuan biota (i.e., Ediacaran biota) in the middle and upper part of the Dengying Formation of Ediacaran system on the western and northern margin of Yangtze Craton in China (Wang et al. 2021), and most of them are symbiotic with vendotaenia. The bottom of the Dengying Formation in the Three Gorges is 551 Ma (Condon et al. 2005; Zhang et al. 2005). Meanwhile, as mentioned before, the *Shannxilithes* and *Charnia* preserved in Zhoujieshan Formation above the Hongtiegou Formation in Qaidam further constrain the age of the Zhoujieshan Formation as 550–539 Ma (Shen et al. 2007; Pang et al. 2021). The Baiyanshao member at the top of the Dengying Formation during the Ediacaran-Cambrian transition period in eastern Yunnan on the western margin of the Yangtze River is rich in macroalgae fossils, small shell fossils appear at the bottom (Xiaoweitoushan section at the bottom of the Cambrian), and suspected *Shannxilithes* can still be seen upwards; Above, the measured U-Pb age of the tuff on the top of the main phosphorus-forming member (Zhongyicun member) is 538 Ma (Jenkins et al. 2002; Sawaki et al. 2008), and the measured U-Pb age of tuff under the initial layer of Chengjiang fauna adjacent to it is 518 Ma (Zhou et al. 2008), but this layer is still lower than Canglangpu Stage in Early Cambrian. At the same time, *Shannxilites* are also more common in late Ediacaran strata in India, Siberia, and China (Zhang et al. 2015).

The trilobite fossil *hsuapis* developed at the lowest layer of the Early Cambrian Xinji Formation overlying the Luoquan diamictite in western Henan indicates the top of the Second Stage of the Early Cambrian (the age of bottom of Second Stage is 521 Ma), which is much higher in the stratigraphic chart than the small shell fauna layer in eastern Yunnan, and the age of which is the middle and late Canglangpu Stage (Liu 1986). Therefore, the upper age limit for Luoquan Formation is 538 Ma.

Micrhystridium-like spinose acritarchs are preserved in the Luoquan Formation and Dongpo Formation above in west Henan, which belong to the *Asteridium–Heliosphaeridium–Comasphaeridium* (AHC) assemblage of lower Meishucunian Stage (Yin and Guan 1999; Ahn and Zhu 2017). The AHC assemblage is widely seen in Lower Cambrian in Canada Avalonia, India, and South Australia, Spain, Yangtze Craton, and Tarim (Ahn and Zhu 2017), therefore, the Luoquan Formation has extended to the Early Cambrian at the latest.

Geochronology

As mentioned earlier, the Luoquan Formation is unconformable on different strata of the Mesoproterozoic. The underlying strata include the Ma'anshan Formation of the Wufoshan Group (~1.8–1.7 Ga) and the Sanjiaotang Formation, Luoyukou Formation (~1.6 Ga), Cuizhuang Formation (~1.4 Ga) of the Luoyu Group (~1.8–1.6 Ga), the

Xunjiansi Formation and Fengjiawan Formation of Guandaokou Group (~1.4 Ga), along with the Baishugou Formation (~1.4 Ga), etc. A long phase of orogenesis and denudation thus occurred before deposition of the Luoquan Formation. For decades, scholars have carried out systematic isotope dating. The age data of the strata above and beneath the Luoquan Formation are as follows:

Age of strata beneath: Luoquan Formation is unconformable on underlying rocks, the youngest of which is the Dongjia Formation. The average K-Ar age of Glauconite in the lower Dongjia Formation is 650 ± 33 Ma (Xing 1989). The underlying volcanic rock of the Hankalchough Formation (equivalent to the Luoquan Formation) in Tarim is 615 Ma (Xu et al. 2009). In addition, beneath diamictite equivalent to Luoquan Formation at the bottom of Baiyanggou Formation of Ediacaran in the North Qilian Mountains, the zircon U-Pb SIMS age of basalt developed in shallow marine carbonate rocks and fine clastic rock series at the top of Zhulongguan Group is 583 ± 3 Ma (Xu et al. 2015), which further defines the minimum age limit of Luoquan Formation. Huang et al. (2023) reported ages of the youngest detrital zircons from Xichangjing Group of the Pochengshan area are 579 and 574 Ma.

Age of strata above: the overlying stratum of the Luoquan Formation is the Dongpo Formation, which is a set of marine greyish green siltstone and fine sandstone mixed with shale of non-glacial origin. The Rb-Sr isochron age of shale is 528 ± 23 Ma (Xing 1989).

Based on the detrital zircon population and the chronological significance of stratigraphic age constraints, meanwhile, combined with the age of tuff from the standard outcrop of the Ediacaran-Cambrian transition period of the Yangtze Craton, the depositional age of the Ediacaran-age diamictite developed in East Asia-Central Asia has roughly limited to 574–551 Ma (Fig. 12), which is later than the age of Gaskiers (580 Ma) in North American.

In recent years, many scholars (Hu et al. 2015; Zhang et al. 2016; Dong et al. 2017; Kang et al. 2018; Li et al. 2018) have systematically studied the detrital zircon population of the Luoquan Formation (and equivalents). The analytical results show that in the Zhengmuguan Formation diamictite or the Lower Cambrian clastic rocks in Helan Mountains, western North China Craton, the age of the youngest zircons in the detrital zircon population are 0.91–0.82 Ga (Dong et al. 2017); On the southern margin of the North China Craton, the youngest zircons from the lower Zhaoling Formation in Liquan, Xi'an are 0.85–0.82 Ga (Zhang et al. 2016), and the youngest zircons from the diamictite of the Luoquan Formation in Shimen, Luonan are 1.1–1.2 Ga (Hu et al. 2015), the youngest zircons of black diamictite of the Fengtai Formation in Huoqiu, Huainan are 0.92–0.87 Ga (Li et al. 2018); The youngest zircons of diamictite of Luoquan Formation in Luonan and Pirangdong Formation in North Korea are 0.88 Ga (Hu et al. 2015), so it can only indicate that the Luoquan Formation in North China Craton and equivalents are younger than 1.0–0.8 Ga. However, there is conclusive geological evidence that the North China Craton lacks Cryogenian and Ediacaran (Lu et al. 2008), and the volcanic rocks under the diamictite of Hankalchough Formation (the equivalent of Luoquan Formation) in Tarim are 615 Ma (Xu et al. 2009), which excludes the possibility that they belong to Cryogenian. Thus, according to the results of the detrital zircon analysis, it is inferred

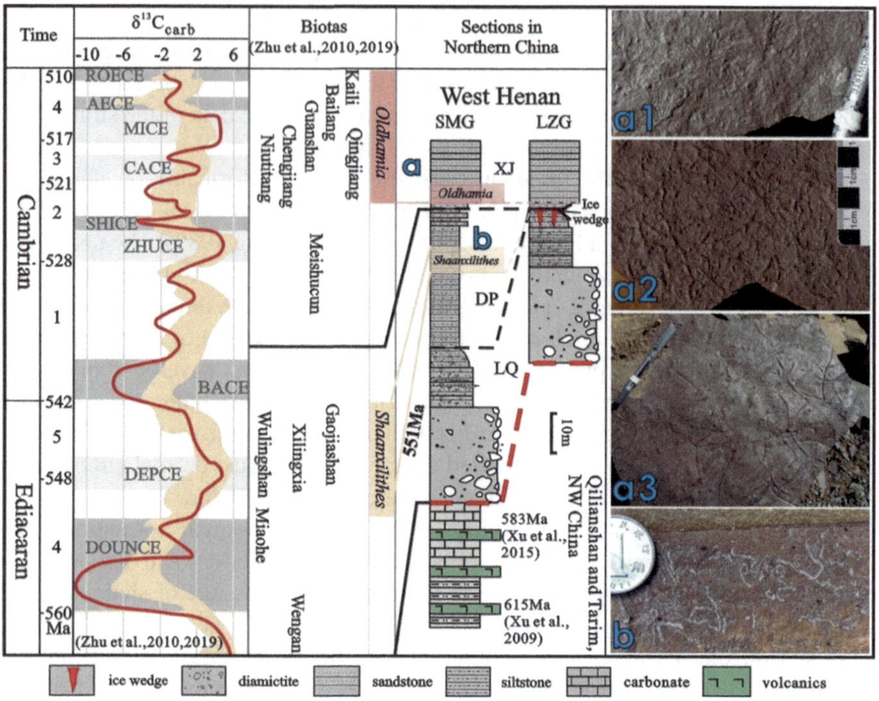

Fig. 12 Comprehensive chart of stratigraphy, biology, and δ^{13}C carbon isotope curve of Ediacaran-Cambrian transition period in West Henan Province (Modified from Zhu 2010; Zhu et al. 2019; panels a1–a3 and b are trace fossils, correspond to the place alongside the log of Shimengou section). Abbreviations: SMG—Shimengou section; LZG—Liuzhanggou section; XJ—Xinji Formation; DP—Dongpo Formation; LQ—Luoquan Formation

that the maximum age limit of the Luoquan Formation is between Ediacaran and the lower Cambrian. The lower part of the log of Shimenggou section is combined from sections from Qilian Mountain and Tarim in Northwest China, along with the ages of volcanics (583 Ma, Xu et al. 2015; 615 Ma, Xu et al. 2009), which further constrained the lowert limit of Luoquan glaciation.

It can be concluded that the chronological framework of the Luoquan Formation in North China is, the maximum age limit >574 Ma and the minimum age limit <551 Ma at the top, which may not only cross the Ediacaran-Cambrian boundary but also significantly later than the Gaskiers glaciation in North America. At the same time, it indicates that there may be two glacial events (~580 Ma, ~551 Ma) in the global Ediacaran-Cambrian transition period, which basically coincides with the development law of δ^{13}C negative drift events during this period. It is also very clear that the duration of the Luoquan Formation (~33 Ma) is significantly shorter than that of the Cryogenian (Sturtian and Marinoan combined).

3.3.2 The Glacial Sedimentary and Erosional Landscapes of Luoquan Formation

Characteristics of Luoquan Formation Diamictite

(1) Rock types: The lithofacies assemblages of the Luoquan Formation are mainly consisting of massive/weakly stratified diamictite, pebbly sandstone, lenticular sandy conglomerate, siltstone, and dropstone-bearing siltstone. Sometimes dropstone-bearing siltstone has a higher content and generally appears in the upper or middle part of the succession, reflecting the characteristics of subaqueous sedimentation of the upper or middle part of the Luoquan Formation.
(2) Fabric: Among the sections of the Luoquan Formation in the investigation area, except for Yangpo and Zhoucun sections where the gravel content of the diamictite can reach 40–50%, the gravel content in the remaining sections is generally relatively low. The gravel composition is mainly sandstone, quartzite, siltstone, and carbonate rock, followed by chert and dolomitic limestone, subangular to subrounded, the composition of gravel generally depends on the extent of erosion of the underlying Mesoproterozoic basement by a glacier. For example, in Yangpo, the massive diamictite at the bottom of the Luoquan Formation is dominated by dolomite and sandstone, the Luoquan Formation is unconformable above the Luoyukou Formation stromatolite-laminated dolomites and argillaceous sandy dolomites. The gravels of Luoquan Formation in Shimengou and Gandanchang are mainly sandstone and quartzite, and are unconformable with the underlying sandstone of the Sanjiaotang Formation of Ruyang Group and the sandstone of Ma'anshan Formation of Luoyu Group. At the junction with Shaanxi and Henan Provinces, such as Zhuyang Town, Lingbao County, the gravels are almost entirely dolomite. In Zhoucun and Shangxuma sections, gravels show some imbrication.
(3) Sedimentary structures: The gravels in all sections have obvious glacial characteristics, ① various diameters and density of gravels; ② abundant striae on the surface; ③ many gravels are in the shape of an electric iron (someone called a slipper-like shape), that is, they have a flat and smooth bottom surface, and saddle-shaped, bullet-shaped, faceted forms that reflect the characteristics of subglacial erosion, as well as pits, cracks, and crescentic fractures; ④ the surface of gravel has a calcareous film, which has a secondary cause, that is, after the glacier melts, the calcium carbonate in the dissolved water precipitates on the surface of the gravel. It is a common phenomenon commonly seen in diamictite and is found in glacial strata around the world; The cement of the diamictite is closely related to the lithology of the underlying strata. Generally, quartz sandstone is the main lithology, followed by carbonate rock; ⑤ dropstone-bearing siltstone constitutes another major sedimentary facies of the Luoquan Formation-ice raft deposition. The lithology of dropstones is diverse, depending on the type of debris carried by the glacier. The range of dropstone diameter also varies widely, ranging from a few centimeters of fine gravels to

Fig. 13 The micro- and meso-scale erosional forms developed on Pavement #1 in the Shimengou section (Chen et al. 2020). **a** Part of the pavement #1 in Shimengou section; **b** Diamictite lodged; **c** Bedrock steps and crossed striae; **d** Glacial cleavage; **e** p forms and striae; **f** Cross striae; **g** Crescentic cracks on the pavement; **h** p-forms and fractures; **i** Cross-cutting striae; **j** Crescentic fractures; **k** Crescentic gouges

dozens of centimeters of boulders. The "upward bending and downward winding" laminae characteristics are developed or not obvious, but the dropstones always exist in the form of a downward gravity center along the vertical line.

Subglacial Erosional Landscapes Beneath the Luoquan Formation

A large number of exquisite glacial erosional landscapes and subglacial erosional forms developed during the Luoquan glaciation, including fossil glacial pavements which are rare in the last glacial period of the Precambrian, have been preserved in the northern and southern sections of the investigation area. In addition, a series of well-preserved meso-and micro-scale subglacial erosional structures can be seen on the surface of pavements (Fig. 13), which are described as follows.

Micro-scale erosional forms:

(1) Striae: The most common structure on the surface of the Mesoproterozoic quartz sandstone and quartzite below the Luoquan Formation, a few millimeters to a few centimeters wide, generally not far extended, and occasionally 4–5 m long. Striae generally appear densely and parallel, and two sets of crossed striae are also very common, and sometimes it can be seen that the striae cut deep into the lodgement tills on the surface of the pavement, indicating two or more periods of glacier movements. On the cross section, the striae are arc-shaped and gradually become shallower toward the direction of the glacier flow, indicating the gradual blunting of the asperities carried under beneath the glacier. As a type of striae, nail head striae are micro-scale and reliable movement indicators, but because of their small scale, they become blurred after weathering erosion. The scale of the nail head can be as small as centimeter-level, or tens of centimeters long. The deep and wide pits in the direction of the nail head represent the sudden cessation of asperities erosion at the bottom of the glacier; In addition, hairpin striae are often developed. The morphology is crescent at the hairpin head, and straight deep striations are on both sides.

(2) Grooves and ridges: A large, deeply abraded group of striae and the bulges between them, up to several centimeters to more than ten centimeters wide and several centimeters deep. Crest-trough patterns are shown on the cross-section, reflecting the large-size asperities carried at the bottom of the glacier, which are eroded on the basement rocks in the form of ploughing under the lubrication of glacier meltwater.

(3) Crescentic fractures and gouges: Crescentic fractures are fractures that appear regularly in groups due to the fracturing of the subglacial bedrock. The width is generally more than 10 cm, and the length is more than ten to tens of centimeters. It is composed of multiple fractures that are closely stacked. Crescentic gouges are horseshoe-shaped pits with an average length of 20–60 cm and a depth of 1–2 cm. The interval is slightly larger and the stacking is not close. The size can also be larger than the crescentic fractures, and there are dense striae around them. Both crescentic fractures and gouges are good signs indicating glacier movement.

(4) The term P-forms (plastically moulded forms): The term P-forms was introduced by Dahl (1965) to describe smoothed depressions eroded into bedrock, which consisted of meltwater or debris-rich basal till at the bottom of a fast-flowing glacier. They are produced in various forms, such as horizontal shell-shaped grooves (*Muschelbruch*), longitudinal spindle flutes (Spindle flute), and ditch-shaped grooves (*Cavetto*) developed on the glacial pavements in the Shimengou section, and are accompanied by abundant striations. By analogy with the erosional landforms of modern glaciers, P-forms are generally formed in the subglacial environment of temperate or cold-temperature polythermal glaciers (Hambrey and Glasser 2012).

Meso-scale erosional forms

(5) Roches moutonnées: When micro-scale erosional forms, such as striae, can only record very local ice flow patterns, meso-scale erosional forms can more clearly reflect the relationship between glaciers and bedrocks. This is because at this time changes in geomorphology and ice flow patterns can be observed more clearly and easily. Different from the micro-scale erosional forms that are eroded downwards, this kind of medium-sized ice erosion structure is dominant in general and partially streamlined. Roches moutonnées developed in the Gandanchang section has striae consistent with the long axis direction on the polished pavements, and horizontally developed joints can be seen at the end.

(6) Bedrock steps: They can be seen on the pavements of the Shimengou and Gandanchang sections. They are in the shape of descending steps. The height of each step is 15–30 cm, and generally 3–4 steps appear continuously. The edge where the step intersects with the interface is sometimes embodied as a perfect circle with striae developed, and the direction is orthogonal to the edge of steps.

Striae generally widen and deepen toward the forward direction of the glacier, which are formed when power is reduced or gravels are unloaded. However, the shape, depth, width, and narrowness of the striae are related to the erosion strength of the overlying ice. Nonetheless, when it is unclear whether the erosion intensity of the glacier will increase or decrease, the use of the striae may not be able to accurately determine the direction of the glacier movement; meanwhile, there is still ambiguity in judging the direction of movement by striae. Therefore, when only striae are not sufficient as proof for judging the direction of glacier movement, the other erosion forms can be used as auxiliary evidence to infer the direction of ancient ice flows.

For example, the horns of crescentic fractures point down-ice, whereas those of crescentic gouges point up-ice; the shape of P-forms can also be used to judge the direction of glacier movement, with sharp convex-upflow rims and indistinct downflow margins; roches moutonnées is a streamlined structure formed on the erosional bedrock under the glacier, with asymmetric characteristics towards the glacier, the extension direction of the long axis is parallel to the movement direction of the glacier, and generally, the stoss face forms an acute angle with the bedrock, down-ice lee faces reflect the sharp section formed by the quarry of the glacier, and nearly vertical to the bedrock; the climbing direction of the bedrock steps also indicates the flow direction of the glacier.

In summary, the range of subglacial bedforms allows insight into not only ice flow direction, but allows insight into subglacial conditions.

Glacial Sedimentary Evolution of Luoquan Formation

Diamictite-bearing deposits, age equivalent to the Luoquan Formation in North China, are widely distributed across large areas of Southeast-Central Asia. The glacial sedimentary succession shows a perfect and complete sedimentary stratigraphic framework from land to glacial lake and even to the shallow sea. Owing to the southward movement of the Luoquan ice mass suggested by the erosional forms (Le Heron et al. 2019; Chen et al. 2020), the basement rocks eroded most severely in the north. As a result, the lower part of the Luoquan Formation is mainly composed of basal tillite and represented by massive diamictites directly piled above the pavements, erratic cobbles also appear (e.g., Shangxuma, Le Heron et al. 2018), probably representing lodgement tills related to glacial advance and meltout tills formed by subsequent glacial retreat. However, in sections with no pavements preserved, such as the Zhoucun section in the north, the Luoquan Formation is directly unconformable on the Mesoproterozoic basement strata, indicating that the lower Luoquan Formation could be periglacial outwash plain deposits, represented by massive/stratified diamictite, and intercalated dropstone-bearing siltstone (e.g., Luoquan village). The upper Luoquan Formation is composed of purplish-red dropstone-bearing siltstone or silty mudstone, conglomerate or sandy lenses can be found in the upper Luoquan Formation in many sections. They are interbedded with purplish-red fine sandy diamictite, interpreted as the product of meltwater channels in a proglacial environment (e.g., Yangpo). In addition, there are sand wedges whose strike is nearly east-west formed at the bottom of Xinji Formation. They directly intrude into the underlying Luoquan Formation, and are interpreted as the fills to ice-wedge casts, representing the permafrost zone in a periglacial environment. In contrast, in the Jiunvdong section in the south, and further southeast in Huoqiu, Anhui, ice-rafted deposits in laminated fine clastic rocks occur on the top of the Luoquan Formation (or equivalent Madian Formation or Fengtai Formation), which suggest that the glacier had advanced in the marine-terrestrial transition zone. That is, the ice sheet reached the shoreline, perhaps extending seaward as an ice shelf. The edges of ice shelves then broke up to form icebergs drifting out into the open sea to produce dropstones. Finally, the extensive glacial retreat-advance process resulted in the deposition of the glauconite-bearing sandstone and siltstone of the Dongpo Formation representing normal shallow coastal deposits above the Luoquan Formation in the investigation area (Fig. 14).

In general, the vertical sedimentary sequence of Luoquan Formation diamictites in different paleogeographic positions exhibit spatiotemporal development characteristics as follows: from the north part of the research area or the base of the sequence, there are massive or stratified diamictites with the sandy-gravel lenses. Upwards they gradually transit into laminated fine sandstone or siltstone rich in dropstones which vary in the grain size and composition, representing the typical glacial lacustrine facies in a periglacial environment. To the top of sequence or south part of the research area, they evolve into siltstone, silty shale, or mudstone, with or without dropstones, indicating an obvious open shallow sea environment. Also, the glacial sedimentary sequence indirectly reflects the paleogeographic shift of the glacial environment from continent to sea, suggesting the process that continental glacier advancing towards the sea, and gradual retreatment of continental glacier subsequently, as well as the sedimentary process. Therefore, the evidences above strongly support that there existed a giant ice sheet during the Luoquan age in Western Henan Province and even in North China.

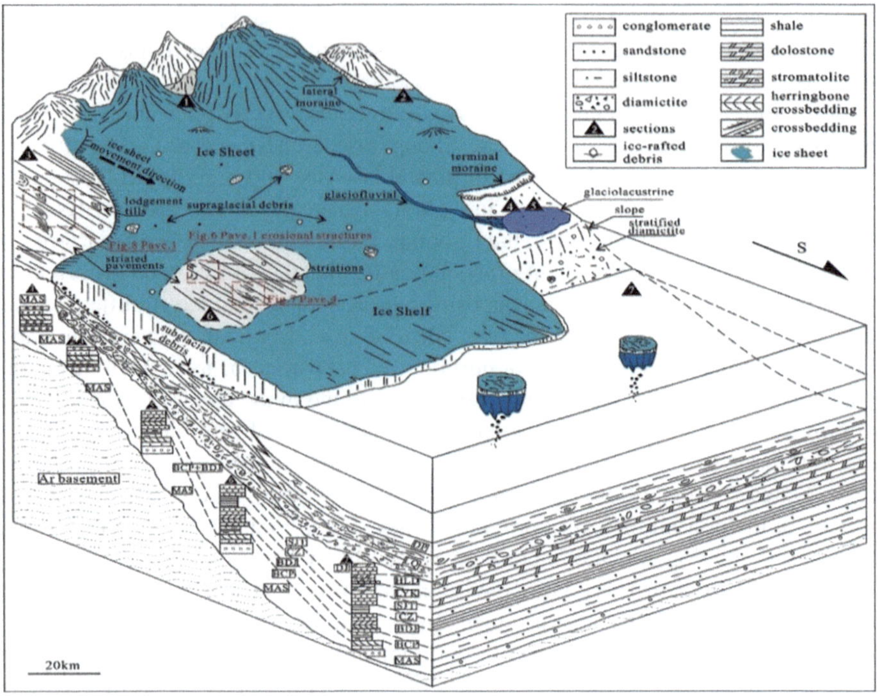

Fig. 14 A reconstruction of the ice sheet developed during the Luoquan period in west Henan, North China. The descriptions of certain pavements are shown in the previous figures (Chen et al. 2020)

3.3.3 Regional Correlation of the Luoquan Formation

In addition to the Luoquan Formation which is widely developed in western Henan, diamictite in the same layer as the Luoquan Formation, with the same sedimentary characteristics and ages are also found in the neighboring Luonan of Shanxi Province, Yongji of Shaanxi Province, and Fengtai of Anhui Province, as well as the Kuruktag of Xinjiang Uygur Autonomous Region, Qaidam of Qinghai Province, and Helan Mountain of Ningxia Hui Autonomous Region in northwestern China. In recent years, many scholars have conducted systematic research and discussion on this topic, most scholars basically hold the view of the glacial origin, except for a few individual objections. Zhang and Tang (1982), Wang et al. (1998) believe that the diamictite in the Shangzhangwan of Luonan, Shaanxi Province is the product of the joint action of glaciers and melting ice water, and the pebbly mudstone with clear and colorful laminae is glacial lacustrine deposits; Zheng and Dou (1980) compared the Fengtai Formation in Huainan and the Luoquan Formation in Western Henan Province, and believed that they belonged to the mountain glacial debris flow in continental rift and offshore ice-rafting. Meanwhile, there are multiple layers of small shell fossils at the bottom of the overlying Cambrian phosphate rock series, such as *Purella* sp., *Cassidina* sp., *Xianfengia* sp., *Spinovennia* sp., *Aanabarella* sp., *Hvolithellas* sp., *Chancelloria* sp., *Allonnia* sp. Li et al. (2018) also based on the characteristics of the element geochemistry and detrital zircon ages of the Fengtai Formation in Huoqiu, Huainan, Anhui Province and proposed that the glacial period had extended to the early Cambrian.

In contrast, Wang and Wang (1993) interpreted the Fengtai Formation diamictite as not only bears glacial characteristics, but also has the characteristics of debris flow and slump deposits. Cao et al. (2006a) not only classified the silty mudstone of Fengtai Formation in Huainan as a Cryogenian product according to the Pb–Pb age (620 Ma), but also pointed out that the development of phosphorus nodules in Fengtai Formation represents a warm climate environment, not an ice age (Cao et al. 2006b). Zhou et al. (2010) also speculated that the "gravel-bearing sandy oolitic dolomite" of Luoquan Formation in Shuiyu, Shanxi Province is the product of transgression or modified by transgression under warm climate, which has nothing to do with glaciation.

Research on the diamictite in northwestern China has also yielded fruitful results. First of all, many scholars consider that equivalent diamictite to the Luoquan Formation in Tarim, Qaidam, and Helan Mountain in Ningxia Hui Autonomous Region is also of glacial origin (Zhang et al. 1979, 1987; Gu 1982; Yang and Zheng 1985; Zhang 1991; Xiao et al. 2004; Zhang et al. 2006; Yang et al. 2007; Shen et al. 2010, 2011; Li et al. 2011; Zhu and Wang 2011; Cai et al. 2013, 2014; Yang et al. 2013); However, although Xiao et al. (2004) considered the diamictite (equivalent to the Luoquan Formation) of the Hankalchough Formation to be of glacial origin, and cap carbonate rocks are also developed with low or extremely low negative values (−4.5 to −6‰ or −12 to −15‰), but the age can only be limited to later than the Marinoan ice age. Shen et al(2010) believed that the diamictite of the Hongtiegou Formation in Qaidam (equivalent to the Luoquan Formation) also developed cap carbonate rock with a thickness of 8 m, but it was later than the Gaskiers glacier (580 Ma) in Newfoundland, Canada, belongs to the late Ediacaran (glacial period) diamictite, and may be compared with the mid-Ediacaran Shuram event (~570 Ma) in Oman and other places. However, there are still many scholars who insist on correlating the Luoquan Formation (and equivalent sediments) with the Gaskiers glacial products in North America (Gao et al. 2010, 2013).

As mentioned above, not only the southern, western, and northwestern margins of the North China Craton, but also the Neoproterozoic Pirangdong Formation (equivalent to the Luoquan Formation) in the Korean Peninsula exposes glacial diamictite belonging to the Ediacaran-Cambrian transition period (Hu et al. 2012b, 2015; Kim et al. 2016). In addition, Kyrgyzstan and Kazakhstan of Central Asia adjacent to the Tarim Craton in northwestern China also developed Ediacaran-Cambrian transition period diamictite, which is locally called the Baykonur glaciation (Chumakov et al. 2011). The glacial outwash sedimentary stratigraphic succession, contact relationship with the upper and lower strata, and rock assemblage characteristics of the Baykonur diamictite can be completely compared with the Luoquan Formation in China, it also shows the characteristics of deposition from a large continental ice sheet, and the spatial distributions of diamictite are wide, and the facies types are abundant and typical.

4 Itinerary

The excursion lasts for 5 days, including Shangxuma, Yiyang in Luoyang city, Shimengou, Yangpo, Luoquan village, and Liuzhanggou in Ruzhou city. The strata involved include Mesoproterozoic Ruyang Group, Luoyu Group-Ediacaran Luoquan-Dongpo-Cambrian Xinji formations, Zhushadong Formation, etc. By inspecting the petrological, sedimentological, and paleontological characteristics of Mesoproterozoic-Neoproterozoic-Cambrian strata, especially the characteristics of Luoquan glacial deposits, we will discuss the stratigraphic succession, sedimentary environment, paleontological characteristics, and chronology of the Luoquan-Dongpo-Cambrian Xinji formations. The scale, nature, and paleogeographic significance of ice masses over the Ediacaran-Cambrian transition will also be discussed.

4.1 DAY ONE: August 27th Route 1 Shangxuma Section

After the journey by taking G653 (7:44 a.m.–11:53 a.m.) (or G671, 8:13 a.m.–12:35 a.m.) from Beijing West Railway Station, we'll arrive at Luoyang Longmen Railway Station at noon, and then go to the Longmen Yijun hotel for lunch and check-in. After a short rest, the first field trip to the Shangxuma section will start at 3:00 p.m. Luoyang city was the capital of the Six Dynasties in ancient China, with a long history and brilliant culture. Particularly, the Buddha Grottoes in Longmen, throughout more than 1500 years history, are the treasures of ancient Chinese stone carving art, which records the development and changes in many fields, such as politics, economy, religion, and culture in ancient China. It is known as one of the four major Grottoes in China.

- **Key word: Glacier accompanied by Buddha Grottoes for thousands of years**

The Shangxuma section starts at 112° 33′ 32″, 34° 31′ 34.72″, and ends at 112° 33′ 30″, 34° 31′ 34″. This section is only about 15 km away from Luoyang Longmen, about 20 min by car (Fig. 15). After lunch, driving southeast to the section at Shangxuma village in Luoyang, the section is just located along the road. At the parking place, one pavilion sits above the hill. The first thing we will see is the grey-white massive thick layered limestone of the Cambrian Zhushadong Formation across the ridge of the mountain, which extends for several kilometers. The local villagers call it the "dragon vein" of Wolonggang (in Chinese means a dragon is circling on the hill). At the foot of the hill, the section is exposed along the road, which is very convenient to reach.

Fig. 15 Route map from the Luoyang Longmen Station to hotel to Shangxuma section (©Google Map)

The glacial pavements and subglacial erosional forms developed on the top of quartz sandstone of the Mesoproterozoic Ma'anshan Formation are the main tasks for observation. Due to the influence of early road construction, the integrity of glacial pavement is damaged and only exposed on one side of the road on a small scale, which has been protected as a geological relic. In addition, another important observation content lies in the sedimentary succession, sedimentary characteristics, and sedimentary structures of the Ediacaran Luoquan-Cambrian Xinji-Zhushadong formations (Figs. 16 and 17).

Fig. 16 Panoramic photo of Shangxuma section. Red line indicates the upper Xinji Formation and the lower Luoquan Formation

Stop 1: Glacial pavement at the top of the Mesoproterozoic Ma'anshan Formation

Layer: Layer 0 of Shangxuma section, quartz sandstone of the Mesoproterozoic Ma'anshan Formation (Figs. 17 and 18a)

Detailed description: Under the Luoquan Formation, the glacial pavement is developed on the top of the quartz sandstone of the Mesoproterozoic Ma'anshan Formation. The occurrence of the Ma'anshan Formation is 326°∠16° (Fig. 17). The unconformity relationship between the bottom of the Luoquan Formation and the glacial pavement can be observed (Fig. 18b). Various types of striae, such as nail head striae and hairpin striae, display two different groups of extension directions. The strike of striae ranges from 110° to 160°, indicating the movement of glacier is in a north–south direction. Besides, the intersection relationship between the striations and the lodgement till which developed on the surface of the glacial pavements (Fig. 18a, c, f).

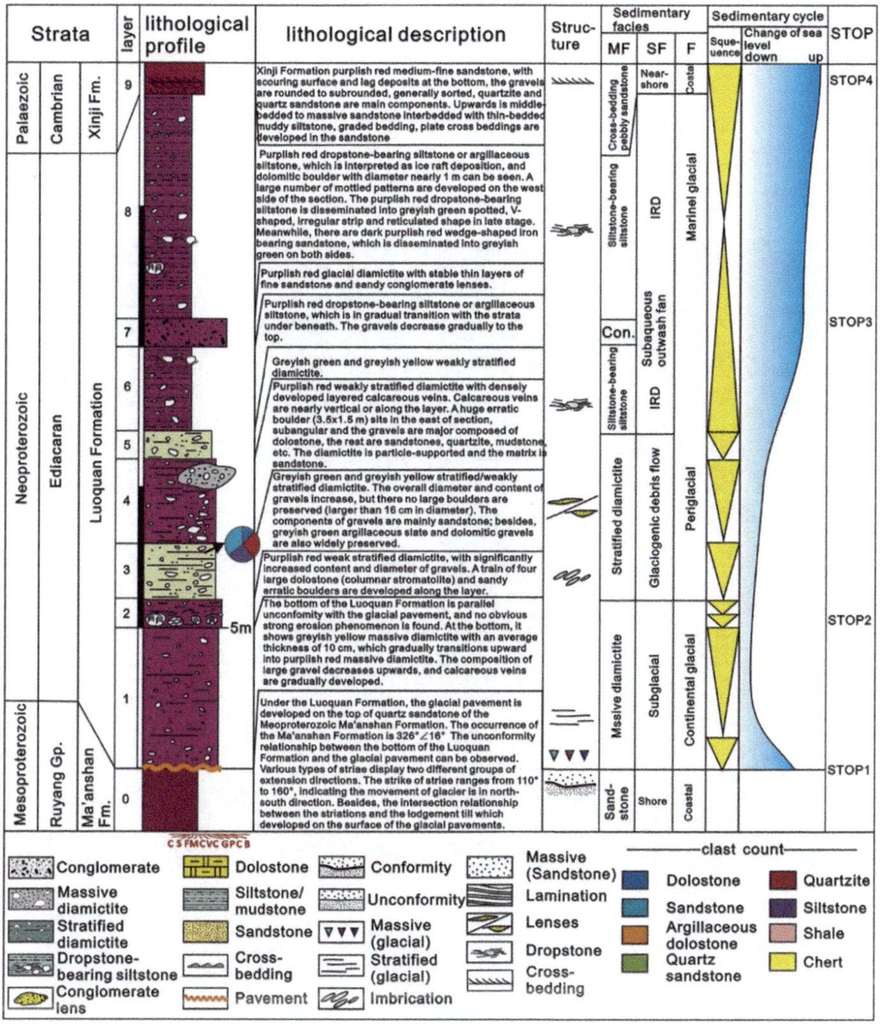

Fig. 17 Comprehensive stratigraphic data from the Shangxuma section

Grooves and ridges are the most common bedforms during the processes of glacier transportation and erosion, they are tightly associated and show wavy fluctuation laterally. Obvious striations are also developed inner the groove (Fig. 18d, e) with strike ranges from 145° to 151°, which are consistent with the main strike of the striations on the glacial pavement. The R shear (Riedel shear) effect of the glacier on the basement leads to a high-low step fluctuation on the glacial pavements (Fig. 18g).

Fig. 18 The striated pavement and associated structures at Shangxuma (Le Heron et al. 2018). **a** Scale drawing of the striated pavement together with measurements of 72 striations. Note the relationship between apparently crosscutting striation sets on the scale drawing. **b** Contact between striated pavement and diamictites lying directly on the contact. **c** View of an area of cross-cutting striae: a simple interpretation, without considering the wider geomorphic evidence, would be that these represent two phases of subglacial abrasion in central China. **d** and **e** Two views of the curvilinear nature of some striations at the outcrop scale, as shown in **a**. **f** Striated pavement draped by diamictite. Some striations are also developed in the very basal deposits (i.e., over a few mm) in this diamictite. **g** Organization of the striated pavement into an upper and a lower step, as also shown in **a**

Stop 2: Glacial diamictite in the middle and lower part of the Luoquan Formation

Layers: Layers 1–7 of the Ediacaran Luoquan Formation (Fig. 17)

Detailed description: Stop 2 focuses on the sedimentary characteristics of the middle and lower part of the Luoquan Formation, which are introduced as follows:

Layer 1, the bottom of the Luoquan Formation is parallel unconformity with the glacial pavement, and no obvious strong erosion phenomenon is found. At the bottom, it shows greyish-yellow massive diamictite with an average thickness of 10 cm, which gradually transitions upward into purplish-red massive diamictite. The composition of large gravel decreases upwards, and calcareous veins are gradually developed (Fig. 18b).

Layer 2, purplish-red weak stratified diamictite, with significantly increased content and diameter of gravels. A train of four large dolostones (columnar stromatolite) and sandy erratic boulders are developed along with the layer (Fig. 19a).

Layer 3, greyish green and greyish-yellow stratified/weakly stratified diamictite. The overall diameter and content of gravels increase, but they're no large boulders are preserved (larger than 16 cm in diameter). The components of gravels are mainly sandstone; besides, greyish green argillaceous slate and dolomitic gravels are also widely preserved (Fig. 19b).

Layer 4, purplish-red weakly stratified diamictite with densely developed layered calcareous veins. Calcareous veins are nearly vertical or along with the layer. A huge erratic boulder (3.5×1.5 m) sits in the east of the section, subangular and the gravels are major composed of dolostone, the rest are sandstones, quartzite, mudstone, etc. The diamictite is particle-supported and the matrix is sandstone (Fig. 19c).

Layer 5, greyish green and greyish-yellow weakly stratified diamictite (Fig. 19d).

Layer 6, purplish-red dropstone-bearing siltstone or argillaceous siltstone, which is in gradual transition with the strata under beneath. The gravels decrease gradually to the top (Fig. 19e).

Layer 7, purplish-red glacial diamictite with stable thin layers of fine sandstone and sandy conglomerate lenses (Fig. 19f).

Stop 3: Sedimentary characteristics of the upper Luoquan Formation-purplish red dropstone-bearing siltstone

Layer: layer 8 of the Luoquan Formation (Fig. 17)

Detailed description: This stop focuses on the upper part of the Luoquan Formation, as described below.

Layer 8, purplish-red dropstone-bearing siltstone or argillaceous siltstone, which is interpreted as ice raft deposition, and dolomitic boulder with a diameter nearly 1 m can be seen. A large number of mottled patterns are developed on the west side of the section. The purplish-red dropstone-bearing siltstone is disseminated into greyish green spotted, V-shaped, irregular strip and reticulated shape in the late stage. Meanwhile, there is dark purplish-red wedge-shaped iron-bearing sandstone, which is disseminated into the greyish green on both sides (Fig. 20a–c).

Fig. 19 Plates of stops 1–2 in Shangxuma section. **a** layer 2, purplish-red weak stratified diamictite, a train of four large dolostones (columnar stromatolite) and sandy erratic boulders are developed along with the layer; **b** layer 3, greyish green and greyish-yellow weakly stratified diamictite; **c** layer 4, purplish-red weakly stratified diamictite with densely developed layered calcareous veins. A huge erratic boulder sits in the east of section; **d** layer 5, greyish green and greyish-yellow weakly stratified diamictite; **e** layer 6, purplish-red dropstone-bearing siltstone or argillaceous siltstone; **f** layer 7, purplish-red glacial diamictite with stable thin layers of fine sandstone and sandy conglomerate lenses

Stop 4: Sedimentary characteristics of the top Luoquan Formation and the boundary with the Cambrian Xinji Formation

Layers: the boundary between Luoquan Formation (layer 8) and the Xinji Formation (layer 9) (Fig. 17)

Detailed description: This stop focuses on the top of the Luoquan Formation, the boundary with Xinji Formation, and sedimentary characteristics, as described below.

Fig. 20 Plates of stops 3–4 in Shangxuma section. **a–c** Layer 8, purplish-red dropstone-bearing siltstone. A large number of mottled patterns are developed on the west side of the section; **d** The relationship between the Luoquan Formation and Xinji Formation is sharp parallel unconformity contact. The top of the Luoquan Formation is purplish-red dropstone-bearing siltstone or argillaceous siltstone. The based part of Xinji Formation is purplish-red medium-fine sandstone, the bottom is scouring surface, with gravel retention deposition; The Xinji Formation are inter-bedded by purplish-red middle-layer massive sandstone and thin-layer mudstone siltstone, developed graded bedding and plate cross-bedding, etc.

Layer 8, purplish-red dropstone-bearing siltstone or argillaceous siltstone, shows a sharp boundary with overlying Xinji Formation sandstone. Most of the gravels are pebbles, which are generally sorted (Fig. 20d).

Layer 9, purplish-red medium-fine sandstone of Xinji Formation, with scouring surface and lag deposits at the bottom, the gravels are rounded to subrounded, generally sorted, quartzite and quartz sandstone are main components. Upwards are middle-bedded to massive sandstone interbedded with thin-bedded muddy siltstone, graded bedding, plate cross beddings are developed in the sandstone (Fig. 20e). The contact relationship between Luoquan Formation and Xinji Formation is a parallel unconformity.

4.2 *DAY TWO: August 28th Route 2 Yiyang Zhoucun Section*

- **Key word: Ice flowing under the windmills**

7:00–8:00 Breakfast at 7:00 a.m. then check out.

8:00–9:30 Heading for the Zhoucun section at Yiyang county by 8:00 a.m., takes 1.5 h to reach the section.

9:30–11:30 Yiyang Zhoucun section observation (Fig. 21). The section locates at the back hill between the Zhoucun village and Miaocun village in east Yiyang county. A spectacular group of wind turbines has been built on the top of the mountain, which greatly improves the accessibility of the road. The section starts at 112° 16′ 47.17″, 34° 31′ 34.72″, and ends at 112° 16′ 49.42″, 34° 31′ 32.33″. Drive westward to the entrance of Zhoucun village, climb the hill along the path, and the contact relationship between the shale of Cuizhuang Formation, Luoyu Group, and the diamictite the basal Luoquan Formation can be observed, mainly focusing on the sedimentary characteristics of the shale of Mesoproterozoic Cuizhuang Formation-Ediacaran Luoquan Formation and Cambrian Xinji Formation (Fig. 22).

11:30–12:30 We can go up the hill directly along the path, or drive up the hill northward to the wind turbine group on the top of the hill for lunch. The scenery here is spectacular, with an impressive bird's-eye view of the Luohe River and the complete Yiyang section.

12:30–14:30 After lunch, we go down the road, and observe the sedimentary succession of Luoquan Formation diamictite-Xinji Formation, the color change and exposure genetic structure within Luoquan Formation, and the contact relationship between Luoquan Formation and the underlying Beidajian Formation sandstone of Mesoproterozoic basement Ruyang Group along the north hillside road.

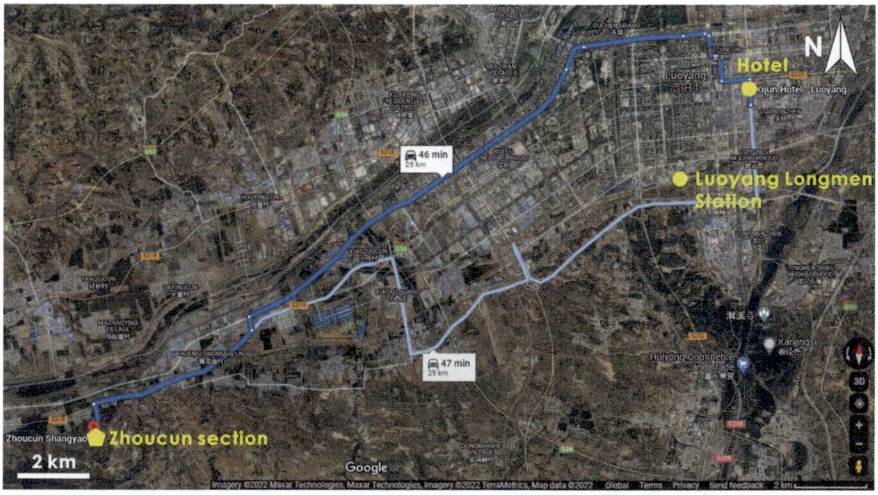

Fig. 21 Route map of hotel-Yiyang Zhoucun section (©Google Map)

Fig. 22 Panoramic photo of Zhoucun section, Yiyang. Yellow line indicates the upper Xinji Formation and the lower Luoquan Formation

Stop 1: The contact relationship between the Cuizhuang Formation of Mesoproterozoic Luoyu Group and the Luoquan Formation

Layer: Layer 0 of Zhoucun section (Fig. 23)

Detailed description: Purplish-red and greyish green thin-bedded argillaceous siltstone intercalated with middle-bedded siltstone of Cuizhuang Formation, small cross-bedding, and parallel bedding are developed (Figs. 23 and 24a, b).

The contact relationship between the argillaceous siltstone of Cuizhuang Formation of Luoyu Group and Ediacaran Luoquan Formation is weak parallel unconformity, and the bottom of Luoquan Formation exhibits dark grey or dark greyish green glacial diamictite containing pebble-cobble (Fig. 24c).

Stop 2: Sedimentary characteristics and succession of Luoquan Formation

Layers: Layers 1–14 of Zhoucun section (Fig. 23)

Detailed description:

Overall characteristics of Ediacaran Luoquan Formation of Zhoucun section, Yiyang county.

(1) In general, the Luoquan Formation develops two groups of fining upward cycles (Fig. 24e): massive diamictite-weakly stratified diamictite-dropstone bearing siltstone, gravels show two groups of imbricate structures in opposite directions locally, and lenticular sandstones are developed in dropstone bearing siltstone and laminated siltstone (Fig. 24f).
(2) The boulder containing massive diamictite in layer 11 falls into the matrix argillaceous siltstone, the diameter of a single boulder can reach more than 1 m. They are mainly subrounded and subangular occasionally, gravels composed of conglomerate can be observed. This massive diamictite represents the beginning of the second glacial advance and retreat cycle of the Luoquan Formation. We can also observe the massive diamictite on the left side of this section offset by fault (Fig. 24k).
(3) The dropstone bearing siltstone of the upper to top of the Luoquan Formation and the purplish-red siltstone intrude downward into the lower greyish green siltstone in a wedge shape (Fig. 24j).
(4) The Luoquan Formation develops two groups of upward fining sequences of massive diamictite- dropstone bearing siltstone, representing two large-scale ice advance and retreat cycles.

This stop mainly focuses on the continuous and complete sedimentary succession and characteristics of the Luoquan Formation (Fig. 24m, n), which are introduced as follows.

Layer 1, dark grey pebble-cobble glacial diamictite, the gravels are mainly composed of dolostone from the Mesoproterozoic basement rocks, such as oolitic dolostone, stromatolite dolostone, laminated dolostone, etc., followed by quartzite; The long axis of single gravel can reach 50 cm, and the gravels show weak orientation in general (Fig. 24d, e).

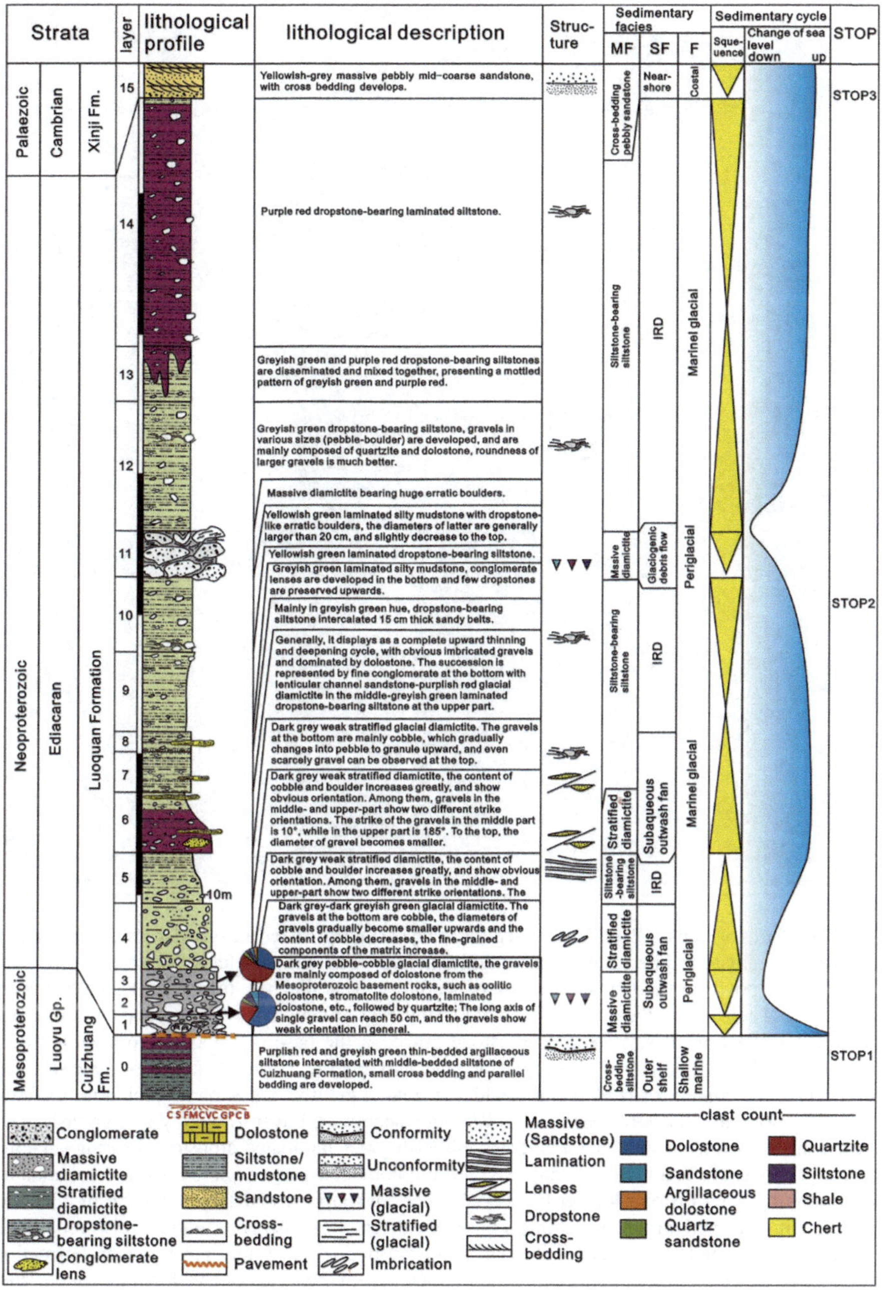

Fig. 23 Comprehensive stratigraphic histogram of Zhoucun section

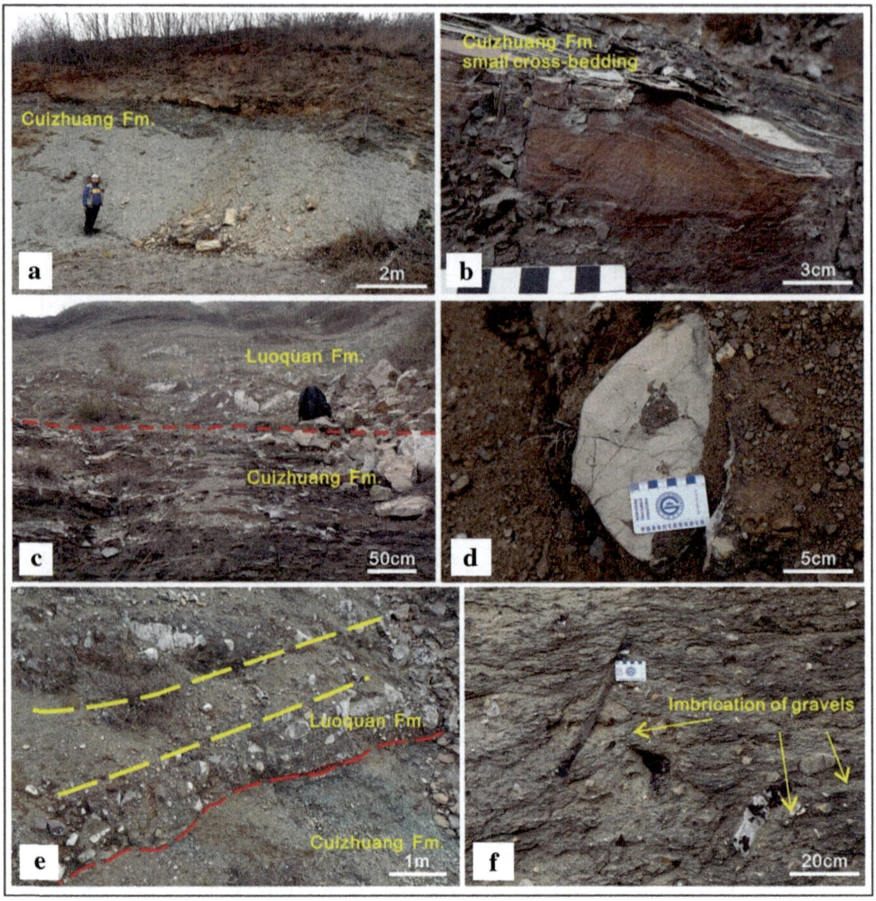

Fig. 24 Plates of stops 1–2 in Zhoucun section. **a** layer 0, purplish red and greyish green thin-bedded argillaceous siltstone intercalated with middle-bedded siltstone of Mesoproterozoic Cuizhuang Formation, Luoyu Group; **b** small-scale cross-bedding and parallel beddings; **c** weak parallel unconformity between the argillaceous siltstone of Cuizhuang Formation of Luoyu Group and Ediacaran Luoquan Formation; **d** stromatolite dolostone gravel in the base of Luoquan Formation glacial diamictite; **e** layer 1, dark grey pebble-cobble glacial diamictite; **f** layer 4, dark grey weak stratified diamictite, the content of cobble and boulder increases greatly, and show two obvious different strike orientations. Plates of stops 1–3 in Zhoucun section. **g** Layer 7, mainly in a greyish-green hue, dropstone-bearing siltstone; **h** layer 9, yellowish-green laminated dropstone-bearing siltstone; **i** layer 11, massive diamictite bearing huge erratic boulders; **j** layer 12, greyish green laminated dropstone-bearing siltstone; **k** layer 13, greyish green, and purplish-red dropstone-bearing siltstones are disseminated and mixed, presenting a mottled pattern of greyish green and purplish-red; **l** layer 15, the sharp boundary between Cambrian Xinji Formation and Ediacaran Luoquan Formation, the bottom of Xinji Formation is pebbly medium-coarse quartz sandstone with parallel and graded beddings. **m** Layers 6–12 of Zhoucun section; **n** layers 1–7 of Zhoucun section

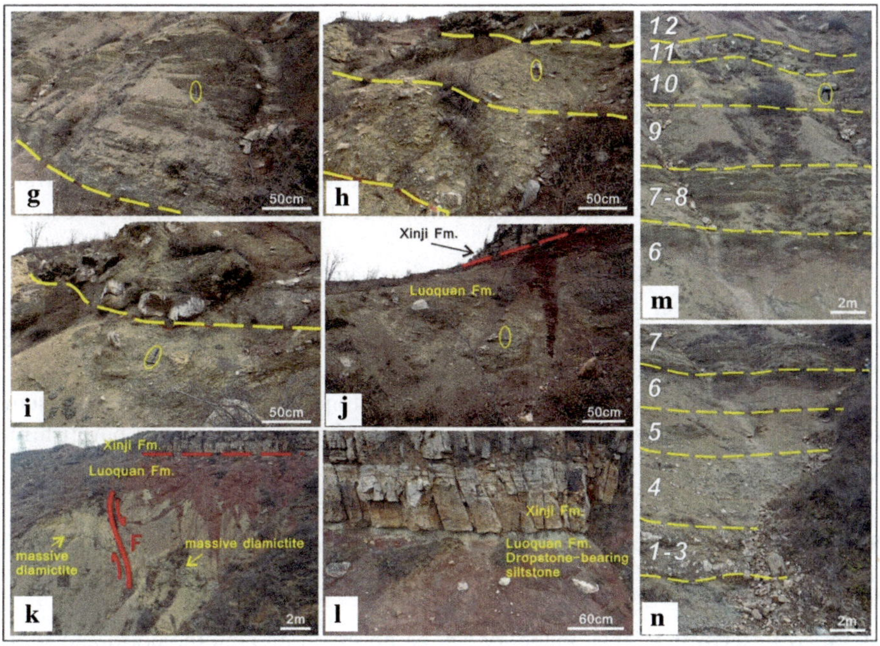

Fig. 24 (continued)

Layer 2, dark grey-dark greyish green glacial diamictite. The gravels at the bottom are cobble, the diameters of gravels gradually become smaller upwards and the content of cobble decreases, the fine-grained components of the matrix increase (Fig. 24e).

Layer 3, the gravels at the bottom are slightly smaller, and saddle grooves and insignificant striae formed by subglacial erosion are mostly seen on the gravel surface. Some gravels even show the characteristics of the desert aeolian prism (Fig. 24e).

Layer 4, dark grey weak stratified diamictite, the content of cobble and boulder increases greatly, and show obvious orientation. Among them, gravels in the middle- and upper-part show two different strike orientations. The strike of the gravels in the middle part is 10°, while in the upper part is 185°. To the top, the diameter of gravel becomes smaller (at ~1 m, Fig. 24f).

Layer 5, dark grey weak stratified glacial diamictite. The gravels at the bottom are mainly cobble, which gradually changes into pebble to granule upward, and even scarcely gravel can be observed at the top (Fig. 24m).

Layer 6, generally, it displays as a complete upward thinning and deepening cycle, with obvious imbricated gravels and dominated by dolostone. The succession is represented by fine conglomerate at the bottom with lenticular channel sandstone-purplish red glacial diamictite in the middle-greyish green laminated dropstone-bearing siltstone at the upper part (Fig. 24m).

Layer 7, mainly in a greyish-green hue, dropstone-bearing siltstone intercalated 15 cm thick sandy belts (Fig. 24g).

Layer 8, Greyish green laminated silty mudstone, conglomerate lenses are developed in the bottom and a few dropstones are preserved upwards (Fig. 24g).

Layer 9, yellowish-green laminated dropstone-bearing siltstone (Fig. 24h).

Layer 10, yellowish-green laminated silty mudstone with dropstone-like erratic boulders, the diameters of latter are generally larger than 20 cm and slightly decrease to the top (Fig. 24h).

Layer 11, massive diamictite bearing huge erratic boulders (Fig. 24i).

Layer 12, greyish green dropstone-bearing siltstone, gravels in various sizes (pebble-boulder) are developed and are mainly composed of quartzite and dolostone, roundness of larger gravels is much better (Fig. 24j).

Layer 13, greyish green and purplish-red dropstone-bearing siltstones are disseminated and mixed, presenting a mottled pattern of greyish green and purplish-red (Fig. 24k).

Layer 14, purplish-red dropstone-bearing laminated siltstone (Fig. 24k).

Stop 3: Boundary between Luoquan Formation and Xinji Formation

Layer: Layer 15 of Zhoucun Section (Fig. 23)

Detailed description: This boundary can be observed on both the north and south sides of the hillside. The section on the south side is precipitous to observe from a close distance. The north side is beside the road, which is very convenient to observe.

(1) In the south, the boundary and contact relationship between the dropstone-bearing siltstone of Luoquan Formation and the above Cambrian Xinji Formation can be observed at the top of the hill, just below the windmill (Fig. 24l). The Xinji Formation is yellowish-brown massive gravelly medium-coarse sandstone with cross-bedding.

(2) The sharp boundary between Luoquan Formation and Cambrian Xinji Formation in the north slope, the Xinji Formation is massive gravelly medium-coarse quartz sandstone with parallel and graded bedding. In ascending order, Luoquan Formation consists of purplish-red massive glacial diamictite and greyish green glacial diamictite (Fig. 25a), the boundary between the two colors exhibits wavy fluctuation.

Stop 4: The exposure structures in the massive glacial diamictite in Luoquan Formation

Detailed description: This stop is located at the roadside of the north slope. We can observe purplish red-greyish green massive glacial diamictite of Luoquan Formation, the interface between the two types of glacial diamictite in different colors exhibits wavy fluctuation (Fig. 25b), which is similar to the ancient weathering crust. The original color is greyish green and the purplish-red is a secondary color, showing an extremely hot environment. The dolomitic gravels in the glacial diamictite have been eroded to form holes, and some gravels only have a layer of crust (Fig. 25c, d),

Fig. 25 Plate of stops 4–5 in Zhoucun section. **a** The sharp boundary between Luoquan Formation and Cambrian Xinji Formation in the north slope; **b** Purplish red-greyish green massive glacial diamictite of Luoquan Formation, the interface between the two types of glacial diamictite in different colors exhibits wavy fluctuation; **c**, **d** The dolomitic gravels in the glacial diamictite have been eroded to form holes; **e** The succession of purplish-red dropstone-bearing siltstone and above greyish green glacial diamictite; **f** The purplish-red glacial diamictite and the sandstone of Xinji Formation are offset by the normal fault in the southwest of this succession

which are in the shape of purplish-red ctenoid, filamentous, reticulated, and crusty. There are also cracks and holes formed by leaching and dissolution in the diamictite, that is, after the karstification of carbonate in an arid environment, the color changed from greyish green to red.

Stop 5: Mesoproterozoic Beidajian Formation basement rocks-Ediacaran Luoquan Formation-Cambrian Xinji Formation section

Detailed description: The exposed sandstone of the Mesoproterozoic Beidajian Formation consists of an integral part of the mountain. The succession of purplish-red dropstone-bearing siltstone and above greyish green glacial diamictite can be seen (Fig. 25e). Due to normal fault in the southwest of this succession, the purplish-red glacial diamictite and the sandstone of Xinji Formation are offset (Fig. 25f). The overall stratigraphic framework is Mesoproterozoic Beidajian Formation sandstone-purplish red glacial diamictite-greyish green glacial diamictite-purplish red dropstone-bearing siltstone-greyish green glacial diamictite-purplish red diamictite of Ediacaran Luoquan Formation-Cambrian Xinji Formation sandstone.

14:30–17:00 After the field trip of the day, leaving Yiyang county by bus, and heading for Ruzhou city, stay in Tianrui Zhongzhou international hotel. All of the rest of field trips in the next few days will be carried out around Ruzhou city.

4.3 *DAY THREE: August 29th Route 3* Lushan Shimengou Section

- **Key words: The fairy pointed the way to the glacier**

7:00–8:00 Breakfast at 7:00 a.m. be packed and on the road by 8:00 a.m.

8:00–15:00 Today we will go to the Shimengou section from Zhongzhou international hotel. The Shimengou section starts at 112° 43′ 14″, 33° 56′ 15″, ends at 112° 16′ 49.42″, 34° 31′ 32.33″. It's a long way and takes about 2 h for one-way travel (Fig. 26). There are not only various geological landscapes but also some hominid sites of the Neolithic age on the top of the cliff. We will be spending the day inspecting this wonderful place.

The main observations in the Shimengou section are as follows: the continuous strata and sedimentary succession of Mesoproterozoic Sanjiaotang Formation-Ediacaran Luoquan Formation-Dongpo Formation, and Cambrian Xinji Formation; exquisite middle-small scale subglacial erosional forms on the glacial pavements of top Sanjiaotang Formation; abundant trace fossils of Early Cambrian at the bottom of Xinji Formation (Figs. 27 and 28).

STOP 1: Sedimentary characteristics and boundary of Ediacaran Luoquan Formation, Dongpo Formation, and Cambrian Xinji Formation

Layers: Layers 1–7 of Shimengou section (Fig. 4.28)

Detailed description: In the Shimengou section, the outcrop of the Luoquan Formation is separated by the road. From here we can see the lower part of Luoquan Formation is yellowish-brown massive glacial diamictite, with 10 cm in diameter subrounded quartzite gravel at the bottommost, upwards is yellowish-brown dropstone-bearing siltstone (Figs. 4.28 and 29a–e). Above the road, outcrops continue

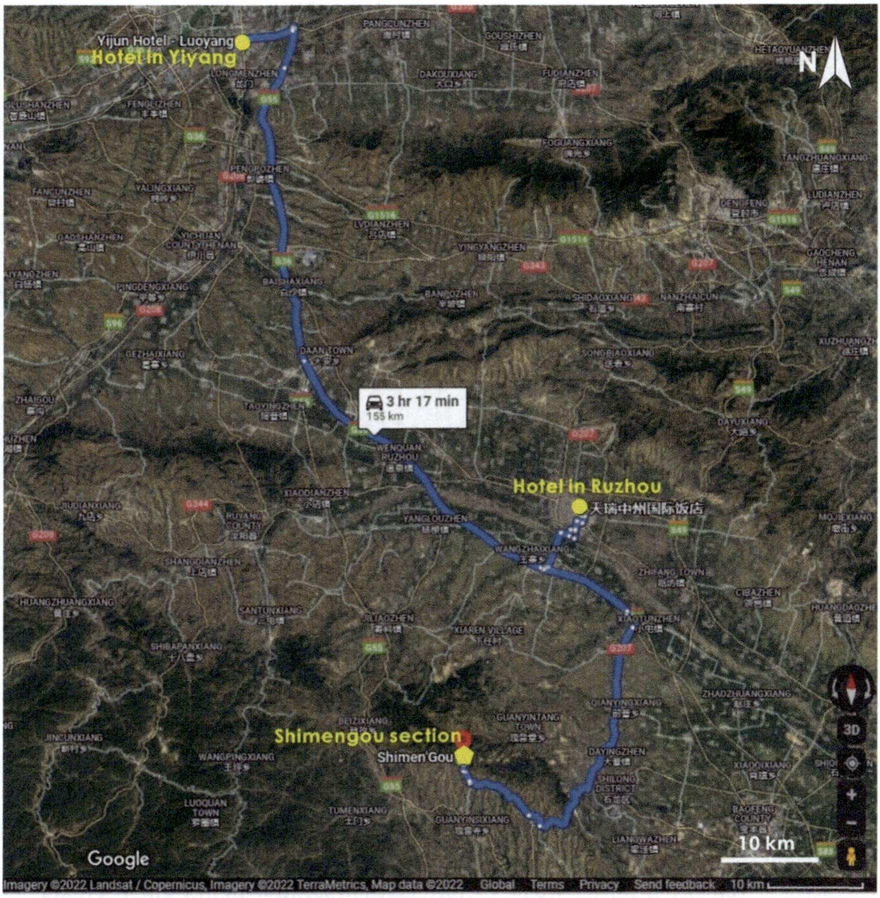

Fig. 26 Route map of hotel-Lushan Shimengou section (©Google Map)

to occur along the hill, there are still dropstone-bearing siltstone with abundant sandstone and conglomerate lenses. The dropstones are centimeter to tens of centimeters in diameter, and are mostly composed of quartz sandstone or quartzite, dominated by subrounded and few subangular (Fig. 29f–i).

The overlying Dongpo Formation are mostly greyish-green thin-bedded siltstone interbedded with thin sandstone layer, scarcely gravels are found, graded bedding and cross-bedding are developed (Fig. 29j, k). The Dongpo Formation gradually transits into purplish-red siltstone or shale upward at the top, which is quite different from the sandstone of Xinji Formation, indicating a parallel unconformity (Fig. 29l). Continuous and complete sedimentary characteristics of the Luoquan Formation-Dongpo Formation are described as follows:

Fig. 27 Panoramic photo of Shimengou section, Lushan. Red dashed line indicates the boundary of Luoquan Formation and the underlying Sanjiaotang Formation

Layer 1, purplish-red massive/weakly stratified glacial conglomerate, directly parallel unconformable overlayed on the glacial pavement (Fig. 29a, b). Scattered massive diamictite can be seen elsewhere on the glacial pavement.

Layer 2, ~50 cm thick dropstone-bearing siltstone, low gravel content and dominated by dolostone and quartz sandstone.

Layer 3, yellowish-brown weakly stratified diamictite with low gravel content but dominated by dolomitic composition (Fig. 29c).

Layer 4, yellowish-brown dropstone-bearing siltstone, silty mudstone. Dolostone boulder with a diameter of ~50 cm can be seen in the middle and upper part. A large number of calcareous layers are preserved discontinuously in the middle part, as well as sandy conglomerate lenses or discontinuous bands (Fig. 29c–e).

Layer 5, yellowish-brown dropstone-bearing siltstone (Fig. 29f), the content of sand increases upward and overall gravel content is significantly increased. The dropstone structure is gradually obvious in the upper part (Fig. 29g). Laterally, sandy conglomerate lenses are developed continuously, and the gravels inside the lenses are arranged horizontally (Fig. 29h, i).

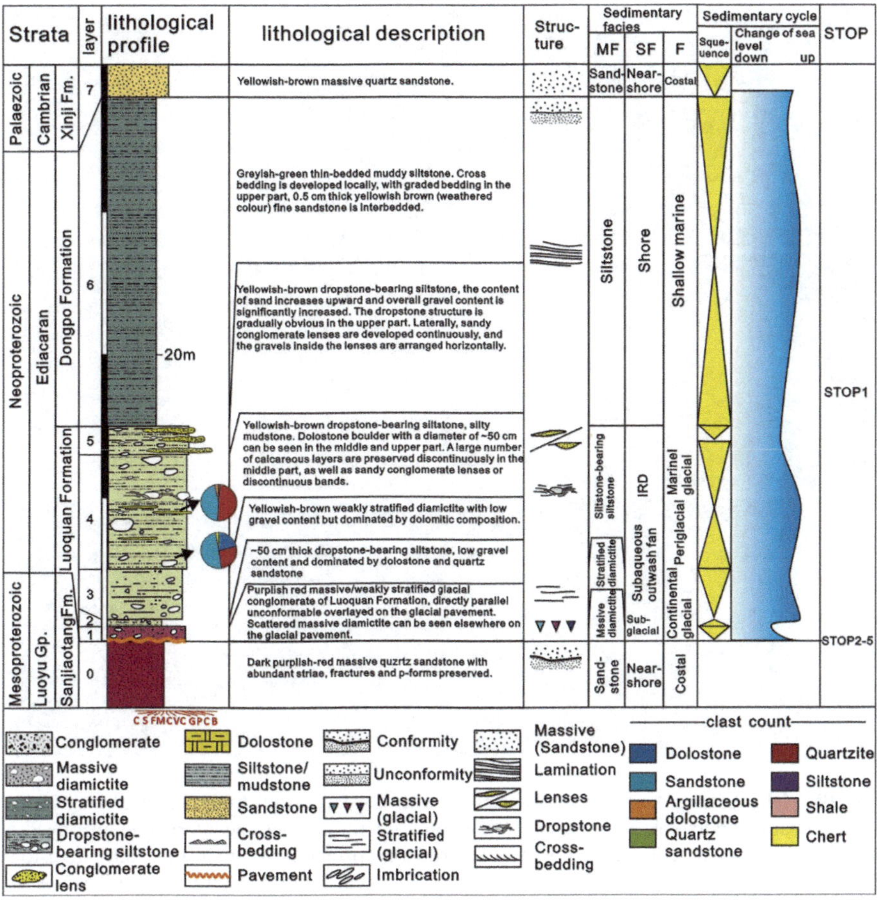

Fig. 28 Comprehensive stratigraphic histogram of Shimengou section

Layer 6, greyish-green thin-bedded muddy siltstone of the Dongpo Fm. Cross bedding is developed locally, with graded bedding in the upper part, 0.5 cm thick yellowish-brown (weathered color) fine sandstone is interbedded (Fig. 29j, k).

Layer 7, The Xinji Formation yellowish-brown massive quartz sandstone (Fig. 29l).

STOP 2–5: Subglacial erosional forms and glacial pavement of the surface of Mesoproterozoic Sanjiaotang Formation, Luoyu Group

Layer: Layer 0 of Shimengou section (Fig. 28)

From south to north, the glacial pavements of the Shimengou section are divided from No. 1 to No. 4 according to the spatial distribution (Fig. 30). Based on the scale of subglacial erosional forms and glacial erosional landscape, micro- and meso-scale subglacial erosional forms are recognized on the glacial pavements of the Shimengou

Fig. 29 Plates of stop 1 in Shimengou section. **a**, **b** Layer1, purplish-red massive/weakly stratified glacial conglomerate of Luoquan Formation, directly parallel unconformable overlayed on the glacial pavement; **c–e** Layer 4, yellowish-brown dropstone-bearing siltstone, silty mudstone, dolomitic erratic boulder, and sandy conglomerate lenses can be seen locally; **f** Layer 5, yellowish-brown dropstone-bearing siltstone; Plates of stops 1 in Shimengou section. **g** Layer 5, yellowish-brown dropstone-bearing siltstone; **h** Layer 5, relatively continuous sandy conglomerate lens developed in upper yellowish-brown dropstone-bearing siltstone. **i** Layer 5, dropstones preserved in the upper part. **j**, **k** Layer 6, Dongpo Formation greyish-green thin-bedded muddy siltstone, siltstone. **l** Layer 7, Xinji Formation yellowish-brown massive quartz sandstone

section. At the same time, further subdivision can be carried out according to the type, morphology, forming mechanism, and glacier kinematic characteristics (Table 2). Micro-scale subglacial erosional forms indicate the direction of glacier movement include glacial striae, crescentic fractures, and plastically moulded forms (P-forms) such as *muschelbruch*, *cavetto*, and spindle flute. Meso-scale subglacial erosional forms are roches moutonnées and bedrock steps, which are usually accompanied by micro-scale subglacial structures, together representing the astounding eroding and reworking ability of glaciers. A total of 562 sets of striae data were collected on the four glacial pavements, with an average trend of 176.2° (Table 3). The comprehensive analyses of the indicative characteristics of these subglacial erosional structures in

Fig. 29 (continued)

different scales show that glacier has a kinematic characteristic that moved from north to south.

Glacial pavements #1 (Stop 2)

Detailed description:

Glacial pavements #1 is the largest one among the four glacial pavements in the Shimengou section (Fig. 31). Micro-scale subglacial erosional forms are abundant. Multiple groups of oblique striae indicate different glacier movement directions. Most of them are preserved on the smooth glacial pavements (Fig. 32), in addition, some can be found in the pits and plastically moulded forms caused by glacier erosion. Even some lodgement tills are cut by the striae.

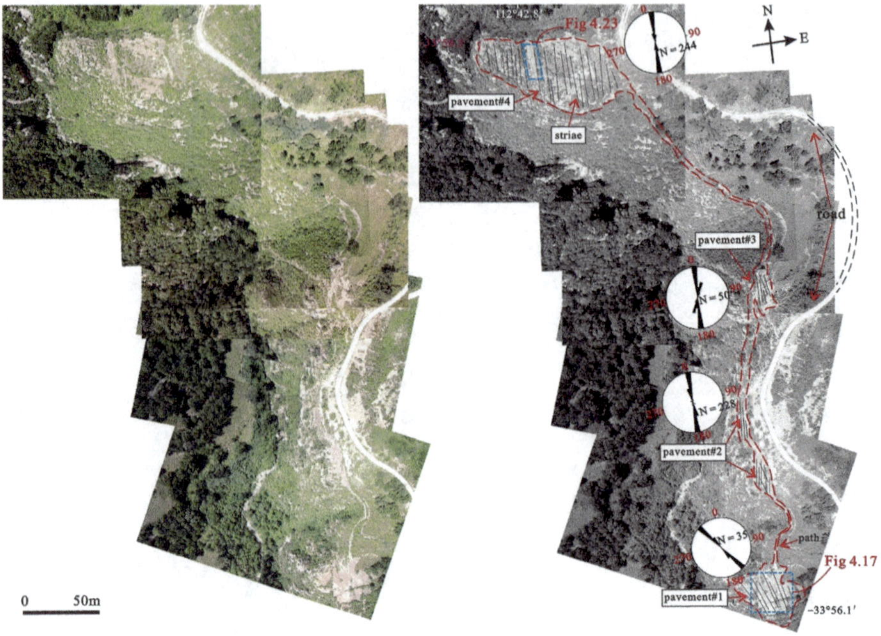

Fig. 30 Composite aerial photographs taken by a drone image of the Shimengou pavements in west Henan province original pictures provided by prof. Le Heron D. P., from approximately south to north, altogether 4 pavements outcrop are alongside the road. The rose diagram refers to striae orientations measured on superposed striated surfaces in four pavements (Chen et al. 2020)

Table 2 Morphology and size classification of subglacial erosional forms and glacial landscapes presented in Shimengou section (Chen et al. 2020)

Morphology	Scale				
	Micro	Meso			Macro
	1 cm	10 cm	1 m	100 m	10 km
Swelling, partly streamlined	Roches moutonnées				
Concave, streamline	Nail head striae	Striae			
	Crescentic fractures				
	P-forms				
Concave, partially streamlined					U-valley

Table 3 The strikes of striations on the studied glacial pavements in the Shimengou section (Chen et al. 2020)

Glacial pavement	#1	#2	#3	#4	Others
Count (set)	35	228	50	244	5
Average trend (°)	147	180	189	185	180
Range of orientation (°)	140–160	135–212	155–219	170–222	178–186

Fig. 31 A bird's eye view of pavement #1 in the Shimengou section (Chen et al. 2020). Details see in Fig. 13

Fig. 32 Glacial striae developed on the glacial pavement #1 in Shimengou. More than two sets of striae are indicated in plates **a–d**. Scale bar is 10 cm and the lens cap is 5 cm

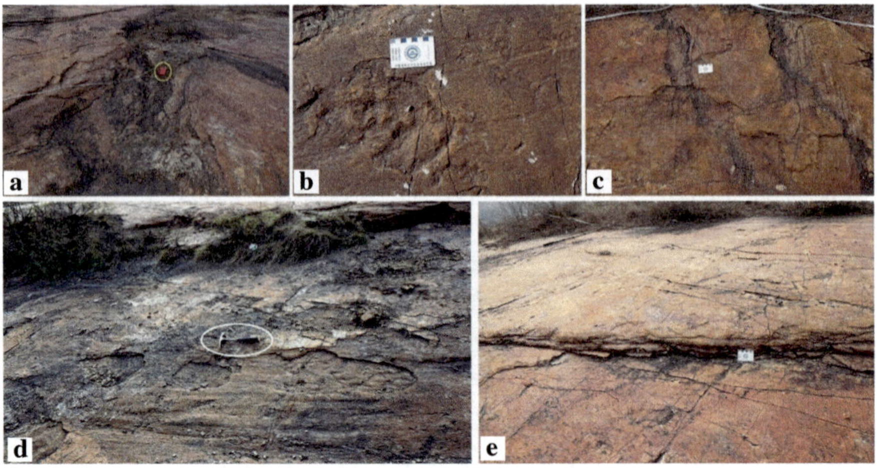

Fig. 33 Glacial structures developed on the glacial pavement #1 in Shimengou. **a** lodgement tills, glacial pits, and striae; **b** glacial pits and striae; **c** torsional cracks; **d** crescentic gouges; E-bedrock steps. Scale bar is 10 cm, the notebook is 20 cm, and the hammer is 30 cm

Other types of medium to small glacial structures, such as glacier erosional pits, crescentic cracks, and gouges, torsional cracks, bedrock steps also developed on glacial pavement #1 (Figs. 13 and 33).

Glacial pavement #2 (Stop 3)

Detailed description:

Glacial pavement #2 is combined with three connected smaller glacial pavements (Fig. 34). It belongs to the layer 0 of the Shimengou section, we can inspect the unconformity relationship between the diamictite of the bottom Luoquan Formation and the Sanjiaotang Formation of the Luoyu Group.

There are a series of continuous and densely developed striations on the glacial pavement #2, some of the striations have obliqued with each other (Fig. 35a). Some striations can reach several meters (Fig. 35b). Hairpin striations (Fig. 35c) which are also indicative of the direction of glacier movement are also developed. Besides, narrow pits (Fig. 35d) or gentle grooves (Fig. 35f) carved by glaciers on the basement rocks are also widely developed.

Glacial pavement #3 (Stop 4)

Detailed description:

Compared with the other glacial pavements in Shimengou, glacial pavement #3 is smaller, and fewer subglacial erosional structures are well preserved, the glacial striae are the major subglacial erosional structure (Fig. 36).

Fig. 34 A bird eye's view of glacial pavement #2 in Shimengou

Fig. 35 Various subglacial erosional structures on glacial pavement #2 in Shimengou. **a**, **b** densely developed striations; **c** hairpin striations; **d**, **e** glacial erosional pits. Scale bar is 10 cm

Fig. 36 Glacial pavement #3 in Shimengou

Glacial pavement #4 (Stop 5)

Detailed description:

The most abundant and delicate subglacial erosional structures are developed on the glacial pavement #4. Moreover, there are various plastically moulded forms (P-forms)

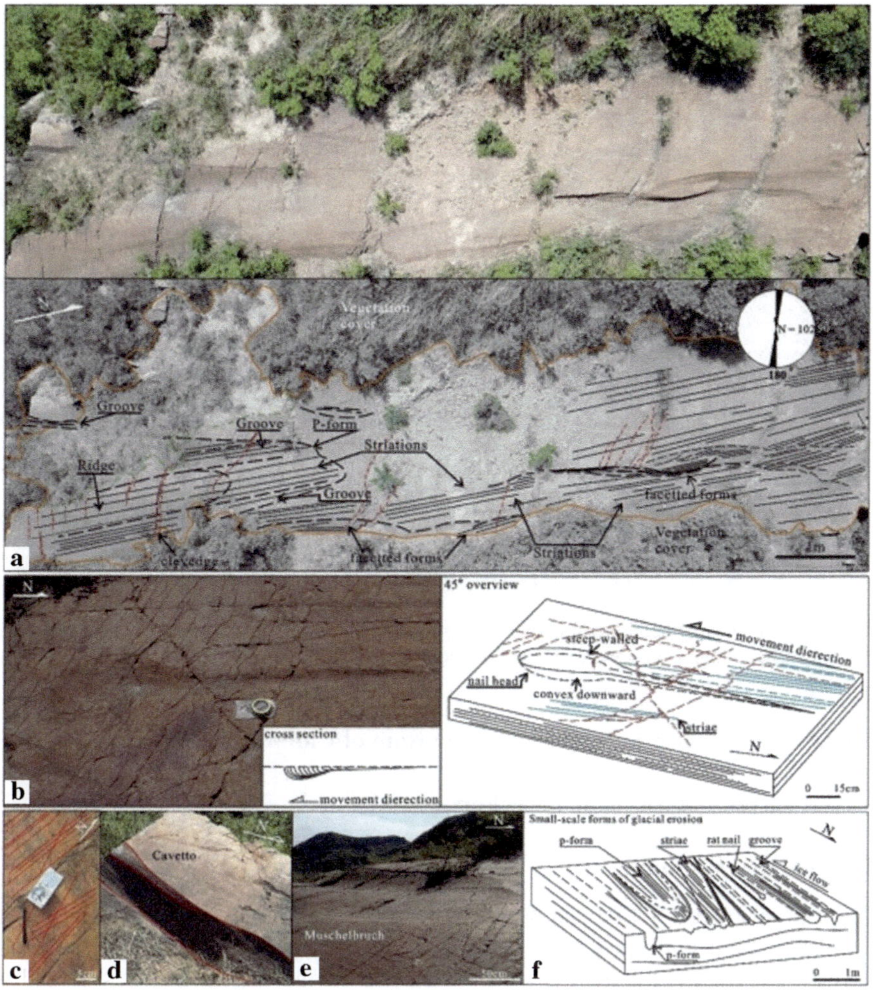

Fig. 37 Bird's-eye view of Shimengou #4 glacial pavement (Chen et al. 2020). **a** Plastically moulded forms (P-forms), glacial striate; **b** nail head striate; **c** obliqued striations; **d** spindle flutes; **e** muschelbruch; **f** cartoon model of subglacial erosional structures

structures, such as spindle flutes, *muschelbruches,* and facetted forms (Fig. 37a, d–f), and associative ridges, gouges, as well as densely distributed striations (Fig. 37a, c). For the latter, the special nail head striae are also developed (Fig. 37b). In addition, crescentic fractures and gouges, and various forms of glacial erosion pits can also be seen (Fig. 38). These subglacial erosional structures are excellent indicators of the direction of glacier movement and the mechanism of glacier thermal regime.

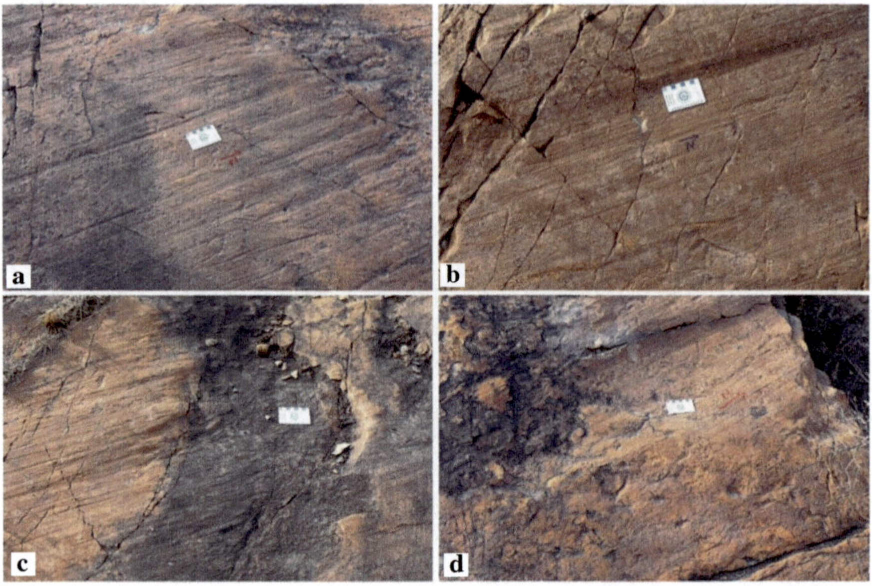

Fig. 38 Various subglacial erosional structures on Shimengou #4 glacial pavement. **a, b** densely developed striations; **c** crescentic gouges; **d** glacial erosion pits and striations. Scale bar is 10 cm

Stop 6: Trace fossils preserved in the bottom of Cambrian Xinji Formation

Layer: Layer 7 of Shimengou section

Detailed description:

Trace fossils (Figs. 39b–d and 40) are well-preserved in the basal sandstone of the Cambrian Xinji Formation (Fig. 39a). The trace fossils are mainly represented by horizontal cylindrical burrows, with 8 ichnogenera and 12 ichnospecies, including *Arenicolites sp., Didymaulichnus Lyelli, Gordia Marina, Helminthopsis, Planolites Montanus, Treptichnus pedum, Treptichnus pollardi,* and 2 unnamed trace fossils. This fossil assemblage belongs to an atypical kind of *Cruziana* ichnofacies and was formed in sandy soft substrata in a subtidal environment with normal salinity. These trace fossils indicate that the lower Xinji Formation is no earlier than the Canglangpu period (Miao and Zhu 2014).

Fig. 39 Well-preserved trace fossils in the bottom sandstone substrata of the Xinji Formation

4.4 DAY FOUR: August 30th Route 4 Luoquan Village Section

- **Key words: Luoquan Formation in Luoquan village**

7:00–8:00 Breakfast at 7:00 a.m...be packed and on the road by 8:00 a.m.

8:00–12:00 Depart from the hotel, we first head for Luoquan village in Mangchuan Town, which locates in the south of Ruzhou City (Fig. 41). The Luoquan Formation was named at Luoquan village 60 years ago, Chinese sedimentologists confirmed the glacial origin of this set of sediments and named it Luoquan Formation, therefore, it is of great commemorative significance. The Luoquan village section has become a glacier geological heritage conservation area. Drive into Luoquan village, and the outcrops are exposed along one side of the road.

The section starts at 112° 45′ 3.29″, 34° 0′ 9.27″, ends at 112° 45′ 31.95″, 33° 59′ 52.90″. The main observation contents are glacial sedimentary characteristics and sedimentary successions such as glacial diamictite, subglacial soft-sediment deformation, and dropstone structure in the middle and lower part of Luoquan Formation (Figs. 42 and 43). Because of the coverage and rural construction, the contact relationship between the bottom of the Luoquan Formation and the quartz sandstone of Mesoproterozoic Beidajian Formation mentioned in the previous paper cannot be seen, the glacial pavements are not found, either. Due to the exposure of the Luoquan Formation being unclear, the bottom and the upper boundary are hard to observe directly. Combined with previous data and field observation, it is speculated

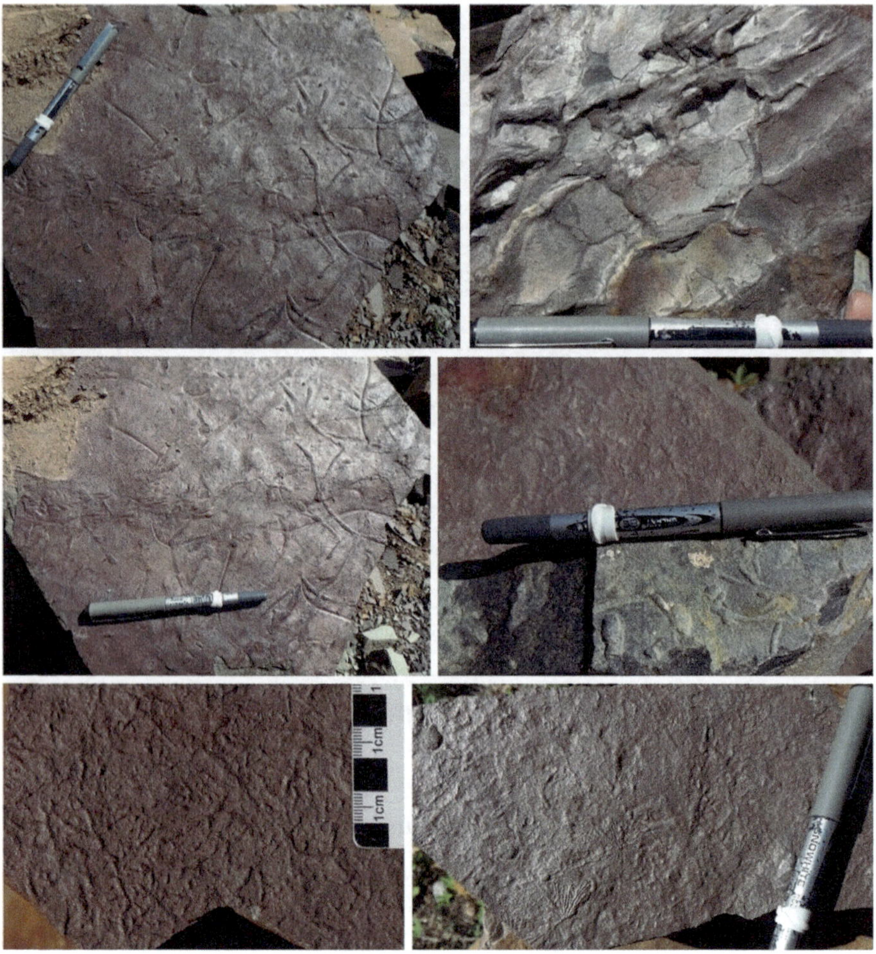

Fig. 40 Well-preserved trace fossils in the bottom sandstone substrata of the Xinji Formation. 15:00–17:00 Finish the field observation and return to the hotel

that the exposed part along the road is the lower to the middle part of the Luoquan Formation.

Stop 1: Sedimentary characteristics of the lower part of Luoquan Formation

Layer: layer 1 of Luoquan village section (Fig. 43**)**

Detailed description: At stop 1, we mainly focus on the sedimentary characteristics of the lower part of the Luoquan Formation, and the introduction are as follows:

Layer 1, light purplish red and reddish-brown glacial diamictite. The boundary between it and the lower Beidajian Formation cannot be seen. It is inferred that it belongs to the bottom-middle part of the Luoquan Formation, representing one

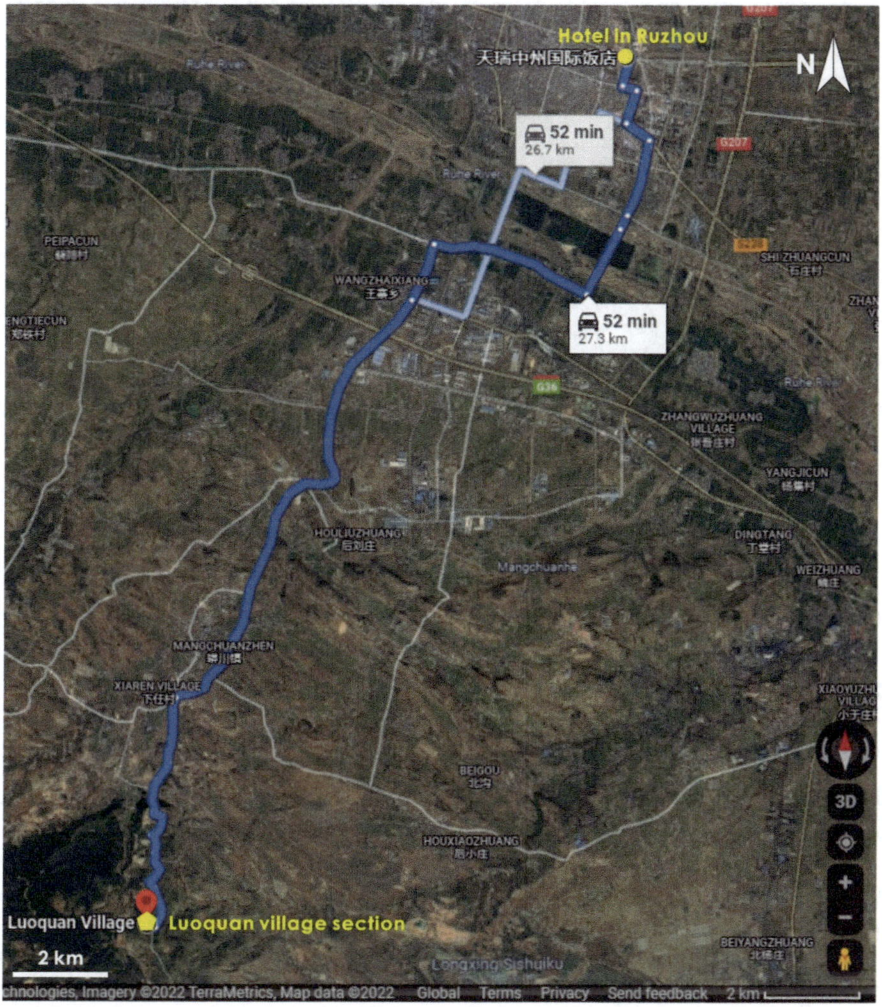

Fig. 41 Route map of the hotel-Luoquan village section (©Google Map)

ice-advance process. The content of gravel is 30–40%, sub-angular to sub-rounded shaped, general sorted. The gravels are mainly dolostone, sandstone, and shale, and the gravel orientation is not obvious. The gravels are mainly developed at the bottom, and the content gradually decreases upward. However, erratic boulders can be seen, and multiple gravel bands or lenses are developed at the upper part (Fig. 44a–d).

Fig. 42 Panoramic photo of Luoquan village section

Stop 2: Dropstone-bearing siltstone of Luoquan Formation

Layers: Layer 2–3 of Luoquan village section (Fig. 43)

Detailed description: Stop 2 focuses on the sedimentary characteristics of the middle and lower part of the Luoquan Formation, and the introductions are as follows:

Layer 2, greyish-green dropstone-bearing siltstone, with sandstone lenses (Fig. 44e, f). Laterally, a 2 m long sandstone lens develops along with the layer and inclines to the lower right, wedges into the underlying diamictite (Fig. 44g).

Layer 3, dark greyish-green stratified diamictite, with coarser gravels and higher gravel content. Two sub-angular quartzite gravels are distributed along with the layer (Fig. 44h, i).

Stop 3: Sedimentary characteristics and soft-sediment deformation structures in the middle of Luoquan Formation

Layers: Layers 4–7 of Luoquan village section (Fig. 43)

Detailed description: This stop focuses on the sedimentary characteristics of the middle part of the Luoquan Formation, and the detailed introduction is as follows:

Layer 4, light purplish red and reddish-brown stratified diamictite (Fig. 44j, k).

Layer 5, light purplish red and reddish-brown diamictite, yellowish-brown leaching color, and greyish-green appear alternately (Fig. 44j, k).

Layer 6, light purplish red and reddish-brown stratified diamictite, calcareous thin layers are developed through the layer, and yellow purple red, and greyish-green appear alternately (Fig. 44j, k).

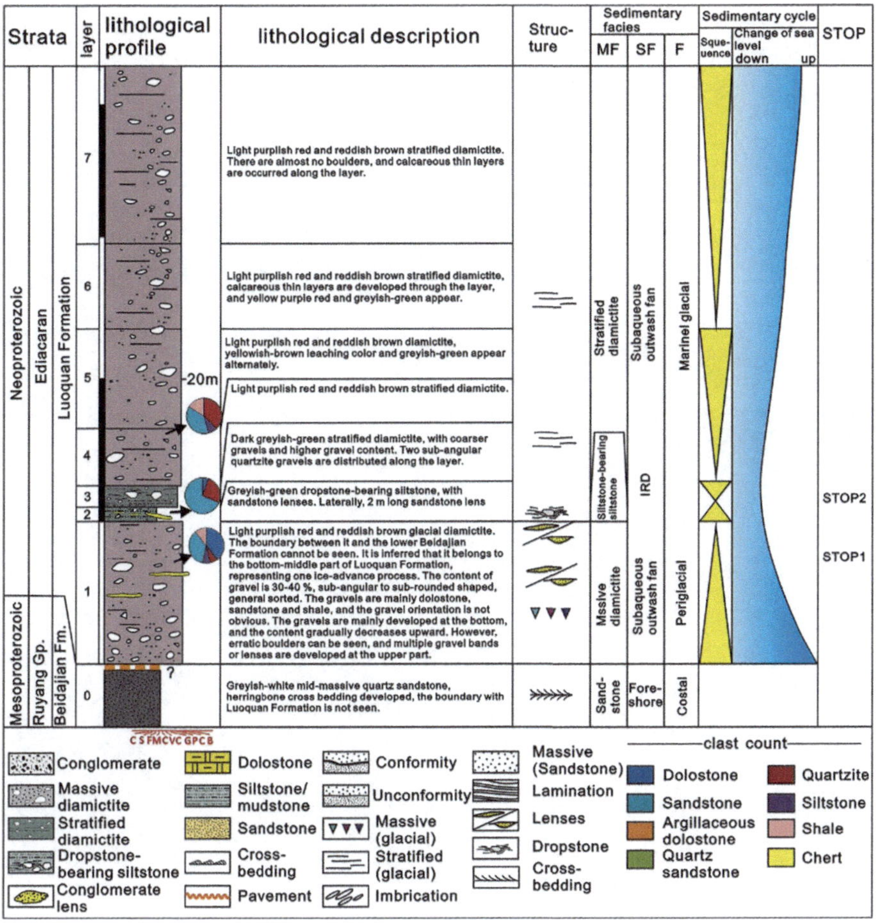

Fig. 43 Comprehensive stratigraphic histogram of Luoquan village section

Layer 7, light purplish red and reddish-brown stratified diamictite. There are almost no boulders, and calcareous thin layers occur along with the layer (Fig. 44j, k).

In the north section, the soft-sediment deformation structures and low angle thrust deformation in massive diamictite of Luoquan Formation can be observed, above is the undisturbed or undeformed nearly horizontal strata (Fig. 45). Due to the outcrop being heavily covered on the hillside, putative ~80 m thick greyish-green and purplish-red shale of Dongpo Formation is exposed above the Luoquan Formation. The boundary between Dongpo Formation and Cambrian Xinji Formation is also difficult to observe, but other sections can be a supplement.

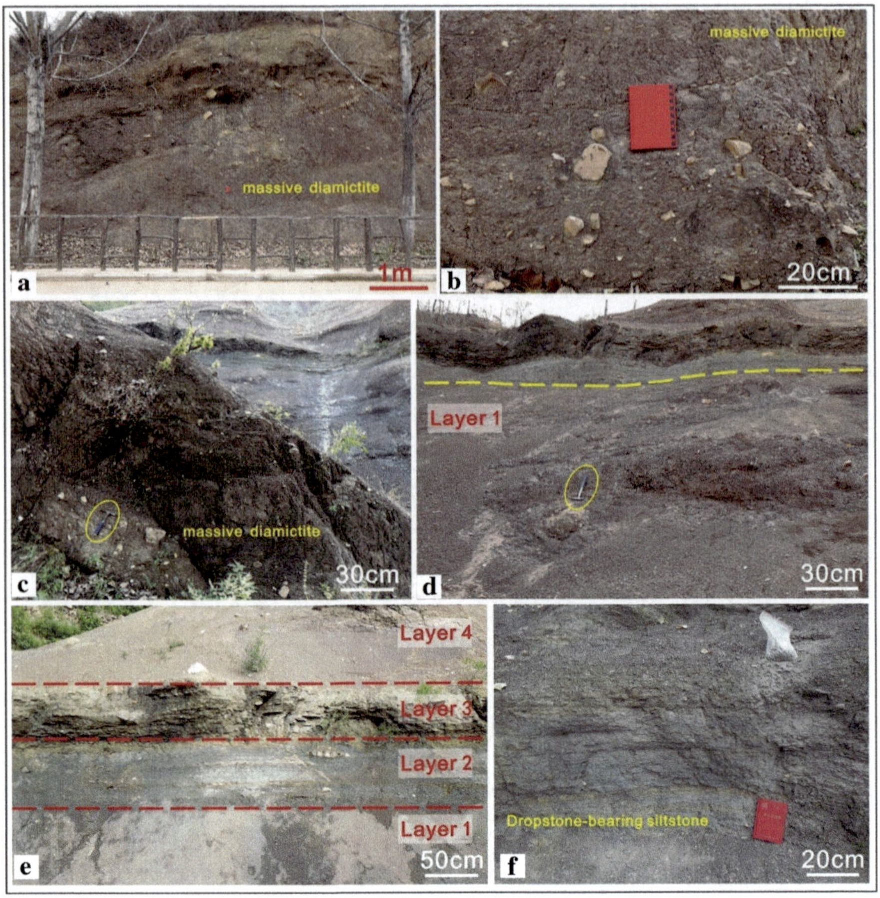

Fig. 44 Plates of stops 1–3 in Luoquan village section; Plates of stops 1–3 in Luoquan village section. **a–d** Layer 1, light purplish red and reddish-brown glacial diamictite. The content of gravel is 30–40%, sub-angular to sub-rounded shaped, general sorted. The gravels are mainly dolostone, sandstone, and shale, and the gravel orientation is not obvious. The content of gravels gradually decreases upward and multiple gravel bands or lenses are developed at the upper part; **e** Layers 1–4; **f** Layer 2, greyish-green dropstone-bearing siltstone, with sandstone lenses; **g** Layer 2, 2 m long sandstone lens develops along with the layer and inclines to the lower right, wedges into the underlying diamictite; **h**, **i** Layer 3, dark greyish-green stratified diamictite; **j** Layers 4–7; **k** Light purplish-red and reddish-brown stratified diamictite of layers 4–7

Stop 4: Glacier geological culture museum in Luoquan village

Detailed description: A series of display boards and monuments set up in Luoquan village relate people about the long history of geological investigation and research of Luoquan Formation glacier in Luoquan village (Fig. 46).

At present, Luoquan village is gradually becoming a "culture village of glacial geology", and a Luoquan glacier geological culture museum in the center of Luoquan

Fig. 44 (continued)

village is planned to build. This is not only a "pilgrimage" in our field trip but also enables more people to understand glaciers from the perspective of science popularization, participate in the protection of glacier deposition, and help promote the science popularization and construction of livable earth.

4.5 DAY FOUR: August 30th Route 5 Yangpo Section

- **Key words: Ice-advance and ice-retreat at the Yangpo section**

12:00–15:30: After finishing the inspection of Luoquan Village in the morning, we will go to the Yangpo section to the west of it (24 km), and it takes about 1 h by car (Figs. 47 and 48). The section starts at 112° 35′ 47.81" E, 34° 3′ 4.33" N, and ends at 112° 35′ 47.54" E, 34° 2′ 55.60" N).

A total of 5 tuff zircon ages were obtained from the middle and top of the Luoyukou Formation of the Yangpo section, which are 1611 ± 8, 1634 ± 10, 1638 ± 9, 1639 ± 13, and 1662 ± 60 Ma, all indicating the Luoyukou Formation is older than 1.6 billion years (Su et al. 2012; Wang 2015; Li et al. 2017; Peng et al. 2018). Therefore, a huge unconformity at least one billion years apart should exist between the Luoquan Formation and the underlying Mesoproterozoic basement. The Luoquan Formation is unique in color, sedimentary structure, and gravel composition (Fig. 48), and can be divided into two distinct members. The iceberg keel plough marks developed on the top surface of the basal massive diamictite; the gravels of the massive diamictite in the lower part were from the erosion of the underlying basement dolostone and quartz sandstone; the middle and upper part is purplish-red bedded dropstone-bearing siltstone, with a large number of channel sandstones (Fig. 50a, d), gravel

Fig. 45 Large-scale soft-sediment deformation structures in the Luoquan Formation in the vicinity of the type section at Luoquan Village (Le Heron et al. 2018). **a** and **b** 30 m amplitude recumbent antiform. Evidence of breaching by low angle thrusts and mantled by comparatively less deformed beds. **c** and **d** Smaller-scale isoclinal fold, truncated by a possible thrust surface and overlain by undeformed subhorizontal diamictite. These soft-sediment deformation structures are interpreted as the products of glaciotectonism

and sandstone lenses (Fig. 51b, d). Small-scale cross-bedding (Fig. 50g) and weakly stratified diamictite (Figs. 4.36h and 51c) are developed; dropstones in different shapes are developed in the upper laminated silty mudstone (Figs. 50j, k, and 51e, i). The Luoquan Formation is unconformably overlain by cross-bedded and layered sandstone of the Xinji Formation (Fig. 50l). The boundary between the Luoquan Formation and the Xinji Formation exposes at the bottom of the ditch.

Stop 1: The boundary between the Mesoproterozoic Luoyukou Formation and the Ediacaran Luoquan Formation

Layer: Layer 0 of Yangpo section (Fig. 49)

Detailed description: The Luoyukou Formation of the Luoyu Group comprises yellowish-brown, light-grey-yellow, thin-bedded argillaceous dolostone, with a

Fig. 46 Luoquan culture village of glacial geology

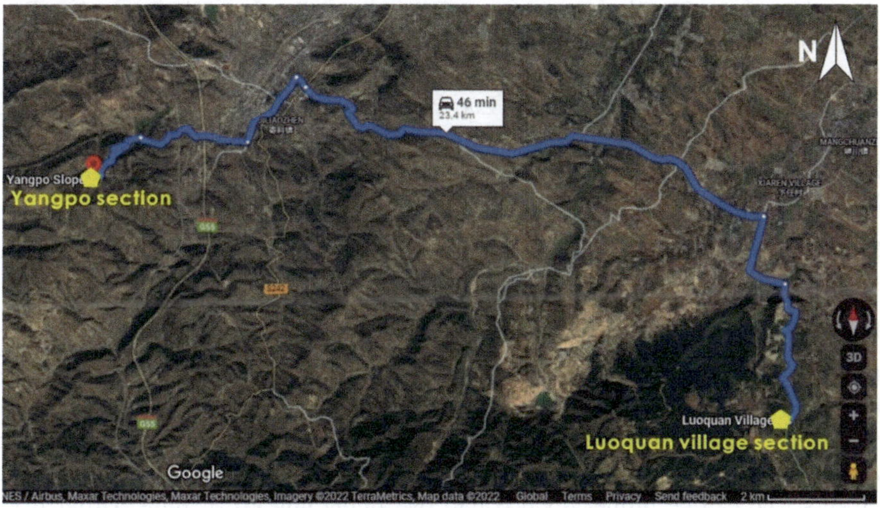

Fig. 47 Route map of Luoquan village section to Yangpo section (©Google Map)

Fig. 48 Panoramic photo of Yangpo section

thickness of 0.1–0.5 cm in a single layer. Parallel bedding, wavy bedding, and transitional layers are developed within the single layer. Purple mudstone and argillaceous siltstone are developed in the dolomites, which are recessed due to weathering (Figs. 50 and 51a).

We will observe the tuff of the Luoyukou Formation and its contact interface with the Luoquan Formation. The massive diamictite at the bottom of the Luoquan Formation unconformably in contact with the layered argillaceous dolomite of the Luoyukou Formation (Fig. 50a).

Stop 2: The diamictite and iceberg keel plough marks at the bottom of the Luoquan Formation

Layer: Layer 1 of Yangpo section (Fig. 49)

Detailed description: This outcrop exposes the sedimentary characteristics of the bottom of the Luoquan Formation. Details are as follows.

Layer 1, purple-red, reddish-brown massive diamictite. The gravels of diamictite are mainly cobbles and boulders, which represents a strongly eroded ice advance process, and can be further divided into two fining upward rhythms. Differ from other sections, the matrix is purple-red siltstone, and the clasts are mainly laminar argillaceous dolomite from the Luoyukou Formation, a small amount of quartz sandstone and quartzite are also observed (Fig. 50b). At the top of the massive diamictite, there are unusual grooves formed (Fig. 50c). An open question remains as to the origins of these, which we shall discuss at the outcrop together.

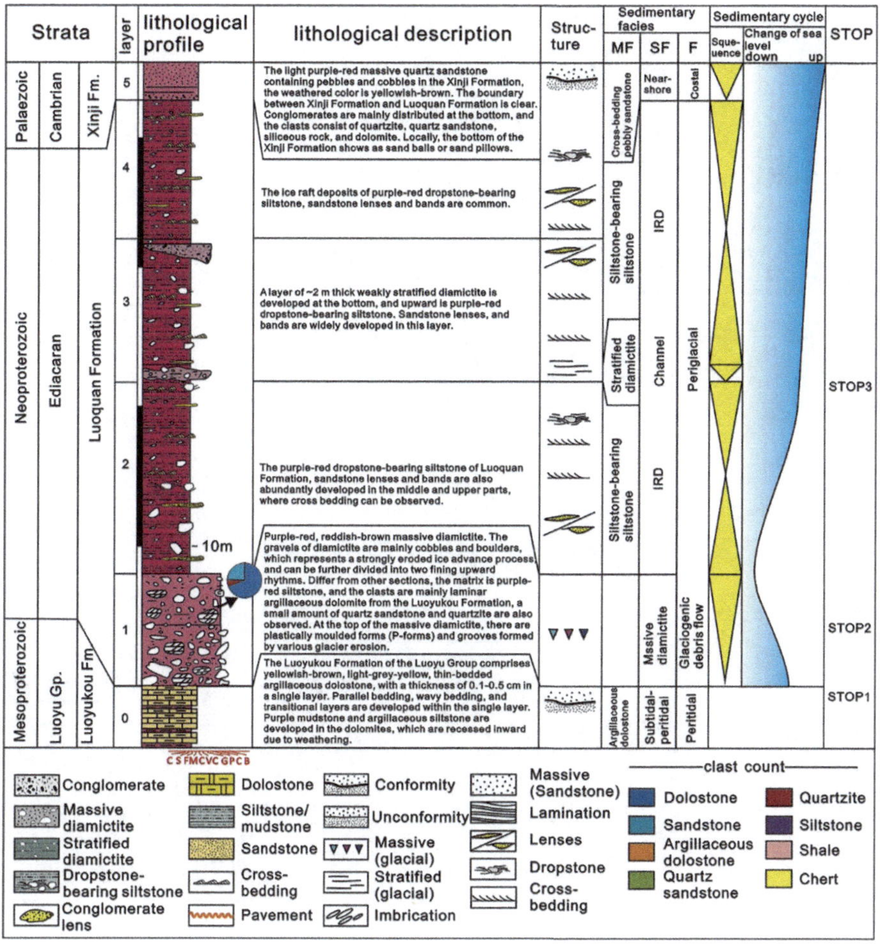

Fig. 49 Comprehensive stratigraphic histogram of Yangpo section

Stop 3: The boundary between the Luoquan Formation and the Xinji Formation and sedimentary characteristics of the middle and upper Luoquan Formation

Layers: Layers 2–5 of Yangpo section (Fig. 49)

Detailed description: We will mainly observe the sedimentary characteristics of the middle and upper Luoquan Formation and its contact relationship of the Xinji Formation, details are as follows.

Layer 2, the purple-red dropstone-bearing siltstone of Luoquan Formation, sandstone lenses, and bands are also abundantly developed in the middle and upper parts, where cross-bedding can be observed (Fig. 50d, e).

Fig. 50 Plates of stops 1–3 in Yangpo section. **a** The Luoyukou Formation of the Luoyu Group comprises yellowish-brown, light-grey-yellow, thin-bedded argillaceous dolostone, with a thickness of 0.1–0.5 cm in a single layer. Parallel bedding, wavy bedding, and transitional layers are developed within the single layer; **b** Layer 1, Light purple-red, reddish-brown massive diamictite; **c** Plastically moulded forms and grooves formed by various glacier erosion and iceberg keel plough at the top of the first layer; **d**, **e** Layer 2, the purple-red dropstone-bearing siltstone, sandstone lenses, and bands are also abundantly developed in the middle and upper parts; **f** Layer 3, purple-red dropstone-bearing siltstone, sandstone lenses, and bands are widely developed in this layer. **g**, **h** Layer 3, a layer of ~2 m thick weakly stratified diamictite is developed at the bottom, and upward is purple-red dropstone-bearing siltstone; **i**, **j** Layer 4, the ice raft deposits of purple-red dropstone-bearing siltstone; **k**, **l** Layer 5, the light purple-red massive quartz sandstone containing pebbles and cobbles in the Xinji Formation, the weathered color is yellowish-brown. The boundary between Xinji Formation and Luoquan Formation is clear

Fig. 50 (continued)

Layer 3, a layer of ~2 m thick weakly stratified diamictite is developed at the bottom, and upward is purple-red dropstone-bearing siltstone. Sandstone lenses and bands are widely developed in this layer (Fig. 50f–h).

Layer 4, the ice raft deposits of purple-red dropstone-bearing siltstone, sandstone lenses, and bands are common (Fig. 50i, j).

Layer 5: The light purple-red massive quartz sandstone containing pebbles and cobbles in the Xinji Formation, the weathered color is yellowish-brown. The boundary between Xinji Formation and Luoquan Formation is clear. Conglomerates are mainly distributed at the bottom, and the clasts consist of quartzite, quartz

Fig. 51 The lens body and the dropstone in the purple-red dropstone bearing siltstone in the upper Luoquan Formation of the Yangpo section, as well as the cobbles with striation (from Le Heron et al. 2018). The striated pavement and associated structures at Shangxuma. **a** Scale drawing of the striated pavement. Note the relationship between apparently crosscutting striation sets on the scale drawing. **b** Contact between striated pavement and diamictites lying directly on the contact. **c** View of an area of cross-cutting striae: a simple interpretation, without considering the wider geomorphic evidence, would be that these represent two phases of subglacial abrasion in central China. **d, e** Two views of the curvilinear nature of some striations at the outcrop scale, as shown in a. **f** Striated pavement draped by diamictite. Some striations are also developed in the very basal deposits (i.e., over a few mm) in this diamictite. **g** Organization of the striate pavement into an upper and a lower step, as also shown in panel A

sandstone, siliceous rock, and dolomite. Locally, the bottom of the Xinji Formation shows sand balls or sand pillows (Fig. 50k, l).

15:00–17:00 End the investigation and return to the hotel.

4.6 DAY FIVE: August 31st Route 6 Liuzhanggou Section

- **Key word: Sand wedges bridge the Precambrian and Cambrian**

7:00–8:00 Breakfast at 7:00 a.m… leave at 8:00 a.m.

8:00–11:30 We will go to the Liuzhanggou section to observe the sand wedge between 8:00 a.m. and 11:30 a.m.

The Liuzhanggou section's distance to the hotel is ~40 km, and the travel time is about 1 h. The section starts at 112° 41′ 39.8″ E, 34° 18′ 59.4″ N and ends at 112° 41′ 39.8″ E, 34° 18′ 59.4″ N. The section is located at the side of a country road, which is divided into two parts: the hillside and the bottom of a ditch (Figs. 52 and 53). The key points of this section are the sedimentary succession of Luoquan Formation diamictite and Xinji Formation and the contact relationship of two Formations. The lower part of the Luoquan Formation is mainly comprising greyish-green dropstone-bearing siltstone. Patches of greyish-green development locally developed in greyish-green dropstone-bearing siltstone. Gravel is the majority, occasionally medium to coarse gravel. Poor sorting and subangular-subrounded gravels are the majority, occasionally medium to coarse gravel. The composition of gravels is greyish-green subangular siltstone, followed by quartzites and quartz sandstone. with sandwiched thin layers of the yellowish-brown sandy conglomerate. Multiple thin layers of yellowish-brown sandy conglomerate extend along with the layer or diagonally, even some lenticular or sausage-shaped (Fig. 55b). The dropstone is abundant and has a common calcareous shell. The upper part of the Luoquan Formation is purplish-red dropstone-bearing siltstone (Fig. 55g–h). Glacial striation can be seen on some dropstone surfaces. Based on the above characteristics, we can conclude that the gravel was transported from the subglacial environment and pushed to the deepwater area by an iceberg raft. The sand wedges at the top of the Luoquan Formation are the most characteristic in the section and detailed in the next.

Fig. 52 Route map of hotel-Liuzhanggou section (©Google Map)

Stop 1: The diamictite of Luoquan Formation

Layers: Layers 1–5 of Liuzhanggou section (Fig. 54)

Detailed description: We mainly observed the sedimentary characteristics of the Luoquan Formation as follows:

Layer 1, light greyish-green, greyer dropstone-bearing siltstone, no obvious orientation. The dropstone is abundant and has a common calcareous shell (Fig. 55a–c).

Layer 2, light purplish red dropstone-bearing siltstone with sandwiched thin layers of the yellowish-brown sandy conglomerate. The yellowish-brown sandy conglomerate extends along with the layer or diagonally, even some lenticular or sausage-shaped. The dropstone-bearing siltstone in some layers is disseminated as greyish-green patches or rings after diagenesis.

Layer 3, yellowish-brown, light greyish-green, light greyish-yellow dropstone-bearing siltstone.

Layer 4, purplish-red dropstone-bearing siltstone. Locally intercalated with greyish-green dropstone-bearing siltstone. There are striations on the surface of the dropstone (Fig. 55d–f).

Layer 5, purplish-red dropstone-bearing siltstone and locally intercalated with imbricated gravel structures. Patches of greyish-green development locally developed in the layer.

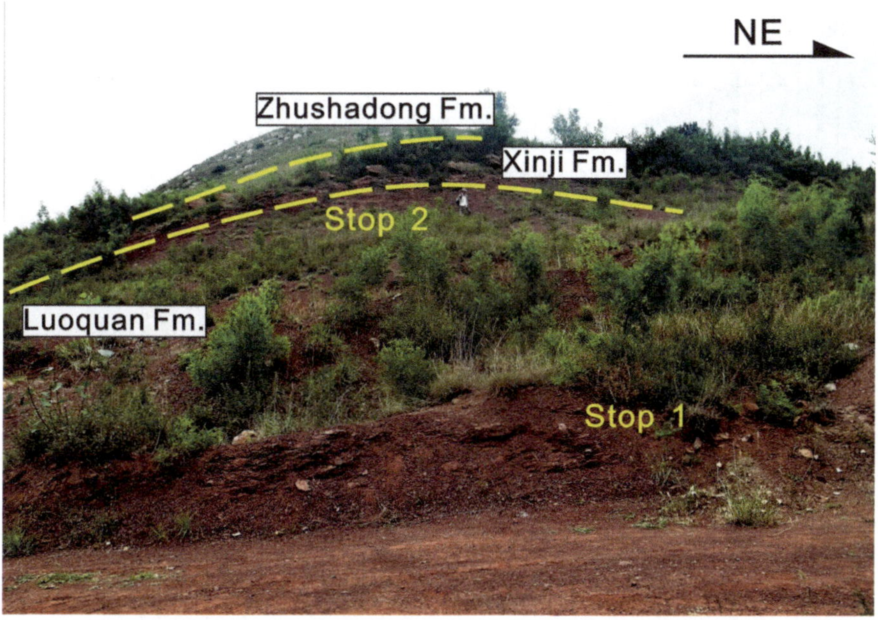

Fig. 53 Panoramic photo of Liuzhanggou section. Yellow dashed lines indicates the boundary of Luoquan Formation and Xinji Formation (lower line), and the boundary of Zhushadong Formation and Xinji Formation (upper line)

Stop 2: The boundary between Luoquan Formation and Cambrian Xinji Formation: Sand wedge is wedged from Xinji Formation into Luoquan Formation

Layers: Layers 6–7 of Liuzhanggou section (Fig. 54)

Detailed description: The characteristics of sand wedge and the boundary between Luoquan Formation and Cambrian Xinji Formation are mainly observed as follows:

Layer 6, purplish-red dropstone-bearing siltstone and sand wedge on the top surface of the diamictite of the Luoquan Formation and covered by sandstone or conglomerate of Xinji Formation.

The characteristics of the sand wedges are as follows:

The Liuzhanggou section displays 51 sand wedges in total (Fig. 56). Sand wedges in the vertical cross-section are 0.2–4 m deep and 3–25 cm wide. Sand wedges show horizontal spacing of 0.3–2 m (Figs. 56 and 57). The middle or upper parts of the sand wedges show a symmetrical narrow V-shape with slightly curved boundaries (Fig. 57b–e). Most sand wedges taper abruptly downward (from approximately 20 cm width to approximately 7–10 cm) in the upper part and continue tapering gradually downward (Fig. 4.43c, d). Some sand wedges display convex-outwards morphologies or occasionally branching, saclike and intersecting morphologies (Fig. 57). A small number of sand wedges have a spindle-shaped appearance in which the upper part

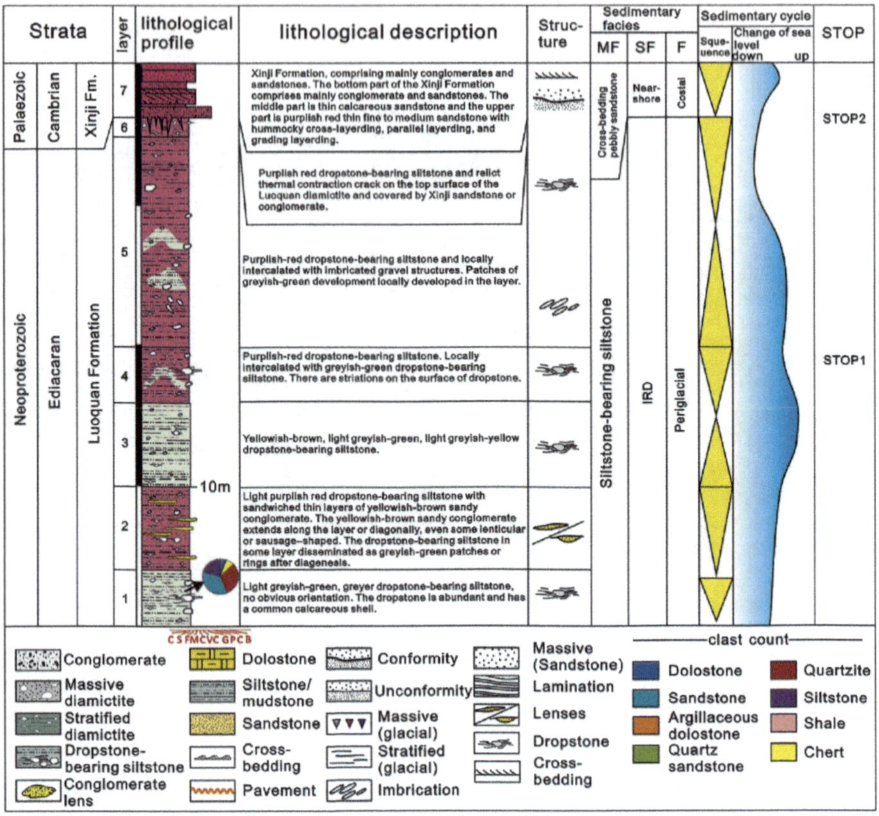

Fig. 54 Stratigraphic column of Liuzhanggou section

is narrower than the middle (Fig. 57b–e). The sand wedges are not exposed in plain view. However, 20 sand wedges in the vertical cross-section show that the directions of the extension are all different, with a trend orientation of 15°–195°, because a wedge parallel to the vertical cross-section does not appear on the section (Fig. 57k) (Wang et al. 2022).

The fillings, which are significantly different from Luoquan Formation diamictite, but have the same color, lithology, and texture as the overlying Xinji Formation sediments, are mainly sands and pebbles (Fig. 57). The composition of grains is chiefly subrounded by quartz and quartzites. Dolomite, sandstone, and mudstone grains are rare in the fillings of sand wedges. Observations under a scanning electron microscope show that 2 kinds of surface textures including dish-shaped impact scars and bulbous-edged impact scars are very conspicuous on the smooth surfaces of rounded quartz grains. Fillings display coarsening upwards, i.e., pebbles are concentrated upwards, but downwards, the fillings exhibit intense heterogeneity and good textural maturity (Fig. 57). In some sand wedges, the particle size often decreases distinctly from one side to the other side in the plane (Fig. 57e, j). Both vertical laminations

Fig. 55 Plates of stop 1 of Liuzhanggou section. **a–c** Layer 1, light greyish-green, greyer dropstone-bearing siltstone; **d** Layer 2, light purplish-red dropstone-bearing siltstone; **e** Layer 4, Purplish red dropstone-bearing siltstone; **f** striated dropstone in layer 4; **g, h** Layer 5, purplish-red dropstone-bearing siltstone; **i** The boundary between Luoquan Formation and Cambrian Xinji Formation (layer 6)

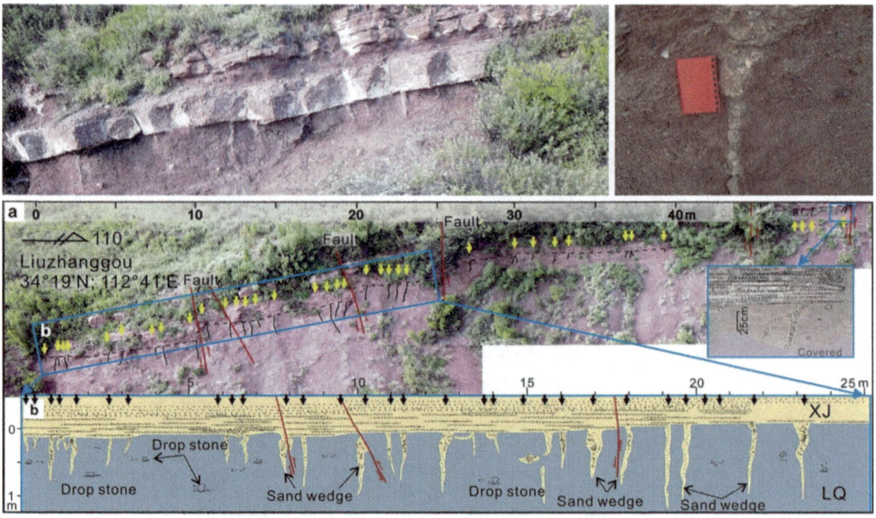

Fig. 56 Outcrop of Liuzhanggou section provide one-dimensional exposure of sand wedges

(mm-scale) of sands (Fig. 57f–j) and the a-axis of pebbles inside sand wedge very distinctly parallel the wedge boundary in some sand wedges (Fig. 57i). However, a massive infill, with a chaotic appearance and lacking systematic vertical trends in particle size, is found in the upper part of the majority of sand wedges (Fig. 57h). Notably, the host rock adjacent to wedges is undeformed in massive diamictites (Wang et al. 2022).

Layer 7, Xinji Formation, comprising mainly conglomerates and sandstones. The bottom part of the Xinji Formation comprises mainly conglomerate and sandstones. The middle part is thin calcareous sandstone and the upper part is purplish-red thin fine to medium sandstone with hummocky cross-layering, parallel layering, and grading layering.

11:30 End field observation and we will come back to the hotel for lunch.

12:00–13:00 Check-out time.

13:00 We will ride to the railway station together and participants depart from Ruzhou.

End of a Very Happy Trip.

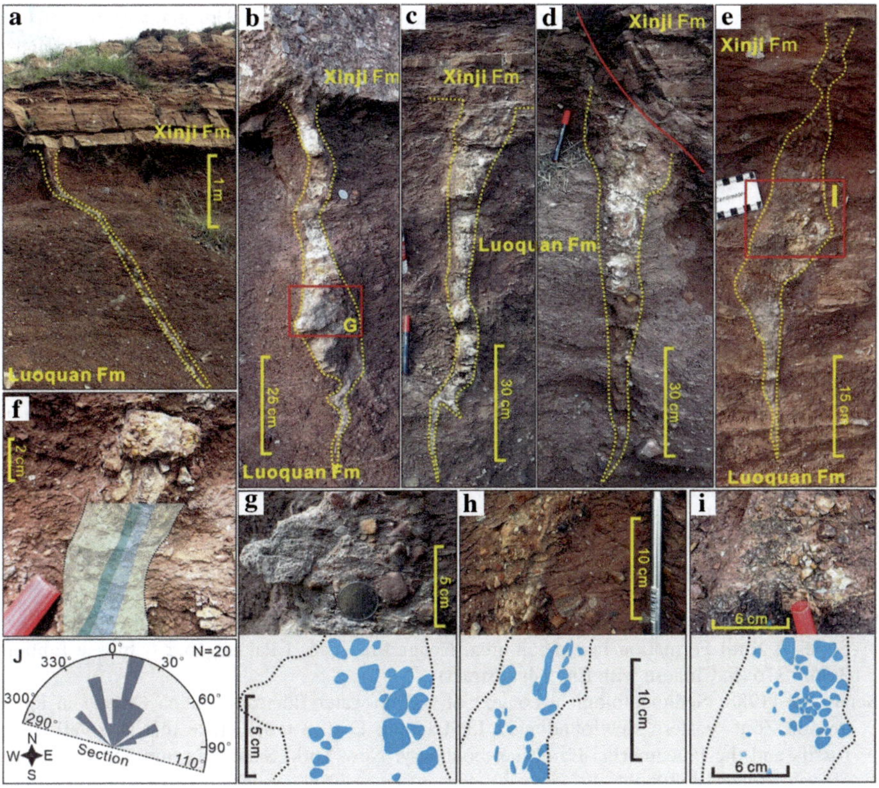

Fig. 57 The characteristics of sand wedges on a macroscopic scale. **a** A sand wedge with a visible depth of 4 m at Shangxuma; **b** V-shaped wedge with the middle bulge is covered by conglomerates of Xinji Formation; **c** The V-shaped wedge with a slightly curved boundary is covered by sandstone of Xinji Formation; **d** The V-shaped sand wedge tapered abruptly downward in the upper part and continued tapering slowly downward; **e, f** The sand wedge with particle size from 2 cm on one side to 0.1 cm on the other side; **g** The filling material of the wedge is mainly sand and pebbles; **h** is a scaled-up section of panel B; **i** At the middle of this wedge, the longest axes (a-axes) of pebbles parallel to the edges; **j** The rose diagram of trend orientation shows the branch, intersection, and may imply the polygonal patterns of sand wedges in the plane

Acknowledgements The study of the Ediacaran Luoquan Glaciation in North China is financially supported by the Peking International Giant Science Program (grant number: Z201100008320007); National Natural Science Foundation of China (grant number: 42072135); China Geological Survey program (grant number: DD20190005; DD20221649); National Key Research and Development Program of China (grant number: 2022YFF0800302; 2016YFC0601001) and the Neoproterozoic stratigraphic framework and sedimentary characteristics study of Shennongjia (AS2101).

The authors are very grateful to chief director Profs. Xingliao Zhang and Jincheng Feng from the Natural Resources of Henan province, chief engineer Prof. Shiyan Wang from Henan Institute of Geological Survey, Prof. Yuansheng Geng from the Institute of Geology, Chinese Academy of Geological Sciences, Prof. Deshun Zheng and Dr. Biyun Zhang from Henan Polytechnic University for great help. Sincere thanks are also given to the administrative institutions of the Department of Natural Resources of Henan Province for providing convenience for our work. Yanwei Zheng,

Xiao Yu, Fengyuan Yu, Hongli Zhang and Yushuang Wang from China University of Geosciences (Beijing) are appreciated for their UAV mapping and model building work. Qingbo Zhao from Henan Provincial Administration of Cultural Heritage for the files of Xianren cave in Lushan Shimengou section.

Last but not the least, the authors owe gratitude to the guidance given by the organizing committee of the 21st International Sedimentological Congress to the route selection of this field trip and the compilation of this guidebook, and we wish the congress a complete success.

References

Ahn SY, Zhu M (2017) Lowermost Cambrian acritarchs from the Yanjiahe Formation, South China: implication for defining the base of the Cambrian in the Yangtze platform. Geol Mag 154(6):1217–1231

Allen PA, Leather J, Brasier MD (2004) The Neoproterozoic Fiq glaciation and its aftermath, Huqf supergroup of Oman. Basin Res 16(4):507–534

Allen PA, Etienne JL (2008) Sedimentary challenge to Snowball Earth. Nat Geosci 1(12):817–825

Bai H, Kuang H, Liu Y et al (2020) Marinoan-aged red beds at Shennongjia, South China: evidence against global-scale glaciation during the Cryogenian. Palaeogeogr Palaeoclimatol Palaeoecol 559:109967

Bai WB, Qi YA, Guo YH et al (2018) Storm deposits and relevant trace fossils from the Cambrian Series 2 Xinji Formation in Lushan area, Henan Province. J Palaeogeogr (Chinese Edition) 3:365–376 (in Chinese with English abstract)

Benus AP (1988) Sedimentological context of a deep-water Ediacaran fauna (Mistaken Point, Avalon Zone, eastern Newfoundland). In: Landing E et al (eds) Trace fossils, small shelly fossils and the Precambrian-Cambrian boundary: New York. State Museum and Geological Survey Bulletin 463:8–9

Bingen B, Griffin WL, Torsvik TH et al (2005) Timing of Late Neoproterozoic glaciation on Baltica constrained by detrital zircon geochronology in the Hedmark Group, south-east Norway. Terra Nova 17(3):250–258

Brasier MD (1990) Phosphogenic events and skeletal preservation across the Precambrian-Cambrian boundary interval. Geol Soc Lond Spec Publ 52(1):289–303

Cai XF, Yang J, He WJ et al (2013) A restudy of the upper Sinian Zhengmuguan and Tuerkeng formations in the Helan Mountains. J Stratigr 37:377–386 (in Chinese with English abstract)

Cai XF, He WJ, Yang J et al (2014) Sedimentary characteristics, regional comparison and implication of the Sinian cap dolomites in the middle Helan Mountain. Resour Surv Environ 35:106–113 (in Chinese with English abstract)

Cao G, Zhang S, Liu ZQ et al (2006a) Discussion on age of the Fengtai Formation along Southeastern margin of the North China Block. Chin J Geol (scientia Geologica Sinica) 41(4):720–728 (in Chinese with English abstract)

Cao G, Zhang S, Sui FG et al (2006b) Analysis on the Lithofacies and Paleogeography of Fengtai Group in the Southeast North China Block. Acta Sedimentol Sin 24(2):210–216 (in Chinese with English abstract)

Cawood PA, Strachan RA, Pisarevsky SA et al (2016) Linking collisional and accretionary orogens during Rodinia assembly and breakup: implications for models of supercontinent cycles. Earth Planet Sci Lett 449:118–126

Chen RJ (1990) Division and correlation of Middle and Upper Proterozoic in western Henan and its adjacent areas. Henan Geol 8(3):28–34 (in Chinese)

Chen X, Kuang H, Liu Y et al (2020) Subglacial bedforms and landscapes formed by an ice sheet of Ediacaran-Cambrian age in west Henan, North China. Precambr Res 344:105727

Chen X, Kuang H, Liu Y et al (2021) Revisiting the Nantuo Formation in Shennongjia, South China: a new depositional model and multiple glacial cycles in the Cryogenian. Precambr Res 356:106132

Chen XS, Kuang HW, Liu YQ et al (2023) Sedimentary characteristics and evolution of Ediacaran glaciation in west Henan, southern North China Craton. Palaeogeography Palaeoclimatology Palaeoecology 632:111866

Chumakov NM (2004) Trends in global climate changes inferred from geological data. Stratigr Geol Correl 12(2):117–138

Chumakov NM (2011) Glacial deposits of the Baykonur Formation, Kazakhstan and Kyrgyzstan. Geol Soc Lond Mem 36(1):303–307

Cohen PA, Macdonald FA (2015) The Proterozoic record of eukaryotes. Paleobiology 41(4):610–632

Condon D, Zhu M, Bowring S et al (2005) U-Pb ages from the neoproterozoic Doushantuo Formation, China. Science 308(5718):95–98

Corkeron ML, George AD (2001) Glacial incursion on a Neoproterozoic carbonate platform in the Kimberley region, Australia. Geol Soc Am Bull 113(9):1121–1132

Corsetti FA, Grotzinger JP (2005) Origin and significance of tube structures in Neoproterozoic post-glacial cap carbonates: example from Noonday Dolomite, Death Valley, United States. Palaios 20(4):348–362

Creveling JR, Bergmann KD, Grotzinger JP (2016) Cap carbonate platform facies model, Noonday Formation, SE California. Bulletin 128(7–8):1249–1269

de Alvarenga CJS, Figureueiredo MF, Babinski M et al (2007) Glacial diamictites of Serra Azul Formation (Ediacaran, Paraguay belt): evidence of the Gaskiers glacial event in Brazil. J S Am Earth Sci 23(2–3):236–241

Dahl R (1965) Plastically sculptured detail forms on rock surfaces in northern Nordland, Norway. Geogr Ann Ser B 47:83–140

Dong XP, Hu JM, Li ZH et al (2017) Provenance of Ediacaran (Sinian) sediments in the Helanshan area, North China Craton: constraints from U-Pb geochronology and Hf isotopes of detrital zircons. Precambr Res 298:490–511

Ernst RE (2014) Large igneous provinces. Cambridge University Press, Cambridge

Ernst RE, Youbi N (2017) How Large Igneous Provinces affect global climate, sometimes cause mass extinctions, and represent natural markers in the geological record. Palaeogeogr Palaeoclimatol Palaeoecol 478:30–52

Etemad-Saeed N, Hosseini-Barzi M, Adabi MH et al (2016) Evidence for ca. 560 Ma Ediacaran glaciation in the Kahar formation, central Alborz Mountains, northern Iran. Gondwana Res 31:164–183

Etienne JL, Allen PA, le Guerroué E et al (2011) The Blaini formation of the lesser Himalaya, NW India. Geol Soc Lond Mem 36(1):347–355

Eyles N, Young GM (1994) Early and Late Paleozoic glaciations. In: Deynoux M, Miller JMG, Domack EW et al (eds) Geodynamic controls on glaciation in Earth history, in Earth's Glacial Record. Cambridge University Press, Cambridge, pp 10–18

Eyles N (2008) Glacio-epochs and the supercontinent cycle after ~3.0 Ga: tectonic boundary conditions for glaciation. Palaeogeogr Palaeoclimatol Palaeoecol 258:89–129

Fike DA, Grotzinger JP, Pratt L M et al (2006) Oxidation of the Ediacaran ocean. Nature 444:744e747

Font E, Trindade RIF, Nédélec A (2006) Remagnetization in bituminous limestones of the Neoproterozoic Araras Group (Amazon craton): hydrocarbon maturation, burial diagenesis, or both? J Geophys Res: Solid Earth 111:B06204

Gao LZ, Wang ZQ, Xu ZQ et al (2010) A new evidence from zircon SHRIMP U-Pb dating of the Neoproterozoic diamictite in Quruqtagh area, Tarim basin, Xinjiang, China. Geol Bull China 29(2/3):205–213 (in Chinese with English abstract)

Gao LZ, Guo XP, Ding XZ et al (2013) Nanhuan glaciation event and its stratigraphic correlation in tarim plate, China. Acta Geosci Sinica 34(1):39–57 (in Chinese with English abstract)

Ge DK, Cui ZJ (1993) A Basal melt-out structure in Luoquan formation and its significance. Acta Geol Sin 4:348–356 (in Chinese with English abstract)

Germs GJB, Gaucher C (2012) Nature and extent of a late Ediacaran (ca. 547 Ma) glacigenic erosion surface in southern Africa. S Afr J Geol 115(1):91–102

Grotzinger JP, Fike DA, Fischer WW (2011) Enigmatic origsin of the largest-known carbon isotope excursion in Earth's history. Nat Geosci 4(5):285–292

Gu QC (1982) Late Precambrian Moraine in Helan Mountain. J Stratigr 6:156–157 (in Chinese)

Guan BD, Geng WC, Rong ZQ et al (1988) Middle-Upper Proterozoic on the north slope of East Qinling. Henan Science and Technology Press, Zhengzhou, Henan

Guan B, Wu R, Hambrey MJ et al (1986) Glacial sediments and erosional pavements near the Cambrian-Precambrian boundary in western Henan Province, China. J Geol Soc 143:311–323

Haldorsen S, Von Brunn V, Maud R et al (2001) A Weichselian deglaciation model applied to the Early Permian glaciation in the northeast Karoo Basin, South Africa. J Quat Sci 16(6):583–593

Hambrey MJ, Glasser NF (2012) Discriminating glacier thermal and dynamic regimes in the sedimentary record. Sed Geol 251–252:1–33

He TG, Pei F (1985) The discovery of bivalves from the lower Cambrian Xinji formation in fangcheng county, Henan province. J Chengdu Coll Geol (1):61–66+113–114 (in Chinese with English abstract)

Hebert CL, Kaufman AJ, Penniston-Dorland SC et al (2010) Radiometric and stratigraphic constraints on terminal Ediacaran (post-Gaskiers) glaciation and metazoan evolution. Precambr Res 182(4):402–412

Hewitt G (2000) The genetic legacy of the quaternary ice ages. Nature 405(6789):907–913

Hoffmann PF, Kaufman AJ, Halverson GP et al (1998) A neoproterozoic snowball earth. Science 281(5381):1342–1346

Hoffman PF, Halverson GP, Domack EW et al (2007) Are basal Ediacaran (635 Ma) post-glacial "cap dolostones" diachronous? Earth Planet Sci Lett 258(1–2):114–131

Hoffman PF, Li ZX (2009) A palaeogeographic context for Neoproterozoic glaciation. Palaeogeogr Palaeoclimatol Palaeoecol 277(3):158–172

Hoffman PF (2009) Pan-glacial-a third state in the climate system. Geol Today 25(3):100–107

Hoffman PF, Macdonald FA (2010) Sheet-crack cements and early regression in Marinoan (635 Ma) cap dolostones: regional benchmarks of vanishing ice-sheets? Earth Planet Sci Lett 300(3–4):374–384

Hofmann M, Linnemann U, Hoffmann KH et al (2015) The four Neoproterozoic glaciations of southern Namibia and their detrital zircon record: the fingerprints of four crustal growth events during two supercontinent cycles. Precambr Res 259:176–188

Högström A, Jensen S, Palacios T et al (2013) New information on the Ediacaran-Cambrian transition in the Vestertana Group. Finnmark, northern Norway, from trace fossils and organic-walled microfossils 93:95–106

Hong QY (1984) Preliminary research on gravity flow sediments of Sinian period in several areas, China. J Southwest Petroleum Inst 4:1–17 (in Chinese with English abstract)

Hu GH, Zhou YY, Zhao TP (2012a) Geochemistry of Proterozoic Wufoshan Group sedimentary rocks in the Songshan area, Henan Province: implications for provenance and tectonic setting. Acta Petrologica Sinica 28(11):3692–3704 (in Chinese with English abstract)

Hu B, Zhai M, Li T et al (2012b) Mesoproterozoic magmatic events in the eastern North China Craton and their tectonic implications: Geochronological evidence from detrital zircons in the Shandong Peninsula and North Korea. Gondwana Res 22(3–4):828–842

Hu B, Zhai MG, Wu JL et al (2015) Zircon LA-ICP-MSU-PB age of Upper Proterozoic moraine clasts in the southern margin of North China block and its geological significance. Acta Mineral Sinica 35(S1):701 (in Chinese)

Hu J, Li Z, Gong W et al (2016) Meso-neoproterozoic stratigraphic and tectonic framework of the North China craton. In: Zhai M, Zhao Y, Zhao T (eds) Main tectonic events and metallogeny of the North China Craton. Springer, Singapore, pp 393–422

Hu J, Wang H, Wang M (2017) Geochemistry and origin of the Neoproterozoic Dahongliutan banded iron formation (BIF) in the Western Kunlun orogenic belt, Xinjiang (NW China). Ore Geol Rev 89:836–857

Hu J, Li C, Tong J et al (2020a) Glacial origin of the Cryogenian Nantuo Formation in eastern Shennongjia area (South China): Implications for macroalgal survival. Precambr Res 351:105969

Hu J, Wang H, Zhang L (2020b) A rare earth element and Nd isotopic investigation into the provenance and deposition of the Dahongliutan banded iron formation and associated carbonates, NW China: Implications on Neoproterozoic seawater compositions. Precambr Res 342:105685

James NP, Narbonne GM, Kyser TK (2001) Late Neoproterozoic cap carbonates: Mackenzie Mountains, northwestern Canada: precipitation and global glacial meltdown. Can J Earth Sci 38(8):1229–1262

Jenkins RJF, Cooper JA, Compston W (2002) Age and biostratigraphy of Early Cambrian tuffs from SE Australia and southern China. J Geol Soc 159(6):645–658

Jiang G, Kennedy MJ, Christie-Blick N et al (2006) Stratigraphy, sedimentary structures, and textures of the late Neoproterozoic Doushantuo cap carbonate in South China. J Sediment Res 76(7):978–995

Kang Y, Chen G, Xia XY et al (2018) Detrial Zircon U-Pb Geochronology and Its Geological Implication of the Nancaode and Zhuanghegou Formations in southern margin of Ordos Basin. Acta Geol Sin 92(9):1829–1842 (in Chinese with English abstract)

Kaufman AJ, Knoll AH (1995) Neoproterozoic variations in the C-isotopic composition of seawater: stratigraphic and biogeochemical implications. Precambr Res 73(1–4):27–49

Keller CB, Husson JM, Mitchell RN et al (2019) Neoproterozoic glacial origin of the Great Unconformity. Proc Natl Acad Sci 116(4):1136–1145

Kennedy MJ (1996) Stratigraphy, sedimentology, and isotopic geochemistry of Australian Neoproterozoic postglacial cap dolostones; deglaciation, delta 13 C excursions, and carbonate precipitation. J Sediment Res 66(6):1050–1064

Kim MC, Yang JH, Peng P et al (2016) Characteristics of carbon isotope of the Ediacaran and Lower Cambrian strata in the Pyongnam basin, DPR Korea. Acta Petrol Sinica 32(10):3180–3186 (in Chinese with English abstract)

Kirschvink JL (1992) Late Proterozoic low-latitude glaciation: the Snowball Earth. In: Schopf JW, Klein C (eds) The proterozoic biosphere. Cambridge University Press, Cambridge, pp 51–52

Kirschvink JL, Gaidos EJ, Bertani LE et al (2000) Paleoproterozoic snowball Earth: extreme climatic and geochemical global change and its biological consequences. Proc Natl Acad Sci 97(4):1400–1405

Kuang HW, Liu YQ, Peng N et al (2019) Carbon isotopic characteristics and sedimentology of Ediacaran cap carbonates: a case study from the Shennongjia area, northern Yangtze craton. Acta Geol Sin 93(9):2139–2157 (in Chinese with English abstract)

Kuang HW, Liu YQ, Peng N et al (2022) Ediacaran cap dolomite of Shennongjia, northern Yangtze Craton, South China. Precambr Res 368(11):106483

Lang X, Chen J, Cui H et al (2018) Cyclic cold climate during the Nantuo glaciation: evidence from the Cryogenian Nantuo Formation in the Yangtze Block, South China. Precambr Res 310:243–255

Le Guerroué E, Allen PA, Cozzi A (2006) Chemostratigraphic and sedimentological framework of the largest negative carbon isotopic excursion in Earth history: the Neoproterozoic Shuram Formation (Nafun Group, Oman). Precambr Res 146:68–92

Le Guerroué E (2010) Duration and synchroneity of the largest negative carbon isotope excursion on Earth: the Shuram/Wonoka anomaly. Comptes Rendus Geoscience 342:204e214

Le Heron DP, Cox G, Trundley A et al (2011) Sea ice-free conditions during the Sturtian glaciation (early Cryogenian), South Australia. Geology 39(1):31–34

Le Heron DP, Busfield ME, Kamona F (2013) An interglacial on snowball Earth? Dynamic ice behaviour revealed in the Chuos Formation, Namibia. Sedimentology 60(2):411–427

Le Heron DP, Vandyk TM, Kuang H et al (2019) Bird's-eye view of an Ediacaran subglacial landscape. Geology 47:705–709

Le Heron DP, Vandyk TM, Wu G et al (2018) New perspectives on the Luoquan Glaciation (Ediacaran-Cambrian) of North China. Depos Rec 4:274–292

Letsch D, Large SJE, Buechi MW et al (2018) Ediacaran glaciations of the west African Craton-Evidence from Morocco. Precambr Res 310:17–38

Li CD, Zhao LG, Chang QS, Xu YW, Wang SY, Xu T (2017) Zircon U-Pb dating of tuff bed from Luoyukou Formation in western Henan Province on the southern margin of the North China Craton and its stratigraphic attribution discussion. Geol China 44(3):511–525 (in Chinese with English abstract)

Li H F (2011) Study of the genesis and significance of glacial rocks of Zhengmuguan formation of Sinian system Helanshan in Ningxia. Dissertation, Shandong University of Science and Technology (in Chinese with English abstract)

Li L, Zhang X, Yun H et al (2016) New occurrence of Cambroclavus absonus from the lowermost Cambrian of North China and its stratigraphical importance. Alcheringa: An Australas J Palaeontol 40(1):45–52

Li M, Vandyk TM, Wu G et al (2020) A window into the great unconformity: insights from geochemistry and geochronology of Ediacaran glaciogenic rocks in the North China Craton. J Asian Earth Sci 194:104327

Li W, Li DY, Zheng SD (1990) The sedimentary characteristics of the alluvial fan of the underwater off-shore slope, Luoquan Formation in the Yiyang area, Henan Province. J Xi'an College Geol 12(3):32–38;102–103 (in Chinese with English abstract)

Li ZS, Zhao ZY, Li JX et al (2018) Provenance analysis of the black clastic rock series at the Cambrian Base in the Southeastern Margin of North China Craton and its constraints on the late Precambrian geologic evolution. Acta Geol Sin 92(9):1803–1828 (in Chinese with English abstract)

Li ZX, Evans DAD, Halverson GP (2013) Neoproterozoic glaciations in a revised global palaeogeography from the breakup of Rodinia to the assembly of Gondwanaland. Sedimentary Geol 294:219–232

Liu CA (1960) Tillite in early Paleozoic in southern North China and its stratigraphic significance. Acad Sci Ministry Geol (in Chinese)

Liu YH (1986) Xinji formation of Low Cambrian in Henan province. Henan Geol 4:30–37 (in Chinese)

Liu YQ, Kuang HW, Chen XS et al (2023) Ediacaran diamictites and glaciation in northern China. Acta Geol Sin 97(12). https://doi.org/10.19762/j.cnki.dizhixuebao.2023158 (in Chinese with English abstract)

Lu SN, Zhao GC, Wang HC et al (2008) Precambrian metamorphic basement and sedimentary cover of the North China Craton: a review. Precambr Res 160(1–2):77–93

Macdonald FA, Wordsworth R (2017) Initiation of Snowball Earth with volcanic sulfur aerosol emissions. Geophys Res Lett 44(4):1938–1946

McCay GA, Prave AR, Alsop GI et al (2006) Glacial trinity: neoproterozoic Earth history within the British-Irish caledonides. Geology 34:909–912

Meng Y, Zuo P, Zheng D et al (2018) The earliest clastic sediments overlying the Xiong'er volcanic rocks: Implications for the Mesoproterozoic tectonics of the southern North China Craton. Precambr Res 305:268–282

Miao LY, Zhu MY (2014) Trace fossils from the basal Cambrian Xinji formation in southern north China plate and its chronological significance. Acta Palaeontol Sinica 53(3):274–289 (in Chinese with English abstract)

Mickala OR, Vidal L, Boudzoumou F et al (2014) Geochemical characterization of the Marinoan "Cap Carbonate" of the Niari-Nyanga basin (central Africa). Precambr Res 255:367–380

Miller N, Johnson PR, Stern RJ (2008) Marine versus non-marine environments for the Jibalah Group, NW Arabian shield: a sedimentologic and geochemical survey and report of possible Metazoa in the Dhaiqa Formation. Arab J Sci Eng 33(1):55–78

Mu YJ (1981) Luoquan Tillite of the Sinian System in China. In: Hambrey MJ, Harland WB (eds) Earth's pre-Pleistocene glacial record. Cambridge University Press, Cambridge, pp 402–413
Mu YJ (1983) Origin and age of Moraine in Louquan formation. Henan Geol 2:35–46 (in Chinese with English abstract)
Narbonne GM, Xiao S, Shields GA et al (2012) The Ediacaran period. In: Gradstein FM, Ogg JG, Schmitz MD et al (eds) The geologic time scale. Elsevier, Netherland, pp 413–435
Ogg JG, Ogg GM, Gradstein FM (2016) A concise geologic time scale: 2016. Elsevier, Netherland
Pan B, Topper TP, Skovsted CB et al (2018) Occurrence of Microdictyon from the lower Cambrian Xinji Formation along the southern margin of the North China Platform. J Paleontol 92(1):59–70
Pang K, Wu C, Sun Y et al (2021) New Ediacara-type fossils and late Ediacaran stratigraphy from the northern Qaidam Basin (China): paleogeographic implications. Geology. https://doi.org/10.1130/G48842.1
Peng N, Kuang H, Liu Y et al (2018) Recognition of geological age for acanthomorphic acritarchs from the Ruyang Group southern margin of North China Craton and its implication for evolution of early eukaryotes. J Palaeogeogr (Chinese Edition) 20(04):595–608
Pu JP, Bowring SA, Ramezani J et al (2016) Dodging snowballs: Geochronology of the Gaskiers glaciation and the first appearance of the Ediacaran biota. Geology 44(11):955–958
Runnegar B (1985) Shell microstructures of Cambrian molluscs replicated by phosphate. Alcheringa 9(4):245–257
Rooney AD, Strauss JV, Brandon AD et al (2015) A Cryogenian chronology: two long-lasting synchronous Neoproterozoic glaciations. Geology 43(5):459–462
Rose CV, Maloof AC (2010) Testing models for post-glacial 'cap dolostone' deposition: Nuccaleena Formation, South Australia. Earth Planet Sci Lett 296(3–4):165–180
Sawaki Y, Nishizawa M, Suo T et al (2008) Internal structures and U-Pb ages of zircons from a tuff layer in the Meishucunian formation, Yunnan Province, South China. Gondwana Res 14(1–2):148–158
Shen B, Xiao S, Dong L et al (2007) Problematic macrofossils from Ediacaran successions in the North China and Chaidam blocks: implications for their evolutionary roots and biostratigraphic significance. J Paleontol 81(6):1396–1411
Shen B, Xiao S, Zhou C et al (2010) Carbon and sulfur isotope chemostratigraphy of the Neoproterozoic Quanji Group of the Chaidam Basin, NW China: Basin stratification in the aftermath of an Ediacaran glaciation postdating the Shuram event? Precambr Res 177:241–252
Shen B, Xiao S, Bao H et al (2011) Carbon, sulfur, and oxygen isotope evidence for a strong depth gradient and oceanic oxidation after the Ediacaran Hankalchough glaciation. Geochim Cosmochim Acta 75:1357–1373
Stouge S, Christiansen JL, Harper DAT et al (2011) Neoproterozoic (Cryogenian–Ediacaran) deposits in East and North-East Greenland. Geol Soc Lond, Mem 36(1):581–592
Su WB, Li HK, Xu L et al (2012) Luoyu and Ruyang Group at the south margin of the North China Craton (NCC) should belong in the Mesoproterozoic Changchengian System: direct constraints from the LA-MC-ICPMS U-Pb age of the tuffite in the Luoyukou Formation, Ruzhou, Henan, China. Geol Surv Res 35(2):96–108 (in Chinese with English abstract)
Torsvik TH, Cocks LRM (2013) Gondwana from top to base in space and time. Gondwana Res 24(3–4):999–1030
Vernhet E, Youbi N, Chellai EH et al (2012) The Bou-Azzer glaciation: evidence for an Ediacaran glaciation on the west African craton (anti-atlas, Morocco). Precambr Res 196:106–112
Vickers-Rich P, Fedonkin MA (2007) The proterozoic (2.5 billion to 542 million years ago). In: Fedonkin MA, Gehling JG, Grey K et al (Eds) The rise of animals. Evolution and diversification of the Kingdom Animalia. The Johns Hopkins University Press, Baltimore, pp 29–51
Wang MZ, Yang TF (1996) Discussion on the development and genesis of Luoquan formation in Yiyang county, Henan province. J xi'an College Geol 2:13–17 (in Chinese with English abstract)
Wang PC (2015) The study on the geological causes and ages of the Luoquan and Dongpo formations in Ruzhou, western of Henan. Dissertation, Guizhou University (in Chinese with English abstract)

Wang PC, Wang Y (2015) The geological cause of the Luoquan formation in Ruzhou, western Henan. J Guizhou Univ (natural Sciences) 32:49–52 (in Chinese with English abstract)

Wang SX, Guo LY, Zhao DH (1998) Sinian system in Luoquan-Gucheng area, Shanxi. Northwest Geosci 2:38–40 (in Chinese with English abstract)

Wang X, Wang Z (1993) Gravity flow and slump deposits of the Fengtai formation in western Anhui. Reg Geol China 2:131–139 (in Chinese with English abstract)

Wang X, Zhang XL, Liu W et al (2021) Biostratigraphic constraints on the age of Neoproterozoic glaciation in North China. J Asian Earth Sci 219:104894

Wang YC, Kuang HW, Liu YQ et al (2022) Composite sand-ice wedge pseudomorphs suggestive of a frosty Ediacaran–Cambrian transition. Precamb Res 382:106873

Wang XF, Li DW, Du YS et al (2015) Zircon U-Pb Dating of the Sub-Volcanic Rocks Dismantled from Gaoshanhe and Yunmengshan Formations in Southern Margin of North China Craton. Geotecton Metallog 39(5):919–932 (in Chinese with English abstract)

Wang ZH (1996) Researching on stratigraphic framework and discussion on the chronolithology of Luoquan age to Dongpo age. Henan Geol 14(4):17–23 (in Chinese with English abstract)

Wu RT, Guan BD (1988a) Glacigenic characteristics of the Luoquan Formation and Sediment gravity flow reworking on It. Acta Geol Sinica-Eng Edn 1:325–339

Wu RT, Guan BD (1988b) On the glacicene characteristics of the Luoquan Formation and its reworking by gravity flows. Acta Geol Sin 1:78–89 (in Chinese with English abstract)

Wu RT, Yang YQ (1991) The indicators for recognizing glacial event. Earth Sci J China Univ Geosci (4):353–359+477 (in Chinese with English abstract)

Xiao S, Bao H, Wang H et al (2004) The Neoproterozoic Quruqtagh Group in eastern Chinese Tianshan: evidence for a post-Marinoan glaciation. Precambr Res 130(1–4):1–26

Xing YS, Gao ZJ, Liu GZ et al (1989) The Upper Precambrian of China. Geological Publishing House, Beijing, China (in Chinese)

Xu B, Xiao S, Zou H et al (2009) SHRIMP zircon U-Pb age constraints on Neoproterozoic Quruqtagh diamictites in NW China. Precambr Res 168(3):247–258

Xu X, Song S, Su L et al (2015) The 600–580 Ma continental rift basalts in North Qilian Shan, northwest China: links between the Qilian-Qaidam block and SE Australia, and the reconstruction of East Gondwana. Precambr Res 257:47–64

Yang YQ, Wu RT (1992) Underlying paleo ice erosion surface and ice erosion structure of Luoquan Formation in western Henan. Hennan Geol 10(4):282–285+321 (in Chinese)

Yang YQ, Wu RT (1994) The sedimentary characteristics and genesis anylsis for complex conglomerate of Luoquan Formation in western Henan. Hennan Geol 2:119–126 (in Chinese with English abstract)

Yang RD, Zhang CL, Luo XR et al (2007) New evidence on Neoproterozoic Hankalchough Glaciation in Quruqtagh, Xinjiang, West China. Geol Rev (2):228–232+298 (in Chinese with English abstract)

Yang J, Zeng ZX, Cai XF et al (2013) Carbon and oxygen isotopes analyses for the Sinian carbonates in the Helan Mountain, North China. Chin Sci Bull 58(32):3943–3955

Yang J, Lyons TW, Zeng Z et al (2019) Geochemical constraints on the origin of Neoproterozoic cap carbonate in the Helan Mountains, North China: implications for mid-late Ediacaran glaciation? Precambr Res 331:105361

Yang SP, Zheng SN (1985) The Trace fossils of Zhengmuguan Foraion in the Helan Mountain, Ningxia. J Earth Sci 10:9–10 (in Chinese)

Yao J, Xiao S, Yin L et al (2005) Basal Cambrian microfossils from the Yurtus and Xishanblaq formations (Tarim, north-west China): systematic revision and biostratigraphic correlation of Micrhystridium-like acritarchs. Palaeontology 48(4):687–708

Ye G (1984) Geological investigation of Luoquan Formation. Geol Sci Technol Inf 1:69–71 (in Chinese)

Yin CY, Gao LZ (1995) The early evolution of the acanthomobphic acritarchs in China and their biostratigraphical implication. Acta Geol Sinica (4):360–371+385–386 (in Chinese with English abstract)

Yin LM, Guan BD (1999) Organic-walled microfossils of Neoproterozoic Dongjia Formation, Lushan County, Henan Province, North China. Precambr Res 94(1–2):121–137

Youbi N, Ernst R E, Söderlund U et al (2020) The Central Iapetus magmatic province: an updated review and link with the ca. 580 Ma Gaskiers glaciation. In: Adatte T, Bond DPG, Keller G et al (eds) Mass extinctions, volcanism, and impacts: new developments. Geol Soc Am Spec Pap 544, Colorado, p 35–66

Young GM, Brunn VV, Gold DJC et al (1998) Earth's oldest reported glaciation: physical and chemical evidence from the Archean Mozaan Group (~2.9 Ga) of South Africa. J Geol 106(5):523–538

Young GM (2017) Precambrian: overview. reference module in earth systems and environmental sciences. Elsevier, Netherland. https://doi.org/10.1016/B978-0-12-409548-9.10498-1

Young GM (2019) Aspects of the Archean-Proterozoic transition: how the great Huronian Glacial Event was initiated by rift-related uplift and terminated at the rift-drift transition during break-up of Lauroscandia. Earth Sci Rev 190:171–189

Yue L, Liu ZL, Ma Y (2018) Sedimentary environment and depositional evolution of the Mesoproterozoic Bingmagou Formation on the southern margin of the North China Craton. Sci Rep 8(1):1–12

Yue L, Liu ZL, Veerle V et al (2020) Glacial sedimentary sequence and evolution of the Fengtai Formation, southeastern margin of the North China Craton. Geol Rev 66(02):307–323 (in Chinese with English abstract)

Zhan S, Chen Y, Xu B et al (2007) Late Neoproterozoic paleomagnetic results from the Sugetbrak Formation of the Aksu area, Tarim basin (NW China) and their implications to paleogeographic reconstructions and the snowball Earth hypothesis. Precambr Res 154(3–4):143–158

Zhang BY (2019) Sedimentary environment analysis of Neoproterozoic Luoquan Formation in Western Henan Province. Dissertation, Henan Polytechnic University (in Chinese with English abstract)

Zhang BY, Zheng DS (2019) Morphological characteristics of quartz grains in the Luoquan formation of the Neoproterozoic in Lushan, Henan province. J Chin Electron Microsc Soc 38(06):657–663 (in Chinese with English abstract)

Zhang BY, Zheng DS, Liu YG (2019) Sedimentary environment analysis of Neoproterozoic Luoquan Formation in Lushan area, Western Henan Province 43(4):417–424 (in Chinese with English abstract)

Zhang B, Zheng D (2021) Sedimentological, geochemical, and geochronological constraints on the origin of the Neoproterozoic Luoquan Formation at the Southern Margin of the North China Craton. Int Geol Rev. https://doi.org/10.1080/00206814.2021.1894609

Zhang HJ, Wang XL, Wang X et al (2016) U-Pb ziron ages of tuff beds from the Hongzaoshan Formation of the Quanji Group in the north margin of the Qaidam Basin, NW China, and their geological significances. Earth Sci Front 23(6):202–218 (in Chinese with English abstract)

Zhang L, Du YS, Zuo JX et al (2008) Negative anomaly of carbon isotope from carbonates of Sinian Dongpo Formation in Ruzhou and Lushan, Henan Province and Its geological significance. Earth Sci J China Univ Geosci 33(4):523–530 (in Chinese with English abstract)

Zhang S, Jiang G, Zhang J et al (2005) U-Pb sensitive high-resolution ion microprobe ages from the Doushantuo Formation in south China: constraints on late Neoproterozoic glaciations. Geology 33(6):473–476

Zhao G (2001) Palaeoproterozoic assembly of the North China craton. Geol Mag 138(1):87–91

Zhang K (1991) Tilloid at the bottom of the sedimentary covers on the margins of the Ordos Basin. Reg Geol China 1:79–85 (in Chinese with English abstract)

Zhang SC, Tang SW (1982) The characters of tillites of Shangzhangwan Formation, Luoquan county at the Northern slope of Qinling range. Sci Geol Sin 4:371–378 (in Chinese with English abstract)

Zhang WZ, Li HK, Wang GF et al (2001) Palaeomegnetic study of neoproterozoic glaciation, eastern china and implication of geology. Prog Precambr Res 1:35–57 (in Chinese with English abstract)

Zhang WT, Zhu ZL, Yuan KX et al (1979) The boundary between Cambrian and upper Precambrian in the South and southwest of North China. Acta Stratigr Sinica 3(1):51–56 (in Chinese)
Zhang Y, Wang PJ, Liu WZ et al (2006) Sedimentary environment of the Neoproterozoic in Quaruqtagh region. Xinjiang Geol (4):365–368+477 (in Chinese with English abstract)
Zhang ZL, Hua H, Zhang ZF (2015) Problematic Ediacaran fossil Shaanxilithes from the Jiucheng Member of Wangjiawan Section in Jijing, Yunnan Province 54(1):12–28 (in Chinese with English abstract)
Zhao TP, Deng XQ, Hu GH et al (2015) The Paleoproterozoic-Mesoproterozoic boundary of the North China Craton and the related geological issues: a review. Acta Petrol Sinica 31(06):1495–1508 (in Chinese with English abstract)
Zheng DS, Wang PX, Sun FB (2016) Sedimentary environment of mesoproterozoic bingmagou formation in Jiyuan, Western Henan province. Geol Sci Technol Inf 35(1):1–7 (in Chinese with English abstract)
Zheng DS, Meng Y, Sun FB et al (2017) REE geochemical characteristics of sandstones of Mesoproterozoic Bingmagou Formation in Yichuan. J Henan Polytech Univ (Natural Science) 36(1):38–45 (in Chinese with English abstract)
Zheng DS, Yang DL, Yang WT et al (2019) Sedimentary Environment and Controlling Factors of the Cambrian Xinji Formation, Lushan Area, Western Henan. Geoscience 33(3):604–614 (in Chinese with English abstract)
Zheng W, Liu SX (2012) Analysis on lithofacies and sedimentary environments evolution of Sinian in Lushan Area, western Henan Province. J Henan Polytech Univ (Natural Science) 31(6):669–673 (in Chinese with English abstract)
Zheng WW, Dou SC (1980) On features of the glacial deposition and their contrast question of fental and Luoquan formation in the North of Anhui. J Hefei Polytech Univ 2:48–74 (in Chinese with English abstract)
Zhou MZ, Li YZ, Lu SN et al (2008) SHRIMP U-Pb zircon age of tuff at the bottom of Niutitang Formation in Zunyi and its geological significance. Chin Sci Bull 53(1):104–110 (in Chinese)
Zhou KK, Xu XS, Mu CL (2010) Transgressive deposits from the Luoquan formation in Shuiyu, Ruicheng, Shanxi. Sediment Geol Ththyan Geol 30(3):39–45 (in Chinese with English abstract)
Zhu MY (2010) The origin and cambrian explosion of animals: fossil evidences from China. Acta Palaeontol Sinica 49(3):269–287
Zhu MY, Wang H (2011) Neoproterozoic glaciogenic diamictites of the Tarim Block, NW China. Geological Society, London, Memoirs 36(1):367–378
Zhu MY, Zhao FC, Yin ZJ (2019) The Cambrian explosion: advances and perspectives from China. Sci China Earth Sci. https://doi.org/10.1007/s11430-019-9508-4
Zuo P, Li Y, Zhang G et al (2019) Reviews of the Mesoproterozoic to Neoproterozoic sedimentary sequences and new constraints on the tectono-sedimentary evolution of the southern margin of the North China Craton. J Asian Earth Sci 179:416–429

Triassic Marine Stratigraphy, Sedimentology and Paleontology of the Xingyi-Guanling Area, Guizhou Province, SW China

Tingshan Zhang, Wei Yang, Mihai Emilian Popa, and Kainan Mao

1 Objectives and Strategy

The Triassic marine strata are well developed and completely preserved in the Xingyi-Guanling area, where sediment differentiations are obvious, and marine reptile fossils are abundant and perfect. They will provide important references for us to better understand the effect of ecological sedimentary processes during the evolution of carbonate platforms and ecological responses and feedback mechanisms during extreme climate processes in the Triassic. We will visit three geological landscape points in this geological field investigation.

(1) **The Guanling-Yongning Triassic Section and Guanling Fossil National Geopark**

This section completely records the formation and evolution process of the carbonate factory during the Lower to Middle Triassic palaeogeographic environment. Furthermore, the Guanling biota, which is discovered in the Upper Triassic part of this section, is a rare marine fossil reservoir around the world.

(2) **The Triassic "S" Shape Sedimentary Facies Transformation Zone**

The facies transformation zone between the Yangtze carbonate platform and the Nanpanjiang Basin. Controlled by paleotopography and sediment sources, the facies

T. Zhang (✉)
School of Geosciences and Technology, Southwest Petroleum University, Chengdu, China
e-mail: zts_3@126.com

W. Yang · K. Mao
School of Resources and Environmental Engineering, Guizhou Institute of Technology, Guiyang, China

M. E. Popa
Department of Geology and Doctoral School of Geology, University of Bucharest, Bucharest, Romania

X. Hu (ed.), *Field Trip Guidebook on Chinese Sedimentary Geology*,
https://doi.org/10.1007/978-981-99-6936-4_9

transformation had an "S"-shaped banding distribution in the Xingyi-Guanling area. It clearly shows the Middle Triassic sedimentary system evolution of "platform-slope-basin" and carbonate factory breakdown during the Carnian Pluvial Episode (CPE).

(3) **The Xingyi Triassic Section and the Museum of Xingyi National Geological Park**

This section is characterized by community successions of the Xingyi biota. Moreover, seismic-tsunamic and storm depositional sequences are well developed in the middle-upper part of the Falang Formation. It offers a unique sedimentological viewpoint to us to systematically understand the climatic, ecological and taphonomic evolution in the middle-upper Triassic.

In addition, the special stratigraphic and tectonic conditions and the unique hydrogeological properties result in the rich combination of karst forms, creating a gorgeous landscape of geological relics in the Xingyi-Guanling area.

2 Field Trip Schedule

This trip (Premeeting) is five days in length and includes four stations. We will focus on the following topics: (1) lithological characteristics of the main stratigraphic units of the Triassic in the Xingyi-Guanling area, (2) primary sedimentary structures and soft sediment deformation structures, (3) the sedimentary sequence and contact relationships between the stratigraphic units, (4) the Xingyi Fossil Reservoir and Guanling Fossil Reservoir, and (5) understanding the formation and collapse process of the Triassic carbonate platform from the perspective of carbonate production. More importantly, we aim to understand the role of the Triassic ecosystem–environment coupling process in the evolution of carbonate platforms (Fig. 1).

Day 0: Guiyang-Guanling

The investigators are requested to assemble in Guiyang, the capital city of Guizhou, in the southwest part of China. The shuttle bus will transport passengers from Guiyang to Guanling County, Anshun City;

Day 1: Geological Survey in Guanling Area

The investigators will focus on the investigation of the Jialingjiang Formation-Falang Formation geological section of the Triassic in Yongning, Guanling and the Fossil In Situ Museum of the Guanling Fossil Group National Geopark;

Day 2: Geological Investigation in Guanling-Xingyi Area

The investigators will travel by bus to Xingyi City, Southwest Guizhou Autonomous Prefecture, and focus on investigating the geological section of the Middle Triassic Series Longtou Formation in Guanling Duanqiao, the facies transition zone between

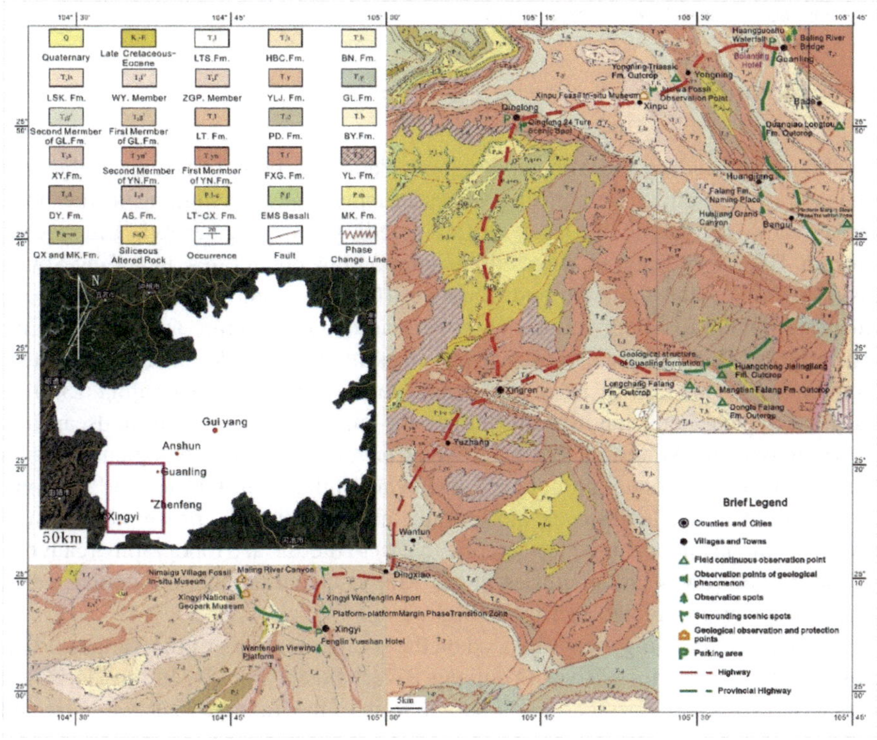

Fig. 1 Location map of the Xingyi and Guanling areas

the Middle Triassic Series Longtou Formation and Bianyang Formation, and the geological section of Zhenfeng Longchang Triassic Yangliujing Formation—Laishike Formation;

Day 3: Geological Survey in the Xingyi Area

The investigators will focus on the investigation of the geological heritage landscape of Wanfenglin, the facies transition zone between the Middle Triassic Series Yangliujing Formation and Longtou Formation, the Guizhou Dinosaur Fossil In situ Museum of Nimaigu Village in Xingyi, and the Xingyi National Geological Museum;

Day 4: Tourism Geological Survey in the Qinglong Area

The investigators will focus on the investigation of Qinglong's "Twenty-Four Bends" Anti-Japanese War Road and related cultural landscapes. In the afternoon, the group will travel by bus back to Guiyang Longdongbao Airport and fly to Beijing.

3 Tectonic and Stratigraphy

3.1 Tectonic

Guizhou is located on the Yunnan-Guizhou Plateau; it borders Hunan to the east, Guangxi to the south, Yunnan to the west, Sichuan and Chongqing to the north and governs six prefecture-level cities, including Guiyang, Zunyi, Liupanshui, Anshun, Bijie, and Tongren, three ethnic autonomous prefectures, including Southeast Guizhou, South Guizhou and Southwest Guizhou, among which Xingyi and Guanling belong to Southwest Guizhou Autonomous Prefecture and Anshun City, respectively. The topography of that region is undulating, dominated by plateau mountains and hills, with an altitude range of 370–2207 m, averaging 1300 m. They are typical karst landform development areas, which include landscapes of the Huangguoshu Waterfall, Maling River Canyon and other geological relics of runoff erosion and mountain collapse, the Wanfenglin Peak Cluster and Maomao Cave and other cave landscapes, etc., completely recording the process of transformation from the underground topography (cave) to the surface topography (canyon). At present, all towns and countries in the Xingyi-Guanling area are connected by expressways or provincial high-grade highways.

The Xingyi-Guanling area is located on the southern margin of the Yangtze Plate, close to the Nanpanjiang River (Youjiang River) Basin (Fig. 2). In the Hercynian Period of the Late Paleozoic, the southern margin of the Yangtze Plate was a passive continental margin; in the Early Triassic, it gradually transitioned from a discrete margin to a convergent margin (Xiao 2005). The Early-Middle Triassic platform-slope-basin bed in the area is well developed and has a drastic facies transition (Fig. 3). It is an important part of the Triassic "S" type facies transition zone in Guizhou (Regional Geology of Guizhou Province, 1987). The Triassic shallow-water carbonate sedimentary system is divided into carbonate ramps and marginal carbonate platforms; the latter mainly developed in the Middle-Late Triassic, with large undulations in the platform and orderly coexistence of reefs and beach facies at the edge of the platform. The deep-water slope area is mainly composed of gravity flow deposits and suspended deposits, and the three intersect and coexist in the facies transition zone area.

The region was controlled by structural extension in the Late Triassic, and "graben horst" patterns of various scales were formed on carbonate platforms; in particular, deep-water platform depressions of different scales were developed inside carbonate platforms. For example, Dingxiao, Xiemi, Nimaigu of Xingyi, Guizhou, Jiyang Mountain, Luopingzhong Mountain, and Banqiao of Fuyuan, Yunnan are all located in deep-water platform depression areas subject to varying degrees of barriers, which developed Xingyi biota of nearshore types, of which Sauropterygia is the most developed. There are also molecules of protosaurus and Cyamodus of placodus and medium-large ichthyosaur fossils. After the death of this group of organisms, they were quickly transported to the lower deep-water hypoxic environment and well buried. The flood events in the Yazhiliang (Carnian) period caused the sea level to

Fig. 2 Regional palaeogeographic setting

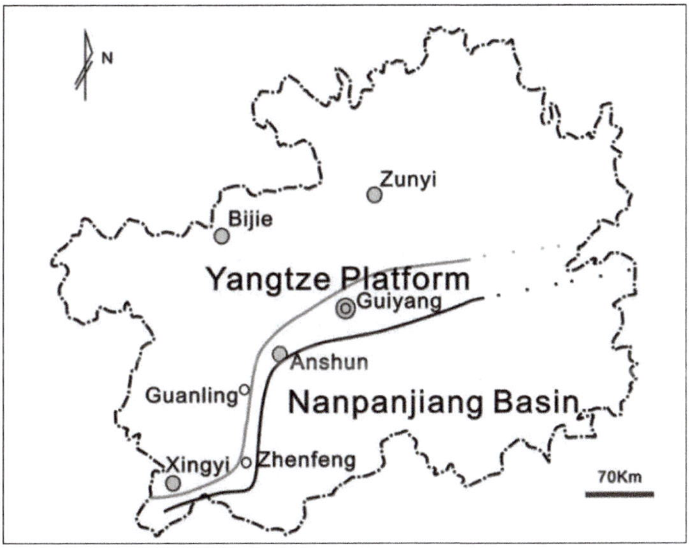

Fig. 3 The Triassic "S" type facies transition zone in Guizhou

rise rapidly (Jin et al. 2015, 2020; Shi et al. 2010). Large carnivorous Ichthyosauria (Guanling biota) were mainly active in the deep-water platform depression area. Earlier reports showed that the Guanling fauna is more widely distributed than the Xingyi fauna (Wang Liting, oral communication), but a large number of prosperous finsosaurus, protosaurus, and other marine reptiles that may live near shore in the Xingyi fauna no longer exist in the Carnian period. This implies that the ecological environment of the two biological groups has changed significantly, and the culprit is related to the Carnian Pluvial episode. With the impact of the Carnian Pluvial episode on the ecological environment, marine reptiles in the Guanling Biota also disappeared in the long river of history as a flash in the pan. Not only that, the Carnian Pluvial episode also completely changed the appearance of the carbonate platform. The turbidite system composed of mudstone, siltstone, and lithic sandstone submerged the carbonate platform and formed a regional deep-water continental shelf deposition.

After the Indosinian movement, most areas of South China were transformed into land or continental depositional areas. The lack of marine stratigraphy was common in the Late Triassic. The Xingyi-Guanling area not only preserved the Upper Triassic Subzhiliang Stage but also produced a rich and complete marine vertebrate Fossil Group. Since the Yazhiliangian Stage in the region and the International Carnian Stage are basically comparable (Tong et al. 2019), the Xingyi-Guanling area has become an important place for studying the extreme climate-deposition-ecological evolution of the Late Triassic (Zhang et al. 2015; Sun et al. 2016).

3.2 *Stratigraphy*

The Triassic in Guizhou is well developed, with a complete sequence and rich biological fossils. The Lower-Middle Triassic Series is widely distributed in the area west of the Zunyi-Kaili line and is mainly composed of marine carbonate and clastic rocks; the Upper Series is fully developed in southwestern Guizhou and belongs to marine-terrestrial deposits (Table 1).

Triassic Jialingjiang Formation (T_1j)

Formerly known as the "Yongningzhen Formation," it is exposed in West Guizhou-North Guizhou and other places in Guizhou, with a thickness of 500–650 m. The lithology is mainly light gray-dark gray thin-thick bedded micritic limestone and dolomite intercalated with oolitic limestone, microbial rock and mudstone, and the top is mostly collapsed breccia dolomite (containing gypsum salt bed) (Fig. 4). It contains bivalves, ammonites, and conodonts and is in integrated contact with the underlying Yelang Formation.

The Jialingjiang Formation in the area can be divided into four sections according to lithology:

Table 1 Regional stratigraphic correlation in the Triassic

Chronostratigraphic unit			Lithostratigraphic unit		
			Upper Yangtze region		Nanpanjiang region
System	International Stage	South China Stage	Xingyi-Zunyi area	Zhengfeng-Guiyang area	
Jurassic			Ziliujing FM.		
Upper Triassic	Rhaetian	Peigucuo	Erqiao FM.		
			Huobachong FM.		
	Norian		Banan FM.	Sanqiao FM.	
	Carnian	Yazhiliang	Laishike FM.	Gaicha FM.	Lashike FM.
			Falang FM.		Heimiaowan FM.
Middle Triassic	Ladinian	Xinpu	Ziliujing FM.	Longtou FM.	Bianyang FM.
	Anisian	Guandao		Poduan FM.	Xinyuan FM.
			Guanling FM.		
Lower Triassic	Olenekian	Chaohu	Jianglingjiang FM.	Anshun FM.	Luolou FM.
	Induan		Yelang FM.	Daye FM.	
Permian	Changhsingian				

gray thin-medium thick bedded micritic limestone, Guanling area

dolomite, gypsum breccia dolomite intercalated with variegated mudstone, Guanling area

Fig. 4 The typical lithologic characteristics in the Jialingjiang FM.

The fourth section: gray–yellowish gray thin-medium thick bedded dolomite, gypsum breccia dolomite intercalated with variegated mudstone, which is often intercalated with vitric tuff on the top.

The third section is gray thin-medium thick bedded micritic limestone intercalated with argillaceous limestone, dolomitic limestone, worm-like limestone, and dolomite.

The second section: Interbeds of purplish red–yellowish green thin-medium thick bedded claystone, silty mudstone and argillaceous limestone, micritic dolomite in unequal thickness, which are rich in bivalves.

The first section: gray–dark gray thin-thick bedded micritic limestone, grain stone, with mud mounds developed.

The characteristics of the lithological sequence of the Jialingjiang Formation indicate that during this period, the sea level decreased, the accommodating space decreased, and the sedimentary environment transitioned from a moderate gentle slope to an inner gentle slope.

Triassic Guanling Formation (T_2g)

It is mainly distributed in the upper Yangtze stratigraphic zone, including Xingyi, Zhenfeng, Guiyang, and other places, with a thickness of approximately 500 m. The lower section (Songzikan section) is characterized by rhythmic interbeds of gray green-purplish red mudstone, gray marl, micritic or endoclastic limestone, and argillaceous dolomite (Fig. 5), with a gray green–yellowish green vitric tuff of 0.3–2 m at the bottom (commonly known as "mung bean rock"), which is a sign of good regional stratigraphic contrast; mud cracks and marine reptile footprints can be seen in the argillaceous dolomite in the middle of this section near Zhenfeng Niuchang and Longchang. The upper section (Shizishan section) is mainly gray thin-medium bedded micritic limestone, oncoid limestone, worm-like argillaceous limestone intercalated with mudstone, and stromatolites are seen on the top of Yongning, Guanling and other places. The Panxian biota fossils were produced near Xinmin, Panxian.

When the Guanling Formation was deposited, tropical shallow-water carbonate factories began to dominate. The second section of the Guanling Formation was

micritic or internal clastic limestone in Songzikan Section, Zhenfeng area	rhythmic interbeds of gray green-purplish red mudstone, gray marl in Songzikan Section, Liuzhi area

Fig. 5 Typical lithologic characteristics in the Guanling FM.

breccia dolomite, Zhenfeng area

stromatolitic limestone, Guanling area

Fig. 6 Typical lithologic characteristics in the Yangliujing FM.

significantly affected by high-frequency sea-level changes, and the resulting meter-scale cycle sequence characteristics were obvious and had a well-ordered pattern of vertical stacking, with a gradual transition from a gentle carbonate slope to a carbonate platform in the area.

Triassic Yangliujing Formation (T_2y)

The distribution range is the same as that of the Guanling Formation, with a thickness variation of 400–1100 m. It is mainly composed of light gray–gray medium-thick bedded massive dolomite and breccia dolomite. The top is transformed into thick bedded lamellar limestone (stromatolitic limestone) at Zhenfeng Longchang, Guanling Duanqiao and other places (Fig. 6). Fossils in this group are rare, and fossils of gastropods and other mollusks are viewed. It is in integrated contact with the underlying Guanling Formation.

The Yangliujing Formation in Yongning of Guanling and other places in the area is dominated by collapsed breccia dolomite, which forms multiple sets of sedimentary cycles with thick bedded crystalline dolomite representing beach facies, reflecting the restricted platform environment on the inner side of the platform margin.

Triassic Longtou Formation (T_2l)

It is mainly distributed in Anshun, Guanling, Huaxi of Guiyang, and other places in Zhenfeng-Guiyang Community, with a thickness of 370–1130 m. The lithology is light gray-off-white thick bedded massive micritic limestone, thrombite, sparite algae bonded limestone, sparite oncoid limestone, etc., including bivalves and brachiopod fossils, which are preserved in fragments and are typical reef beach facies deposits on the margin of the platform.

Limestones of this formation are in obvious tooth-like interspersed facies transition relationships with dolomite of the Yangliujing Formation and sand shale of the Bianyang Formation. Among them, the facies transition between the limestone of the Longtou Formation and the dolomite of the Yangliujing Formation is clear in Zhaojiadu, Wanfenglin, etc., of Xingyi, and the topography is obvious (Fig. 7).

Fig. 7 Cone-shaped stone pinnacles (Yangliujing FM.) and tower-shaped stone pinnacles (Longtou FM.) combination, Xingyi area

Triassic Bianyang Formation (T_2b)

It is mainly distributed in the Youjiang stratigraphic zone, with a thickness of 137–1568 m. Generally, they are terrigenous clastic sandstones and claystones of deep-water continental shelves integrated above the Xinyuan Formation and below the Heimiaowan Formation, containing a small number of bivalves and ammonites. In Bangui of Guanling, Nidang of Xingyi and other places, the facies transition relationship between the limestone of the Longtou Formation and the sand shale of the Bianyang Formation can be clearly observed (Fig. 8).

Triassic Falang Group ($T_{2\text{-}3}f$)

It is mainly distributed in Wusha, Dingxiao of Xingyi, Longchang of Zhenfeng, Duanqiao, Yongning of Guanling, Langdai of Liuzhi and other places, with a thickness of 0–684 m, which is generally 200–300 m. The age belongs to the late Xinpu (Latin)-Yazhiliang (Carnian) period.

The lower section of the Falang Formation is called the Zhuganpo section, and the lithology is gray-dark gray thin-medium bedded bioclast-bearing micritic limestone, calcirudyte, and nodular limestone, partially intercalated with flint micritic limestone

Fig. 8 The landscape of the facies transition

Fig. 9 The typical lithologic characteristics in Falang FM.

(Fig. 9), including bivalves *Halobia*, brachiopod *Koninckina*, crinoid *Traumatocrinus* and so on. Dingxiao, Wusha of Xingyi and Luoping of Yunnan and other places are important producing areas of the Xingyi biota represented by the *Keichousaurus Hui*, which are produced in the lower part of this section. The main reptiles are *Kueichousaurus*, *Nerhesaurus*, *Anshunsaurus*, etc. There are also abundant fish fossils, such as *Guizhouniscus*, *Brachysomus*, *Xingyia*, etc., which are in integrated contact with the underlying Yangliujing Formation and Longtou Formation.

The upper section of the Falang Formation is called the Wayao section, which is mainly composed of dark gray–gray black thin-medium bedded micritic limestone, marl intercalated with calcareous mudstone, and calcareous sandstone (Fig. 9). The bottom of this section is often rich in iron and manganese, which can form local concentrations of ore. In the Xinpu area of Guanling, the lower part of this section produces the Guanling biota, which is famous for marine reptiles, including *Syngnathus*, *Ichthyosauria*, and *Placodontia*. It is also rich in ammonites *Protrachyceras*, *Trachyceras*, bivalve *Daonella*, petrified wood *Dadoxylon* and so on.

Triassic Laishike Formation (T_3ls)

It is mainly distributed in areas such as Xingyi, Zhenfeng, Zhenning, Guanling, and Liuzhi, with thicknesses of 125–850 m. The main lithology is composed of rhythmic interbeds of gray–gray green thin-medium bedded mudstone, silty mudstone, and fine-medium-grained lithic quartz sandstone of unequal thickness interbedded with argillaceous siltstone, and the sandstone often contains features of turbidites such as load cast, flute cast and Bouma sequences. It is rich in ammonite *Trachyceras*. It is continuously deposited with the underlying Falang Formation and Heimiaowan Formation.

4 Field Stop Description

4.1 Geological Survey in Guanling Area (Stop 1)

The scope of the geological survey in the Guanling area is concentrated in Yongning Town of Guanling County and Duanqiao Town of Guanling County. Among them, the section of Yongning of Guanling is located approximately 10 km west of Guanling County. The outcrop is on National Highway 320. The total length is over 7 km. The section is 1588 m thick. The Jialingjiang Formation, Guanling Formation, Yangliujing Formation and Falang Formation are exposed from bottom to top. A brief description of the measured section is as follows:

The Wayao section of the Falang Formation, 20.17 m thick, no peak is seen (beds 96–100). Interbeds of dark gray–black gray medium-thick bedded argillaceous limestone and black gray calcareous shale of unequal thickness, with well-developed argillaceous lamellae, and calcium balls, bivalves or crinoids with a high abundance arranged in an oriented arrangement are commonly seen under the microscope, with obvious handling characteristics. The lower part of the section produces Changuanling biota, which is characterized by vertebrates such as *Ichthyosaurus* and *Syngnathus* and invertebrates such as ammonites and crinoids.

- - - - - - Conformity - - - - - -

The Zhuganpo section of the Falang Formation, 168.62 m thick, (beds 85–95).

Gray thin-medium bedded bioclast-bearing micritic limestone, sand-clastic micritic limestone, argillaceous-silty limestone intercalated with yellowish gray–gray thin bedded calcareous shale. The limestone beds are mostly undulating. The top has increased argillaceous matter, which are interbeds of gray calcareous shale and gray-rust red thin bedded bioclast-bearing micritic limestone of unequal thickness. Among them, the limestone bed has developed a large number of branched traces of foraging, and the surface of limestone is rusty red, as it is often affected by iron and manganese oxidation.

- - - - - - Conformity - - - - - -

Yangliujing Formation, 484.79 m thick (beds 78–84).

Gray medium-thick bedded granular micritic limestone intercalated with off-white silty limestone, locally dolomitized, with dissolved pores with an oriented arrangement developed. It has a thickness of 35.14 m.

Light gray–gray medium bedded-massive granular silty dolomite, the grains are mainly sand-sized and gravel-sized grains, the latter is angular, poorly rounded, and has the characteristics of typical collapse breccia. It has a thickness of 354.64 m.

In light gray-off-white medium-thick bedded mud-sparite granular dolomite, the grains are mainly sand-sized and gravel-sized grains and bioclasts, and the later dissolution reformation is strong. It has a thickness of 95.01 m.

- - - - - - Conformity - - - - - -

Guanling Formation, 564.34 m thick.

The second section of Guanling Formation (beds 51–77).

Light gray–gray thin-medium thick bedded micritic limestone, micritic grainstone, sparite grainstone, locally dolomitized, and the grains include sand-sized and gravel-sized grains and ooids. It has storm graded bedding, and the thickness is 99.69 m.

Gray thin-medium bedded bioclast-bearing micritic limestone and micritic limestone, with high argillaceous content at the bottom, often develop argillaceous laminae or bands, with a thickness of 212.65 m.

The first section of Guanling Formation (beds 30–51).

Gray–grayish yellow thin-medium bedded siltstone intercalated with thin bedded silty mudstone, which is 127.91 m thick.

Gray green-purplish red thin bedded argillaceous siltstone and silty mudstone are 10.67 m thick.

Interbeds of gray–gray green thin-medium bedded argillaceous siltstone and silty mudstone of unequal thickness, with a thin bedded micritic limestone bed at the bottom, and it is 89.76 m thick.

Gray thin-medium bedded micritic limestone intercalated with argillaceous siltstone; the bottom is collapse breccia, has a scour surface and is 22.6 m thick.

The lower part is vitric tuff (mung bean rock), the upper part is gray thin-medium bedded argillaceous siltstone, and it is 1.5 m thick.

- - - - - - Conformity - - - - - -

Jialingjiang Formation, 350.08 m thick.

The fourth section of the Jialingjiang Formation (beds 25–29).

Grayish yellow–black gray silty mudstone, argillaceous dolomite, shale intercalated with collapse breccia, 62.16 m thick.

The third section of the Jialingjiang Formation (beds 20–24).

Gray micritic dolomite, micritic limestone intercalated with marl, and worm-like structures developed in micritic limestone have thicknesses of 90.36 m.

The second section of the Jialingjiang Formation (bed 19).

Yellowish green thin bedded silty mudstone, 13.1 m thick.

The first section of the Jialingjiang Formation (beds 1–18).

Gray thin-medium bedded micritic limestone intercalated with thin bedded marl and micritic dolomite, with a thickness of 35.61 m.

Off-white thin-thick bedded massive algal oolitic limestone, micritic limestone, and oolitic limestone. This bed has developed two sets of thrombolite mounds. The reef-building organisms are mainly *Tubiphytes*, with bed 11 being the most typical. The reef base reef core reef cap is fully developed, and it is representative of the earliest organisms that revived on the carbonate ramp in the Early Triassic. The thickness is 22.82 m.

Gray medium-thick bedded micritic limestone, worm-like limestone intercalated with algal oolitic limestone, with developed argillaceous bands at the bottom, with a thickness of 118.34 m.

The main visit points for the geological section of the Jialingjiang Formation-Falang Formation of the Triassic in Yongning of Guanling are: (1) Jialingjiang Formation Mud mound; (2) Gypsum breccia at the top of Jialingjiang Formation-cave breccia deposits at the bottom of Guanling Formation; (3) The "red bed" deposition of the first section of the Guanling Formation; (4) the thin bedded carbonate sedimentary sequence of the second section of the Guanling Formation; (5) the stratigraphic boundary between the Yangliujing Formation and the Falang Formation; (6) the stratigraphic boundary between Zhuganpo section-Wayao section of the Falang Formation.

Stop 1.1: Mud Mound of the Jialingjiang Formation in the Guanling Yongning Section

The Jialingjiang Formation's depositional age belongs to the Olenekian stage of the Early Triassic Series and belongs to the biological recovery period after the Permian–Triassic biological extinction. In the middle and lower part of the first section of the Jialingjiang Formation in the Guanling Yongning section, there are mud mounds (bed 11) with a height of 8 m. The reef base is composed of oolitic limestone, and the reef core and reef wings are mainly composed of *Tubiphytes* and thrombites (Fig. 10). There is an overlying reef cover composed of oolitic limestone, constituting a complete biological reef development cycle. Further investigation showed that there were mud mounds of different sizes in Yongning of Guanling, Longchang of Zhenfeng and other places in the same period, indicating that the main carbonate builders during this period were mainly autotrophic microorganisms such as cyanobacteria. The mud mound factory often played the role of a precursor to post disaster flooding and recovery during the major transition period of geological history. Current research shows that the mud mound is representative of the early recovery of biological construction on the carbonate ramp of the Early Triassic in Guizhou.

Stop 1.2: Jialingjiang Formation Gypsum Breccia-Guanling Formation Cave Breccia in the Yongning Guanling Section

The breccias at the top of the Jialingjiang Formation-bottom of the Guanling Formation are concentrated. The formation of breccia dolomite at the top of the Jialingjiang Formation is related to the dissolution of the underlying gypsum leading to collapse of

Fig. 10 Mud mounds of the Jialingjiang Formation

the overlying strata (Fig. 11). The cave breccia at the bottom of the Guanling Formation appears as plate-like argillaceous gravel, which is highly angular, unsorted and single in composition. The bottom has an obvious scouring surface.

Stop 1.3: Red Beds of the Guanling Formation in the Guanling Yongning Section

The upper part of the first section of the Guanling Formation developed a set of tidal flat deposits dominated by gray green-purplish red silty mudstone-argillaceous dolomite. Mud fractures are common in the bedding. Reptile footprints are still seen near Niuchang and Longchang of Zhenfeng, representing the peritidal environment formed by the decline in sea level in the early Middle Triassic (Fig. 12).

Stop 1.4: Stratigraphic Boundary Line of the Yangliujing Formation-Falang Formation in the Guanling Yongnings Section

The upper part of the Yangliujing Formation in the Guanling Yongning section is characterized by typical collapse breccia-like dolomite, which often constitutes a thick bedded-massive granular dolomite-collapse breccia-like dolomite cycle sequence,

Fig. 11 The breccias at the top of the Jialingjiang Formation-bottom of the Guanling Formation

Fig. 12 Red beds of the Guanling Formation

which is restricted platform deposits subject to the barrier. However, the overlying Falang Formation is a dark gray thin bedded micritic limestone intercalated with mudstone, in which the organic matter content is relatively high. The bioclasts are mainly thin-shelled bivalves, showing obvious deep-water sedimentary characteristics, which is different from the "Indosinian retreat" in many places in southern China. This stratigraphic boundary is an integrated contact.

Studies have shown that at the end of the Triassic Latin period in the Xingyi-Guanling area, a series of growth faults and graben horst structures were formed under the influence of regional extension, which controlled the depositional appearance of the Falang Formation. The Franco Formation is formed in deep-water depressions inside the platform, distributed along the inner side of the "S" type facies transition zone, and controls the distribution of the Xingyi biota.

Stop 1.5: Stratum Boundary Line Between Zhuganpo Section and Wayao Section of Falang Formation in Guanling Yongning Section

The top of the Zhuganpo section of the Falang Formation is a thin bedded micritic limestone intercalated with calcareous mudstone. Rich deep-water facies trace fossils are often distributed on the calcareous mudstone bed, with low differentiation, but show multiple branches and level-oblique crossing tracing features along the bed (Fig. 13), accompanying crinoids and bivalves. The top bed of the Zhuganpo section is often rich in iron and manganese minerals, and their formation time is comparable to that of the position of the upper manganese deposits of the Gejiu Formation in eastern Yunnan and other places.

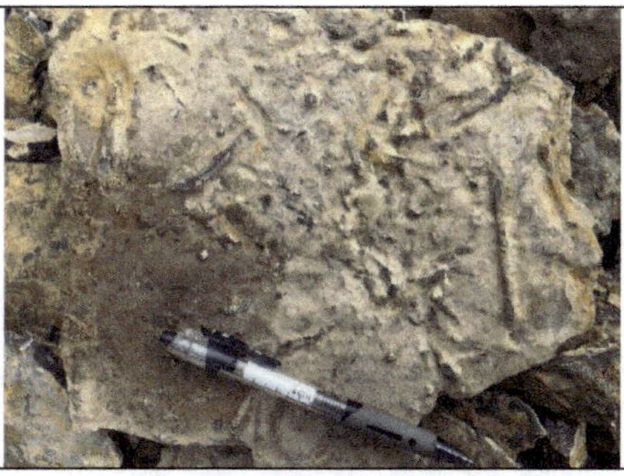	
The trace fossil in Zhuganpo Section	The Crinoidea in the bottom of Wayao section

Fig. 13 The features of fossils in Falang FM.

The Wayao section of the Falang Formation is in integrated contact with the underlying Zhuganpo Section (Fig. 14). The bottom of the Wayao Section is composed of dark gray thin bedded laminar limestone, sandy limestone, bivalve limestone, etc. The lower part at 5–10 m produced Guanling biota, which is represented by invertebrates such as crinoids and marine reptiles. Terrigenous quartz can be seen in the Wayao section of the area, and the phase is transitioned into calcareous fine-medium sandstone northward to the area of Langdai of Liuzhi. This is the sedimentary response to the Carnian Pluvial episode (Fig. 15).

Stop 1.6: Guanling Biota

The Guanling Fossil Library features a large number of exquisitely preserved fossils of marine reptiles and crinoids, which is unique in the world. The Guanling biota is mainly distributed in the 6–8 m condensed section from the bottom of the Wayao section of the Falang Formation, mainly including bivalves (*Halobia* and *Daonella*, etc.), ammonites (Trachyceras and Sirenites senticosus), well-preserved pseudoplanktonic traumatocrinus, *Ichthyosauria* (*Shastasaurus* such as *Guizhou Ichthyosauria* and *Guanling Ichthyosauria*), and *Syngnathus* (*Xinpulong Syngnathus* and *Anshun Syngnathus*, etc.) (Fig. 16). Sedimentary-paleoecological studies have shown that the ecological-burial environment of the Guanling biota is clearly related to the Carnian Pluvial episode. The lack of oxygen in the ocean caused by the Carnian Pluvial episode caused the final demise of the Guanling biota.

4.2 Xingyi-Guanling Area Geological Survey (Stop 2)

The scope of the geological survey in the Xingyi-Guanling area is concentrated between Duanqiao town in Guanling County, Bangui County in Guanling County, and Longchang town in Zhenfeng County. Among them, Bangui County of Guanling County is located at the boundary between Guanxing Highway and Huajiang Grand

Fig. 14 Stratum boundary line between the Zhuganpo section and Wayao section of the Falang Formation

Canyon, a famous tourist route in the province. It is an important distribution area of the Triassic "S" type facies transition zone in Guizhou. It is the facies transition zone between the Longtou Formation and the Bianyang Formation. At the same time, the area passes through Huajiang Gorge, with rich fluvial-karst landform types.

Stop 2.1: Limestone of Longtou Formation

The Longtou Formation was first created near the Longtou Formation Village in Qingyan, Guizhou, representing the Middle Triassic Latin Period platform margin reef (beach) facies deposits in Central Guizhou. It is distributed along the "S" type facies transition zone with a bandwidth of approximately 10 km. The geological section of the Longtou Formation of the Triassic in Duanqiao, Guanling County, is located at Road X478, Bade Country, Duanqiao Town, Guanling County. The thick bedded-massive sparite oncoid limestone and algal reef limestone of the Longtou Formation are mainly exposed. The Longtou Formation can be divided into three sections according to rock types and assemblage characteristics: the main body of the lower section is an oncoid limestone-stromatolite sedimentary cycle, the middle section is composed of a sparite oolitic limestone-algal reef limestone cycle, the upper section is composed of an oncoid limestone-pelletoidal limestone cycle (Fig. 17), and the sedimentary sequence of the Longtou Formation is closely related to the Latinian regression event.

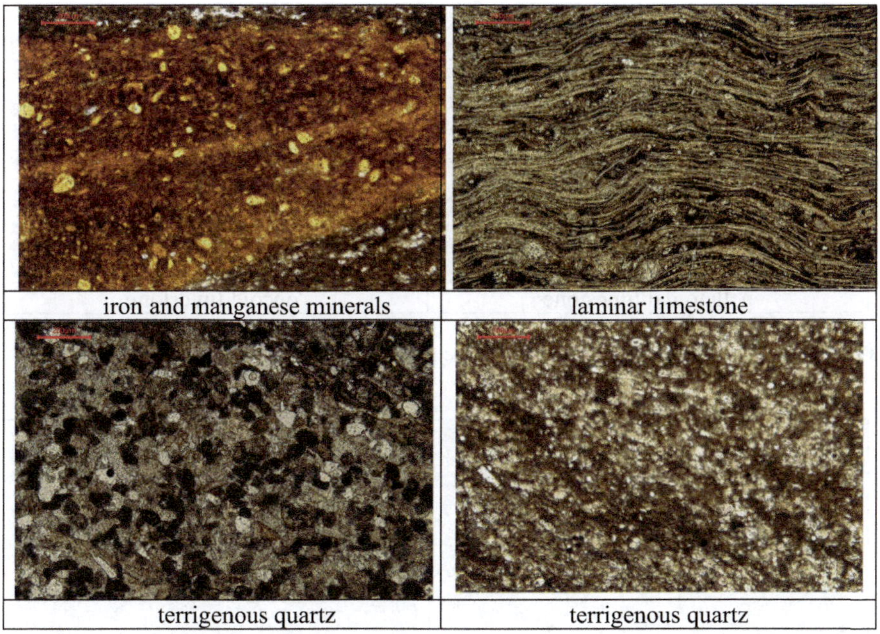

Fig. 15 Typical lithologic characteristics in Falang FM.

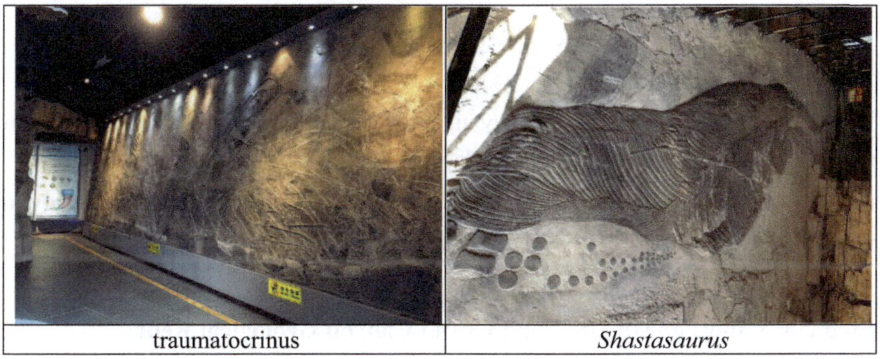

Fig. 16 Guanling fossil library

Stop 2.2: Stratigraphic Boundary Between the Longtou Formation and Falang Formation

The Longtou Formation is in integrated contact with the Falang Formation in the area of Duanqiao, Guanling. At the observation point, dark-colored thin bedded micritic limestone and bioclastic micritic limestone of the Zhuganpo section of the Falang Formation overlie the Longtou Formation (Fig. 18), which has formed a large difference from the stratigraphic boundary between the Longtou Formation-Gaicha

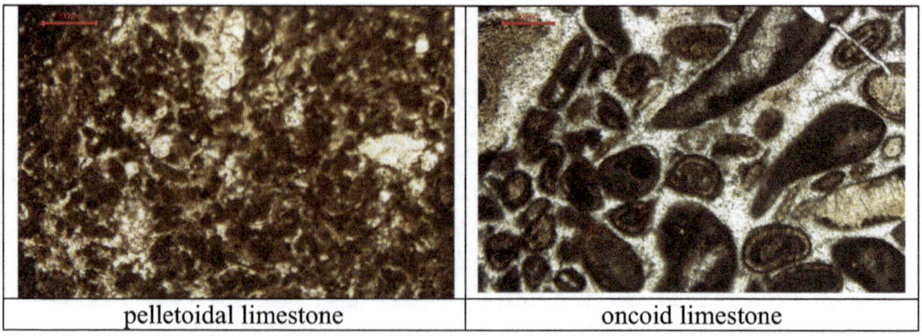

Fig. 17 Typical lithologic characteristics in the Longtou FM.

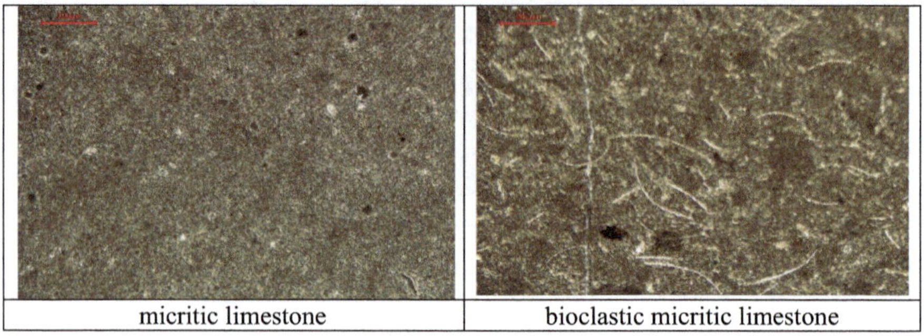

Fig. 18 Typical lithologic characteristics in the Fanglang FM.

Formation of Longtou Village, Qingyan, Guiyang, indicating the strong sedimentary differentiation of the Middle-Late Triassic in central Guizhou-southwest Guizhou.

Stop 2.3: Guanling Bangui Middle Triassic Series Longtou Formation-Bianyang Formation Facies Transition Zone

See Fig. 19.

Stop 2.4: Landscape of Huajiang Grand Canyon Geological Relics

It is distributed in the Guangshan-Hongyan area of Bangui County of Guanxing Highway. The middle Triassic Series Longtou Formation off-white thick bedded-massive sparite oolitic limestone and algal reef limestone are exposed on the north side of the road, representing platform margin reef facies deposits. The occurrence is 123°∠16°. At the south side of the road is the yellowish gray thin-medium bedded argillaceous siltstone and fine sandstone of the Middle Triassic Series Bianyang Formation, with the development of the Bouma sequence, which represents the deep-water slope turbidites (Fig. 19). The occurrence is 120°∠15°, and the two intersect in a tooth shape in this area and form a unique carbonate-clastic rock landform type.

Fig. 19 Facies transition zone

The Huajiang Grand Canyon is located at the junction of Guanling and Zhenfeng in Guizhou, with a total length of approximately 80 km and a total area of 300 km^2. It is a typical representative of the evolution of underground rivers to surface canyons. The canyon is divided into two phases, a wide valley and a narrow valley, with different directions. On both sides of the Huajiang Grand Canyon, there are a large number of mysterious ancient rock paintings and cliff rock carvings, including the Ma Ma Cliff Murals, Niujiaojing Mural Group, and Han Yuan Cave wall paintings and murals, which have high historical, artistic, and scientific value (Fig. 20).

The Beipanjiang Grand Bridge is the affiliated bridge of the Guanling-Xingyi high-grade highway. The main span is 388 m. When completed, it is a suspension bridge designed with two-way prestressed concrete stiffened slab beams for the largest span in China. It spans the Huajiang Grand Canyon. The distance from the bridge surface to the river is 486 m, and it is one of the tallest bridges in the world.

Stop 2.5: Stratigraphic sequence of the Falang Formation in the Zhenfeng Longchang section

The section of Longchang, Zhenfeng is located 2 km south of Longchang Town, Zhenfeng County, Southwest Guizhou Autonomous Prefecture, with good traffic

The Huajiang Grand Canyon	Ma Ma Cliff Murals

Fig. 20 Huajiang Grand Canyon geological relic landscape

conditions. The section exposes the Middle Triassic Series Yangliujing Formation-Upper Triassic Series Laishike Formation (Fig. 21), with continuous stratum exposure and a high degree of research.

The measured stratum is as follows:

In the Triassic Laishike Formation, no peaking is observed.

9.2 m thick

18. Gray medium bedded argillaceous fine sandstone, the weathered color is gray–yellow, and the joints are extremely developed.

2.74 m thick

17. Gray yellow thin bedded marl, thin bedded earthy yellow fine sandstone, the total thickness of the sandstone is approximately 60 cm, the middle part is intercalated with gray thin bedded limestone, and the thickness of a single bed is approximately 6 cm.

6.46 m thick

------ Conformity ------

Wayao section of the Triassic Franco Formation

16. The lower part is gray thin bedded argillaceous limestone, the argillaceous content increases upward, it is thin bedded argillaceous limestone intercalated with gray thin slab-like mudstone, with a large number of fossil fragments on the surface, and the upper part is gray black thin bedded argillaceous shale intercalated with marl, argillaceous sandstone, horizontal bedding is developed, and a large number of biological fossils are observed.

9.2 m thick

15. Dark gray thin-medium bedded argillaceous limestone, with rhythmic beds visible.

1.78 m thick

14. Earthy yellow argillaceous shale, severely weathered.

3.49 m thick

13. Gray thin-medium bedded argillaceous limestone, the argillaceous content decreases upward, and it transitions to gray medium bedded micritic limestone, partially intercalated with thin slab-like micritic limestone.

1.61 m thick

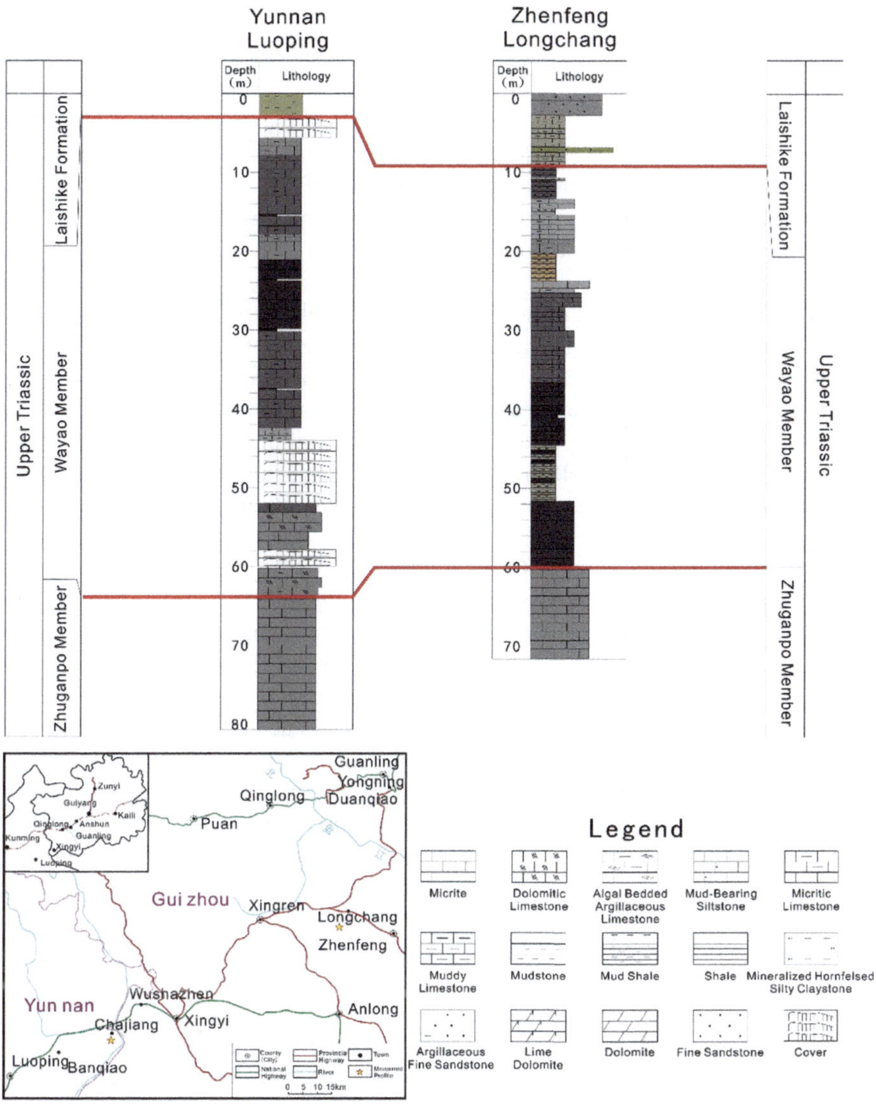

Fig. 21 Lithostratigraphic correlation of the Wayao Section

12. Gray–black thin bedded argillaceous limestone, where the argillaceous content decreases upward, is interbedded with thin slab-like limestone.

1.75 m thick

11. Gray black thin bedded marl, with visible argillaceous laminae, gradually disappears upward.

3.04 m thick

10. Gray black thin-medium bedded argillaceous limestone, a single thin bed is approximately 8 cm thick, the rhythmic bed is visible, and the rhythmic bed gradually disappears upward.

2.01 m thick

9. Black thin-medium bedded marl intercalated with black shale. A single thin bed is approximately 6 cm thick. A large number of fossils of crinoid stems and ammonite can be seen on the surface. The color becomes lighter upward, and it is a gray black thin-medium bedded marl.

12.45 m thick

8. The bottom is black thin-medium bedded argillaceous limestone with high argillaceous content. A single thin bed is approximately 8 cm thick, intercalated with black argillaceous shale. A large number of bivalve and ammonite fossils are seen on the surface. The middle part is yellowish gray argillaceous shale intercalated with black argillaceous shale, and a large number of ammonites and bivalve fossils are seen. The upper part is yellowish gray argillaceous shale interbedded with black argillaceous shale, and a large number of fossil fragments are seen.

15.5 m thick

- - - - - - Conformity - - - - - -

Zhuganpo section of the Triassic Franco Formation

7. Dark gray medium-thick bedded micritic limestone and fossils of crinoid stems are seen.

11.45 m thick

6. The bottom is gray thin bedded marlstone intercalated with thin bedded argillaceous limestone, a single thin bed is approximately 6–10 cm thick, and the marl content increases upward; the middle and upper parts are dark gray medium bedded micritic limestone, and a single medium bed is approximately 30 cm thick, partially intercalated with 1–2 cm argillaceous bands.

10.24 m thick

5. Dark gray medium-thick bedded micritic limestone, with coral fossils on the surface, and a single medium bed is 30–40 cm thick.

11.54 m thick

4. Dark gray medium bedded micritic limestone, partially intercalated with argilla ceous bands, approximately 1–2 cm thick. Fossils of ammonites, corals, crinoid stems, and bivalves can be seen on the surface. Corals are metasomatized by calcite, and pyrite crystals can be seen. Approximately 1.4 m of the upper part is covered by vegetation, sporadic outcrops can be seen, and the exposed lithology is dark gray medium bedded micritic limestone.

14.16 m thick

3. Dark gray thin-medium bedded micritic limestone, the thin bed is approximately 9 cm thick, with calcareous nodules with a diameter of approximately 5–20 cm, and a crust-like structure formed by recrystallization can be seen on the top.

2.09 m thick

2. Dark gray thin-medium bedded micritic limestone, with a single thin bed approximately 5–8 cm thick.

0.8 m thick

- - - - - - Conformity - - - - - -

The top of the Triassic Yangliujing Formation, the bottom of which is not seen.

1. Off-white thick bedded dolomite intercalated with medium bedded dark gray micritic limestone on the top.

13.98 m thick

The lithology of the Zhuganpo section of the Falang Formation is dominated by dark gray micritic limestone, partially intercalated with gray thin bedded marl, and a small number of fossils of ammonites, corals, crinoid stems, and bivalves can be seen in the middle and upper parts. The lithology of the Wayao section of the Falang Formation is mainly black thin bedded argillaceous limestone, gray black argillaceous shale, gray black thin bedded silty limestone, partially intercalated with thin plate like-lens like micritic limestone, mudstone and medium bedded micritic limestone. The top is intercalated with argillaceous sandstone. Since the deposition of the Wayao section, terrigenous fragments began to increase, and benthic biological fossils began to become increasingly deficient until they were totally depleted, reflecting that the carbonate platform in the middle and late stages of the Falang Formation has been subject to climate-ecological threats, and the corresponding sedimentary response has appeared.

Stop 2.6: Stratigraphic boundary between the Falang Formation and Laishike Formation in the Zhenfeng Longchang section

The lithology of the Laishike Formation is dominated by gray-earthy yellow thin-medium bedded siltstone, with a turbidite sequence developed, and it is in integrated contact with the underlying thin bedded argillaceous limestone intercalated with argillaceous shale in the Wayao section of the Falang Formation (Fig. 22). The Laishike Formation clastic rock deposits overlie the deep-water intracarbonate platform depressions in the Falang Formation, representing the complete termination of the Late Triassic carbonate factories in the Upper Yangtze Region.

4.3 Geological Survey in the Xingyi Area (Stop 3)

The scope of the geological survey in the Xingyi area is concentrated in the Xingyi Fengdu-Wanfenglin Subdistrict and Xingyi Wusha Town. Xingyi is located in an important distribution area of the Triassic "S" type facies transition zone in Guizhou. It not only produced the world-famous Xingyi biota in the Franco Formation but also formed typical and diverse peak cluster landforms in the Xingyi area due to strong structural folds and transformations in the later period.

Stop 3.1: Facies transition zone between the Yangliujing Formation and Longtou Formation in Zhaojiadu, Xingyi

Xingyi Zhaojiadu is the distribution area of the Maling River canyon. The west side of the canyon is the peak cluster area, which mainly exposes gray thin-medium bedded argillaceous dolomite and micritic dolomite of the Middle Triassic Series Yangliujing

Fig. 22 Stratigraphic boundary between the Falang Formation and Laishike Formation

Formation, featuring cone-shaped stone pinnacles. The rock occurrence is gentle, and the joints and fissures are densely developed, but limited by the bed, the vertical extensibility is poor. On the east side of the canyon is the distribution area of tower-shaped stone pinnacles, which mainly exposes the thick bedded-massive sparite oolitic limestone and algal reef limestone of the Middle Triassic Series Longtou Formation. The rock bed has a gentle dip angle, and vertical joints and fissures are sparsely developed, but the vertical extensibility is good, and there are collapse residues at the bottom of the peak. The two form a jagged, interlocking pattern on the plane and vertically constitute the landform of Daimao Mountain (Fig. 23).

Stop 3.2: Xingyi Wanfenglin peak cluster depression geological relics landscape

Wanfenglin is located in southeastern Xingyi city, Guizhou Province, bordering Guangxi to the south, reaching the convergence of three rivers at the junction of Yunnan, Guangxi and Guizhou Provinces in the west and connecting to the main peak of Wumeng Mountain in the north. It has a total area of 2,000 square kilometers. It is the largest and most typical karst peak forest in China (Fig. 24) and is well-known with the saying "of so many mountain peaks in the world, only the peaks here become forests."

Xu Xiake, a geographer from the Ming Dynasty in China, visited Xingyi twice. He wrote in his diary: "among the mountains of western Guangdong, some are purely composed of stones, some are partially composed of stones, they stand separately

Fig. 23 Facies transition zone between the Yangliujing Formation and Longtou Formation

Fig. 24 Wanfenglin landscape

without mixing with each other. Mountains in southern Yunnan all feature earthy peaks, and some are decorated with stones, but there are few stones, so ring-shaped depressions are more common. The mountains in southern Qiannan are in a state between the two, and they are unique in that they are extremely high." "It is speculated that this cluster of peaks begins here in the southwest and ends in Daozhou in the northeast (today's Didao of Hunan). It is majestic for thousands of miles and is a spectacular view in the southwest, and this is the best in the southwest." The peak cluster described by Xu Xiake as "majestic for thousands of miles" back then is the Wanfenglin forest mainly located in Xingyi. After hundreds of years, people are surprised to realize that it is the "spectacular view in the southwest."

At present, the Xingyi Wanfenglin Scenic Area is represented by Dongfenglin in Nahui Village, Wanfenglin Subdistrict, which often forms a combination of peak clusters and depressions, with sinkholes at the bottom of depressions; from the upper part of Nahui Village to the lower part of Nahui Village, a sloping valley is developed and formed along the fault zone in the east–west direction, the center gathers surface runoffs, and the valleys on both sides are flat, on which rice is often planted, which constitutes beautiful natural and pastoral scenery.

Stop 3.3: Xingyi Fossil Library

The Guizhou Dinosaur Fossil In-Situ Museum is located in northwestern Nimaigu village, Wusha town, Xingyi city. The excavation site of the Guizhou Dinosaur Fossil In-Situ Museum covers an area of 1,500 square meters, and the work area for fossil excavation is 600 square meters. A total of 779 specimens of various vertebrate fossils were excavated, which are mainly composed of *Keichousaurus*, *Syngnathus*, *Nothosaurus*, *Lariosaurus*, *Ichthyosauria*, *Potanichthys xingyiensis*, *Birgeria*, crinoids, ammonites, bivalves, etc. The Guizhou Dinosaur Fossil In-Situ Museum displays the continuous and complete Triassic Falang Formation Zhuganpo section and part of the Wayao section and systematically reveals important data such as the fossil assemblage, output characteristics and resource scale of the Xingyi biota. It provides important data for an in-depth understanding of the diversity and the sedimentary-ecological evolution process of the Xingyi biota.

Measured data of the section of Nimaigu, Wusha, Xingyi (Fig. 25):

In the Wayao section of the Triassic Falang Formation, no peak is observed.

25–30: The lower part is gray medium bedded lamellar limestone, the upper part is gray medium bedded micritic limestone, the "eyelid-eyeball" structure is observed, the argillaceous content is higher than the lower 24 beds, of which the lamellar limestone reflects the interstorm period, and "eyelid-eyeball" structured micritic limestone reflects the storm period.

8.63 m thick

24. The lower part is gray thin-medium bedded lamellar limestone. There are many crinoline stems and ammonites in the limestone. The laminae are mainly argillaceous laminae, 20 cm diameter flint masses can be seen, and mound-like cross bedding is developed upward. The "eyelid-eyeball" structure can be seen. The upper part is gray medium bedded argillaceous-silty limestone, with undulating bedding, containing argillaceous laminae. A clear pungent odor can be smelted after hitting it, which "smells like rotten egg."

5.1 m thick

Gray medium bedded argillaceous limestone and bioclast-bearing silty limestone intercalated with thin bedded dark gray marl. A large number of complete bivalves are seen on the surface of the marl, and the shell decoration is clear.

1.97 m thick

- - - - - - Conformity - - - - - -

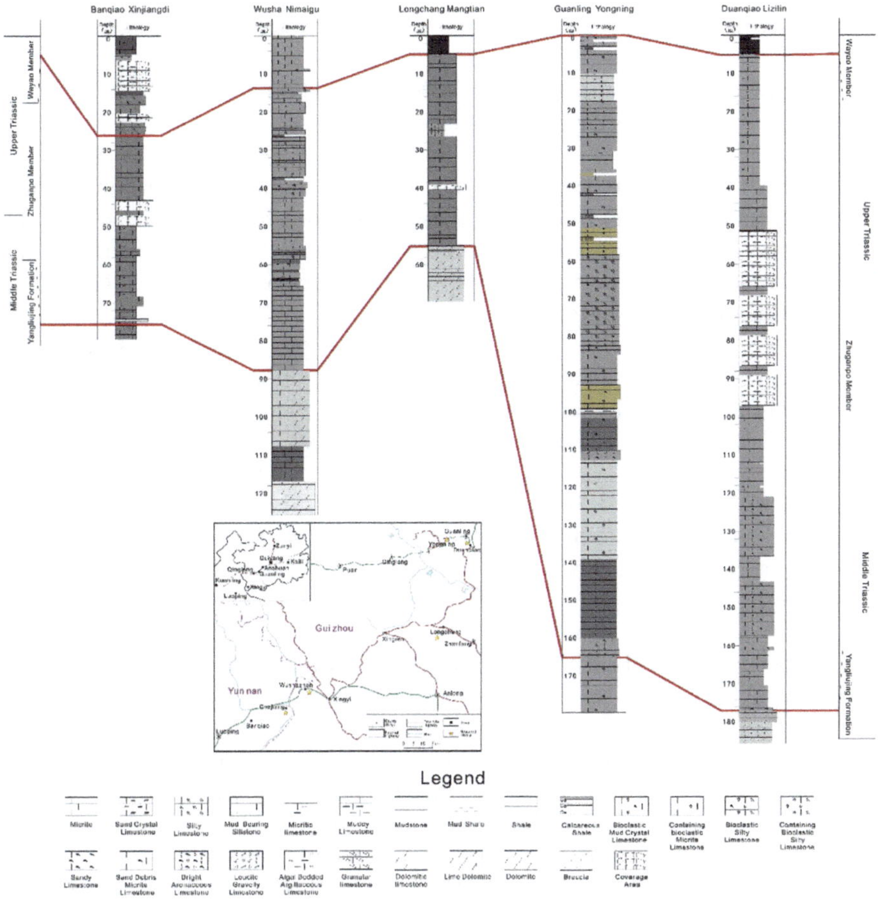

Fig. 25 Lithostratigraphic correlation of the Zhuganpo Section

Zhuganpo section of the Triassic Falang Formation

22. The gray medium bedded argillaceous limestone.

1.77 m thick

21. Gray thin-medium bedded soft sediment deformation layer, the lithology is bioclast-bearing micritic limestone intercalated with thin marl, argillaceous laminas are seen.

1.53 m thick

20. Gray medium bedded bioclast-bearing micritic limestone interbedded with gray thin bedded-shale marl in unequal thickness. There are a large number of well-preserved bivalves on the surface of marl, with clear shell decoration, and no obvious undulations are seen in the bedding.

1.29 m thick

19. Gray medium bedded micritic limestone intercalated with argillaceous bands.
1.16 m thick

18. Gray thin-medium bedded bed soft sediment deformation layer, the lithology is micritic limestone, argillaceous laminae are seen.
3.35 m thick

17. Gray thin-medium bedded (bioclast-bearing) micritic limestone intercalated with breccia (autobreccia)-bearing argillaceous bedding, breccia diameter is 5 mm–6 cm, sorting is poor, bivalve bioclasts are seen, the top is intercalated with dark gray calcareous shale, a large number of bivalves can be seen on the surface of shale, and shell decoration is clear and complete.
4.02 m thick

16. Gray medium bedded bioclast-bearing micritic limestone and argillaceous gravel bed are seen in the layer. Ammonites with a diameter of 4 cm are seen in natural bed 188, and the bedding is gently undulating.
1.82 m thick

15. Gray thin bedded bioclast-bearing micritic limestone intercalated with medium bedded bioclastic micritic limestone. There is an argillaceous gravel bed in the layer. There are many bioclasts near the gravel bed, and the bed is gently undulating.
2.64 m thick

14. Gray medium bedded bioclast-bearing micritic limestone, and there are argillaceous gravel beds in the bed. There are two types of gravel beds. The first type of gravel is formed in situ, with poor rounding, poor sorting, and a large gravel diameter. The second type of gravel has obvious transport characteristics. The rounding is medium-good, it is subround-subangular, and the sorting is good. The two gravel beds mostly appear alternately. In natural bed 170, the two gravel types can appear at the same time. The bedding is gently undulating, with en-line expansion joints partially are seen, with the top limestone intercalated with dark gray calcareous shale.
3.5 m thick

13. Gray medium-thin bedded bioclastic micritic limestone, bioclast-bearing micritic limestone, with undulations in the bedding which gradually slows upward. There is an argillaceous gravel bed in the bed with a gravel diameter of 1–4 cm, which is subrounded-subangular, and the top is thin bedded micritic limestone intercalated with dark gray calcareous shale.
3.16 m thick

12. Gray thin-medium bedded bioclast-bearing micritic limestone and gravel-bearing argillaceous beds are seen, the gravel diameter is less than 1 cm, flint masses are seen, and no obvious undulations are seen in the bedding.
2.15 m thick

11. Gray medium-thick bedded micritic limestone, with an argillaceous gravel bed within the bed, with a gravel diameter of 1–4 cm, which is subrounded-subangular, gentle undulations are seen in the bedding, and flint masses are seen in some parts.
5 m thick

10. Gray thin bedded micritic limestone intercalated with medium bedded micritic limestone, with bivalves as the main bioclasts. A pungent odor can be smelled upon

striking it, which "smells like rotten egg," and there are no obvious undulations in the bedding.

1.57 m thick

9. Gray medium bedded micritic limestone with argillaceous gravel bed in it. The gravel at the bottom has a diameter of approximately 3 mm, which is subround-subangular. The sorting is good, and an obvious erosion surface structure can be seen. The gravel diameter increases upward, the gravel diameter is 3–13 cm, with poor sorting and good rounding, and it is in a subround-subangular shape and shows a flat distribution as a whole. Then, the gravel diameter decreases further upward, the gravel diameter is 3 mm–1 cm, the sorting is good, and it is subround-subangular, with undulations on the bedding.

4.13 m thick

8. Gray thin bedded (bioclast-bearing) micritic limestone. The limestone contains argillaceous laminae, with an oriented distribution of flint masses developed in the bottom layer. Undulating argillaceous patterns are seen between beds, and gentle undulations are seen in the bedding. Gravel-bearing argillaceous beds can be seen upward.

4.56 m thick

7. Gray medium bedded bioclast-bearing micritic limestone, with scattered bioclast distribution, which is dominated by bivalves and echinoderms. Gravel-bearing argillaceous bands and argillaceous laminae are developed in the limestone, and autobreccia is seen in the argillaceous bed. The bioclasts show an oriented arrangement, and the gravel is poorly rounded and distributed in a chain shape. The bottom shows scattered flint masses, 15 cm in diameter, and the bedding is slightly undulating.

1.25 m thick

6. Dark gray medium bedded argillaceous limestone intercalated with dark gray thin slab-like marl. The marl has horizontal laminae. The surface of the limestone is stained with no obvious undulations. It contains fossils of *Ichthyosauria, Syngnathus* and crinoids.

1.21 m thick

5. Gray–gray black thin-medium bedded argillaceous limestone intercalated with black gray thin bedded limestone, thin bedded marl, thin slab-like argillaceous limestone, with developed surface laminae. It contains fossils of *Keichousaurus, Ichthyosauria, Nothosaurus, Lariosaurus*, fishes, shrimps, bivalves, ammonites, and crinoids.

5.96 m thick

4. Gray medium bedded limestone with medium-thin bedded limestone interbedded with argillaceous limestone on the top.

21.3 m thick

------ Conformity ------

At the top of the Middle Triassic Series Yangliujing Formation, the bottom is not seen.

3. White gray–gray thick bedded-massive dolomitic limestone.

19.9 m thick

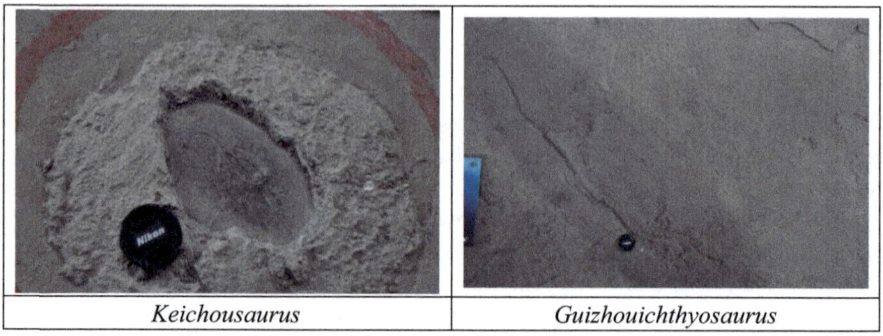

Fig. 26 Typical fossils in Xingyi biota

2. Dark gray thick bedded-massive limestone.

6.1 m thick

1. Off-white–gray thick bedded-massive dolomite with developed knife-cut shaped karrens.

7.3 m thick

The Xingyi biota is mainly composed of marine reptiles, fishes, ammonites, bivalves and other fossil groups represented by *Keichousaurus hui young*. According to current research, the Xingyi biota is mainly produced at the bottom of the Zhuganpo section of the Falang Formation and is divided into two fossil groups from bottom to top. The lower fossil section is produced from the section of gray black thin bedded marl, marked by the nearshore *Sauropterygia*, showing the affinity to the West Tethys; the upper Fossil Group is produced from the section of gray medium bedded limestone intercalated with marl, featuring *Guizhouichthyosaurus*, and they show affinity to North America in paleobiogeography (Fig. 26).

Stop 3.4: Tsunami-Seismic Deposits in the Zhuganpo Section of the Falang Formation in the Nimaigu Section of Xingyi

The middle and upper parts of the Zhuganpo section of the Falang Formation in the Xingyi-Guanling area (marked as above bed 52) are mainly characterized by interbeds of calcirudyte limestone and micritic limestone of unequal thickness, and the bedding often has wave-like undulations. The gravel-sized grains have two characteristics: quasi-in situ formation and off-site transportation (Fig. 27).

The deformation structure of the soft sediments in the upper part of the Zhuganpo section of the Falang Formation is well developed and is characterized by hydroplastic folds. The axial plane is in the preferred orientation, the core of the anticline is significantly thicker, and the folds are cut by the overlying sedimentary bed (Fig. 28). Frequent earthquakes and the resulting tsunamis (?) may not be conducive to the preservation of marine reptiles.

Fig. 27 Calcirudyte limestone

Fig. 28 The deformation structure of the soft sediments

Stop 3.5: Storm sedimentary sequence of the Wayao section of the Falang Formation of the Nimaigu section in Xingyi

A typical near-source storm sequence is developed at the bottom of the Wayao section of the Falang Formation. The sequence of graded bedding-parallel lamella bedding-mound shaped crossbedding is the most typical, forming multiple cycle sequences of storm period-interstorm period. A large number of crinoids, bivalves and other bioclasts with oriented arrangements can be seen in the graded bedding (Fig. 29).

Stop 3.6: Xingyi biota

The Xingyi National Geological Park Museum mainly displays Middle Triassic paleontological fossils and sedimentary processes in the Xingyi area. Due to the structural differentiation of the Middle Triassic Late Latin Period in the area, the Xingyi area shows obvious tectonic subsidence. The water body in the Xingyi area changes from shallow to deep, and its biological appearance is that the lower fossil bed is dominated by shallow sea creatures, while the upper fossil bed is dominated by deep-sea *Ichthyosauria* (Fig. 30), transforming into a deep-sea ecological type. The biological diversity of Xingyi biota shows the transformation of marine reptiles from near-shore to far-shore types. It also reflects the disappearance of the confines of

Fig. 29 The storm sedimentary sequence of the Wayao section

Fig. 30 Xingyi National Geological Park Museum

intraplatform depressions caused by sea level rise and the process by which far-shore type organisms quickly occupy the original ecological niche of *Keichousaurus.*

4.4 Qinglong Area Tourist Geological Survey Route (Stop 4)

The Qinglong area is located in the transition zone between platform facies and slope facies from the middle section of the Yunnan-Guizhou Plateau to Guangxi. In the Ming Dynasty, the Qinglong area was designated Annan Garrison due to its important geographical location featuring "multiple checkpoints and overlapping mountain ridges, locking Yunnan and Guizhou." It was changed to Annan County in the Kangxi Period of the Qing Dynasty and later renamed Qinglong County. Qinglong County is rich in geological heritage resources and natural and cultural landscapes. Among them, the "Twenty-Four Bends" anti-Japanese road is the most famous. It has an important strategic position and played an active historical role in the history of the Chinese Anti-Japanese War and the World Anti-Fascist War. Qinglong has a unique climate, diversified soil types and rich plant resources. It is also the place where Pleistocene tea seed fossils were discovered. In Qinglong, there are mainly metal mineral resources such as gold, antimony, lead and zinc and nonmetallic mineral resources such as coal, fluorite, Guizhou jadeite, marble, limestone, calcite, and porcelain clay.

Stop 4.1: Ming and Qing Garrison System and Annan Ancient City

The garrison system is designed to transfer military households from all over the country to support local areas. It undertakes the main purposes of reclamation under the military footprint and defense and is the main military system in the Ming and Qing Dynasties. Usually, the garrisons set up in areas not subject to formal administrative divisions are called real land garrisons, which have the function of the prefecture's center and once had a significant impact on the construction of administrative divisions in Guizhou.

Due to the strategic location of the Qinglong area, in the 23rd year of Hongwu (1390), the commander Liang Hai built the Annan Garrison City (today's Qinglong County). The main task was to control the Yunnan-Guizhou Post Road and guard the section from Guiyang to Yunnan Pingyi. In the 25th year of Hongwu (1392), Annan Garrison City was completed. The perimeter of the city wall was 743 Zhang (a unit of length (= 3.33 m)), with gates in each direction. There were 36 city shops inside and several temples outside (Fig. 31). The garrisons in Guizhou have the dual functions of administrative management and military defense. Therefore, the garrisons set for the purpose of guarding are essentially "military-administered governments." During the mid-Ming Dynasty-the Yongzheng period of the Qing Dynasty, due to the continuous migration of the military households and the end of the surrounding wars, the garrisons that belonged to the military system were gradually transferred to the administrative systems of prefectures, Zhou (an ancient administrative division), and counties. The site of the garrison gradually changed from a military function to a town that is a political, cultural, economic, and educational center. In the 26th year of Kangxi, Annan garrison was dissolved and changed to Annan County. Because the name of the county was the same as the original name of Vietnam, it was renamed Qinglong in 1941.

Stop 4.2: "Twenty-Four Bends" Anti-Japanese War Road

The "Twenty-Four Bends" Anti-Japanese War Road is located 1 km southwest of Qinglong County, Guizhou Province, between the Qinglong Mountains and Mopan Mountain, with a total length of 3.98 km and an effective road width of 7.5 m. The height difference between the starting and ending points of this section of road is 265 m. It is characterized by 24 turning curves (curved turns). In addition, it also has a minimum curve radius of 8.125 m, a horizontal curve ratio of 52.788%, an average longitudinal slope of 6.67%, a largest longitudinal slope of 14.7% and other characteristics (Fig. 32). The drainage system of the entire road is well designed. The horizontal ditches and longitudinal ditches are connected with six stone arch culverts to achieve the greatest stability of the drainage system.

The survey for the road started in 1928, and construction fully began in 1935. It was completed and opened to traffic in September 1936. During the Anti-Japanese War, a large number of materials and persons aiding China on the Yunnan-Burma Road, Hump Route, and Ledo Road had to be transported to the rear of Chongqing and

Fig. 31 Annan Garrison City (repaired)

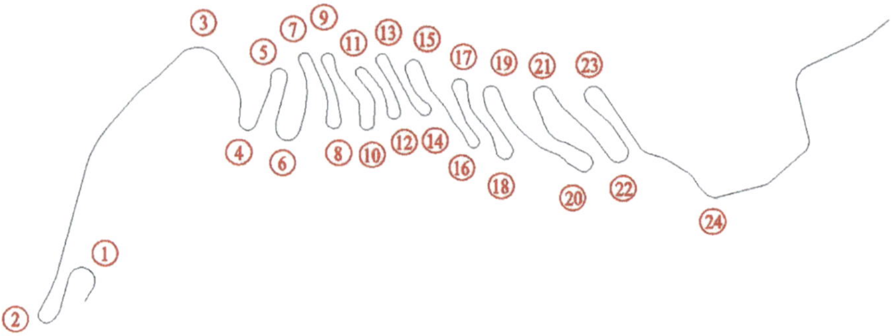

Fig. 32 "Twenty-Four Bends" pattern

the front lines of the Anti-Japanese War through Qinglong's "Twenty-Four Bends" traffic throat; at the same time, the Second Army, Fifth Army, and Eighth Army of the National Revolutionary Army successively included the series of Chinese Expeditionary Forces stationed in Qinglong for training. After the outbreak of the Pacific War, U.S. Army General Joseph Stilwell served as the commander-in-chief of the U.S. Army's China-Burma-India Theater and Chief of Staff of Allied Forces for the Chinese Theater. To meet the monthly transportation needs of more than 15,000 tons of strategic materials, he strengthened maintenance on Qinglong's "Twenty-Four Bends" strategic node and dispatched the 1880 Aviation Engineering Battalion

Fig. 33 Gas stations (repaired)

Technical Company of the U.S. Army aiding China to Qinglong to assist in the refurbishment of the road. The U.S. Army Wartime Production Department of the Chinese Theater Command also established vehicle maintenance stations and gas stations successively in Qinglong County (Fig. 33), the 105th Rear Hospital of the Ministry of Military and Political Affairs stationed at Qinglong was responsible for the rescue.

After the opening of Ledo Road, the American military journalist John Albert took the lead in shooting the "Twenty-Four Bends" panorama in Qinglong and published it with *Stilwell Road—Land Route to China* as the title on *National Geographic*. It was later published and reprinted countless times by major media in Europe, America, Japan, Southeast Asia, and the National Government. However, since the above publications did not indicate the specific location, the road sections in the photos have long been interpreted as a certain section of the Stilwell Road by various media and experts. In 2002, Mr. Ge Shuya, an expert on the history of Yunnan and Burma, confirmed that it is located in Qinglong, Guizhou. In May 2006, the "Twenty-Four Bends" Anti-Japanese War Road was listed in the sixth batch of national key cultural relic protection units upon approval by the State Council (Fig. 34).

At present, Qinglong's "Twenty-Four Bends" Anti-Japanese War Road is well preserved, still in normal operation, and there are few traffic accidents. The stone arch culverts and retaining walls on the road are also intact after 80 years of wind and rain, all reflecting the scientific design of wartime engineers of the road and strict control of the engineering quality. At the same time, as a cultural landmark of the Ledo Road (Stilwell Road), the road has made an important contribution to China's victory in the Anti-Japanese War. It is also a symbol of the unity of the Chinese and American allied forces in the War of Resistance and an important historical testimony of the friendship and cultural exchanges between the Chinese and American people.

Fig. 34 "Twenty-Four Bends" (From 1945 to 2006)

Stop 4.3: Qinglong ancient tea tree and tea seed fossils

Since ancient times, Qinglong has not only been an important tea trading hub but also an important tea producing area suitable for the growth and development of shade-tolerant plants such as tea trees. Qinglong has a 600-year recorded history of tea planting. There are a large number of primitive wild-type, evolutionary and semievolved, tree-type, half-tree-type and shrub-type ancient tea tree resource communities and cultivated large tea tree resources totaling more than 50,000 plants in the county. They present the characteristics of community diversity. At 594 m above sea level in Guangzhao Town, there is golden camellia; at 1236 m above sea level in Zima County, there is *Camellia rubituberculata*; at 1475 m above sea level in Zhongyun Town, there is Palu tea; at 1680 m, there is red medicine red camellia; and at 2025 m, there are *Camelliatetracocca*, *Banpo Camellia arborescens*, and *ilex latifolia thumb,* etc., and many ancient tea trees have a history of more than a thousand years.

In 1980, tea seed fossils were discovered in the Pleistocene series residual bed of the virgin forest of Yuntou Mountain in Qinglong (Fig. 35). The researcher Guo Shuangxing from the Nanjing Institute of Geology and Paleontology of the Chinese Academy of Sciences believed that features such as that there were dents on the side of the umbilicus of the fossil and that the top of the seed was flat or slightly protruded are very similar to the seeds of modern *Camelliatetracocca*, so it can be classified as *Camelliatetracocca.* The discovery of Qinglong tea seed fossils puts an end to the

Fig. 35 Yuntou Mountain in Qinglong

long-term disputes over the origin of tea trees in the tea industry and at the same time provides valuable evidence for the study of tea history and the development and evolution of tea families (plants) in China.

Key References

Jin X, Shi ZQ, Wang YY, Duan X, Cheng M (2015) Mid-Carnian (Late Triassic) extreme climate event: advances and unsolved problems. Acta Sedimentol Sin 33:105–115

Jin X, Gianolla P, Shi ZQ, Franceschi M, Caggiati M, Du YX, Preto N (2020) Synchronized changes in shallow water carbonate production during the Carnian Pluvial episode (Late Triassic) throughout Tethys. Glob Planet Change 184:103035

Shi ZQ, Qian LJ, Xiong ZJ, Zeng DY (2010) Carnian crisis occurring in SW China and its ideational origin. Bull Mineral Petrol Geochem 29:227–232

Sun Y, Wignall P, Joachimski MM, Bond DPG, Grasby SE, Lai XL, Wang LN, Zhang ZT, Sun S (2016) Climate warming, euxinia and carbon isotope perturbations during the Carnian (Triassic) crisis in South China. Earth Planet Sci Lett 444:88–100

Tong JN, Chu DL, Liang L, Shu WC, Song HJ, Song T, Song HY, Wu YY (2019) Triassic integrative stratigraphy and timescale of China. Sci China Earth Sci 62:189–222

Xiao JF (2005) Sedimentary-volcanic tuffs during early Middle Triassic volcanic event in Guizhou province and their stratigraphic significance. Chin J Geochem 24(4):338–344

Zhang Y, Li MS, Ogg JG, Montgomery P, Huang CJ, Chen ZQ, Shi ZQ, Enos P, Lehrmann DJ (2015) Cycle-calibrated magnetostratigraphy of middle Carnian from South China: implications for Late Triassic time scale and termination of the Yangtze platform. Palaeogeogr Palaeoclimatol Palaeoecol 436:135–166

Safety

Guizhou belongs to the subtropical humid monsoon climate zone. The average temperature of the hottest month (July) is generally 22–25 °C; it is a typical area with a cool summer, with much precipitation that is obvious in the rainy season, many cloudy days and little sunshine. Therefore, the investigators need to prepare clothing and utensils for various weather conditions.

The karst landforms are typical in Guizhou, and there may be sudden geological disasters. A primary triggering factor for disasters is rainfall, and the period of occurrence is mainly from May to September. The investigators must pay attention to the warning signs for hidden geological hazards near the inspection site and observe the time schedule provided by the explainers. In addition, because most of the inspection sites are located on the side of the highway, investigators must pay attention to the traffic on the highway.

The investigators must actively cooperate with the epidemic prevention requirements of the conference affairs team and conform to the requirements of the health code, itinerary code presentation, temperature measurement, wearing masks, etc. If fever, fatigue, cough, sore throat, sneezing, diarrhea, vomiting, jaundice, conjunctival congestion, rash, and other suspected symptoms occur, they should be reported to the conference affairs team immediately, and medical attention should be sought to assess the condition as soon as possible.

For the above precautions, all field investigators are requested to actively cooperate. If they refuse to cooperate, all consequences should be borne by the offender.

The emergency telephone number for police is 110. The emergency telephone number for ambulance is 120.

Field Leaders

Tingshan Zhang

Tingshan Zhang is a full professor and doctoral supervisor of Southwest Petroleum University. His main research activity focuses on Palaeozoic reef and Late Triassic palaeoenvironmentpalaeoclimate and palaeoecology.
Mobile: +8613982001379
Email: zts_3@126.com

Mihai Popa

Mihai Popa is a professor at the University of Bucharest. His main research is focused on paleoecology, stratigraphy and continental sedimentology.
Mobile: +40213053817
Email: mihai@mepopa.com

Wei Yang

His research activity is primarily devoted to Paleozoic carbonate platform evolution and paleoenvironmental evolution.

Late Permian Reef Buildups and the Puguang Gas Field in the Northeastern Sichuan Basin, SW China

Qianglu Chen, Zongquan Hu, Taizhong Duan, Kun Yuan, and Tao Li

1 Introduction

At the turn of Paleozoic and Mesozoic, the South China Plate was composed of several small plates in the Tethys Ocean at low latitude (Yin et al. 1999). The Yangtze region is one of regions with the best stratigraphic record and the most abundant fossils in the East Tethys and even in the world in the Permian–Triassic transitional period. It records the Emei tafrogenesis, the evolution of the Paleotethys, PTB mass extinction and other important geological events. Thus, it is the best option for studying geological catastrophe at the turn of Paleozoic and Mesozoic, as well as being the focus and hotspot of international researches.

Reef is a type of carbonate buildup formed by a large number of in-situ growing sessile organisms in a carbonate deposit system (Dunham 1970; Fan and Zhang 1985a, b; Riding 2002). After diagenetic transformation, the associated reef and grain shoal sedimentary bodies often contain pores and cavities, and become good oil and gas reservoirs, which is one of the critical reservoir types of giant marine carbonate oil and gas fields in the world. Permian was a key reef-forming period with widespread

Q. Chen (✉) · Z. Hu · T. Duan · K. Yuan
Sinopec Petroleum Exploration and Production Research Institute, Beijing, China
e-mail: chenql.syky@sinopec.com

Z. Hu
e-mail: huzongquan.syky@sinopec.com; huxm@nju.edu.cn

T. Duan
e-mail: duantz.syky@sinopec.com

K. Yuan
e-mail: yuankun.syky@sinopec.com

T. Li
Sinopec Zhongyuan Oilfield Company, Beijing, China
e-mail: ltao.zyyt@sinopec.com

X. Hu (ed.), *Field Trip Guidebook on Chinese Sedimentary Geology*,
https://doi.org/10.1007/978-981-99-6936-4_10

reefs around the world. And those reefs developed in the Upper Yangtze region of China were the most typical reefs in Late Permian. Since 1970s, the composition, biological species, sedimentary environment and ecology, basic characteristics, type and diagenesis, architectures, development model and distribution pattern of reefs have been studied in depth (Zhang 1985, 1992; Liu and Rigby 1992; Wu et al. 2003; Mu et al. 2003, 2005; Wu et al. 2012; Hu et al. 2012a, b; Rigby et al. 1989; Li et al. 2018; Chen et al. 2018). Reef in the Jiantianba outcrop (or Jiantianba reef) in the Yangtze region is one of the most typical reefs of Permian and one of the reefs that have been most investigated. It includes many calcisponge species, most of which are well preserved. Common species residing in Europe and North America and endemic species only found in this region jointly form a distinctive Lichuan sponge fauna (Fan and Zhang 1986, 1987; Li et al. 2018).

With the deepening of basic geological researches including sedimentary outcrops, it is gradually recognized that the alternation of platform and shelf/trough developed on the northeast of the upper Yangtze Platform (now the East Sichuan Province-West Hubei Province) with reef and shoal sedimentary systems at the margin during the Late Permian. These basic geological cognitions have triggered the new ideas for oil and gas exploration in the northern part of the Upper Yangtze region, and played a role of geological guidance for the discovery of Puguang and other giant gas fields (Ma et al. 2005, 2007a, b, c, 2010; He et al. 2015). The discovery of Puguang gas field and the confirmation of presence of deep and ultra-deep reservoirs with high porosity and permeability further led to the research upsurge of sedimentary paleogeography, sequence stratigraphy, platform margin structure, diagenesis, reservoir characteristics, formation mechanism and related aspects in the northern part of the Upper Yangtze region and even China marine basins (Ma 2006; Ma et al. 2005, 2007a, b, c, 2010; Wang et al. 2006, 2010; Zhao et al. 2010; Hu et al. 2012a, b; Zhao et al. 2011; Li et al. 2015, 2018; Wu et al. 2015; Xing et al. 2017, 2018). During this period, the *Integration of Outcrop and Modern Analogs in Reservoir Modeling* edited by G. M. Grammer (2008) and the *Outcrop and Reservoir Modeling of Carbonate Platform Margin Sedimentary Systems* edited by Li Sitian (2014) emphasized the model of sedimentary environment based on modern sediments and large sections on outcrops to identify sedimentary system and reservoir architecture, which can be referred in the exploration of complex marine carbonate subtle reservoirs in ultra-deep strata. The proposed investigation route in Upper Yangtze region, i.e., the marginal reef of Jiantianba platform in Lichuan—the marginal reef and shoal of Panlongdong platform in Xuanhan—Puguang gas field, demonstrates the Late Permian platform margin sedimentary system, development of typical Late Permian reefs in the world, development of the Early Triassic grain shoals, and the successful practice that the sedimentological study serves oil and gas exploration. Moreover, the route also includes the sedimentary response to the PTB mass extinction event (the largest in Phanerozoic), that is, the widely developed reefs in the Permian were covered by grain shoal deposits in the Early Triassic, in other words, the Reef Gap in epeiric sea (Flügel 2002). This change of sedimentary facies is closely related to the paleoclimate and paleo-environment of the earth's surface systems.

1.1 Tectonic Evolution and Sedimentary Response

The study area is located in the northeast of the present-day Sichuan Basin. Sichuan Basin belongs to the Middle–Upper Yangtze Craton, and lies in the western part of the Yangtze plate in terms of tectonic setting (Fig. 1).

The Middle–Upper Yangtze craton evolved with the alternation of extension and compression during geological history. On the basis of Jinning basement (formed in the Middle Neoproterozoic), it is composed of Sinian–Middle Triassic marine

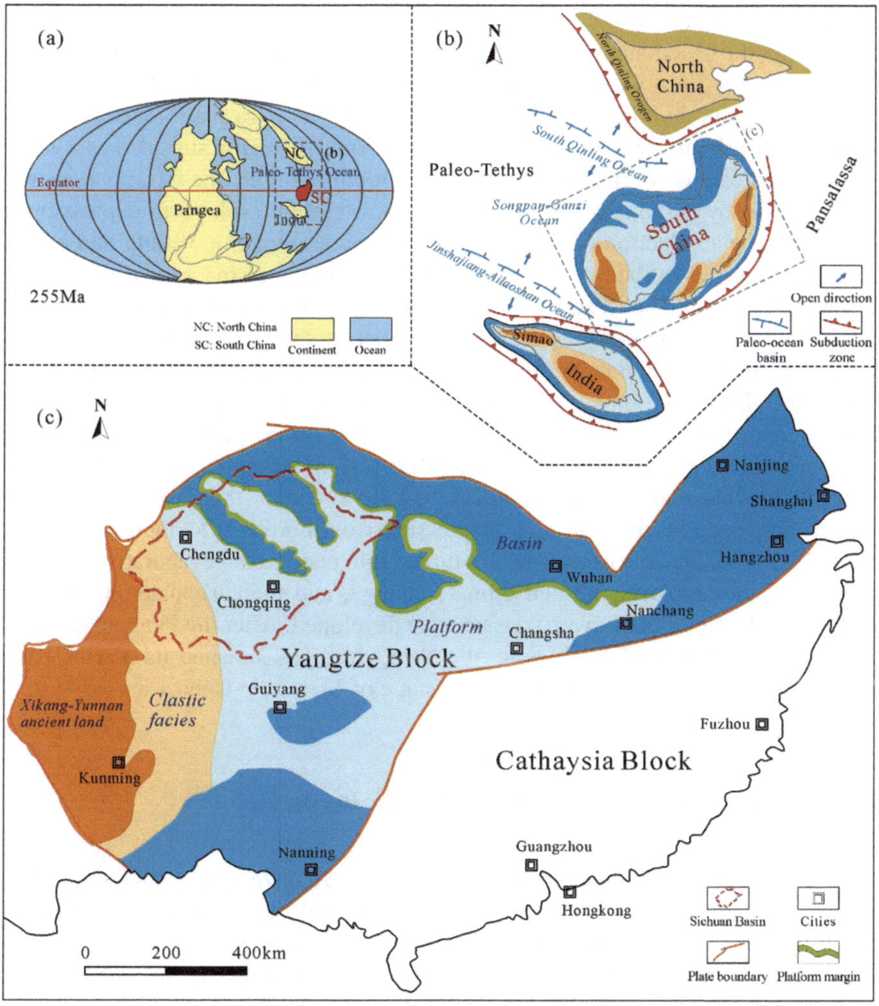

Fig. 1 Paleogeographic pattern of Yangtze Plate during Late Permian. (**a**) and (**c**) are modified from Wu et al. (2021); (**b**) is modified from Huang et al. (2021)

strata and Upper Triassic–Quaternary continental strata. From the breakup of South China ancient land in Early Neoproterozoic to the formation of a new unified South China continent in Silurian, the basin has experienced 3 evolution stages which were rift basin stage in Nanhua (Cryogenian)–Sinian period, passive continental margin and cratonic basin stage during Cambrian–Middle Ordovician, and intracratonic depression stage during Late Ordovician–Silurian. Accordingly, the sedimentary assemblage of clastic–carbonate rocks in the Nanhua–Sinian systems, the marine shale–carbonate during Cambrian–Middle Ordovician, and the shale and siltstone of deep-water shelf to shallow-water shelf facies during Late Ordovician–Silurian were deposited.

From Devonian to Middle Triassic the basin experienced marine basin evolution stage. After the Caledonian movement, the Yangtze region was a vast erosional area. In the Devonian–Early Carboniferous, the Upper Yangtze region was denudated as an ancient land, and the Middle Yangtze region developed intra-platform depressions. In the Late Carboniferous–Middle Permian, the Yangtze region subsided as a whole and suffered extensive transgression. In the Middle Permian, the basin pattern was characterized by carbonate platform in the south and intra-platform depression in the north. The subduction of the Paleotethys, mantle plume upwelling and crustal uplifting occurred in the western Upper Yangtze region at the end of the Middle Permian, resulting in uplift and denudation of the upper strata of the Middle Permian (Maokou Formation) to various degrees. Influenced by the tectonic movement at the end of the Middle Permian and the Emei tafrogenesis in the Late Permian (Fig. 2) (Luo 1981, Luo and Long 1992), large-scaled magmatic eruption occurred in the western margin of the Yangtze plate, formed mafic-intermediate volcanic rocks with a thickness of 1000–2000 m. And the eastern part went through a relative subsidence which led to clastic-carbonate deposits. In Late Permian, the carbonate platform further developed because of the transgression from northeast to southwest. In the end of Early Triassic, the open platform turned into restricted—evaporative platform because of the sea level fall and climate change, and the carbonate and gypsum deposits of the Middle—Lower Triassic were developed. After the Indosinian movement in the end of Middle Triassic, the Yangtze region finished its marine basin history by stepping into continental basin stage of Mesozoic-Cenozoic.

1.2 Palaeogeographic Pattern of Late Permian

After Caledonian movement (end-Silurian), the Yangtze and Cathaysian block merged into the South China Plate (SC). Global paleogeographic reconstruction shows that during the Late Permian, the South China Plate was in the north near the equator, surrounded by the Paleotethys and Panthalassa in the east and west, respectively (Fig. 1a). It was separated from the North China Plate (NC) in the north by the Qinling Micro-plate. From the Late Devonian to the early Middle Triassic, the Qinling Micro-plate became a deep-sea basin connecting the Panthalassa in the east and opening westward to the Paleotethys, and joined by scissor pattern from east to

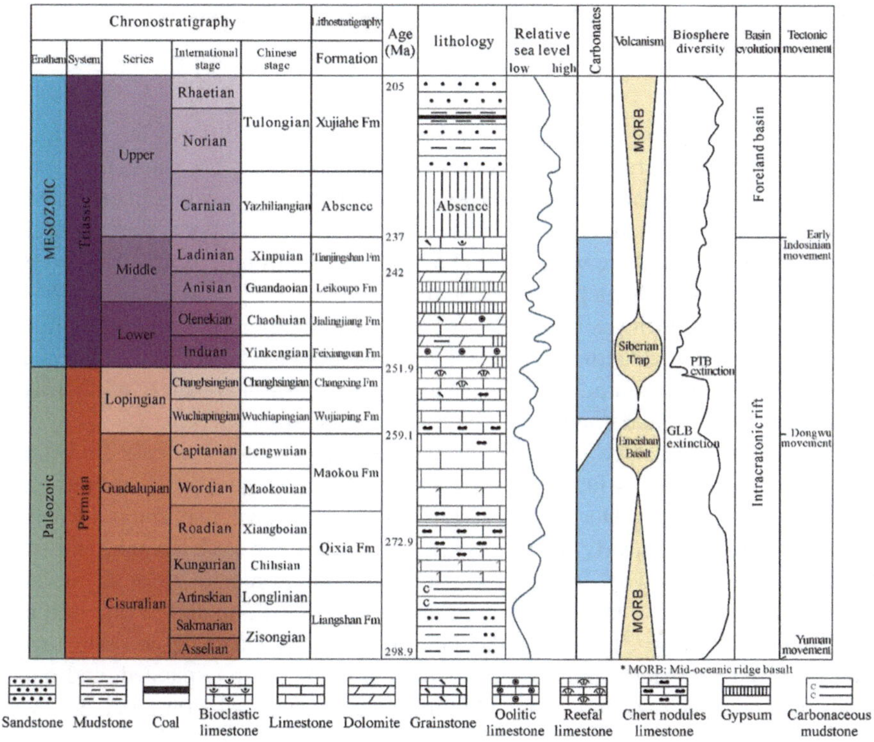

Fig. 2 Permian–Triassic strata and sedimentary events in the Middle–Upper Yangtze region

west. It survived until the Ladinian of Middle Triassic (Yin et al. 1999). During this period, the northern margin of the Yangtze Plate was composed of passive continental margin and deep shelf deposits. The Simao block in the south split from the Yangtze Plate since the Early Devonian and reached the peak in the Late Carboniferous–Early Permian, forming a back-arc spreading ocean basin. Since Permian, the Tethys Ocean subducted eastward to form Permian–Triassic volcanic magmatic arcs in Lancang River and Jiangda-Weixi area, and recombined in the Yangtze margin in the Early Triassic. During the Indosinian movement at the end of the Middle Triassic, the South China plate converged with the North China plate, which led to the closure of the Qinling Ocean (Fig. 1b).

The Dongwu movement in the end of the Middle Periman and the Emei tafrogenesis in the Late Permian led to a topographic trend sloping from southwest to northeast in the Upper Yangtze region. During the Late Permian, seawater trangressed from northeast to southwest, resulting in the palaeogeographic distribution pattern which developed Xikang-Yunnan antient land, Sichuan-Yunnan flood plain, littoral clastic platform, carbonate platform, slope and basin from west to east (Fig. 1c).

1.3 Characteristics of Sedimentary Facies Near the P–T Boundry

During the Changxing stage of Late Permian, tectonic differential subsidence occurred within the cratonic basin due to the NW-trending basement fault extension under the regional extension background. There was obvious platform–shelf sedimentary differentiation on the carbonate ramp in the Upper Yangtze region. From north to south, there were Chengkou-Exi marine trough, Kaijiang-Liangping shelf and Pengxi-Wusheng platform trough, which generally struck nearly NW–SE. It stretched to the open sea in the northwest and gradually narrowed and shallowed to the platform in the southeast. The sedimentary thickness increased to the southeast while the sea becoming shallower, forming a depositional pattern of alternative carbonate platform facies and deep–shallow water shelf facies (Fig. 3).

The Feixianguan stage of the Early Triassic represented a process from transgression to regression. Affected by regression and arid climate, the shoreline shrank northward as a whole. During the early Feixianguan stage, the Pengxi-Wusheng platform trough was nearly filled, the Kaijiang-Liangping shelf shrank, and water depth became shallower. The platform margin shoals on both sides of the shelf were developed, and a large set of micrite, oolitic limestone and oolitic dolomite were deposited (Fig. 3). During the late Feixianguan stage, large-scale regression occurred. Restricted platform and evaporative platform expanded, and purple mudstone, argillaceous limestone, dolomitic limestone and dolomite, etc. were deposited in the restricted platform. From the end of Early to Middle Triassic, with the relative decline of sea level, the restricted platform and evaporative platform were developed continuously with gypsum in the Jialingjiang Formation–Leikoupo Formation (Fig. 3).

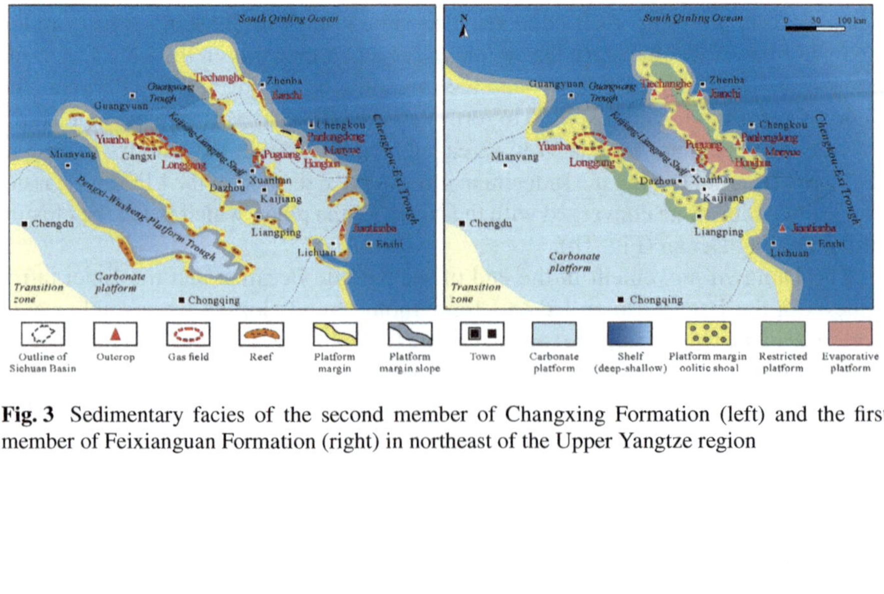

Fig. 3 Sedimentary facies of the second member of Changxing Formation (left) and the first member of Feixianguan Formation (right) in northeast of the Upper Yangtze region

The formation, evolution and extinction of the trough or deep-water shelf controlled the evolution of the platform margin. During Changxing and early Feixianguan stage in north Upper Yangtze region, two platforms were formed in the east and the west respectively, bounded by the deep-water shelf in the Kaijiang-Guangyuan area. The east platform was connected with the Chengkou-Exi trough in the east. Under this structure and paleogeographic background, three rows of nearly parallel NW–SE trending carbonate platform marginal zones were formed in North Upper Yangtze region during the Changxing–early Feixianguan stage. From west to east, there were Yuanba-Longgang-Liangping platform margin zone, Tiechanghe-Puguang-Luojiazhai platform margin zone, and Jianchi-Panlongdong-Honghua-Lichuan platform margin zone. The development of these platform margin zones spanned the great extinction event at the turn of Permian and Triassic. It formed a unique superimposed development pattern of platform margin reef and shoal during Changxing stage and oolitic shoals during Feixianguan stage, with inherited evolution characteristics in space (Fig. 3) (Xing et al. 2017).

1.4 Characteristics of Late Permian Reefs

1.4.1 Brief Review of Reef Researches

Reef is a kind of special carbonate anti-wave structure composed of various organisms in carbonate areas. According to their age, reefs can be divided into modern coral reefs and ancient reefs. The modern coral reefs are mainly composed of corals and red algae that cover the coral skeleton. As to ancient reefs, there are other kinds of organisms that could build up the framework besides corals. In order to be distinguished from modern coral reefs, ancient reefs are called organic reefs (Fan and Zhang 1985a, b).

As to reef researches, in the 1950s, most scholars emphasized the role of organisms, and strictly followed the principle that reefs are anti-wave structures formed by frame-building organisms. Since the 1970s, the concept of reefs has been divided or expanded. Many researchers generally classified carbonate rocks buildup to be reefs regardless of whether they have anti-wave ability and organic framework. Therefore, Dunham put forward the concepts of stratigraphic reef and ecologic reef (Dunham 1970). Stanton (1967) and Heckel (1974) proposed the term of buildup, emphasizing the positive geomorphic uplift, which mainly includes “reef”, “organic mound” and “organic reef”. Toomey (1981) thought that only the buildup formed after the Middle Triassic by hexacorals as the frame-builders can be named as reef, while the buildup composed of other frame-building organisms can be named as mound. This definition is obviously different from the typical concept of lime-mud mound. Fan and Zhang (1985a, b) indicated that reef could be composed of organisms or few organisms, emphasizing that it is a three-dimensional carbonate geometry. Riding (2002) proposed that reefs can be divided into three types (matrix-supported reefs, skeleton-supported reefs and cement-supported reefs) according to the differences of textural components and sediments in lithofacies, on the basis of matrix, skeletons and cavity/

cement (MSC). “Reef” discussed here refers to the ecologic reef in the narrower sense (Dunham 1970), which emphasizes the carbonate rock geometry in 3D space mainly composed of in-situ organisms. The organisms consist of frame-building organisms, joint coating organisms and reef attaching organisms. The main marks are as follows (Fan and Zhang 1985a, b): (1) there are obvious facies belt assemblages, such as reef front, slope, reef core, reef back, or reef core and reef flank in lateral direction; (2) there are obvious sedimentary landform of mound uplift; (3) there is no bedding in reef core generally; (4) frame-building organisms and joint organisms are well developed; (5) the frame-building organisms grew in situ and did not undergo transportation; (6) there is a pectinate structure composed of early seafloor cements (fibrous calcite, foliated calcite and columnar calcite) (Liu and Rigby 1992); (7) the original biological framework pores and inter-biological framework pores are well developed, but most of them are cemented by sparry calcite; and (8) dolomitization is easy to occur. Reef rocks can be divided into two types: in-situ deposits and allogene secondary deposits (Embry and Klovan 1971; Wu 1992, 1997). The in-situ sedimentary carbonate rocks are mainly composed of framestone formed by frame-building organisms, bindstone formed by binding organisms (cyanobacteria) adsorbing and capturing lime mud, and bafflestone formed by dendritic organisms that act as encroachment to resist lime mud. The allogene secondary deposits include reef breccia that is formed by sliding along the reef front slope. They are floatstone and rudstone etc. (Dunham 1962; Embry and Klovan 1971; Fan and Zhang 1985a).

1.4.2 Distribution Characteristics

The Permian marine strata are widely distributed in the world. Permian is an important period of reef development in geological history. Fan and Wu (2005) systematically summarized the paleogeographic distribution of Permian reefs. The early Permian reefs are mainly distributed in the northern shelf of the Pangaea, the foreland basins of the Ural Mountains and the Caspian Basin of Kazakhstan. Globally, the reefs during Qixia stage of Middle Permian are limited, while the reefs in Maokou stage are widely distributed. The Kapitan reef in the Guadalupe Mountains of North America is the most typical one in the world and is regarded as a reef model. The reefs in the Maokou stage are characterized by a large number of sponges-archaeolithoporella. The Late Permian Wujiaping reefs are represented by the reefs in the Zechstein Basin of Western Europe, which are mainly stromatolites or bryozoan reefs. South China has the most typical reefs of Changxing stage, with various types and abundant organisms. Sponge is the main frame-buildup organisms, which constitutes a bright spot in the distribution of the world Permian reefs (Fan and Wu 2005).

Calcisponges took part in reef-building during Maokou stage, and developed rapidly during Late Permian. They reached its peak and ended in Changxing stage. So far, no reef with calcisponges as the main frame-building organism has been found in Mesozoic in China. In Europe, calcisponges had obvious evolution discontinuity between Permian and Triassic, but flourished again after the Middle Triassic, and developed rapidly in the Late Triassic and Jurassic. It is inferred that calcisponges might occur in the eastern part of the Paleotethys, that is, the Yangtze region. At that time, it was in the low latitude area near the equator, which was very favorable for the survival of calcisponges. The development of calcisponges reached its peak during the Late Permian and spread westward, and continued to the Mesozoic. The abundance of calcisponges determine the distribution of reefs in the Permian–Triassic Paleotethys. During Maokou–Changxing stage, reefs with calcisponges as frame-building organisms were widely developed in the Upper Yangtze region, but rarely in Europe. However, during the Triassic, reefs with calcisponges as frame-building organisms were widely distributed, while absent in China (Zhang 1992).

Studies found that there were no reefs in the world during the Early Triassic. The extinction of epigenetic reef-building animals caused by PTB extinction event is the main reason for the absence of reefs in neritic area.

In the Upper Yangtze region, the Permian reefs are widely distributed in Lichuan of Hubei Province, Longlin of Guangxi Province, Ziyun of Guizhou Province, Cili of Hunan Province and eastern Sichuan Province, etc. Reefs include intra-platform patch reef, platform margin reef, shallow ramp reef, and platform margin slope reef. The patch reefs are mainly distributed in the interior of platform, characterized by small scale, such as Laolongdong of Beipei in Chongqing, etc. The platform margin reefs are distributed as discontinuous barrier reef in the Chengkou-Exi trough and the east side of Kaijiang-Liangping shelf etc., such as Jiantianba in Lichuan of Hubei, Honghua in Kaixian County of Chongqing, Panlongdong in Xuanhan County of Sichuan, etc., and Puguang, Huanglongchang and other places on the eastern platform margin of the Kaijiang-Liangping shelf (e.g. the reefs in the Changxing Formation drilled by Wells PG8 and PG6). They are also distributed on the margin of carbonate platform in Southeast Yunnan. The shallow ramp reefs are mainly distributed in the shallow ramp zone of the western margin of Kaijiang-Liangping shelf, such as the reefs drilled by Wells YB3 and YB2 in Yuanba area, with the characteristics of small single-stage thickness and multiple development stages (Zhao et al. 2010). The platform margin slope reefs are chiefly distributed in the middle–lower part of the margin slope of the E-xi trough, such as Huangnitang, Huajiaoping and Manyue. The reefs in gentle mound shape are mainly composed of bafflestone.

1.4.3 Reef-Building Organisms—Calcisponges

The reef-building organisms of the Late Permian reefs include calcisponge, hydroid and bryozoan, coating organisms including calcareous alga species (e.g. archaeolithoporella and tubiphytes), and the reef-attached organisms including brachiopods, echinoderms, gastropods, lamellibranchia and foraminifera, etc. Calcisponges are the main reef-building organisms of Late Permian in the Yangtze region of China, including sphinetozoa, inozoa and Sclerospongiae. About 20 genera of sphinetozoa have been found and described in southern China, accounting for nearly 2/3 of all genera in the sphinetozoa suborder. The inozoa have 8–10 genera, which are more than half of the total genera of this kind of sponges. In the sclerospongiae, two basic types of the plate sponge have been found in China. Zhang (1992) summarized that the distribution of calcisponges is characterized by (1) considerable quantity: calcisponges takes more than 30% in most Permian reefs in the Upper Yangtze region, and the content of sphinetozoa is up to 30–40% in the Jiantianba reef of Lichuan City, Hubei Province; (2) highly differentiated genera and species of sponges: calcisponges from reef cores in different places have various shapes, such as twisted, grape, bag, long column and thin plate. The populations are mostly branched or clumped, with various types; (3) ecological characteristics in relation to hydrodynamic conditions; and (4) a close symbiotic relationship: all kinds of calcisponges often live together, forming the dominant organisms in reef. In addition, sponges are closely related to other species of organisms.

Calcisponges can participate in the formation of reefs in different ways, thus forming different types of reef limestone. Particularly, framestone is characterized by calcisponges with maximum supporting framework. Bafflestone is characterized by calcisponges with coarse branch and fine branch shape and capable of encroaching lime mudstone. Bindstone is characterized by shell or plate-shaped calcisponges and other organisms with the ability of coating and bonding. From a macroscopic point of view, the same reef may contain different types of reef limestone, but it is limited to a certain facies zone, and may be dominated by only one type of reef limestone, which mainly depends on the dominant reef-building organisms.

1.5 Oil and Gas Reservoirs in Reef-Shoal Facies: A Case Study of Puguang Gas Field

Reef and shoal facies play an important role in carbonate oil and gas exploration. Reef and shoal bodies are one of the important reservoir types of large marine carbonate oil and gas fields in the world. Puguang gas field is the first discovered super-giant monolithic marine gas field with the largest burial depth and the highest resource abundance in China. The reservoir of Puguang gas field is typical carbonate reef-shoal deposits.

The present-day tectonic location of Puguang gas field is in a nose structural belt of an anticline belt in the northeast fault-fold belt of the Sichuan Basin (Fig. 4). The type of gas reservoirs belongs to structural-lithologic trap. The reservoirs are reef dolomite in the Changxing Formation and oolitic shoal dolomite in the Feixianguan Formation. The gypsum and marl in the 4th Member of the Feixianguan Formation compose the direct caprock of the gas reservoirs in the 1st–3rd Members of the Feixianguan Formation. The thick gypsum strata in the Triassic Jialingjiang Formation and Leikoupo Formation are the direct caprock and regional caprock of the gas reservoirs. Many sets of source rocks are developed in this region, including gray-black shale in the Lower Cambrian, black shale and mudstone in the Lower Silurian Longmaxi Formation, and coal measures and dark mudstone in the Upper Permian Longtan Formation. Among them, the source rocks in the Longmaxi Formation and Longtan Formation are the main gas sources of Puguang gas field.

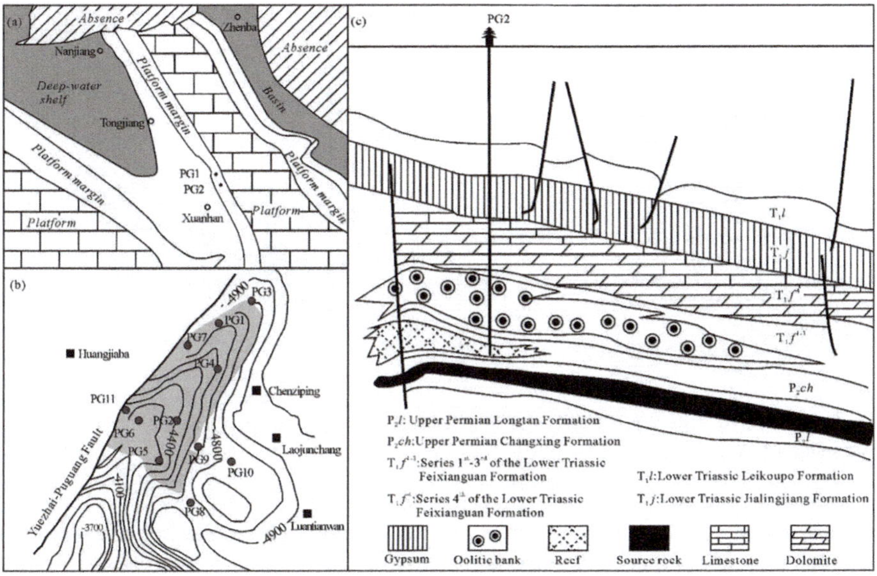

Fig. 4 **a** Sedimentary facies of initial Changxing Formation in Puguang area, Sichuan Basin; **b** tectonic position of Puguang gas field; **c** profile of Puguang gas reservoir (Ma et al. 2007a, b, c)

The reservoirs of Puguang gas field are the Permian Changxing Formation and the Lower Triassic Feixianguan Formation, with buried depth of 4776–6169 m. The main lithologies of high-quality reservoirs are oolitic/residual oolitic dolomite, crystalline dolomite, bioclastic dolomite, dolorudite and sponge reefal dolomite, etc. The oolitic/residual oolitic dolomites have well-developed intragranular dissolution pores, with pinhole dissolution pores which can be observed in core (Fig. 5).

The maximum porosity of dolomite is 28.8%, with an average of 8.17% (according to 780 cores). The maximum permeability is up to 3354 $\times$ 10^{-3} μm^2, with an

Fig. 5 Pore characteristics of reservoir in Feixianguan Formation in Puguang gas field. **a** Oolitic and residual oolitic dolomite in Puguang gas field; **b** core with pinhole dissolution pores

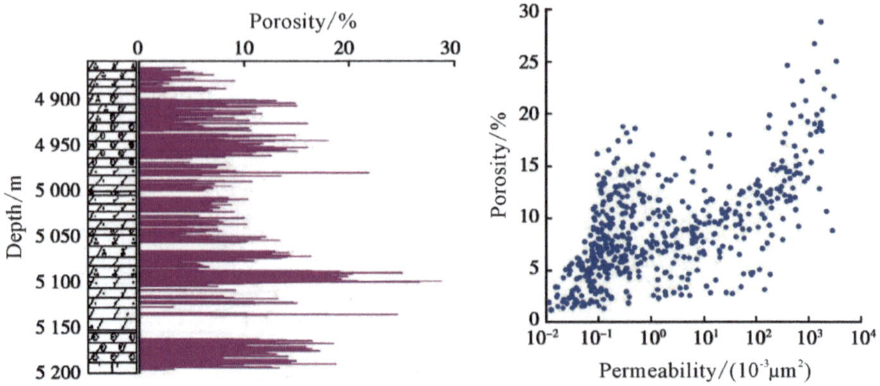

Fig. 6 Vertical distribution of porosity and its relationship with permeability in Feixianguan Formation of Well PG2 (Ma et al. 2007a, b, c)

average of 94×10^{-3} μm^2 (according to 684 cores). There is a better correlation between porosity and permeability, and reservoirs are dominated by porous (dissolution pore) type (Fig. 6). According to the burial history analysis, the burial depth of the Feixianguan Formation could reach 8000 m at the end of Cretaceous. High-quality porous reservoirs can be developed at such a deep burial depth, which indicates that the average porosity of carbonate rocks tends to increase with depth under certain geological conditions. This exploration practice breaks through the understanding that the reservoir pores gradually decrease with the increase of burial depth, and the reservoir pores die out when the burial depth reaches 3500 m, which has important guiding significance for the global deep and ultra-deep carbonate exploration (Ma et al. 2010).

Statistics show that the reservoir physical properties of Puguang gas field are closely related to carbonate sedimentary facies and subfacies (Tables 1 and 2) (Ma et al. 2010). Hereinto, the platform margin shoal facies is the best, the platform margin reef and platform evaporite facies are the second, and the slope facies is the worst. Exposed shoal, tidal channel, bafflestone and shoal subfacies are the best reservoirs. The sedimentary environment is an important controlling factor for the development of high-quality reservoirs. The best reservoirs are mainly developed in the sedimentary environment with strong hydrodynamic force, repeated wash by fluids and exposure environment (Fig. 7).

1.6 Field Trips Route and Stops

The field trips are carried out at the junction of Chongqing Municipality, Hubei Province and Sichuan Province in Southwest China. The investigation route and stops are shown in Fig. 8. There are 3 investigation sites: (1) Jiantianba outcrop in

Table 1 Porosity and permeability of different sedimentary facies in Changxing-Feixianguan Formations, Puguang area (Ma et al. 2010)

Sedimentary facies	Porosity (%)			Permeability ($\times 10^{-3}$ μm^2)		
	Sample amount	Min–Max	Mean	Sample amount	Min–Max	Mean
Slope	21	0.93–1.76	141	2	0.0103–0.432	0.0565
Restricted platform	84	1.26–20.94	313	57	0.004–41.5414	1.4876
Open platform	17	1.32–1.95	162	17	0.0143–1.075	0.0846
Platform margin shoal	744	1.11–28.86	924	664	0.0163–7973.7685	174.811
Platform margin reef	275	1.12–14.51	448	274	0.0116–1391.2082	16.6226
Platform evaporites	591	0.45–17.24	454	526	0.0001–9664.887	81.9397

Table 2 Porosity and permeability of different sedimentary subfacies in Changxing-Feixianguan Formations, Puguang area (Ma et al. 2010)

Sedimentary subfacies	Porosity (%)			Permeability ($\times 10^{-3}$ μm^2)		
	Sample amount	Min–Max	Mean	Sample amount	Min–Max	Mean
Exposed shoal	925	1.11–28.86	8.38	842	0.0163–7973.7685	157.8231
Framestone	70	1.12–8.38	2.19	70	0.0116–129.1227	2.1079
Bafflestone	109	1.27–14.51	6.50	108	0.0148–223.2907	7.1924
Tidal channel	65	3.12–23.05	8.47	63	0.0599–9664.887	265.8044
Evaporation flat	465	0.45–17.24	3.76	402	0.0001–5418.85	46.1705
Shoal	277	1.11–23.05	5.83	260	0.0191–7973.7685	177.902
Shallow ramp	44	0.47–1.83	1.02	44	0.0101–0.3777	0.0492
Lagoon	46	0.26–8.91	2.11	28	0.0108–0.7155	0.095

Lichuan City, western Hubei Province, including 3 stops; (2) Panlongdong outcrop in Xuanhan County, Dazhou City, northeastern Sichuan Province, including 2 stops; and (3) Puguang gas field in Xuanhan County, Dazhou City, Sichuan Province, including 2 stops. The investigation in the Jiantianba outcrop focuses on the sedimentary sequence, architectural units in reef core, and sedimentary microfacies of carbonate platform margin, and the Late Permian reef-building organisms (the calcisponges with abundant species). The investigation in the Panlongdong outcrop focuses on the evolution sequence of reefs in the Changxing Formation of the Late Permian and the development characteristics of grain shoal in the Feixianguan Formation of the Early

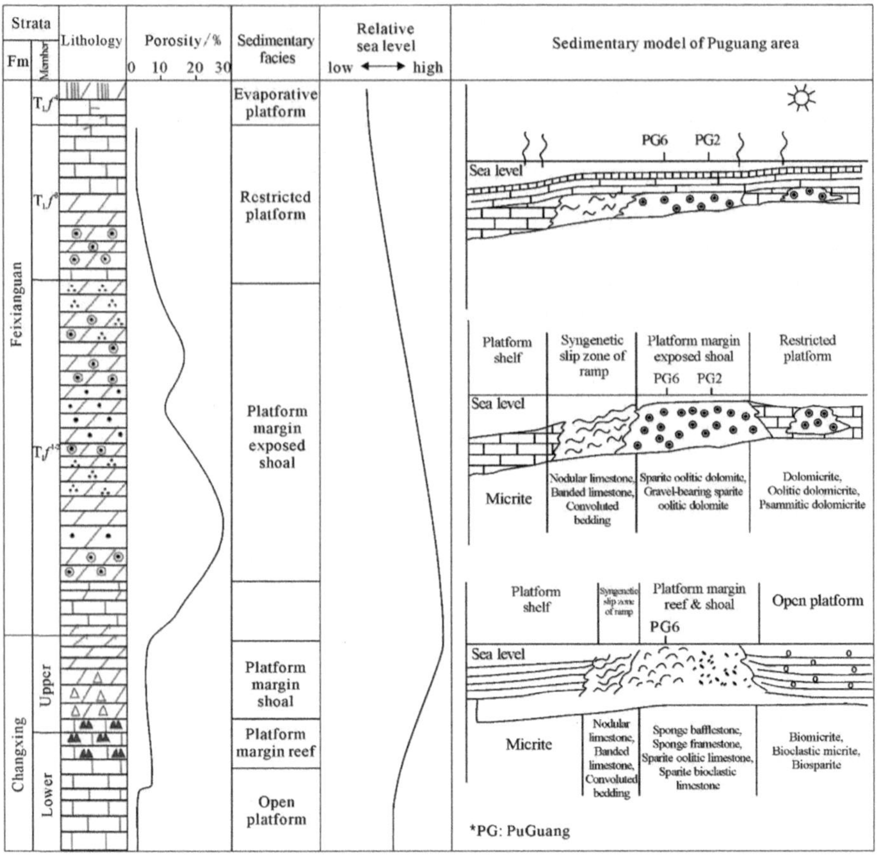

Fig. 7 Sedimentary evolution and sedimentary model of Changxing-Feixianguan Formations in Puguang area (Ma et al. 2007a, b, c)

Triassic. The investigation in Puguang gas field focuses on the successful application of reef-shoal sedimentation and reservoir research in oil and gas exploration and development.

The investigation schedule is arranged as follows:

Day 1: Fly from Beijing (Capital/Daxing International Airport) to Chongqing Jiangbei Airport, and then take bus to Lichuan City, Hubei Province.

Day 2: Investigate the Jiantianba outcrop, and then go to Dukou Township, Dazhou City, Sichuan Province by bus.

Day 3: Investigate the Panlongdong outcrop and Puguang gas field.

Day 4: Go to Chongqing by bus, and the investigation is over.

Chongqing is one of the important central cities in China. It is a comprehensive transportation hub in Southwest China, where the Yangtze River and Jialing River meet. It has beautiful scenery and numerous cultural relics.

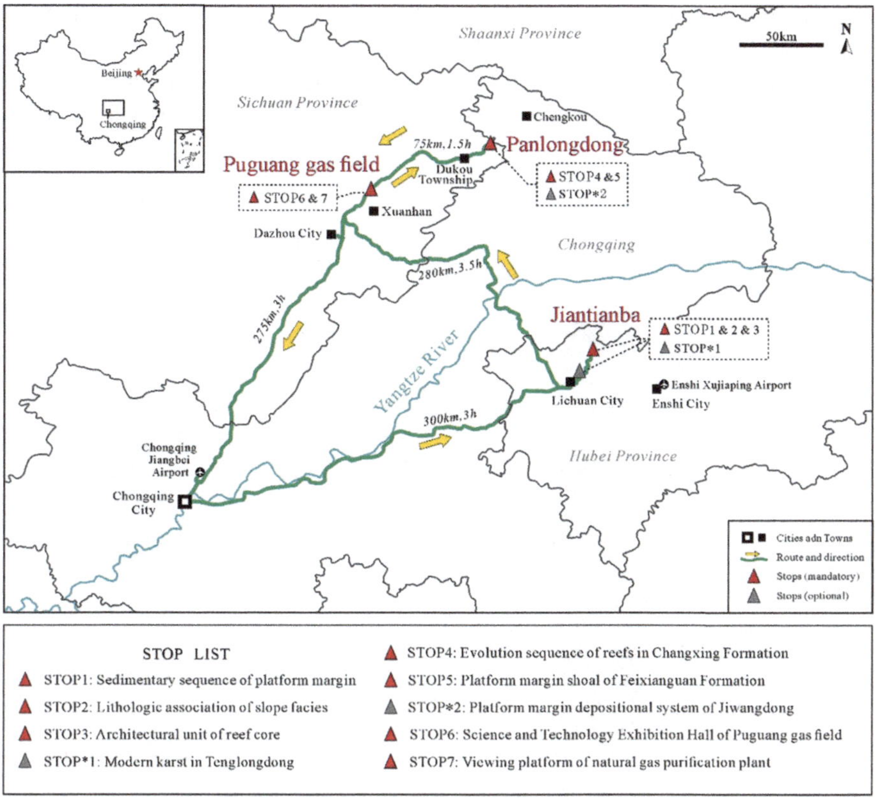

Fig. 8 Investigation route and stops

2 Field Itinerary

2.1 Late Permian Reefs in Lichuan City, Hubei Province

The Jiantianba reef in Lichuan City is a typical outcrop of Late Permian sponge reef in the Upper Yangtze region. Since the 1980s, extensive and in-depth researches have been carried out on it by scholars (Fan et al. 1982; Fan and Zhang 1985b, 1986, 1987; Zhang 1985; Liu and Rigby 1992; Guo and Riding 2010; Zhang 1992; Hu et al. 2012a, b; Wu et al. 2015; Chen et al. 2018; Li et al. 2018). It is one of the most studied reefs.

The Jiantianba outcrop is located in Jiantian village, Lichuan City, western Hubei Province, about 25 km northeast of Baiyang Town, Lichuan City. Structurally, it is located in the north-central part of Lichuan synclinorium in East Sichuan fold belt. The Upper Permian–Lower Triassic strata are exposed on the Jiantianba outcrop, including (from bottom to top): Wujiaping Formation–Changxing Formation–Daye Formation (Feixianguan Formation) (Fig. 9).

The Jiantianba reef outcrop has convenient transportation conditions. We can observe the Wujiaping Formation–Changxing Formation–Daye Formation as a whole

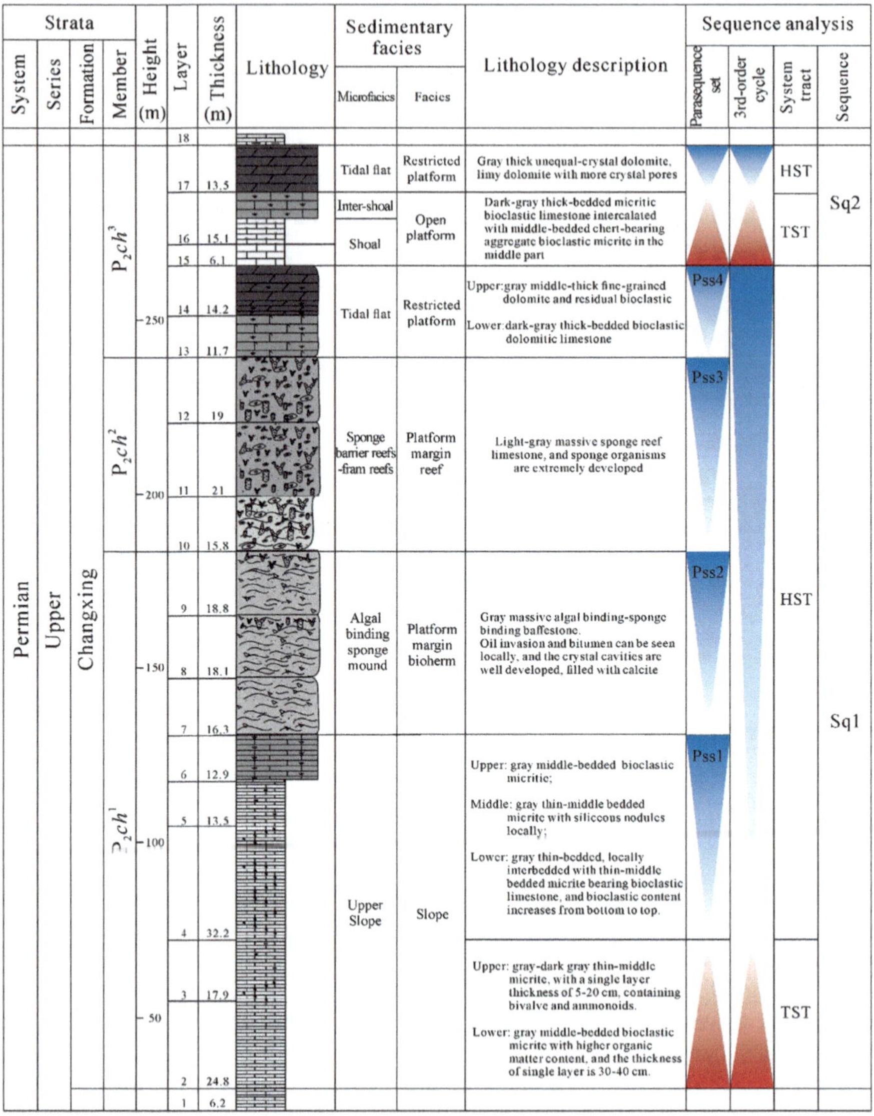

Fig. 9 Stratigraphic column of Changxing Formation in Jiantianba outcrop (modified from Hu et al. 2012a, b)

along village road, and understand the sedimentary sequence of shelf margin–slope–platform margin reef–restricted platform–open platform macroscopically. In recent years, due to quarrying activities, the reef core was mechanically cut to form a three-stage stone mining step, which clearly shows a variety of microfacies in the reef core, providing a good platform for observing the characteristics of the internal structure unit combination of the reef.

Stop 1: Platform margin sedimentary sequence in Upper Permian

At the end of Late Permian (Changxing stage), with the rise of regional sea level, most areas in the northeast of the Upper Yangtze region were mainly open platform deposits (Fig. 3), while in Chengkou-Exi area deep-water basin deposits dominated by siliceous rocks were developed. Jiantianba in Lichuan City was located in the platform margin facies belt between the deep-water basin and the carbonate platform. The deep-water slope facies was deposited during the early stage, and the water gradually became shallow upward which resulted in deposits of the platform margin reef facies in Changxing Formation. The restricted platform facies–open platform facies were developed at the deposition end of the Changxing Formation. The Wujiaping Formation underlying the Changxing Formation was in conformity with the overlying Daye Formation.

The Changxing Formation can be divided into three members from bottom to top (Fig. 9): (1) the lower member is composed of dark-gray thin–middle bedded micrite and bioclastic micrite, and is early deep-water slope facies deposit; (2) the middle member is platform margin reef deposits composed of light-gray massive bindstone, bafflestone and framestone; and (3) the lower part of the upper member is light gray-gray middle-thick fine-grained dolomite, residual bioclastic limy dolomite, gray thin limestone interbedded with black shale. The upper part is interbedding gray middle-thick micritic bioclastic limestone and limy dolomite with various thickness, which represents the restricted platform-lagoon-open platform deposit at the end of the Changxing Formation.

Stop 1 is located in Jiantian Village Cultural Square. The panoramic view of the carbonate platform margin depositional system of the Late Permian Changxing stage can be macroscopically observed. We can have an overall understanding of the sedimentary sequence, and the extension and distribution of sponge reef cores can be macroscopically observed (Fig. 10).

Stop 2: Lithological combination of the slope and the lateral distribution of the reefs

The Changxing Formation and overlying Wujiaping Formation contact in conformity, and their interface is a lithologic-lithofacies transition interface. Under the interface, the Wujiaping Formation is composed of black thin carbonaceous shale with thin siliceous interlayers, gray-black thin carbonaceous shale interbedded with gray-black thin–medium layered marl, and dark gray thin–medium micrite of shelf facies. Above the interface, the lithology is dark-gray thin–medium layered bioclastic micrite deposit at the bottom of the Changxing Formation of slope facies. The lower member of the Changxing Formation is composed of several parasequence cycles

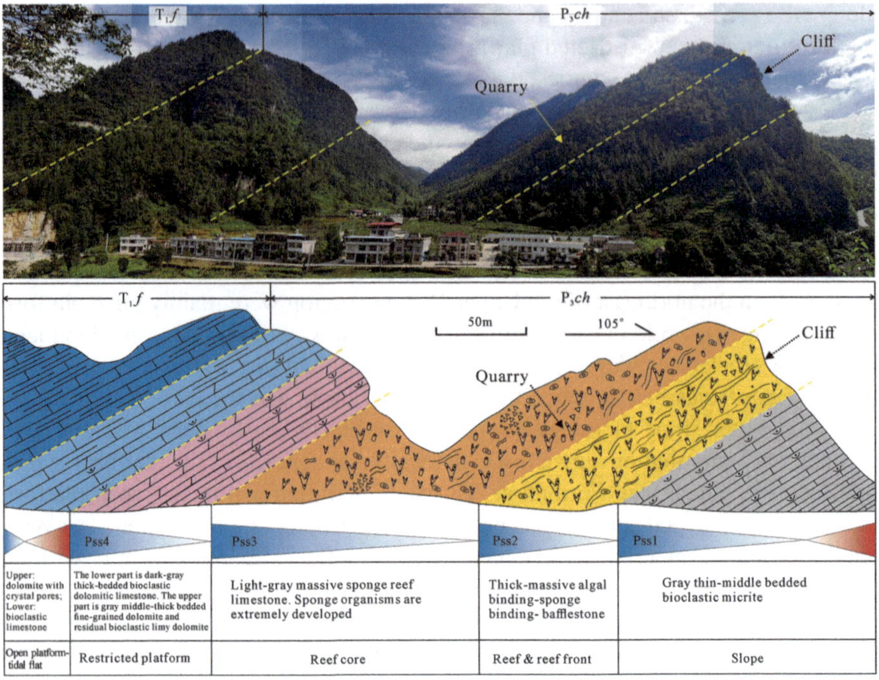

Fig. 10 Panoramic view of platform margin sedimentary sequence of Changxing Formation in Jiantianba outcrop (modified from Hu et al. 2012a, b)

formed by gray thin–medium layered micrite to gray medium layered micrite and gray middle micritic bioclastic limestone. Locally, there are gravel limestone and thin-medium layered micrite intercalated with siliceous strip deposits, which is a set of low-energy slope facies deposits. Jiantianba reef is developed based on micrite and micritic bioclastic limestone. Along the Jiantian village highway, we can observe the reef section and the slope facies deposition formed during the foundation stage of reef development (Fig. 11).

Stop 3: Characteristics of architectural units in reef core

As the core of Jiantianba reef is cut into many step sections by quarrying, the architecture and evolution sequence of the reef can be observed on one plane or three planes stereoscopically. According to the classification and naming scheme of Riding (2002) (Fig. 12), at least four typical reef microfacies can be identified (Fig. 13), namely, filled-skeleton sponge framestone, open-skeleton sponge framestone, binding-skeleton sponge framestone, segment or cluster bafflestone (Chen et al. 2018), which represent different hydrodynamic and paleoecological environments.

Fig. 11 Panoramic view of cliff section of Jiantianba reef

Fig. 12 Reef fabric classification. The MSC triangle shows the relationship among matrix, skeleton and cement in pore and cavity (Riding 2002)

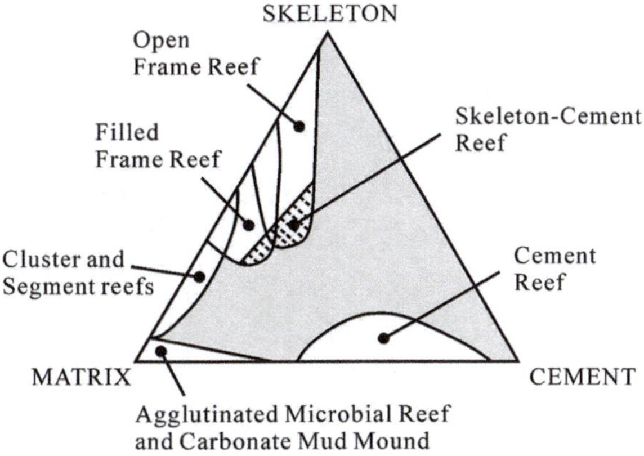

① Filled-skeleton sponge framestone

It is located in the lower part of reef core (right side of the quarry step). The main reef-building organism is sphinetozoa, whose individual length can reach 20–40 cm and diameter can be 2–3 cm. On the section facing the step, the calcisponges generally appear in transverse, longitudinal or oblique sections, showing two conditions (upright and lodging) of the sphinetozoa. The sponge skeletons are rarely in contact with each other, and there is stable interval between individuals (Fig. 13①). The matrix is composed of marl, and there are some biological debris, such as echinoderm, stem of crinoids, planktonic foraminifera and so on. The rapid accumulation of marl fills the gap between sponges, resulting in the rapid burial of sponges in the process of sponge growth. In this microfacies, there is a lack of algal laminar incrust and tubiphytes around the sponge skeleton, and the calcite content is lower.

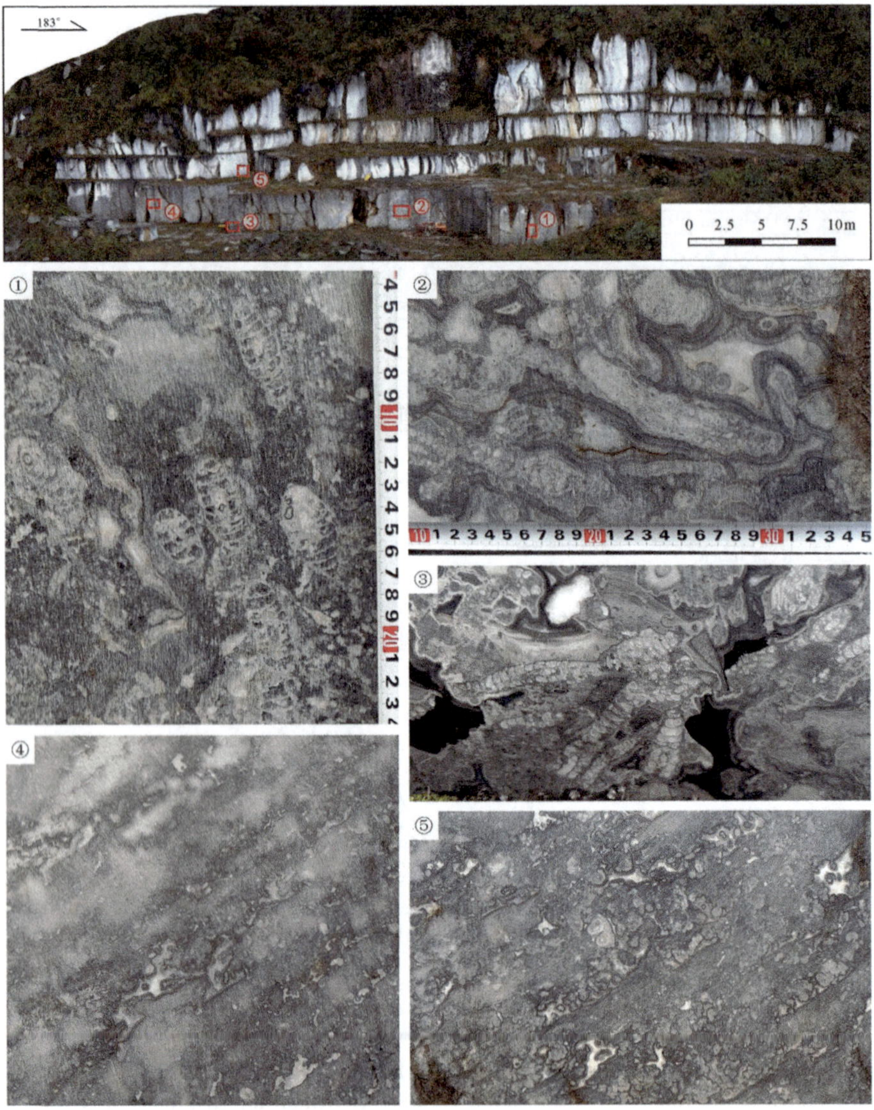

Fig. 13 Four typical types of reef facies in Jiantianba quarry. ① Filled-skeleton sponge framestone, ② open-skeleton sponge framestone, ③ binding-skeleton sponge framestone, ④ and ⑤ segment or cluster bafflestone

② Open-skeleton sponge framestone

It is mainly located in the middle of the quarrying step (the middle zone of the reef core), and sphinetozoa is the main reef-building organism. The individual can be 10–30 cm in length and 1–2 cm in diameter. A large number of dense sponge skeletons are developed alternately and contacted each other (Fig. 13②). There are

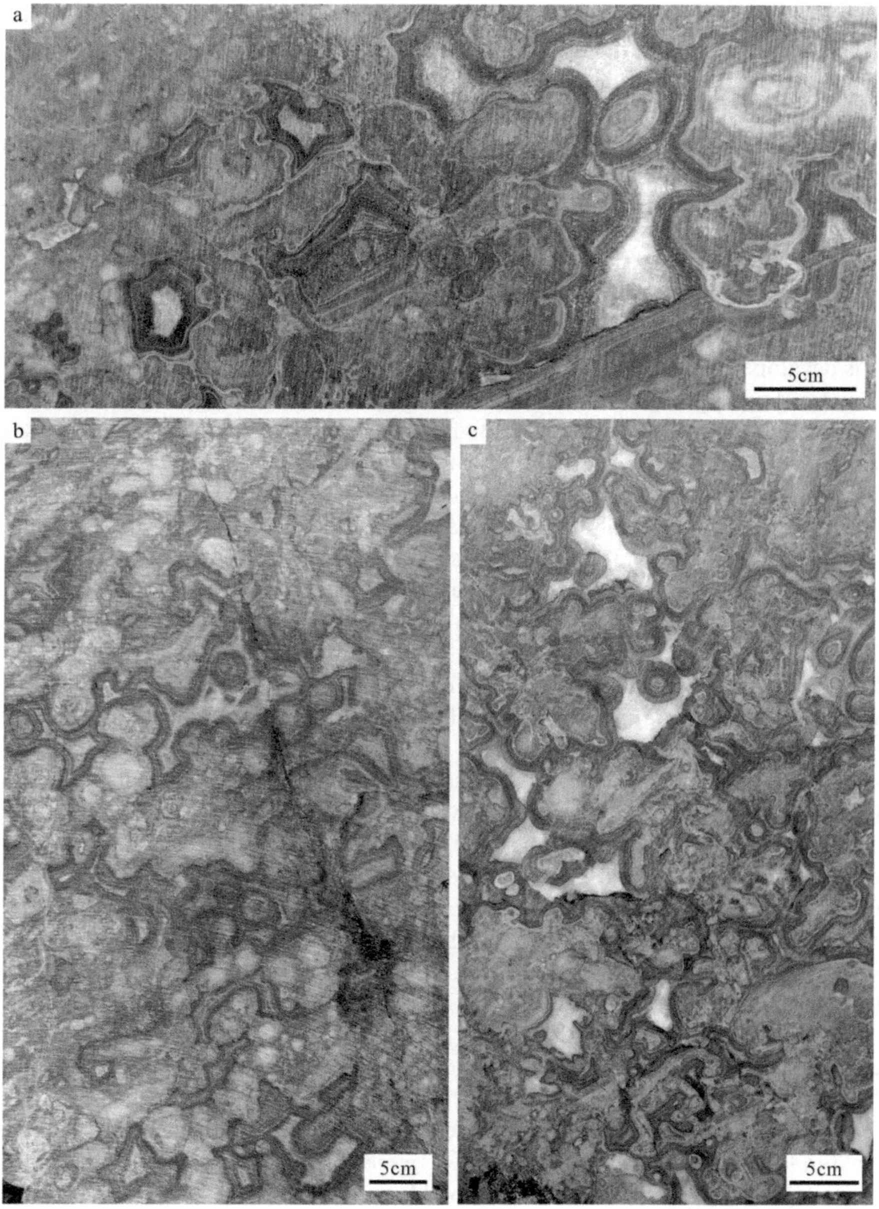

Fig. 14 Macroscopic characteristics of binding-skeleton sponge framestone

a lot of algal laminar incrusts on the outside of the sponge skeletons. In the macroscopic hand specimens, the incrusts twins and wraps the sponge skeletons, showing concentric circle. It is generally believed that the algal laminar incrusts make the sponges have a supporting role and become a reef frame with anti-wave function. The development of the open sponge framework limestone is mostly in a relatively clean water environment. The pores between the sponge frameworks are filled with spar calcite. Early fibrous and radial seafloor cements and massive calcite filling can be seen. The appearance of the open-skeleton sponge framestone indicates a shallow water and strong hydrodynamic sedimentary environment.

③ Binding-skeleton sponge framestone

It is developed in the upper part of the reef core (the middle-left of the quarry step). The reef-building organisms mainly include sphinetozoa, inozoa, sclerospongiae, hydra and tubiphytes. The sponge individual is shorter, and the diameter is thicker, which is suitable for stronger hydrodynamic environment. During the development of reefs, the association of microorganisms makes the sponge wrapped into a massive pattern. A large number of algal laminae and early seafloor cements are laminated, wrapped and grow around the sponge mass. White rhombic calcite cement is common between sponge aggregates (Figs. 13③ and 14).

④ Segment or cluster bafflestone

It is mainly developed at the bottom of reef cap unit (the left of the quarrying step), with more sphinetozoa and inozoa, showing a lodging pattern. Algal laminar incrusts are well developed outside the sponge visceral cavities, with fibrous, radial and massive calcite cements. In the macroscopic observation of lithology, the lithology is dense, showing the characteristics of overlapping and layering between the lodging sponge section and bioclastic limestone. In this microfacies, stromatactis is developed, with flat bottom and finger-shaped structure at the top (Fig. 13④, ⑤), thus it is generally considered as the sedimentary structure of intertidal zone. The origin of the stromatactis may be related to the shielding of sponge and the activity of microorganism. The jointed sponge bafflestone mainly constitutes the reef crest unit (Fig. 15).

By overall observation of the sponge reefs in Lichuan, it is found that the size of calcareous sponge varies little from bottom to top, but there are obvious differences in lithologic structure. The lower part (near the right side of the section) is filled framestone, which is characterized by lack of algal incrust. The upper part (from the right to the middle of the section) is open framestone and jointed framestone, with circle layer of algal incrusts, and reef-building organisms are coated or jointed to form a mass shape. Although calcareous sponges have a certain anti-wave ability (Li et al. 2018), the reef-forming mode of the calcareous sponges is different from that of modern corals. Corals constantly secrete calcareous skeletons and build strong anti-wave structures through their own in-situ proliferation. However, most genera and species of the calcareous sponges need the assistance of other organisms to build reefs. The calcareous sponges proliferate in large quantity in the favorable environment suitable for survival. After death or lodging, they are winded and jointed

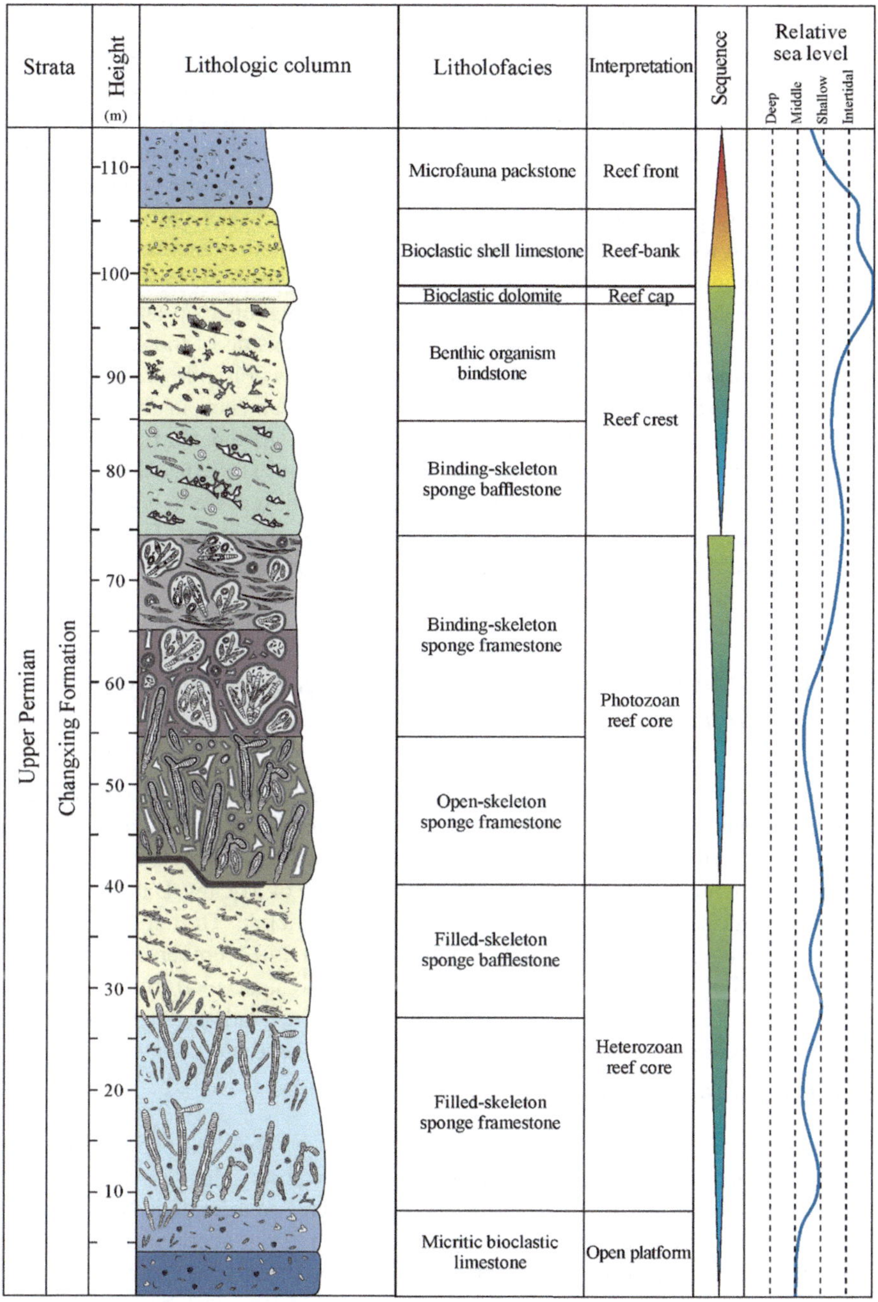

Fig. 15 Development pattern of Jiantianba reef (modified from Chen et al. 2018)

by calcareous algae and other binding organisms, and form a new bottom material after fixation. The second generation of calcareous sponges grow and develop on them, and are bound and fixed by bonding organisms after death or lodging. After repeat and accumulation for generations, a thick anti-wave structure can be formed (Zhang 1992). The difference of various rock structures of the Jiantianba reef is the reflection of reef-forming conditions. It can be observed in the outcrop that the content of micritic matrix decreases obviously from the lower part of the sponge reef limestone to the upper part, while the spar cementation increases obviously, reflecting the enhancement of hydrodynamic conditions from the lower to the upper part. In addition, the anti-wave ability of reef-building organisms is enhanced by the coating effect of archaeolithoporella. At the top (the left side of the section), dark-gray algal binding limestone, bafflestone and calcareous dolomite are interbedded, and geopetal structure is developed. Stromatactis can be seen in the bafflestone, which indicates the subtidal and intertidal environment. The fenestral structure in the bindstone may indicates the intertidal and supratidal zone. The existence of dolomite may demonstrate the development of restricted lagoon, represents the low-energy sedimentary environment, and also marks the end of reef buildup. The architecture and evolution of the Jiantianba reef can be illustrated as Fig. 15.

2.2 *Reef and Grain Shoal in Panlongdong, Xuanhan County, Sichuan Province*

The Panlongdong outcrop of reef and grain shoal is located in Bashan Grand Canyon, Xuanhan County, northeastern Sichuan Province, along S029 provincial road, about 80 km away from Puguang gas field. The regional structure is located in the northeast edge of the Sichuan Basin, belonging to the front zone of Micang-Daba mount, and is located in the southwest wing of the Panlongdong anticline. The exposed strata are mainly Permian Changxing Formation and the 1st and 2nd Members of the Triassic Feixianguan Formation. It was in the northwest of the same platform margin zone with Lichuan reef (Fig. 3). Strata development on the Panlongdong outcrop and platform margin reef-shoal reservoirs are highly correlated with Puguang gas field.

Stop 4: Platform margin reef of Panlongdong

The Panlongdong outcrop section is along S209 highway. There is a parking lot beside the road, which is convenient for parking and observation. The start coordinate of the outcrop is N31° 45′ 15″, E108° 27′ 49″, and its end coordinate is N31° 45′ 15″, E108° 27′ 29″. The outcrop is well exposed, being a set of accretion-progradation platform margin reef, it develops two stages of reefs. The first stage represents the initial formation of the reef ecosystem and constitutes the reef base of the main reef-forming cycle. It contains complete association of reef base, reef core and reef cap with smaller scale. The second stage is the main reef-forming stage. In the cycle of secondary sea level becoming shallower, on the basis of the first stage of small reefs, reef-building organisms (such as sphinetozoa and inozoa) and reef attached organisms (such as brachiopods, bivalves, echinoides and foraminifera) continued to develop and flourish, forming large dome-shaped reefs of accretion-prograding type. The reef base corresponds to the palaeogeomorphic highland formed by the first stage of reefs. The reef core developed sponge bafflestone, bindstone and framestone. In addition, the proportion of framestone increases upwards, and finally ends. Sphinetozoa, inozoa and sclerospongiae are developed in each layer of the reef, which constitute the main part of reef-building organisms. The protoliths of reef cap are mainly bioclastic and echino-clastic packstone and wackestone. All these in reef cap has been dolomitized to fine crystal and powder crystal dolomite. These two reef-forming stages constitute a large reef cycle, composed of reef base, reef core and reef cap unit from bottom to top (Fig. 16).

Reef base: It is the growth base of reef, consisted of layers 1–15 (Fig. 17). The lower part is composed of dark-grey moderate-thick bedded-massive wackestone and grainstone in the bioclastic shoal and inter-shoal facies. The upper part of layer 9 begins to develop moderate-bedded sponge bafflestone with packstone and lime mudstone. It belongs to a small patch reef. The reef base of the first stage is layers 1-lower 9, mainly composed of grainstone, wackestone and limy mudstone. The reef core consists of layered sponge bafflestone interbedded with packstone and

Fig. 16 Panoramic view of Panlongdong reef outcrop section in Xuanhan

limy mudstone in layers upper 9-lower 14. The sponges include sphinetozoa, inozoa and sclerospongiae. The reef cap corresponds to the grainstone, dolomitic mudstone and dolomite wackestone in layers upper 14–15. This small reef can represent the ecosystem of the reef has been initially established, and formed the reef base of the second main reef-forming cycle (Fig. 17).

Reef core: It is the main part of the wave-resistant body of reef, composed of layers 16–23 (Fig. 18). The lithology is light-grey sponge bafflestone, sponge framestone

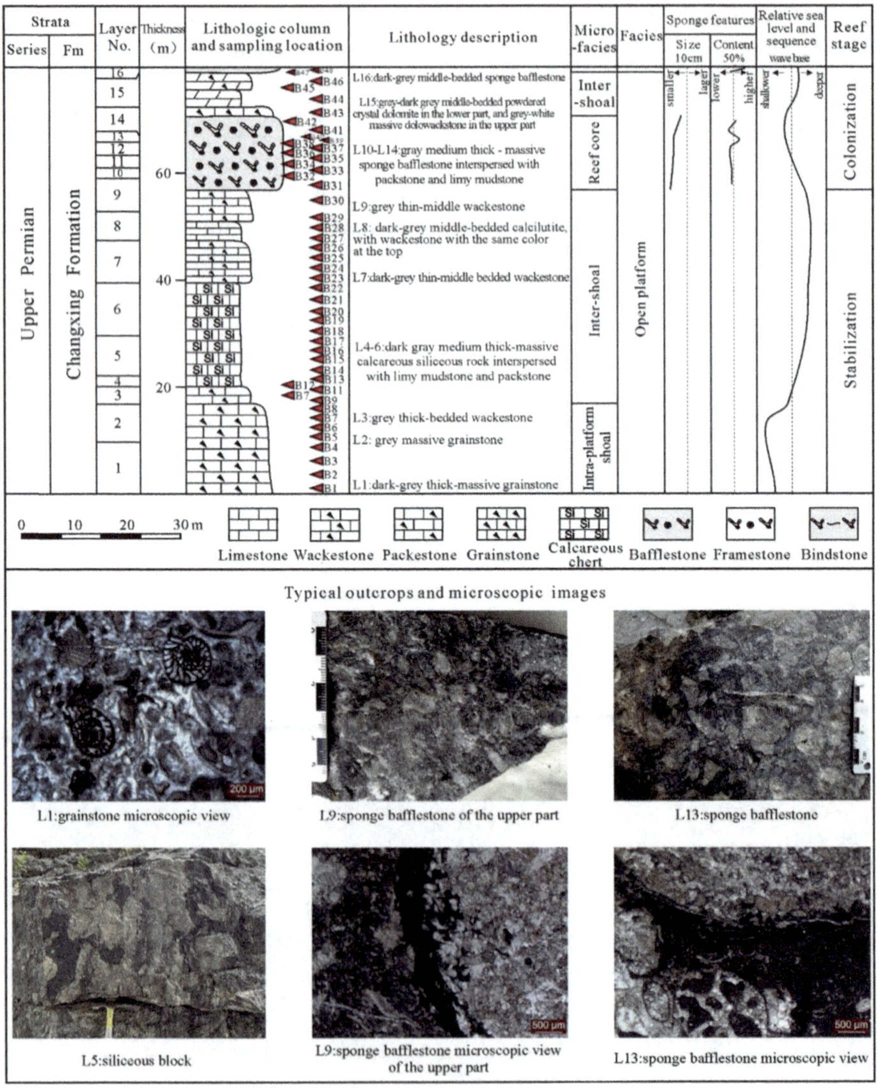

Fig. 17 Comprehensive column of reef base on Panlongdong section

and bindstone, dominated by sponge bafflestone. Hereinto, layers 16–17 are sponge bafflestone; layers 18–19 are sponge baffling dolomite; layer 20 is chiefly alternative sponge bafflestone and framestone, and layers 21–23 are mainly sponge bafflestone intercalated with bindstone. Reef-building organisms of the reef core are mainly sphinetozoa, inozoa and few sclerospongiae. The content of the reef-building organisms is 20–50%, among which layers 18–19 have the highest organism content, up to 45–50%. Most sponges grow vertically and a few obliquely to bedding surface, which reflects the upright and pitching growth and burial state of sponges. Coating organisms (e.g., archaeolithoporella, tubiphytes and dasyclads) surround the reef-building organisms, such as sponges and bryozoans, and play a role of fixation and adhesion. Reef attaching organisms include brachiopod, lamellibranch, crinoid, fusulinid and foram etc. Most of the reef rocks have been dolomized to be reef dolomite. Pores and caves are well developed in the reef, some of which are filled with calcite or bitumen.

Reef cap: It is developed above the reef core, composed of Layers 24–25. Due to the strong dolomitization, the current lithology is fine crystal dolomite and powder crystal dolomite. Wu et al. (2015) classified it as bioclastic limestone of open platform through protolith restoration.

Stop 5: Platform margin grain shoal of Panlongdong

On the Panlongdong section, the platform margin shoal is developed above the reef facies, which is a set of superimposed continuous deposition of reef and shoal. Platform margin shoals are developed in the background of relative sea level decrease and are composed of platform margin shoals and inter-shoal subfacies (Fig. 19).

Platform margin shoal is mainly oolitic shoal and less psammitic shoal with tabular and wedge crossbedding. Grain shoal is mainly composed of sparry oolitic dolomite, psammitic limestone, powder crystal psammitic dolomite, sandy and intraclast oolitic limestone and sandy dolomite. The solution pores are well developed in the grain shoal deposits, and a large number of pinhole solution pores can be seen. Some of the solution pores are filled with bitumen (Fig. 19).

To sum up, Panlongdong region has experienced 4 platform margin reef evolution stages and 2 platform margin shoal evolution stages. The sedimentary processes of reefs and shoals can be summarized as follows (Mu et al. 2003, 2004, 2005; Hu et al. 2012a, b; Li et al. 2015; Wu et al. 2015).

Compared with the reef development model established by James and Mountjoy (1983), the development of the Panlongdong reefs can be divided into four stages: ① Stabilization stage (Layers 1-middle 9): the early deeper water environment evolved near the wave base, and a large amount of bioclastic accumulation was formed. Organisms (such as gastropod, brachiopod, bivalve and echinodermata) mainly reproduced in large numbers, with a small number of fixed growing sponges. In this period, the water depth, terrain, topography and hydrodynamic force created conditions for the formation of later reefs. ② Colonization stage (Layers upper 9–15): the reef-building organisms dominated by sponges began to grow, and the reef

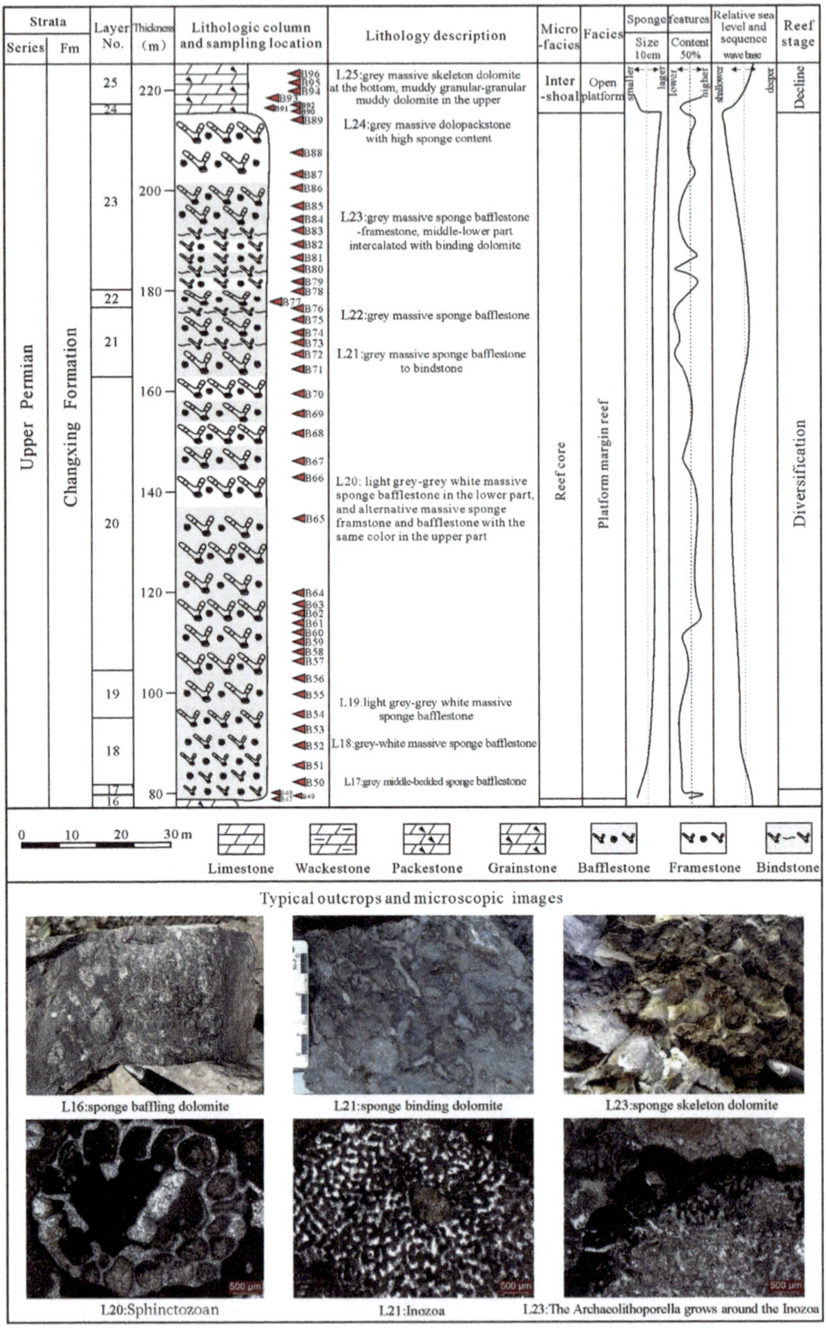

Fig. 18 Comprehensive column of reef core on Panlongdong section

Fig. 19 Development characteristics of grain shoal on platform margin of Panlongdong section

system began to be established. The reefs were in the embryonic stage of development, and the accumulation rate of the reefs was relatively slow. The layered sponge bafflestone was mainly developed, often transited to be inter-reef deposition. It was the initial reef-forming stage. ③ Diversification stage (Layers 16–23): all kinds of reef-building organisms entered the flourishing period, and the reef ecosystem was stable. The reefs were developed continuously with the rise of relative sea level, forming large dome-shaped reef, and the framework reefs were developed more and more upwards, indicating that the water became shallower and the energy became higher. It was the main reef-forming period. ④ Decline stage (Layers 24–25): the decline of relative sea level led to the exposure and erosion of the reefs, resulting in the decline of the reefs and the end of the large-scale reef-building process.

Based on the paleo-geomorphology of reef, the prograding platform margin shoal assemblage was developed when the relative sea level decline. The grain shoal was deposited in platform margin shoal and inter-shoal environments. These sediments were strongly dolomitized. At the end of the Permian, due to the changes in the physical and chemical conditions of the ocean, the biological features changed greatly, which is also the reason why the widely developed reefs were replaced by grain shoal deposits.

2.3 Puguang Gas Field

Puguang gas field is located in Puguang Town of Xuanhan County, Sichuan Province. It is the first integrated super-giant high-sulfur-containing marine gas field developed successfully in China, and the main gas source of the energy artery of "Sichuan-East Gas Transmission" project in China's "11th Five-Year Plan" period. The discovery of the gas field fully demonstrates the guiding role of basic sedimentological research in oil and gas exploration industry. The study of reef and shoal facies on the Late Permian platform margin in the Upper Yangtze region of China has not only important academic value, but also important industrial application value.

Puguang gas field was discovered in a new round of exploration under the condition that the pioneers had done a lot of work with less outcome and the exploration had been stagnated for 10 years. The discovery of Puguang gas field is the innovation result of exploration ideas, geology cognitions and exploration technologies (Ma et al. 2010), also a successful practice of sedimentology research in oil and gas exploration and geology. Based on the basic researches on sedimentary evolution and facies belt distribution including sedimentary outcrop, it is recognized that Xuanhan-Daxian area had the basic geological conditions to form reef-shoal porous dolomite reservoir during the Changxing-Feixianguan stage, which may be a favorable area for the development of structural-lithologic traps. The main factors affecting the gas reservoir scale and reserves abundance are the thickness, distribution area and pore development degree of the reservoir (Ma et al. 2005, 2010). Under the guidance of this idea, the lithofacies paleogeography and distribution law of reef-shoal facies belt during the deposit period of the Changxing-Feixianguan Formations were studied, the sedimentary facies model was established, and the main reservoir types were determined using the methods of sedimentology and reservoir geology. The multi-disciplinary information (geology, drilling, well logging, seismic and so on) were integrated to predict lateral reservoir distribution. Fine reef-shoal development model and seismic facies model were established (Fig. 20). On this basis, fine reservoir prediction was carried out. Well PG1 was deployed at the position 1300 m lower than the high point of Puguang structure. This well was completed in 2003, and obtained 42.37×10^4 m^3 of absolute open flow of natural gas in testing. Great breakthrough was obtained in the first drilling, leading to the discovery of Puguang gas field. This has promoted the deep exploration in Yuanba and Heba in Northeast Sichuan and the Sichuan Basin, expanded the exploration space of the Sichuan Basin,

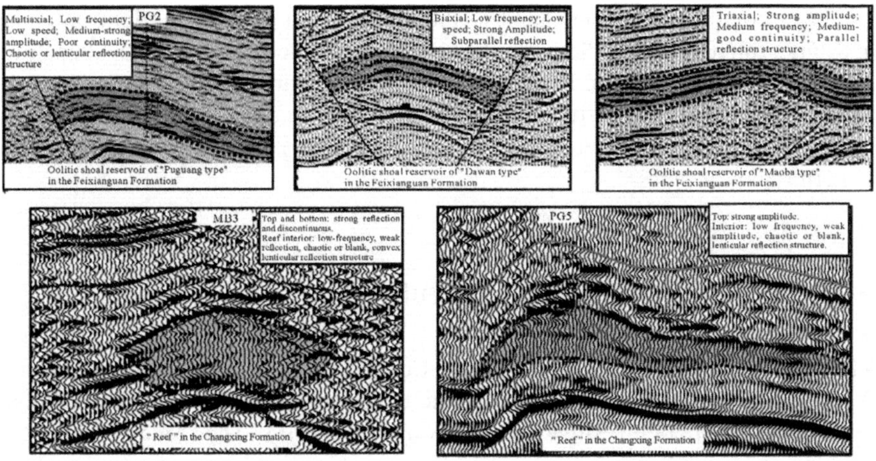

Fig. 20 Seismic prediction model of reef-shoal reservoir in Puguang area (Ma et al. 2010)

and is also a beneficial reference for oil and gas exploration in the world (Ma et al. 2010).

Stop 6: Science and Technology Exhibition Hall of Puguang gas field

Puguang gas field has the characteristics of ultra-deep burial, high pressure, high production, high hydrogen sulfide and carbon dioxide content. Large-scaled development of this kind of gas fields is a world-class problem. The Science and Technology Exhibition Hall of Puguang gas field applies cores, models and various records to show the exploration and development history of the Upper Yangtze region, as well as the exploration ideas of structural-lithologic traps dominated by reef-shoal porous dolomite reservoirs, typical reservoir rocks and cores of Changxing-Feixianguan Formations, fine description of reef-shoal reservoirs, gas field development and construction, and gas field safe and standardized operation etc. Moreover, the pore development of oolitic dolomite can also be observed by polarizing microscope in the exhibition hall.

Stop 7: Natural gas purification plant

The natural gas purification plant is the first super-large (10 billion cubic meters) high-sulfur natural gas purification plant in China, and is a key part of the major project of "Sichuan-East Gas Transmission" in the "11th Five-Year Plan". It is located in Puguang Town, Xuanhan County, Sichuan Province, covering an area of 2.27 km^2 with 6 sets of purification devices and supporting public, storage and transportation facilities. It has an annual processing capacity of 12 billion cubic meters of high-sulfur natural gas and an annual sulfur production capacity of 2.4 million tons. It is the largest high-sulfur natural gas purification plant in Asia and the second in the world. Since it was put into commercial operation on August 31, 2010, it has maintained safe, green and efficient operation for 11 consecutive years. Commercial gas has been continuously transported to thousands of enterprises in eight provinces and cities along the middle and lower reaches of the Yangtze River, benefiting nearly 200 million people. It has made important contributions to optimizing and adjusting China's energy structure, improving domestic natural gas supply capacity, and promoting the coordinated economic development of central-western region and eastern region.

3 Proposed Investigation Schedule

The 21st International Sedimentological Congress will be held in Beijing, China, from August 22 to August 26, 2022. The conference includes oral presentations, poster displays, field investigation and short-term courses. The "Late Permian Reef Buildup in Upper Yangtze Region and Puguang Gas Field" investigation route, as one of the post-conference investigation routes, is scheduled to begin in the morning of August 27, 2022, Beijing time. Please set your clock to Beijing time in advance (UTC/GMT + 08:00).

3.1 Dates and Location

Investigation dates: August 27–30, 2022.

Departure place: Beijing Capital/Daxing International Airport (the flight schedule for August 2022 is not available now).

Ending place: Chongqing City.

3.2 Travel Arrangements

The participants are requested to arrive at Beijing Capital/Daxing International Airport before 7 AM on August 27, 2022, and take flight CA1429 to Chongqing Jiangbei Airport. The following schedule is:

(1) August 27: Take bus in the afternoon from Chongqing to Lichuan City, Hubei Province, and stay in Lichuan City.
(2) August 28: (Morning) Field investigation in Jiantianba, Lichuan City; (Afternoon) go to Dukou Township, Xuanhan County, Dazhou City, Sichuan Province by bus and stay in Dukou Township.
(3) August 29: (Morning) Field investigation in Panlongdong, Xuanhan County; (Afternoon) visit Puguang gas field, then stay in Dazhou City.

(4) August 30: Take bus in the morning from Dazhou to Chongqing, and finish the field investigation.
(5) August 30: Return to Beijing from Chongqing or continue sightseeing in Chongqing in the afternoon.

Table 3 shows the specific schedule of "Late Permian Reef Buildup in Upper Yangtze Region and Puguang Gas Field" investigation route.

Table 3 Schedule of investigation route in north Upper Yangtze region

Date	Time	Trip	Transportation mode	Contents	Note
August 27 (day 1)	09:00–12:05	Beijing-Chongqing Jiangbei	Plane	Inbound journey	Lunch: Chongqing
	13:00–17:30	Chongqing-Lichuan City	Bus	Journey	Accommodation: Lichuan
August 28 (day 2)	07:30–09:00	Lichuan-Jiantian Village	Bus	Journey	
	09:00–12:00	Jiantianba section		Field investigation in Jiantianba	Lunch: Jiantian Village
	13:00–19:00	Jiantian Village-Dukou Township	Bus	Journey	Accommodation: Dukou Township
August 29 (day 3)	07:00–08:00	Dukou Township-Panlongdong	Bus	Journey	
	08:00–11:30	Panlongdong section		Field investigation in Panlongdong	
	11:30–12:30	Panlongdong-Dukou Township	Bus	Journey	Lunch: Dukou Township
	13:00–14:30	Dukou Township-Puguang Town	Bus	Journey	
	14:30–16:30	Puguang gas field		Visit exhibition hall and viewing platform	
	16:30–17:30	Puguang-Dazhou	Bus	Journey	Accommodation: Dazhou
August 30 (day 4)	08:00–11:00	Dazhou-Chongqing	Bus	Journey	Lunch: Chongqing
	14:25–17:30	Chongqing Jiangbei-Beijing Daxing	Plane	Return journey	

4 Logistics

The following arrangements for this field trip are provided for your reference.

4.1 Accommodation and Meals

Participants in this investigation will be arranged with the accommodation from the night of August 27 to the night of August 29 (in Lichuan City, Hubei Province on August 27; Dukou Township, Xuanhan County, Dazhou City, Sichuan Province on August 28; Dazhou City, Sichuan Province on August 29). All the reserved hotels are safe, clean and comfortable. Breakfasts and dinners will be at the hotels where you stay, lunches will be at the restaurants along the way or bento. Necessary meals/ drinks will be arranged during the investigation.

4.2 Transport

The investigation stops are all located within 50 km of your hotels, and the TOYOTA COASTER minibus will be used for the transportation.

4.3 Weather

Dazhou City of Sichuan Province and Lichuan City of Hubei Province are located in southern China, which belong to typical subtropical monsoon climate. Please prepare your rainproof tools for any possible rainfall in the rainy August. During the investigation period, in Lichuan with the altitude of about 1400 m, the forecast temperature is about 20–28 °C; in Dazhou with the altitude of about 550 m, the forecast temperature is about 24–33 °C.

Moreover, the participants are recommended to prepare sunshade cap, sunscreen cream, antiskid and rainproof outdoor shoes, as well as a thin coat, in order to prevent any possible sun exposure, rainfall and temperature changes during the investigation.

4.4 Safety

(1) Be careful to avoid vehicles, bad road conditions, rivers, and falls.
(2) Be careful during investigation at high and steep locations, in order to avoid any possible damage by falling rocks. Observe the signals of geological disasters,

such as abnormal noise, loosening and collapse. Wear the helmet at or near the crags.

(3) The investigation arrangement will be delayed or cancelled in case of rainstorm, gale, thunderstorm, hail and other adverse weather conditions, or the possibility of natural disasters.
(4) Follow the arrangement of the organizer during the investigation. DO NOT leave the team and act alone.
(5) Inform the organizer before the trip if you have any health problems (e.g. hypertension, heart disease, asthma, diabetes, etc.). Common medicines will be available at the vehicles.
(6) When necessary, call 110 for police support, 120 for emergency medical assistance, or 112 where there is no mobile phone signal for field emergency help.

Acknowledgements We appreciate Professor Hu Xiumian of Nanjing University for his assistance in the scheduling for the investigation, Professor Lin Changsong of China University of Geosciences (Beijing) and Professor Xie Xinong of China University of Geosciences (Wuhan) for their guidance and suggestions on the arrangement of the investigation stops, and Professor Lu Yongchao of China University of Geosciences (Wuhan) for his valuable opinions on the investigation. Professor Xing Fengcun of Chengdu University of Technology worked with us in examining the sections in hot summer. Dr. Chen Beichen shared the research results of Jiantianban reef selflessly. Many officials and staffs of Sinopec Zhongyuan Oilfield Company and Puguang Company also provided their great support to our work. The preparation of this guide book has referred to and used the research results of many scholars, some of which may not be listed in the references in time. Thanks to all the researchers.

References

Chen B, Xie X, Al-Aasm IS et al (2018) Depositional architecture and facies of a complete reef complex succession: a case study of the Permian Jiantianba Reefs, Western Hubei, South China. Minerals 8(11):533

Dunham R J (1962) Classification of Carbonate rocks according to depositional texture. AAPG Memoir 1:62–84, 108–121

Dunham RJ (1970) Stratigraphic reefs versus ecologic reefs. AAPG Bull 50(10):1931–1950

Embry AF, Klovan JE (1971) A late Devonian reef tract on northeastern Banks Island, N.W.T. Bull Canad Petrol Geol 19(4):730–781

Fan J, Wu Y (2005) Main features of the Permian reefs of world and their palaeogeographic distribution. J Palaeogeogr 7(003):287–304

Fan J, Zhang W (1985a) Basic concept, classification and identification characteristics of reefs. Acta Petrologica Sinica 1(003):47–61+99

Fan J, Zhang W (1985b) Sphinctozoans from late Permian reefs of Lichuan, West Hubei, China. Facies 13(1):1–43

Fan J, Zhang W (1986) Late Permian sphictozoan sponges in Lichuan, Western Hubei. Chin J Geol 2:151–160

Fan J, Zhang W (1987) Fibrous sponges and tabulozoan in the Late Permian reefs in Lichuan, Western Hubei. Chin J Geol 4:326–333

Fan J, Zhang W, Ma X et al (1982) Basic characteristics and development law of Permian reefs in Western Hubei. Chin J Geol 3:274–282
Flügel E (2002) Triassic reef patterns. In: Kiessling W, Flügel E and Golonka J (eds) Phanerozoic reef patterns. Tulsa, SEPM Special Publication 72:391–463
Guo L, Riding R (2010) Microbial micritic carbonates in uppermost Permian reefs, Sichuan Basin, southern China: some similarities with recent travertines. Sedimentology 39(1):37–53
He D, Li D, Wang C (2015) Miracles due to one sedimentary facies map: on mobilistic tectono-paleogeography. Chin J Geol 50(02):391–409
Heckel PH (1974) Carbonate buildups in the geological record: a review. SEPM Special Publication Reefs in Time and Space 18:90–155
Hu M, Wei H, Qiu X et al (2012a) Reef composition and their forming models of Changxing Formation in Jiantianba section of Lichuan, western Hubei. Acta Sedimentol Sin 30(1):33–42
Hu M, Hu Z, Qiu X et al (2012b) Platform edge reef and bank structure and depositional model of Changxing formation in Panlongdong section, Xuanhan, northeastern Sichuan. J Earth Sci 4:431–441
Huang H, He D, Li Y et al (2021) Late Permian tectono-sedimentary setting and basin evolution in the Upper Yangtze region. South China: Implications for the formation mechanism of intra-platform depressions. J Asian Earth Sci 205:104599
James NP, Mountjoy EW (1983) Shelf-Slope break in fossil carbonate platforms: an overview. SPEM 33:189–206
Li Q, Miao S, Jiang Q (2015) Sedimentary characteristics and reef-forming model of Changxing Formation in Panlongdong Section of Xuanhan Sichuan. J Jilin Univ: Earth Sci Ed 45(5):1322–1331
Li Y, Wu Y, Jiang H (2018) Taphonomic characteristics of a Permian calcisponge reef in Lichuan, Hubei province and its paleoenvironmental significance. Acta Palaeontologica Sinica 57(2):212–227
Liu H, Rigby JK (1992) Diagenesis of the Upper Permian Jiantianba Reef, West Hubei, China. J Sediment Res 62(3):367–381
Luo Z (1981) The influence of tafrogeny on the formation of petroleum and other minerals since Late Paleozoic in Southwest China. Acta Geologica Sichuan 00:4–25
Luo ZL, Long XM (1992) The uplifting of the longmenshan orogenic zone and the subsidence of the west sichuan foreland basin. Acta Geologica Sichuan 12(1):1–17
Ma Y (2006) Cases of discovery and exploration of marine fields in China (Part 6): Puguang Gas Field in Sichuan Basin. Mar Orig Pet Geol 11(02):35–40
Ma Y, Guo X, Guo T et al (2005) Discovery of the large-scale Puguang Gas Field in the Sichuan Basin and its enlightenment for hydrocarbon prospecting. Geol Rev 51(4):477–480
Ma Y, Guo T, Zhao X et al (2007a) Formation mechanism of deep high-quality dolomite reservoirs in Puguang Gas Field. Sci China (Ser D: Earth Sci) 037(0z2):43–52
Ma Y, Guo X, Guo T et al (2007b) The Puguang gas field: new giant discovery in the mature Sichuan Basin, southwest China. AAPG Bull 91(5):627–643
Ma Y, Mu C, Tan Q et al (2007c) Reef-bank features and their constraint to reservoirs of natural gas, from Permian Changxing formation to Triassic Feixianguan formation in Daxian-Xuanhan area of Sichuan province, South China. Earth Sci Front 14(1):182–192
Ma Y, Cai X, Guo X et al (2010) The discovery of Puguang Gas Field. Eng Sci 12(10):14–23
Mu C, Tan Q, Wang L et al (2003) Late Permian bioherm oil paleo-pool discovered in Panlongdong, Xuanhan County, Sichuan. Geol Rev 49(3):60–64
Mu C, Tan Q, Yu Q et al (2004) The organic reefs and their reef-forming model for the Upper Permian Changxing Formation in northeastern Sichuan. Sediment Geol Tethyan Geol 24(003):65–71
Mu C, Ma Y, Wang R et al (2005) Diagenesis of the Upper Permian Panlongdong organic reefs in northeastern Sichuan. Sediment Geol Tethyan Geol 25(1):198–202
Riding R (2002) Structure and composition of organic reefs and carbonate mud mounds: concepts and categories. Earth Sci Rev 58:163–213

Rigby JK, Fan J, Zhang W (1989) Sphinctozoan sponges from the Permian reefs of south China. J Paleontol 63(4):404–439
Stanton RJ (1967) Factors controlling shape and internal facies distribution of organic carbonate buildups. AAPG Bull 51(12):2462–2467
Toomey DF (1981) European fossil reef models. SPEM Special Publication 30:1–7
Wang Y, Wen Y, Hong H et al (2006) Petroleum geological characteristics of deep water deposits in Upper Permian-Lower Triassic trough in Sichuan basin and adjacent areas. Oil Gas Geol 27(005):702–714
Wang S, Jiang X, Guan H et al (2010) Origin of intragranular dissolution pores of oolite dolomite reservoirs in Puguan Gasfield, Northeastern Sichuan Province. Acta Sedimentologica Sinica 01:13–19
Wu Y (1992) Structural rock types and structural facies of reefs. Sci China (Ser B) 3:304–310
Wu Y (1997) Classification of reef rocks. Geol Rev 43(003):281–289
Wu YS, Fan JS, Jin YY (2003) Emergence of the late Permian Changhsingian Reefs at the end of the Permian. Acta Geologica Sinica 77(3):289–296
Wu L, Jiao Y, Rong H et al (2012) Reef types and sedimentation characteristics of Changxing Formation in Manyue-Honghua section of Kaixian, Northeastern Sichuan Basin. J Earth Sci 23(4):490–505
Wu Y, Jiang H, Liu L et al (2015) Revised sedimentary facies and lithostratigraphy of the Upper Permian Changxing Formation at Panlongdong, Xuanhan, Sichuan Province. ACAT Geologica Sinica 89(2):412–424
Wu W, Liu W, Mou C et al (2021). Organic-rich siliceous rocks in the upper Permian Dalong Formation (NW middle Yangtze): provenance, paleoclimate and paleoenvironment. Marine Pet Geol 123:104728
Xing F, Lu Y, Guo T et al (2017) Sedimentary texture diversity of different carbonate platform margins and it its significance for petroleum exploration: a case study of carbonate platform margins in Feixianguan period of the Early Triassic, NE Sichuan basin, China. Acta Petrologica Sinica 33(4):1305–1316
Xing F, Hu H, Hou M et al (2018) Carbonate reservoirs cycles and assemblages under the tectonic and palaeogeography control: a case study from Sichuan Basin. Earth Sci 43(10):3540–3552
Yin H, Wu S, Du Y et al (1999) South China defined as part of Tethyan archipelagic ocean system. Earth Sci J China Univ Geosci 24(1):1–12
Zhang W (1985) A new discovery of Sphinctozoa fossils from Late Permian reefs in Lichuan, Western Hubei. Chin J Geol 4:364–374
Zhang W (1992) Study on Permian reefs and paleoecology of reef-building organisms. Adv Earth Sci 7(6):31
Zhao W, Guo T, Cai Z et al (2010) Reef types and control factors of Changxing Formation of Permian in Northeast Sichuan. Geoscience 5:951–956
Zhao W, Xu C, Wang T et al (2011) Comparative study and significance of natural gas accumulation in reef shoal bodies of Changxing-Feixianguan Formations of Permian and Triassic in Longgang and Luojiazhai-Puguang areas of Sichuan Basin. Chin Sci Bull 56(Z2):2404

Unconventional Gas Resources from the Upper Ordovician Wufeng-Lower Silurian Longmaxi Shales in the Fuling Shale Gas Field of Southern Chongqing City, SW China

Long Shengxiang, Du Wei, Li Maowen, Cao Wei, Ma Jun, and Tian Jingchun

1 Objectives and Strategy

Participants will examine (1) the main facies and physical stratigraphy of the organic-rich shales of the Upper Ordovician Wufeng Formation and the first member of the Lower Silurian Longmaxi Formation; (2) differences in the lateral distribution of these shales within or around the eastern margin of the Sichuan Basin as a result of the changes in tectonic uplift, volcanism and glacio-eustatic sea level; (3) great karst landscape; and (4) Fuling Shale Gas Field, a great milestone of China's shale gas industry. We will discuss the vertical variations in the Wufeng-Longmaxi Formations and their relationships with tectonics, sedimentation, and paleontological features. A history of shale gas exploration and development in China will be introduced.

2 Introduction

Shale gas exploration and development in China have made significant breakthroughs in southeastern and southern Sichuan Basin, with multiple shale gas fields such as Fuling, Changning, Weiyuan, Zhaotong, Weirong and south Pingqiao-Dongsheng being built. The Wulong-Qijiang block (in southern Chongqing city) adjacent to the Fuling and south Pingqiao-Dongsheng gas fields displays strong structural deformation and erosion at different positions due to intensive tectonic activity since the Middle-Late Yanshanian period, leading to the exposure of the Lower Paleozoic to Mesozoic strata to the surface. These include the organic-rich shales of the Upper

L. Shengxiang (✉) · D. Wei · L. Maowen · C. Wei · M. Jun · T. Jingchun
SINOPEC Petroleum Exploration and Production Research Institute, SINOPEC Jianghan Company, SINOPEC East China Company, Chengdu University of Technology, Chengdu, China
e-mail: huxm@nju.edu.cn

X. Hu (ed.), *Field Trip Guidebook on Chinese Sedimentary Geology*,
https://doi.org/10.1007/978-981-99-6936-4_11

Ordovician Wufeng Formation and the first member of the Lower Silurian Longmaxi Formation, the target horizons of the above shale gas fields. This field trip will observe three well-preserved outcrops and 1–2 alternate outcrops for a great opportunity to see the stratigraphy, lithology, paleontology and vertical variation within the Upper Ordovician Wufeng Formation and Long-1 Member, as well as the changes in shale quality from the marginal facies to the depocenter based on geological outcrops and stratigraphic correlations. Great karst landforms will be enjoyed en route at the Natural Bridges and the surrounding area. Toward the end of the trip, we will visit the Fuling Shale Gas field, view the spectacle of shale gas field construction and independent innovation technology, and share the stories of shale gas discovery and achievements in China.

Period: 4 days.

Field Trip Leaders:

Li Maowen (SINOPEC Petroleum Exploration & Production Research Institute), limw.syky@sinopec.com;
Long Shengxiang (SINOPEC Petroleum Exploration & Production Research Institute), lsx.syky@sinopec.com;
Du Wei (SINOPEC Petroleum Exploration & Production Research Institute), duwei.syky@sinopec.com;
Cao Weisheng (SINOPEC Jianghan Oilfield Company), caows.jhyt@sinopec.com;
Ma Jun (SINOPEC East China Oil & Gas Company), dzgcmj@163.com;
Tian Jingchun (Chengdu University of Technology), tjc@cdut.edu.cn.

3 Proposed Investigation Schedule

3.1 Field Trip Route and Logistics

DAY 1: Beijing Chognqing-Nanchuan
13:45 Arrival at Chonqing by airplane;
14:00–16:30 Leave for Anwen town, Qijiang District;
16:30–17:00 Visit Guanyinqiao outcrop;
17:00–19:20 Leave for Yuejing Hotel in Nanchuan District;
19:20–20:30 Dinner at hotel.

DAY 2: Nanchuan-Daozhen-Wulong
07:30–08:00 Breakfast at hotel;
08:00–9:30 Leave for Daozhen County, Zunyi City, Guizhou Province;
9:30–11:00 Visit Bayu outcrop;
11:00–12:30 Lunch break;
12:30–15:30 Leave for Huangying town, Wulong District;
15:30–17:00 Visit Huangying outcrop;

17:00–17:30 Leave for Yuzhu Garden Hotel in Wulong;
17:30–18:30 Dinner at hotel.

DAY 3: Wulong-Zibotang-Three Natural Bridges-Wulong
07:30–08:00 Breakfast at hotel;
08:00–10:00 Leave for Zibotang, Wulong District;
10:00–11:30 Visit Tiansheng outcrop;
11:30–12:30 Leave for Three Natural Bridges;
12:30–13:30 Lunch break;
13:30–17:00 Visit carbonate rock corrosion at Three Natural Bridges;
17:00–19:00 Leave for hotel (undetermined) at Fuling District;
19:00–20:00 Dinner at hotel.

DAY 4: Nanchuan-Fuling Shale Gas Field-Chognqing
07:30–08:00 Breakfast at hotel;
08:00–09:30 Leave for Fuling Shale Gas Field;
09:30–12:00 Visit Fuling Shale Gas Field;
12:00–13:00 Lunch break;
13:00–16:00 Back to Chongqing, happy ending (Fig. 1).

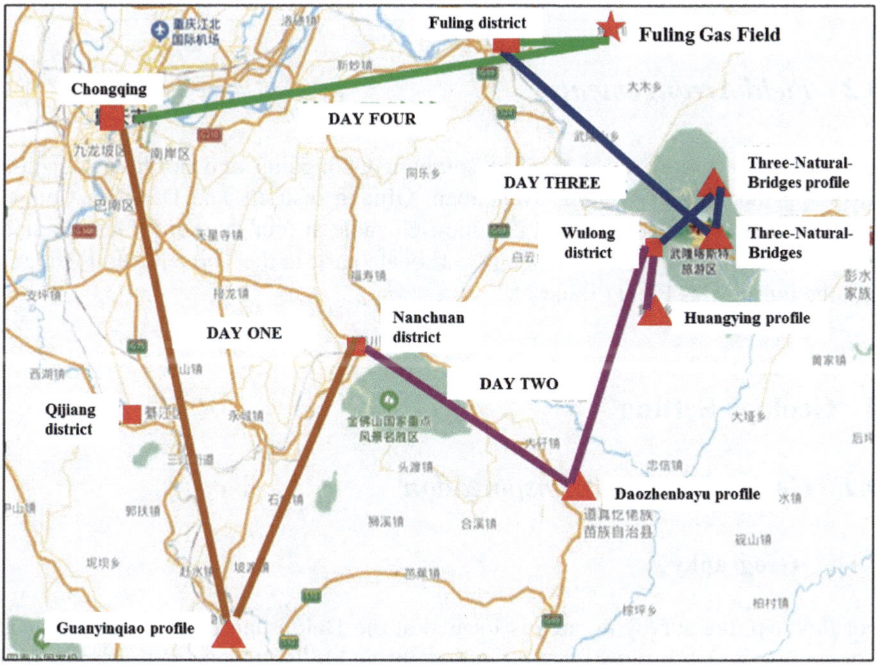

Fig. 1 Outcrop locations and traffic map (from Baidu.com)

Table 1 Travel arrangements and field observations

Date	Stops	Observations	Locations	Time (h)
DAY 1	1.1	Wufeng-Longmaxi shale at Guanyinqiao outcrop	Beside highway near Anwen town, Qijiang district	1.0
DAY 2	2.1	Wufeng-Longmaxi shale at Bayu outcrop (sublayer ⑤–⑨)	Beside highway near Daozhen county, Zunyi city, Guizhou province	1.5
	2.2	Wufeng-Longmaxi shale at Bayu outcrop (sublayer ①–⑤)	Beside Road S203, Shuanghe village, Huangying town, Wulong district	1.5
DAY 3	3.1	Wufeng-Longmaxi shale at Tiansheng outcrop (sublayer ①–④)	Beside road near Zibotang, Jielong town, Wulong district	1.5
	3.2	Carbonate rock corrosion at Three-Natural- Bridges	Three-Natural- Bridges, Wulong district	3.5
DAY 4	4.1	Brief introduction about Jiaoye 1 well and Fuling Gas Field	Jiaoshi town, Fuling district	1.0
	4.2	Gathering and transportation station	Jiaoshi town, Fuling district	0.5

3.2 Field Arrangements

Our field trip will spend 4 days in southern Chongqing and northern Guizhou, involving the Fuling, Wulong, Nanchuan, Qijiang districts and Daozhen County. We will visit Wufeng-Longmaxi organic-rich shale at four outcrops (Guanyinqiao, Daozhen, Huangying and Tiansheng), carbonate karst at the Three Natural Bridges, and the Fuling Gas Field (Table 1).

4 Geology Setting

4.1 Geography and Transportation

4.1.1 Geography

For this trip, the surveying area is located at the Daloushan fold belt in southeast Chongqing, which is dominated by mountains and hills. It is characterized by high mountain ridges, thousands of hills and valleys, numerous rivers and caves created by long-term dissolution (Fig. 2).

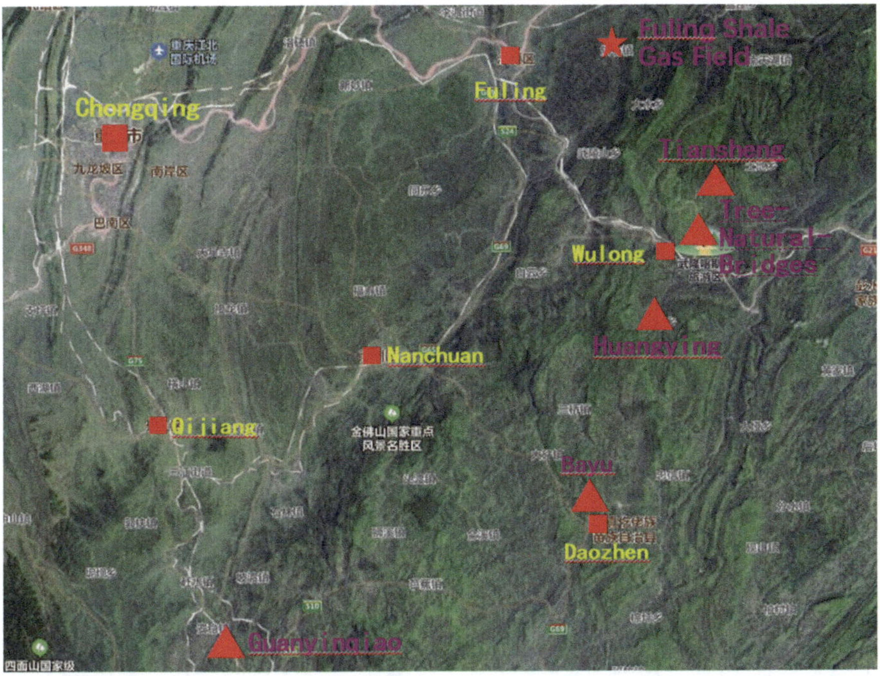

Fig. 2 Geographical map showing the Pengshui-Nanchuan region, Chongqing (from Baidu.com)

4.1.2 Traffic Routes

Local transportation, including railways, expressways and highways, is well developed in the surveying area. Vehicles are able to reach villages and households (Figs. 1 and 2).

All stops are close to highways and can be reached by vehicles, except in rare situations affected by natural disasters (e.g., heavy rain, highway freezing or collapse, traffic accidents, or force majeure).

4.2 Geological History

The surveying area is located in the Upper Yangtze, bounded by the Qiyueshan fault in the north and adjacent to the Sichuan Basin. Similar to the Sichuan Basin, it also experienced six primary tectonic cycles, e.g., the Yangtze, Caledonian, Hercynian, Indosinian, Yanshan and Xishan cycles (Fig. 3). The surveying area was integrated with the Sichuan Basin in the first four cycles, with similar geological characteristics, while it was gradually differentiated from the Sichuan Basin in the last two cycles. Intensive tectonic deformation, uplift and erosion resulted in complex and

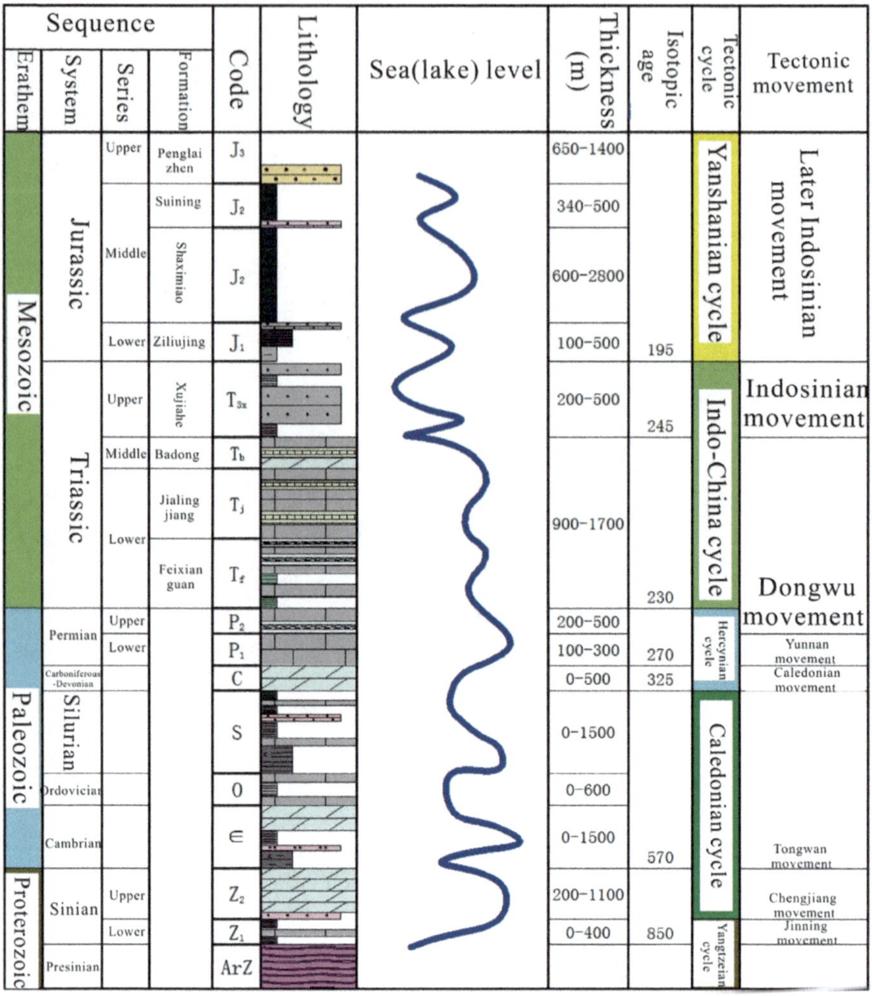

Fig. 3 Comprehensive tectonic and stratigraphic column in the survey area

diverse structural patterns, where organic-rich shale was preserved sporadically. As a consequence, shale gas reservoirs with normal and abnormally low pressures have been widely developed.

Based on previous investigations on the geology and tectonic movements in and around the Sichuan Basin, the regional geological history of the surveying area is briefly summarized as follows.

The most important tectonic event in the Yangtze cycle was the Jinning movement, which occurred before the Sinian period, resulting in reoccurrence of the pre-Sinian geosynclines and folds as well as the integration of the Yangtze paraplatform. After that, the late Early Sinian (NANTUO period) entered the great glacial period, where the Yibin and eastern Chongqing regions were covered by moraine rock, with a thickness of 50–500 m, and up to 1000–3000 m locally in the Rongjiang and Huaihua regions.

The Caledonian cycle began in the late Sinian, with a relatively stable passive margin being formed in the Doushantuo period. Two sets of lithologic assemblages, i.e., black shale mixed with argillaceous dolomite and phosphorite, purplish red and yellow gray sand-mudstone and argillaceous dolomite mixed with phosphorite, were deposited on an uneven seafloor during the transgressive cycle. The sea basin floor in the Dengying period was high in the west and low in the east, where provenance was located in the west. The surveying area and adjacent areas were restricted sea platforms, depositing dolomite, argillaceous dolomite, and algal dolomite mixed with gypsum dolomite, with a residual thickness of 500–1000 m. Gypsum from a brackish lagoon was deposited near Changning. The Zhijin movement (or Tongwan movement) at the end of the Sinian was characterized by large-scale uplift, and the upper part of the Dengying Formation was widely eroded. In the early Cambrian, it was still a epeiric sea with high in the west and low in the east, where the black carbonaceous shale and silty shale in the Niutitang Formation, the primary source rocks in this area, were deposited, with a thickness of 50–200 m. It was gradually transited into sandy shale and sand-shale interbeds in the middle and late stages and limestone in the late stage. A semirestricted environment to the lagoon occurred widely in the middle and late Cambrian, with a relatively shallow water body. A NE-trending underwater uplift developed in the surveying area during the early Ordovician. From the middle and late Ordovician to the Silurian, the westward subduction of the ancient Pacific plate activated the relatively stable passive margin, uplifting the southern part of the Upper Yangtze Platform as a whole and eroding the Middle and Upper Ordovician and the Lower Ordovician Meitan (Dawan) Formation on a large scale. The Honghuayuan Formation was eroded locally. The Leshan-longnusi uplift, Central Guizhou to East Yunnan uplift and southern Xuefeng uplift all rose rapidly at this time. During this period, the Sichuan-Guizhou Depression appeared in southeastern Sichuan, depositing sequences in a relatively closed and anoxic intraplatform basin during the Wufeng–Longmaxi period, thus forming another high-quality source rock.

The Guangxi movement at the end of the Silurian joined the South China and Yangtze Platform to an integrated South China block, indicating the end of the Caledonian cycle. Most areas from southeastern Sichuan to the surveying area were in a paleocontinental environment during the Hercynian cycle, after the Guangxi movement, and at the end of the Carboniferous, the Yunnan movement uplifted and eroded Devonian and Carboniferous deposits. Rising global sea level in the early Permian swamped the whole Yangtze from south to north, depositing the lower Permian shallow water carbonate rocks with a thickness of 350–500 m. Strong extension due to the opening of the ancient Tethys in the later stage resulted in a regional Dongwu movement, which exhibits four characteristics in this region: (1) the NE-trending Huayingshan fault was activated again, resulting in basalt overflow along the fault systems; (2) horst-graben structure derived from rifting was characterized by the "Bainikuang Layer" in the Maokou period; (3) an intensive eroded area occurred with Yanggaosi as the erosion center; (4) a nearly EW extending fault occurred with thick deposits in the south and thin deposits in the north. Changes in water depth during the Late Permian Wujiaping period resulted in alluvial sandstone and mudstone in the west and coal-bearing sandstone and mudstone deposited in coastal plains in the east. The coal bearing sequence, named the Longtan Formation, can also be found in the Chishui region. Its time equivalent in the Fuling and east Guiyang regions, named the Wujiaping Formation, consists mainly of limestone. This was followed by the Changxing Formation, where shallow marine limestone occurred widely between eastern Yibin and Wanxian, while sand mudstone from coastal plain facies developed west of Gusong and Zigong.

The Indosinian cycle began in the Early Triassic. In the early stage of the Early Triassic (Feixianguan Formation), shale and siltstone were widely deposited in coastal plain facies in southeastern Sichuan. Subtidal carbonate and sandy mudstones were deposited in the Luzhou and Chishui areas, along with limestone and oolitic limestone. Approximately 450–500 m of dolomite was deposited in the east. Limestone, dolomite and gypsum were deposited in the late stage of the Early Triassic (Jialingjiang Formation). The marine basin was relatively closed in the Middle Triassic (Leikoupo period), characterized by the deposition of dolomite, argillaceous dolomite and mudstone. It formed a regressive cycle dominated by sandstone, mudstone, limestone, dolomite and gypsum.

At the end of the Middle Triassic, with the rapid closure of the paleo-Tethys, the western and northern edges of the Yangtze block collided with the Qiangtang block from northwest to southeast and collided with the North China block at a high angle from north to south, resulting in strong compression, deflection and subsidence to form a "peripheral foreland basin" in the Songpan-Ganzi on the west margin and the south Qinling on the north margin of the Yangtze block (Zheng et al. 2003). At this time, the "Luzhou ancient uplift" in Luzhou and the "Kaijiang ancient uplift" west of Wanxian County developed in the NE direction, which was the core of the foreland uplift. Therefore, strong erosion lasting for 20–25 Ma in southeastern Sichuan was thought to be the cause for the removal of the Lower Triassic Maantang and Xiaotangzi formations. During the late Indosinian in the late stage of the Late Triassic (Norian-Rhaetian period), the comprehensive reoccurrence of the Ganzi-Aba geosyncline resulted in a strong depression in front of the Longmen Mountains in the northwest Sichuan Basin, thus initiating the stage of the Western Sichuan foreland-like basin. Westward subsidence forced the foreland uplift to migrate rapidly toward the southeast. As a result, southeast Sichuan became a foreland slope gently inclining to the northwest, and immediately, overlapping occurred from northwest to southeast during the Late Triassic Xujiahe period, where the sedimentary thickness increased rapidly from east (250–50 m in Chishui and Fuling) to the west (~500 m in Luzhou and 2000 m in Chengdu). A large foreland continental lake was developed, with coasts, lakes and rivers (coal-bearing clastic facies) occurring from west to east, ending the marine sedimentary history of the entire region. In South China, the marine sedimentary history lasted almost 2300 Ma from the Luliang period to the Late Triassic. Westward subduction of the Pacific plate in the Yanshanian cycle since the Jurassic shifted the tectonic activity zone from the front of Longmen Mountain to the southeast of the Sichuan Basin. The rapid overthrust of the Jiangnan uplift developed an intracontinental depression in the northwest, which was strongly connected with the depression in the south of the Dabashan uplift, forming depressions along Wanxian-Nanchuan-Gulin and an uplift zone along Luzhou-Chishui in the west. The latter extended to the Jiangnan uplift in the east, Longmenshan uplift in the west and Dabashan uplift in the north. Under the background of the western Sichuan foreland basin, the depression was more like a backbulge zone after foreland uplift (Luzhou-Dachuan), which lasted until the end of the Jurassic period.

During the Late Yanshan to Himalayan cycle, oblique subduction of the Pacific plate from the end of the Jurassic to the end of the Early Cretaceous further compressed the Yangtze plate, forming a primary episode of the Yanshan movement, where the ancient land southeast of the Qiyueshan fault greatly advanced to the Sichuan Basin, shrinking the deposition zone of the foreland basin. Strong folding in eastern Sichuan from the end of the Middle Eocene to the early Tertiary and strong thrusting in western Sichuan during Neogene to Paleogene collectively referred to the Sichuan movement, which was the I episode of the Himalayan movement. It occurred in the middle and late Eocene in the east and in the Neogene to Paleogene in the west. The Sichuan movement ended the sedimentary history of the large continental lake basin, which was an important period for the formation of the current tectonic pattern in the Sichuan Basin. The folds in the southeastern Sichuan Basin were partially formed at this stage, which was further enhanced during episodes II and III of the Himalayan movement.

The regional geological history suggests that two sets of organic-rich shales are developed in the surveying area: one is the Lower Cambrian Niutitang organic-rich shale, e.g., black carbonaceous shale and silty shale, with thicknesses of 50–200 m. Currently, no major breakthrough in shale gas exploration has been made in this area. Another is the Upper Ordovician Wufeng and Lower Silurian Longmaxi organic-rich shale, e.g., black, gray black siliceous shale, carbonaceous shale and silty shale, with thicknesses of approximately 100 m.

Although multiple tectonic movements occurred after the Silurian period, the weak compressive strength did not lead to intensive folding or local structures. The whole rise and fall in the early stage resulted in the absence of Carboniferous and Devonian strata in this area. It experienced intensive erosion and deplanation for 20–25 Ma in the later stage, absenting the Ma'antang Formation and Xiaotangzi Formation.

This area began to be intensively compressed during the Yanshanian cycle, especially the Middle Yanshanian movement. A large number of NNE-trending folds were formed, and faults were well developed after the late Yanshanian and Himalayan movements (Fig. 4). Meanwhile, the eroded thickness was large and varied greatly laterally, e.g., 3000–5000 m at the Pengshui and Wulong synclines but 5000–7000 m at anticlines. Wufeng to Long-1 organic-rich shale was significantly eroded, especially at the anticline, which can only be observed at residual synclines, e.g., Wulong, Pengshui, Daozhen (Fig. 5). Therefore, a large number of outcrops of Wufeng to Long-1 organic-rich shale can be found at the edge of the syncline, where outcrops near highways are generally continuously exposed, and various geological

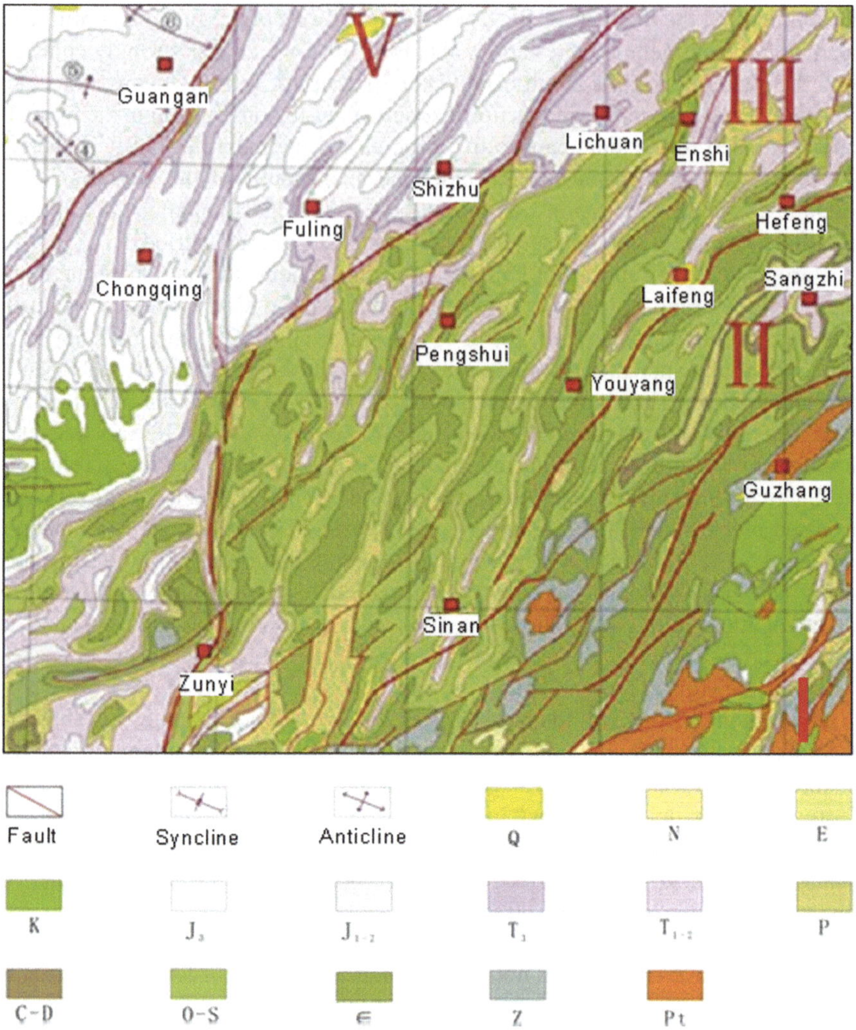

Fig. 4 Himalayan tectonic outline at the surveying and surrounding areas. I—Xuefeng Mountain uplift; II—Extened western margin of Xuefeng Mountain; III—Trough-like faulted and folded belt in Hubei, Chongqing, Hunan and Guizhou; IV—Sichuan Basin

phenomena are clearly visible, which is suitable for scientific investigation and in-depth research. However, the Lower Cambrian Niutitang organic-rich shale is not included in this field trip because of its limited exposure.

The denudation was also accompanied by strong dissolution. The Permian to Middle Triassic carbonate rocks are exposed on a large scale in the survey area. Karst features, also known as karst topography, have been formed under long-term filtration and dissolution. It is a belisted karst in southern China together with Libo

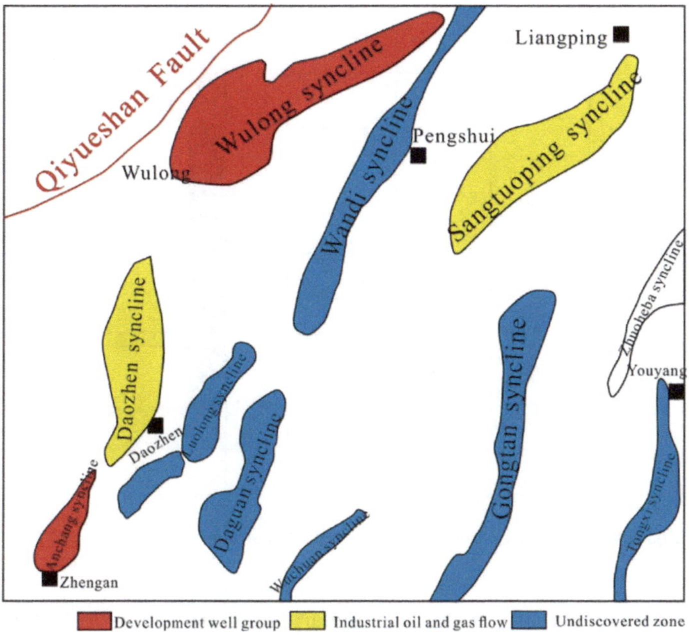

Fig. 5 Schematic diagram of residual syncline and shale gas exploration and development in southern Chongqing and northern Guizhou

in Guizhou and Shilin in Yunnan and has been listed in the "World Natural Heritage List" by UNESCO. The dissolution at the Tiansheng three bridges will be visited on this fieldtrip.

4.3 Geological Characteristics of the Wufeng to Long-1 Member

4.3.1 Stratigraphy and Sedimentology

The Upper Ordovician Wufeng Formation and Lower Silurian Longmaxi Formation can be divided into 5 parts: Wufeng shale, Guanyinqiao, Long-1, Long-2 and Long-3 members, while the corresponding strata are the Upper Katian and Hirnantian of the Upper Ordovician, Rhuddanian, Aeronian and Lower Telychian of Llandovery, with a time span of 447.62–438.13 Ma (Fig. 6). The target interval of this survey is the Wufeng to Long-1 Member, while the Long-1 Member can be further divided into three subunits.

Sedimentary microfacies of the Wufeng Formation to Long-1 Member in wells and typical outcrops were investigated based on lithological association, proportion

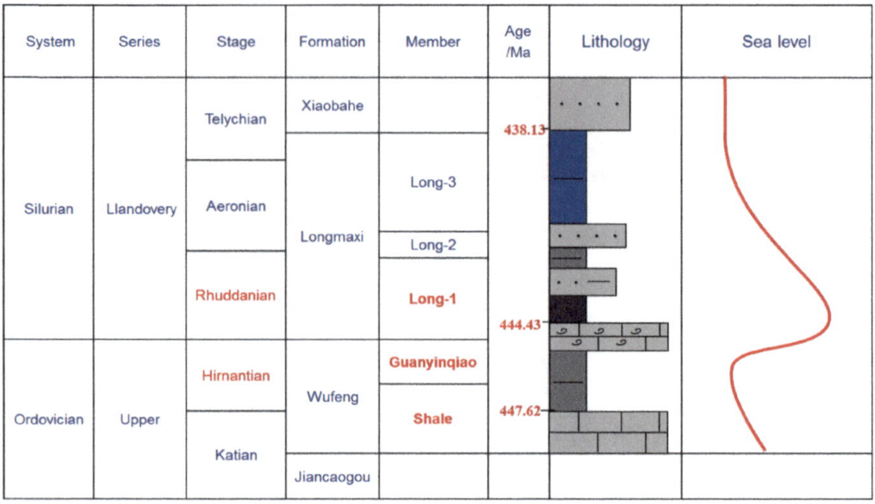

Fig. 6 Stratigraphy of upper Ordovician-Lower Silurian strata

of siliceous shale and silty shale, sedimentary structure and hydrodynamics, storm flow and turbidity current far from source, fossil type and ecology, TOC, TH/U ratio, log facies and seismic facies. Two subfacies and eight microfacies were identified (Table 2).

A semi closed and wide bay with openings in the north appeared in the Middle and Upper Yangtze during the Wufeng period. A paleogeographic pattern of three uplifts and one depression was formed in the present Sichuan Basin and its surrounding areas. Approximately 10 m of deep-water siliceous shales developed in the eastern and northern Sichuan Basin, which were under a siliceous deep-water shelf as a whole with stratigraphic centers near the Liye 1 and Enye 1 blocks. The thickness of the Wufeng Formation decreased from the center to the surroundings, with approximately 5–6 m in the survey area (Fig. 7).

The Sichuan Basin and its surrounding areas during the Guanyinqiao period inherited the paleogeographic pattern of the early Wufeng period, i.e., the central Guizhou Uplift was uplifted into a paleocontinent, and the central Sichuan-Hannan Uplift was uplifted continuously but had not yet been exposed to water. The western Hunan and Hubei were uplifted into an underwater highland, and the south Sichuan Basin—northeast Sichuan Basin were flat and deep depressions. Eastern Yunnan, northern Guizhou, southeastern Chongqing and western Hubei were basically at the slope of the depression margin. This area was located in the argillaceous shallow water shelf because of sea-level decline, while a set of thin calcareous mudstones with shells was formed due to storms.

The warming climate and global ice sheet melting during the deposition of Submember I of the Long-1 Member rose sea level rapidly and formed the largest transgression in the Silurian area, forming a wide semi closed deep-water bay in the northeast-east-southern Sichuan Basin. The sediments in the Fuling area were

Table 2 Sedimentary facies of the Wufeng Formation–Longmaxi Formation in the Jiaoshiba and surrounding areas

Subfacies	Shallow water shelf				Deep water shelf			
Microfacies	Sandy-muddy	Muddy	Storm	Turbidite	Calcareous	Muddy	Sandy-muddy	Siliceous
Water depth	Between normal wave base and storm wave base, 50–100 m				Storm wave bottom to shelf break, 100–200 m			
Lithological association	Interbedded dark gray siltstone and mudstone	Gray–green argillaceous shale	Shell limestone, Calcareous mudstone with shell	Interbedded siltstone and mudstone, fine sandstone	Calcareous shale and black–gray argillaceous shale	Mainly black–gray argillaceous shale	Black–graySilty shale and argillaceous shale	Mainly gray black siliceous shale
Sedimentary structure	Horizontal bedding and wavy bedding	Horizontal bedding	Massive bedding, shell fragment varying gradually	Cross-bedding, parallel bedding	Parallel bedding	Massive bedding, Horizontal bedding	Horizontal bedding, wavy bedding, rhythmic bedding	Massive bedding
Fossil	Graptolite in local position	Foraminiferan	Brachiopod fossils	None	Minor graptolite	Graptolite, sponge spicule, hornblende fossils	Mainly graptolite, followed by sponge spicule and hornblende	Radiolaria, sponge spicule et cl., abundant graptolite
Siliceous content/%	40–60, avg.47	25–34, avg.28	Approximately 30	70–80, avg. 75	18–30, avg. 22	25–34, avg. 28	23–40, avg. 33	40–70, avg. 47
Pyrite/%	/	Approximately 2	Approximately 2	Approximately 1.3	/	0.5–3, avg.2	1.5–6, avg.3.5	2–8, avg. 5
TOC/%	<0.5			<0.2	About 2.0	0.5–3.2, avg. 1.6	1–3.5, avg. 2.0	3–6, avg. 3.6

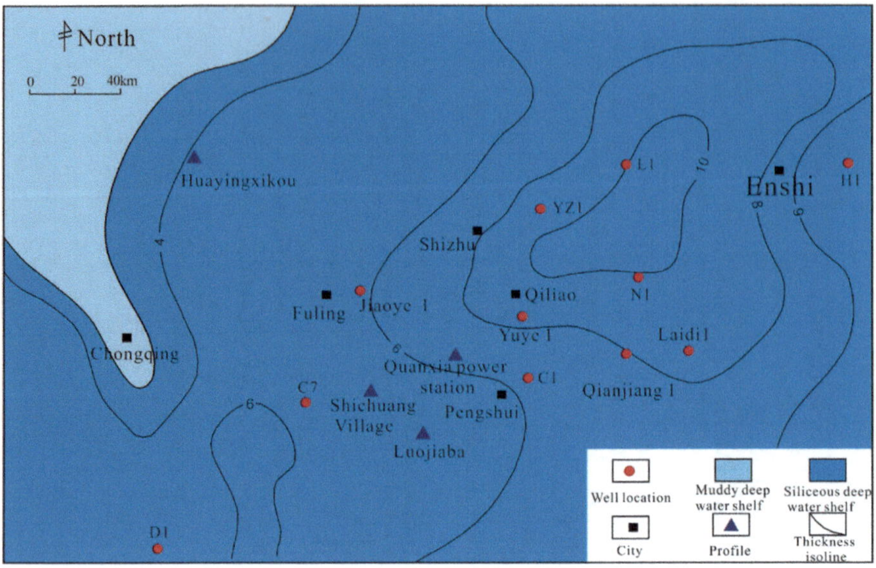

Fig. 7 Sedimentary microfacies of the Wufeng Formation in the survey area

deposited dominantly in a deep-water siliceous shelf, with a thickness of 30–40 m, while depocenters with a thickness >50 m developed in the north and southwest directions, respectively. Controlled by the Jiangnan-Xuefeng archicontinent, muddy deep-water shelf microfacies were developed in the southeast. Uplifting at the present central Sichuan Basin led to the development of a shore argillaceous shallow water shelf microfacies around the uplift and a siliceous deep water shelf microfacies in the perimeters (Fig. 8).

The fall in sea level and regression northward during the deposition of the II subunit shifted the depo-center to the middle and northern Sichuan Basin. Consequently, the deep-water zone shrank significantly toward the east in the Qianjiang-Lichuan shelf, which was dominated by calcareous deep-water shelf microfacies. In contrast, the Fuling area was dominated by sandy-muddy deep-water shelf microfacies, with a thickness of 20–30 m, where the laminar silty shale was the typical lithofacies. Additionally, uplift at the central Sichuan Basin was responsible for the development of shore argillaceous shallow water shelf microfacies around the uplift and argillaceous and sandy-muddy deep-water shelf microfacies away from the uplift (Fig. 9).

A further drop in sea level northward during the deposition of the III subunit continuously migrated the depo-center to the middle and northern Sichuan Basin. The water body east of Lichuan-Enshi-Qianjiang was shallow, where gray dark gray silty mudstone and argillaceous siltstone were developed under sandy-muddy shallow water shelf microfacies. The Fuling area is located in the muddy deep-water shelf, with sediments dominated by 20–30 m of argillaceous shale. The sedimentary patterns around the uplift in central Sichuan were similar during the deposition of the II and III submembers (Fig. 10).

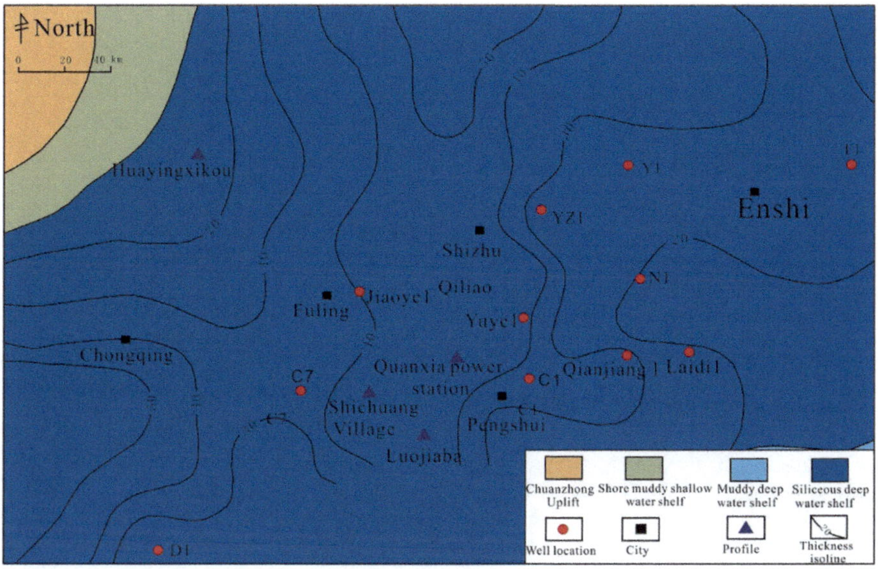

Fig. 8 Sedimentary microfacies of the I subunit of the Long-1 Member in the survey area

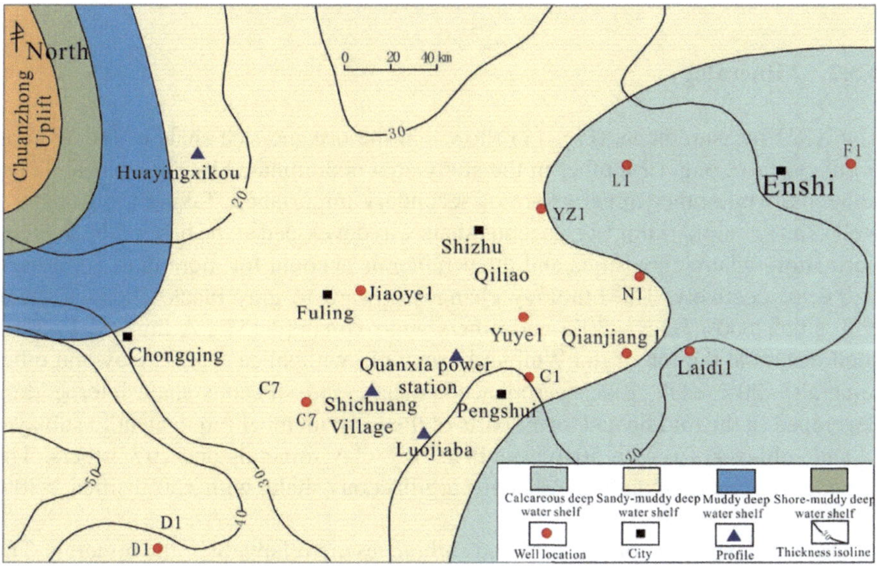

Fig. 9 Sedimentary microfacies of the II submember of the Long-1 Member in the surveying area and surrounding areas

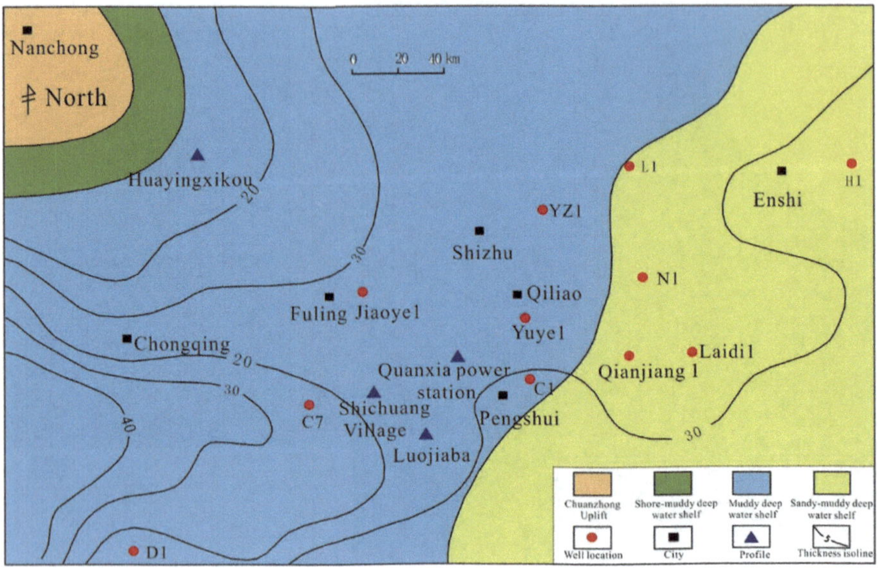

Fig. 10 Sedimentary microfacies of the III submember of the Long-1 Member in the surveying area and surrounding areas

4.3.2 Mineralogy

The XRD measurements (Fig. 11) show that the organic-rich shale of the Wufeng Formation to Long-1 Member in the study area is dominated by siliceous and clay minerals, while other minerals are of secondary importance. Taking the Longye 1 well as an example, a thin argillaceous shale was developed at the base of the Wufeng Formation, where clay, silica and other minerals account for more than 50, 30 and <20%, respectively. The lithology changes upward to gray black siliceous shale, with silica >60%, clay <30% and others approximately 15%. A thin marl with a shell occurs at the top of the Wufeng Formation, with silica > 50%, clay and other minerals ~20% each. Black carbonaceous shale and siliceous shale interbedding developed in the middle and lower parts of the I sub member (equivalent to sublayer ③ and sublayer ④), with 50–65% silica, 20% clay minerals and 20% others. The upper part consists mainly of siliceous argillaceous shale, with <50% silica, >30% clay, and 20–30% others.

Quartz, K-feldspar, plagioclase and carbonates are relatively brittle minerals. The brittle mineral content, mainly quartz, in the high-quality shales of the Longye 1 well is 79.1%. The quartz content increases from sublayer ⑤ to sublayer ① from top to bottom, while the feldspar, carbonate, pyrite and other brittle mineral contents decrease (Table 3). Although brittleness is overall high, the relatively higher brittleness at the lower part leads to better fracturing performance.

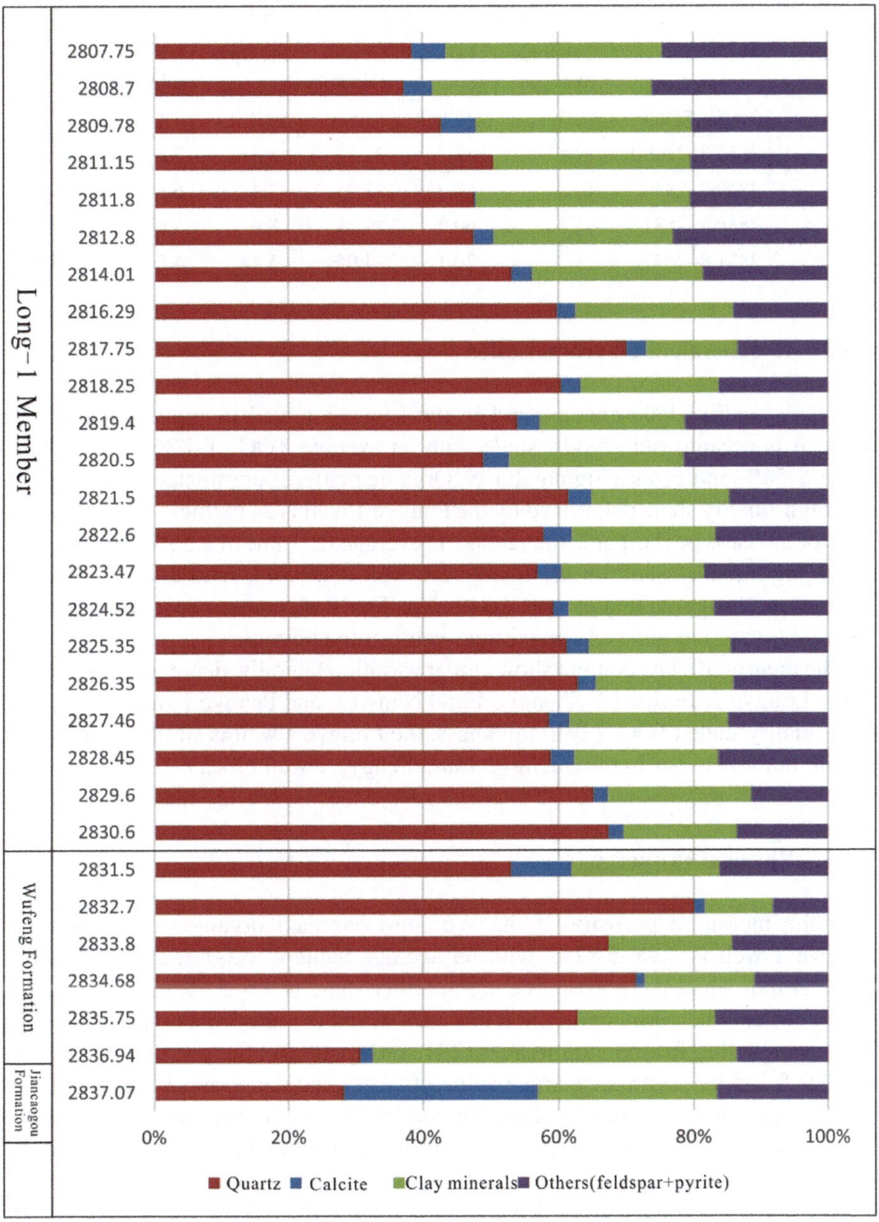

Fig. 11 Shale mineral composition at the base of the Wufeng Formation-Longmaxi Formation in the Longye 1 well

Table 3 Brittle mineral content in the Longye 1 well (From SINOPEC East China Company 2015)

Sublayer	Depth (m)	Thickness (m)	Quartz	Clay mineral	Feldspar	Carbonate mineral	Pyrite
⑤	2805–2812.3	7.3	42.8	31.6	10.75	7.1	4.45
④	2812.3–2819.8	7.5	57.28	21.32	8.56	7.64	4.34
③	2819.8–2830.1	10.3	60.24	21.3	7.4	6.83	3.48
②	2830.1–2831.8	1.7	60.2	19.35	8.9	8.6	2.95
①	2831.8–2837	5.2	70.1	24.04	5.44	4.5	3.82

4.3.3 Geochemical Characteristics

The high-quality shale encountered in the Longye 1 well is approximately 32 m thick. It is organic-rich, type I shale with an average TOC of 4.36%. Its high Ro value, 2.54%, indicates a high maturity. Organic matter is dominated by sapropelite. The high-quality shale penetrated by the Pengye 1 well is 24 m thick, with an average TOC of 3.05% type II organic matter and an average Ro value of 2.55%. The thickness of the high-quality shale in the Shengye 1 well is 27.5 m, with a TOC content of 3.2%, type II organic matter, and 2.54% Ro. The TOC content decreases vertically from bottom to top and has a maximum value at the bottom.

The measured TOC content shows great variation laterally, decreasing in the order of the Longye 1, Jiaoye 1 1, Shengye 1 and Nanye 1, and Pengye 1 wells. In terms of high-quality shale (TOC $\geq$ 2%) thickness, the Longye 1 well is similar to the Jiaoye 1 well, but it is lower in the Shengye 1 and Pengye 1 well blocks.

4.3.4 Physical Property

That the measured porosity of the Wufeng-Longmaxi organic-rich shale in the Longye 1 well is 2.98–5.87%, with an average value of 5.00%; the permeability is 0.034 md, while the porosity of high-quality shale is 4.6–5.5%, with an average of 4.95%, and the permeability is 0.034 md, which is regarded as type I shale. The log interpretation from Schlumberger (2015) suggests that the porosity of sublayers ①–③ is 5.1–5.9%, with an average value of 5.5%, and the permeability is 360–448 nD. The porosity is vertically characterized by "high in upper and lower units and low in middle unit"; however, the permeability is lower in the upper unit and high in the lower unit".

The porosity of the Wufeng-Longmaxi organic-rich shale in the Pengye 1 well is 2.3–11.7%, with an average value of 5.8%; the permeability is 0.1–1.559 mD, with an average value of 0.042 mD. The porosity of high-quality shale is 2.05%, which is regarded as type I shale.

The porosity in the Shengye 1 well is 0.12–3.79%, with an average value of 2.05%. The porosity of high-quality shale is 2.33%, which is regarded as type II shale.

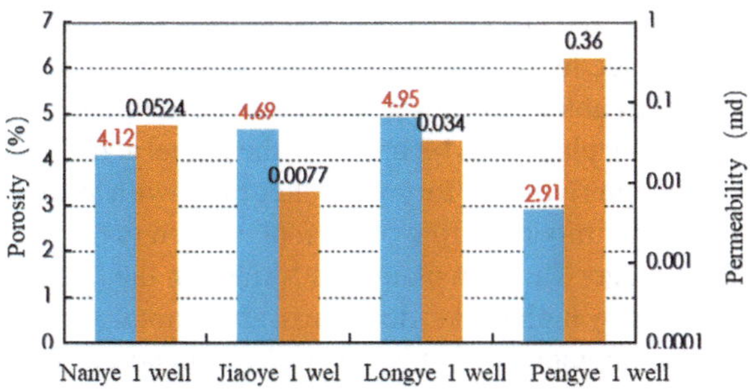

Fig. 12 Porosity and permeability of high-quality shale from different wells (data from Sinopec Co., Ltd. Huadong Oil Field Branch 2015)

Although the permeability is generally low, it is within the range measured from the commercially developed North American shales.

The comparison suggests (Fig. 12) that the high-quality shales in the Longye 1, Nanye 1 and Jiaoye 1 wells have similar porosities and permeabilities, while the high-quality shale in the Pengye 1 well has the lowest porosity. The Pengye 1 well has the highest permeability, and the Jiaoye 1 well has the lowest permeability, while the abnormally high permeability in the Pengye 1 well may be affected by faults.

4.4 *Karst at Wulong*

Carbonate dissolution is common in the Wulong district of Chongqing city. Zhang et al. (2012) reported 26 karst caves with lengths >10 km in China, three of which occur in the Wulong district, ranking in the top five, while eight of the 20 karst caves with depths >400 m in China are in the Wulong district. Dissolution not only aids in developing karst caves underground but also forms sinkholes and natural bridges on the surface and peak-cluster depressions, all of which are referred to as karst systems. The Chongqing Tourism Administration concludes that Wulong district has three independent karst systems, namely, Furong Cave, Three Natural Bridges, and Qingkou Tiankeng, which can be called "Three Spectacular Karst Landforms in southern China".

Zhang et al. (2012) suggests that the development of a large karst cave system requires ① pure, thick and continuously distributed carbonate rocks, ② rising units of neotectonics, ③ a concentrated supply of abundant rainfall, surface runoff and allogenic water, ④ lowlands for drainage, and ⑤ slope zones and other geological conditions.

Xia et al. (2010) suggested that a complex structure pattern has gradually developed in the region since the Yanshanian movement. The planation surface in the western Hubei period was formed at the end of the Paleogene after long-term erosion and planation, with gentle and low peak clusters and shallow sags, reflecting karsts at an old age. The crust was strongly uplifted during episode B of the Himalayan movement, and the planation surface in the western Hubei period disintegrated in the Paleogene. The crust was stable for a long time in the Neogene, and the planation surface in the Shanyuan period was developed. At this time, a warm and humid climate enabled strong karstification, developing red weathering crust, which was a sign of late planation. Meanwhile, the gravel at the top of Jishijian Mountain indicates that a large river was formed at that time, with strong transportation capacity, representing an important role of normal erosion in the formation of the planation surface at this stage. The peneplain was formed in the late Pliocene, while karsts were characterized by high peaks, deep sags, abundant wave-rift valleys and karst caves, which was the geomorphic landscape of karst in adulthood. From the Late Pliocene to the Early Quaternary, the mountains created by episode C of the Himalayan movement rose strongly, and the planation surface in the Shanyuan stage disintegrated, entering a short structure-stable period. Under a warm and humid climate, strong karstification formed new sinkholes, funnels, sags, etc. At the same time, the migrated river gravel in Jiaoshiba Guanyin cave and large surface rivers and many underground rivers on the current ground show that lateral inhibition was responsible for the denudation surface of this level. However, no wide and gentle planation surface was formed in this area, where only one denudation surface was formed in a short period. The landform of this denudation surface was mainly karst peak-cluster mountains and valleys, where typical peak-cluster depressions were relatively rare and formed before 2 Ma. The neotectonics have become more intense since the Quaternary, where crusts rose sharply and rivers in this region cut down rapidly. Under these conditions, the Wujiang River and its tributaries formed the magnificent Wujiang Canyon, Furong River Canyon, etc., which was called the Wujiang period. Three short pauses with unequal times occurred during this period, forming three-level terraces in the wide valley. They caused multiple variations in the direction of groundwater, developed caves with three layers, and finally formed natural bridges on the surface and cave systems with complex underground morphology.

5 Field Itinerary

5.1 *DAY 1*

Stop 1.1 Guanyinqiao Outcrop

The Guanyinqiao outcrop is located next to Anwen town, Qijiang District, Chongqing, with a length of 62.6 m (Fig. 13). The Upper Ordovician Linxiang Formation (O_3l), Wufeng Formation (O_3l), Guanyinqiao Member and Long-1 Member are exposed at the Guanyinqiao outcrop.

The Linxiang Formation is a medium-sized to thick massive limestone with a gross thickness of 2–3 m, which has a clear boundary with the overlying Wufeng Formation (Fig. 14).

The middle and lower parts of the Wufeng Formation are black carbonaceous shale (3.5 m thick, thin single layer). The shear joints perpendicular to the bedding plane are

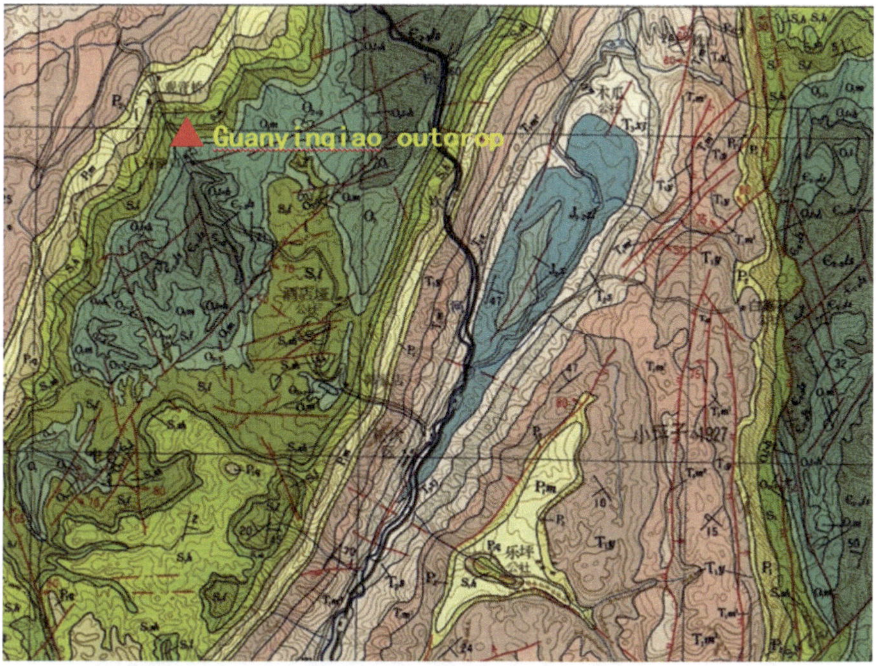

Fig. 13 Location of the Guanyinqiao outcrop in Qijiang district

Fig. 14 Boundary between the Linxiang Formation and Wufeng Formation at the Guanyinqiao outcrop

arranged parallel, with joint widths <0.2 cm, which are commonly filled by calcite and mudstone. The upper part of the Wufeng Formation is limestone (commonly known as the Guanyinqiao Member), with a thickness <1 m. It conformably contacts the overlying Longmaxi Formation (S_1l) (Fig. 15).

The Longmaxi Formation (S_1l) in this area is well developed. The silty shale is broken, and what is below it is gray black to black carbonaceous shale, carbonaceous mudstone, and argillaceous siltstone interbedded with calcareous mudstone. Graptolites occur in large numbers, and the organic matter content is high. The upper part of the Longmaxi Formation is mainly composed of silty shale and carbonaceous shale. The joint in this part has a high density compared with the lower part, which is more prone to breakage. The weathering products are usually sawdust-like (Fig. 16), and spherical weathering can be observed due to differential weathering (Fig. 17).

Fig. 15 Guanyinqiao member at the Guanyinqiao outcrop

Fig. 16 Joint growth in the Upper Longmaxi Formation

Fig. 17 Spherical weathering in the Longmaxi shale

5.2 *DAY 2*

Stop 2.1 Bayu Outcrop

The Bayu outcrop is located beside a reservoir at Shaba village beside S207 road, Yuxi town, Daozhen County, Guizhou Province. The Daozhen area in northern Guizhou Province is situated at the Wuling-Xiangexi folded and faulted zone in the Middle-Upper Yangtze block (Fig. 18), which is bounded by Xuefeng Uplift in the east, Yunnan Guizhou Guangxi Uplift in the southwest, and shale gas production construction in the southeast edge of Sichuan Basin in the northwest. The outcrop is distributed along the road with a length of approximately 400 m.

The Jian-1 Member (O_3j), Wufeng Formation (O_3w) and Long-1 Member (S_1l) are exposed at the Bayu outcrop, which can be divided into nearly 50 layers (Fig. 19). The lithology of the Jiancaogou Formation is mainly gray to dark gray eyeball-shaped marl. The Wufeng Formation is 7.23 m thick and is mainly composed of black argillaceous siliceous shale and siliceous shale, and no Guanyinqiao Member is found at the top. The Long-1 Member is 99.85 m thick and consists of black to gray black siliceous shale, gray black argillaceous siliceous shale, gray silty shale and gray black argillaceous shale.

Nine sublayers can be divided from the Wufeng Formation to the Longmaxi Formation at the Bayu outcrop, which are sublayer ① to sublayer ⑨, based on lithological variation, color, TOC and the graptolite zone (Figs. 20 and 21). The sublayer division scheme matches that of shale-gas reservoirs in the southeastern Sichuan Basin. All 9 sublayers can be observed in the dry season, and only sublayers ⑤-⑨ can be observed in the high-flow period.

The Nos. 1–6 layers at the outcrop correspond to the Wufeng shale (sublayer ①), with a gross thickness of 7.23 m. The sublayer contacts conformably with the underlying Jiancaogou nodular limestone. A thin argillaceous siliceous shale is developed at the bottom, which transitions upward to siliceous shale and is locally weathered to brown siliceous shale. There is no shell-bearing marl in the Guanyinqiao

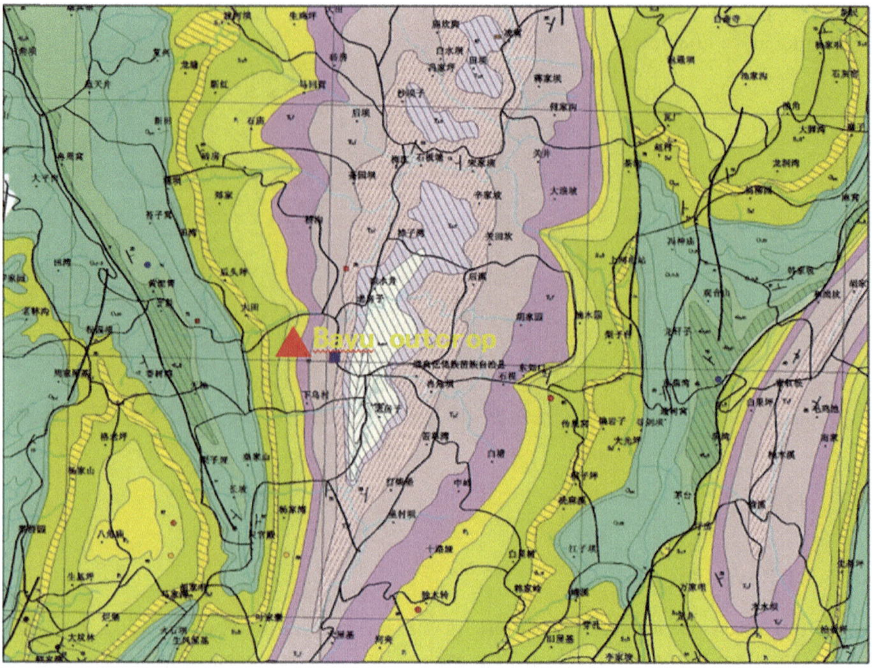

Fig. 18 Regional structure and geographical location of the Bayu outcrop

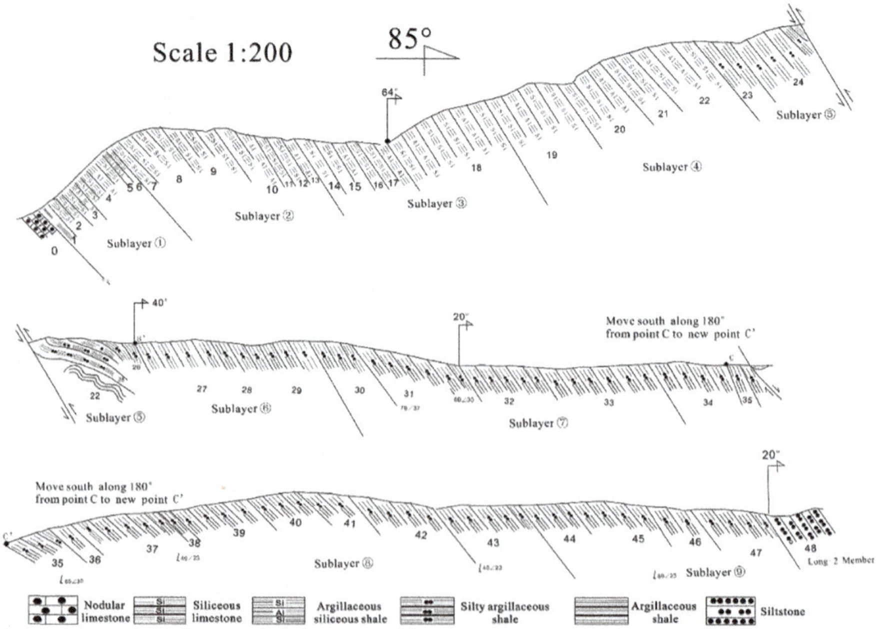

Fig. 19 Measured Wufeng-Longmaxi sequences at the Bayu outcrop

Fig. 20 Tops and bottoms of sublayers ①–⑦ at the Bayu outcrop

Fig. 21 Boundary between the Long-1 Member and Long-2 Member at the Bayu outcrop

Member. Siliceous shale is massive interbedded with gray thin bentonite (1 cm) due to intensive cleavage and weathering. A large number of graptolites are developed in shale, including *Dicellogaraptus complexus*, *Dicellograptus Hopkinson, Smplexograptus, Orthograptus, Amplexograptus latus, Rectograptus* sp., *Paraorthograptus pacificus* and *Diplograptus*. WF2–WF3 graptolite zones can be observed from bottom to top in sublayer ①, where no WF1 and WF4 graptolite zones can be found. *Dicellograptus* and *Dicellogaraptus complexus* occur at the top of sublayer ②, while *Paraorthograptus pacificus* can be found at sublayer ④.

The Nos. 7–13 layers at the bottom of the Longmaxi Formation are sublayers ②–③, with thicknesses of 12.07 m. They are mainly black carbonaceous shale interbedded with siliceous shale. The bottom is in conformable contact with the Wufeng siliceous shale and gradually transitions upward to argillaceous siliceous shale. A minor group of discontinuous silty laminae and calcite laminae can be found, with high-angle network-like calcite veins. Abundant graptolites can be found in long and fine shapes, e.g., *Glyptograptus* and *Agetograptus* sp. in sublayer ②, *Primar Akidograptus, Tip-shaped climacograptus, Normalo graptolite, Hamatograptus* sp., *Coronograptus* sp. and *Pribylograptus acuminatus* in sublayer ③. Sublayers ②–③ and the lower part of sublayer ④ are the LM1 (*Normalograptus persculptus*)-LM4 (*Cystograptus vesiculosus*) graptolite zone.

The Nos. 14–18 layers are sublayer ④ of the Longmaxi Formation, with a thickness of 12.69 m, and the bottom contacts conformably with the underlying carbonaceous shale. The layer is mainly composed of black carbonaceous shale interbedded with siliceous carbonaceous shale, and network-like calcite veins are developed in siliceous carbonaceous shale. Graptolites are in high density, mostly in long and fine shapes, e.g., *Coronograptus gregarius, Pernerograptus revolutus kurek, Demirastrites, Orthograptus, Pristiograptus* and *Glyptograptus tamariscus* linearis perner. The upper part of this blayer is the LM5 (*Coronograptus cyphus*) graptolite zone.

The Nos. 19–22 layers are sublayer ⑤ at the Long-1 Member, with a thickness of 13.93 m. It is dominated by gray black carbonaceous silty mudstone interbedded with black shale, which is a typical reverse cycle deposition order. The silty content increases upward with a small group of discontinuous millimeter-scale silty laminations and spherical weathering. The graptolites on cores are mainly *Coronograptus gregarius, Climacograptus, Glyptograptus* and *Orthograptus*. The upper part is pristiograptus, which is the LM6 graptolite zone.

The Nos. 25–29 layers are the sublayer ⑥ at the Longmaxi Formation, with a thickness of 11.06 m. It is mainly dark gray laminated silty shale interbedded with thin argillaceous shale, which conformably contacts the sublayer ⑤. High-density millimeter-scale silt, pyrite and calcite laminae and millimeter-scale pyrite laminae can be observed with lamina thicknesses of 2 cm and spacings between laminae of 1–5 mm, respectively. Jagged pastrites or janograptus arcuatus, *Demirastrites* and *Amplexograptus* can be found on cores, which is also the LM6 graptolite zone.

The Nos. 30–32 layers are sublayer ⑦ at the Longmaxi Formation, with a thickness of 9.79 m. It is mainly dark gray massive shale with silty shale. Low-density laminae are composed of silt, pyrite and calcite, with local calcite laminae and bedding nodular pyrite. The cross-section graptolites include rastrites, pristiograptus, *Pseudorthograptus* sp., and *Monograptus*, which is the LM7 graptolite zone.

The Nos. 33–39 layers are the ⑧ at the Longmaxi Formation, with a thickness of 16.48 m. Deep gray to gray black medium-sized to thin argillaceous shale interbedded with medium-sized silty shale are developed in this layer, with a small group of bedding nodular pyrite. Rastrites, Orthograptus, Amplexograptus, *Climacograptus*? sp., *Torquigraptus* sp. and a small amount of Monograptus are found in this layer, which is the LM7 graptolite zone.

The Nos. 40–47 layers are sublayer ⑨ at the Long-1 Member, with a thickness of 16.61 m. It is mainly deep gray to gray black medium-sized to thin silty shale and black argillaceous shale. Climacograptus, suspected pristiograptus and monograptus can be observed on cores, which is the LM7 graptolite zone. It is overlain by gray siltstone at the Long-2 Member.

The high-quality shale at sublayers ①–④ of the Bayu outcrop is 2.88% in TOC and 26.1 m in thickness, which is similar to that of the Pengye 1 well in Southeast Sichuan Bain and is lower than that of the Longye 1 well and Jiaoye 1 well. The TOC in sublayers ⑤–⑨ is <1.0% (Fig. 22). The bitumen reflectance of Longmaxi shale at Bayu outcrop is 2.13–2.28%, with the equivalent vitrinite reflectance of 1.73–1.83% (avg. 1.78%), which is lower than that in Longye 1 well and Pengye 1 well.

The Wufeng-Longmaxi shale in the Bayu outcrop is dominated by quartz and clay minerals, with carbonate minerals of secondary importance <20% (Fig. 22).

Stop 2.2 Huangying Outcrop

The Huangying outcrop is located beside S203 Road, Jianshe West Road, Shuanghe village, Huangying Township. It is structurally located on the south wing of the Wulong syncline (Fig. 23). The Jurassic strata are exposed at the core of the Wulong syncline, and the Triassic-Cambrian strata are successively exposed at the south wing. The Huangying outcrop clearly shows the Wufeng Formation and the lower part of the Long-1 Member.

The measured thickness of the Huangying outcrop is 60 m, with a true thickness of 32.7 m. The strata are the Middle Ordovician Baota Formation (no exposed bottom), Upper Ordovician Linxiang Formation, Wufeng Formation, Guanyinqiao Member, and lower part of the Lower Silurian Long-1 Member from bottom to top (Fig. 24). The Wufeng black shale is in parallel unconformity contact with the Linxiang Formation, where gray and white bentonite with a thickness of 1 cm occurs at the contact (Fig. 25a). The Longmaxi Formation and Wufeng Formation were successively deposited with conformable contacts. The boundary is marked by dark gray shell argillaceous limestone (Guanyinqiao Member) with a thickness of 0.32 m at the top of the Wufeng Formation (Fig. 25b). Obvious bubbling can be observed after hydrochloric acid treatment.

The Wufeng Formation is a black siliceous shale deposited in the Katian-Hernantian deep-water shelf, with bedding, joints and high-angle fractures, which is correlated with sublayer ① in wells (Figs. 26 and 27). Dicellograptus complanatus lapworth can be observed. The graptolite content is high, generally 15–40%, and increases to the middle and upper parts. This graptolite zone is defined as WF2–3.

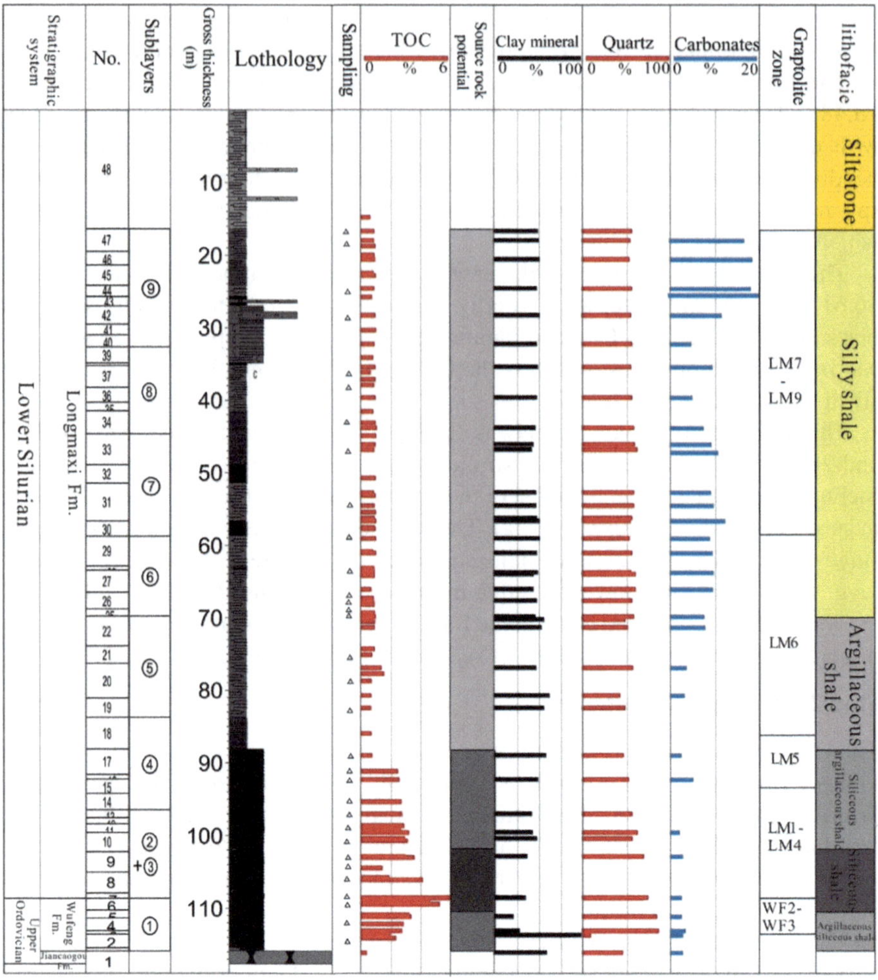

Fig. 22 Mineral variation at the Wufeng to Long-1 Member at the Bayu outcrop

Hirnantia fossils can be seen in the Guanyinqiao Member at the top, whose graptolite zone is defined as WF4-LM1.

The lower part of the Long-1 Member was deposited at the Rhuddanian deep-water shelf. It is gray black siliceous shale in sublayers ②–⑤ (Figs. 26 and 27). The thickness of sublayer ② at the bottom is 2.3 m, where persculptograptus persculptus can be observed. Graptolite is extremely abundant, accounting for 25-75% of the plane porosity in cores. The graptolite zone is defined as LM1-2. This indicates that graptolites began to grow and accelerated, while a large number of other organisms died out after the short glaciation in the Hernantian at the end of the Ordovician.

Sublayer ③, mainly black siliceous shale and black shale, was deposited at the deep-water shelf during the early-middle transgression in Ludan, with a thickness

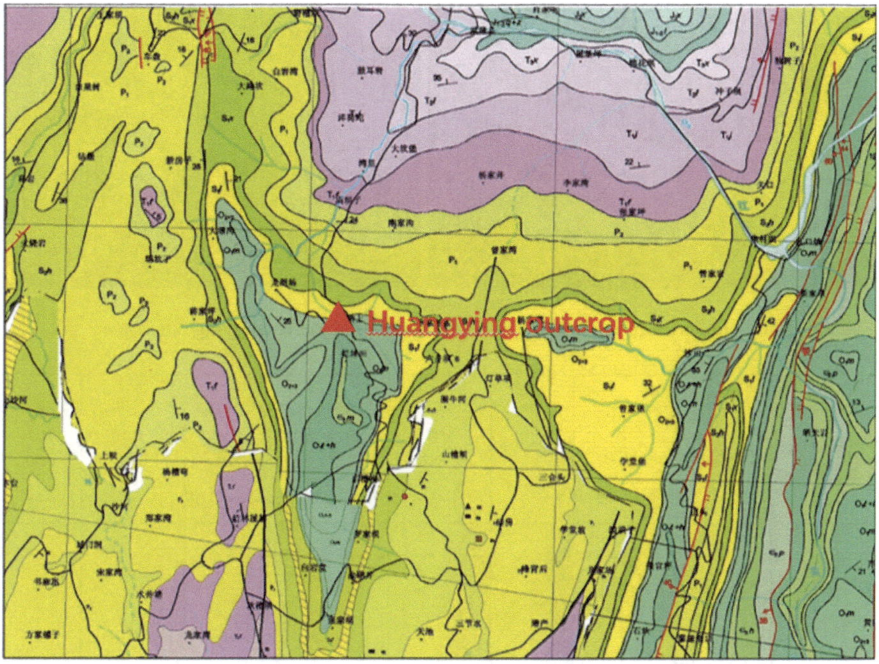

Fig. 23 Geological map of the Wulong syncline and location of the Huangying outcrop

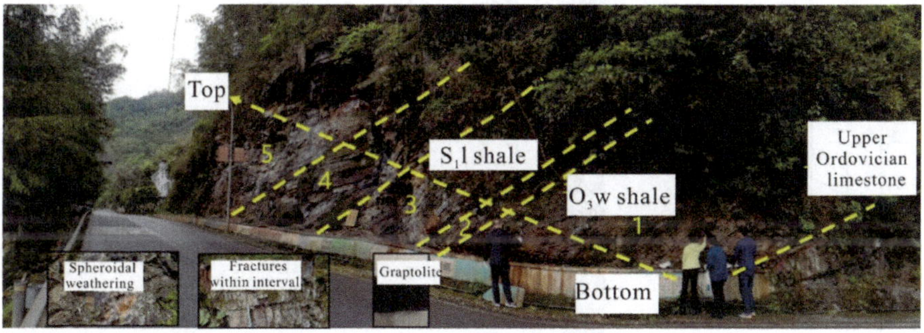

Fig. 24 Observed Wufeng-Longmaxi outcrop at Huangying town

of 11.4 m. Multidecollement, lamellation and joints can be found, and a microfault can be observed in the middle, with a fault distance <1 m. Three groups of high-angle fractures can be observed, which are half-filled or filled by calcite and pyrite. Persculptograptus persculptus and Akidograptus ascensus davies can be found, with graptolite contents of 10–40%. This graptolite zone is LM3–5.

Sublayer ④ was deposited on a deep-water shelf during the later transgression in Ludan, with a thickness of 5.8 m, primarily black shale and black carbonaceous shale. Graptolites, e.g., Normalo- graptus angustus and Clinoclimacograptus, are

Fig. 25 Two boundaries of the Wufeng–Long-1 Member at the Huangying outcrop

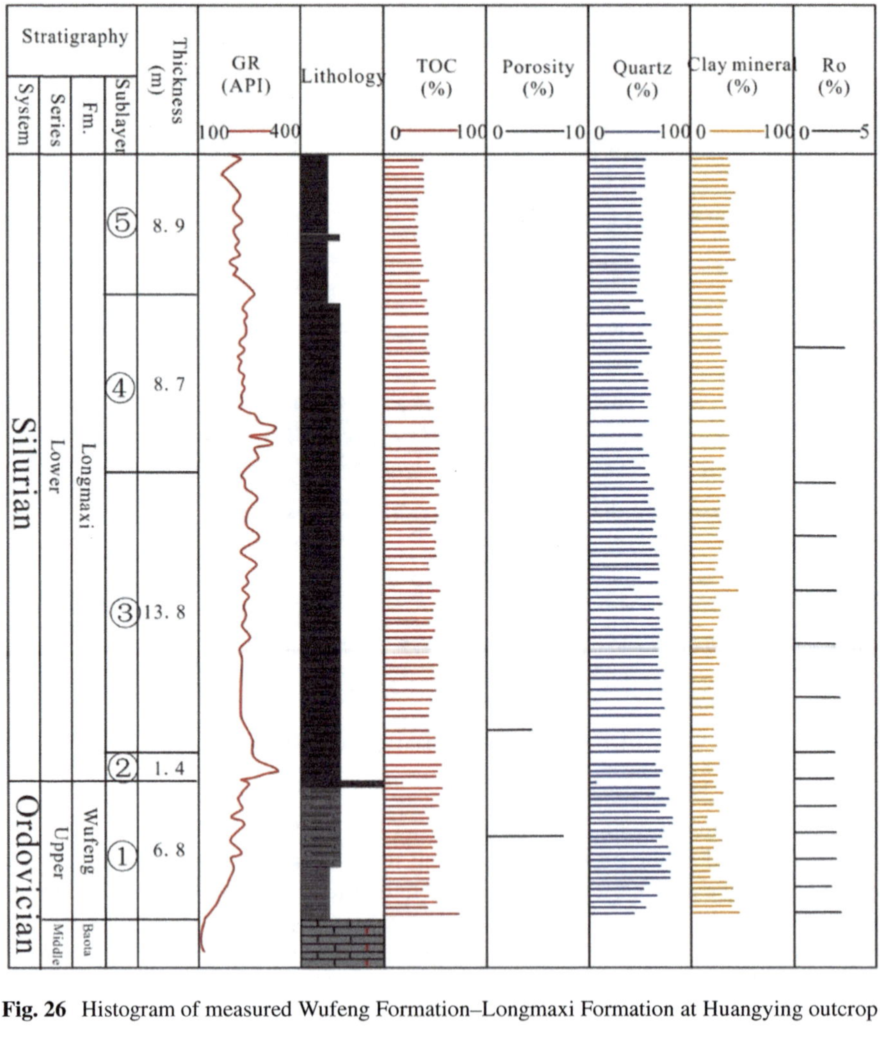

Fig. 26 Histogram of measured Wufeng Formation–Longmaxi Formation at Huangying outcrop

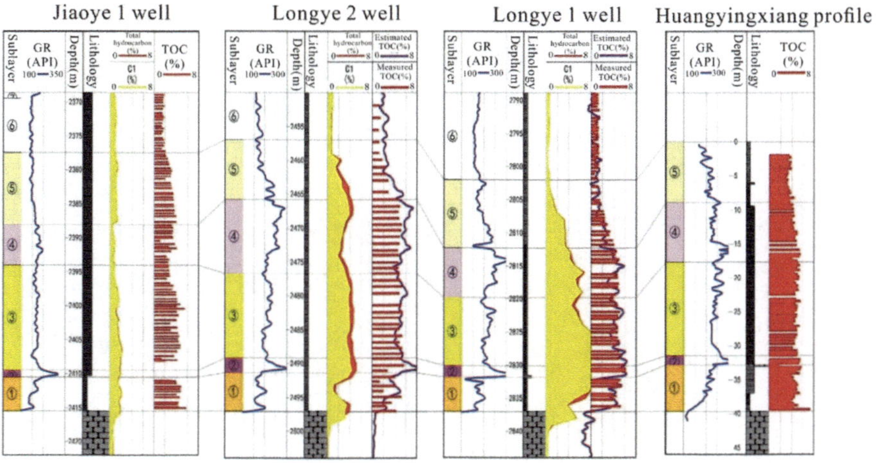

Fig. 27 Crosswell section showing Wufeng Formation–Longmaxi Formation in Chongqing

widely developed. Graptolite individuals are larger than those in the lower part. This graptolite zone is named LM5.

Sublayer ⑤, lamellated black shale, was deposited at the transition zone from deep-water shelf to semideep-water shelf during the high stand—early regression of the ludanian, with a thickness of 6.8 m. Large Normagraptus angustus, Clinoclimacograptus, etc., can be found, with a content of 10–30%, which is defined as LM5.

The high-quality shale at the Huangying outcrop is 32.7 m thick. TOC is in the range of 1.74–7.38%, with an average value of 4.55%. Ro is approximately 2.37–3.06%, with an average value of 2.60%. The effective porosity is 4.41–7.49%, averaging 5.95%. The quartz content is 6.10–82.00%, with an average value of 61.02%. The clay content is 8.20–45.90%, with an average value of 28.70%. In summary, the high-quality shale in the Wulong syncline has a large thickness, high TOC, large porosity, moderate thermal maturity and high hydrocarbon generation intensity, which is a favorable normal-pressure shale gas exploration target in the residual syncline outside the basin.

5.3 *DAY 3*

Stop 3.1 Tiansheng Outcrop

The Tiansheng outcrop is located beside the highway near Zibotang, Jielong Town, Wulong District, Chongqing, 32 km away from the Fuling Shale Gas Field (Fig. 28). It is structurally situated at the north wing of the Wulong syncline. Each sublayer has a clear boundary and key symbols, as well as abundant graptolites. At present,

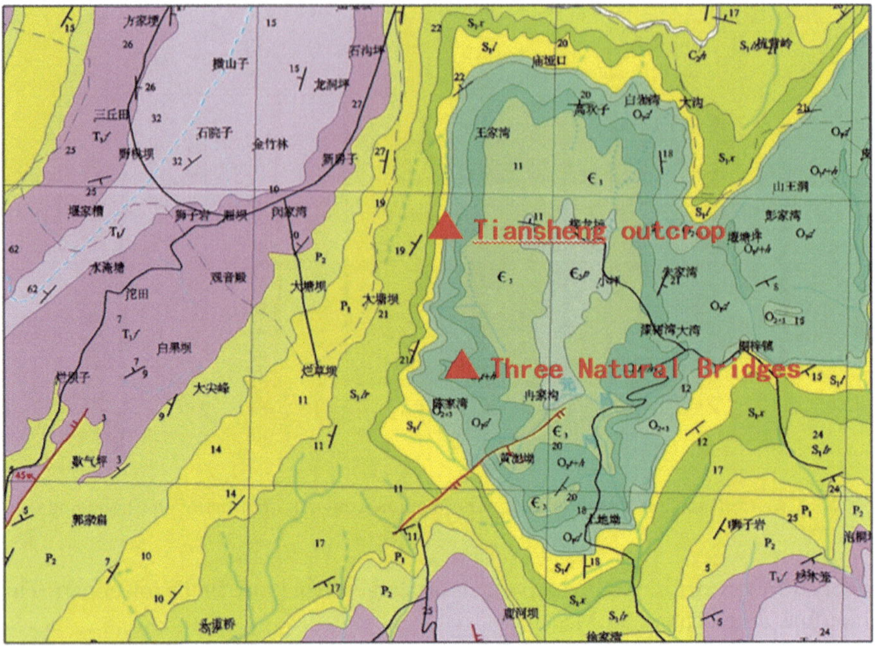

Fig. 28 Location of Tiansheng otcrop and three natural bridges

boundaries have been clearly marked, and general geological characteristics can be described (Fig. 29).

Fig. 29 Perspective view of the measured Wufeng Formation–Longmaxi Formation at the Tiansheng outcrop

The Upper Ordovician Wufeng Formation and Lower Silurian Long-1 Member in the Tiansheng outcrop can be divided into 16 sublayers from bottom to top (Table 4). A comparison between the outcrop and adjacent wells can calibrate the 16 sublayers with sublayers ①–④ in the area. The specific division is as follows.

Sublayer 0: Jiancaogou Formation, gray thick massive micritic and fine-grained dolomitic limestone, containing grained clasts and pyrite crystal aggregates of 1–3 mm. The dolomite content increases from bottom to top.

Sublayer ①: main section of the Wufeng Formation, including layers ①–⑧ at the outcrop, gray black medium-sized siltstone-bearing carbonaceous siliceous shale. An obvious lithologic boundary and conformable contact can be observed between gray black shale at the bottom of the ① sublayer and carbonate rock (dolomitic limestone) of the Jiancaogou Formation (Fig. 30).

Guanyin Member: layer ⑨ at the outcrop, at the top of the Wufeng Formation with a thickness of 0.75 m (Fig. 31), gray black medium-sized limestone with shells, calcareous shale, abundant Hirnantia fossils (Fig. 32), the division between the Wufeng Formation and the overlying strata.

Sublayers ②–③: at the lower section of the Long-1 Member, there is gray black massive to medium-sized siliceous shale, which can be divided into 7 layers at the outcrop from bottom to top. The bottom contacts conformably with the Guanyinqiao Member, and the top contacts conformably with the sublayer ④ (Fig. 33).

Layer ⑩: gray black medium-sized silty-bearing carbonaceous siliceous shale, 1.40 m.

Layer ⑪: gray black massive to medium-sized siliceous shale, 1.38 m.

Layer ⑫: gray black medium-sized siliceous silty shale, 1.42 m.

Layer ⑬: gray black medium-sized to laminar silty-bearing carbonaceous siliceous shale.

Layer ⑭: gray black thin argillaceous silty shale, 0.80 m.

Layer ⑮: gray black medium-sized silty-bearing siliceous mudstone, silty increasing from bottom to top, 3.73 m.

Layer ⑯: gray black silty-bearing carbonaceous siliceous shale, medium-sized and thin layers at the top with a thickness of 5.28 m and an accumulated thickness of 15.03 m.

Sublayers ④ and above: seriously weathered and covered, inferred gray black shale.

The Upper Ordovician Wufeng Formation and Lower Silurian Long-1 Member in the Tiansheng outcrop are dominated by quartz, with an average content of 67.58%. Clay minerals are of secondary importance, with an average content of 26.82%. The potassium feldspar, plagioclase, siderite and pyrite contents are all lower than 5%. Mineral contents vary greatly among different sections (Table 5). The Upper Ordovician Wufeng Formation has a higher quartz content, an average of 71.86%, than the Longmaxi Formation (62.78%).

The measured TOC at the Tiansheng outcrop is between 1.10 and 5.94%, with an average value of 4.14%. The TOC in the Wufeng Formation ranges from 3.26 to 5.20%, with an average value of 4.03%, while the TOC in the Long-1 Member ranges from 1.10–5.94%, with an average value of 4.28%. TOC gradually decreases

Table 4 Brief geological stratification of the Tiansheng outcrop

Stratigraphic system					Division			Lithology description
Erathem	System	Series	Formation	Code	No.	Depth from bottom (m)	Thickness (m)	
Paleozoic	Silurian	Lower	Longmaxi	S_1l	⑯	26.69	5.28	Gray black siliceous carbonaceous mudstone, medium-thin layer at the top, 1–2 mm brown silty horizontal lamina on core intersection
					⑮	21.41	3.73	Gray black medium-sized mudstone with silt and silica, silt content increasing from bottom to top, radish-like weathering products and abundant graptolites
					⑭	17.68	0.80	Grayish black lamellar silt- and silica-bearing shale
					⑬	16.88	1.02	Grayish black medium-sized to lamellar mud- and silt- bearing siliceous shale, weathering surface with obvious rust color, some spherical weathering, fine-columnar weathering products
					⑫	15.86	1.42	Gray black, medium-sized, muddy and silty siliceous argillaceous shale, no obvious hand blacking and angular weathering surface
					⑪	14.44	1.38	Grayish black thick medium-sized argillaceous silty siliceous shale, angular weathering surface, abundant graptolite fossils
					⑩	13.06	1.40	Grayish black medium-sized carbonaceous siliceous shale, angular weathering surface

(continued)

Table 4 (continued)

Stratigraphic system					Division			Lithology description
Erathem	System	Series	Formation	Code	No.	Depth from bottom (m)	Thickness (m)	
	Ordovician	Upper	Wufeng	O_3w	⑨	11.66	0.75	The lower part: gray black medium-sized carbonaceous siltstone, spherical weathering in local position. The upper part: gray black medium-sized shell limestone
					⑧	10.91	1.09	Gray black medium-sized carbonaceous siliceous shale, angular weathering surface
					⑦	9.82	0.42	Gray black lamellar carbonaceous siliceous shale
					⑥	9.40	1.32	Gray black medium-sized and lamellar carbonaceous siliceous shale
					⑤	8.08	0.72	Gray black lamellar carbonaceous siliceous shale interbedded with gray and gray black bentonite
					①	7.36	2.00	Grayish black medium-sized and lamellar carbonaceous siliceous shale interbedded with bentonite, and the color of bentonite varying from gray to black from bottom to top
					③	5.36	0.71	Grayish black medium-sized carbonaceous siliceous shale with grayish white bentonite

(continued)

Table 4 (continued)

Stratigraphic system					Division			Lithology description
Erathem	System	Series	Formation	Code	No.	Depth from bottom (m)	Thickness (m)	
					②	4.65	1.10	Gray black medium-sized carbonaceous siliceous rocks interbedded with gray gray white bentonite
					①	3.55	1.55	Gray black massive carbonaceous siliceous limestone, obvious spherical weathering, broken weathering products, no obvious hand blacking
			Jiancaogou	O_3j	0	2.00	2.00	Gray massive calcareous dolomite with fine-grained structure, containing debris and pyrite crystal aggregates of 1–3 mm, limestone decreasing and dolomite increasing from bottom to top

Fig. 30 Contact between sublayer ① and the Jiancaogou Formation

Fig. 31 Contact between the Guanyinqiao Member and its upper and lower layers

from bottom to top (Fig. 34). The organic matter types are dominated by amorphous sapropelic, sapropelic fragments, collinite and vitrodetrinite. The calculated kerogen type index (TI) is between 11 and 40, indicating type II_2 kerogen. The vitrinite reflectance determined by measured bitumen reflectance (Rb) is approximately 2% or higher, which is at the over thermal maturity stage.

Fig. 32 Hirnantia fossils in limestone with shells in the Guanyinqiao Member

Fig. 33 Images of sublayers ②–③ as well as the boundary at the top and bottom

Table 5 Mineral components at the Tiansheng outcrop

Formation	Average mineral content/%					
	Clay mineral	Quartz	K feldspar	Plagioclase	Siderite	Pyrite
Long-1 Member	30.16	62.78	2.55	4.13		3.10
Wufeng	23.86	71.86	1.05	2.54	0.80	4.27

Stop 3.2 Three Natural Bridges

Three natural bridges are located 20 km southeast of Wulong District, Chongqing. It is an AAAAA Scenic Spot in Chongqing. It is a typical karst topography and a rare geological wonder eco-tourism in China (Fig. 35). In terms of geological background,

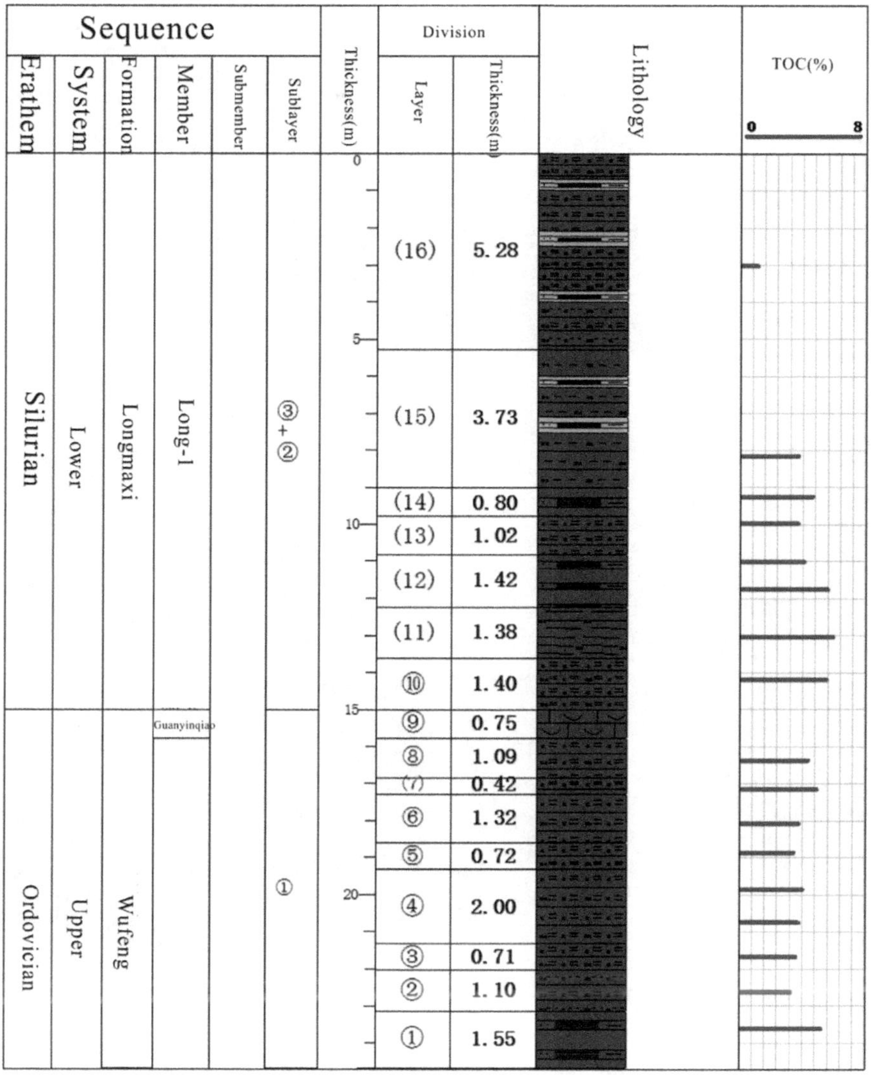

Fig. 34 Variation in measured TOC at the Tiansheng outcrop

the Three Natural Bridges are situated between the Zhongliangshan anticline and Wulong syncline, where strata exposure is controlled by these two structures, e.g., successively from old to new from northwest to southeast: the Silurian, Permian, Triassic and Jurassic. The strata are inclined to the southeast with dip angles of 10°–35°. Three natural bridges are developed in the Lower Triassic Feixianguan and Jialingjiang carbonate rocks.

Fig. 35 Images of some scenic spots of three natural bridges (from Baidu.com)

The karst system in this area is characterized by abundant colorful, rare and typical valleys, which are mainly composed of canyons, undercurrents, natural bridges, sinkholes, caves, karst depressions, karst springs, eroded rock peaks, etc. Three large-scale and great bridges, e.g., the Tianlong bridge, Qinglong Bridge and Heilong bridge, are the most attractive and largest bridge groups in the world.

Karst topography is developed by water dissolution on soluble rocks (mostly limestone). It is widely developed in China, while southern Chongqing is one of the largest karst clusters.

Karst caves are underground spaces derived from the dissolution and erosion of undercurrents along various fractures on soluble rocks. Sinkholes or even canyons can be developed when expanded karst caves cannot support overburden. Natural bridges can be developed in some narrow zone, where collapsed overlying rocks are supported by sinkholes and canyons, such as bridges.

The sinkhole development in the Three-Natural-Bridges can be explained by the following: several sets of Carboniferous to Middle Triassic carbonate strata were originally deposited on the Wufeng-Longmaxi shale in southern Chongqing and northern Guizhou Province. Intensive compression in this area is due to tectonic movement, especially the Yanshan movement and Himalayan movement uplifted and severely eroded strata, where Permian to Middle Triassic carbonate strata only exist in the syncline and other lower parts. These residual strata were also cut by a large number of faults and fractures of different scales. The long period and large volume of rain in southern China caused surface water (fresh water) to flow along faults and fractures, eroding carbonate rocks, especially limestone, on both sides for tens of millions of years. Dissolution can further occur along soluble rock strata, gradually developing undercurrent (the so-called underground river), and the created space is the karst cave. As mentioned above, sinkholes and canyons can eventually form when karst caves develop gradually. Of course, they can be further developed and may eventually evolve into clusters and stone forests until they die out.

The primary attractions in three natural bridges are as follows.

Tianlong Bridge: the most lifelike artificial bridge, and the bridge opening is a regular rectangle. The bridge is 235 m in height, 150 m in thickness, and 147 in width, with an average arch and span of 96 and 34 m, respectively.

Tianfu official residence: it is an ancient official post station that was first built in the Tang Dynasty (619 A.D.). The current one was rebuilt in a later period.

Qinglong Bridge: This bridge is the most magnificent and dangerous of the three bridges, as it is similar to an open gate. Rainbow can be observed through the gate after rain. The bridge is 281 m in height, 168 m in thickness, and 124 m in width, with an average arch and span of 103 m and 31 m, making it the most beautiful natural bridge in the world. Cliffs are on both sides of the cave, where birds wander frequently. Cloud and fog occurrence makes it like a wonderland.

Heilong Bridge: it is named after the deep and dark inside the cave, whose top is similar to a black dragon hiding here. The bridge is 223 m in height, 107 m in thickness, and 193 m in width, with an average arch and span of 1116 and 28 m, respectively. The cave is high, narrow and dark. Four springs can be found in the cave with different shapes and postures.

5.4 DAY 4

The Fuling Gas Field is near Jiaoshiba, and our visiting schedules are as follows.

Stop 4.1 Jiaoye 1 Well Platform and Fuling Gas Field Achievements

Jiaoye 1 well platform and Fuling Gas Field achievements: a brief understanding of Sinopec shale gas exploration history, Jiaoye 1 well and its significance, Fuling Gas Field development and primary achievements.

Stop 4.2 Gathering and Transportation Station

Gathering and transportation station: visiting a station near Jiaoshiba to briefly learn about technologies and current gas output.

6 Important Notes and Personal Safety

6.1 Important Notes

ISC, the sponsoring organizations or field trip leaders and their employers maintain insurance covering illness or injury for individuals.

(1) It is strictly forbidden to conduct a survey alone.
(2) For field trips, transportation with the trip phone numbers and email addresses of hotels and trip leaders prior to the event date. Personal communication tools should be unblocked to keep in touch with others during field operation.
(3) Cooperation is encouraged to ensure safety when passing cliffs.
(4) Proper clothing and supplies are needed for the outdoors (umbrella, hat, sturdy footwear and backpack).
(5) Depending on location, temperatures can range from 25 to 40 °C with or without rain. Participants are advised to check local weather forecasts for the latest updates.
(6) Insect repellent and sun protection (sunscreen and hats) are recommended.
(7) Pay attention to possible snakes, insects, animals, plants, etc., when passing through areas covered by vegetation, e.g., forest and bush, etc. Try to keep away as much as possible. In case of emergency in the forest, please remain calm and call for help.
(8) It is advised to shut off mobile phones, radios, walkie-talkies and other electrical appliances when passing through high voltage transmission lines and towers.
(9) Emergency measures for thunderstorms in field operations: ① have rain gear available and ready before going up the mountain; ② do not stay at places prone to flash floods or collapse as well areas prone to lightning, e.g., at the top of the mountain, under large trees, near ditches, and under unstable scarps; ③ return to designated stops after thunderstorms, or contact team leaders to report return times and routes; and ④ take care to no slip in rain or after rain to ensure safety.
(10) If you use a prescription drug, please inform a team leader of the primary function, usage and dosage so that it can be administered properly in the event of an emergency.
(11) Emergency plans are necessary for any accidents.

6.2 *Guidance for COVID-19 Prevention*

(1) Know how COVID-19 spreads and take steps to protect yourself and others.
(2) Wash your hands frequently and maintain hygiene, avoid touching your mouth, eyes and nose. Practice respiratory hygiene, cover coughs and sneezes.
(3) Maintain 6 feet of distance (approximately 2 arm lengths) from other people and wear a mask.
(4) Monitor your health daily. Medical treatment should be sought and reported to the epidemic prevention group when the temperature is > 37.3 °C.
(5) Report to the field epidemic prevention group when personnel or residents had fever, cough, shortness of breath, or other symptoms of COVID-19.

Acknowledgements The author would like to give sincerest thanks to Professor Huang Rong'an from the Chongqing Institute of Geology and Mineral Resources, Professor Bao Shujing from the China Geological Survey, Professor Zhijinchuan from the China University of Geosciences (Beijing), senior engineer Wei Xiangfeng and senior engineer Xia Weishu from the Sinopec Exploration Company for their great support in making this trip possible.

References

Chen X, Rong JY (1999) From biostratigraphy to tectonics with Ordovician and Silurian of South China as example. Geoscience 13(4):385–389
Chen X, Rong JY, Li Y et al (2004a) Facies patterns and geography of the Yangtze region, South China, through the Ordovician and Silurian transition. Palaeogeogr Palaeoclimatol Palaeoecol 204:353–372
Chen X, Fan JX, Melchin MJ, et al (2004b) The process and mechanism of the extinction and recovery of graptolites in South China at the end of Ordovician. In: Jiayu R, Zongjie F (eds) Mass extinction and its subsequent recovery: evidence from the Paleozoic in South China. China University of science and Technology Press, Hefei, pp 9–54, 1037–1038
Chen X, Melchin MJ, Sheets HD et al (2005) Patterns and processes of latest Ordovician graptolite extinction and recovery based on data from South China. J Paleontol 79:842–861
Chen X, Rong JY, Fan JX et al (2006) The Global boundary Stratotype Section and Point (GSSP) for the base of the Hirnantian stage (the uppermost of the Ordovician system). Episodes 29:183–196
Chen X, Zhang YD, Fan JX et al (2012) The process of Guangxi movement: evidence from biological and lithofacies. Chin Sci: Geosci 42(11):1617–1626
Feng ZZ, Peng Yong M, Jin ZK, et al. Lithofacies paleogeography of Cambrian and Ordovician in China. Geological Publishing House, Beijing, pp 1–276
Guo ZW, Deng KL, Han YH et al (1996) Formation and evolution of Sichuan Basin. The Geological Publishing House, Beijing, pp 163–175
Li YF, Fan TL, Gao ZQ et al (2012) Sequence stratigraphy of Silurian black shale in Southeast Chongqing. Nat Gas Geosci 23(2):299–306
Liang DG, Guo TL, Chen JP et al (2009) Some progresses on studies of hydrocarbon generation and accumulation in marine sedimentary regions, Southern China (Part 3): controlling factors on the sedimentary facies and development of Palaeozoic marine source rocks. Marine Petrol Geol 14(2):1–19
Liu BJ, Xu XS. Lithofacies paleogeography Atlas of South China (Sinian–Triassic). Science Press, Beijing, pp 1–239

Ma YS, Chen HD, Wang GL (2009) Sequence stratigraphy and paleogeography of southern China. Science Press, Beijing, pp 254–280

Rong JY, Chen X, Wang Y et al (2011) The evolution of Qianzhong ancient land at the turn of Ordovician and Silurian: evidence and enlightenment. Chinese Sci: Geosci 4(10):1407–1415

Rong JY, Melchin MJ, Williams SH et al (2008) Report of the restudy of the defined global stratotype of the base of the Silurian system. Episodes 31:315–318

Sheehan PM (2001) The Late Ordovician mass extinction. Ann Rev Earth Planet Sci 29:331–364

Tang G. The "Three Spectacular Karst Landforms" in Wulong. The World and Chongqing, pp 50–57

Wang HZ (1985) Ancient geographic Atlas of China. The Geological Publishing House, Beijing

Xia KS, Yuan DX, Xie SY (2010) Preliminary study on morphology features of karst in the lower reaches of Wujiang River—a case in Wulong, Chongqing and the surrounding area. Carsologica Sinica 29(2):196–204

Xu KD, Ma L et al (2002) New progress and basic framework of tectonic in South China (the first series). Petroleum Industry Press, Beijing, pp 180–194

Zhang YH, Zhu DH (2012) Large Karst Caves distribution and development in China. J GuiLin Univ Technol 32(1):20–28

Zheng HR, Gao B, Peng YM et al (2013) Sedimentary evolution and shale gas exploration direction of Lower Silurian in Middle and Upper Yangtze. Acta Palaeogeographic Sinica 15(5):645–656

Mesozoic-Cenozoic Sedimentary Geology in the Tibetan Himalaya: Continental Collision, Palaeoceanography and Hyperthermal Events

Xiumian Hu, Xi Chen, Wei An, Jiangang Wang, Jingen Dai, Juan Li, Qun Liu, and Zhong Han

1 Geological Introduction

The Himalayan–Tibetan orogen is the youngest and most spectacular of all the continent–continent collisional belts on Earth, which occupies high-altitude Himalaya and Karakorum ranges in the south and the Tibetan plateau to the north (Yin and Harrison 2000). This orogenic system was largely created by the Indo-Asian collision over the past ~60 Ma (Hu et al. 2016; An et al. 2021). In the southern Tibet of

X. Hu (✉)
Nanjing University, Nanjing, China
e-mail: bobliu@pku.edu.cn

X. Chen · J. Dai
China University of Geosciences, Beijing, China
e-mail: xichen@cugb.edu.cn

J. Dai
e-mail: daijingen@cugb.edu.cn

W. An
Hefei University of Technology, Hefei, China
e-mail: anwei@hfut.edu.cn

J. Wang
Institute of Geology and Geophysics, CAS, Beijing, China
e-mail: wangjiangang@mail.iggcas.ac.cn

J. Li
Nanjing Institute of Geology and Paleontology, Nanjing, China

Q. Liu
Nanjing Xiaozhuang University, Nanjing, China

Z. Han
Chengdu University of Technology, Chengdu, China

X. Hu (ed.), *Field Trip Guidebook on Chinese Sedimentary Geology*,
https://doi.org/10.1007/978-981-99-6936-4_12

China, this orogen belt is east–west trending and extends more than 1000 km. From north to south, it can be divided into 5 tectonic units including the Gangdese Arc, Xigaze Forearc Basin, Yarlung Zangbo Suture Zone and Tethys Himalayas (Tibetan Himalayas). In this field trip, the geological characteristics of typical outcrops of different units will be investigated. Each of the unit experienced complicated history of evolution related to the subduction and closure of Neo-Tethys Ocean. Additionally, the paleoceanographic conditions of eastern Tethys Ocean are archived in the marine sedimentary successions and will be investigated including several hyperthermal events.

1.1 Gangdese Arc

The Gangdese magmatic arc refers to the Mesozoic–Palaeogene igneous rocks developed along the southern margin of the Lhasa block that were formed as a result of northward subduction of the Neo-Tethyan oceanic lithosphere. It is the middle part of the ~ 6,000-km-long Neo-Tethyan arc belt that extends from Kohistan-Ladakh, through southern Tibet and western Burma, to Sumatra in southeastern Asia (Figs. 1 and 2) (Zhang et al. 2019a, b).

Intrusive rocks of the Gangdese arc (Gangdese batholith) are mainly granodiorite and monzogranite, with subordinate gabbro, gabbro-diorite, diorite, tonalite and granite. Volcanic rocks spread across the Gangdese arc also show a wide range of composition, including basalt, andesite, dacites and rhyolite (e.g., Zhu et al. 2008; Kang et al. 2014). Pre-Cenozoic volcanic rocks had experienced green-schist facies metamorphism, and were intruded by later granitoids. Cenozoic volcanic rocks represented by the Linzizong volcanic successions are unmetamorphosed, only mildly

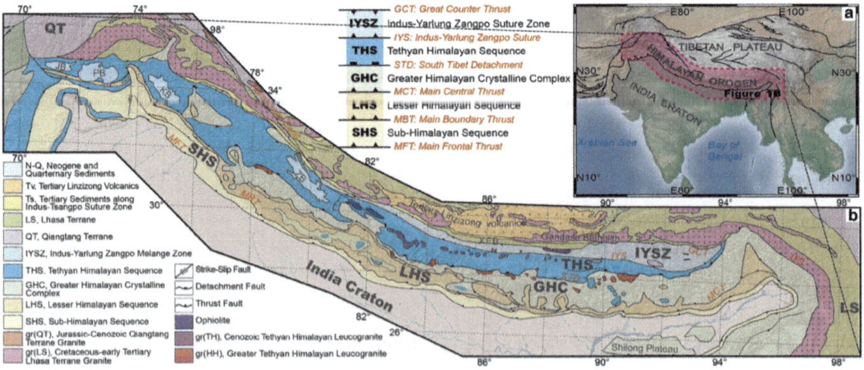

Fig. 1 **a** Tectonic and topographic map of the Tibetan Plateau, adapted from (Tapponnier et al. 2001) and (Yin 2006). The pink box indicates the simplified regional geologic map of the Himalayan Orogen (**b**), modified from Yin and Harrison (2000), Yin (2006), and Sciunnach and Garzanti (2012)

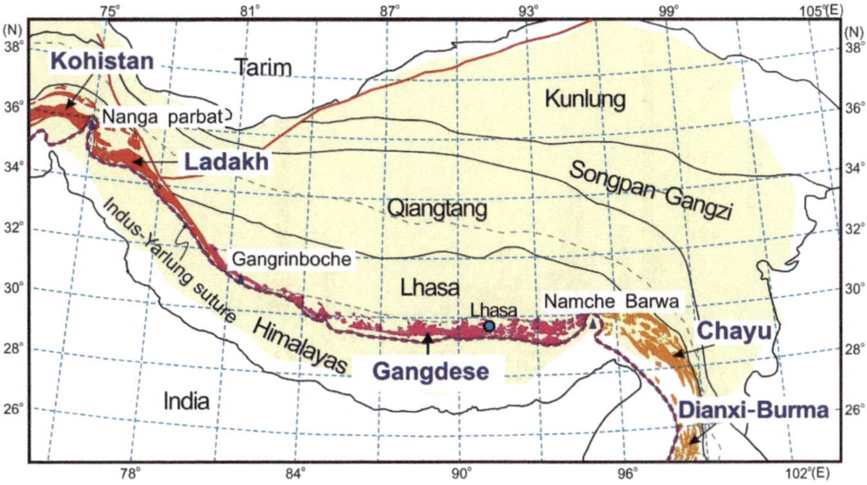

Fig. 2 Location and distribution of the Gangdese arc (Modified after Ji et al. 2009)

deformed, and overlain older rocks above a regional unconformity (Mo et al. 2008; Lee et al. 2009; Zhu et al. 2015). Sedimentary covers of the Gangdese arc are only patchily preserved, including Upper Jurassic–Lower Cretaceous carbonates and shallow-marine to nonmarine clastic rocks (Leier et al. 2007a, b; Wang et al. 2020).

Geochronological study indicated the Gangdese arc had a long-lived magmatic history from Early Mesozoic to Neogene (Fig. 3) (Ji et al. 2009; Hou et al. 2015). Magmatic activity in the Gangdese arc was started as early as Middle Triassic (~237 Ma, Wang et al. 2016a, b), although the Neo-Tethyan subduction zone was fully developed only since Early Cretaceous as indicated by tectonic and sedimentary evidences. Tectonic setting and petrogenesis of the pre-Cretaceous igneous rocks are still puzzled and controversial. Arc magmatism was mainly active during Cretaceous to Palaeogene, and shows two magmatic peaks at ~90 Ma and ~51 Ma, respectively (Ji et al. 2009). Postcollisional magmatic activity were represented by the Miocene potassic and ultrapotassic rocks, which commonly occur as isolated complexes and intruded into the Gangdese batholiths and volcano-sedimentary covers (Chung et al. 2003; Hou et al. 2004; Zhao et al. 2009).

Igneous rocks of the Gangdese arc are characterized by extremely depleted isotopic features, indicating magmatic derivation from a juvenile source (Ji et al. 2009; Zhu et al. 2011). Zircon crystals from these rocks are thereby characterized by very positive $\varepsilon_{Hf}(t)$ values (Ji et al. 2009; Hou et al. 2015), which make them highly recognizable during provenance study. Gangdese zircons have been widely used as sedimentary fingerprint to trace Asian active margin sediments in the foreland basin, remnant oceanic basin as well as synorogenic conglomerate, to constraint the timing of Indian-Asian continental collision, uplift and erosion of Gangdese-Himalayan orogenic belt and evolution of major drainages originated from the Tibetan plateau (e.g., Hu et al. 2015; Najman et al. 2017; Colleps et al. 2020).

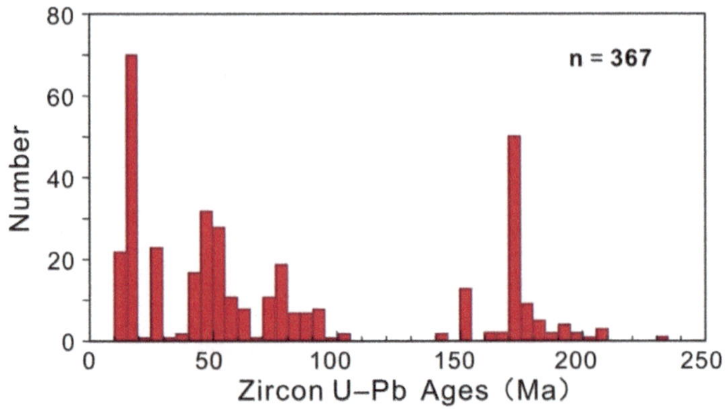

Fig. 3 A summary of published zircon U–Pb ages from the Gangdese arc

Structural analyses have shown that notable (~60%) crustal shortening had taken place on the Lhasa block during the Late Cretaceous (e.g., Murphy et al. 1997; Volkmer et al. 2007). As a result, the Gangdese belt might have reached elevations of up to 4 km in the Early Paleogene (Kapp et al. 2007), i.e., an Andean-type mountain has already been established stretching the Gangdese arc well before the India–Asian collision in Early Cenozoic (Ding et al. 2014). By contrast, crustal shortening of southern Asian margin after the India–Asian collision is limited. Recent rapid exhumation occurred in the Miocene, which might have related to the delamination of the subducting Indian slab or incise of the Yarlung-Zangbo (e.g., Harrison et al. 1992; Carrapa et al. 2014).

1.2 Xigaze Forearc Basin

The Xigaze forearc basin (XFB) is located to the south of the Gangdese arc in an east–west oriented belt extending over 500 km along the Yarlung Zangbo suture zone. With a width ranging from 50 to 22 km (Wang et al. 2012), the forearc basin is composed of flysch-dominated Xigaze Group and shallow marine succession including the Padana and Qubeiya formations and the overlying Palaeogene Cuojiangding Group (Fig. 4) (Liu et al. 1988; Wan et al. 2001).

The Xigaze Group is composed of the Chongdoi, Sangzugang and Ngamring Formations. The Chongdoi Formation comprises two members, the lower member of purplish-red radiolarian chert, siliceous mudstone interbedded with ophiolitic breccia and the upper of volcaniclastic turbidites, with numerous tuff layers throughout both members. The lower member is deposited directly upon the pillow basalts of the Xigaze ophiolite and its upper contact is conformable with the overlying Ngamring Formation (Girardeau et al. 1984; An et al. 2014; Wang et al. 2017a, b). The lower member contains a late Barremian–late Aptian radiolarian association (Ziabrev et al.

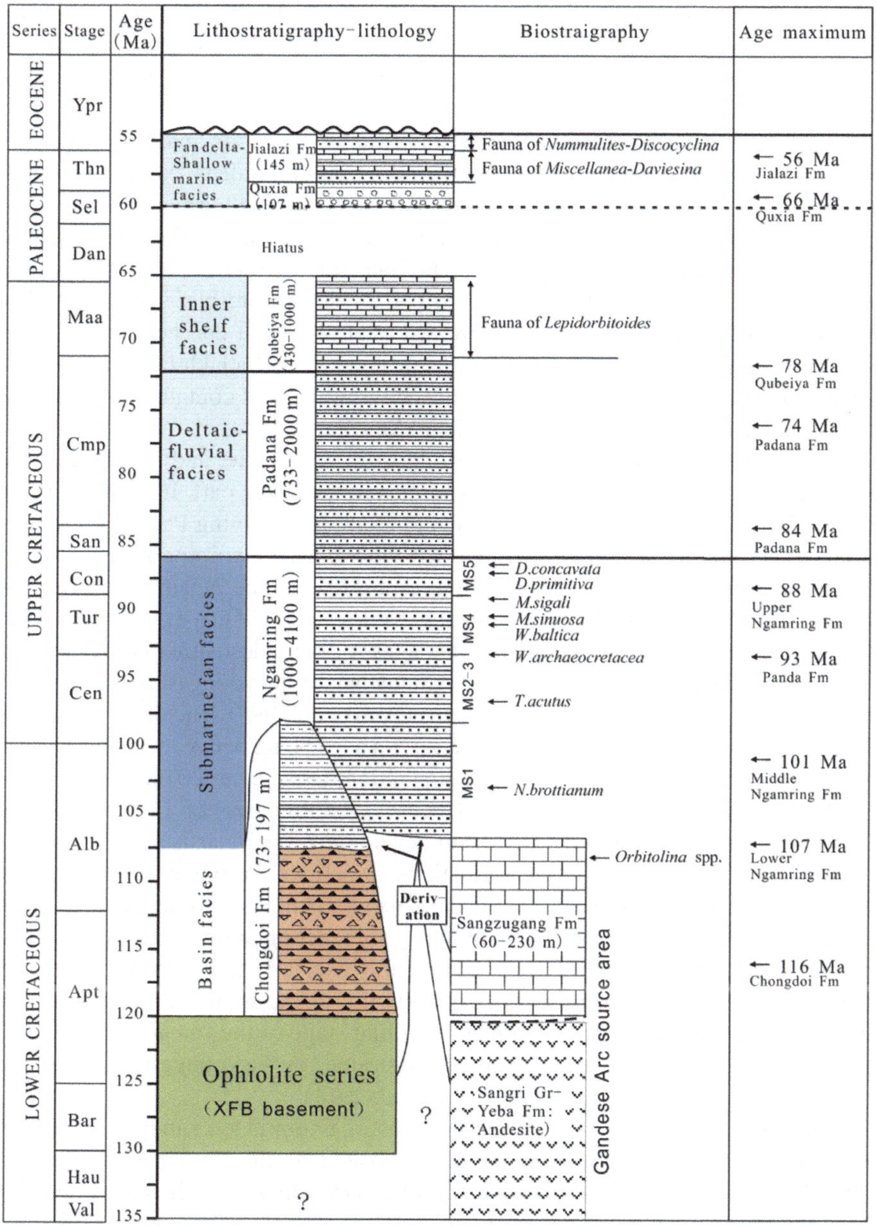

Fig. 4 Diagram showing the stratigraphic framework of the XFB with data of biostratigraphy, maximum age of detrital zircon U–Pb isotopes, and lithofacies (modified after Wang et al. 2012). In the column of the biostratigraphy, the fauna of the Qubeiya and Jialazi formations are from Wan et al. (2001), and the biozones of the Ngamring Formation are cited from Wan et al. (1998). The column of the age maximum is from Wu et al. (2010) and Hu et al. (2016)

2003) and tuffaceous zircon ages from ~119 to 113 Ma (Wang et al. 2017a, b), whereas the upper member contains detrital zircons with the youngest peak age of 116 Ma and tuffaceous zircon ages from ~113 to 110 Ma (late Aptian; Wu et al. 2010; Wang et al. 2017a, b).

The Sangzugang Formation consists of thick-bedded or massive dark-gray bioclastic limestones with abundant benthic foraminifera, rudists, corals, and minor bivalves. This unit is in fault contact with either the Ngamring Formation or the latest Oligocene-Early Miocene Kailas Formation, and its original stratigraphic relationships are uncertain (Wang et al. 2012). The Sangzugang Formation was deposited in reefal environments along the northern part of the forearc basin during the late Aptian to early Albian (Liu et al. 1988; An et al. 2014). It can be considered an equivalent to the rudistid-bearing Khalsi Limestone of the Ladakh Himalaya, which is paraconformably overlain by volcaniclastic turbidites that contain ammonoids of the late Albian age at the base (Garzanti and Van Haver 1988).

The Ngamring Formation consists of sandstone and shale, and is characterized by a series of large channelized conglomerates in the lower part. It is conformable with both the underlying Chongdoi Formation and the overlying Padana Formation in the south, and locally in fault contac with the Xigaze Ophiolite in the south or the Kailas Formation in the north (Wang et al. 2012). The Ngamring Formation forms a synclinorium striking W-E in the Xigaze area (Wang et al. 2012), with Aptian–Albian and Cenomanian–Turonian ages for the oldest strata in the northern and southern flanks respectively, and late Coniacian for the youngest strata along the axis indicated by foraminiferal associations (Wan et al. 1998). The depositional age is also supported by the youngest detrital-zircon ages being between 84 and 107 Ma (Albian–Santonian; Wu et al. 2010; An et al. 2014; Orme et al. 2014, 2016).

The shallow-water parts of the XFB with a total thickness of 1472 m provides auxiliary information about forearc basin development and consists of the Padana, Qubeiya, Quxia, and the Gyalaze formations, which were cropped out to the west of Ngamring county. The Santonian-Campanian Padana Formation is composed of varicolored sandstone interbedded with shales, minor conglomerates and limestones. Best exposed in Sangsang, this unit conformably overlies the Ngamring Formation and is in fault contact with the Xigaze Ophiolite to the south. It was interpreted to be shelf-deltaic-fluvial deposition at the filling stage of the forearc basin, with a depositional age of Santonian-Campanian constrained by the youngest detrital zircon ages and some fossils (Wu et al. 2010; An et al. 2014).

The overlying Qubeiya Formation consists of fine-grained sandy wackestones intercalated with fine-grained sandstones deposited on the inner shelf (Liu et al. 1988; Wan et al. 2001). A deposition age of Maastrichtian is indicated by benthic foraminiferal assemblages (Liu et al. 1988; Hu et al. 2016). The Quxia Formation contains a lower part of shales with thick-bedded sandstones, and an upper part of conglomerates, sandstones and shales arranged in fining-upward cycles. This unit was deposited in a fluvial-fan delta environment at the early Paleocene (Hu et al. 2016). The Jialazi Formation mainly consists of sandy foraminiferal limestones and interbedded sandstones deposited in shallow-marine, distal fan-delta environments.

The depositional age is constrained as Paleocene to earliest Eocene by larger benthic foraminifera and tuffaceous zircon ages (Hu et al. 2016; Jiang et al. 2021).

Evolution of the Xigaze forearc includes an early underfilled stage represented by abyssal cherts and turbidites of the Chongdui and Ngamring Formations and an overlying filled stage represented by the Padana, Qubeiya, Quxia and Jialazi formations, which matches well with a typical upward-shoaling sequence from deep-water turbidites to slope, shelfal, and eventually deltaic or even fluvial deposits of forearc basin. Based on both sedimentological (Wang et al. 2012) and provenance analysis, five successive stages are identified in the evolution of the Xigaze forearc basin (Fig. 5). (1) Late Barremian–Aptian (127–117 Ma): The Chongdui Formation records abyssal siliceous sedimentation in the initial phases of Neo-tethyan subduction, when negligible sediment was produced because the nascent arc was still largely submerged. The upper Aptian to lower Albian Sangzugang Formation developed subsequently (117–110 Ma), on the shallow-marine northern edge of the Xigaze forearc basin. (2) Albian–early Cenomanian (104–99 Ma): The Chongdui turbidites and lower Ngamring turbidites were deposited in deep-sea fan environments. Detritus was largely supplied by Lower Cretaceous magmatic rocks of the Gangdese arc, with minor contributions from its basement (as documented by the subordinate population of pre-Mesozoic zircons) and from erosion of the Sangzugang Formation (Fig. 5a). (3) Cenomanian–early Coniacian (99–88 Ma): The middle Ngamring turbidites were derived from both Cretaceous (ca. 100 Ma) and Jurassic (ca. 157 Ma) magmatic rocks of the Gangdese arc, and subordinately from the Sangzugang Formation (Fig. 5b). (4) Coniacian–Santonian (88–83 Ma): The upper Ngamring turbidites were derived from both the Gangdese arc and the central Lhasa terrane (Fig. 5c). (5) Early Campanian (83–76 Ma): The Padana Formation records coastal to fluvio-deltaic sedimentation with a slight increase in quartz, possibly due in part to the recycling of older siliciclastic sediments (Fig. 5d).

1.3 Xiukang Complex and Trench Basin

The Xiukang Complex, currently 10–20 km-wide and in tectonic contacts with the Yarlung-Zangbo Ophiolite in the north and with Tethyan Himalayan strata in the south, has been considered to be products of the Neotethyan subduction and subsequent India-Asia collision (Tapponnier et al. 1981; Cai et al. 2012; An et al. 2017; Metcalf and Kapp 2019). The mélange, including exotic blocks of sandstone, limestone, chert and basalt with various sizes and ages, displays a typical block-in-matrix fabric. The generally deformed matrix is represented by upper Triassic–lower Aptian abyssal chert, siliceous mudstone and local turbidites showing N-S trending stretching lineation and top-to-south S-C fabrics (Cai et al. 2012). The occurrence of *Globotruncanas* within the limestone blocks suggested a post-Late Cretaceous age for the formation of the mélange (Tapponnier et al. 1981). Three groups of sandstone blocks with different provenance and depositional setting are distinguished by their petrographic, geochronological and isotopic fingerprints (An et al. 2017). Blocks of

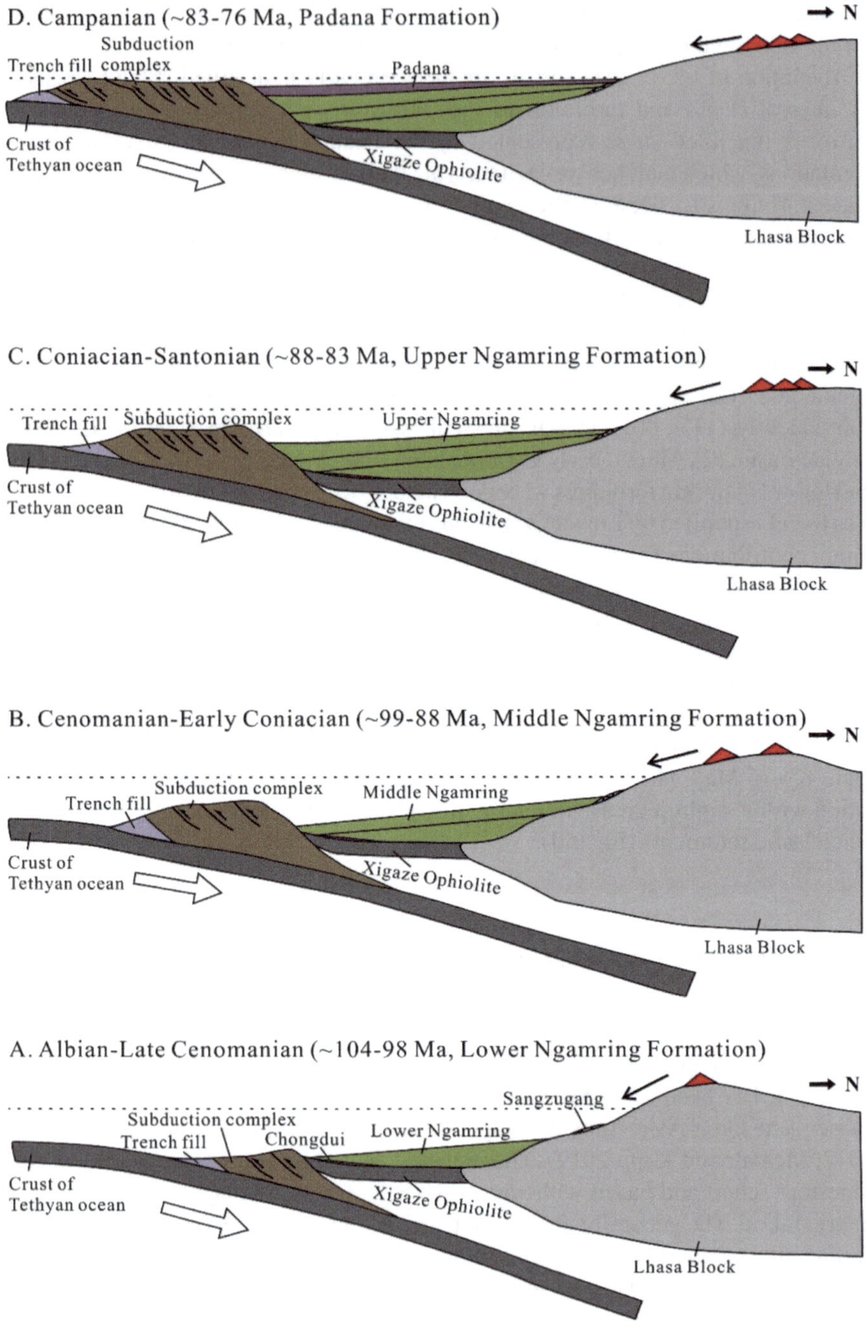

Fig. 5 Schematic sections illustrating the envisaged evolution of the Xigaze forearc basin from Albian to Campanian times (An et al. 2014)

turbiditic quartzarenite originally sourced from the Indian continent were deposited in pre-Cretaceous time on the northernmost edge of the Indian passive margin and eventually accreted at the early stage of the India–Asia collision. Two distinct groups of volcaniclastic-sandstone blocks were derived from the central Lhasa block and Gangdese magmatic arc. One group was deposited in the trench and/or on the trench slope of the Asian margin during the early Late Cretaceous (Cai et al. 2012; An et al. 2017), and the other group in a syn-collisional basin just after the onset of the India–Asia collision in the Early Eocene (An et al. 2017).

The abundant limestone blocks, varying in size from cm to km, are mostly bioclastic, with Middle–Late Permian crinoids and bryozoans, and Early Triassic, Early Jurassic and Late Cretaceous foraminifera (Tapponnier et al. 1981; Jin et al. 2015 and references therein). These exotic blocks have been variously interpreted to represent either remnants of seamounts within the Neotethys Ocean (Wang and Mu 1980), or of a carbonate platform situated on the northern fringe of Gondwana (Guo et al. 1991). Subsequent studies of the brachiopod fauna have documented transitional/mixed characters between Cathaysian and Gondwanan faunas, favoring deposition on the outer peri-Gondwanan shelf or on seamounts within Neotethys (Shen et al. 2003a, b). Blocks from seamounts within the Neotethys Ocean, as indicated by limestones and cherts directly deposited on OIB-type basalt, have been reported recently in Zhongba (Dai et al. 2012). Decimetric to metric basaltic blocks appear occasionally. They are mainly porphyritic and/or amygdaloidal, have experienced low-grade hydrothermal metamorphism, and display within-plate-basalt geochemical affinity, similar to Indian-Reunion "hotspot" lavas (Dupuis et al. 2005). Metric blocks of radiolarian chert have been identified as two groups, Middle–Upper Triassic ones with continental-margin affinity and Upper Jurassic–Lower Cretaceous ones with oceanic-basin affinity (Zhu et al. 2006).

Late Cretaceous and Paleocene trench basins have been recently reported along the central part of the Yarlung Zangbo suture zone, including the Jiachala Formation in Gyangze area (Fu et al. 2018), the Rongmawa Formation in the Dênggar area (Cai et al. 2012), and the Luogangcuo Formation and Sangdanlin-Zheya succession exposed in Saga area (An et al. 2018a, b; Hu et al. 2020). The Jiachala, Rongmawa, and Luogangcuo formations accumulated in Late Cretaceous during Neo-Tethyan ocean subduction whereas the Sangdanlin and Zheya formations accumulated in the middle Paleocene during the onset of the India-Asia collision, marking a transitional stage starting with pre-collisional sedimentation and ending with syn-collisional deformation.

The Jiachala Formation is composed of sandstone interbedded with shale deposited in a submarine-fan lobe environment (Fu et al. 2018). The Rongmawa Formation consists of lower and upper members with thin-bedded sandstone interlayered with siltstone, separated by a middle member of mud dominated interval interbedded with chert and sandstone. The Luogangcuo Formation is composed of conglomerate interlayered with sandstone and shale. Conglomerates was deposited by debris flows while sandstone by turbidity currents in a distal submarine fan (An et al. 2018a, b). Provenance analysis suggests that detritus of sandstones was derived mainly from the Gangdese arc and central Lhasa terrane while the conglomerates were sourced from the Xiukang Complex.

In the Saga area, The Sangdanlin Formation is mainly composed of chert and siliceous shale with turbiditic sandstone beds increasing upwards. The Zheya Formation is composed of turbiditic sandstone intercalated with shale (Wang et al. 2011; DeCelles et al. 2014). The depositional age is precisely constrained by coccolith and radiolarian biostratigraphy coupled with zircon chronostratigraphy as middle Paleocene (59 ± 1 Ma; Hu et al. 2015). Turbiditic sandstones in the lower Sangdanlin Formation are either quartzose derived from India or feldspatholitho-quartzose volcaniclastic derived from the Gangdese arc and central Lhasa terrane (Wang et al. 2011; DeCelles et al. 2014). In the upper Sangdanlin and Zheya formations, instead, all turbidites are sourced transversally from the Gangdese arc and central Lhasa (Wang et al. 2011; DeCelles et al. 2014; Wu et al. 2014a, b; Hu et al. 2015, 2016; An et al. 2021).

1.4 Yarlung Zangbo Suture Zone

The Yarlung Zangbo Suture Zone (YZSZ) marks the southernmost and youngest suture in Tibetan Plateau (Yin and Harrison 2000; Fig. 6 inset). This suture includes remnants of the Neo–Tethyan oceanic lithosphere accreted to Eurasia (Allègre et al. 1984; Dai et al. 2011; Hébert et al. 2012). The Yarlung Zangbo ophiolite (YZO) comprises well preserved to dismembered ophiolitic massifs. Ophiolites are fragments of ancient oceanic lithosphere that are emplaced at convergent plate margins. Previous studies document that the YZO formed during 130–120 Ma (Dai et al. 2012, 2013, 2021; Li et al. 2009; Liu et al. 2016; Wang et al. 2006; Wu et al. 2014a, b; Zhang et al. 2016a, b; Zhang et al. 2016a, b; Fig. 6a, b). In the south, the YZO thrusts over a narrow belt of ophiolitic mélange which is composed of blocks of serpentinized peridotite, mafic rock, amphibolite, chert, limestone, and sandstone (Guilmette et al. 2009); while in the north, the YZO is overlain by flysch of the Xigaze Group (Dürr 1996; Wang et al. 2012). The nonconformable depositional contact between YZO lava and sediment of Chongdoi Formation defines the YZO as the base of the Xigaze forearc basin (An et al. 2014; Dai et al. 2015; Wang et al. 2017a, b).

1.4.1 Tectonic Settings of the Yarlung Zangbo Ophiolite (YZO)

Slow-Spreading Mid-Ocean-Ridge Model

Several tectonic models have been proposed for the YZO. In the early 1980s, detailed geological mapping was conducted in the Xigaze Ophiolite in the central YZO including the Dazhuqu to the Jiding massifs (Fig. 6b). These studies revealed that the YZO did not form at a magma-rich mid-oceanic spreading ridge because of the paucity of mafic rocks, but was generated at a slow-spreading ridge (Girardeau et al.

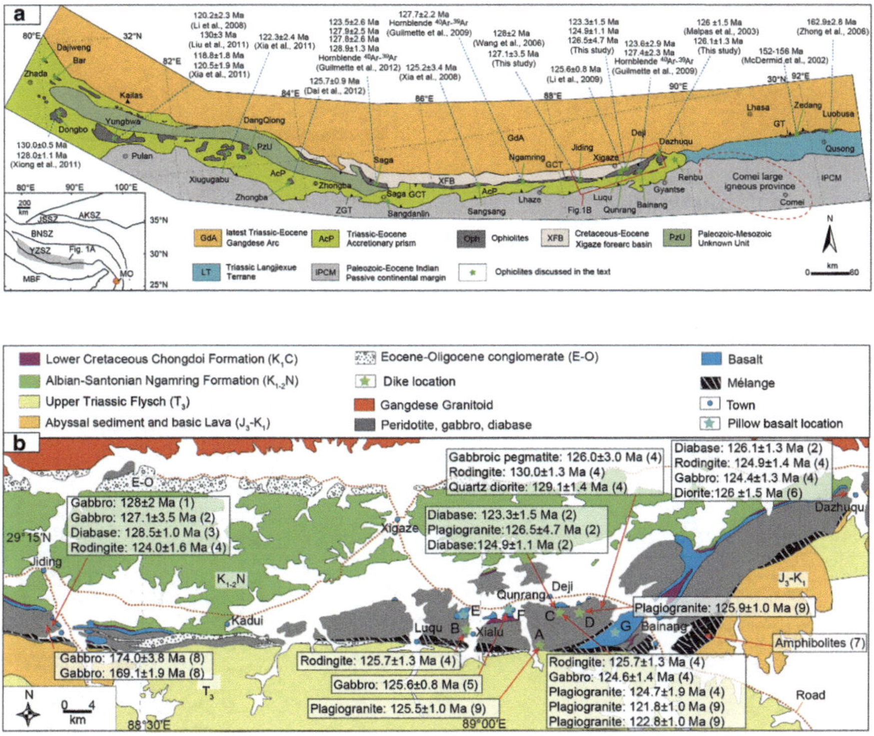

Fig. 6 **a** Schematic tectonic map of southern Tibet showing the ophiolitic massifs within the YZSZ (after Dai et al. 2013). Zircon U–Pb ages of the mafic rocks and hornblende $^{40}Ar/^{39}Ar$ ages of the amphibolites across the whole YZSZ are also presented. Major faults: GCT, Great Counter thrust; ZGT: Zhongba-Gyangze thrust; GT: Gangdese thrust. Inset is the simplified tectonic map of Tibetan Plateau showing major sutures. The major sutures are: AKSZ, A'nemaqin-Kunlun suture zone; JSSZ, Jinshajiang suture zone; BNSZ, Bangong-Nujiang suture zone; YZSZ, Yurlung Zangbo suture zone. MO, Myitkyina ophiolite, Myanmar. **b** Geological map of the Xigaze area (Wang et al. 1987) showing the locations of dated samples (after Dai et al. 2021). References: (1) Wang et al. (2006); (2) Dai et al. (2013); (3) Bao et al. (2013); (4) Zhang et al. (2016a, b); (5) Li et al. (2009); (6) Malpas et al. (2003); (7) Guilmette et al. (2009); (8) Wang et al. (2018); (9) Dai et al. (2021)

1985a; Nicolas et al. 1981). Limited thin crustal outcrops might be related to exhumation after ophiolite emplacement (Burg et al. 1987). This possibility was confirmed by thermochronological data from the Oligocene sediments within the YZSZ which reveal about 6 km (vertical) of rock removal (Carrapa et al. 2014).

The slow-spreading ridge tectonic model was developed by Wu et al. (2014a, b) and his group (Fig. 7a). Zhang et al. (2016a, b) argued that Xigaze Ophiolite basalts are typical N–MORBs without influence from subduction–related fluids. However, the following lines of evidence argues against this conclusion: (1) most mafic rocks show negative Nb anomalies and positive U and Pb anomalies, indicating that their source was affected by subduction–related fluids (Hébert et al. 2012); (2) although at least 85% of sedimentary Nd and Hf are recycled into the mantle, the Hf and

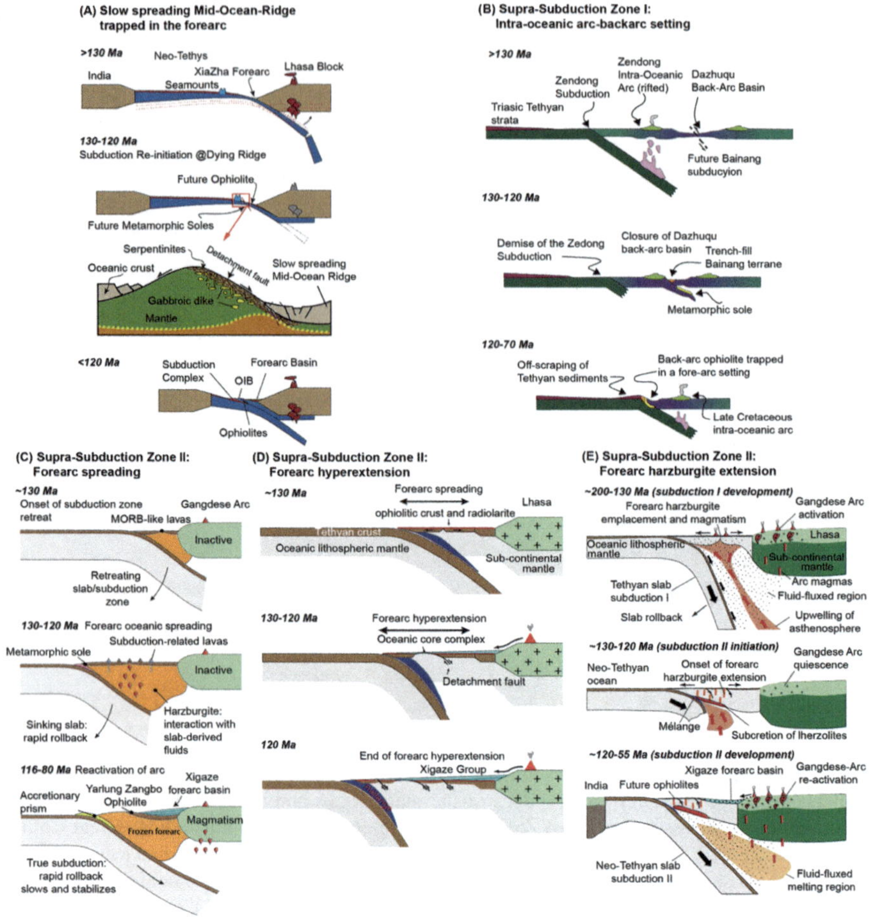

Fig. 7 Proposed tectonic models for the Yarlung Zangbo ophiolite (after Dai et al. 2021). **a** Slow spreading mid-ocean-ridge trapped in the forearc (Liu et al. 2014; Zhang et al. 2019a, b); **b** Supra–Subduction Zone I: complex intra–oceanic arc-backarc setting (Hébert et al. 2012; Guilmette et al. 2009); **c** Supra–Subduction Zone II: Forearc spreading (Dai et al. 2013, 2021); **d** Forearc hyper-extension (Maffione et al. 2015); **e** Forearc harzburgite extension and lherzolite subcretion (Xiong et al. 2016)

Nd isotopic compositions of the Izu–Mariana arc lavas are not markedly different from those of MORBs (Chauvel et al. 2009). Therefore, the MORB–like Nd–Hf isotopic compositions do not rule out the possibility that mafic rocks witnessed subduction–related fluids.

Liu et al. (2016) proposed that Xigaze Ophiolite gabbroic rocks were generated beneath a fossil slow–spreading ridge because gabbro pods or dikes are associated with diabase dikes or sills, suggesting that the gabbros were exhumed when the diabases were generated. Rapid exhumation of mantle peridotites and gabbroic rocks of the Xigaze Ophiolite may have allowed intrusion of the diabase dikes and sills. The above arguments do not constrain the tectonic setting of the Xigaze Ophiolite. If the Xigaze Ophiolite was formed in an SSZ environment, the mantle peridotites and gabbroic rocks could also be intruded by diabase dikes. In addition, Liu et al. (2018) observed that gabbroic intrusions in the Jiding sequence are more evolved than Dazhuqu and Bainang gabbroic rocks which are characterized by higher Cr_2O_3 but lower TiO_2 and REE contents in both clinopyroxene and bulk compositions. They proposed that the lower oceanic crust within the Xigaze Ophiolite is highly segmented and discontinuously distributed, similar to that exposed at modern slow– and ultra–slow–spreading ridges. Such variations may be related to both magma evolution and magma source, since they also proposed that melt–crystal reaction played a key role in magma generation. More recently, Liu et al. 2019 proposed that Dazhuqu mantle peridotites experienced 0–6% garnet–facies melting followed by 10–18% melting in the spinel stability field based on clinopyroxene trace element modeling. This is similar to the degree of garnet–facies melting inferred for many MOR peridotites rather than fore–arc peridotites. The occurrence of abyssal peridotite is not strong evidence that the Xigaze Ophiolite was generated in the slow–spreading MOR because a previous study revealed the occurrence of both fertile and depleted mantle in the Zedang area (Xiong et al. 2016). Specifically, previous studies revealed that more than half of the Luqu peridotites (Fig. 7b) have ultra–refractory compositions indicated by both whole rock and clinopyroxene compositions (Zhang et al. 2017).

The same group proposed a subduction re–initiation model at a dying mid–ocean–ridge (Zhang et al. 2019a, b; Fig. 7a) following previous studies of Guilmette et al. (2012). This interpretation was based on the high geothermal gradient recorded by the metamorphic sole in the Lhaze ophiolitic mélange which was also reported in other mélange massifs including Bainang, Buma and Saga (Guilmette et al. 2012). This model emphasized that the location of the subduction zone was close to the Eurasian continental margin (Zhang et al. 2019a, b), while previous models proposed that subduction was intra-oceanic (Guilmette et al. 2012). This model proposed (1) the entire YZO was a trapped remnant of MORB lithosphere that was generated in a slow–spreading MOR; and (2) the metamorphic sole originated from the MOR crust in the lower plate. However, the majority of MOR was generally subducted, not accreted, just as illustrated by the currently subducting Nazca Ridge beneath the northern South American margin (Hampel et al. 2004). All the above observations can be well explained in a subduction initiation model.

Supra-Subduction Zone Model

Most studies of mineral and whole–rock geochemical data from the YZO reveal a heterogeneous mantle that possesses refractory whole rock major element compositions and U–shape REE patterns (Dai et al. 2013; Zhang, et al. 2017; Xiong et al. 2017), as well as the high Cr# spinels for most YZO peridotites (Hébert et al. 2003; Zhang, et al. 2017; Xiong et al. 2017), indicating that they underwent hydrous melting. Mafic rocks show slight depletion in light REE, and have a small negative

Ta–Nb anomaly, suggesting the influence of a subduction component. The abundances of incompatible elements in the mafic rocks are similar to back–arc–basin basalts (BABB; Bédard et al. 2009; Hébert et al. 2012). Therefore, the YZO was proposed to belong to a complex intra-oceanic arc–back-arc setting, analogous to the Lau basin (Aitchison et al. 2000; Hébert et al. 2003; Guilmette et al. 2009; Fig. 7b). However, there is no evidence for a slightly older intra-oceanic arc along the YZSZ that a BAB could form behind (Wu et al. 2014b). Instead, the associated Gangdese magmatic arc lies to the north, separated from the ophiolites by the Xigaze forearc basin (Fig. 6b).

Based on geochemical similarities between basalts and late–stage mafic dikes in the YZO and forearc basalts and boninites in the IBM, Dai et al. (2013) was first to propose that the YZO formed in a forearc spreading setting (Fig. 7c). This forearc spreading model was further developed in the following two models: (1) Forearc hyperextension: paleomagnetic and field geological observations reveal the occurrence of extensional detachment faults in the YZO (Maffione et al. 2015). These exhumed mantle rocks and mélange on the seafloor and were subsequently covered by forearc basin strata (Fig. 7d). Forearc hyperextension might help explain dismemberment of the YZO. (2) Forearc harzburgite extension and lherzolite subcretion: both harzburgite and lherzolite outcrop in the Zedang Ophiolite of the eastern YZO (Xiong et al. 2016; Fig. 7e). The equilibration temperature of lherzolite is 250–150 °C higher than that of the harzburgite, suggesting that harzburgites formed earlier (possibly at 200–130 Ma during an earlier episode of subduction) and at shallower levels than lherzolite. The authors proposed that the subcretion of lherzolites, intrusion of dolerite dykes, and doming and extension of overlying harzburgitic lithosphere occurred coevally during 130–120 Ma of the second stage of subduction (Fig. 7e).

1.4.2 Geological Settings of the Xigaze Ophiolite (XO)

The Xigaze Ophiolite (XO), located in the central part of the Yarlung Zangbo suture zone (Dai et al. 2021), displays well preserved sections from mantle to crustal rocks, together with overlying sedimentary strata (Fig. 6b). Marine sedimentary rocks consist of mudstone, chert, felsic tuffs and fine–grained volcaniclastic deposits and are exposed along the northern margin of the XO, with Late Barremian to Late Aptian depositional ages (Ziabrev et al. 2003; Figs. 6b and 7b). The Xigaze Ophiolite mafic sequence is thin (approximately 2 km) (Girardeau et al. 1985b). A well–preserved sequence of pillowed lava occurs in several places such as Qunrang, Deji, and Luqu. Minor cumulate gabbro has only been reported at Dazhuqu, Bainang and Jiding massifs (Liu et al. 2018; Fig. 6b). Sheeted diabase is rarely observed except at Deji and Bainang massifs (Wang et al. 1987). The Xigaze Ophiolite is dominated by mantle peridotite (Figs. 6b and 7b). These are variably serpentinized (Hébert et al. 2003) and are intruded by various dikes, such as diabase, rodingite, plagiogranite, and gabbroic pegmatite (Fig. 8b).

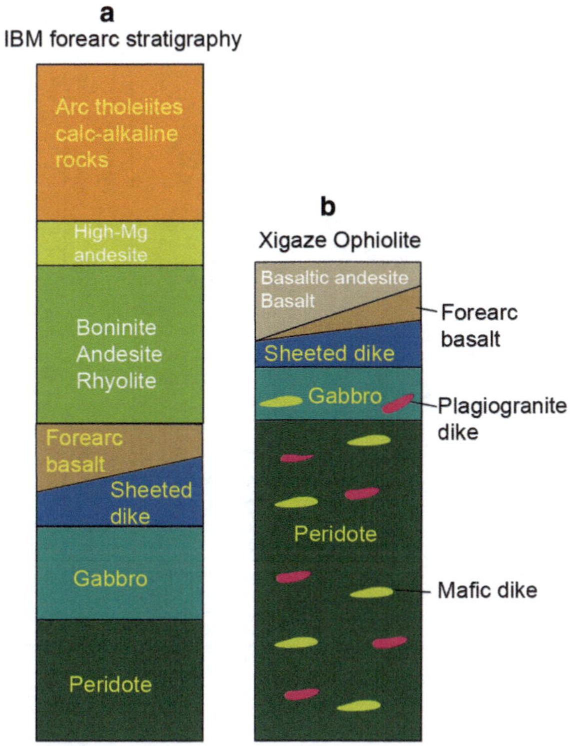

Fig. 8 Schematic columnar sections of Izu–Bonin–Mariana (IBM) forearc (**a**, modified from Ishizuka et al. 2011) and Xigaze Ophiolite (**b**, modified from Girardeau et al. 1985a)

In the XO, there are plenty of radiometric ages from mafic rocks including isotropic gabbro, diabase sill and various gabbroic/diabase dikes intruded into both the crustal and mantle section. These ages range from 124 ± 1.6 Ma to 130 ± 1.3 Ma (Fig. 7b; Bao et al. 2013; Dai et al. 2013, 2021; Liu et al. 2016; Zhang et al. 2016a, b), which are comparable to those of the plagiogranites (Four plagiogranite samples from three different locations of a, c, d in Fig. 6b yielded zircon U–Pb ages of 121.8 ± 1.0 Ma, 122.8 ± 1.0 Ma, 125.5 ± 1.0 Ma, 125.9 ± 1.0 Ma, respectively, Fig. 9a–d). These ages indicate that both mafic and felsic intrusive rocks in the Xigaze Ophiolite were generated in the Early Cretaceous, i.e., 121.8–130.0 Ma (Fig. 9e). However, Middle–Late Jurassic ages (174.0 ± 3.8 Ma and 169.1 ± 1.9 Ma) from two gabbro samples from the Dazhuqu and Jiding massifs have been reported recently in the Xigaze Ophiolite (Wang et al. 2018), products of an earlier stage of igneous activity in the Neo–Tethyan Ocean, unrelated to the Early Cretaceous magmatism. Most ages suggest that the duration of Xigaze Ophiolite magmatic activity during the Early Cretaceous stage was about 10 Ma, consistent with the conclusions of Hu and Stern (2020).

The main samples and field observations in this field guide book were focused around several massifs (a–d for gabbroic pegmatite, plagiogranite, diabase dike in the mantle and basaltic dike in the crust, e–g for pillow basalt; Fig. 6b).

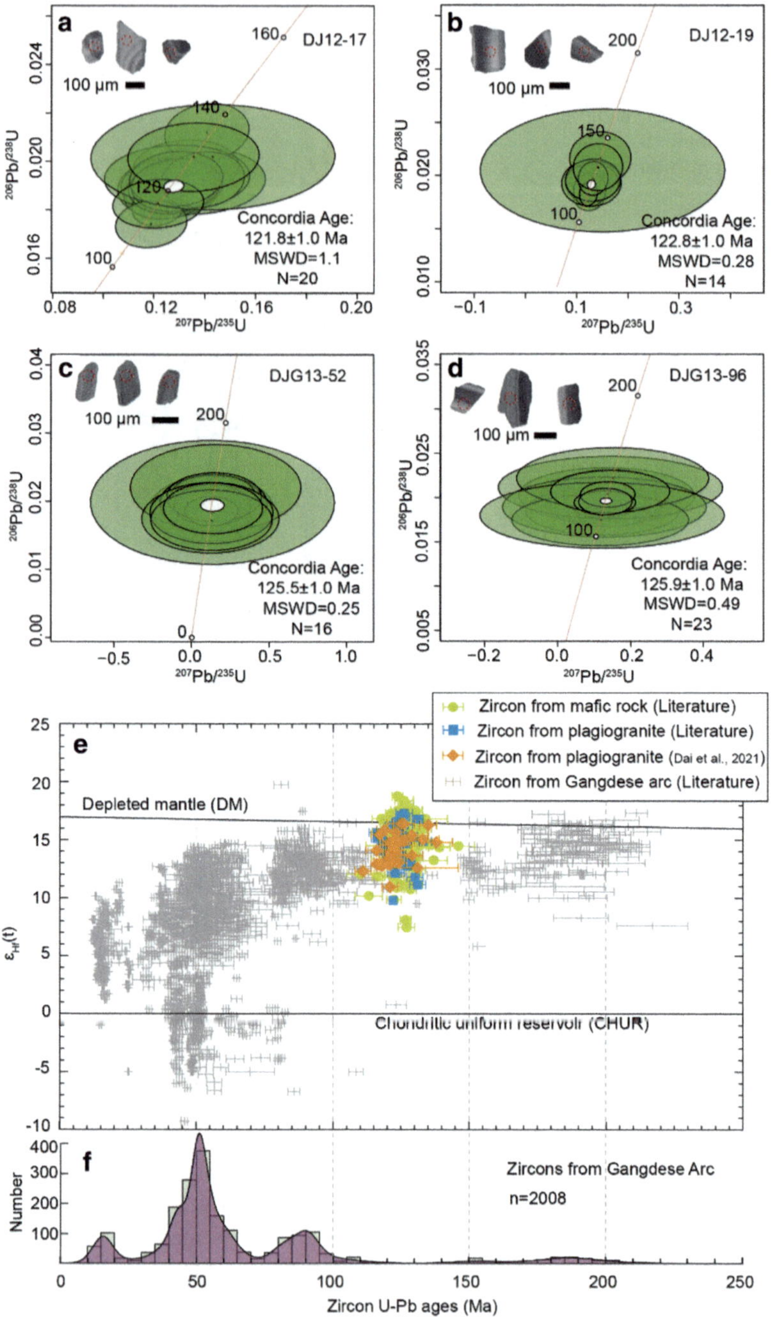

a
DJ12-17
100 μm
160
140
120
100
Concordia Age:
121.8±1.0 Ma
MSWD=1.1
N=20
0.024
0.020
0.016
$^{206}Pb/^{238}U$
0.08
0.12
0.16
0.20
$^{207}Pb/^{235}U$
b
DJ12-19
100 μm
200
150
100
Concordia Age:
122.8±1.0 Ma
MSWD=0.28
N=14
0.030
0.020
0.010
−0.1
0.1
0.3
c
DJG13-52
100 μm
200
0
Concordia Age:
125.5±1.0 Ma
MSWD=0.25
N=16
0.04
0.03
0.02
0.01
0.00
−0.5
0.0
0.5
1.0
d
DJG13-96
100 μm
200
100
Concordia Age:
125.9±1.0 Ma
MSWD=0.49
N=23
0.035
0.025
0.015
0.005
−0.2
0.0
0.2
0.4
e
Zircon from mafic rock (Literature)
Zircon from plagiogranite (Literature)
Zircon from plagiogranite (Dai et al., 2021)
Zircon from Gangdese arc (Literature)
Depleted mantle (DM)
Chondritic uniform reservoir (CHUR)
$\varepsilon_{Hf}(t)$
25
20
15
10
5
0
-5
-10
f
Zircons from Gangdese Arc
n=2008
Number
400
300
200
100
0
50
100
150
200
250
Zircon U-Pb ages (Ma)

◀**Fig. 9** Zircon U–Pb concordia plots of four plagiogranite samples (**a**–**d**: DJ12–17, DJ12–19, DJG13–52, DJG13–96) and representative cathodoluminescence images of zircons. Plots of $\varepsilon_{Hf}(t)$ versus U–Pb ages of zircons from plagiogranite and mafic rock in the Xigaze Ophiolite (**e**) and zircon U–Pb ages of Gangdese Arc (**f**). Data source of mafic rock and plagiogranite from Dai et al. 2013; Liu et al. 2016; Zhang et al. 2016a, b; Please see Dai et al. (2021) for the detailed data source of the Gangdese arc

1.5 Syncollisional and Post-Collisional Basins

1.5.1 Syncollisional Basin

After initial continent–continent collision (i.e., the first contact of two continents after the end of an oceanic subduction), an underfilled seaway still exists between the two continents. The marine basin was inherited, and developed above former forearc and passive marginal basins. This syncollisional basin will be gradually filled until final retreat of seawater from the collision zone. Then, the landmasses were completely integrated and connected by fluvial systems (Hu et al. 2016b). In some literature, the syncollisional basin was called as an underfilled peripheral foreland basin (Sinclair 1997). A remarkable feature of syncollisional basin is that sediments from the active margin could be transported across the trench and deposited on the former passive margin.

In the Gangdese forearc, syncollisonal basin developed above the Xigaze forearc basin (Figs. 9 and 10). Syncollisional strata including the Quxia Formation and the Jialazi Formation were well preserved and exposed in the Cuojiangding area of Zhongba County (Fig. 10). The Quxia Formation is a suite of (~105-m-thick) coarse-grained clastic rocks, consists of conglomerates, sandstones and shales that were deposited in fluvial to fan-delta environments. The underlying latest Cretaceous sandy wackestones and fine-grained sandstones of the Qubeiya Fomration were deposited in deltaic to inner shelf environments, representing the final the final filling of the Xigaze forearc basin. The overlying Jialazi Formation was conformably contact with the Quxia Formation, and consists of ~240-m-thick sandy foraminiferal limestones and interbedded sandstones, deposition in shallow marine, distal fan-delta environments (Hu et al. 2016b). The Paleocene-Eocene Thermal Maximum (PETM) event was recognized in the Jialazi Fomration, as indicated by carbon-isotope excursion and facies change. It was accompanied by increasing siliciclastic supply, indicating enhanced hydrological circulation (Jiang et al. 2021).

Larger benthic foraminifera from the Qubeiya Formation indicates a late Maastrichtian depositional age, whereas those from the Jialazi Formation indicate late Paleocene-early Eocene age (SBZ3-SBZ5, ~58–54 Ma). Considering the limited thickness of the Quxia Formation and rapid depositional rate in the fan-delta environment, there should be a hiatus of ~7–8 Ma between the Qubeiya Formation and the Quxia Formation (Ding et al. 2005; Hu et al. 2016). Although this unconformity was not directly observed in the field because of poor outcrop conditions, abrupt facies change between the two formations support this speculation. Dramatic change

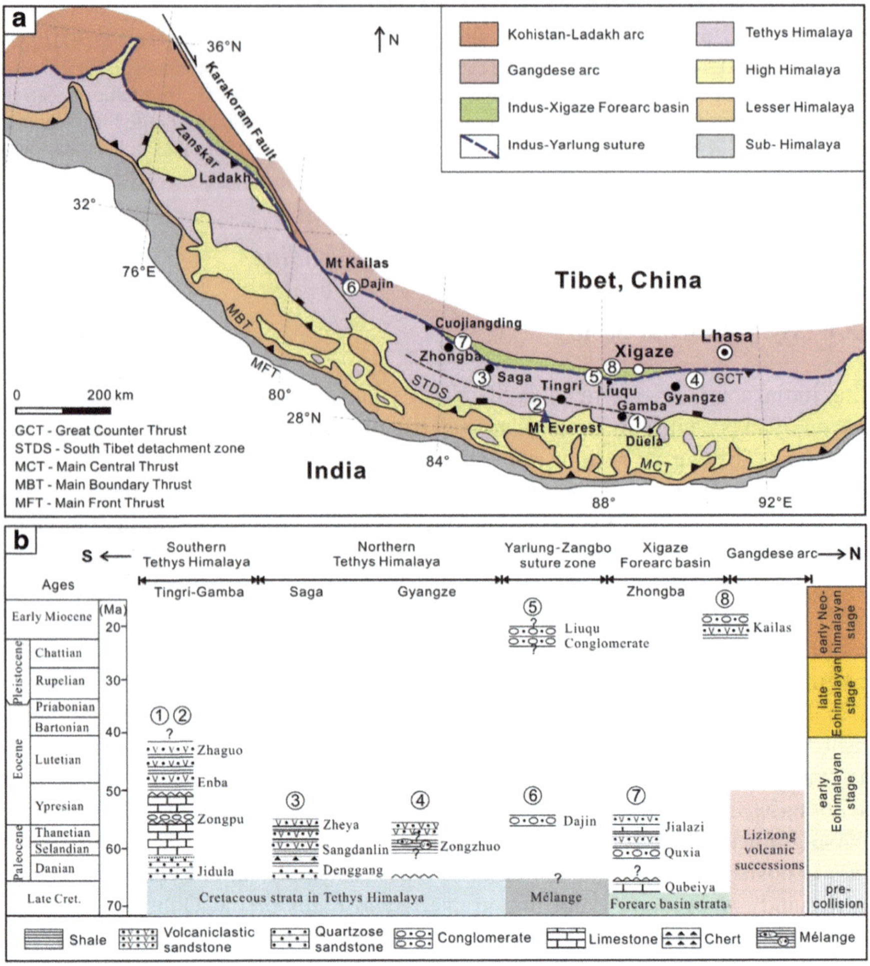

Fig. 10 **a** Simplified geological map shows location of major syncollisional and post-collisional sedimentary units along the Indian-Asian collision belt. **b** Stratigraphic correlation of syncollisional and post-collisional sedimentary records (Hu et al. 2017)

of depositional environments and the possible depositional hiatus were interpreted as a consequence of incipient collision between India and Asia. If this is the truth, sedimentary response to the collision onset would be constrained between 65 and ~58 Ma in the Gangdese forearc. Subsequently, the southern margin of Asia was uplifted and rapid exhumation ensued. Petrography pf sandstones changes from 'undissected arc' to 'dissected arc' through the stratigraphic sequence (Hu et al. 2016).

Further west in the Kailas region (Fig. 10), the earliest Eocene coarse-grained shallow-marine clastic rocks (Dajin Formation) contain large amounts of recycled sedimentary rock fragments. U–Pb age and Hf isotopic signatures of detrital zircons indicate ultimate provenance of these rocks was the Gangdese arc. Sedimentary recycling of forearc-basin strata documented in the Dajin Formation may be also a consequence to the India-Asia collision (Wang et al. 2015).

South of the Yarlung suture in the Tethyan Himalaya, syncollisional basin could be divided into two belts, analogous to the Indian passive margin strata. The northern zone consists of deep-marine turbidites, chert and olistostromes, while the southern zone was filled shallow marine carbonates, and shelf to fluvial clastic rocks.

Syncolisional strata of the northern zone were well documented in the Saga area (Figs. 9 and 10). The Sangdanlin Formation and the conformably overlying Zheya Formation consist of hundreds-m-thick interbedded volcaniclastic sandstones, black mudstones, purple-reddish chert and siliceous shales. The appearance of chert and siliceous shale, and lack of carbonate sediments indicate a palaeo-water depth deeper than the carbonate compensation depth (CCD). Sandstone beds commonly display sharp contact with underlying mudrocks, and are arranged in thicken- and coarsen-upward trends. Sedimentary structures of sandstone layers include normal grading, flute casts, ditch moulds, and slump structures, indicating deposition by turbidity currents. Slumping deposition with exotic blocks of conglomerate, sandstone, and carbonate is common throughout the succession, indicating an unstable setting (Wang et al. 2011; DeCelles et al. 2014). Provenance analysis indicated that volcaniclastic sandstones in the Sangdanlin and Zheya formations were derived from the Gangdese arc to the north. By contrast, the underlying strata (the Denagng Formation) consist of quartzose sandstones, silty sandstone and silty shale. Provenance study suggested an exclusive Indian source (Wang et al. 2011; Wu et al. 2014a, b; DeCelles et al. 2014; Hu et al. 2015). Arrival of Asian sediments onto the distal margin of Indian continent provides a solid constraint on the timing of Initial India-Asia collision. The age for provenance changes was precisely dated by radiolarian fossils, calcareous nanofossils and zircon U–Pb age of tuffs to be 59 $\pm$ 1 Ma (Hu et al. 2015).

In Gyangze area to the east, syncollisional sediments were represented by the Zongzhuo mélange (Figs. 9 and 11). The Zongzhuo mélange shows typical 'blocks in matrix' texture. The blocks are several centimeters to tens of meters in size, and include sandstone, chert, basalt, and basaltic conglomerate. The matrix is mainly black shale and siltstone. Provenance analysis indicated that clastic sediments of the mélange were derived from the Asian active margin (Wu et al. 2014a, b). It is suggested that the Zongzhuo mélange had been formed in the trench after the initial collision between Indian and Asian continents (Sun et al. 2011).

Syncolisional strata of the southern Tethyan Himalayan zone were documented in Tingri-Gamba areas (Figs. 9 and 10). Although initial India-Asia collision occurred as early as middle Paleocene (59 $\pm$ 1 Ma), carbonate platform (The Zongpu Fomration) continued to early Eocene (~50 Ma) in Tingri-Gamba area (Hu et al. 2012; Zhang et al. 2012; Li et al. 2015a, b). Notably, a suite of conglomerate occurs at the Paleocene-Eocene boundary. The clasts in the conglomerate are all carbonate rocks and were derived from underlying strata. Microfacies analysis indicates that Paleocene carbonates below the conglomerate were deposited in a carbonate mid-ramp environment, whereas the Eocene carbonates above the conglomerate were deposited in a restricted lagoon (Li et al. 2015a, b). Therefore, occurrence of the conglomerate records a regression event. The conglomerate was interpreted by some scholars as a record of forebulge unconformity associated with India-Asia collision, but others suggested it is a response to the PETM event (Li et al. 2020a, b). The first appearance of Asian detritus occurs in the overlying Enba formation, which consists of calcareous shales and sandstones deposited in shelf to delta-front environments (Najman et al. 2010; Hu et al. 2012). The youngest strata in Tingri-Gamba areas are the fluvial Zhaguo Formation, deposition of which indicates the final retreat of seawater from this region.

1.5.2 Syn-Orogenic Conglomerates

Syn-orogenic conglomerate, also called as continental 'molasses' in some literature, is generally held to be a typical product of continental collision, documenting the stage of rapid topographic growth of a foreland fold-thrust belt. In the Himalayan orogen, typical molasses was deposited in the sub-Himalaya along the south frontier of the orogenic belt in Neogene time. This age is much later than initial continental collision as well as the final filling of the syncollisional basin. Syn-orogenic conglomerates had also been deposited and preserved along the Yarlung-Zangbo suture zone north of the Himalayan orogenic belt.

There are two syn-orogenic conglomerate belts along the Yarlung-Zangbo suture zone: one is the Kailas Conglomerate (or Gangrinboche Conglomerate), exposed along the southern foothills of the Gangdese arc; and the other is the Liuqu conglomerate, exposed to the south of the Kailas Conglomerate along the southern side of the Yarlung-Zangbo ophiolite (Figs. 9, 10, 11 and 12). The Kailas Conglomerate was deposited unconformably above the Gangdese magmatic rocks, and was commonly buried by the Great Counter Thrust with the Xigaze forearc strata in the hanging wall. The sedimentary environment of the Kailas Conglomerate evolved from lakes and meandering rivers at the base to braided rivers and alluvial fans at the top (DeCelles et al. 2011; Wang et al. 2013). Sediments of the base strata were derived entirely from the Gangdese arc in the north, whereas detritus from the Tethys Himalaya, Xigaze forearc basin strata, and Yarlung-Zangbo ophiolites in the south increase upward and become dominant at the top. The zircon ages of tuff layers and pollen constrain the age of the Kailas Conglomerate as late Oligocene-early Miocene (26–18 Ma; Aitchison et al. 2009; Li et al. 2010; DeCelles et al. 2011; Wang et al. 2013). Deposition of the Kailas Conglomerate was considered as a consequence of short-lived extension along the suture followed by the development of the Great Counter Thrust, which were geodynamically induced by the rollback and breakoff of the subducting Indian lithosphere (DeCelles et al. 2011).

The Liuqu Conglomerate was exposed between accretionary mélanges (to the south) and ophiolites, forearc deposits (to the north) along a central, 150-km-long segment of the India-Asia suture zone in southern Tibet. The age of the Liuqu conglomerate has been the object of a long-term controversy. Fossil plants indicated that it could be mid-late Eocene in age (Tao 1988; Fang et al. 2004), and sporopollen fossils indicated a late Eocene to Oligocene age (Wei et al. 2009). However, low

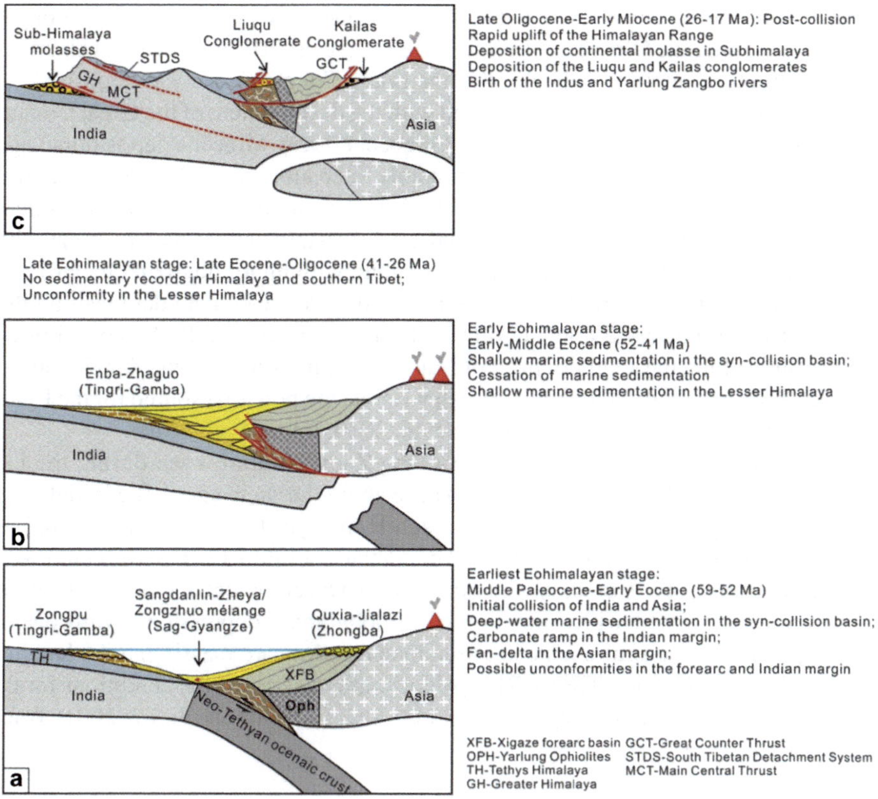

Fig. 11 Syncollisional to post-collisional basin evolution along the Indian-Asian collision belt (Hu et al. 2017)

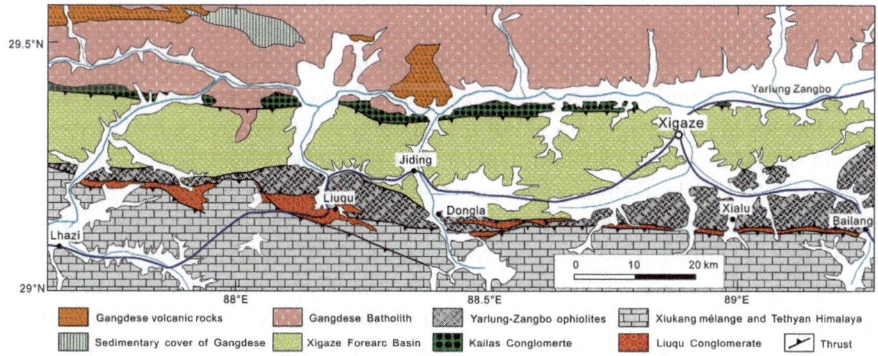

Fig. 12 Geological map shows the distribution of the Kailas Conglomerate and the Liuqu Conglomerate in the Xigaze area, southern Tibet (Hu et al. 2017)

temperature thermochronology data suggested instead that it is notably younger, probably during Late Oligocene-Early Miocene (Li et al. 2015a, b; Leary et al. 2016). Recent structural analysis indicates that the Liuqu Conglomerate was deposited in a contractional setting, probably associated with the development of the Great Counter Thrust system, thus supporting the Late Oligocene-Early Miocene depositional age. The Liuqu conglomerate was deposited mainly in an alluvial fan-braided river environment (Davis et al. 2002). Clasts of the Liuqu Conglomerate are of polymictic composition, varying among outcrops, and are dominantly quartz-arenite, litharenite, basalt, radiolarian chert, phyllite, and slate, along with minor quartzite, gabbro, and serpentinized ultramafic rocks. Paleocurrent and provenance data demonstrate that sediment was transported north-northwest from the hanging wall of a coeval thrust fault system along the southern limit of Liuqu outcrops. The Yarlung-Zangbo ophiolite, Xiukang Mélange, and Tethys Himalaya strata are potential source for the Liuqu Conglomerate (Davis et al. 2002; Wang et al. 2010; Leary et al. 2016).

Conglomerates deposited near the Yarlung-Zangbo suture zone during the late Oligocene to early Miocene (26–18 Ma) have about the same age as the initial stage of "molasse"-type sedimentation on the sub-Himalaya. During this time window, both the collisional suture zone and the Himalayan fold-thrust belt to the south were subjected to rapid uplift, erosion, and exhumation. Development of the Great Counter Thrust was associated with formation of a topographic depression (syn-orogenic basin) parallel to the suture zone between the Gangdese arc massif and the hanging wall of the Great Counter Thrust. This paleogeography created the conditions for the formation of the Yarlung-Zangbo and Indus rivers (Sinclair and Jaffey 2001; Wang et al. 2013).

1.6 Tethys Himalaya

The Tethys Himalaya is one of the major tectonic zones within the Himalaya Orogen (Gansser 1964). It is boarded by the Greater Himalaya Complex in the south and the Indus-Yarlung Zangbo Suture Zone in the north. The Tethys Himalaya (THS) has been located at the passive margin of Indian continent from the birth of Neo-Tethys Ocean till Indian-Asia collision, happened in the early Paleogene (Hu et al. 2015; An et al. 2021). The sedimentary succession in this area archives the tectonic and oceanographic evolution history of Neo-Tethys Ocean. Sedimentary rocks deposited in the Tethys Himalaya are exposed from Zanskar (NW India) to southern Tibet (SW China; Fig. 1; Sciunnach and Garzanti 2012), stretching for about 1500 km. In south Tibet, the hyper-extended THS margin is traditionally subdivided into the northern/outer (nTHS) and southern/inner (sTHS) subzones (Brookfield 1993; Liu and Einsele 1994; Yin and Harrison 2000; Yin 2006), separated by the Gyirong-Kangmar Thrust (Ratschbacher et al. 1994) or the Gamba-Tingri Fault (Wang et al. 1996). Lithologies of Mesozoic in nTHS are mainly deep-sea deposits. By contrast, in sTHS the Mesozoic sedimentary rocks were deposited on shallow marine environments consisting coastline to outer shelf.

The most widely spread sediments and spectacular outcrops in the Tethys Himalaya are of Jurassic to Eocene marine deposits (Fig. 13). Owing to few vegetation cover, the sediment successions are well cropped out along main roads in this area. In sTHS, the Jurassic succession is composed of carbonates in the Lower to Middle Jurassic, which is overlain by siliceous sediments, i.e., the black shales and quartz arenites of Menkadun Fm. The depositional environments of the carbonates were documented as tidal flats, reef rimed platform in the Hettangian-Pliensbachian and carbonate ramp in the Toarcian-Aalenian (Jadoul et al. 1998; Han et al. 2016). Between the carbonates and black shales, an interval of iron oolites developed, which is a few meters thick and is correspondence with global transgressive initiated in late Middle Jurassic. Black shales of the Menkadun Fm. are also called the 'Spiti Shale', which contains abundant well preserved ammonites. The Lower Cretaceous (?Berriasian-Aptian) in sTHS is characterized by volcanoclastic sandstones and black shales with clastic dykes, which recorded reactivation of the passive margin. From Albian to the latest Cretaceous the carbonate composition increased upward. The lithologies in this interval are mainly marl/limestone couplets and limestones deposited in ramp environments. A disconformity of ~10 Myr lies between early Campanian pelagic mudstones or foraminiferal wackstones and early Maastrichtian? intraclast packstones can be observed (Hu et al. 2012). This disconformity is interpreted as the regional uplift by the Deccan hotspot (Li et al. 2019a, b). Cretaceous in nTHS is composed mainly deep sea shales, turbidites and exotic blocks. The Early Cretaceous are volcanclastics and dark greenish gray shales overlain by Upper Cretaceous calcareous shales, radiolarian cherts, red shales with slumped limestones and olistostrome (Hu et al. 2008).

Recent years, the equivalent levels of global paleoceanographic events including Toarcian Oceanic Anoxic Event (T-OAE), Cretaceous OAEs, Paleocene-Eocene Thermal Maximum (PETM), and Cretaceous Oceanic Red Beds has been studied in detail. Moreover, the regional tectonic evolution, includes the reactivation of the passive margin in Early Cretaceous, the interaction between Deccan Trap and the Tethys Himalaya in Late Cretaceous, closure of Neo-Tethys Ocean revealed by latest marine horizon and provenance changes, have also been documented in last decades.

2 Schedule of the Field Trip

The field trip is mainly along the Yarlung Zangbo river in the south Tibet and the northside of the Himalaya. It takes 9 days to travel across four tectonic units of the Himalaya Orogen (Table 1), i.e., Gangdese Arc, Suture Zone, Xigaze Forearc Basin and Tethys Himalaya. We will investigate 27 typical outcrops (Fig. 14) related to closure history, the sedimentation process and paleoceanographic events in the Neo-Tethys Ocean. The field trip is mainly focused on the sediments of Jurassic to Paleogene. The igneous of Gangdese Arc and massif volcanic rocks of ophiolites will be observed as well. Additionally, one stop about the Permian–Triassic boundary event is involved. Therefore, the topics of the field trip are broad, which include not

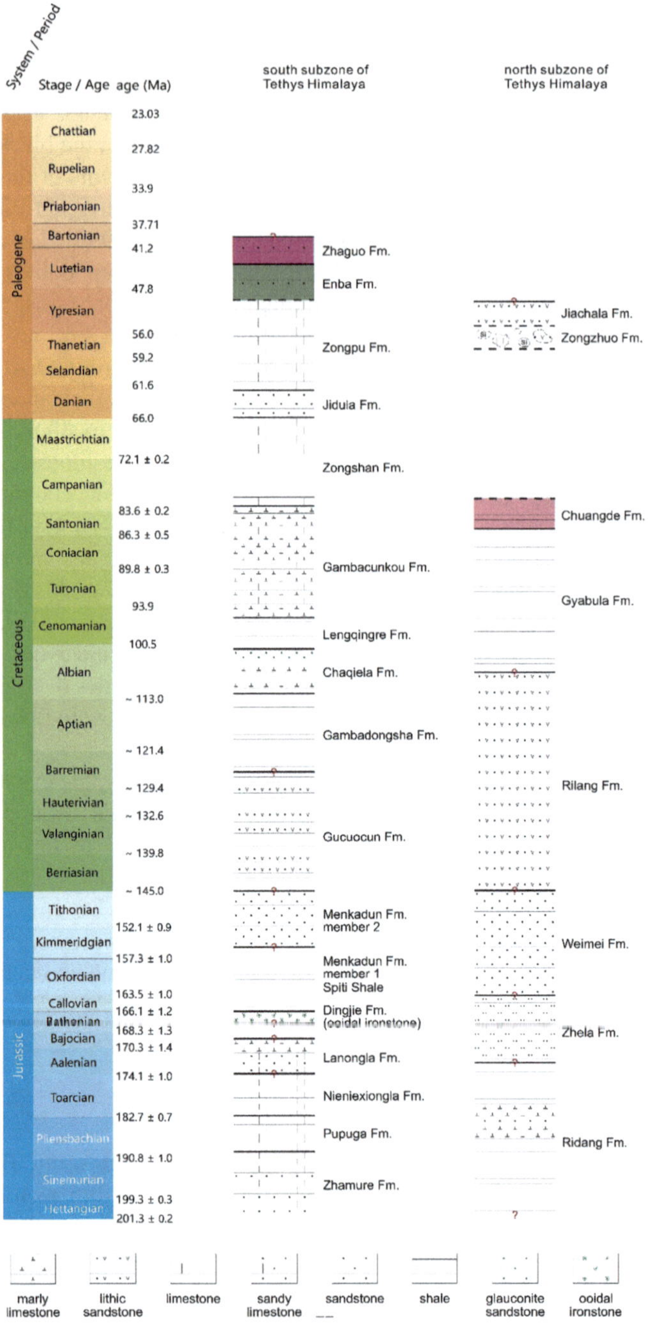

Fig. 13 Lithostratigraphy of Jurassic-Cretaceous in the south Tethys Himalaya and north Tethys Himalaya

Table 1 General schedule of the field trip

Date	Route	Stops	Observations
Day 1	Beijing to Lhasa		
Day 2	Lhasa to Xigaze	1–4	• Gangdese Batholith • Kailas Conglomerate • Channel conglomerate of the lower Nagmiring Formation in the Xigaze forearc basin
Day 3	Xigaze to Lhatse	5–9	• Sediments and tuff layer in Nagmiring Formation, Xigaze forearc basin • Liuqu Conglomerate • Sedimentary blocks of Xiukang Complex
Day 4	Lhatse to Old Tingri	10–12	• Upper Triassic deep-water sediments of northern subzone of Tethys Himalaya • Cretaceous-Paleogene shallow-water succession and hyperthermal events of southern subzone of Tethys Himalaya
Day 5	Old Tingri	13–17	• Latest marine horizon of Neo-Tethys • Early Cretaceous clastic dykes and cold seep carbonates • Lower Jurassic shallow-water succession and hyperthermal event • Triassic-Jurassic boundary sedimentation of southern subzone of Tethys Himalaya • Cenozoic N-S striking rift
Day 6	Old Tingri to Saga	18–19	• Middle-Upper Jurassic succession: transition from carbonate platform to siliceous clastic shelf • Permian–Triassic boundary sedimentation of south Tethys Himalaya
Day 7	Saga	20–21	• Late Cretaceous trench sediments of Neo-Tethys, Luogangcuo Formation • Cretaceous-Paleocene abyssal Sangdanlin section
Day 8	Saga to Xigaze	22–23	• Indian Asian provenance reversal recorded in Mubala section • Tectonic contact between the Xigaze forearc basin and Yarlung Zangbo Suture Zone
Day 9	Xigaze to Gyangze	24–26	• Ophiolites of Yarlung Zangbo Suture Zone • Sedimentary contact between the Xigaze forearc basin and Yarlung Zangbo Suture Zone
Day 10	Gyangze to Lhasa	27–29	• Cretaceous succession in north Tethys Himalaya • Triassic deep-water Langjiexue group
Day 11	Lhasa to Beijing		

only the tectonic sedimentation, but also the sedimentation of carbonate factory, submarine canyon, slope apron and of hyperthermal events. Most of the scheduled outcrops have been well studied and distributed in the last decades.

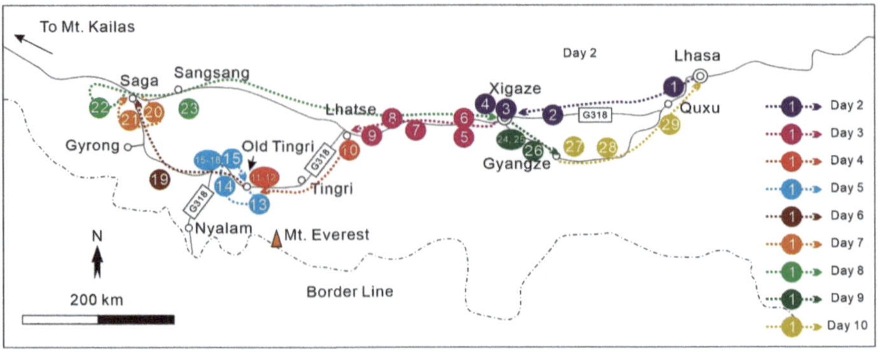

Fig. 14 Sketch shows the localities of the visiting stops

3 Description of Stops

Stop 1 Gangdese Batholith Near the Nietang Village, Quxu County (GPS: 29°29′57″N, 90°56′19″E)

This is a typical outcrop of the Gangdese batholith, and in literature it was called Quxu Batholith. Porphyritic monzogranite and coarse-grained granodiorite contain large amount of fine-grained gabbroic–dioritic enclaves (Fig. 15). Zircon U–Pb geochronology indicates the granitic hosts and the mafic microgranular enclaves have a same age of ~51 Ma (Ji et al. 2009). Co-occurrence of felsic and mafic rocks indicate significant magma mixing during emplacement of these intrusions. Hf isotopes from both the host rocks and the enclaves are also the same, with εHf(t) values of 11.0 ± 0.5 (Ji et al. 2009). Consistent Hf isotopic signatures indicate the mantle-derived gabbro and diorite and crustal-derived granitoids had become isotopically homogeneous during magma mixing, or they share a common isotopic source.

Emplacement of the Quxu pluton corresponds to the Paleocene- Eocene (~65–41 Ma) magmatic flare-up in the Gangdese arc. It is suggested that this magmatic active period was caused by rollback of the subducting Neotethyan oceanic slab and then culminated by the slab break-off that occurred around 50–45 Ma (Chung et al. 2005; Zhu et al. 2015). The intrusive and volcanic rocks (i.e., the Linzizong Volcanic Successions) formed during this period make up the main part of the present Gangdese arc.

On this outcrop, we can also see a Miocene (16.8 ± 0.4 Ma) granodioritic dyke intruded into the Eocene pluton (Fig. 15). The dyke show clear chilled margins and is characterized by lots of potassium feldspar phenocryst. These porphyry intrusions record an important post-collisional magmatic stage in the Gangdese arc, and are related to the formation of the Gangdese porphyry copper belt (Hou et al. 2009).

Fig. 15 Eocene monzogranite with dioritic enclaves (**a**, **b**); Miocene granodioritic dyke with potassium feldspar phenocryst (**c**, **d**). Photos are provided by Wei-Qiang Ji

Stop 2 Kailas Conglomerate at Dazhuka Bridge (GPS: 29°17′30.1″N, 89°47′54.4″E)

An exposure of the Kailas Conglomerate is observed. The conglomerate consists of gravels of various size and sandy matrix. Clasts are subangular to well rounded, poorly sorted. Rude normal grading and cross bedding are occasionally observed. Sedimentary textures indicate a braided river environment (Fig. 16). Gravel clasts are of polymictic composition, including chert, mafic and ultra-mafic rocks, sandstones, etc. The Yarlung-zangbo ophiolites and the accretionary mélange are the main source of the sediments.

Interlayer of felsic tuffs have been reported from nearby outcrops. $^{40}Ar/^{39}Ar$ dating of biotite and plagioclase separates from the tuffs gave ages of 20–22 Ma (Aitchison et al. 2009).

Fig. 16 Field photos of the Kailas Conglomerate in Dazhuka. Photos are provided by Jian-Gang Wang

Stop 3 Channelized Conglomerates of the Ngamring Formation in Xigaze Forearc Basin, Xigaze City (GPS: 29°16′01.4″N, 88°51′49.5″E)

At west of the Tashilhunpo Monastery in Xigaze city, five set of channel systems were identified within a west–east oriented zone extending 120 km long, interpreted to be channeled submarine fan megasequence (Fig. 17). In the Xigaze area, the submarine fan was outcropped 10–20 km west and east of the town. Five sandstone–conglomerate bodies as high, steep relief in the field were identified west of the Xigaze town: Quarry Ridge, Minor Ridge, Castle Ridge, Monastery Ridge, and West Ridge (Wang et al. 2012). These strata were developed during Cenomanian based on benthic foraminifera from siltstone along strike (Wan et al. 1998), with detritus mainly derived from the Gangdese magmatic arc, the central Lhasa terrane to the north and the underlying Sangzugang Formation (Wu et al. 2010; An et al. 2014).

At this stop, the Monastery Ridge system is composed by superimposed channel fills and overlying overbank/slope sediments, with a thickness of > 200 m. Channel fills are represented by conglomerates fining upwards to massive turbiditic sandstones with lenses of conglomerate and mudstone while overbank/slope sediments are characterized by mudstone intercalated with thin-bedded sandstone. Clastic-supported conglomerates contain well-rounded pebbles to cobbles of volcanics, granites, limestone, mudstone and chert. Bouma sequence Ta-c, scoured bases and shale intraclasts

Fig. 17 Panoramic view of the Monastery Rigde system of the Ngamring Formation. Photo by Wei An

were developed in sandstone beds (Wang et al. 2012). One rhyolitic cobble from its counterpart west of the Xigaze town was dated at ~95 Ma and one andesitic cobble from the southern limb of the forearc basin was dated at ~105 Ma (Dai et al. 2015).

Stop 4 Thrust Contact of the Kailas Conglomerate with the Ngamring Fm. At Tunqiong Village (GPS: 29°19′41.2″N, 88°47′35.5″E)

The top of the Kailas Conglomerate and its fault contact with the Xigaze forearc strata are observed (Fig. 18). The Kailas Conglomerate consists of purple-red, massive or laminated mudrock with carbonate nodules, and thin-bedded to massive siltstone, sandstone and fine-grained conglomerate. The red mudrocks were deposited in a floodplain, whereas sandstones and conglomerates represent braided river channels. Clasts are mainly various colored chert, with subordinate quartz sandstone, litharenite, mafic and ultramafic rocks, granite, and volcanic rocks. These sediments were derived from both sides of the basin, including the Gangdese arc to the north and the Xigaze forearc basin, the Yarlung-Zangbo ophiolite, the accretionary mélange and even the Tethyan Himalayan strata to the south. Deposition of the Kailas Conglomerate indicates a paleo-Yarlung-Zangbo River was established at the very start of the Miocene (Wang et al. 2013). On this outcrop, we can also see a lamprophyre dyke (~18 Ma; Yin et al. 1994) cross-cutting the stratigraphic beds of the Kailas Conglomerate. To the south, the Xigaze forearc strata (the Ngamring Formation turbidites) thrust northward over the Kailas Conglomerate by the Great Counter Thrust (Fig. 18).

Stop 5 Cold Seep Carbonates and Turbidites of the Ngamring Fm. Of the Xigaze Forearc Basin at Kadui Village (GPS: 29°09′06″N, 88°37′56″E)

This outcrop is located at the Kardoi village, which is ~20 km west to the Xigaze city and ~500 m south to the China-Nepal road. Nodular concretions are developed within the sandstone and shale intercalations of the Nagmiring Formation in Kardoi area (Fig. 19). Their sizes are much variable from a few centimeters to tens of centimeter.

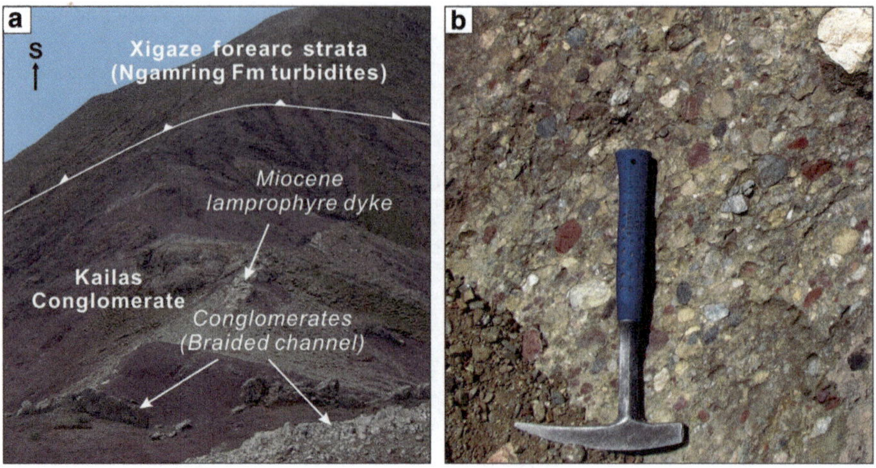

Fig. 18 Field photos of the Kailas Conglomerate in Tunqiong area, Xigaze. Photos are provided by Jian-Gang Wang

TOC content (0.05–0.31%) in the concretion is less than their host sediments (0.10–1.00%). The δ ^{13}C values range from –26.68‰ to –6.07‰ (PDB) (Liang et al. 2016). Within the nodular concretions, the δ ^{13}C values show a gradual decrease from the periphery to the center (Liang et al. 2016). These systematic variations reflect the diagenetic changes that occurred in the host sediment and indicate that the growth of the nodular concretions was sequential, from an early formed center to a later-formed periphery. The carbon source of the nodular concretions was derived from a mixture of anaerobic methane oxidation, methanogenic CO_2, and seawater-dissolved inorganic carbon (Liang et al. 2016).

At this stop, the sediments of submarine fan system can be observed (Fig. 20). The succession belongs to the megasequence 3, which is dominated by middle-outer fan to basin facies, e.g., gravity-flow sandstones, lobe and interchannel sandstone-shale couplets and basinal/pelagic limestones (Wang et al. 2012). It represents an assemblage of interchannel, overbank and channel sediments, which are clearly shown in the Fig. 20. At the base of sandstone unit, flute casts of ~0.5 m in size developed and indicat the sediments sourced from north (Fig. 21).

Stop 6 Tuff Layer in the Ngamring Formation of Xigaze Forearc Basin (GPS: 29°09′05″N, 88°37′13″E)

This stop will observe one tuff layer within the Ngamring Formation near the Kadui village. This tuff layer with a thickness of ~ 0.2 m was discovered within the upper Ngamring Formation (Fig. 22a, b) and one sample (ART01) was collected from this tuff layer. Zircons from this sample show relatively large size with a length of 100–200 μm, and a width of 80–120 μm. They show sector and oscillatory zoning and euhedral, long-prismatic shape. Their Th/U ratios are between 0.71 and 1.54. Fifteen zircon grains yield a weighted mean age of 91.3 ± 0.9 Ma (MSWD = 2.7; Fig. 22c, d;

Fig. 19 The cold seep carbonates, which are nodular concretions enclosed by dark gray shales and fine sandstones of the Nagmiring Formation in Kardoi area. Professor Chengshan Wang (left) and Qiyu Huang (Right) were standing in front of the outcrop. Photo by Xi Chen

Fig. 20 The submarine fan deposits of the Cretaceous Nagmiring Formation. The scour surface between the sandstone and the turbidites is clearly shown. At the base of the sandstone unit, flute casts were developed. Photo by Xi Chen

Fig. 21 Enlarged view of the flute casts show the paleo-current orientated from right (north) to left (south). The flute casts are of ~ a half meter in size. Photo by Xi Chen

Dai et al. 2015). This age provides more precise constraints on the deposition timing of the Ngamring Formation. Given that the tuff layer is conformably interbedded with the flysch, this age represents the timing of the air fall tuff and thus the age of deposition for the Ngamring Formation. Twenty analyses of tuff sample ART01 yield initial $^{176}Hf/^{177}Hf$ ratios of 0.282890–0.283005. These analyses show positive $\varepsilon_{Hf}(t)$ values of + 6.0 to + 10.2 and young $T_{DM}{}^{C}$ ages of 472–716 Ma, implying that they were also associated with an origin of juvenile crust. This interpretation is consistent with the extensive granitoid magmatism with large positive $\varepsilon_{Hf}(t)$ values in the Gangdese arc (e.g., Ji et al. 2009).

Stop 7 Liuqu Conglomerate Near Liuqu Village (GPS: 29°8′34.7″N, 88°9′25.1″E)

Exposure of the Liuqu Conglomerate and its structure relationship with surrounding tectonic domains are observed. The Liuqu Conglomerate here is folded into a several-kilometer-wide syncline by movement of faults along both its northern and southern boundaries. Along its northern outcrop boundary, the Liuqu Conglomerate is overthrust by ophiolitic rocks, whereas to the south, it is overthrust by the Xiukang Complex (Leary et al. 2016) (Fig. 23).

Lithofacies of the Liuqu Conglomerate are mainly clast-supported, massive or horizontally stratified, pebble-cobble conglomerate; red or green, massive mudstones; and massive and horizontally laminated siltstone. Massive, horizontally stratified, or trough cross-bedded sandstone; Matrix-supported, boulder conglomerate are occasionally presented. Sedimentological analysis shows the Liuqu Conglomerate is composed of mixed fluvial and sediment-gravity flow lithofacies assemblages locally interbedded with mature paleosols, suggesting a

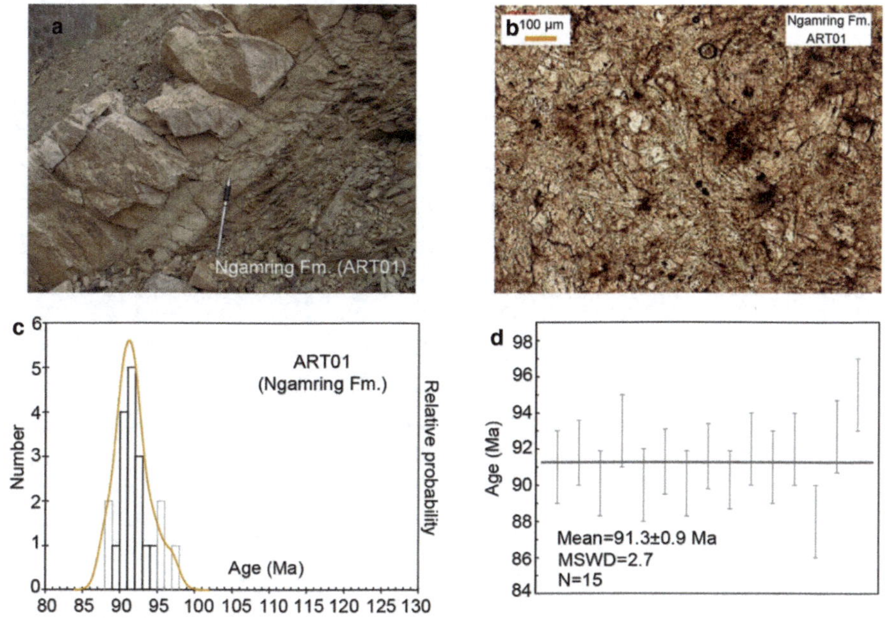

Fig. 22 Field photographs (**a**) and microstructure photomicrographs (**b**) of the Tuff in the Chongdoi Formation. Probability and weighted average of $^{206}Pb/^{238}U$ ages for sample ART01. Gray box in the probability diagram represents samples that plot outside of a Gaussian distribution or the highly discordant analyses, and they were excluded from the weighted mean age calculation (After Dai et al. 2015). Abbreviations: Fm., Formation

stream-dominated alluvial-fan environment (Davis et al. 2002; Leary et al. 2016) (Fig. 23).

Lithofacies and sedimentary texture indicate the Liuqu Conglomerate was deposited proximal to the source. Paleocurrent measured from clast imbrications suggests northeastward-northwestward sediment transport, indicating detritus were derived mainly from source terranes to the south. Basalt and serpentinized clasts were derived from the ophiolites; Green lithic sandstone and slate were most likely derived from the mélange complex; quartz sandstone, quartzite, and limestone were also likely derived from the mélange, with a probable ultimate origin in the Tethys Himalayan; chert clasts could be derived either from the ophiolites or from the mélange (Wang et al. 2010). Erosion of ophiolites and mélange could be a result of development of the Great Counter Thrust (Leary et al. 2016).

Stop 8 Xiukang Complex with Permian Exotic Limestone Blocks Near Quxia Village (GPS: 29°7′2.8″N, 88°2′57″E)

This stop is located at south of Quxia village to the west of G318 Road near the 5000 km milestone in Lhatse county.

The sedimentary-matrix mélange is composed of exotic limestone, sandstone, basalt and chert blocks within strongly deformed phyllitic matrix in the Zhongbei

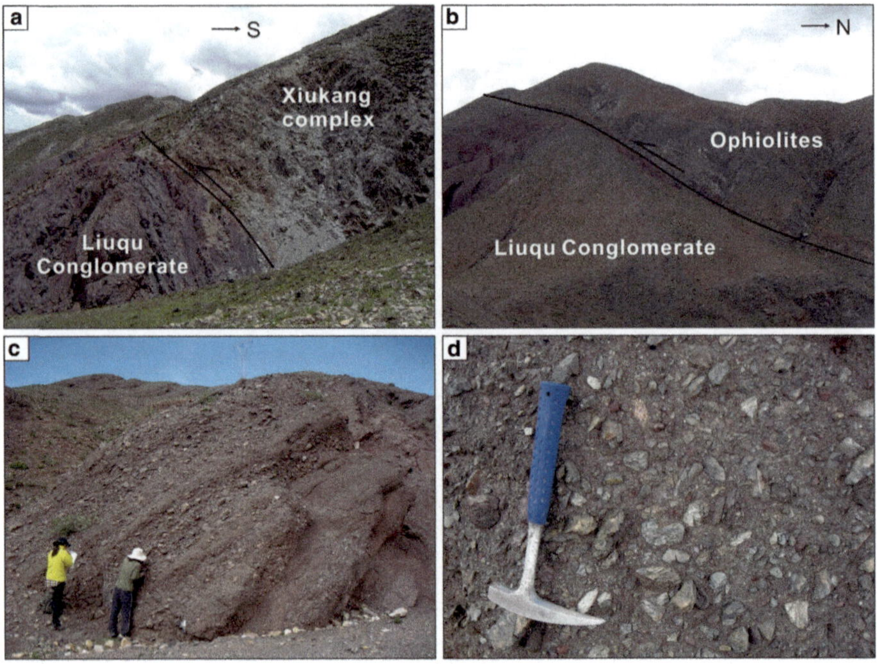

Fig. 23 Field photos of the Liuqu Conglomerate. Photos are provided by Jian-Gang Wang

area, Lhatse county. Numerous limestone blocks size from meter to kilometer, among which large ones form ridges of mountains and small ones often half buried or floating within matrix (Shen et al. 2003b; Jin et al. 2015). At this stop, the limestone blocks extend up to 3 km and for about 1 km along the road, in a wavy contact with the phyllitic matrix (Fig. 24). Brown, purplish red and greenish gray bioclastic limestones are intercalated with thin-bedded mudstone (Fig. 25). Fragments of crinoid and bryozoan, brachiopod shells and corals with peri-Gondwanan and cathaysian affinity are reported, indicating a middle-late Guadalupian age (Jin et al. 2015). The bioclastic limestone generally contains terrigenous clasts and/or intercalated with mudstone, suggesting a proximal depositional environment. Generally, these blocks are considered as dismembered units of middle Permian carbonate successions broken during the rifting process in the Late Triassic (Shen et al. 2003b; Jin et al. 2015).

Stop 9 The Xiukang Complex with Sandstone Blocks at Colha Pass Along G318 in Lhatse County (GPS 29°04′08.0″N, 87°59′29.7″E)

Small sandstone blocks and peridotite blocks are preserved within mudstone matrix at this stop (Figs. 26 and 27). Quartoze sandstones (Fig. 27c) are outcropped up to

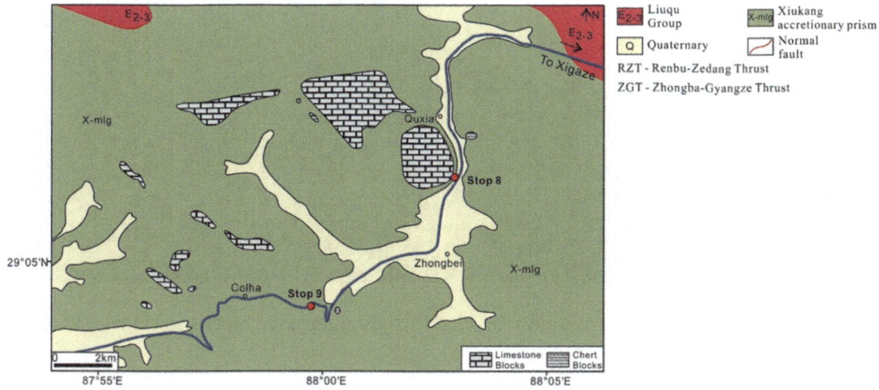

Fig. 24 Geological map of the Zhongbei area, Lhatse county (After An et al. 2017)

Fig. 25 Limestone blocks at Stop 9 (left: red bioclastic limestone intercalated with marls; right: bioclastic limestone). Photo by Wei An

a few kilometers in size, consist of thick-bedded or lenticular quartzarenite intercalated with thin-bedded gray siliceous mudstone. These blocks were derived from the Indian craton, deposited as Indian passive margin strata and incorporated into the accretionary prism during India-Asia collision (Fig. 28). Meanwhile, feldspatho-litho-quartzose volcaniclastic sandstones are Late Cretaceous (Fig. 27d) or Paleogene (Fig. 27e, f) based on the youngest detrital zircon U–Pb ages and sourced from the Lhasa terrane (Fig. 28). The Late Cretaceous ones are rhythmically-interbedded with mudstone in m-size outcrops while the Paleogene ones are exposed sparsely along the G318. Therefore, these blocks were interpreted as deposited in trench/trench-slope basin accreted into the Xiukang Complex during Neo-tethyan subduction or syn-collisional basin and dismembered during the India-Asia collision (An et al. 2017). Also, a panoramic view of the Xiukang Complex will be observed at this stop (Fig. 26).

Fig. 26 Outcrops of Stop 8 (left: sandstone blocks; right: limestone blocks as ridges). Photo by Wei An

Stop 10 Upper Triassic Deep-Water Nieru Formation of the Northern Subzone of Tethyan Himalaya Near Lhatse County (GPS 29°00′15.8″N, 87°39′14.4″E)

The Nieru Formation consists slate and quartzo-sandstone with occasionally thin-bedded limestone and chert deposited in a deep marine environment on the Indian passive margin. The depositional age of Carnian-Norian is indicated by bivalve fossils (Yu and Zhen, 1979). Provenance studies indicate a continental block source from the Indian continent (Cai et al. 2016). In the Lhatse area, the Nieru Formation was intruded by Early Cretaceous bimodal suite of mafic and felsic rocks as part of the Tethyan Himalaya igneous province occurring across the whole Tethyan Himalaya (Chen et al. 2018). We will observe grey slate intruded by mafic dikes at this stop (Fig. 29), which display porphyritic textures with phenocrysts of olivine, augite and plagioclase, yielding a zircon age of 117–118 Ma (Chen et al. 2018).

Stop 11 The Cretaceous Nirang Section of Southern Tethys Himalaya Near Old Tingri (GPS: 28°39′56″N, 87°02′10.14″E)

At this stop we will observe the Cretaceous succession (Albian-Maastrichtian) at Nirang section in the Tingri area, which is ~40 km west of the city of Tingri and is located on the southern slope of the Zhepure mountain (Fig. 30).

In the Tingri area, the Cretaceous to Paleogene are mainly exposed in the Zhepure Mountain and form a syncline, which extends ~50 km from the New Tingri City in the east to the Old Tingri City in the west. Previous studies on the Cretaceous are mainly focused on the north flank of the syncline, e.g., Tingri section (Willems et al. 1996) and Gongzha section (Hao and Wan, 1985; Li et al. 2006; Hu et al. 2012). However, the strata before middle Cenomanian are not exposed in the north of Zhepure Mountain. Therefore, the Albian strata has not been identified in the Tingri area before. In last two years, we found a new section, namely Nirang section located in the south flank of the Zhepure Mountain, which contains Cretaceous of ~1,200 m thick (Fig. 31, Zhang et al. in review). The succession of the Upper Cretaceous in the upper half of the Nirang section is quite similar to the published

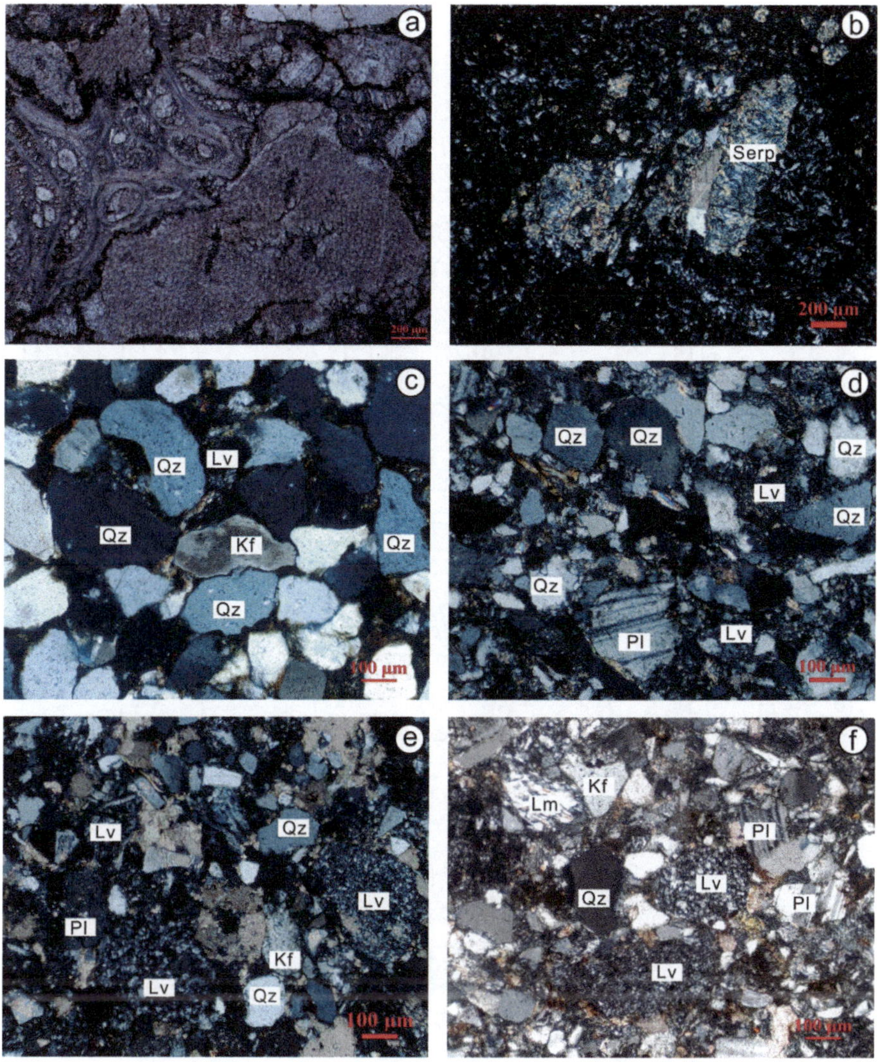

Fig. 27 Microphotographs from An et al. (2017): **a** bioclastic limestone with crinoids and bryozoans (limestone block at Stop 8); **b** serpentinized peridotite (near Stop 9); **c** Group 1 quartzarenite; **d** Group 2 volcaniclastic sandstone; **e** and **f** Group 3 volcaniclastic sandstone. (Qz = quartz; Kf = K-feldspar; Pl = plagioclase; Lm = metamorphic lithic fragments; Lv = volcanic lithic fragments)

Tingri and Gongzha sections. In this section, we can investigate the most complete succession of Albian-Campanian in the Tingri area: Albian fine siliciclastic rocks and marls with glauconitic sandstones in the top (Chaqiela Fm.), Cenomanian gray shales (Lengqingre Fm.), late Cenomanian to Santonian marls and marly limestone couplets (Gamba Cunkou Fm.), and limestones of Campanian to ?Maastrichtian

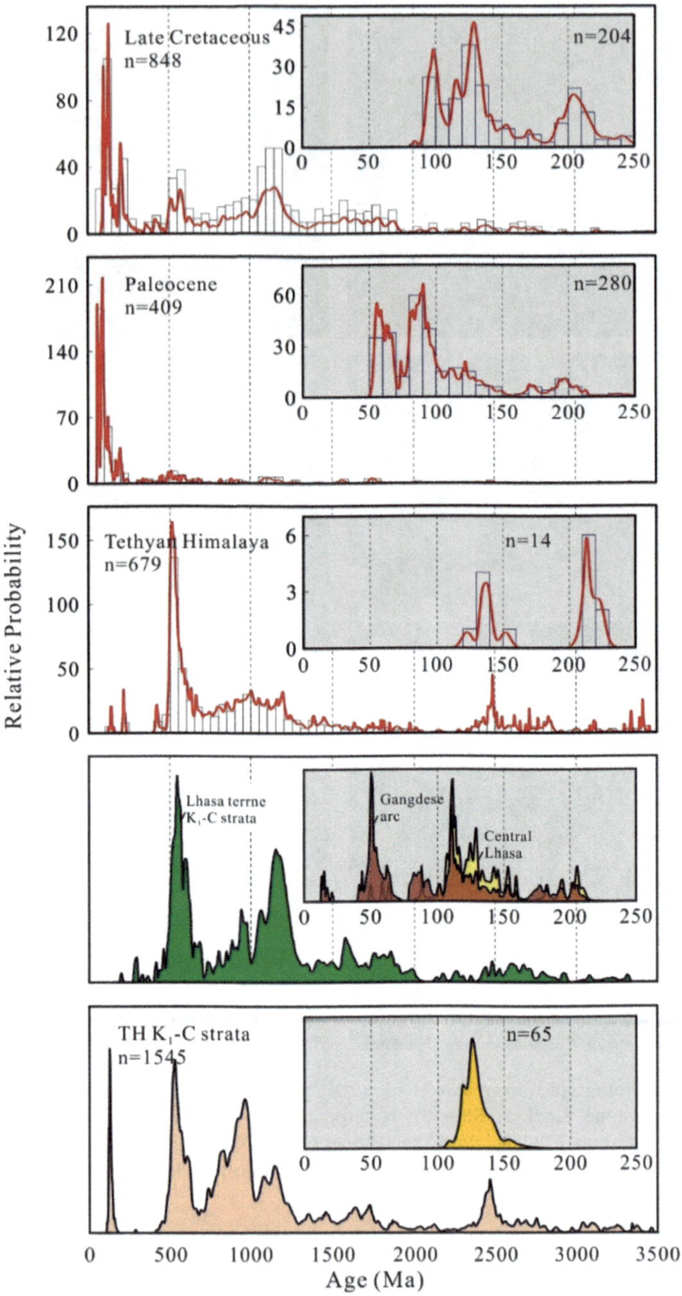

Fig. 28 U–Pb age spectra of detrital zircons in Group 1–3 sandstone blocks from the Cola pass, data from An et al. (2017) and Metcalf and Kapp (2019). Shown for comparison are data from Gangdese arc (Ji et al. 2009 and references therein), the Lhasa block (Leier et al. 2007a, b), Tethyan Himalaya successions (Zhu et al. 2011; Wu et al. 2014a, b)

Fig. 29 Outcrop of Stop 10 (left: slate with lens of limestone; right: diabase block within the Nieru Formation). Photo by Wei An

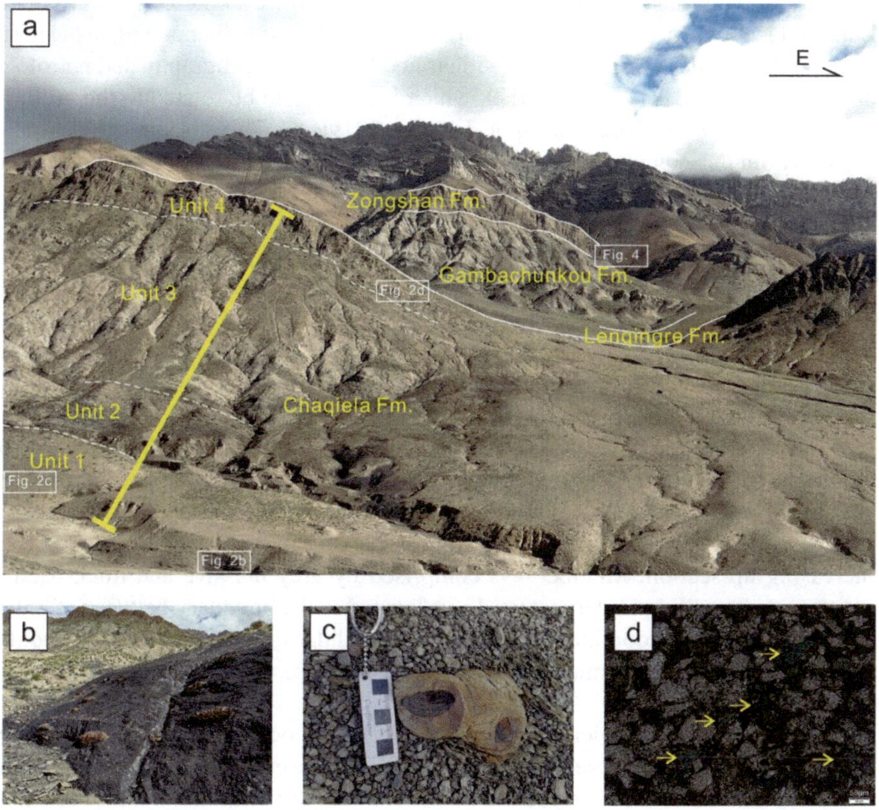

Fig. 30 Field photo of the Nirang section (**a**), underlying black shales of Gamba Dongshan Formation (**b**), seep carbonate concretions in the unit 1 of the Albian (**c**) and Glauconitic sandstone in thin section (**d**), glauconites are indicated by yellow arrows (Liu et al. in review)

(Zongshan Fm.) (Fig. 30, Liu et al. in review). Abundant trace fossils including *Zoophycos*, and other fossils e.g., shark teeth, can be found in the Albian glauconitic sandstones. Moreover, seep carbonate concretions were developed in the silts in the lower ~80 m of the section. Nannofossil biostratigraphy and chemostratigraphy studies have been submitted to international journals. Carbon isotope curve of Albian can be well correlated with the presentative sections in east Pacific, western Tethys and other areas. Furthermore, spectral analysis on the carbon isotope curve, magnetic susceptibility and TOC contents revealed that orbital-paced regional hydrological processes caused changes in terrigenous inputs and organic matter burial, which are responsible for periodically fluctuations of carbon isotope values (Liu et al. in review). Oceanic Anoxic Event 1b, 1c and 2 are identified in this section with biostratigraphy, lithostratigraphy and chemostratigraphy studies. OAE 1b and 1c is identified with the chemostratigrpahy (Liu et al. in review). Disconformity in the Late Cretaceous limestonesbetween Jiubao Fm. and Zhepure Shanpo Fm. in the north slope of the Zhepure Mountain were identified by the biostratigraphy and petrology observation (Fig. 31, Hu et al. 2012), which is inferred to be related to the uplift of Tethys Himalaya by Deccan hotspot (Li et al. 2019b, a). This disconformity is also found within the Zongshan Fm. of Nirang section. According to study of calcareous nannofossils, the spanning of this disconformity is less than 10 myr (Figs. 31 and 32.

Stop 12 Paleocene-Eocene Thermal Maximum (PETM) in Shallow-Water Carbonate Platform Environment in the Tethys Himalaya at Shenkezha Section in Old Tingri (GPS 28°26′39.1″N, 86°40′22.5″).

At this stop, we will observe the Paleocene-Eocene boundary section in shallow-water carbonate platform environment. The PE boundary here was located in the upper part of Zongpu Formation. The Zongpu Formation can be subdivided into four members, from bottom to top: Member 1 are dominated by thin- to medium-bedded yellowish-grey limestones and marly limestones with green algae and foraminifera; Member 2 to 3 is a dark grey nodular limestone with marly interbeds, the marly is increasing up-section; Member 4 is comprised by grey massive and thick-bedded limestones. The PE boundary interval was located in the boundary between member 3 and member 4, it is dominated by 18 m-thick nodular limestone and marly nodular limestone in the lower to middle part, and 6 m -thick massive limestone in the upper part (Fig. 33).

Detailed sedimentological, paleontological, and geochemical analysis indicates that a relative rise in sea level coincided with PETM onset, continued through PETM core, and terminated with a regression at PETM recovery (Fig. 34). At PETM onset, corresponding to the SBZ4/SBZ5 boundary, no obvious impact on biota and specifically on larger benthic foraminifera (LBF) is observed. The major biotic change occurs later on at PETM recovery, corresponding to the SBZ5/SBZ6 boundary. Our data suggest that relative sea level, rather than temperature, exerted the main control on benthic biota during the PETM. The $\delta^{13}C_{carb}$ value of bulk carbonates are significantly ^{13}C-depleted, which we attribute to environmental change driven by relative sea-level fluctuations.

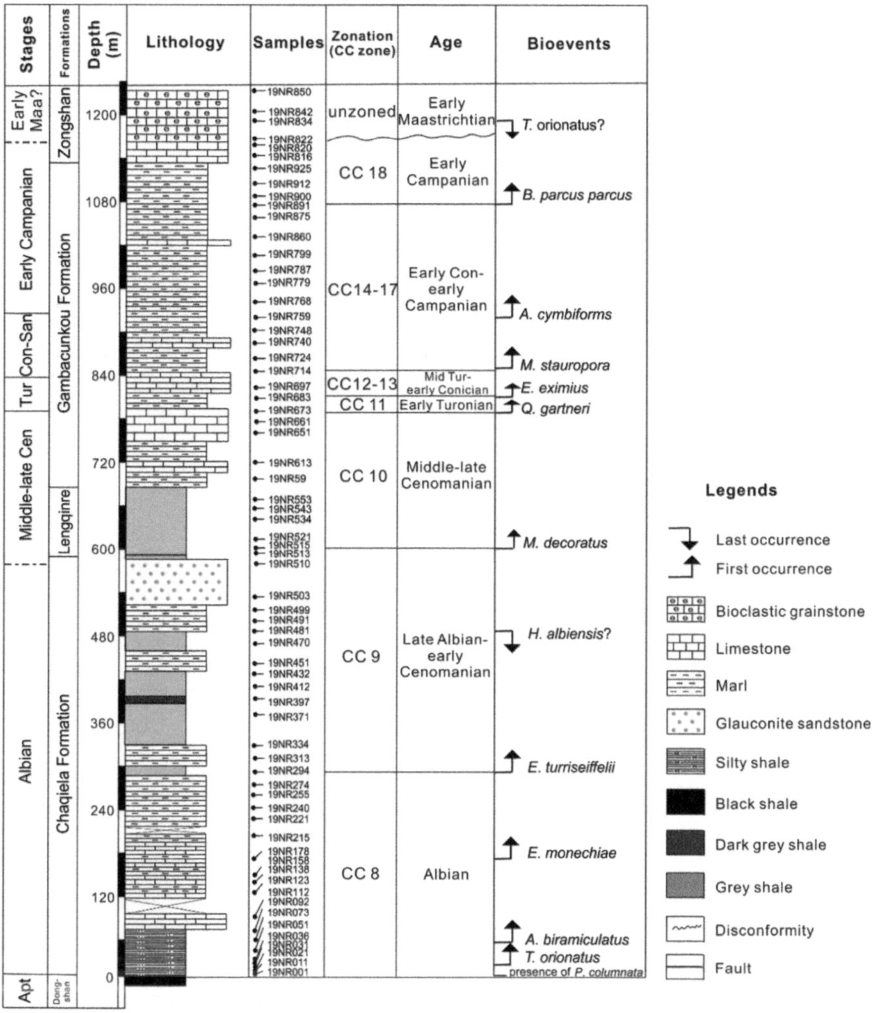

Fig. 31 Calcareous nannofossils biostratigraphy of the Nirang section (Zhang et al. in review)

Stop 13 Early Cenozoic Succession and Latest Marine Horizon in the Tethys Himalaya at Longjiang Section Near Old Tingri (GPS 28°26′39.1″N, 86°40′22.5″)

The Paleogene Zongpu, Enba and Zago formations are well exposed in the Longjiang section, about 5 km on the eastern side of the Old Tingri to Rungbulk monastery. At this location, the Upper Cretaceous gray marl (Gambacunkou Formation) transitions upward into sandy marl and marl with unequal thickness, and then upward into a marl layer (Zhepureshanpo Formation). The Palaeogene foraminifer-rich Zongpu Formation is sitting on top of the Zhepureshanpo Formation with fault contact which

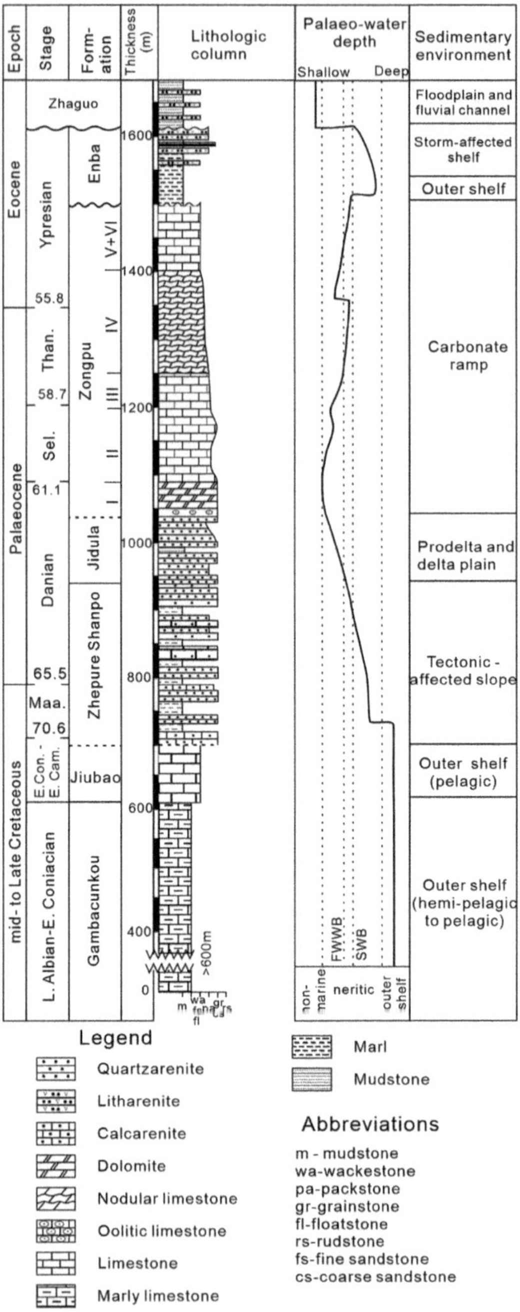

Fig. 32 Composite stratigraphic log of the Cretaceous—Palaeogene strata at the Zhepure Mountain area showing interpreted palaeo-water depth trend and sedimentary environments (Hu et al. 2012 and references there in). FWWB–fare weather wave base; SWB–storm wave base

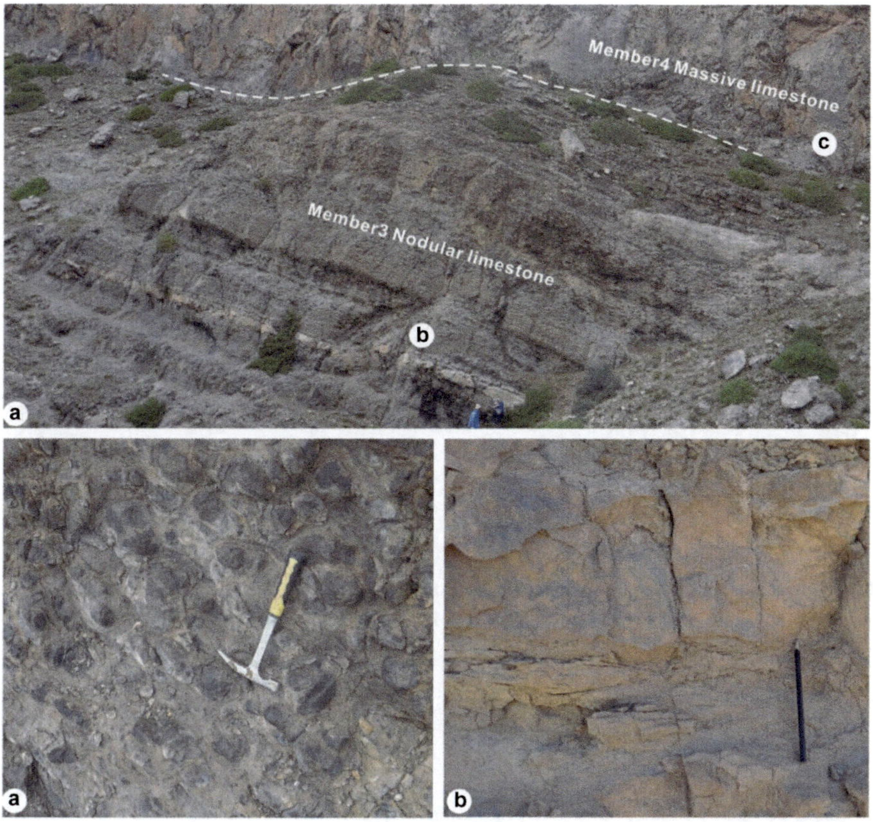

Fig. 33 Outcrop photographs of PE boundary interval in the Shenkezha section (**a**) showing: (**b**) Member3 nodular limestone in the lower and middle part; and, (**c**) Member4 thick-bedded to massive limestone in the upper part (Li et al. 2020a, b)

is thrusted from north to south dipping to the north. The uppermost of the Zongpu Formation and the overlying Enba and Zhaguo formations are nicely exposed in the Longjiang section representing the latest marine deposits in the Himalayan area (Fig. 35) (Wang et al. 2002; Hu et al. 2012).

Larger benthic foraminifers were distributed in the limestones of the Zongpu Formation indicating Eocene SBZ7-SBZ12 (Fig. 36). The PETM record in shallow-marine carbonates of the Zongpu Formation (Li et al. 2021) reported in the Zhepure Mountain is not outcropped due to the thrusting. There is an erosional interface on the top of the Zongpu Formation, which is in parallel or slightly angular unconformity with the overlying Enba Formation. The main body of the Enba Formation is dark gray calcareous shale sandwiched by thin lithic sandstone, with a visible thickness greater than 40 m. The Enba Formation is in sedimentary contact with the Zhaguo Formation. The main body of the Zhaguo Formation is purple-red shale with a few

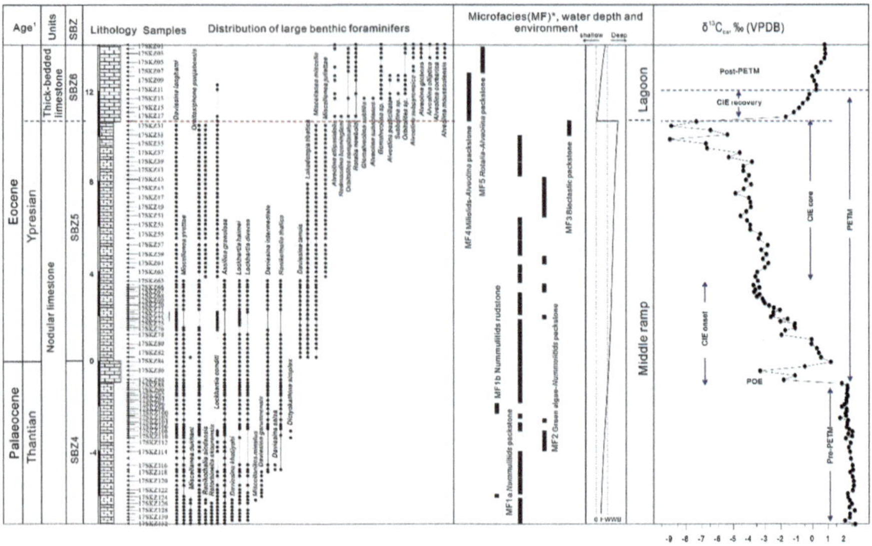

Fig. 34 Lithological log of the Shenkezha section (Tibet) showing the distribution of large benthic foraminifera, carbonate microfacies, interpreted paleowater depths, sedimentary environments and carbon-isotopes across PETM (Li et al. 2020a, b)

Fig. 35 Panoramic photo of the Longjiang section (southeast of the camera, photo by Xiumian Hu)

sandstone layers. Green nodular conglomerate is seen in the red sandstone, all of which are carbonate nodules, formed by the redeposition of palaeosoil nodules.

Provenance analysis from the sandstones in the Enba and Zhaguo formations including detrital modal, detrital zircon U–Pb age dating and Hf isotopes indicated that the sedimentary rocks of both the Enba and Zhaguo formations have similar provenance, with the main sources being rocks of the Lhasa terrane (Fig. 37). The Enba Formation records the first appearance of detritus from the Asian margin (Wang

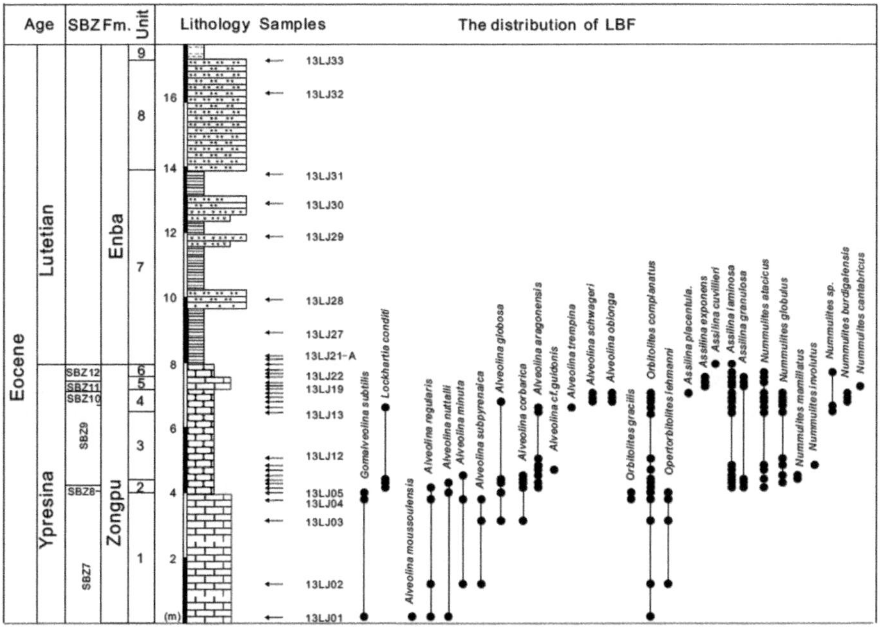

Fig. 36 The larger benthic foraminiferal distribution in the Longjiang section. Foraminiferal identification by Prof. Dr. Marcelle BouDagher-Fadel from University College London

et al. 2002; Zhu et al. 2005; Najman et al. 2010; Hu et al. 2012) in the southern Tethyan Himalaya in shallow-water environment in the northern Indian margin.

Stop 14 Lower Cretaceous Clastic Dykes and Cold Seep Carbonates Near Gucuo Village (GPS: 28°44′58″N, 86°18′38″E)

We will investigate field characteristics of Lower Cretaceous clastic dykes and cold seep carbonates in the southern Tethys Himalaya. This outcrop is located approximately 30 km northwest of Old Tingri and approximately 5 km south of the village of Gucuo. The clastic dykes and cold seep carbonates are logged at two sections positioned next to each other: Gucuo East (GPS: 28°44′36″N, 86°18′39″E) and Gucuo II (GPS: 28°44′58″N, 86°18′38″E) (Fig. 38). These two sections, which have a total thickness of approximately 480 m, correspond to Gucuo Formation and Gamba Dongshan Formation. Both sections consist largely of dark grey shales intercalated with volcaniclastic (Chen et al. 2017; Guo et al. 2021).

Clastic dykes composed of lithic-rich volcaniclastic sandstone are widely exposed in the lower Aptian shales. Most clastic dykes are subvertical. The length and width of clastic dykes range from 5 to 50 m and from 20 to 150 cm, respectively. The dykes exhibit sharp and undeformed contacts with the surrounding shales. The majority of the dykes display a sheet-like geometry and generally they do not thicken or thin significantly. Many dykes are distributed along a series of normal faults, and some display cross-cutting and drag relationships with each other (Fig. 39). The U–Pb age

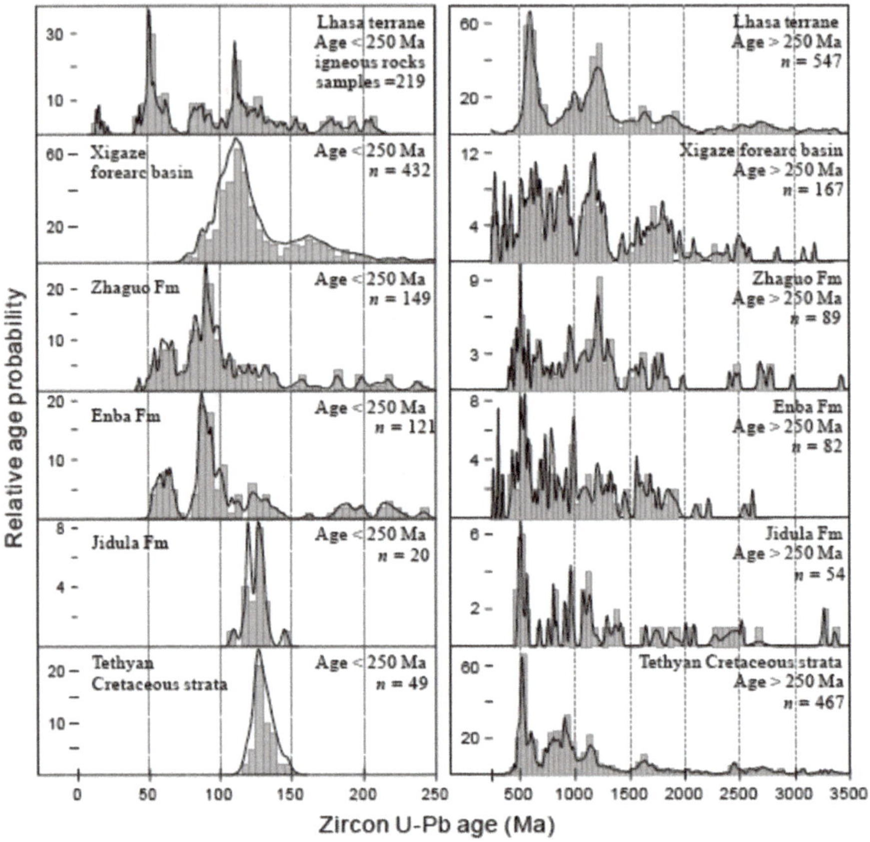

Fig. 37 Relative probability plots of analysed U–Pb data in Enba and Zhaguo formations (Hu et al. 2012)

Fig. 38 Outcrop of Gucuo II section showing clastic dykes and cold seep carbonates (Chen et al. 2017)

spectra of zircon grains contained in clastic dykes display a youngest peak at 124.3 ± 2 Ma. Petrographic analyses show that the dykes were derived from underlying strata. The liquefaction of underlying strata was induced by earthquakes associated with the extensive magmatic activities and regional extension that affected the Indian passive margin of Neotethys during the Early Cretaceous (Guo et al. 2021).

A suite of carbonate nodules with a thickness of about 1.38 m developed under the oceanic anoxic events OAE1a of the Gucuocun Formation. The carbonate nodule layer can be subdivided into three layers: the lower layer is small pipe-shaped nodules

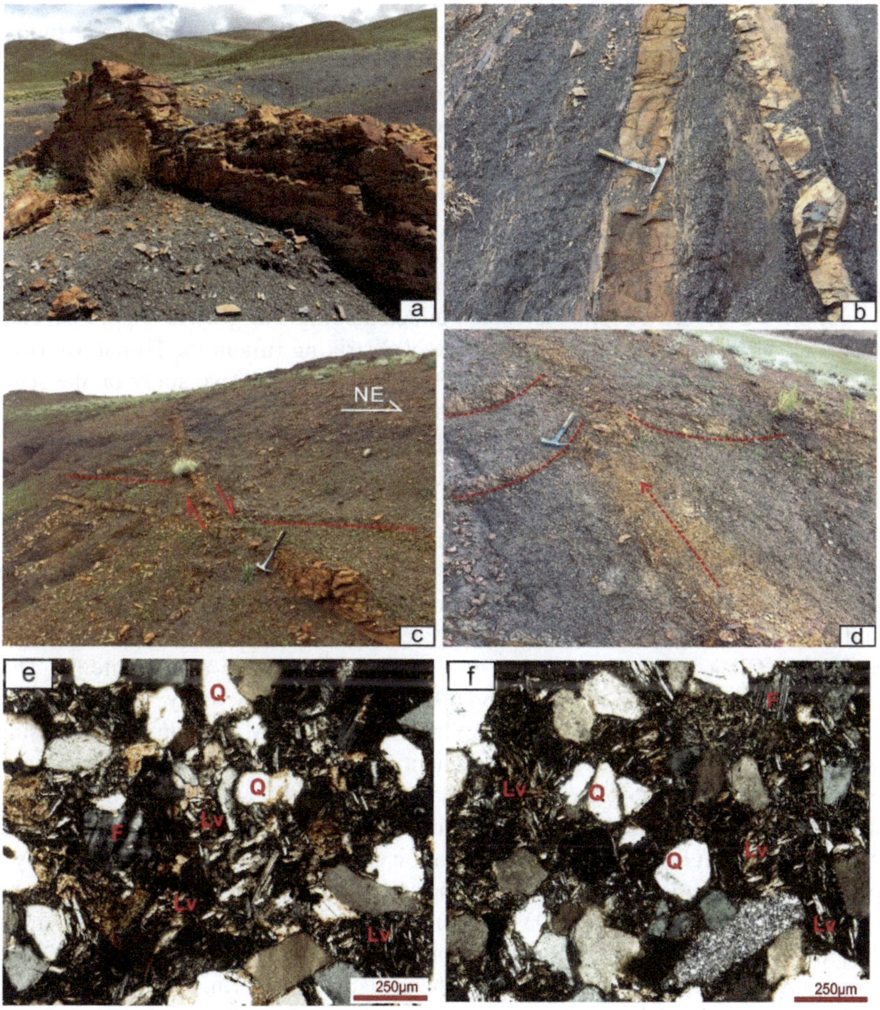

Fig. 39 Clastic dykes in outcrops and photomicrographs. (**a, b**) sharp undeformed contacts; (**c**) dykes distributed along normal faults; (**d**) drag relationship between clastic dykes; (**e–f**) sandstones within underlying volcaniclastic sandstones clastic dykes (Guo et al. 2021)

with a length-to-width ratio of 2.4 cm/1.8 cm, which are produced in parallel or obliquely across the layers. and the middle layer is composed of ellipsoidal and plate-shaped nodules. The length-to-width ratio of nodules is 17 cm/11.5 cm. The upper layer is breccia, and the breccia is composed of limestone. It is a sandy composition, and the nodules are mainly composed of authigenic carbonate minerals and detrital materials, with the development of banded and botryoidal cements (bbc), framboidal pyrite and other common sedimentary fabrics of modern cold spring carbonate rocks (Fig. 40, Fan, 2021)). The nodules contain trace fossils, foraminifera, cephalopods, bivalves and other biological fossils. The δ ^{13}C value of carbonate rocks is −23.26–−7.14‰ (Fan, 2021). Sulfur isotope values of carbonate associated sulfates are 17.3–19.0‰ (Chen unpublished data). The morphological, mineralogical and geochemical analysis of the carbonate nodules all show that the nodules are cold seep carbonate rocks, indicating the occurrence of cold spring leakage under the sediment–water interface. The underlying clastic dykes provide a channel for upward seepage.

Stop 15 Kung Cuo Half Graben in Tingri (a Far View from the Location of GPS: 28°46′11.3″N, 86°20′53.1″E)

The formation of the Cenozoic N-S striking rifts in the Tibetan Plateau is the consequence of continuous contraction after the India-Asia collision. Its formation and evolution are of great significance for understanding the growth of the Tibetan Plateau. The main extension period of the N-S striking rifts in the Himalaya-Tibet Plateau is Miocene. Garzione et al. (2003) did a comprehensive study on the sediments in the Thakkhola graben, central Nepal and showed east–west extension and environmental change at that time in southern Tibet.

The Kung Co half graben is bounded by an active N–S normal fault with a minimum vertical offset of 1600 m. Mahéo et al. (2007) estimated the onset of normal faulting by combining high and medium temperature (U–Pb, Ar/Ar) and low temperature ((U–Th)/He) thermochronometry of the Kung Co pluton, a two-mica granite of the northern Himalayan granitic belt that outcrop in the footwall of the fault. Biotite and muscovite Ar/Ar ages, are close from each other [~16 ± 0. Ma 2 (Ms) and ~15 ± 0.4 Ma (Bt)], which is typical of fast cooling. The zircon and apatite (U–Th)/He ages range from 11.3 to 9.6 Ma and 9.9 to 3.7 Ma respectively. These He ages are indicative of (1) fast initial cooling, from 11.3 to ~9 Ma, gradually decreasing with time and (2) a high geothermal gradient (~400 °C/km), close to the surface at ~10 Ma. The Kung Co pluton was emplaced at about 22 Ma (U–Pb on zircon) at less than 10 km depth and 520–545 °C. Subsequent to its shallow emplacement, the pluton underwent fast thermal re-equilibration ending around 7.5 Ma, followed by a period of slow cooling caused either by the end of the thermal re-equilibration or by very slow exhumation (0.02–0.03 mm/yr) from ~7.5 Ma to at least 4 Ma. In either case the data suggest that the exhumation rate increased after 4 Ma, in which this increase was interpreted to be related to the initiation of the Kung Co normal fault.

Looking northward into the distance from Kung Co village, we have a view of Quaternary landforms and distribution features of recently active normal faults (Fig. 41). A series of triangular facets truncate the ridges between U-shaped glacial valleys. The lower part of each facet is hard and composed of shattered granite

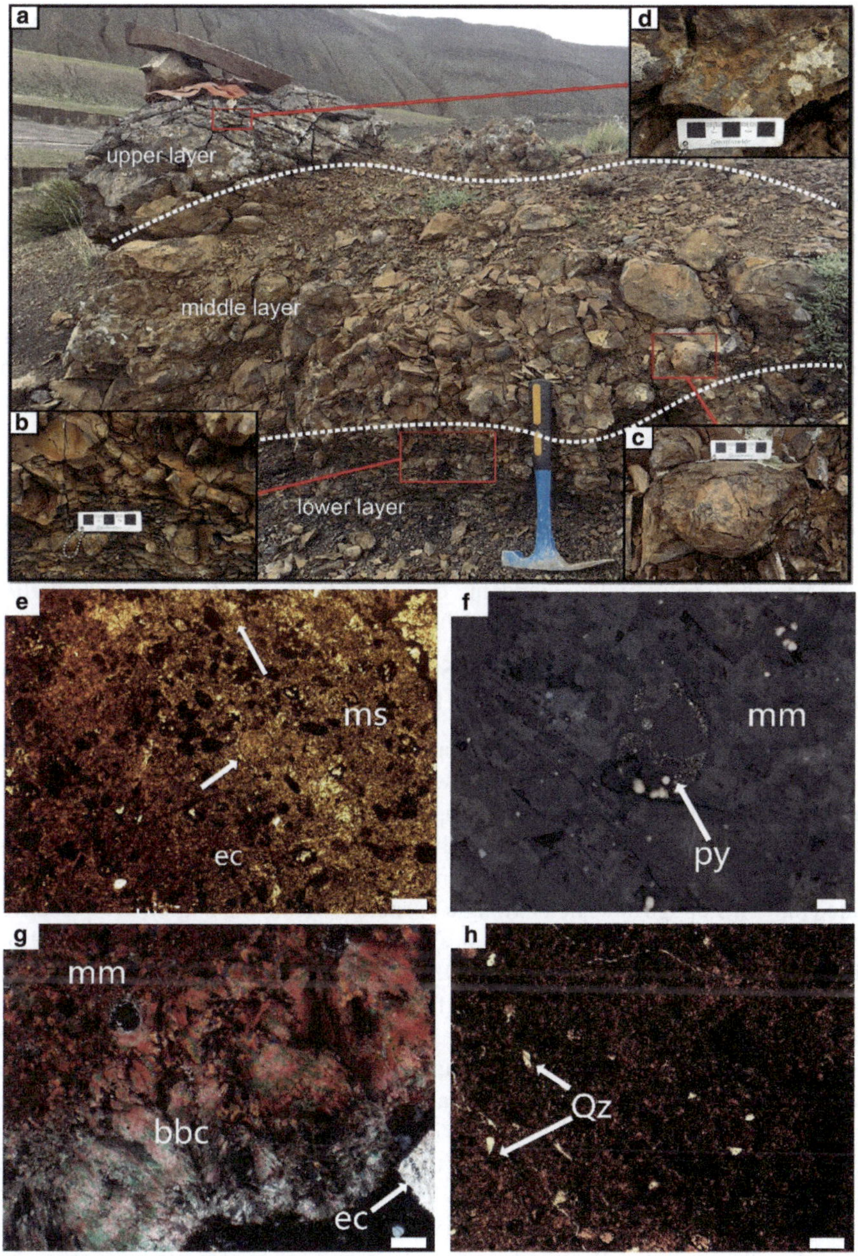

Fig. 40 Field photographs and photomicrographs of the cold seep carbonate (Fan, 2021). **a** the morphology of nodules in section (**b**) pipe-shaped nodules in the lower layer (**c**) ellipsoidal and plate-shaped nodules in the middle layer (**d**) breccia in the upper layer (**e–h**) thin section micrographs of representative authigenic carbonates. py: pyrite; mm: carbonate with matrix micrite; bbc: banded and botryoidal calcite; ec: equant calcite; ms: microspar

containing striations. The glacial valleys are "hanging" and horizontal breaks in slope are clearly displayed across the range front. Total vertical displacement on the major normal fault is larger than the relief of the bedrock landform which is more than 800 m.

Stop 16 Triassic-Jurassic Boundary Sedimentation at Gemig Section (GPS: 28°42.179′N, 86°08.623′E)

The Gemig section (Fig. 42) in the Nyalam area is located at the 5259 milestone of the Tingri–Nyalam Highway, about 2 km south of the small village of Gemig. The section provides a continuous exposure of ammonoid-bearing, uppermost Triassic and basal Jurassic strata. In total 11 lithological units have been recognized in this section which is composed of the Menkadun formation (unit 0), Zhamure formation (unit 1) and the Germig formation (units 2–10) and the overlying quartz-sandstone series (Fig. 43) (Yin et al. 1999).

The T-J boundary falls within the the Germig formation, which consists of alternations of fine calcareous sandstones and subordinate siltstones. In thin section the sandstones are seen to be well sorted, calcite-cemented quartz arenites. The highest bed contains abundant grains of glauconite. In the T-J boundary beds pyrite aggregates become locally highly abundant. This is stratigraphically close to the earliest recorded *Psiloceras* (Hallam et al. 2000).

Stop 17 Lower Jurassic Carbonate Succession and Outcrop of the Toarcian Oceanic Anoxic Event at Nianduo Section (GPS: 28°40′52″N, 86°08′7″E)

The Nianduo section is located near Nianduo village between the 5263 and 5264 km milepost on the Tingri–Nyalam Highway in the Nyalam County. This section preserves the shallow-water carbonate deposits on the so-called Kioto Platform which extensively developed during the Pliensbachian to Toarcian interval in the southern zone of the Tethys Himalaya. The lithostratigraphic unit has been traditionally subdivided into the Pupuga and Nieniexiongla Formations from the bottom to top and well studied in the Nianduo (Nyalam County) and the Wölong (Tingri County) sections (Jadoul et al. 1998; Han et al. 2016, 2021a, 2021b, 2022). The Pupuga Formation represents shallow-water carbonate platform, dominated by bioclastic grainstones/packstones and yielding *Lithiotis* bivalves. The overlying Nieniexiongla Formation represents a deeper water carbonate ramp and is mainly composed of micrites alternating with coarser grained storm-generated layers. The larger benthic foraminifera, the occurrence of *Lithiotis* bivalves and carbon-isotope chemostratigraphy date the Pupuga Formation as the Pliensbachian to lowest Toarcian (Jadoul et al. 1998; Han et al. 2016, 2021a, b; Jiang et al. 2020). The T-OAE records has been placed at the Pupuga–Nieniexiongla transitional interval, characterized by the occurrence of the characteristic negative CIE, significant sea-level rise, carbonate-platform crisis, and abundant storm-generated layers (Han et al. 2016, 2021a, b).

In the field, the Pupuga Formation is dominated by thick- and medium-bedded, offwhite grainstones (Fig. 44a–g), whereas the Nieniexiongla Formation is characterized by thin-bedded, dark grey micrites (Fig. 44a, d, i, j) (Han et al. 2016).

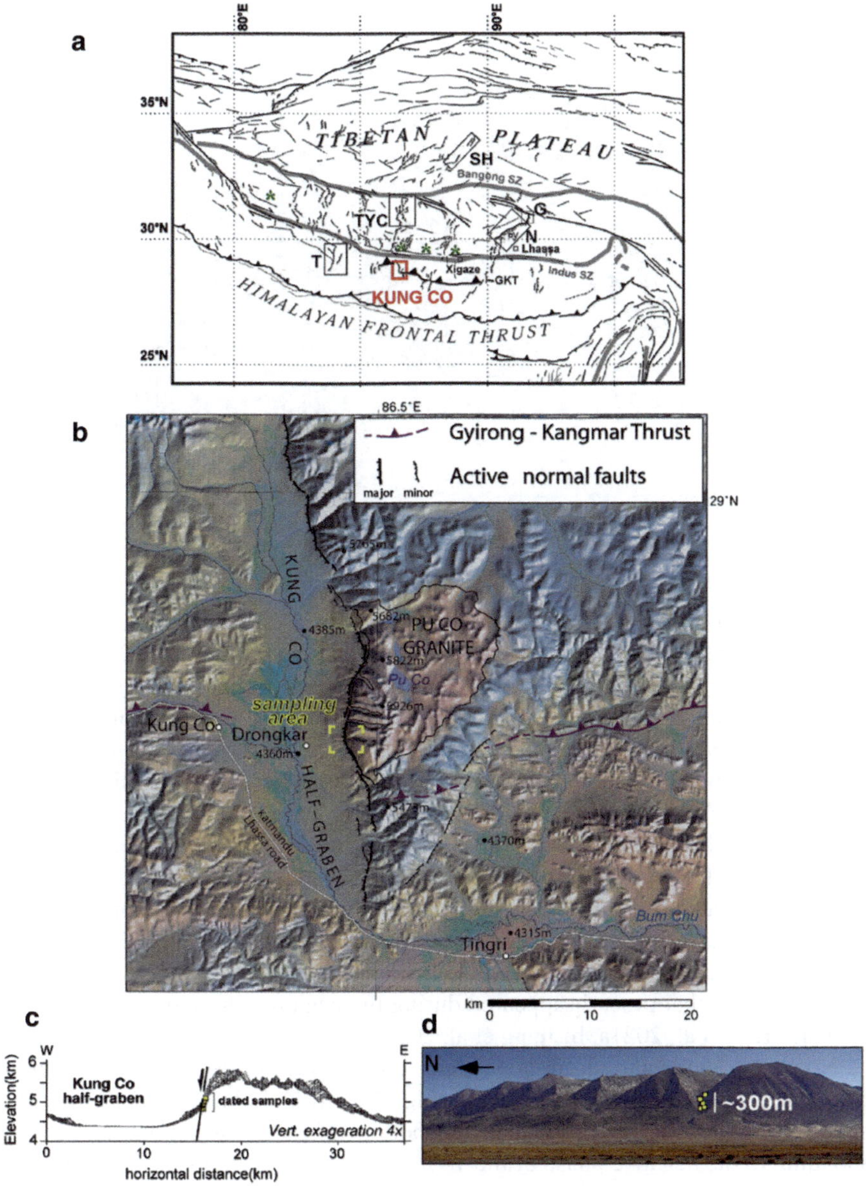

Fig. 41 Structural setting of the Kung Co half graben and sample location (Mahéo et al. 2007). **a** Main active faults (black) and suture zones (grey) of the India–Asia collision zone. Green *: N–S trending andesitic dykes. G: Gulu rift, N: Nyainqentangla graben, T: Thakkhola graben, TYC: Tangra Yum Co rift, SH: Shuang Hu graben, y: Yangbajing. **b** Landsat satellite image and Digital Elevation Model (SRTM data) of Kung Co area and main tectonic features (Giyrong–Kangmar Thrust location). **c** E–W topographic profiles across Kung Co half graben and normal fault (SRTM data) and sampling locations. **d** Photograph of Kung Co normal fault and sampling location. View towards the east from valley floor

Fig. 42 Field photo of the Gemig section. Photo by Zhong Han

The formation boundary is obvious and shown in Fig. 2a. Based on the biostratigraphy, the negative excursion of ~–2.5‰ recorded in organic matter (–2.6 to 28.4 m) was confirmed to be the T-OAE CIE (Fig. 45). The microfacies analysis show that the platform-top grainstones deposited under high-energy conditions of the Pupuga Formation are abruptly overlain by middle/outer ramp micrites reflecting low-energy conditions of the Nieniexiongla Formation during the onset of the T-OAE CIE (Fig. 3; Han et al. 2016). This change also records the early Toarcian global sea-level rise, as envisaged in other sites (Hallam, 1981; Hesselbo and Jenkyns 1998; Han et al. 2016). The Kioto Carbonate Platform suffered the carbonate productivity crisis, which is directly illustrated by biotic extinction and obvious reduction of carbonate content. Based on the data recently obtained from the Kioto Platform, the foraminiferal diversity severely decreased and the *Lithiotis* bivalves, prominent framework builders in reef systems at that time, disappeared during the negative phase of the T-OAE CIE (Fig. 45; Han et al. 2021a, b; Jiang et al. 2020).

Notably, abundant storm-generated coarse-grained interlayers, including carbonate, e.g. ooid (Fig. 45i), peloid (Fig. 45f, h) and clastic grains (Fig. 45g), occurred in the micrites formed in the bottom Nieniexiongla Formation, corresponding to the negative phase of the T-OAE CIE (Han et al. 2016). The associated sedimentary structures are dominated by hummocky and swaley cross-stratification (HCS and SCS; Fig. 45a), U/V-shaped gutter casts (Fig. 45b), parallel laminations (Fig. 45c), climbing ripples (Fig. 45d), current ripple (Fig. 45e), graded bedding (Fig. 45c) and sharp erosive bases (Fig. 45f) in field and microfacies.

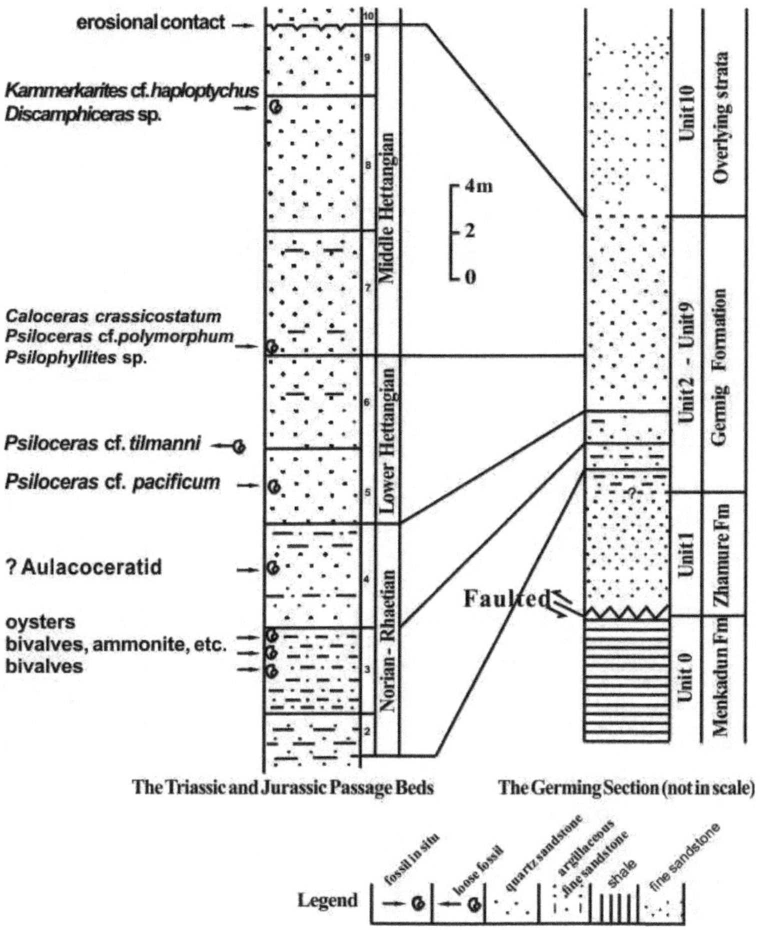

Fig. 43 The Triassic and Jurassic boundary passage beds in the Germig section (Yin et al. 1999)

Stop 18 Middle Jurassic Iron Oolites and Late Jurassic Spiti Shales at Kongcuo Section (GPS: 28°29′56″N, 87°06′32″E)

The Kongcuo section is in a valley of the Menbu area, near the Nylam county in the south subzone of the Tethyan Himalaya (Fig. 46). At this section, the Bajocian-Callovian Ferruginous Oolite Formation (FOF) (Dingjie Fm.) (Yin, 2010) and Callovian black shales (member 1 of Menkadun Fm.) are well exposed. The member 1 of the Menkadun Fm. is equivalent to "Spiti Shale", which is described across the Tethyan Himalaya from west to east: Zanskar, Lahaul, western Spiti (Losar area),

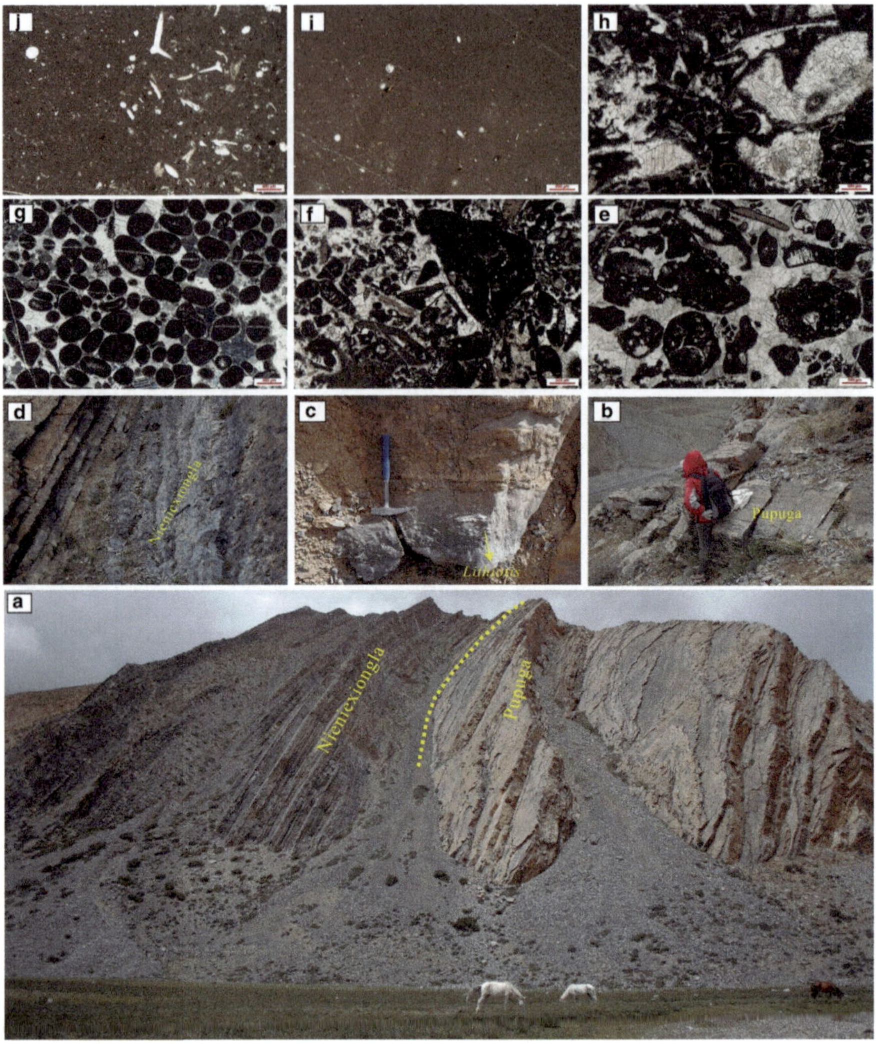

Fig. 44 Field (**a–d**) and carbonate microfacies (**b–j**) photographs of the Nianduo section. The Pupuga Formation is dominated by thick- and medium grainstone/packstone (**a–c** and **e–h**) with *Lithiotis* bivalve (**c**). The Nieniexiongla mainly consists of thin-bedded micrites (**a, d, i** and **j**) intercalated with abundant storm-generated coarse-grained beds (Fig. 38). **e**: Lump grainstone; **f**: Bioclastic grainstone; **g**: Ooid grainstone; **h**: Bioclastic packstone; **i**: Micrite; **j**: Spiculitic micrite. Figure 4b, d are from Han et al. (2022) and others from Zhong Han

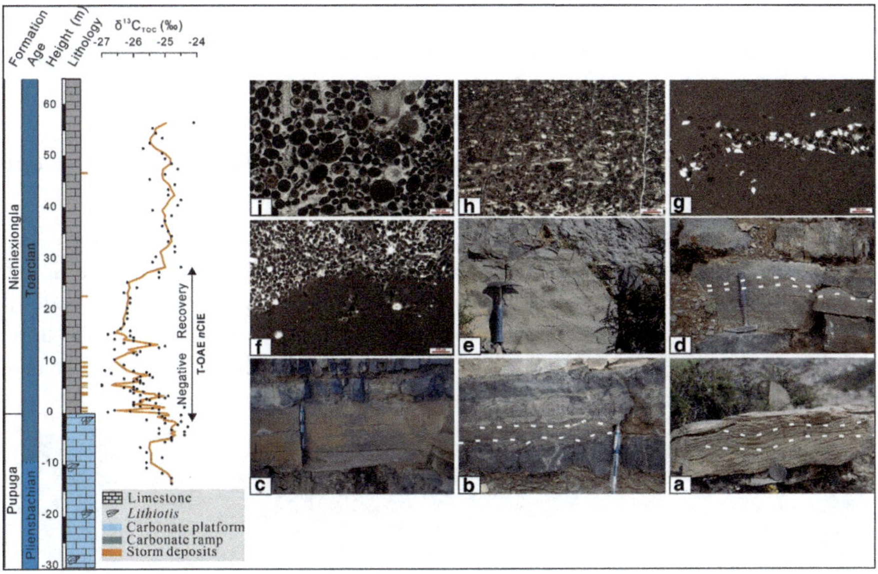

Fig. 45 Photographs of storm-generated deposits and associated sedimentary structures in the field and microfacies. **a** Storm beds with HCS/SCS; **b** V-shaped gutter casts filled with mixed coarse carbonate and terrestrial grains showing HCS/SCS; **c** Storm beds with parallel lamination; **d** Sandy beds within micrite. **e** Current ripple marks; **g** Micrites with an intercalated lamina enriched in siliciclastic minerals; **h** Intraclastic bioclastic packstone with graded bedding; **i** Ooid grainstone within the micrites. Figure 4a–e, h are from Han et al. (2016) and others from Zhong Han

central Spiti (Pin Valley), Dolpo, Thakkhola, Manang, and south Tibet (Everest-Shishapangma area) (Sciunnach and Garzanti, 2012). The Kongcuo section starts at the middle of the Lalongla Fm. and ends in the black shales of the Menkadun Fm. (Fig. 46a). The measured succession is ~55 m thick (Fig. 47). From 0 to ~27 m, the sediments consist mainly of marl and marly limestone couplets with a few medium bedded bioclastic limestones in the top, which belongs to the Lanongla Fm. The contact between FOF and the Lalongla Fm. is an erosive base (Fig. 46b). FOF spans from ~27 m to 32 m (Fig. 47). The ooidal ironstone of FOF macroscopically shows reddish to yellow, some parts are disseminated by Fe through weathering, and thus it was described as "Fe-cap" in previous studies (Fig. 46a). The FOF can be divided into two lithology units based on the content of bioclasts. Unit A is from the bottom to 1.4 m of FOF. It is composed of medium bedded ironstones with abundant fossils including corals, bivalves, foraminifera and echinoderms (Fig. 46b). At ~1 m of the FOF, abundant belemnites arranged parallel to the bedding surface were found (Fig. 46c). An ammonite rich layer was found at ~1.3 m from the base of FOF (Fig. 46d). Unit B, from 1.4 m to the top of the FOF, is characterized by ooidal ironstones, which consist of a large amount of ferruginous ooids and fine quartz grains cemented by brown ferruginous minerals. Only few fossils can be observed in this unit. The ferruginous ooids are green colored. However, the color of cements is dark brown resulting in the brown color of the weathered surface. The deposit of ooidal ironstone was coincided with the onset of a global transgression. Upwards, FOF is overlain by dark shales of the Menkadun Fm. (Han et al. under review), where abundant well-preserved ammonites can be found.

Stop 19 Permian–Triassic Boundary Sedimentation at Selong Xishan Section, Nyalam County (GPS: 28°40′15″N 85°49′36″E)

The Selong Xishan section (Fig. 48a) is located ~1 km northwest of Selong village by the road from Nyalam to Gyirong, with an altitude of ~4500 m. In the Selong area, the uppermost Selong Group was subdivided into two parts: the lower is dominated by bioclastic rudstone intercalated with dark grey silty shale and the upper mainly consist of massive crinoid grainstone (Fig. 48B-E; Shen et al. 2000; An et al.

Fig. 46 **a** Photo of the Kongcuo section. **b** The erosive base between FOF and the Lalongla Fm., fossils are marked by white circles. **c** Abundant belemnites arranged parallel to the bedding surface. **d** The ammonite rich layer (Han et al. under review)

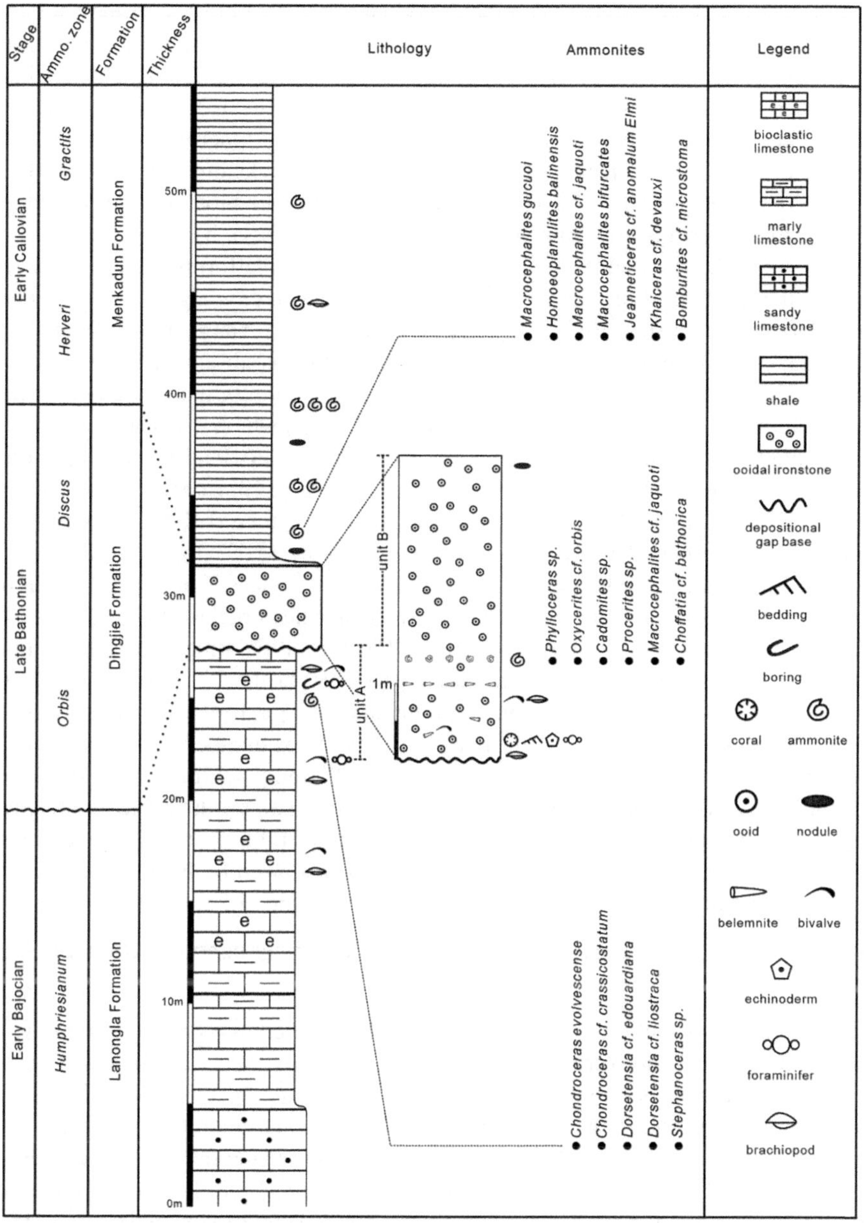

Fig. 47 Lithologic log of the Kongcuo section (Han et al. under review)

2018a, b; Li et al. 2019a, b). The brachiopod and conodont biostratigraphy suggest a Wuchiapingian of the lower part and latest Changhsingian age of the upper part, respectively, and therefore a Lopingian age for the uppermost Selong Group (Shen et al. 2000; Orchard et al. 1994; Wang et al. 2017a, b; Yuan et al. 2018). The overlying Kangshare Formation is in conformable contact with the uppermost Selong Group (Fig. 48a), and characterized by thin-bedded bioclastic packstone/wackestone (Shen et al. 2000; An et al. 2018a, b). The conodont and ammonite biostratigraphy indicate an Induan age for this unit (Shen et al. 2006; An et al. 2018a, b).

For the PTB, the demise of low-latitude carbonate platforms widely occurred where the reef-building corals and symbiotic calcareous algae and fusulinid completely lost during this biotic crisis (Stanley and Yang, 1994; Wang and Sugiyama, 2000). Notably, the Selong Xishan section is absent of large mounds/reefs but yields carbonate producers of metazoan community including bryozoans, echinoderms and brachiopods during the Late Permian, which suggest a cool-water carbonate factory at that time (Shen et al. 2006; Li et al. 2019a, b). The Uppermost Permian Selong Group consists of four microfacies (Fig. 49; Li et al. 2019a, b): the lower part is dominated by bioclastic wackestone/floatstone (MF6); the upper part consists of calcrete (MF1; Fig. 49A), crinoid grainstone (MF2; Fig. 49B) and coral-bearing grainstone (MF3; Fig. 49C). By contrast, the lower Kangshare Formation develops densely packed bivalve grainstone with micritic intraclasts (MF4; Fig. 49D), bioclastic packstone with diverse fossils (MF5; Fig. 49E), bioclastic wackestone/floatstone (MF6; Fig. 49F) and lime-mudstone (MF11). The temporal microfacies changes suggest: (1) a sedimentary environmental change from lower outer ramp to upper inner ramp beneath the PTB, marked by the dolostone bed; (2) subsequently, the inner ramp was abruptly succeeded by a middle ramp setting. The microfacies results also show that the loss of most cold-water carbonate-producing assemblage of bryozoans, brachiopods and echinoderms across the PTB biotic crisis led to collapse of the Late Permian carbonate factory, which was followed by benthic automicrite factory with only minor amounts of calcifying metazoans (Li et al. 2019a, b).

Stop 20 Late Cretaceous Trench Sediments at Luogangcuo Section in Saga County (GPS 29°15′40.8″N, 85°20′35.8″E)

The Luogangcuo section is located at about 12 km southeast of Saga county, at the south bank of the Yarlung Zangbo river (Fig. 50). The Luogangcuo Formation at this section is composed of thick-bedded or amalgamated conglomerates with minor sandstone in the lower part and massive conglomerate in the upper part (Fig. 51). Graded bedding and oblique lamination occur in the section. The conglomerates are poorly sorted, with blocks of chert, sandstone, and limestone. Granules to cobbles consist of radiolarian chert (70%–85%), subordinate shale (3%–23%), sandstone (1%–20%), and minor limestone, indicating a proximal provenance from the adjacent accretionary prism. Meanwhile, the lateral limited sandstone beds are feldspatho-quartzo-lithic volcaniclastic sourced from the Gangdese magmatic arc and the central Lhasa terrane. The depositional age was constrained as ca. 88–81 Ma by youngest detrital zircon ages. Therefore, the section was interpreted to

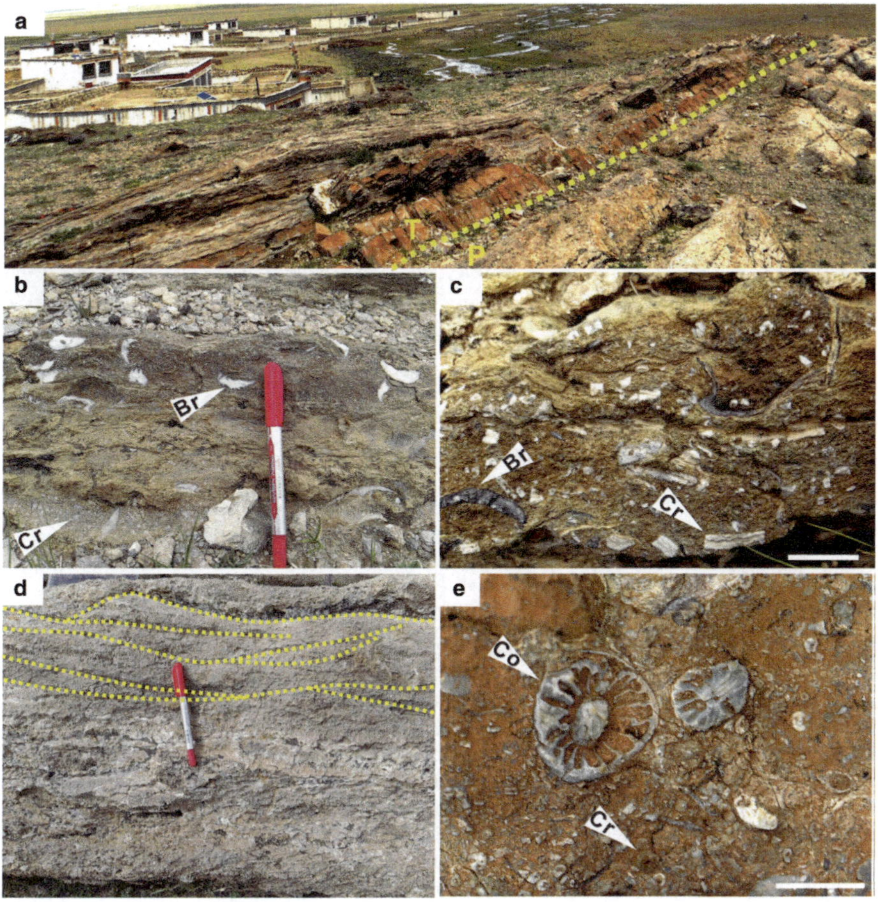

Fig. 48 Field photographs of the Selong Xishan section, showing the feature of the Permian–Triassic Boundary (yellow dashed line). Field of view is *ca* 2.5 m. **b** Thin-bedded, fossiliferous limestone containing abundant fragments of crinoids (Cr) and brachiopods (Br), −5.9 m. **c** Bryozoan–echinoderm rudstone containing abundant fragments of bryozoans, crinoids (Cr) and brachiopods (Br), −4.5 m. Scale bar = 1 cm. **d** Crinoid grainstone showing cross-stratification (denoted by dashed yellow lines), −1.5 m. **e** Coral-bearing grainstone that contains occasional corals (Co) floating in crinoid (Cr) grainstone, −0.25 m. Scale bar = 1 cm. Photographs and descriptions are from Li et al. (2019a, b)

be deposited by proximal submarine fan in the Neo-tethyan trench during Late Cretaceous.

Stop 21 The Sangdanlin Section, Near Sangdanlin Village (GPS 29°15′28N″, 85°14′52″E), About 5 km South to the Saga. Near the Highway from Saga to Gyirong

The Sangdanlin section (Fig. 52) preserves the most complete succession of Cretaceous-Paleocene abyssal strata documented so far in the Himalayas, and has been consequently the focus of research to date.

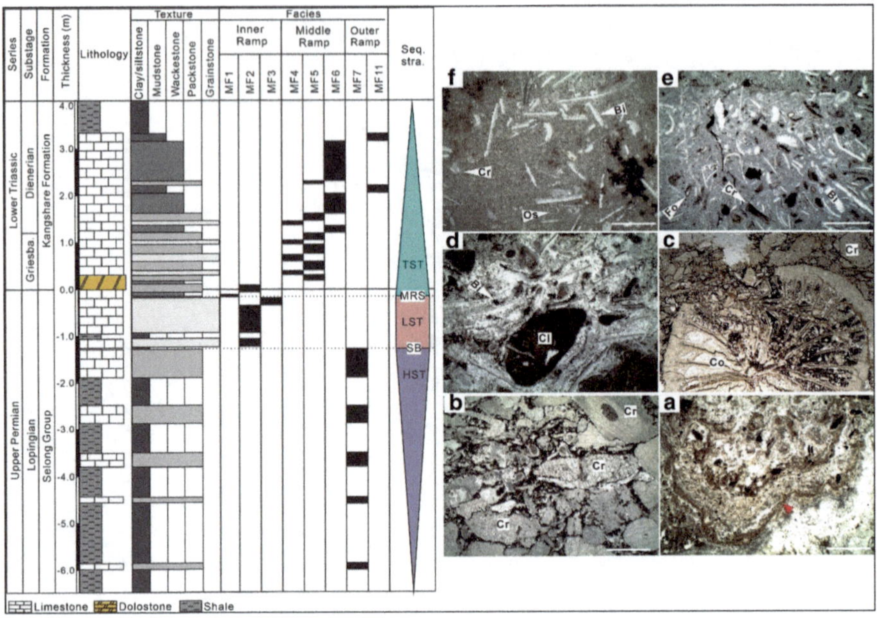

Fig. 49 Microfacies distribution (left) and photographs (right) of the Selong Xishan section during the Permian–Triassic boundary. A. Calcrete (MF1) withdownward-growing, crust-like structure (red arrow) containing amorphous brown pigments, −0.15 m; B. Crinoid grainstone (MF2) exhibiting densely packed crinoid (Cr) fragments, −0.70 m; C. Coral-bearing grainstone containing slightly abraded rugose corals (Co) and abundant crinoid (Cr) fragments, −0.25 m; D. Shelly tempestite (MF4) consisting of densely packed bivalve shells (Bi) with rounded micritic intraclasts (Cl), + 0.30 m; E. Bioclastic packstone (MF5) containing abundant crinoid (Cr) and occasional bivalve (Bi) fragments and foraminifera (Fo), +1.25 m; F. Bioclastic wackestone containing diverse fossils (MF6), including bivalves (Bi), crinoids (Cr) and ostracods (Os), +1.8 m. Note that the scale bar = 1 mm. except for the Fig. 3 C with scale bar = 1 cm. Photographs and descriptions are from Li et al. (2019a, b)

The Sangdanlin section can be divided into three units from base to top (Fig. 53): the Denggang, Sangdanlin and Zheya formations (Wang et al. 2011; DeCelles et al. 2014). The Denggang Formation (units 1–10) is approximately 105 m thick and composed of light yellow fine to coarse quartzose sandstones interbedded with grey-green (weathering to light yellow) silty sandstone and silty shale. It is capped by a 60 cm-thick, green volcaniclastic sandstone bed (unit 10). The overlying Sangdanlin Formation (units 11–21) is approximately 95 m thick, and composed of purple-red siliceous shale, chert, quartzose sandstone, and feldspatho-litho-quartzose volcaniclastic sandstone. Quartzose sandstones are found in unit 13, lower part of units 16, 17, and 19. Volcaniclastic sandstones are found in units 14 and 15, upper part of units 16, 18 and 22. The alternating quartzose and volcaniclastic sandstones are unique of the Sangdanlin Formation. The overlying Zheya Formation (units 22–49) is more than 500 m-thick and composed of grey-black silty shale, thick to massive layers of

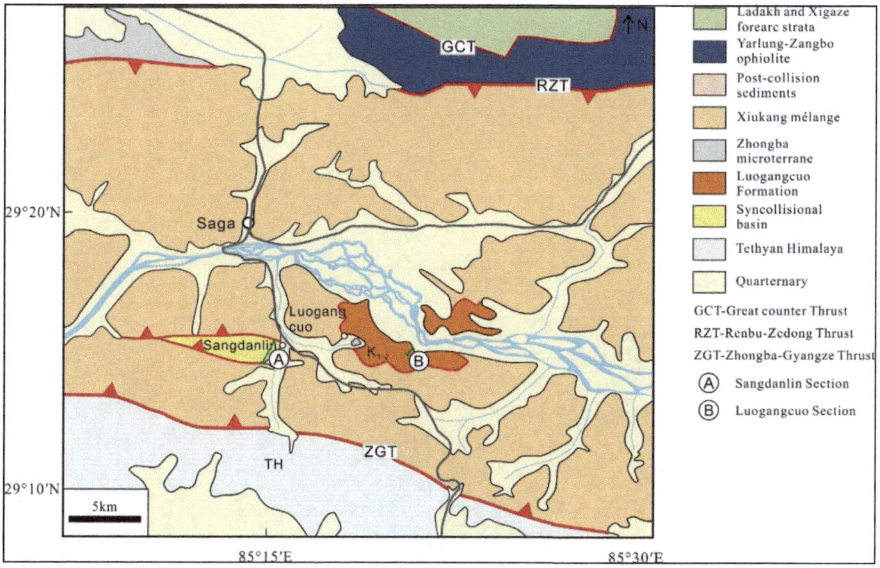

Fig. 50 Geological map of the Sangdanlin and Luogangcuo sections near Saga

Fig. 51 Luogangcuo section at Stop 19 (left: Panorama of the Luogangcuo section; right: sandstone and chert blocks within conglomerate). Photo by Xiumian Hu

grey-green feldspatho-litho-quartzose volcaniclastic sandstone and a few intervals of mottled chert.

The thickness of sandstone beds in the Denggang Formation varies from several centimetres to several metres, and the layers exhibit sharp contact with the underlying and overlying mudrocks. Sedimentary structures include cross-lamination, parallel-lamination, normal grading, and flute casts. The Denggang Formation is interpreted as deposited at the toe of the Indian continental slope, and thicker sandstone beds as deposited by high-density turbidity currents (Wang et al. 2011; DeCelles et al. 2014). The appearance of purple-red siliceous shale in the Sangdanlin Formation indicates

Fig. 52 Field photographs of the Sangdanlin section (Hu et al. 2017). **a** Panorama of the section, looking to the northwest; stratigraphic units are indicated; **b** sandstone bed of unit 8 (Denggang Formation); **c** boundary between quartzose sandstone of unit 17 and volcaniclastic sandstone of unit 18 (Sangdanlin Formation); **d** mudrocks of unit 26 yielding calcareous nannofossils; **e** turbiditic volcaniclastic sandstones of unit 27 (Zheya Formation); **f** sedimentary olistoliths in unit 30; **g** interbeded tuff in unit 47 (Zheya Formation)

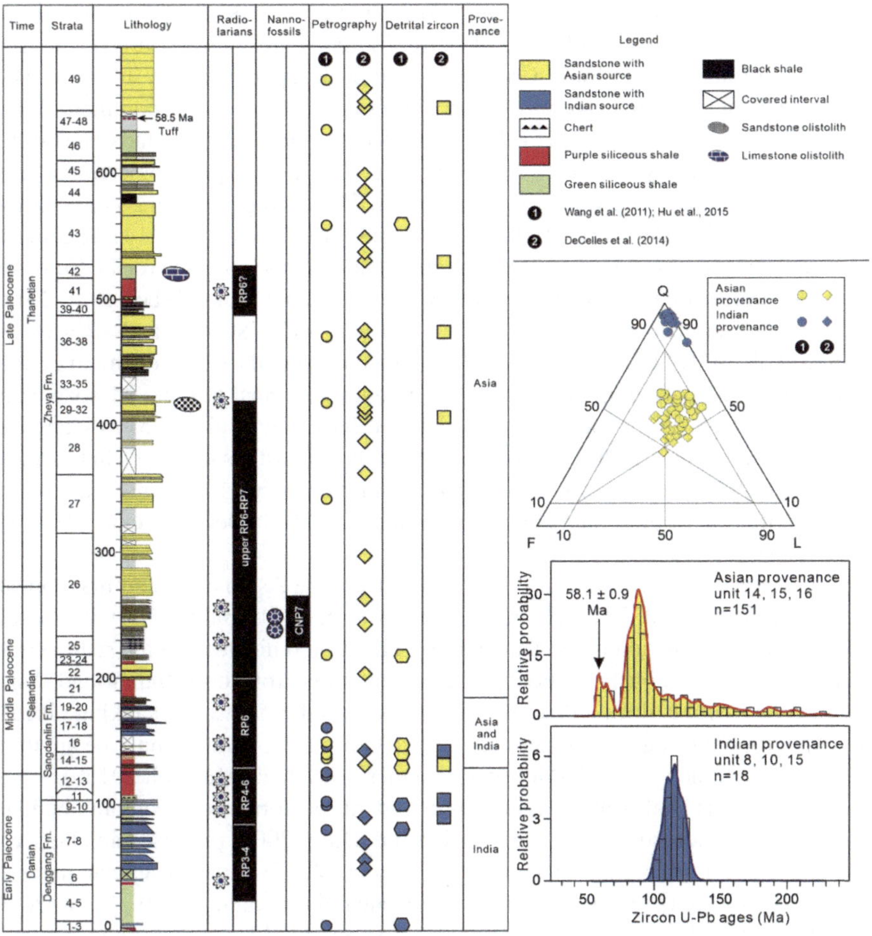

Fig. 53 Integrated bio- and chrono-stratigraphy, sandstone petrography, detrital-zircon chronostratigraphy and provenance change in the Sangdanlin section. Detrital-zircon U–Pb age patterns highlight the sharp difference between Indian-derived quartzose turbiditic sandstones and Asian-derived volcano-plutonic turbiditic sandstones. After Hu et al. (2016)

a palaeo-water depth deeper than the carbonate compensation depth (CCD). Volcaniclastic sandstones in the Sangdanlin Formation mostly display lenticular shape and sharp contact with underlying mudrocks. Slump the Zheya Formation sedimentary structures of sandstone layers include normal grading, flute casts, ditch moulds, and slump structures, indicating deposition by turbidity currents. Background sedimentation is represented by siliceous shale deposited in abyssal environment (Wang et al. 2011; DeCelles et al. 2014). Exotic blocks of conglomerate, sandstone, and carbonate in silty shale are common in the upper part of the Zheya Formation.

The quartz content of the Denggang Formation exceeds 98%, and most particles are well rounded. The U–Pb ages of detrital zircons are mainly Early Palaeozoic to Precambrian, with an Early Cretaceous age peak (approximately 140–120 Ma). These characteristics indicates provenance from the northern Indian continent (Wang et al. 2011; DeCelles et al. 2014; Hu et al. 2015). Volcaniclastic sandstones in the Sangdanlin and Zheya formations have drastically different petrologic characteristics from the underlying Denggang Formation. The average detrital composition is Qm:F:Lt = 41:9:50 (Wang et al. 2011) or 44:25:31 (DeCelles et al. 2014). Textures of lithic fragments indicate origin from intermediate to felsic volcanic rocks. Embayed monocrystalline quartz was derived from felsic volcanic rocks (Wang et al. 2011). Ages of detrital zircons cluster at ~196–180 Ma, ~125–80 Ma, and ~70–54 Ma (Wang et al. 2011; DeCelles et al. 2014), which correspond to the three main periods of magmatic activity in the Gangdese arc. In addition, most of these Mesozoic and Cenozoic zircon grains yield positive εHf(t) values (Hu et al. 2016) as Gangdese zircons. These data conclusively indicate that volcaniclastic sandstones in the Sangdanlin and Zheya formations were derived from the Gangdese magmatic arc on the southern margin of the Asian continent (Wang et al. 2011; DeCelles et al. 2014; Hu et al. 2016).

Provenance analysis indicates that detritus from the Asian continent first appears in unit 14 (Wang et al. 2011; Hu et al. 2016). Alternating turbidite beds with sharply contrasting composition in the Sangdanlin Formation indicates that the sedimentary basin was synchronously receiving clastic materials from both Indian and Asian continental margins during this time period. Accumulation was thus taking place in the topographic trough that separated the two continents, which cannot be anything else than the trench where India began to underthrust Asia (Hu et al. 2016b).

The siliceous shales in the Sangdanlin and Zheya formations contain abundant radiolaria, but their stratigraphic age is debated. Ding (2003) proposed a Palaeocene age (radiolarian zones RP1-RP7), whereas Li et al. (2007) found a Campanian radiolarian assemblage in unit 11 and an early Eocene radiolarian in unit 16. Chan (2006) found RP5 radiolaria in layer S4 (equivalent to units 11, 12), a mid-Cretaceous radiolarian assemblage in layers S5 (equivalent to units 31, 32) and S7 (equivalent to unit 41), and Palaeocene radiolaria of zones RP6 and RP7 near unit 29. They concluded that the Sangdanlin section may contain multiple structural repetitions. A careful biostratigraphic study was carried by Prof. Ted Moore on the Sangdanlin section (Hu et al. 2015), where radiolarian fossils are rare and not well preserved. Moreover, radiolarian assemblages include species of different age in the same sample. Depositional ages have thus to be inferred from the youngest fossils of the assemblage, whereas older fossils, representing more than half of the assemblage in most samples, are interpreted as redeposited (Hu et al. 2015). With the limitations posed by extensive recycling, poor preservation and imperfect definition of the Paleogene radiolaria time scale, we concluded that (Hu et al. 2015): radiolaria in the siliceous shales of the Sangdanlin Formation (units 11–20) are Palaeocene in age (RP4-RP6 zones), while radiolaria in the lower part of the Zheya Formation (units 25 and 32) may extend to zone RP7. Calcareous nanofossils were also found for the first time at the base of the Zheya Formation (unit 26), indicating an age corresponding to calcareous nannozone CNP7 (Hu et al. 2015).

DeCelles et al. (2014) found a tuff layer at the top of the Zheya Formation (unit 48). The weighted average of 38 U–Pb ages obtained by LA-MC-ICPMS on zircon crystals is 58.5 ± 0.6 Ma (2σ), which provides a precise age constraint for this layer, located more than 500 m above unit 14 recording the first appearance of Asian derived detritus. Combined with evidence mentioned above this supports rapid sedimentation in the trench during the very first stage of the India-Asia collision.

Stop 22 The Mubala Section, About 60 km West to the Saga (GPS 29°19′41N″, 84°41′26″E)

The Mubala section, ~60 km west of the Sangdanlin section (Fig. 54) is a newly discovered, ~2 km-thick siliciclastic succession that conformably overlies interbedded chert and dark grey siliceous shale equivalent to the Cretaceous Zhela Formation deposited on the distal part of the Indian shelf within the Tethyan Himalaya (An et al. 2021). The Mubala section preserves the same stratigraphy as the Sangdanlin section, from Denggang Formation to Zheya Formation (An et al. 2021). However, the name of Formations has been revised according to the recent biostratigraphic work (Li et al. 2022). The Seven Formations preserved in the Mubala section are now called Zhela, Weimei, Rilang, Duobeng, Chuangde, Sangdanlin and Zheya (Fig. 55, Li et al. 2022).

The c. 300 m thick Zhela Formation are quartzose and sedimentary structures indicate a submarine lobe environment. The overlying Rilang Formation is sublitharenite and sedimentary structures also indicate a submarine lobe environment. Then Rilang Formation is capped by a sand injection complex intruding into a chert dominated interval. The lithology of the sandstone is glaucony-bearing feldspatho-quartzose. Immediate above the sand injection interval is a red shale dominated interval, with folded chert layers, folded carbonate layers, chert blocks and quartzose blocks, indicating a mass transport deposit (MTD I). Then another mass transport deposit (namely MTD II), with folded and shingled sandstone, immediately sitting above MTD I (Liu

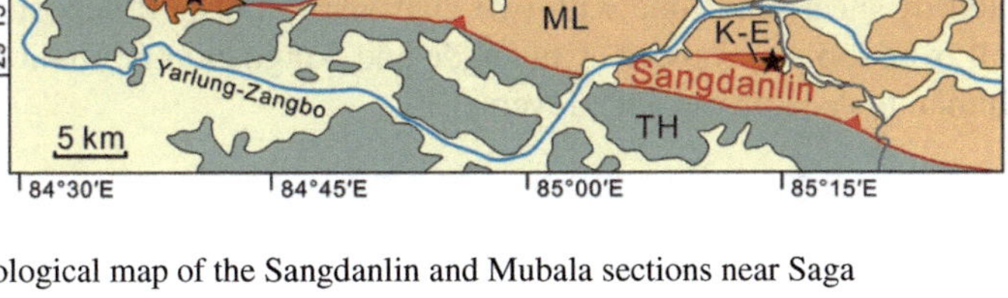

Fig. 54 Geological map of the Sangdanlin and Mubala sections near Saga

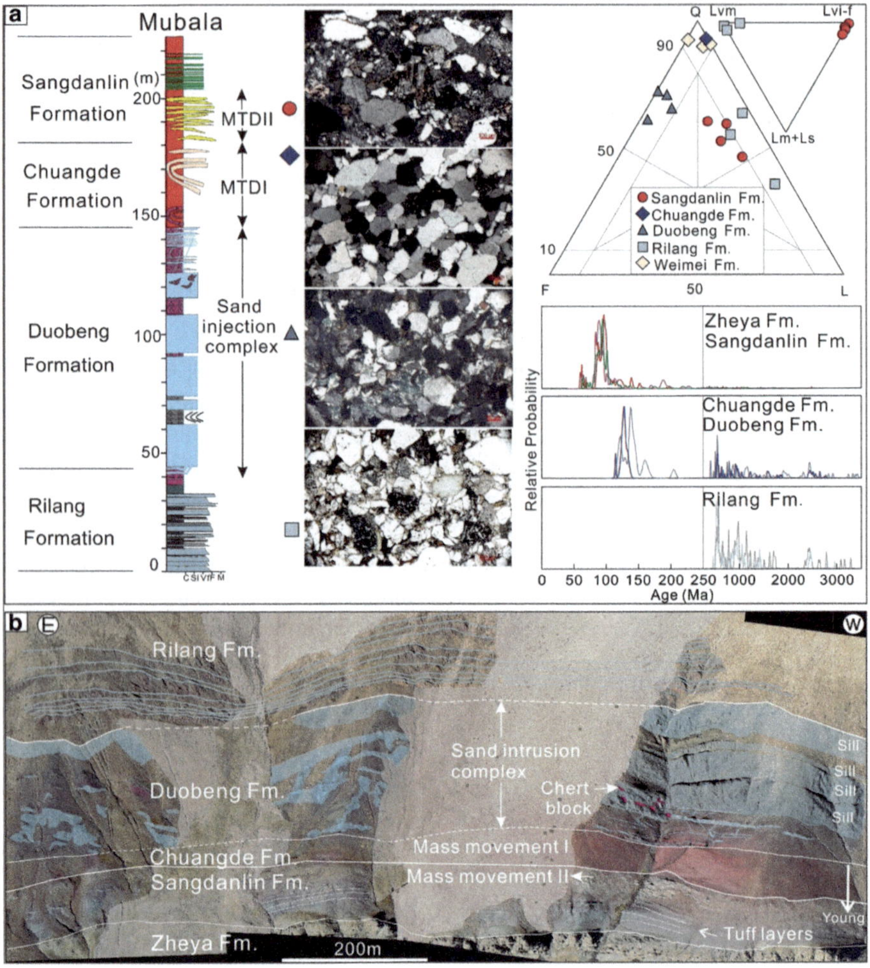

Fig. 55 The **a** Sedimentary column of Mubala section (left), lithology of Rilang to Sangdanlin Formations (middle), provenance data of the Mubala section (right); **b** Outcrop interpretation of Mubala section

et al. 2021). The lithology of this sandstone is volcaniclastic sublitharenite. By provenance study, the Sangdanlin sandstone is the first Asia derived sandstone, and before which the sandstones are all coming from Indian (An et al. 2021). Zheya Formation is represented by sheet sandstone with two giant mass transport deposits, the sediments are all from Indian. Thus, the Chuangde and Sangdanlin Formation represent an abrupt Indian to Asia provenance reversal.

This Indian to Asian provenance reversal (IAPR) is dated by a tuff layer immediately above it at 61 ± 0.3 Ma (zircon U–Pb age using Secondary Ion Mass Spectrometry; Fig. 56, An et al. 2021). Also, radiolarian fossils identified from the same layers give an age of around 61 Ma (Fig. 56, Li et al. 2022). Thus, the Indian to Asian provenance reversal has been tightly constrained at 61 ± 0.3 Ma, and the well exposed outcrop offer a great opportunity to disclose the sedimentary processes occurring before and during continental collision.

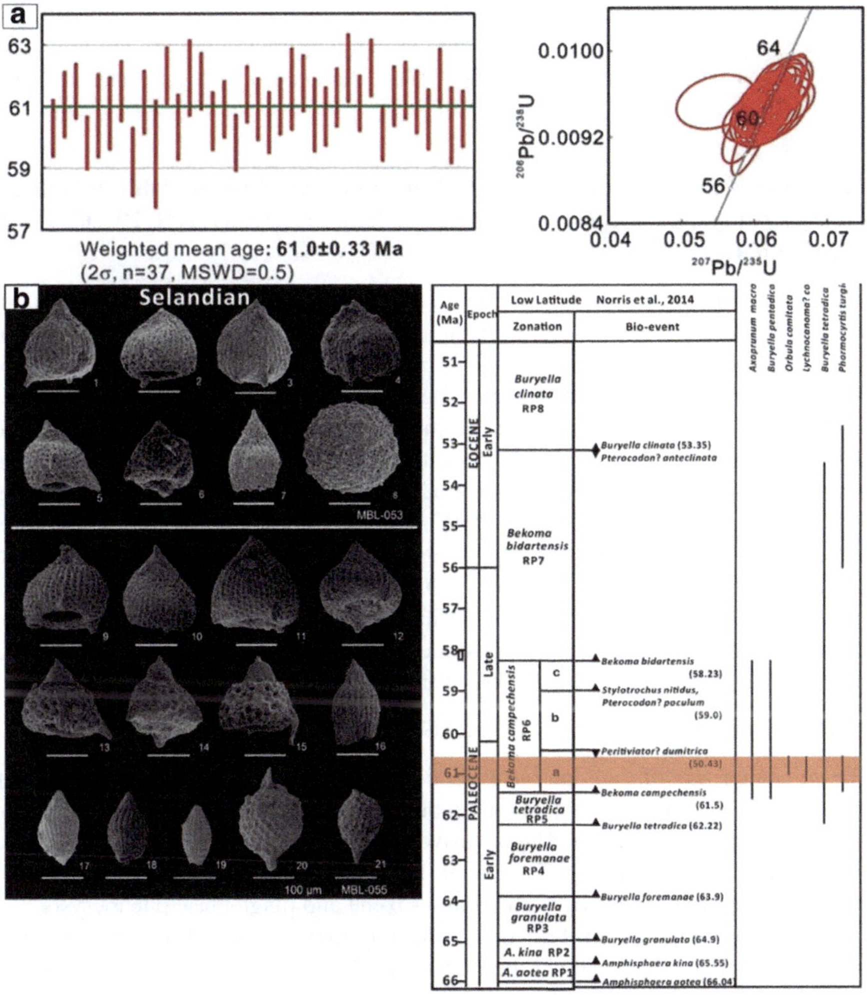

Fig. 56 The **a** Age plots for the tuff immediately above the IAPR; **b** Radiolarian fossil identified from the chert layer immediately above the IAPR with an indicative age

Fig. 57 Outcrops of Stop 21 (left: purple mudstone interbedded within thin-bedded sandstone) and Stop 19b (right: the fault contact between the Padana Formation and the Xigaze ophiolite). Photo by Wei An

Stop 23: Tectonic Contact Between the Xigaze Forearc Basin and the Ophiolite of Yarlung Zangbo Suture Zone at Riwuqi in Sangsang Area (GPS: 29°20′37.6″N, 86°36′59.6″E)

The Xigaze ophiolite lies under a southward thrust with the Padana Formation deposited in a deltaic-fluvial environment in the Xigaze forearc basin. In the Sangsang area, the depositional age of the Padana Formation is constrained between ca. 83–76 Ma by the youngest detrital zircon ages (An et al. 2014). Provenance studies suggests the sandstone were derived from the Lhasa terrane in the north (An et al. 2014; Huang et al. 2015). This stop was interpreted to a direct contact between the Ngamring Formation with highly sheared ophiolitic mantle or ophiolitic mélange (Huang et al. 2015; Maffione et al. 2015). However, the red sandstone interbedded within purple and grey mudstone represents part of the fluvial-deltaic Padana Formation (Wang et al. 2017a, b). Therefore, this stop was reinterpreted to be a tectonic contact between the Xigaze forearc basin and the ophiolite consisting of serpentinized peridotites (Fig. 57). Red sandstone of the Padana Formation, the fault gouge and fractured serpentinites can be found at this stop.

Stop 24 Qunrang Pillow-Lava Basalt (GPS: 29°09′17.7″N, 89°02′40.2″E)

The volcanic rocks of the top crustal section are mainly massive and pillow basalts (Fig. 58a, b). Secondary veins and amygdules of carbonate, quartz, and chlorite are observed in the pillow basalt. The intrusive section in the crust consisting of gabbro, diabase and sheeted sill (Fig. 58c) is cut by basaltic and diabase dikes (Fig. 58d). Basaltic dikes are porphyritic with clinopyroxene and plagioclase phenocrysts.

Most Xigaze Ophiolite pillow basalts display LREE depleted patterns similar to that of N–MORB (Fig. 59a). In the plots of Zr/Yb versus Sm/Yb and Zr versus Y, all our and published basalts from the Xigaze Ophiolite plot in the "MORB array" (Fig. 59d) and overlap with global MORB (Fig. 59c). However, pillow basalts and other published basalts display pronounced depletion of Nb and Ta, and some enrichment of LILE, resembling those of IBM FABs (Fig. 60a, b). In this context, the generation mechanism of the Xigaze Ophiolite basalts is similar to IBM FABs, that

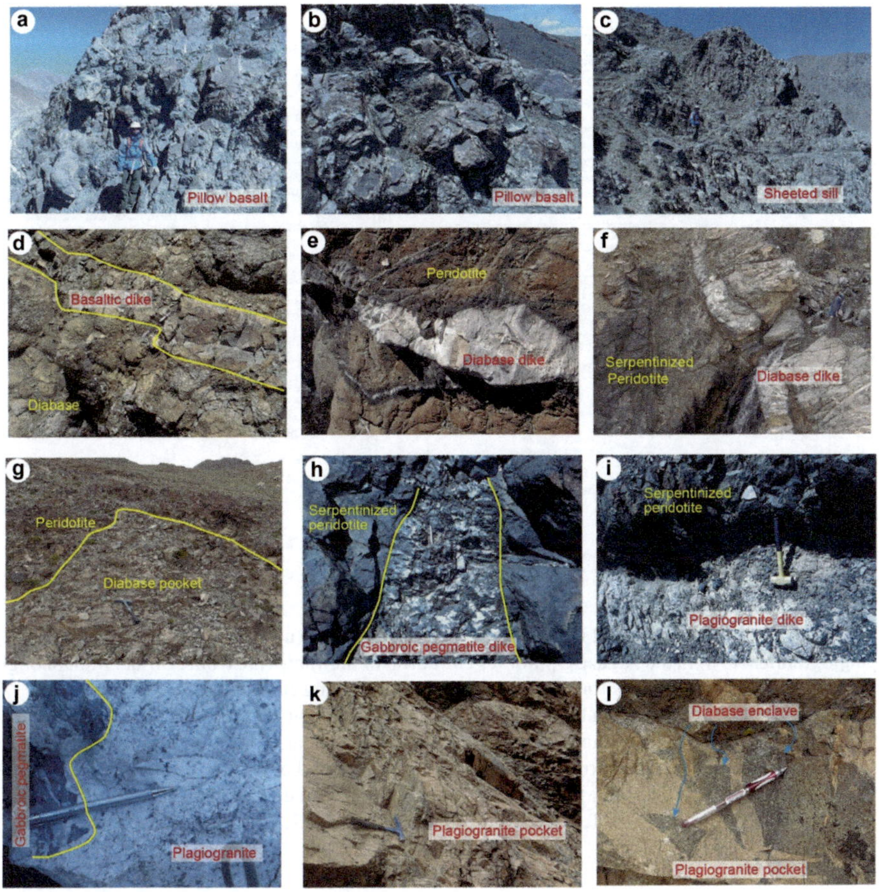

Fig. 58 Field photographs of the Xigaze Ophiolite showing pillow basalt, sheeted sill and basaltic dike in the crust section, and various mafic and felsic dikes in the mantle. Pillow basalt (**a**, **b**), sheeted sill (**c**), basaltic dike (**d**) in the crust; various dikes including diabase dike/pocket (**e**–**g**), gabbroic pegmatite (**h**), plagiogranite (**i**–**k**) in the mantle, and the association of gabbroic pegmatite and plagiogranite (**j**), diabase enclave in the plagiogranite pocket (**m**)

is, the partial melting was mainly caused by decompression as the contribution of fluids derived from the sinking slab during this stage was limited. Xigaze Ophiolite basalts have higher Zr/Y and Zr/Yb ratios than those of IBM FABs, indicating that they may have been generated from a more fertile mantle source compared to IBM FABs (Fig. 59c, d).

Stop 25 Luqu Ophiolite Massif (Mantle Rocks and Various Dikes, GPS: 29°09′09.28″N, 88°53′35.12″E)

The mantle peridotite is dominated in the XO, and it is mainly composed of variably serpentinized harzburgite associated with dunite and lherzolite (Girardeau et al.

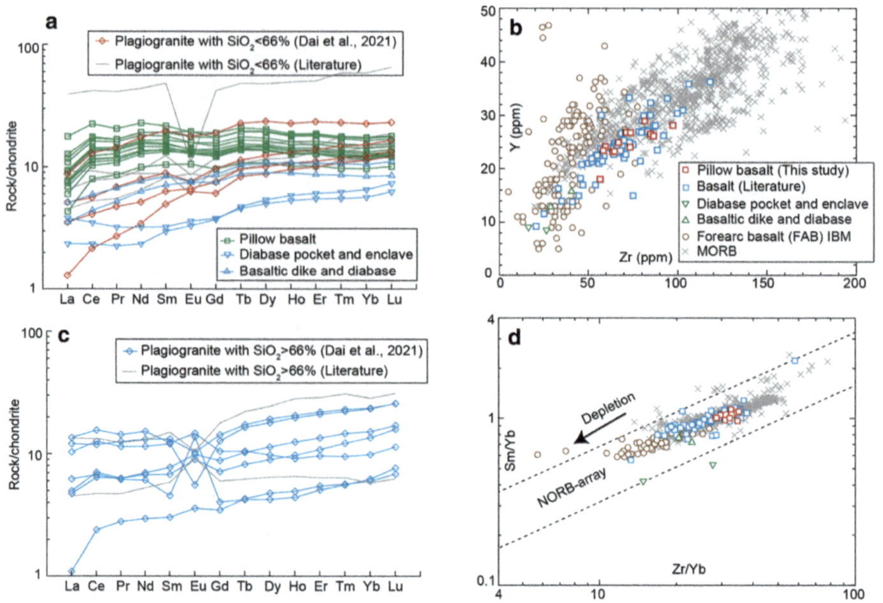

Fig. 59 Chondrite–normalized REE patterns of pillow basalts (**a**), basaltic and diabase dikes (**a**), plagiogranites (**a, b**) in the Xigaze Ophiolite. The normalizing values are from Sun and McDonough, 1989. Compared data sources for plagiogranites are from Zhang et al. 2016a, b. Variations of (**c**) Zr versus Y, (**d**) Zr/Yb versus Sm/Yb for basalt and diabase. Note that Zr/Y and Zr/Yb ratios for Xigaze Ophiolite basalts are higher than IBM forearc basalts, but similar to MORB, indicating derivation from a more fertile mantle source. Basalts of the XO from Bao et al. 2013; Chen and Xia, 2008; Dubois–Côté et al. 2005; Malpas et al. 2003; Li et al. 2012; Niu et al. 2006; Zhang et al. 2016a, b. FABs from Ishizuka et al. 2011, 2018; Reagan et al. 2010, 2017; global MORBs from Gale et al. 2013

1985a; Hébert et al. 2003; Nicolas et al. 1981). These mantle rocks are crosscut by abundant mafic dikes and sporadic plagiogranite dikes (Dai et al. 2013, 2021). Diabase dikes/pockets (for brevity, in some places we use dikes for both) are widespread in the peridotite and are 0.3–4.5 m wide (Fig. 58e–g). They are composed of plagioclase and clinopyroxene and sometimes are intensely altered. Gabbroic pegmatites occur as 3–4–m–wide pods or 0.1–0.3–m–wide dikes (Fig. 58h). Gabbroic pegmatites are dominated by clinopyroxene megacrysts with coarse–grained plagioclase and amphibole. Plagiogranite usually occurs as dikes from 0.3 m to 1 m wide (Fig. 58i), or pockets with an area of ~40 × 45 m (Fig. 58h, i; Location C in Fig. 6). At Location A in Fig. 6, an association of gabbroic pegmatites and plagiogranite was observed, and their contact is diffuse (Fig. 58J). Most diabase enclaves present sharp contacts with the host plagiogranite (Fig. 58m). One sample (DJ12–20) was collected from these enclaves. The plagiogranite is medium–grained, and mainly consists of plagioclase and quartz, with minor amphibole, biotite and orthopyroxene, as well as common accessory minerals such as zircon, titanite. It is traversed by pumpellyite veins. Plagioclase is subhedral to euhedral and is partly altered into secondary

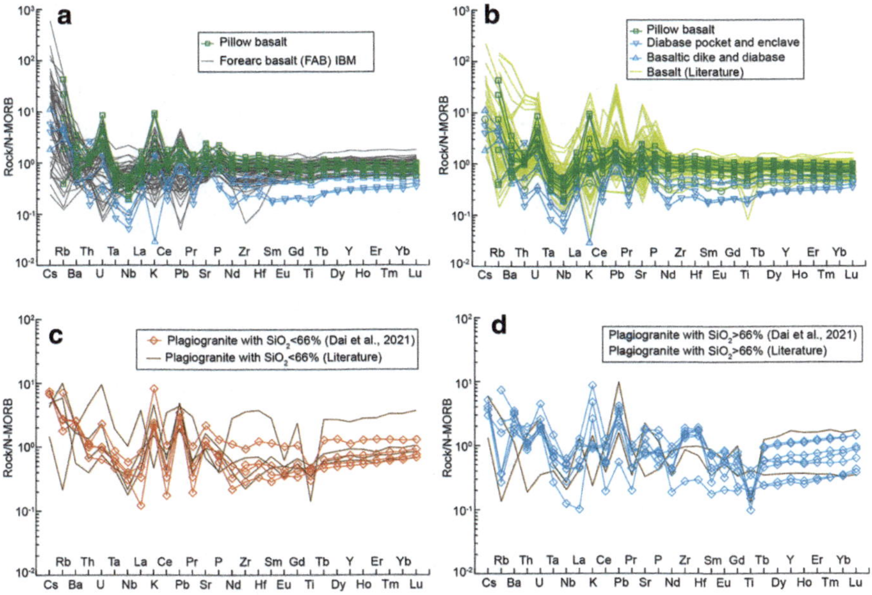

Fig. 60 N–MORB normalized trace element diagrams for pillow basalts (**a, b**), basaltic and diabase dikes (**a, b**) plagiogranites (**c, d**). N–MORB normalizing values are from Sun and McDonough (1989). Plagiogranites from Zhang et al. 2016a, b; basalts from Bao et al. 2013; Chen and Xia, 2008; Dubois–Côté et al. 2005; Malpas et al. 2003; Li et al. 2012; Niu et al. 2006; Zhang et al. 2016a, b

minerals, mainly epidote and sericite. Amphibole is euhedral and ranges in color from brown to green. Orthopyroxene occurs both as megacrysts and as small grains. Quartz is commonly medium– and fine–grained, anhedral to subhedral.

Basaltic dike and diabase enclave/pocket samples are characterized by LILE enrichment and significant HFSE depletion, especially negative Nb and Ta anomalies (Fig. 60b), suggesting that they reflect flux melting of a depleted mantle. Their depleted mantle source is also supported by low Zr and Y contents (Fig. 59c).

Stop 26 Chongdui Formation of the Xigaze Forearc Basin at Qunrang Village (GPS: 29°09′17.7″N, 89°02′40.2″E)

Base strata of the Xigaze forearc basin (the Chongdui Formation) are observed in this section (Fig. 61). The Chongdui Formation here consists of ~95 m thick, steeply north-dipping strata, deposited on top of pillow basalts of the Xigaze ophiolite. The strata could be subdivided into two members: a lower member comprising purplish red chert (6.4 m thick) overlain by siliceous mudrocks (12.6 m thick), and a conformably overlying upper member dominated by sandstones and shales (76 m thick). The top of the succession is faulted against a different sliver of ophiolitic basalt to the north (Wang et al. 2017a, b).

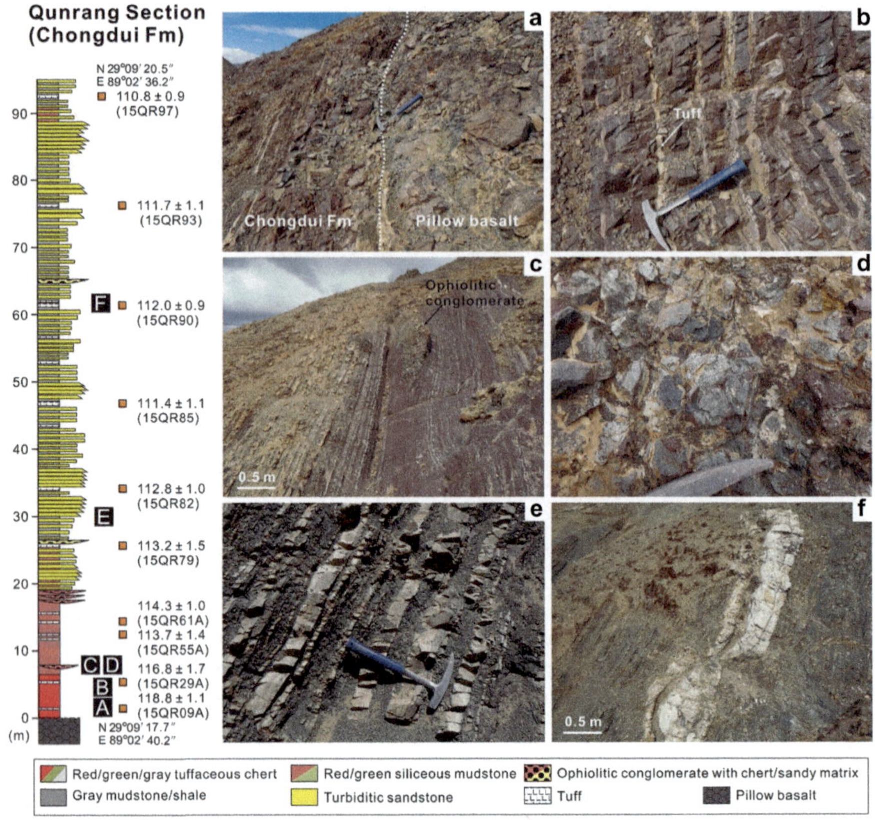

Fig. 61 Stratigraphy (left) and field photos (right) of the Chongdui Formation in the Qunrang section (After Wang et al. 2017a, b). **a**, Boundary between the Chongdui Formation and pillow basalts of the Yaarlung-Zangbo ophiolites; **b**, Chert from the lower member; **c**, conformably contact between the lower and upper members, also show a lenticular conglomerate bed in the lower member; **d**, Ophiolitic conglomerate; **e**, turbidites; **f**, white tuff layer in the upper member

Chert beds in the lower member are commonly 5–10 cm thick and rich in radiolarian fossils, whereas siliceous mudrocks are massive. Tuffaceous laminaes and thin tuff layers (0.5–3 cm) are common throughout the lower member, increasing in frequency and thickness (up to 10 cm) up-section. The upper member comprises thin-bedded, fine-grained sandstones, greenish gray shales, minor purplish radiolarian mudrocks, and medium to thick-bedded (10–60 cm) white tuffs. Sandstone beds, commonly 5–20 cm thick, laterally continuousand arranged in thickening-upward cycles, include massive, plane-parallel-laminated and rippled sandstones. Four lenticular conglomerate beds with siliceous matrix are intercalated in the siliceous mudrocks of the lower member, and two conglomerate beds with sandy-mud matrix occur in the upper member. Conglomerate clasts are mostly subangular to an-gular basalt and minor diabase, gabbro and serpentinite. Radiolarian chert, siliceous shale and interbedded tuffs of the lower member represent abyssal sediments deposited onto oceanic crust with in-tercalated fallout tephra. Sandstones of the upper member were deposited by low-density turbiditic flows on a lower deep-marine fan, whereas the conglomerates with mafic and ultramafic gravel were deposited by submarine debris flows sourced from nearby ophiolitic highs (Wang et al. 2017a, b).

Radiolarian faunas and Zircon SIMS U–Pb ages of tuffs gave accurate constraints on the depositional age of the Chong-dui Formation (Ziabrev et al. 2003; Wang et al. 2017a, b). Chronostratigraphic results suggested that initial deposition of the Chongdui Formation at late Barremian, immediately after the formation of the Yarlang-Zangbo ophiolite (~131–124 Ma). Deposition of turbidites started at ~113 Ma. Detritus of turbiditic rocks are from the Gangdese arc.

Stratigraphy of the Chongdui Formation indicates (1) The Yarlung-Zangbo ophiolite were formed adjacent to the Lhasa terrane, in the Gangdese forearc; (2) The Xigaze forearc basin had experienced a starved stage of ~10–15 Ma before turbiditic filling; (3) The Asian active margin of Neotethys was fully established only after Early Cretaceous.

Stop 27 Upper Cretaceous Oceanic Red Beds of the North Tethys Himalaya at Chuangde Section, Gyangze County (GPS 28°58′00″N, 89°44′05″E)

The Cretaceous marine strata of north Tethys Himalaya can be observed in the Chuangde section in Gyangze area (Fig. 62). The volcanoclastics of Rilang Fm., shales of Gyabula Fm., red shale with exotic limestones of Chuangde Fm. and olistostrome of Zongzhuo Fm. are well exposed in this section. The Chuangde section is one of the representatives for Cretaceous Oceanic Red Beds (CORBs), which is considered as a major change of the Cretaceous oceanic condition from anoxic to oxic in deep-sea environments (Wang et al. 2005a, b). The CORBs in the Gyangze basin mainly consist of shales, thin-bedded marls, re-sedimented limestones and clasts supported conglomerates-breccias (e.g., Chen et al. 2011).

CORBs in Chuangde section is mainly composed of ca. 30 m thick red siliceous shales intercalated with thin bedded marls, caused by turbidites (Hu et al. 2006). Medium to thick bedded variegated limestones were found at the base and in the upper part of the CORB succession. The limestone beds are commonly deformed and in sharp contact with the shales. At the base of the succession, a single limestone bed (limestone unit of Chuangde 1, Lc1; Chen et al. 2011) is out-cropping with an average thickness of 0.5 m. The thickness of the bed varies laterally from tens of centimeters to about 1.5 m. In contrast, the medium bedded limestones are interbedded with shales and are discrete laterally in the middle part of the succession (Lc2; Chen et al. 2011). Mudstone fragments paralleling the dip of the sequence are distributed widely in the limestone beds. The upper 10 m of the formation is characterized by the occurrence of lenticular and round limestone olistoliths enclosed by red shales (Lc3; Chen et al. 2011). The limestones are mainly gray in color. The planktonic foraminiferal result indicates that the onset of the Chuangde CORBs of early Campanian age (Chen et al. 2011), which may be slightly younger than previously documented (Li et al. 2005; Wan et al. 2005).

Fig. 62 Field photo of the Chuangde section showing the quartz arenites of latest Jurassic Weimei Fm., volcanoclastics of Rilang Fm., shales of Gyabula Fm., red shale with exotic limestones of Chuangde Fm. and olistostrome of Zongzhuo Fm

Stop 28 Longma, Zongzhuo Mélange Aside the Main Road from Gyangze-Ngyangze (GPS: N28°57′47.0″, E89°44′14.0″)

Here near the Baisa village, the Zongzhuo mélange is comprised by exotic blocks and unmetamorphic shale and mudstone 'matrix' (Fig. 63). Exotic blocks are several centimeters to tens of meters in size, and comprise a variety of lithologies including sandstones, cherts, basalts, limestones and conglomerates. Detrital modes of the sandstones bocks are composed of dominantly volcaniclastic grains and subordinately quartzes and feldspars, similar to sandstones from the Xigaze forearc basin (Sun et al. 2011). Detrital zircon U–Pb ages from an exotic sandstone block are mainly clustered at ~77–129 Ma, with two peaks at ~90 Ma and 110 Ma. Occurrence of abundant volcanic detritus and Mesozoic detrital zircons suggest that the Zongzhuo mélange contain significant amount of materials originally derived from the Gangdese volcanic arc.

Stop 29 Triassic Langjiexue Group, Nanggarze Area (GPS: 29°13′42.7″N, 90°37′58.2″E)

The Upper Triassic (roughly Carnian to Norian) Langjiexue Group is a suite of turbidites exposed south of the Yarlung-Zangbo suture zone in the eastern Himalaya (Fig. 64). It tectonically belongs to the Tethys Himalayan zone, but its provenance signatures significantly different from those of other Tethys Himalayan sandstones. These features include: (1) Southward to southwestward paleocurrent direction; (2)

Fig. 63 Panoramic view on the Zongzhuo mélange near the Baisa village

'Recycled Orogen Provenance' detrital composition; (3) Depleted whole-rock Nd isotopes; (4) Late Paleozoic to early Mesozoic (400–200 Ma, mostly clustered at 280–220 Ma with peak at ~245 Ma) detrital zircons; and (5) Arc-basalt or mantle-peridotite-derived chrome spinels (Meng et al. 2019 and references therein). As the Langjiexue Group is everywhere in fault contact with Tethys Himalayan strata, its depositional setting and provenance had been for a long time. Some scholars proposed that the Langjiexue Group was part of an exotic terrane deposited in the forearc region of either the Lhasa block or an intra-oceanic arc, and accreted to the northern Indian margin during the convergence and collision between India and Asia, but others insist it was originally deposited on the distal Indian margin only with unique depositional process (see Wang et al. 2016a, b and Liu et al. 2020 for summary).

A recent study by Meng et al. (2019) found that the Qulonggongba Formation in the southern Tethyan Himalayan zone and the Langjiexue Group had similar sandstone petrography. The presence of Late Paleozoic to early Mesozoic detrital

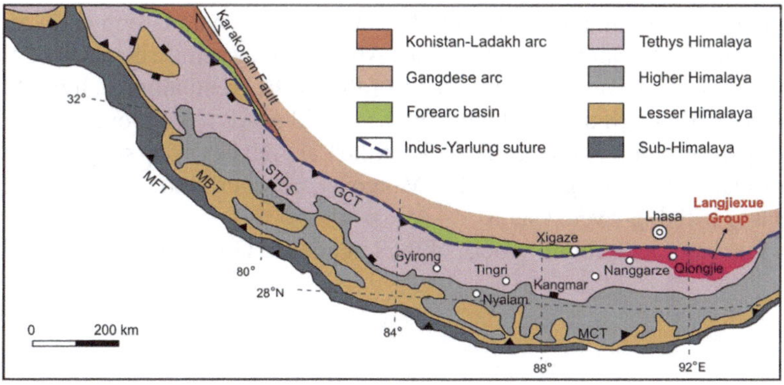

Fig. 64 Tectonic map shows distribution of the Langjixue Group

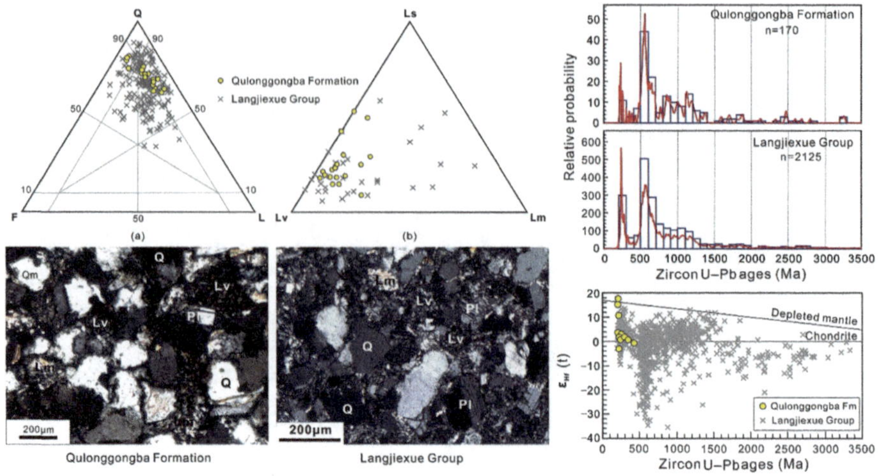

Fig. 65 A comparison of provenance feature between the Langjixue Group and the Qulonggongba Formation (Data are from Meng et al. 2019)

zircons in the Qulonggongba Fomration further proves that it has consistent provenance with the Langjiexue Group (Fig. 65). Because the Qulonggongba Formation certainly belongs to the Tethys Himalayan zone, the provenance similarity with the Langjiexue Group indicates that the latter is also an in situ Tethys Himalayan sedimentary sequence. That is to say, the Langjiexue Group is deep marine sediments of the north Indian margin.

Unique provenance features of the Langjiexue Group would be attributed to the special paleogeography during it was deposition. Wang et al. (2016a, b) suggested that Late Triassic rifting along the northwestern margin of astern Gondwana might have changed the original paleogeography. Thermal uplift rift would have formed a long relief that forced northwestward-draining river systems to be deflected southwestward, carrying detritus from the Gondwanide orogen to as far as the northern Indian margin and filling the huge turbiditic fan of the Langjiexue Group (Fig. 66). This model is somewhat analogous to the East African rift system, which forced the Nile River to flow northward along the strike of the rift to eventually reach the Mediterranean Sea. Although this working hypothesis still needs to be proven, discovery of syn-depositional bimodal volcanic rocks in the Langjiexue Group supports the model (Huang et al. 2018; Meng et al. 2021). Moreover, huge turbiditic fans are necessarily associated with big-river systems.

On this outcrop, turbiditic rocks of the Langjiexue Group are observed (Fig. 67).

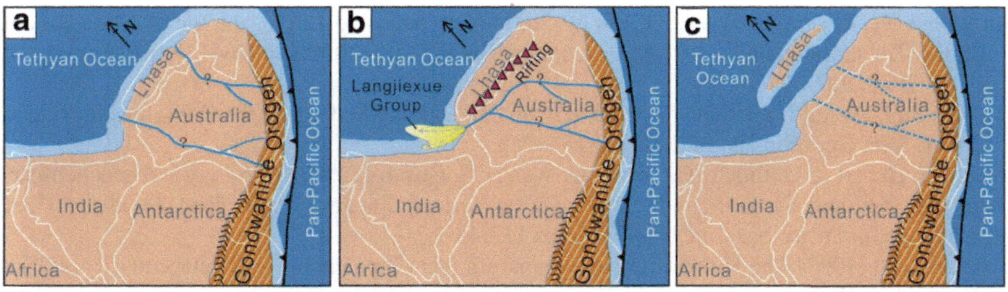

Fig. 66 Late Triassic paleogeographic evolution of northwestern margin of eastern Gondwana, showing depositional process of the Langjiexue Group (After Meng et al. 2019)

Fig. 67 Field photos of the Langjixue Group in Nanggarze area. Photos are provided by Jian-Gang Wang

References

Aitchison J, Badengzhu DA, Liu J, Luo H, Malpas J, McDermid I, Wu H, Ziabrev S, Zhou MF (2000) Remnants of a Cretaceous intra-oceanic subduction system within the Yarlung-Zangbo suture (southern Tibet). Earth Planet Sci Lett 183:231–244

Aitchison JC, Ali JR, Chan A, Davis AM, Lo CH (2009) Tectonic implications of felsic tuffs within the Lower Miocene Gangrinboche conglomerates, southern Tibet. J Asian Earth Sci 34:287–297

Allègre CJ, Courtillot V, Tapponnier P, Hirn A, Mattauer M, Coulon C, Jaeger JJ, Achache J, Schärer U, Marcoux J, Burg JP, Girardeau J, Armijo R, Gariépy C, Göpel C, Li TD, Xiao XC, Chang CF, Li GQ, Lin BY, Teng JW, Wang NW, Chen GM, Han TL, Wang XB, Den WM, Sheng HB, Cao YG, Zhou J, Qiu HR, Bao PS, Wang SC, Wang BX, Zhou YX, Xu RH (1984) Structure and evolution of the Himalaya-Tibet orogenic belt. Nature 307:17–22

An W, Hu X, Garzanti E, BouDagher-Fadel MK, Wang J, Sun G (2014) Xigaze forearc basin revisited (South Tibet): provenance changes and origin of the Xigaze Ophiolite. Geol Soc Am Bull 126(11–12):1595–1613

An W, Hu X, Garzanti E (2017) Sandstone provenance and tectonic evolution of the Xiukang Mélange from Neotethyan subduction to India–Asia collision (Yarlung-Zangbo suture, south Tibet). Gondwana Res 41:222–234

An W, Hu X, Garzanti E (2018a) Discovery of Upper Cretaceous Neo-Tethyan trench deposits in south Tibet (Luogangcuo Formation). Lithosphere 10(3):446–459

An X, Zhang Y, Zhu T, Zhang Y, Yuan D (2018b) Stable carbon isotope characteristics of Permian-Triassic boundary at the Selong Xishan Section. Earth Sci 43:2848–2857

An W, Hu X, Garzanti E, Wang JG, Liu Q (2021) New precise dating of the India–Asia collision in the Tibetan Himalaya at 61 Ma. Geophys Res Lett 48:e2020GL090641. https://doi.org/10.1029/2020GL090641

Bao PS, Su L, Wang J, Zhai QG (2013) Study on the Tectonic Setting for the Ophiolites in Xigaze, Tibet. Acta Geol Sin-Engl Ed 87(2):395–425

Bedard E, Hebert R, Guilmette C, Lesage G, Wang CS, Dostal J (2009) Petrology and geochemistry of the Saga and Sangsang ophiolitic massifs, Yarlung Zangbo Suture Zone, Southern Tibet: evidence for an arc-back-arc origin. Lithos 113:48–67

Brookfield M (1993) The Himalayan passive margin from Precambrian to Cretaceous Times. Sed Geol 84:1–35

Burg JP, Leyreloup A, Girardeau J, Chen GM (1987) Structure and Metamorphism of a Tectonically Thickened Continental Crust: the Yalu Tsangpo Suture Zone (Tibet): Philosophical Transactions of the Royal Society of London. Ser A, Math Phys Sci 321(1557):67–86

Cai F, Ding L, Leary RJ, Wang H, Xu Q, Zhang L, Yue Y (2012) Tectonostratigraphy and provenance of an accretionary complex within the Yarlung-Zangpo suture zone, southern Tibet: insights into subduction–accretion processes in the Neo-Tethys. Tectonophysics 574–575:181–192

Cai F, Ding L, Laskowski AK, Kapp P, Wang H, Xu Q, Zhang L (2016) Late Triassic paleogeographic reconstruction along the Neo-Tethyan Ocean margins, southern Tibet. Earth Planet Sci Lett 435:105–114

Carrapa B, Orme DA, DeCelles PG, Kapp P, Cosca MA, Waldrip R (2014) Miocene burial and exhumation of the India-Asia collision zone in southern Tibet: response to slab dynamics and erosion. Geology 42(5):443–446

Chauvel C, Marini JC, Plank T, Ludden JN (2009) Hf-Nd input flux in the Izu-Mariana subduction zone and recycling of subducted material in the mantle. Geochem Geophys Geosyst 10(1):Q01001

Chen GW, Xia B (2008) Platinum-group elemental geochemistry of mafic and ultramafic rocks from the Xigaze ophiolite, southern Tibet. J Asian Earth Sci 32(5–6):406–422

Chen X, Wang CS, Kuhnt W, Holbourn A, Huang YJ, Ma C (2011) Lithofacies, microfacies and depositional environments of Upper Cretaceous Oceanic red beds (Chuangde Formation) in southern Tibet. Sed Geol 235:100–110

Chen X, Idakieva V, Stoykova K, Liang H, Yao H, Wang C (2017) Ammonite biostratigraphy and organic carbon isotope chemostratigraphy of the Early Aptian oceanic anoxic event (OAE 1a) in the Tethyan Himalaya of southern Tibet. Palaeogeogr Palaeoclim Paleoecol 485:531–542

Chen SS, Fan WM, Shi RD, Liu XH, Zhou XJ (2018) 118–115 Ma magmatism in the Tethyan Himalaya igneous province: constraints on Early Cretaceous rifting of the northern margin of Greater India. Earth Planet Sci Lett 491:21–33

Chu MF, Chung SL, O'Reilly SY, Pearson NJ, Wu FY, Li XH, Liu D, Ji J, Chu CH, Lee HY (2011) India's hidden inputs to Tibetan orogeny revealed by Hf isotopes of Transhimalayan zircons and host rocks. Earth Planet Sci Lett 307(3–4):479–486

Chung SL, Liu D, Ji J, Chu MF, Lee HY, Wen DJ, Lo CH, Lee TY, Qian Q, Zhang Q (2003) Adakites from continental collision zones: Melting of thickened lower crust beneath southern Tibet. Geology 31:1021–1024

Chung SL, Chu MF, Zhang YQ, Xie YW, Lo CH, Lee TY, Lan CY, Li XH, Zhang Q, Wang YZ (2005) Tibetan tectonic evolution inferred from spatial and temporal variations in post-collisional magmatism. Earth-Sci Rev 68:173–196

Colleps CL, McKenzie NR, Horton BK, Webb AAG, Ng YW, Singh BP (2020) Sediment provenance of pre- and post-collisional Cretaceous-Paleogene strata from the frontal Himalaya of northwest India. Earth Planet Sci Lett 534:116079

Dai JG, Wang CS, Hébert R, Santosh M, Li YL, Xu JY (2011) Petrology and geochemistry of peridotites in the Zhongba ophiolite, Yarlung Zangbo Suture Zone: implications for the Early Cretaceous intra-oceanic subduction zone within the Neo-Tethys. Chem Geol 288(3–4):133–148

Dai J, Wang C, Li Y (2012) Relicts of the Early Cretaceous seamounts in the central-western Yarlung Zangbo Suture Zone, southern Tibet. J Asian Earth Sci 53:25–37

Dai J, Wang C, Polat A, Santosh M, Li Y, Ge Y (2013) Rapid forearc spreading between 130 and 120 Ma: evidence from geochronology and geochemistry of the Xigaze ophiolite, southern Tibet. Lithos 172–173:1–16

Dai JG, Wang CS, Zhu DC, Li YL, Zhong HT, Ge YK (2015) Multi-stage volcanic activities and geodynamic evolution of the Lhasa terrane during the Cretaceous: insights from the Xigaze forearc basin. Lithos 218:127–140

Dai JG, Wang CS, Stern RJ, Yang K, Shen J (2021) Forearc magmatic evolution during subduction initiation: insights from an Early Cretaceous Tibetan ophiolite and comparison with the Izu-Bonin-Mariana forearc. GSA Bull 133:753–776

Davis AM, Aitchison JC, Luo H, Zyabrev S (2002) Paleogene island arc collision-related conglomerates, Yarlung-Tsangpo suture zone, Tibet. Sedimentary Geology 150(3):247–273

DeCelles PG, Kapp P, Quade J, Gehrels GE (2011) Oligocene-Miocene Kailas basin, southwestern Tibet: record of postcollisional upper-plate extension in the Indus-Yarlung suture zone. Geol Soc Am Bull 123(7–8):1337–1362

DeCelles P, Kapp P, Gehrels G, Ding L (2014) Paleocene-Eocene foreland basin evolution in the Himalaya of southern Tibet and Nepal: implications for the age of initial India–Asia collision. Tectonics 33(5):824–849

Ding L, Xu Q, Yue YH, Wang HQ, Cai FL, Li S (2014) The Andean-type Gangdese Mountains: Paleoelevation record from the Paleocene-Eocene Linzhou Basin. Earth Planet Sci Lett 392:250–264

Ding L, Kapp P, Wan XQ (2005) Paleocene–Eocene record of ophiolite obduction and initial India-Asia collision, south central Tibet. Tectonics 24:TC3001

Ding L (2003) Paleocene deep-water sediments and radiolarian faunas: implications for evolution of Yarlung-Zangbo foreland basin, southern Tibet. Sci China Ser D: Earth Sci 46

Dubois-Côté V, Hébert R, Dupuis C, Wang CS, Li YL, Dostal J (2005) Petrological and geochemical evidence for the origin of the Yarlung Zangbo ophiolites, southern Tibet. Chem Geol 214(3–4):265–286

Dupuis C, Hébert R, Dubois-Côté V, Guilmette C, Wang CS, Li YL, Li ZJ (2005) The Yarlung Zangbo Suture Zone ophiolitic mélange (southern Tibet): new insights from geochemistry of ultramafic rocks. J Asian Earth Sci 25(6):937–960

Dürr SB (1996) Provenance of Xigaze fore-arc basin clastic rocks (Cretaceous, south Tibet). GSA Bull 108(6):669–684

Fan HH (2021) Early Cretaceous cold seep carbonates in Dingri, Southern Tibet: the feather and formation model. Master thesis, China University of Geosciences (Beijing) (in Chinese with English abstract)

Fang AM, Yan Z, Liu XH, Pan YS, Li JL, Yu LJ, Tao JR (2004) The age of the plant fossil assemblage in the Liuqu Conglomerate of southern Tibet and its tectonic significance (in Chinese with English abstract). Prog Nat Sci 14:1419–1427

Fu H, Hu X, Crouch EM, An W, Wang J, Garzanti E (2018) Upper Cretaceous trench deposits of the Neo-Tethyan subduction zone: Jiachala Formation from Yarlung Zangbo suture zone in Tibet, China. Sci China Earth Sci 61(9):1204–1220

Gale A, Dalton CA, Langmuir CH, Su Y, Schilling JG (2013) The mean composition of ocean ridge basalts. Geochem Geophys Geosyst 14(3):489–518

Gansser (1964) Geology of the Himalayas. Interscience Publishers John Wiley and Sons, New York, pp 1–289

Garzanti E, Van Haver T (1988) The Indus clastics: forearc basin sedimentation in the Ladakh Himalaya (India). Sed Geol 59(3):237–249

Garzione CN, DeCelles PG, Hodkinson DG, Ojha TP, Upreti BN (2003) East-west extension and Miocene environmental change in the southern Tibetan plateau: Thakkhola graben, central Nepal. Geol Soc Am Bull 115(1):3–20

Girardeau J, Marcoux J, Zao Y (1984) Lithologic and tectonic environment of the Xigaze ophiolite (Yarlung Zangbo suture zone, southern Tibet, China), and kinematics of its emplacement. Eclogae Geol Helv 77:153–170

Girardeau J, Mercier JCC, Xibin W (1985a) Petrology of the mafic rocks of the Xigaze ophiolite, Tibet: implications for the genesis of the oceanic lithosphere. Contrib Miner Petrol 90(4):309–321

Girardeau J, Mercier JCC, Yougong Z (1985b) Origin of the Xigaze ophiolite, Yarlung Zangbo suture zone, southern Tibet. Tectonophysics 119(1–4):407–433

Guilmette C, Hébert R, Wang C, Villeneuve M (2009) Geochemistry and geochronology of the metamorphic sole underlying the Xigaze Ophiolite, Yarlung Zangbo Suture Zone, South Tibet. Lithos 112(1–2):149–162

Guilmette C, Hebert R, Dostal J, Indares A, Ullrich T, Bedard E, Wang CS (2012) Discovery of a dismembered metamorphic sole in the Saga ophiolitic melange, South Tibet: assessing an early cretaceous disruption of the Neo-Tethyan suprasubduction zone and consequences on basin closing. Gondwana Res 22(2):398–414

Guo T, Liang DY, Zhang YZ, Zhao CH (1991) Geology of Nagri, Tibet (Xizang). China University of Geosciences Press, Wuhan, p 464

Guo H, Chen X, Yao H, Han K, Garzanti E, Liu X, Fan H (2021) Lower Cretaceous clastic dykes in southern Tibet: characteristics and palaeogeographic significance. Geol J 56:1726–1739

Hallam A (1981) A revised sea-level curve for the early Jurassic. J Geol Soc 138:735–743

Hallam A, Wignall PB, Yin J, Riding JB (2000) An investigation into possible facies changes across the triassic–jurassic boundary in southern tibet. Sed Geol 137:101–106

Hampel A, Adam J, Kukowski N (2004) Response of the tectonically erosive south Peruvian forearc to subduction of the Nazca Ridge: analysis of three-dimensional analogue experiments. Tectonics 23

Han Z, Hu X, Li J, Garzanti E (2016) Jurassic carbonate microfacies and relative sea-level changes in the Tethys Himalaya (southern Tibet). Palaeogeogr Palaeoclimatol Palaeoecol 456:1–20

Han Z, Hu X, BouDagher-Fadel M, Jenkyns HC, Franceschi M (2021a) Early Jurassic carbon-isotope perturbations in a shallow water succession from the Tethys Himalaya, southern hemisphere. Newsl Stratigr 54:461–481

Han Z, Hu X, Boudagher-Fadel MK, Jenkyns HC, Franceschi M (2021b) Early Jurassic carbon-isotope perturbations in a shallow-water succession from the Tethys Himalaya, southern hemisphere. Newsl Stratigr 54(4):461–481

Han Z, Hu X, He T, Newton RJ, Jenkyns HC, Jamieson RA, Franceschi M (2022) Early Jurassic long-term oceanic sulfur-cycle perturbations in the Tibetan Himalaya. Earth Planet Sci Lett 578:117261

Han Z, Hu X, Hu Z, Jenkyns H, Su T, Geochemical evidence from the Kioto Carbonate Platform (Tibet) reveals enhanced terrigenous input and deoxygenation during the early Toarcian. Glob Planet Chang, In review

Han KB, Chen X, Garzanti E, Yao HW, Guo HF, Liu X, Zhu SP, Formation process of the Mid-Jurassic oolitic ironstone in Nyalam area, southern Tibet: implications for the paleoceanography in eastern Tethys margin. Unpublished

Hao Y, Wan X (1985) The marine Cretaceous and Tertiary strata of Tingri, Xizang (Tibet). Contribution to the Geology of the Qinghai-Xizang (TIBET) Plateau 17:227–332 (in Chinese with English abstract)

Harrison TM, Copeland P, Kidd WSF, Yin A (1992) Raising Tibet. Science 255:1663–1670

Hébert R, Bezard R, Guilmette C, Dostal J, Wang CS, Liu ZF (2012) The Indus-Yarlung Zangbo ophiolites from Nanga Parbat to Namche Barwa syntaxes, southern Tibet: first synthesis of petrology, geochemistry, and geochronology with incidences on geodynamic reconstructions of Neo-Tethys. Gondwana Res 22(2):377–397

Hébert R, Huot F, Wang CS, Liu ZF (2003) Yarlung Zangbo ophiolites (Southern Tibet) revisited: geodynamic implications from the mineral record. In: Dilek Y, Robinson PT (eds) Ophiolites in earth history, vol 218. Geological Society, London, Special Publications, 165–190

Hesselbo S, Jenkyns H (1998) British Lower Jurassic Sequence Stratigraphy, pp 561–581

Hou ZQ, Gao YF, Qu XM, Rui ZY, Mo XX (2004) Origin of adakitic intrusives generated during mid-Miocene east-west extension in southern Tibet. Earth Planet Sci Lett 220:139–155

Hou ZQ, Yang ZM, Qu XM, Meng XJ, Li ZQ, Beaudoin G, Rui ZY, Gao YF, Zaw K (2009) The Miocene Gangdese Porphyry copper belt generated during post-collisional extension in the Tibetan orogeny. Ore Geol Rev 36:25–51

Hou ZQ, Duan LF, Lu YJ, Zheng YC, Zhu DC, Yang ZM, Yang ZS, Wang BD, Pei YR, Zhao ZD, McCuaig TC (2015) Lithospheric architecture of the Lhasa Terrane and its control on ore deposits in the Himalayan-Tibetan Orogen. Econ Geol 110:1541–1575

Hu H, Stern RJ (2020) Early Cretaceous subduction initiation beneath southern Tibet caused the northward flight of India. Geosci Front 11(4):1123–1131

Hu XM, Luba J, Sarti M (2006) Mid-Cretaceous oceanic red beds in the Umbria–Marche Basin, central Italy: constraints on paleoceanography and paleoclimate. Palaeogeogr Palaeoclimatol Palaeoecol 233:163–186

Hu XM, Jansa L, Wang CS (2008) Upper Jurassic-Lower Cretaceous stratigraphy in south-eastern Tibet: a comparison with the western Himalayas. Cretac Res 29:301–315

Hu XM, Sinclair HD, Wang JG, Jiang HH, Wu FY (2012) Late Cretaceous-Palaeogene stratigraphic and basin evolution in the Zhepure Mountain of southern Tibet: implications for the timing of India–Asia initial collision. Basin Res 24(5):520–543

Hu XM, Garzanti E, Moore T, Raffi I (2015) Direct stratigraphic dating of India-Asia collision onset at the Selandian (middle Paleocene, 59±1Ma). Geology 43(10):859–862

Hu XM, Wang JG, BouDagher-Fadel M, Garzanti E, An W (2016b) New insights into the timing of the India–Asia collision from the Paleogene Quxia and Jialazi Formations of the Xigaze forearc basin, South Tibet. Gondwana Res 32:76–92

Hu XM, An W, Garzanti E, Liu Q (2020) Recognition of trench basins in collisional orogens: Insights from the Yarlung Zangbo suture zone in southern Tibet. Sci China Earth Sci 63:2017–2028

Hu XM, Garzant IE, Wang JG, Huang WT, An W, Webb A (2016a) The timing of India–Asia collision onset—Facts, theories, controversies. Earth-Sci Rev 160:264–299

Hu X, Wang J, An W, Garzanti E, Li J (2017) Constraining the timing of the India–Asia continental collision by the sedimentary record. Sci China Earth Sci 60

Huang WT, van Hinsbergen DJJ, Maffione M, Orme DA, Dupont-Nivet G, Guilmette C, Ding L, Guo ZJ, Kapp P (2015) Lower Cretaceous Xigaze ophiolites formed in the Gangdese forearc: evidence from paleomagnetism, sediment provenance, and stratigraphy. Earth Planet Sci Lett 415:142–153

Huang Y, Cao HW, Li GM, Brueckner SM, Zhang Z, Dong L, Dai ZW, Lu L, Li YB (2018) Middle-Late Triassic bimodal intrusive rocks from the Tethyan Himalaya in South Tibet: geochronology, petrogenesis and tectonic implications. Lithos 318:78–90

Ishizuka O, Tani K, Reagan MK, Kanayama K, Umino S, Harigane Y, Sakamoto I, Miyajima Y, Yuasa M, Dunkley DJ (2011) The timescales of subduction initiation and subsequent evolution of an oceanic island arc. Earth Planet Sci Lett 306(3–4):229–240

Ishizuka O, Hickey-Vargas R, Arculus RJ, Yogodzinski GM, Savov IP, Kusano Y, McCarthy A, Brandl PA, Sudo M (2018) Age of Izu-BoninMariana arc basement. Earth Planet Sci Lett 481:80–90

Jadoul F, Berra F, Garzanti E (1998) The Tethys Himalayan passive margin from Late Triassic to Early Cretaceous (South Tibet). J Asian Earth Sci 16:173–194

Ji WQ, Wu FY, Chung SL, Li JX, Liu CZ (2009) Zircon U-Pb geochronology and Hf isotopic constraints on petrogenesis of the Gangdese batholith, southern Tibet. Chem Geol 262:229–245

Jiang S, Song H, Kemp DB, Dai X, Liu X (2020) Two pulses of extinction of larger benthic foraminifera during the Pliensbachian–Toarcian and early Toarcian environmental crises. Palaeogeogr Palaeoclimatol Palaeoecol 560:109998

Jiang JX, Hu XM, Li J, BouDagher-Fadel, Garzanti E (2021) Discovery of the Paleocene-Eocene thermal maximum in shallow-marine sediments of the Xigaze forearc basin, Tibet: a record of enhanced extreme precipitation and siliciclastic sediment flux. Palaeogeogr Palaeoclimatol Palaeoecol 562:110095

Jin X, Huang H, Shi Y, Zhan L (2015) Origin of Permian exotic limestone blocks in the Yarlung Zangbo Suture Zone, Southern Tibet, China: with biostratigraphic, sedimentary and regional geological constraints. J Asian Earth Sci 104:22–38

Kang ZQ, Xu JF, Wilde SA, Feng ZH, Chen JL, Wang BD, Fu WC, Pan HB (2014) Geochronology and geochemistry of the Sangri Group Volcanic Rocks, Southern Lhasa Terrane: implications for the early subduction history of the Neo-Tethys and Gangdese Magmatic Arc. Lithos 200–201:157–168

Kapp P, DeCelles PG, Leier AL, Fabijanic JM, He S, Pullen A, Gehrels GE (2007) The Gangdese retroarc thrust belt revealed. GSA Today 17(7):4–9

Leary RJ, DeCelles PG, Quade J, Gehrels G, Waanders G (2016) The Liuqu Conglomerate, southern Tibet: early Miocene basin development related to deformation within the Great Counter Thrust system. Lithosphere 8(5):427–450

Lee HY, Chung SL, Lo CH, Ji J, Lee TY, Qian Q, Zhang Q (2009) Eocene Neotethyan slab breakoff in southern Tibet inferred from the Linzizong volcanicrecord. Tectonophysics 477:20–35

Leier AL, DeCelles PG, Kapp P, Ding L (2007a) The Takena Formation of the Lhasa terrane, southern Tibet: the record of a Late Cretaceous retroarc foreland basin. Geol Soc Am Bull 119:31–48

Leier AL, DeCelles PG, Kapp P, Gehrels GE (2007b) Lower Cretaceous strata in the Lhasa Terrane, Tibet, with implications for understanding the early tectonic history of the Tibetan Plateau. J Sediment Res 77:809–825

Li XH, Wang CS, Hu XM (2005) Stratigraphy of deep-water Cretaceous deposits in Gyangze, southern Tibet, China. Cretac Res 26:33–41

Li XH, Jenkyns HC, Wang CS, Hu XM, Chen X, Wev Y, Huang YJ, Cui J (2006) Upper Cretaceous carbon- and oxygen-isotope stratigraphy of hemipelagic carbonate facies from southern Tibet, China. J Geol Soc 163:375–382

Li YL, Wang CS, Hu XM, Bak M, Wang JG, Chen L (2007) Characteristics of early Eocene radiolarian assemblages of the Saga area, southern Tibet and their constraint on the closure history of the Tethys. Chin Sci Bull 52:2108–2114

Li JF, Xia B, Liu LW, Xu LF, He GS, Wang H, Zhang YQ, Yang ZQ (2009) SHRIMP U-Pb dating for the gabbro in the Qunrang ophiolite, Tibet: the geochronology constraint for the development of eastern Tethys basin. Geotecton Metallog 33(2):294–298

Li JG, Guo Z, Batten DJ, Cai H, Zhang Y (2010) Palynological stratigraphy of the Late Cretaceous and Cenozoic collision-related conglomerates at Qiabulin, Xigaze, Xizang (Tibet) and its bearing on palaeoenvironmental development. J Asian Earth Sci 38:86–95

Li WX, Zhao ZD, Zhu DC, Dung GC, Zhou S, Mo XX, DePaolo D, Dilek Y (2012) Geochemical discrimination of tectonic environments of the Yarlung Zangpo ophiolite in southern Tibet. Acta Petrol Sin (Yanshi Xuebao) 28(5):1663–1673

Li GW, Kohn B, Sandiford M, Xu ZQ, Wei L (2015a) Constraining the age of Liuqu Conglomerate, southern Tibet: Implications for evolution of the India–Asia collision zone. Earth Planet Sci Lett 426:259–266

Li J, Hu X, Garzanti E, An W, Wang J (2015b) Paleogene carbonate microfacies and sandstone provenance (Gamba area, South Tibet): stratigraphic response to initial India–Asia continental collision. J Asian Earth Sci 104:39–54

Li M, Song H, Woods AD, Dai X, Wignall PB (2019b) Facies and evolution of the carbonate factory during the Permian-Triassic crisis in South Tibet, China. Sedimentology 66:3008–3028

Li J, Hu X, Garzanti E, Banerjee S, BouDagher-Fadel M (2020a) Late Cretaceous topographic doming caused by initial upwelling of Deccan magmas: Stratigraphic and sedimentological evidence. Geol Soc Am Bull 132:835–849

Li J, Hu X, Garzanti E, BouDagher-Fadel M (2021) Climate-driven hydrological change and carbonate platform demise induced by the Paleocene-Eocene thermal maximum (southern Pyrenees). Palaeogeogr Palaeoclimatol Palaeoecol 567:110250

Li J, Hu X, Garzanti E, Banerjee S, Boudagher-Fadel M (2019) Late cretaceous topographic doming caused by initial upwelling of deccan magmas: stratigraphic and sedimentological evidence. Geol Soc Am Bull 132(3–4)
Li J, Hu XM, Zachos JC, Garzanti E, BouDagher-Fadel M (2020) Sea level, biotic and carbon-isotope response to the Paleocene–Eocene thermal maximum in Tibetan Himalayan platform carbonates. Glob Planet Chang 103316
Li X, Hu X, An W, Liu Q, Garzanti E, Jun M (2022) From Neo-Tethyan convergence to India–Asia collision: radiolarian biostratigraphy of the Cretaceous to Paleocene deep-water Tethys Himalaya. Newsletters Strat
Liang H, Chen X, Wang C, Zhao D, Weissert H (2016) Methane-derived authigenic carbonates of mid-Cretaceous age in southern Tibet: types of carbonate concretions, carbon sources, and formation processes. J Asian Earth Sci 115:153–169
Liu G, Einsele G (1994) Sedimentary history of the Tethyan basin in the Tibetan Himalayas. Geol Rundsch 83(1):32–61
Liu CJ, Yin JX, Sun XX, Sun YY (1988) Marine Late Cretacous–early Tertiary sequences—the non-flysch deposits of the Xigaze Forearc Basin in the south Xizang. Mem Inst Geol, Acad Sin 3:130–157 (in Chinese)
Liu CZ, Zhang C, Yang LY, Zhang LL, Ji WQ, Wu FY (2014) Formation of gabbronorites in the Purang ophiolite (SW Tibet) through melting of hydrothermally altered mantle along a detachment fault. Lithos 205:127–141
Liu T, Wu FY, Zhang LL, Zhai QG, Liu CZ, Ji WB, Zhang C, Xu Y (2016) Zircon U-Pb geochronological constraints on rapid exhumation of the mantle peridotite of the Xigaze ophiolite, southern Tibet. Chem Geol 443:67–86
Liu T, Wu FY, Liu CZ, Tribuzio R, Ji WB, Zhang C, Xu Y, Zhang WQ (2018) Variably evolved gabbroic intrusions within the Xigaze ophiolite (Tibet): new insights into the origin of ophiolite diversity. Contrib Miner Petrol 173(11):91
Liu T, Wu FY, Liu CZ, Zhang C, Ji WB, Xu Y (2019) Reconsideration of Neo-Tethys evolution constrained from the nature of the Dazhuqu ophiolitic mantle, southern Tibet. Contrib Miner Petrol 174(3):23
Liu YM, Dai JG, Wang CS, Li HA, Wang Q, Zhang LL (2020) Provenance and tectonic setting of Upper Triassic turbidites in the eastern Tethyan Himalaya: Implications for early-stage evolution of the Neo-Tethys. Earth Sci Rev 200:103030
Liu Q, Kneller B, An W, Hu X (2021) Sedimentological responses to initial continental collision: triggering of sand injection and onset of mass movement in a syn-collisional trench basin, Saga, southern Tibet. J Geol Soc 178(6)
Liu X, Zhang Y, Han KB, Batenburg SJ, Guo HF, Ma C, Yao HW, Fan HH, Wu Q, Chen X, Chemo- and Cyclostratigraphic records of the Albian from the Tethyan Himalaya of southern Tibet, China. In review
Maffione M, van Hinsbergen DJJ, Koornneef LMT, Guilmette C, Hodges K, Borneman N, Huang W, Ding L, Kapp P (2015) Forearc hyperextension dismembered the south Tibetan ophiolites. Geology 43(6):475–478
Mahéo G, Leloup PH, Valli F, Lacassin R, Arnaud N, Paquette JL, Fernandez A, Haibing L, Farley KA, Tapponnier P (2007) Post 4 Ma initiation of normal faulting in southern Tibet. Constraints from the Kung Co half graben. Earth Planet Sci Lett 256(1):233–243
Malpas J, Zhou MF, Robinson PT, Reynolds PH (2003) Geochemical and geochronological constraints on the origin and emplacement of the Yarlung Zangbo ophiolites, southern Tibet. In Dilek Y, Robinson PT (eds) Ophiolites in Earth History: Geological Society [London] Special Publication, vol 218, pp 191–206
Meng ZY, Wang JG, Ji WQ, Zhang H, Wu FY, Garzanti E (2019) The Langjiexue Group is an in situ sedimentary sequences rather than an exotic block: constraints from coeval Upper Triassic strata of the Tethys Himalaya (Qulonggongba Formation). Sci China Earth Sci 62:83–79

Meng ZY, Wang JG, Garzanti E, Han Z, Chen G (2021) Late Triassic rifting and volcanism on the northeastern Indian margin: a new phase of Neo-Tethyan seafloor spreading and its paleogeographic implications. Palaeogeogr Palaeoclimatol Palaeoecol 570:110367

Metcalf K, Kapp P (2019) History of subduction erosion and accretion recorded in the Yarlung Suture Zone, southern Tibet. Geol Soc, Lond, Spec Publ 483(1):517–554

Mo XX, Niu YL, Dong GC, Zhao ZD, Hou ZQ, Zhou S, Ke S (2008) Contribution of syncollisional felsic magmatism to continental crust growth: a case study of the Paleocene Linzizong Volcanic Succession in southern Tibet. Chem Geol 250:49–67

Murphy MA, Yin A, Harrison TM, Dürr SB, Chen Z, Ryerson FJ, Kidd WSF, Wang X, Zhou X (1997) Did the Indo–Asian collision alone create the Tibetan Plateau? Geology 25(8):719–722

Najman Y, Appel E, Boudagher-Fadel M, Bown P, Carter A, Garzanti E, Godin L, Han J, Liebke U, Oliver G, Parrish R, Vezzoli G (2010) Timing of India–Asia collision: geological, biostratigraphic, and palaeomagnetic constraints. J Geophys Res, Solid Earth 115(B12):B12416

Najman Y, Jenks D, Godin L, Boudagher-Fadel M, Millar I, Garzanti E, Horstwood M, Bracciali L (2017) The Tethyan Himalayan detrital record shows that India–Asia terminal collision occurred by 54 Ma in the Western Himalaya. Earth Planet Sci Lett 459:301–310

Nicolas A, Girardeau J, Marcoux J, Dupre B, Wang XB, Cao YG, Zheng HX, Xiao XC (1981) The Xigaze ophiolite (Tibet): a peculiar oceanic lithosphere. Nature 294:414–417

Niu XL, Zhao ZD, Mo XX, Depaolo DJ, Dong G, Zhang SQ, Zhu DC, Guo TY (2006) Elemental and Sr-Nd-Pb isotopic geochemistry for basic rocks from Decun–Angren ophiolites in Xigaze area, Tibet: implications for the characteristics of the Tethyan upper mantle domain. Acta Petrol Sin (Yanshi Xuebao) 22(12):2875–2888

Orchard M, Nassichuk W, Rui L (1994) Conodonts from the lower Griesbachian Otoceras Latilobatum Bed of Selong, Tibet and the position of the Permian—Triassic Boundary

Orme DA, Carrapa B, Kapp P (2014) Sedimentology, provenance and geochronology of the upper Cretaceous–lower Eocene western Xigaze forearc basin, southern Tibet. Basin Res 387–411

Orme DA, Laskowski AK (2016) Basin analysis of the Albian-Santonian Xigaze Forearc, Lazi Region, South-Central Tibet. J Sediment Res 86(8):894–913

Ratschbacher L, Frisch W, Liu G, Chen C (1994) Distributed deformation in southern and western Tibet during and after the India–Asia collision. J Geophys Res Atmos 99:19917–19945

Reagan MK, Pearce JA, Petronotis K, Almeev RR, Avery AJ, Carvallo C, Chapman T, Christeson GL, Ferré EC, Godard M, Heaton DE, Kirchenbaur M, Kurz W, Kutterolf S, Li H, Li Y, Michibayashi K, Morgan S, Nelson WR, Prytulak J, Python M, Robertson AHF, Ryan JG, Sager WW, Sakuyama T, Shervais JW, Shimizu K, Whattam SA (2017) Subduction initiation and ophiolite crust: new insights from IODP drilling. Int Geol Rev 59(11):1439–1450

Reagan MK, Ishizuka O, Stern RJ, Kelley KA, Ohara Y, Blichert-Toft J, Bloomer SH, Cash J, Fryer P, Hanan BB, Hickey-Vargas R, Ishii T, Kimura J-I, Peate DW, Rowe MC, Woods M (2010) Fore-arc basalts and subduction initiation in the Izu-Bonin-Mariana system. Geochem Geophys Geosystems 11(3):Q03X12

Sciunnach D, Garzanti E (2012) Subsidence history of the Tethys Himalaya. Earth-Sci Rev 111:179–198

Shen SZ, Archbold NW, Shi GR, Zhong-Qiang C (2000) Permian brachiopods from the SelongXishan section, Xizang (Tibet), China Part 1: stratigraphy, strophomenida, productida and rhynchonellida. Geobios-Lyon 33:725–752

Shen S, Dongli S, Shi GR (2003a) A biogeographically mixed late Guadalupian (late Middle Permian) brachiopod fauna from an exotic limestone block at Xiukang in Lhaze county, Tibet. Journal of Asian Earth Sciences 21(10):1125–1137

Shen S, Shi GR, Archbold NW (2003b) A Wuchiapingian (Late Permian) brachiopod fauna from an exotic block in the Indus–Tsangpo suture zone, southern Tibet, and its palaeobiogeographical and tectonic implications. Palaeontology 46(2):225–256

Shen SZ, Cao CQ, Henderson CM, Wang XD, Shi GR, Wang Y, Wang W (2006) End-Permian mass extinction pattern in the northern peri-Gondwanan region. Palaeoworld 15:3–30

Sinclair H (1997) Tectonostratigraphic model for underfilled peripheral foreland basins: an Alpine perspective. Geol Soc Am Bull 109(3):324–346
Sinclair HD, Jaffey N (2001) Sedimentology of the Indus Group, Ladakh, northern India: implications for the timing of initiation of the palaeo-Indus River. J Geol Soc Lond 158:151–162
Stanley SM, Yang X (1994) A double mass extinction at the end of the Paleozoic era. Science 266:1340–1344
Sun GY, Hu XM, Wang JG (2011) Petrologic and provenance analysis of the Zongzhuo mélange in Baisha area, Gyangze, Southern Tibet. Acta Geol Sin 85(8):1343–1351
Sun S, McDonough W (1989) Chemical and isotopic systematics of oceanic basalts: implications for mantle composition and processes. In Saunders AD, Norry MJ (eds) Magmatism in the Ocean Basins. Geological Society [London] Special Publication 42:313–345
Tao J (1988) Plant fossils from the Liuqu Formation in Lhaze County, Xizang and their paleoclimatological significances. Memoirs of the Institute of Geology, Chinese Academy of Science (in Chinesewith English abstract). Science Press, Beijing, pp 223–238
Tapponnier P, Mercier JL, Proust F, Andrieux J, Armijo R, Bassoullet JP, Brunel M, Burg JP, Colchen M, Dupre B, Girardeau J, Marcoux J, Mascle G, Matte P, Nicolas A, Li T, Xiao X, Chang C, Lin B, Li G, Wang N, Chen G, Han T, Wang X, Den W, Zhen H, Sheng H, Cao Y, Zhou J, Qiu H (1981) The Tibetan side of the India–Eurasia collision. Nature 294(5840):405–410
Tapponnier P, Zhiqin X, Roger F, Meyer B, Arnaud N, Wittlinger G, Jingsui Y (2001) Oblique stepwise rise and growth of the Tibet Plateau. Science 294:1671–1677
Volkmer JE, Kapp P, Guynn JH, Lai Q (2007) Cretaceous-Tertiary structural evolution of the north central Lhasa terrane, Tibet. Tectonics 26:TC6007
Wan X, Wang L, Wang C, Jansa L (1998) Discovery and significance of Cretaceous fossils from the Xigaze forearc basin, Tibet. J Asian Earth Sci 16(2):217–224
Wan XQ, Ding L, Li JG, Cai HW (2001) Latest Cretaceous to Early Eocenemarine strata in the Zhongba region, Tibet. J Stratigr 25(4):267–272
Wan XQ, Lamolda MA, Si JL, Li GB (2005) Foraminiferal stratigraphy of Late Cretaceous red beds in southern Tibet. Cretac Res 26:43–48
Wang C, Liu Z (1999) Xigaze forearc basin and Yarlung Zangbo suture zone. Geological Publishing House, Tibet
Wang YJ, Mu XN (1980) Some new observation on the Permian biostratigraphy of the Himalayan province in southern Tibet. J Stratigr 4(2):145–151
Wang XD, Sugiyama T (2000) Diversity and extinction patterns of Permian coral faunas of China. Lethaia 33:285–294
Wang XB, Bao PS, Xiao XC (1987) Ophiolites of the Yarlung Zangbo River, Xizang (Tibet). Publishing House of Surveying and Mapping, Beijing, p 117
Wang C, Xia D, Zhou X, Chen J, Liu Z (1996) Field Trip Guide: T121/T387 geology between the Indus–Yarlung Zangbo Suture Zone and the Himalaya Mountains (Xizang), China. Geological Publishing House, Beijing, pp 49–156
Wang CS, Li XH, Hu XM, Jansa LF (2002) Latest marine horizon north of Qomolangma (Mt Everest): implications for closure of Tethys seaway and collision tectonics. Terra Nova 14:114–120
Wang CS, Hu XM, Sartid M, Scotte RW, Li XH (2005a) Upper Cretaceous oceanic red beds in southern Tibet: a major change from anoxic to oxic, deep-sea environments. Cretac Res 26:21–32
Wang R, Xia B, Zhou G, Zhang Y, Yang Z, Li W, Wei D, Zhong L, Xu L (2006) SHRIMP zircon U-Pb dating for gabbro from the Tiding ophiolite in Tibet. Chin Sci Bull 51(14):1776–1779
Wang JG, Hu XM, Wu FY, Jansa L (2010) Provenance of the Liuqu Conglomerate in southern Tibet: a Paleogene erosional record of the Himalayan–Tibetan orogen. Sed Geol 231(3–4):74–84
Wang JG, Hu XM, Jansa L, Huang ZC (2011) Provenance of the Upper Cretaceous-Eocene deep-water sandstones in Sangdanlin, southern Tibet: constraints on the timing of initial India–Asia collision. J Geol 119(3):293–309

Wang C, Li X, Liu Z, Li Y, Jansa L, Dai J, Wei Y (2012) Revision of the Cretaceous-Paleogene stratigraphic framework, facies architecture and provenance of the Xigaze forearc basin along the Yarlung Zangbo suture zone. Gondwana Res 22:415–433

Wang JG, Hu XM, Garzanti E, Wu FY (2013) Upper Oligocene-Lower Miocene Gangrinboche Conglomerate in the Xigaze area, southern Tibet: implications for Himalayan uplift and paleo-Yarlung-Zangbo initiation. J Geol 121(4):425–444

Wang JG, Hu X, BouDagher-Fadel M, Wu FY, Sun GY (2015) Early Eocene sedimentary recycling in the Kailas area, southwestern Tibet: implications for the initial India–Asia collision. Sed Geol 315:1–13

Wang C, Ding L, Zhang LY, Kapp P, Pullen A, Yue YH (2016a) Petrogenesis of Middle-Late Triassic volcanic rocks from the Gangdese belt, southern Lhasa terrane: Implications for early subduction of Neo-Tethyan oceanic lithosphere. Lithos 262:320–333

Wang JG, Wu FY, Garzanti E, Hu X, Ji WQ, Liu ZC, Liu XC (2016b) Upper Triassic turbidites of the northern Tethyan Himalaya (Langjiexue Group): the terminal of a sediment-routing system sourced in the Gondwanide Orogen. Gondwana Res 34:84–98

Wang JG, Hu X, Garzanti E, An W, Liu XC (2017a) The birth of the Xigaze forearc basin in southern Tibet. Earth Planet Sci Lett 465:38–47

Wang L, Wignall PB, Sun Y, Yan C, Zhang Z, Lai X (2017b) New Permian-Triassic conodont data from Selong (Tibet) and the youngest occurrence of Vjalovognathus. J Asian Earth Sci 146:152–167

Wang XC, Liu WL, Zhong Y, Hu XC, Xia B, Huang W (2018) Geochemical and zircon U-Pb age constraints on the origin of the Mesozoic Xigaze ophiolite, Yarlung Zangbo suture zone, SW China. Int Geol Rev 60(10):1267–1289

Wang JG, Hu X, Garzanti E, BouDagher-Fadel MK, Liu ZC, Li J, Wu FY (2020) From extension to tectonic inversion: Mid-Cretaceous onset of Andean-type orogeny in the Lhasa block and early topographic growth of Tibet. Geol Soc Am Bull 132(11–12):2432–2454

Wang CS, Li XH, Hu XM, Wan XQ, Yin JR, Huang YJ, Huang SJ, Li GB (2005) Tethyan Himalayan sedimentary geology and continental paleoceanography. Geological Publishing House, Beijing, p 373 (in Chinese)

Wei L, Liu X, Yan F, Mai X, Zhou X, Li G, Liu X (2009) Discovery and preliminary study on palynofossils from the Paleogene Liuqu Conglomerates in southern Xizang (Tibet). Acta Microbiol Sin 26:249–260

Willems H, Zhou Z, Zhang B, Gräfe KU (1996) Stratigraphy of the Upper Cretaceous and Lower Tertiary strata in the Tethyan Himalayas of Tibet (Tingri area, China). Geol Rundsch 85:723–754

Wu FY, Ji WQ, Liu CZ, Chung SL (2010) Detrital zircon U-Pb and Hf isotopic data from the Xigaze fore-arc basin: constraints on Transhimalayan magmatic evolution in southern Tibet. Chem Geol 271(1–2):13–25

Wu F, Ji W, Wang J, Liu C, Chung S, Clift PD (2014a) Zircon U-Pb and Hf isotopic constraints on the onset time of India–Asia collision. Am J Sci 314(2):548–579

Wu FY, Liu CZ, Zhang LL, Zhang C, Wang JG, Ji WQ, Liu XC (2014b) Yarlung Zangbo ophiolite: a critical updated view. Acta Petrol Sin (yanshi Xuebao) 30(2):293–325

Xiong Q, Griffin WL, Zheng JP, O'Reilly SY, Pearson NJ, Xu B, Belousova EA (2016) Southward trench migration at ~130–120 Ma caused accretion of the Neo-Tethyan forearc lithosphere in Tibetan ophiolites. Earth Planet Sci Lett 438:57–65

Xiong F, Yang J, Robinson PT, Gao J, Chen Y, Lai S (2017) Petrology and geochemistry of peridotites and podiform chromitite in the Xigaze ophiolite, Tibet: implications for a suprasubduction zone origin. J Asian Earth Sci 146(supplement C), 56–75

Yin A (2006) Cenozoic tectonic evolution of the Himalayan orogen as constrained by along-strike variation of structural geometry, exhumation history, and foreland sedimentation. Earth Sci Rev 76:1–131

Yin A, Harrison TM (2000) Geologic evolution of the Himalayan–Tibetan orogen. Annu Rev Earth Planet Sci 28:211–280

Yin A, Harrison TM, Ryerson FJ, Chen WJ, Kidd WSF, Copeland P (1994) Tertiary structural evolution of the Gangdese thrust system in southeastern Tibet. J Geophys Res 99:18175–18201

Yin J, Enay R, Wan X (1999) The first report of the late Triassic-Early Jurassic passage beds in the eastern tethyan himalaya. Comptes Rendus De Lacademie Des Sciences Serie II A Sciences De La Terre Et Des Planetes, 329, 0–133

Yin JR (2010) Jurassic ammonites of Tibet. Geological Publishing House, Beijing, pp 227–235 (in Chinese with English abstract)

Yu ZX, Zhen AZ (1979) Geologic map of the Lhasa region at a scale of 1:1000000. Geologic Publishing House

Yuan DX, Zhang YC, Shen SZ (2018) Conodont succession and reassessment of major events around the Permian–Triassic boundary at the Selong Xishan section, southern Tibet, China. Global Planet Change 161:194–210

Zhang QH, Willems H, Ding L, Gräfe KU, Appel E (2012) Initial India–Asia continental collision and foreland basin evolution in the Tethyan Himalaya of Tibet: evidence from stratigraphy and paleontology. J Geol 120:175–189

Zhang C, Liu CZ, Wu FY, Zhang LL, Ji WQ (2016a) Geochemistry and geochronology of mafic rocks from the Luobusa ophiolite, south Tibet. Lithos 245:93–108

Zhang LL, Liu CZ, Wu FY, Zhang C, Ji WQ, Wang JG (2016b) Sr-Nd-Hf isotopes of the intrusive rocks in the Cretaceous Xigaze ophiolite, southern Tibet: constraints on its formation setting. Lithos 258–259:133–148

Zhang XR, Chung SL, Lai YM, Ghani AA, Murtadha S, Lee HY, Hsu CC (2019a) A 6000-km-long Neo-Tethyan arc system with coherent magmatic flare-ups and lulls in South Asia. Geology 47:573–576

Zhang C, Liu CZ, Xu Y, Ji WB, Wang JM, Wu FY, Liu T, Zhang ZY, Zhang WQ (2019b) Subduction re-initiation at dying ridge of Neo-Tethys: insights from mafic and metamafic rocks in Lhaze ophiolitic mélange, Yarlung–Tsangbo suture zone. Earth Planet Sci Lett 523:115707

Zhang C, Liu CZ, Wu FY, Ji WB, Liu T, Xu Y (2017) Ultra-refractory mantle domains in the Luqu ophiolite (Tibet): petrology and tectonic setting. Lithos, 286–287(supplement C), 252–263

Zhang Y, Liu X, Watkins DK, Bruno MDR, Yao HW, Han KB, Guo HF, Chen X, The Cretaceous (early Albian to early 1 Campanian) biostratigraphy and paleotemperature reconstruction of the eastern Tethys: calcareous nannofossil evidences from southern Tibet, China. In review

Zhao ZD, Mo XX, Dilek Y, Niu YL, DePaolo DJ, Robinson P, Zhu DC, Sun CG, Dong GC, Zhou S, Luo ZH, Hou ZQ (2009) Geochemical and Sr-Nd-Pb-O isotopic compositions of the post-collisional ultrapotassic magmatism in SW Tibet: petrogenesis and implications for India intra-continental subduction beneath southern Tibet. Lithos 113:190–212

Zhu B, Kidd WSF, Rowley DB, Currie BS, Shafique N (2005) Age of initiation of the India-Asia collision in the east-central Himalaya. J Geol 113(3):265–285

Zhu J, Du Y, Liu Z, Feng Q, Tian W, Li J, Wang C (2006) Mesozoic radiolarian chert from the middle sector of the Yarlung Zangbo suture zone, Tibet and its tectonic implications. Sci China Ser D 49(4):348–357

Zhu DC, Pan GT, Chung SL, Liao ZL, Wang LQ, Li GM (2008) SHRIMP zircon age and geochemical constraints on the origin of Lower Jurassic volcanic rocks from the Yeba Formation, southern Gangdese, South Tibet. Int Geol Rev 50:442–471

Zhu DC, Zhao ZD, Niu YL, Mo XX, Chung SL, Hou ZQ, Wang LQ, Wu FY (2011) The Lhasa Terrane: record of a microcontinent and its histories of drift and growth. Earth Planet Sci Lett 301(1–2):241–255

Zhu DC, Wang Q, Zhao ZD, Chuang SL, Cawood PA, Niu YL, Liu S-A, Wu FY, Mo XX (2015) Magmatic record of India–Asia collision. Sci Rep 5:14289

Ziabrev S, Aitchison J, Abrajevitch A, Davis A, Luo H (2003) Precise radiolarian age constraints on the timing of ophiolite generation and sedimentation in the Dazhuqu terrane, Yarlung–Tsangpo suture zone, Tibet. J Geol Soc, London 160(4):591–599

A Flooding-Dominated Lacustrine Basin: Upper Triassic Fluvial-Crevasse to Fluvial-Delta Successions in the Yanhe and Shiwanghe Sections, Ordos Basin, China

Xiangbo Li, Rukai Zhu, Carlos Zavala, Xianyang Liu, Tian Yang, Huaqing Liu, Ping Luo, Zhiyong Gao, Bin Hao, and Zhanlong Yang

1 Geological Setting

(1) **Present location and regional tectonic characteristics of the Ordos Basin**
The Ordos Basin is located in the western part of the North China Plateau (Fig. 1). This basin extends along five provinces of China (Shaanxi, Gansu, Ningxia, Shanxi, and Inner Mongolia), and is currently limited by the Yinshan Mountains in the north, the Qinling Mountains in the south, the Luliang Mountains in the east, and the Helan-Liupan Mountains in the west. The Ordos Basin covers an area of more than 37 × 10^4km^2. This basin went through 5 evolutionary phases including Middle-Late Proterozoic aulacogen, Early Paleozoic

X. Li (✉) · R. Zhu · H. Liu · P. Luo · Z. Gao · B. Hao · Z. Yang
Exploration and Development-Northwest Branch (NWGI), Research Institute of Petroleum, PetroChina, Beijing, China
e-mail: lixiangbo911@sina.com

R. Zhu
e-mail: zrk@petrochina.com.cn

H. Liu
e-mail: 782352398@qq.com

P. Luo
e-mail: luoping@petrochina.com.cn

Z. Gao
e-mail: gzy@petrochina.com.cn; gzybox@163.com

B. Hao
e-mail: haobin86@petrochina.com.cn

Z. Yang
e-mail: yang_zl@petrochina.com.cn

C. Zavala
Universidad Nacional del Sur, Buenos Aires, Argentina
e-mail: czavala@uns.edu.ar

X. Hu (ed.), *Field Trip Guidebook on Chinese Sedimentary Geology*,
https://doi.org/10.1007/978-981-99-6936-4_13

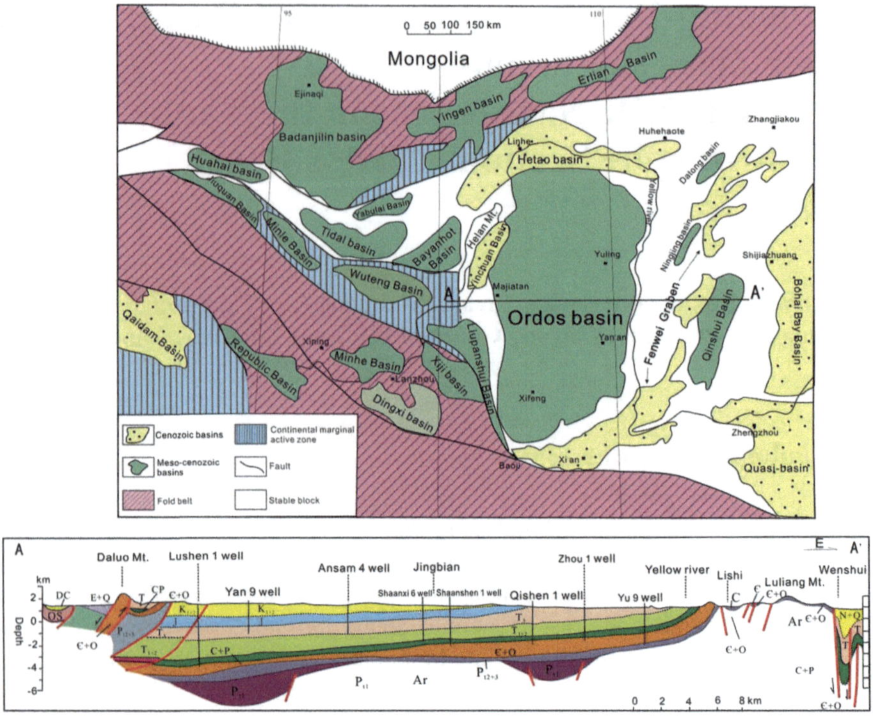

Fig. 1 Structural location of the Ordos Basin (above) and EW geologic cross-section (below)

carbonate platform, Late Paleozoic coastal plain, Mesozoic nonmarine basin and Cenozoic—to date rifted basin. There is a gentle geological structure in the basin area, that is a regional monocline tilting and dipping towards the west with a structural gradient of about 1 degree.

(2) **Tectonic location and prototype basin characteristics in Late Triassic**
During the Middle and Late Triassic, the Ordos Basin was located in the inland area of the northern margin of the Tethys Ocean (Fig. 2). It was an inland foreland lacustrine basin. The original sedimentary area of the basin was much larger than the present residual area. It is a super-large asymmetrical inland craton depression basin boundary by the Hexi Corridor in the west, the Southern North China region in the east, the Qinling-Qilian orogenic belt in the south, and the foot of the Yinshan Mountains in the north (Fig. 3).

X. Liu
National Engineering Laboratory for Exploration and Development of Low-Permeability Oil & Gas Fields, PetroChina Changqing Oilfield Company, Beijing, China
e-mail: lxy3_cq@petrochina.com.cn

T. Yang
Chengdu University of Technology, Chengdu, China

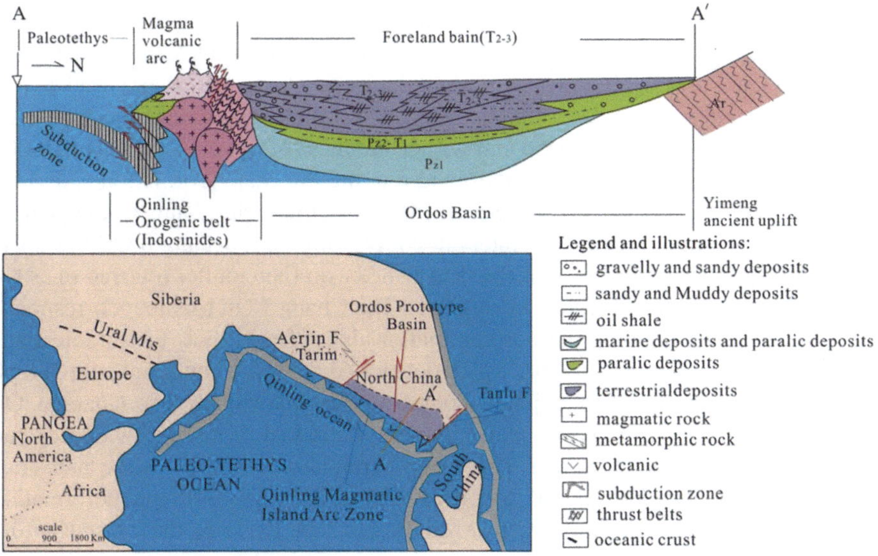

Fig. 2 Regional profile and tectonic setting of the Ordos Basin during the Middle to Late Triassic (Ladinian–Norian). The upper cross-section shows a back-arc foreland basin resulting from the subduction of the Qinling oceanic plate. The lower map shows the location of the Ordos Basin in the regional context (after Bai and Ma 2019)

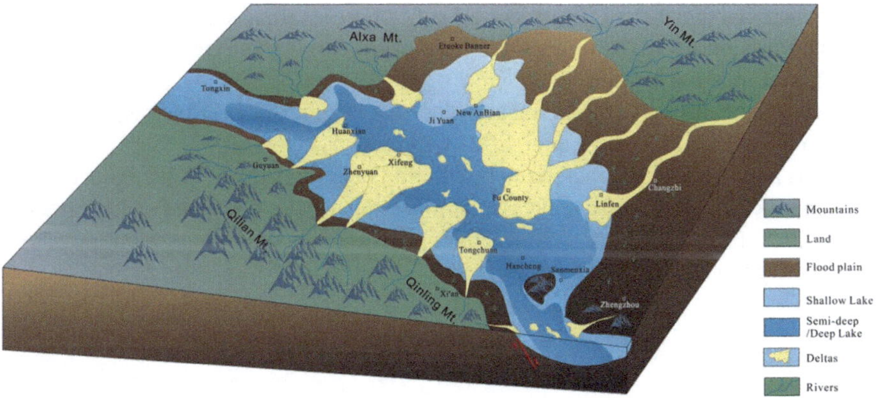

Fig. 3 Summary of late Triassic Maximum Lake flooding sedimentation in Ordos Basin

(3) **Sedimentary characteristics and traditional stratigraphic division of the Middle and Upper Triassic in the Ordos Basin**

The Triassic constitutes one of the thickest sedimentary successions in the basin, and is composed of Lower, Middle, and Upper Series. The Liujiagou, Heshanggou, and Zhifang formations from the Lower and Middle Series are composed of red beds accumulated in different fluvial and lacustrine environments developed in arid conditions. These deposits constitute a thick succession associated to a small and very shallow lacustrine basin. No source rocks were recognized in this succession. The Upper Series is composed of the Yanchang Formation, which is a clastic unit with a thickness of more than 1000 m (1000–1500 m) (Yang 2002). From bottom to top, the Yanchang Formation is divided into ten members (sections), termed from base to top as Chang 10 to Chang 1 (Table 1). Among them, the interval from Chang 10 to Chang 8 Oil Members comprises the initial stage of the lacustrine basin (hereinafter referred to as the early stage of the Yanchang Formation). The Chang 7 Oil Member corresponds to the largest flooding period (hereinafter also referred to as the middle stage of the Yanchang Formation), which developed the most important source rock series in the basin (Zhangjiatan Shale). During the Chang 6 to Chang 1 Oil Members, the basin experienced a process of gradual shrinking, swamping and gradually closing of the lacustrine basin (hereinafter referred to as the later stage of the Yanchang Formation). Since the basin underwent uplift and deformation during the Indosinian movements (Middle and Late Triassic), the top and bottom of the Yanchang Formation were both characterized by tectonically-induced unconformities.

(4) **Paleoclimate during the Late Triassic**
The middle-late Triassic climate can be divided into three stages (Li et al. 2022). In the Ladinian Age (Chang 10 to Chang 8), the climate is dry. In the Carnian period (Chang 7 to Chang 4 + 5), the climate is warm and humid. In the Norian-Rhaetian Ages (Chang 3 to Chang 1), the climate turned semi-dry again (Fig. 4).

(5) **Depositional environments and evolution during the accumulation of the Yanchang Formation**
Two depositional models for arid and wet conditions in Yanchang Formation have been proposed (Li et al. 2021a, b) (Fig. 5).
The arid model is characterized by a fluvial fan system (e.g., flood plain, crevasse channels, and crevasse splays) developed during the low stand system tract.
The wet model is characterized by a lacustrine fluvial-delta system developed during the transgressive period.

Table 1 Subdivision of the Yanchang Fm. in Ordos Basin

Strata system				Thickness (m)		Lithology
System	Formation	Sub-FM	Oil member			
Triassic	Yanchang	T_{3y5}	Chang 1	0–240		Dark mudstones, muddy siltstones, fine-medium grained sandstones, interbedded with coal seams
		T_{3y4}	Chang 2	Chang 2_1	40–45	Grayish green massive fine-grained sandstones interbedded with gray mudstones
				Chang 2_2	40–45	Light-gray fine-grained sandstones interbedded with gray mudstones
				Chang 2_3	40–50	Gray, light-gray fine-grained sandstones interbedded with gray mudstones
			Chang 3	100–110		Gray, grayish-brown fine-grained sandstones interbedded with dark mudstones
		T_{3y3}	Chang 4 + 5	Chang 4 + 5_1	30–50	Dark mudstones, limy mudstones, coal seams interbedded with thin-bedded siltstones and fine-grained sandstones
				Chang 4 + 5_2	30–50	Light gray silt-fine-grained sandstones interbedded with dark mudstones
			Chang 6	Chang 6_1	35–45	Brownish gray, greenish-gray fine-grained sandstones interbedded with dark mudstones
				Chang 6_2	35–45	Light grayish-green silt-fine-grained sandstones interbedded with dark mudstones
				Chang 6_3	35–40	Grayish black mudstones, muddy siltstones, silt-fine-grained sandstones interbedded with thin tuff levels
			Chang 7	80–100		Dark mudstones, limy mudstones, oil shales interbedded with thin silt-fine-grained sandstones
		T_{3y2}	Chang 8	70–85		Dark mudstones, sandy mudstones interbedded with gray silt-fine-grained sandstones
			Chang 9	90–120		Dark mudstones, shales interbedded with gray silt-fine-grained sandstones
		T_{3y1}	Chang 10	280		Gray thick-massive medium-fine-grained sandstones with coarse-grained sandstones at the bottom
	Zhifang FM					Grayish purple mudstones, sandy mudstones interbedded with purplish-red medium-fine-grained sandstones

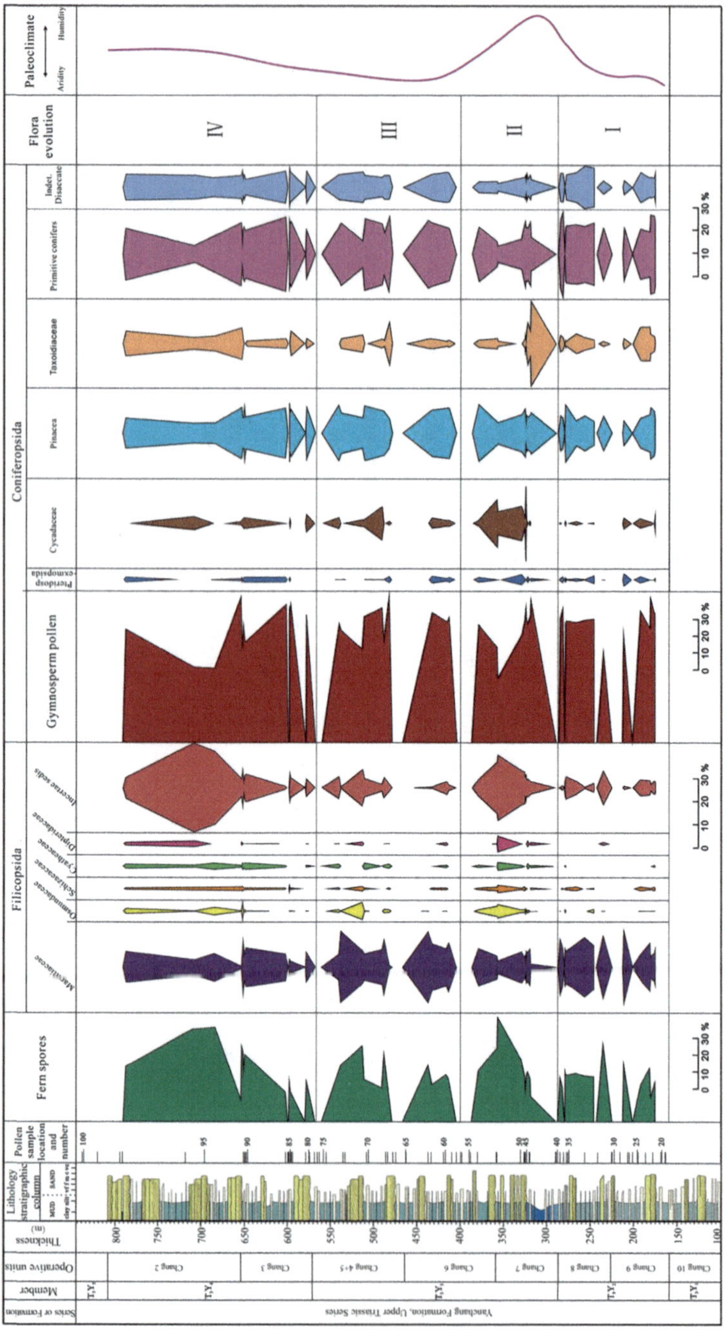

Fig. 4 Distribution of main palynoflora types and palaeoclimatic evolution at the depositional stage of the Yanchang Formation (*Note* no palynoflora found in Chang 10 Member and Zhifang Formation)

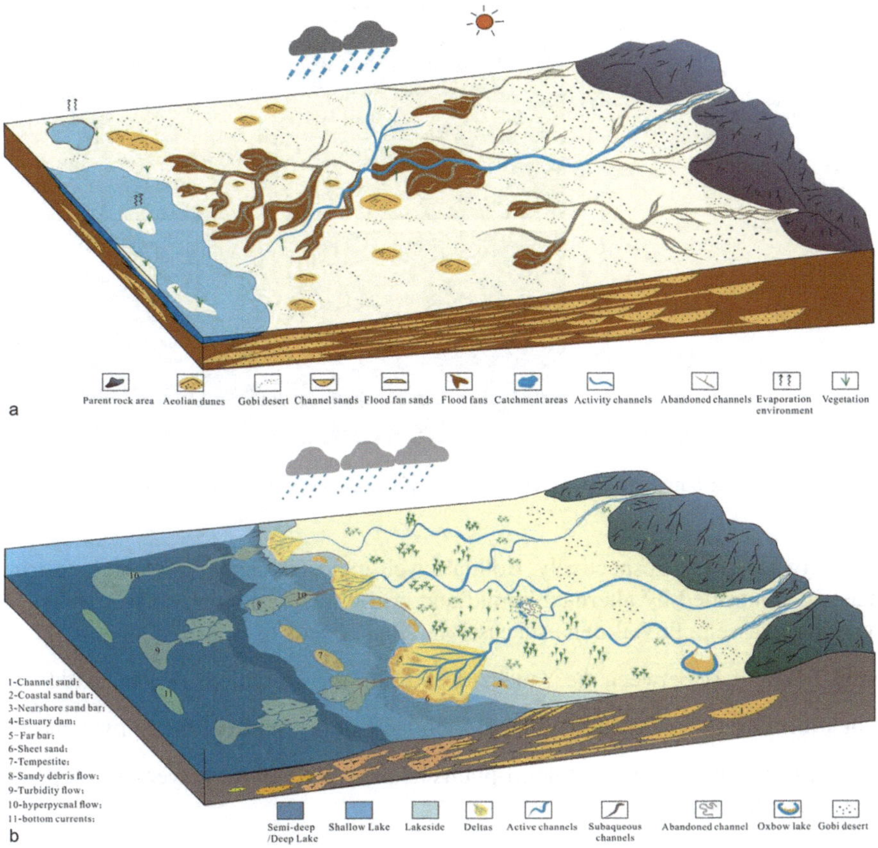

Fig. 5 Depositional model for the Yanchang Formation based on alternate humid and arid climate evolution. **a** Depositional model for arid periods. The figure shows that in the arid environment the catchment area of the continental basin is small, and the main depositional units are flood plain, channels, and crevasse fans. Due to frequent channel diversions, sedimentation is usually dominated by aggradation. **b** Depositional model of fluvio-deltaic, and lacustrine systems affected by gravity flows (mainly hyperpycnal flows) developed during humid periods. The figure shows that in a humid environment, the catchment area of the continental basin is large, and the main depositional units are fluvio-deltaic, deep to semi-deep lake, and various kinds of gravity-flow deposits (mostly hyperpycnal flows), with foreset deposits developed at the delta front

2 Stop Descriptions

The upper Triassic Yanchang Formation outcrop profiles introduced here are mainly located in the surroundings of the Yichuan and Yanchang cities in the northern Shaanxi province. They include two outcrop profiles, (1) the Yanhe section in the Yanchang County and (2) the Shiwanghe section in the Yichuan County. In outcrops, this formation is about 800 m thick with a gentle dip to the west of less than 1°. The outcrops of Chang 10 to Chang 3 Oil Member are incised by the Yanhe

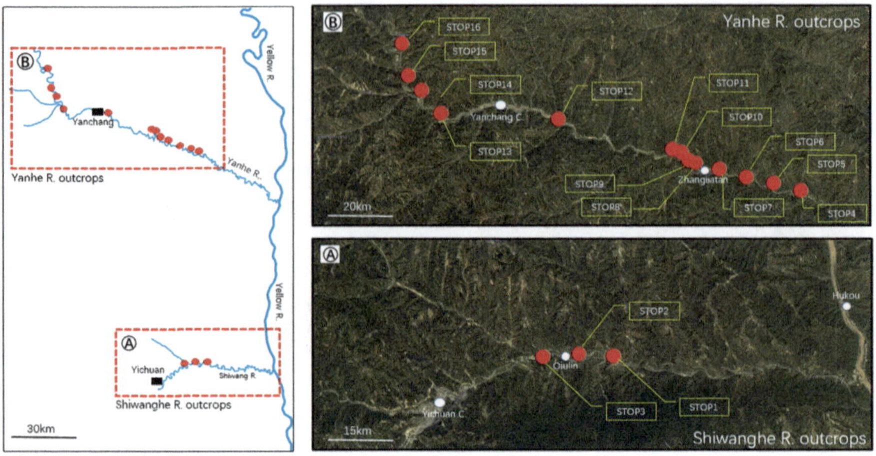

Fig. 6 Google Earth image showing the locations of the Yanhe and Shiwanghe Sections

and Shiwang rivers, respectively. Both are well-exposed, showing inland lacustrine deposits controlled by alternate humid and arid climate settings, corresponding to fluvial-crevasse to fluvial-delta successions (Figs. 6 and 7).

Stop 1 Boundary between the Yanchang and Zhifang formations (Figs. 8 and 9).

Location East of the Xiantou Village, Yichuan County, north of the Shiwanghe.

Content Note the presence of flood-plain channels and paleosol layers developed in arid environments.

Stop 2 Flood plain and sand body architectures in the middle of the Chang 10 Member (Figs. 10 and 11).

Location East of the Qiulin Town, Yichuan County, north of the Shiwanghe.

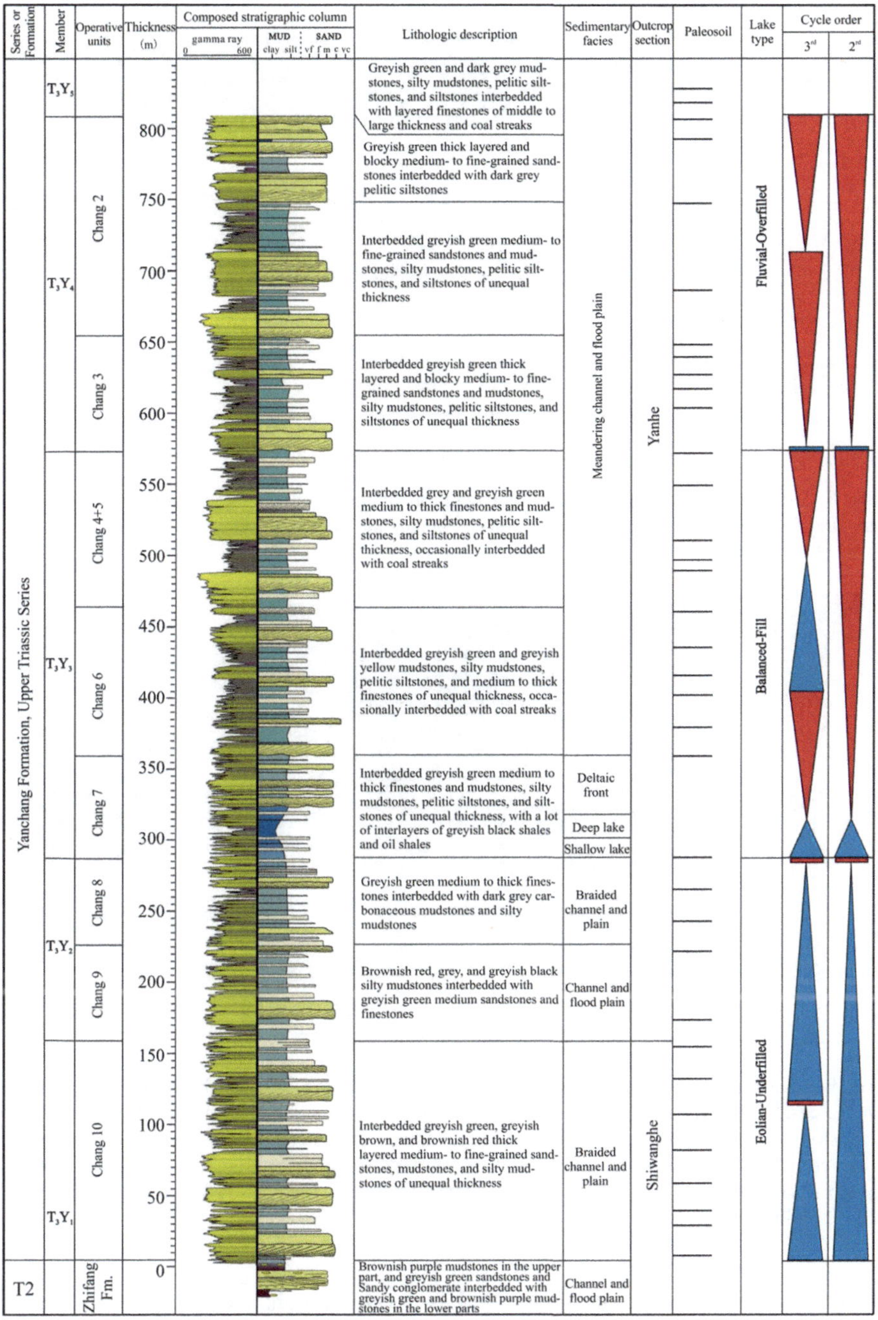

Fig. 7 Sedimentary log of the Yanchang Fm at the Yanhe and Shiwanghe Sections

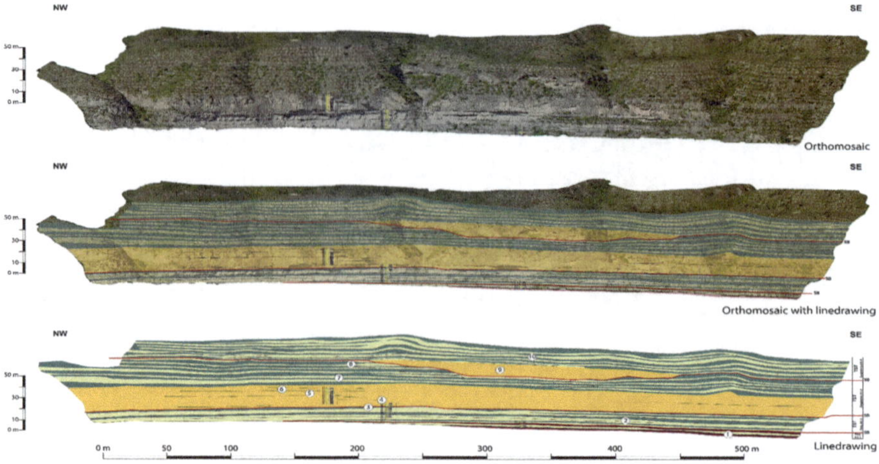

Fig. 8 Drone photos and geological interpretation of the boundary between the Yanchang and Zhifang formations at the east of the Xiantou Village, Yichuan County

Fig. 9 Measured sedimentary log and detailed pictures showing typical sedimentary characteristics in the east of the Xiantou Village. **a** Measured log showing a section dominated by fine-grained sandstones. This section starts with reddish claystone with calcareous nodules. The upper part is characterized by sandy mudstones exhibiting wavy heterolithic bedding. **b** Macroscopic outcrop photo; **c**, **d** Partial zoom in picture b; *Note* CCH—Crevasse channel, CVS—crevasse splay, FP—Flood-plain deposit

Content These deposits are interpreted as flood-plain channels and paleosol layers Developed in arid environments.

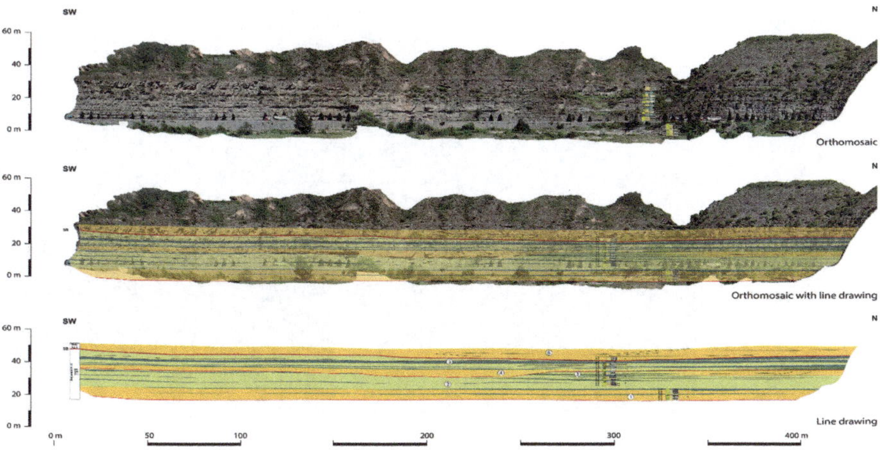

Fig. 10 Drone photos and geological interpretation of Chang 10 Member at the east of the Qiulin Town

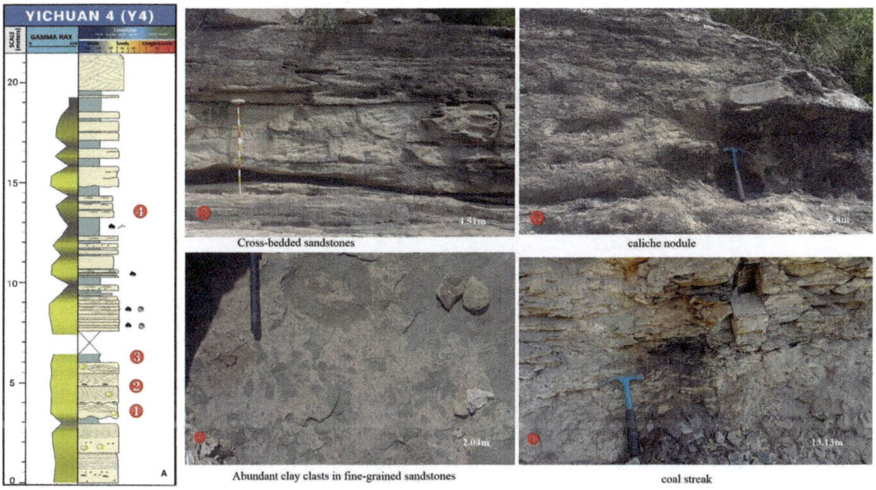

Fig. 11 Measured stratigraphic log and detailed pictures of typical sedimentary characteristics in the east of the Qiulin Town (detailed picture). The measured log shows an alternation between medium- to fine-grained massive or cross-bedded sandstones with abundant clay clasts and thin siltstone layers. *Note* the red circle with the numbers shows the location of detailed pictures

Stop 3 Characteristics of the sequence boundary between the Chang 10 and Chang 9 Member of Yanchang Fm (Fig. 12).

Location North of the Xiayangwan Village, Yichuan County, north of the Shiwanghe.

Content Flood-plain crevasse channels and paleosol layers developed in arid environments.

Fig. 12 Measured stratigraphic log and detailed pictures of typical sedimentary characteristics in the north of the Xiayangwan Village. **a** Sedimentary log showing massive, laminated, tabular fine-grained sandstone beds (many of them with sharp-based), and poorly sorted siltstones with abundant soil structures, caliche nodules, root traces, plant debris, and coal seams. **b** Macroscopic outcrop photo; **c**, **d** Partial zoom in picture b, indicating paleosols in arid environments; *Note* CCH—Crevasse channel; CVS—crevasse splay

Stop 4 Deposition of aggradational flood-plain deposits in the lower middle part of the Chang 9 Member from the Yanchang Formation (Figs. 13 and 14).

Location Baijiahe Village, Yanchang County, north of the Yanhe.

Content Deposits interpreted as of flood-plain crevass splays, crevass channels, and mud cracks developed in arid environment with flood events.

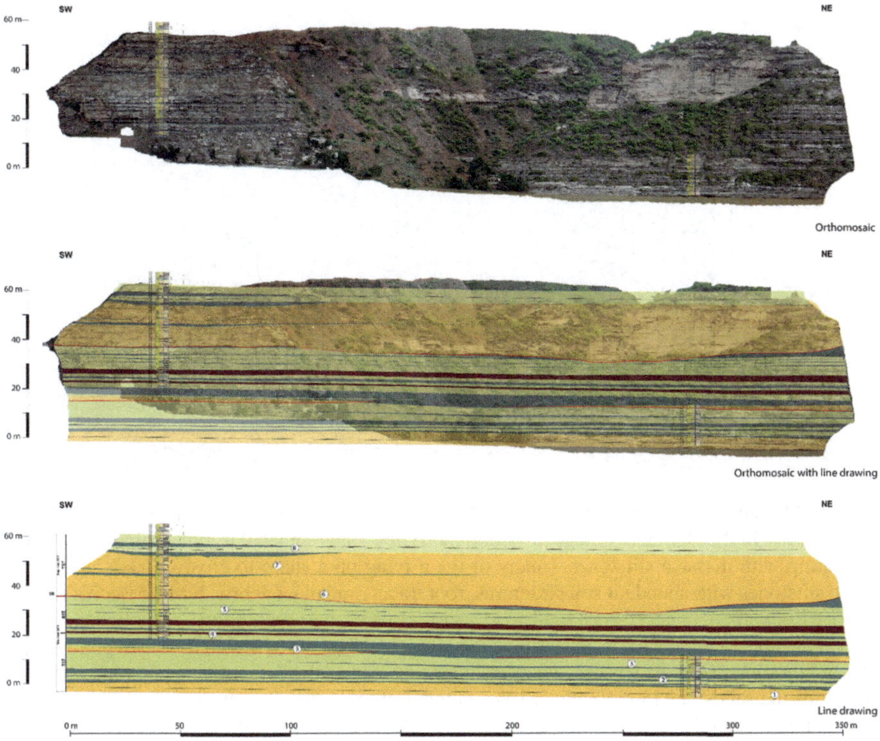

Fig. 13 Drone photos and geological interpretation of Chang 9 from the Baijiahe Village

Stop 5 Characteristics of the sequence boundary between the Chang 9 and Chang 8 members from the Yanchang Formation (Figs. 15 and 16).

Location West of the Laohugou Village, Yanchang County, north of the Yanhe.

Content Paleosol layers, iron oxide layers, aeolian deposits.

Fig. 14 Measured stratigraphic log and detailed pictures showing typical sedimentary characteristics observed in the Baijiahe Village. **a** Sedimentary log showing tabular, fine-grained sandstones alternating with massive siltstones, (many of them lying on a sharp-to erosive base), and poorly sorted siltstones with abundant soil structures, root traces, and plant debris. **b** Macroscopic outcrop photo; **c** Rockfall, light gray-green sandstone, its lower part is massive, slightly cross-bedding developed in the upper part, the location is shown in picture b; **d** The top of the sandstone in picture c, see a lot of "tortoise shell" similar polygonal cracks, suspected of the dry cracking phenomenon; **e** the bottom of the sandstone in picture c, a large number of carbonized plant stems and leaves can be seen, indicating flood events. *Note* CVS—crevasse splay; FP—Floodplain deposit; CCH—Crevasse channel

Fig. 15 Measured stratigraphic log and detailed pictures of typical sedimentary characteristics shown in the west of the Laohugou Village. **a** Detailed sedimentary log. The lower part shows a basal mudstone succession bearing root traces and oxide beds, followed in a sharp boundary by cross-bedded, fine-grained sandstones (maybe aeolian in origin) bearing carbonate concretions. The upper part is mainly composed of fine-grained sandstones exhibiting a thickening-upwards trend. **b** Macroscopic outcrop photo; **c**, **d** Partial zoom in picture b, showing plant roots and iron oxide layers

Fig. 16 Detailed photo of aeolian deposits at the west side of the Laohugou Village. These deposits are composed of laminated and cross-bedded fine-grained sandstones with sedimentary structures considered diagnostic of aeolian origin, such as climbing translated strata, granular ripples, and grain-fall dunes

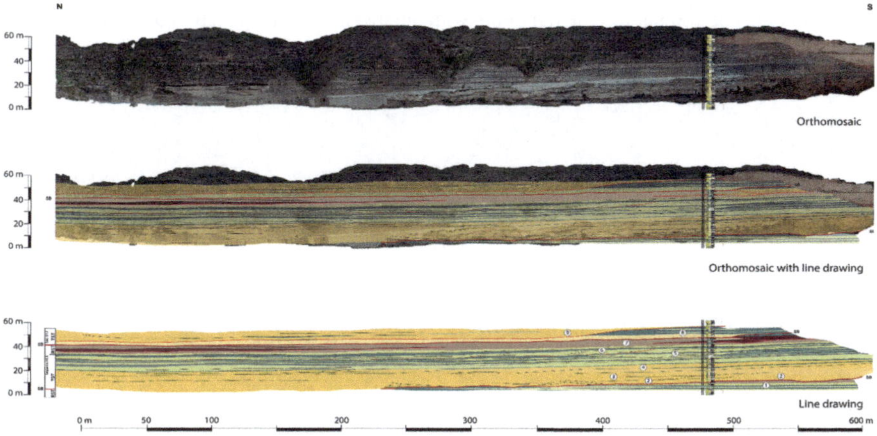

Fig. 17 Drone photos and geological interpretation of Chang 8 Member from the Qianwangjiahe Village

Stop 6 Boundary, sand body architecture, and sedimentary environments in the Chang 8_1 to Chang 8_2 members (Figs. 17, 18 and 19).

Location Qian Wangjiahe Village, Yanchang County, north of the Yanhe River.

Content Flood-plain branch channels, interchannel.

Fig. 18 Measured stratigraphic log and detailed pictures showing typical sedimentary characteristics at the Qianwangjiahe Village. **a** Measured log showing an alternation between fine-grained sandstones and laminated siltstones, or a rhythmic alternation between massive sandstone and siltstone deposits. Frequently, erosive-based, medium-grained sandstone with mud clasts over bed boundaries are recognized. Abundant plant debris, coal streak, root traces within siltstone deposits. **b** Macroscopic outcrop photo. The red circle with numbers shows the location of detailed pictures

Fig. 19 Inter-distributary (**a** and **b**) and distributary channels (**c**) of Chang 8_1 delta/flood plain, Qianwangjiahe locality. *Note* CH—Channel; CVS—crevasse splay; FP—Flood-plain deposit; ACH—Abandoned channel deposits; CCH—Crevasse channel

Stop 7 Sedimentary facies change at the abrupt boundary between Chang 7 and Chang 8 Member (Figs. 20 and 21).

Location Erlipan Village, Yanchang County, north of the Yanhe.

Content Paleosol layers, shallow-water or subaerial bioturbations, deep-water oil shales.

Stop 8 Shale deposits accumulated during the maximum flooding period of the Chang 7 Member (Fig. 22).

Location Located in the Zhangjiatan Shale Monument.

Fig. 20 Picture near the abrupt change in sedimentary facies between Chang 7 and Chang 8 Oil Member in the Erlipan Village, Yanhe Section. **a**—Macroscopic photo; **b**, **c**—Partial zoom in picture **a**

Fig. 21 Sedimentary phenomena near the abrupt boundary of sedimentary facies between Chang 7 and Chang 8 Member in the Erlipan Village, Yanhe Profile (see Fig. 20 for the location of the picture)

Fig. 22 The Maximum Flooding Surface (MFS), composed of oil shales and delta front deposits at the Zhangjiatan Shale Monument in the Chang 7 Member. **a** Measured log showing a fine-grained succession dominated by argillaceous mudstones (oil shale), fine-grained sandstones with an abrupt base ranging in thickness from 0.2 to 0.8 m. Sandy and muddy deposits are moderately bioturbated. In places, siltstones exhibit abundant plant debris. **b** Macroscopic outcrop photo; **c**, **d**—Partial zoom in picture **b**

Content Semi-deep lake shale (major source rock in the Mesozoic) of the Chang 7 MFS.

Stop 9 Frequent changes of the sedimentary sequence between deep lake, delta front, and delta plain in the Chang 7 Member (Figs. 23, 24 and 25).

Location 4 km west of the Zhangjiatan Shale Monument, north of the Yanhe.

Content shallow-water or subaerial bioturbations, deep-water oil shale.

Stop 10 Chang 7 ~ Chang 6_3 prograding delta and distributary architecture (Figs. 26 and 27).

Location North side of Yanhe River, Tanjiahe Village, Yanchang County.

Content Depositional characteristics of Chang 7 delta-front distributary, inter-distributary, Chang 6_3 delta-plain distributary and incised channel deposits.

Fig. 23 Drone photos and geological interpretation of Chang 7 from the west side of the Zhangjiatan Shale Monument

Fig. 24 Measured stratigraphic log and detailed pictures of typical sedimentary characteristics at the west side of the Zhangjiatan Shale Monument. **a** Measured log showing a fine-grained succession dominated by argillaceous mudstones (oil shale), fine-grained sandstones over an abrupt base, which frequently exhibits abundant plant debris. Occasionally, very fine-grained sandstones bearing erect trunks and root traces are recognized. **b** Macroscopic outcrop photo; **c** Very fine-grained sandstones bearing trunks; **d** Thinly bedded sandstone succession with parallel-laminated, climbing ripple cross-lamination and massive sandstone; **e** oil shale

Fig. 25 Detailed picture of typical sedimentary characteristics of the Chang 7 Member on the west side of the Zhangjiatan Shale Monument

Fig. 26 Drone photos and geological interpretation of Chang 7 ~ Chang 6_3 from Tanjiahe Village

Fig. 27 Measured stratigraphic log and detailed pictures showing typical sedimentary characteristics of the Chang 7 to Chang 6_3 Oil Member in the Tanjiahe Village. **a** The measured log comprises Chang 6 and Chang 8 Oil Members, the former showing a fine-grained, thick-bedded sandstone succession consisting of sharp- to erosive-based cross-bedded sandstone beds. Plant debris and trunks are very common within sandstone deposits. The latter has the same lithology as stop 9; **b** Macroscopic picture of outcrop; the red circle with the number represents the location of the detailed pictures, including ❶ Heterolithic deposits overlaid by very fine-grained sandstones exhibiting soft-sediment deformation structures; ❷ Horizontal worm traces of silty mudstone bedding; ❸ Paleosol layers; ❹ Cross-bedded fine-grained sandstones. Note the presence of trunks in the upper cross-bedded sandstone bed (24.42 m)

Stop 11 Chang 7–Chang 6_3 prograding delta and distributary architecture (Figs. 28 and 29).

Location North side of Yanhe River, Tanjiahe Village, Yanchang County.

Content Chang 6_3 delta-plain distributary and insised valley deposits.

Stop 12 Features of the sequence boundary within Chang 6 member (Figs. 30 and 31).

Location North side of Yanhe River, Zhujiawan Village, Yanchang County.

Content Depositional characteristics of flooding plain underlying, architecture of the channel sand bodies overlying the sequence boundary.

Fig. 28 Drone photos and geological interpretation of Chang 7–Chang 6_3 from Tanjiahe Village

Fig. 29 Delta-front distributary channels at the top of Chang 7 and insised valleys at the bottom of Chang 6_3 from Tanjiahe Village. *Note* DCH—Distributary channel; CVS—crevasse splay; FP—Flood-plain deposit; ACH—Abandoned channel deposition; CCH—Crevasse channel

Fig. 30 Measured stratigraphic log and detailed pictures of typical sedimentary characteristics of the Chang 6 Oil Member, Zhujiawan Village. **a** Measured log showing mainly fine-grained sandstones, consisting of thick, sharp-based, tabular sandstone bodies. Most sandstones are laminated, except basal massive sandstone beds. Plant debris and caliche nodule are common in sandstone beds; **b** Macroscopic outcrop photo; **c**, **d** Partial zoom in picture b

Fig. 31 Panorama of Chang 6_1 fluvial sandbody showing its internal architecture. *Note* CH—channel; FP—Floodplain deposit; SB—Sequence boundary

Stop 13 Flood plain deposits. Oil seepage in the Chang 4 + 5—Chang 6_1 Member (Figs. 32, 33 and 34).

Location North of the Fenjia Village, Yanchang County.

Content Oil seepage, paleosol, sand body architecture.

Stop 14 Flood plain deposits. Oil seepage in the Chang 4 + 5 Member (Fig. 35).

Location Hujia Village, Yanchang County.

Content Oil seepage, paleosol, sand body architecture.

Fig. 32 Drone photos and geological interpretation of Chang 4 + 5–Chang 6_1 from the north of the Fenjia Village

Fig. 33 Measured stratigraphic log and detailed pictures of typical sedimentary characteristics in the Chang 4 + 5 to Chang 61 Oil Member in the north of the Fenjia Village. **a** Measured log showing the characterisics of Chang 6_1 and Chang 4 + 5 Oil Members. Chang 6_1 Oil Member shows an alternation of fine-grained sandstones and massive siltstone deposits, with common root traces and plant debris. Occasionally, huge trunks are recognized in life position. Chang 4 + 5 Oil Member consists of sharp-based, tabular sandstone bodies ranging in thickness from 0.2 to 1 m. Sandstones are characterized by abundant trunks and partly carbonized plant debris, and are fine-to medium-grained in size, displaying trough and asymptotic cross-bedding as well as parallel-lamination. **b** Macroscopic outcrop photo; **c**, **d**, **e** Partial zoom in picture **b**

Fig. 34 Detailed pictures of typical sedimentary characteristics in the Chang 4 + 5—Chang 6_1 Member in the north of the Fenjia Village

Fig. 35 Measured stratigraphic log and detailed pictures of typical sedimentary characteristics in the Chang 4 + 5 Member in the Hujia Village. **a** Measured log showing a sandy interval. Sandstones are mainly composed of massive sand bodies exhibiting flat, sharp-based beds (less erosive). In places, sandstones are intercalated by heterolithic deposits consisting of an alternation between tabular-to lenticular sandstone layers and massive siltstones. Occasionally, sandstones are highly bioturbated and show abundant root traces and plant debris. Huge trunks were disposed of on life position within massive sand; **b** Macroscopic outcrop photo; **c**, **d**, Partial zoom in picture b; *Note*: CVS—crevasse splay

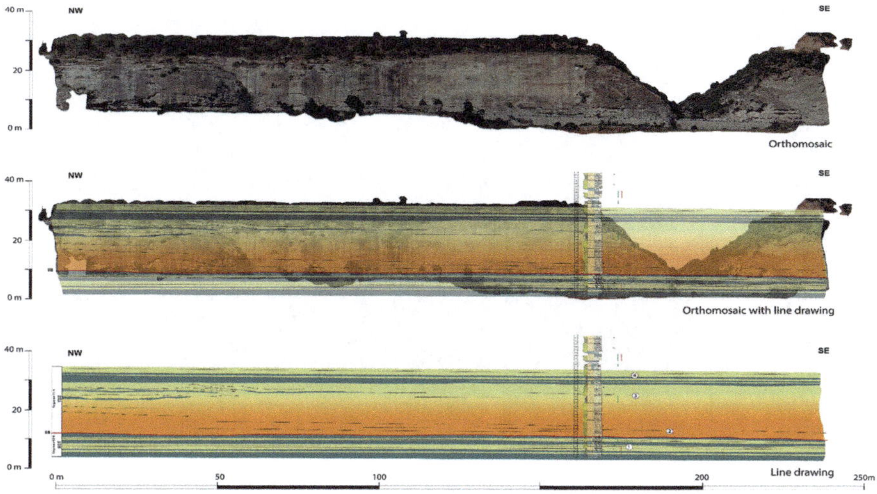

Fig. 36 Drone photos and geological interpretation of Chang 4 + 5–Chang 6_1 from the Zhangjiawan Village

Stop 15 Meandering channel deposits and sequence boundary between Chang 3 and Chang 4 + 5 Member (Figs. 36 and 37).

Location Zhangjiawan Village, Yanchang County.

Content Sedimentary environment, sand body architecture.

Stop 16 Flood events and the accumulation of flood plain deposits in the Upper of the Chang 3 Member (Figs. 38, 39 and 40).

Location East of Heijiabao Village, Yanchang County.

Content Fossil trees, paleosol, sand body architecture.

Fig. 37 Measured stratigraphic log and detailed pictures of typical sedimentary characteristics from the Chang 3 to Chang 4 + 5 Oil Member in the Zhangjiawan Village. **a** Measured log showing two parts: the lower part consists of an alternation of fine-grained sandstone bodies and moderately to highly bioturbated siltstone/heterolithic deposits with abundant plant debris and trunks. The upper part is mainly composed of medium-to fine-grained sandstone bodies exhibiting large-scale asymptotic and trough cross-bedding. **b** Macroscopic outcrop photo; the red circle with numbers show the location of detailed pictures, including ❶ Very fine sandstones with trunks; ❷ Highly bioturbated heterolithic deposits; ❸ Cross-bedding within fine-grained sandstones

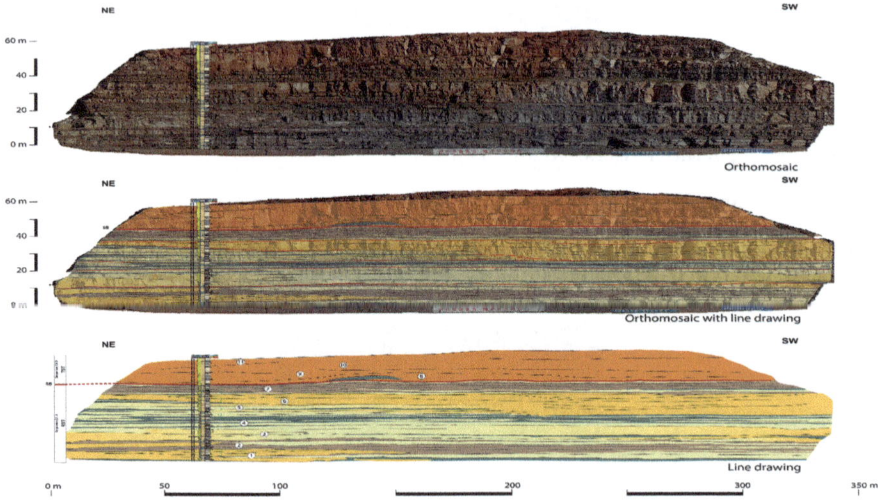

Fig. 38 Drone photos and geological interpretation of Chang 3 from the East of the Heijiabao Village

Fig. 39 Sedimentary log and detailed pictures of typical sedimentary characteristics in Chang 3 Oil Member in Heijiabao Village. **a** Measured log showing an alternation between moderately to highly bioturbated sandy mudstone bodies and fine-grained sandstone bodies. Sandstones are mainly composed of massive sand bodies exhibiting flat, sharp-based beds (less erosive). In places, huge trunks disposed in life position within massive sandstones are recognized. **b** Macroscopic outcrop photo; **c**–**l** Upright fossil trees; see picture B for their location; *Note* CVS—crevasse splay; FP—Flood-plain deposit

Fig. 40 Flood events in Chang 3, Yanchang Formation, East of Heijiabao Village. **a** A tree in life position. Its roots in host rocks of dark red silty mudstones with a mass of lebensspur fossils, indicating flood plain microfacies. A stem in host rocks of blocky argilliferous silt-finestones with a flat bottom and no bedding, indicating crevasse-splay deposits caused by the high rate of sedimentation. It was inferred that the tree was located in the low-lying area at the end of the river; it was buried and changed into fossils due to floods. The location of photo a is shown in the blue box in Fig. 39b; **b** Interpretation of photo a; **c** Local zoom of the photo. *Note* CVS—crevasse splay; FP—Floodplain deposit

3 Yan 1 Well-The First Oil Well on the Mainland of China

See Fig. 41.

4 Principal Investigator of Section Construction

Leaders:

Xiangbo Li (Research Institute of Petroleum Exploration and Development-Northwest Branch (NWGI), PetroChina, lixiangbo911@sina.com);

Rukai Zhu (Research Institute of Petroleum Exploration and Development, PetroChina, zrk@petrochina.com.cn);

Carlos Zavala (Universidad Nacional del Sur, Argentina, czavala@uns.edu.ar);

Xianyang Liu (National Engineering Laboratory for Exploration and Development of Low-Permeability Oil & Gas Fields; PetroChina Changqing Oilfield Company, lxy3_cq@petrochina.com.cn);

Tian Yang (Chengdu University of Technology, tianyang9645@gmail.com);

Huaqing Liu (Research Institute of Petroleum Exploration and Development-Northwest Branch (NWGI), PetroChina, 782,352,398@qq.com);

Ping Luo (Research Institute of Petroleum Exploration and Development, PetroChina, luoping@petrochina.com.cn);

Zhiyong Gao (Research Institute of Petroleum Exploration and Development, PetroChina, gzy@petrochina.com.cn, gzybox@163.com);

Bin Hao (Research Institute of Petroleum Exploration and Development-Northwest Branch (NWGI), PetroChina, haobin86@petrochina.com.cn);

Zhanlong Yang (Research Institute of Petroleum Exploration and Development-Northwest Branch (NWGI), PetroChina, yang_zl@petrochina.com.cn).

Assistant:

Zhangyan (Research Institute of Petroleum Exploration and Development-Northwest Branch (NWGI), PetroChina).

Fig. 41 A full view of Well Yan 1, inside the CNPC Hope Primary School, Yanchang

Guojinyi (Research Institute of Petroleum Exploration and Development-Northwest Branch (NWGI), PetroChina).

Huangjunping (Research Institute of Petroleum Exploration and Development-Northwest Branch (NWGI), PetroChina).

Wangjing (Research Institute of Petroleum Exploration and Development-Northwest Branch (NWGI), PetroChina).

Dengxiuqing (PetroChina Changqing Oilfield Company).

Yangweiwei (PetroChina Changqing Oilfield Company).

References

Bai Y, Ma Y (2019) Geology of the Chang 7 member oil shale of the Yanchang Formation of the Ordos Basin in central North China. Pet Geosci 26(2):petgeo2018-091

Li X, Liu H, Deng X, et al (2021a) The concept of fluvial fans in an arid environment: a new explanation of the origin of "sand-filled basins" in the Yanchang Formation, Ordos Basin. Acta Sedimentol Sin 39(5):1208–1221 (in Chinese)

Li X, Liu H, Huang J et al (2021b) Alternation of arid-humid climate and formation and distribution of Fluvial fan sand in the central area of Inland Lake basin-taking Yanchang Formation in Ordos basin as an example. Acta Geol Sin 95:1–17 (in Chinese)

Li X, Zhu R, Hui X, et al (2022) Sedimentological response of late Triassic Carnian Pluvial Episode (CPE) in lacustrine basin: a case study from Yanchang Formation of Ordos Basin. Acta Geolog Sin 2022 (online publishing, in Chinese)

Precambrian to Lower Paleozoic Depositional Strata of the Western Margin of the North China Block

Jitao Chen, Wenjie Li, and Paul M. Myrow

1 Brief Description

This field excursion will show you superbly exposed Precambrian to lower Paleozoic depositional strata in the western margin of the North China Block. The bedrock consists of Paleoproterozoic gneiss, Mesoproterozoic siliciclastic facies, Cambrian mixed siliciclastic and carbonate facies, and Ordovician platform and slope carbonate (and siliciclastic facies), all of which are separated by remarkable unconformities. We will emphasize the lower Paleozoic strata and will see a large variety of textbook-level sedimentary structures and facies.

The lower Paleozoic record a significant stratigraphic hiatus (Furongian–Lower Ordovician), the timing and magnitude of which are similar to those of the northern Indian continental margin of the Himalaya, suggesting a linkage between northwest China Block and northern Indian, which were potentially influenced by same orogeny. On the other hand, the Middle Ordovician records an abrupt facies change from platform carbonate (Zhuozishan Formation) to slope carbonate and overlying shale (the Kelimoli Formation), the latter with spectacular, and locally unique, slumping structures and mass transport deposits. This facies change reflects a tectonic transition from passive margin to foreland basin, and associated platform foundering and steepening of the slope.

J. Chen (✉) · W. Li
Nanjing Institute of Geology and Palaeontology, CAS, Nanjing, China
e-mail: jtchen@nigpas.ac.cn

W. Li
e-mail: wjli@nigpas.ac.cn

P. M. Myrow
Colorado College, CO Springs, USA
e-mail: pmyrow@coloradocollege.edu

X. Hu (ed.), *Field Trip Guidebook on Chinese Sedimentary Geology*,
https://doi.org/10.1007/978-981-99-6936-4_14

The field excursion will cover many aspects of sedimentary geology, including classical sedimentary features and lithofacies to basinal evolution.

- **Duration**: 2 + 4 days
- **Transportation**: Airplane + 4WD vehicles
- **Difficulties**: Moderate to high difficulty; can be steep climbing
- **Climate, temperature, recommended field equipment, passport**: In late August, the daytime temperature in Wuhai is usually between + 20 °C and + 31 °C. The climate of Wuhai area is arid, so bring water bottles (we can also provide bottled water). Hiking boots, hat, sunblock, and sunglasses (up to yourself) are necessary for the field. Hiking canes are helpful. Passports and visas are needed to entire China.

The SBG Section in the Subaigou Valley (Photo Taken in 2014 by Jitao Chen)

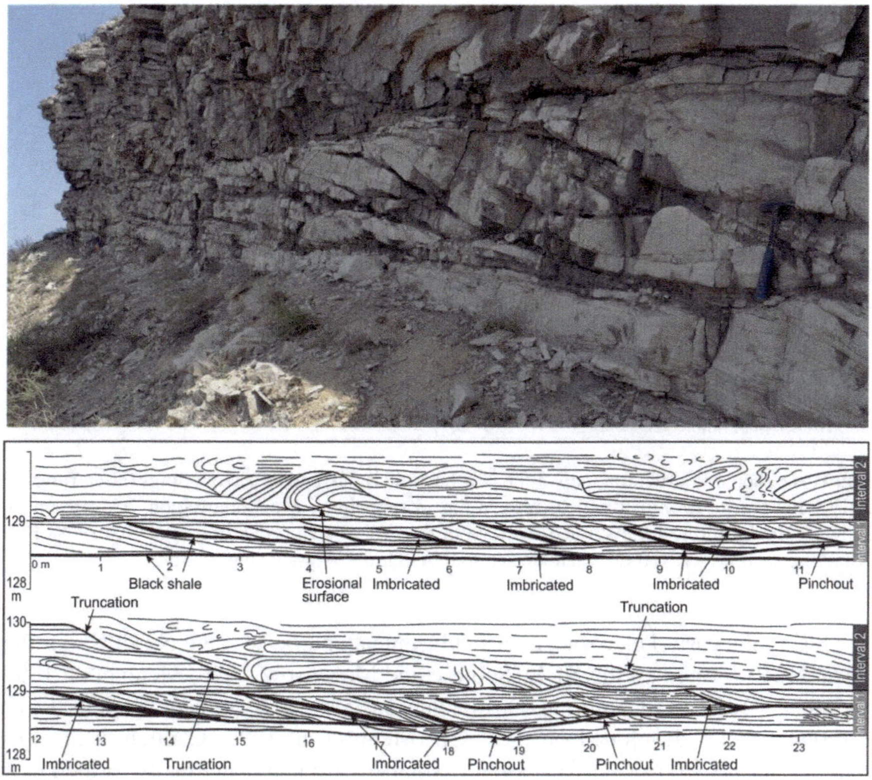

Imbricated Bed and Highly Folded Beds in the XLF Section (Li Et Al., 2022 Sedimentology)

2 Introduction

The North China Block (NCB) is one of the oldest cratons in the world, containing Archean–Paleoproterozoic (3.85 to 1.6 Ga) metamorphic basement rocks (Bian et al. 2001). The NCB is bound by the North Qilian Orogen to the west and southwest, the Central Asian Orogen to the north, and the Qinling–Dabie-SuLu (ultral) high–pressure metamorphic belts to the south and southeast (Xiao et al. 2003; Shi et al. 2013; Song et al. 2013; Dong and Santosh 2016; Fig. 1). Our field excursion is located in Wuhai City, Inner Mongolia, close to the western margin of the NCB. This area is near the north–south striking Zhuozishan–Gangdeershan Mountain region, which is connected to the Helanshan Moutain to the south, the Alax landmass to the northwest, the North Qilian Orogen to the southwest, and the North China Platform to the east (Darby and Ritts 2002; Zhang et al. 2011).

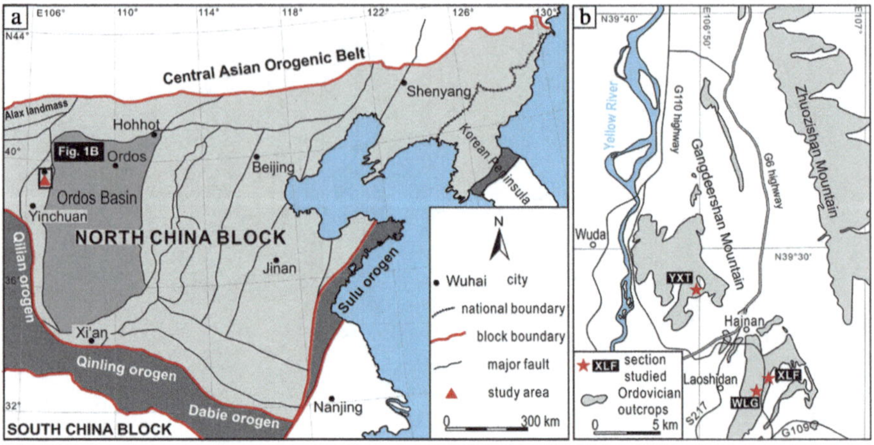

Fig. 1 **a** Schematic structural map of the North China Block and its surroundings (after Yang et al. 2019), and the location of the field area in the western Inner Mongolia region. **b** Close-up map showing the location of the studied sections in the Wuhai area (after Jing et al. 2016). XLF: the Xilaifeng section; YXT: the Yixiantian section; WLG: the Wolonggang section

During the early Paleozoic, a vast subtropical epeiric sea (the North China Platform, NCP) formed on the NCB (Scotese and McKerrow, 1990; Meng et al. 1997). A thick (~1800 m) Cambrian–Ordovician succession of mixed siliciclastic and carbonate facies developed on the NCP (Meng et al. 1997; Chough et al. 2010; Myrow et al. 2015; Fig. 2). Slope and basinal environments gradually formed in the study area during the Middle Ordovician (Feng et al. 1998, 2004). During the early Katian (Late Ordovician), regional uplift related to the Huaiyuan Epeirogeny led to extensive erosion, and formation of a substantial unconformity between the uppermost Ordovician and the lowermost Carboniferous (Chen et al. 2013; Zhen et al. 2016).

Chronostratigraphy		Formation	Major lithology
Penn-sylvanian	Bashkirian	Yanghugou	Sandstone, coal, shale, limestone
Upper Ordovician	Katian	Sheshan	Turbidite sandstone, siltstone, shale, bioclastic grainstone
	Sandbian	Gongwusu	Greenish-grey mudstone, thin-bedded limestone, sandstone
		Lashenzhong	Turbidite sandstone, siltstone, shale
		Wulalike	Slump-debris flow deposits, black shale
Middle Ordovician	Dariwillian	Kelimoli	Graptolitic black shale, thin-bedded limestone, turbidite
		Zhuozishan	Bioturbated and bioclastic limestone, microbialites
		Sandaokan	Quartzose sandstone, bioclastic dolostone, limestone
Furongian	Fengshanian	Abuqiehai	Dolostone, carbonate conglomerate
	Changshanian		Thin-bedded limestone & dolostone, conglomerate
Cambrian Series 3	Kushanian		Bedded limestone, oolite, conglomerate
	Changhian	Zhangxia	Limestone-shale alternations, oolite, conglomerate
	Hsuchuangian	Hulusitai	Thin-bedded limestone, greenish-grey & purple shales
	Maochuangian	Taosigou	Sandstone, purple & greenish-grey shale, bedded limestone
Cambrian Series 2	Lungwang-miaoan		Sandstone, greenish-grey shale, dolostone
	Canglangpuan	Wudaotang	Dark gray dolostone, dolomitic limestone, limestone
		Suyukou	Phosphatic sandstone, glauconitic arkose sandstone
Ediacaran/Sinian		Zhengmuguan	Conglomerate, breccia, and gravelly sandstone; tillite
Meso-proterozoic	Jixian Sys.	Wangquankou	Minor sandstone, thick-bedded dolostone with stromatolite
	Chang-cheng Sys.	Huangqikou	Quartz sandstone, slate, and minor dolostone

Fig. 2 Proterozoic to Paleozoic stratigraphy of the western Inner Mongolia and Ningxia regions (i.e., Helanshan and Zhuozishan mountains) of the North China Block

3 Excursion Route and Road Log

DAY 1 (Tuesday, August 16, 2022): reception (pickup at Yinchuan airport) Arrival: at Yinchuan from all over the world Stay: Yinchuan Hotel Welcome banquet at the Yinchuan Hotel
DAY 2 (Wednesday, August 17, 2022): Start: Yinchuan (8 am) → Wuhai: 4WD for 2.5 H Overview: Cambrian–Ordovician succession at Gandeer (GDE) Mountain, with giant Genghis Khan statue (88.98 m high) on the top. We can see various typical Cambrian–Ordovician facies including ribbon limestone, flat-pebble conglomerate, and maceriate microbialite. Overview of the Cambrian–Ordovician succession and brief tour to the Genghis Khan statue at the top (Fig. 3). Stops: Stop 1. Various gutter casts and maceriate microbialites Stop 2. Overview of the Cambrian–Ordovician succession at the top Stop 3. Genghis Khan Statue Park Stop 4. Yellow River Park Welcome banquet at the Wuhai hotel
DAY 3 (Thursday, August 18, 2022): Start: Hotel Overview: Paleoproterozoic gneiss/Mesoproterozoic siliciclastic facies/Cambrian–Ordovician carbonate at the Subaigou section (SBG) (Fig. 3). Stops: Stop 1. Nonconformity between Paleoproterozoic and Mesoproterozoic Stop 2. Unconformity between Mesoproterozoic and Cambrian Stop 3. Mixed siliciclastic and carbontate facies Stop 4. The Abuqiehai Formation at the SBG section Stop 5. Maceriate microbialite Stop 6. Sandstone facies in Ordovician strata Stop 7. Ordovician paleokarst Dinner at the Wuhai lamb BBQ
DAY 4 Friday, August 19, 2022): Start: Hotel Overview: Ordovician carbonate slope deposits with various soft-sediment deformation features at the Xilaifeng (XLF) section (Fig. 3). Stops: Stop 1. Overview of the XLF section Stop 2. Undeformed layers of the middle part of the Lower Kelimoli Formation Stop 3. Slump sheet laterally truncated by debrite (99.5 m) Stop 4. Folds, chaotic bed, and debrites (117-119 m) Stop 5. Imbricated beds (128.5 m) Stop 6. Chaotic beds (139 m) Stop 7. Carbonate breccia and grainstone (175 m) Dinner at the Wuhai Hansen winery

(continued)

(continued)

DAY 5 (Saturday, August 20, 2022): Start: Hotel Overview: Ordovician slope deposits with remarkable channelized carbonate breccia at the Yixiantian (YXT) section (Fig. 3). Stops: Stop 1. Lower Kelimoli Formation Stop 2. Upper Kelimoli Formation Stop 3. Lenticular carbonate breccia Stop 4. Carbonate breccia, mud-armored breccia ball and irregular slump block Stop 5. Various outcrops of the breccia-filled channel Farewell dinner at the Wuhai hotel
DAY 6 (Sunday, August 21, 2022): Departure for Beijing 21ISC meeting Start: Wuhai (8:30 am) → Yinchuan (10:00 am) 4WD for 2.5 H Yinchuan → Beijing: preliminary by airplane CA1282

4 Geological Itinerary and Excursion

4.1 *DAY 1 (Wednesday, August 17, 2022): Gandeer (GDE) Mountain*

Overview:

Cambrian–Ordovician succession at Gandeer (GDE) Mountain, with giant Genghis Khan statue (88.98 m high) on the top (Fig. 4). We can see various typical Cambrian–Ordovician facies including ribbon limestone, flat-pebble conglomerate, and maceriate microbialite. Overview of the Cambrian–Ordovician succession and a brief tour to the Genghis Khan statue at the top.

Stop 1. Various Gutter Casts and Maceriate Microbialites

Gutter casts form by rapid filling of storm current incised gullies and depressions (Myrow 1992). Abundant gutter casts of flat-pebble conglomerates and bioclastic grainstone are present in the lime mudstone and marlstone alternations (Figs. 5, 6, 7, 8). They show high variety in shape, lateral continuity, and infill. Some contain coarse-grained sediment in the lower part and marly sediment in the upper part of the gutter casts (Fig. 6). Sandy infills display hummocky cross-stratification.

The presence of gutter casts with coarse-grained infill and basal erosion surfaces suggests initial erosion of ribbon limestone by powerful bottom currents that were most likely generated by storms. The hummocky cross stratification resulted from storm-induced combined flows.

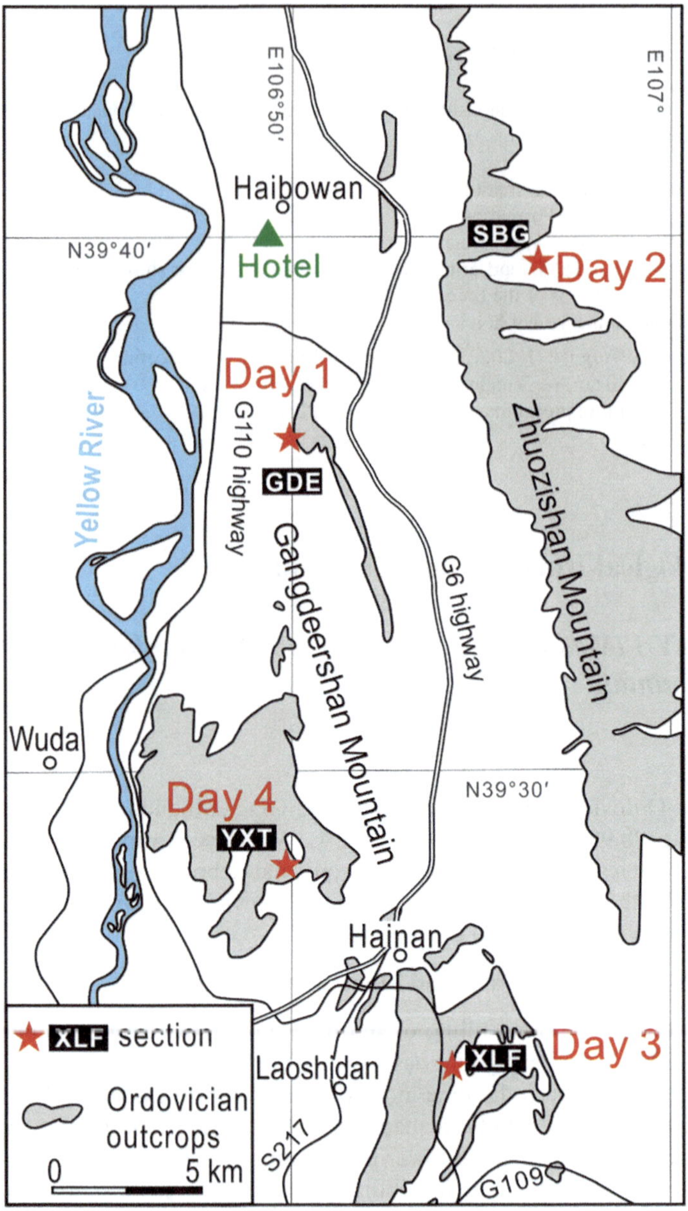

Fig. 3 Excursion route of Wuhai area

Fig. 4 Cambrian–Ordovician strata with giant Genghis Khan statue on the top Photo from internet (https://www.sohu.com/a/155227351_645084)

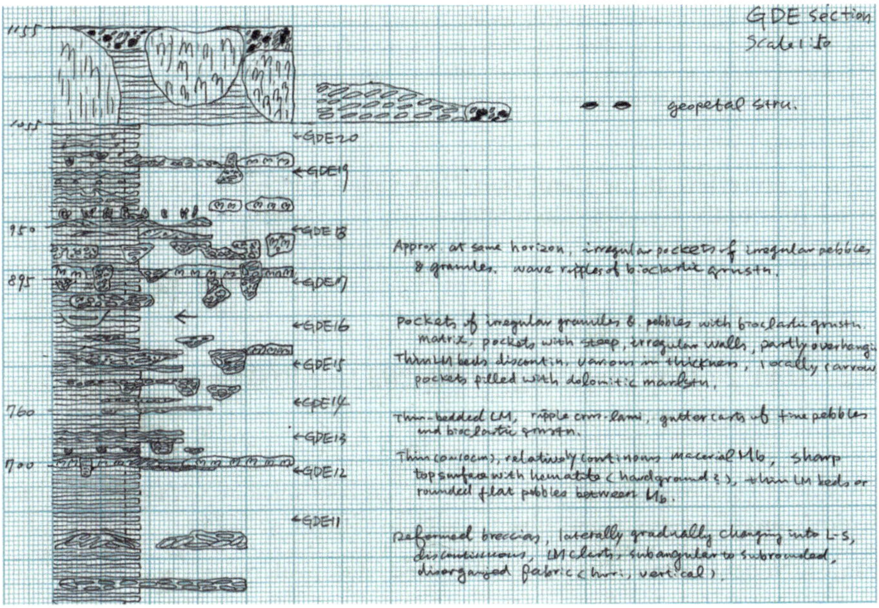

Fig. 5 Field note and sketch of the measured GDE section (by Jitao Chen)

Fig. 6 Ribbon limestone dominated facies in the Cambrian Abuqiehai Formation, showing several levels of gutter casts, mostly filled with intraclasts and fossil fragments. The gutter cast in the middle shows a relatively narrow lower part, filled with intraclasts and fossil fragments, and a wider upper part, filled with dolomitic marlstone and a few thin limestone layers. Scale bar is in cm

Stop 2. Overview of the Cambrian–Ordovician Succession at the Top (and a Brief Tour to the Genghis Khan Statue)

See Figs. 9, 10.

4.2 *DAY 2 (Thursday, August 18, 2022): SBG, Zhuozishan Mountain*

Overview: Paleoproterozoic gneiss/Mesoproterozoic siliciclastic facies/Cambrian–Ordovician carbonates at the Subaigou section (SBG; Fig. 11).

Stop 1. Nonconformity Between Paleo- and Mesoproterozoic

See Figs. 12, 13, 14, 15.

Stop 2. Unconformity Between Mesoproterozoic and Cambrian

The upper part of the Mesoproterozoic Huangqikou Formation is characterized by thick (tens of meters up to more than 100 m to the South; e.g., Zhang et al. 2020), white, cross-stratified quartzose sandstone. The exact age of the sandstone is

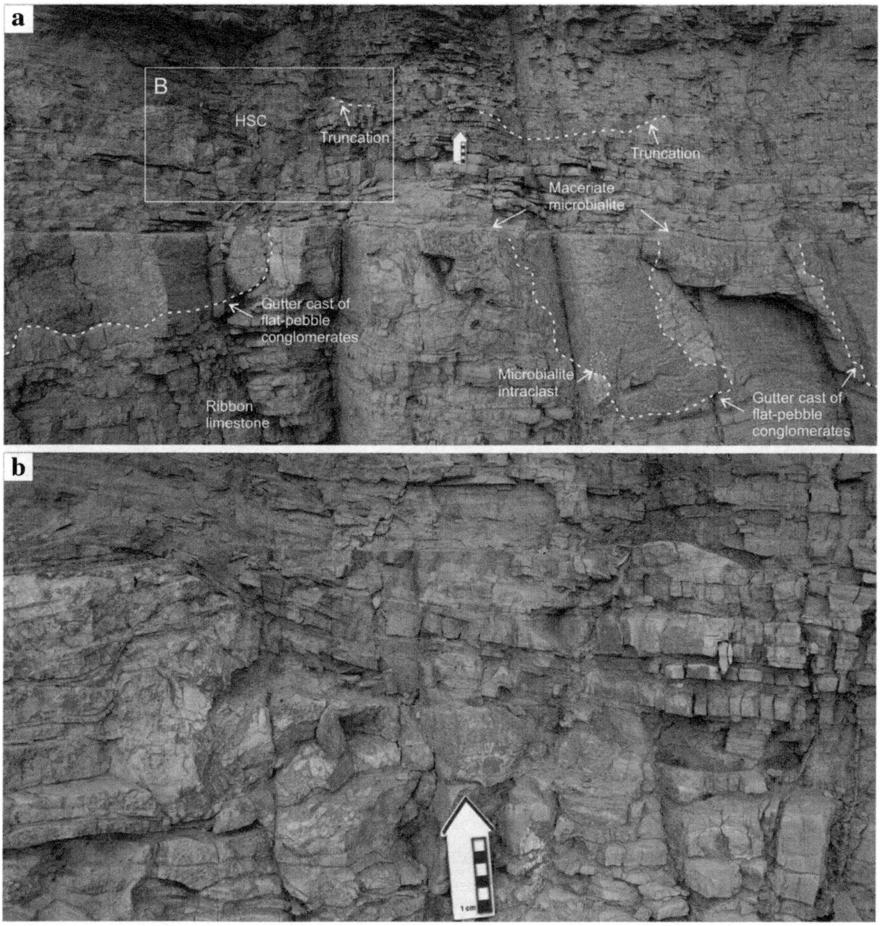

Fig. 7 Ribbon limestone dominated facies in the Cambrian Abuqiehai Formation, showing various structures such as truncation surfaces, hummocky and swaley cross stratification (HSC), and deeply incised gutter casts. There are also thin and wide lenses (or channels) of flat-pebble conglomerate across the outcrop. These structures collectively suggest frequent reworking and deposition by storm-induced processes in a setting that generally experienced low-energy, background sedimentation of ribbon limestone. Scale bar is in cm

unknown. Some workers argue for a lower Cambrian age (like the quartzose sandstone of the basal Cambrian Liguan Formation to the southeast of the North China Platform). Like most researchers, we regard it as Mesoproterozoic, given that no fossils or bioturbation were found in the sandstone.

The unconformity between Mesoproterozoic and Cambrian here (Fig. 16) is equivalent to the global "Great Unconformity", as in other parts of the North China Platform, where Cambrian strata sit above the Precambrian basement. To the south (e.g.,

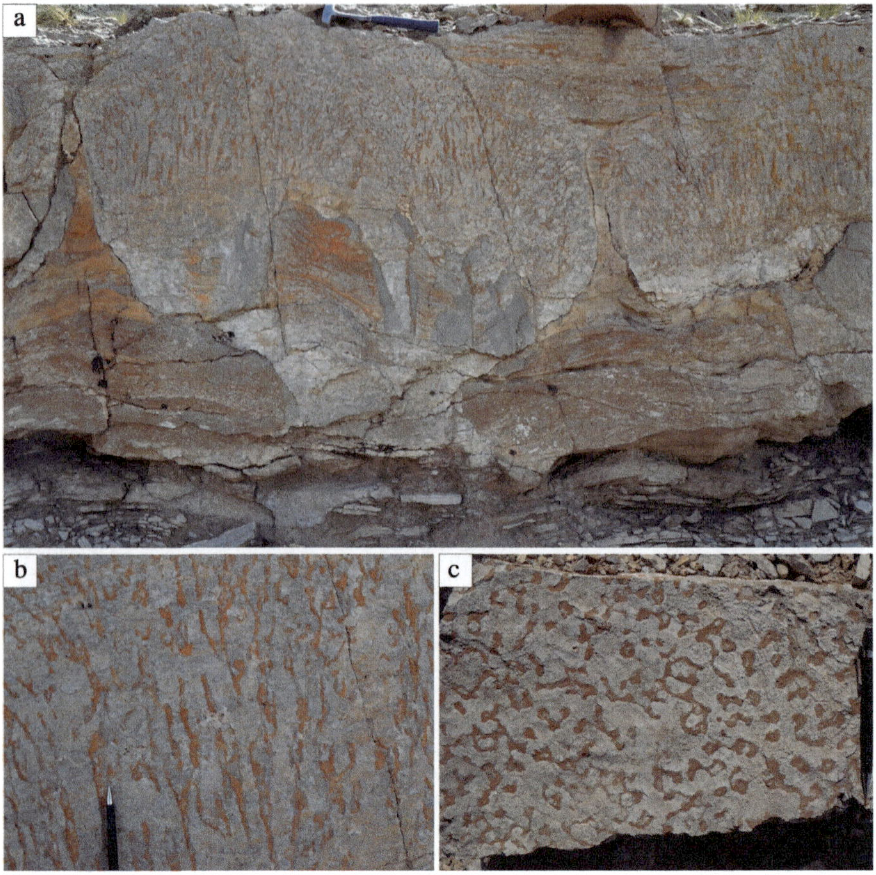

Fig. 8 Typical features of maceriate microbialites in the middle–upper Cambrian of the North China Platform (Lee et al. 2010, 2014) and Laurentia (Shapiro et al. 2006). This type of microbialites does not show either laminated or clotted structures, and thus does not fall into the four common categories of microbialites (i.e., stromatolite, thrombolite, dendrolite, and leiolite; Riding 2000). They show interconnected digitate structures in the vertical plane (B) and irregular honeycomb-like structures on the bedding plane **c**, and maze-like (maceriate) meso-structures in 3D (Shapiro et al. 2006; Lee et al. 2010), which thus are call maceriate microbialites (Figs. 9, 10)

the Suyukou section of the Helan Mountain in Yinchuan), the Cambrian strata overlie the Neoproterozoic diamictite of the Zhengmuguan Formation.

The Cambrian Taosigou Formation is characterized by sandy dolostone, fenestral dolostone with oncoids and stromatolitic laminae, stratified and bioturbated sandstone, and mudstone with microbially induced sedimentary structures, as well as paleokarst and mud cracks. Collectively, these suggest very shallow marine settings, frequently subaerially exposed due to base-level fall (Fig. 17). Various trace fossils and trilobites are present in shale, suggesting an early Miaolingian (3rd epoch of the Cambrian) age.

Fig. 9 The GDE section with the Genghis Khan statue being constructed in 2012, photo by Jitao Chen

Fig. 10 The Genghis Khan statue (88.98 m high), photo from internet (https://www.sohu.com/a/155227351_645084)

Fig. 11 The SBG section in the Subaigou Valley (photo taken in 2014 by Jitao Chen)

Fig. 12 The group photo for the first excursion to the SBG section in 2012, by the sign of the Paleoproterozoic and Mesoproterozoic boundary

Fig. 13 Nonconformity between the Paleoproterozoic basement Zhaochigou/Helan Group gneiss and the overlying basal red beds of conglomerate and sandstone of the Mesoproterozoic Huangqikou Formation

Fig. 14 The Paleoproterozoic Zhaochigou/Helan Group consists of gneiss with dark purple color and high-angle (~80°) foliation. Above the nonconformity is the basal conglomerates of the Mesoproterozoic Huangqikou Formation. Conglomerate is composed of well-rounded cobbles (up to 15 cm diameter) and poorly sorted matrix. Conglomerate bed varies in thickness from 2.5 m to 20 cm across ~ 10 m of outcrop. Conglomerate changes upward into cross-stratified sandstone

Fig. 15 Planar/trough cross-stratified and parallel-stratified sandstone of the Mesoproterozoic Huangqikou Formation

Fig. 16 Unconformity between the Mesoproterozoic Huangqikou Formation of white, cross-stratified quartzose sandstone and the overlying Cambrian Taosigou Formation of mixed siliciclastic and carbonate facies

Stop 3. Mixed Siliciclastic and Carbontate Facies

The Cambrian Taosigou Formation is characterized by mixed siliciclastic and carbonate facies of the lower Cambrian, typical of the entire North China Platform (e.g., Shandong region in the eastern part of the NCP; Lee and Chough 2011). The succession yields the oldest trilobites of the Maochuangian Stage (~lower Wuliuan Stage of the Miaolingian, Cambrian). Trace fossils are also common in most facies (Fig. 18).

Stop 4. The Abuqiehai Formation at the SBG Section

The Abuqiehai Formation (Guzhangian Stage, Miaolingrian) is dominated by ribbon limestone, microbialites, and plat-pebble conglomerate. The Furongian is absent in the SBG section (Fig. 19).

Stop 5. Maceriate Microbialite

The microbialite is characterized by digitate or maceriate (maze-like) vaguely stromatolitic and thrombolitic structures (Fig. 20a, b), which is composed of microstromatolites with associated patches of sponge spicules (Fig. 20c, d). They commonly show mounded or highly irregular morphologies. Thick microbialite mounds rest on a conspicuous erosional surface that gradually cuts out approximately 94 cm

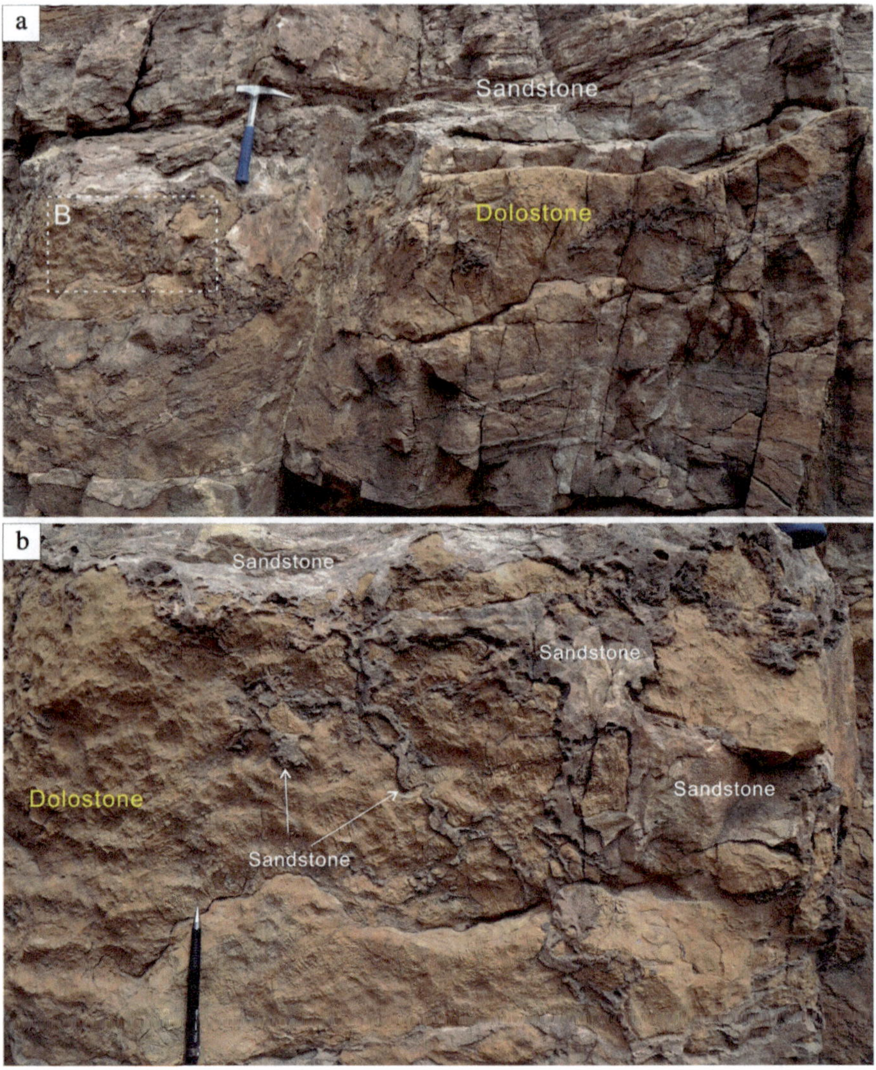

Fig. 17 Several horizons of paleokarst up to 1 m deep developed in dolostone of the lower Taosigou Formation (Cambrian). The paleokarst consists of a network of irregular branching sand-filled cavities that cut the dolostone host rock and were filled with quartz sands. They have highly irregular and sharp margins. The presence of paleokarst in the dolostone indicates that it was subaerially exposed and chemically weathered after cementation. The paleokarst developed at one stratigraphic level after significant lowering of base level at that time to expose previously deposited subtidalintertidal marine strata

Fig. 18 Mixed siliciclastic and carbonate facies of the Cambrian Taosigou Formaiton at the SBG section

of underlying strata (Fig. 20e). These microbialite mounds are surrounded by flat-pebble conglomerate at the base, and by ribbon mudstone and calcareous shale at the top.

Stop 6. Sandstone Facies in Ordovician Strata

This facies consists of white, tan, and brown, very fine to very coarse quartz sandstone (Fig. 21). Internal structure includes massive, bioturbated, and burrowed textures, the latter including *Skolithos* (Fig. 21a–c). Other beds contain parallel lamination and trough cross stratification (Fig. 21d). The sandstone is relatively well sorted, but some beds are poorly sorted with grain sizes ranging either from fine to coarse or medium to very coarse sand. Trough cross-stratified sets are 3 to 12 cm thick. Several beds contain small (1 to 2 mm) floating intraclasts of carbonate mudstone.

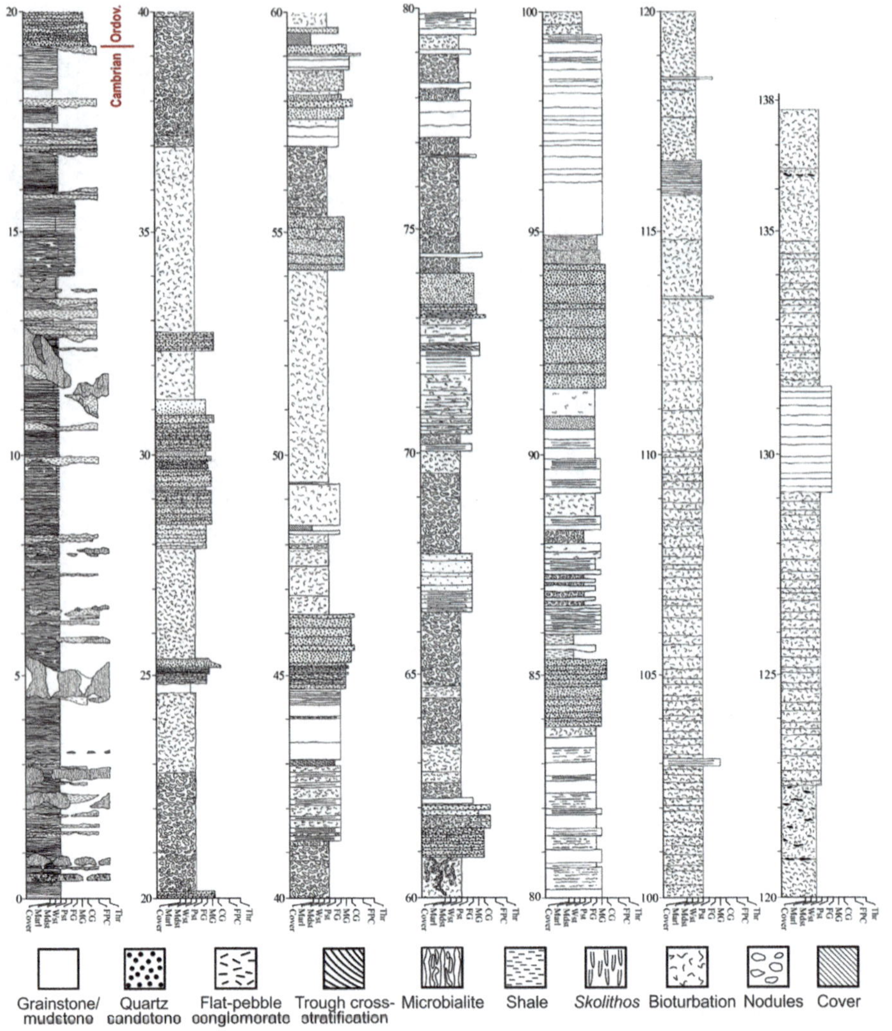

Fig. 19 Detailed stratigraphic columns of the upper Abuqiehai Formation at the Subaiyingou section (SBG). Mdst: lime mudstone; Wst: wackestone; Pst: packstone; FG: fine grainstone; MG: medium grainstone; CG: coarse grainstone; FPC: flat-pebble conglomerate; Thr: microbialite

Stop 7. Ordovician Paleokarst

An ~ 90 cm deep paleokarst is present in bioturbated wackestone at 60.91 m of the SBG section, and is overlain by a 1.28 m thick unit of quartz sandstone (Fig. 22). The paloeokarst consists of a network of irregular branching sand-filled cavities that cut the wackestone host rock. They have highly irregular, sharp, and scalloped margins, and contain a fill of quartz grains ranging in size from medium to coarse sand up to granules (Fig. 22a and b). Several 3–7 cm thick coarse sandstone beds that overlie

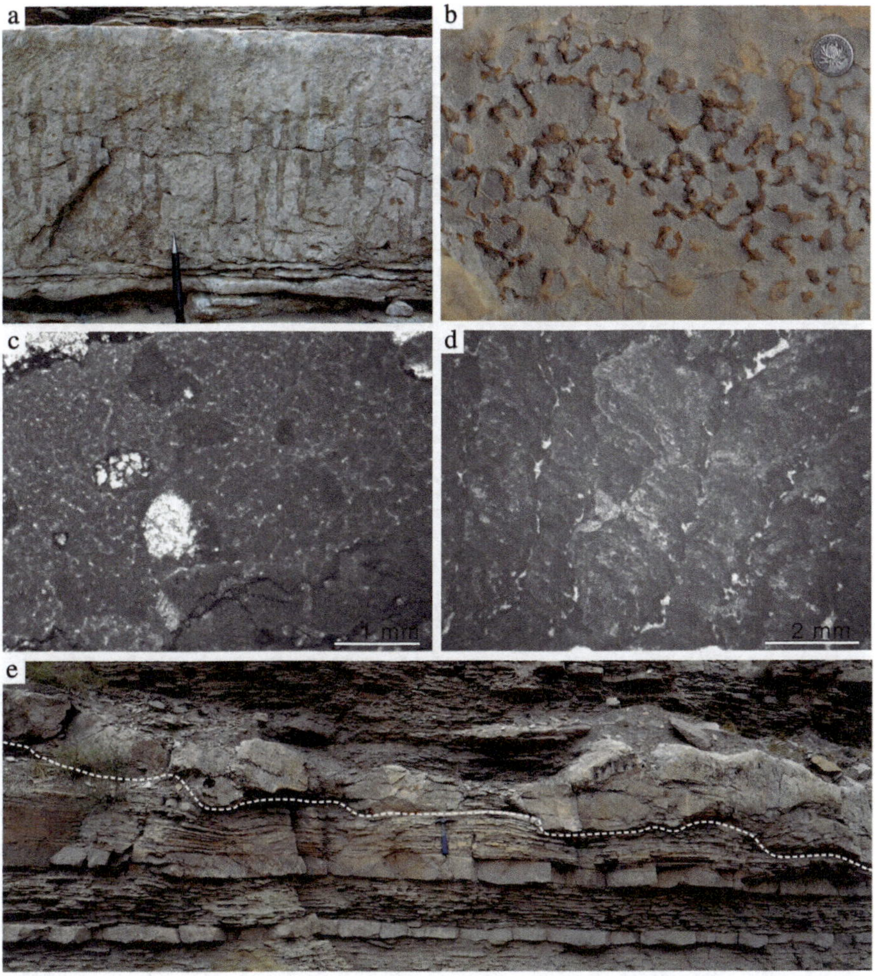

Fig. 20 **a** Vertical section, showing digitate, branching structures. **b** Bedding plane, showing rambling, maze-like structures. Coin is 25 mm in diameter. **c** Patch of sponge spicules in a maceriate microbialite. **d** Microstromatolite in a maceriate microbialite. **e** A maceriate, thrombolitic microbialite and flat-pebble and grainstone complex, down-cutting the underlying strata via an irregular surface

a paleokarst (described below) are deformed with downwarping of the sandstone laminae into the underlying depression in the wackestone (Fig. 22c and d).

Fig. 21 **a** Contact between orange-weathering uppermost Cambrian dolostone and Ordovician sandstone and bioturbated mudstone. **b** White and brown weathering sandstone with bed containing abundant *Skolithos* trace fossils. Hammer is 28 cm long. **c** Bioturbated sandstone. Pencil tip in the lower left for scale. **d** Trough cross-stratified sandstone. Scale bar is in cm. Hammer is 30 cm long

4.3 DAY 3 (Friday, August 19, 2022): Xilaifeng (XLF) Section

Overview: The XLF section exposes the Kelimoli and Wulalike formations (middle to late Darriwilian; Figs. 23 and 24). The Kelimoli Formation is characterized by ribbon lime mudstone and marl (Lower Kelimoli), and interbedded graptolitic black shale and thin-bedded lime mudstone (Upper Kelimoli; Fig. 23a). Slumps with

Fig. 22 Examples of the paleokarst between meters 60 and 61 of the SBG section. **a** Bioturbated wackestone with sandstone filled in paleokarst. **b** Closeup of a sandstone-filled paleokarst in wackestone. **c** Bioturbated wackestone bed with sandstone-filled paleokarst, overlain by partly deformed sandstone. **d** Closeup of sandstone-filled paleokarst in wackestone. Pencil is 14 cm and scale bar is in cm

various internal SSDS are extensively exposed in the upper part of Lower Kelimoli Formation (Fig. 23b, c and d). The Wulalike Formation is composed mainly of graptolitic black shale, and layers of lenticular carbonate breccia and grainstone at the base (Figs. 23e and 24).

Stop 1. Overview of the XLF Section

At the foot of the XLF section is the typical ribbon lime mudstone and marlstone facies of the Lower Kelimoli Formation (Fig. 25). Intact trilobite fossils can be found on the bedding planes (Fig. 26).

Stop 2. Undeformed layers of the middle part of the Lower Kelimoli Formation

In the process of going up the mountain, the typical banded mud and mud rocks in the central part of the Kelimoli group (55.5 ~ 85 m) were revealed. This facies consists of the plug -in plate -shaped mud rock layer and 10% to 15% of the hidden

Fig. 23 **a** Overview of the major outcrop of the Xilaifeng (XLF) section. **b–d** Major deformed intervals (slumps) of the Lower Kelimoli Formation. **e** Carbonate breccia and grainstone of the base of the Wulalike Formation

mud rock layer, which is the main phase of the Krimoli group. Lime mud rocks are dark gray and the climate is light gray. Lime mud rocks are either uniform or thin layers (Figs. 27 and 28). Limin mud layer is 2 ~ 15 cm thick, with an average of 10 cm, and the thickness of the lithium rock layer is generally less than 1 cm.

The lime mudstone contains many intact trilobites (nileid fauna) (e.g. Zhou et al. 1989). Conodonts and radiolarians (Fig. 29) are also rich in the ribbon limestone (e.g. Wang et al. 2013; Jing et al. 2015). Trace fossils are relatively scarce in this facies, with only subvertical and subhorizontal dwelling and/or feeding traces. This facies is interpreted to indicate a low-energy deep marine depositional condition.

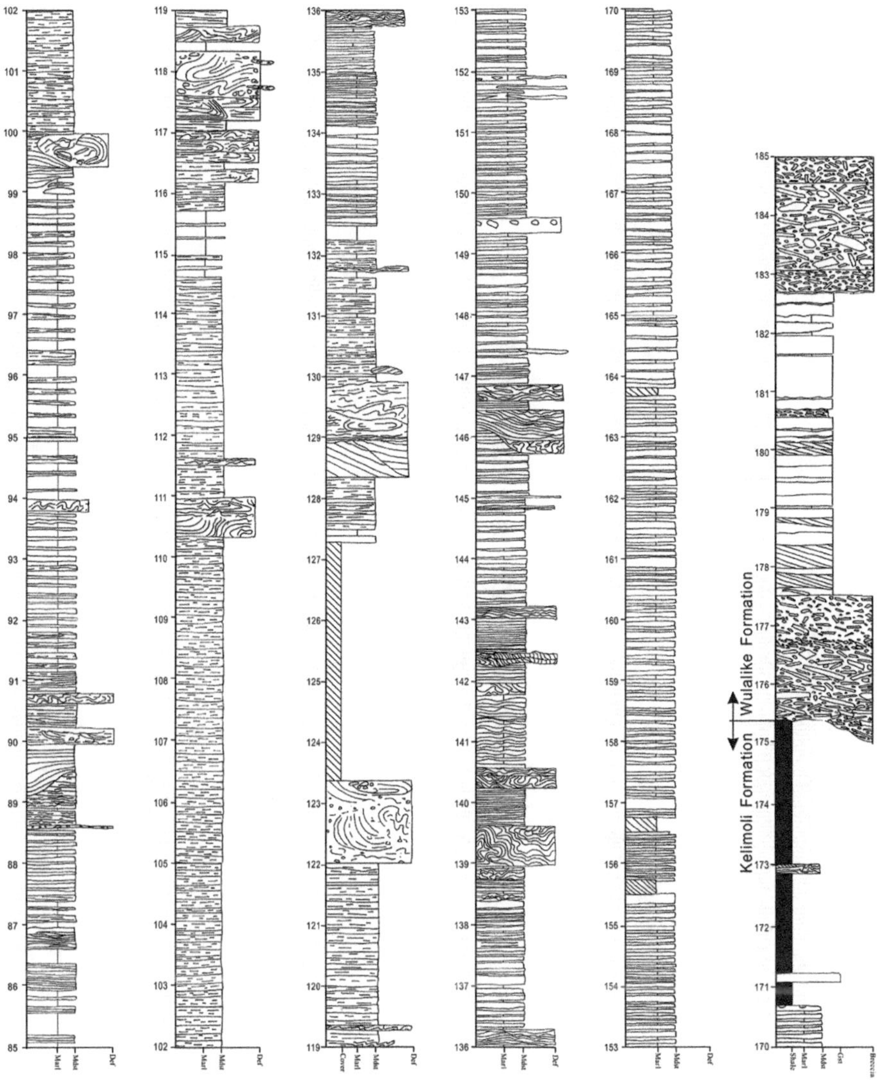

Fig. 24 Datailed sedimentologic log of the XLF Sect. (85–185 m) by Jitao Chen and Paul Myrow. Def: deformed interval; Mdst: lime mudstone; Gst: grainstone

Stop 3. Slump Sheet Laterally Truncated by Debrite (99.5 m)

(For general description of Stop 3, see next page.) (Fig. 30).

General Description of Stop 3:

There is a significant grapestone in the XLF part (Figs. 30 and 31). Dedication is exposed to more than 15 m along the face of 330° along the hit 330°. Its maximum

Fig. 25 Thin-bedded, homogeneous to crudely laminated lime mudstone facies

Fig. 26 Relatively intact trilobite fossils on bedding planes of the ribbon lime mudstone

Fig. 27 Ribbon lime mudstone (homogeneous) and marlstone

Fig. 28 Lime mudstone, showing finely, horizontal- to slightly undulating lamination

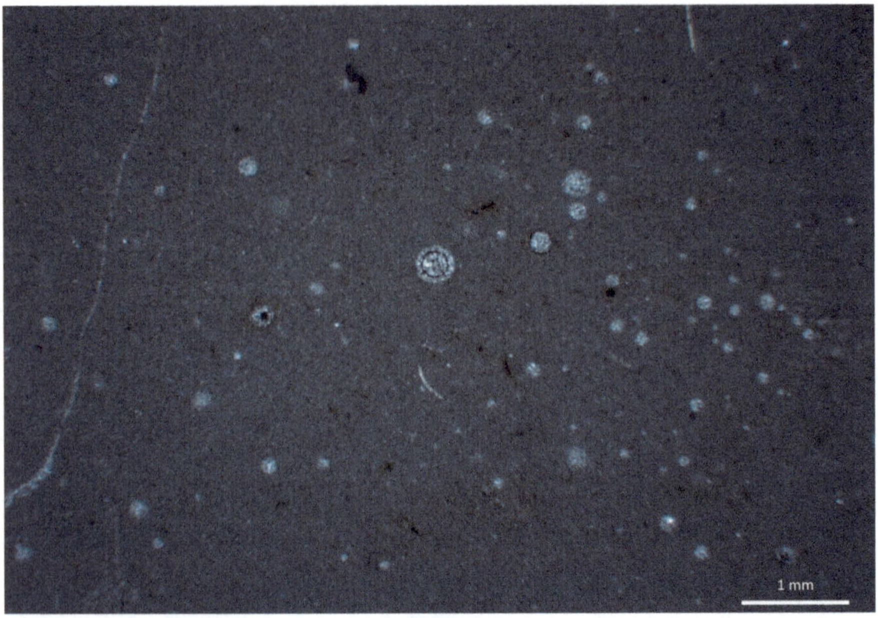

Fig. 29 Lime mudstone containing radiolarian fossils

Fig. 30 Debrite and slump sheet at ~ 99.5 m in the XLF section. **a** Debrite and chaotic bed contacted by a detaching surface. The "100 m" with a white bar in **a** and **b** indicates the stratigraphic height. **b** Lenticular debrites extended from **a** to ~330° direction

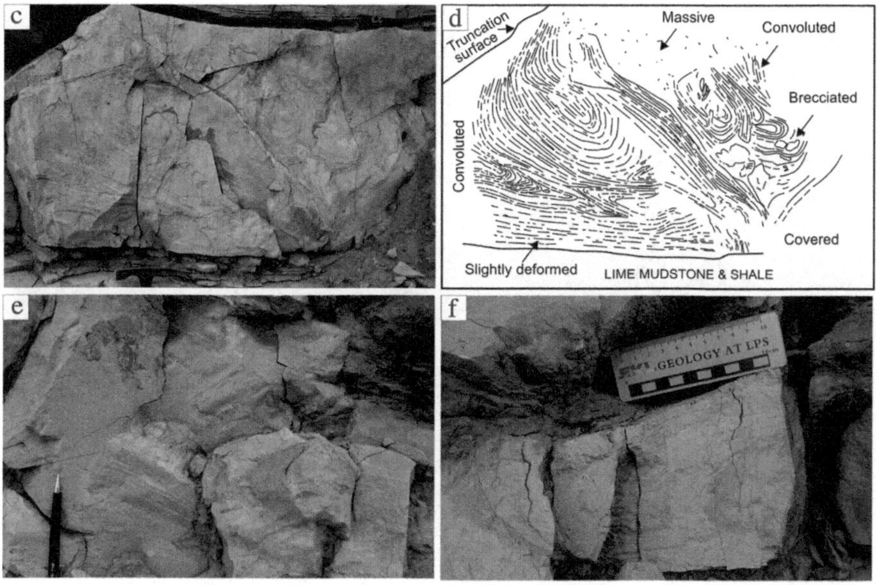

Fig. 31 **c** Detailed deformational feature of the chaotic bed beneath the detaching surface. **d** Line drawing of **c** by Jitao Chen. **e** Tabular, laminated carbonate clasts within debrites. Pencil is 14 cm long. **f** Recumbent folded lamination within the matrix of the lower part of debrite. Scale bar is in cm. Hammer in **a**, **b** and **c** is 28 cm long

thickness is about 35 cm, and it is squeezed on the side (Fig. 30a and b). Grape stones have a fluctuation foundation that can cover thin layer of limestone and shale (Fig. 30a and b). The lower part has a horizontal folding plate layer (Fig. 31e and f) with floating debris and CM scale.

Portuguese beds are terminated on sliding steep slopes. The title wall is a bed with a corrected layer of the layer (Figs. 30a, 31c and d). The steep geometric shape of the slide is 35°. A bed with a horizontal and disagreement folds, the supplies and corner clams of the memo bed are adjacent to and cut off as a steep slope of the slide, facing ~ 330° (Fig. 31c and d). These features may record spatial transition and interaction between multi -phase deformation and signs and debris flows (in printing).

Stop 4. Folds, Chaotic Bed, and Debrites (117–119 m)

The folds and chaos beds of various scale (MM-to M scale) are the most common SSDs in the middle to upper Kelimoli strata. The fold geometric shape has variable geometric shapes: tilt, flip, horizontal and disagreement (Fig. 32). Some horizontal folds are separated by each other and have obvious opposite distances in the cross section (Fig. 33e).

Some folds are tight, S-or Z-shaped, with continuous beds on the surface, and can be tracked by folds (Fig. 33b). Many horizontal folds have narrow hinge areas and relatively straight folding limbs, showing Z -shaped glyphs. A spectacular example

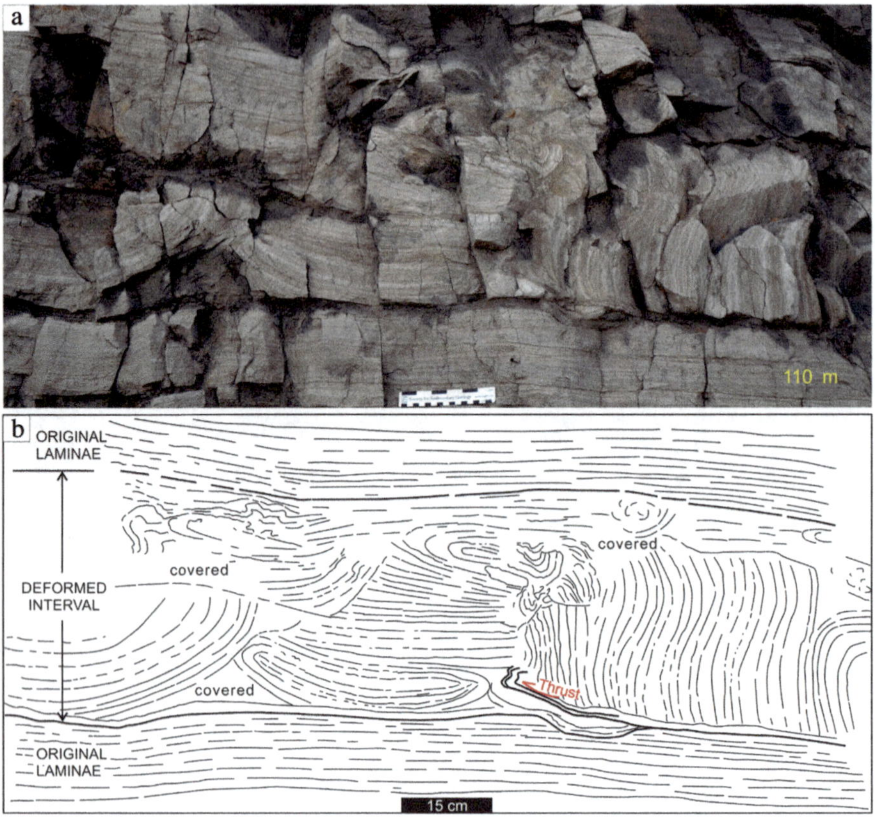

Fig. 32 Chaotic bed at ~110 m in the XLF section. **a** Highly folded and convoluted deformed interval. **b** Line drawing of **a** by Jitao Chen. Scale bar is in cm in **a**

of 118 m in the XLF cross -section shows the horizontal change along the display (Fig. 33), the height deformation, the 30–50 cm thick snow VR folds, divide the side into 10 cm thick tilt folds, and then at a certain distance at a certain distance Slightly drooping layer, and then slightly paved within a certain distance. A few meters northwest (Fig. 33a, b and c). A series of thrust rests above the basis of Mary limestone (Fig. 33a). Following the ups and southeast of the southeast, these deformed beds have become large -scale, staircases and thick beds (tens to 1 m) sleeper and flip folds (Fig. 33e). These thick horizontal folds are usually S -shaped and are preserved or separated by the fault plane. In most cases, when the sharp contact with the upper cover and the undesirable bed below, the folds are cut off (Fig. 33d and e). In addition, the bottom layer of these folding beds shows the local preservation of debris (Fig. 33a and d).

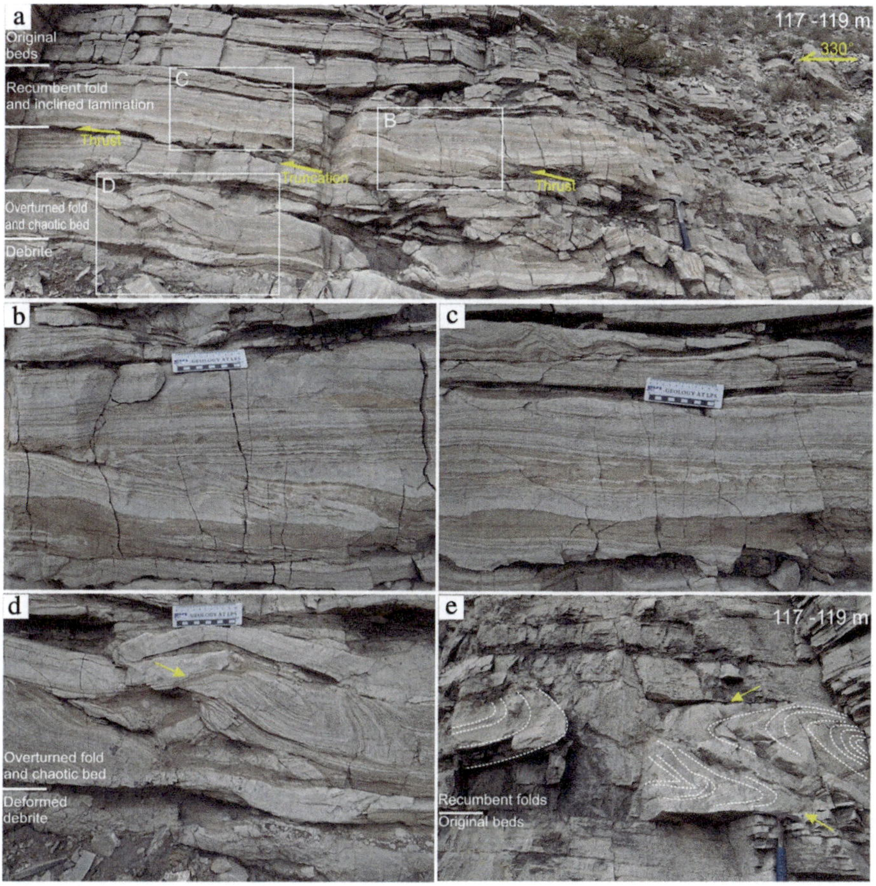

Fig. 33 Folds and chaotic beds at 117–119 m in the XLF section. **a** Outcrop features of folds, inclined lamination, chaotic bed and debrites. Note the thrust and truncation features at the bottom of recumbent folds. **b** Z-shaped recumbent chevron folds. **c** Inclined lamination. Note the lateral variation trend from recumbent folds **b** to inclined lamination **c**. **d** Overturned fold (eroded by the overlying layer, arrow) within a chaotic bed, overlying a deformed debrites. **e** Stepped recumbent folds, present several meters from **a** to 150° direction, separated by a décollement (arrows) between the overlying and underlying layers. Hammer in **a** and **e** is 28 cm long. Scale bar is in cm in **b**, **c** and **d**

Stop 5. Imbricated Beds (128.5 m)

At 128.5–130 m in the XLF section, there are two deformed intervals that are exposed laterally for over 20 m (Figs. 34 and 35). The lower interval (128.5–129 m, ~ 0.5 m thick, interval 1 in Figs. 34 and 35) and the upper interval (129–130 m, ~ 1 m thick, interval 2 in Figs. 34 and 35) are separated by an undulatory surface. Beneath the lower interval, there is a 10 cm thick black shale, which, in turn, rests on minimally deformed, thin-bedded, finely laminated lime mudstone (Figs. 34a and 35d).

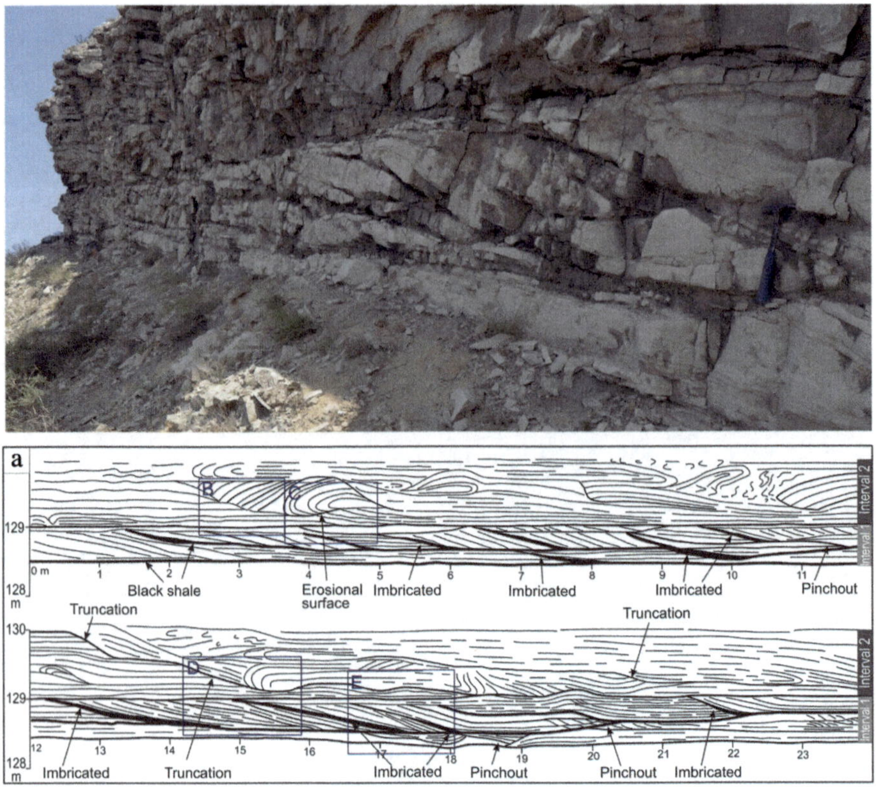

Fig. 34 Imbricated bed and highly folded beds at 128–130 m in the XLF section. **a** Line drawing of the imbricated bed (interval 1) and the overlying stepped, strongly folded bed (interval 2)

The lower section is mainly composed of a wedge -shaped body. Each wedge -shaped width exceeds 2 m, and the maximum thickness is about 20 cm. The interior tattoo layer of the wedge in this interval usually fluctuates slightly. Through the measurement of the 20 layer of tile layer, after the tilt -angle correction is corrected, the direction of the tile layer is 92° ~ 139°, and the vector average is 109.5°. B, C, D). The stacking layer section of about 50 cm in the lower section of the lower part is in the southeast (150°) unable to deform layer (Fig. 35f) with a thick 20 cm. On the contrary, when describing in the same direction, the upper part (layer Sect. 2) fold deformation is weak, and some folds are replaced by stacked tile layer (Fig. 35f).

The stacking layer may be formed by a series of thrust of a 20 cm thick layer. The horizontal changes of the height of the height of the overlapping unit to the upper cover unit (interval 2 in Fig. 34) indicates that the stacking tiles are likely to be formed during the slump of the sedimentation block and develop with the folds Wait, in the pressure machine).

Stop 6. Chaotic Beds (139 m)

The chaotic layer of the XLF contains folds and rotary patterns (Fig. 36), but there is no space and strata trend, which is disorderly layer. The main thickness of the chaos bed is about 50 cm ~ 1 m, which is located on the sliding layer of the unresolved bed and the folds. XLF Sect. 139–139.5 m chaotic bed layer (Fig. 36a and b) is located on the bed layer (Fig. 36c) with tilted folds and associated flame structures at 50 cm; these horizontal folds form a small direction in one direction. The size of the flat folds and curling layers (Fig. 36a) forms reverse wrinkles in another direction (Fig. 36d). The contact between the sloping layer and the chaos layer is suddenly (Fig. 36a). Although the structure stored in the sliding layer has obvious horizontal changes, from breaking to wrinkles and rotary stratums, the upper chaos layer is quite uniform along the outsiders (Fig. 36).

Significant shear stress and uneven squeezing stress related to sliding may cause the formation of chaos (Allen and Banks 1972; Owen 1996; Rosetti 1999). The chaotic bed layer on the fold sliding tablet may represent two independent quality handling events. The height deformed sliding layer (chaotic layer) covers the previous sliding layer along the surface of the Décollement, which marks the contact between the layers and the layer. The differences in the two layers of internal structure may reflect the two destructive flow behaviors, which may be determined by the degree of water incorporation or the rock nature of the original sedimen (li et al., In present).

Stop 7. Carbonate Breccia and Grainstone (175 m)

The thickness of the carbonate and crystal rock at the XLF section is about 10 m, which is composed of the lower and upper part of the corner gravel (175 and 182.6 m) and the coarse stone (about 0.3–1 mm in diameter) in the middle unit (Fig. 37a). The thickness of the lower and upper corners of the corner arms is about 2.4 m, but it has an erosive base surface, which can alleviate up to 41 cm. Corner gravel is embedded, table, angle, white lime mudstone, dark gray grain, and gray Calcisiltite.) Similar ingredients and textures (Kelimoli strata), as well as medium -gray, rough oil crystal rock matrix (Figs. 37b, c, 38b and d). At least 10 cm long of these corners of these corner karma units show the uniform direction of the ancient flow in the west.

The intermediate unit consists of a huge 5.1 m and the grooves, and the very rough biological debris particles (Fig. 38a, b and c). The thickness of the bed is 10 to 70 cm. In some places, it is inserted with 2 to 3 cm thick, and the discontinuous lime mud rock bed (Fig. 38a). The groove horizontal beds account for about 40% of the unit, with a thickness of 10 to 37 cm (Fig. 37c and 38b); the rest of the unit consist of large -scale stone beds. The cross layer in the grain stone is local complex or overturned (Fig. 38c). In some places, there are floating fragments of lime mudstone.

In view of the abundance of sand debris, delations and normal grading, carbonate corner gravel sedimentation is limited to the concentrated density flow of the channel (Mulder and Alexander (2001). The intermediate unit of rough biofidella particles records the transportation of carbonate bone sand in the low speed flow of turbulence, which flows than the speed of the caulmeter transport, but it is powerful and lasting enough to maintain the dunes migrated in coarse sand (Nordin 1971; Dinehart 1989).

Fig. 35 **b** Truncated overturned fold (dash line). **c** Recumbent fold and underlying imbricated bed. Note the internal detachingassociated with the thrust (dash line). Scale bar is in cm. **d** Undulated sliding surface (dash line) between the chaotic bed and the normal beds and underlying imbricated bed. **e** Imbricated bed complex. **f** Imbricated bed (interval 2) interbedded with undeformed layers (interval 1). Hammer in **b**, **d**, **e** and **f** is 28 cm long

The thinness of lime mud rock indicates that the flow is unstable, and sometimes the mud is deposited from the suspension solution in the channel. The deformation granular stone indicates that a soft sink accumulation failure occurs in the channel, just like the remaining carbonate slope, this article is recorded in this article (li et al., Printing).

4.4 DAY 4 (Saturday, August 20, 2022)

Overview: Ordovician platform to slope deposits with remarkable channelized carbonate breccia at the Yixiantian (YXT) section (Figs. 39 and 40).

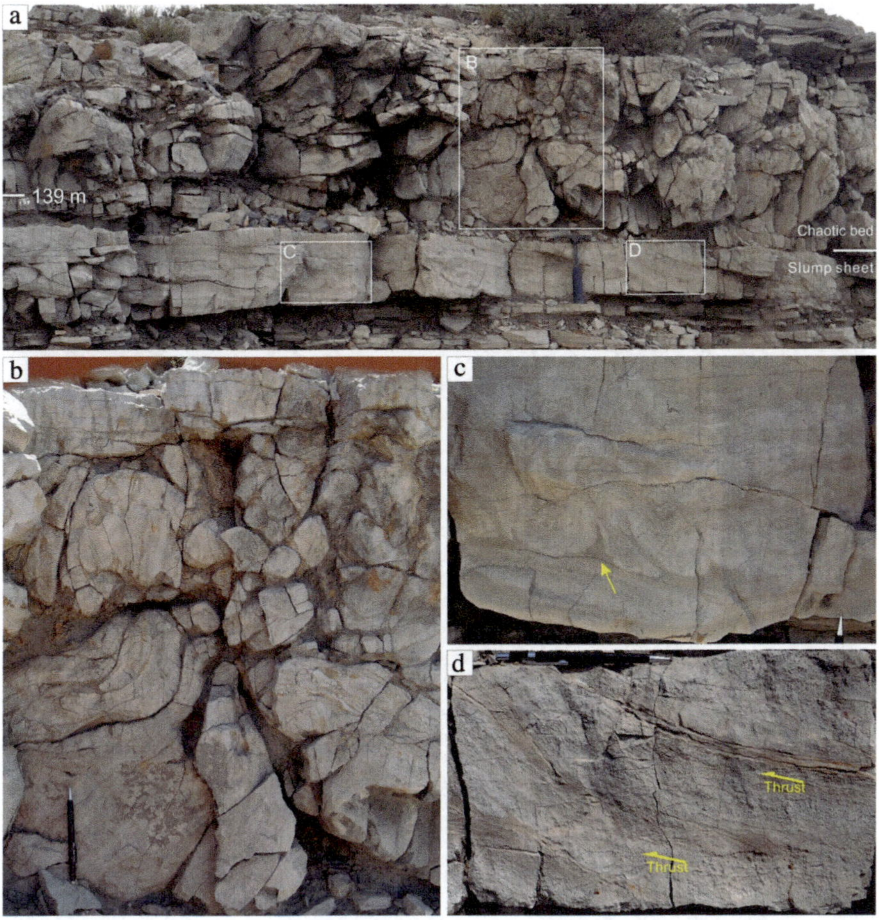

Fig. 36 Chaotic bed at ~139 m in the XLF section. **a** Chaotic bed and the underlying slump sheet. Hammer is 28 cm long. **b** Highly deformed chaotic bed. **c** Inclined flamed structure (arrow) and undulating lamination. **d** Thrust folds. Pencil is 14 cm long in **b**, **c** and **d**

Stop 1. Lower Kelimoli Formation

Here we can see the bioturbated and bioclastic limestone-dominated Zhuozishan Formation and the interbedded lime mudstone and marl of the Lower Kelimoli Formation (Figs. 40 and 41). The Lower Kelimoli Formation at the YXT section is very similar to that of the XLF section, indicating a possibly similar deep water condition.

Stop 2. Upper Kelimoli Formation

The upper part (graptolite-bearing black shale) of the Kelimoli Formation at the YXT section is remarkably thicker than that of the XLF Sect. (16 m vs 5 m). In

Fig. 37 Carbonate breccia and grainstone at the XLF section. **a** Outcrop of the carbonate breccia and grainstone layer at the base of the Wulalike Formation. **b** Imbricated, tabular carbonate breccia and trough cross-stratified, very coarse bioclastic grainstone. **c** Different texture features between the upper, middle and lower beds of the lower carbonate breccia unit

Fig. 38 **a** Discontinuous lime mudstone beds within the middle unit. **b** Cross-bed of the middle unit. The thin lower unit is located at the edge of the channel. **c** Convoluted and overturned stratification of the middle unit. **d** Bedding surface of the upper unit showing carbonate breccia and large grainstone intraclasts

consideration of the inferred paleoslope orientation (WNW), the black shale and limestone at YXT most likely reflect a deeper slope-basin environment (Fig. 41).

Stop 3. Lenticular Carbonate Breccia

We correlate the breccia at the base of Wulalike Formation at the YXT section to carbonate breccia and grainstone of the XLF section (Fig. 42). This breccia is sporadically exposed at Stop 3, but there is one 7 m long and 0.6–1.3 m thick lens, with a generally flat bottom and an undulatory top surface. The breccia pinches out to the both ends, and one end is slightly deformed. There are two kinds of lime mudstone intraclasts: (1) small (mm to several cm in diameter) and subangular to subrounded; and (2) flat and longer (several cm to tens of cm in diameter), and subangular to angular. The breccia is clast-supported, with marl or calcisiltite matrix. It is noteworthy that four small patches of carbonate breccia are present 2.3 m below the level of the breccia described above (Fig. 43). Further features of the breccia are exposed at the next stops.

Stop 4. Carbonate Breccia, Mud-Armored Breccia Ball and Irregular Slump Block

YXT has a large number of lens carbonate rock angle arges, almost all of them appear on a strata level near the bottom of the Urarik group (Fig. 44a, d and e), but there are below the primary layer of the above coarse grains. Two structures (Fig. 44a). First,

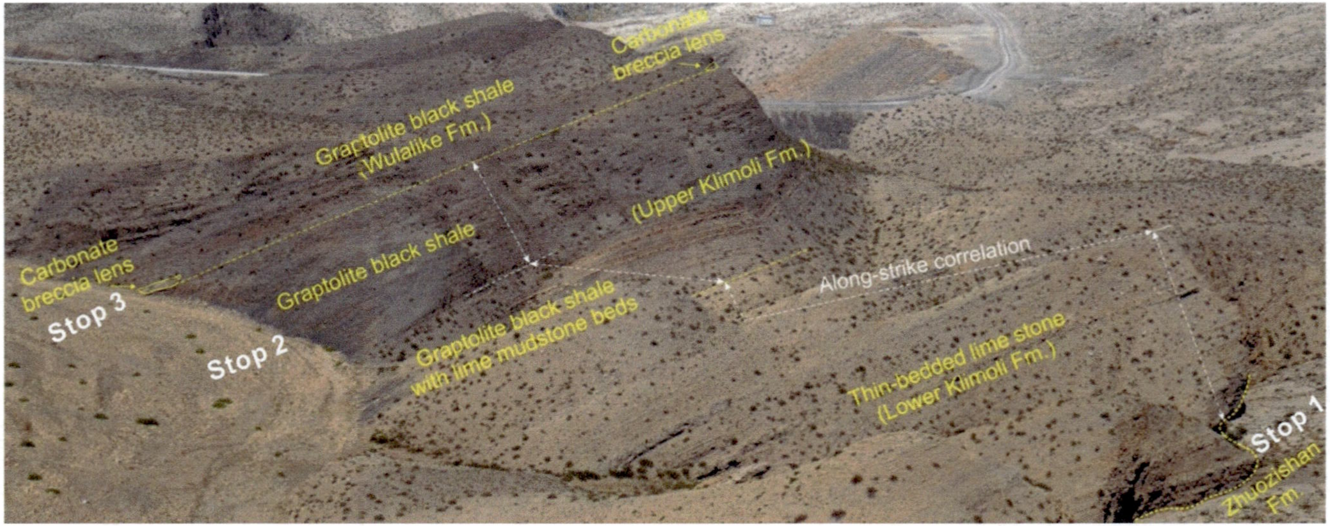

Fig. 39 Overview of the YXT section, with major lithostratigraphic units. The YXT section is located approximately 20 km south of Wuhai and ~ 10.5 km from the XLF section

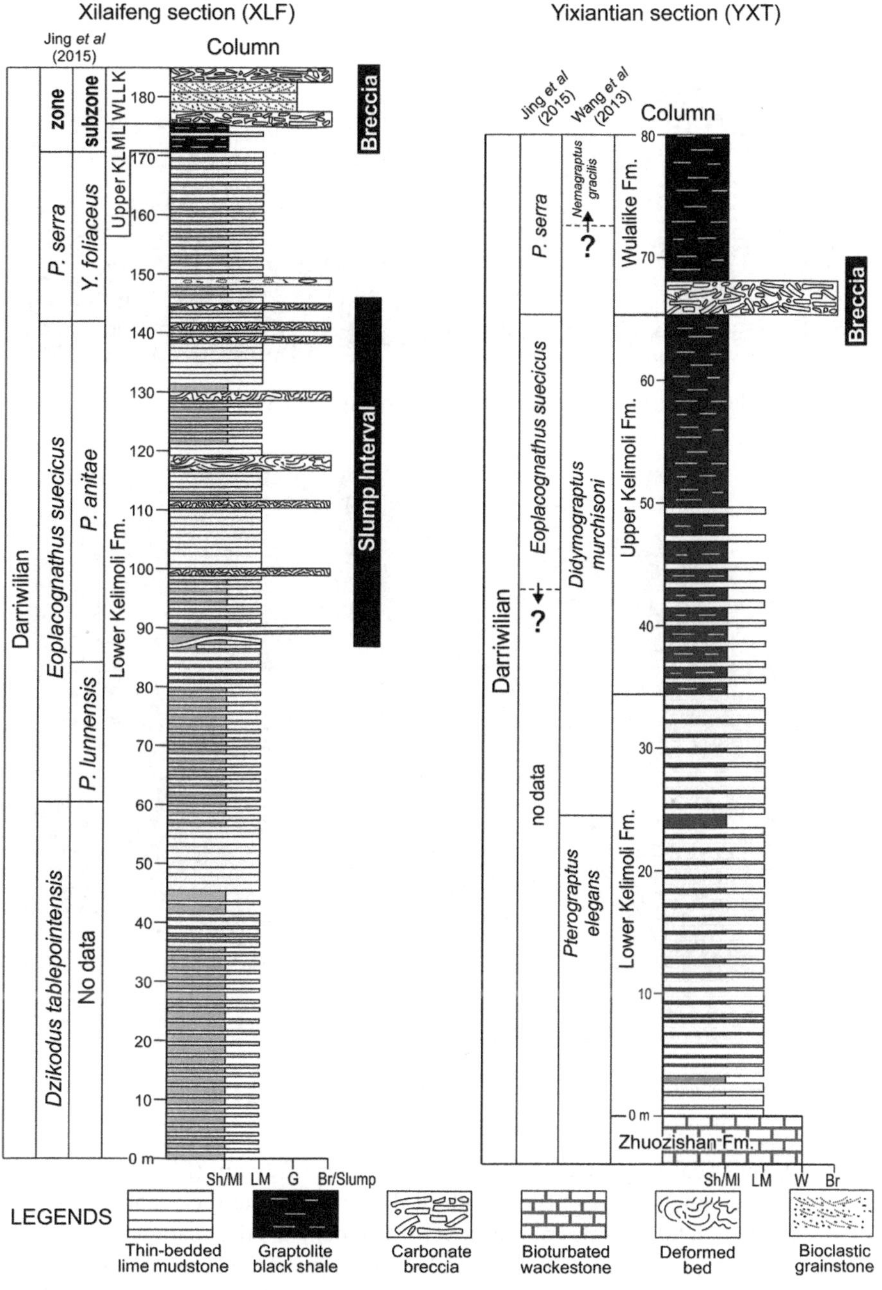

Fig. 40 Stratigraphic sequences of the Xilaifeng and Yixiantian sections. Conodont and graptolite zonation are from Jing et al. (2015) and Wang et al. (2013) respectively. Conodont zones: *P. serra*—*Pygodus serra*; *P. lunnensis*—*Pygodus lunnensis*; *P. anitae*—*Pygodus anitae*; *Y. foliaceus*—*Yangtzeplacognathus foliaceus*. Formations: KLML—Kelimoli Formation; WLLK—Wulalike Formation

Fig. 41 The Zhuozishan and Lower Kelimoli formations of the YXT section

Fig. 42 Lenticular carbonate breccia at the YXT section

there is a spherical fragrance block with a diameter of 29 cm (Fig. 44b) at 60 cm below the protagonist's args. The length of the debris in the rock ranges from 1 to 7 cm. About 1 cm thick shale coating covers the block, and the coating is a relatively continuous block (non -layered) edge, and it is obvious at the layer -like shale in the block and the enclosure (Fig. 44b). Followed by a small hill -shaped mining bed with a width of about 50 cm, a thickness of about 20 cm, and height, the base is about 30 cm below the spherical block, consisting of white tattoo small crumbs and mud

Fig. 43 Interbedded graptolitic black shale and thin- to medium-bedded lime mudstone of the Upper Kelimoli Formation

rock matrix (Fig. 44c). The top of the bed is highly irregular, and some of them are protruding, forming more plate -shaped accumulation.

The overall geometry shape of the corner clams shows that it may scroll on the muddy bottom on the slope and accumulate about 1 cm thick mud coating (now shale), that is, it is a mud armor debris, and typical typical typical The contrary to the armor of the armor (CF. Van de Mer Merwe et al. 2011; Maier et al. 2020). Second, small, highly irregularly irregularized tattoo dick -shaped sediments have irregular geometric shapes (li et al., In press).

Stop 5. Various Outcrops of the Breccia-Filled Channel

There are usually two types of corner arges of YXT profile, although there are examples of gradients between these two scale. 0.7–2 m thick large-scale strata horizontal distribution can reach tens of meters (Figs. 45a and 46a). These show the complex filling of multiple partitions, and each partition shows the difference in alignment, size, and circles of debris. The arrangement of debris on the plane (2D) vertical surface from random (Fig. 46b and e) to horizontal (Fig. 46d), and then steep (Fig. 45d) (extensive in some cases) (Fig. 45e). Multiple exposure inspections have shown that about the display of about the SE-NW direction (average ~150° -330°) display bureau stacking tiles (Fig. 45a and e). Almost orthogonal NE-SW exposure (average ~40°–220°) shows the obvious horizontal direction of debris (Fig. 46d). Displays in other directions are not neatly arranged. The debris is composed of debris in the lone rock, but there are also a small amount of black shale blocks (Fig. 44d and 46b).

Smaller corner arges (Figs. 44 and 47) are steep channel-filling forms, either attached to a more continuous upper overflow or isolated in the shale. The latter is usually only a few to tens of cm thick (Figs. 44e, 47a, d, and e). Although a significant

Fig. 44 Breccia-filled channel, mud-armored breccia ball and slumped breccia mass at the YXT section. **a** Different types of carbonate breccia deposits. **b** Mud-armored spherical breccia block. **c** Irregular mounded carbonate breccia complex. **d** Breccia-filled channel with black shale intraclast (arrow). **e** Breccia-filled channel with a laminated lime mudstone cap (arrow). Pencil is 14 cm long

example is 1.9 m thick, the width from 0.5 m at the bottom to 1.8 m (Fig. 47c). The wide and depth ratio of these channels filled with corners is 0.5 ~2.6 (average 1.4). Some rock layers display a series of channels, which are connected from a relatively thin rock layer (Fig. 47d) in the horizontal direction. The thicker river sediment (generally and gt; 1 m) is often wider (10 m), and usually the top of the mild mound (Fig. 45a).

The isolated and connected narrow groove charging bodies show uniform debris or normal grades (Fig. 46d and 47b) in the entire corner arms partition. Many layers include laminar (partial cross-layer) calcium mud to block-weak layer-like lime mudstone (Fig. 44e, 45a–c, 47b and f). A fine -grained partition is usually a few centimeters thick, although they can reach 21 cm thick. In some cases, the division of calcium pink rock/mud rock is uniform thickness and is limited to the width of the lens -like corner grave layer. The top is a hill -shaped cornea layer that usually

Fig. 45 Features of channels crop out along NW–SE at the YXT section. **a** Longitudinal profile (along 150°-330° direction) of a widely extended channel (20 m) with mounded top, and lime mud cap and wing. **b** Close-up of **a**, arrow shows the pinch out of the lime mud cap and its irregular base. **c** Photomicrograph of fine-grained caps, composed mainly of lime mud and cross laminatd peloidal calcisiltite with tiny bioclasts. **d** Steeply orientated, flat clasts within a channel. **e**. Imbricated clasts within a channel along NW–SE (140°–320°). Hammer is 28 cm long. Pencil is 14 cm long

becomes thinner to the thickness of the corner arges. In a few cases, they will squeeze at the high point of the corner arges (Fig. 45b).

Like the breccia of XLF, the ordinary plate -shaped geometry of debris reflects them from thin to medium -sized band -shaped mud stones, which is the characteristic of the Kelimoli group. The geometric shapes of these tablets and the very large debris size will greatly increase the internal friction angle of the sediment, which requires the quality loss event to occur on a relatively steep slope. There are no debris in the shale outside the river. Unless there is a complete downhill by -to -wing by the thick sediments (corner arges) under the deep flow of deep flow, this will indicate

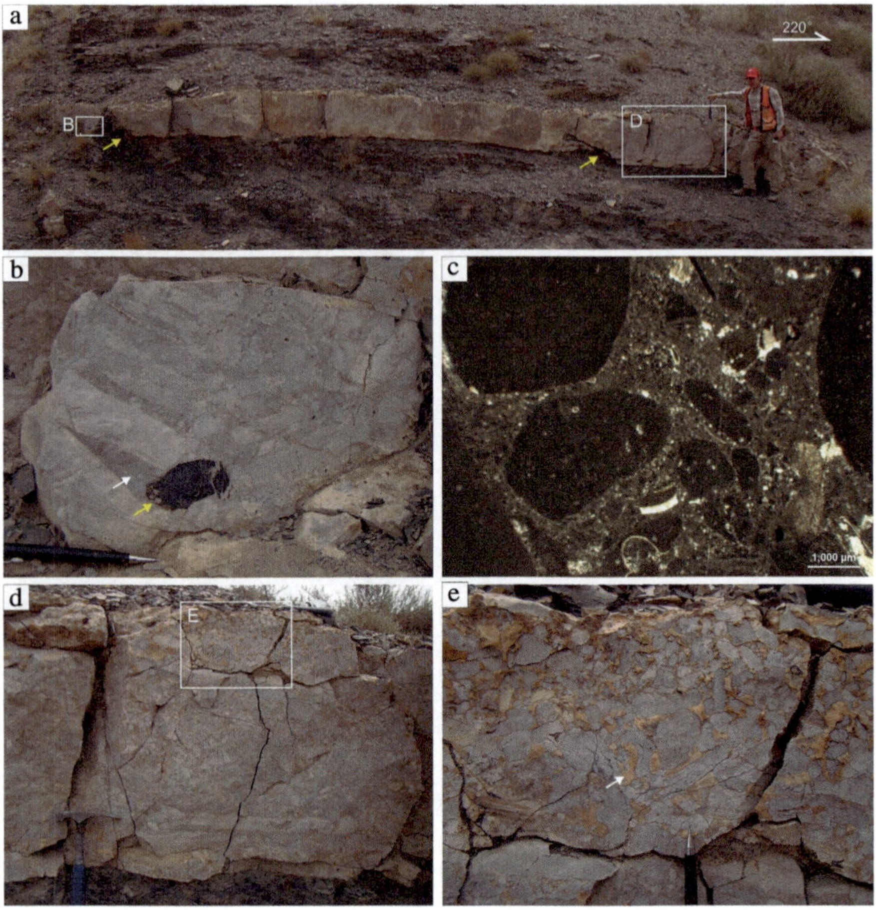

Fig. 46 Features of a channel crop out along NE-SW at the YXT section. **a** A ten-meter-wide cross-section of a channel along 40°–220° direction, with erosive features (arrows). **b** Close-up of **a**, showing a black mudstone clast (yellow arrow) and calcisiltite matrix (white arrow). **c** Photomicrograph of breccia with calcisiltite matrix, composed of silty quartz and μm scale bioclasts. **d** Close-up of **a**, showing normal grading of the breccia and horizontal flat intraclasts at the bottom. **e** Close-up of **d**. Disorganized intraclasts, with dolomitic marlstone matrix (arrow). Hammer is 28 cm long. Pencil is 14 cm long

that the water flow is limited to the river channel. The dominant position of the thick granular matrix indicates that the flow is non -sticky, so it is not a mudslide. Large debris (tens of centimeters), and has no leveling, indicating that the debris is carried by high density flow (ie, sand debris flow) (Mulder and Alexander 2001). However, the steep carbonate rock angle rinsered layer shows a strong flow of erosion. The weak erosion properties of high -density flows, and the extremely rare of debris in black mud rocks, indicate that the fluid flows excessively open and forms a deep channel formed before the corners of the corners. The mud cover of the YXT corner sedimentary material represents a dense mud dispersed body, and it is formed at the same time as the high concentration gravel flow filled with the narrow channel (li, etc., which is squeezed) (Fig. 47).

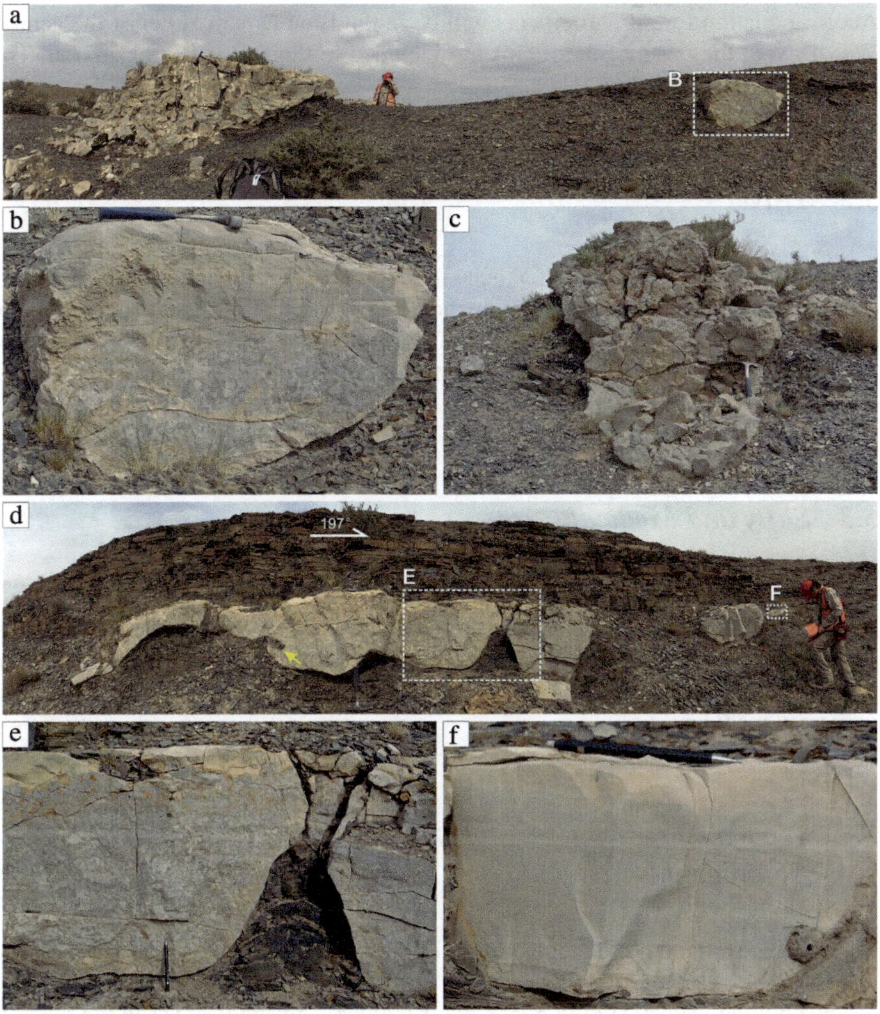

Fig. 47 Breccia-filled channels at the YXT section. **a** Isolated channels within black shale. **b** Normal graded, small-scale breccia-filled channel, capped by weakly laminated lime mudstone. **c** Deep-incised breccia-filled channel. **d** Laterally linked channels by relatively thin beds above and black shale intraclast (arrow). **e** Breccia cuts downsection into black shale. **f** Lime mudstone wing in black shale. Hammer is 28 cm long. Pencil is 14 cm long

5 Field Logistics

5.1 Accommodation

Accommodation for all participants will be in the Blue Dream Hotel of the central Wuhai. Note the costs of the rooms are included in the fee of this excursion. The hotel will provide breakfast.

5.2 Travel

All participants should arrive at Yinchuan Airport on August 16. We will pick you up in the Yinchuan Airport, and stay in the Yinchuan hotel on August 16. During the excursion, we will take minibus and 4WD vehicles to the field sections. On August 21, 2022 (Sunday), we will take vehicles from Wuhai to Yinchuan Airport and fly to Beijing.

5.3 Safty and Medical Care

The field is moderate to high difficulty, and can be steep climbing. The participants should have health insurance for the journey. All foreign participants are required to bring with them health insurance contracts, for the time of the trip, from the insurance company that provide international insurance policy program. You should take along necessary medicine.

5.4 Temperature, Climate and Recommended Field Equipment

In late August, the daytime temperature in Wuhai is usually between + 20 °C and + 31 °C. The climate of Wuhai area is arid, so bring water bottles (we can also provide bottled water). Hiking boots, hat, sunblock, and sunglasses (up to yourself) are necessary for the field. Hiking canes are helpful.

5.5 Emergency Numbers

Police: 110;
First aid: 120;
Fire alarm: 119.

Acknowledgements The integrated sedimentologic and stratigraphic study on the Precambrian (mostly the Mesoproterozic Huangqikou and Wangquankou formations) through to the Cambrian–Ordovician successions in the western margin of the North China Block was dated back to a decade ago, when the two senior authors of this field guide (Paul Myrow and Jitao Chen, and their Chinese and U.S. colleagues and students) first visited the SBG and GDE sections in 2012. Jitao was a postdoctoral fellow, back then, with Prof. Sung Kwun Chough at the Seoul National University, South Korea. Wenjie joined the team in 2016 and started leading the study on the carbonate slumps and breccias in the Kelimoli and Wulalike formations (middle Ordovician) from 2019. Over the ten years, we have been visiting and working in these outcrop sections every year. Unfortunately, Paul and his students were not able to travel to China due to break out of the COVID-19 in 2020.

At this occasion, we would like to thanks our (former) students, who have joined us in the fieldtrip over the past ten years. They are: Zachary Snyder, Anne J. Hakim, Lea Linse, Tianran Zhang, Xiaoxu Tian, Yuze Huang, Peter Mow, Wenli Yang, Hao Xin, and Zhixin Sun. We have many sweet memories together in the field. We still remember that we've been hiking in the 40 °C weather over 2 hours in order to reach the WQG section in 2016 when the roads were destroyed by floods and our field vehicles were stuck in the valley. We also remember that Tianran carried a huge watermelon (~15 kg) to the waist of the Sheperd House section in the SBG valley…

We highly appreciate the fruitful discussion in the field with Jinliang Yuan, Xuejian Zhu, Justin V. Strauss, and Xiuchun Jing. We thank Stephen Leslie, David A. Fike, C. Mark Fanning, and N. Ryan McKenzie for contribution of their expertise to the studies that were successfully published. We thank Xu Chen and Yuandong Zhang from the Naning Insitute of Geology and Palaentology, Chinese Adademy of Sciences, for identification of graptolites and discussion on the Ordovician stratigraphy.

We would like to thank Xiumian Hu from the Nanjing Unversity for his continuous support and encouragement ever since from the beginning in organizing the 21the ISC fieldtrips. We are grateful to finantial support from National Natural Science Foundation of China, Chinese Academy of Sciences, State Key Laboratory of Palaeobiology and Stratigraphy, and National Science Foundation of U.S.

References

Allen JRL, Banks NL (1972) An interpretation and analysis of recumbent-folded deformed cross-bedding. Sedimentol 19:257–283

Bian Q, Gao S, Li D, Ye Z, Chang C, Luo X (2001) A Study of the Kunlun-Qilian-Qinling Suture System. Acta Geol Sinica—English Ed 75:364–374

Chen X, Bergström SM, Zhang Y, Wang Z (2013) A regional tectonic event of Katian (Late Ordovician) age across three major blocks of China. Chin Sci Bull 58:4292–4299

Chough SK, Lee HS, Woo J, Chen J, Choi DK, Lee SB, Kang I, Park TY, Han Z (2010) Cambrian stratigraphy of the North China Platform: revisiting principal sections in Shandong Province, China. Geosci J 14:235–268

Darby BJ, Ritts BD (2002) Mesozoic contractional deformation in the middle of the Asian tectonic collage: the intraplate Western Ordos fold–thrust belt, China. Earth Planet Sci Lett 205:13–24

Dinehart RL (1989) Dune migration in a steep, coarse-bedded stream. Water Resour Res 25:911–923
Dong Y, Santosh M (2016) Tectonic architecture and multiple orogeny of the Qinling Orogenic Belt, Central China. Gondwana Res 29:1–40
Feng Z, Bao Z, Zhang Y (1998) Lithofacies paleography of Ordovician carbonatite formation in Ordos Basin. Geol Publ House, Beijing
Feng Z, Peng Y, Jin Z, Bao Z (2004) Lithofacies paleogeography of the Cambrian and Ordovician in China. Pet Ind Press, Beijing
Jing X, Zhou H, Wang X (2015) Ordovician (middle Darriwilian-earliest Sandbian) conodonts from the Wuhai area of Inner Mongolia, North China. J Paleontol 89:768–790
Jing X, Zhou H, Wang X (2016) Biostratigraphy and biofacies of the Middle Darriwilian (Ordovician) conodonts from the Laoshidan section in the western margin of the North China Craton. Mar Micropaleontol 125:51–65
Lee HS, Chough SK (2011) Depositional processes of the Zhushadong and Mantou Formations (Early to Middle Cambrian), Shandong Province, China: Roles of archipelago and mixed carbonate–siliciclastic sedimentation on cycle genesis during initial flooding of the North China Platform. Sedimentol 58:1530–1572
Lee JH, Chen J, Chough SK (2010) Paleoenvironmental implications of an extensive maceriate microbialite bed in the Furongian Chaomidian Formation, Shandong Province, China. Palaeogeogr Palaeoclimatol Palaeoecol 297:621–632
Lee JH, Chen J, Choh SJ, Lee DJ, Han Z, Chough SK (2014) Furongian (Late Cambrian) sponge–microbial maze-like reefs in the North China Platform. Palaios 29:27–37
Li W, Chen J, Anne JH, Myrow PM (in press) Middle Ordovician mass-transport deposits from western Inner Mongolia, China: mechanisms and implications for basin evolution, Sedimentol
Maier KL, Paull CK, Caress DW, Anderson K, Nieminski NM, Lundsten E, Erwin BE, Gwiazda R, Fildani A (2020) Submarine-fan development revealed by integrated high-resolution datasets from La Jolla Fan, offshore California, USA. J Sediment Res 90:468–479
Mulder T, Alexander J (2001) The physical character of subaqueous sedimentary density flows and their deposits. Sedimentol 48:269–299
Meng X, Ge M, Tucker ME (1997) Sequence Sequence stratigraphy, sea-level changes and depositional systems in the Cambro-Ordovician of the North China carbonate platform. Sed Geol 114:189–222
Myrow PM (1992) Pot and gutter casts from the Chapel Island Formation, Southeast Newfoundland. J Sediment Res 62:992–1007
Myrow PM, Chen J, Snyder Z, Leslie S, Fike DA, Fanning CM, Yuan J, Tang P (2015) Depositional history, tectonics, and provenance of the Cambrian-Ordovician boundary interval in the western margin of the North China block. GSA Bull 127:1174–1193
Nordin CF (1971) Statistical properties of dune profiles. Prof. Paper US Geol. Surv., 562-F, p 41
Owen G (1996) Experimental soft-sediment deformation: structures formed by the liquefaction of unconsolidated sands and some ancient examples. Sedimentol 43:279–293
Riding R (2000) Microbial carbonates: the geological record of calcified bacterial–algal mats and biofilms. Sedimentol 47:179–214
Rossetti DDF (1999) Soft-sediment deformation structures in late Albian to Cenomanian deposits, São Luís Basin, northern Brazil: evidence for palaeoseismicity. Sedimentol 46:1065–1081
Scotese CR, McKerrow WS (1990) Revised World maps and introduction. Geol Soc, London
Shapiro RS, Awramik SM (2006) Favosamaceria cooperi new group and form: A widely dispersed, timerestricted thrombolite. J Paleontol 80:411–422
Shi Y, Yu JH, Santosh M (2013) Tectonic evolution of the Qinling orogenic belt, Central China: New evidence from geochemical, zircon U-Pb geochronology and Hf isotopes. Precambr Res 231:19–60
Song S, Niu Y, Su L, Xia X (2013) Tectonics of the North Qilian orogen, NW China. Gondwana Res 23:1378–1401

Van Der Merwe WC, Hodgson DM, Flint SS (2011) Origin and terminal architecture of a submarine slide: a case study from the Permian Vischkuil Formation, Karoo Basin, South Africa. Sedimentol 58:2012–2038

Wang Z, Bergström SM, Zhen YY, Chen X, Zhang Y (2013) On the integration of Ordovician conodont and graptolite biostratigraphy: new examples from Gansu and Inner Mongolia in China. Alcheringa: Australas J Palaeontol 37:510–528

Xiao W, Windley BF, Hao J, Zhai M (2003) Accretion leading to collision and the Permian Solonker suture, Inner Mongolia, China: Termination of the central Asian orogenic belt. Tectonics 22:1–20

Yang R, van Loon AJ, Jin X, Jin Z, Han Z, Fan A, Liu Q (2019) From divergent to convergent plates: Resulting facies shifts along the southern and western margins of the Sino-Korean Plate during the Ordovician. J Geodyn 129:149–161

Zhang J, Li J, Liu J, Feng Q (2011) Detrital zircon U-Pb ages of Middle Ordovician flysch sandstones in the western Ordos margin: new constraints on their provenances, and tectonic implications. J Asian Earth Sci 42:1030–1047

Zhang T, Myrow PM, Fike DA, Mckenzie NR, Yuan J, Zhu X, Li W, Chen J (2020) Sedimentology, stratigraphy, and detrital zircon geochronology of Mesoproterozoic strata in the northern Helan Mountains, western margin of the North China Block. Precambr Res 343:105730

Zhen YY, Zhang Y, Wang Z, Percival IG (2016) Huaiyuan Epeirogeny—Shaping Ordovician stratigraphy and sedimentation on the North China Platform. Palaeogeogr Palaeoclimatol Palaeoecol 448:363–370

Zhou ZY, Zhou ZQ, Zhang J (1989) Ordovician trilobite biofacies of North China Platform and its western marginal area. Acta Palaeontologica Sinica 28:296–313

Devonian-Carboniferous Carbonates in Guilin, South China: Depositional Records of Platform-Basin Complex and Major Biocrisises

Chen Dai-zhao and Qing Hai-ruo

1 Summary

Guilin is a well-known tourist city in China and abroad, marked by the breathtaking scenery of spectacular karstic landforms. These landforms dominantly comprise the Devonian carbonates with subordinate Lower Carboniferous carbonates deposited principally in the fault-controlled platform-basin system in which mass extinction events of Late Devonian (Frasnian-Famennian and Devonian-Carboniferous transitions), also known as Kellwasser and Hangenberg events, were well recorded as well. Besides, fault-controlled hydrothermal dolostones were also developed in some carbonate successions such as the basal successions of Devonian carbonates (Givetian) and the lower part of Carboniferous (Visean).

This excursion will show you the diversified depositional facies (restricted carbonate platform interior, open platform-platform margin, and marginal slope-basinal facies), time- and facies-specific faunal assemblages (i.e., stromatoporoids, tentaculitids), and the stratigraphic-geochemical records of Late Devonian biocrisises, as well as the hydrothermal dolostones in the Givetian and Visean carbonates in Guilin area. As a bonus, this trip will also show you how the primary (intraplatform) facies had largely controlled the development and distribution of the most spectacular karstic landforms including the cave (speleothem) systems in Guilin.

C. Dai-zhao (✉)
Institute of Geology and Geophysics, Chinese Academy of Sciences, Beijing, China
e-mail: dzh-chen@mail.iggcas.ac.cn

Q. Hai-ruo
Department of Geology, University of Regina, Regina, Canada
e-mail: hairuo.qing@urgina.ca

X. Hu (ed.), *Field Trip Guidebook on Chinese Sedimentary Geology*,
https://doi.org/10.1007/978-981-99-6936-4_15

2 Paleogeographic and Tectonic Contexts

During the Devonian Period, the South China Block (coupled Yangtze and Cathaysian blocks) was located close the equator in the eastern flank of the Paleo-Tethys Ocean, and to the west of the Gondwanaland (Fig. 1).

From the late Early Devonian (Emsian), especially from the Middle Devonian Givetian (Middle *varcus* Zone), the South China block was initially fragmented and subsided along the antecedent coupling zone between Yangtze and Cathaysia blocks from Guangxi to Hunan in the context of a progressive transtensional tectonic activity, resulting in a transtensional (strike-slip) rift basin domain there, notably from the Late Devonian (Chen et al. 2001a, b; Fig. 2).

Carbonate platforms (or shelf in the Early Devonian) developed on the palaeo-highs were commonly bounded by curvilinearly narrow interplatform basins (or troughs) extending locally for several 100s' km long. The specific platform-basin configuration as shown in Figure 2 was interpreted to having resulted from the major NNE–SSW-trending, sinistral strike-slip faulting along the deep-seated basement zones, inducing a series of secondary fault zones (i.e., *en echelon*, synthetic and antithetic strike-slip fault zones, and extensional (pull-apart process) fault zones), thereby leading to a extremely complicated platform-basin configuration which complex could continue into the Early Carboniferous (Fig. 2a, b; Chen et al. 2001a, b, 2006).

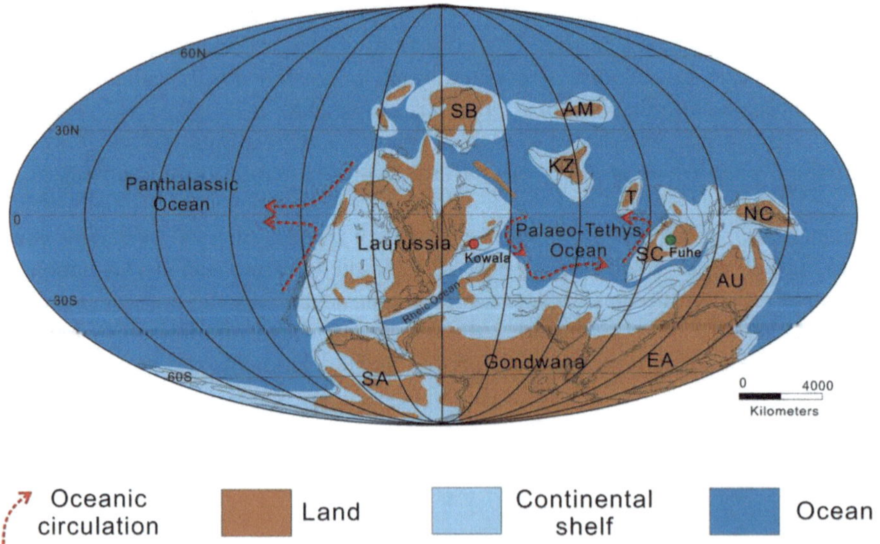

Fig. 1 Global paleogeographic map of the latest Frasnian of Devonian. Abbreviations: SB, Siberia; AM, Amuria; KZ, Kazakhstan; T, Tarim; NC, North China; SC, South China; AU, Australia; EA, East Antarctica; SA, South America (modified after Kiessling et al. 2002). Note the position of South China Block (SC)

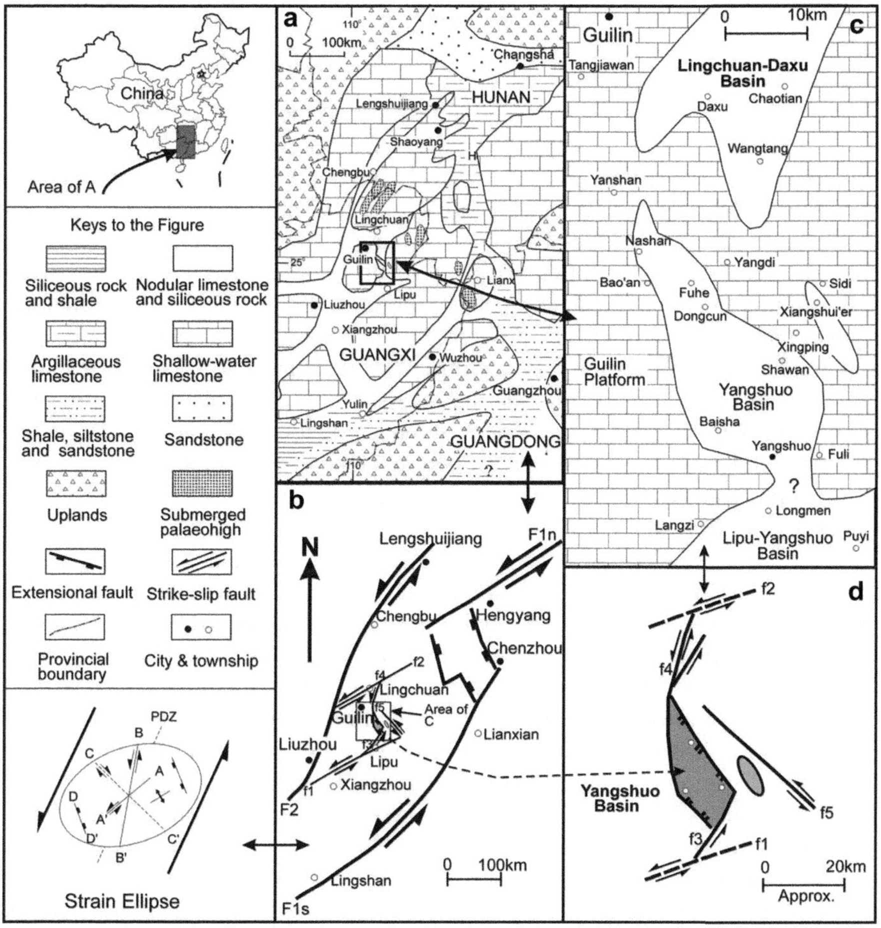

Fig. 2 Late Devonian lithofacies palaeogeography and tectonic setting in South China (after Chen et al. 2001b, 2006)

The Devonian Guilin Carbonate Platform was nearly 40 km wide from east to west and about 80 km from south to north. It was almost surrounded by deep-water basins or interplatform basins except two shallow-water corridors that had linked to a larger platform to the east so it looked like a half-detached platform (Fig. 2a, c). Between the two platforms, and further delimited by the two shallow-water corridors, was a spindle- to rhomb-shaped basin created mostly in northern Yangshuo (named as Yangshuo Basin). This unique configuration likely resulted from pull-apart process induced by the secondary strike-slip faulting activity (Fig. 2d; Chen et al. 2001b, 2002a, b). Such a complicated platform-basin complex resulted in diversified depositional facies (or lithofacies) across different platform-basin profiles due to different topographic reliefs. Stromatoporoid reefs initially colonized the platform margins from the early Givetian, then rapidly evolved to stromatoporoid-microbial and microbial reefs in the early Frasnian, whilst microbial buildups predominated over the Famennian.

The Devonian marine transgression initiated from the Qingfang remnant trough in the southern Gunagxi and expanded progressively northwards in the South China (Fig. 2), in response to the basin-scale, intense basement faulting at depth (e.g., Wu Yi in Zhong et al. 1992; Chen et al. 2001a). During the Early Devonian, marine transgression only reached the central and western parts of Guangxi, including the Liujing and Dale areas. At that time, frequent sea level changes resulted in frequent progradations/retrogradations of siliciclastic shorelines, leading to the frequent lithologic changes that were represented by different lithologic units (formations/members) in stratigraphic columns. By the very late Early Devonian and early Middle Devonian, the sea gradually covered the northeastern part of Guangxi (including Guilin) and southern Hunan Province. It was not until the Eifelian or Givetian that the marine transgression largely expanded into the central Hunan (e.g., the Xikuangshan area; Fig. 2), represented by the Tiaomajian Formation (Eifelian or Givetian age uncertain, probably early Givetian in age).

According to Wu Yi (in Zhong et al. 1992, p 286–288), a short-term regression occurred by the late Eifelian and mid-Givetian, respectively, whereas transgression persisted from the late Givetian through Frasnian, although episodically. However, during the Middle and Late Devonian transitional period, it seems that water depth changed in different ways in different depositional settings. In some areas it became deeper as that is represented by the deposition of the Liujiang Formation and the Baqi Formation (see the Stratigraphy section below; Fig. 3); the former is characterized by bedded cherts, and the latter, by thin- to medium-bedded limestones intercalated with bedded cherts. In the platform settings, the strata are characterized by regressive carbonate successions (decreasing stacked cycles/cycle sets) or even missing (Chen et al. 2001a); those deposited proximal to terrigenous provenances, i.e., in central Hunan, the lower Frasnian is mainly composed of sandstones. Tectonic controls might have played an important role, as stressed by Chen et al. (2001a, b).

In the later Frasnian, there were two important deepening events: the *hassi* Zone transgression and the Lower *rhenana* Zone transgression, respectively corresponding to the Lower Rhinestreet Event, and the *semichatovae* transgression recognised world-wide, during which the ammonoid *Mesobeloceras*? fauna and *Manticoceras* fauna appeared, respectively in central Hunan Province (Ma et al. 2004).

During the Frasnian-Famenian transition, sea level dropped substantially in platform facies evidenced by occurrence of paleokarst in platform interior in Guilin city center (Chen and Tucker 2004), occurrence of large-scale sandstones proximal to the terriginous sources in the Leimingdong section of central Hunan, occurrence of thick dolomite successions in the Dushan section of Guizhou and Longmenshan section of Sichuan. In deeper water settings, increasing occurrences of gravity-flow deposits (calcidebrites/megabreccias and calciturbidites) on the slope-basin settings (Chen et al. 2001a, 2002a, b; Chen and Tucker 2003). However, the picture seems somehow complicated due to the uncertain in biostratigraphic division such that both slightly shallower and deeper trends have been reported (Wang and Ziegler 2002).

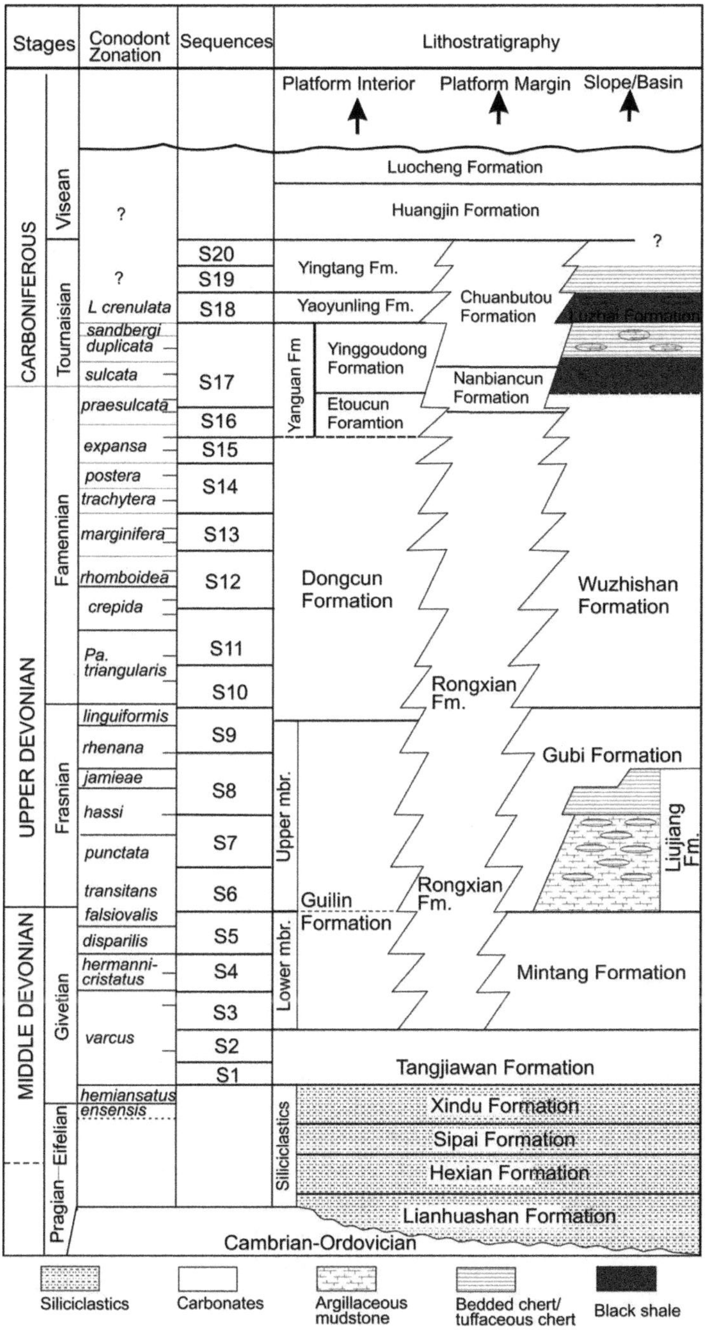

Fig. 3 Devonian-Lower Carboniferous stratigraphic successions and relationship between the platform and basin in Guilin area. Correlation with the international stages are derived from various sources (Zhong et al. 1992; GBGM, 1988, 1994; Shen 1995)

During the Famennian, the lithological succession is rather simple in a specific locality in Guangxi, but the sea level fluctuations can be generally identified through the cycle stacking patterns and are fairly comparable, notably the larger-scale sea level drop during the mid-late Famennian (corresponding to the conodont *marginifera* and *postera* zones) as evidenced by occurrence of paleokarst/paleosoil horizons in several platforms (Chen et al. 2016). Although the sea level generally dropped somewhat compared with the Frasnian, the basic paleogeographic setting was little changed. This setting persisted into the very early Carboniferous (Tournasian), but the platforms tended to be non-rimmed as ramps. The platform-basin differentiation further became waning in the Visean, so that carbonate ramps predominated over the shallow-water regimes that time.

3 Stratigraphy

Due to diverse paleogeographic and depositional settings of Devonian in South China and Guilin area as well (Fig. 2), there are abundant stratigraphic nomenclatures (formations and members) coined for the strata deposited in different settings, although some of which have only minor differences. Relatively, the platform-basin differentiation in the early Carboniferous (i.e., Visean) tended to be waning, so that the lateral variations in stratigraphy were also less variable.

In Guilin area, the stratigraphic units (nomenclatures) of Devonian and Lower Carboniferous, vertical successions, and lateral relationship between the platform and basin are illustrated in Fig. 3. The stratigraphic correlation of Devonian and Lower Carboniferous in the Guilin area (Fig. 3) is largely based on the conodont data (Yu 1988; Ji 1989a, b; GBGM 1994; Shen 1995).

The basal Tangjiawan Formation is the first carbonate horizon, which overlies the siliciclastic Xindu Formation; it is mainly made of stromatoporoid biostromal facies representing the initial stage of platform-basin evolution. Further upwards, the stratigraphic units vary in different depositional settings. In platform interior, the stratigraphic units, in ascending order, include Guilin, Dongcun and Etoucun formations, spanning from Givetian to Frasnian and Famennian in age. The overlying Lower Carboniferous strata include Yingoudong, Yaoyunling, Yingtang formations of Tournaisian Stage, and Huangjin and Luocheng formations of Visean. Of these, the Guilin Formation is characterized by cyclic stromatoporoid biostromes internating with microbial laminites (meter-scale cycles). The Dongcun Formation mainly comprises cyclic fenestral limestones bearing only euryohaline fauna (gastropod, ostracod, foraminifer and extensive microbes-cyanobacteria) and bioclastic wackestone bearing minor stromatoporoids and corals in the upper part. Etoucun Formation (or Member #1 of Yanguan Formation) is the uppermost horizon of Devonian and mainly composed of dark grey thin- to medium-bedded bioclastic mudstones/wackestones bearing extensive foraminifera (typically *Quasiendothyra*) and minor stromatoporoids and corals (*Cystophrentis* sp.) and brachiopods (*Cyrtospirifer* sp.).

The overlying Carboniferous starts with the **Yinggoudong Formation** (or Member #2 of Yanguan Formation which is characterized by the dark grey, platy to thin-bedded lenticular (ribbon) bioclastic wackestone/mudstone in the base, passing upward into medium- to thick bedded bioclastic wackestone/packstone in which foraminifers are abundant but generally megafauna absent. The top is commonly marked by an erosive surface covered by a phophatic-iron crust. The **Yaoyunling Formation** commences with grey to dark grey thin- to medium-bedded bioturbated argillaceous wackestone/mudstone (or marlstones), intercalated with crinoidal grainstones/packstones, grading upward into thicker-bedded crinoidal packstones/grainstones intercalated with cherty bands or shaley partings. Except for the widespread crinoids, faunal assemlages also include corals (*Pseudouralinia tangpakouensis*) and brachiopods (*Eoshoristites transversa*) and foraminifers. The **Yingtang Formation** is mainly composed of grey to dark grey, medium- to thick-bedded limestones and finely crystalline dolostones demarcated from the Yaoyunlin Formation by the occurrence of dolostone. Faunal assemblages include corals (*Pseudouralinia sp.*) and foraminifers. The overlying **Huangjin Formation** comprises three lithological members: The first member, demarcated from the underlying Yingtang dolostones, comprises grey to dark grey medium-bedded bioclastic-peloidal and foraminiferal wackestones intercalated with argillaceous mudstone and cherty bands in the middle part. Locally (Mopanshan, Guiju), fault-controlled dolostones occur in the top horizon. The second member mainly comprises dark grey medium-bedded bioclastic mudstone/wackestone intercalated thin-bedded marlstone, bioclastic mudstone and stratiform coral boundstone; locally thick-bedded dolostone and interbedded chert/limestone occur in the middle part (Guiju). The third member is only partially cropped out, comprises grey to dark grey medium- to thick-bedded lime mudstome, bioclastic wackestone/mudstone intercalated argillaceous mudstone and cherty bands/concretions, passing to thin-bedded laminated marlstones and mudstones intercalated bioclastic wakestones/packstones in deeper setting (Guiju). Abundant Visean fossil assemblages include foraminifers, corals, brachiopods and ostracods. The **Luocheng Formation** is only locally exposed on the peak of Mopanshan, and demarcated from the dolostone of the uppermost Huangjin Formation; it comprises grey to light grey medium- to thick-bedded bioclastic wackestones/packstones and mudstones, locally intercalated with cherty bands/concretions and argillaceous strips.

In platform margin setting, stratigraphic unit mainly comprises Rongxian Formation ranging from the upper Givetian through Famennian, which is overlain by Nanbiancun or Etoucun Formation as well (Fig. 3). It is characterized by ecological buildups and shoal deposits and generally can be divided into two members. The lower member comprises dark grey thick-bedded to massive grainstones/packstones composed of peloids, ooids, oncoids and microbial fragments or mixed components, and reefs/mounds of stromatoporoid-dominant boundstones (coral- and *Chaetetes*-stromatoporoid, *Smithiphyllum* bafflestone) and microbial boundstones (*receptaculitid*-microbial mounds). The upper member is distinguished from the lower member by lighter color (light grey mostly), few stromatoporoid and dominance of microbial components, and spans mostly the Famennian Stage. It generally comprises light grey thick-bedded to massive grainstones/packstones of peloids, ooids, oncoids and microbial fragments, notably mixture of multiple grains, and microbial reefs/mounds dominated by variable cyanobacterial elements, notably the *Epiphyton-Renalcis* groups. Vertically it can sit on either the lower member of Rongxian Formation or the Gubi Formation; laterally it commonly passes basinwards into the Wuzhishan Formation (nodular limestone). Further upwards is the Nanbiancun Formation proposed by Yu (1988), it is characterized by brachiopod shell-dominant limestones (coquina) of a couple of meters, although bearing other taxa of fossils, such as crinoid. This formation straddles across the Devonian/Carboniferous (D/C) boundary. The overlying **Chuanbutou Formation** is a laterally variable lithostratigraphic unit. Where it was deposited near the upper marginal slope zone, medium- to thick-bedded grainstones/packstones mixed with skeletal and nonskeletal grains predominate over, as observed at Nanbiancun section. Passing basinwards, thin- to medium-bedded wackestones/packstones predominated over the formation in which bedded cherts and black shales are intercalated or interfingered, It can be further subdivided into three members, which approximately correspond to the Yinggoudong, Yaoyunling and Yingtang formations in the platform interior. Similar to the platform interior, this unit was further overlain by Huangjin and Luocheng formations as well.

In marginal slope to basinal setting, the stratigraphic succession includes Mintang, Gubi, Liujiang and Wuzhishan formations (Fig. 3). The **Mintang Formation** is characterized by the alternation of laminated microbialites and deep-water carbonates (thin- to medium-bedded Tentaculitid-bearing mudstones), and ranges mainly within the upper Givetian. The **Liujiang Formation** is characterized by starved basinal deposits of siliceous rocks (bedded cherts) intercalated with tuffaceous and shaley interbeds, with minor tentaculitid argillaceous or cherty limestones in the base. The **Gubi Formation** (mainly Frasnian) is chiefly composed of gravity-flow deposits at the platform margin and pelagic limestones (e.g. nodular limestone) with minor gravity-flow intercalations near the basin centre. The **Wuzhishan Formation** represents distal slope and basin deposits dominated by pelagic nodular limestones with minor calciturbidites, mainly of Famennian age (Fig. 3). The overlying Carboniferous strata (Tournasian) is the **Luzhai Formation** which is dominated by basinal deposits composed of dark grey to black black shale, bedded chert and argillaceous limestones/marlstones, and can be further subdived into three members. The first member starts with black shales with muddy siltstone and silty shale interbeds, grading upwards into thin-bedded cherts intercalated with siliceous shales, marlstones and cherty limestones in the mid-upper part. The second member commences with black shales with marlstone intercalations in the base, grading into black shales with laminated limestone concretions in the middle part, and dark grey to black thin-bedded lime mudstones intercalated with marlstones and bioclastic mudstones/wackestones. The third member is only partially exposed, and dominated by thin-bedded cherts intercalated with siliceous shale.

4 Itinerary Arrangement

South China is the most important region for the Devonian researches in China on various aspects, including paleontology, biostratigaphy, lithostratigraphy, sedimentology, and Guangxi is one of the most important areas for the study of the Devonian geology in South China (Guangxi Bureau of Geology and Mineral Resources, 1987), where many typical marine Devonian sections deposited in different facies domains were successively cropped out.

Guilin is an area characterized by diversified (mainly Devonian) depositional facies, including restricted carbonate platform interior, open platform-platform margin, and marginal slope-basinal facies (Figs. 2 and 4) in which the Late Devonian mass extinctions events, notably the F-F and D-C biocrisis events were well preserved in different depositional facies. Therefore, Guilin is one of the highly potential areas suitable for neritic-pelagic stratigraphic correlation and reconstruction for the paleoenvironmental and paleoclimatic changes that may have caused these critical biotic crises in the Phanerozoic. Moreover, rapid, even sharp facies changes in space and time provide a good opportunity to examine the critical controls on platform development and evolution on aspects of tectonic activities and eustatic fluctuations.

Additionally, fault-related dolostones present in the Givetian and Lower Carboniferous carbonates (originally limestones) provides useful examples to understand the fault-controlled hydrothermal dolomitization and mineralization, which could be very common in the deep-buried (or even deep-buried) carbonates interesting for the hydrocarbon reservoir sedimentologists and strata-bound ore geologists.

Lastly, from the angle of sedimentologists, we also have the chance to examine how the primary (Devonian) depositional facies had controlled the development/distribution of the spectacular karstic landscapes (karstic tower hill/pillar forests; See the front cover photo) gaining the reputation of the "peerless Guilin's Sceneries" which are dominantly distributed along two platform facies-influenced scenery corridors of Li River from Mopanshan wharf to Xingping town and the western flank of Guilin-Yangsuo Highway, respectively (Fig. 4). The visiting spots (stops) of this field excursion are mostly arranged along the two facies-associated scenery corridors wit minors in basinal domains between them in three days (Fig. 4).

It is our hope that, after this field excursion, you will not only gain some basic information/knowledge of typical Chinese Devonian strata on aspects of stratigraphy, sedimentology, event sedimentation, dolomitization, but also enjoy the spectacular karstic sceneries around Guilin. Participants themselves will arrange to arrive and leave Guilin where are easily got accessible through domestic and international air flights and high-speed trains (Fig. 5).

The Day of Arrival

Participants will arrange by themselves to arrive Guilin through either air flight or Express train (Fig. 5) and to visit any localities of interesting in or around the city.

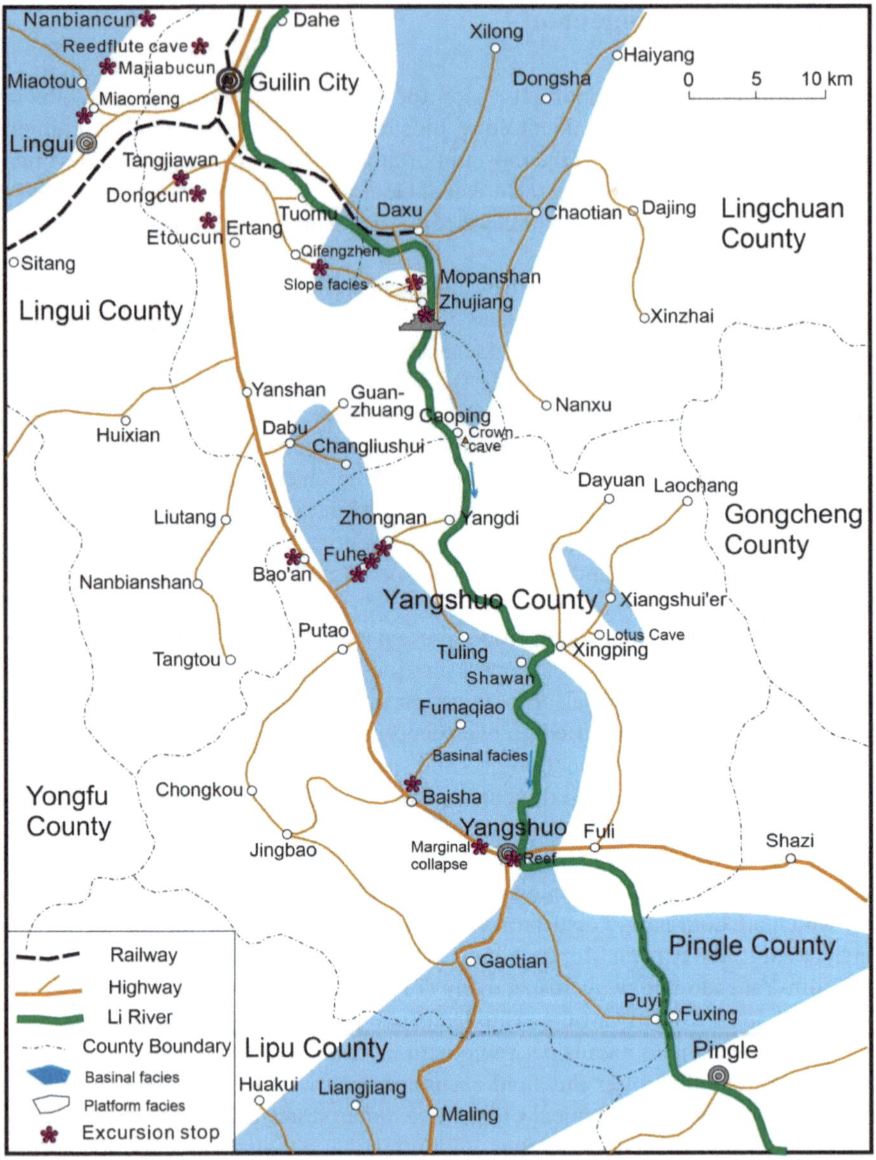

Fig. 4 Map showing facies distribution/pattern of the Late Devonian around Guilin-Yangshuo area, and locations of visiting stops

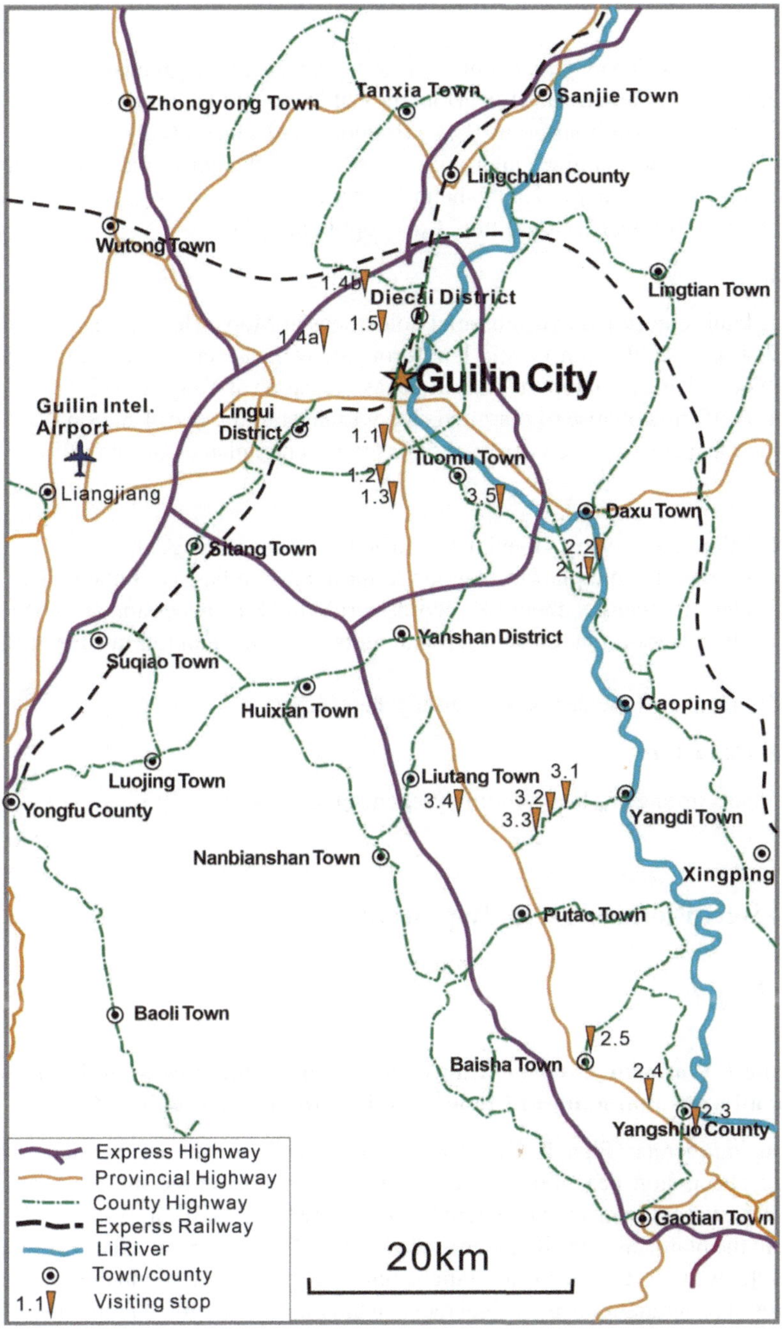

Fig. 5 Map showing transportation networks around Guilin and locations of excursion stops (1.1–3.5)

Day 1

Stop 1: Givetian-Frasnian platform interior deposits near Tangjiawan (1.1);
Stop 2: Frasnian-Famennian (F/F) Boundary at Detergent Factory (1.2);
Stop 3: Devonian-Carboniferous (D/C) Boundary at Etoucun (1.3)
Stop4: Famennian marginal microbial reefs at Majiabucun (1.4a) and Devonian-Carboniferous boundary at Nanbiancun (1.4b)
Stop 5: Cave carbonates (Reed-flute Cave, optional; 1.5)

Day 2

Stop 1: Fault-controlled hydrothermal dolostones at Mopanshan (2.1);
Stop 2: Facies control on karstic landforms along Li River (Boat excursion; 2.2);
Stop 3: Givetian stromatoporoid reef to Frasnian slope at Yangshuo (2.3);
Stop 4: Platform collapse of marginal slope, Liangshuijing petrol station (2.4); Stop 5: Slope-basinal succession across the Frasnian-Famennian at Baisha (2.5);

Day 3

Stop 1: Microbial reefs of Givetian to Frasnian at Zhongnan (3.1);
Stop 2: Evolution from platform lagoon to intraplatform basin at Fuhe (3.2);
Stop 3: Depositional-geochemical records across the F/F succession at Fuhe (3.3);
Stop 4: Stromatoporoid reefs to microbial reefs from Givetian to Frasnian at Bao'an (3.4)
Stop 5: Frasnian slope deposits at Qifengzhen (3.5)

Day of Departure

Participants arrange to leave Guilin by themselves.

5 Descriptions of Field Trip Stops

5.1 Day 1

Devonian Platform Interior Succession Near Tangjiawan and Devonian-Carboniferous Boundary at Etoucun and Nanbiancun, Guilin City

In today's itinerary (Figs. 4–6), we will firstly visit the Givetian-Frasnian platform interior succession near Tangjiawan (Stop 1.1), then examine the platformal F/F boundary succession at the Detergent Factory and Shengwanli (Stop 1.2), and further move to the plaformal D/C boundary succession at Etoucun (Stop 1.3). On the way to Stop 1.4b, we will stop shortly at Majiabucun to see the geometry and internal fabrics/textures of progradational microbial reefs on the western margin of Famennian Guilin platform (1.4a), and finally stop at Nanbiancun to look the D/C boundary succession on the marginal slope setting (Stop 1.4b). After finishing these visits, an alternative

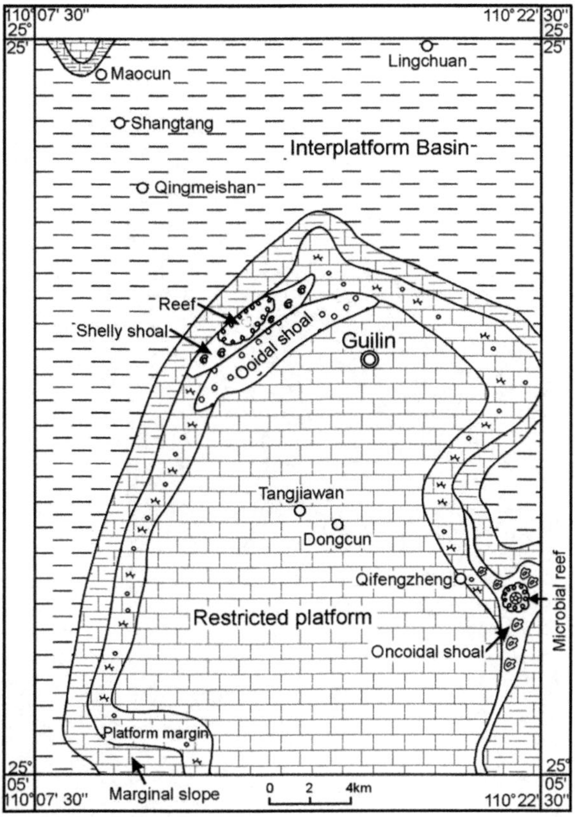

Fig. 6 Palaeogeographic setting of the northern portion of the Guilin Platform during the early Late Devonian

option to visit the Reed-flute Cave is tentatively planned on the way back (Stop 1.5), if time is available (Figs. 4, 5).

These localities of outcrops are located near the downtown of Guilin City and make them easily accessible by coach or bus.

5.1.1 Stop 1

Platform Interior Deposits from Givetian to Frasnian at Tangjiawan

The outcrop starts from Tangjiawan village and extends eastwards along the hill (see Figs. 4, 5 for location). The first carbonate horizon that overlies the siliciclastic succession (Xindu Foration) is named as the Tangjiawan Formation; it is mainly composed of thick stromatoporoid biostromal facies (Fig. 7a, b), grading upwards into microbial laminites, thus constituting very thick shallowing-upward cycles (up to 20 m thick) (Fig. 7c; Chen et al. 2001b). This stratal unit was locally strongly dolomitized, in which some of biotic skeletons were dissolved as mouldic vugs and plugged by saddle dolomite cements (Fig. 7d) (Chen et al. 2004), however, the

Fig. 7 **a** Very thick to massive biostromal facies composed of stromatoporoid boundstone, overlying the microbial laminite (standing person (ME Tucker) for scale). **b** Close-up view for the stromatoporoid rudstone/bafflestone dominated by tabular and hemispheric stromatoporoid members. Hammer head for scale (15 cm wide). **c** Shallowing-upward cycle (arrow) commencing with tabular to hemispheric stromatoporid bafflestone/wackestone, passing upwards into the microbial laminite. Hammer for scale (30 cm long). **d** Strongly dolomitized stromatoporoid wackestone/bafflestone, mouldic skeletal vugs plugged by saddle dolomite and later megacalcite crystals (arrows). Mark pen for scale (14 cm)

dolomitization was generally inhomogeneous, more intensive along fracture zones, so it was likely induced by structurally-controlled hydrothermal activity.

The overlying Guilin Formation is characterized by more extensive cyclicity and less dolomitization. Extensive meter-scale shallowing-upward cycles commonly start with stromatoporid bafflestones or wackestones, grading upwards into branching stromatoporoid-dominated facies capped by microbial laminites (also see Stop

2). Several karstic surfaces characterized by cinder-like rubble horizons with argillaceous seams occur in the upper part of this formation (Chen et al. 2001b).

5.1.2 Stop 2

Frasnian-Famennian (F/F) Boundary at Detergent Factory (Near Dongcun)

This outcrop is well cropped out at the Detergent Factory near Dongcun, about 6 km SW away from the downtown Guilin. Palaeogeographically, this locality was located in the platform interior of Guilin Platform (Figs. 2, 4, 6).

At this locality, we will examine the sharp variations in lithofacies and biofacies across the F/F boundary succession, across which two lithostratigraphic units are present: Guilin and Dongcun formations (Fig. 8). The Guilin Formation is characterized by cyclic subtidal-dominant facies associations containing abundant fossils, i.e., spherical to massive stromatoporoids, and branching stromatoporoids (*Amphipora* or *Paramphipora*) (Fig. 9). The Dongcun Formation is characterized by intertidal-supratidal cyclic fenestral limestone-dominant facies associations (Figs. 8 and 10), in which normal marine benthic fauna decline sharply with extremely rare branching *Amphipora* in the base (~10 m thick). By way of contrast, gastropods and ostracods are present continously across the boundary and further upwards. These fenestral limestones mostly consist of variable microbial members and non-radiospheric calcispheres, probably a kind of phytoplanktons in the surface water, and algal fragments (Chen and Tucker 2003; Chen et al. 2016) (Figs. 8 and 10). Although the boundary between Guilin and Dongcun formations is very close to the F/F boundary, it cannot be definitely determined by the conodont data. With this simple stratigraphic framework, detailed high-resolution cyclostratigraphic analysis were carried out and correlated with the deep-water counterparts (Chen and Tucker, 2003).

Cyclostratigraphic analysis (vertical stacking patterns) on the F/F boundary successions revealed two third-order depositional sequences (SFr, SFa), in which numerous generally meter-scale, upward-shallowing cycles are extensive. Apparent decrease in accommodation space (or relative sea-level) is unraveled in the base of Dongcun Formation, in parallel with the massive declining of benthic fauna (Fig. 10). Note the last occurrence of stromatoporoids likely represents the F/F boundary in the shallow-water carbonate succession (Figs. 8, 10 and 11). Another small-scale cycle above marks the upper sequence boundary of Fa. The subsequent three thickening cycles (with thick bioturbated lime mudstone bases) in the base of SFa (Figs. 10 and 11) represent an apparent deepening during the transgression of sequence SFa; these deepening cycles can be correlated well with their equivalents in basinal successions (Wuzhishan Formation) one by one.

It is a little pity that, due to the construction of city expansion, the Detergent Factory F/F boundary succession is not accessible now. An alternative section at Shenwanli Quarry (about 200–300 m south of the backhill) is available for visiting where the Guilin Formation is more visible by excavation (Fig. 12a) and the F/F boundary can be observed by climbing along the hillslope (Fig. 12b). At this locality,

Fig. 8 **a** Panoramic view of Frasnian/Famennian boundary in the lower Dongcun Formation (above the Dongcun/Guilin Formation boundary) in the platform succession at the Detergent Factory of Guilin. **b** Close-up view for the Frasnian/Famennian boundary (see the box in A for location), the top surface of the mottled mudstone showing the last appearance of stromatoporoid **c** likely represents the Frasnian/Famennian boundary. Vertical yellow arrows mark the meter-scale upward-shallowing cycles

except for the typical lithofacies of Guilin Formation such as the *Amiphipora* floatstone, bioturbated (mottled) mudstone, an apparent karstic surface that marks the sequence boundary of sequence SFr was excavated out in the quarry (Fig. 12c); it was characterized by cinder-like rubble horizon developed on the top of subtidal *Amphipora* floatstone/wackestone, and paved with greenish claystone (palaeosol) and dark grey shale/mudrock (Fig. 12c, d; Chen et al. 2001a). This surface seemingly was caused by the tectonic-induced abrupt uplift, rather than by a progressive shallowing, which was temporally linked to the abrupt deepening in the basin (Chen et al. 2001a, b).

Fig. 9 Typical lithofacies and faunal assemblage in Guilin Formation in the upper Guilin Formation. **a** Stromatoporoid rudstone/floatstone bearing overturned bulbous stromatoporoid and broken *Amphipora*. **b** Spaghetti-like *Amphipora* bafflestones/floatstone

5.1.3 Stop 3

Devonian-Carboniferous (D/C) Boundary at Etoucun

This outcrop is located at Etoucun-Luojiacun, southern Guilin, about 1 km southeast of the F/F boundary at the Detergent Factory (or Shengwanli). Similarly, this locality was also located in the platform interior (Figs. 1, 4, 5 and 13).

At this locality, we will examine the lithofacies and faunal changes across the D/C boundary in platform successions (Figs. 14 and 15). The Etoucun Formation (or No. 1 Member of Yanguan Fm.) is demarcated from the Dongcun Formation by the topmost fenestral limestones, and mainly comprise dark grey medium-bedded to massive bedded bioclastic-peloidal wackestones/mudstones bearing abundant foraminifers with minor stromatoporoids and brachiopods (Fig. 16c, d). The basal Carboniferous succession (Yingoudong Fm.) starts with dark grey lenticular (or boudinage-like) limestones, grading into medium- to thick-bedded bioclastic wackestones bearing abundant foraminifers and ostracods (Fig. 16e, f).

The D-C boundary defined by the foramineral fauna assemblages (Fig. 15) are generally consistent with that constrained with conodont data. More precise placement of the D/C boundary can be made based on the positive carbon isotope excursion climax to near the top of first upward-shallowing cycle in the base of Yinggoudong Formation correlated with that of para-stratotype GSSP section for the D-C boundary at Nanbiancun (Fig. 17).

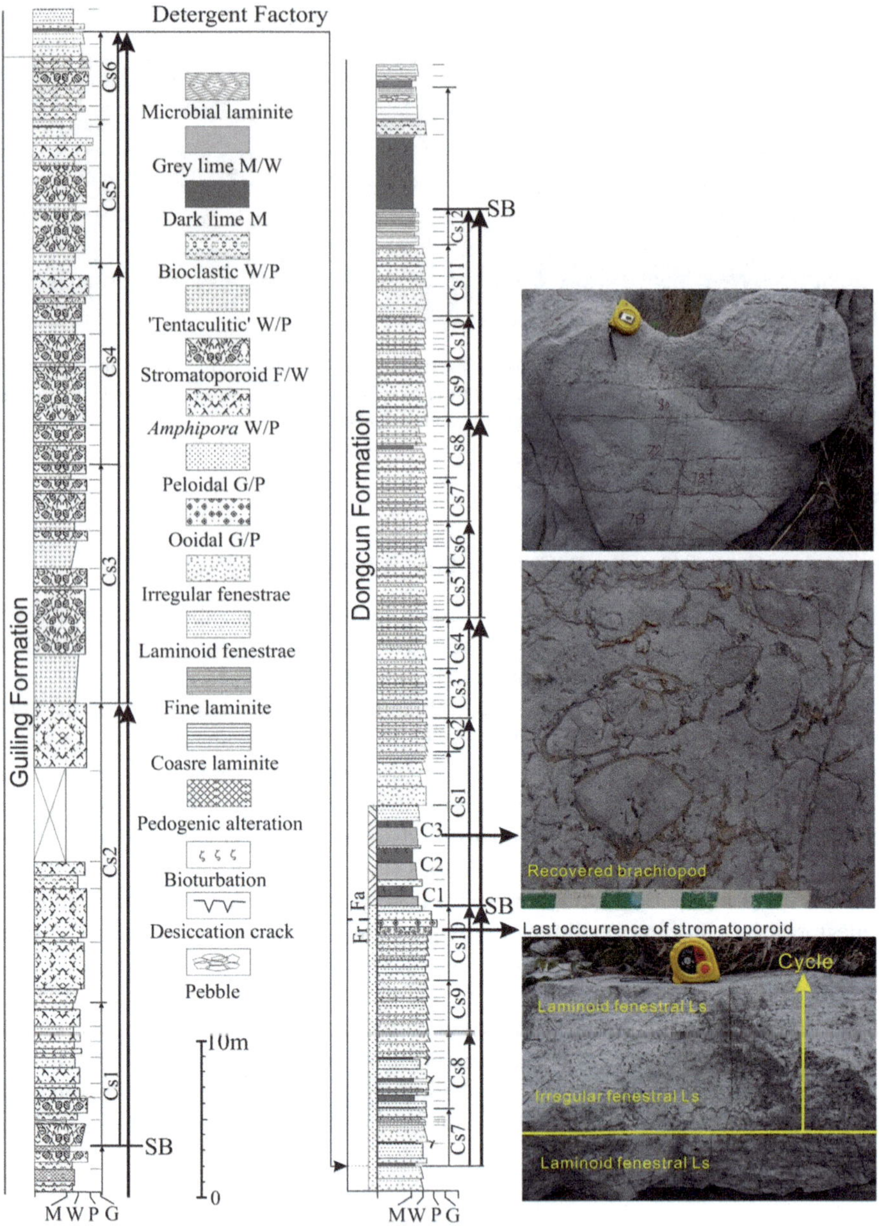
Detergent Factory
Microbial laminite
Grey lime M/W
Dark lime M
Bioclastic W/P
'Tentaculitic' W/P
Stromatoporoid F/W
Amphipora W/P
Peloidal G/P
Ooidal G/P
Irregular fenestrae
Laminoid fenestrae
Fine laminite
Coasre laminite
Pedogenic alteration
Bioturbation
Desiccation crack
Pebble
10m
0
Guiling Formation
Dongcun Formation
Cs1
Cs2
Cs3
Cs4
Cs5
Cs6
Cs7
Cs8
Cs9
Cs10
Cs11
Cs12
C1
C2
C3
SB
Fa
Fr
M W P G
Recovered brachiopod
Last occurrence of stromatoporoid
Cycle
Laminoid fenestral Ls
Irregular fenestral Ls
Laminoid fenestral Ls

◀**Fig. 10** High-resolution cyclostratigraphy in the platform succession across the F/F boundary. Note the internal facies variations within cycles and cycle stacking patterns. Short horizontal lines on the right of the logs mark bounding surfaces of dm- to m-scale shallowing-upward cycles. Vertical thin, medium and thick line arrows define cycle sets (Cs), mesocycle sets and megacycle sets, respectively (modified after Chen and Tucker 2003). Compared with previous placement of the F/F boundary on the top of first apparent deepening cycle (C1) (Chen and Tucker 2003), the F/F boundary is newly placed on the top of the oolite/mottled wackestone bearing the last occurrence of the stromatoporoid in the lower part of Dongcun Formation (see Fig. 7b, c) in view of the similar cycle patterns in basinal successions constrained by new conodont zonation (Chen et al. 2013; Huang et al. 2018). The photos on the right side of the log are typical lithofacies around the F/F boundary which are characterized by cyclic (from irregular to laminoid) fenestral limestones. The earliest recovered Rhynchonellid brachiopods (i.e., *Yuananellina* sp.) appear in the third deepening (thickening) cycle (C3) above the F/F boundary

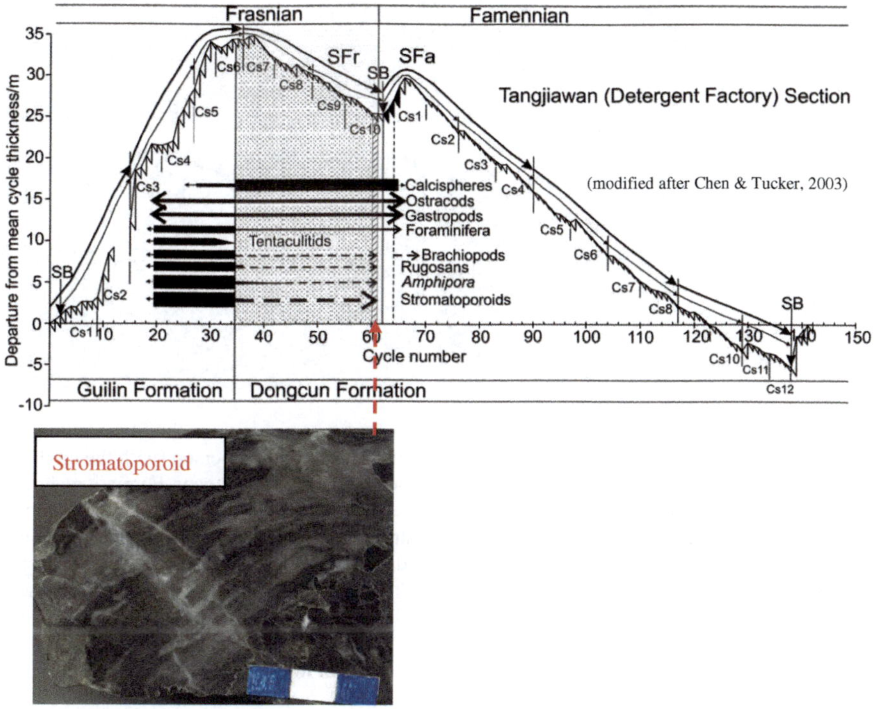

Fig. 11 Fischer-plots showing cycle stacking patterns, long-term changes in accommodation space (or sea-level changes) and biotic declining patterns patterns across the F/F boundary at Tangjiawan. Thin line arrows and thick arrows define the ranges of mesocyle sets and megacycle sets, respectively. Dotted interval represents the horizon of massive decline of normal marine fauna coincident with a significant decrease in accommodation space; slashed bar represents the highest occurrence of marine benthos (stromatoporoid) at this locality above which is the F/F boundary suggested. Black triangles below the curve represent three thick deepening cycles corresponding to the commencement of the Wuzhishan Formation of basinal successions (modified after Chen and Tucker 2003)

Fig. 12 **a** Panoramic view of Frasnian/Famennian boundary succession at the Shenwanli quarry on the backhill of the Detergent Factory section southwards. Standing person (with the circle) for scale (1.7 m). **b** Frasnian/Famennian boundary in the lower part of Dongcun Formation showing the identical facies succession at Detergent Facotory section. Thick-bedded stromatoporoid-bearing bioturbated mudstone marks the last occurrence of stromatoporoid (hammer posited). See the inset box in A for location. **c** Irregular karstic surface (dashed line) overlain with greenish claystone (paleosol) (arrows) in the upper Guilin Formation. Standing person for scale (1.75 m). **d** Close-up view on the karstified horizon characterized by cinder-like rubble horizon with fractures/fissures filled with paleosol (yellow arrow). Original rock is *Amphipora* wackestone/rudstone with minor hemispheric stromatoporoids (blue arrows). See the inset box in C. Hammer is 28 cm long

5.1.4 Stop 4

Famennian Marginal Microbial Reefs at Majiabucun and Devonian-Carboniferous Boundary at Nanbiancun

On the way to the Nanbiancun D/C boundary section (Figs. 4, 5), we will have a short stop at Majiabucun (or Xingcheng) to visit the geometry and internal fabrics/ textures of progradation microbial reefs of Famennian on the western margin of Guilin platform (Fig. 18). At this locality, at least four prograding-westward microbial buildups can be seen on the cliff of the hill in which collapsed megabreccia body

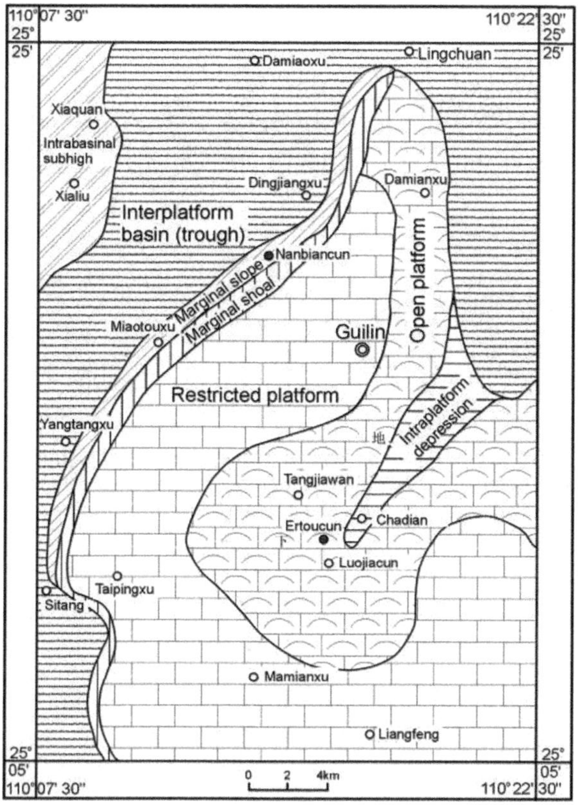

Fig. 13 Palaeogeographic setting of the northern portion of Guilin Platform during the Early Carboniferous (modified from GBGMR, 1988)

was developed in the foreslope of one reef. On the other side of the main road (Zhongying Road), we will check the internal fabrics/textures of microbial buildups common in the Famennian time, e.g., built up typically by cyanobacterial assemblage of *Epiphyton* and *Renalcis* (Shen and Webb 2004).

Afterwards, we will go ahead for the Nanbiancun D/C section. This outcrop is located at Nanbiancun, about 4 km northwest of Guilin city, and now is protected as a Geopark (Fig. 19). Palaeogeographically, it was located at the marginal slope of the Guilin Platform during the D/C transition (Fig. 13). Integrated studies of lithostratigraphy, biostratigraphy, chronostratigraphy, lithofacies, geochemistry and magnetostratigraphy have been conducted on the D/C boundary (Yu et al., 1988), which was approved as the para-stratotype section candidate of the D/C boundary (Figs. 20 and 21).

At this locality, we will examine the litho- and biostratigraphic variations across the D/C boundary, which is placed along the boundary between beds 55 and 56 in the top of Rongxian Formation (Fig. 22a–c), based on the evolution of conodont

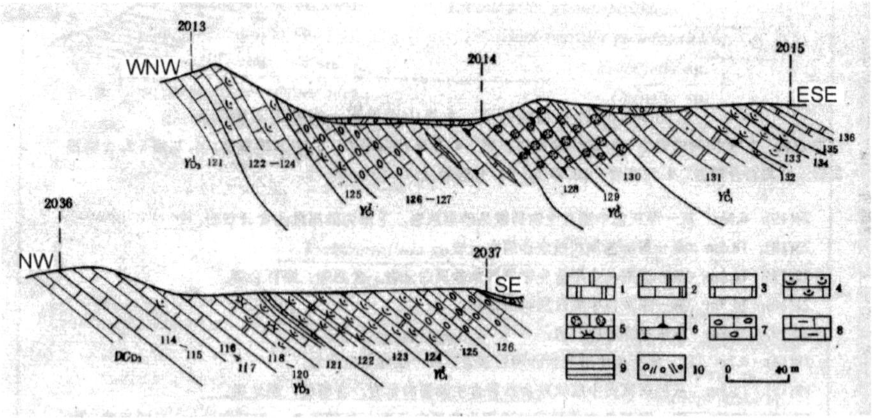

Fig. 14 Outcrop cross-section showing general features across D/C boundary at Ertoucun-Luojiacun villages. 1-dolomite, 2-dolomitic limestone, 3-limestone, 4-bioclastic limestone, 5-crinoidal limestone, algal limestone, 7-nodular limestone, 8-argillaceous limestone, 9-shale, 10-regolith (after GBGMR 1988)

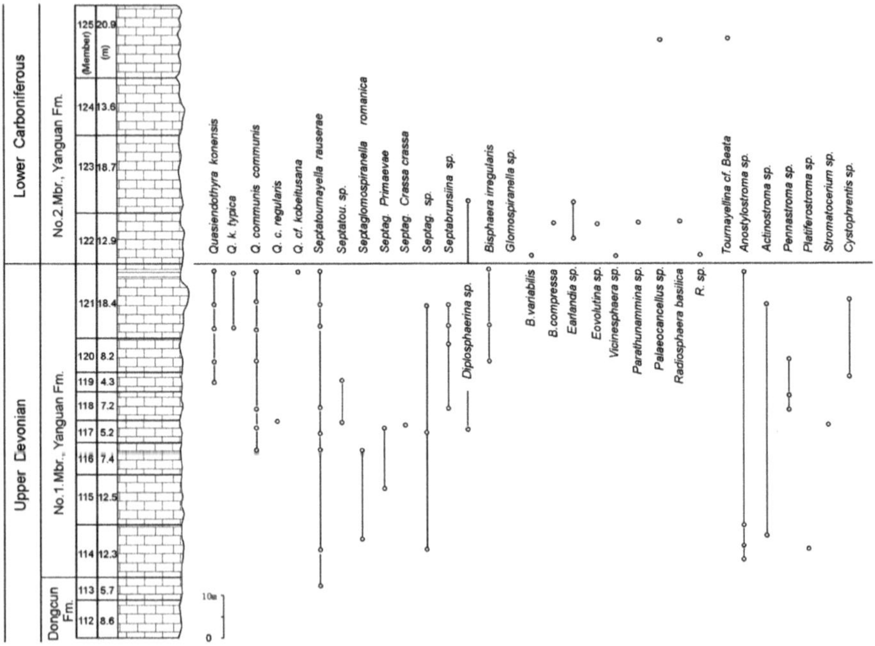

Fig. 15 Main faunal (foraminera) variations across the D/C boundary in the platform succession at Etoucun (after GBGMR 1988)

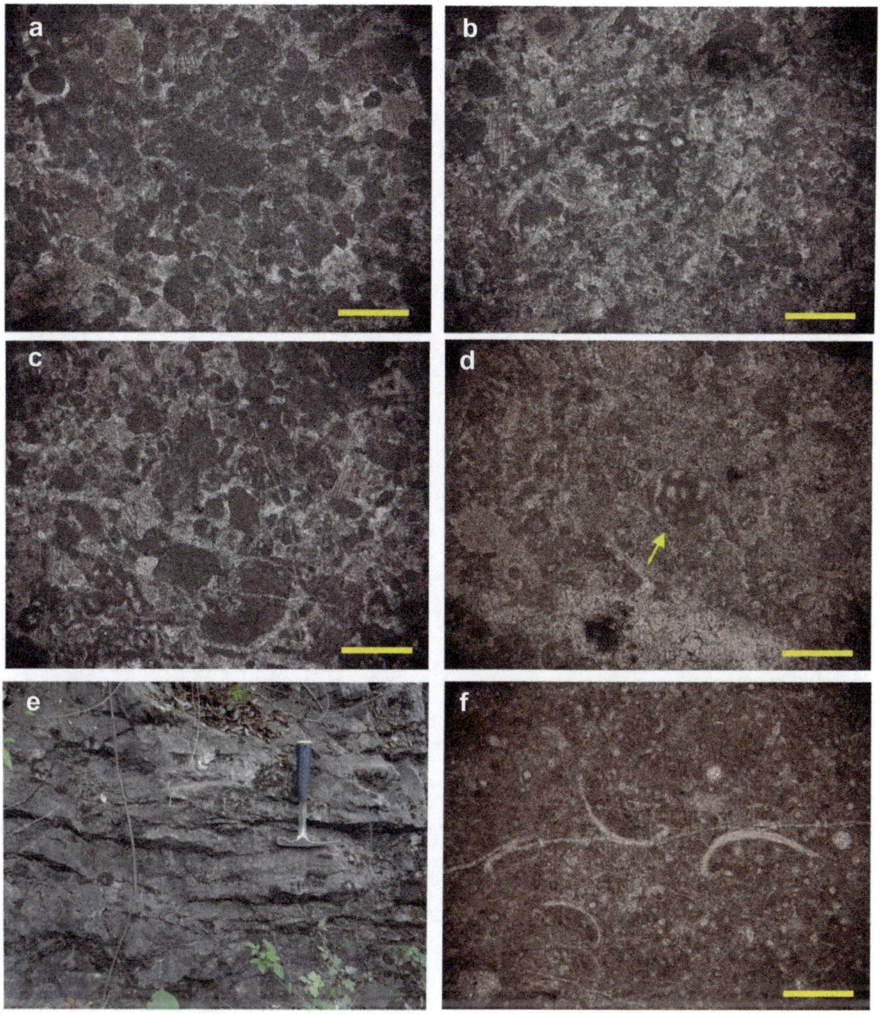

Fig. 16 Lithological features across the D-C boundary succession at Etoucun. All photomicrophs were taken under plane polarized light. Scale bar = 0.5 mm. Hammer = 28 cm. **a**, **b** Photomicrographs for the microfacies of uppermost Dongcun Formation, composed of foraminiferal-peloidal packstone **a**, and peloidal-foramineral wackestone/mudstone **b**. **c**, **d** Main microfacies in the Etoucun Formation., including foramineral-peloidal packstone/grainstone **c** and foraminiferal wackestone **d**. Note the typical foraminifer *Quasiendothyra* sp. (arrow) in the Etoucun Formation. Paired panels for macrofacies of lenticular limestone **e** and microfacies ostracod-calcispheric mudstone **f** in the base of Yinggoutong Formation

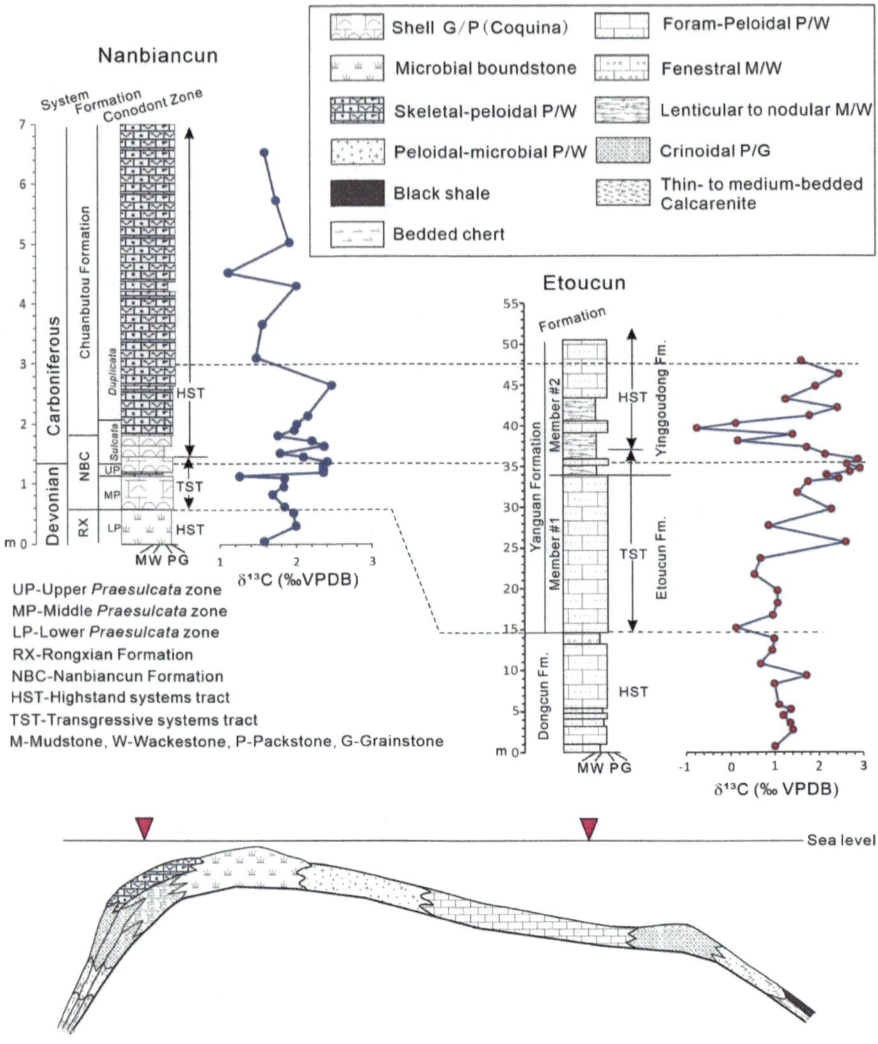

Fig. 17 Depositional facies successions and ^{13}Ccarb variations in different depositional environments across the asymmetric Guilin platform geometry (relative steep slope on western margin vs. mild slope on the eastern flank) during the Devonian/Carboniferous (D/C) transition at Etoucun and Nanbiancun, respectively (see Fig. 12 for the depositional setting and location). The D/C boundary coincides with the climax of positive ^{13}Ccarb excursion slightly above the Yinggoudong and Nanbiancun formations

Siphonodella praesulcata—*S. sulcata* community (Figs. 22 and 23). An apparent event could have occurred between beds 48 and 50, expressed by a serrated surface on the top of Bed 48 overlain by a thin layer (Bed 49) of black shale (0.4 cm) or directly by the coquina limestone (Bed 50) (Fig. 21a, d), above which coquina limestones are intercalated with several thin layers of dark grey shale or argillaceous strips which are

Fig. 18 Panoramic view on the geometry of progradational microbial reefs of Famennian on the western margin of Guilin platform at Majiabucun (entrance to the Xingcheng Community)

Fig. 19 A monument erected in the Geopark of Nanbiancun Para-stratotype section for the D/C GSSP, Linchuan District, Guilin

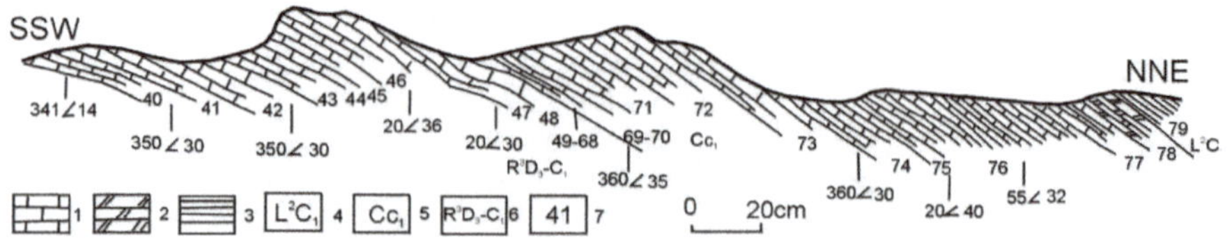

Fig. 20 Outcrop cross-section of D/C boundary at Nanbiancun. 1-limestone, 2-siliceous rock, 3-shale, 4-No. 2 Member of Luzhai Formation, 5-Chuanbutou Formation, 6-No. 3 Member of Rongxian Formation, 7-Bed number (after GBGMR 1988)

Fig. 21 Panoramic view of D/C boundary succession at Nanbiancun (Village on the south). Note the lateral variations in strata forms and facies

generally fitted with bean-shaped limestone bands (Fig. 23). This coquina limestone-dominant succession (beds 49—67) was named as the Nanbiancun Formation by some researchers (Fig. 17, 21; Yu et al. 1988).

Carbon isotopic variations across the conondont-constrained D/C boundary succession at Nanbiancun show that there is an overall positive excursion ranging from the upper *praesulcata* to the lower *duplicata zones*, which are further composed three sub-spikes. The D/C boundary (between beds 55 and 56) precisely coincides with the climax of the basal carbon isotope shift which starts simultaneously with the rapid transgression (sea-level rise) (Fig. 17).

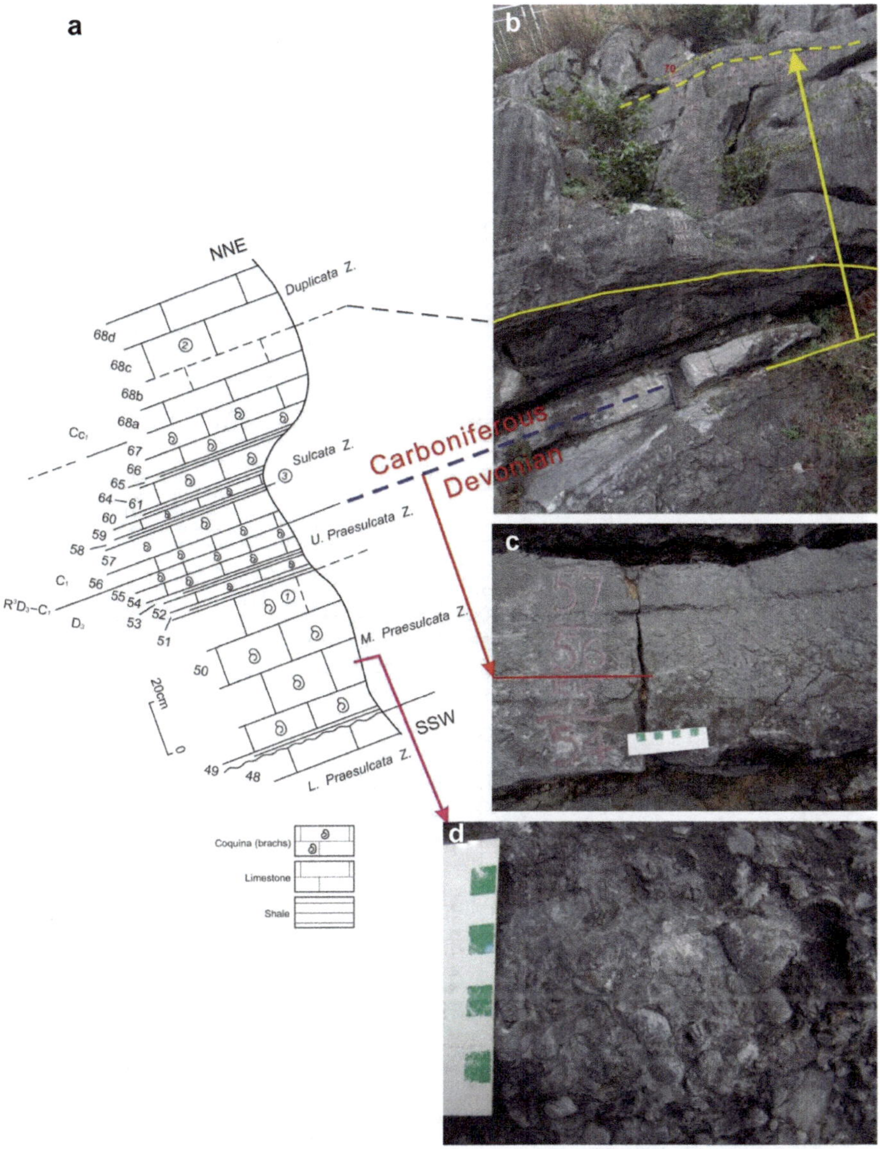

Fig. 22 **a** Detailed lithofacies (coquina-dominant) successions across the D/C boundary at Nanbiancun (after GBGMR). **b** Overall view of the D/C boundary succession. The Up-dip arrow marks the upward-shallowing cycle. **c** Close-up view of the D/C boundary with the shelly beds in which the brachiopods are mostly disarticulated. **d** Close-up view of the coquina (Bed 50) in which the rhynchonellid brachiopods are mostly articulated

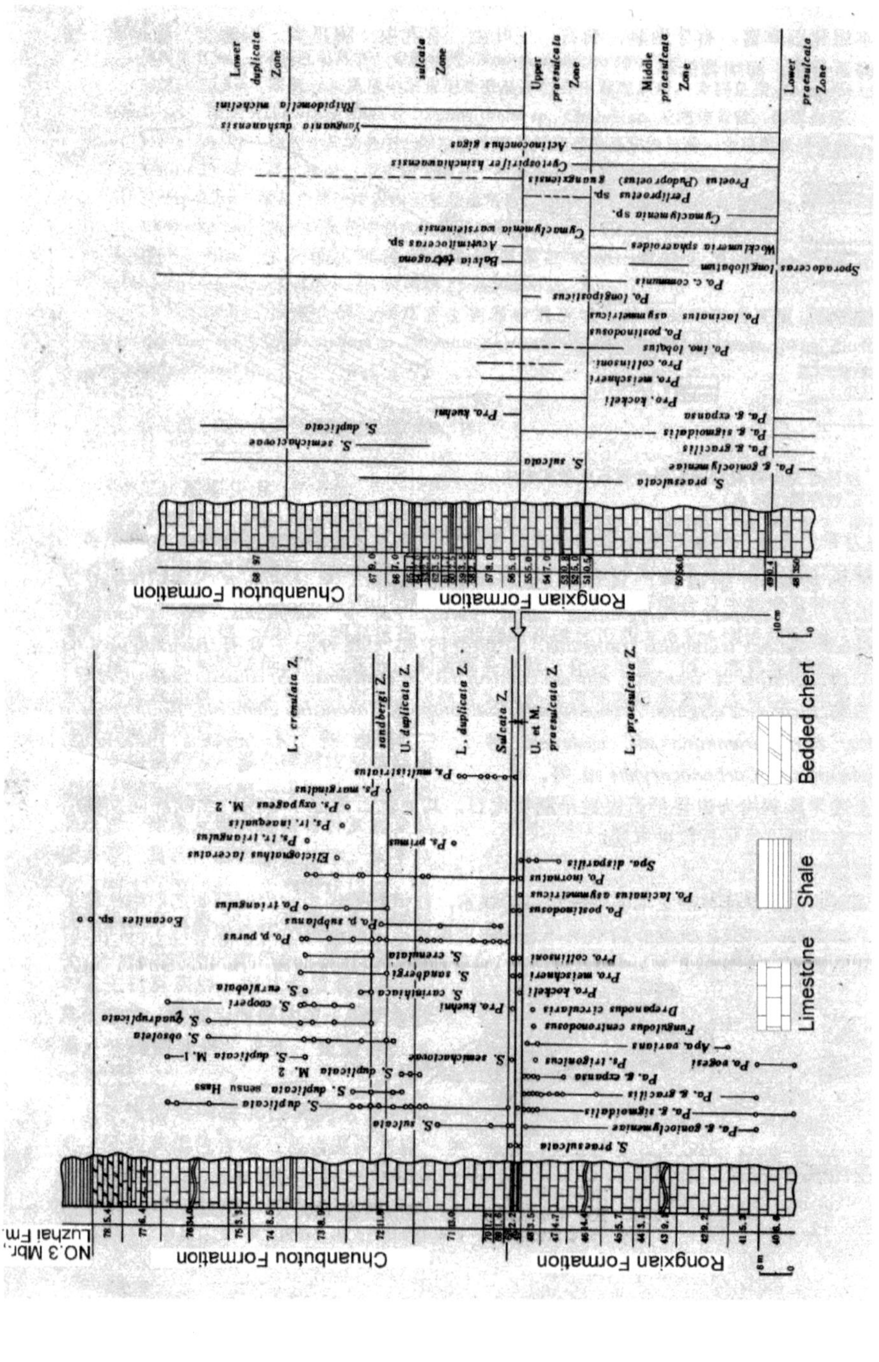

Fig. 23 Conodont biostratigraphy across the D/C boundary at Nanbiancun (Yu 1988; GBGMR 1988)

On aspect of sedimentary facies, near the top of Rongxian Formation (Bed 48), the carbonate rocks mainly comprise microbial packstones/wackestones (Fig. 24a) and microbial boundstones in which traces of Epiphyton-Renalcis assemblages still can be identified (Fig. 24b), indicating they were deposits of microbial buildup (reef). In the overlying articulate coquina bed (Bed 50), the matrix or framework was still made up by the microbial elements (boundstones) (Fig. 24c, d), so these articulated brachiopods (*Rhychonellids*) could be the fauna attached onto the microbial buildups colonizing the platform margin (Fig. 17), gnawing the microbes as the foods. This ecological behavior of *Rhychonellids* was also reported in the Middle Devonian microbial buildups (Chen et al. 2002b, a). Farther upwards in the thinner-bedded shelly beds of Nanbiancun Formation, coquinas dominantly comprise disarticulated brachiopod shell fragments (Figs. 22c, 24e) showing bed-parallel fabrics, indicating transportation from other places, the i.e., rhychonellid-dwelling forereef of microbial reefs on the platform margin and deposition on the slope. In the overlying Chuanbutou Formation, thick-bedded to massive peloidal-microbial packstones/wackestones with subordinate skeletal fragments (crinoid, brachiopod, ostracod) predominate over (Fig. 24f), likely representing the deposits of foreslope of the microbial buildups (Figs. 17 and 21).

Moreover, down to the toe of the hill along the bank of creek (<500 m in distance), we can see a rapid lithofacies change to the deeper criquina-dominated succession across the D/C boundary (Fig. 25a, b). Above the massive crinoidal grainstone interval, deposits tend to be cyclic (I-III). In cycle I, crinoidal shoal was colonized by microbial mound (MM) (Fig. 25a, c), implying shallowing or environmental deterioration. The subsequent stratal onlap and thinning beds of cycle II indicate the marine transgression. The occurrence of thinner platy to lenticular ostracod wackestone (Fig. 25a, d) in the base of cycle III suggests an accelerating deepening process, the subsequent demise of crinoidal fauna and colonization of microbialite (Fig. 25a, e) probably represent the biotic crisis across the D/C transition.

Farther basinward to west (~500 m), basinal black shales (Member #3 of Chuanbutou Formation) was excavated (Figs. 23, 26). Farthest to west about 1 km, bedded cherts and black shales comprise almost all of this formation (Li 1992).

If looking southwards, a magnificent contrast in landform comes into the eyes: characteristic karstic towers/hills developed on the previous carbonate platform on the left (east) and very mild and flat-lying plains developed on the slope-basinal setting on the right (west), exemplifying the primary depositional facies controls on the later geomorphologic development and evolution.

5.1.5 Stop 5 (Optional)

Cave (Speleothem) Carbonates

On the way back from the Stop 4, we will visit the Reed-flute Cave, a typical karstic cave system, also a very popular scenic interest in Guilin area, where we can see the spectacular stalagmites and stalactites whose diversified shapes had inspired

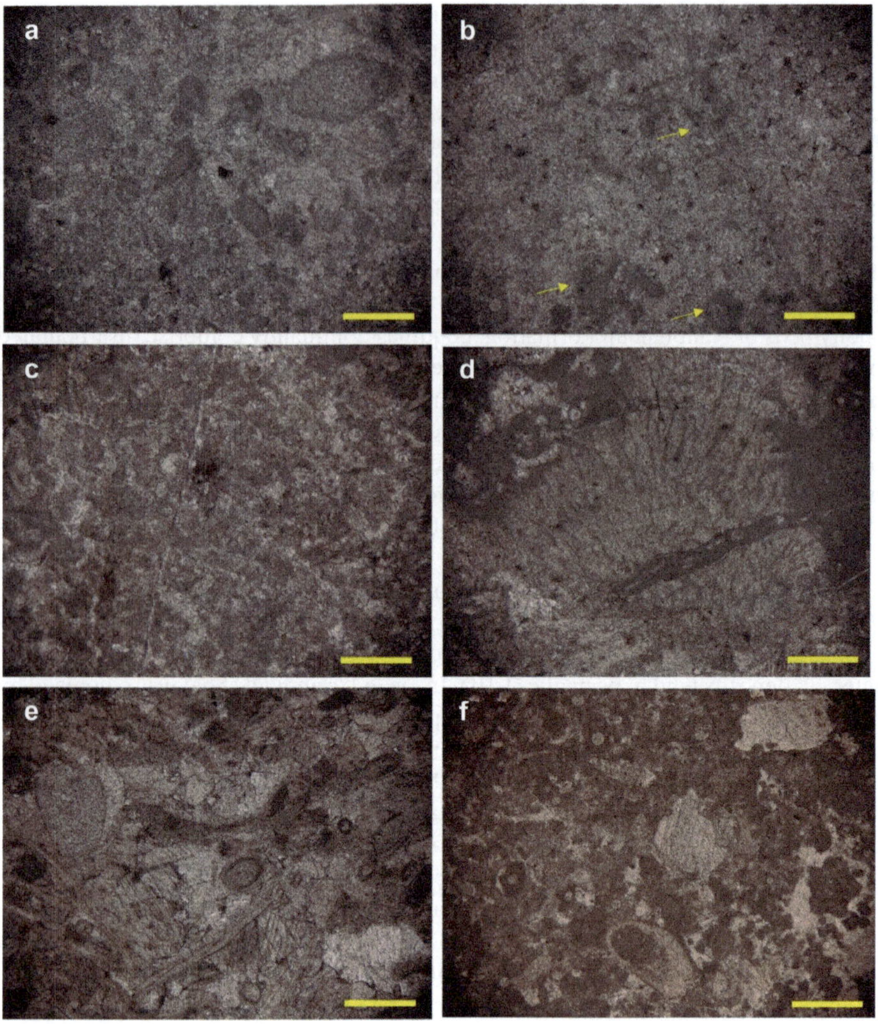

Fig. 24 Photomicrographs showing main microfacies across the D/C boundary at Nanbiancun. Plane polarized light. Scale bars = 0.5 mm. **a** Microbial-peloidal packstone/wackestone. Bed 48. **b** *Epiphyton*-like (arrows) bindstone. Bed 48. **c** Microbial bindstone in the articulated coquina, Bed 50. **d** Overgrown calcified *Rivularia* (*Ortonella*) in the microbial bindstone in the articulated coquina bed, Note the micritized burrows and intense micritized rims surrounding the microbe. Bed 50. **e** Brachiopod shell packstone with minor crinoids. Bed 57. **f** Skeletal-microbial packstone/ grainstone in which skeletal fragments mainly include crinoids, brachiopod and foraminifer. Bed 70. See Fig. 20 for detailed strata position

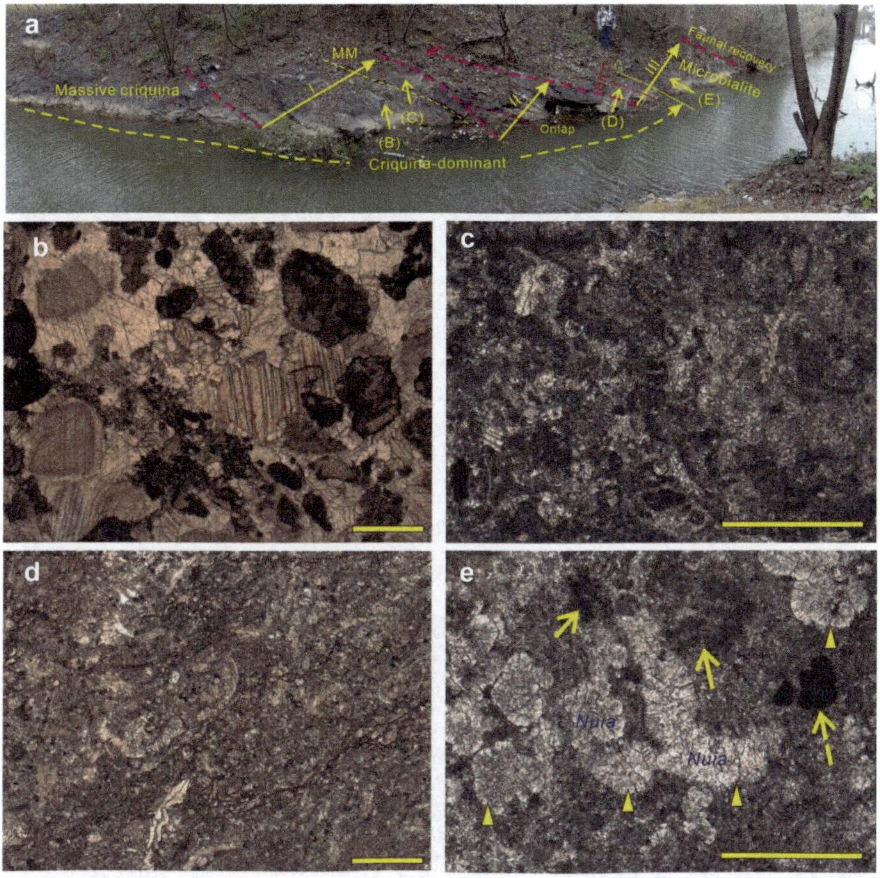

Fig. 25 **a** Panoramic view on the D/C boundary succession along the creek at Nanbiancun which is overwhelmingly composed of criquinas (crinoidal limestones) evolving from massive to cyclic (thin- to thick-bedded) successions (I-III). In cycle I, the crinoidal shoal aggraded to microbial mound (MM), indicating a shallowing process. In cycle II, basal strata lap out onto the top of MM, indicating a transgressive process onto the slope. The basal occurrence of platy to lenticular limestone in cycle III indicates an accelerating marine incursion. Crinoidal fauna can extend upward to the basal horizon of cycle III (below the hammer (28 cm; circled)) above which microbialites dominated, likely caused by the biotic crisis, so that the D/C boundary is tentatively placed within or on the top of the basal horizon composed of platy bedded bioclastic wackestone/packstone. **b** Photomicrograph showing the criquina dominated by crinoids with subordinate microbial fragments. **c** Microscopic lithology (bindstone) of microbial mound (MM) colonizing the crinoidal shoal. **d** Photomicrograph of platy stylolized limestone comprising ostracod wackestone/mudstone with minor microbial particle. **e** Photomicrograph of microbialite (bindstone) composed of micritized bush-like *Epiphyton* (arrows), *Renalcis* (dashed arrow) and *Nuia* (triangles) and other unidentifiable elements in the upper part of cycle III. See **a** for the locations of **b–e**. Photomicrographs of **b-e** were taken under plane-polarized light. Scale bars in (B-E) are 1 mm long

Fig. 26 Black shales in the Member #3 of Chuanbutou Formation about 500 m west of the Nanbiancun section

Fig. 27 Photo showing the splendid stalactites and stalagmites in the karstic cave (Reed-flute Cave)

numerous legendary stories (Fig. 27). To make your own stories using your wonderful imaginations from what you have seen!

5.2 *Day 2*

On today's trip, on the way to boat wharf, we will stop at Mopanshan to look the fault-controlled hydrothermal dolostones in the Lower Carboniferous Huangjin Formation, then we will have a boat trip on the Li River to see the spectacular karstic landscapes. When taking off the boat at Yangshuo, we will have a look on the Middle Devonian stromatoporoid reefs. On the way back from Yangshuo, we will have a short stop at a petrol station, ~ 1 km away from the Yangshuo, to look the platform collapse near the F/F transition, and have another stop at Baisha to look the basinal facies succession across the F/F boundary (Figs. 4 and 5).

5.2.1 Stop 1

Fault-Related Dolomitization at Mopanshan

On the way to the Mopanshan (or Zhujiang nearby) pier, we will firstly stop at Mopanshan shortly to see the fault-controlled hydrothermal dolostone along the highway roadside. At this locality, we can see the direct evidence on how the faulting had controlled the hydrothermal dolomitization (Fig. 28a) and related petrographic indicators such as zebra texture and brecciation in which vugs/veins were extensively cemented initially by the saddle dolomites (Fig. 28b, c) which generally followed by other mineral phases (i.e., calcite mostly).

5.2.2 Stop 2

Facies Controls on the Karstic Landscape Along the Li River

After Stop 1, We will take the coach directly to the Zhujiang (or Mopanshan) pier, then board on the two-storey sightseeing boat to start the excursion along the Li River (Lijiang) down to Yangshuo (Figs. 4 and 5). A boat trip on the peaceful *Li River* is the absolutely high point of any visit to the Guilin. A cruise from Guilin to Yangshuo, visitors will experience the winding and twisting Li River, pass many bizarre tower hills whose shapes have inspired and fired numerous imaginations and legendary stories (Figs. 29 and 30). If you are luck, you can experience the peaceful and harmonic life of local people along the river: cormorant fishermen in narrow bamboo boats (rafts), bathing children, water buffaloes, small tranquil villages and women doing their washing on banks of the river along the way.

Except for the sightseeing interests, the geology related to the formation of spectacular karstic landscapes could be definitely a dessert for this trip. The most breathtaking karstic landscapes occur in the Devonian Rongxian Formation (Upper Devonian) along the river between Caoping and Xingping towns, along which a narrow shallow-water "bridge" or corridor (platform) occurred, linking the Guilin Platform

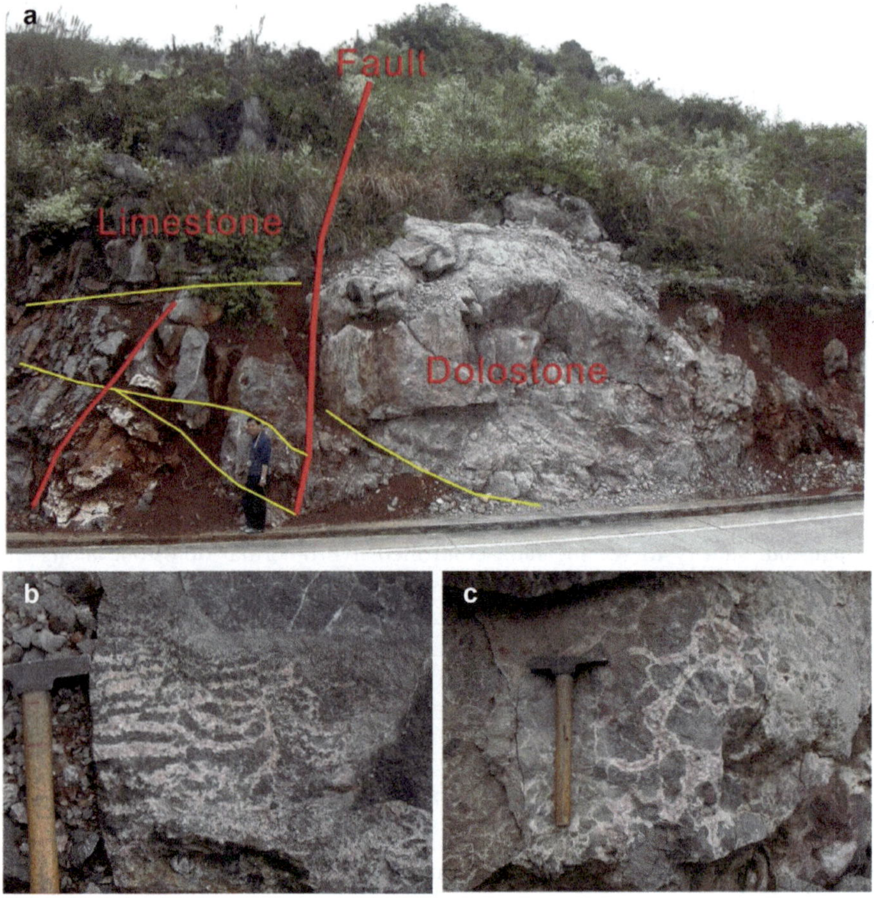

Fig. 28 Fault-controlled hydrothermal dolostones at Mopanshan, eastern Guilin. **a** Panoramic view on the fault-controlled hydrothermal dolomitization. Note the dolostones on the right side of fault. **b** Zebra texture in the dolostone in which the zebra vugs were cemented initially by saddle dolomites. Hammer for scale (30 cm long). **c** Brecciation in the dolostone in which the breccias were lined initially by saddle dolomites then megacrystalline calcites. Hammer for scale (30 cm)

to a larger platform to the east in the Devonian (Figs. 2 and 4). The Rongxian Formation is generally composed of chemically pure and physically brittle microbial limestones, readily resulting in fracturing in the course of later folding and uplift at the onset of the Mesozoic. Progressive fracturing, leaching and dissolution by meteoric waters in the course of exhumation of the specific limestones finally evolved into the spectacular karstic tower hills as that we would see. By contrast, no typical karstic towers but generally mild hills commonly occur in the more argillacous and thinner-bedded basinal deposits as we see along the river from the Xingping down to the Yangshuo (Fig. 31). When approaching Yangshuo city, the typical karstic landscapes

Fig. 29 Photo showing the well-known scenery of painted (or nine-horse) hill along the Li River near Yangdi

Fig. 30 Photo showing the famous scenery of picturesque karstic landscapes along the Li River near Xingping town

once again come into our eyes due to facies transition to platform domains again (cf. Fig. 4).

Fig. 31 The apparent contrast in landscape controlled by the primary depositional facies between platform (behind the dashed line) and basinal facies domains across the winding Li River at Shawancun near Xingping town (see Fig. 4 for location and facies). Characteristic karstic towers/ pillars were developed in the platform facies regime (behind the dashed line) while densely-vegetated mild hills/mounded hills developed in the deep-water basinal facies regime (in front of dashed line)

5.2.3 Stop 3

Middle Devonian Stromatoporoid Reefs to the Slope Setting

As taking off the boat at Yangshuo pier, we will gather again and go to the Yangshuo Bridge (about 500 m by walk) to visit the Givetian stromatoporoid reefs in the Tangjiawan Formation (the earliest carbonates in Guilin area) (Fig. 32), focusing on reef builders and architecture of the basal one due to time restriction. Afterwards, we will take on the coach and stop at Liangshuijing (~1 km apart) along the Yangshuo-Guilin Highway (be careful for the traffic) where we will check the facies transition to the slope setting in the Mintang Formation, characterized by thin-bedded lime mudstones, calciturbidites and stratal onlap termination with deep-water faunal assemblages, notably tentaculitids (Fig. 32). It should be noted, there also occur cyclic finely laminated microbialites overlying the calciturbidites, which are not common in deep-water setting in previous studies. Going upward further, there occurs the bedded chert succession of Liujiang Formation (Frasnian) bearing tentaculitids and radiolarians, indicating a significant deepening into the basinal setting.

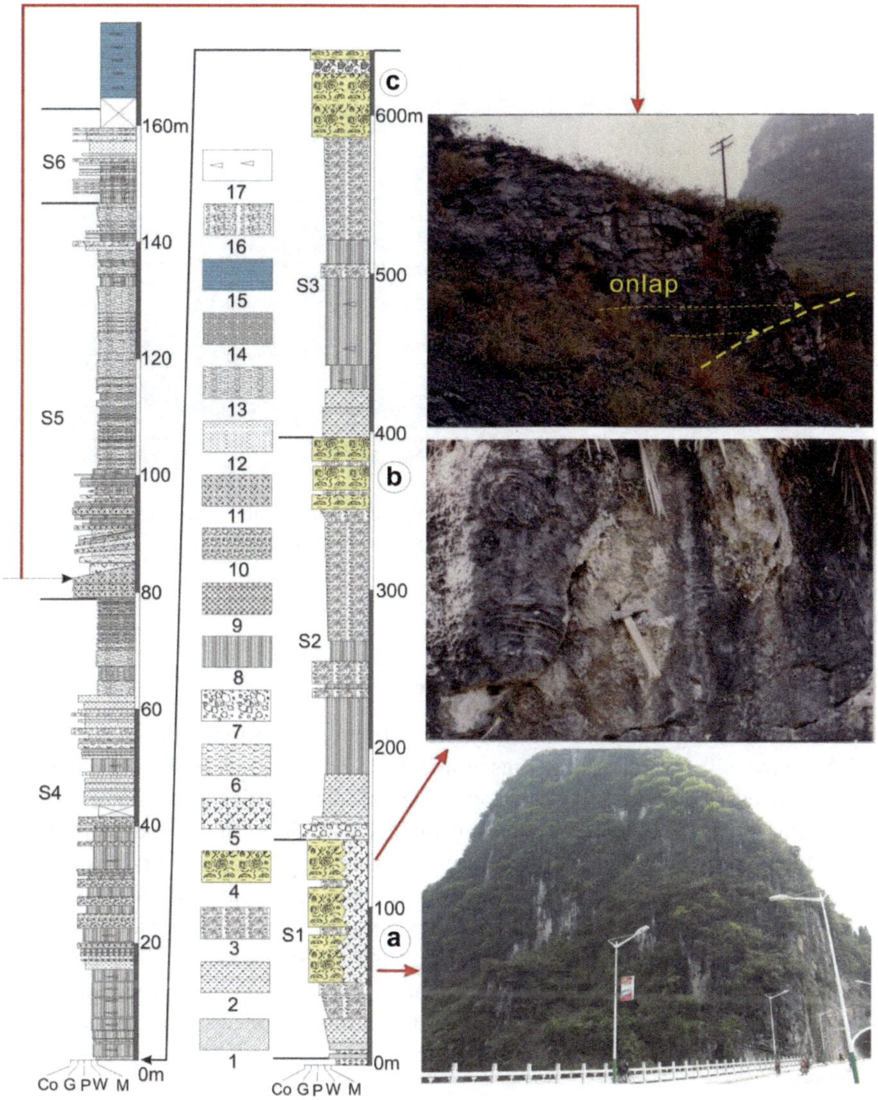

Fig. 32 Lithological succession from Tangjiawan Formation (S1-S3) to Mintang Formation (S4-6) from Yangshuo Bridge to Liangshuijing. Three phases of stromatoporoid reefs were developed at this locality compared with two phases of reefs (or buildups) elsewhere in general. S1-S6 are depositional sequences (Modified from Chen et al. 2001b). M-lime mudstone, W-wackestone, P-packstone, G-grainstone, Co-conglomerate, B-bindstone, F-floatstone. 1-Calcareous shale, 2-Thin-bedded lime M, 3-Skeletal W/P, 4-Platy stromatoporoid F/P, 5-Stromatoporoid B/F, 7-*Amphipora* F/P, 8-Coquina, 9-Breecia/megabreccia, 10-Peloidal-oncoidal P/G, 11-Stromatoporoid rudstone, 12-Intraclastic-skeletal P/G, 13-Calciturbidite, 14-Deep water microbialite, 15-Bedded chert/siliceous shale, 16-Pebbly conglomerate, 17-Tentaculitid

Fig. 33 Photo showing the large-scale udulatory slide and fault escarpment, overlain by megabreccia and translational blocks at the petrol station near Yangshuo

5.2.4 Stop 4

Platform Collapse on the Marginal Slope

By further walking along the highway, we will have a short stop at a petrol station ~ 1.5 km away from Yangshuo (Figs. 1 and 4), to have a quick look on the platform collapse occurring in the F/F transitional interval. A magnificent undulatory slide to fault escarpment is overlain by thick megabreccias (~50 m thick) and translationational blocks (Fig. 33). These gravity-flow deposits are considered as the equivalents to the calciturbidites in deeper basinal successions (i.e., at Fuhe) (Chen et al. 2001b; Chen et al. 2002a, b; Chen and Tucker 2003). Afterwards, we will head for the next stop at Baisha (about 8 km in distance) through the highway along which typical karstic tower hills/pillars frequently come into your eyes on the left side (platform facies) whereas only mild (or mounded) hills dominate the landforms on the eastern side (basinal facies).

5.2.5 Stop 5

Slope-Basinal Succession Across the F/F Boundary at Baisha

This locality is about 10 km north of Yangshuo, on the eastern of Baisha Town along the Guilin-Yangshuo highway (Figs. 2 and 4). At this locality, we will stop to examine the lithological variation across the F/F boundary in the basinal succession, from the Liujiang, Gubi to Wuzhishan formations.

Fig. 34 **a** Bedded chert intercalated with brownish tuffaceous layers in Liujiang Formation at Baisha. Hammer for scale (35 cm long). **b** Nodular limestones in Wuzhishan Formation at Baisha. Pen for scale (10 cm)

The Liujiang Formation consists of thin-bedded chert interbedded with brownish and grey tuffaceous layers (Fig. 34a), which is overlain by Gubi Formation comprising calcareous gravity flow deposits (debrites and turbidites) (Fig. 35a, b) and nodular (or lenticular) limestones. It is further overlain by the Wuzhishan Formation composed overwhelmingly of nodular limestones (Fig. 34b) with minor calciturbidites (Fig. 36). The boundary between Gubi and Wuzhishan formations is characterized by the thin-bedded calciturbidites (Gubi Formation) overlain by three shale-based nodular limestone cycles, grading upwards into overall nodular limestones (Fig. 36), identical with the F/F succession at Fuhe as we will see on next day. Two third-order depositional sequences are identified across the F/F boundary: the lower sequence Fra and the upper sequence Fa (Chen and Tucker 2003), in which smaller-scale cycle sets (Sc) consisting of (generally 6–8) bundling cycles are well illustrated (Fig. 35c). Therefore, a sea-level fall followed by a rapid sea-level rise occurred during the F/F transition (Chen et al. 2002a, b; Chen and Tucker 2003). An overall positive carbon isotope excursion was revealed across the F/F boundary (Chen et al. 2002a, b,2005).

5.3 *Day 3*

In today's excursion, we will firstly go ahead to Yangdi to visit the Middle-Upper Devonian shallow-water carbonate succession and the gorgeous karstic landscapes, then back to the Zhongnan village to see the depositional facies, stratal pattern and geometry of microbial buildup in the platform marginal zone during Givetian-Frasnian interval. Afterwards we will be back further to Fuhe to look the facies transition from a lagoon to basinal environment recorded in the Tangjiawan, Mintang, and Liujiang formations along a rural highway. A detailed examination on the F/F boundary event will be carried out in the Gubi and Wuzhishan formations at the

Fig. 35 Gravity flow deposits in the Gubi Formation at Baisha. **a** Flat pebble conglomerate derived from the unsolidated slope deposits near the base of Gubi Formation. **b** Flat pebble-dominated conglomerate with skeletal fragments (arrows) in the upper part of the conglomerate horizon. Br-brachiopod, Cr-Coral, St-Stromatoporoid. **c** Calciturbidites intercalated with lenticular mudstones. Note the bundling cycles (arrows) each composed of eight stratification beds

Fuhe reservoir. Then we will transfer to Bao'an where we will look the Givetian stromatoporoid reefs and overlying Frasnian shoal-microbial complexes. Finally, we will go to Qifengzhen in southeastern Guilin to visit the slope deposits dominated by calciturbidites, megabreccias and intraformational truncations (see Figs. 2 and 4 for location and depositional settings).

5.3.1 Stop 1

Microbial Reef During the Givetian-Frasnian Interval

The outcrop is located near Zhongnan village, on the way between the main highway (Guilin-Yangshuo Road) and Yangdi. Geologically, it was located at the transitional zone from the platform margin to marginal slope (Figs. 2, 4 and 37).

The original stratal pattern and geometry from platform margin to marginal slope (horizontal to declined strata) are likely preserved from Zhongnan to Wulibei. During the Givetian, the platform margin was rimmed with microbial buildups (Figs. 38, 39), which was transferred to a sandy shoal-rimmed carbonate platform system in the early Frasnian (Figs. 40 and 41) (Chen et al. 2002b, a).

This situation occurred in accompanying with a large-scale platform collapse, such that massive megabreccias and slump blocks being scraped from the mid-upper slope and foreslope moved downdip to the lower slope (Figs. 40–42), likely resulting

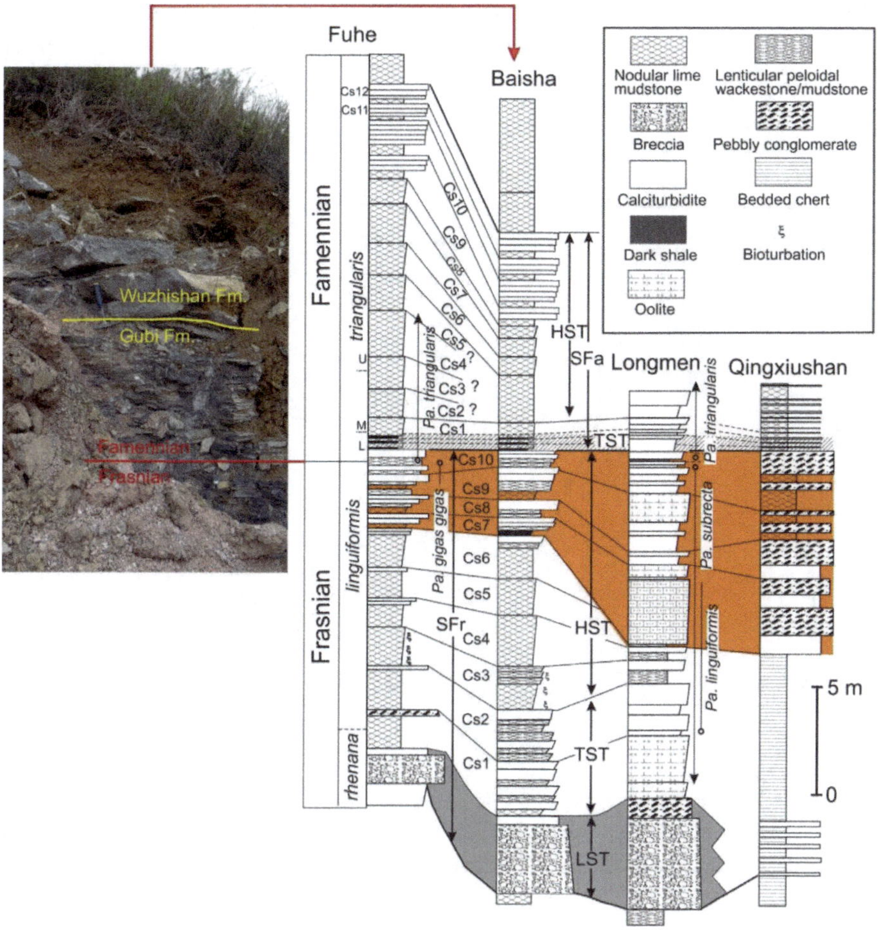

Fig. 36 Depositional (third-order) sequences in deep-water successions across the F/F boundary in Guangxi area. Sequences: SFr, top Frasnian sequence; SFa, lowermost Famennian sequence. Cs1—Cs12, cycle sets (parasequence sets) within sequences. LST, TST and TST: lowstand, transgressive and highstand systems tracts, respectively. Dotted horizon is the calciturbidites of late HST deposits in SFr; slashed horizon is the transgressive deposits of SFa (modified from Chen and Tucker, 2003, with additional data). The F/F boundary is moved down to the top of calciturbidite horizon of Gubi Formation with new conodont biostratigraphic constraints (Chen et al. 2013; Huang et al. 2018). Note the lithological variations across Gubi and Wuzhishan formations and the F/F boundary placement at Baisha

from block-tilting at the platform margin in response to a jerky faulting. In the latest Frasnian, small-scale microbial buildups colonized once again on the slope, which further thrived on the platform margin during the F/F transitional interval and later on (Chen et al. 2002a, b).

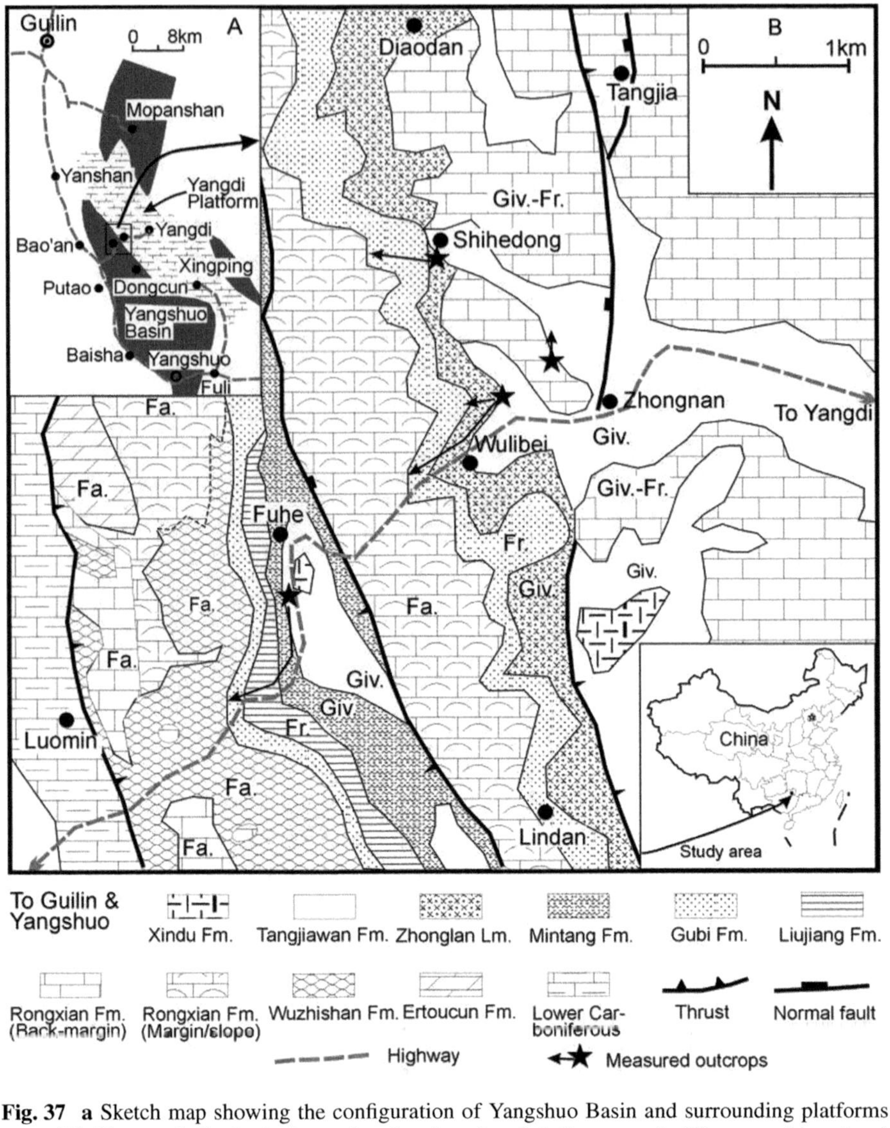

Fig. 37 **a** Sketch map showing the configuration of Yangshuo Basin and surrounding platforms around Guilin area. **b** Geological map showing deposits of platform margin (Zhangnan Limestone) to slope and basin (Mintang, Gubi and Liujiang formations) crop out from Fuhe to Zhongnan along the rural highway. Giv.-Givetian, Fr.-Frasnian, Fa.-Famennian

5.3.2 Stop 2

Evolution from Restricted Lagoon to Intraplatform Basin

The stop starts from Fuhe to the Fuhe Reservoir, about 4 km away from the main highway to Guilin (see Figs. 2 and 4 for location and setting). At this stop, we can see

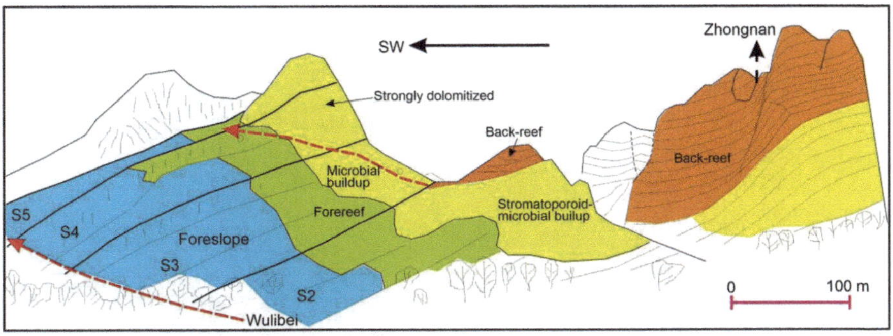

Fig. 38 A panoramic sketch drawing of photo mosaics showing the stratal pattern transition from the horizontal platform strata (the hill right back in distance) to the inclined foreslope strata of Givetian. Note the aggradational architecture of microbial buildup on the slope. Sequences (S2-S5) are separated by the thick lines. Box at the bottom of hill is 3.5 m high (after Chen et al. 2002a, b)

Fig. 39 **a** Branching tubular microbes built up the framework in the reef core. **b**, **c** Forereef microbial bindstones showing in-situ growth forms and cavities filled by radiaxial fibrous calcite cements and later equant calcite. **c** Small-sized articulated brachiopods (*Leirorhychus* sp.) attached in the microbial buildups (fore-reef) (after Chen et al. 2002a, b)

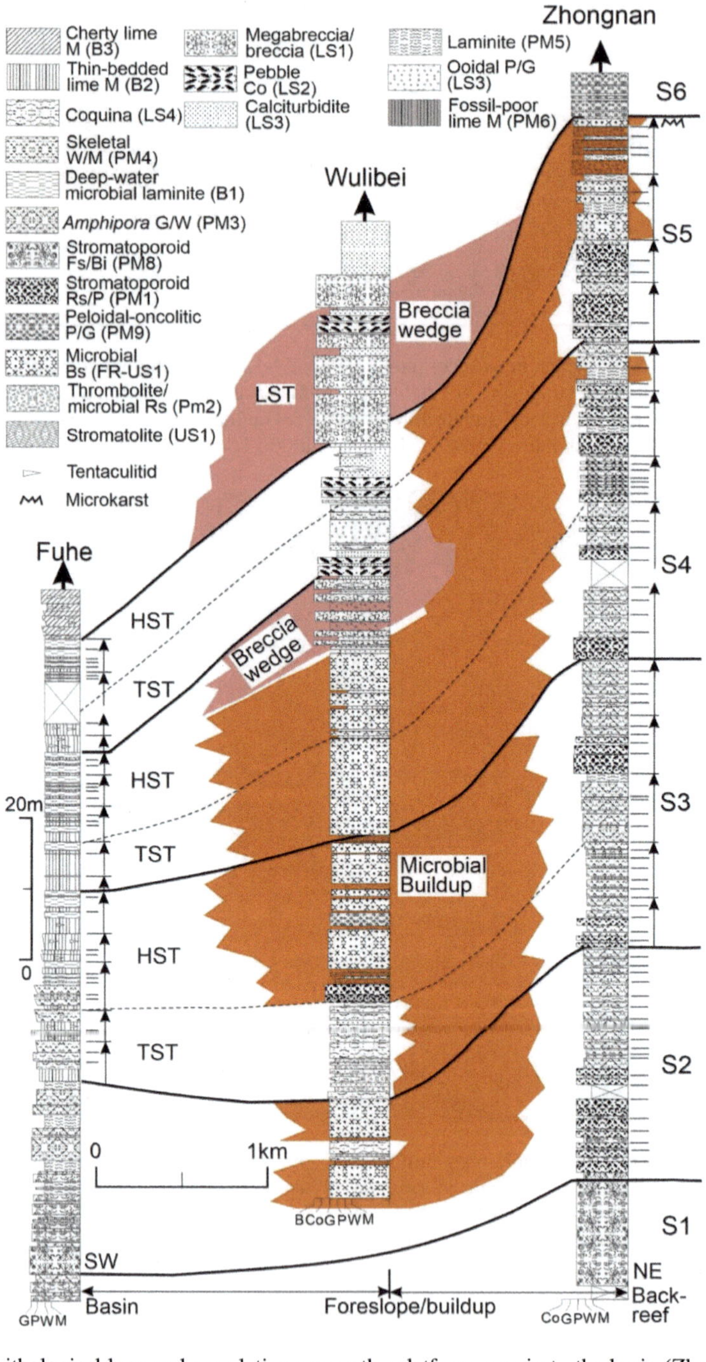

Fig. 40 Lithological logs and correlation across the platform margin to the basin (Zhongnan via Wulibei to Fuhe) during Givetian time. The marigin was rimmed by microbial buildups. S1-S6 = sequences (after Chen et al. 2002a, b)

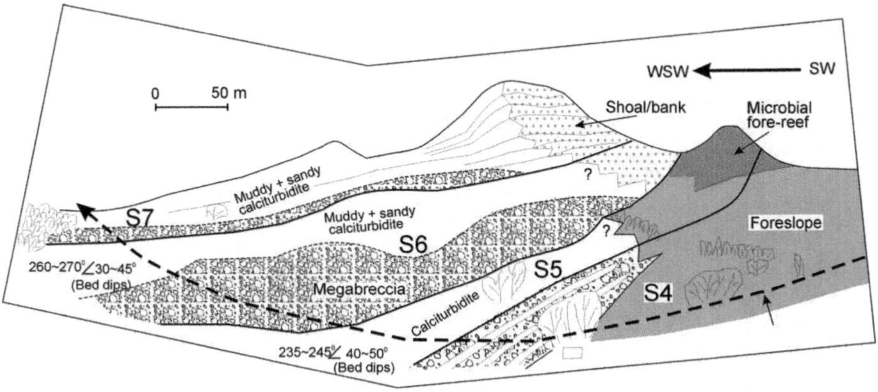

Fig. 41 Sketch drawing of photo mosaics showing the general depositional succession and geometry changes from microbial reef-rimmed platform to sand-shoal platform system from the uppermost Givetian to the Frasnian. The two lower megabreccia/breccia units are composed of microbialites derived from the updip microbial buildups. The upper breccia unit at the base of S7 is mainly composed of stromatoporoid buildup clasts. This diagram is the westward-extending part of Fig. 37 approximately in the dipping direction of slope. Bush-like patterns mean heavy vegetation. Box at the bottom is 3.5 m high

the complete stratigraphic succession from Tangjiawan to Mintang, Liujiang, Gubi and Wuzhishan formations, spanning from Givetian to Famennian in age (Figs. 3 and 37), experiencing the evolution from a restricted lagoon to the intraplatform basin. Firstly, we will see the basal carbonate unit of Tangjiawan Formation, which is dominated by *Amphipora* floatstone/bafflestones deposited in the restricted platform lagoon. However, this formation was totally dolomitized, with extensive mouldic vugs of skeletons and zebra fractures cemented by saddle dolomite and calcite crystals (Fig. 43) as extensively seen at Tangjiawan and elsewhere, suggesting intense hydrothermal dolomitization during burial (Chen et al. 2004).

The Mintang Formation is mainly composed of bioclastic (including tentaculitid) mudstones/wackestone intercalated with deep-water laminated microbialites (Fig. 37; Chen et al. 2001b, 2002a, b), indicating a deepening process after deposition of the Tangjiawan Formation. Liujiang Formation comprises basal chert limestones and bedded cherts upwards in the lower Frasnian (Fig. 44). The siliceous succession, 20–25 m thick, consists of black thin-bedded chert interbedded with brownish tuffaceous layers (Fig. 44b, c) in which pumice clasts and pelagic tentaculitids and radiolarian are present (Fig. 44d, e). The widespread deposition of bedded cherts in basins/troughs points to the strongest tectonic-induced differentiation of Devonian platform-basin complexes in South China: deepening in the basin regime vs. shallowing in the platform regime (Chen et al. 2001a, b).

REE geochemistry of the siliceous deposits (Fig. 45) suggests these chert deposits were mainly precipitated from silica-rich seawaters likely mixed with hydrothermal fluids vented through the basement faults at depths (cf., Fig. 2 and 4), as indicated by convex-shapped REE patterns and positive Eu anomalies (Chen et al. 2006).

Fig. 42 Slope deposits of Frasnian at Wulibei. **a** Megreccias at the base of S7. Hammer for scale (37 cm). Gubi Formation. **b** Polished slab of megabreccia **a**. Clasts comprise stromatoporoid (St), *Amphipora* (Am), brachiopods (Br), echinoid (Ec) and ostracod. **c** platy to boudinage-bedded hemipleagic cherty limestones with slight slump folds. Thin-bedded calciturbidites intercalated. **d** Strongly deformed slump folds in the hemiplegic limestones

In this light, relatively, the hydrothermal activities were the strongest at Shawancun, intermediate at Fuhe and weakest at Baisha, which were closely related to the intensity of basement faulting (i.e., relative strong close to the master fault zone).

5.3.3 Stop 3

Depositional-Geochemical Records Across the Basinal F/F Succession

The visit point is located along the Fuhe Reservoir where the F/F boundary succession includes the Gubi and Wuzhishan formations in ascending order. The Gubi Formation comprises gravity-flow deposits (calcidebrites and calciturbidites) in the lower part, nodular limestones in the middle part and calciturbidite horizon in the upper part. The Wuzhishan Formation is dominantly composed of nodular limestones intercalated with minor calciturbidites (Fig. 36). Recent conodont data suggest to placing the F/

Fig. 43 Strongly dolomitized limestones of Tangjiawan Formation by hydrothermal activity. Note the saddle dolomite-filling mouldic vugs of *Amphipora* and brecciation

F boundary across near the top of calciturbidite horizon (top of second last bundling cycle), moving down 0.83 m from the previous boundary (Figs. 46–48; Chen et al. 2013; Huang et al. 2018).

Across the F/F boundary succession, different hierarchies of depositional cyclicity are extensive (Chen and Tucker 2003). Two third-order depositional sequences (SFr, SFa) are identified across the F/F boundary (Figs. 36, 47, 48), in which higher-order cycles are further recognized. The boundary between this two sequences is very close to, but a little above the biostratigraphic boundary between the Frasnian-Famennian boundary constrained by the conodont zonation (Figs. 47 and 48; Chen et al. 2013; Huang et al. 2018). Therefore, a relative sea-level fall and a subsequent sea-level rise are evident across the F/F boundary.

The smallest-scale cycle is the stratification bed; 6–8 stratification beds commonly bundle into a cycle (bundling cycle), which are further grouped into cycle sets (with 6–8 cycles), mesocycle sets (with two cycle sets) and megacycle sets (with four cycle sets), respectively (Figs. 48, 49). Such a cycle stacking pattern suggests a Milankovitch orbital forcing on the cycle stacking patterns and cyclicity. In this case, the bundling cycles, cycle sets, mesocycle sets and megacycle sets were formed in response to the perturbations of precession, eccentricity, intermediate eccentricity and long eccentricity, respectively (see Chen and Tucker, 2003), thereby each spanning 16–18, 100, 200 and 400 kyrs. Cyclostragraphic analysis revealed four cycle sets in the uppermost horizon of SFr related to sea-level fall (~400 kyr), and three bundling cycles in the lowermost of SFa in response to a rapid sea-level rise (~50 kyr).

Fig. 44 **a** The boundary between Liujiang and Mintang formations demarcating the organic-rich dark grey thin-bedded lime mudstones of the basal Liujiang Formation from the microbial laminite of underlying Mintang Formation. **b** Thin-bedded cherts interbedded with tuffites in the upper Liujiang Formation. **c** Close-up view of the siliceous rock showing the rhythmic black cherts and light-colored porous tuffites. Scale bar = 2 cm. **d** Abundant tentaculitid shell fragments in the bedded chert. **e** Radiolarians present in bedded chert

High-resolution C (δ^{13}Ccarb–δ^{13}Corg) and S (SCAS—^{34}SPyrite) isotopic studies from identical sample sets were carried out within the high-resolution sequence-cycle stratigraphic frameworks across the F/F boundary (Fig. 50). An overall positive excursion of δ^{13}Ccarb–δ^{13}Corg pairs (Chen et al. 2005), which is independent on the abundance of organic matter, starts from the base of calciturbidite horizons (upper *linguiformis* zone) (four eccentricity-forcing cycles), and ends in the base of nodular limestone successions (two eccentricity-forcing cycles), thereby spanning ~600 ka in duration (Fig. 50).

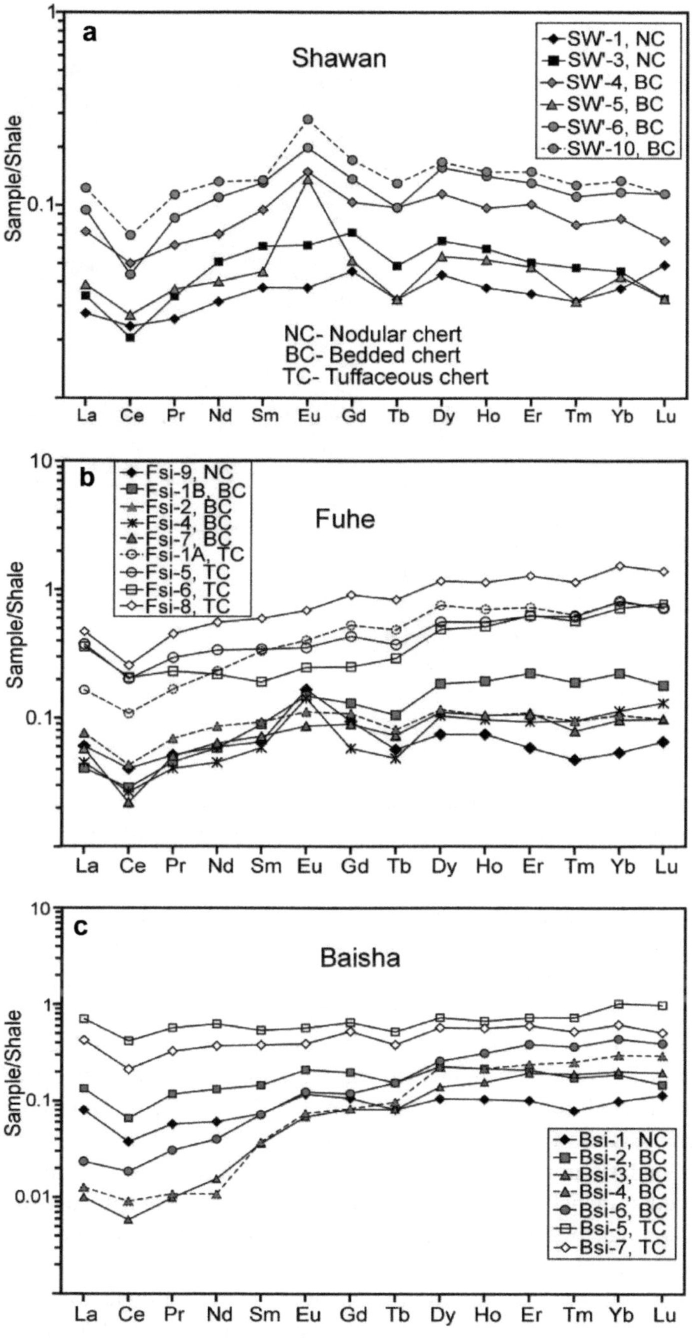

Fig. 45 Shale-normalized REE abundance patterns of different types of cherts (nodular, bedded and tuffaceous cherts), respectively from Shawan **a**, Fuhe **b** and Baisha **c** (see Figs. 2 and 4 for location and tectonic settings) (After Chen et al. 2006)

Fig. 46 The new placement of F/F boundary across near the top of Gubi Formation (calciturbide horizon) compared with previous placement across near the base of Wuzhishan Formation

Moreover, the magnitude of the maximum δ^{13}Corg excursion is larger (~4.5‰ VPDB) than that of δ^{13}Ccarb excursion (~2.5‰ VPDB), but the timing of the δ^{13}Corg excursion is later (~ 100 ka). Within this overall positive excursion, three shorter-term isotopic perturbations, each spanning ~200 ka, can be further identified, which are better expressed by the variations in Δ^{13}C values (= δ^{13}Ccarb—δ^{13}Corg) (Figs. 50 and 51) which are largely related to the variations in atmospheric *P*CO2 levels, thereby frequent climate fluctuations (Chen et al. 2005). High-resolution cyclostratigraphic analysis shows a strong obliquity power of the δ^{13}Ccarb time series across the F/F boundary, notably at Fuhe (Fig. 52; De Vleeschouwer et al. 2017), indicating a predominance of high-latitude orbital influence on the global carbon cycling that time, similar to the cases found in other critical transitions, i.e., the OAE2 in Cretaceous (Fig. 52).

For the S isotopic dataset, ^{34}Spyrite changes are more frequent than the ^{34}SCAS changes, implying a relative low sulfate concentration during the F/F transition (Chen et al. 2013). Although a positive excursion in ^{34}SCAS is also revealed, an apparent

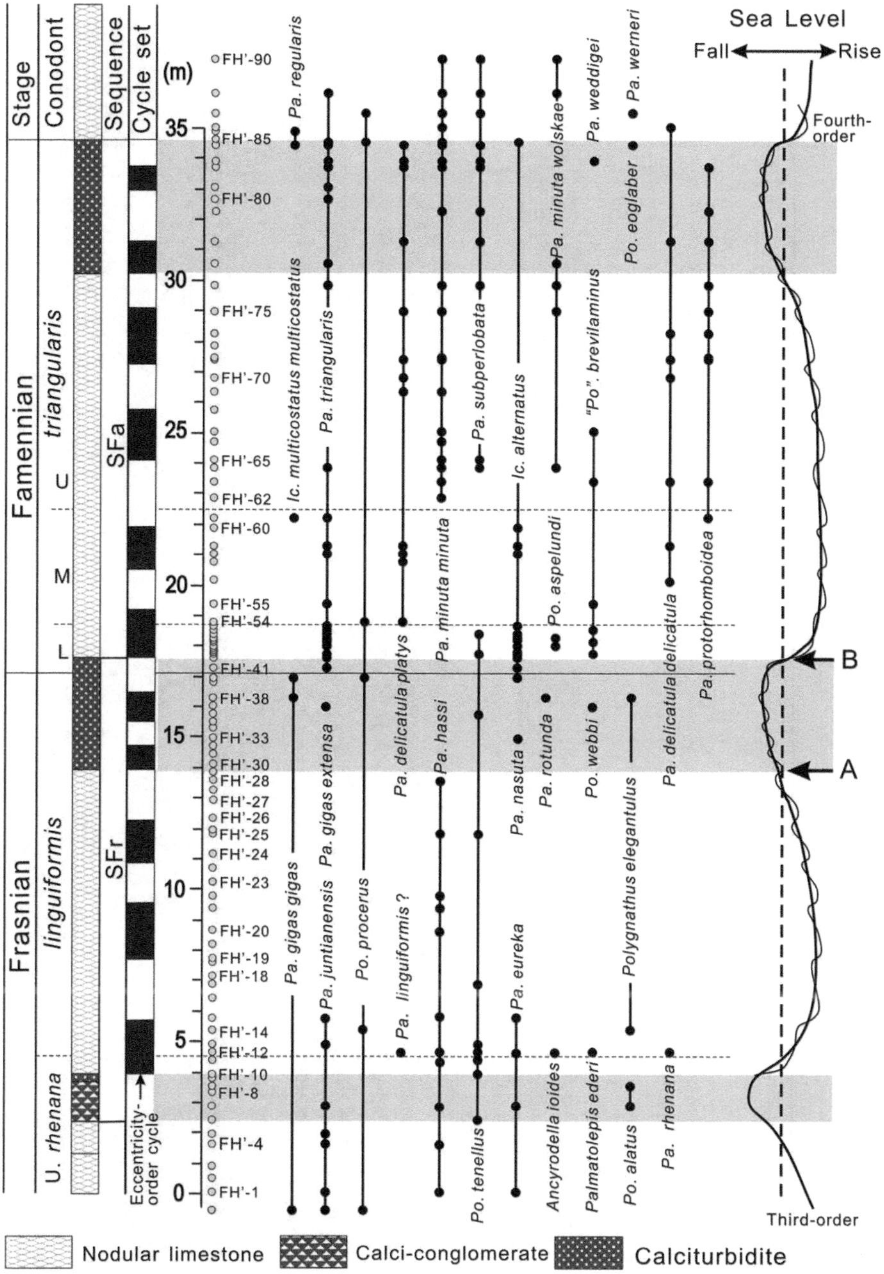

Fig. 47 New conodont bisotratigraphy integrated with cyclostratigraphy and sea level changes across the F-F boundary at Fuhe. The new data suggest a placement of the F-F boundary through the uppermost calciturbidite horizon, lower than previous position (Chen et al. 2013)

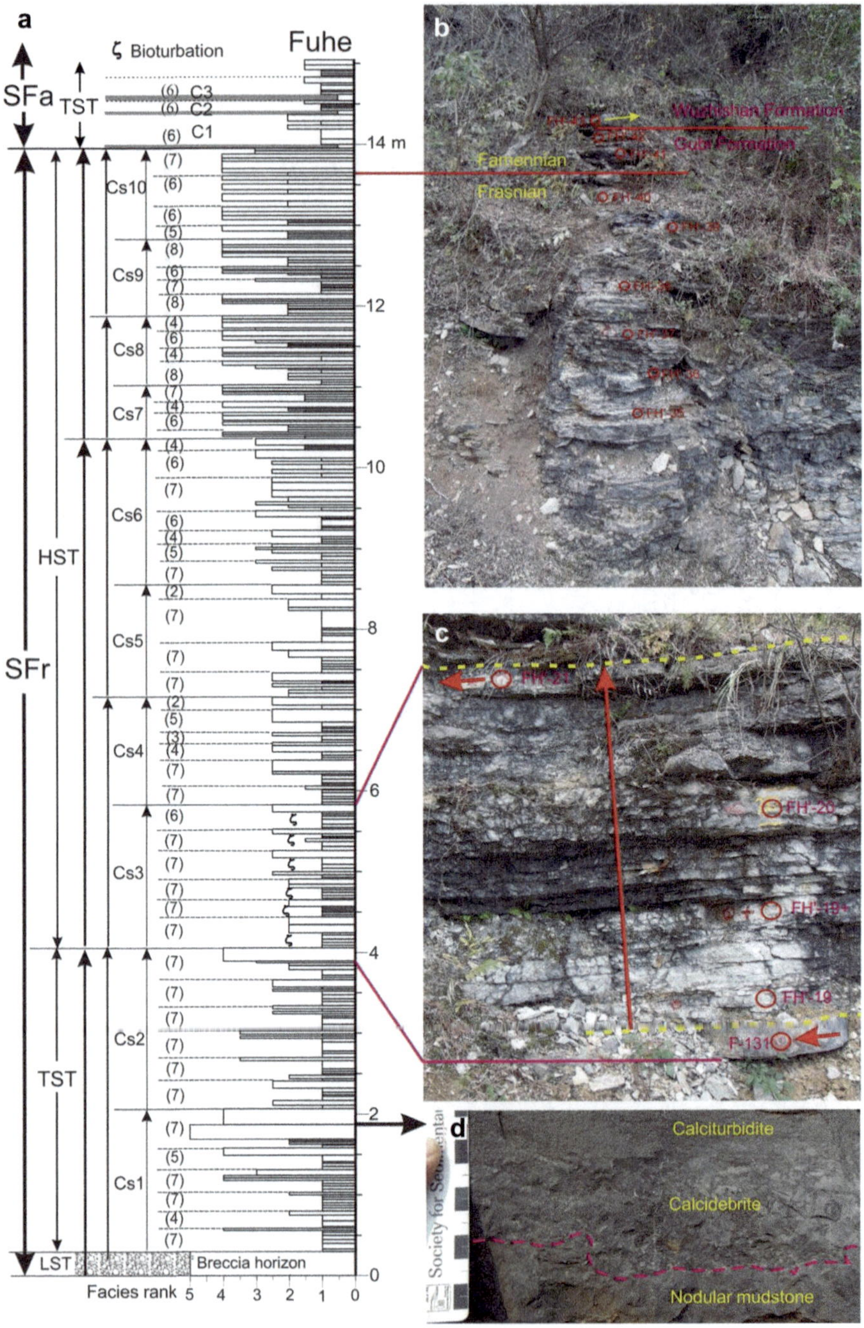
a
ζ Bioturbation
Fuhe
SFa
TST
(6) C3
(6) C2
(6) C1
14 m
SFr
HST
TST
LST
Cs10
Cs9
Cs8
Cs7
Cs6
Cs5
Cs4
Cs3
Cs2
Cs1
Breccia horizon
Facies rank 5 4 3 2 1 0
12
10
8
6
4
2
0
b
Wuzhishan Formation
Gubi Formation
Famennian
Frasnian
c
FH-21
FH-20
FH-19
F-131
d
Calciturbidite
Calcidebrite
Nodular mudstone

◀**Fig. 48** Details of stacking patterns of higher-order cycles in SFr at Fuhe. Bundling cycles **c** (numbers within brackets: numbers of stratification beds of each cycle) are grouped into cycle sets (Cs1—Cs10), mesocycle sets (marked by the second line of arrows from the right), and megacycle sets (marked by the third line of arrows from the right). Facies ranks: 0.5—shale, 1—nodular lime mudstone, 2—lenticular/boundinage peloidal wackestone/mudstone, 3—cm-bedded distal calciturbidite, 4—calciturbidite, 5—breccia (modified from Chen and Tucker 2003)

Fig. 49 Bundling cycles in the nodular limestones of Gubi Formation, Fuhe. Hammer for scale (35 cm long)

timing offset occurs relative to the C isotope excursion, reaching the climax at the onset of calciturbidite horizon (or sea-level fall).

Recent high-resolution conodont Sr isotopic data demonstrate three distinctive variation cycles (I, II and III), notably at Fuhe, fluctuating from low (negative) to higher Sr isotope ratios (positive excursion) (Zhang et al. 2020a, b; Fig. 51). Temporally, fluctuations I and II occurred immediately before the F-F boundary and fluctuation III started almost from the onset of the Famennian age. These conodont $^{87}Sr/^{86}Sr$ (low-to-high) fluctuations temporally match well the paces of conodont ^{18}O (low to high) and ^{13}Ccarb-org (high to low variations) fluctuations from the identical section published previously. These Sr isotopic fluctuations thus corroborate the cyclic variations in volcanic activity/flux and weathering intensity/riverine flux in parallel with climate warm-to-cool fluctuations. Mass balance modeling shows the first exceptionally high Sr isotope spike (I) could have been only yielded by a dramatic increase in terrestrial flux in a somewhat restricted Paleo-Tethys, likely due to closure of the oceanic domain between Laurussia and *Gondwana,* and sea-level fall

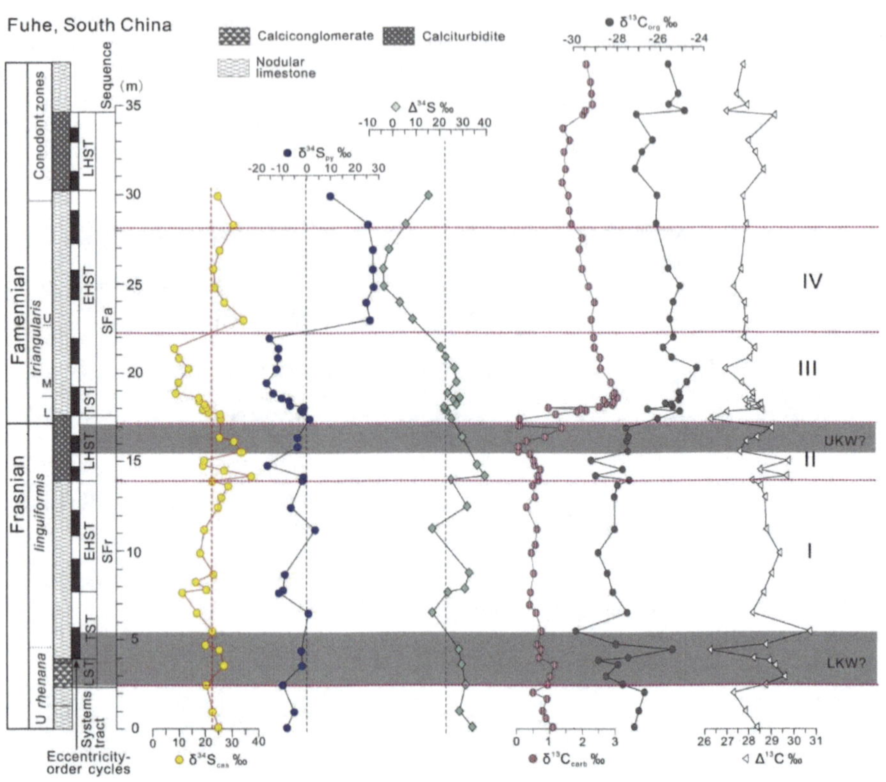

Fig. 50 **C-S** isotopic systematics from micrites across the F/F boundary at Fuhe, Guilin. $\Delta^{13}C = \delta^{13}C_{carb} - \delta^{13}C_{org}$. ^{34}S = SCAS—^{34}SPyrite. Carbon and sulfur isotopic compositions are reported in per mil with respect to VPDB and VCDT standards, respectively. Shaded horizons roughly correspond to the upper and lower Kellwasser horizons (Integrated from Chen et al. 2005, 2013)

(Zhang et al. 2020; Fig. 1). This scenario would enhance the seawater eutrophication/anoxia in the pericontinental basins along the Paleo-Tethys, causing initial strikes on the low-latitude shallow marine benthos. Subsequently, the frequent climatic fluctuations may have further caused escalating detrimental ecological pressures, leading to their massive demise.

5.3.4 Stop 4

Stromatoporoid Reef-to Microbial Reef-Rimmed Platform Margin from Givetian to Frasnian at Bao'an

This visiting locality (Bao'an) was located on the platform margin where two-phase stratomatoporoid reefs were developed during the Givetian, and evolved to microbial buildups and backreef shoals during the Frasnian (Figs. 4 and 53).

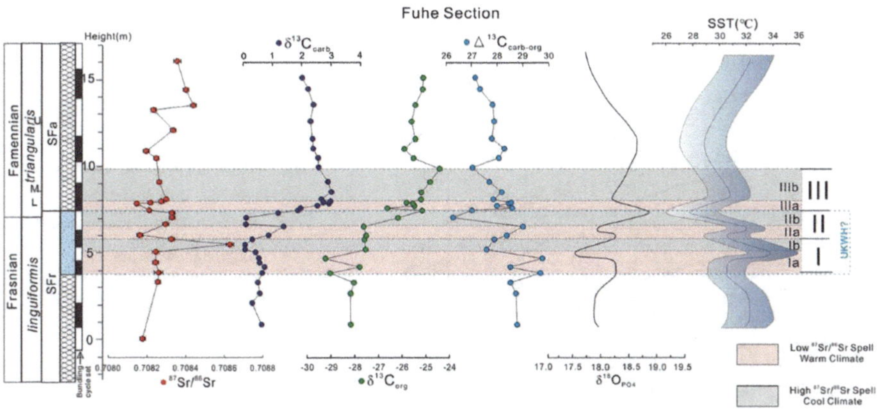

Fig. 51 High-resolution variations in Sr, O isotope compositions of condonts and estimated seawater surface temperature (SST) compared with the C isotope variations across the F/F boundary at Fuhe (Zhang et al. 2020a, b). The bulk carbonate C isotopic data are derived from Chen et al. (2005). O isotopic data and SST values are from Huang et al. (2018). Horizontal shaded intervals (I-III) mark three distinctive isotopic cycles

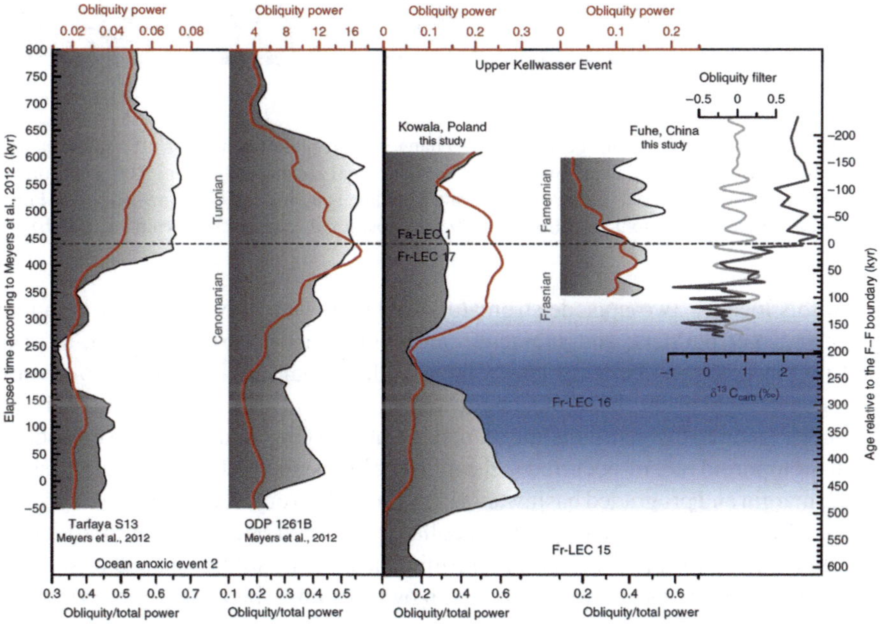

Fig. 52 Comparison of signals of obliquity forcing across the OAE-2 and UKW. Red lines indicate total power in the obliquity band (0.015–0.04 per kyr), and the black lines indicate the fraction of the power in the obliquity band, as compared to the total power at frequencies less than 0.1 per kyr. The Fuhe δ^{13}Ccarb time series and its obliquity-filter visualize the notable imprint of the obliquity on the course of the UKE positive δ^{13}Ccarb excursion (after De Vleeschouwer et al. 2017)

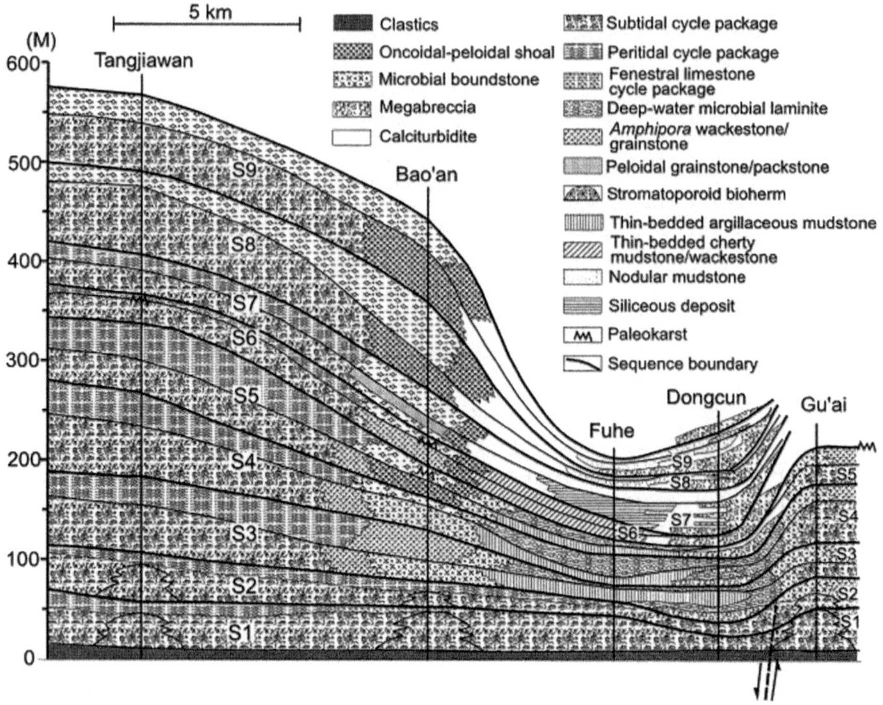

Fig. 53 Cross-section showing general facies changes from platform interior (Tangjiawana and Guilin formations) at Tangjiawan through platform marginal zone (Tangjiawan, Rongxian formations) at Bao'an to the basinal setting (Tangjiawan, Mintang, Liujiang and Gubi formations) at Fuhe. See Fig. 4 for paleogeographic locations. S1-S9 = depositional sequences (Chen et al. 2001b)

At this locality, two-episode stromatoporoid reefs were developed in which reef geometry, reef builders, and subfacies can be identified clearly (Fig. 53a, b). The first reef system was built up into a large-scale domal architecture up to several 10's meter high in the base of Tangjaiwan Formation, evolving to fore-reef and foreslope featured by olistoliths/megabreccias later on (Fig. 53a). The next episode of stromatotoporoid reef re-occurred in the mid-late highstand systems tract (HST) stage on the margin and prograded basinwards in the late HST (or forced regressive systems tract). Afterwards, large-scale platform collapse occurred once again in fore-reef (or foreslope) along the platform margin, forming large-scale wedge-shaped olistoliths/megabreccias (or lowstand wedge) overlain by subsiding (thickening-platformward) reef bodies; Fig. 53a).

In the overlying Rongxian Formation, microbialites predominated over (Fig. 53). In the lower part, in general, the microbial buildups/mounds are mainly composed of thrombolite-like microbialtes with laminated microbiate patches (Fig. 54c); locally *receptaculitids* are present (Fig. 54d). Microbes cover a wide spectrum of elements, but mostly of cyanobacteria.

Fig. 54 **a** Panoramic view on the two-episode stromatoporoid reefs and foreslope collapse wedges in the Tangjiawan Formation (Givetian) at Bao'an. LST-lowstand systems tract, TST-transgressive systems tract, HST-highstand systems tract. **b** Reef builders of in-situ stromatoporoids in growth position. Hammer for scale (25 cm). **c** Thrombolite-like microbial buildup with laminated microbial patches and framework cavities locally filled with submarine fibrous calcite cements (arrows). Mark pen = 14 cm. **d** *Receptaculitid* present locally in the microbialtes

In the mid-upper part, sedimentary succession is mainly composed of skeletal-oncoidal packstones, *Amphipora* grainstone and fenestral limestone (Fig. 55a–d), indicating the lithofacies deposited slightly backwards from the microbial buildups. In the fenestral limestone-dominated intervals, some subaerial exposure indicators such as sheet-like cracks (Fig. 55d), tepee structures (Fig. 55e) and microkarstic surface (Fig. 55f) are present, indicating a very shallow water condition subject to intermittent subaerial exposure. As a whole, all these features generally support a platform margin rimmed with stromatoporoid reefs, notably in the Givetian interval and microbial reefs in the Frasnian time.

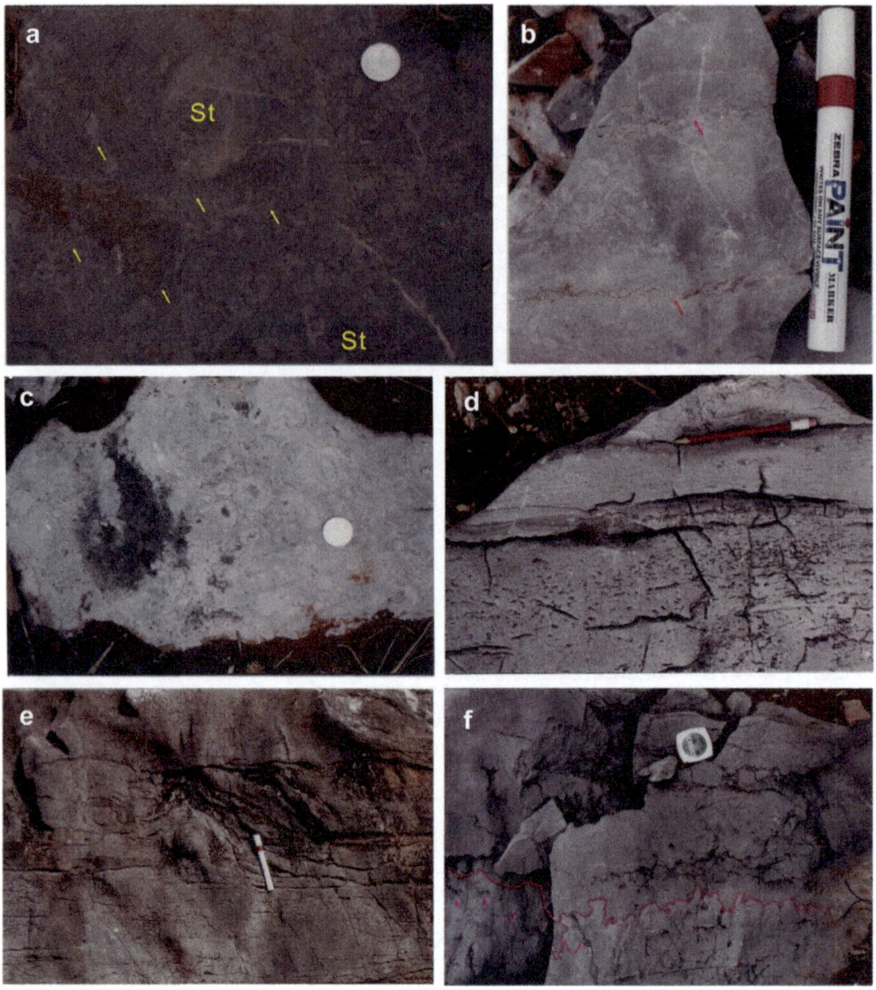

Fig. 55 Backreef lithofacies present mainly in the Rongxian Formation (Frasnian). **a** skeletal-oncoidal packstone/grainstone composed mainly of oncoids, bulbous stromatoporoid (St), *Amphipora* clast (arrows) and peloids. The skeletal grains commonly served as cortexes of the oncoids. Coin for scale (2.5 cm across). **b** *Amphipora* grainstone/packstone with multiple internal erosive surfaces (arrows). Mark pen for scale (14 cm). **c** Oncolite composed dominantly of oncoids and other fragments (microbial fragments, peloids, ostracods), Coin for scale (2.5 cm). **d** Fenestral limestone composed of alternating irregular to laminoid fenestral limestone. Note the sheet-crack filled with submarine fibrous calcite cements in the laminoid fenestral horizon. Pen for scale (12.5 cm). **e** Tepee structure locally occurred in the laminoid fenestral horizons. **f** Serrated microkarstic surface overlain by peloidal grainstone/packstone, under which dissolution pipes/vugs filled by either the overlying peloidal grainstone/packstone (the larger ones; arrows) or submarine fibrous calcite cements

5.3.5 Stop 5

Frasnian Slope Deposits at Qifengzhen, Southeastern Guilin

Qifengzhen, gaining its name for the peculiar picturesque karstic tower hills/pillars, is about 10 km southeast of Guilin. Paleogeographically, it was located on the eastern slope of Guilin Platform, notably during the Frasnian (Fig. 4), so that a slope succession of Frasnian (~300m thick) was deposited, used to be named as the Qifengzhen Formation roughly equivalent to the Gubi Formation (For the convenience, I would temporarily use this nomenclature).

Qifengzhen Formation (~300 m thick) is characterized by dark grey thin-bedded calciturbidites, lenticular (boudinage-shaped) wackestones/mudstones in the lower part and thicker-bedded oncolites, microbialites (stromatolites, thrombolite) in the upper part. It overlies conformably the Guilin Formation dominated by light-coloured thick-bedded grainstones/packstones bearing ooids, peloids, microbial fragments and other intraclasts, indicating platform marginal facies. Upwards, Qifengzhen Formation is overlain by Famennian Rongxian Formation composed of ligh grey thick-bedded limestones (microbial buildups, grainstones/packstones).

In the lower part of the section, the carbonates are dominated by platy to thin-bedded calciturbidites and lenticular wackestones/mudstones with shale partings (Fig. 56a). In the mid-lower part, we can observe the onlap surface and downward-curving intraformational truncation (Fig. 56a), which commonly occur on the slope setting. In the middle part, two wedge-shaped megabraccia bodies were developed (Fig. 56a, b) which contain various lithoclasts of fenestral limestone, oncolite, elongate pebble of calciturbidite, indicating derivation from the upper slope-marginal zone. Locally, soft sediment deformation structures (e.g., slump folds) can be observed. In the upper part, microbial-bearing limestones (oncolite, stromatolite and thrombolite) are intercalated (Fig. 56c).

6 Other Related Information

As stated before, the Guilin city is dispersed on the karstic lowlands/plains surrounded by numerous karstic tower hills dominantly composed of Devonian carbonates, Except for the scenery spots and historic sites, there are also many localities of sedimentological and paleontological interests just in walking distance, such as the Elephant Hill, the symbol of Guilin, showing the paleokarst features across the F/F boundary (Fig. 57), Laoren Shan (Old Man Hill) exhibiting accumulation of *Amphipora*, coral and brachiopod in Frasnian, as well as the Diecai Shan (Rising Rose Hill). They are worthy of visiting.

Fig. 56 Slope deposits of Frasnian at Qifengzhen. **a** Panoramic view of the slope deposits composed mainly of thin-bedded calciturbidites in which onlap termination (OL) and intraformation truncation (downward-concaving) surface and megabreccias (M1, M2) were developed, locally soft deformation (e.g., slump fold) occurred. Note the differences in the intraformational truncation surface and onlap surface. **b** Close-up view of the megabreccia (M2), ca. 3 m thick. **c** Oncoidal floatstone in the thicker-bedded strata in upper part. Key for scale (~5 cm long)

7 Logistic Information

Guilin is located in the subtropical area, so it generally hot and somewhat humid at the end of August (25–35 °C), and not uncommon raining as well. We would recommend participants to prepare themselves umbrellas and/or other water-proof coverings. In addition, Guilin is a popular tourist city well known for its peculiar scenic beauty, so easily got accessible by air flight and Express railway system (Fig. 5). We would like to arrange the pick-up services at the Guilin Liangjiang International Airport on the arriving and departure days. Those who take the Express trains (or other transportation vehicles) are encouraged to take the Taxi or walk to the reserved hotel by themselves. All participants will be arranged to stay in a same hotel (4 stars) near Guilin city centre. During field trips, a lunch box will be provided, and an evening buffet be prepared in the hotel for all participants. Travelling in Guilin area is safe. If there is any emergency, please let us know immediately or call 110 (police) and 120 (health help) directly.

Fig. 57 Elephant Nose Hill comprising the Dongcun Formation in city center, symbol of Guilin City. Note the paleokarst surface under the Nose (arrows), the undulatory paleokarstic surface (even sinkholes) on the light grey fenestral limestones was covered by transgressive black bioclastic (ostracod mainly) mudstones/wackestones deposited in deeper-water (lagoon) condition. This subaerial exposure surface and related karstification were supposed to have been formed in response to sea-level fall at the end of Frasnian (Chen et al. 2004)

Acknowledgements The depositional records of Devonian-Carboniferous carbonates presented here are the main research outcomes of the senior guide achieved since 1990's. We sincerely thank the Chinese National Natural Science Foundation (NSFC) for the long-term financial support, enabling these researches possible. During field works, numerous invaluable helps were provided by senior geologists Zhenliang Li, Bao'an Yin, Hongjin Lu from the Regional Geological Survey and Research Institute of Guangxi based in Guilin.

References

Chen DZ, Guo ZH, Jiang MS, Guo C, Ding Y (2016) Dynamics of cyclic carbonate deposition and biotic recovery on platforms during the Famennian of Late Devonian in Guangxi, South China: Constraints from high-resolution cycle and sequence stratigraphy. Palaeogeog Palaeoclimat Palaeoecol 448:245–265

Chen DZ, Wang JG, Racki G, Li H, Wang CY, Ma XP, Whalen, M (2013) Large sulphur isotopic perturbations and oceanic changes during the Frasnian-Famennian transition of the Late Devonian. J Geol Soc Lond 465–476

Chen DZ, Qing HR, Yan X, Li H (2006) Hydrothermal venting and basin evolution (Devonian, South China): Constraints from rare earth element geochemistry of chert. Sediment Geol 183:203–216

Chen DZ, Qing HR, Li RW (2005) The Late Devonian Frasnian-Famennian (F/F) biotic crisis: Insights from δ^{13}Ccarb, δ^{13}Corg and ^{87}Sr/^{86}S isotopic systematics. Earth Planet Sci Lett 235:151–166

Chen DZ, Qing HR, Yang C (2004) Multistage hydrothermal dolomites in the Middle Devonian (Givetian) carbonates from the Guilin area, South China. Sedimentology 51:1029–1051

Chen DZ, Tucker ME (2004) Palaeokarst at the Frasnian-Famennian transition, Guilin, South China and its implication. J Geol Soc Lond 161:895–899

Chen DZ, Tucker ME (2003) The Frasnian-Famennian mass extinction: insights from high-resolution sequence stratigrphy and cyclostratigrphy in South China. Palaeogeogr Palaeoclimat Palaeoecol 193:87–111

Chen DZ, Tucker ME, Shen YA, Yans J, Preat A (2002a) Carbon isotope excursions and sea-level change: implications for the Frasnian-Famennian biotic crisis. J Geol Soc Lond 159:623–626

Chen DZ, Tucker ME, Zhu JQ, Jiang MS (2002b) Carbonate platform evolution: from a bioconstructed platform margin to a sand-shoal system (Devonian, Guilin, South China). Sedimentology 49:737–764

Chen D, Tucker ME, Zhu J, Jiang M (2001a) Carbonate sedimentation in a starved pull-apart basin, Middle to Late Devonian, southern Guilin, South China. Basin Res 13:141–167

Chen DZ, Tucker ME, Jiang MS, Zhu JQ (2001b) Long-distance correlation between tectonic-controlled, isolated carbonate platforms by cyclostratigraphy and sequence stratigraphy in the Devonian of South China. Sedimentol 48:57–78

De Vleeschouwer D, da Silva A-C, Sinnesael M, Chen DZ, Day J, Whalen M, Guo ZH, Claeys P (2017) Timing and pacing of the Late Devonian mass extinction event regulated by eccentricity and obliquity. Nature Com 8:226. https://doi.org/10.1038/s41467-017-02407-1

Guangxi Bureau of Geology and Mineral Resources (GBGMR) (1987) The Sedimentary Facies Palaeogeography and Relatively Mineral Deposits of Devonian in Guangxi. Guangxi People's Publishing House, 292 pp (in Chinese with English abstract)

Guangxi Bureau of Geology and Mineral Resources (GBGMR) (1988) Report of regional geological survey for Guilin area (scale: 1: 50, 000)

Guangxi Bureau of Geology and Mineral Resources (GBGMR) (1994) Report of regional geological survey for Guilin-Yangshuo area (scale: 1: 50, 000)

Huang C, Gong YM (2016) Timing and patterns of the Frasnian-Famennian event: Evidences from high-resolution conodont biostratigraphy and event stratigraphy at the Yangdi section, Guangxi, South China. Palaeogeogr Palaeoclimat Palaeoecol 448:317–338

Huang C, Joachimski MM, Gong Y (2018) Did climate changes trigger the Late Devonian Kellwasser Crisis? Evidence from a high-resolution conodont δ^{18}OPO4 record from South China. Earth Planet Sci Lett 495:174–184

Ji Q (1989) On the Frasnian-Famennian mass extinction event in South China. Courier Forsch-Inst Senckenberg 117:275–301

Li ZL (1992) The Devonian-Carboniferous boundary of different sedimentary facies in Guilin. Regnal Geol China 2:97–110

Ma XP, Bai SL (2002) Biological, depositional, microspherule, and geochemical records of the Frasnian/Famennian boundary beds, South China. Palaeogeogr Palaeoclimatol Palaeoecol 181:325–346

Ma XP, Sun YL, Hao WC, Liao WH (2002) Rugose corals and brachiopods across the Frasnian-Famennian boundary in central Hunan, South China. Acta Palaeontol Pol 47:373–396

Ma XP, Gong YM, Chen DZ, Racki G, Chen XQ, Liao WH (2016) The Late Devonian Frasnian-Famennian Event in South China—Patterns and causes of extinctions, sea level changes, and isotope variations. Palaeogeogr Paleoclimat Palaeoecol 448:224–244

Ma XP, Sun YL, Bai ZQ, Wang SQ (2004) New advances in the study of the Upper Devonian Frasnian of the Shetiaqiao section, central Hunan, south China. J Stratigr 28(4):367–374 (in Chinese with English abstracts)

Qie WK, Liu JS, Chen JT, Wang XD, Mii H-S, Zhang XH, Huang X, Yao Y, Algeo TJ, Luo GM (2015) Local overprints on the global carbonate d13C signal in the Devonian-Carboniferous boundary successions of South China. Palaeogeogr Paleoclimat Palaeoecol 418:290–303

Shen JW (1995) Middle-Upper Devonian conodont succession and sea-level change in Guilin. Acta Micropaleont Sinica 12:251–273

Shen JW, Zhang SL (1997) A Late Devonian (Frasnian) coral-bafflestone reef at Houshan in Guilin, South China. Facies 37:85–98

Shen JW, Yu CM, Bao HM (1997) A Late-Devonian (Famennian) Renalcis-Ephiphyton reef at Zhaijiang, Guilin, South China. Facies 37:43–46

Shen JW, Webb GE (2004) Famennian (Upper Devonian) clacimicrobial (Renalcis) reef at Miaomeng, Guilin, Guangxi. South China. Palaeogeogr Paleoclimat Palaeoecol 204:394

Song H, Song H, Algeo TJ, Tong J, Romaniello SJ, Zhu Y, Chu D, Anbar AD (2017) Uranium and carbon isotopes document global-ocean redox-productivity relationships linked to cooling during the Frasnian-Famennian mass extinction. Geology 45:887–890

Wang C-Y, Ziegler W (2002) The Frasnian-Famennian conodont mass extintion and recovery in South China. Senckenbergiana Lathaea 82:463–493

Wang X, Liu SA, Wang ZR, Chen DZ, Zhang LY (2018) Zinc and strontium isotope evidence for climate cooling and constraints on the Frasnian-Famennian (~372 Ma) mass extinction. Palaeogeogr Paleoclimat Palaeoecol 498:68–82

Whalen MT, Sliwinski MG, Payne J, Day JE, Chen D, Da Silva A-C (2015) Chemostratigraphy and magnetic susceptibility of the Late Devonian Frasnian-Famennian transition in western Canada and southern China: Implications for carbon and nutrient cycling and mass extinction. In: Da Silva AC, Whalen M, Hladil J, Chadimova L, Chen D, Spassov S, Boulvain F, Devleeschouwer X (Eds) Magnetic Susceptibility Application: A window onto Ancient Environments and Climatic Variations. Geological Society (London) Special Publication 414: 37–72, 2015

Yu CM (1988) Devonian-Carboniferous Boundary in Nanbiancun, Guilin, China—Aspects and Records. Science Press, Beijing, 390pp. with 87 pls

Yu CM (1989) Comprehensive evaluation of the Devonian-Carboniferous boundary section in Nanbiancun, Guilin. Guangxi. Acta Geologica Sinica 3:246–254

Zhang LY, Chen DZ, Kuang GD, Guo ZH, Zhang GJ, Wang X (2020a) Persistent oxic deep ocean conditions and frequent volcanic activities during the Frasnian-Famennian transition recorded in South China. Global Planet Change 195:103350. https://doi.org/10.1016/j.gloplacha.2020.103350

Zhang LY, Chen DZ, Huang TY, Yu H, Zhou XQ, Wang JG (2020b) An abrupt oceanic change and frequent climate fluctuations across the Frasnian-Famennian transition of Late Devonian: Constraints from conodont Sr isotope. Geol J 55(6):4479–4492. https://doi.org/10.1002/gj.3657

Zhong K, Wu Y, Yin BA, Liang YL, Yao ZG, Peng JL (1992) Stratigraphy of Guangxi, China. Part 1: Devonian of Guangxi. China Univ Geosci Press, 384 pp (in Chinese with English abstract)

Stratigraphy and Sedimentology of the Proterozoic at the Jixian Section, Tianjin

Mei Ming-Xiang

1 Introduction

Jixian County, an ancient county (Yuyang County) in North China, is situated in the southern foot of the Yanshan Range and in the northern outer suburbs of Tianjin municipality. It is about 90 and 100 km from Beijing and Tianjin respectively. The Proterozoic strata (~1800 Ma to ~850 Ma) are well exposed in the north of the Jixian area (Fig. 1).

In this area, a syncline with folding axis extending in a NW direction is observed in the Hongshuizhuang to Jingeryu villages with the Cambrian rocks developed as the core of the syncline. A thrust following the strike of strata spread out from Yangzhuang to Chuancangyu villages where a stratigraphic repetition is found. In addition, a big N-S trending fault occurs from Huangyaguan to Yangzhuang and a set of thrusts with a NEE trend is seen along the front of the hill, to the north of the county town.

The Jixian Proterozoic succession overlies Archaean gneiss and schist near the Great Wall, and extends south for ca 20 km to near Jixian city where it is overlain by Cambrian rocks. The Proterozoic geology of Jixian area and its vicinity was first introduced by Gao Zhenxi (C. C. Kao) et al. in 1934. They described the lithology and stromatolites there and made the lithostratigraphic subdivision. Since then, the Jixian section has been regarded as a typical section for the Middle and Upper Proterozoic of North China. This is the classic 'Jixian Section' of North China (Zhu 1993; Zhu et al. 1994; Wu and Chen 1997; Su 2014).

After 1960's, the Proterozoic successions around Jixian and Yanshan Range have been studied by many geologists on lithology, palaeontology, geochemistry, geochronology, pleomagnetism and plaeogeography, and a lot of papers and some

M. Ming-Xiang (✉)
China University of Geosciences, Beijing, China
e-mail: meimingxiang@263.net

X. Hu (ed.), *Field Trip Guidebook on Chinese Sedimentary Geology*,
https://doi.org/10.1007/978-981-99-6936-4_16

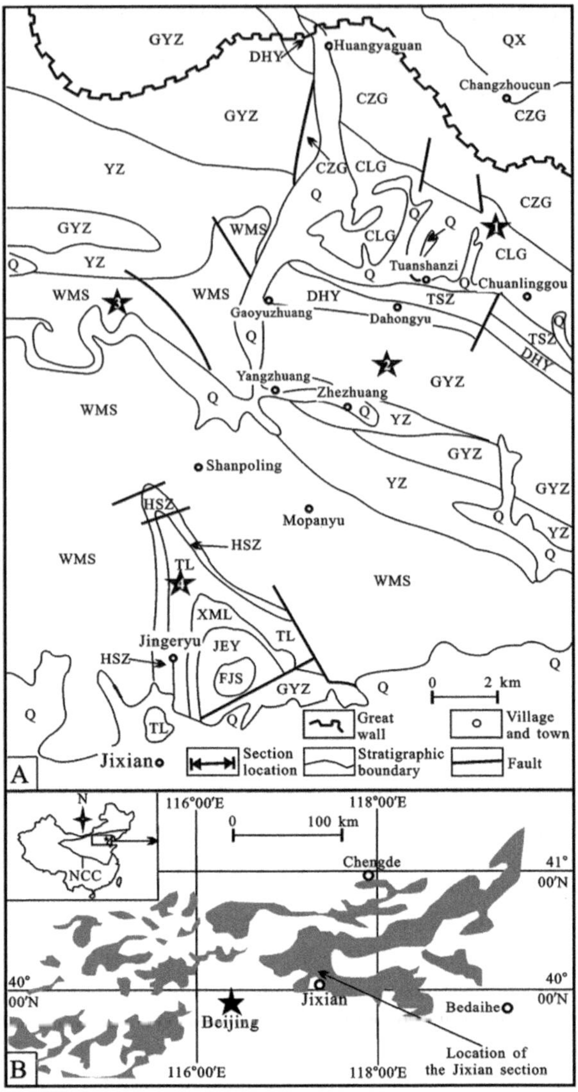

Fig. 1 Simplified geological map (**a**) and sketch map (**b**) for the Jixian area in Tianjin, North China. The units in **a** are as follows: Changzhougou (CZG), Chuanlinggou (CLG), Tuanshanzi (TSZ), and Dahongyu (DHY) formations of the Paleoproterozoic; Gaoyuzhuang (GYZ), Yangzhuang (YZ), Wumishan (WMS), Hongshuizhuang (HSZ), Tieling (TL) and Xiamaling (XML) formations of the Mesoproterozoic; Changlongshang (CLS), and Jingeryu (JEY) formations of the Neoproterozoic; Fujunshan Formation (FJS) of the Cambrian; Q indicates modern alluvium; and stars marked by 1 to 4 refer to four stops. NCC in **b** means the North-China Craton, and the gray shadings refer to the outcrops of strata from the Paleo-to Mesoproterpozoic

monographs have been published. Previously, the Changzhougou-Xiamaling succession as a whole was regarded as ~1800–950 Ma, and the Jixian Group (with the Tieling Formation at its top) as ~1400–1000 Ma (Zhu 1993; Zhu et al. 1994; Wu and Chen 1997). However, the new data indicate that the Tieling Formation is ~1.4 Ga rather than ~1.0 Ga in age. Generally, the Jixian Proterozoic succession can be subdivided into two parts (as summarized by Tosti and Riding (2017)). The lower part (ca 1650 to 1320 Ma) is generally divided into the Changcheng Group that had been traditionally defined as the Changchengian System (Changzhougou, Chuanlinggou, Tuanshanzi and Dahongyu formations), the Jixian Group that had been traditionally defined as the Jixianian System (Gaoyuzhuang, Yangzhuang, Wumishan, Hongshuizhuang and Tieling formations) and the Xiamaling Formation. The upper part (ca 1000 to 800 Ma) consists of the Qingbaikou Group that had been traditionally defined as the Qingbaikouan System (Changlongshan and Jingeryu formations) (Su et al. 2014). These ages are based on SHRIMP data (Gao et al. 2007; Li et al. 2013; Su et al. 2014). Thus, these new data suggest that the Changzhougou Group is Late Palaeoproterozoic, whereas the Jixian Group and Xiamaling Formation are ca 1600 Myr to ca 1320 Myr old (early to mid-Mesoproterozoic), and that the remainder of the Mesoproterozoic is absent (Gao et al. 2007; Li et al. 2013; Su et al. 2014; Su 2014) due to the long-term hiatus as that between the Mesoproterozoic Xiamaling and the Neoproterozoic Changlongshan Formation within the current Proterozoic time scale (Fig. 2).

Era	Age	Period	Description
Neoproterozoic	541(Ma) – 635	Ediacaran	Ediacara refers to a name from Australian Aboriginal term for place near water, so the Edicaran Period delegates the evolution of first megascopic marine animals, and the GSSP bottom boundary of the Edicaran is marked by the end of both the global glaciation and the excursion of carbon isotope.
	635 – 850	Cryogenian	"Cryos" means ice, and the Cryogenian Period represents the period of global glaciations, i.e. glacial deposits that typify the late Proterozoic are most abundant during this interval.
	850 – 1000	Tonian	"Tonas" means stretch, thus, the Tonian Period delegates an interval of the further major platform cover expansion (e.g., Upper Riphean, Russia.; Qingbaikou, China; basins of Northwest Africa), following final cratonization of polymetamorphic mobile belts, below.
Mesoproterozoic	1000 – 1200	Stenian	"Stenos" means narrow, so the Stenian further indicates an interval of the narrow belts of intense metamorphism & deformation, that is, narrow polymetamorphic belts, characteristic of the mid-Proterozoic, separated the abundant platforms and were orogenically active at about this time (e.g., Grenville, Central Australia)
	1200 – 1400	Ectasian	"Ectsis" means extension, so the Ectasian Period delegates a period of the continued expansion of platform covers, that is, platforms continue to be prominent components of most shields.
	1400 – 1600 Ma	Calymmian	"Calymma" means cover, so the Calymmian Period refers to an interval that is marked by the development of platform covers and is characterized by expansion of existing platform covers, or by new platforms on recently cratonized basement (e.g. Riphean of Russia).
Palaeoproterozoic	1600 Ma – 1800	Statherian	"Statheros" means stable and firm, so the Statherian Period represents an interval of stabilization of cratons or cratonization, and this period is characterized on most continents by either new platforms (e.g., North China, North Australia) or final cratonization of fold belts (e.g. Baltic Shield, North America).
	1800 – 2050	Orosirian	"Orosira" means mountain range, so the Orosirian Period refers to a global orogenic period, and the interval between about 1900 Ma and 1850 Ma was an episode of orogeny on virtually all continents.
	2050 – 2300	Rhyacian	"Rhyax" means stream of lava, so the Rhyacian Period represent an interval of the injection of layered complexes, that is, the Bushveld Complex (and similar layered intrusions) is an outstanding event of this time, but the age of the Bushveld seems unlikely to change dramatically.
	2300 – 2500 Ma	Siderian	"Sideros" means iron, so the Siderian Period delegates a period of the widespread development of the banded iron formations (BIF), and the earliest Proterozoic is widely recognized for an abundance of BIF, which peaked just after the Archeane-Proterozoic boundary.

Fig. 2 The current Proterozoic time scale. From the International Commission on Stratigraphy, based on Plumb (1991)

More interestingly, under a fully revised Proterozoic time scale proposed by Van Kranendonk et al. (2012), the Changcheng Group, the Jixian Group and the Qingbaikou Group at the Jixian section are ca. 1800 Myr (or 1600 Ma) to ca 850 Myr old (the revised Mesoproterozoic or the Rodinian Period; Fig. 3). A revised Mesoproterozoic Era would extend from ca. 1780 Ma to ~850 Ma and encompass the period of environmental stability marked by sulfidic deep oceans, the slow diversification of eukaryotes, and the disassembly of supercontinent Columbia and subsequent aggregation of supercontinent Rodinia (1300 ~ 900 Ma), since no subdivisions of this era into periods have yet been recognized. Thus, a revised Mesoproterozoic Era is temporarily grouped into one period, i.e. the Rodinian Period (Van Kranendonk et al. 2012). Further, since the disappearance of the molar-tooth structure at ~720 Ma that has been used as a mark to revise the bottom boundary in "International Chronostratigraphic Chart (2015)" (Fan et al. 2015), a revised Mesoproterozoic Era extending from ca. 1780 Ma to ~850 Ma that was proposed by Van Kranendonk et al. (2012) should be extend from ca. 1780 Ma to ~720 Ma.

The interpretation of the position of the NCC within Nuna/Columbia and subsequently in Rodinia (Bogdanova et al. 2009; Pesonen et al. 2012; Wilde 2014) is

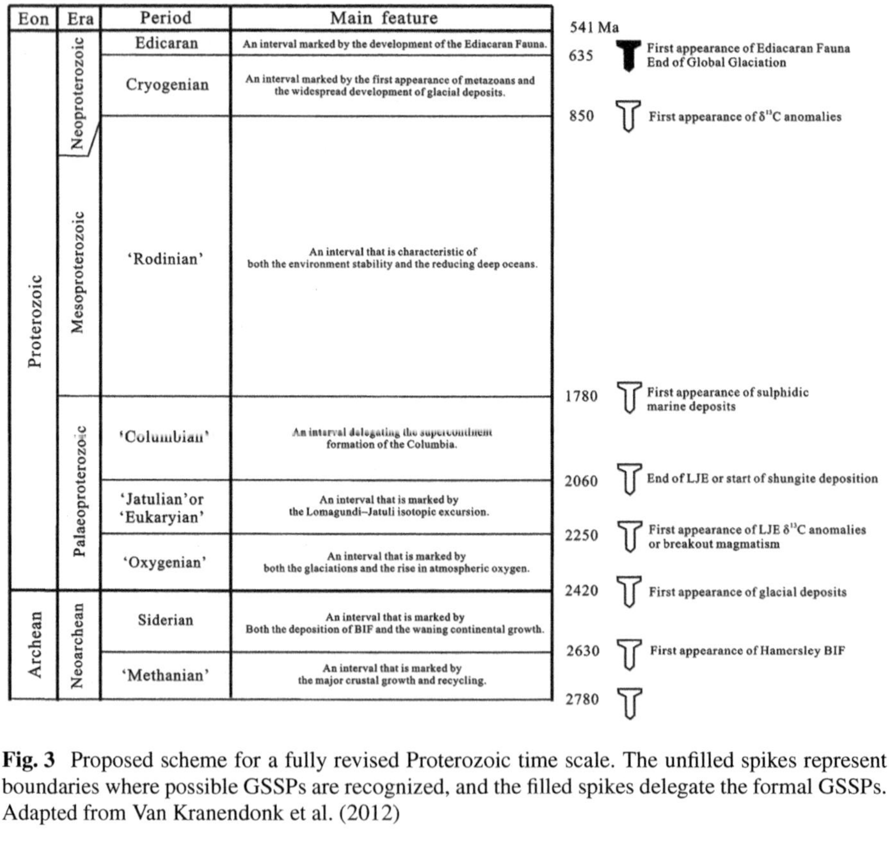

Fig. 3 Proposed scheme for a fully revised Proterozoic time scale. The unfilled spikes represent boundaries where possible GSSPs are recognized, and the filled spikes delegate the formal GSSPs. Adapted from Van Kranendonk et al. (2012)

profited from the following researches. The ca. 1650 to 1400 Myr old Changcheng–Jixian succession, with the Tieling Formation at its top (Li et al. 2013), is currently thought to have accumulated in an extensive ensialic basin, the Yanliao Aulacogen on the Eastern Block of the North-China Craton (NCC) that could be part of Nuna/Columbia (Zhang et al. 2012; Zhao 2014), prior to rifting and emplacement of diabase sills at ca 1.35 Ga that marks 'final breakup of the NCC from the Columbia supercontinent' (Zhao and Cawood 2012; Zhao 2014). The global continental reconstructions have placed the NCC external to India and to portions of Australia on the south-east side of Nuna at ca. 1.6 Ga (Zhang et al. 2012), and external to Siberia and portions of Australia on the eastern side of Nuna at 1.45 Ga. Further, Further, Chen et al. (2013) placed North China close to India and Siberia, and Wang et al. (2015) placed it between Laurentia and Baltica. Zhang et al. (2012) proposed that from ca 1.80 to 1.4 Ma, North China was in an equatorial to low-latitude location, consistent with red beds, reefs (presumably the Tieling stromatolites) and dolomites in these deposits, and that the Jixian area was approximately 25 degrees north during Tieling deposition.

Much of the Changcheng Group and Jixian Group succession (ca 1650 to 1400 Ma) is dominated by carbonates (Zhu 1993; Zhu et al. 1994; Wu and Chen 1997). Stromatolites are locally abundant, particularly in the Wumishan, and Tieling formations of the Jixian Group. They include microdigital forms (for example, *Pseudogymnosolen*) and large *Jacutophyton* in the Wumishan, and numerous columnar and branched forms (attributed to *Anabaria*, *Baicalia* and *Chihsienella*) in the Tieling Formation (Zhu 1993; Cao and Yuan 2006). In addition to the stromatolite, the microbial-induced sedimentary structure (MISS) within the Chuanlingou Formation, the Milakovitch-driven cycle within the Wumishan Formation and the molar-tooth structure within the Gaoyuzhuang Formation are also interesting sedimentological phenomena during the revised Mesoproterozoic or the Rodinian Period. Furthermore, the greater time span of the Mesoporteroic succession (the Changcheng Group, the Jixian Group and the Qingbaikou Group) at the Jixian section that are ca 1800 Myr (or 1600 Ma) to ca 850 Myr old (the revised Mesoproterozoic or the Rodinian Period; Fig. 3), provides the opportunity for observing Earth's processes at a broader scale, gaining an improved understanding on issues such as the mechanisms governing stratigraphic cyclicity and the variability thereof (Catuneanu et al. 2005) like the Milakovitch-driven cycle and its orderly stacking pattern in the third-order depositional sequence within the Wumishan Formation. Therefore, it provides better insights into some key issues of sequence stratigraphy.

2 Stratigraphy

2.1 *The Changcheng Group (or the Changchengian System)*

The Late Palaeoproterozoic Changcheng Group (or the Changchengian System) includes four formations with a total thickness of 2762 m. The Changzhougou Formation is constituted of sandstoes. The second, Chuanlinggou Formation, is composed mostly of neritic shales passing transitionally upward into the dolomitic Tuanshanzi Formation. In conformable contact with the Tuanshanzi Formation, the fourth Dahongyu Formation consists of littoral and neritic sandstone, shale and potassium-rich trachyte with cherty dolostone in the upper part.

At the Jixian section in Tianjin (Fig. 1), a four-formation division of the Changcheng Group is employed, given the existence of a clear unconformity between the Dahongyu and Gaoyuzhuang formations (Zhu et al. 2005; Lamb et al. 2007) that represents a hiatus in deposition from 50 to 100 Ma (e.g., Chen he Wu 1997; Zhu et al. 1994, 2005). Furthermore, the upper and lower boundaries of the Dahongyu Formation are marked by unconformities that are thought to indicate crustal uplift and that are termed the Xingcheng Uplift (lower boundary) and Qinglong Uplift (upper boundary) (e.g., Zhu et al. 1994). The cyclicity of sedimentary facies (e.g., Mei et al. 2000; Catuneanu et al. 2005; Eriksson et al. 2005) in the Changcheng Group at the Jixian section makes up one first-order sequence that can be subdivided into 4 s-order sequences and 14 third-order sequences (Fig. 4).

The Changzhougou Formation, the lowermost unit of the Changcheng Group, consists of grayish purple, massive to thick-bedded conglomerate and coarse-grained sandstone, pinkish and white quartz sandstone, sedimentary quartzite and some thin-bedded sandstone, fine-grained sandstone, clayey and feldspathic sandstone with a thickness of 859 m. The lowermost Changzhougou Formation rests unconformably upon the Archean metamorphic Zunhua Complex-Group (Fig. 5a) and is chiefly composed of conglomerate, pebble-bearing sandstone and arkosic sandstones (Fig. 5b–c) of near-shore facies in the lower portion and massive fine-grained sandstone (Fig. 5d) of far-shore marine facies in the middle-upper part, although the lower portion was traditionally interpreted as the fluvial deposits. Thus, the Changzhougou Formation makes up a second-order sequence, which can be subdivided into 4 third-order depostional sequences with the abrupt changing surfaces of sedimentary facies from the restricted mid-to supratidal muddy sandstones to the massive subtidal sadstones (Fig. 5e). Well-preserved acritarchs had been observed within shales and silty shales of the ~1800 million-year-old Changzhougou Formation (lowermost Changcheng Group) in the Pangjiapu Region of North China (the western Yanshan area; Lamb et al. 2009). These microbial body fossils, although they are sphaeromorphs, have a size and fine-scale morphological complexity that would attribute them to the eukaryotic domain on the tree of life. These results suggest that eukaryotic microbial life was well established by the end of the Paleoproterozoic and had already undergone a moderate amount of diversification (Lamb et al. 2007, 2009; Zhang et al. 2021).

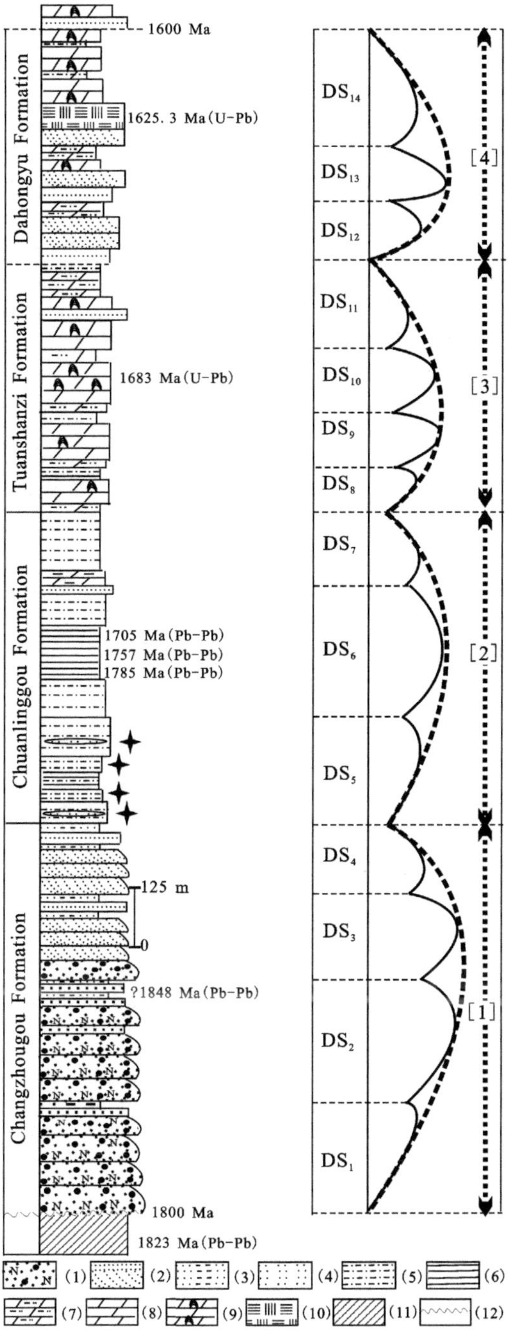

Fig. 4 Sequence stratigraphic division of the Paleoproterozoic Changcheng Group at the Jixian section, also showing age data. DS_1 to DS_{14} refer to 14 third-order sequences identified in the Changcheng Group at the Jixian section, which are grouped into 4 s-order sequences ([1] to [4]). Lithological codes (in the key at the bottom of the figure) are as follows: (1) = conglomerate sandstone with feldspar sand, (2) = cross-bedded sandstone, (3) = muddy siltstone, (4) = siltstone, (5) = silty shale, (6) = shale, (7) = sandy and muddy dolomite, (8) = dolomite, (9) = stromatolitic dolomite, (10) = volcanic rock, (11) = metamorphic rock, (12) = angular unconformity. Stars indicate the stratigraphic locations of siltstone dykes

Fig. 5 Chief features of sedimentology and stratigraphy for the Changzhougou Formation at the Jixian section. Image **a** depicts the unconformity (the arrow) between the Changzhougou Fomation (ca. 1800 Ma) and the Archean metamorphic Zunhua Complex-Group (ca. 2500 Ma), which is marked by weathering crust of allitization; images **b** and **c** express the scouring surfaces (the arrow) within thick-bedded to massive sandstones with cross-beddings; image **d** shows the typical massive fine sandstones of the near-shore facies in the upper part of the Changzhougou Formation; and image **e** indicates the possible sequence boundary of the top boundary of the third third-order depositional sequence within the Changzhougou Formation (the DS_3 in Fig. 4)

The Chuanlinggou Formation consists chiefly of silty and illitic lenses. Its thickness is 889 m at the Jixian section, but it thins markedly towards the eastern and western part of the Yanshan Range. Acritarchs are abundant in this formation and they can be represented by following groups:

Sphaeromorphida (e.g. *Leiominuscula, Margominuscula, Dictgosphaera, Kildinella* and *Leiosphaerida*), Scaphomorphida (e.g. *Dictgosphaera, Kildinella* and *Schizovalvia*), Versimorphida including blade-like, lobe-like, valve-like and band-like forms (e.g. *Pholimorphaba, Lobomorpha* and *Valvimorpha*); and Nematomorphida (e.g. *Qingshania* and *Changchengonema*). In addition, some macroscopic fossil algae of Chuaria-like, ribbon-shaped remains were ever reported. The Chuanlinggou Formation at the Jixian section is composed mainly of peritidal silty shales and can be subdivided into 3 third-order sequences (DS_5 to DS_7; Fig. 4) that are markedly different from sandstones of the underlying Changzhougou Formation and dolomites of the overlying Tuanshanzi Formation. Siltstone dykes occur mainly in fine peritidal deposits with lenses and lenticular beds of siltstones in the lower part of the Chuanlinggou Formation, which might represent a type of microbial induced sedimentary structures (MISSs) that will be introduced in detail in the following context. These fine-grained deposits belonging to the restricted supratidal facies make up the transgressive and early high-stand system tracts of the third-order sequence, and silty shales of the restricted lagoon facies constitute the late high-stand system tracts. There are no stromatolites in this formation at Jixian section, but two horizons containing stromatolites have been found in the western Yanshan Range.

The Tuanshanzi Formation is mainly composed of muddy and silty dolomicrite with a total thickness of 518 m (Fig. 6). The lower member of this formation is rich in muddy, silty and carbonaceous lithologies, with pyrite impregnations and black carbonaceous films on the bedding planes. In the upper member, the content of clastics increases considerably upward and dolomcrite sandstone, sandy dolostone, thin-bedded sandstone and small stromatolite bioherms are common. The lower member of the Tuanshanzi Formation is characterized by even bedding with local quiet, weakly reducing, subtidal environment (lagoonal facies), whereas its upper part is formed in an unstable environment from the intertidal zone to the supratidal zone. Previously, the Tuanshanzi Formation was regarded as a formation only containing a few of stromatolites, microfossils and doubtful carbonaceous megafossils dominated by *Tyrasotaenia* (Hofmann and Chen 1981). However, a large collection of carbonaceous megafossils represented by Iongfenshanids (*Antiqufolia gixianfolium* and others; Zhu and Chen 1995) have been found at the lower part of this fommtion in recent years, and they have been regarded as the most sufficient palaeoproterozoic megascopic eukaryotes or metaphytes. Generally, the Tuanshanzi Formation makes up a second-order sequence that can be subdivided into four thir-order depositional sequences (Fig. 4) in term of following features, i.e. the transgressive system tracts are composed of thick-bedded to massive dolostones with stromatolites from the subtidale to intertidal facies (Fig. 6a) and the high-stand system tracts of third-order sequences are constituted of sily and muddy dolostones with mud-cracks from the intertidal to supratidal facies (Figs. 6b–d). Furthermore, both the central development of purple sandy and silty mudstones belonging to the supratidal facies and an

Fig. 6 Chief features of sedimentology and stratigraphy for the Tuanshanzi Formation at the Jixian section. Image **a** shows the thick-bedded dolostones with stromatolites generally making up the transgressive system tract of third-order sequences; image **b** depictes the thin-to middle-bedded silty and sandy dolostones making up the high-stand system tract of third-order sequences; images **c** and **d** express mud-crack structures within silty and sandy dolostones; and image **d** indicates the boundary between the Tuanshanzi and the Dahongyu formations (the course arrow), under which are marked by the central development of purple silty and sandy mudstones of the supretidal facies of the Tushanzi Formation, over which are characteristic of the vertical stacking pattern of milakovitch-driven meter-scale cycles (from (1) to (4)) of the Dahongyu Formation

upward thinning succession of meter-scale cycles belonging to the tidal-dynamic type (Fig. 6d) mark a third-to second-order sequence boundary. These mud-cracks (Figs. 6b–d) show evidences of unusual cohesive behavior (cracks resulting from shrinking or degassing; resistance to erosion) that is reflected both the big width and the small depth, which can be grouped into the microbial induced sedimentary structures (Schieber 1999; Noffke et al. 2001).

The Dahongyu Formation can generally be subdivided into two parts. The lower part consists mainly of white quartz sandstone and sandy dolostone, intercalated with purple and green tuffeceous siltstone or tuffaceous silty shale, some beds of volcanic breccia, tuffaceous sandstone and high potassium lava. The upper part comprises dolomicrite and cherty dolomicrite with abundant stromatolites. According to the following features, i.e. one retrogradation succession (Figs. 6d and 7a) of meter-scale cycles belonging to the tidal-dynamic type in the transgressive system tract of the third-order sequences (Figs. 6d and 7b), and high-energy subtidal massive sandstones (Fig. 7c and d) make up the middle part (the position of the condensed section) of the third-order sequences, the Dahongyu Formation can be grouped into one second-order sequence that can further be subdivided into three third-order sequences (Fig. 4).

2.2 The Jixian Group (or the Jixiannian System)

Disconformably deposited on the uppermost formation of the Changcheng Group, this system consists of six conformable formations. The lowermost Gaoyuzhuang Formation is composed of lithoral-neritic dolostone and dolomitic limestone. The Yangzhuang Formation is principally made up of marine alternating red-white dolostone with minor limestone. The overlying Wumishan Formation is composed of thick beds of siliceous dolostone intercalated with abundant cherty bands and goes upward into the Hongshuizhuang Formation of dominant black shale. Traditionally, the Tieling Formation was thought as the uppermost part of this group, which consists of shale and limestone with abundant stromatolites. New data suggest that the Xiamaling Formation are ca 1320 Myr old (Gao et al. 2007; Su 2014; Su et al. 2014), which should be grouped into the uppermost part of the Mesoproterozoic Jixian Group. Thus, the Jixian Group is ca 1600 Myr to ca 1320 Myr old.

The Jixian Group is composed of two successions that might represent two first-order sequences: one is constituted of the succession from the Gaoyuzhuang to the Yangzhuang Formation (Fig. 8), another succession is made up of the succession from the Wumishan to the Xiamaling Formation (Fig. 9). The succession from the ca. 1600 m thick Gaoyuzhuang of tidal-flat carbonate rocks to the ca. 700 m thick Yangzhuang Formation (Fig. 8) of red beds with dolostones make up one first-order sequence that can further subdivide into 5 s-order sequences and 17 third-order depositional sequences. Further, the ca. 1600 m thick Gaoyuzhuang of tidal-flat carbonate rocks can be subdivided into 4 s-order sequences and 13 third-order sequences, and the ca.

(a)
165 cm
(b)
9.0 cm
(c)
35.0 cm
(d)
2.5 cm

◀**Fig. 7** Chief features of sedimentology and stratigraphy for the Dahongyu Formation at the Jixian section. Image **a** shows one retrogradation succession (the arrow) of meter-scale cycles belonging to the tidal-dynamic type in the transgressive system tract of the third-order sequences; image **b** indicates meter-scale cycles (the arrow) belonging to the tidal-dynamic type composed of intertidal muddy sandstones and the supratidal purple silty and sandy mudstones in the lower part of the Dahongyu Formation; image **c** depicts meter-scale cycles (the arrow) belonging to the tidal-dynamic type composed of massive subtidal sandstones and thin-bedded intertidal to supratidal muddy sandstone; and image **d** expresses the thin-bedded intertidal to supratidal muddy sandstone (the arrow) making up the upper part of a meter-scale cycle belonging to the tidal-dynamic type

700 m thick Yangzhuang Formation (Fig. 8) of red beds with dolostones make up one second-order sequence that can further subdivide into 4 third-order sequences.

Together with the Tieling, the Hongshuizhuang and the Xiamaling formations, the Wumishan Formation with a thickness of 3336 m make up one first-order sequence that can further be subdivided into 7 s-drder sequences and 31 third-order sequences (Fig. 9). Like the Gaoyuzhuang Formation with a thickness of 1600 m, the Wumishan Formation with a thickness of 3336 m can further be subdivided into 6 s-order sequence and 26 third-order sequences (Fig. 9) reflecting a particular stratigraphical and sedimentological phenomena of the Precambrian which are different from the Phanerozoic.

The Gaoyuzhuang Formation has traditionally been thought as the top of the Changcheng Group by the current subdivision of China (Xinget al. 1989; Song et al. 1991; Zhu et al. 1994; Chen and Wu 1997). However, it is referred to the bottom of the Jixian Group due to its unconformable contact with the underlying Dahongyu Formation and its identical sedimentary and palaeotological characteristics with the "Jixian System (Jixian Group)" (Zhu et al. 2005). This formation is primarily composed of carbonate rocks, subdivided into four members (or subformations) as follows:

The first Member (or the Guandi Subformation) is mainly characterized by dark-grey cherty dolomicrites, lenticular layers of dololutite and dolorudite, and is intercalated with thin-bedded clayed dolomicrites and shales or silty shales. In addition, a bed of sandstone (Fig. 10a–b) with ripple marks (Fig. 10c), about three meters thick, where an erosional surface occurs on the top of its underlying bed, constitutes the transgressive deposits of the Gaoyuzhuang Formation.

According to the cyclicity reflected by succession of sedimentary facies, i.e. the transgressive system tracts (TSTs) are generally made up of thick-bedded dolostones with straomatolites of the subtidal to intertidal facies (Fig. 11a–b) and the high-stand system tracts (HSTs) are mainly composed of muddy dolostone of the supratidal facies (or the lagoon facies; Fig. 11c), the first Member of Gaoyuzhuang Formation makes up one second-order sequence that can further be subdivided into six third-quences (Fig. 8).

The second Member (or the Sangshuan subformation) of the Gaoyuzhuang Formation may be subdivided into two third-order sequences. Thick-bedded to massive manganiferous dolostones of the intertidal to subtidal facies (Fig. 12a) make up the TST unit of third-order sequences, which represent the response of the upward deepening process of sedimentary environments resulting from the sea-level rising.

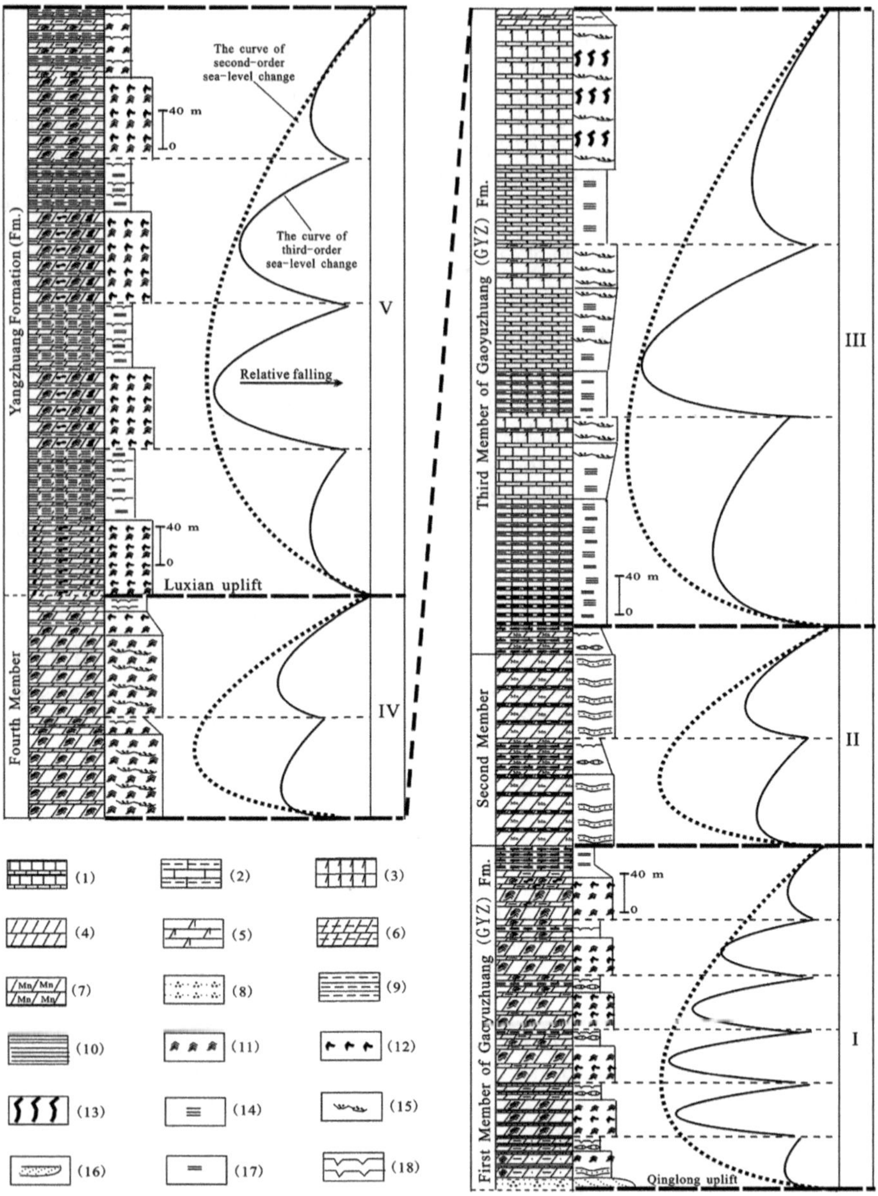

Fig. 8 Sequence stratigraphic division for a succession from the Gaoyuzhuang to the Yangzhuang Formation in the lower part of the Jixian Group at the Jixian section. I to V refer to 5 s-order sequences identified in the succession from the Gaoyuzhuang to the Yangzhuang Formation in the lower part of the Jixian Group at the Jixian section, which can further subdivided into 17 third-order sequences. Lithological codes (in the key at the bottom of the figure) are as follows: (1) = limestone, (2) = marl, (3) = dolomitic limestone, (4) = dolostone, (5) = lime dolostone, (6) muddy dolostone, (7) = manganese dolostone, (8) = sandstone, (9) = mudstone, (10) = shale, (11) = stromatolite, (12) = chert, (13) = molar-tooth structure, (14) = lamilite, (15) = scouring surface, (16) = cross bedding, (17) = horizontal bedding, and (18) = mud crack

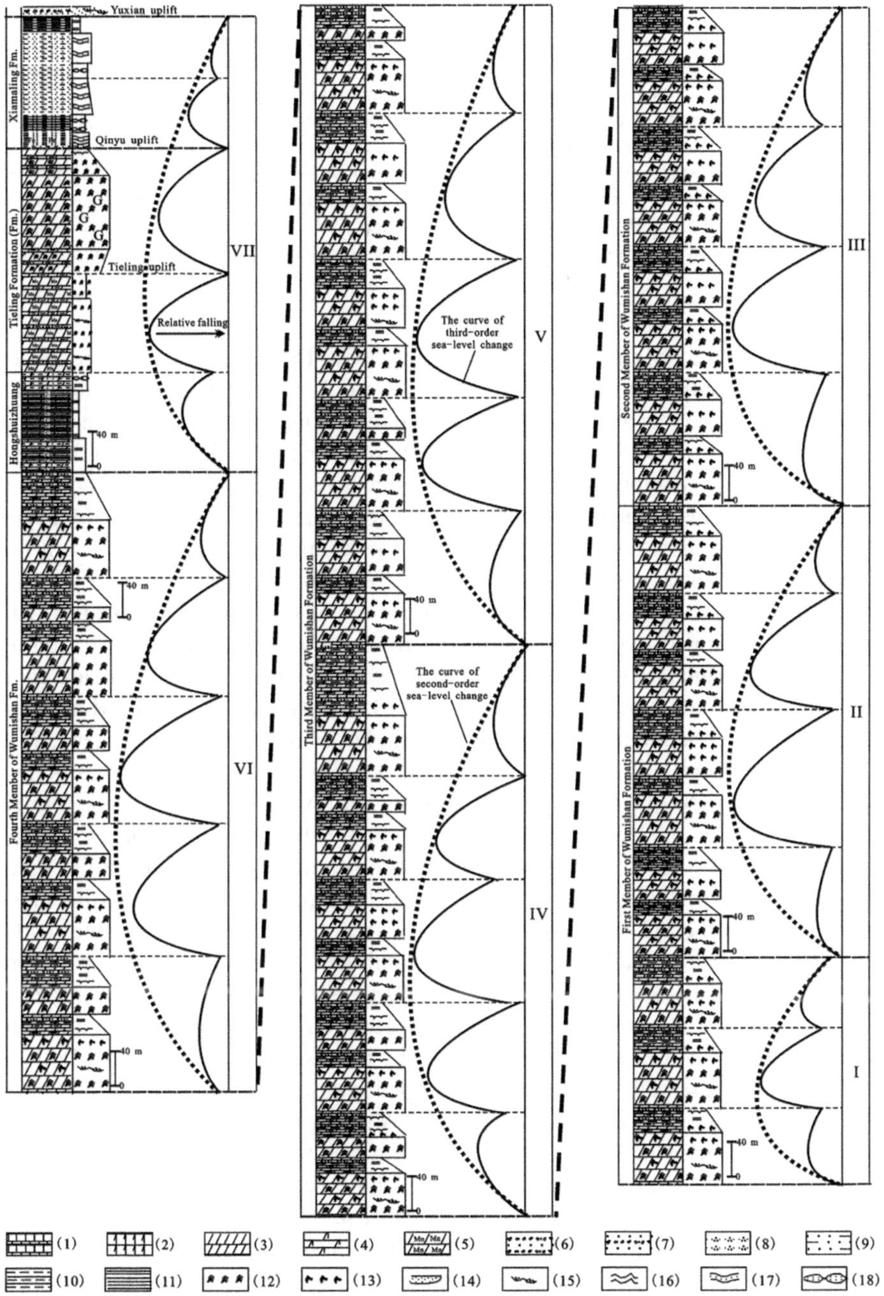
Yuxian uplift
Xiamaling Fm.
Qinyu uplift
Tieling Formation (Fm.)
Tieling uplift
Relative falling
Hongshuizhuang
40 m
0
VII
Fourth Member of Wumishan Fm.
VI
Third Member of Wumishan Formation
The curve of third-order sea-level change
The curve of second-order sea-level change
V
IV
Second Member of Wumishan Formation
First Member of Wumishan Formation
III
II
I
(1)
(2)
(3)
(4)
(5)
(6)
(7)
(8)
(9)
(10)
(11)
(12)
(13)
(14)
(15)
(16)
(17)
(18)

◂**Fig. 9** Sequence stratigraphic division for a succession from the Wumishan to the Xiamaling Formation in the upper part of the Jixian Group at the Jixian section. I to VII refer to 7 s-order sequences identified in the succession from the Wumishan to the Xiamaling Formation in the upper part of the Jixian Group at the Jixian section, which can further subdivided into 31 third-order sequences. Lithological codes (in the key at the bottom of the figure) are as follows: (1) = limestone, (2) = dolomitic limestone, (3) = dolostone, (4) = lime dolostone, (5) = manganese dolostone, (6) = sandy conglomerate, (7) = gravel sandstone, (8) = sandstone, (9) = muddy sandstone, (10) = mudstone, (11) = shale, (12) = stromatolite, (13) = chert, (14) = cross bedding, (15) = scouring surface, (16) = tidal vein stratification,, (17) = tidal wavy stratification, and (18) = tidal lenticular stratification

The HSTs (or forced regressive system tracts (FRSTS)) of third-order sequences are generally composed of muddy dolostones or dolomitic shales of the supratidal facies (or the lagoon facies; Fig. 11b–c), which represents the response of the upward shoaling process of sedimentary environments resulting from the sea-level falling.

Most of the Precambrian carbonates are made up of dolomites, but a set of limestones with the development of carbonate meter-scale cycles of the subtidal type (Fig. 13c) constitutes the third Member (or the Zhangjiayu Subformation) of the Gaoyuzhuang Fm. and makes up a particular non-stromatolitic succession at the Jixian section, which is mainly made up by a set of 681m thick limestones with laminae and molar-tooth (MT) structures that make up one second sequence including 3 third-order sequences (Fig. 8). These carbonate meter-scale cycles are made up of marls (Fig. 13a) and limestones (Fig. 13b and d). Two types of primary sedimentary strcutures, which are laminae (Fig. 13c) and molar-tooth (MT) structures (Fig. 2, 3, 4, 5, 6, 7, 8, 9 and 10d), are widely developed in this non-stromatolitic carbonate succession; and the limestones can be grouped into two categories, i.e. the leiolite (Riding 2000) and the laminate (Fig. 2, 3, 4, 5, 6, 7, 8, 9, 10 and 11). That is, the MT structures are developed within the thick-bedded to medium-bedded leiolite limestones from the middle ramp to shallow ramp. Therefore, the forming sedimentary environment of the MT structures is generally deeper than that of stromtolites (tidal flat) and is relatively shallower than that of laminites (deep-to middle ramp). More importantly, lots of macroscopic fossils of algae ("Ma" in Fig. 2, 3, 4, 5, 6, 7, 8, 9, 10 and 11) are frequently paragenesis with the MT structures, and are different from the spheroidal MT structures described by Furniss et al. (1998) and by Pollock et al. (2006). These body fossils can be grouped into several genuses, i.e. the *Genus Chuaria*, the *Genus Shouhsienia* (Xing et al. 1989), the *Genus Tawuia* and the *Genus Phascolites* (Xing et al. 1989; Du 1992), all of which make up an assemblage of the *Chuaria-Shouhsienia.* Although similar fossils have been found in the strata that are older than the Gaoyuzhuang Formation, most of them are marked by the carbonaceous compressions but not by body fossils. The biological attribute to these body fossils of megascopic algae that are found in limestone beds of the Gaoyuzhuang Formation at the Jixian section remain uncertain, thus it can be concluded that the further studies for these body fossils of the Mesoproterozoic Gaoyuzhuang Formation are important and meaningful to search for both the evolutionary history and the origin of the metaphyta, and are useful to solve other relative problems. And

Fig. 10 The trasgressive deposits that are constituted of sandstone with the thickness of 3 m in the bottom part of the Gaoyuzhuang Formation at the Jixian section. Image **a** refers to the boundary (the arrow) between the Dahongyu and the Gaoyuzhuang formations, which had been defined as the unconformity representing the Qinglong uplift; image **b** shows the thick-bedded to massive sandstone with the thickness of 3 m in the bottom part of the Gaoyuzhuang Formation; and image **c** expresses the ripple within the tragsressive sandstone

these body fossils of megascopic algae become the important contents of the non-stromatolitic carbonates of the third Member of the Gaoyuzhuang Formation at the Jixian section.

Fig. 11 Chief features of sedimentology and stratigraphy for the first Member of the Gaoyuzhuang Formation at the Jixian section. Images **a** and **b** show the thick-bedded dolostones with stromatolites making up the TSTs of the third-order sequence; and image **c** expresses the boundary (the arrow) between the first and the second Member of the Gaoyuzhuang Formation, which is a typical sequence boundary that is marked by muddy dolostones of the supratidal facies (or lagoon facies)

The fourth Member of the Gaoyuzhuang Formation (or the Huanxiusi Subformation) begins with thick-bedded to massive dark stromatolite dolostones (Figs. 15a–b) and is intercalated with light-yellow micritic dolostone without stromatolites (Fig. 15c). Within thick-bedded to massive dark stromatolite dolostones, there some large-scale scouring surfaces filled by cm-scale oncoids (Fig. 15b), which means that they are formed in an high-energy subtidal environment. According the cyclicity

Fig. 12 Chief features of sedimentology and stratigraphy for the second Member of the Gaoyuzhuang Formation at the Jixian section. Images **a** shows the thick-bedded manganese dolostones with stromatolites making up the TSTs of the third-order sequence; and images **b** and **c** express the sequence boundary (the arrow) that are marked by muddy dolostones or dolomitic shales of the supratidal facies (or lagoon facies)

reflected by the succession from the thick-bedded to massive dark stromatolite dolostones (Fig. 15a–b) of the subtidal facies making up the TST to the micritic dolostone without stromatolites (Fig. 15c) of the inter-to supratidal facies making up the HST, two third-order sequences can be discerned within the fourth Member of the Gaoyuzhuang Formation, which can be grouped into one second-order sequence (Fig. 8).

Fig. 13 Chief features of sedimentology and stratigraphy for the third Member of the Gaoyuzhuang Formation at the Jixian section. Images **a** shows the marl; image **b** indicates the lamilite; image **c** expresses one subtidal meter-scale cycle that is made up the marl (the lower part) and the thick-bedded limestone (the upper part); and image **d** depicts the leiolite limestone with development of both the macroscopic fossils of algae (MA) and the molar-tooth structures (MT)

Fig. 14 Laminate (Lm) and leiolite (Le) with MT structures in the top part of the third Member of the Gaoyuzhuang Formation at the Jixian section. Images **a** shows the meter-scale cycle that id made up of the laminate (Lm) and the leiolite (Le) with MT structures; image **b** indicates the morphological features of MT atructures (MT) on the bedding plane; image **c** expresses the morphologically intrastratal features of MT atructures (MT)

The Gaoyuzhuang Formation contains various fossils. The stromatolites occur mainly in the first Member, which are represented by *Confusoconophyton multiangulum, Gaoyuzhuangia crassibrevis, G. Bulbosa, G. gaoyuzhuangensis, Tabuloconigera paraepiphyta, Conophyton garganicum* and *C. cylindricum.* The microfossils referred to microscopic algae have been also found mainly in the cherty rocks of the first Member. Although they are preserved badly in the Jixian area, a great quantity of this kind of fossils has been found in the same horizon in adjacent areas. The

Fig. 15 Chief stratigraphic and sedimentological features of the fourth Member of the Gaoyuzhuang Formation at the Jixian section. Images **a** shows the massive stromatolitic dolostones of subtidal facies; image **b** indicates the scouring surface (the arrow) filled by cm-scale oncoids; and image **c** expresses the typical sequence boundary that is marked by an abrupt changing surface from the micritic dolostones of the intertidal to supratidal facies to the massive stromatolitic dolostones of subtidal facies. Notes the scouring surface filled by oncoids (the arrow in image (**b**)) might means that oncoids are resulted from atrong currents

acritarchs in this formation are poor in general and only a small number of them has been found in the first and third Members. The carbonaceous macroscopic fossils of algae are represented by *G.rvpania spiralis (namely Sangshuania spiralis, S. sangsuanensis* and *S. buccinata*) and they occur only in the lower part of the third Member. Although it is needed to be further researched, body fossils of macroscopic fossils of algae as shown in Fig. 13c do represent interesting and important phenomena.

The obvious peculiarity for the Yangzhuang Formation is constituted of the argillite-like, silt-bearing dolomitic mudstones (Fig. 16a), red and brick-red in colour, which represents a typical red beds of the Mesoproteroic that are resulted from the sedimentation in arid lagoon. In addition, there are a lot of thick-bedded siliceous dolostones with stromatolites (Fig. 16b) of the intertidal to supratidal facies. In term of the abrupt changing surfaces from the red beds of the arid lagoon facies to the dolostones with stromatolites of the intertidal to supratidal facies (Fig. 16b–c), which mark the sequence boundary in the Yangzhuang Formation, one second-order sequences that can further be subdivided into four third-order sequences can be discerned within the Yangzhuang Formation (Fig. 8).

The Wumishan Formation is the largest unit with a thickness of 3336 m that makes up one third of the total thickness of the Proterozic in the Jixian Section. The rocks of this formation are deposited in a tidal environment and characterized by a distinct cyclicity (Figs. 17 and 18). This formation traditionally had been subdivided into four members or (subformations) as follows; (1) The first Member is Lozhuang Subformation; (2) The second Member equals to the Mopanyu Subformation; (3) The third Member can be defined as the Ershilipu Subformation; (4) The fourth Member refers to the Shanpoling Subformation.

The Wumishan Formation contains abundant stromatolites and microscopic fossils of algae (Liang et al. 1985; Xing et al. 1989; Zhu 1993; Zhu et al. 1994). In addition to the visible taxa in common, the stromatolites in this formation are characterized by many kinds of very small-sized ones, namely ministromatolites (Liang et al. 1985), which are represented by *Psoudogvmnosollen moponyueuse, S'chuphus parvas* and *Mlcrostylus deusus*. These ministromatolites occur mainly in the Lozhuang (lst) Subformation and the lower part of the Ershilipu (3rd) Subformation, while the others do mainly develope in the Mopanyu (2nd) subformation and the upper part of the Ershilipu (3rd) subformation, and are represented by big-sized taxa, such as *Conophytes lituum, C. shanpolingese, Jacutophyton furcatum* and *Colonnella discreta.* Meanwhile, there are some medium-sized stromatolites at the top of the Shanpoling (4th) Subformation, which are *Pseudochihsienela, Wumishanella* and *Paraconophyton.*

A multitude of fossil algae have been found in black cherty layers in this formation. The main elements include *Palaeolynbya, Sphonophycus, Eoentophysalis, Oscillatoriopsis, Sacdyngbya, Hyxococcodes, Harythrix* and *Archeaoellipsotdes* etc. Moreover, some eukaryotic algae have also been reported and they can be represented by *Pandorinepsis, Uronemopsi,, Hyalothecopsis* and *Templuma.*

Carbonate strata of the Mesoproterozoic Wumishan Formation in the Jixian area near Tianjin are ~3300 m thick and were deposited over some 100 million years. Metre-scale cycles (parasequences) dominate the succession, which are described

Fig. 16 Chief features of sedimentology and stratigraphy for the Yangzhuang Formation at the Jixian section. Images **a** shows the mudstones with red and brick-red in color making up the red bed; image **b** indicates a abrupt changing surface from the red bed to the light yellow dolostones with stromatolites, which represents a typical sequence boundary (the arrow); and image **c** depicts the sequence boundary (the arrow) between the underlying Yangzhuang and the overlying Wumishan Formation, which is also marked by a abrupt surface from the red bed to the light yellow dolostones with stromatolites

as the Wumishan cycles (Mei and Tucker 2013; Figs. 17 and 18). They are generally carbonate type rock types before tide, and they are mostly alternative to similar symmetrical rocks. The central part is a thicker layer of stone biological layer and smaller thrombosis-tumor biological reef. Lallings-Baiyun Rock Shaw and ancient

Fig. 17 Chief features of the Wumishan cycle dominating the Wumishan Formation. Image **a** expresses a typical Wumishan cycle that made up of an approximately symmetrical lithofacies succession with a thickness of about 15 m; image **b** indicates the bottom boundary (the arrow) of the Wumishan cycle; image **b** shows the stromatolite (**a**) and the thrombolite (**b**) of the subtidal facies making up the central part of of the Wumishan cycle; image **d** refers to the thrombolite (**b**); and image **e** expresses the top boundary (the arrow) of the Wumishan cycle, which can also dose both the base composed of micritic dolostone **d** of the Wumishan cycle and an upward shoaling succession of "siliceous microbial laminate (**c**)-micritic dolostone (**d**)-muddy dolostone **e**-dolomitic shales (**f**) and palaeosoil (**g**)" in the upper part of the Wumishan cycle

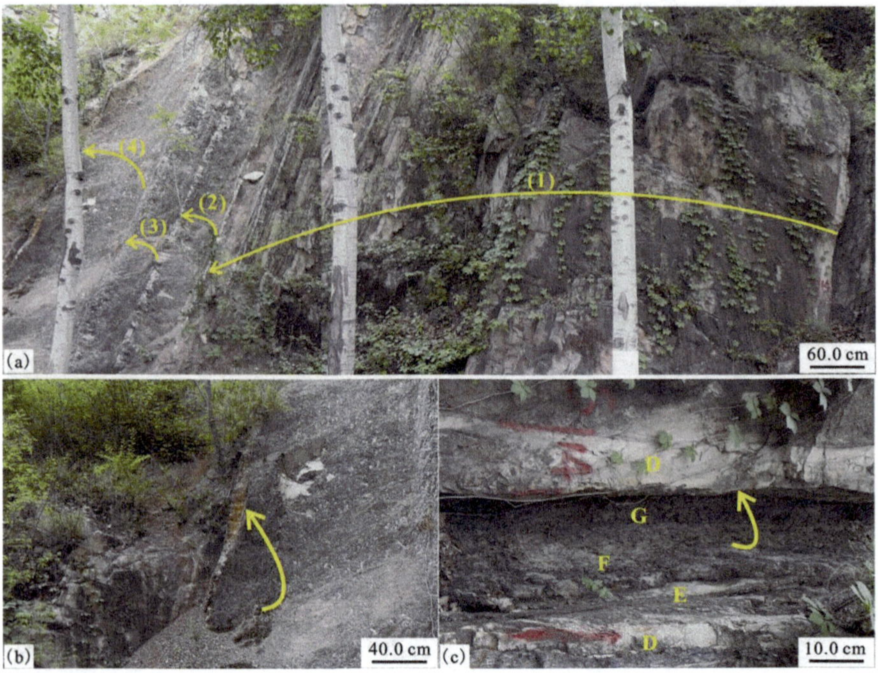

Fig. 18 Images showing the widespread 1:4 stacking pattern and some details in the cyclic succession of the Wumishan cycles. Image **a** expresses the cyclic succession of the Wumishan cycles, which consists of 4 cycles ((1) to (4); image **b** shows a boundary between cycles that can be used as the sequence boundary (the arrow); and image **c** indicates a the top boundary (the arrow) of the Wumishan cycle, that also dose show both the base composed of micritic dolostone (**d**) of the Wumishan cycle and an upward shoaling succession of "siliceous microbial laminate. **c**-micritic dolostone **d**-muddy dolostone **e**-dolomitic shales (**f**) and palaeosoil (**g**)" in the upper part of the Wumishan cycle

soil hat constitute the top layer. Umishan Rotary Betting Boundary is a typical expo sure, with rich evidence of freshwater into rocks. The universal 1: 4 stack mode shows that a single fogmine mountain rotation is the sixth-order sub-layer sequence, and the four sub-layer sequences form a fifth-order sub-layer set. There are 5–8 beds or couplets locally, which can be recognized. The layer sequence in the sixth-order sub-layer sequence is vertical, forming a fifth-order sub-layer set, which is interpreted as the result of the environmental fluctuations controlled by Milan Civic's rhythm control, that is, progress, short-term eccentricity, and long-term eccentric superposition. The 1: 4 superposition mode (Fig. 18) in the rotation sequence shows that the rhythm of the Milan Kovich in ancient China and the Yuan Dynasty is similar to that of the Milan Civic's rhythm. According to the rotation superposition mode, 26 three-level sequences can be divided into 6 s-level sea invasion-regression large sequences (or sequence sets), which reflects the composite sequence (Fig. 9) that is relative to the changes in the sea level (Fig. 9).

The Hongshuizhuang Formation is only 131 m thick and is subdivided into two parts. The lower part is composed of greyish or dark yellow, thin bedded, muddy dolostones with some intercalations of black and green shales (Fig. 19a). The upper part is dominated by black and green shales with fine and flat silty laminations (Fig. 19b). This formation is mainly the sediments formed in a lagoonal (or bay) facies. A lot of acritarchs with distinct features are found through the maceration of shales (Xing et al. 1989) which are characterized by new types of quadrangular, triangular, global and scaffold forms, such as *Quadratimorpha, Triangumarpha, Leiofusa, Scaphida, Trachysphaeridium* and *Pseudezonosphera* etc. In addition, a few of filamental forms and carbonaceous megafossils of *Chuaria* are also found. The Hongshuizhuang Formation dominated by dark-shales makes up a third-order sequence, which represents a large-scale condensed succession between the Wumishan Formation and the Tieling Formation.

The Tieling Formation, which generally and tranditionally thought as the uppermost stratigraphic unit of the Mesoproterozoic Jixian Group (Xing et al. 1989; Zhu et al. 1994; Chen and Wu 1997), consists of two members: The lower manganese-dolostone member (Fig. 20) and the upper limestone member (Figs. 21 and 22), totally about 330m in thickness. Based on the existence of a layer of "ferriferous sandstone" or palaeokarst breccias (Fig. 20d), a disconformity was formerly established between the two members. Thus, the Tieling Formation makes up two third-order sequences (Fig. 9).

A number of microscopic acritarchs and stromatolites (Fig. 22) have been found in the lower shale member and upper limestone beds respectively. The stromatolites are very well developed and constitute a "stromatolitic palace" in the Jixian area (Zhu 1993; Zhu et al. 1994; Cao and Yuan 2006). They are represented by *Kussiella, Scopulimorpla, Anabaria, Baicalia, Tielingella* and *Chihsienella.* The Tieling Formation is disconformably overlain by the Xiamaling Formation, which tanditionally thought as the bottom unit of the Neoproterozoic Qingbaikouan System (or the Qingbaikou Group) (Xing et al. 1989; Zhu et al. 1994; Chen and Wu 1997).

Further, glauconites developed in limestones of the stromatolites (Fig. 22d) in the second member of the Mesoproterozoic Tieling Formation at the Jixian section, are mainly rich in the boundary between the stack stacks and the grains is a colloidal form that reflects the characteristics of sea greenstone formed in the high-energy shallow sedimentation environment. High KO2 content (& gt; 8%) indicates that these sea greenstones can be classified as high-evolved sea greenstone. These sea greenstones have the characteristics of deployment environment with high energy shallow water, a special form that appears at the junction of stack stones and matrix, and the earlier era formed in the late period of the ancient Chinese and Yuan Dynasty. Greenstone is very different. In terms of deposits, the ancient Tieling group sea greenstone in the ancient Tieling group in Jixian County reflects normal high energy shallow water is in the mild reduction state of low oxygen in the late China and Yuan Dynasty. Obviously, the sea greenstone in the lintest layer of the ancient Tieling group of the China-Yuan Dynasty cannot be used as a identification sign of the concentrated section of the layers, nor can it be classified as a product formed by long-term cracking holes. Therefore, the Tieling Group Sea Greenstone is the representative

Fig. 19 Chief features of sedimentology and stratigraphy for the Hongshuizhuang Formation at the Jixian section. Images **a** shows the dark shales; and image **b** indicates a abrupt changing surface from the shales to the manganese dolostone of the first Member of the Tieling Formation, which represents a sequence boundary (the arrow)

Fig. 20 Chief features of sedimentology and stratigraphy for the first Member of the Tieling Formation at the Jixian section. Images **a** and **b** show the manganiferous dolostones with intercalations of green shales; and images **c** and **d** indice the exposure surface between the first and the second members of the Tieling Formation (the arrow), which is marked by the palaeokarst breccias underling boundary

of the former Hanwu Ji, which indirectly reflects the diversity of the environmental environment in the long-evolved geological history.

A 168m-thick suit of yellowish green to greyish green shally siltstone and lenticular sandstone, disconformably underlain by the Tieling Formation, is named as the Xiamaling Formation after its original locality in the western hills of Beijing. This sequence is deposited in supratidal to intertidal environments with current marks in the bedding plane of sandstone, such as furrow casts, groove casts and mud cracks

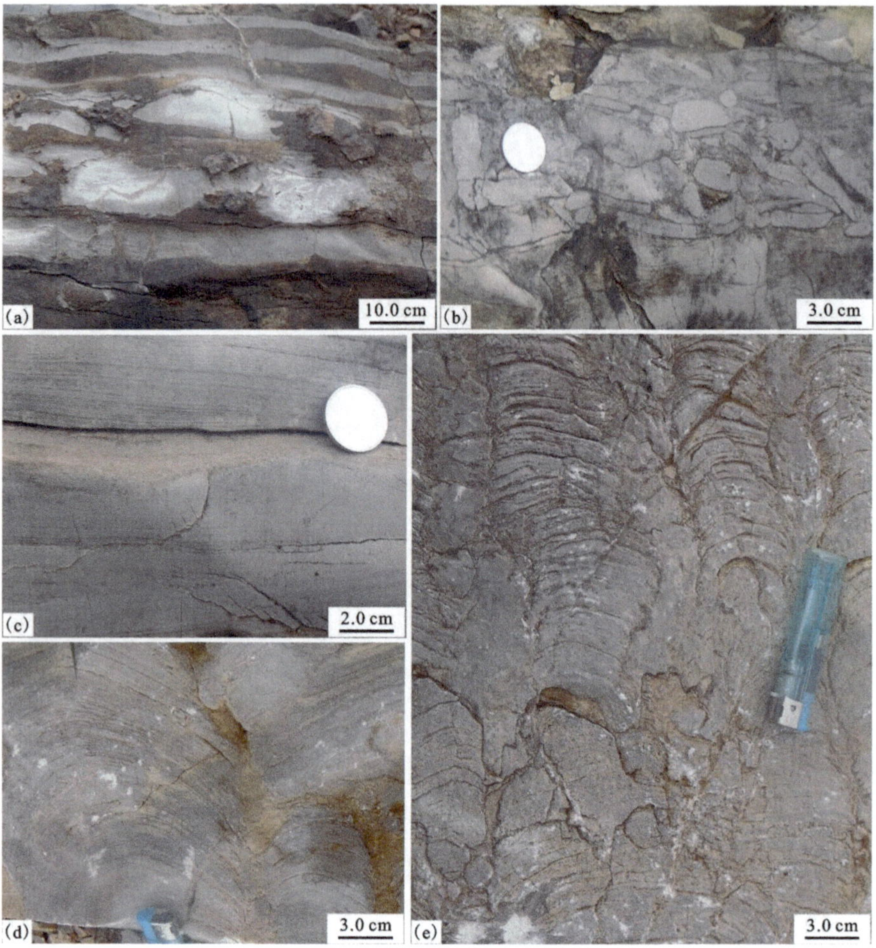

Fig. 21 Chief features of sedimentology and stratigraphy for the bottom part of the second Member of the Tieling Formation at the Jixian section. Image **a** shows the banded micritic limestone; image **b** indicates the lens of edgewise conglomerates within the banded micritic limestone in image (**a**); image **c** refers to the laminate; image **d** depicts the wavy stromatolite; and image **e** expresses the column stromatolite. All of them make up a succession with the thickness of about 20 m from the laminate to the stromatolite in the bottom part of the second Member of the Tieling Formation

etc. Abundant acritarchs, such as *Microconcentrica, Gloeocapsomorpha, Symplassosphaerdium* and *Tasmanites,* were obtained from the silty shales by maceration. They show the particular shapes and ornamentations, which distinguished them from those in the underlying strata. Like an obvious unconformity that has been thought as the Tieling Uplift between the first and the second Members of the Tieling Formation (Figs. 9 and 20c–d), both the bottom and the top boundaries of the Xiamaling Formation are marked by unconformities, the bottom boundary represents the Qinyu Uplift (Figs. 9 and 23a) and the top boundary do the Yuxian Uplift (Figs. 9, 24, and 25a).

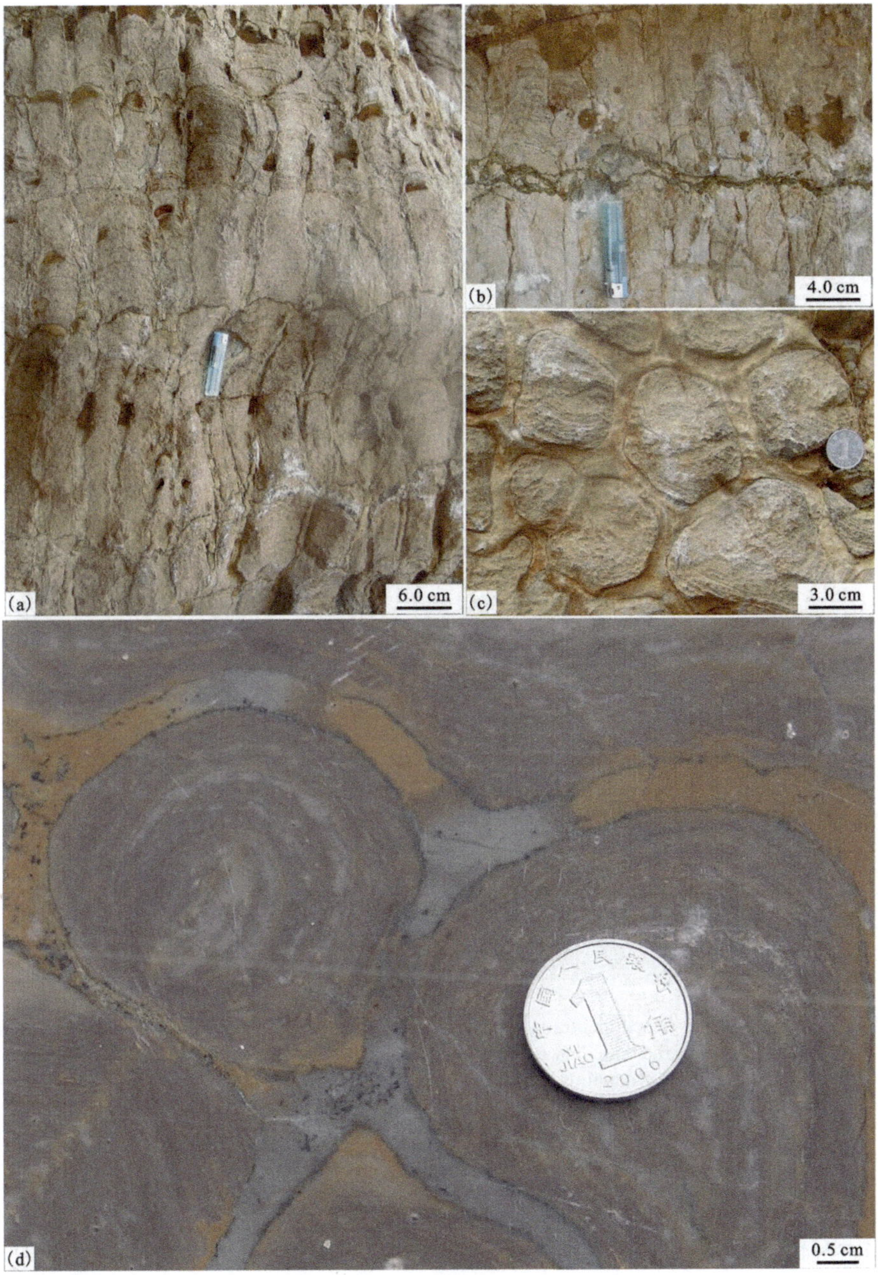

Fig. 22 Column stromatolites making up the central part of the second Member of the Tieling Formation at the Jixian section. Images **a** and **b** express the giant column stromatolites with growing hiatus surfaces (the arrow); image **c** and **d** depict the morphological features of stromatolites in the bedding plane, and image **d** shows glauconites as a fashion of colloids in the boundary between stromatolites and their matrix

Fig. 23 Chief features of sedimentology and stratigraphy for the Xiamaling Formation at the Jixian section. Image **a** shows the boundary (the arrow) between the underlying Tieling Formation (dolostones) and the overlying Xiamaling Formation (sandy shales); and images **b** and **c** indicate shales making up the Xiamaling Formation

Further, new data suggest that the long-term hiatus between the Mesoproterozic Xiamaling (~1320 Ma) and the Neoproterozoic Changlongshan Formation (~900 Ma) of the unconformity of the Yuxian Uplift (Figs. 9, 24, and 25a).

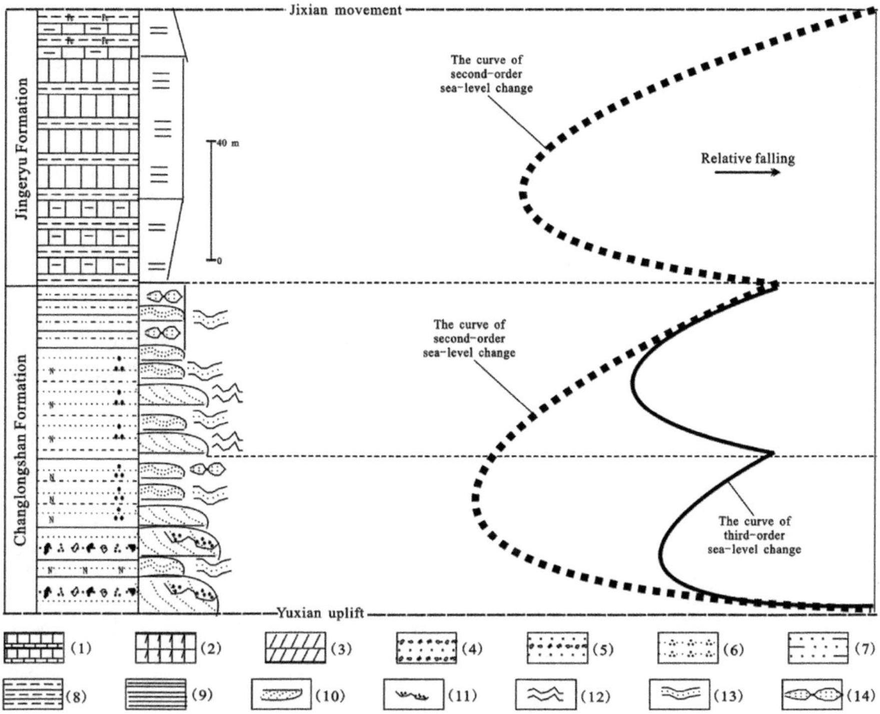

Fig. 24 Sequence stratigraphic division of the Neoproterozoic Qingbaikou Group at the Jixian section. Lithological codes (in the key at the bottom of the figure) are as follows: (1) = limestone, (2) = dolomitic limestone, (3) = dolostone, (4) = conglomerate, (5) = gravel dolostone, (6) = sandstone, (7) = muddy sandstone, (8) = mudstone, (9) = shale, (10) = cross bedding, (11) = scouring surface, (12) = flaser bedding, (13) = wavy bedding, and (14) = lenticular bedding

2.3 The Neoproterozoic Qinbaikou Group (or the Qingbaikouan System)

The Qinbaikou Group (or the Qingbaikouan System) includes two formations, which constitute two independent and integrated sedimentary successions (Fig. 24). In disconformable contact with the underlying the Xiamaling Formation that chiefly consists of sandy shales and shale, the lower unit is the Changlongshan Formation of conglomerate, arkosic sandstone, glauconitic sandstone and shale. The upper unit refers to the Jingeryu Formation that is mainly made up of limestone. The Qingbaikou Group is unconformably overlain by the Fujunshan Formation belonging to the late Epoch 2 of the Cambrian. Therefore, both the bottom and the top boundaries of the Qinbaikou Group (or the Qingbaikouan System) are typical unconformities representing long-term hiatus of more than several hundreds of millions years.

Fig. 25 Chief features of sedimentology and stratigraphy for the Changlongshan Formation at the Jixian section. Image **a** shows the boundary (the arrow) between the underlying Xiamaling Formation (Shales) and the overlying Changlongshan Formation (sandstone with lens of conglomerates), i.e. the unconformity of the Yuxian uplift; image **b** expresses sandy conglomerates making up the bottom of the Changlongshan Formation; and image **c** indicates course sandstones with two-direction cross-beddings making up the bottom of the Changlongshan Formation

The Changlongshan (or Luotuoling) Formation was ever referred to the lower part of the Jingeryu Formation and was first named as the Changlongshan Subformation. However, other suggested later that a new formation named as the Changlongshan Formation, and consequently the name Luotuoling Formation was given by Xing et al. in 1989 according to the name of its typical section in Jixian area. The Changlongshan Formation begins with basal conglomerate followed successively by arkosic-yellowish green and green shale and overlain by glauconite-bearing arkosic sandstone (Fig. 25). The upper part of this formation is characterized by purple, grey and black shale (Fig. 26a). There are two types of glauconites in sandstone: the first is granular and distributed along the cross-bedding planes as sand grains; the second constitutes the cements of interstitial fillings of the sand grains. This Formation is 118 m thick. It unconformably overlies the Xiamaling Formation in Jixian area (Fig. 25a) or other older horizons including the Archean in adjacent area, and conformably underlies the Jingeryu Formation.

The Jingeryu Formation was once regarded as the upper part of the "Jingeryu Formation", but many researches including this book regard it as an independent formation. This formation is only 100 m thick in Jixian and consists mainly of limestone in various colours, which varies from red, greyish green, light green, greyish brown to green on the top. According to the sedimentary structures, these rocks have been deposited in subtidal environment.The Jingeryu Formation is unconformably overlain by the Fujunshan Formation belonging to the late Epoch 2 of the Cambrian, which is marked by an unconformity of the Jixian Movement (Figs. 24 and 26c). The Fujunshan Formation is largely composed of dolostone, limestone and dolomitized limestone, and contains a fauna of *Redlichia chinensls* (Walcott) and *Megapalaeolenus fengyangensis* (Chu) of trilobite in the dolomitized limestone. Particularly, a set of massive breccias with the thickness of about 40 m makes up the bottom part of the Fujunshan Formation at the Jixian section, which might be the tsunamite that needs further research.

In summarization, within the framework of the current Proterozoic time scale, there was a long–term (more than 3000 Ma) hiatus between the Mesoproterozoic and the Neoproterozoic (an unconformity of the Yuxian Uplift; Figs. 9, 24, 25a) at the Jixian section. However, the Changcheng Group, the Jixian Group with the top unit of the Xiamaling Formation, and the Qingbaikou Group at the Jixian section, do represent a stratigraphical record during a particularly geological time in the lengthy earth history, i.e., a revised Mesoproterozoic Era that is temporarily grouped into one period, i.e. the Rodinian Period (Van Kranendonk et al. 2012) or a period from ~1.85 to 0.85 Ga has been referred to as the 'boring billion' in term of the stability in the environment recorded by marine $\delta^{13}C$ (Holland 2006; Roberts 2013). Therefore, there are certainly lots of sedimentological mysteries as well as straigraphical problems at the Jixian section that are needed to be further researched in the future.

3 Some Important Phenomena of Sedimentology

Among of lots of interested phenamena of strtigraphy and sedimentology at the Jixian section, four important phenamena of strtigraphy and sedimentology, i.e. (1) microbial mat-related siltstone dykes in the lower part of the Chuanlinggou Formation, (2) the molar-tooth (MT) structures in the upper part of the third Member of the Gaoyuzhuang Formation, (3) Wumishan cycles and their vertical stacking pattern in long-term sequences of the Wumishan Formation and (4) the glauconite within the

(a)
46.0 cm
(b)
16.0 cm
(c)
Canon
7.0 cm
(d)
4.0 cm
(e)
26.0 cm

◀**Fig. 26** Chief features of sedimentology and stratigraphy for both the Jingeryu Formation and the bottom part of the Cambrian Fujunshan Formation at the Jixian section. Image **a** shows the boundary (the arrow) between the underlying Changlongshan Formation (Shales) and the overlying Jingeryu Formation (limestones); image **b** expresses muddy limestones making up the Jingeryu Formation; image **c** indicates the boundary (the arrow) between the underlying Jingeryu Formation (muddy limestones) and the overlying Fujunshan Formation (breccias), i.e. the unconformity of the Jixian movement; images **d** and **e** further express the fundamental features of breccias making up the bottom part of the Cambrian Fujunshan

diversified column stromatolites of the second Member of the Tieling Formation are remarkable sedimentary deposits for their sophisticated origin. Their locations are shown in Fig. 1 that are marked by four stars in Fig. 1a.

Stop 1: Microbial Mat-Related Siltstone Dykes in the Lower Part of the Chuanlinggou Formation

Because of the common absence of direct evidence for microbial existence, the identification of microbial mat-related sedimentary structures or features in siliciclastic deposits poses a difficult challenge, and the development of microbial films and mats on sedimentary surfaces can only be inferred from the abnormal glutinosity and plasticity of sediments prior to lithification. Schieber summarized four groups of microbial mat-related features preserved within siliciclastic rocks that indicate the occurrence of microbial mats in the depositional paleoenvironment. Similar features were still defined as the matground structure (Pflüger 1999) or as the microbial induced sedimentary structure (MISS; Noffke et al. 2001) that was further grouped into the primary sedimentary structure of a fifth category based on the classification for the primary sedimentary structure finished by Pettijohn and Potter.

The Chuanlinggou Formation at this section is composed mainly of peritidal silty shales (thickness > 900 m) and can be subdivided into 3 third-order sequences (DS_5 to DS_7; Fig. 4) that are markedly different from sandstones of the underlying Changzhougou Formation and dolomites of the overlying Tuanshanzi Formation. Siltstone dykes occur mainly in fine peritidal deposits with lenses and lenticular beds of siltstones in the lower part of the Chuanlinggou Formation. These fine-grained deposits make up the transgressive and high-stand system tracts of the third-order sequence DS_5 (Fig. 4).

Numerous siltstone dykes are found in peritidal silty shales in the lower part of the Paleoproterozoic Chuanlinggou Formation at the Jixian section. These dykes have been interpreted as (1) the oldest known trace fossils, based on their burrow-like geometry and orientation parallel and perpendicular to bedding (e.g., Liu and Liu 1992); and (2) earthquake-related veins of liquefied sand, based on the occurrence of macro-and microscopic sedimentary fabrics that are indicative of liquation (e.g., Qiao and Gao 2007). Thus, these siltstone dykes represent an important phenomenon for the origin of these dykes remains debated. This phenomenon can be examined at the small road nearby Chuanlinggou village (40. 1973°N, 117.4827°E to 40. 1945°N,

117.4780°E), ca. 5 km south of Changzhougou village or ca. 35 km north of Jixian city (the first star in Fig. 1a).

3.1 Macroscopic Features of Siltstone Dykes

The Chuanlinggou Formation at the Jixian section is markedly different from the underlying Changzhougou Formation. The Changzhougou Formation consists of high-energy sandstone strata (thickness > 800 m) that possibly formed in a tidal beach environment during a slow transgression, whereas the Chuanlinggou Formation contains fine-grained deposits (total thickness > 900 m) that represent peritidal to lagoonal deposits in a back-barrier environment. The lower part of the Chuanlinggou Formation is mainly silty shales with lenses and lenticular beds of siltstone (Fig. 27) that contain features of tidal bedding.

The lower part of the Chuanling Formation (thickness > 200 m) at the Jixian section contains silty shales with lenses and lenticular beds of sedimentary siltstones; these rocks contain numerous siltstone dykes (Figs. 28 and 29). Along bedding planes, the dykes occur as spindle-shaped sandy cracks that narrow toward each end, resembling biological burrows (Fig. 27), with lengths of 3–5 cm and widths of 2–4 mm in the central parts of the dykes. In surfaces oriented perpendicular to bedding planes, the dykes are sinuous and are variable in length and width (Fig. 28): the larger dykes have widths of 2–4 mm and lengths exceeding 10 cm, whereas smaller dykes have widths less than 1 mm and lengths of 3–5 cm. The dykes may cut across several lenses and lenticular beds of siltstone; however, the jumping-off points (marked "P" in Fig. 29) for most of the dykes coincide with the lenses and lenticular beds of sedimentary siltstones, which indicates that the silts making up

Fig. 27 Images of the macroscopic features of silty shale with lenses and lenticular beds of sedimentary siltstones in the lower part of the Chuanlinggou Formation at the Jixian section

Fig. 28 Images showing the macroscopic feature in the bedding plane of siltstone dykes within the lower part of the Chuanlinggou Formation. Image **a** and **b** show centimeter-scale spindle-shaped siltstone dykes (the arrow) upon a bedding plane in silty shale within the lower part of the Chuanlinggou Formation

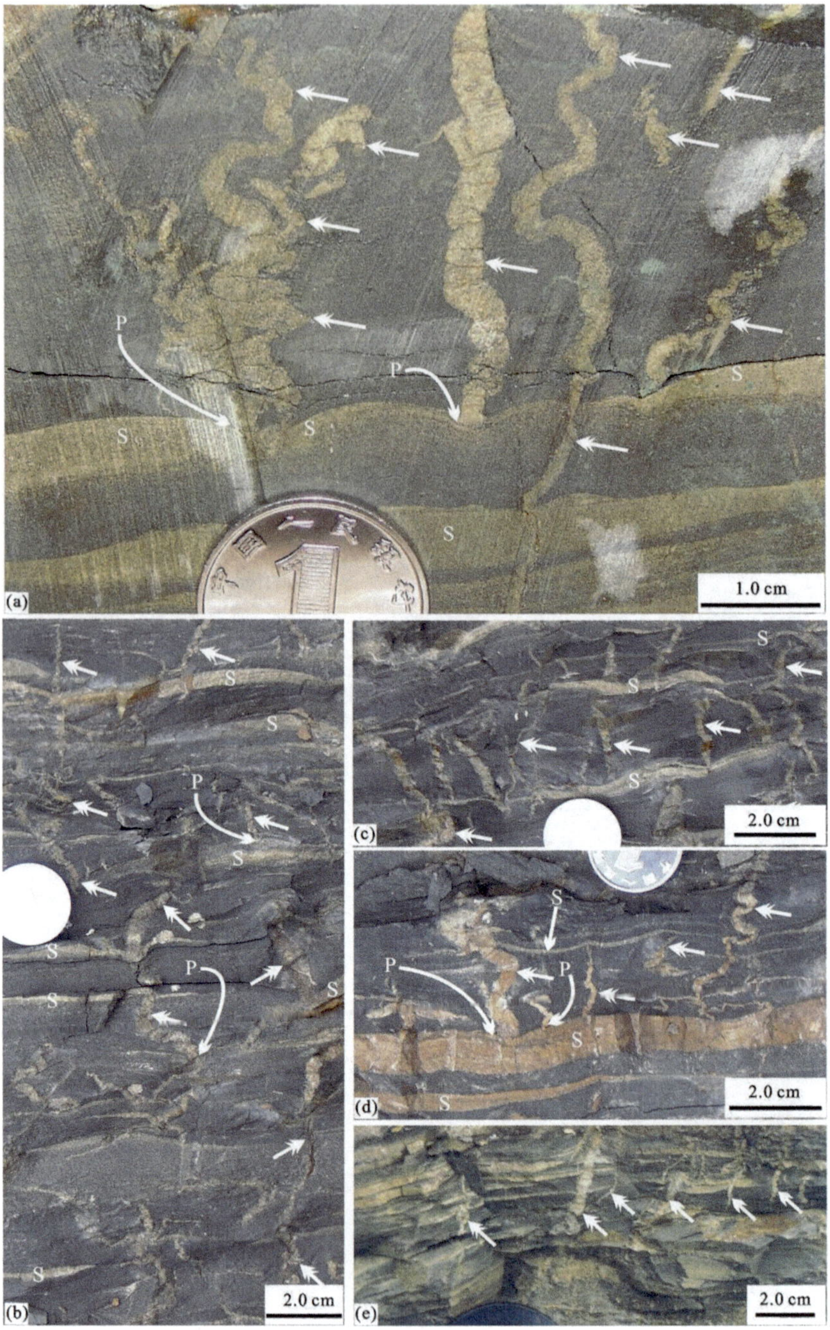

P
P
S
S
S
S
S
(a)
1.0 cm
S
S
P
S
S
S
S
P
S
(b)
2.0 cm
S
S
S
(c)
2.0 cm
S
P
P
S
(d)
S
2.0 cm
(e)
2.0 cm

◀**Fig. 29** Images showing the fundamental features of siltstone dykes within silty shales in surfaces oriented perpendicular to bedding in the lower part of the Chuanlinggou Formation at the Jixan section. Within images **a** to **e**, the extension lengths of siltstone dykes are as much as 10 cm or more (the arrow); "S" indicates sedimentary lenses and lenticular beds of sedimentary siltstone. Some dykes cut across the lenses and lenticular beds, which represent the jumping-off point (P) for most of the dykes

the lenses and lenticular beds of sedimentary siltstones become the source of silts making up the siltstone dukes.

The main features of the siltstone dykes are as follows: (1) along bedding planes, most of the dykes have a spindle-like morphology consisting of silt-filled cracks that narrow toward each end (Fig. 28); (2) in surfaces oriented perpendicular to bedding, most of the dykes are sinuous or tabular, although some have irregular shapes (Fig. 29); (3) some siltstone dykes cut across several lenses and lenticular beds of siltstone, and the fact that the jumping-off points of some dykes ("P" in Fig. 29) coincide with lenses and lenticular beds of siltstone indicates that the silt grains within the dykes originated from deformation of the lenses and lenticular beds of siltstone; (4) there exists a clear interface between the dykes and host rocks (silty shales); and (5) most of the dykes are tabular, with thicknesses of 2–4 mm, lengths of 3–5 cm, and heights greater than 10 cm (Figs. 27 and 28).

3.2 *Microfabrics of Siltstone Dykes*

Silt grains make up 65%–70% of the siltstone dykes in the lower part of the Chuanlinggou Formation (Fig. 31a–d), which is a higher proportion than that in the lenses and lenticular beds (Fig. 30a–d). Most of the silt grains within the dykes are subangular and some are well rounded, showing greater rounding than silt grains within the sedimentary siltstone. The dykes are characterized by the following features: (1) a clear interface between the dyke and host rock (silty shale), marked by residual organic material; (2) compared with the lenses and lenticular beds (Fig. 30a–d), the dykes (Fig. 31a–d) have a higher content of silt grains, a more localized distribution of silt grains, and silt grains are more rounded; and (3) the edges of silt grains are obscure (Fig. 31d), possibly indicating an origin via precipitation or liquation.

A clear interface is observed between the dykes and host silty shale (Figs. 31a, 32, and 33), marked by residual organic material. The walls of siltstone dykes show a thread-like (widths of ~0.05 mm) network of residual organic material that also occurs within the host rock close to the dyke margins (Fig. 32a–c); the side walls of siltstone dykes are 0.25–0.30 mm wide. Importantly, the dykes have an obvious upper interface (Figs. 32d–f and 33) that is more pronounced that the side walls (Figs. 31a and 32a–c). The upper interface also consists of residual organic material (widths of 0.10–0.15 mm). The upper surfaces of the dykes are convex, and horizontal laminae within the host rock become sinuous toward the upper surface, and are parallel to

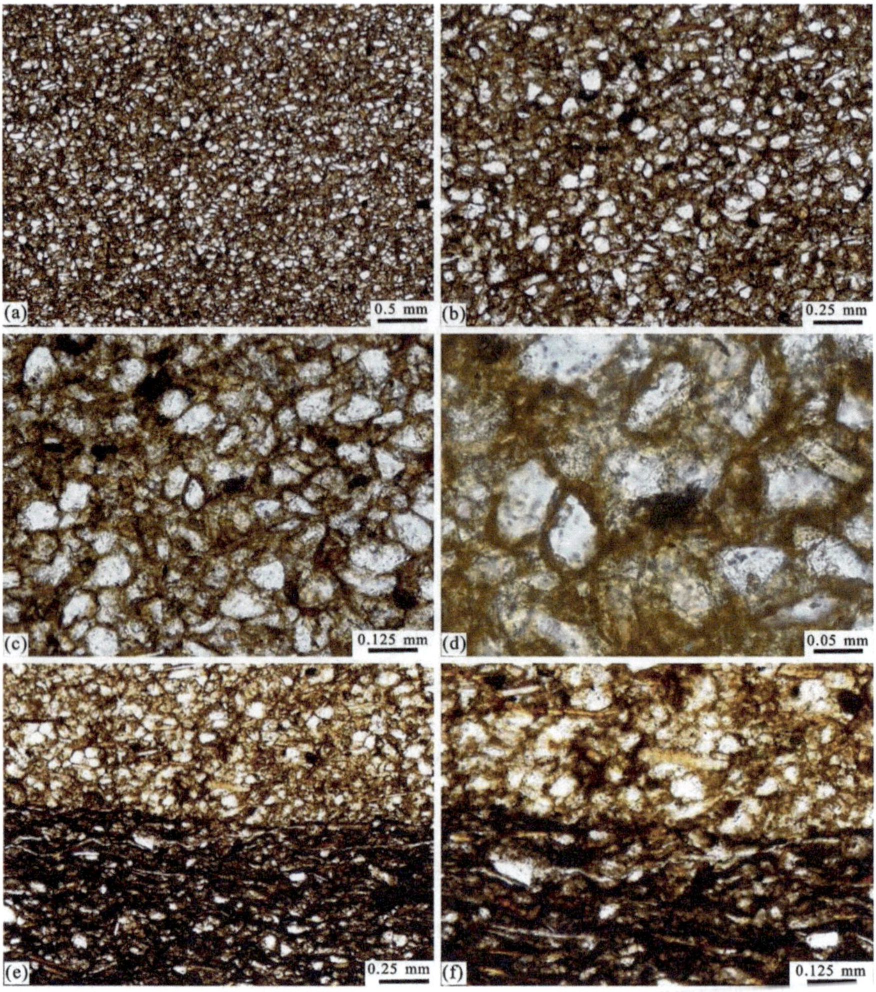

Fig. 30 Images of sedimentary siltstone that occurs as lenses and lenticular beds within silty shale in the lower part of the Chuanlinggou Formation at the Jixian section. Images **a** to **d** express the sedimentary siltstone at various magnifications; images **e** to **f** depict the interface between the host rock (silty shale; lower part of each image) and sedimentary siltstone (upper part). All images were taken under plane-polarized light

the surface (Figs. 32d–f and 33). These observations indicate that the siltstone dykes formed during early-stage diagenesis and are different structures to the lenses and lenticular beds of sedimentary siltstone that occur within the host rock.

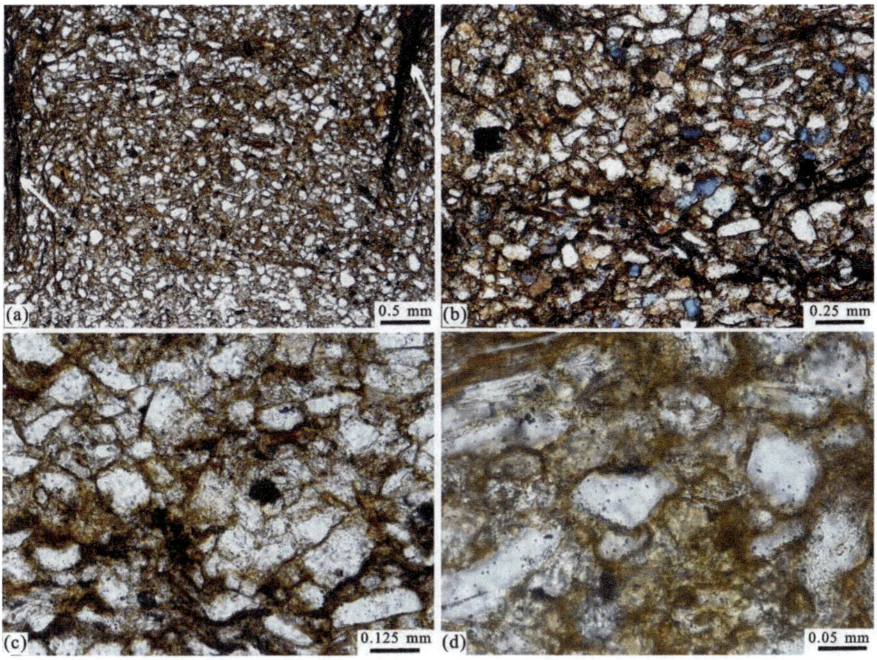

Fig. 31 Images of siltstone dykes in the lower part of the Chuanling Formation at the Jixian section. Images **a** to **d** show siltstone at various magnifications. Arrows in **a** indicate the interface between the siltstone dyke and host rock. All images were taken under plane-polarized light

3.3 Sedimentary Structures Contemporaneous with the Siltstone Dykes

The siltstone dykes coexist with widespread wrinkle structures (Noffke et al. 2001; Porada and Bouougri 2007) and local palimpsest ripples (Noffke 1998; Noffke et al. 2001). As seen in Fig. 34, the palimpsest ripples that developed on the depositional surface of silty shales are marked by a low ripple index and variable orientations, with wavelengths of 10–12 cm and amplitudes of 1–2 mm. These observations indicate weak current action during formation of the ripples.

Wrinkle structures are widely developed upon bedding planes within the silty shale that contains lenses and lenticular beds of sedimentary siltstone (Fig. 35), at the following scales: (1) large-scale wrinkle structures, which are similar to small-scale ripples (Fig. 35b), have wavelengths of 3–6 mm and amplitudes of 0.5–1.0 mm, and occur on the depositional surfaces of muddy siltstone; most of the wave crests are smooth, although some are angular; (2) medium-scale wrinkle structures, which have variably oriented wave crests with widths of 0.5–1.0 mm and amplitudes of 0.1–0.2 mm, and occur mainly on the depositional surfaces of silty shale; most of the wave crests are smooth, although some have are angular (Fig. 35a, c, e, f); and (3) small-scale wrinkle structures, which are chaotically distributed upon the

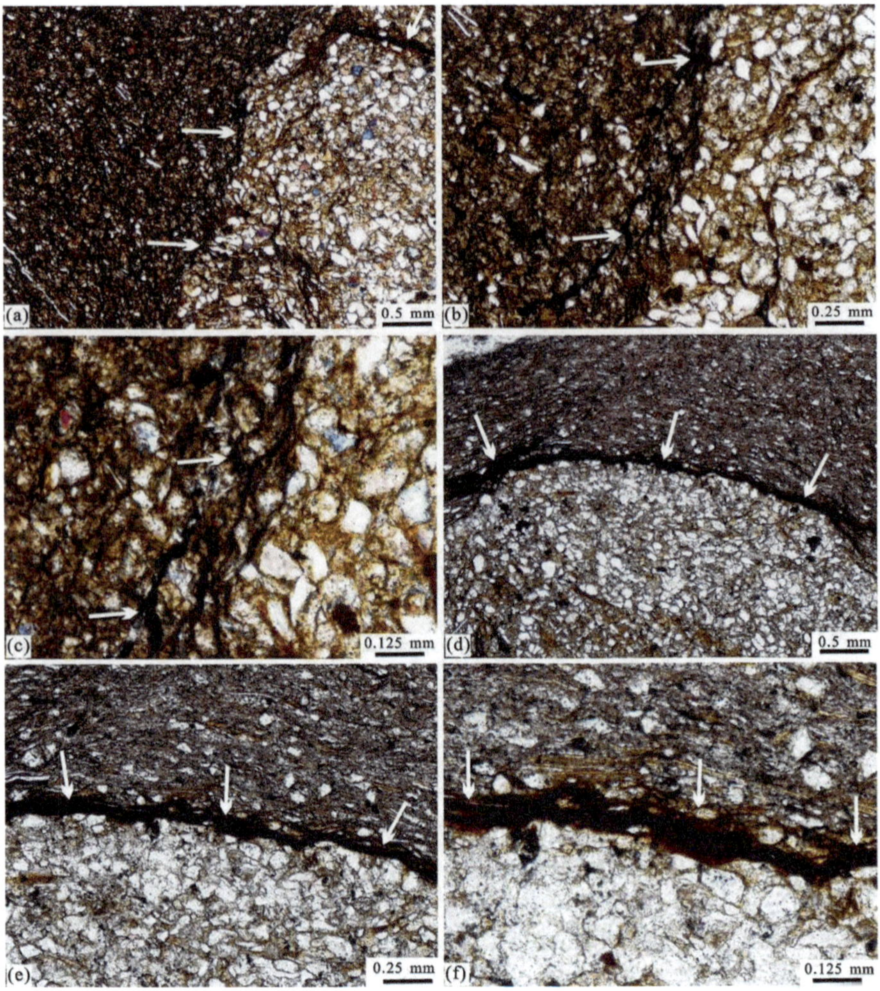

Fig. 32 Images of the margins of siltstone dykes in the lower part of the Chuanlinggou Formation at the Jixian section. Images **a** to **c** express the side wall of siltstone dykes (arrows); and images **d** to **f** show the upper interface of siltstone dykes. All images were taken under plane-polarized light

depositional surfaces of silty shale without obvious wave-like crests (Fig. 35d). These wrinkle structures are paragenetic with the silty dykes (Fig. 3, 4, 5, 6, 7, 8 and 9a, c, and e). Therefore, together with the siltstone dykes, palimpsest ripples and wrinkle structures make up an association of sedimentary structures in the peritidal silty shales with lenses and lenticular beds.

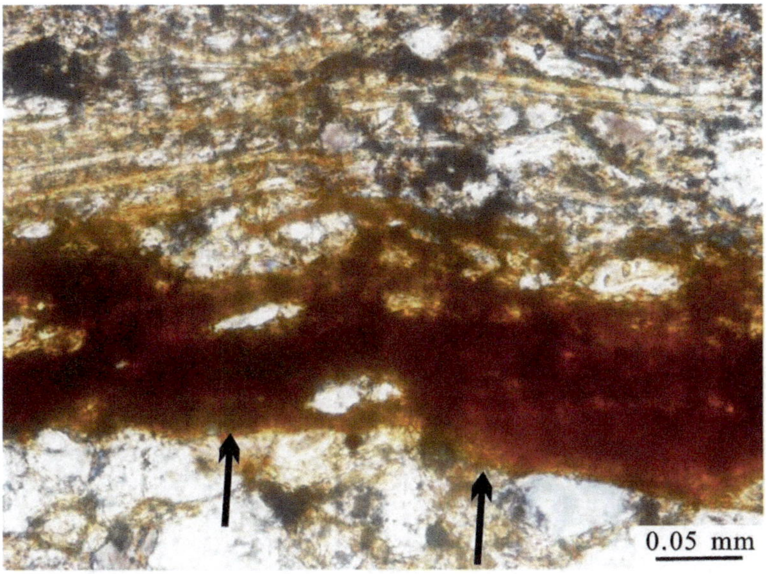

Fig. 33 Image of the upper boundary of a siltstone dyke in the lower part of the Chuanlinggou Formation at the Jixian section. The upper interface (width of dyke wall is 0.10–0.15 mm; arrows) consists of residual organic material and silt-sized grains. Plane-polarized light

Fig. 34 Palimpsest ripples upon a bedding surface in silty shale from the lower part of the Chuanlinggou Formation at the Jixian section. The low wave index values and variable orientations are characteristic features of the ripples

Fig. 35 Wrinkle structures upon bedding planes in silty shale that contains lenses and lenticular beds of sedimentary siltstone in the lower part of the Chuanlinggou Formation at the Jixian section. Images **a** to **f** express the wrinkle structures (large arrows) of various scales; image **b** indicates the large-scale wrinkle structures developed on the depositional surface of muddy siltstone; image **d** shows the small-scale wrinkle structures; and image **a**, **c** and **e** to **f** depict the medium-scale wrinkle structures (small arrows indicate siltstone dykes)

3.4 Possible Origin of Siltstone Dykes

The silty shales with lenses and lenticular beds are characterized by a relatively high content of residual organic material and a distinct association of sedimentary structures comprising tidal lenticular and wave bedding, siltstone dykes (Figs. 28 and 29), wrinkle structures (Fig. 35), and palimpsest ripples (Fig. 34).

The burrow-like shapes of the siltstone dykes (Fig. 28) led to a biological explanation of their origin; i.e., they were interpreted as the oldest trace fossils of metazoans

(Liu and Liu 1992) and grouped into the *Planolites* (Liu and Liu 1992). Zhang and Hua (2003) argued against such an interpretation, based on the following features: (1) a lack of bifurcating and jointing phenomena in dykes oriented perpendicular to bedding (Fig. 29); (2) the occurrence of clear liquefaction features (Fig. 29); (3) the occurrence of a clear interface between dykes and the host rock, including the side walls of dykes (Fig. 31a–c), and a clear upper interface (Figs. 32d–f and 33); and (4) obvious features of early-stage diagenesis of the dykes, including the convex shape of the upper surfaces and the sinuous nature of horizontal lamina within the host rock, oriented parallel to the upper surface of dykes (Fig. 32d–f).

Based the obvious liquefaction and early-stage diagenesis features of siltstone dykes, as observed macroscopically (Figs. 28 and 29) and microscopically (Figs. 31, 32 and 33), the dykes have been interpreted as earthquake-induced veins filled by liquefied sand (Pratt 1998a, b; Qiao and Gao 2007). However, apart from the convex upper surfaces of the dykes and related sinuous horizontal lamina within the host rock (Fig. 32d–f), there are few indications of earthquake-related deformation, such as small-scale extensional and compressional faults, and disharmonic soft-sediment folds that might have been expected to have developed within the host silty shales and even the dykes themselves. Furthermore, it is difficult to estimate how many earthquakes might have occurred during deposition of the silty shale, which is greater than 200 m thick. Importantly, the sedimentary structures that coexist with siltstone dykes include wrinkle structures and palimpsest ripples, which are not earthquake-related structures. Thus, there exists some doubt regarding the validity of an interpretation of the dykes as earthquake-related veins filled with liquefied sand (e.g., Pratt 1998a, b; Qiao and Gao 2007).

The observed wrinkle structures (Fig. 35) and palimpsest ripples (Fig. 34) are important indicators of the development of microbial mats on the depositional surface (e.g., Noffke 1998; Noffke et al. 2001; Porada and Bouougri 2007). Furthermore, these structures are representative of those that result from microbial mat-related leveling of the depositional surface, and are considered as typical microbial sedimentary structures (cf. Noffke 1998; Schieber 1998; Noffke et al. 2001; Mei et al. 2006; Porada and Bouougri 2007). Therefore, the wrinkle structures and palimpsest ripples that coexist with the siltstone dykes indicate that microbial mats developed on the sediment surface during deposition (Noffke 1998; Schieber 1998; Pflüger 1999; Noffke et al. 2001; Mei et al. 2006; Porada and Bouougri 2007), and may represent a key in understanding the formation of the siltstone dykes.

Similar to the air-escaping process that generated sandy pillar structure in modern and Precambrian fluvial deposits (see McManus and Bajabaa 1998), and the expansion and migration process that forms molar-tooth structures (e.g., Furniss et al. 1998; Pollock et al. 2006), the siltstone dykes analyzed in the present study are interpreted to have formed via the following complex process during early-stage diagenesis: (1) degassing caused by the splitting of organic material within silty shales that were enriched in residual organic material (Fig. 28) that originated from microbes; (2) dewatering of fine deposits (silty shales; Fig. 27) that may have been enriched with diagenetic water during early-stage diagenesis; (3) liquefaction of unconsolidated silty sediments within the sedimentary siltstones that represent the lenses and lenticular-bedding shapes observed within the silty shales (Figs. 28 and 29), which was genetically related to the above degassing and dewatering processes; and (4) formation of the interface of siltstone dykes (Figs. 32 and 33) resulting from the concentration of residual organic material.

The source of silt within the siltstone dykes was the sedimentary siltstone that occurs as lenses and lenticular beds within the silty shales (see Figs. 28 and 29), and the pressure required for liquefaction associated with degassing and dewatering was generated by sealing of the sediment surface by microbial mats. Thus, the development of microbial mats on the depositional surface and the enrichment in organic matter within the silty shale resulted from microbial activity in the Paleoproterozoic; furthermore, the siltstone dykes can be interpreted as a microbial mat-related sedimentary structure. If the wrinkle structures and palimpsest ripples that coexist with the siltstone dykes are interpreted as microbial mat-induced structures that developed upon the sediment surface (i.e., parallel to bedding) as a result of the bio-leveling effect of microbial mats (cf. Noffke 1998; Schieber 1998; Noffke et al. 2001; Porada and Bouougri 2007), the siltstone dykes could be interpreted as microbial mat-induced sedimentary structures that formed perpendicular to bedding (cf. Noffke et al. 2001; Mei et al. 2006). Although the microbial-mat related structures and features preserved within the siliciclastic rock record are viewed as sedimentary structures, and their similar features and structures were still defined as the matground structure (Pflüger 1999) or as the microbial induced sedimentary structure (MISS; Noffke et al. 2001) that was further grouped into the primary sedimentary structure of a fifth category based on the classification for the primary sedimentary structure that was subdivided into four categories, most of them actually are of a complex organophysico-chemical origin that frequently leads to multiple explanations (Schieber 1998). Therefore, the siltstone dykes in the lower part of the Chuanlinggou Formation provide an example that had result in the multiple explanations and a type of the microbial mat-related structure formed during a complex diagenesis process that include the liquation of sedimentary grains.

Stop 2: Molar-Tooth (MT) Structures in the in the Upper Part of the Third Member of the Gaoyuzhuang Formation

After the first description of molar-tooth (MT) structures by Bauerman, many mechanisms for MT genesis have been proposed, including subaqueous shrinkage or syneresis, microbial mat growth, replacement of evaporites, microbially-induced gas-bubble expansion and migration (Furniss et al. 1998), and earthquake-induced dewatering (Qiao et al. 1994; Pratt 1998a, b). Some geologists have suggested that the MT structures are a type of biogenic sedimentary structure with a particular palaeoenvironmental significance (Meng and Ge 2002; Kuang et al. 2006). Pollock et al. (2006) suggested a process of microscale gas-sediment interaction, which is similar to the microbially-induced gas-bubble expansion / migration model. Bishop and Sumner (2006) and Bishop et al. (2006) proposed a wave-induced fluid-flow model for the precipitation of the microspar in the MT cracks, brought about by both the generation of gas from organic decay and fluid movement due to wave loading. Since there have always been earthquakes and waves/storms, and microbes too have had a long geological history, it is difficult to explain the scarcity of MT structures in the Phanerozoic. Most geologists now agree that the key to interpreting MT structures lies in requires understanding both the mechanism for the formation of the MT cracks and the generation of the microcrystalline calcite (microspar) filling the cracks (Pollock et al. 2006). MT structure is well developed in limestones of the upper part of the third Member of the Gaoyuzhuang Formation at the Jixian section, and provides insights into the origin of MT structures in general. MT structures can be examined at the small road from Dahongyu to Zhezhuang village (40. 1535°N, 117.4795°E to 40. 1522°N, 117.4781°E), ca. 5 km south of Dahongyu village or ca. 30 km north of Jixian city (the location that is marked by the second star in Fig. 1a).

3.5 The Third Member of the Gaoyuzhuang Formation at the Jixian Section

At the Jixian section in Tianjing City, the Mesoproterozoic Gaoyuzhuang Formation is a ~1600m thick carbonate succession of Clymmian age (1600 to 1400 Ma; Xing et al. 1989; Du 1992; Zhu et al. 1994; Chen and Wu 1997; Mei 2006, 2007), in which the third member (or the Zhangjiayu Subformation) is mainly composed of ~681m of limestone with lamination and well-developed MT structures (Fig. 36).

The third member contains shelf shales (marls), micritic limestones and peritdidal dolomitic limestones. Micritic limestones comprise two main types: massive micrites and laminated micrites. The common sedimentary structures are the laminates and MT structures (cf. Mei 2006, 2007). Some scoured-surfaces overlain by oncolite grainstone or silty microbial-intraclastic wackestone are present in thicker-bedded aphanitic micrites. The MT structures are chiefly developed in massive micrites.

Three types of meter-scale cycles occur in the third member of the Gaoyuzhuang succession. The first type (L-M cycle) is made up of alternating beds of limestone and marl or shale; this type formed in deeper-water, such as an outer ramp or open shelf (cf. Mei et al. 2000; "a" in Fig. 36b). The second type is a subtidal carbonate cycle, characterised by upward-increasing grain-content, upward-coarsening grain-size, upward-thickening beds, and upward-shallowing facies (cf. Mei et al. 2000; Osleger and Read 1991; "b to d" in Fig. 36b). The third type is a peritidal carbonate cycle with upward-decreasing grain-content, upward-fining grain-size, upward-thinning beds and upward-shallowing facies (cf. Mei et al. 2000; "e" in Fig. 36b).

There are many difficulties in identifying higher third-order sequences within very thick Precambrian successions, such as that represented by the Gaoyuzhuang Fm., because of the poor biostratigraphic and chronostratigraphic control (Gao et al. 1996; Mei 2006, 2007). However, many Meso-Neoproterozoic carbonates are cyclic, with regular, repeated patterns of lithofacies defining meter-scale cycles (e.g. Song et al. 1991). The cycle stacking patterns can be used to recognise third-order sequences. The longer-term patterns of facies, thickness changes and sequences, and the extent of regional unconformities, are then the basis for defining second-order sequences.

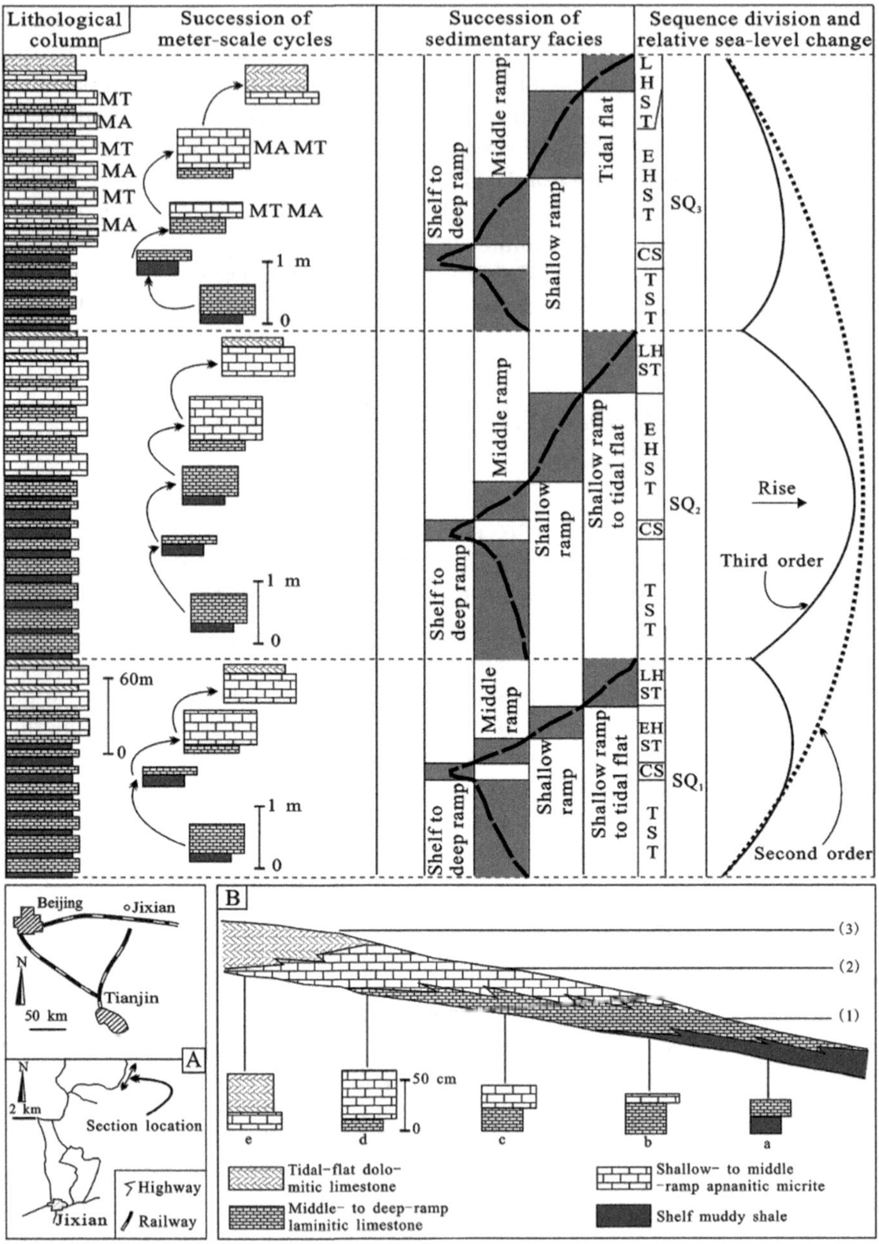

Lithological column
Succession of meter-scale cycles
Succession of sedimentary facies
Sequence division and relative sea-level change
MT
MA
MT
MA
MT
MA
MA MT
MT MA
1 m
0
60m
0
Shelf to deep ramp
Middle ramp
Shallow ramp
Tidal flat
Shallow ramp to tidal flat
LHST
EHST
CS
TST
SQ3
SQ2
SQ1
Rise
Third order
Second order
Beijing
Jixian
Tianjin
N
50 km
A
2 km
Section location
Highway
Railway
B
(3)
(2)
(1)
50 cm
e
d
c
b
a
Tidal-flat dolomitic limestone
Middle- to deep-ramp laminitic limestone
Shallow- to middle-ramp apnanitic micrite
Shelf muddy shale

◂**Fig. 36** Sedimentary succession and sequence-stratigraphic divisions for the third member of the Mesoproterozoic Gaoyuzhuang Fm. at Jixian section. Three third-order sequences, SQ_1 to SQ_3, can be discerned in the third member of the Gaoyuzhuang Fm. These third-order sequences, which are marked by regularly vertical stacking pattern of meter-scale cycles ("a" to "e" in "B") and the particular succession of sedimentary facies, can be grouped into one second-order sequences. TST = transgressive systems tract; HST = highstand systems tract; CS = condensed section; EHST and LHST = early and the late highstand systems tract respectively; MT = MT structure, and MA = possible fossils of megascopic algae, both of which are developed in ramp limestones that are marked by aphanitic micrites in the EHST of SQ_{11}. Note that a non-strmatolitic carbonate succession makes up the third member of the Gaoyuzhuang Formation. "A" shows the section location. "B" shows the different types of meter-scale cycles ("a" to "e") and their changing spectrum of sedimentary environment; (1) sea level, (2) fair wave base, (3) storm wave base

Using this approach, the four members of the Gaoyuzhuang Fm. are seen to represent four distinct depositional periods. According to the vertical stacking patterns of the meter-scale cycles, the third member of the Gaoyuzhuang Fm. at Jixian can be divided into 3 third-order sequences (SQ_1 to SQ_3 in Fig. 36) and can then be grouped into one second-order sequence.

In the subtidal meter-scale cycles, the laminated micrite facies generally passes up into massive micrite ("b to d" in Fig. 36b; sea following Fig. 37a–b). This suggests that the depositional environment of the massive micrites with the MT structure was shallower than that of the laminated micrites. MT structures occur in medium- to thick-bedded massive micrites deposited in a mid-to shallow-ramp environment. Thus, the depositional environment of the MT-bearing facies is generally deeper than that of the stromatolitic dolomites (tidal flat), but shallower than that of the laminated micrites (outer-to mid ramp). Thus, most MT limestones are developed in the early high-stand system of the SQ_3 (Fig. 36).

It is perhaps significant that there are many structures interpreted as macroscopic algal fossils associated with the MT structures (c.f. Mei 2006, 2007) as shown in Fig. 13d in above context. These are quite different from the spheroidal MT structures described by Pollock et al. (2006) and are different from the three-dimensional fossils filled with MT microspars described by Hofmann. The algal fossils are three-dimensional and can be grouped into several genera, i.e. the genus *Chuaria*, with a spheroidal-subspheroidal shape, the ellipsoidal *Shouhsienia*, *Tawuia* an elongate straight tube, and the podiform *Phascolites* (Du 1992), together constituting a *Chuaria-Shouhsienia* assemblage (c.f. Mei 2006, 2007). Although similar fossils have been found in strata older than the Gaoyuzhuang Fm., most of those are carbonaceous impressions rather than three-dimensional fossils. The precise biological affinites of these megascopic algae remains uncertain, and further studies would clearly be useful.

Fig. 37 Morphology of MT structures in limestones of the third member of the Gaoyuzhuang Fm. at Jixian. **a** to **b** meter-scale cycles (MSC) made up of laminated micritic limestone (**a**) and aphanitic micritic limestone (**b**); **c** to **f** morphology of MT structures in plan view, and there are some three-dimentional fossils of macroscopic algae (e.g. the genus *Chuaria*; indicated by black arrows) in MT limestone **c**. Rectangles in **c** and **d** refer to the parts that are enlarged as shown in Fig. 3–12. Note that one long linear crack (with loneness of more than 30 cm; indicated by white arrows) can cross and break MT cracks of both the large-scale ribbon with spindle shape and the medium-scale ribbon with spindle shape (**c**)

3.6 Morphologies of Molar-Tooth Cracks

The MT structures developed in the limestone beds of the third Member of the Gaoyuzhuang Fm. are distinguished by their irregular configuration. Complexity and irregularity of MT structures (Fig. 37c–f) increase upward within cycles, concomitant with shallowing of the environment (Fig. 37a–b).

In plan view, MT structures have the regular to irregular shape of sheet-like ribbons that are tapered at the ends, and the ends are gradually pinched out (Fig. 37c–f). They typically measure a few millimeters in width, and a few centimeters to tens of centimeters in length.

MT cracks can be subdivided into two types based on morphology: MTa, MTb (Fig. 38a–b). Large-scale ribbon MT cracks (MTa in Fig. 38a–b) are spindle-shaped with a width of 1 to 2 cm in the middle part and tens of centimeters in length. The medium-scale MT cracks (MTb in Figs. 38a–b) are also spindle-shaped with a width of several millimeters in the middle part and a length of several tens of centimeters. These two types of MT crack can be interconnected or quite separate. Importantly, one long linear form of microfault with the length of more than 30 cm (indicated by white arrows in Figs. 37c and 38a) can cross and break MT cracks of both the large-scale ribbon MT cracks (MTa in Fig. 38a) or the medium-scale MT cracks (MTb in Fig. 38a). MT structures in the shape of sheet-like ribbons chiefly occur as vertical structures that are straight to crenulated to sinusoidal in cross-section (Fig. 39). Numerous spheroidal bodies (Figs. 37c and 38a) are scattered within the aphanitic micrite host rock, but, as noted above, they are three-dimensional fossils of macroscopic algae, i.e. the *Chuaria.*

3.7 Molar-Tooth Intraclasts

Scour structures and lenses of packstone-wackestone occur within the mid-to shallow-ramp aphanitic micritic limestones and are interpreted as the result of storms. The clasts in the storm beds are mainly intraclasts of MT limestone (Fig. 40). Various grain sizes show that these intraclasts of MT limestone include rudites, arenites and siltites (Fig. 41). And most of these intraclasts are sharp-edged, and matrices are micrites. Pratt (1998a, b) termed MT intraclast beds as molar-tooth coquinas, and explained their formation by multiple phases of dilation, filling and cementation, related to seismic shocks or shaking. Bishop et al. (2006), on the other hand, suggested that MT intraclasts indicate calcite microspar precipitation very soon after deposition and cracking of host-sediments, and that the pebbly lags (Fig. 40), demonstrating the brittle nature of crack fills, were simply the result of storm action. Thus, the particular MT intraclasts have special meanings for the understanding of forming process of the MT structure.

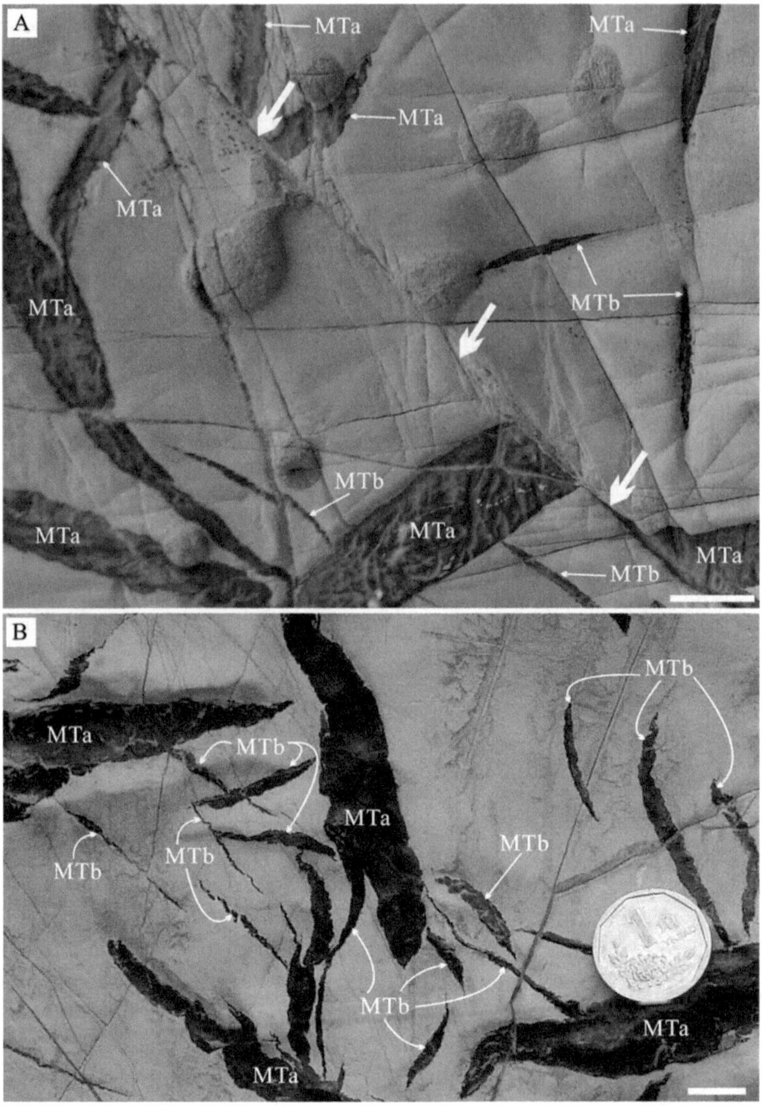

Fig. 38 Complex and irregular configuration of MT cracks in plan view. **a** an enlargement of the parts indicated by the rectangle in Fig. 2C; **b** an enlargement of the parts indicated by rectangle in Fig. 2D. Large and small sets of MT cracks: MTa = large-scale ribbon with spindle shape; MTb = medium-scale ribbon with spindle shape. Note that the spheroidal bodies that are possible fossils of macroscopic algae (e.g. the genus *Chuaria*) are scattered within the massive micrites (**a**). The arrowed in "A" refer to the long linear crack that is the same to that as shown in Fig. 2C, which is a microfault. Scale bars, 1 cm

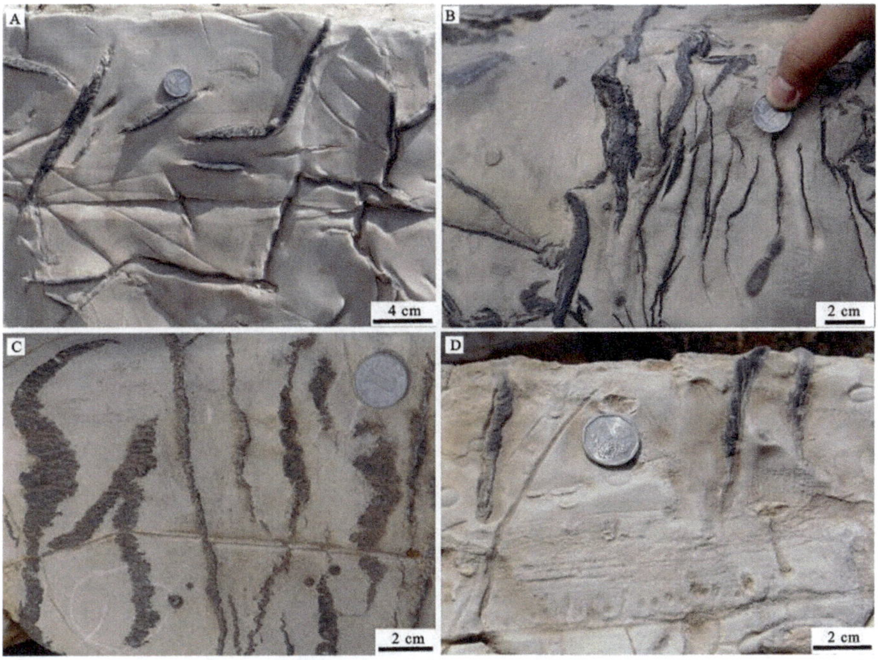

Fig. 39 Morphology of MT structures on the surface perpendicular to bedding

3.8 Molar-Tooth Microspars and Host Massive Micrites

The host rock of the MT structure is composed of massive micrite. Scours and storm beds, as shown in Fig. 40, indicate that this aphanitic micrite was deposited above storm wave surface. Macroscopically, compared with the host rocks that are composed of grey aphanitic micrite, the MT structures themselves have a darker colour (Figs. 37, 38 and 39).

The calcite microspar filling the cracks is composed of uniform, equant, limpid crystals, 25 to 50 microns in diameter, with planar crystalline boundaries (Fig. 42). In thin-section, the aphanitic micrite crystals (<4 μm in diameter) are cloudy in appearance and darker than the MT microspar (Fig. 42a).

More importantly, the interface between the MT microspar and the micritic host rock is a black zone, width from 0.2 to 0.5 mm, which is composed of organic material with pyrite crystals (Fig. 42a). Within the microspar filling the MT cracks, there are many pyrite crystals and organic remnants that are seen as black 'microblobs' (Fig. 42a–d). The massive micrite itself has a relatively structureless microfabric and lacks any clear lamination ("AM" in Fig. 42a), clots or dendritic fabric (Riding 2000). However, the cloudy appearance of the massive micrite crystals suggests high organic matter content.

Fig. 40 Intraclasts (arrowed) derived from MT structures. Lag deposit consisting of fragments of MT structure, with fining-up of clast size, derived from storm erosion

Interestingly and importantly, the organic-rich interface at the margin of the MT cracks has more pyrite crystals than the calcite microspar of the MT structure and the massive micrite (arrowed in Fig. 42a) of host rock. Partly, there are some remnants of host massive micrites within the MT calcite microspar (Fig. 42c). And there are many larger pyrite crystals (about 50 microns) within these remnants of host massive micrites (Figs. 3–16d). Microscopically, calcite microspars filling the MT crack are quite different from the interior fillings within the spheroidal body (Figs. 3–11C), i.e. the *Chuaria*. The interior fillings within the spheroidal body have possibly biological natures (Figs. 3–17).

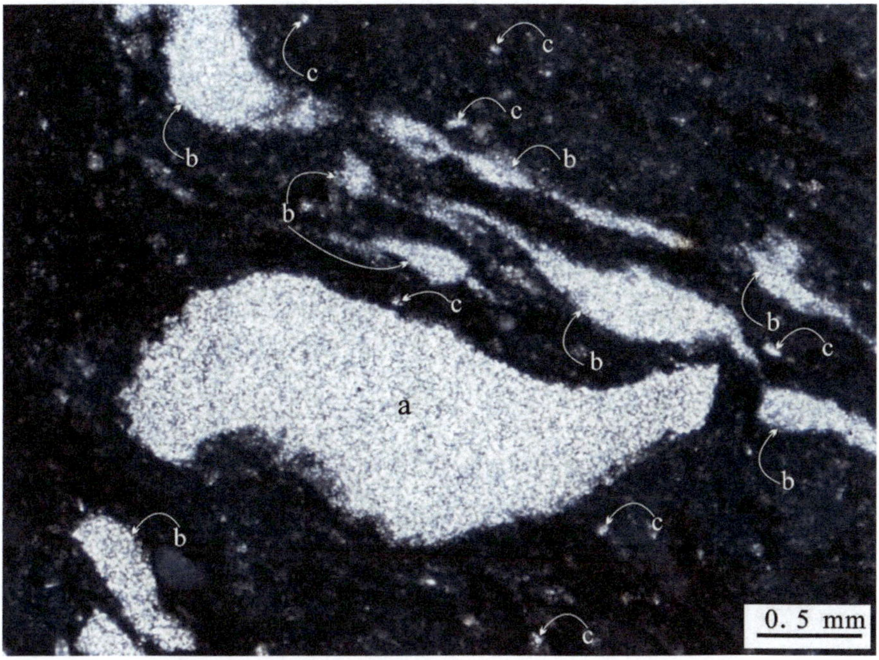

Fig. 41 Microscopic images of MT intraclasts. These various intraclasts include the rudites (**a**), arenites (**b**) and siltites (**c**). Most of them are sharp-edged, and matrixes are massive micrites that are the same to host rocks

3.9 Chemical Composition of MT Structure and Its Host Rock

There are strong differences in component and geochemistry among the calcite microspar filling MT cracks, the fillings of three-dimensionally preserved fossils and the massive micrite of the host rock (Table 1). The organic carbon content of calcite microspars filling the MT crack is lower (about 0.03%; Table 1) than that of the massive micrites of the host rock (0.1%; Table 1) and that of of three-dimensionally preserved fossils (0.09% to 0.1%; Table 1). That is, the organic carbon content of massive micrite of the host rock is similar to that of three-dimensionally preserved fossils. Geochemistrically, the calcite microspars filling MT cracks has higher carbon isotopic value ($\delta^{13}C_{PDB}$ is 0.03‰; Table 1) than that of three-dimensionally preserved fossils ($\delta^{13}C_{PDB}$ is from –1.3 to –1.8‰; Table 1) and that of the massive micrites of the host rock ($\delta^{13}C_{PDB}$ is from –0.2 to –0.5‰; Table 1). And the three-dimensionally preserved fossil has more negative carbon-isotopic value. These three materials, have similar oxygen-isotopic value ($\delta^{18}0_{PDB}$ is from –5.3 to –6.2‰; Table 1).

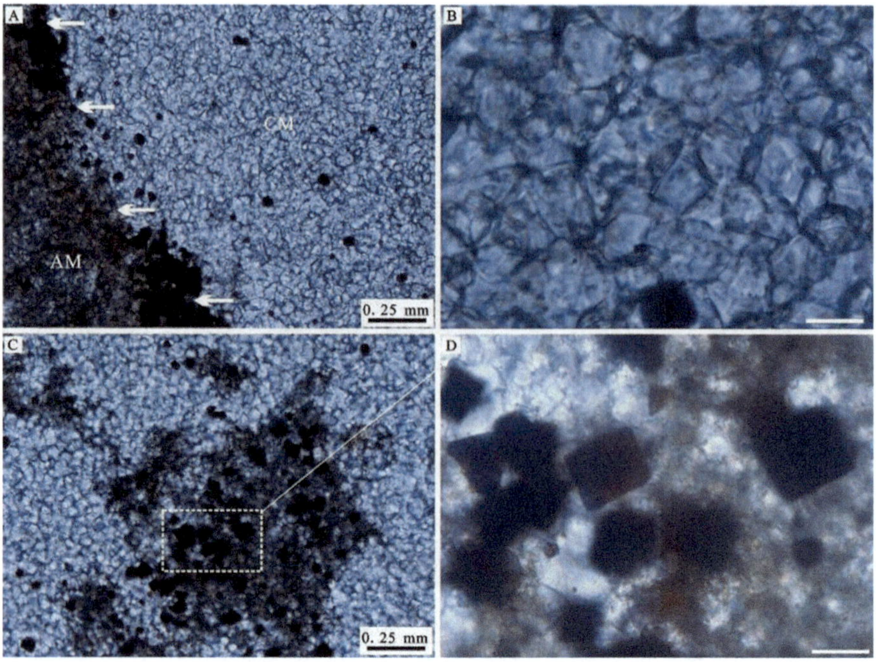

Fig. 42 MT microspar and host rock. **a** MT microspar (CM) and its host rock of aphanitic micrite (AM), the interface between them is a black zone rich in organic matter and pyrite (arrowed); **b** an enlargement of MT microspar, scale bar is 50μm; **d** remnant of host aphanitic micrites with many larger pyrite crystals within the MT microspar; **d** an enlargement of remnant of host aphanitic micrites with many larger pyrite crystals within the MT microspar, scale bar is 50μm

3.10 The Origin of Molar-Tooth Structures

Any satisfactory hypothesis for the origin of MT structures must take into account the following phenomena: (1) the mechanism for generating the MT cracks; (2) the process of calcite microspar precipitation and filling of the MT cracks; (3) the generation of the organic-rich interface with pyrite between the calcite microspar of the MT cracks and aphanitic micrite of the host limestone, and (4) the temporal restriction of MT structures to Precambrian strata.

In a similar manner to Qiao et al. (1994) and Fairchild et al. (1997), Pratt (1998a, b) envisaged MT structures as forming in an original terrigenous clay-lime mud mixture during dewatering caused by earthquake shocks. According to this model, (1), a variety of sheet-like and spindle-shaped cracks were generated during sediment compaction caused by seismic shaking; (2) a complex process of sediment sorting was caused by the seismic shaking, i.e. the lime mud, granular in character, segregated from the clay, and was carried into the fissures and cracks as a slurry; (3) the host sediments were consolidated and sheared by the continued seismic shaking, and during this process the lime mud in the fissures started to lithify almost immediately

Fig. 43 Microscopic images of the three-dimensional spheroidal fossils of macroscopic algal. The microstructure of the fossil with possibly biological origin, including the silicificational part (MA_1), and the calcificational part (MA_2), which is strongly different from the massive micrites of the host rock (AM) and from the MT calcite microspar as shown in Fig. 42

Table 1 Geochemical and carbon-organic analysis results for the aphanitic micrite of host rock (HR), the molar-tooth calcite calcite microspar (MT) and the algal megafossil (MA) in the third member of the Gaoyuzhuang Formation at the Jixian section

Sample	Organic-carbon content (%)	Carbon-isotopic value ($\delta^{13}C_{PDB}$; ‰)	Oxygen-isotopic value ($\delta^{18}O_{PDB}$; ‰)
HR-1	0.1	−0.5	−6.0
HR-2	0.1	−0.2	−5.8
HR-3	0.1	−0.2	−5.8
MT-1	0.03	+0.3	−5.5
MT-2	0.03	+.3	−5.5
MT-3	0.03	+0.3	−5.5
MA-1	0.1	−1.8	−6.2
MA-2	0.09	−1.3	−5.3
MA-3	0.09	−1.3	−5.3

by grain-growth, with $CaCO_3$ cement being precipitated from seawater. However, as noted by Furniss et al. (1998) for Belt-Purcell MT structures, the MT beds in the Gaoyuzhuang Fm. lack all the typical seismic deformational structures such as small-scale extensional and compressional faults, and disharmonic soft-sediment folds. More importantly, Pratt (1998a, b) ascribed the absence of MT structures in the Phanerozoic to changes in sediment rheology brought about by increased organic binding due to diversification of the microbiota, i.e. an increased abundance of organic matter in Phanerozoic sediments and the water column. It is thought this led to an essentially irreversible change in the geotechnical properties of marine sediments which prevented the generation of MT structures. However, Pratt's model does not seem to apply to the MT structures in the Gaoyuzhuang Fm. since the host aphanitic micrite is itself rich in organic matter. On the other hand, Pratt's explanation for the formation of calcite microspar filling MT cracks, through near-spontaneous direct physicochemical precipitatation that is probably as spontaneous "whitings" in the water column, may also be incorrect.

Based on innovative experiments, Furniss et al. (1998) visualized the formation of MT structures as a two-stage process. The first stage involved the generation of biogenic gas from decaying organic matter, creating a series of cracks, fissures and bubbles, filled with H_2S, CO_2 and CH_4, within a meter or so of the depositional surface. The gas could not escape because the surface was sealed, probably by microbial mats. In the second stage, precipitation of microspar in the open cracks prior to compaction was probably mediated by microbes. Pollock et al. (2006) supported this model of microbially-induced gas-bubble expansion and migration, and invoked microscale gas-sediment interaction to interpret crack morphology in terms of gas pressure and substrate shear strength. Pollock et al. (2006) suggested that precipitation of microspar was linked to gas production during organic matter decomposition within the host substrate. Generally, the model of gas migration and expansion is the best interpretation on the MT structure of the Gaoyuzhuang Formation.

As suggested by James et al., molar-tooth may reflect a secular set of global conditions that promoted early precipitation of calcite in the Proterozoic. Pollock et al. (2006) speculated that gas production would have resulted in pore-fluids with an elevated concentration of dissolved organic molecules, which may have induced precipitation of MT microspar when the pore fluids came into contact with $CaCO_3$ supersaturated seawater. Thus, the restriction of MT structures largely to Mesoproterozoic through Neoproterozoic strata probably reflects a critical level of carbonate saturation that could have limited early substrate lithification, thereby allowing crack production, but remained high enough so that organic catalysts were able to initiate precipitation of the MT microspar.

The lack of microbial laminae and the ptygmatic cracks filled with microspar suggest that the MT structure is not simply a result of microbial mat growth. This is supported by the occurrence of MT structures in facies of deeper-water than peritidal dolomites containing stromatolites (Fig. 36). An origin through evaporite replacement is also ruled out by facies considerations; evaporites and / or their pseudomorphs do occur in the tidal-flat facies but not in mid-outer ramp facies where the MT structures are concentrated.

MT structures of the Gaoyuzhuang Formation are chiefly developed in massive micrites that are the same as *leiolites*, as described by Riding (2000), which are homogeneous microcrystalline carbonates of probable microbial origin. Since the massive micrites are a kind of microbial carbonate (Riding 2000), the closely associated laminated micrites may be a type of mat-ground deposit (Pflüger 1999; Noffke et al. 2001; Mei et al. 2006), i.e. also a type of microbial carbonate. The formation of massive micrite could have been the result of various metabolic processes, such as synthetic uptake of CO_2 and/or release of $HCO_3^-$by cyanobacteria, and ammonification, denitrification and sulphate reduction by other bacteria, which would have increased alkalinity and stimulated carbonate precipitation (Riding 2000). These processes, promoted by microbial activity, could have led to rapid nucleation and lithification of the massive micrite and so the development of the MT structures.

As shown in Fig. 42, several features of the Gaoyuzhuang Fm. MT structures, such as the obvious interface rich in pyrite crystals and organic matter, the larger pyrite crystals within the microspar than within the host sediments, and the irregular, complex morphology of the MT cracks, as shown in Figs. 38, 39 and 40, are consistent with the microbially-induced gas-bubble expansion and migration model. However, the following aspects should be emphasized and discussed further to interpret the sedimentological features of these MT structures.

Four viable models exist for the formation of the Gaoyuzhuang MT cracks themselves: syneresis cracking, seismic cracking, gas-expansion cracking and wave-induced cracking. However, only gas-expansion and wave-induced cracking are consistent with the formation of the MT cracks in the limestone of the third Member of the Gaoyuzhuang Formation at Jixian (Figs. 37, 38 and 39).

As shown in Figs. 37, 38 and 39, various MT cracks, i.e. the larger ribbon (MTa in Fig. 38), the smaller ribbon (MTb in Fig. 38). And thin fracture cracks can cross and break the former types of the MT cracks. Thus, they all appear to have formed as multiple generations.

Many scoured surfaces are developed within the host rocks (aphanitic micrite) of the MT structures (c.f. Mei 2006, 2007), which suggest that most of these structures formed on an active seafloor. Thus their formation could be explained by the wave-induced fluid-flow model (Bishop et al. 2006); an active seafloor with storms and waves would have resulted in the dewatering and degassing to generate MT structures.

Ultimately, the MT cracks of the Gaoyuzhuang Formation can be interpreted as the early-diagenetic bodies of the degassing process resulting from the decomposition of organic substance of the host aphanitic micrites enriched with organic substance that provided by microbe activity. This conclusion is based on the following features of the MT cracks of the Gaoyuzhuang Formation as shown in Figs. 37 and 38: (1) the larger width of the larger ribbon MT cracks (the maximum value is about 2 cm; Fig. 37c–f; MTa in Fig. 38) that is disproportionate to the length (the maximum value of the length is about 20 cm; Fig. 37c–f; MTa in Fig. 38), which is difficult to be interpreted with normal physical forming mechanism of the void crack; (2) the complex and irregular configuration of MT cracks in plan view (Figs. 37 and 38) that shows they all appear to have formed as multiple generations; (3) as shown in Fig. 37e, the overall volume of the MT structures can locally reach to 40% of the host rock, which is also difficult to be interpreted with normal physical forming mechanism of the void crack.

As that as noted by Mei (2006, 2007), the large ribbon-shaped MT bodies of the Gaoyuzhuang Formation are distinct from the vein-shaped MT bodies of the latest Neoproterozoic strata in the Shandong-Liaoning-Jilin-Anhui area in the eastern part of China (Meng and Ge 2002; Qiao et al. 1994). Doing not come singly but in pairs, an unusually dolomitic nodules with abundant cement-filled MT structures that are formed in mid-ramp carbonates and are not found in surrounding thinly bedded dolomicrites in the early Neoproterozoic Burovaya Formation of the Turukhansky Uplift, Siberia provide an interesting example to interpret the formation of the MT structure. Therefore, various morphologies of MT bodies can be interpreted by Furniss et al. (1998) model of the gas-bubble migration and expansion.

The microspar filling the MT cracks is marked by its uniform crystal size and equant shape (Fig. 42). The interface composed of organic matter with pyrite crystals between the microspar of the MT structure and the aphanitic micrite of the host rock (Fig. 42) suggests that the MT cracks acted as fluid conduits and mixing sites for pore-water and sea-water where organic catalysts were able to initiate precipitation of calcite and so cause rapid lithification.

Lag deposits of MT coquina (Fig. 40) reflect the rapid lithification of the MT structure. This phenomenon indicates that the MT structures were lithified rapidly prior to erosion by storm currents. Furthermore, some intraclasts are entirely the MT crack fills (Fig. 41), showing that in situ precipitation of the carbonate was penecontemporaneous with or soon after the formation of the MT crack itself and even before the lithification of the host aphanitic micrite. This also shows clearly that the formation of the MT cracks and their filling by calcite microspar took place before the action of strong storm currents.

Some special remnants of host aphanitic micrites within MT microspars and with many larger pyrite crystals (Fig. 42c–d), indicate that the recrystallization and replacement of the host micrites cannot be eliminated during the forming of the MT microspar.

Thus, the formation of the MT microspar is a complex process: (1) a special pore-water rich in dissolved organic matter was first formed during the decomposition of organic matter in the presence of calcite-supersaturated Proterozoic seawater; (2) the dynamics provided by wave-induced fluid-flow or degassing and dewatering processes of host aphanitic micrites made this special pore-water migrate into the MT cracks; (3) the presence of calcite-supersaturated Proterozoic seawater resulted in the precipitation of microcrystalline calcite and its rapid lithification. The sealing effect of microbial mats or scums on the bedding surface of the host sediment would have prevented gas from escaping out of the sediment; the interface between the MT microspar and the host sediment, as shown in Fig. 42, formed a lining to the MT crack and so a favoured place for the precipitation of the calcite.

The organic material clearly formed a lining to the MT crack, and it is possible that this created a barrier, so that the calcite microspar filling the MT cracks precipitated within a sealed cavity. This interface could be the result of one or several mechanisms, such as dewatering, degassing, and decomposition of organic material in the aphanitic micrite. Furthermore, this distinct interface demonstrates that the MT cracks might not have been empty during their formation, but may have been filled with a particular diagenetic gas-rich fluid, resulting from the deformational processes of the host aphanitic micrite. This idea is different from the hypotheses of many researchers, who have suggested that the MT cracks formed early but were empty pores, before the calcite microspar fill was precipitated there sometime later (Furniss et al. 1998; Pratt 1998a, b).

Microscopically, calcite microspars filling the MT crack are quite different from the fillings within the spheroidal body (Figs. 37c and 38a). O'Connor referred to similar structures as 'blobs' and Furniss et al. (1998) and Pollock et al. (2006) had a spheroidal MT category; probably their structures are also remains of macroalgae. The microscopic features of the spheroidal bodies like the *Chuaria* (Figs. 37c and 38a) in limestones of the Gaoyuzhuang Formation are different from the calcite microspar filling the MT cracks. These spheroidal algae megafossils have particular infillings with possible cellular microstructure (Fig. 38). This particularity might reflect in vivo calcification of cellular substance that is similar to in vivo cyanobacteria sheath calcification and the excellent preservation of organic material within the spheroidal algae megafossils (Fig. 38) may reflect the extremely rapid lithification, perhaps on timescales of days to weeks.

Geochemical and carbon-organic analysis (Table 1) reflects the difference among the host rock (HR_1 to HR_3 in Table 1), the molar-tooth microspar (MT_1 to MT_3 in Table 1) and the algal megafossil (MA_1 to MA_3 in Table 1). For the molar-tooth calcite microspar, the particularly positive value of carbon-isotopic value ($\delta^{13}C_{PDB}$; 0.3 ‰) and the relatively small value of organic-carbon content (0.03%) indicate that some alterations resulting from the early diagenesis that forms the molar-tooth structure, and further express the basic depositional and diagenetic features of the molar-tooth structure. For the host rock composed of the aphanitic micrites, the relatively negative values of carbon-isotopic value ($\delta^{13}C_{PDB}$; –0.2 to –0.5 ‰) and the relatively large value of organic-carbon content (0.1%) indicate that this kind of microbial limestone can provide the organic substance for the formation of the molar-tooth structure. For the algal megafossil, the relatively much more negative carbon-isotopic values ($\delta^{13}C_{PDB}$; −1.3 to-1.8 ‰) and the relatively large values of organic-carbon content (0.09% to 0.1%) indicate that algal megafossils had enriched organic substance and have the fundamental biological features. Further, the similar value of the oxygen-isotopic value ($\delta^{18}0_{PDB}$; from –5.3 to 6.2 ‰; Table 1) among the host rock, the molar-tooth calcite microspar and the algal megafossil shows that the fossilization of megascopic algae was formed in the depositional period, the forming of molar-tooth structures was generated in the early diagenetic period and all of them have undergone same buried diagenesis. Furthermore, according to some important features of MT microspars of the Gaoyuzhuang Foration, i.e. the larger calcite crystals that is from 25 to 50 μm (Fig. 42b), the particular interface composed of organic matter with pyrite crystals between the microspar of the MT structure and the aphanitic micrite of the host rock (Fig. 42a), the lower organic-carbon content and the more positive value of $\delta^{13}C_{PDB}$ than the host micrites and the algae fossils (Table 1) and the special remnant of host micrites within the MT micospars (Fig. 42c–d), it can be inferred that the precipitation of MT calcite microspars has undergone two key processes. The first process is a solution of the host micrites in an acid diagenetic environment within the gas-migrating conduit resulting from the decomposition of organic substance of host rocks. The second process refers to the rapid precipitation of calcite microspars in an alkali diagenetic environment within the gas-migrating conduit resulting from microbial sulphate reduction.

Ultimately, although the MT structures cannot be interpreted as direct microbial deposit, the formation of the MT structures could well be genetically related to microbial activities. The formation of MT crack-like bodies is genetically related the degassing process of host aphanitic micrites enriched with organic substance provided by microbes. And the calcite microspars filling the MT crack-like bodies are resulted from a complex precipitation within a special pore-water rich in dissolved organic matter and was first formed during the decomposition of organic matter in the presence of calcite-supersaturated Proterozoic seawater. Sedimentologically, the MT structure could be grouped as a type of mat-ground structure (Noffke 1998; Pflüger 1999) and further into the microbially-induced primary sedimentary structure category (fifth category of Noffke et al. 2001; Mei et al. 2006). This idea is similar to that of the concept, i.e. "MT structures are the enigmatic microbially influenced structures in Proterozoic Carbonates".

Stop 3: Wumishan Cycles and Their Vertical Stacking Pattern in Long-Term Sequences of the Wumishan Formation

As that as mentioned in above context, carbonate strata of the Mesoproterozoic Wumishan Formation in the Jixian area near Tianjin are ~3300 m thick and were deposited over some 100 million years. Metre-scale cycles (parasequences) dominate the succession, which are described as the Wumishan cycles (Mei and Tucker 2013; Fig. 17). The widespread 1:4 stacking pattern in the cyclic succession (Fig. 18) suggest that the Milankovitch rhythms had similar ratios in the Mesoproterozoic as in the Phanerozoic. Based on the cycle stacking patterns, 26 third-order sequences can be distinguished and these group into 6 s-order, transgressive–regressive megasequences (or sequence sets), all reflecting a composite, hierarchical succession of relative sea-level changes (Fig. 9). Importantly, 7 types of lithofacies and about 1000 cycles make up a cyclic succession with the thickness of about 3300 m, which is marked by a set of remarkable sedimentary deposit (Hinnov 2013). Good outcrops for the Wumishan cycles can be examined at the highway from Jixian city of Tianjin to Pinggu city of Beijing (40. 1510°N, 117.4980°E to 40. 1620°N, 117.3794°E), ca. 35 km north of Jixian city (the location that is marked by the third star in Fig. 1a).

3.11 The Basic Model for the Wumishan Cycles

In sharp contrast to most meters of carbonate carbonate rock rotation in the typical carbonate rock platform in geological records (such as Brett et al. 2006) believes that Umishan Rotary has similar symmetry to Rock phase replacement characteristics. In the middle is a stack of stone biological layers and thrombosis-tumor rock biological reefs, the lower part is Chaoping Baiyun Rock, and the upper part is the lagoon with ancient soil hat-the white cloud rock shale. Umishan Rotary has obvious first deep and then shallow special signs. It is different from the "sequence sequence" defined by Vail et al. (1977). component.

There are four main types of rock phase types in the Umishan group in Tianjin: (1) Tide down layer of stone biological strata, (2) tide blocks-tumor rock biological reef, (3) tide flat cloud rock, (4) lagoon-the trendy white cloud rock page Rock (Figs. 44, 45, 46 and 47 section a ~ g).

Dolomite stack stone biological layer under the tide. The biological layer is usually less than 1 m thick and can follow dozens of meters. There are inner corners and gravel in the layer, and there are limest mud rocks between the stacked stones. The Umishan group has a variety of stack stones, including the Conophyton-Pseudogymnosolen combination, which can be divided into two sub-combinations of grinding jade stack stacks and hillside stack stones.

The combination of Musch stacks is mainly composed of small (CM-scale) stacked stone, such as PSEUDOGYMNOSOLEN MOPANYUENSE LIANG, TSAO, Scyphus Parvus Liang, etc., which are common in the lower part of the Umishan group. As shown in Fig. 44a, these small micro-finger stack stones are explained as a product of direct precipitation of carbonic acid in the sea. The hillside stacked stone sub-combination is mainly giant stack stone, with a height of up to 2 m (Fig. 44c), including Conophyton Litum Maslov, C. Coniforme Liang, etc., C. ShanPolingense Liang, etc. wait. These are more common in the upper part of the Umishan group. The biological layer is deposited in the low-energy environment.

B. Subtidal thrombolitic-oncolite bioherms. These structures can reach 2 m and 1 m high, consisting of microlachets. The sand particles-sized particles (condensate or thrombo particles) and pebbles composed of thrombolytic stone fragments, as well as thrombolytic stones, appear in a thin layer in the lens body and surface. They may be generated by storms. Thromer-tumor rock biological reefs are deposited in a medium-to-high-energy shallow environment. Catagraphs refers to thrombosis particles, and thrombosis and tumorite are classified as microtonic rocks. As shown in Fig. 44e, the rock phase unit lacks a real stack stone (layered structure). The block structure of the rock phase is interpreted as a secondary product of microorganism-organic degradation and calcification, as RIDING (2000) was proposed in his term "thrombosis after sedimentation". According to the detailed study of ZHAO (1992), some thrombosis particles with vague coating may be the embryo shape of tumor stone. The rock phase belongs to the type of non-superimposed stone microbial carbonate rock, which is worth further research. Interestingly, if the condensation of this rock phase is the sedimentary thrombosis (c.f. Riding 2000) or early tumor rocks (c.f. Zhao 1992), then it is likely to be the most shown in Fig. 45a. One of the ancient thrombite (Riding 2000). In addition, the micro-numbers stacking stack of the Musch-layer Stone combination may be the youngest stack layer formed by the stack of carbonate directly on the bottom of the sea to sedimentary sedimentation One of the stones, because this form of stack stone is mainly developed in the ancient and ancient Yuan Dynasty.

C. Intertidal silicified microbial laminated dolomites. Microbial laminites are marked by wavy, hummocky and planar stromatolites. As shown in Figs. 44g and

A
(a)
4 cm
A
D
E
G
F
E
(b)
30 cm
A
(c)
4 cm
D
G
F
E
D
(d)
40 cm
B
(e)
6 cm
A
D
G
F
E
D
C
A
(f)
30 cm
S
S
C
B
(g)
2 cm
SB
(h)
300 cm
D
(1)
2 cm
(j)
200 cm

◄**Fig. 44** The lithofacies units and the general characteristics of the Wumishan cycles. Symbols from A to G refer to the lithofacies units of the Wumishan cycles described in the text. SB refers to the boundary of the third-order sequences. S is a scoured surface. The arrows represent the boundaries of the Wumishan cycles. Image a shows a stromatolitic biostrome made up of *Pseudogymnosolen*; image b shows the boundary of a Wumishan cycle (the arrowed), which is expressed as the abrupt change from the shallowing-up succession of "E–F-G" in the upper part of the underlying cycle to the deepening-up succession of "E-D-A" in the lower part of the overlying cycle; image c shows a stromatolitic biostrome made up of *Conophyton*; image d shows the cycle boundary expressed as the abrupt change from the unit G at the top of the underlying cycle to the unit D in the lower part of the overlying cycle; image e shows a unit B facies without stromatolites; image f shows a cycle boundary expressed as an abrupt change, similar to those in images b and d; image g shows a unit C facies with a scoured surface (S); image h shows the sequence boundary that separates a succession of asymmetrical cycles without stromatolitic biostromes and thrombolitic-oncolitic bioherms in the top part of the underlying sequence from a succession of approximately symmetrical cycles with thick-bedded to massive stromatolitic biostromes; image i shows a unit D facies with mud cracks; image j shows one Wumishan cycle with a thickness more than 10 m

45b, interbedded tabular calcirudites on scoured surfaces with cross-bedding are interpreted as tempestites, although this.

Scholars such as Qiao and Li, Ettensohn et al. (2011), Su and Sun interpreted special types of plate-shaped calcium rocks as a soft sink deformation structure induced by earthquakes. Most of pebbles are mud released by dryness. Strong siliconization and white cloudization are the basic features of the rock phase.

D. Intertidal quartz-sand-rich dolomicrites. The rock-phase unit is mainly mud cracks (Fig. 44i), silicate, calcium, and rock salt pseudo crystals, especially Baiyun Rock (Fig. 44b, d, f) that forms the upper part of Umia. The quartz particles are round and can be eroded by the wind. Silicization and Baiyunhua are common.

E. Supratidal muddy and sandy dolomites. This unit developed mud cracks, silicate and siliconized rocks. Irregular micro-rock-soluble surfaces are covered with lens-like rock-like capacity. Baiyunhua and siliconization are common.

F. Lagoonal-supratidal dolomitic shales. The clay minerals of these shale are mainly sea bubbles, monsters, Illi stone, slopes, and talcs, and a small amount of kaolin stone (Huang et al. 1996). This special clay correlation indicates that the shale is formed by the high-silicon magnesium content and low aluminum content of carbonate rocks in high-silicon magnesium content. The rock phase unit F is interpreted as a rock shale (such as Song et al. 1991). However, from the B, D, and F figures in Fig. 44, it can be seen that most of the base parts of Mount Wumi Mountain are composed of D rock phase units (Baiyun Rock containing mud fissures), and the D rock phase unit is also It constitutes the upper part of the Wumi Mountain. Therefore, between the upper rock phase unit D (Baiyun Rock) and the lower rock phase unit E (tide upper mud and sandy white cloud rock), the rock phase unit F (laxa-tide-on the tide-on the tide-on the tide Baiyun Rock Shark). In the end, the shallow performance from Unity D to G Unity constitutes the upper part of the Umishan Rotate, and the depth of the depth from Union D or E-E-E-A unit A. In Figs. 44b, d, f, and 46c, the

Fig. 45 Images showing the sedimentary features for the unit B and C of the Wumishan cycles. Image **a** shows the thrombolitic grain with different size and various shapes that make up the unit B of the Wumishan cycles (the arrowed); image **b** shows the interbedded tabular calcirudites on scoured surfaces within unit C of the Wumishan cycles, which reflects an important action feature by strong current

boundary of Umeshan's rotation back is always from the mutation of the G Unit (the ancient soil that forms the F unit) to the D (or E) to Sign. These features indicate that the Fylite is unlikely to be the rock of the rock.

G. Palaeosoils. The rock phase is usually composed of 5 ~ 20 cm thick brown shale, which is considered to be a product of ground exposure and weathering effects of the lagallow-upper shale (Union F). It is the cover unit of Umishan rotation, with a few CM ~ 20 cm thick. As shown in Fig. 46c, high-content iron is the main reason for this special color to distinguish the unit from the lagoon-Baiyun Rock Shale below. Glee G (ancient soil) shows that the border of Umeshan Ruck is characterized by exposure-dot-shaped surfaces (Tipper 2000). In addition, the upper part of the Umeshan rotation back to the upper part of the D ~ G rock phase unit is superficial and shallow, which has a strong characteristics of atmospheric rocks, which is related to the formation of the exposed appearance. The revealing appearance of the Wumi Mountain also shows that when evaluating the amplitude of the third-order relative to the sea plane from the Fischer diagram, there is no need to consider the effects of water depth and sediments.) The proposed (see the latter part of this article).

Most of the above-mentioned rock-phase units appear in a specific sequence and are similar to symmetry (Figs. 47 and 48). In the typical rotation, the stacked stone biological layer, thrombosis-tumor bi reef and microbial layer accumulate white cloud rocks to form the middle part; the lower and upper parts are Chaoping Baiyun Rock, and the upper part is the white cloud rock shale with ancient soil hat.

The upper boundary of Umeshan Rotation is characterized by showing appearances. In the upper part of the Baiyun Rock and the tide of Baiyun Rock, it develops calcium, silicon arges, and rock-soluble grave lenses in Baiyun Rock Shale.

In the third-level sequence, Umishan Rotation has a regular superposition rule, which is reflected in the changing form of the rock phase layer order, forming the spectrum shown in Fig. 48. The long-term deepening of the sedimentation environment is a response to the rising level of the three levels of the sea level, forming a thick-like-like layer of stone biological strata and thrombosis-tumor biological reefs. From Figs. 44j, 46a–b, it can be seen that the rock phase sequence of Umeshan's rotation is always symmetrical (as shown in Fig. 48). On the other hand, in response to the sedimentation environment of the three levels of the sea level decline, the sedimentation environment has been shallower for a long time, and the shallow water and exposure are more frequent, the rock phase sequence of the Umeshan rotation is transformed from symmetrical to asymmetric (E in Fig. 48f); stacking stone biological layers and thrombosis-tumor biological reefs thinner, and the lagoon-white cloud rock shale becomes thicker.

G
F
D
E
C
A
B
D
A
B
A
(a)
50 cm
A
D
E+F+G
D
C
D
G
F
E
D
(b)
50 cm
A
JX-89
D
E
G
F
(c)

◀**Fig. 46** Images showing some detail characteristics of the Wumishan cycles. Symbols from A to G refer to the lithofacies units of the Wumishan cycles described in the text. The arrows represent the boundaries of the Wumishan cycles. Image a shows one Wumishan cycle with a thickness more than 10 m that is composed of a succession of "D-A-B-A-B-A-C-D-E–F-G"; Image b shows one Wumishan cycle with a thickness more than 1 m that is composed of a succession of "D-C-D-E–F-G"; Image c shows the cycle boundary expressed as the abrupt change from the unit G at the top of the underlying cycle to the unit E in the lower part of the overlying cycle, in which a high content of iron for the unit G is the main cause of the special colour, marking this unit off from the underlying lagoonal-supratidal dolomitic shale

3.12 Vertical Stacking Patterns Related to Milankovitch Rhythms

In recent years, the deposition and strata have a great interest in the recovery of the rotation of Milan Civic's rhythm. In the year; (Van der Zwan 2002; Schwarzacher 2005; Brett et al. 2006; Tucker and Galan 2010; Zecchin 2007, 2010; Catuneanu and Zecchin 2013). The loop superposition mode is often used to represent this orbit forced control; for example, the 1: 4 arrangement of the North American Cambrian carbonate carbonate rock is explained as the deposition is controlled by short-handed rhythm stacks in long eccentric rhythm. Goldhammer et al., Fischer and Bottjer, Schwarzacher and others believed 1: 5 (the so-called "five ring" in the trilayer) from Chen and Tucker (2003) to 1: 7/8, because the early rotation rhythm was shorter. North China Cambrian Stranger is similar to the stack of rotation. Many studies have shown that the extensive appearance of the Mi-grade sedimentation has a genetic relationship with the rhythm of Milan Civic. However, it must be remembered that the self-loop process also runs at a similar time scale, and the random process can obviously produce similar patterns (Van Der Zwan 2002; Schwarzacher 2005; Brett et al. 2006; Tucker and Galan 2010).

In the Umishan group, there are 4 different types of rotation, as shown in Fig. 48, they generally form a regular sequence, with systematic changes in thickness and phase (Figs. 18 and 49). To. Common 1: 4 superposition mode shows that each basic cycle fluctuates by the environment-related environment of genetic and short-handed rhythms, and the unit or package consisting of four cycles is genetically genetically related to the long-term eccentric rhythmic rhythm. Caused by environmental fluctuations. According to Brett and MEI, a Umi Mountain Rotation can be called the six-order sub-layer order, the combination of four sub-layer prefaces can be Called the fifth-order sub-layer set. In some cases, a sequence can be divided into 5–8 layers (Fig. 50). Although this superposition mode is not common, such as Berger et al. and Schwarzacher, one of which may be the result of the rotation deposition control of the rhythm of the spring equinox, overlapping produce a sub-sequence order to produce sub-sequences Short-eccentric rhythm.

Establishing an astronomical control theory for the establishment of the Cambrian sedimentation is much more difficult than establishing an astronomical control theory that updates the "ice cellar" sedimentary material, because as discussed by MIALL

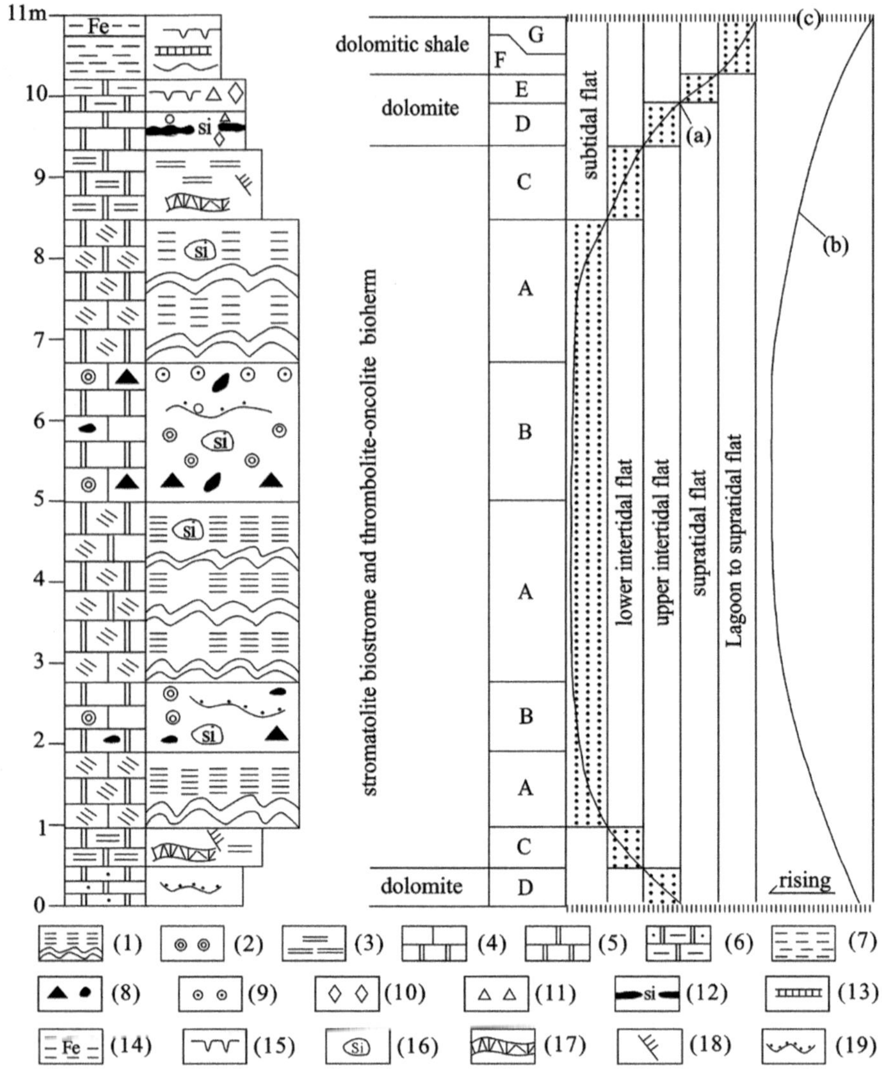

Fig. 47 A general model for the Wumishan cycles. A to G indicates the lithofacies units described in the text. The general characters are shown from A to G, as in Fig. 2. Explanations of (**a**) to (**c**) are as follows: **a** the changing curve of water depth; **b** the curve of high-frequency relative sea-level change; **c** a short-lived exposure surface. Interpretations of (1) to (19) are as follows: (1) stromatolite, (2) oncolite, (3) microbial laminite, (4) dolomitic limestone, (5) dolomite, (6) sandy and muddy dolomite, (7) dolomitic shale, (8) thrombolite, (9) oolite, (10) halite pseudomorphs, (11) karst breccia, (12) silcrete, (13) calcrete, (14) palaeosoil, (15) mud cracks, (16) chert nodule, (17) tabular rudite, (18) cross-bedding, (19) scoured surface

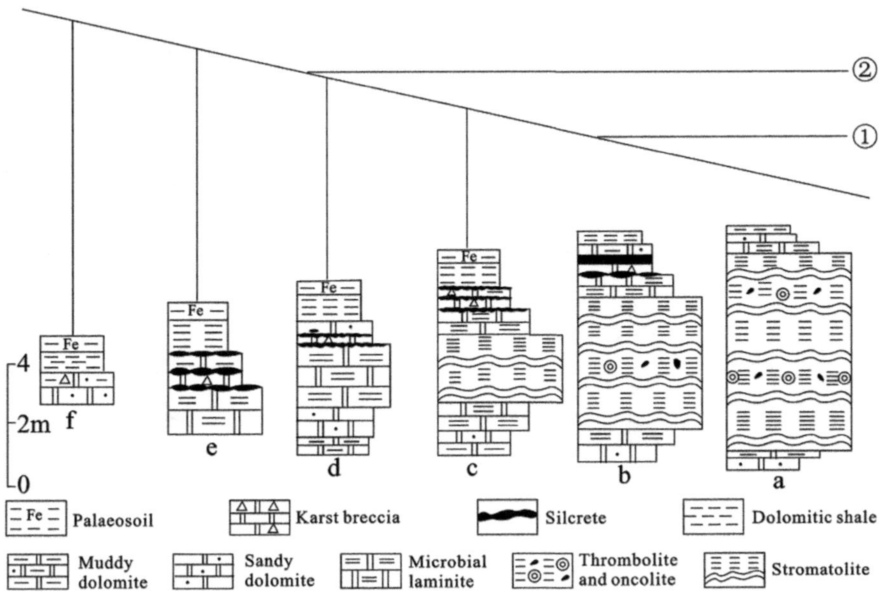

Fig. 48 The general environmental spectrum of the Wumishan cycles. 1 indicates the mean low water (MLW), and 2 indicates the high water (MHW); a to f indicate different kinds of Wumishan cycle

(2005), ERIKSSON and Catuneanu discussed As in the ancient greenhouse conditions in the ancient and Yuan Dynasty, most of the climate change is not so obvious. In addition, the absolute time of the former Han Wuji is more difficult to determine and the accuracy is much lower. Although there is no accurate radiation testing year, the accumulation law of Umishan's rotation back does indicate the control of long-term eccentricity, short eccentricity and progressive rhythm.

3.13 *Vertical Stacking Patterns of the Wumishan Cycles into Long-Term Sequences*

The Umishan group and the upper red water village group, Tieling group, and Xiamaling group may form a first-level rotary sequence, as shown in Fig. 9. This may be related to the genes of the Galaxy Year cycle (GAO et al. 1996) or the typical Wilson rotation of the consequences of the continent of the universe. The formation time is about 300 million years (1500 ma ~ 1300 ± 10 ma, Zhu et al. 1994).

From the vertical superposition mode of Wumi Mountain, the third-order sequence (Fig. 51) can be seen (Fig. 51). In the sea invasion system domain (TST), the fifth-level sub-layer group is usually composed of 4 rotation, which is marked with the upper stone biological layer and thrombosis-tumor rock biological reef as the upward.

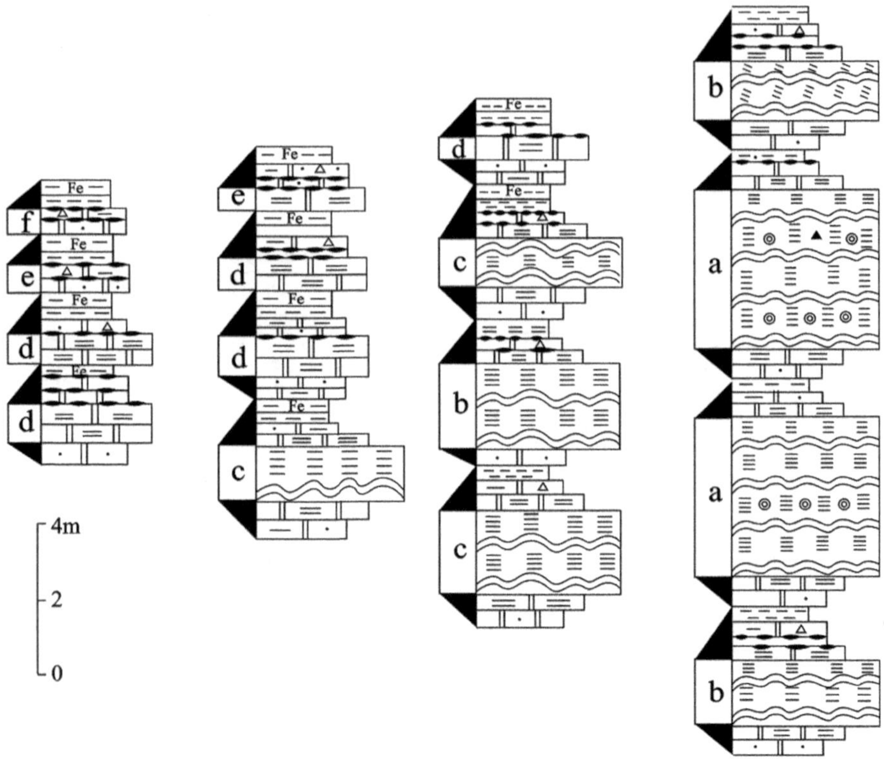

Fig. 49 The common 1:4 stacking pattern of the Wumishan cycles. This stacking pattern is interpreted as the result of environmental fluctuations controlled by the short-eccentricity rhythms superimposed on the long-eccentricity rhythm. The lithological marks are the same as those shown in Fig. 3. "a" to "f" indicate different kinds of Wumishan cycle as shown in Fig. 4

Chaoping Baiyun Rock and Lag Lake-Chao Cloud Rock Shale is the cover layer. In this cycle sequence, obvious concentration (CS) cannot be identified. In the high-ground system (HST) of the third-level sequence (HST), the five-level sub-layer distribution or six-level rotation can be identified. Rock, as well as stacking stone biological layers and thrombosis-tumor biological reefs, the overall spin-upward rotation. At the top of the third-level preface, the six-order rotation appears, and the rock phase sequence is asymmetric. As shown in the "E" or "F" in Fig. 49, the exposure of the characteristics is good, which may be the result of "missing beating". It can be seen that the boundary of the three—level sequence of the Umishan group is the main appearance, similar to the type 1 or 2 layer boundary boundary defined by Vail et al. (1977). As shown in Fig. 51, the vertical superposition mode of Wumishan rotation recovery is the key to identifying the third-class preface.

Figure 51 shows that from the regular vertical superposition mode of Wumi Mountain in the Jixian-Tianjin area, 26 tertiary sequences can be identified, which can be

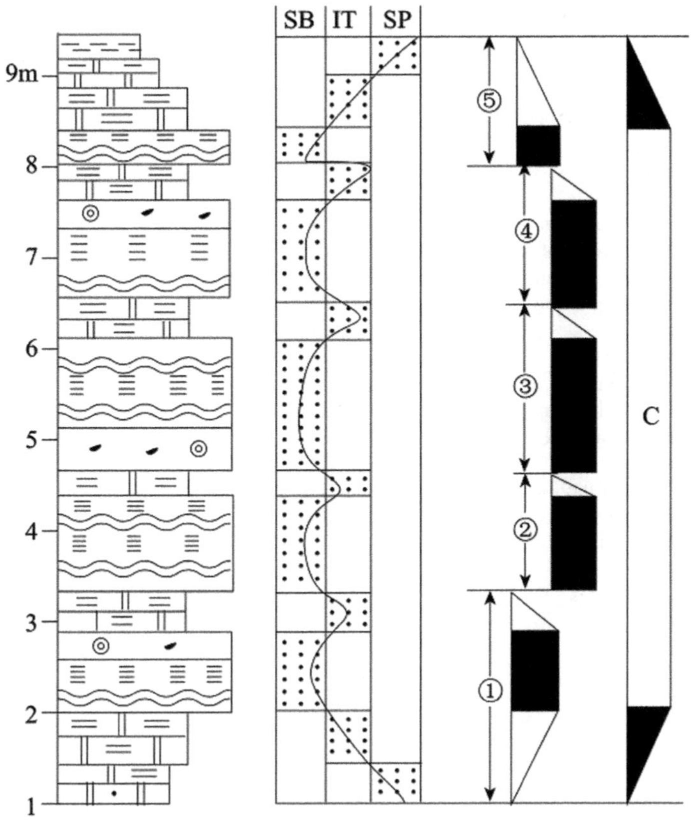

Fig. 50 One single Wumishan cycle may consist of 5 to 8 beds. Although this kind of vertical stacking pattern is rare, it is interpreted as the result of the precession rhythms superimposed on the short-eccentricity rhythm. 1 to 5 represent the beds; C represents one single Wumishan cycle; SB = subtidal facies, IT = tidal-flat facies, SP = supratidal-flat facies. The curve is the changing pattern of water depth. Lithofacies marks are the same as those in Fig. 48

divided into 6 secondary large layers (or layers of serial sets; 3–26). In order to estimate the changing amplitude of third-order relative sea-level change and the duration of the third-order sequences, 626 well-exposed cycles in the middle and upper part of the Wumishan Formation (sequences SQ_8 to SQ_{26} shown in Fig. 52) were studied in detail. The thickness of the cycles ranges from 0.6 to 14.5 m with an average of 4.2 m (Table 2).

The thickness of the Umeshan group is about 3300 m, and the deposition speed is about 100 m/year. During this sedimentary period, the crustal drop rate was about 3 m/100 kyr. On the basis of these assumptions, according to the same method proposed by Read and Goldhammer (1988), the Fischer diagram can be used to estimate the amplitude of the third—order sea plane change. The cycle of short—handed rhythm is equivalent. This is basically in line with the actual situation.

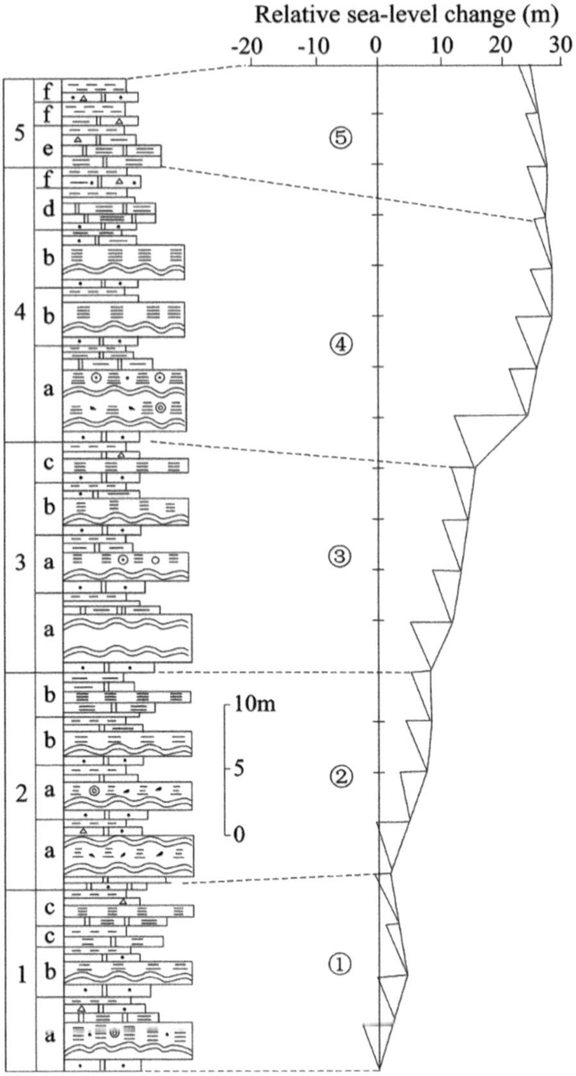

Fig. 51 Facies succession of the third-order sequences of the Wumishan Formation, an example from the 12th third-order sequence (SQ_{12}). Five fifth-order parasequence sets are indicated by 1 to 5, the same as described in Fig. 3–23. In the figure, a to f represent different kinds of Wumishan cycle as shown in Fig. 3–22. According to the chronostratigraphic interpretation of the Wumishan cycles, the duration of this third-order sequence can be estimated tentatively as 2 m.y. The curve on the right of the figure is the Fischer plot, showing that the changing amplitude of relative sea level of this sequence is about 25 m

ENOS Et Al. (1998) pointed out that the well—known three-stacking tide week LOFER rotation usually lacks significant ground exposure evidence, and rarely shows a significant deepening or shallow trend. They believe that the alternation of River's rotation is only the alternation of the tide week (paragraph B) and the sub-(C section), which is better described as … BCBCB … rhythm instead of the real rotation. Therefore, in a similar way, the Umishan Rotation with a typical symmetrical rock sequence can represent another rule, rhythmic alternation.

Most of the three—order sequences of the Umeshan group can be divided into 2 to 3 fourth—order sequences. For example, the third-level sequence of the 12th

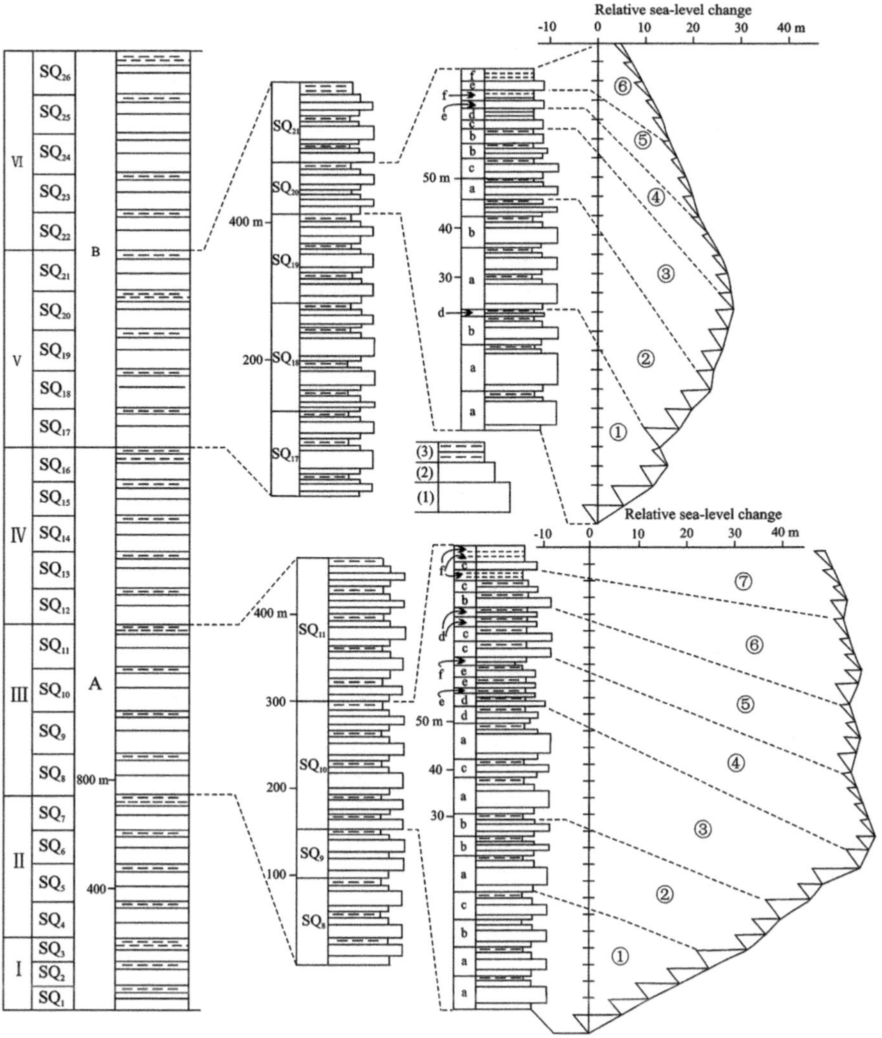

Fig. 52 Simplified schematic depiction of the composite cyclic succession of the Wumishan Formation at the Jixian section in Tianjin, including the 26 third-order sequences (SQ_1 to SQ_{26}) that can be grouped into 6 s-order megasequences (I to VI). Regular vertical stacking patterns of the Wumishan cycles (a to f, which are the same as those described in Fig. 48) and fifth-order parasequence sets (1 to 7 which are the same as those described in Fig. 49) in the long-term third-order sequences, show the nature of the Milankovitch rhythms in the stratigraphic record. Fischer plots are shown for Wumishan sequences SQ_{10} and SQ_{20}

Table 2 Listing of the number of the Wumishan cycles with thickness in metres

Nr[m]	Nr[m]	Nr[m]	Nr[m]	Nr[m]	Nr[m]	Nr[m]	Nr[m]	Nr[m]	Nr[m]	Nr[m]	Nr[m]	Nr[m]	Nr[m]
1 3.15	46 5.00	91 2.50	136 3.30	181 1.95	226 1.10	271 2.85	316 3.10	361 3.10	406 0.80	453 1.70	498 6.05	544 6.55	590 6.90
2 3.85	47 3.50	92 7.00	137 1.10	182 1.60	227 1.60	72 3.25	317 5.10	362 4.90	407 0.80	454 3.30	499 7.10	545 1.70	591 5.50
3 3.80	48 1.25	93 6.50	138 0.65	183 2.00	228 0.55	273 2.40	318 7.95	363 4.45	408 4.85	455 3.95	500 6.55	546 4.80	592 5.00
4 4.40	49 1.10	94 4.50	139 3.40	184 1.40	229 0.70	274 2.10	319 7.55	364 2.70	409 9.30	456 2.50	501 4.90	547 8.30	593 12.50
5 5.80	50 1.05	95 2.00	140 4.90	185 6.35	230 0.60	275 3.90	320 7.00	365 2.65	410 6.25	457 2.30	502 5.00	548 6.00	594 2.35
6 6.45	51 1.15	96 4.00	141 3.20	186 6.25	231 0.90	276 3.85	321 5.95	366 2.15	411 6.60	458 7.10	503 6.30	549 3.50	595 2.10
7 6.00	52 4.00	97 3.50	142 3.00	187 7.65	232 1.70	277 3.70	322 8.30	367 0.60	412 2.70	459 7.80	504 5.60	550 5.45	596 2.15
8 3.75	53 3.50	98 1.50	143 4.70	188 8.10	233 2.95	278 3.30	323 7.30	368 0.80	413 6.20	460 6.90	505 4.50	551 6.80	597 2.45
9 6.80	54 1.40	99 1.10	144 2.85	189 3.80	234 0.90	279 4.30	324 6.10	369 0.80	414 2.20	461 6.10	506 6.95	552 7.30	598 4.75
10 3.40	55 1.20	100 0.70	145 2.40	190 5.90	235 1.60	280 4.85	325 1.05	370 0.60	415 1.90	462 7.77	507 5.80	553 6.30	599 10.40
11 8.90	56 4.95	101 0.85	146 2.10	191 1.80	236 1.55	281 4.50	326 7.15	371 0.50	416 4.20	463 6.80	508 4.60	554 6.10	600 2.40

(continued)

Table 2 (continued)

Nr[m]	Nr[m]	Nr[m]	Nr[m]	Nr[m]	Nr[m]	Nr[m]	Nr[m]	Nr[m]	Nr[m]	Nr[m]	Nr[m]	Nr[m]	Nr[m]
12 1.30	57 3.55	102 0.90	147 3.30	192 4.75	237 1.70	282 3.90	*327* 8.70	372 0.70	417 3.80	464 6.40	509 4.40	555 4.20	601 3.85
13 1.50	58 1.65	103 1.15	148 2.40	193 4.55	238 1.10	283 3.70	328 7.10	373 0.50	418 2.90	465 5.60	510 2.42	556 3.20	602 4.20
14 13.70	59 0.80	104 0.60	149 1.80	194 6.60	239 1.00	284 2.70	329 6.80	374 0.50	419 2.60	466 2.90	511 2.20	557 5.40	603 9.20
15 8.40	60 1.40	105 0.96	150 0.80	195 *7.75*	240 5.50	285 3.60	330 1.05	375 8.35	420 6.30	467 2.30	512 1.20	558 4.60	604 5.95
16 9.30	61 3.70	106 0.90	151 5.90	196 5.90	241 2.60	286 3.00	331 0.85	376 9.40	421 4.80	468 2.90	513 0.80	559 7.40	605 2.32
17 2.00	*62* 1.00	107 0.95	152 9.50	197 6.60	242 2.70	287 2.60	332 0.75	377 5.80	472 3.90	469 2.20	514 1.20	560 5.90	606 3.40
18 3.70	63 1.10	108 6.00	153 4.80	198 3.00	243 1.30	288 2.90	333 0.70	378 0.60	423 3.30	470 3.00	515 1.20	561 6.90	607 4.40
19 7.45	64 1.00	109 4.00	154 3.25	199 6.30	244 1.90	289 39.0	334 0.95	379 7.25	424 4.75	471 3.45	516 1.10	562 7.95	608 3.80
20 0.80	65 3.90	110 1.50	155 3.65	200 3.35	245 2.00	290 3.60	335 1.00	380 5.60	425 4.20	472 2.65	517 1.00	563 6.00	609 3.20
21 0.80	*66* *3.80*	111 2.50	156 3.60	201 5.45	246 1.30	291 1.90	336 0.80	381 6.50	426 3.50	473 2.10	518 7.50	564 5.15	610 3.10
22 0.80	67 3.50	112 5.35	157 1.70	202 1.10	247 1.00	292 2.30	337 0.80	382 3.80	427 2.70	474 1.30	519 8.25	565 7.90	612 7.20
23 2.45	68 3.30	113 4.50	158 1.25	203 4.20	248 1.90	293 1.70	338 0.90	383 4.50	428 0.80	475 1.70	520 5.80	567 6.00	613 6.55
24 7.25	69 6.50	114 4.00	159 11.25	204 3.40	249 2.60	294 1.40	339 0.95	384 4.20	429 0.90	476 1.35	521 4.15	568 4.10	614 10.40

(continued)

Table 2 (continued)

Nr[m]	Nr[m]	Nr[m]	Nr[m]	Nr[m]	Nr[m]	Nr[m]	Nr[m]	Nr[m]	Nr[m]	Nr[m]	Nr[m]	Nr[m]	Nr[m]
25 6.27	70 6.15	115 3.10	160 5.90	205 1.80	250 2.60	295 1.30	340 1.10	385 3.55	430 0.90	477 0.90	523 10.05	569 3.00	615 11.60
26 7.52	71 2.20	116 6.15	161 1.95	206 1.80	251 1.10	296 0.70	341 0.70	386 2.90	431 0.90	478 1.25	524 7.60	570 3.50	616 10.80
27 7.70	72 5.80	117 4.50	162 6.10	207 14.50	252 2.80	297 0.75	342 2.90	387 1.90	432 0.70	479 1.10	525 6.70	571 3.10	617 6.80
28 6.80	73 3.50	118 4.25	163 4.10	208 2.15	253 3.45	298 0.70	343 9.60	388 0.90	434 0.90	480 1.15	526 6.60	572 2.00	618 1.10
29 5.45	74 3.00	119 3.45	164 1.10	209 6.30	254 4.15	299 0.65	344 9.75	389 03.90	435 0.80	481 1.10	527 5.90	573 4.55	619 1.05
30 4.60	75 2.50	120 7.25	165 2.75	210 4.40	255 2.15	300 0.90	345 2.35	390 0.90	436 0.80	482 4.00	528 7.40	574 4.00	620 1.15
31 0.65	76 6.50	121 4.35	166 6.60	211 7.20	256 2.10	301 1.25	346 7.40	391 1.40	437 0.90	483 4.55	529 4.80	575 2.60	621 0.85
32 1.65	77 6.00	122 4.85	167 6.25	212 6.10	257 5.05	302 1.70	347 8.60	392 1.20	438 0.90	484 5.90	530 6.40	576 2.25	622 1.20
33 1.00	78 4.00	123 3.35	168 6.20	213 4.70	258 3.50	303 1.70	348 6.70	393 0.90	439 0.85	485 4.60	531 4.60	577 1.00	623 1.15
34 0.90	79 3.30	124 0.85	169 5.10	214 3.50	259 3.50	304 3.10	349 4.70	394 0.80	440 0.80	486 6.90	532 3.70	578 0.80	624 0.80
35 0.80	80 6.20	125 2.65	170 5.00	215 3.80	260 2.60	305 2.15	350 7.50	395 0.80	441 0.80	487 7.20	533 1.50	579 0.85	625 0.80
36 10.45	81 3.50	126 0.70	171 2.35	216 2.80	261 3.20	306 2.90	351 8.90	396 1.10	442 3.90	488 5.90	534 8.50	580 0.70	626 0.70

(continued)

Table 2 (continued)

Nr[m]	Nr[m]	Nr[m]	Nr[m]	Nr[m]	Nr[m]	Nr[m]	Nr[m]	Nr[m]	Nr[m]	Nr[m]	Nr[m]	Nr[m]	Nr[m]
37 9.30	82 3.00	127 0.70	172 5.70	217 2.50	262 3.30	307 3.05	352 7.50	397 0.90	444 3.35	489 4.60	535 9.30	581 0.95	
38 9.80	83 2.50	128 3.10	173 3.15	218 4.50	263 2.90	308 5.05	353 4.85	398 0.80	445 2.90	490 7.25	536 5.05	582 0.85	
39 9.75	84 6.40	129 2.20	174 0.70	219 3.50	264 2.10	309 2.50	354 8.65	399 0.80	446 3.90	491 8.80	537 4.20	583 0.80	
40 10.75	85 4.40	130 2.20	175 0.50	220 3.00	265 4.70	310 1.40	355 5.80	400 3.60	447 2.25	492 93.30	538 2.00	584 0.80	
41 6.70	86 4.50	131 2.25	176 3.70	221 2.10	266 5.50	311 4.35	356 2.20	401 2.60	448 1.60	493 6.50	539 2.10	585 6.10	
42 6.60	87 3.80	132 2.25	177 5.70	222 0.90	267 4.90	312 5.20	357 1.00	402 4.10	449 6.85	494 6.80	540 2.00	586 6.30	
43 11.40	88 5.50	133 4.30	178 3.50	223 1.30	268 4.40	313 2.10	358 3.60	403 3.25	450 4.90	495 8.80	541 1.50	587 7.25	
44 5.25	89 4.40	134 3.35	179 2.10	224 1.60	269 2.95	314 7.55	359 4.65	404 1.10	451 2.60	496 7.35	542 2.10	588 5.45	
45 11.20	90 3.50	135 3.95	180 1.40	225 0.90	270 2.50	315 7.75	360 3.20	405 3.85	452 2.10	497 5.50	543 2.20	589 5.80	

stage of the Umeshan group contains 2 sub-sequences (Fig. 51), and the 10 and 20th sections can also be divided into two sequences (Fig. 52). In addition, 26 third—order sequences can be further divided into second—order serial packages that are more invasive and regained. There are 6 such second-order sequences (or sequence sets; Fig. 52 I to VI).

It can be seen that the rotation of the Umishan group can be divided into 6 levels, namely the second—level large sequence (or sequence set), the third—level sequence, the fourth—level subsequent sequence, the fifth—level sub—sequence set, and the sixth grade sequence. The latter consists of layers. All these show that the hierarchical control of the Umishan deposition is under changing by the sea level composite. The sedimentation mode of the three—level sequence of the Umeshan group is obviously different from the layer mode defined by Vail et al. (1977). Due to the lack of clear concentrated section and maximum flooding surfaces, the darts of the six—level preface are obvious. However, it is closer to Read and Goldhammer (1988). If there is no precise timing, the regularity of the cycle of different steps will become a very important argument for supporting astronomical control (Figs. 49, 50, 51 and 52).

The 6—level rotary nature found in the Umishan group and its regular vertical superposition can be used to estimate the general time span of the 3—layer sequence developed by the Umishan group. Table 3 lists parameters such as the thickness, number of layers and inference time spans of the Umishan layer. From Figs. 53 and 54, the Fischer Plot results can estimate the maximum and average values of the third-order relative to the sea level.

As shown in Table 3, the above estimates are based on the thickness of 626 cycles (Table 2). These rotations constitute 19 tertiary sequences (SQ8 ~ SQ26), which can be further divided into 4 s—level large—layer order (or layer distribution), distributed in the middle and upper part of the Umishan group. The statistical results of the Fischer diagram show that the sequence thickness of the third-class is 79.1 ~ 196.0 m, with an average of 132.3 m. The maximum three-level sea level change range is 6 ~ 88 m, and the average value is 45 m. The general range of three-level relative to sea level changes is 6 ~ 73 m, with an average of 28 m. The number of cycles (quasi-sequences) of the third-order sequence ranges from 12 to 41, with an average value of 33. Therefore, assuming that the duration of the sequence sequence is equal, the time span of the third order is 1.2 ~ 4.1 m.y, and the average value is 3.3 m.y. The average duration and amplitude of the third-order relative to the sea level rotation are similar to the recorded amplitude in the stratium in the universe (such as Osleger and Read 1991; Brett et al. 2006; Tucker and Galan 2010; Catuneanu and ZECCHIN 2013).

The Wumishan cycles with their approximately symmetrical lithofacies succession might represent a classic case for true sedimentary cycles. The superimposed relationship of four cycles/parasequences constituting one fifth-order parasequence set is interpreted as reflecting environmental fluctuations genetically related to short eccentricity rhythms superimposed on long eccentricity rhythms. Some sub-layers can be further divided into 5 ~ 8 layers, reflecting the environmental fluctuations controlled by the short-term eccentric rhythm of the progressive rhythm. Although

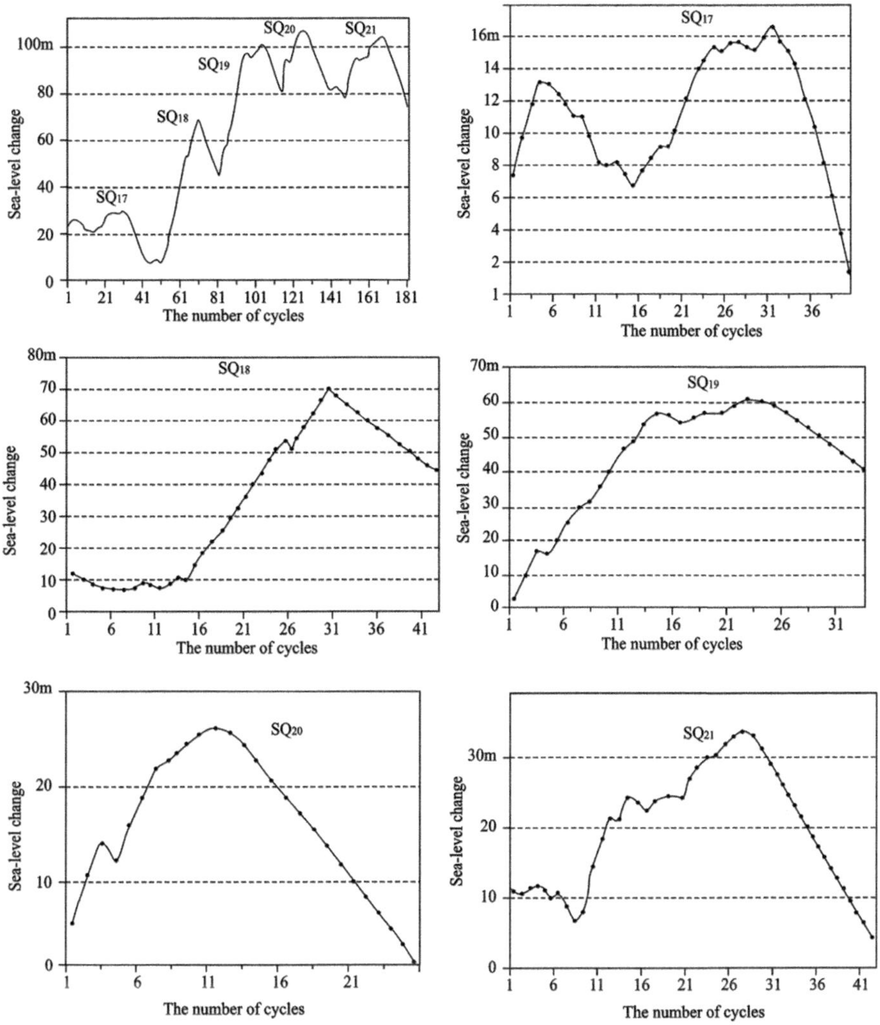

Fig. 53 Fischer plot from the fifth second-order megasequence of the Wumishan Formation composed of 5 third-order sequences from SQ_{17} to SQ_{21} as shown in Fig. 52. The general range of third-order relative sea-level change can be deduced from this plot

the absolute time span of the rhythm of the former Cambrian Milan Kivic is difficult to determine, the proportion of the Milan Civic rhythm reflected from the regular vertical overlapping mode from Umishan has similar to the rendezvous universe. This shows that the duration of the preferential rhythm of Milan Kovich in the former Cambrian, although the rhythm of the progress was short.

According to the vertical overlapping law of Wumei Mountain, 26 three-level sequences are identified, and it is divided into 6 secondary large layers (or sequence sets). A total of 628 rotations were found in the middle and upper part of the Misty

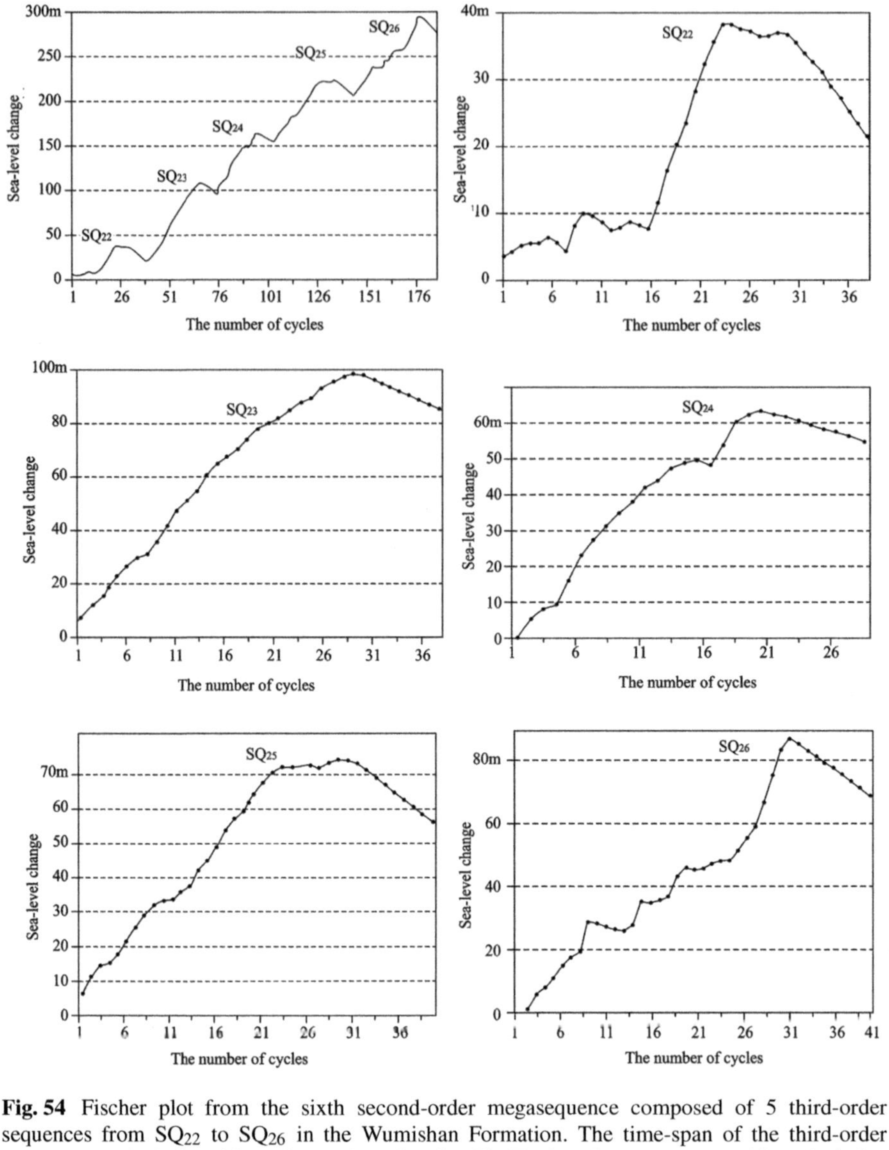

Fig. 54 Fischer plot from the sixth second-order megasequence composed of 5 third-order sequences from SQ_{22} to SQ_{26} in the Wumishan Formation. The time-span of the third-order sequences can be inferred from the number of cycles. The Fischer plot serves as a tool for calculating the amplitude of the third-order sea-level changes

Mountain Group (SQ8 ~ SQ26). These cycles (sequences) are the basis for Fisherto analysis to determine the long-term changes that can accommodate the space or the third-order relative to the sea level change. The Fischer diagram shows that the average maximum amplitude of the third-order sea plane changes is about 45 m,

Table 3 Numbers of the Wumishan cycles of the third-order sequences (SQ_8 to SQ_{26}) according to the Fischer plots shown in Figs. 10 and 11

Serial number of sequence	Thickness of sequence (m)	Number of cycle in each sequence	Inferred time-span of sequence (Ma)	Maximum rise and fall of relative sea-level changes(m)	General rise of relative sea-level changes (m)
SQ_8	106.1	22	2.2	47	40
SQ_9	79.1	12	1.2	15	22
SQ_{10}	145.3	29	2.9	64	47
SQ_{11}	164.0	42	4.2	18	24
SQ_{12}	94.0	20	2.0	21	−14
SQ_{13}	83.3	23	2.3	6	−6
SQ_{14}	131.7	34	3.4	38	30
SQ_{15}	177.5	37	3.7	67	67
SQ_{16}	89.7	43	4.3	40	−40
SQ_{17}	124.9	40	4.0	10	−4
SQ_{18}	154.7	42	4.2	62	29
SQ_{19}	134.9	33	3.3	57	50
SQ_{20}	85.8	25	2.5	27	1
SQ_{21}	119.8	42	4.2	30	−7
SQ_{22}	137.8	40	4.0	34	18
SQ_{23}	182.9	36	3.6	88	75
SQ_{24}	142.7	28	2.8	67	58
SQ_{25}	164.3	39	3.9	69	51
SQ_{26}	196.0	41	4.1	88	73
Mean	132.32	33.05	3.31	44.63	27.79

the average amplitude range is about 28 m, and the average duration of the third-order sequence is about 3.3 m.y. The data shows that the duration of the third-order sequence and the range of the third-order relative to the sea level change are similar to that of the show. The six rotation levels of the Misty Mountain Group and its regular vertical superposition mode reflect the level of the transformation of the Cambrian relative to the sea level.

In the context of Milan Kivic's rhythm control, further study the impact of Umishan Rotary Return on the curatorial mechanism of former Cambrian carbonate white clouds and siliconization, as well as rock global chemistry.

Stop 4: The Glauconite Within the Diversified Column Stromatolites of the Second Member of the Tieling Formation

Sea greenstone is a water—containing silicate—shaped silicate mineral, with different liquid or dark green as the host rock. This clay mineral is an iron-rich mineral, and there is a three-price iron and binary iron at the same time. It can be seen that the necessary condition for the formation of sea greenstone is from the weak oxidation of the medium to the weak restore. Under this special condition, both three-price iron and two-price iron can enter the dissolved state. When the pH value is greater than 2 to 3, the three-price iron starts to precipitate. When the pH value is greater than 5.5 ~ 7, the dose of the binary iron increases; Another necessary condition for the formation of sea greenstone is the alkaline conditions of the medium [1]. In addition to the favorable media environment, the formation of sea greenstone must also have a source of mixed sources. The potassium content is the key factor. The concentration of potassium in normal seawater is generally about 380ppm. The necessary condition for the formation of sea greenstone is that the potassium concentration is greater than 560ppm. Therefore, sea greenstone can be generated in a suitable environment. It is characterized by the alternation of water power alternation, feasible oxidation and reduction state, and suitable acid-alkali medium. Rich aluminum is enriched in the near-shore belt (ordered Matter 1981).

Sea greenstone is the representative of minerals on the surface of the earth, which is widely distributed in marine sediments and modern sediments. In strata records, the output mode of sea greenstone is diversified (Amorosi 1995, 1997; Banerjee et al. 2016). According to the formation mechanism of sea greenstones kind. First of all, the original sea greenstone refers to the green stone that is formed in place without being transferred from the same place. Secondly, heteristic sea greenstone refers to the sea greenstone that is separated from the original formation site by transmitting, which can be divided into heterogeneous green stones and debris sea green stones. The vice-original sea greenstone refers to the sea greenstone particles that are far from the original formal place due to the action of storms, tidal or gravity flow; sea greenstone is mainly stored in the mining bed during the same period. Dedicated sea greenstone refers to the sea greenstone particles deformation of the ancient strata. Therefore, sea greenstone is often transmitted from its original position in the form of particles or debris (Ordin and Matter 1981; Banerjee et al. 2016). The formation of sea greenstone can be classified into heterogeneous sea greenstone.

According to the sea greenstone formation model proposed by Ordin and Matter (1981), sea greenstone minerals are first referred to the micropores of about 5 ~ 10 μm, and then evolved into sea greenstone with the corrosion of the host mineral. The initial precipitation has the characteristics of rich iron, rich potassium-rich sea green rocks that are vertical on the ground; then, these initial sedents have gradually transformed into pane fabrics. With the increase of sea greenstone littering, the residual space of small spores is gradually filled, and the initial precipitation of the sea green stone mask is deformation due to mature (rone effect), which leads to the initial analysis of the sea green stone monsters. Increased content. However, only the potassium content gradually increases with the maturity period. With the enhancement of maturity, more sea green stone mask will be crystal and transform into sea green stone clouds; therefore, the content of K2O can be as high as 9%. Therefore, sea greenstone minerals can be interpreted as products of early corrosion and advanced maturity, rather than alternating products of pre-minerals. More importantly, the potassium content reflected by the surface color of the sea green stone becomes an indicator of the maturity. According to maturity indicators, sea greenstone can be divided into the following types: poverty-stricken potassium type (K2O = 2 ~ 4%); (2) micro-advance evolution (k2o = 4 ~ 6%); (3) evolutionary (k2o = 6 ~ 8%); (4) High evolution (K2O & GT; 8%).

A general view is that the formation of sea greenstone minerals reflects a slow gathering environment, the depth of water is 50 ~ 500 m, currently 100 ~ 200 m, water temperature 15 ~ 20 °C, pH value of 7.5 ~ 8.5. The higher the maturity of the sea greenstone, the longer the sedimentary interval. Therefore, the K2O content is more than 7% of the sea greenstone that can represent geological sedimentation cracks. According to this concept, the sea greenstone, especially the heterogeneous sea green stone, has become a slowdown in the sedimentary environment, weak restore, and large water depth. Emphasized signs for concentrated section (Ordin and Matter 1981; ChaFetz and Reid 2000; Banerjee et al. 2016).

In strata records, there are some exceptions of the output form of sea greenstone. For example, some native sea green stones in the Southern Western Cambrian to the Ova formation develop in the shallow tide beach sediment, rather than developing in the slow deployment deep-water environment (Chafetz and Reid 2000) Essence Similarly, the seized stone greenstone developed in the second section of the ancient Tieling Group in the ancient Tieling Group in the Jixian section is characterized by the colloidal membrane developed by stacked stones and mud crystal (grains) interface, which reflects the native sea. The basic characteristics of green stones are different from modern sea greenstone. Tieling group sea greenstone is mainly developed on the outer wall of the stack stone cylindrical body.

The special output method of Tieling Group Hilogishi represents a special stratigraphy record, which is difficult to explain its causes with the formation of modern sea greenstone; in addition, the Tieling group sea greenstone may reflect the ancient China and Yuan Dynasty (~ 1400 ma; Fig. 45d) is still marked by the hypoxia environment, which is different from modern, which makes the sea greenstone developed in Tieling group stacked stone reef has special significance. The good display of the second section of the Tieling Group is distributed in the old quarry section near Tieling Village (40. 0759° north latitude and 117.4319° east longitude position).

3.14 The Tieling Formation

The ancient Tieling group in the Jixian section can be divided into two sections (or sub-groups); Fig. 55) (Chen and Wu 1997; Chen et al. 1999). The first paragraph (Daizhuangzi group) is a set of manganese-containing white cloud rocks, and the embedded layer is a thin layer containing manganese and potassium mud shale; the second section (that is, the old Fuding Asian group) is a set of linsees, the lower and upper parts, the lower and upper parts The developmental part of Baiyun Rock or Baiyun Rock. The development of ancient rock-soluble and ancient soil on the two sections of the Tieling Group indicates that there is a long-term fracture zone in the stratum; the non-integrated belt represented by the long-term stratum cracks is named

Tieling bulge and green jade (Chen and Wu 1997; Chen et al. 1999). According to MEI et al. 2000), the Central and Yuan Dynasty Tieling Group was divided into two tertiary prefaces (DS1 and DS2 in Fig. 55). The layers of layers divide methods, that is, the intercourse of the rock phase is the basis of identifying the rice-level rotation, and the extent of the sedimentation is the basis of dividing the three-level sequence. More regularity (ERIKSSON et al. 2005; Catuneanu et al. 2005).

From Figs. 9 and 55, it can be seen that the first section of the Tieling Group is a third-level order, that is, the DS1 layer and the front-type carbonate carbonate rocky-grade (MEI et al. 2000) It reflects the basic characteristics of DS1, which is composed of the middle thick layer-like manganese white cloud rock and thin-layer white cloud rock shale (Fig. 20a-b). In the early DS1 high-altitude system domain (EHST), the lower unit of the rice-level rotation was marked by block to thick-layered manganese-containing white cloud rock, representing the basic characteristics of the sedimentation under the tide. The DS1 Sea Invasion System Domain (TST) is mainly composed of the upper tide belt containing manganese-containing white cloud rock. The important thing is that a shallow sedimentary sequence constitutes a late high altitude system domain (LHST), that is, from the upper part of the tidal room with manganese white cloud rock to the ancient rock-soluble sequences composed of hair quality and white cloud rock rocky rock. The sequence reflects the top boundary of the DS1 is an obvious exposure point surface, that is, the non-combined surface of Tieling's bulging (Fig. 20c–d; Chen and Wu 1997; Chen et al. 1999).

The second section of the Tieling Group forms a second-level and third-level order, that is, the DS2 layer order, as shown in Fig. 55. This layer of the sequence is deepened from the tide phase white cloud rock to the lower layer of limestone to deepen the upper layer of limestone; on the contrary, LHST is replaced by the upward layer of the white cloud rock on the tide, the ancient rock dissolves The cover layer is bran-like mudstone. In addition, considering the deformation of ancient rock-soluble substances in the early stages of sea invasion, a set of aluminum soil-braning rock system with a thickness of more than 20m in the bottom of the Xiamaling Group, the border of the DS2 top is obviously incompatible. –6 and 3–29; Chen and Wu 1997; Chen et al. 1999).

What is important is that the EHST of the DS2 is composed of a large group of high—energy stack stone reefs. The thickness of these stacking stone biological layers is generally above 10 m (Figs. 21, 22, 56a). Various types of stack stones form a combination, that is, "Tieling-Chihienella" Combination (zhu 1993). The diverse stack of the stacking stone combination is formed before the decrease of important stack stones at about 1,000 mA (RIDING 1997, 20,000). The outer wall of the cylindrical stacks is only developed with special sea turquoise membranes, namely Chihsienella Chihsiensis and Tielingella Tielingnsis. Therefore, the DS2's EHST unit floor layer is exactly the output layer of the ancient Tieling group in the ancient Tieling group, which is characterized by the high-energy deposition of the shallow environment.

Note that the formation time of the Tieling group exceeds 100 mA, and the group can only be subdivided into two three-level sequences with a thickness of about 200m thick. This actual phenomenon cannot be used as a evidence that the formation time

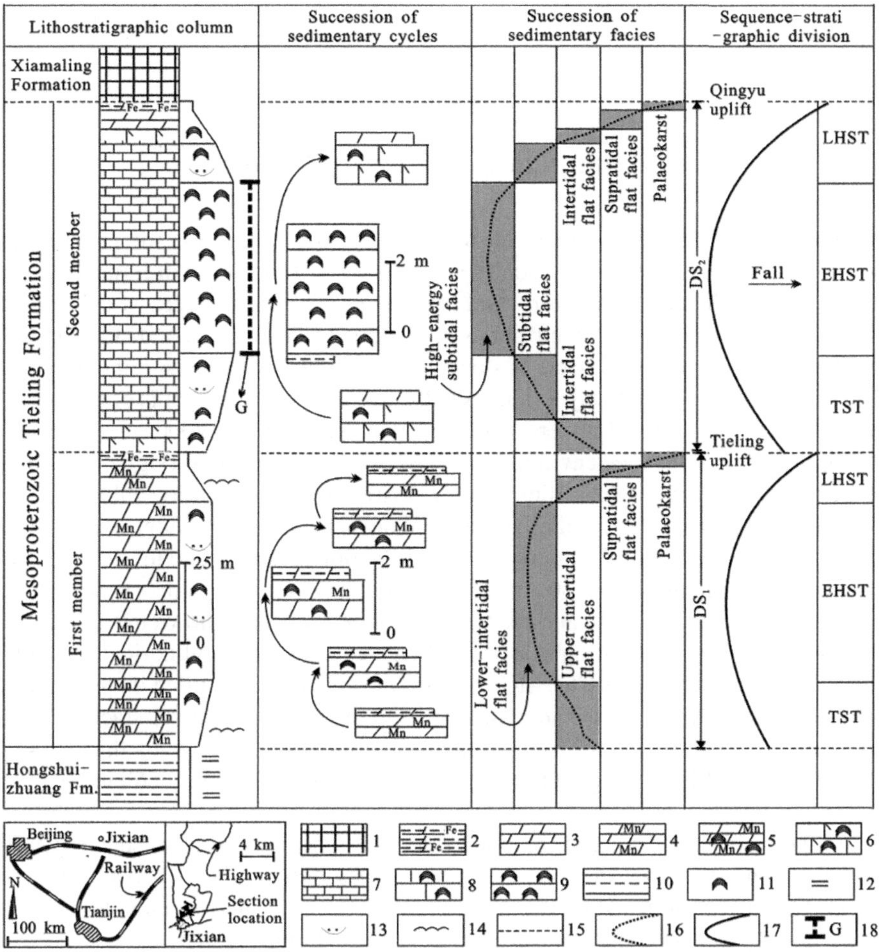

Fig. 55 Sequence-stratigraphic divisions for the Mesoproterozoic Tieling Formation at Jixian section. In the figure, DS_1 to DS_2 refer to two third-order sequences that are discerned in the Tieling Formation, and their component units are as follows: TST, trangressive system tract, and EHST and LHST refer to the early and the late high-stand system tracts, respectively. The meanings of the legends are: 1. bauxitic-ferruginous rock system; 2. Palaeokarst; 3. dolomite; 4. manganiferous dolomite; 5. manganiferous stromatolitic dolomite; 6. stromatolitic dolomitic limestone; 7. limestone; 8. stromatolitic limestone; 9. limestone of stromatolitic biostrome; 10. muddy shale; 11. stromatolite; 12. horizontal lamina; 13. scouring surface; 14. ripple; 15. sequence boundary; 16. changing curve of water depth; 17. changing curve of relative sea-level; 18. outputting horizon of glauconite. Map in the left-lower indicates the section location

of the third-order sequence of the former Cambrian may be compared to the third-order sequence of the universe. The confirmation of the non-integrated band of the non-integrated band reflected by the bulge is still uncertain. It can be inferred that the Tieling Group of the Jixian County is significantly shorter than the punctuation time.

3.15 Macroscopic Outputting Features of Glauconites

As shown in Fig. 56a, a few blocks of stone rock biological reefs or biological layers developed in the middle of the second section of the Tieling Group in Jixian County, that is, the EHST unit of the third-level sequence DS2. Several types of cylindrical stacks, such as Chihsienella Chihsiennsis, Tielingella Tielingnsis, and Baicalia CF. BAICALICA, constitute a stack of stone rock biological hills or biological layers. The thickness of the single layer of the stack stone rock biological hills or biological layers is generally several meters; for a few people, it is up to 2000 m. Therefore, the high-energy turbulence environment of the downward beach is a sign of stacking stone biological reefs or the sedimentation environment of the biological layer.

From 3-30A, it can be seen that a stacked stone reef composed of large overlapping stones such as Tielingella Tielingensis can reach 20m thickness, and the width of the stacked stone column is 10 ~ 30 cm. Crystal, and the outer wall of the stacked stone cylindrical body is generally enriched with sea greenstone. As shown in Fig. 56b, in the stacking stone combination from the hill-shaped stack to the cylindrical and split-shaped stacks (BAICALIA CFBAICALICA), the mud crystal is filled in the pillar of the stacked stone (such as the smooth stone; On the outer wall of the stacked stone column shape with irregular bits, the sea green stone membrane with a thickness of about 0.5 mm ("LE“" in Fig. 56b) and the sea green stone film (Fig. 56b in Fig. 56b "G").

As shown in Fig. 56c, large columnar stacking stone layer surfaces such as Chihsienella Chihsiennsis are irregular columnar shapes, and the size is between 15 and 20cm; Greenstone film with a thickness of 0.2 ~ 0.5 cm. The special colors of sea green liquid green are different from stack stones and mud crystals.

Large cylindrical stacked stone forms such as Tielingella Tielings, which are vertical on the surface, as shown in Fig. 56d: (1) The width of the stacking stone columnar is 15 ~ 20 cm; The width is 5 ~ 10 cm; (3) The outer wall of the layer of stone is generally developed.

In the end, in the Tieling Group's waist rocks, sea green stones are always developed with columnar stacks (Chihsienella Chihsiensisis, Tielingella Tielingnsis, and Baicalia CF. BAICALICA) outer wall glue film, indicating that these sea greenstones should be native sea greenstone. These large cylindrical stacks generally form a stacking stone reef, indicating that its sedimentary environment is a high-energy turbulence environment similar to the tide. Another basic feature is the high deposition rate. Therefore, the Tieling Group Sea Greenstone is different from the modern sea greenstone. It is formed in deep water, and the sedimentation is slow. It is a typical record of former Cambrian fossils.

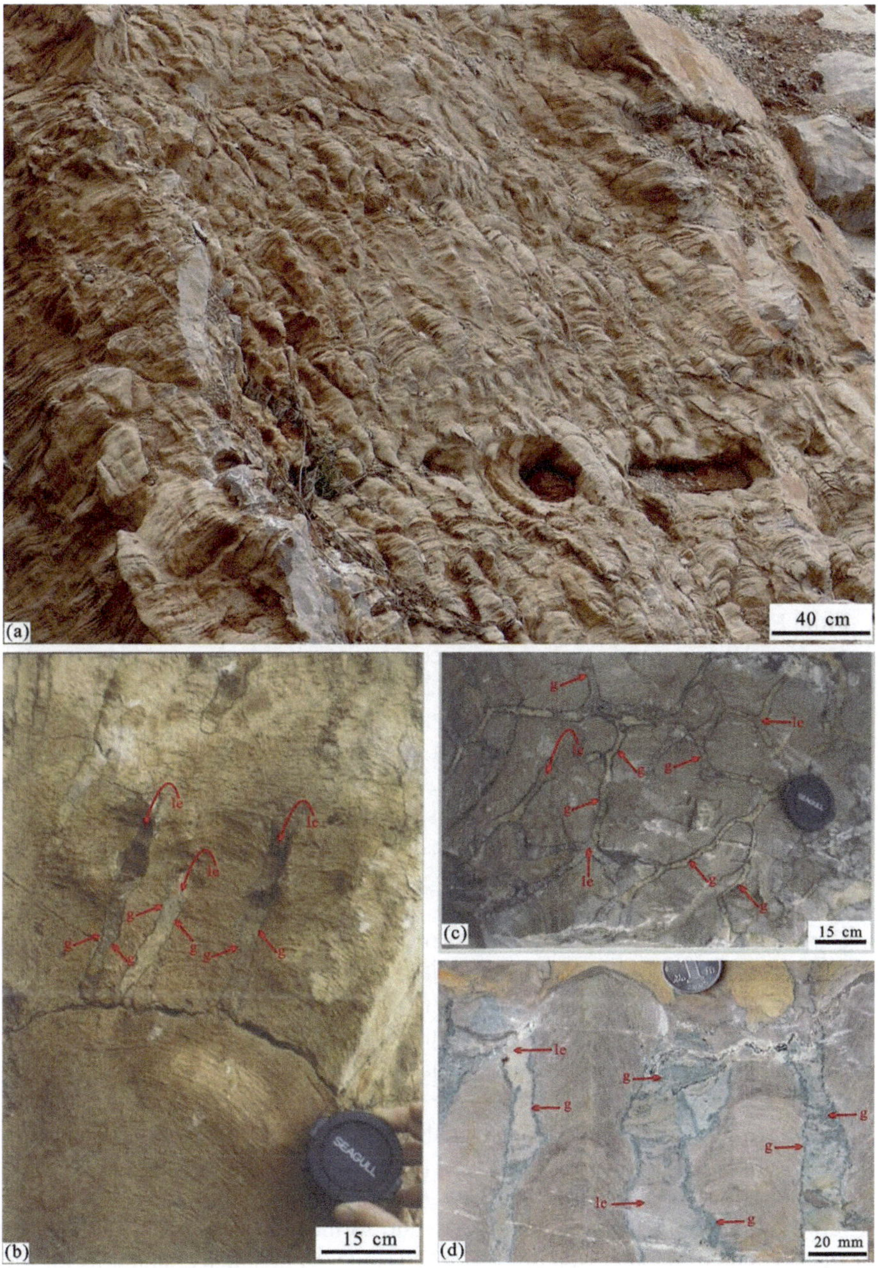
(a)
40 cm
(b)
15 cm
SEAGULL
(c)
15 cm
(d)
20 mm

◀**Fig. 56** Images showing the macroscopic features of the stromatolitic limestones in the second member of the Tieling Formation at the Jixian section. Image **a** refers to a stromatolitic biostrome with thickness of more than 20m, which is composed of the columnar *Tielingella tielingensis*; image **b** refers to an assemble of stromatolites made up by the vertical changes from the large-scale hillock-shaped stromatolites to the bifurcate shaped and the columnar stromatolites (*Baicalia cf. baicalica*) with glauconite film in their exterior walls; image **c** demonstrates the morphology in the bedding plane of the columnar stromatoltes (*Chihsienella chihsienensis*) with glauconite films in their exterior walls; and image **d** shows the morphology of the columnar stromatolites (*Tielingella tielingensis*) perpendicular to the bedding plane and the glauconite films in their exterior walls. From "(**b**)" to "(**d**)", "g" refers to the glauconite, and "le" refers to micrites within stromatolites which is the same as the leiolite

3.16 Microscopic Outputting Features of Glauconites

In macro perspective, Tieling Group Sea Greenstone is mainly developed as colon-shaped stacks, such as Chihsienella Chihsienensis, Tielingella Tielingensis, Baicalia CF. BAICALICA (Fig. 56). Under the mirror, the colloidal membrane of the sea green stone is concentrated in the transition band or the inside of the mud crystal (Fig. 57).

From Figs. 57 and 58, it can be seen that stack stones such as Chihsienella Chihsiennsis, which constitute stacked stone reefs, are stacked by layered black crystals (Fig. 57 "ST", Fig. 3—It is composed of 32A-B). It is different from the mud crystal filled with a stacked stone column-like space ("Le" in Fig. 57). There are obvious layer-like structures in the stack stone itself, with deep layers and shallow alternatives ("ST" in Fig. 57 and 58a-b), corresponding to the macro layered structure, constitute irregular belts, irregularities, irregularities, and irregularities. Or a micro weaving such as the rules condensate. The thickness of the stacking stone wall is 0.1 ~ 0.2 mm, which is composed of dark glittering litter rich in the residual organic matter ("B" in Fig. 57).

Carbonate rock mud crystals (Figs. 56b, d, 57c-f in Figs. 56b, d, 57c–f) may experience re-crystallization deformation. The crystal is larger and the size is between 0.05 and 0.1mm. From a macro perspective, these mud crystals are very small, which can be classified as one category with the "leiolite" that constitutes the growth substrate of the reef rock; it can be seen that the flat layer stone and the overlapping stone form a stack stone combination.

The sea greenstone ("G" in Fig. 57) in the stack stone reef is mainly concentrated or decentralized on the outside of "leiolite", which shows that the macro output mode of sea greenstone is the main It is distributed on the outer wall of the stacked stone, as shown in Fig. 56, which is actually a pseudo-form. These sea greenstone colloids are distributed on the outer wall of the stacked stone cylinder in a granular form, with a size between 0.05 and 0.2mm (Fig. 57b–c and e). The outer part of the mud crystal is above 1mm (Fig. 57a, d and f). Obviously, these colloidal sea greenstone should belong to the original sea greenstone; in addition, these colloidal sea greenstone shows the significant characteristics of the early corrosion of carbonate mud crystal and the late maturity of the sea greenstone, that is, the sea greenstone first fills in

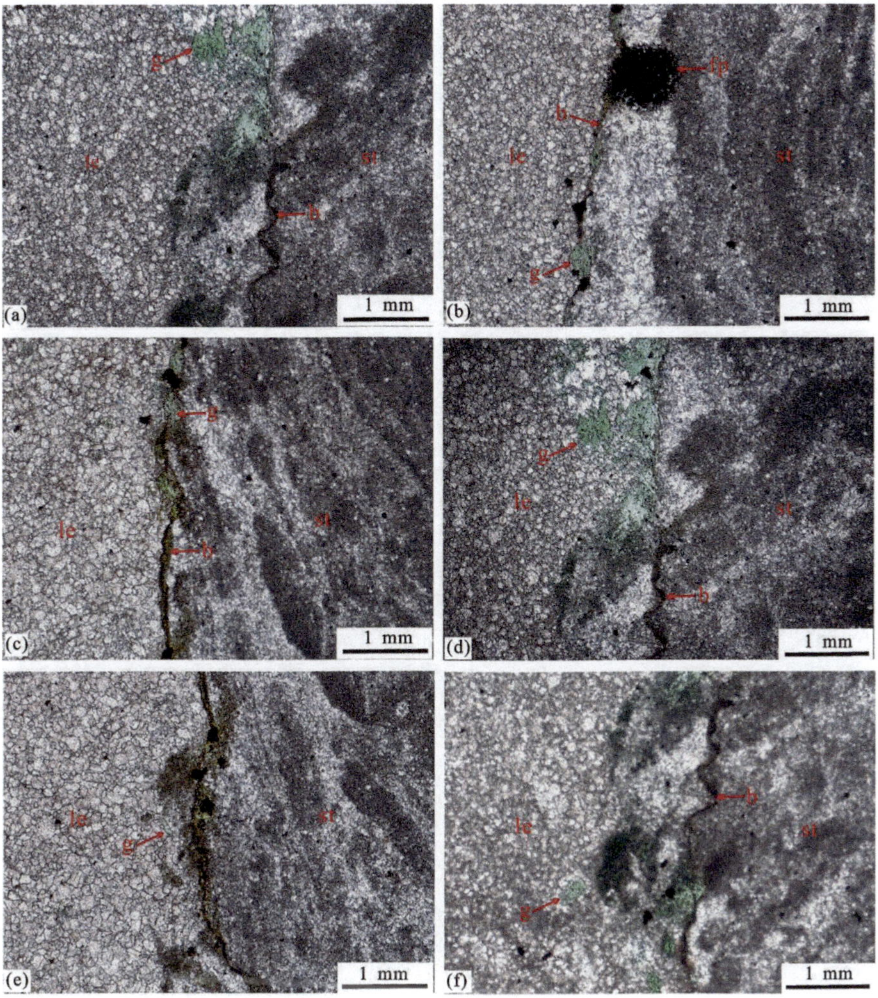

Fig. 57 Images showing the microscopic occurrence modes of glauconites in stromatolitic limestones of the Tieling Formation. All images (**a** to **f** show the distribution diversity of glauconitic colloids; most of these distribute in the boundary between the stromatolite (st) and the micrite (le) and mainly distribute in the sides of micrites. In the images, "g" represents the glauconite, and "b" refers to the boundary between the micrite (le) and the stromatolite (st). The stromatolite type that constitutes the stromatolitic limestone is the *Chihsienella chihsienensis*, and the black spots with the size ranging from 0.01 to 0.1mm are the crystals of pyrite. Note that one possible strawberry congeries of pyrites marked by the "fp" in the exterior part of the stromatolite in image B. Scale bars are 1 mm

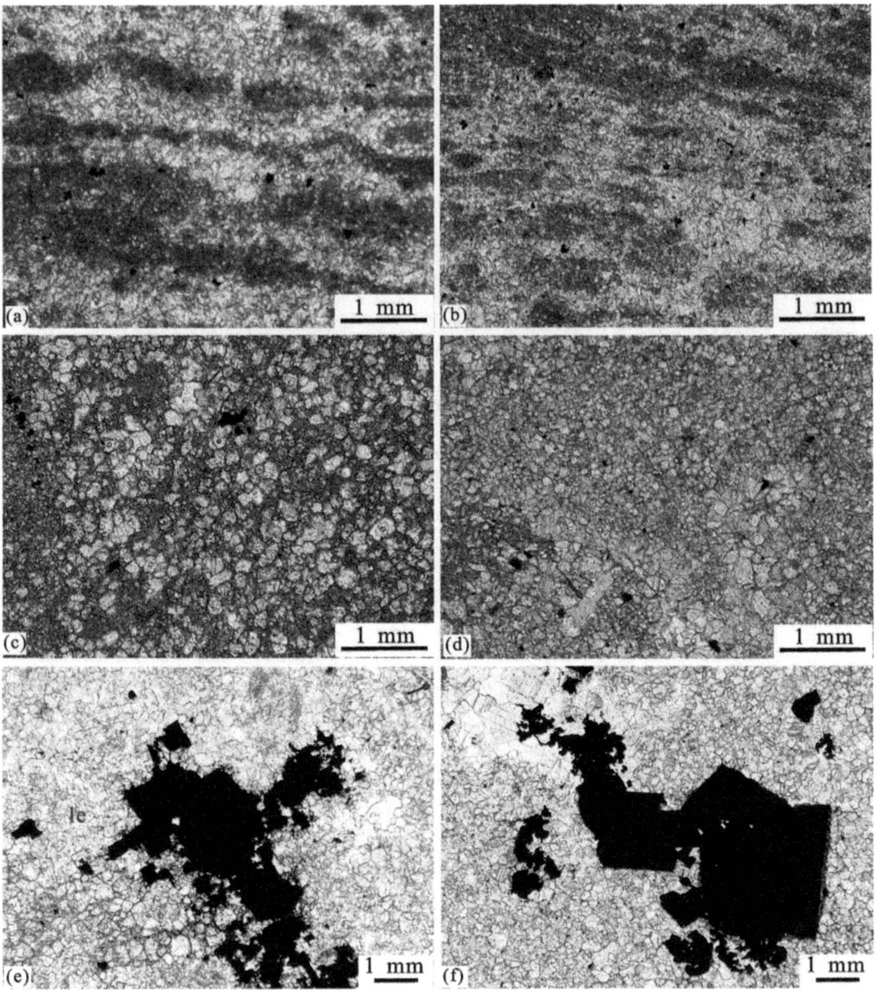

Fig. 58 Imagess showing the microscopic features of the stromatolitic and the leiolitic limestones. Images **a** and **b** show the irregular laminated and irregular clotted microstructures of the stromatolite like the *Chihsienella chihsienensis*, in which the small black blobs are pyrite crystals. Images **c** and **d** demonstrate the micrites of the leiolites filling the spaces among the stromatolitic columnar bodies, where the arrowed are glauconites and the black blobs are pyrites. Note the special congeries of pyrite crystals in micrites of the leiolites in the central parts of images **e** and **f** Scale bars are 1 mm

the first filling of the sea greenstone. The micro-space inside the mud crystal, and then gradually strengthened with the corrosion and maturity, forming irregular sheet shapes.

It can be seen from Fig. 58 that there are several large and small pyrite crystals inside the stacked stone cylindrical body, and there are also a number of flat rock crystals in the special space between the stacked stone cylinder. These yellow iron ore crystals are usually scattered in stacked stones in the form of black spots, with a size between 0.01 and 0.02 mm (Fig. 58a–b). Compared with the stacked stones, the pyrine crystals in the granules are relatively large, between 0.04 ~ 0.1 mm (Fig. 58c–d). More specifically, some several millimeter-sized pyrite polymers have cube's self-forming crystals, and some mud crystals are obviously replaced by sea greenstone. There are almost no sea green stones inside the stack stones. These pyrite particles or crystals, especially in Fig. 57b, may be the polymer of strawberry-shaped pyrite, which may reflect the stack stone itself itself itself The combined stack stone consisting of pyramids is a typical microbial carbonate, which is caused by complex microbial deposition. To. These yellow iron ore may indicate that the high-energy shallow seas environment in the late China and Yuan Dynasty was in a weak restoration state, and further reflected the oxygen level in the atmosphere different from the current. Therefore, the sea greenstone in the limestone of the Tieling group and the relative to the pyrite indicate that the high-energy shallow environment in the late ancient and late period of the mid-to-see in the late ancient and late period is characterized by weak restore state.

The micro-existence characteristics of Tieling Group's Lifesturine Greenstone is a typical record of former Cambrian Fossil: (1) Some features show that these special colloidal green stones are native sea greenstone, that is, irregular colloids are distributed in the grains of the grains grains On the outside of the quilt, the special colloid is concentrated at the junction of stacked stones and mud crystal, and a small amount of stacked stones; (2) a small amount of pyrite in stack stones and mud crystals may reflect the special environment of the shallow seas, and it is in a weak restoration The state is conducive to the generation of sea greenstone, and it may also be a manifestation of insufficient oxygen content in the ancient and atmospheric atmosphere.

3.17 Componential Features of Glauconites

Using EPMA-1600 electronic probe and relative quantitative analysis (GB/T 15,074–94), the ingredients characteristics of the sea greenstone, stacked stones, and granular quilt in Tieling arrays are measured Essence The analysis results are shown in Tables 4 and 5, which reflects the main elements of sea green stones, mud crystals and stacked stones and the component characteristics of the main oxides.

There is almost no sea greenstone in the stack stones. The main elements of stacking limestone are oxygen and calcium, and the corresponding oxides are mainly calcium oxides, which reflect the basic characteristics of pure lithium rocks. It can be seen from Table 4 that TL-9–3 and TL-8–3 sampling points in the stacked stones are 40.13% (TL-9–3) and 40.30% (TL-8– (TL-9–3) and 40.30%. 3), calcium oxide content is 56.57% (TL-9–3) and 56.39% (TL-8–3), respectively. Note that there are some erosion phenomena in the outer part of the stacked stone. The content of potassium (0.13%) and potassium oxide (0.15%) in Tables 1 and 2 in Tables 1 and 2 is low, that is, the outer part of the stack stone There are some new sea green stones.

Table 4 Results of the EMPA analysis for the stromatolitic limestones of the Tieling Fm

Sampling points	Lithological types	Content of the main elements (%)										
		Na	Mg	Al	Si	P	K	Ca	Ti	Mn	Fe	O
TL-9–1	Glauconitic micrite	0.02	0.02	0.09	0.30	0.08	0.29	39.53	0.16	0.19	0.20	16.59
TL-8–1	Glauconitic film	0.04	2.58	6.64	22.64		7.20	4.18	0.04	0.09	6.29	38.14
TL-9–3	Stromatolite				0.14		0.13	40.13		0.09	0.26	16.42
TL-8–3	Stromatolite		0.24	0.04	0.14			40.30		0.13	0.41	16.60
TL-9–2	Micrite		0.21		0.10	0.08		39.61			0.18	16.21
TL-8–2	Micrite		0.14	0.00			0.09	40.96				16.46
TL-GM-1	Glauconitic micrite	0.16	3.00	5.41	23.46		7.10	0.40	0.13		7.89	38.66
TL-GM-2	Glauconitic micrite	0.40	3.15	7.29	23.33		7.18	0.56	0.23		6.01	39.71
TL-GM-3	Glauconitic micrite	0.22	2.74	6.24	23.53		7.42	0.77	0.06		6.73	38.99
TL-GM-4	Glauconitic micrite	0.15	3.06	6.55	23.25		6.98	0.58	0.21		6.69	39.05
TL-G-1	Glauconitic film	0.29	3.01	7.11	23.58		7.27	0.52	0.13	0.07	6.01	39.70
TL-G-2	Glauconitic film	0.35	2.77	8.02	23.43		6.74	1.06			4.71	39.60
TL-G-3	Glauconitic film	0.31	3.14	6.83	23.21		7.27	1.19	0.09	0.17	5.97	39.33
TL-G-4	Glauconitic film	0.35	3.06	8.37	23.28		7.14	0.44			5.27	39.91
TL-G-5	Glauconitic film	0.23	3.01	7.21	23.44		7.70	0.25	0.26	0.30	5.53	39.49
TL-G-6	Glauconitic film	0.14	3.06	9.14	22.99		7.64	0.38	0.16	0.36	3.87	39.97

For the mud crystal filled with stacked stone cylindrical gaps, that is, the grains, the main elements and corresponding of TL-9–2 and TL-8–2 in Tables 3, 4 and 5 The main oxide content is similar to the stack stones, which also illustrates the basic characteristics of pure waters. The calcium content of the TL-9–2 sampling point is 39.61%, and the corresponding calcium oxide content is 55.42%; the relative values of the TL-8–2 sampling point are 40.96%and 57.31%, respectively. In the transition zone near the outer wall of the stacked stone and the overlapping stone, the TL-8–2 sampling point reflects some new green stone, namely small potassium content (0.09%) and the corresponding potassium oxide content (0.11111%).

For the sea greenstone in Tieling group, TL-9–1 and TL-8–1 are the two sampling points of their characteristics. TL-9–1 reflects the characteristics of sea green limbs.

Table 5 Results of the EMPA analysis for the stromatolitic limestones of the Tieling Fm

Sampling points	Lithological types	Content of the main oxides (%)									
		Na_2O	MgO	Al_2O_3	SiO_2	P_2O_5	K_2O	CaO	TiO_2	MnO	FeO
TL-9–1	Glauconitic micrite	0.02	0.04	0.17	0.64	0.17	0.28	55.31	0.27	0.24	0.25
TL-8–1	Glauconitic film	0.06	4.28	12.54	48.43		8.68	5.85	0.07	0.12	8.09
TL-9–3	Stromatolite				0.30		0.15	56.57		0.12	0.33
TL-8–3	Stromatolite		0.40	0.08	0.30			56.39		0.17	0.53
TL-9–2	Micrite		0.35		0.21			55.42			0.23
TL-8–2	Micrite		0.23	0.01			0.11	57.31			
TL-GM-1	Glauconitic micrite	0.22	4.97	10.21	50.97		8.55	0.55	0.22		11.28
TL-GM-2	Glauconitic micrite	0.54	5.22	13.77	49.92		8.65	0.78	0.39		8.60
TL-GM-3	Glauconitic micrite	0.30	4.54	11.79	50.34		8.94	1.07	0.10		9.63
TL-GM-4	Glauconitic micrite	0.20	5.07	12.38	49.74		8.41	0.82	0.35		9.56
TL-G-1	Glauconitic film	0.39	4.99	13.44	50.45		8.75	0.73	0.22	0.09	8.72
TL-G-2	Glauconitic film	0.47	4.59	15.15	50.13		8.12	1.48			6.73
TL-G-3	Glauconitic film	0.42	5.21	12.90	49.64		8.76	1.66	0.15	0.22	8.54
TL-G-4	Glauconitic film	0.38	5.08	15.62	49.79		8.59	0.61			7.53
TL-G-5	Glauconitic film	0.31	4.99	13.63	50.13		9.28	0.35	0.43	0.39	7.91
TL-G-6	Glauconitic film	0.19	5.07	17.28	49.18		9.21	0.53	0.26	0.46	5.54

Its calcium content is 39.53%, and the calcium oxide content is 55.31%; its potassium content is 0.29%, and the corresponding potassium oxide content is 0.28%; there are NA_2O (0.02%), MGO (0.04%), Al_2O_3 (0.17%), SIO_2 (0.64%), P_2O_5 (0.17%), TIO_2 (0.27%), MNO (0.24%), FEO (0.25%) and other oxides. These characteristics of TL-9–1 reflect the basic characteristics of new sea greenstone. TL-8–1 test results show the typical characteristics of highly evolved sea greenstone, corresponding to the dark green film corresponding to sea greenstone. The composition of TL-8–1 is AL (6.64%), SI (22.64%), K (7.20%) and other elements such as high content. The corresponding oxides Al_2O_3 (12.54%), SIO2 (48.43%), K_2O (8.6888 (8.6888) %) High content. The K_2O content is 8.68%, and the TL-8–1 shows the basic characteristics of high-evolved sea greenstone.

In order to further study the characteristics of the highly evolved sea greenstone in Tieling group stacking layer of limestone, the sea greenstone is divided into two categories, namely the glue sea greenstone (Fig. 58c–d) and the membrane-shaped in the replacing Mo Changshi Sea green stones (Figs. 56 and 57). Tables 4 and 5 TL-GM-1 ~ TL-GM-4 sampling point is still called mud crystal sea green stone, Tables 4 and 5 TL-G-1 ~ TL-G-6 sampling point is mud crystal green stone. The potassium content of mud crystal sea greenstone is 6.98 ~ 7.42%, and the content of potassium oxide is 8.41 ~ 8.94% (TL-GM-1 ~ TL-GM-4; Tables 4 and 5). The potassium content of membrane-shaped sea greenstone is 6.74 ~ 8.94%, and the content of potassium oxide is 8.12 ~ 9.28% (TL-G-1 ~ TL-G-6; Tables 4 and 5).

In summary, an electronic probe test was performed on the rich output of the tattoo of Tieling Group, including mud crystal green stones and membrane-like greenstones, indicating that they were highly evolved in sea greenstone. As shown in Table 5, TL-G-5 (typical thin film-like greenstone) samples of potassium oxide are as high as 9.28%, and the basic concepts of these highly evolved Tieling groups and modern sea green stones are not It is too consistent, that is, sea greenstone is formed in a relatively deeper sedimentation environment. In addition, the Tieling Group Sea Greenstone cannot be used as a sign of certain special geological phenomena, such as slow deposits, geological cracks, concentrated sections in layers of layers, etc. Like the Cambrian Haiki green stone introduced by CHFETZ and Reid (2000), the Tieling Group Hai Greenstone is also the representative of the former Cambrian Haihe, which reflects the diversity of sea greenstone in geological history.

In addition, according to the characteristics of low fluctuations in the second section of the Tieling Group, the cross-cutting layer of the cross-cutting section, the obvious thickness of the thickness, the fast fluctuations of the pillar width, the frequency of the columnar branch, the similarity of the adjacent matrix and the stack stone column structure. It is seen that this is a novel formation mode of diverse column-shaped stacks (Fig. 21 and 22). Figures 57 and 58 belong to the Lower Stromatolite Unit proposed by Tosti and Riding (2017). In other words, they are condensed stacks formed by the closed stone formed by carbonate mud, rather than precipitated stacks formed by the precipitation of in situ carbonate. The dynamic interaction between the growth of the cushion, the flow of water and the supply of sediments determines the shape of the Tieling column, and forms the formal diversity of the shape of the ancient branch of the ancient and late-mid-to-premium branches. The rich mud is attributed to the water column that stimulates the stimulating water columns of cyanobacteria, and the cushion composed of microorganisms such as multi-haired cyanobae can capture this rich carbonate rock mud. Signs of calcification filament or other microbial precipitation. However, adjacent substrates different from the micro-structures (Figs. 57 and 58), which are different from the micro-structured stone column, and the sea green stone membranes are mainly distributed in the outer wall of the stack stone, as well as the interior of the stack stone column body The mud crystal in the pyrite and the granules fills the special gap between the stacked stone pillars. These features may reflect some characteristics of the sedimentation of carbonate rocks to form a layer of stone. Therefore, there are many problems that need to be further studied.

Acknowledgements This work benefited from field discussions with Professor Zhu Shixing, Gao Lingzhi and Huang Xueguang. The funding for this research into the middle and upper Proterozoic strata of North China was provided by the Natural Sciences Foundation of China (grant 49802012, 40472065). Acknowledgements are also due to Professor Wang Hongzhen and Shi Xiaoying, the hosts of Special Project (SSER), for their support in this research. We would like to thank Professor Zhou Hongrui for his significant fieldwork contributions.

References

Amorosi A (1995) Glaucony and sequence stratigraphy: a conceptual framework of distribution in siliclastic sequences. J Sedi Res B65:419–425
Amorosi A (1997) Detecting compositional, spatial, and temporal attributies of glaucony: a tool for provenance research. Sed Geol 109:135–153
Awramik SM, Sprinkle J (1999) Proterozoic stromatolites: the first marine evolutionary biota. History Bio 13:241–253
Banerjee S, Mondal S, Chakraborty PP, Meena SS (2015) Distinctive compositional characteristics and evolutionary trend of Precambrian glaucony: example from Bhalukona formation, Chhattisgarh basin, India. Precam Res 271:33–48
Banerjee S, Bansal U, Thorat AV (2016) A review on palaeogeographic implications and temporal variation in glaucony composition. J Palaeog 5(1):43–71
Bishop JW, Sumner DY (2006) Molar tooth structures of the neoarchean montevielle formation, transvaal supergroup, South Africa, I: constraints on microcrystalline $CaCO_3$ precipitation. Sedmentology 53(5):1049–1068
Bishop JW, Sumner DY, Huerta NJ (2006) Molar tooth structures of the neoarchean montevielle formation, transvaal supergroup, South Africa, II: a wave-induced fluid flow model. Sedmentology 53(5):1069–1082
Bogdanova SV, Sà P, Li ZX (2009) Assembly and breakup of Rodinia (some results of IGCP Project 440). Strati Geol Correlation 17:259–274
Bosak T, Knoll AH, Petroff AP (2013) The meaning of stromatolites. Ann Rev Earth and Plane Sci 41:21–44
Brett CE, Allison P, Tsujita C et al (2006) Sedimentology, taphonomy, and paleoecology of meter-scale cycles from the upper Ordovician of Ontario. Palaios 21:530–547
Burne RV, Moore L (1987) Microbialites: organosedimentary deposits of benthic microbial communities. Palaios 2:241–254
Cao R, Yuan X (2006) Stromatolites. Press of China University of Science and Technology, Hefei of China (in Chinese)
Catuneanu O, Martins-Neto MA, Eriksson PG (2005) Precambrian Sequence Stratigraphy. Sedi Geo 176:67–95
Catuneanu O, Zecchin M (2013) High-resolution sequence stratigraphy of clastic shelves II: controls on sequence development. Mar Petrol Geo 39:26–38
Chafetz HS, Reid A (2000) Syndepositional shallow-water precipitation of glauconitic minerals. Sedi Geo 136:29–42
Chen L, Huang B, Yi Z et al (2013) Paleomagnetism of ca. 1.35 Ga sills in northern North China Craton and implications for paleogeographic reconstruction of the Mesoproterozoic supercontinent. Precam Res 228:36–47
Chen J, Wu T (1997) Regional stratigraphy in North China. China University of Geosciences Press, Wuhan of China (in Chinese)
Chen J, Zhang P, Gao Z et al (1999) Chinese Stratigraphic lexicon (Mesoproterozoic). Geological Publishing House, Beijing

Du R (1992) Earth history and paleontology in Precambrian. Geological Publishing House, Beijing (in Chinese)
Eriksson PG, Catuneanu O, Nelson DR et al (2005) Controls on Precambrian sea-level change and sedimentary cyclicity. Sedi Geo 176:43–65
Ettensohn FR, Zhang C, Gao L et al (2011) Soft-sediment deformation in epicontinental carbonates as evidence of paleoseismicity with evidence for a possible new seismogenic indicator: accordion folds. Sedi Geo 235:222–233
Fan J, Peng S, Hou X et al (2015) Official website of the international commission on stratigraphy and the realize of the international chronostratigaphic chart (V2015/01). J Strati 39(2):125–134 (In Chinese with English abstract)
Furniss G, Rittle JF, Winston D (1998) Gas bubble and expansion crack origin of "MT" structure from the Late Precambrian Belt-Purcell supergroup. J Sedi Res 68:104–114
Gao L, Zhang Y, Wang C et al (1996) Preliminary study of the sequence stratigraphy of the middle and upper Proterozoic in Jixian of Tianjin. Regi Geo China 56:64–74 (in Chinese with English Abstract)
Gao L, Zhang C, Shi X et al (2007) A new SHRIMP age of the Xiamaling formation in the North China Plate and its geological significance. Acta Geol Sinica (english Edition) 81(6):1103–1109
Hinnov LA (2013) Cyclostratigraphy and its revolutionizing applications in the earth and planetary sciences. GSA Bull 125:1703–1734
Holland HD (2006) The oxygenation of the atmosphere and oceans. Philosop Trans Roval Soc B 361:903–915
Huang X, He Y, Wang Y (1996) Sepiolite deposits in North China: their occurrence. Geological Publishing House, Beijing (in Chinese), Genesis and Use
Kah LC, Knoll AH (1996) Microbenthic distribution of Proterozoic tidal flats: environmental and taphonomic considerations. Geology 24:79–82
Kuang H, Meng X, Ge M (2006) Disscussion on origin for Molar Tooth carbonate rocks: an example from the Neoproterozoic in Jilin-Liaoning area. J Palaeog 8(2):63–74 (in Chinese with English abstract)
Lamb DM, Awramik SM, Zhu S (2007) Paleoproterozoic compression-like structures from the Changzhougou Formation, China: Eukaryotes or clasts? Precam Res 154:236–247
Lamb DM, Awramika SM, Chapmanb DJ et al (2009) Evidence for eukaryotic diversification in the ~1800 million-year-old Changzhougou formation, North China. Precam Res 173:93–104
Li H, Lu S, Su W, Xiang Z et al (2013) Recent advances in the study of the Mesoproterozoic geochronology in the North China Craton. J Asian Earth Sci 72:216–227
Liang Y, Zhu S, Zhang L et al (1985) Stromatolite assemblages of the late Precambrian in China. Precam Res 29:15–32
Liu H, Liu C (1992) The oldest trace fossils from the Chuanlinggou formation of the Changchengian system at the Jixian section. J Northwest Uni (natural Science) 22:208 (in Chinese)
Liu Y, Liu X, Li Y (1997) Tectonic cyclic sequences in the Mesoproterozoic and Neoprotrozoic aulacogen of Yanshan: a concept of aulacogen tectonic cycle and its hierarychy. Acta Geos Sinica 18(2):142–149 (in Chinese with English abstract)
Lu S, Li H (1991) A precise U-Pb single zircon age determination for the volcanics of the Dahongyu formation, Changcheng system in Jixian. Bull China Aca Geol Sci 22:137–146 (in Chinese with English abstract)
Miall AD (2005) Testing for eustatic sea-level control in the Precambrian sedimentary record. Sedi Geo 176:9–16
Mei M, Zhou H, Du B et al (2000) Meso-and neoproterozoic sedimentary sequences in Jixian, Tianjin, Northern China: division of the first-order sequences of the Precambrian Strata (1800Ma-600Ma) and their correlation with the Phanerozoic strata. Sedi Geo Tethyan Geo 20:48–59 (in Chinese with English abstract)
Mei M (2006) Origin of molar-tooth structure based on sequence-stratigraphic position and macroscopic features: example from Mesoproterozoic Gaoyuzhuang Formation at Jixian section, Tianjin, North China. J China Uni of Geos 17(3):201–208

Mei M, Gao J, Meng Q (2006) From the matground structure to the primary sedimentary structure of a fifth category: significant concepts on sedimentology. Geosciences 20(3):413–422 (in Chinese with English abstract)
Mei M (2007) Implications of the Precambrian non-stromatolitic carbonate succession making up the third member of Mesoproterozoic Gaoyuzhuang Formation in Yanshan Area of North China. J China Uni Geos 18(3):191–209
Mei M (2010) Research on forming mechanism of long-term sequences: the second advance in sequence stratigraphy. J Palaeog 12(6):711–728 (In Chinese with English abstract)
Mei M, Gao J, Qn M, Liu Z (2010) Sedimentary features and their implications of microdigital stromatolites from the Mesoproterozoic Wumishan formation at the Jixian section in North China. Acta Geol Sinica (english Edition) 82(3):483–496
Mei M, Tucker ME (2011) Molar tooth structure: a contribution from the mesoproterozoic Gaoyuzhuang formation, Tianjin City. North China. Acta Geol Sinica (english Edition) 85(5):1084–1109
Mei M, Tucker ME (2013) Milankovitch-driven cycles in the Precambrian of china: the Wumishan formation. J Palaeog 2(4):369–389
Meng X, Ge M (2002) The sedimentary features of Proterozoic microspar (MT) carbonates in China and their significance. Episodes 25:185–195
Meng X, Ge M (2004) Sequences. Science Press, Beijing (in Chinese), Events and evolution of the Sino-Korean plate
Noffke N (1998) Multidirected ripple marks rising from biological and sedimentological processes in modern lower supratidal deposits (Mellum Island, southern North Sea). Geology 26:879–882
Noffke N, Gerdes G, Klenke T et al (2001) Microbially-induced sedimentary structures: a new category within the classification of primary sedimentary structures. J Sedi Res 71:649–656
Osleger DA, Read JF (1991) Relation of eustasy to stacking patterns of meter-scale carbonate cycles, late Cambrian, U.S.A. J Sedi Petro 61:1225–1552
Odin GS, Matter A (1981) De Glauconiarum Origine. Sedimentology 28:611–641
Pesonen LJ, Mertanen S, Veikkolainen T (2012) Paleo-Mesoproterozoic supercontinents—a paleomagnetic view. Geophysica 48:5–47
Pflüger F (1999) Matground structures and redox facies. Palaios 14:25–39
Plumb KA (1991) New Precambrian time scale. Episodes 14:139–140
Pollock MD, Kah LC, Bartley JK (2006) Morphology of MT structures in Precambrian carbonates: influence of substrate rheology and implications for genesis. Jof Sedi Res 76:310–323
Porada HP, Bouougri EH (2007) Wrinkle structure: a critical review. Earth-Sci Rev 81:199–215
Pratt BR (1998a) Syneresis cracks: subaqueous shrinkage in argillaceous sediments caused by earthquake-induced dewatering. Sedi Geo 117:1–10
Pratt BR (1998b) Molar-tooth structures in Proterozoic carbonates: origin from synsedimentary earthquakes, and implication for the nature and evolution of basinal and marine sediments. GAS Bull 110:1028–1045
Qiao X, Song T, Gao L et al (1994) Seismic sequence in carbonate rocks by vibarational liquefaction. Acta Geol Sinica (english Edition) 7(3):243–265
Qiao X, Gao L (2007) Mesoproterozoic palaeogeoearthquake and palaeogeography in Yan-Liao Aulacogen. J Palaeog 9(4):336–352 (in Chinese with English abstract)
Read JF, Goldhammer RK (1988) Use of Fischer plots to define third-order sea-level curves in Ordovician peritidal cyclic carbonates, appalachians. Geology 16:895–899
Riding R (2000) Microbial carbonates: the geological record of calcified bacterial-algal mats and biofilms. Sedimentology 47(1):179–214
Roberts NMW (2013) The boring billion? Lid tectonics, continental growth and environmental change associated with the Columbia supercontinent. Geos Front 4:681–691
Sà P, Sà E, Pesonen LJ et al (2014) Mesoproterozoic paleogeography: supercontinent and beyond. Precam Res 244:207–225
Schieber J (1998) Possible indicators of microbial mat deposits in shales and sandstones: examples from the Mid-Proterozoic Belt Supergroup, Montana USA. Sedi Geo 120:105–124

Shi X, Jiang G (2011) Geomicrobiological action and evolution of the epigeosphere system in the Precambrian. In: Xie S, Ying H, Shi X et al (eds) Geobiology: interaction and synergetic evolution between the Earth' s environment and the life. Science Press, Beijing, pp 190–236 (In Chinese)
Shields GA (2002) Molar tooth microspar: a chemical explanation for its disppearance similar to 750 Ma. Terra Nova 14:108–113
Song T, Gao J (1985) The oldest metazoa fossils: research on the filling-tube structures for the Precambrian Changzhougou Formation in Ming Tombs district. Acta Sedi Sinica 3(2):85–96 (in Chinese with English abstract)
Song T, Zhao Z, Wang C (1991) Proterozoic sedimentary rocks in North-China. Beijing Sciences and Technology Press, Beijing (in Chinese)
Su W, Li H, Huff WD et al (2010) SHRIMP U-Pb dating for a K-bentonite bed in the Tieling formation, North China. Chin Sci Bull 55:2197–2206
Su W (2014) A review of the rexised Precambrian Time Scale (GST 2012) and the research of the Mesoproterozoic chronostratigraphy of China. Earth Sci Fronti 21(2):119–138 (In Chinese with English abstract)
Schwarzacher W (2005) The stratification and cyclicity of the Dachstein Limestone in Lofer, Leogang and Steinernes Meer (Northern Calcareous Alps, Austria). Sedi Geo 181:93–106
Tang D, Shi X, Ma J et al (2016) Mesoproterozoic glaucony as a potential mineral proxy for shallow chemocline in the Precambrian ocean. Earth Sci Fron 23(6):219–235 (In Chinese with English abstract)
Tipper JC (2000) Patterns of stratigraphic cyclicity. J Sedi Res 70:1262–1279
Tosti F, Riding R (2017) Fine-grained agglutinated elongate columnar stromatolites: Tieling formation, ca 1420 Ma, North China. Sedimentology 64:871–902
Tucker ME, Garland J (2010) High-frequency cycles and their sequence stratigraphic context: orbital forcing and tectonic controls on Devonian cyclicity, Belgium. Geol Belgica 13(3):213–240
Vail PR, Mitchum RM, Thompson S (1977) Seismic stratigraphy and global changes of sea level, part 3 and part 4. AAPG Mem 26:63–98
Van der Zwan CJ (2002) The impact of Milankovitch-scale climatic forcing on sediment supply. Sedi Geo 147:271–294
Van Kranendonk MJ, Altermann W, Beard BL et al (2012) A Chronostratigraphic division of the Precambrian. In: Gradstein FM, Ogg JG, Schmitz MD et al (eds) The geologic time scale 2012. Elsevier, Amsterdam, pp 299–392
Wang W, Liu S, Santosh M et al (2015) 1.23 Ga mafic dykes in the North China Craton and their implications for the reconstruction of the Columbia supercontinent. Gond Res 27:1407–1418
Wilde SA (2014) The Precambrian geology of the North China craton: a review and update of the key issues. In: Dilek Y, Furnes H (ed) Evolution of Archean Crust and Early Life Modern Approaches. Solid Earth Sciences, 7, pp 149–178
Xing Y, Gao Z, Liu G et al (1989) The upper precambrian in China. Geological Publishing House, Beijing (in Chinese)
Zecchin M (2010) Towards the standardization of sequence stratigraphy: is the parasequence concept to be redefined or abandoned? Earth-Sci Rev 102:17–119
Zhang S, Li Z-X, DàD E et al (2012) Pre-Rodinia supercontinent Nuna shaping up: a global synthesis with new paleomagnetic results from North China. Earth Plane Sci Lett 353–354:145–155
Zhang F, Wang J, Zhang S et al (2021) Evolution of Proterozoic eukaryotic algae and environmental constraint. Acta Geol Sinica 95(5):1334–1355 (in Chinese with English abstract)
Zhang L, Hua H (2003) A primary discussion on the authentic of the Changchengian macroscopic trace fossils. In: Hao D, Liu Y, Feng Y (eds) Tracing the origin of lives: congratulate professor Du Rulin Won a Lisiguang geological scientific prize. Seismological Press, Beijing, pp 205–210 (in Chinese)
Zhao G, Pà C (2012) Precambrian geology of China. Precam Res 222–223:13–54
Zhao G (2014) Precambrian evolution of the North China Craton. Elsevier, Oxford

Zhao Z (1992) Dispersal grain: the embryonic form of oncolite. Acta Sedi Sinica 10(2):19–27 (in Chinese with English abstract)
Zhu S (1993) Stromatolites in China. Geological Publishing House, Beijing, 1993 (in Chinese).
Zhu S, Xing Y, Zhang P (1994) Biostratigraphic sequence of the middle-upper Proterozoic in North-China platform. Geological Publishing House, Beijing (in Chinese)
Zhu S, Huang X, Sun S (2005) New progress in the research of the Mesoproterozoic Changcheng system (1800~1400 ma) in the yanshan range, north china. J Strati 29(Supple):437–448 (In Chinese with English abstract)

Large-Scale Formation of Manganese, Phosphorus, and Aluminum Deposits in Guizhou, China and Global Deep-Time Changes

Wen-chao Yu, Jiang-hai Yang, and Yuan-sheng Du

1 Introduction

1.1 Snowball Earth Event and Large-Scale Manganese Ore Deposition

Cryogenian (720–635 Ma) is the critical transition period in the geological history. During this period, the break-up of Rodinia supercontinent, the Snowball Earth event, the Neoproterozoic oxidization event (NOE), and the evolution of metazoan. Neoproterozoic is also the peak period for sedimentary manganese deposits, among the widespread Neoproterozoic manganese deposits, that in South China is the most important one for its large reserve and high-grade. Formation of Neoproterozoic manganese deposits was linked with these major geological events and the interactions between the lithosphere, hydrosphere, atmosphere, and biosphere. During the interval of 825 and 740 Ma, the break-up of Rodinia supercontinent led to developing of rift basin worldwide and relevant igneous activities. Area of passive continent margin increased promoted the weathering of aluminosilicate and bury of organic carbon. Both of these processes would consume CO_2 in atmosphere, thus resulted in decreasing temperature and expansion of ice-sheets. Low albedo of ice would reflect the solar energy and finally caused the whole or partial ocean freezing. During the Snowball Earth event, exchanges of material and energy were almost blocked, continental weathering was slow and marine carbonate was vanished.

W. Yu (✉) · J. Yang · Y. Du
China University of Geosciences, Wuhan, China
e-mail: yuwenchaocug@163.com

X. Hu (ed.), *Field Trip Guidebook on Chinese Sedimentary Geology*,
https://doi.org/10.1007/978-981-99-6936-4_17

Neoproterozoic ice age contained four stages in the world. Kaigas and Gaskiers glaciations were probably the continental/apline glaciers, show short-lived and regional characteristics. Sturtian (718–660 Ma) and Marinoan (650–635 Ma) were global glaciations, Sturtian glaciation may contain two episodes.

Manganese ore of the Datangpo Formation rapidly deposited after the termination of Sturtian glaciation. The metallogenic process was the result of the coupling paleoclimate, post-glacial biological recovery, watermass structure, and seawater geochemisty. According to recent study, after the meltdown of ice-sheets, the watermass in the Nanhua Basin was strongly physic and redox stratified, hydrothermal activities input manganese ions into the basin, and they were then fixed by microbially-mediation processes (Fig. 1).

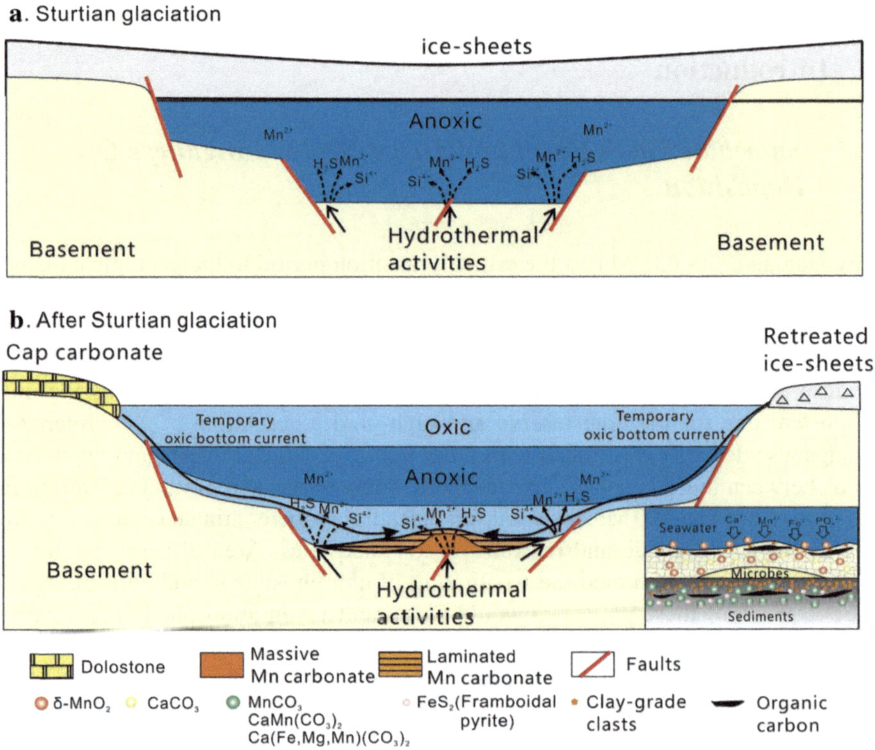

Fig. 1 Model for manganese deposits in the Datangpo Formation, South China (Yu et al. 2019a)

1.2 The Neoproterozoic Oxygenation Event, Ecological Evolution and Phosphogenic Episodes

The phosphorites are interpreted as evented deposits in the geological history and the Phosphogenic Events are commonly closely associated with global ice age, atmosphere and ocean oxygenation and other important geological events (Pufahl and Groat 2017) (Fig. 2). The Neoproterozoic Phosphogenic Event is the first true phosphorite giant on earth, and the phosphorites of Ediacaran Doushantuo Formation accumulated on the Yangtze Block are the one of the most important phosphorus-rich deposits in this phosphogenic event. Phosphogenesis was restricted to photosynthetically produced nearshore oxygen oases during the Paleoproterozoic Phosphogenic Episode, whereas phosphorite deposits developed in the Neoproterozoic Phosphogenic Episode have a larger scale and a more extensive distribution in the world (Pufahl and Hiatt 2012). The Neoproterozoic Phosphogenic Event is interpreted as reflecting massive chemical weathering of P from primitive mafic crust influenced by the Neoproterozoic Oxygenation Event after the "snowball earth" (Och and Shields-Zhou 2012). The resultant delivery of P to the oceans is thought to have stimulated high surface ocean productivities, generating oxygen that contributed to the oxygenation event and life evolution, and then producing the global phosphorite giant (She et al. 2014; Pufahl and Groat 2017).

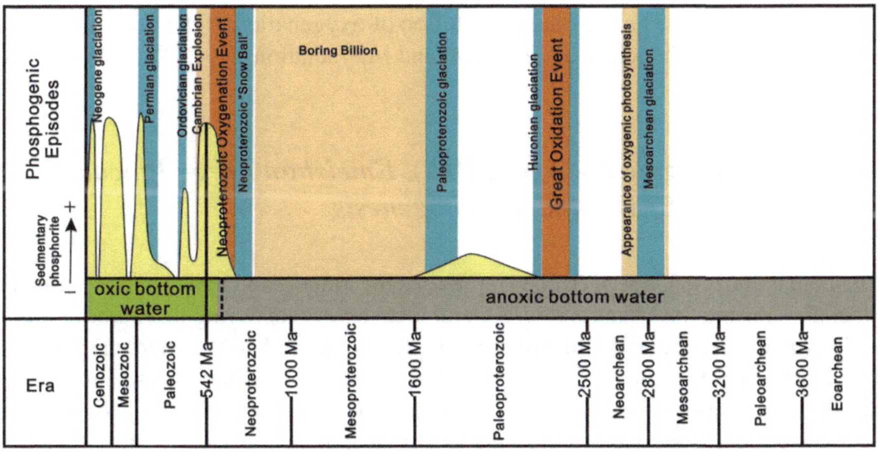

Fig. 2 Temporal distribution of sedimentary phosphorite and important geological events in the Earth history

The sedimentation of Doushantuo phosphorite deposits in central Guizou Province records the coupling relationship between the Neoproterozoic Oxygenation Event and Phosphogenic Event (Zhang et al. 2019). The Doushantuo phosphorite formed as the result of the concentration of phosphate near the near the redox interface through microbial mediation and Fe-redox pumping (She et al. 2014). The upwelling of nutrients to the photic zone of the continental shelf supported the growth of phytoplankton which stimulated the production of photosynthetic oxygen in surface water. The generation of nearshore oxygen oases created suitable redox conditions near the sediment-water interface for chemosynthetic bacteria or Fe-redox pumping to concentrate phosphate and stimulate francolite precipitation. The accumulation of shallow shore phosphorite through the Doushantuo period suggests that the benthic cycling of bioavailable P was a persistent occurrence in coastal oxygen oases. The REE characteristics of Weng'an Doushantuo Fomation indicates the gradually increase of oxygen content in shallow seawater environment. In deeper environment, the anoxic condition in the seafloor could not sequester P to sediments and led to recycling of P directly back to the seawater. This concentration process of P in the deeper water column was likely widespread and may have stimulated primary production in surface seawater environment, contributing to the rise in oxygen and creating the nutrient-rich environment for metazoan animals evolution. As oxygen levels increased, the environments hosting eukaryotes also expanded. The evolution of metazoan life drove the progressive oxygenation of deep oceans through the quick settlement of eukaryotes and reducing consumption of oxygen in the surface ocean. As oxygen content gradually increased in deeper water, the microbial P redox system is established and the phosphorite factory can operate. The distribution of phosphorite deposits with eukaryotic multicellular fossils from shallow-water shoal to the shelf margin-to-slope records the coevolution of oxygenation of the deep ocean, the expansion of phosphogensis environment and life radiation.

1.3 Late Paleozoic Ice Age (LPIA), Emeishan Large Igneous Province (ELIP) and Metallogenesis

The Late Paleozoic Ice Age (LPIA) represents the longest and most intense period of glaciation during the Phanerozoic. Several tectonic events during the Carboniferous and Permian contributed to the initiation of glaciation, such as the southward drift of Gondwana over the polar regions, the uplift of Central Pangean Mountains beyond the alpine equilibrium line altitude (ELA), the closure of the Rheic Seaway between Euramerica and Gondwana, and a consequent change in ocean-circulation patterns. Recent research emphasizes atmospheric CO_2 variations as a driving force for the initiation and evolution of the LPIA. The history of the LPIA is documented in both near-field (glacial sedimentary) and far-field (e.g., eustatic, O-isotopic) records. Near-field records show that it commenced with short-lived alpine glaciation in southern Gondwana (present South America) and southern Euramerica (Appalachian Orogen,

U.S.) during the Devonian-Carboniferous transition. A second glacial episode during the mid to late Tournaisian resulted in deposition of tillites in the Solimões and Amazon Basins of Brazil and possibly in the Eastern Cordillera of northern Chile and Argentina. A third glacial episode during the late Visean left a glacial sedimentary record in Brazil and western Argentina. Above the Visean-Serpukhovian boundary, deposits in Gondwana document continuous widespread glaciation into the Lower Permian (Bishop et al. 2009). Far-field sedimentary records in the U.S., U.K., and central Asia record cyclic marine or marine-terrestrial successions ("cyclothems") that record large-amplitude glacio-eustasy beginning in the late Visean, and coeval chemostratigraphic records ($\delta^{13}C$, $\delta^{18}O$, and $^{87}Sr/^{86}Sr$) show climate-related shifts or fluctuations. Although LPIA glaciation continued unabated from the late Visean (mid-Mississippian) through the mid-Sakmarian (Early Permian), recent studies have inferred that continental ice mass peaked during the Early-Middle Pennsylvanian and early Early Permian, with a relative ice-mass minimum during the Late Pennsylvanian. This pattern has been linked to evidence for multiple glacial centers in Gondwana, with long-term waxing and waning of ice mass triggered by plate motions favoring growth or decay of specific icesheets. In the Northern Hemisphere, although the existence of a Permian North Polar icesheet has been proposed, recent studies have re-interpreted diamictites in northeastern Asia as evidence of Bashkirian-Moscovian sea-ice and Capitanian volcano-sedimentary gravity-flow deposits. From the end of the Moscovian to the Gzhelian (Late Pennsylvanian), a ~9-Myr-long global warming trend resulted in partial deglaciation of Gondwana. After the mid-Sakmarian (mid-Early Permian), deposition of glacial deposits terminated in most Gondwanan basins, except in eastern and western Australia and Tasmania, where glaciation persisted until the Late Permian.

Appropriate paleoclimatic conditions are the basic requirement for the formation of Permo-Carboniferous bauxite in South China. Although the glacial stage of the LPIA might lead to the regional cooling in western Pangea, both CIA and paleo-temperature indicators are testimonies for intense chemical weathering and relatively high temperature (MAT = ~20 °C) in the tropical East Tethys (Yu et al. 2019b). Located in the equatorial East Tethys, the South China Craton was an island block far away from the main continent during Carboniferous and Permian. Major bauxite deposits formed in the coastal plain and/or in coastal karstic depressions, coal or coal-bearing clastic rocks cover bauxite deposits and are occasionally interbedded with them. The former is the sign of the post-glacial transgression and the latter indicates fluctuations of groundwater table during bauxitization. Fluctuations of groundwater table in the coastal plain and coastal karstic depression environment are linked with the sea-level changes. During the LPIA, glacio-eustasy caused by the glacial-interglacial transitions or waxing and waning of icesheets is common. High pCO_2, high sea-level, high groundwater table and low precipitation, low vegetation cover conditions in interglacial stage meet the requirements for bauxitization. Low pCO_2, low sea-level, low groundwater table and high precipitation, high vegetation cover conditions in glacial stage promotes the bauxitization (Fig. 3).

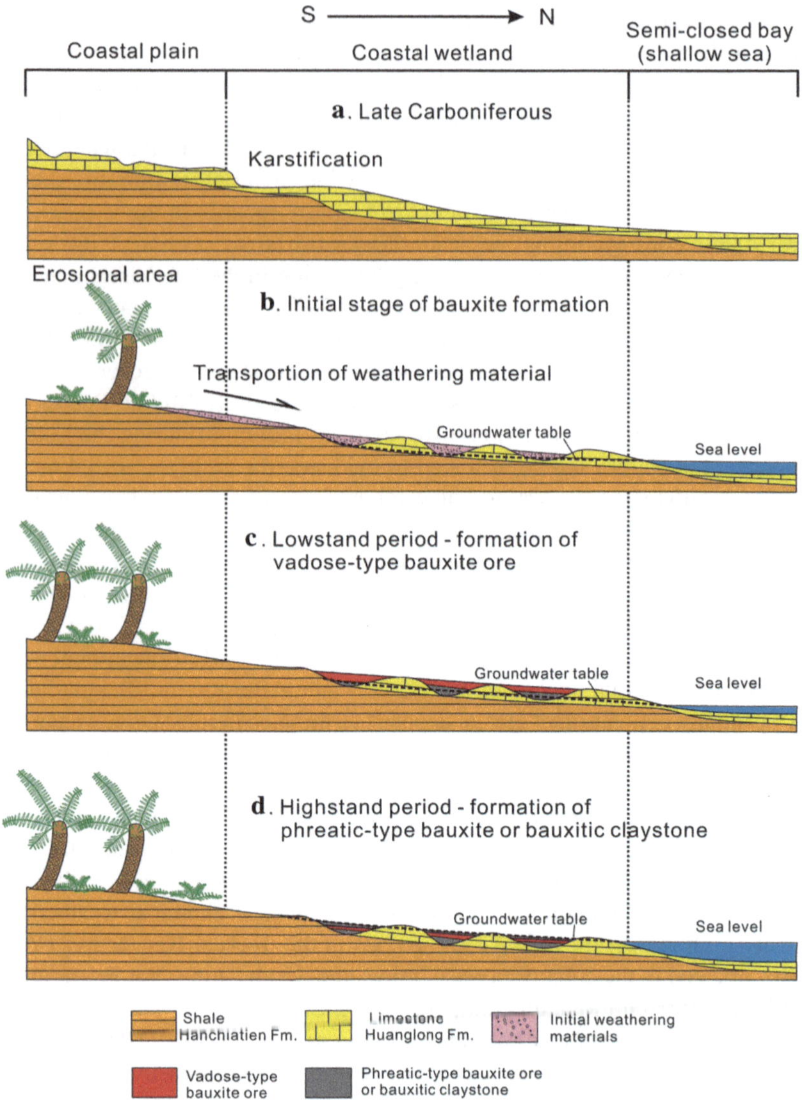

Fig. 3 Model for bauxite formation in South China during Late Paleozoic (Li et al. 2020)

After the Capitanian stage of the Middle Permian, the changes in sea level over the South China Block resulted in a marine transgression event and a marine regression event. The interval between these events was relatively short, 5 Myr or less, and both of these sea-level-change events were recorded in the Maokou Formation. The cause of the marine transgression event could be complex, as it may include (1) climate factors, i.e., the weakening of the Kamura cooling event during the Capitanian age; (2) tectonic factors, i.e., regional tectonic subsidence induced by the initial period of activity in the Emeishan large igneous province (ELIP); and (3) biotic factors, i.e., rapid deterioration of the oceanic ecological conditions, which severely damaged photosynthetic organisms and thus inhibited or extinguished carbonate production factories in the shallow tropical seas that were present at the time.

2 Excursion Route and Road Log

DAY One: Beijing Capital International Airport (PEK) to Tongren Fenghuang Airport (TEN).

DAY Two: Field trip of Cryogenian manganese deposit, the Datangpo section in Songtao County, Tongren City (from the Liangjiehe Formation to the Datangpo Formation). Move to Weng'an County in the afternoon.

DAY Three: Field trip of the Ediacaran phosphorite deposit, the Beidoushan section in Weng'an County. Move to Zunyi.

DAY Four: Field trip of the Carboniferous bauxite deposit and the Permian manganese deposit. Forenoon: section and drillcore of the bauxite deposit in the Jiujialu Formation in the Tuanxi town. Afternoon: section of the manganese deposit in the Maokou Formation in the Sangji town (Fig. 4).

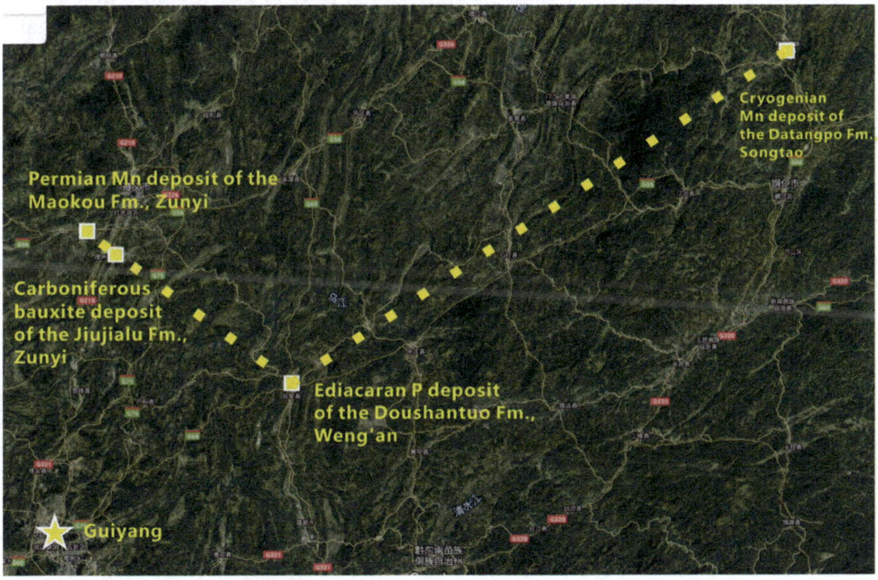

Fig. 4 Excursion Route in Guizhou Province, South China

3 Geological Itinerary and Excursion

3.1 DAY TWO: Field Trip of Cryogenian Mn Deposit

Stop 1

Site No. 001 the Cryogenian Liangjiehe and Tiesi'ao Formations

Location: Maomaoyan section, Mountain road of the Datanpog Village, Songtao County, Tongren City

Please see Figs. 5 and 6 for reginal geology and stratigraphic column. Strata in this area are controlled by NE trend syncline, Cryogenian strata are well preserved.

Length of this section is about 300 m (Fig. 7). The Liangjiehe Formation is mainly composed of thick middle-thick sandstone, it unconformably covers the purple and/or green slate and sandy slate, and it is covered by diamictite of the Tiesi'ao Formation. Dolostone lenses appear in the basal part of the Liangjiehe Formation, laminates can be observed in the dolostone lenses (Fig. 8a). Gravel-bearing sandstone and arkose are shown in the lower and upper Liangjiehe Formation (Fig. 8b), the middle Liangjiehe Formation is dominated by quartz sandstone, parallel beddings and wedge-shaped cross-beddings can be found in the some layers (Fig. 8c, d).

The Tiesi'ao Formation is mainly composed of gray or black diamictite, sizes of gravels are serval centimeters to millimeters, poorly sorting and rounding, matrix supported, high TOC content, pyrites are common. The Tiesi'ao Formation is considered as the glacial marine deposit.

Stop 2

Site No. 002 the Cryogenian Datangpo Formation

Location: Tiekuangping section, Datanpog Village, Songtao County, Tongren City

There are two sections in this site (No. 1 section in Fig. 9 and No. 2 section in Fig. 10). Both of them contain the 1st Member of the Datangpo Formation, including manganese carbonate and black shale deposits (Fig. 9a). Geochronological data from the tuff layers within the Datangpo Formation constrain its initial deposition age as ~659 Ma and its ending age after 654 Ma. Thickness of manganese ore bodies are ranging from tens of centimeters to serval meters, most manganese ores show massive and laminated structures. Due to long time mining, most ore bodies in these sections are empty, leaving the cavities which previously were filled with manganese carbonates. Black shales surround the manganese ore bodies, containing depositional deformation (Fig. 9b, c). Cracks within the black shales are partly filled by manganese carbonate (Fig. 9d, e).

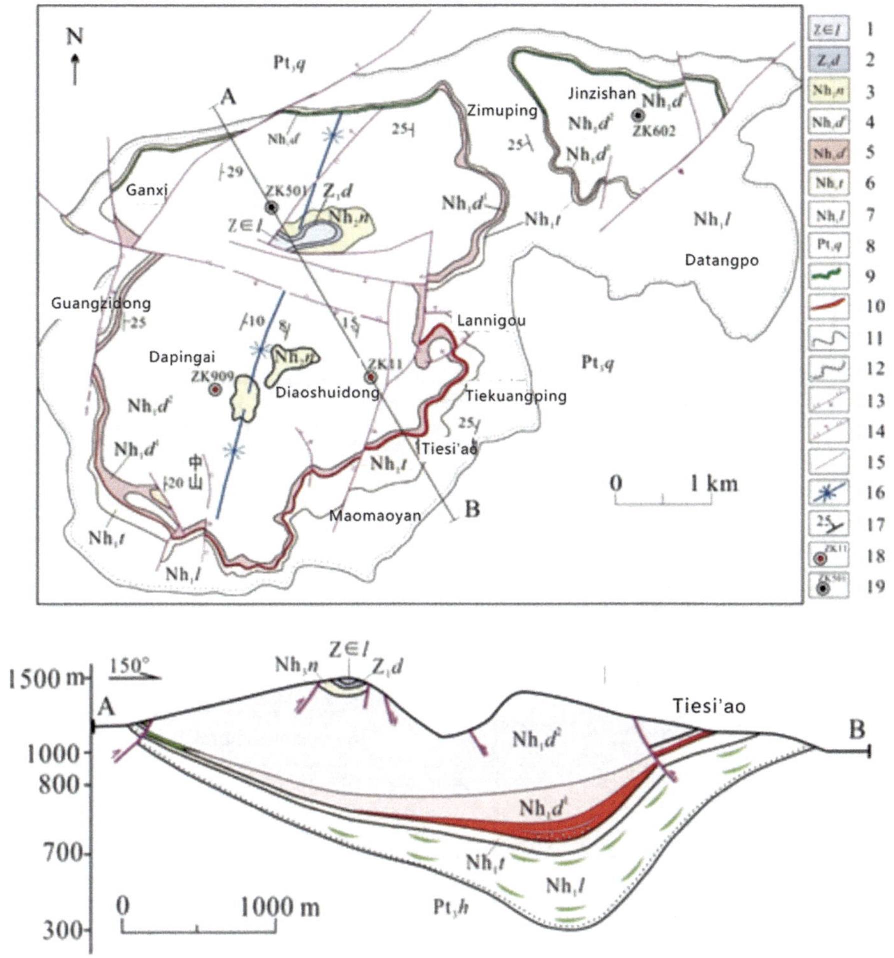

Fig. 5 Geological map and cross-section of the Datangpo village, Songtao County, Tongren City, Guizhou Province

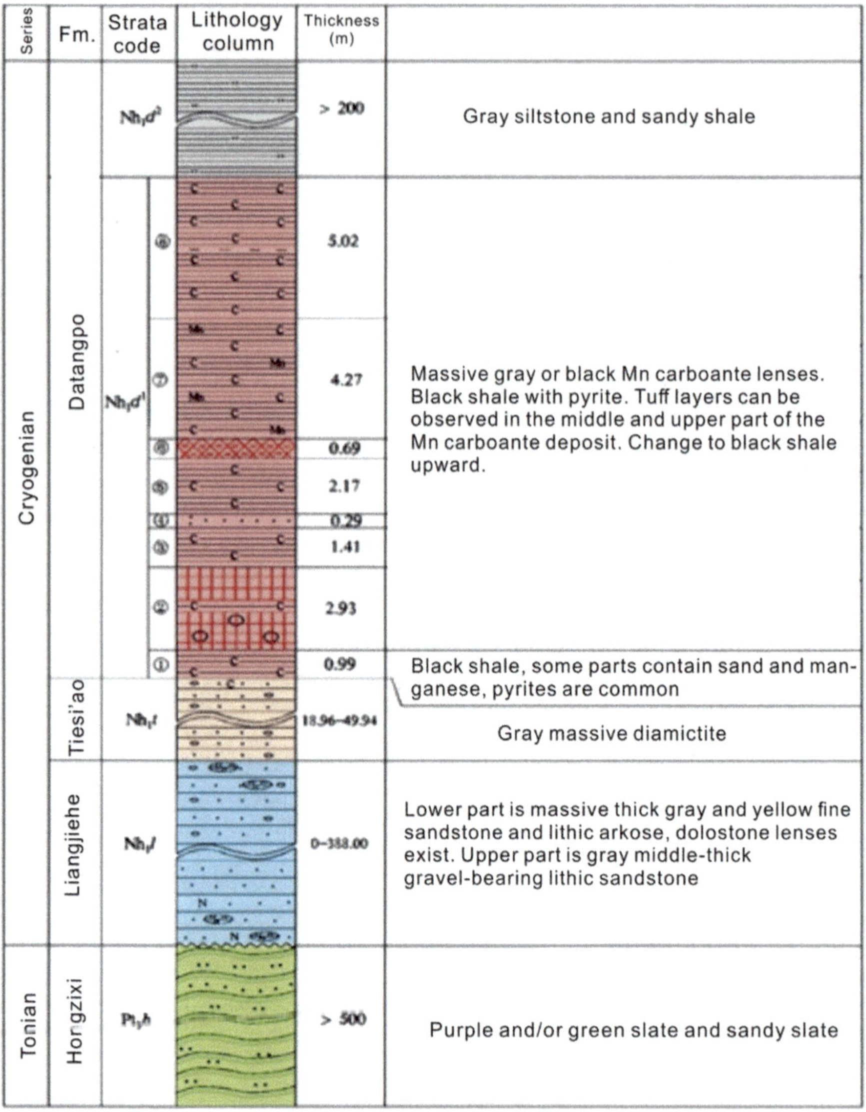

Fig. 6 Regional lithology column in the Datangpo Village, Songtao County, Tongren City, Guizhou Province

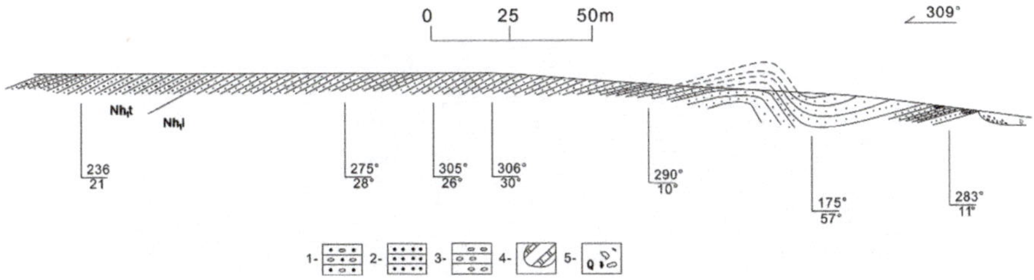

Fig. 7 The Maomaoyan section in the Datangpo Village, Songtao County, Guizhou Province. 1-gravel-bearing sandstone, 2-quartz sandstone, 3-diamictite, 4-dolostone lense, 5-Quarternary sediments

Fig. 8 Sedimentary structures in the Liangjiehe Formation of the Maomaoyan section in the Datangpo Village, Songtao County, Guizhou Province

Fig. 9 Sedimentary structures in the 1st Member of the Datangpo Formation, Tiekuangping No.1 section, Datangpo Village, Songtao County, Guizhou Province. **a** Whole scene of No. 1 section; **b**, **c** Depositional deformations in the basal Datangpo Formation; **d**, **e** Cracks in black shales are filled by Mn carbonate

Fig. 10 Sedimentary structures in the 1st Member of the Datangpo Formation, Tiekuangping No. 2 section, Datangpo Village, Songtao County, Guizhou Province

3.2 *DAY THREE: Field Trip of Late Neoproterozoic Phosphorite Deposits*

Stop 3

Site No. 003 Cap carbonate and upper phosphorite bed in Ediacaran Doushantuo Formation

Location: 300 m west of a three-way intersection in Datang mining area, Weng'an County

Regional geology of Weng'an mining area and stratigraphic column of Doushantuo Formation are showed in Fig. 11a, b. Qingshuijiang Formation, Nantuo Formation Doushantuo Formation and Dengying Formation are present and Ordovician, Silurian, Carboniferous and other Phanerozoic strata are absent in Weng'an area. All of the phosphatic layers developed in Doushantuo Formation, which occurred in both wings of Baiyan-Gaoping Anticline along the direction of south-north.

Strata from top part of the Qingshuijiang Formation to lower and middle part of the Doushantuo Formation could be observed in Site No. 001. The Qingshuijiao Formation, which is unconformably covered by the Doushantuo Formation,

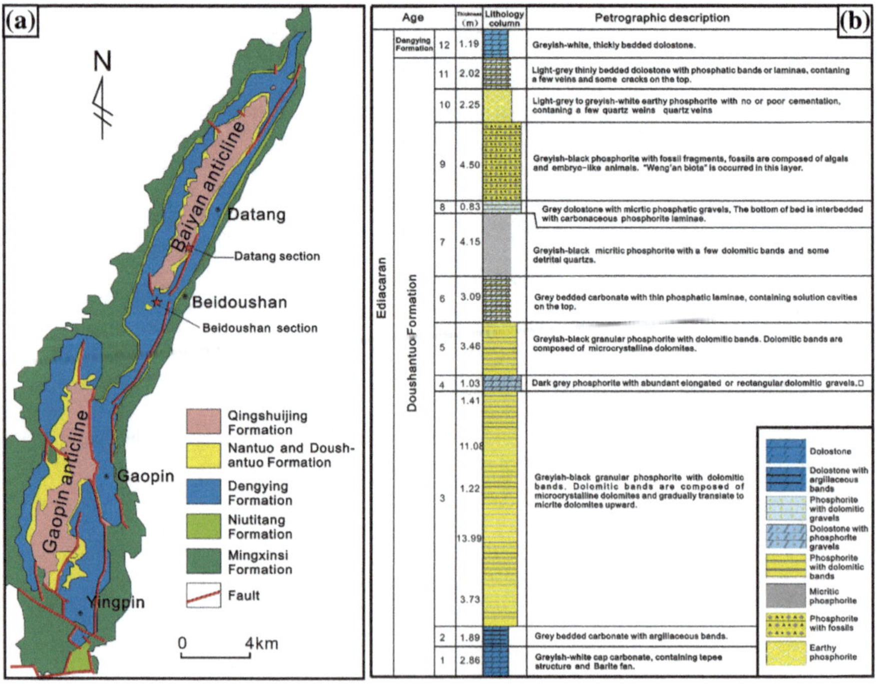

Fig. 11 Geological map (**a**) and regional lithology column (**b**) in Weng'an area, Guizhou Province

is mainly composed of greyish-green siltstone and dark grey shale and slate. The greyish-white, thickly bedded cap carbonate occurred in the bottom of Doushantuo Formation (Fig. 12a). The cap carbonate layer is two meters thick and developed sheet fissures and tepee structures. Most of fissures distribute along the layers but some of them show a vertical distribution. These fissures were filled with siliceous rocks and a few calcites. Light-grey, thickly bedded microcrystalline dolostones (3–5 m thick) covers the cap dolostone, showing some tubercles on the top (Fig. 12b). The interior tubercles are distributed in layers, containing one or two radial and flabellate barite layers (Fig. 12c). The overlying Lower Phosphorite Bed (10 m thick) developed massive granular phosphorite, granular phosphorite interbedded with dolomitic bands (Fig. 12d) and granular phosphorite with micritic or argillaceous laminae from bottom to the top.

Fig. 12 Sedimentary structures of the Doushantuo Formation in the Datang mining area, Weng'an Country. **a** Cap carbonate in the bottom of Doushantuo Formation, **b** tubercles in cap carbonate, **c** flabellate barite in tubercles, **d** granular phosphorite with dolomitic bands in Lower Phosphorite Bed

Stop 4

Site No. 004 Lower and Upper Phosphorite Bed in Ediacaran Doushantuo Formation

Location: Beidoushan mining area in Weng'an Country

The phosphatic succession from Lower to Upper Phosphorite Bed is welled developed in Beidoushan mining area (Fig. 13a). The Lower Phosphorite Bed is 7 m thick and mainly composed of granular phosphorite with dolomitic bands. A bedded dolostone layer containing horizontally arranged ellipsoidal phosphatic gravels are present in the top of the Lower Phosphorite Bed. The thickly bedded dolostone layer covers the Lower Phosphorite Bed. It is 1.5–2 m thick and heavily silicified. The overlying Upper Phosphorite Bed developed thinly bedded carbonaceous, micritic phosphorite layer (1.5 m thick) in the bottom of the Upper Phosphorite Bed. Above it is granular phosphorite bed (>3 m) with abundant biological fossils that could be observed as white spherical particles in macroscopic scale (Fig. 13b). The famous "Weng'an biota" is occurred in this layer, which contains metazoans including embryo-like fossils (Fig. 13c–f).

3.3 DAY FOUR: Field Trip of Carboniferous—Permian Bauxite and Manganese Deposits

Stop 5

Site No. 005 Upper Paleozoic Carboniferous Jiujialu Formation bauxite deposit

Location: Bauxite mining field in Tuanxi Town, Bozhou District, Zunyi City

Cambrian to Cretaceous strata are present near Tuanxi Town, Bozhou District, Zunyi City, Silurian and Devonian strata are absent, and Carboniferous is partly preserved (Fig. 14). On the top of the Cambrian Loushanguan Formation, paleokarstic surface developed, representing the uplift and exposure of the South China Block caused by the Guangxi Movement (late Ordovician to early Devonian). In Zunyi area, the exposure could last to early Permian.

Zunyi City is one of the earliest area that was explored bauxite deposits in China. In this site, we will focus on the bauxite deposit in the Jiujialu Formation. Underlying strata of the Jiujialu Formation is the Cambrian Loushanguan Formation dolostone, and the overlying strata is the mudstone of the Permian Liangshan Formation or the limestone of the Qixia Formation (Fig. 15). Two features are shown in the bauxite

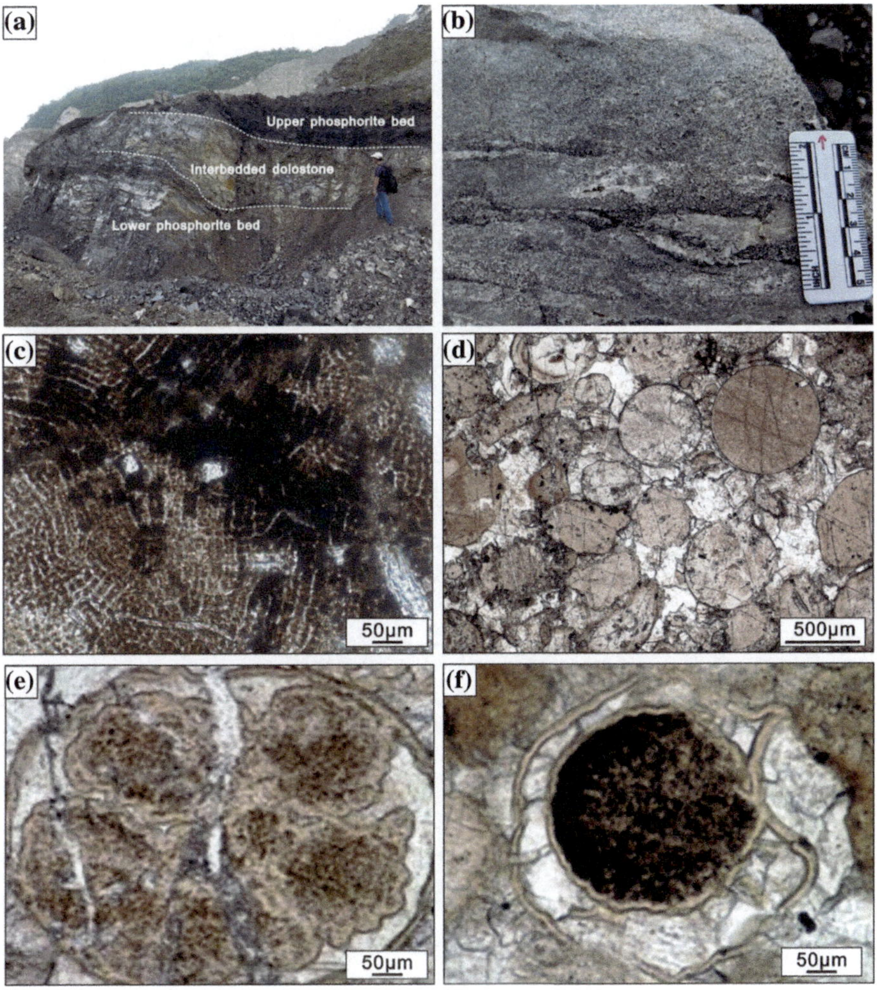

Fig. 13 Sedimentary structures and microstructures present in the Doushantuo Formation of Beidoushan mining area, Weng'an Country. **a** Boundaries between the Lower Phosphorite Bed, interlayer dolostone and Upper Phosphorite Bed, **b** Granular phosphorite with abundant spherical fossils in the Upper Phosphorite Bed, **c** Algal fossils in the Upper Phosphorite Bed, **d**–**f**, embryo-like fossils in the Upper Phosphorite Bed

deposits of Zunyi: (1) dramatic variation show in the thickness of bauxite, from several meters to more than one hundred meters; (2) cycles of bauxite ore layers and bauxitic claystone/coal layers show in the bauxite deposit, the thick deposit present more cycles (Fig. 16).

The bauxite mining filed near the Tuanxi Town, Zunyi City provides us a good spot to observe the distribution and shape of bauxite ore body. Bauxite deposit was filled

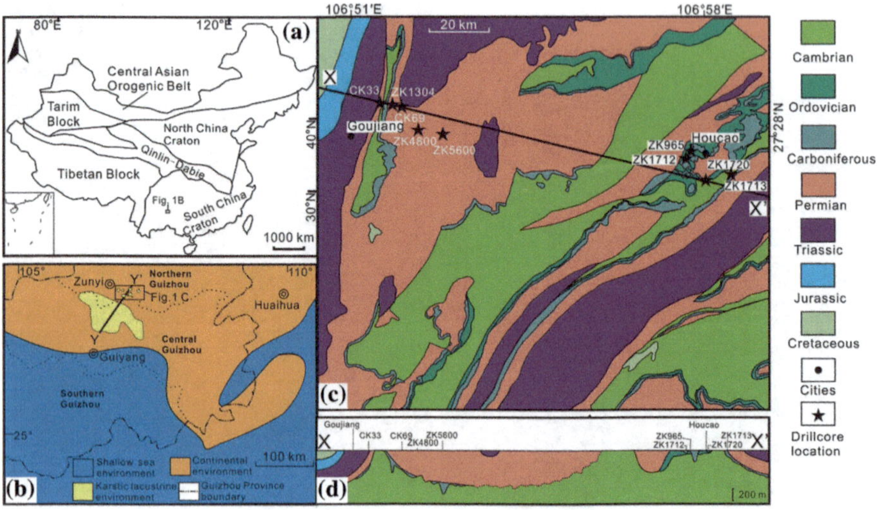

Fig. 14 **a** Location of Zunyi city in China; **b** Paleogeography map of Guizhou during Early Carboniferous; **c**, **d** Geological map and cross-section of studying area

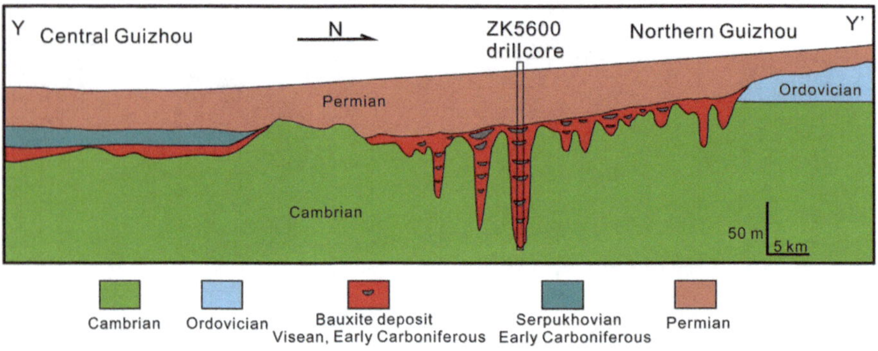

Fig. 15 Cross-section of stratigraphy from central Guizhou to northern Guizhou (Weng et al. 2019)

within the paleokarstic doline, now most bauxite deposit was stoped out and leaving the hiatus (Fig. 17). For detailed investigation and observation of the bauxite deposit, we prepare one complete drillcore (ZK5600, Fig. 18), total thickness of the bauxite deposit in this drillcore reaches 110 m. Low part of the bauxite deposit consist of Fe-rich bauxitic claystone, middle-upper part contain 5 bauxite ore layer—bauxitic claystone cycles. According to sedimentology, mineralogy, and geochemistry study, bauxite ore contain massive, clastic and porous structures, clasts are mostly composed of redepositional bauxite ore. Diaspore, clay minerals (kaolinite, illite, and smectite), pyrite, hematite, and some heavy minerals (zircon and rutile). Bauxitic claystone

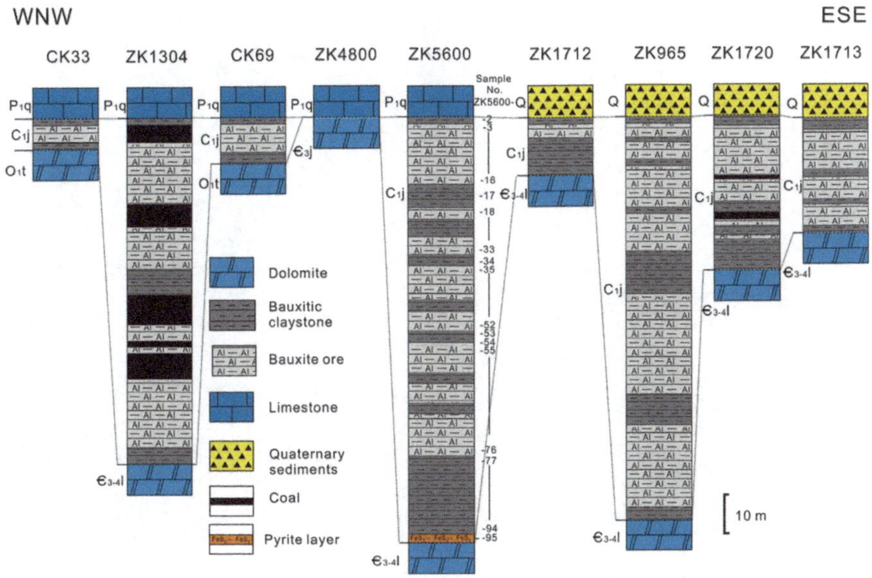

Fig. 16 Correlation of lithology columns in Zunyi

Fig. 17 Mining field near Tuanxi Town, Bozhou District, Zunyi City

usually presents massive structure, containing dominated kaolinite and illite and minor diaspore and pyrite (Fig. 19).

Fig. 18 Bauxite deposit in drillcore ZK5600 near Tuanxi Town, Bozhou District, Zunyi City, total thickness of bauxite deposit is 110 m

Stop 6

Site No. 006 Upper Paleozoic Permian Maokou Foramtion manganese deposit

Location: Sangji Town, Bozhou District, Zunyi City

The study area is located on the western margin of the South China Block and is part of the northern Guizhou stratigraphic region. Strata in this area consist mainly of Cambrian, Permian, and Triassic.

Owing to the Dongwu Movement, the top of the Maokou Formation has undergone varying degrees of denudation in the study area. It is unconformable with the overlying Permian Longtan Formation, and argillaceous mudstone and ferromanganese crust have developed around the interface between these formations; these features are consistent with those of a paleokarstic surface.

The Maokou Formation can be divided into three members. *The first Member* *(P_2m^1)* is distributed throughout the study area, and its lithology consists of thick-bedded to massive light-gray or gray bioclast-bearing limestone intercalated with wavy-/lenticular-bedded, banded argillaceous limestone. Certain parts of this member are bedded with dolomitic limestone and flint bands/nodules. The thickness of this member is very stable (typically 110 m), and it is in conformable contact with the underlying Qixia Formation. *The Second Member (P_2m^2).* The lithology of the second member varies with location in the study area, and it can be categorized into two types. (1) Carbonaceous-siliceous limestone (P_2m^{2a}): its lithology consists mainly of dark-gray thin-bedded silicon- and carbon-bearing limestone, with localized intercalations of gray thin- to medium-bedded silicon-bearing limestone and

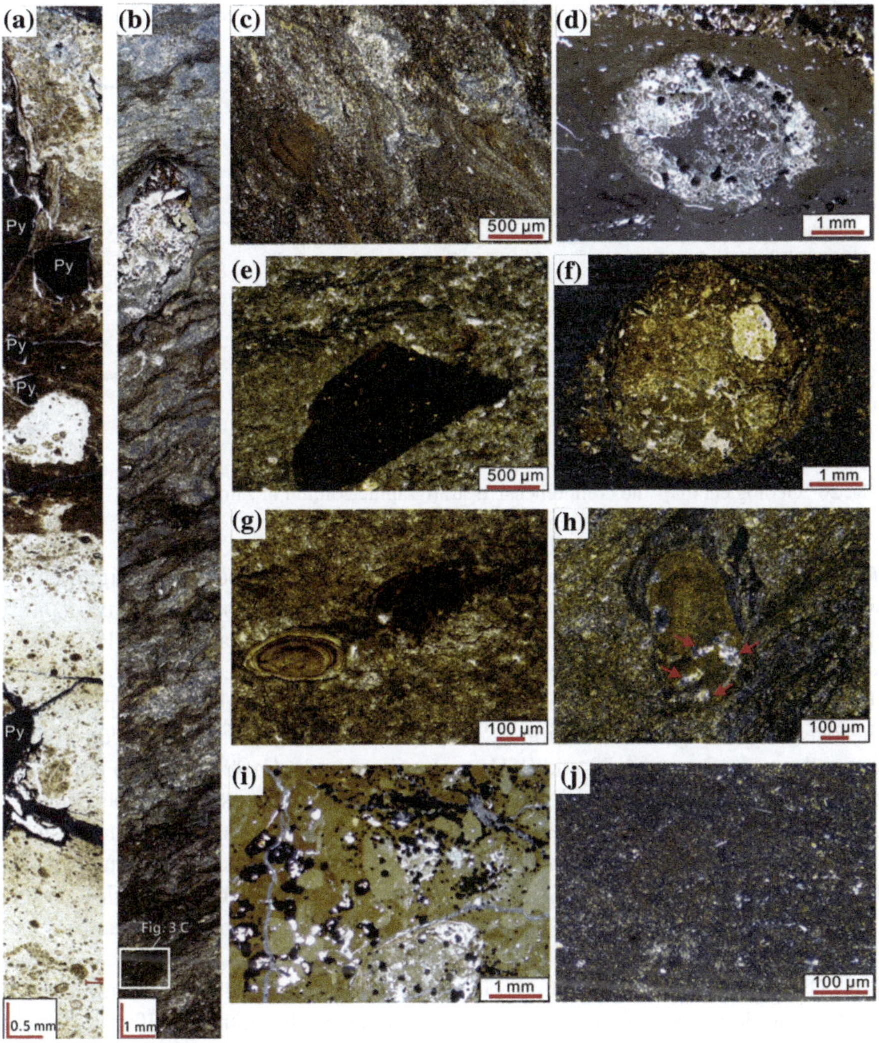

Fig. 19 Structure (**a**, **b**), minerals and textures (**c**–**j**) in bauxite ore of drillcore ZK5600 near Tuanxi Town, Bozhou District, Zunyi City

siliceous rock bands. The silicon and manganese contents of this member steadily increase from the bottom layer to the top layer. The thickness of this Member varies between 40 and 68 m, and the average thickness is 50 m; it is conformable with the First Member of the underlying Maokou Formation. (2) Flint-bearing limestone (P_2m^{2b}): This lithological type consists of gray-black or dark-gray medium-bedded limestone with flint nodules intercalated with dark-gray micritic limestone. The thickness of these rocks varies between 60 and 140 m, and the average thickness is 70 m; this layer is conformable with the First Member of the underlying Maokou Formation.

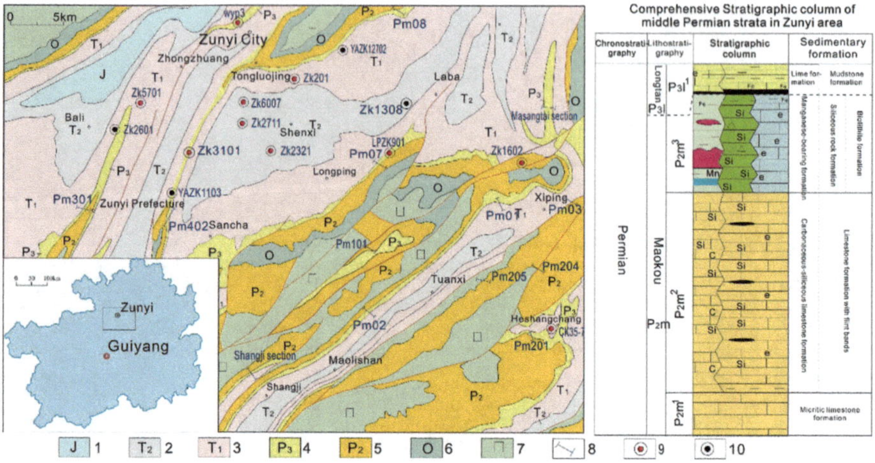

Fig. 20 Geological map and comprehensive stratigraphic column of Zunyi

The Third Member (P_2m^3): the lithology of the Third Member varies significantly with location and includes the following three types. (1) Bioclastic limestone (P_2m^{3a}): its lithology consists of light-gray and gray thick-bedded to massive bioclastic limestone, and its thickness varies between 40 and 50 m. This layer is conformable with the second member of the underlying Maokou Formation and unconformable with the overlying Longtan Formation. (2) Silicified rocks (P_2m^{3b}): its silicon content typically varies between 70 and 95%; pure siliceous rock (99% silicon) appears in certain parts of this layer, as well as linear brecciated siliceous rocks. The thickness of this layer varies between 30 and 50 m. It is conformable with the second member of the underlying Maokou Formation and unconformable with the overlying Longtan Formation. (3) Manganese deposits (P_2m^{3c}): It is a manganese-bearing series of rocks that consists of light-gray, gray, gray-green, and gray-black pyrite-containing clay rocks, silty mudstone, manganese-bearing clay rocks, tuff, and manganese carbonate ore. The thickness of this layer varies between 1.96 and 5.95 m. Commercially exploitable deposits of manganese are often found in the bottom half of this layer. This layer is in direct contact with siliceous limestone from the second member of the underlying Maokou Formation and is unconformable with the overlying Longtan Formation (Figs. 20 and 21).

4 Field Logistics

Since the stops of this route is near the road, the physical difficulty of this route is easy. However, safety in the field is closely related to awareness of potential difficulties, fitness and use of appropriate equipment.

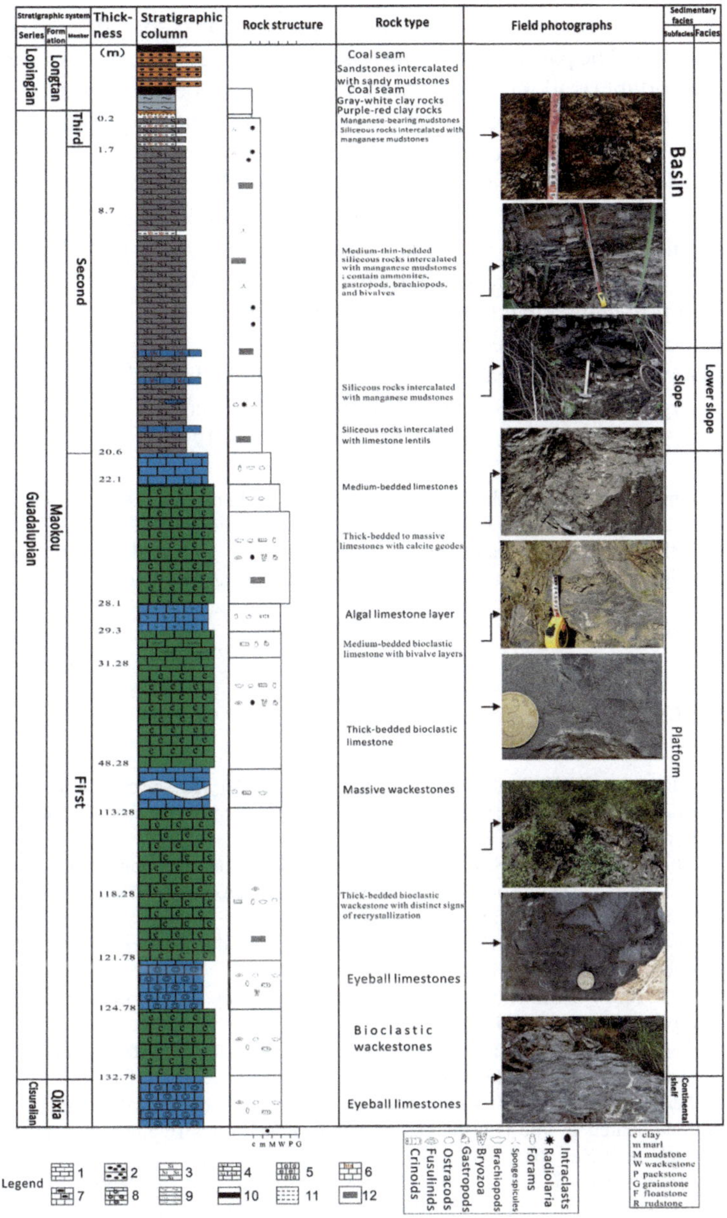

Fig. 21 Lithology column of the Jisang section, Zunyi

Safety is a personal responsibility and all participants should be aware of the following issues. The participants will be carried with one or two bus. All participants require comfortable walking boots. Trainers or running shoes are unsuitable footwear in the field. A waterproof coat/jacket is essential. In August, weather conditions in Guizhou are usually good; anyway, an umbrella is recommended as well as a bottle of water and a backpack. Sun protection can be useful as well a hat or a headscarf. Participants should inform the excursion leaders (in confidence) of any physical or mental condition, which may affect performance in the field (e.g., asthma, diabetes, epilepsy, vertigo, heart condition, back problem, ear disorder, lung disease, allergies etc.), and of problems related to particular diets or food.

The vehicle will carry one basic first aid kit.

Mobile/cellular phone coverage is good but could be limited in some places.

The emergency telephone number for police is 110 and 119.

The emergency telephone number for ambulance is 120.

Acknowledgements We would thank Dr. Yaguan Zhang and Dr. Chao Deng for providing helps to route organizations. Innovation Center of Ore Resources Exploration Technology in the Region of Bedrock, 102 Geological Brigade, 103 Geological Brigade, 106 Geological Brigade and 115 Geological Brigade are greatly thanked for providing helps during the field works.

References

Bishop JW, Montañez IP, Gulbranson EL, Brenckle PL (2009) The onset of mid Carboniferous glacio eustasy: sedimentologic and diagenetic constraints, Arrow Canyon, Nevada. Palaeogeogr Palaeoclimatol Palaeoecol pp. 276–243

Li P, Yu W, Du Y, Lai X, Weng S, Pang D, Xiong G, Lei Z, Zhao S, Yang S (2020) Influence of geomorphology and leaching on the formation of Permian bauxite in northern Guizhou Province, South China. J Geochem Explor 210:106–446

Och L, Shields-Zhou G (2012) The Neoproterozoic oxygenation event: environmental perturbations and biogeochemical cycling. Earth Sci Rev 110:26–57

Pufahl P, Groat L (2017) Sedimentary and igneous phosphate deposits: formation and exploration: an invited paper. Econ Geol 112:483–516

Pufahl P, Hiatt E (2012) Oxygenation of the Earth's atmosphere–ocean system: a review of physical and chemical sedimentologic responses. Mar Pet Geol 32:1–20

She Z, Strother P, Papineau D (2014) Terminal Proterozoic cyanobacterial blooms and phosphogenesis documented by the Doushantuo granular phosphorites II: microbial diversity and C isotopes. Precambr Res 251:62–79

Weng S, Yu W, Algeo T, Du Y, Li P, Lei Z, Zhao S (2019) Giant bauxite deposits of South China: multistage formation linked to Late Paleozoic Ice Age (LPIA) eustatic fluctuations. Ore Geol Rev 104:1–13

Yu W, Algeo T, Yan J, Yang J, Du Y, Huang X, Weng S (2019a) Climatic and hydrologic controls on upper Paleozoic bauxite deposits in South China. Earth Sci Rev 189:159–176

Yu W, Polgari M, Gyollai I, Fintor K, Szabo M, Kovacs I, Fekete J, Du Y, Zhou Q (2019b) Microbial metallogenesis of Cryogenian manganese ore deposits in South China. Precambr Res 322:122–135

Zhang Y, Pufahl P, Du Y, Chen G, Liu J, Chen Q, Wang Z, Yu W (2019) Economic phosphorite from the Ediacaran Doushantuo Formation, South China, and the Neoproterozoic-Cambrian Phosphogenic Event. Sed Geol 388:1–19

GPSR Compliance

The European Union's (EU) General Product Safety Regulation (GPSR) is a set of rules that requires consumer products to be safe and our obligations to ensure this.

If you have any concerns about our products, you can contact us on ProductSafety@springernature.com

In case Publisher is established outside the EU, the EU authorized representative is:

Springer Nature Customer Service Center GmbH
Europaplatz 3
69115 Heidelberg, Germany

Batch number: 10725816

Printed by Printforce, the Netherlands